CIVIL
Engineer's
ILLUSTRATED SOURCEBOOK

CIVIL Engineer's
ILLUSTRATED SOURCEBOOK

ROBERT O. PARMLEY, P.E.

McGraw-Hill
New York Chicago San Francisco Lisbon London Madrid
Mexico City Milan New Delhi San Juan Seoul
Singapore Sydney Toronto

The McGraw-Hill Companies

Cataloging-in-Publication Data is on file with the Library of Congress.

1 2 3 4 5 6 7 8 9 0 KGP/KGP 0 9 8 7 6 5 4 3

ISBN 0-07-137607-0

The sponsoring editorfor this book was Larry S. Hager, the editing supervisor was David E Fogarty, and the production supervisor was Pamela A. Pelton.

Design by Wayne C. Parmley, Parmley Design.

Printed and bound by QuebecorlKingsport.

This book was printed on acid-free paper.

To Harold (Jim) Cleary, P.E.
who encouraged me to pursue
a career in professional engineering

CONTENTS

PREFACE

Any engineer entering the profession, seeking a career in the area of private practice, should be mentored by a well-seasoned professional. But, in today's world of specialists, there are fewer general practioners in civil engineering who are well versed in the full aspect of consulting engineering. When you do find one, that individual most often is too busy to devote much time to training entry-level personnel. Therefore, the novice engineer is at a basic disadvantage.

Being well trained and educated is not sufficient to pursue a career in consulting engineering. One must also have adequate, practical experience. The licensing laws of each state, territory and province have experience requirements mandated before issuing a P.E. (Professional Engineer) registration. However, these requirements are broad based. This sourcebook is not intended to supplant any licensing requirement, but rather designed to augment this segment of the engineer's training.

There are literally scores, if not hundreds, of excellent books on every technical subject, including planning, design, inspection, estimating, report writing, construction management and related topics scattered throughout the engineering literature. Unfortunately, this material is so detailed and topic specific that a single volume, containing an overview or road map of project engineering, appears to be non-existent. It is the intent of this sourcebook to remedy this void in the literature.

Civil Engineer's Illustrated Sourcebook was designed to walk the reader through the entire process of a project; from the initial idea to completion of construction. The project milestones have been presented in chronological sequence with a major focus on preparation of construction plans for typical civil engineering projects. These plans, as well as related documents, have been painstakingly selected from successful projects ranging over a forty-year span.

While it is obvious that this sourcebook cannot cover all facets of a project within these limited pages, let it suffice to say that the presentation is ample and should be a useful guide and reference to the reader.

It must be noted that before undertaking any engineering commission, both the engineer and client should have a clear understanding of the scope of services to be performed, a reasonable timetable for execution of the work and an agreement for monetary compensation for said services. Usually these agreements are formal written documents, properly signed by all parties and in-place, prior to commencing any work.

There are numerous standard contract forms in use for contracting engineering services available from various professional organizations. The leading source for these standard forms of agreement is the Engineer's Joint Contract Documents Committee (consisting of representatives from the American Consulting Engineers Council, the American Society of Civil Engineers, the Construction Specifications Institute and the National Society of Professional Engineers). Additionally, many granting/loan agencies and some regulatory agencies, have either their own typical agreement forms or mandated special requirements. Therefore, whether it is a signed proposal, letter of agreement or an official contract for services, this sourcebook assumes that an engineering contract has been properly executed so the engineer may begin work. With this understanding, the reader is now ready to travel through the following pages. I hope the journey will be rewarding and of lasting value.

All sources of material and documents have been listed where known. A special thank-you to Lana and Ethne for their word processing assistance. As always, Wayne did an excellent job on layout and design of the final format. It was a pleasure working with him again on a major project.

Robert O. Parmley, P.E.
Editor-in-Chief
Ladysmith, Wisconsin

INTRODUCTION

The purpose of this sourcebook is to provide the user with an all-inclusive overview of the general pattern of civil engineering as it is practiced by private consulting firms.

Scope of Sourcebook

The general aim of this sourcebook is to afford those interested in the private practice of civil engineering, whether they are students, teachers, novice engineers, municipal officials, technicians, planners, designers or the public in general, an outline of the discipline with graphic references of actual projects that were successfully completed. While this information is not new, it is widely scattered throughout the engineering literature and no single volume, to date, contains all this information in a logical sequence. Our goal is to remedy this void with a user friendly sourcebook that will have lasting value to the profession.

Most current literature on these subjects are mainly in written form. Since this sourcebook is meant to be all-inclusive and instructive, the following pages are extensively illustrated. Textural material is included only where necessary to supplement the drawings, charts, tables and outlines.

One of the major problems of developing this manuscript was the distilling of available material so the data could fit into the boundary of a single volume. This difficulty was further compounded by the anticipated needs of the future readers. Their concerns certainly will range in magnitude from the very basic to those of great complexity. In an attempt to solve these problems, it would be paramount to select material that would balance the presentation. Therefore, selections were made from a reservoir of over forty years of projects. The emphasis is on basic or general areas of practice.

The Editor-in-Chief cautions the users that the material presented includes general applications, basic patterns and generic examples. It will be noticed by astute readers that many of the sample plans shown were designed under past rules that may no longer apply. We ask the reader to grant us this artistic license to include these plans because they were successfully used to construct facilities, structures and systems that continue to this day to serve their original purpose; many of which were unique and innovative designs. However, for specific projects, the users must consult applicable current codes, rules and regulations, as well as conform to the current requirements of specific regulatory agencies having jurisdiction over the project parameters.

Most civil engineering projects are not large, highly publicized or historical in nature. Rather, they are relatively small, routine, utilitarian and essential to the continuing improvement of modern civilization. The majority of communities are less than 10,000 in population, but still have the fundamental needs of large metropolitan cities; i.e., sanitary sewers, potable water, transportation networks, fire protection, waste treatment and disposal, municipal buildings and related facilities. With this in mind, we have assembled a broad range of relatively small, but fundamental plans to illustrate the wide range of projects that are common to the majority of municipalities and thus basic to the practice of civil engineering.

As previously stated, material for this sourcebook was extracted from decades of files. Therefore, the reader will notice a variation in plan formats and drafting techniques. Trends change and technical methods evolve. Most of the plans within the following pages were hand drafted with conventional tools on either vellum or mylar paper using ink pens or lead drafting pencils. Within recent years, this method of drafting is rapidly being replaced by computer aided drafting (CAD). I fear that some valuable techniques will be sacrificed by totally relying on this electronic process. It was with this thought in mind that we were prompted to include a large volume of plans that were manually drawn. The reader should study the skill and organization of these previous plans and consider incorporating applicable concepts into their future layouts. However, it must be noted that we are not implying that the CAD system is inferior or significantly lacking in scope, but rather warning of its potential for program restrictions and graphic limitations. Let the CAD system serve, rather than the designer conform to its limitations. For a balance, several sets of plans have been included that were entirely prepared by CAD which demonstrates its compatibility with established procedures, if properly programmed.

The format of this sourcebook follows the general and logical sequence of typical projects; i.e., planning, design, bidding process and construction. The following discussion of each category will flesh out their respective divisions.

Planning

All projects begin with a thought that germinates into an idea. As the idea takes root and starts to evolve, a logical planning process must be followed or the project may be subjected to numerous delays, technical pitfalls, possible controversy, lack of support, public misconceptions, budget problems, environment conflict and regulatory resistance.

Each project has its own respective set of unique elements that must be addressed. This sourcebook certainly can not anticipate all of them. However, most elements of civil engineering practice are common to all projects. Therefore, this sourcebook attempts to include these basics in an orderly, logical and chronological sequence so the reader can see the progression of a project to its successful conclusion.

As the project's concept emerges from the individual's mind or group's collective input, notes should be recorded and an outline developed for preparation of an engineering or technical report. This report must be complete and accurate in describing the project or sub-part of the project. The various types of technical reports are discussed in Section 1 and sample outlines provided.

Project scheduling is presented in Section 2 with example outlines and formats extracted from past projects that were successfully completed.

As the project continues, field reconnaissance, surveying and mapping usually will be required. Sample maps, data and information are illustrated in Section 3 to familiarize the reader with the variety of material needed in this phase of project development.

Section 4 discusses the public meeting component of a project. This phase can be extremely critical. Many projects have been terminated or retarded because of improperly administered public meetings. Whether it is an informational meeting, assessment hearing, regular board session, or formal proceeding, one must be fully prepared and understand the purpose to avoid failure.

While regulatory approvals are not obtained until final plans and specifications are submitted, it is wise to understand which agencies have jurisdiction over the proposed project. A full and complete knowledge of the applicable rules and regulations should be known early in the process to avoid potential design conflicts. This area is highlighted in Section 5.

A wise man always counts the cost of a project prior to committing his financial assets. Section 6 details some typical estimating patterns used in civil engineering projects to assist the reader, as well as some basic tables to aid in calculating costs. Bear in mind that second only to protecting the health and safety of the public, the civil engineer should be morally bound to providing their client with the most cost-effective and environmentally sound project.

Design

After the conceptual planning, financial arrangements and the client's notice to proceed, the engineer can commence the design phase. This segment of the project is certainly the most detailed and technically intense.

Preparation of the construction plans and accompanying specifications must be controlled and continuously supervised by a licensed design professional. While there can be a wide variety of technical personnel involved with the preparation of the plans and specifications, it is the responsibility of the supervising engineer to provide the final documents. The final plans and specifications generally require the supervising design professional to seal, date and sign the documents which are then submitted to the applicable regulatory agencies for approval.

The general format and sheet arrangement of plans vary between types of projects, regulatory agency requirements and the preference of each respective design firm. However, the title page of most plans have some basic parameters. Section 7 of this sourcebook attempts to describe some typical examples.

Sections 8 through 20 contain examples of actual plans to provide the reader with a broad range of projects from which to get the feel of plan arrangement. These reduced plans are from files that date back over the years. Most were hand drafted, but a few were prepared with CAD. There was not an attempt made to standardize the layouts because the editor-in-chief wanted to provide the readers with as wide a variety of layouts and drafting techniques as possible. CAD certainly has become the preferred method of drafting, but original plans of existing projects that predate use of computers still must be consulted. Therefore, the young engineer should be somewhat attuned to prior practices. The reader is encouraged to review these examples in hope of discovering techniques and graphic styles that could be incorporated into future plan preparation.

Section 21 illustrates some examples of typical, standard details that are often used in similar projects.

Section 22 describes briefly the process of preparing technical specifications for construction projects. These written documents are designed to further describe the proposed project in a form that drafting can not fully detail. While a picture can be worth a thousand words, sometimes a word or phrase can describe a detail that can not be adequately drawn. Therefore, the plans and specifications must complement each other to ensure a complete understanding of the proposed project.

Bidding Process

Following completion of the construction plans and specifications, the regulatory approval process begins. Since the regulatory agencies are so numerous and varied in their areas of authority, no attempt is made in this sourcebook to list or categorize them. It is sufficient to say that it is the responsibility of the design professional to be fully aware of the applicable codes, rules, regulations and required approvals needed for each respective project.

Normally, following the receipt of all necessary approvals and signed easements, the client authorizes the engineer to commence the bidding process. Each proposed project has its own special parameters. Some may have state or federal wage rate requirements, special minority regulations, federal nondiscrimination clauses, etc. All these special provisions and regulations must be incorporated into the bidding documents that accompany the plans and specifications. Section 23 address this phase and provides examples of generic documents, as well as discussion of the bid proposal.

After the bidding packet; i.e., bidding documents, plans and specifications have been assembled, the advertisement for bids is prepared and properly published in the local newspapers and other applicable publications such as trade journals, builders exchanges, etc. (see Section 24). At this time it is customary to send bidding packets to the client, builders exchanges and appropriate contractors, sub-contractors and suppliers. During the period from first publication date of the Ad for Bids until the bid opening, the engineer must be reasonably available to respond to questions from prospective bidders (contractors) concerning any item in the bidding documents, plans and specifications. Sometimes a pre-bid conference is held to answer questions from contractors. If a question or concern is significant, the engineer prepares a written addendum and issues it to all plan holders, prior to the bid opening. This document is intended to clarify any misunderstandings or correct any discovered error well in advance of bid opening so all prospective bidders are on a level and fair playing field.

When the time arrives for the bid opening, it is customary for the client or the engineer to declare the bidding "closed". This is a formal process and should be a public event with an attendance sign-up sheet to document who were present. All bids must be delivered in a sealed envelope prior to the time of closing and should be initialed by the receiving clerk. Following the bid closing, the engineer and client open each bid in order of receipt and read aloud each respective bidder's proposal.

Following the reading, those in the audience are excused and the engineer is assigned the task of reviewing each bid proposal in detail. Checking the math, bid bond and other applicable documents required. Usually, within a few days after bid opening, the engineer issues an official bid tabulation to all plan holders and the client. At this point in time, the engineer makes a recommendation to the client. If the low bidder has a bona fide bid bond, reasonably balanced bid, good references and has complied with all specific bidding requirements, he is recommended by the engineer. If the client agrees, a Notice of Award is issued to the successful contractor, usually through the engineer. Within a ten to fifteen day period, the contractor supplies a Performance Bond, Payment Bond, Insurance Certificate and other relevant contract documents to the engineer who assembles them into the contract for the client's attorney to review. Following the attorney's certification or the contract, the engineer notifies the contractor and sets a date for the Pre-Construction Conference. Refer to Section 25 for construction contracts.

Construction

The construction phase is where the project finally takes physical form. After those long weeks, months and generally years of effort, the project comes into existence.

Section 26 describes the general format of the Pre-Construction Conference. At this meeting, all players are formally assembled, i.e., owner's representatives, engineer, inspector, contractor, sub-contractors, utility representatives, and other interested parties. The construction contract is formally issued to the contractor with a Notice to Proceed. The plans and specifications are reviewed and a schedule of construction is developed. Chain of authority is discussed and a timetable for various items formatted. Installation of the project sign can be executed at this time and a groundbreaking ceremony sometimes occurs.

Within two to three weeks, shop drawings begin to arrive at the engineer's office for review, comments and approval. Shop drawings are normally required from manufacturers and job shops prior to fabrication. It is the engineer's responsibility to review these drawings to insure they conform to the project plans and specifications. These components are items that are not built at the project site where an inspector can view their construction. Examples of some of these components are: manholes, pumps, electrical panels, structural steel, specific equipment, control units, pre-stress concrete, steel tanks, cabinetry, doors, windows, roof trusses, special hardware, etc. See Section 27 for some exhibits of actual shop drawings.

Safety at the construction site is a major item. It is usually discussed at the pre-construction conference, but should be a continuous area of concern of all people connected to the construction phase. Some contractors have weekly meetings with their personnel to reinforce the need for a safety first mentality among its personnel. Section 28 address some of the facets of construction safety.

The project inspector is the engineer's representative on the site and is responsible for visually inspecting construction. It is mandatory that the inspector keep accurate, detailed, daily written logs of the project. Sketches and photographs of key elements of construction are also paramount for the historical record. While the inspector is not responsible for the work, it is essential that flaws and poor quality workmanship be brought to the engineer's attention as soon as possible so corrective measures can be taken in a timely manner. A good contractor will cooperate with a reasonable inspector to insure that the plans and specifications are followed. Section 29 describes some of the duties of an inspector.

Testing of materials, soils, concrete and systems of a project generally come under the tasks of the inspector. Collecting of concrete samples and other material can be coordinated and cataloged by the inspector. Testing of water systems and similar piping networks is usually executed by the engineer with assistance of the inspector.

Construction staking of the project is performed by the engineer and inspector or by the engineer's survey crew. The inspector normally accompanies the survey crew so he knows the method of staking and has control over the issuance of grade and/or cut sheets to the construction foreman. Refer to Section 30 for further discussion.

The project close-out can be time-consuming and intense. The last 5% of a project is the toughest. The contractor and his subcontractors are eager to move on to their next project and have minimal crew personnel on site. At this point in time, the engineer, with the assistance of the resident inspector, are assigned the task of reviewing the completed project from stem to stern to insure to the client (owner) that the construction, in deed, complies with the plans and specifications, as well as meets all of the requirements of the regulatory approvals. The document that evolves from this process is known as the "Punch List". All unfinished items, flaws, problems, etc. found during inspection of the project when it is at substantial completion are listed on the punch list, which is officially delivered to the prime contractor for his execution. All of the items on the punch list must be satisfied prior to the engineer certifying to project completion. In addition, the contractor must submit lien waivers to the owner before final payment is made. Section 31 contains generic examples of these documents.

Supplemental

The last section in this sourcebook is entitled, Technical Reference. Its contents include information on a wide variety of technical data that is not normally contained in broad based engineering handbooks. This material is condensed for easy usage and avoids lengthy discussion.

The final eight pages of the section are devoted to metric measurement. Basic SI units are described followed by detailed descriptions and conversion factors.

It is hoped that the readers add this reference material to their personal technical files or use this data to commence developing one.

Planning

TECHNICAL REPORTS

Section 1

The dictionary defines a report as an account of some topic specifically investigated or an official statement of facts.

Written technical reports are prepared and presented to inform the reader on a matter which has a special or specific subject.

This section of the sourcebook will briefly discuss a variety of technical reports and provide typical outline examples.

TABLE OF CONTENTS

General Overview

There are several excellent books in circulation that discuss, in detail, the many facets and stages of technical writing. This sourcebook will not attempt to compete with them. However, our intent is to briefly summarize the process and provide, where applicable, some helpful hints and sample tables of contents from actual projects.

It has been stated, as a communicator, mankind is lazy and very inefficient. While natively blessed with a highly sophisticated communication system, we consistently fail to use it properly. This biological communication system consists of a transmitter and receiver coupled into a single unit that is controlled by a computer; all housed within each of our respective heads. Transmission is delivered in three forms; i.e., physical action, speech and writing (in many forms). It is the latter form that this section will briefly discuss.

Whatever the topic of the technical report, consideration for the reader or readers is of supreme importance. The reader must be able to understand the contents of the report and its intended purpose. All facts must be accurate and well documented. Conclusions must be soundly based on those facts and recommendations founded on sound, professional judgement, supported by the logical evaluation of those relevant facts.

There are many facets that enter into the preparation of a useful and well-executed technical report. While it is impractical to list all of them, this writer will mention the key elements that have been of value throughout the years. They are:

1 - Know your subject

2 - Prepare an outline

3 - Document all facts and organize them in logical sequence

4 - Develop a rough draft

5 - Be brief and to the point

6 - Assemble applicable documents as supporting exhibits

7 - Acquire top quality illustrations, charts, maps and photographs

8 - Fine tune

9 - Proofread

10 - Insist on good word processing; i.e., text style, page design and heading highlighting

11 - Final read and pier review

12 - Professional cover and binder

It must be noted, for the record, that technical reports must be grammatically correct and follow accepted writing principles. However, space and purpose does not allow inclusion of this material. The Editor-in-Chief strongly advises that the reader obtain a good technical writing text book for reference.

Engineering Reports

Engineering reports are the primary documents that launch projects. Therefore, they must be properly organized, technically accurate and have a logical sequence to fully demonstrate their proposal. However, the document must be flexible in its format to allow reasonable modification as the project evolves to fruition. Since most times the Engineering Report is the main focus of a public meeting or official hearing, it is prudent to have the final draft reviewed by qualified associates. Additionally, the Engineering Report is a key component of submittals to regulatory agencies for plan review and their approval.

A typical example of the Table of Content of a Engineering Report follows.

M & P Project No. 01-115

-ENGINEERING REPORT-
for
MUNICIPAL UTILITY REHABILITATION PROJECT
W. THIRD AVENUE

Village
of
Sheldon, Wisconsin
February, 2002

Prepared by:

MORGAN & PARMLEY, LTD.
115 West 2nd Street, South
Ladysmith, Wisconsin 54848

Source: Morgan & Parmley, Ltd.

-TABLE OF CONTENTS-

-LIST OF ILLUSTRATIONS-

-LIST OF EXHIBITS-

M & P Project No. 93-159

ENGINEERING REPORT
for
PROPOSED
MUNICIPAL WATER SYSTEM
Village
of
Tony, Wisconsin
February, 1994

prepared by:

MORGAN & PARMLEY LTD.
Professional Consulting Engineers
115 West 2nd Street South
Ladysmith, Wisconsin 54848

-TABLE OF CONTENTS-

-i-

Source: Morgan & Parmley, Ltd.

-ii-

-LIST OF ILLUSTRATIONS-

-LIST OF TABLES-

-iii-

-LIST OF REFERENCE EXHIBITS-

-ABBREVIATIONS USED IN REPORT-

SYMBOL		DESCRIPTION
DNR	-	Department of Natural Resources
Fe	-	Iron
FmHA	-	Farmers Home Administration
gpcd	-	gallons per capita per day
gpd	-	gallons per day
gpm	-	gallons per minute
ISO	-	Insurance Services Office of Wisconsin
MG	-	million gallons
MGD	-	million gallons per day
mg/l	-	milligrams per liter
Mn	-	Manganense
ppm	-	Parts per Million
PSC	-	Public Service Commission of Wisconsin
psi	-	pounds per square inch (gauge)
USGS	-	U.S. Geological Survey (All elevations referred to in this report are USGS)
VOCs	-	Volatile Organic Compounds

Preliminary Reports

Preliminary Reports are just that; they are "preliminary." They come into play when a final report can not be written because all of the facts, testing and/or parameters are not currently available. Therefore, final conclusions and recommendations are held in abeyance, however, the subject can be discussed at the level required. Generally, preliminary reports are utilized to keep a proposed project flowing by providing concepts and details to the appropriate audience and thus avoid delays in the process. The following Table of Contents is presented for guidance. Note that the general format is similar to a final engineering report, with some supportive exhibits and documentation absent.

M & P Project No. 90-139

PRELIMINARY
-ENGINEERING REPORT-

MUNICIPAL STREET IMPROVEMENT PROJECT
(PHASE ONE)

Village
of
Gilman, Wisconsin

JAN 8 1992

prepared by:

MORGAN & PARMLEY LTD.
Professional Consulting Engineers
115 West 2nd Street South
Ladysmith, Wisconsin 54848

Source: Morgan & Parmley, Ltd.

TABLE OF CONTENTS

LIST OF ILLUSTRATIONS

-iii-

LIST OF TABLES

LIST OF EXHIBITS

-iv-

Feasibility Studies

A Feasibility Study or Evaluation Report begins with an idea or concept and then development occurs, followed by an analysis to assess whether it is technically and/or economically sound.

The feasibility document may be either a letter or a full-blown formal report. When a letter is chosen, the project is usually small and the audience (individual or committee) has adequate background to quickly assimilate the material. The formal feasibility study is used for more complex or comprehensive projects. A sample Table of Contents for the formal presentation follows.

M & P Project No. 90-123

ENGINEERING REPORT

-PRELIMINARY FEASIBILE CONCEPT-
for
COMPOSTING FACILITY

City
of
Ladysmith, Wisconsin
October 15, 1990

prepared by:

MORGAN & PARMLEY, LTD.
Professional Consulting Engineers
115 West 2nd Street South
Ladysmith, Wisconsin 54848

Source: Morgan & Parmley, Ltd.

-TABLE OF CONTENTS-

-i-

-LISTING OF ILLUSTRATIONS-

-LISTINGS OF EXHIBITS-

-ii-

MAJOR TECHNICAL REFERENCES

1-Applicable sections of Wisconsin Administrative Code.

2-Final Report: Wisconsin Co-Composting Demonstration Study, by Dr. Aga S. Razvi (March, 1987)

3-The Utilization of Solid Waste Composts, Co-Composts, and Shredded Refuse on Agricultural Lands (Literature Review)

4-Solid Waste Composting in the U.S., by Nora Goldstein, Bio-Cycle-Nov., 1989

5-Solid Waste Composting Facilities, by Goldstein & Spencer, BioCycle-Jan. 1990

6-Fillmore County, Minn. Resource Recovery Center, by Tim. L. Goodman, Sept. 29, 1989

7-Portions of: Pollution Control Agency Solid Waste Management Rules-Printed Jan., 1989

8-Wisconsin Act 335-The Recycling Law-DNR Publ-IE-041 Rev 6/90

9-Grants Provided by the Recycling Law-Published by DNR, Publ-IE-046-90 Rev.

10-Accounting for and Spending Recycling Grants-Published by DNR, Publ-IE-048-90 REV

11-Forming Responsible Units-Published by DNR, PUBL-IE 044-90 REV 8/24/90

12-How New Units Can Receive Recycling Grants, Published by DNR, Publ-IE-047-90 REV

13-Small Town Designs for Big Impact, by Mark Selby & Joe Carruth, BioCycle, Aug., 1989

14-Solid Waste Management, by D. Joseph Hagerty, et al, Van Norstrand Reinhold 1973

15-Environmental Engineers' Handbook, by Liptak, editor, Chilton Book Co., 1974

Facility Plan Documents

A Facility Plan is an all-inclusive, generally massive, document that is used for major projects such as a new municipal wastewater treatment facility. It is the paramount purpose of a Facility Plan to provide a solution to an existing environmental problem by recommending a solution that is not only environmentally sound, but the most cost-effective.

This is without a doubt the most costly, time consuming and technically demanding report to prepare. The multitude of areas that must be properly addressed and the ultimate exposure to public hearings, regulatory agencies' reviews and challenges from special interest groups places this at the top of technical report writing. Typically, this document can take 18 to 24 months to finalize.

The following Table of Contents, extracted from an actual Facility Plan, reveals the many facets that are required to be incorporated into the final document that can be accepted by all parties. The following pages reveal the complex and varied tasks needed to successfully complete the document.

M & P Project No. 96-113

-AMENDMENT-
to
FACILITIES PLAN
for
MUNICIPAL WASTEWATER TREATMENT FACILITY

Village
of
Sheldon, Wisconsin
July, 1997

Prepared by:

MORGAN & PARMLEY, LTD.
Professional Consulting Engineers
115 West 2nd Street, South
Ladysmith, Wisconsin 54848

Source: Morgan & Parmley, Ltd.

-TABLE OF CONTENTS-

-LIST OF ILLUSTRATIONS-

-LIST OF TABLES-

Sewer Use Ordinance

While a Sewer Use Ordinance is not truly a technical report, it was felt that some discussion is warranted in this sourcebook to fully round out our coverage.

The Sewer Use Ordinance (SUO) is a requirement of most state regulatory agencies that have authority over municipal wastewater collection and treatment facilities (WWCTF). This document contains rules and regulations that govern the local sanitary sewer utility. In addition, this document, or ordinance, contains a User Charge System (UCS) which establishes the financial operation; i.e., user fees, operation & maintenance costs and related expenses. Refer to the following Table of Contents of a typical format.

M & P Project No. 98-109

-MUNICIPAL WWCTF-
SEWER USE ORDINANCE
and
USER CHARGE SYSTEM

Village
of
Sheldon, Wisconsin
May, 2001

Prepared by:

MORGAN & PARMLEY, LTD.
Professional Consulting Engineers
115 West 2nd Street, South
Ladysmith, Wisconsin 54848

Source: Morgan & Parmley, Ltd.

TABLE OF CONTENTS

– i –

– ii –

Wellhead Protection Plan

The Wellhead Protection Plan (WHPP) is a relatively recent document promoted by the US/EPA and mandated by most states. This document's purpose is an attempt to protect the potable water sources of public wells by regulating environmental activity within the surrounding area of active wells.

As with the proceeding topic, it was felt by the Editor-in-Chief that a brief discussion of the WHPP document was in order. The limited quantity of good quality drinking water is a very serious threat to our population, and it must be protected. Increased demands, both private and industrial, certainly make this a major priority. Therefore, the reader is directed to the following Table of Contents of a recent WHPP for a brief review of its topics.

M & P Project No. 96-172

-WELLHEAD PROTECTION PLAN-

Village
of
Gilman, Wisconsin
April, 1998

Prepared by:

MORGAN & PARMLEY, LTD.
Professional Consulting Engineers
115 West 2nd Street, South
Ladysmith, Wisconsin 54848

Source: Morgan & Parmley, Ltd.

TABLE OF CONTENTS

-ILLUSTRATIONS-

-APPENDIX-

Planning

PROJECT SCHEDULING

Section 2

In order for an engineer to be successful, it is essential to be well organized. An overall master schedule is mandatory in every project. Without one, the project will suffer and result in a crippled effort.

TABLE OF CONTENTS

The Project Engineer

As previously stated, project engineers must be well organized to be successful. In addition, they must be technically competent and very disciplined. Incorrect decisions and improper actions can, in most cases, prove extremely costly and adversely affect the project progress and jeopardize the budget.

A good project engineer should never lose self-control and be continuously in charge of the project. Without a well-designed schedule, the project engineer can not steer a logical course. Therefore, from square one, the engineer must begin to outline the project schedule and formulate the sequence of key elements.

Throughout the project the engineer must, of course, interact with a wide variety of technical personnel, construction tradesmen, utility representatives, regulatory staff, environmental groups, elected officials, private citizens, media personnel, legal authorities, public exposure, regulatory inspectors, accounting agents, grant administration, political officials, and subordinate engineering staff within the consulting engineering firm. Certainly, in order to be an effective project engineer, one must have good communication skills and leadership qualities to successfully complete a project. Time does not allow, nor is it the primary purpose, to list all of the other non-engineering characteristics that are molded into a top quality project engineer. However, it should be noted that these qualities are blended and mixed from an understanding of business, professional ethics, moral fiber, experience, devotion to serve, goal orientation and a genuine respect for the profession.

In summary, a proficient project engineer must possess a wide range of skills, but without following a well-designed schedule, a project will not be successfully executed.

Event Calendar–Summary of Basic Tasks

Item No.	Description	Projected Dates	Notes
i	Define Project Scope		
ii	Municipality Authorizes Engineer to Begin		
1	Initial Conference: Municipality/Engineer		
2	Research Records & Files		
3	Field Survey & Reconnaissance		
4	Video Photograph Project Site (prior to Commencing Construction)		
5	Buried Utilities Located & Flagged		
6	Research Property Survey Records		
7	Soil Borings, Test Drilling & Subsurface Investigation		
8	Public Works Committee Meeting-Refine Project Scope		
9	Reduce Field Notes		
10	Preliminary Budget-Construction Cost Estimate		
11	Utility Committee Meeting-Construction Cost Estimate		
12	Finalize Budget-Construction Cost Estimate		
13	Preliminary Plans & Specifications		
14	Municipality Meeting – Review & Approve Prel. P & S		
15	Easement Survey & Legal Documents (If Req'd)		
16	Final Plans & Specifications		
17	Engineering Report (Summary of Work)		
18	Proposal & Bidding Documents		
19	Obtain Concurrence from Municipality for Final P & S		
20	Submit Plans & Specifications to Regulatory Agencies		
21	Obtain Approvals from Regulatory Agencies:		
22	DNR (If Req'd.)		
23	PSC (If Req'd.)		
24	DOT (If Req'd.)		
25	Corps of Engineers (If Req'd.)		
26	Private Utility Company Approvals:		
27	Cable TV		
28	Electrical & Gas		
29	Telephone		
30	Public Hearing (Assessments to Property Owners) (If Req'd.)		

Source: Morgan & Parmley, Ltd.

Item No.	Description	Projected Dates	Notes
31	Organize Bidding Process (Commence)		
32	Obtain Wage Rates (State and/or Federal)		
33	Advertisement for Bids		
34	Supply Plans & Specs to Builders Exchanges		
35	Send Plans & Specs to Contractors		
36	Pre-Bid Meeting (If Req'd.)		
37	Bid Opening:		
38	Minutes, Bid Tabulation & Analysis of Proposals		
39	Recommendation to Municipality		
40	Municipality gives Notice of Award		
41	Prepare Construction Contract Documents		
42	Secure Municipality Atty's Certification of Construction Contract		
43	Obtain Municipality's Resolution to Sign Construction Contract		
44	Supervise Signing of Construction Contract		
45	Pre-Construction Conference		
46	Municipality to Sign Notice to Proceed		
47	Shop Drawing Review		
48	Construction Staking		
49	General Inspection Services, Daily Log, etc.		
50	Construction Records		
51	Contractor Payment Requests Review & Certification		
52	Status Reports (Periodic)		
53	Inspect Testing of Installation		
54	Final Inspection & Certification		
55	Preparation of Construction Record Drawings		
56	Final Payment Request Review & Close-Out Certification		

Note: The projected dates are targets and may vary, due to conditions beyond the control of the Municipality.

Summary of Engineering Services

(Post Plan Approval)

1. Apply for State Wage Rates
2. Prepare Advertisement for Bids
3. Submit Ad to Ladysmith News & Western Builder
4. Prepare Bidding Documents
5. Print & Bind Plans, Specifications & Bidding Documents
6. Send Plans, Specs. & Bidding Documents to Builders Exchanges
7. Distribute Plans, Specs & Bidding Documents to Prospective Bidders
8. Tabulate Plan Holders List
9. Prepare Addendas (if necessary)
10. Interpret Plans & Specs; i.e. answer technical questions from Prospective Bidders
11. Supervise Bid Opening
12. Prepare Bid Tabulation & Distribute to Plan Holders
13. Review Bids & Investigate Qualifications of Low Bidder
14. Prepare Summary for Village Board
15. Prepare Notice of Award for Village to Execute
16. Prepare Construction Contract
17. Review Construction Contract w/Village Attorney for Certification
18. Coordinate & Manage Pre-Construction Conference
19. Attend Groundbreaking Ceremony
20. Review Shop Drawings
21. Coordinate Plan Submittal for Roof Trusses and Precast Concrete
22. Provide Construction Staking
23. Periodic Inspection Visits to Site @ Key Events
24. Perform Soil & Concrete Sampling & Lab Services
25. Process Change Orders (if required)
26. Review & Certify Pay Requests from Contractor
27. Perform Final Inspection
28. Prepare Punch List
29. Supervise Close-Out Process
30. Obtain Lien Waivers & Consent of Surety for Final Payment
31. Verify that Contractor demonstrate operation of HVAC equipment to Village Personnel
32. Verify that O & M Manuals for Equipment are Supplied
33. Brief Status Reports to Village Board at Regular Meetings

Please Note: This does not include any resident inspector services.

Source: Morgan & Parmley, Ltd.

Planning

FIELD RECONNAISSANCE, SURVEYING & MAPPING

Section 3

Initially, a proposed site must be investigated to determine its general suitability for the intended use. Environmental issues must be resolved. Archeological and historical clearance must be obtained. Wildlife and endangered species impacts must be fully addressed. Safety issues and concerns of citizens must be evaluated. Sensitive environmental areas must be protected. A vegetational survey should be conducted, if applicable. Exploration of existing soils and groundwater should be accomplished to determine suitability for the proposed project.

Following completion of the foregoing elements and proper clearance, the final field reconnaissance, surveying and mapping are executed. This section provides samples of a wide variety of these layouts showing project site maps, topographic mapping, sub-divisions, plats, land title survey, composite layouts, certified survey maps, property survey, investigation identification mapping, municipal mapping, right-of-way mapping, easement mapping, and service area mapping.

TABLE OF CONTENTS

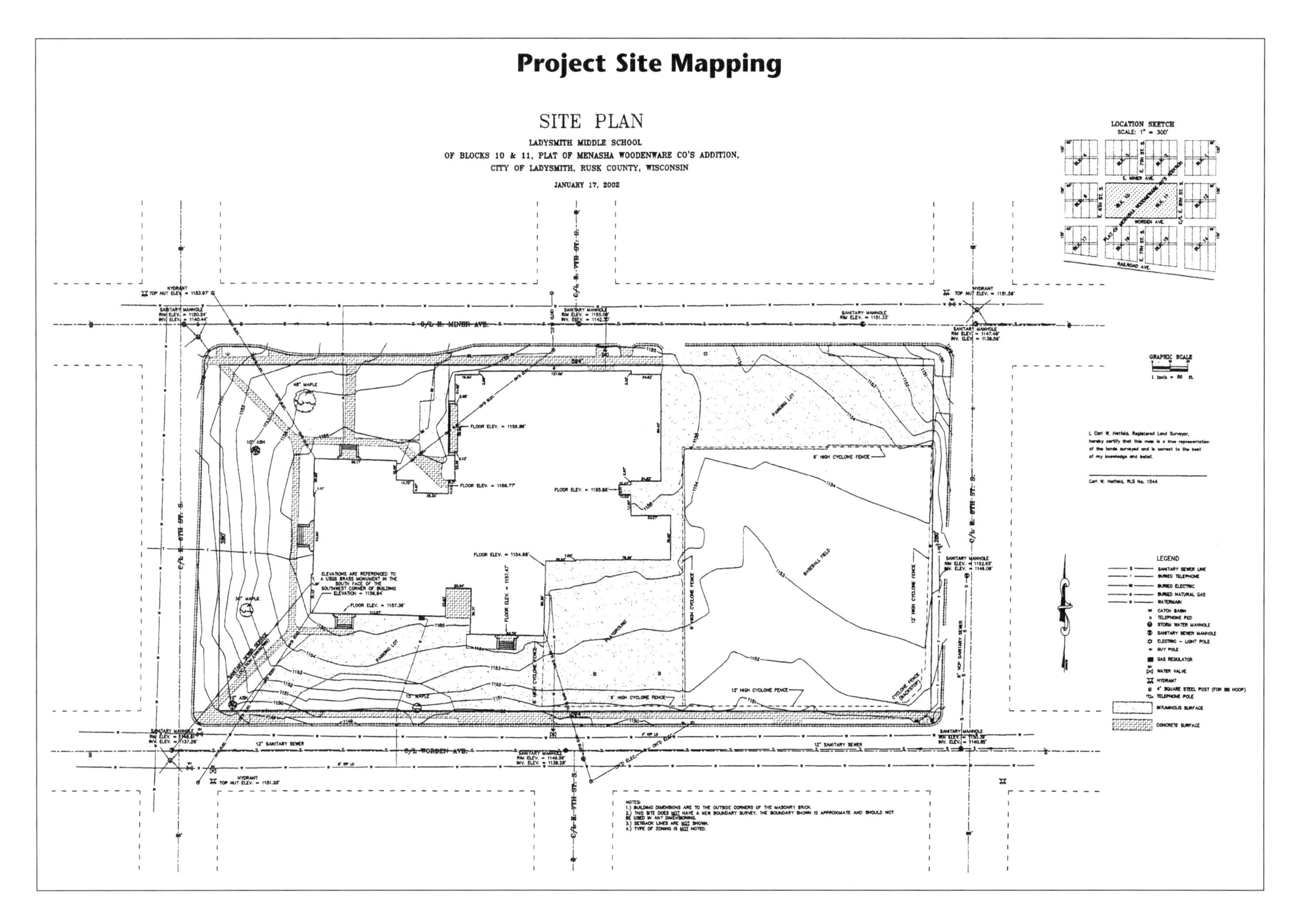

Source: Polk County Land Surveying

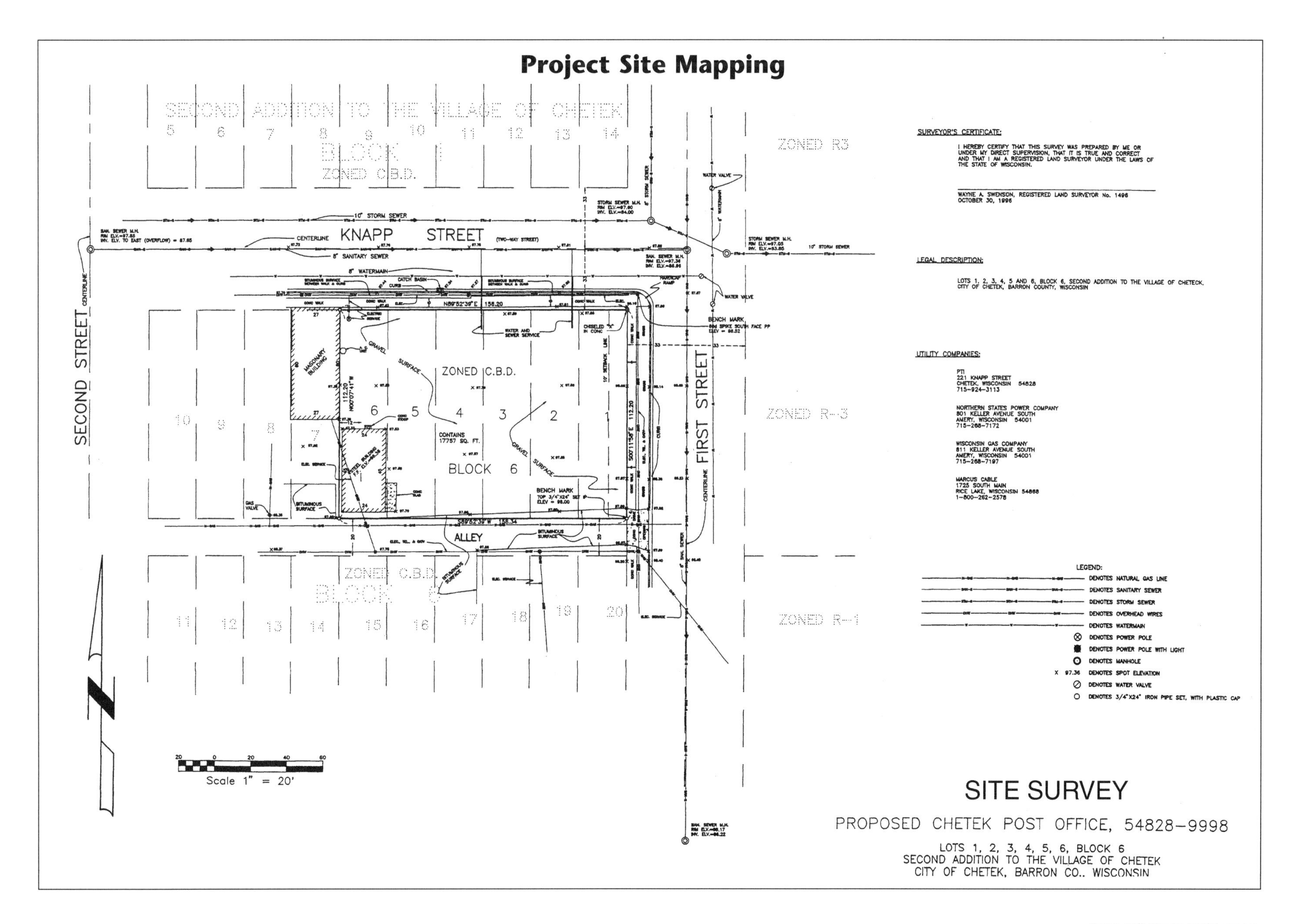

Source: Polk County Land Surveying

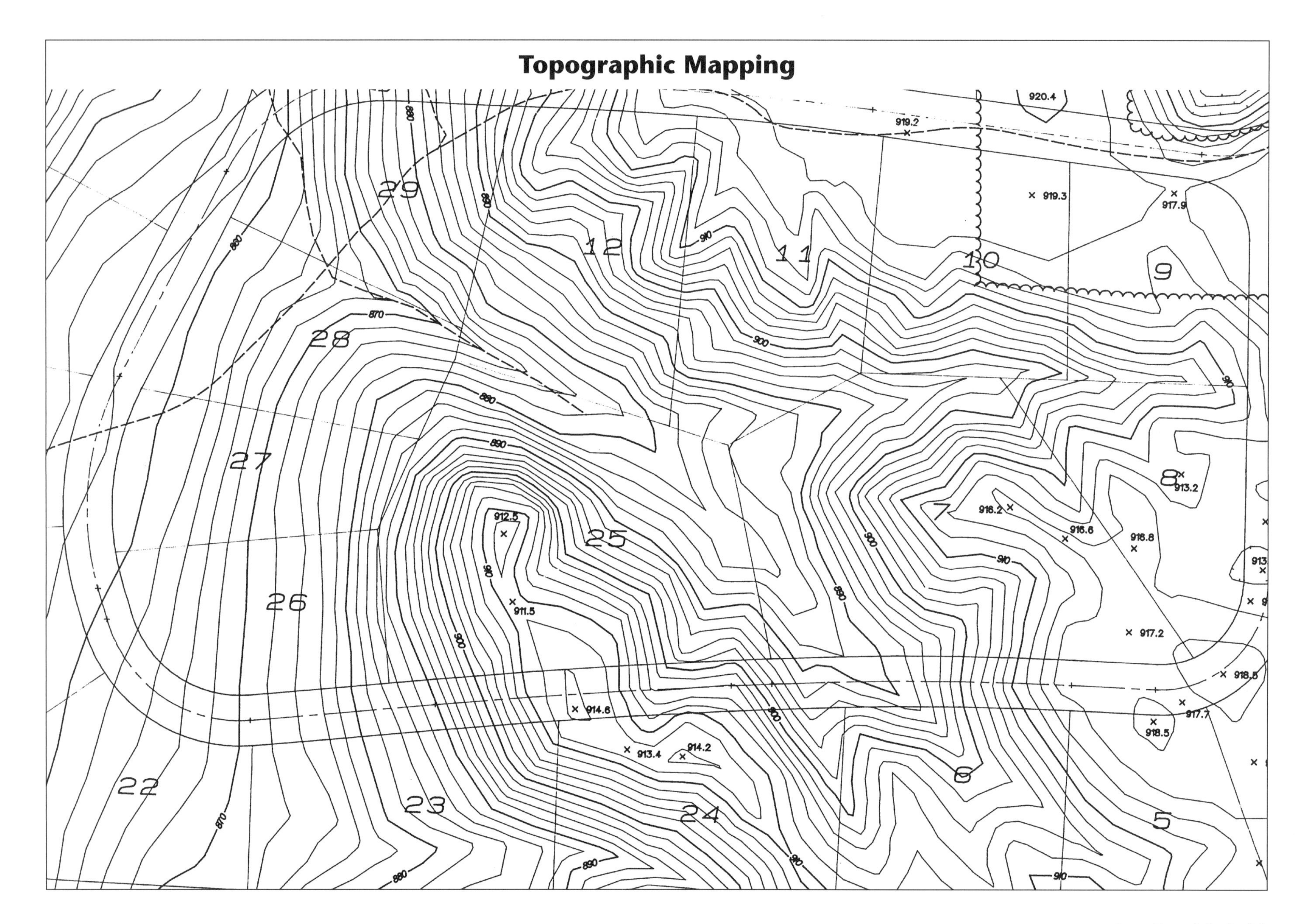

Source: Morgan & Parmley, Ltd.

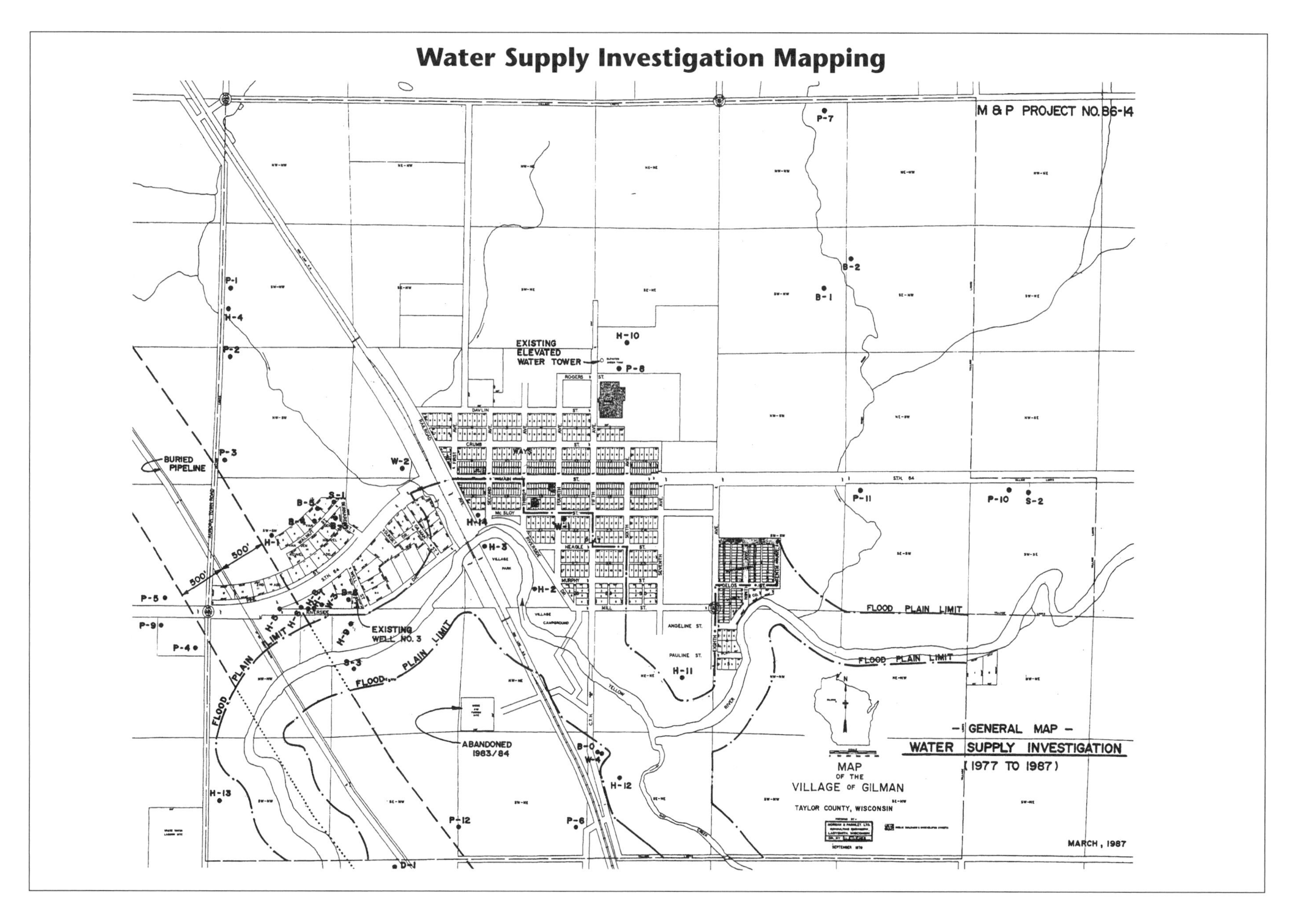

Source: Morgan & Parmley, Ltd.

Municipal Mapping

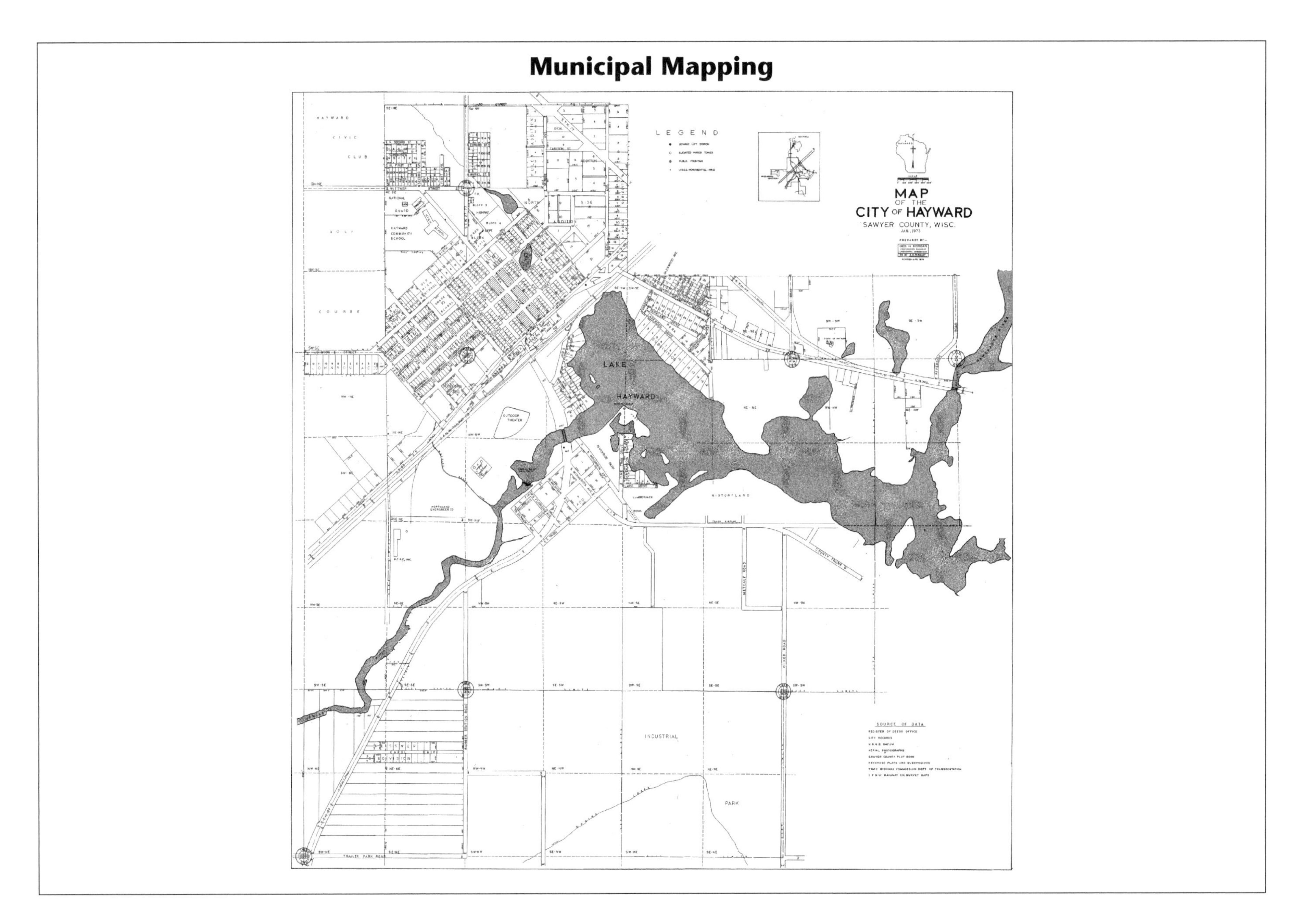

Source: Morgan & Parmley, Ltd.

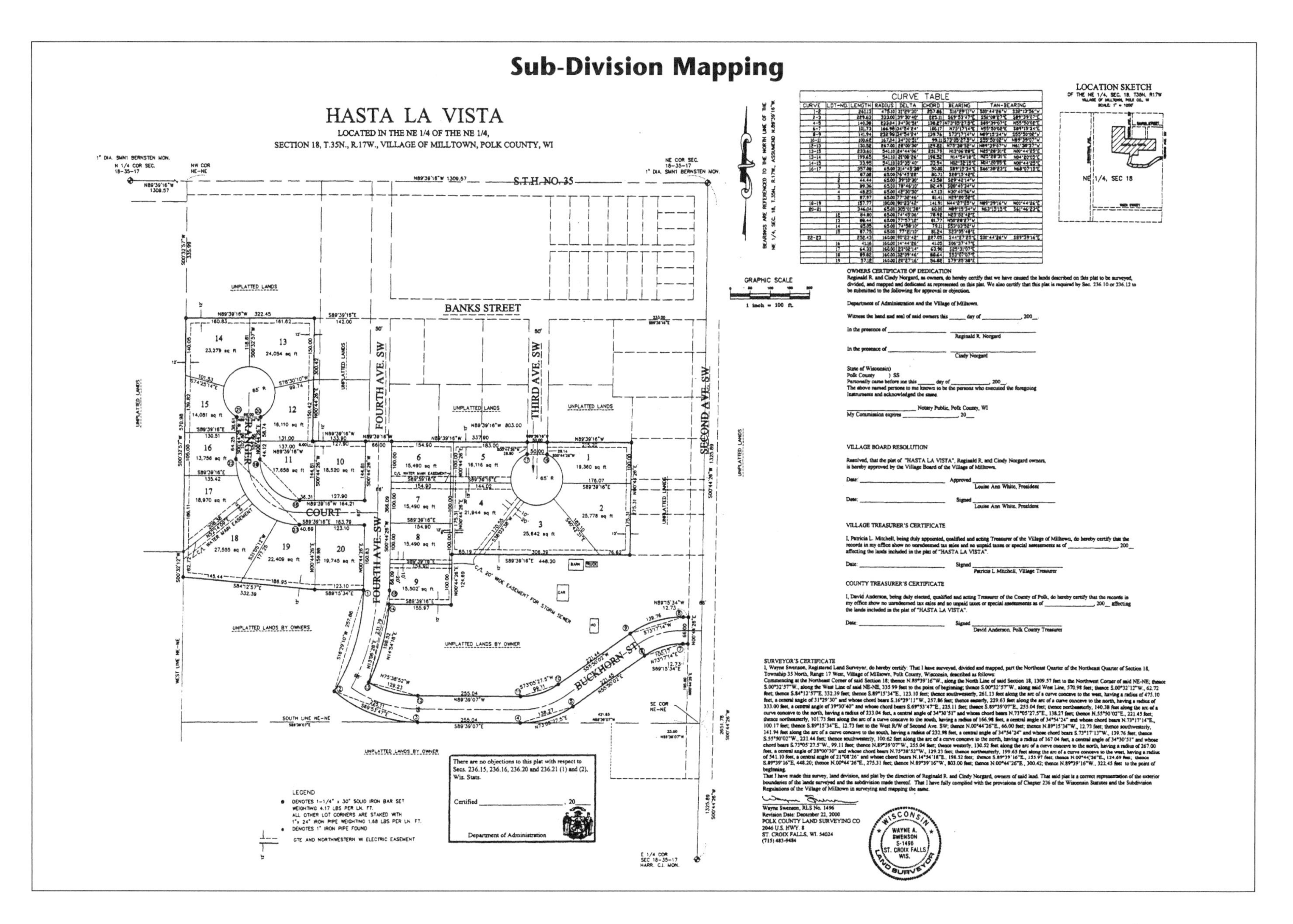

Source: Polk County Land Surveying

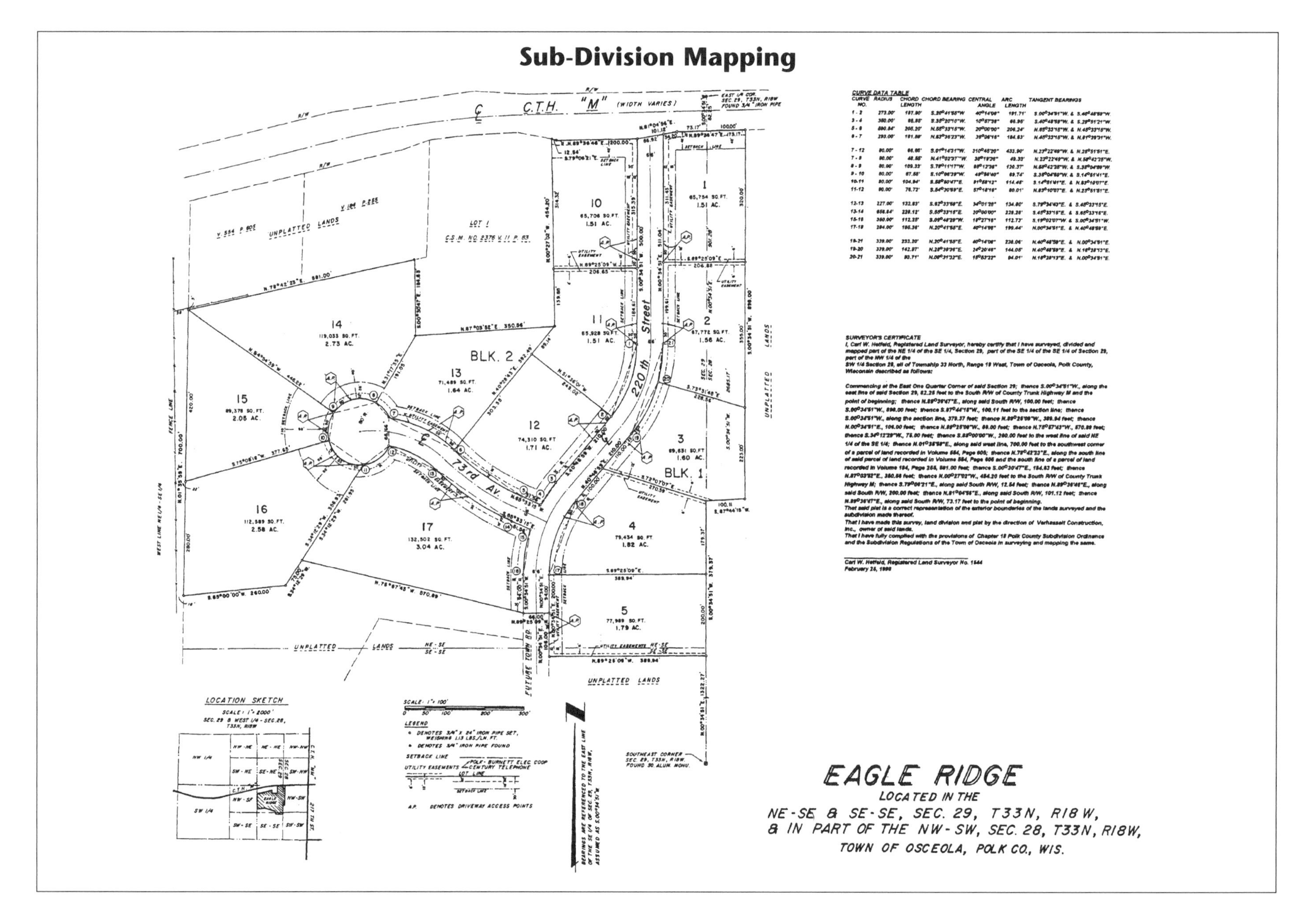

Source: Polk County Land Surveying

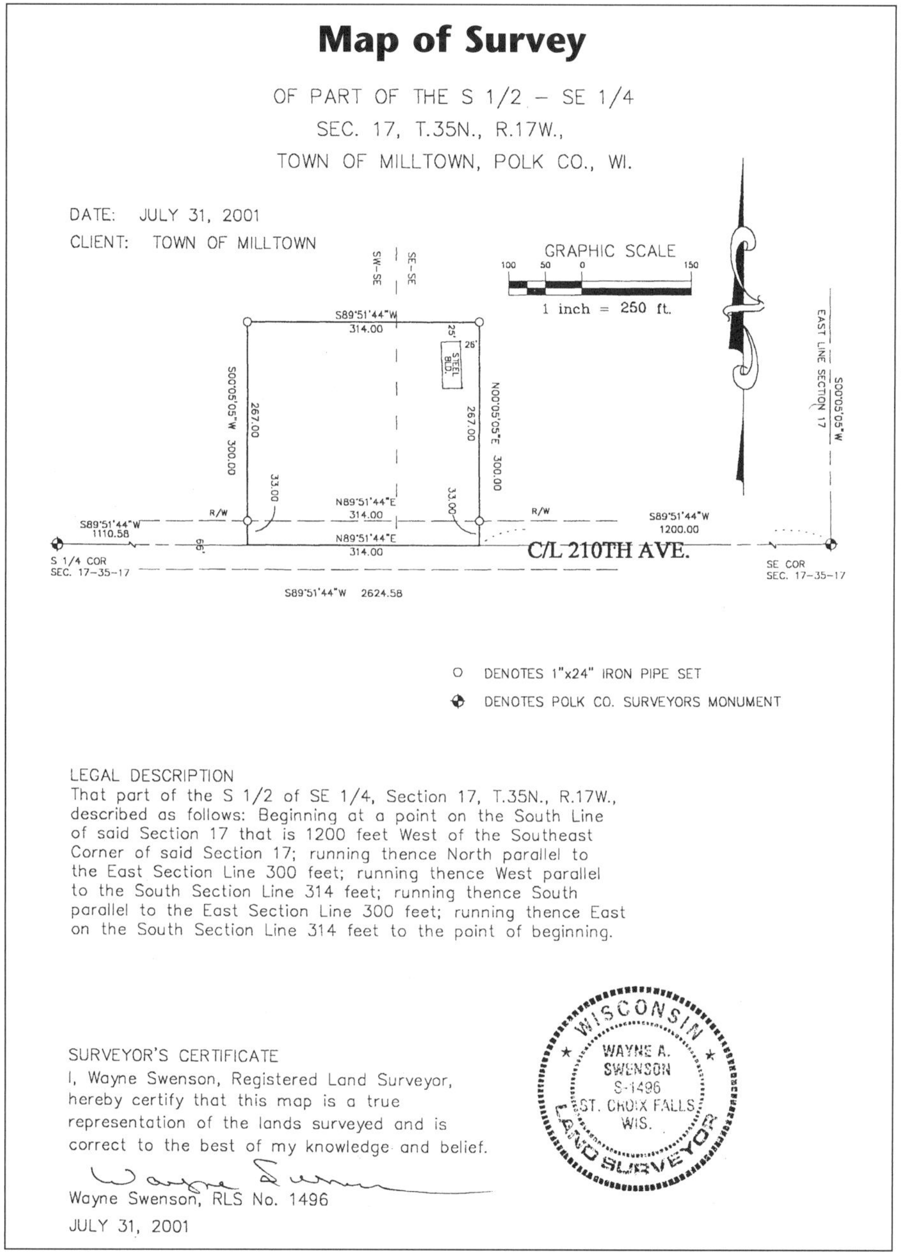

Map of Survey

OF PART OF THE S 1/2 – SE 1/4
SEC. 17, T.35N., R.17W.,
TOWN OF MILLTOWN, POLK CO., WI.

DATE: JULY 31, 2001
CLIENT: TOWN OF MILLTOWN

GRAPHIC SCALE
100 50 0 150
1 inch = 250 ft.

SW-SE | SE-SE

S89°51'44"W 314.00
25' 26' STEEL BLD.
S00°05'05"W 300.00
267.00
N00°05'05"E 300.00
267.00
33.00
33.00
N89°51'44"E 314.00
N89°51'44"E 314.00
R/W
R/W
S89°51'44"W 1110.58
S89°51'44"W 1200.00
66'
S 1/4 COR SEC. 17–35–17
C/L 210TH AVE.
SE COR SEC. 17–35–17
S89°51'44"W 2624.58
EAST LINE SECTION 17
S00°05'05"W

O DENOTES 1"x24" IRON PIPE SET
⊕ DENOTES POLK CO. SURVEYORS MONUMENT

LEGAL DESCRIPTION
That part of the S 1/2 of SE 1/4, Section 17, T.35N., R.17W., described as follows: Beginning at a point on the South Line of said Section 17 that is 1200 feet West of the Southeast Corner of said Section 17; running thence North parallel to the East Section Line 300 feet; running thence West parallel to the South Section Line 314 feet; running thence South parallel to the East Section Line 300 feet; running thence East on the South Section Line 314 feet to the point of beginning.

SURVEYOR'S CERTIFICATE
I, Wayne Swenson, Registered Land Surveyor, hereby certify that this map is a true representation of the lands surveyed and is correct to the best of my knowledge and belief.

Wayne Swenson, RLS No. 1496
JULY 31, 2001

WISCONSIN LAND SURVEYOR
WAYNE A. SWENSON
S-1496
ST. CROIX FALLS
WIS.

Source: Polk County Land Surveying

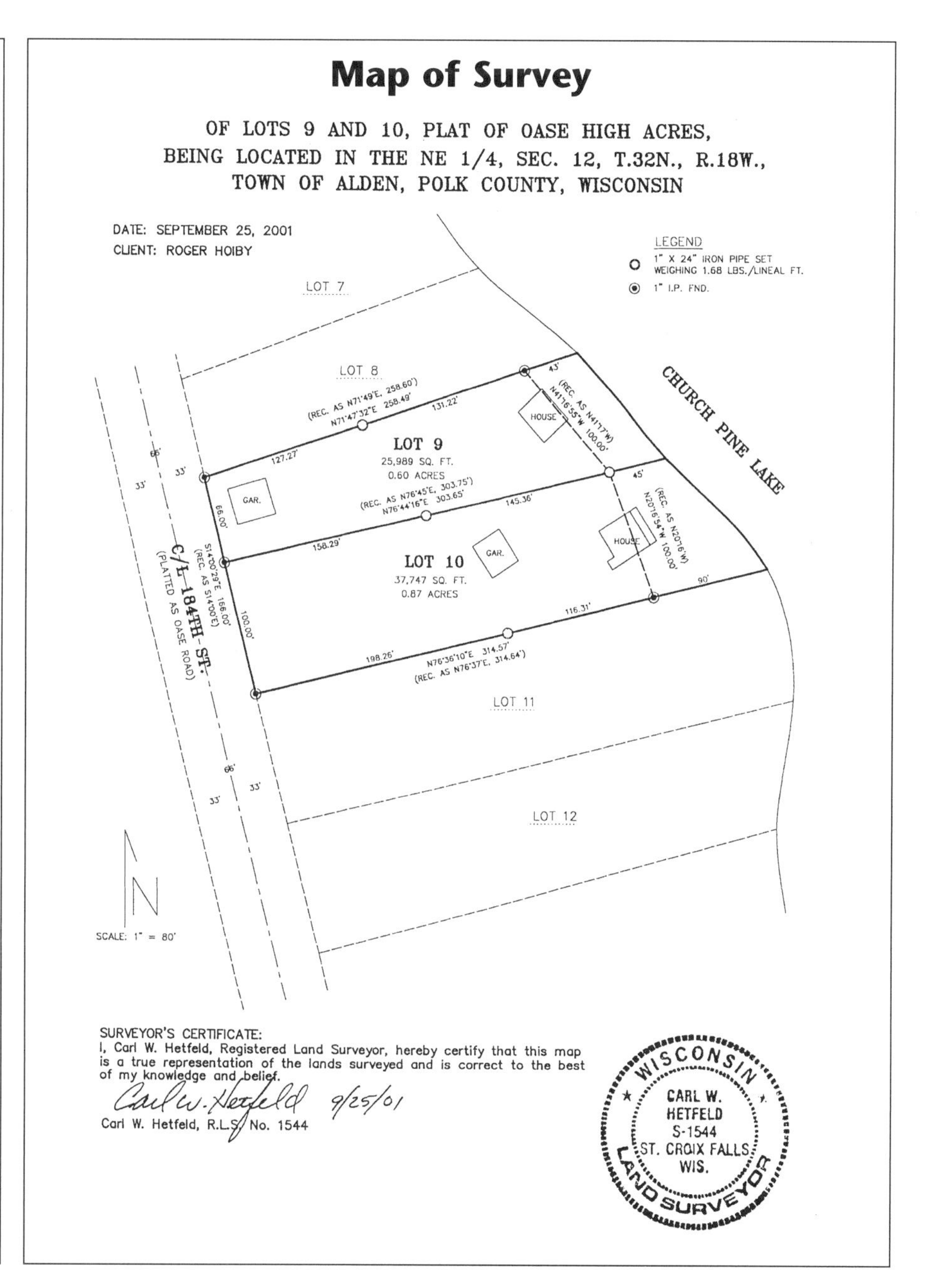

Source: Polk County Land Surveying

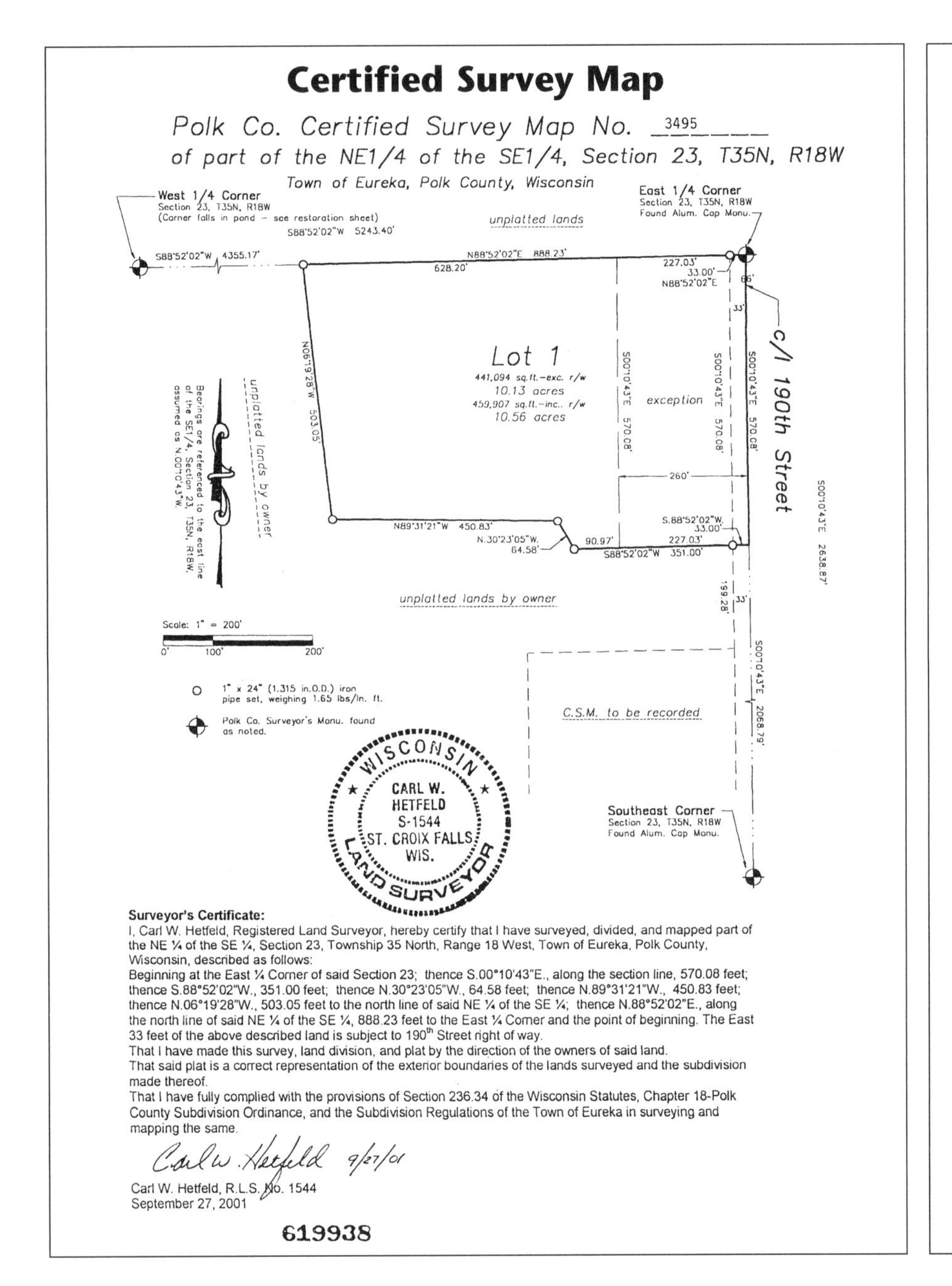

Surveyor's Certificate:

I, Carl W. Hetfeld, Registered Land Surveyor, hereby certify that I have surveyed, divided, and mapped part of the NE ¼ of the SE ¼, Section 23, Township 35 North, Range 18 West, Town of Eureka, Polk County, Wisconsin, described as follows:

Beginning at the East ¼ Corner of said Section 23; thence S.00°10'43"E., along the section line, 570.08 feet; thence S.88°52'02"W., 351.00 feet; thence N.30°23'05"W., 64.58 feet; thence N.89°31'21"W., 450.83 feet; thence N.06°19'28"W., 503.05 feet to the north line of said NE ¼ of the SE ¼; thence N.88°52'02"E., along the north line of said NE ¼ of the SE ¼, 888.23 feet to the East ¼ Corner and the point of beginning. The East 33 feet of the above described land is subject to 190th Street right of way.

That I have made this survey, land division, and plat by the direction of the owners of said land.

That said plat is a correct representation of the exterior boundaries of the lands surveyed and the subdivision made thereof.

That I have fully complied with the provisions of Section 236.34 of the Wisconsin Statutes, Chapter 18-Polk County Subdivision Ordinance, and the Subdivision Regulations of the Town of Eureka in surveying and mapping the same.

Carl W. Hetfeld 9/27/01

Carl W. Hetfeld, R.L.S. No. 1544
September 27, 2001

619938

Source: Polk County Land Surveying

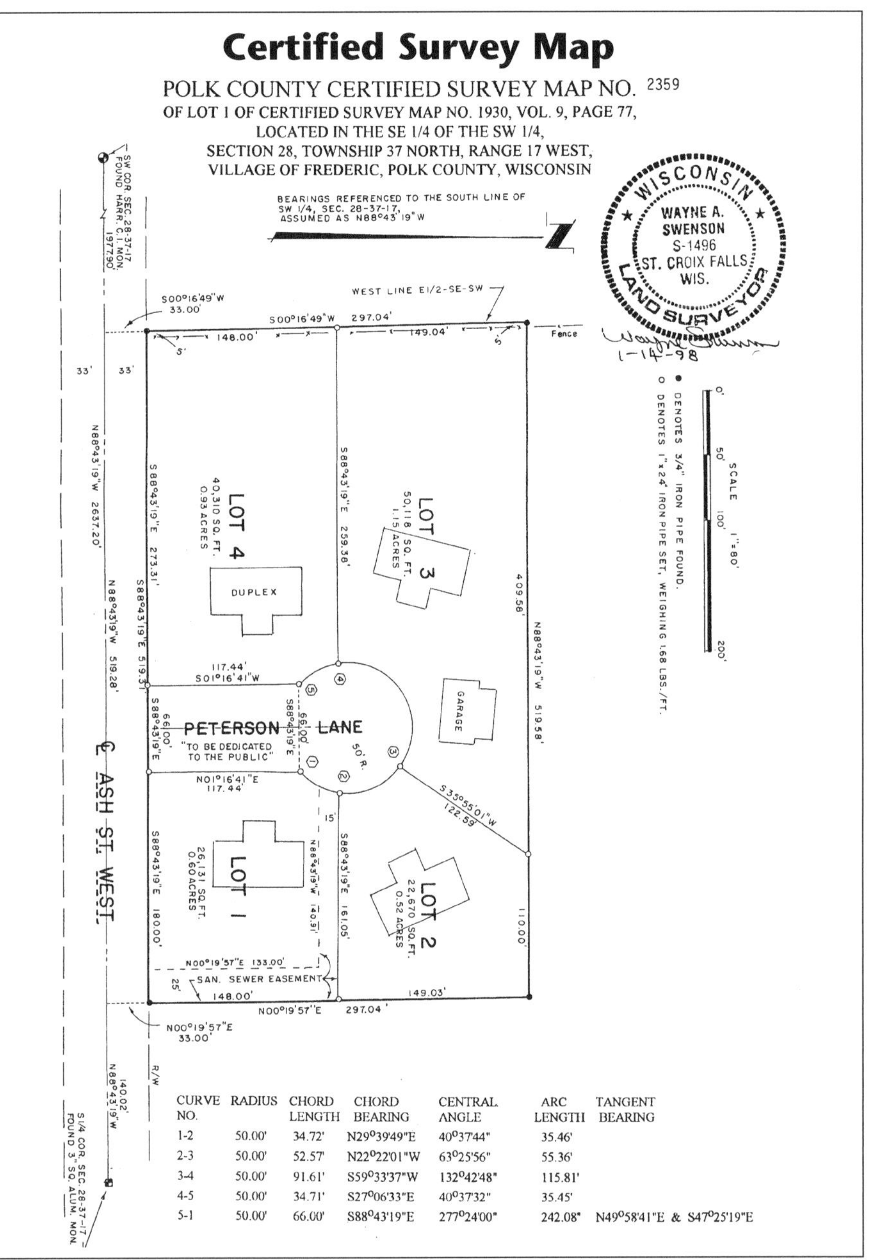

CURVE NO.	RADIUS	CHORD LENGTH	CHORD BEARING	CENTRAL ANGLE	ARC LENGTH	TANGENT BEARING
1-2	50.00'	34.72'	N29°39'49"E	40°37'44"	35.46'	
2-3	50.00'	52.57'	N22°22'01"W	63°25'56"	55.36'	
3-4	50.00'	91.61'	S59°33'37"W	132°42'48"	115.81'	
4-5	50.00'	34.71'	S27°06'33"E	40°37'32"	35.45'	
5-1	50.00'	66.00'	S88°43'19"E	277°24'00"	242.08'	N49°58'41"E & S47°25'19"E

Source: Polk County Land Surveying

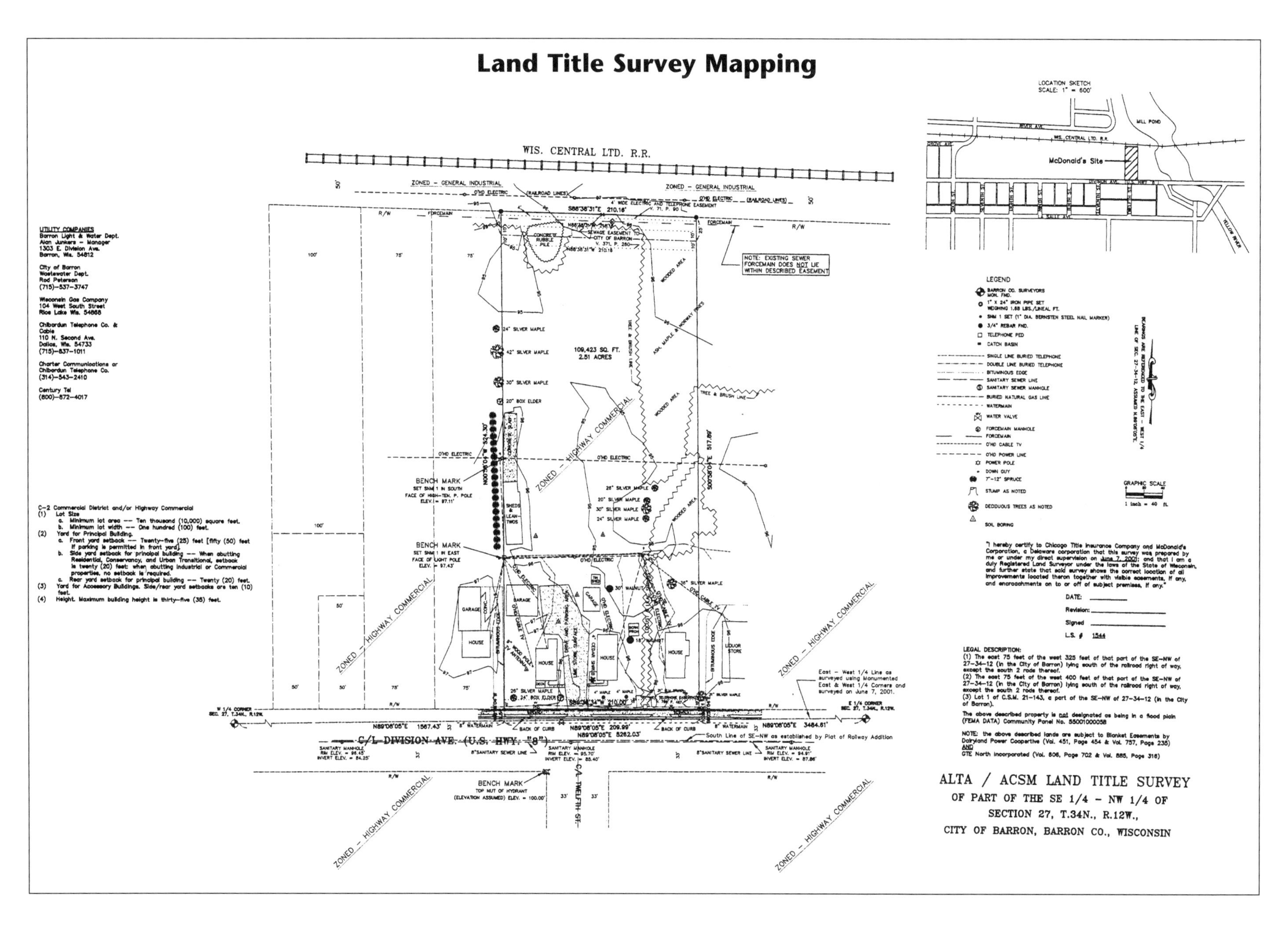

Source: Polk County Land Surveying

Important Facts About Land Descriptions

LAND MEASUREMENTS, TOWNSHIPS, SECTIONS, MEANDERED WATER, GOVERNMENT LOTS, ETC.

WHAT IS A LAND DESCRIPTION?

A land description is a description of a tract of land in legally acceptable terms, so as to show exactly where it is located and how many acres it contains.

TABLE OF LAND MEASUREMENTS

LINEAR MEASURE				SQUARE MEASURE			
1 inch	.0833 foot	16½ feet	1 rod	144 sq. in.	1 sq. ft.	43560 sq. ft.	1 acre
7.92 inches	1 link	5½ yards	1 rod	9 sq. ft.	1 sq. yd.	640 acres	1 sq. mile
12 inches	1 foot	4 rods	100 links	30¼ sq. yds	1 sq. rod	1 sq. mile	1 section
1 vara	33 inches	66 feet	1 chain	16 sq. rods	1 sq. chain	36 sq. miles	1 township
2¾ feet	1 vara	80 chains	1 mile	1 sq. rod	272¼ sq. ft.	6 miles sq.	1 township
3 feet	1 yard	320 rods	1 mile	1 sq. chain	4356 sq. ft.	208 ft. 8 in. sq.	1 acre
25 links	16½ feet	8000 links	1 mile	10 sq. chains	1 acre	80 rods sq.	40 acres
25 links	1 rod	5280 feet	1 mile	160 sq. rods	1 acre	160 rods sq.	160 acres
100 links	1 chain	1760 yards	1 mile	4840 sq. yds.	1 acre		

In non-rectangular land descriptions, distance is usually described in terms of either feet or rods (this is especially true in surveying today), and square measure in terms of acres. Such descriptions are called Metes and Bounds descriptions and will be explained in detail later.

In rectangular land descriptions, square measure is again in terms of acres, and the location of the land in such terms as N½ (north one-half), SE¼ (south east one-fourth or quarter), etc. as shown in Figures 2, 3, 4 and 5.

MEANDERED WATER & GOVERNMENT LOTS

A meandered lake or stream is water, next to which the adjoining landowner pays taxes on the land only. Such land is divided into divisions of land called government lots. The location, acreage and lot number of each such a tract of land, was determined, surveyed and platted by the original government surveyors.

The original survey of your county (complete maps of each township, meandered lakes, government lots, etc.) is in your courthouse, and this original survey is the basis for all land descriptions in your county (see figure 1).

IMPORTANT
THE GOVERNMENT LOT NUMBER GIVEN TO A PIECE OF LAND, IS THE LEGAL DESCRIPTION OF THAT TRACT OF LAND.

HOW CAN YOU TELL WHETHER WATER IS MEANDERED OR PRIVATELY OWNED?

On our township maps, if you find government lots adjoining a body of water or stream, those waters are meandered. If there are no government lots surrounding water, that water is privately owned, the owner is paying taxes on the land under the water, and the owner controls the hunting, fishing, trapping rights, etc., on that water, within the regulations of the State and Federal laws, EXCEPT where such water is deemed navigable, other rulings may sometimes pertain.

As a generality (but not always), meandered water is public water which the public may use for recreational purposes, fishing, hunting, trapping, etc., provided that there is legal access to the water, or in other words, if the public can get to such waters without trespassing. There still is much litigation concerning the same to be decided by the courts.

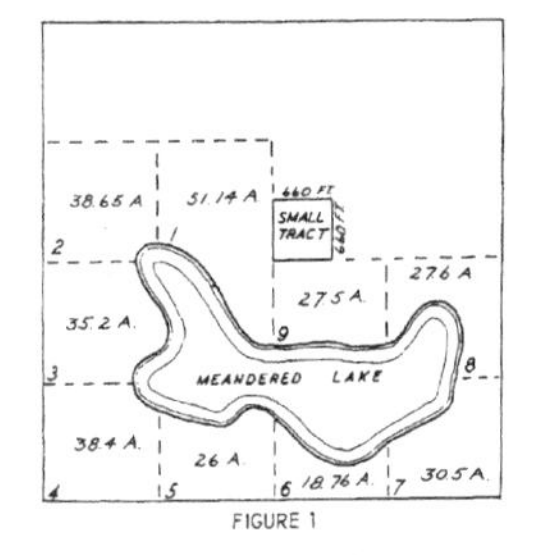

FIGURE 1

SAMPLE SECTIONS SHOWING RECTANGULAR LAND DESCRIPTIONS, ACREAGES AND DISTANCES

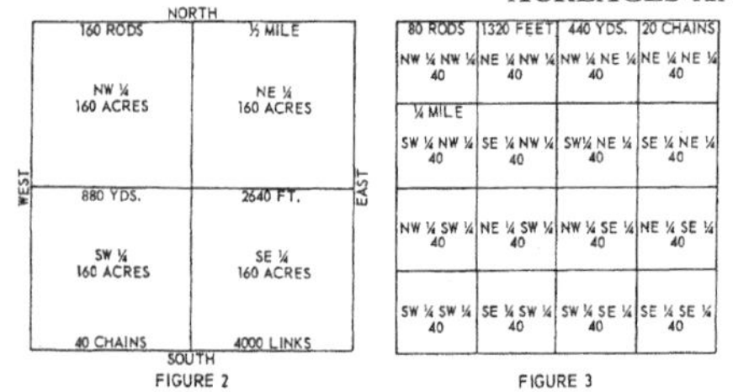

FIGURE 2

80 RODS	1320 FEET	440 YDS.	20 CHAINS
NW ¼ NW ¼ 40	NE ¼ NW ¼ 40	NW ¼ NE ¼ 40	NE ¼ NE ¼ 40
¼ MILE			
SW ¼ NW ¼ 40	SE ¼ NW ¼ 40	SW¼ NE ¼ 40	SE ¼ NE ¼ 40
NW ¼ SW ¼ 40	NE ¼ SW ¼ 40	NW ¼ SE ¼ 40	NE ¼ SE ¼ 40
SW ¼ SW ¼ 40	SE ¼ SW ¼ 40	SW ¼ SE ¼ 40	SE ¼ SE ¼ 40

FIGURE 3

1 MILE - 320 RODS - 5280 FEET - 80 CHAINS
N ½ 320
N ½ N ½ SW ¼ 40
W ½ NW ¼ SE ¼ 20
E ½ NW ¼ SE ¼ 20
N½NE¼SE¼ 20
S ½ N ½ SW ¼ 40
S½NE¼SE¼ 20
W ½ SW ¼ SW ¼ 20
NE ¼ SW ¼ SW ¼
SE ¼ SW ¼ 40
S ½ SE ¼ 80
E½SE¼SW¼SW¼-5 ACRES
W½SE¼SW¼SW¼-5 ACRES
10

FIGURE 4

W ½ NW ¼ 80
E ½ NW ¼ 80
E ½ 320
N ½ SW ¼ 80
N½SW¼SW¼ 20
NW ¼ SE ¼ SW ¼
NE ¼ SE ¼ SW ¼
SE ¼ SW ¼ SW ¼
SW ¼ SE ¼ SW ¼
SE ¼ SE ¼ SW ¼
N½SW¼SW¼SW¼-5 ACRES
S½SW¼SW¼SW¼-5 ACRES
10 10

FIGURE 5

THE BEST WAY TO READ LAND DESCRIPTIONS IS FROM THE REAR OR BACKWARDS

Descriptions of land always read FIRST from either the North or the South. In figures 2, 3, 4 and 5, notice that they all start with N (north), S (south), such as NW, SE, etc. They are never WN (west north), ES (east south) etc.
IMPORTANT: It is comparatively simple for anyone to understand a description, that is, determine where a tract of land is located, from even a long description. The SECRET is to read or analyze the description from the rear or backwards.

EXAMPLE: Under figure 4, the first description reads E½, SE¼, SW¼, SW¼. The last part of the description reads SW¼, which means that the tract of land we are looking for is somewhere in that quarter (as shown in figure 2). Next back we find SW¼, which means the tract we are after is somewhere in the SW¼ SW¼ (as shown in figure 3). Next back, we find the SE¼, which means that the tract is in the SE¼ SW¼ SW¼ (as shown in figure 5). Next back and our last part to look up, is the E½ of the above, which is the location of the tract described by the whole description (as shown in figure 4).

TO INTERPRET A LAND DESCRIPTION - LOCATE THE AREA ON YOUR TOWNSHIP PLAT, THEN ANALYZE THE DESCRIPTION & FOLLOW IT ON THE PLAT MAP.

TOWNSHIP SURVEY INFORMATION

A CONGRESSIONAL TOWNSHIP CONTAINS 36 SECTIONS OF LAND 1 MILE SQUARE

A CIVIL OR POLITICAL TOWN MAY BE LARGER OR SMALLER THAN A CONGRESSIONAL TOWNSHIP.

DIAGRAM SHOWING HOW SECTIONS ARE NUMBERED IN A TOWNSHIP

6	5	4	3	2	1
7	8	9	10	11	12
18	17	16	15	14	13
19	20	21	22	23	24
30	29	28	27	26	25
31	32	33	34	35	36

FIGURE 6

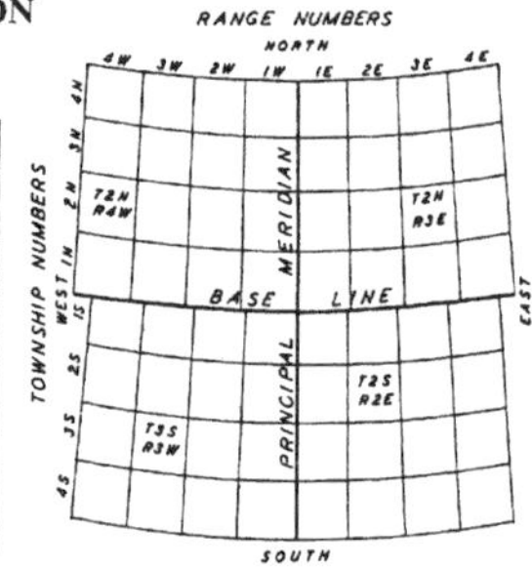

FIGURE 7

TOWNSHIPS

Theoretically, a township is a square tract of land with sides of six miles each, and containing 36 sections of land. Actually this is not the case. Years ago, when the original survey of this state was made by the government engineers, they knew that it was impossible to keep a true north and south direction of township lines, and still keep getting township squares of 36 square miles. As they surveyed toward the north pole, they were constantly running out of land, because the township lines were converging toward the north pole.

If you will turn to one of the township maps in this plat book, you will notice that on the north and on the west of each township, there are divisions of land which show odd acreages. In some townships, these odd acreages are called government lots (because they were given a lot number), and at other times left as FRACTIONAL FORTIES OR EIGHTIES. It was at the option of the original government surveyors as to whether they would call these odd acreages government lots, or fractional forties and eighties.

The reason for these odd acreages is that the government surveyors adjusted for shortages of land which developed as they went north, by making fractional forties, eighties or government lots out of the land on the west side of a township, and the same for the land on the north side of a township to keep east and west lines running parallel. In other words it was impossible to fit full squares into a circle.

Townships sometimes vary in size from the regularly laid-out township (see figure 6). Suppose that the dotted line in figure 6 is a river separating two counties. The land north and west of the river could be a township in one county, the land south and east could be a township in another county. Whichever county the land is in, it still retains the same section, township and range numbers for purposes of land descriptions.

Each township has a township number and also a range number (sometimes more than one of each if the township is oversized, or a combination of more than one township and range).

Government surveying of townships is run from starting lines called base lines and principal meridians. Each township has a township number. This number is the number of rows or tiers of townships that a township is either north or south of the base line. Also each township has a range number. This number is the number of rows or tiers of townships that a township is either east or west of the principal meridian (see figure 7). EVERY DESCRIPTION OF LAND SHOULD SHOW THE SECTION, TOWNSHIP AND RANGE IT IS LOCATED IN.

**TOWNSHIPS MAY BE EITHER NORTH OR SOUTH OF THE BASE LINE
RANGES MAY BE EITHER EAST OR WEST OF THE PRINCIPAL MERIDIAN.**

METES AND BOUNDS DESCRIPTIONS
AND EXPLANATION OF DIRECTION IN TERMS OF DEGREES

WHAT IS A METES AND BOUNDS DESCRIPTION? It is a description of a tract of land by starting at a given point, running so many feet a certain direction, so many feet another direction etc., back to the point of beginning. EXAMPLE: In figure 1 notice the small tract of land outlined. The following would be a typical metes and bounds description of that tract of land. "Begin at the center of the section, thence north 660 feet, thence east 660 feet, thence south 660 feet, thence west 660 feet, back to the point of beginning, and containing 10 acres, being a part of Sec. No. etc."

IMPORTANT: To locate a tract of land from a metes and bounds description, start from the point of beginning, and follow it out (do not read it backwards as in the case of a rectangular description).

The small tract of land just located by the above metes and bounds description could also be described as the SW¼ SW¼ NE¼ of the section. In most cases, the same tract of land may be described in different ways. The rectangular system of describing and locating land as shown in figures 2, 3, 4 and 5 is the most simple and almost always used when possible.

A circle contains 360 degrees. Explanation: If you start at the center of a circle and run 360 straight lines an equal angle apart to the edge of the circle, so as to divide the circle into 360 equal parts, THE DIFFERENCE OF DIRECTION BETWEEN EACH LINE IS ONE DEGREE.

In land descriptions, degree readings are not a measure of distance. They are combined with either North or South, to show the direction a line runs from a given point.

HOW TO READ DESCRIPTIONS WHICH SHOW DIRECTIONS IN TERMS OF DEGREES

In figure 8, the north-south line, and the east-west line divide the circle into 4 equal parts, which means that each part contains 90 degrees as shown. Several different direction lines are shown in this diagram, with the number of degrees each varies east or west from the north and south starting points (remember again that all descriptions read from the north or south).

We all know what north-west is. It is a direction which is half-way between North and West. In terms of degrees the direction north-west would read, north 45 degrees west (see figure 8).

EXAMPLE OF A LAND DESCRIPTION IN TERMS OF DEGREES

At this time, study figure 8 for a minute or two.

In figure 8, notice the small tract. The following metes and bounds description will locate this small tract. "Begin at the beginning point, thence N 20 degrees west — 200 feet, thence N 75 degrees east — 1190 feet, thence S 30 degrees east — 240 feet, thence S 45 degrees west — 420 feet, thence west — 900 feet back to the point of beginning, containing so many acres, etc."

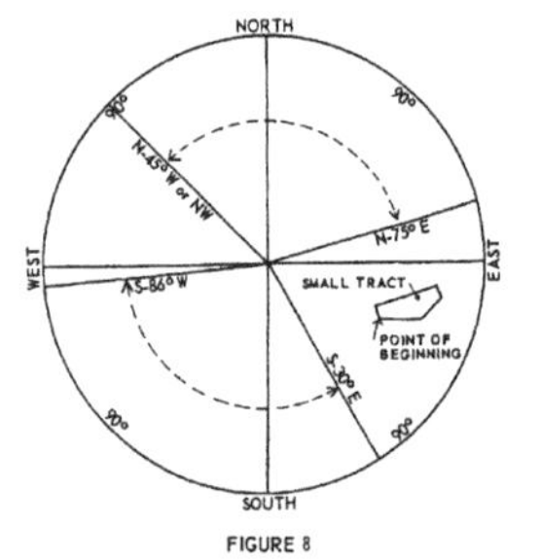

FIGURE 8

Source: Rockford Map Publishing, Inc.

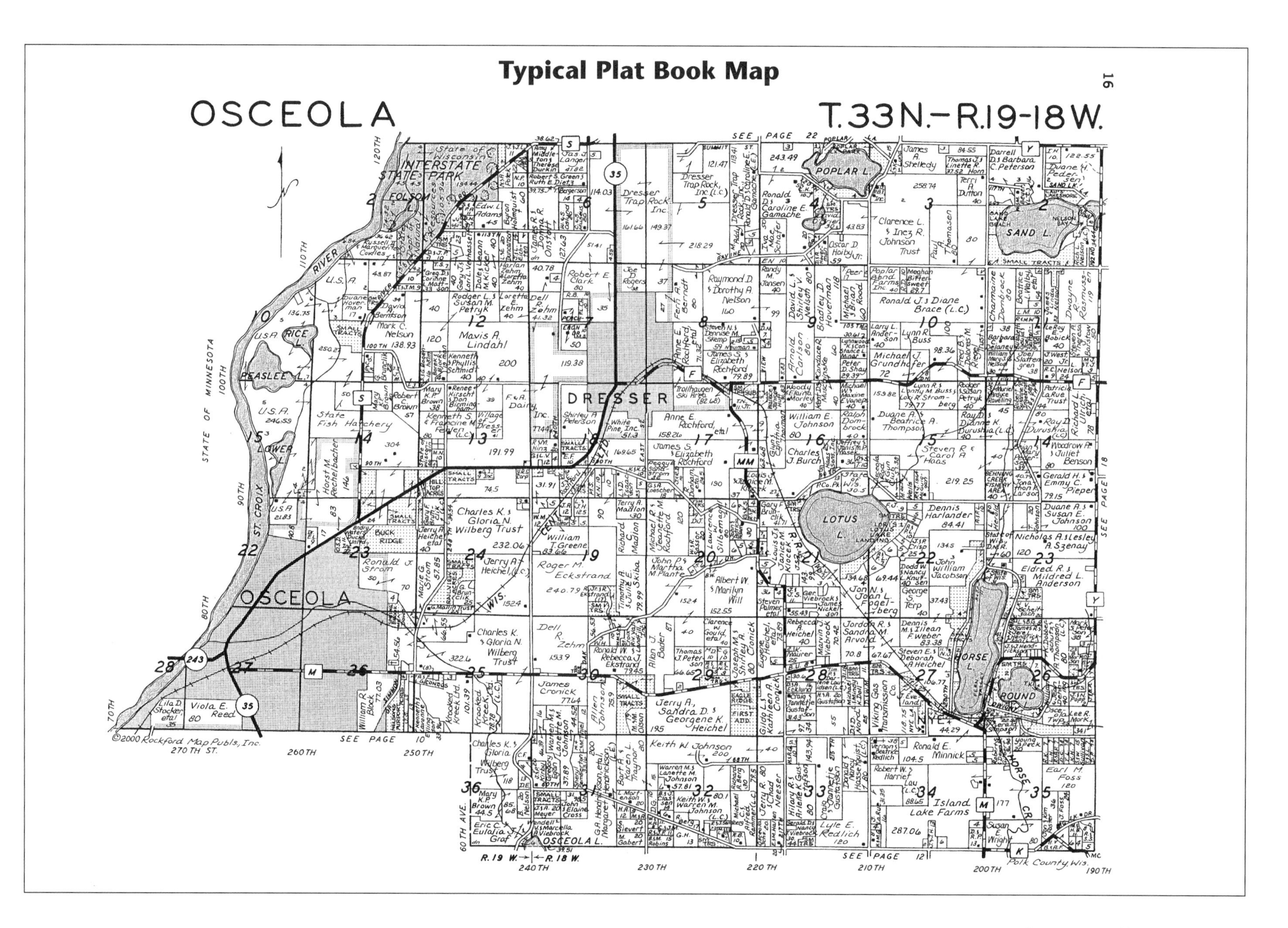

Source: © 2000 Rockford Map Publishing, Inc.

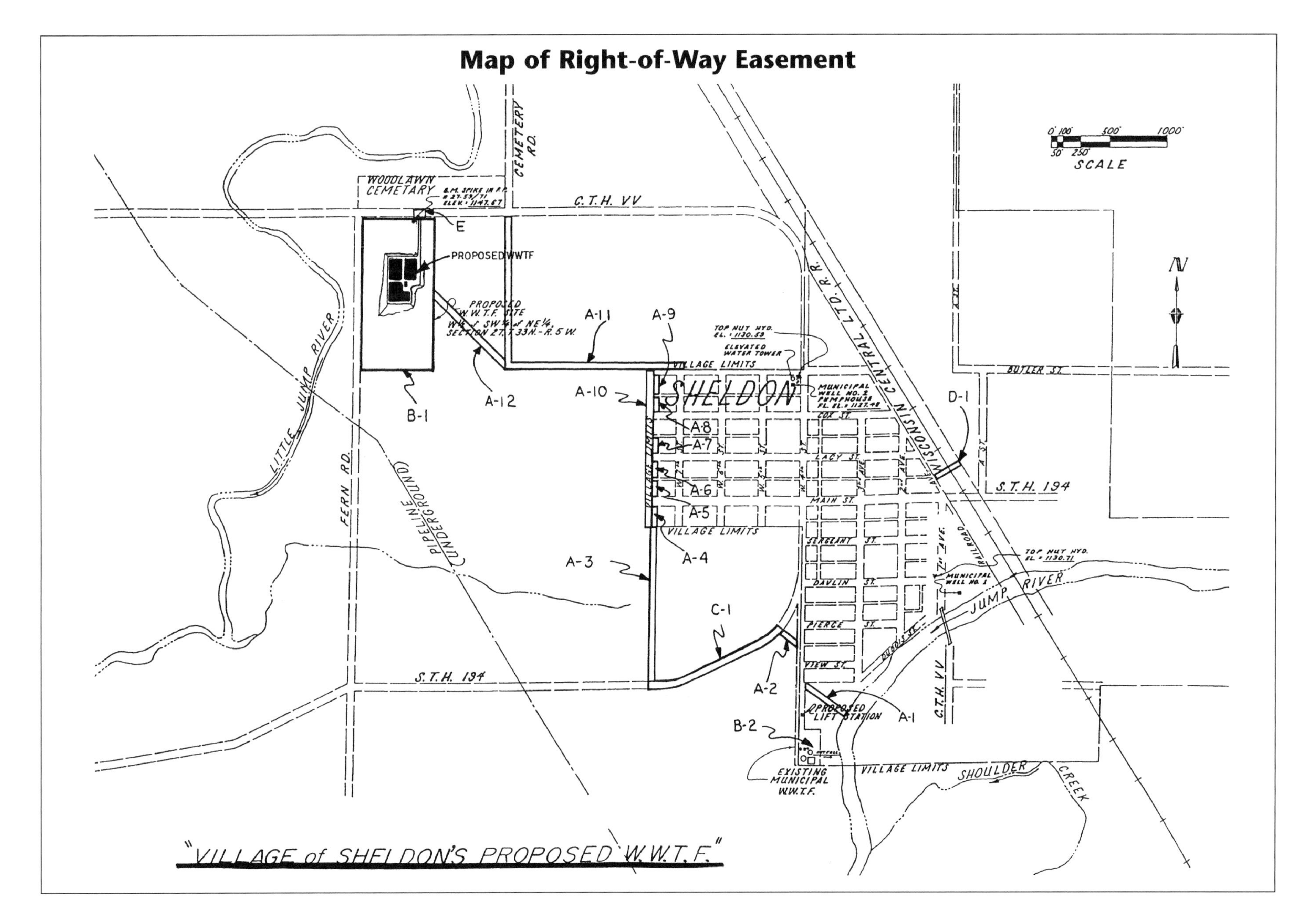

Source: Morgan & Parmley, Ltd.

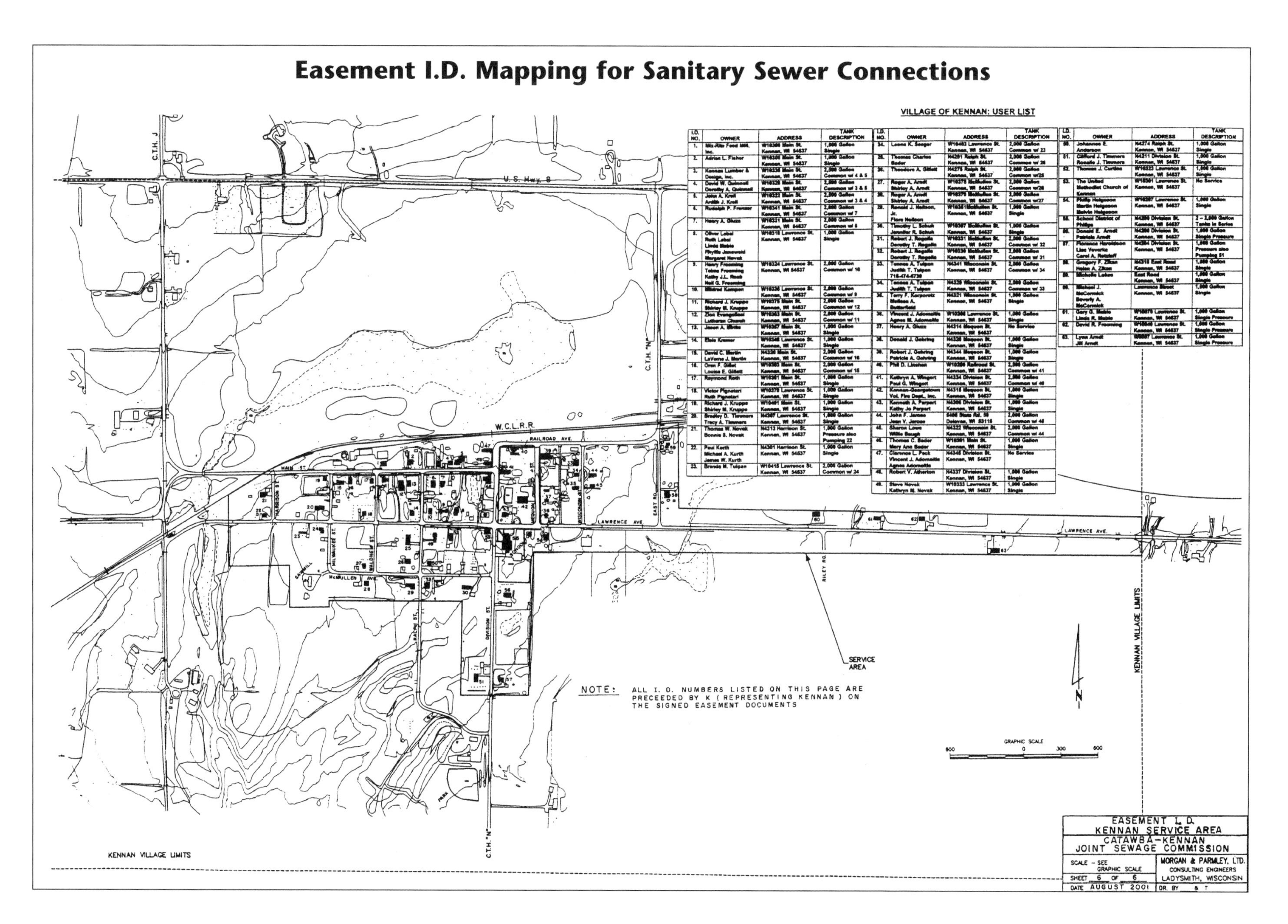

VILLAGE OF KENNAN: USER LIST

I.D. NO.	OWNER	ADDRESS	TANK DESCRIPTION
1.	Miz-Rite Food Mill, Inc.	W16369 Main St. Kennan, WI 54637	1,000 Gallon Single
2.	Adrian L. Fisher	W16356 Main St. Kennan, WI 54637	1,000 Gallon Single
3.	Kennan Lumber & Design, Inc.	W16336 Main St. Kennan, WI 54637	3,000 Gallon Common w/ 4 & 5
4.	David W. Quinnell Dorothy A. Quinnell	W16328 Main St. Kennan, WI 54637	2,000 Gallon Common w/ 3 & 5
5.	John A. Kroll Ardith J. Kroll	W16322 Main St. Kennan, WI 54637	2,000 Gallon Common w/ 3 & 4
6.	Rudolph P. Franzer	W16341 Main St. Kennan, WI 54637	2,000 Gallon Common w/ 7
7.	Henry A. Glaza	W16331 Main St. Kennan, WI 54637	2,000 Gallon Common w/ 6
8.	Oliver Lobel Ruth Lobel Linda Mable Phyllis Janowski Margaret Novak	W16318 Lawrence St. Kennan, WI 54637	1,000 Gallon Single
9.	Henry Froeming Telma Froeming Kathy J.L. Raab Neil G. Froeming	W16324 Lawrence St. Kennan, WI 54637	2,000 Gallon Common w/ 10
10.	Mildred Kampen	W16336 Lawrence St. Kennan, WI 54637	2,000 Gallon Common w/ 9
11.	Richard J. Kruppe Shirley M. Kruppe	W16375 Main St. Kennan, WI 54637	2,000 Gallon Common w/ 12
12.	Zion Evangelical Lutheran Church	W16363 Main St. Kennan, WI 54637	2,000 Gallon Common w/ 11
13.	Jason A. Mintz	W16387 Main St. Kennan, WI 54637	1,000 Gallon Single
14.	Elsie Kramer	W16348 Lawrence St. Kennan, WI 54637	1,000 Gallon Single
15.	David C. Martin LaVerne J. Martin	N4326 Main St. Kennan, WI 54637	2,000 Gallon Common w/ 16
16.	Oren F. Gillet Louise E. Gillett	W16383 Main St. Kennan, WI 54637	2,000 Gallon Common w/ 15
17.	Raymond Roth	W16391 Main St. Kennan, WI 54637	1,000 Gallon Single
18.	Victor Pignatari Ruth Pignatari	W16378 Lawrence St. Kennan, WI 54637	1,000 Gallon Single
19.	Richard J. Kruppe Shirley M. Kruppe	W16401 Main St. Kennan, WI 54637	1,000 Gallon Single
20.	Bradley D. Timmers Tracy A. Timmers	N4387 Lawrence St. Kennan, WI 54637	1,000 Gallon Single
21.	Thomas W. Novak Bonnie S. Novak	N4313 Harrison St. Kennan, WI 54637	1,000 Gallon Pressure also Pumping 22
22.	Paul Kurth Michael A. Kurth James W. Kurth	N4301 Harrison St. Kennan, WI 54637	1,000 Gallon Single
23.	Brenda M. Tulpan	W16415 Lawrence St. Kennan, WI 54637	2,000 Gallon Common w/ 24
24.	Leona K. Seeger	W16403 Lawrence St. Kennan, WI 54637	2,000 Gallon Common w/ 23
25.	Thomas Charles Bader	N4291 Ralph St. Kennan, WI 54637	2,000 Gallon Common w/ 26
26.	Theodore A. Gillett	N4275 Ralph St. Kennan, WI 54637	2,000 Gallon Common w/25
27.	Roger A. Arndt Shirley A. Arndt	W16378 McMullen St. Kennan, WI 54637	2,000 Gallon Common w/28
28.	Roger A. Arndt Shirley A. Arndt	W16376 McMullen St. Kennan, WI 54637	2,000 Gallon Common w/27
29.	Ronald J. Neilson, Jr. Flora Neilson	W16361 McMullen St. Kennan, WI 54637	1,000 Gallon Single
30.	Timothy L. Schuh Jennifer R. Schuh	W16367 McMullen St. Kennan, WI 54637	1,000 Gallon Single
31.	Robert J. Rogalla Dorothy T. Rogalla	W16331 McMullen St. Kennan, WI 54637	2,000 Gallon Common w/ 32
32.	Robert J. Rogalla Dorothy T. Rogalla	W16339 McMullen St. Kennan, WI 54637	2,000 Gallon Common w/ 31
33.	Tommas A. Tulpan Judith T. Tulpan 715-474-6730	N4341 Wisconsin St. Kennan, WI 54637	2,000 Gallon Common w/ 34
34.	Tommas A. Tulpan Judith T. Tulpan	N4329 Wisconsin St. Kennan, WI 54637	2,000 Gallon Common w/ 33
35.	Terry F. Korporatz Melissa A. Butterfield	N4321 Wisconsin St. Kennan, WI 54637	1,000 Gallon Single
36.	Vincent J. Adomaitis Agnes M. Adomaitis	W16386 Lawrence St. Kennan, WI 54637	1,000 Gallon Single
37.	Henry A. Glaza	N4314 Maquon St. Kennan, WI 54637	No Service
38.	Donald J. Gehring	N4328 Maquon St. Kennan, WI 54637	1,000 Gallon Single
39.	Robert J. Gehring Patricia A. Gehring	N4344 Maquon St. Kennan, WI 54637	1,000 Gallon Single
40.	Phil D. Linehan	W16289 Railroad St. Kennan, WI 54637	2,000 Gallon Common w/ 41
41.	Kathryn A. Wingert Paul G. Wingert	N4334 Division St. Kennan, WI 54637	2,000 Gallon Common w/ 40
42.	Kennan-Georgetown Vol. Fire Dept., Inc.	N4316 Maquon St. Kennan, WI 54637	1,000 Gallon Single
43.	Kenneth A. Parpart Kathy Jo Parpart	N4306 Division St. Kennan, WI 54637	1,000 Gallon Single
44.	John F. Jarosa Jean V. Jarosa	5486 State Rd. 59 Delavan, WI 53115	2,000 Gallon Common w/ 45
45.	Sharon Lowe Willie Baugh	N4322 Wisconsin St. Kennan, WI 54637	2,000 Gallon Common w/ 44
46.	Thomas C. Bader Mary Ann Bader	W16301 Main St. Kennan, WI 54637	1,000 Gallon Single
47.	Clarence L. Peck Vincent J. Adomaitis Agnes Adomaitis	N4348 Division St. Kennan, WI 54637	No Service
48.	Robert V. Atherton	N4337 Division St. Kennan, WI 54637	1,000 Gallon Single
49.	Steve Novak Kathryn M. Novak	W16333 Lawrence St. Kennan, WI 54637	1,000 Gallon Single
50.	Johannes E. Anderson	N4274 Ralph St. Kennan, WI 54637	1,000 Gallon Single
51.	Clifford J. Timmers Rosalie J. Timmers	N4211 Division St. Kennan, WI 54637	1,000 Gallon Single
52.	Thomas J. Curtiss	W16323 Lawrence St. Kennan, WI 54637	1,000 Gallon Single
53.	The United Methodist Church of Kennan	W16361 Lawrence St. Kennan, WI 54637	No Service
54.	Philip Helgeson Martin Helgeson Melvin Helgeson	W16287 Lawrence St. Kennan, WI 54637	1,000 Gallon Single
55.	School District of Phillips	N4280 Division St. Kennan, WI 54637	2 – 2,000 Gallon Tanks in Series
56.	Donald E. Arndt Patricia Arndt	N4280 Division St. Kennan, WI 54637	1,000 Gallon Single Pressure
57.	Florence Haroldson Lisa Vevarka Carol A. Retzlaff	N4284 Division St. Kennan, WI 54637	1,000 Gallon Pressure also Pumping 51
58.	Gregory F. Zikan Helen A. Zikan	N4318 East Road Kennan, WI 54637	1,000 Gallon Single
59.	Michelle Lukes	East Road Kennan, WI 54637	1,000 Gallon Single
60.	Michael J. McCormick Beverly A. McCormick	Lawrence Street Kennan, WI 54637	1,000 Gallon Single
61.	Gary G. Mable Linda R. Mable	W16075 Lawrence St. Kennan, WI 54637	1,000 Gallon Single Pressure
62.	David R. Froeming	W16040 Lawrence St. Kennan, WI 54637	1,000 Gallon Single Pressure
63.	Lynn Arndt Jill Arndt	W16007 Lawrence St. Kennan, WI 54637	1,000 Gallon Single Pressure

Source: Morgan & Parmley, Ltd.

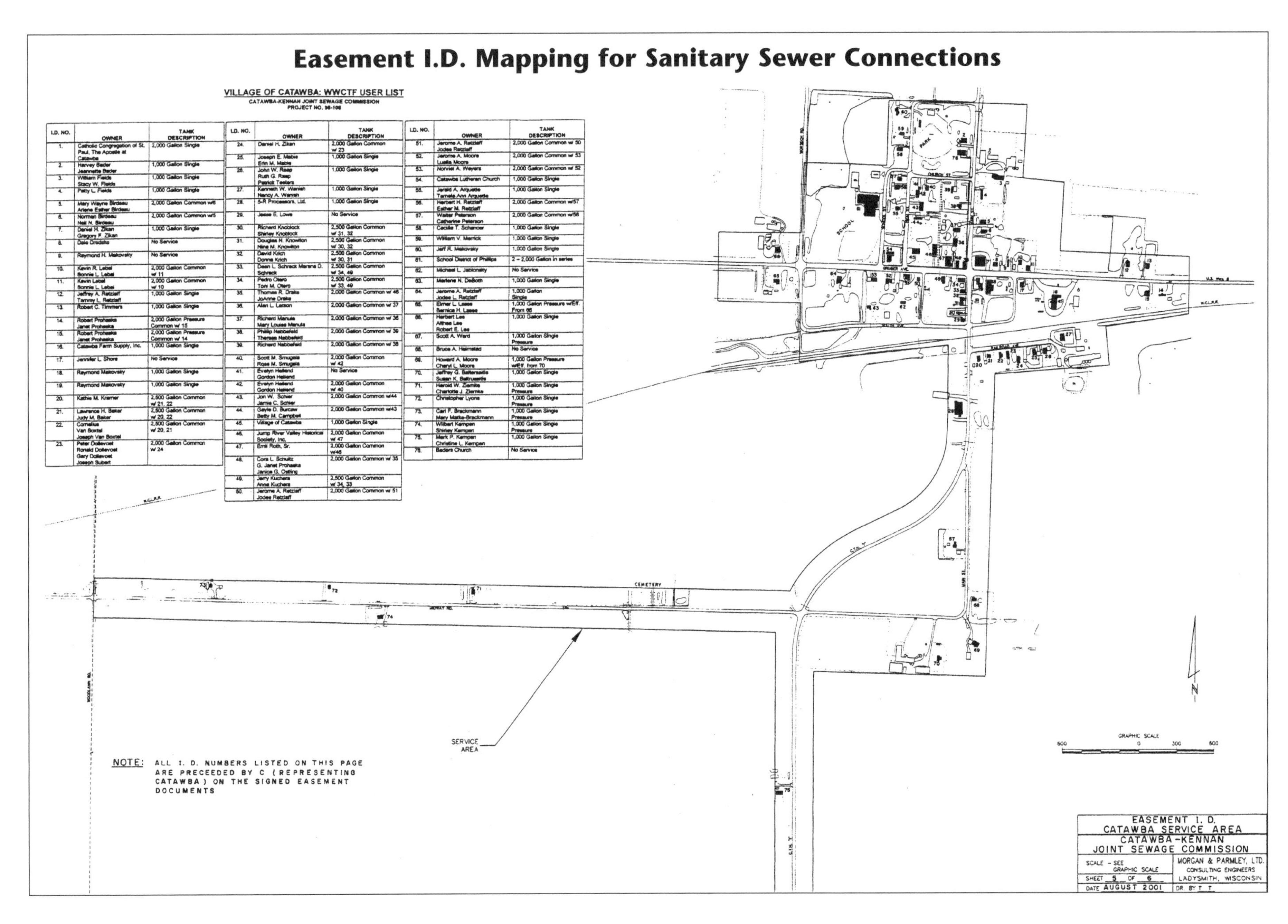

Easement I.D. Mapping for Sanitary Sewer Connections

VILLAGE OF CATAWBA: WWCTF USER LIST
CATAWBA-KENNAN JOINT SEWAGE COMMISSION
PROJECT NO. 98-106

I.D. NO.	OWNER	TANK DESCRIPTION
1.	Catholic Congregation of St. Paul, The Apostle at Catawba	2,000 Gallon Single
2.	Harvey Bader Jeannette Bader	1,000 Gallon Single
3.	William Fields Stacy W. Fields	1,000 Gallon Single
4.	Patty L. Fields	1,000 Gallon Single
5.	Mary Wayne Birdeau Arlene Esther Birdeau	2,000 Gallon Common w/6
6.	Norman Birdeau Neil N. Birdeau	2,000 Gallon Common w/5
7.	Daniel H. Zikan Gregory F. Zikan	1,000 Gallon Single
8.	Dale Dredske	No Service
9.	Raymond H. Makovsky	No Service
10.	Kevin R. Lebal Bonnie L. Lebal	2,000 Gallon Common w/ 11
11.	Kevin Lebal Bonnie L. Lebal	2,000 Gallon Common w/ 10
12.	Jeffrey A. Retzlaff Tammy L. Retzlaff	1,000 Gallon Single
13.	Robert C. Timmers	1,000 Gallon Single
14.	Robert Prohaska Janet Prohaska	2,000 Gallon Pressure Common w/ 15
15.	Robert Prohaska Janet Prohaska	2,000 Gallon Pressure Common w/ 14
16.	Catawba Farm Supply, Inc.	1,000 Gallon Single
17.	Jennifer L. Shore	No Service
18.	Raymond Makovsky	1,000 Gallon Single
19.	Raymond Makovsky	1,000 Gallon Single
20.	Kathie M. Kramer	2,500 Gallon Common w/ 21, 22
21.	Lawrence H. Baker Judy M. Baker	2,500 Gallon Common w/ 20, 22
22.	Cornelius Van Boxtel Joseph Van Boxtel	2,500 Gallon Common w/ 20, 21
23.	Peter Dollevoet Ronald Dollevoet Gary Dollevoet Joseph Subert	2,000 Gallon Common w/ 24
24.	Daniel H. Zikan	2,000 Gallon Common w/ 23
25.	Joseph E. Mabie Erin M. Mabie	1,000 Gallon Single
26.	John W. Raap Ruth G. Raap Patrick Teeters	1,000 Gallon Single
27.	Kenneth W. Wanish Nancy A. Wanish	1,000 Gallon Single
28.	5-R Processors, Ltd.	1,000 Gallon Single
29.	Jesse E. Lowe	No Service
30.	Richard Knoblock Shirley Knoblock	2,500 Gallon Common w/ 31, 32
31.	Douglas N. Knowlton Nina M. Knowlton	2,500 Gallon Common w/ 30, 32
32.	David Krich Donna Krich	2,500 Gallon Common w/ 30, 31
33.	Dean L. Schreck Marena D. Schreck	2,500 Gallon Common w/ 34, 49
34.	Pedro Otero Toni M. Otero	2,500 Gallon Common w/ 33, 49
35.	Thomas R. Drake JoAnne Drake	2,000 Gallon Common w/ 48
36.	Alan L. Larson	2,000 Gallon Common w/ 37
37.	Richard Manula Mary Louise Manula	2,000 Gallon Common w/ 36
38.	Phillip Nabbefeld Theresa Nabbefeld	2,000 Gallon Common w/ 39
39.	Richard Nabbefeld	2,000 Gallon Common w/ 38
40.	Scott M. Smugala Rose M. Smugala	2,000 Gallon Common w/ 42
41.	Evelyn Helland Gordon Helland	No Service
42.	Evelyn Helland Gordon Helland	2,000 Gallon Common w/ 40
43.	Jon W. Schier Jamie C. Schier	2,000 Gallon Common w/44
44.	Gayle D. Burcaw Betty M. Campbell	2,000 Gallon Common w/43
45.	Village of Catawba	1,000 Gallon Single
46.	Jump River Valley Historical Society, Inc.	2,000 Gallon Common w/ 47
47.	Emil Roth, Sr.	2,000 Gallon Common w/46
48.	Cora L. Schultz G. Janet Prohaska Janice G. Ostling	2,000 Gallon Common w/ 35
49.	Jerry Kuchera Anna Kuchera	2,500 Gallon Common w/ 34, 33
50.	Jerome A. Retzlaff Jodee Retzlaff	2,000 Gallon Common w/ 51
51.	Jerome A. Retzlaff Jodee Retzlaff	2,000 Gallon Common w/ 50
52.	Jerome A. Moore Luella Moore	2,000 Gallon Common w/ 53
53.	Norviel A. Weyers	2,000 Gallon Common w/ 52
54.	Catawba Lutheran Church	1,000 Gallon Single
55.	Jerald A. Arquette Tamela Ann Arquette	1,000 Gallon Single
56.	Herbert H. Retzlaff Esther M. Retzlaff	2,000 Gallon Common w/57
57.	Walter Peterson Catherine Peterson	2,000 Gallon Common w/56
58.	Cecilia T. Schancer	1,000 Gallon Single
59.	William V. Merrick	1,000 Gallon Single
60.	Jeff R. Makovsky	1,000 Gallon Single
61.	School District of Phillips	2 – 2,000 Gallon in series
62.	Michael L. Jablonsky	No Service
63.	Marlene N. DeBoth	1,000 Gallon Single
64.	Jerome A. Retzlaff Jodee L. Retzlaff	1,000 Gallon Single
65.	Elmer L. Lasee Bernice H. Lasee	1,000 Gallon Pressure w/Eff. From 66
66.	Herbert Lee Althea Lee Robert E. Lee	1,000 Gallon Single
67.	Scott A. Ward	1,000 Gallon Single Pressure
68.	Bruce A. Halmstad	No Service
69.	Howard A. Moore Cheryl L. Moore	1,000 Gallon Pressure w/Eff. from 70
70.	Jeffrey G. Baltersaitis Susan K. Baltrusaitis	1,000 Gallon Single
71.	Harold W. Ziemke Charlotte J. Ziemke	1,000 Gallon Single Pressure
72.	Christopher Lyons	1,000 Gallon Single Pressure
73.	Carl F. Brackmann Mary Matke-Brackmann	1,000 Gallon Single Pressure
74.	Wilbert Kempen Shirley Kempen	1,000 Gallon Single Pressure
75.	Mark P. Kempen Christine L. Kempen	1,000 Gallon Single
76.	Baders Church	No Service

Source: Morgan & Parmley, Ltd.

Flood Plain Mapping

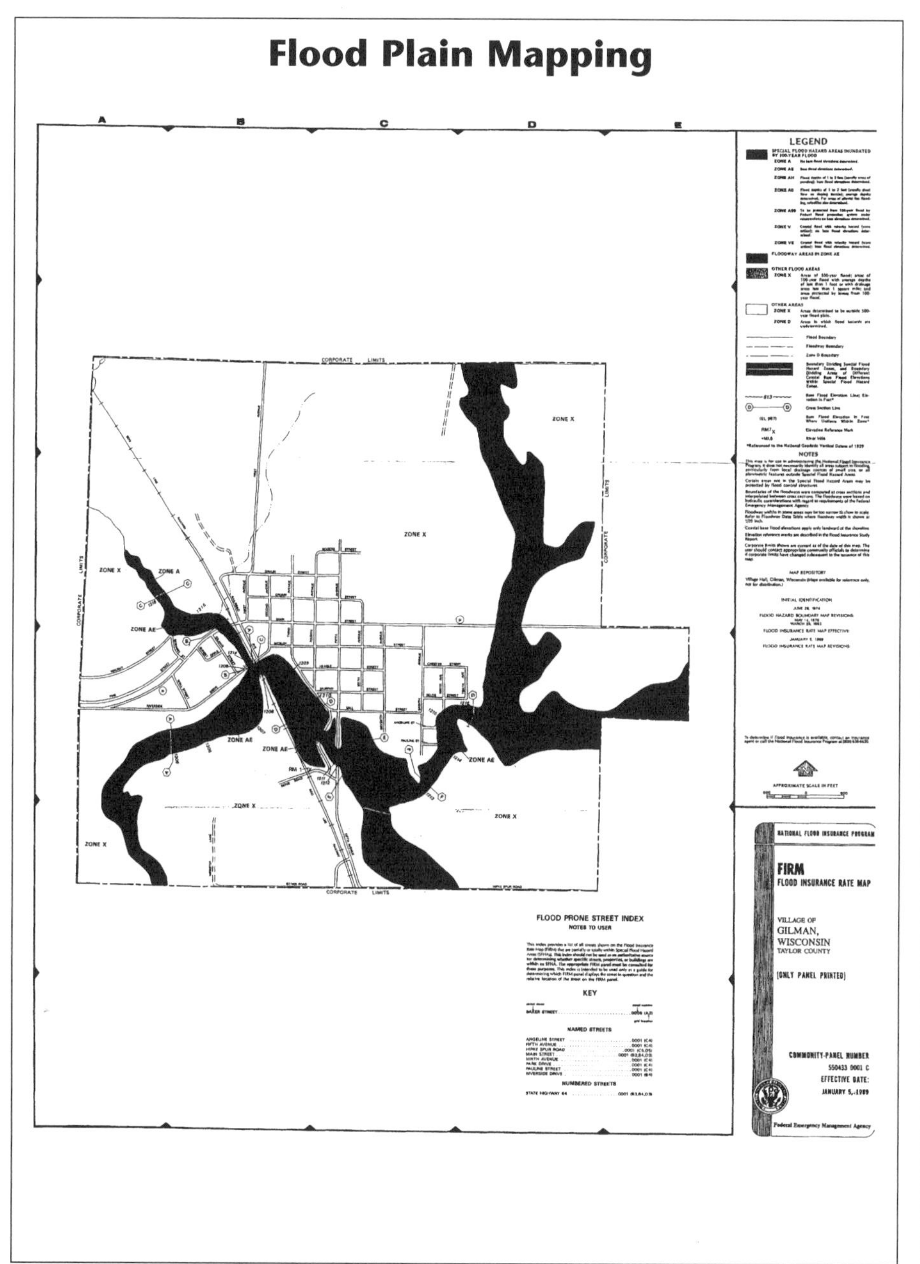

Source: FEMA

Planning Area Boundary Mapping

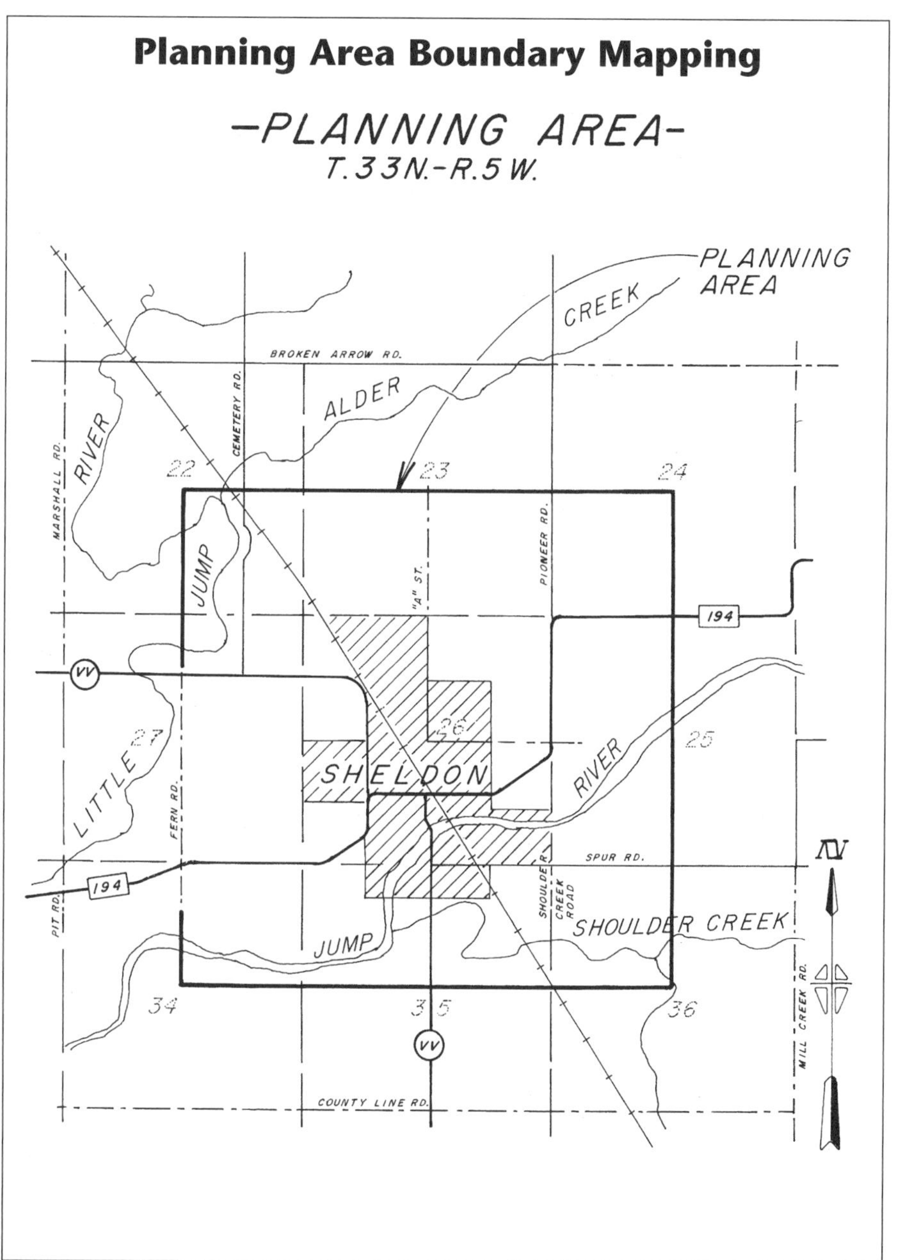

Source: Morgan & Parmley, Ltd.

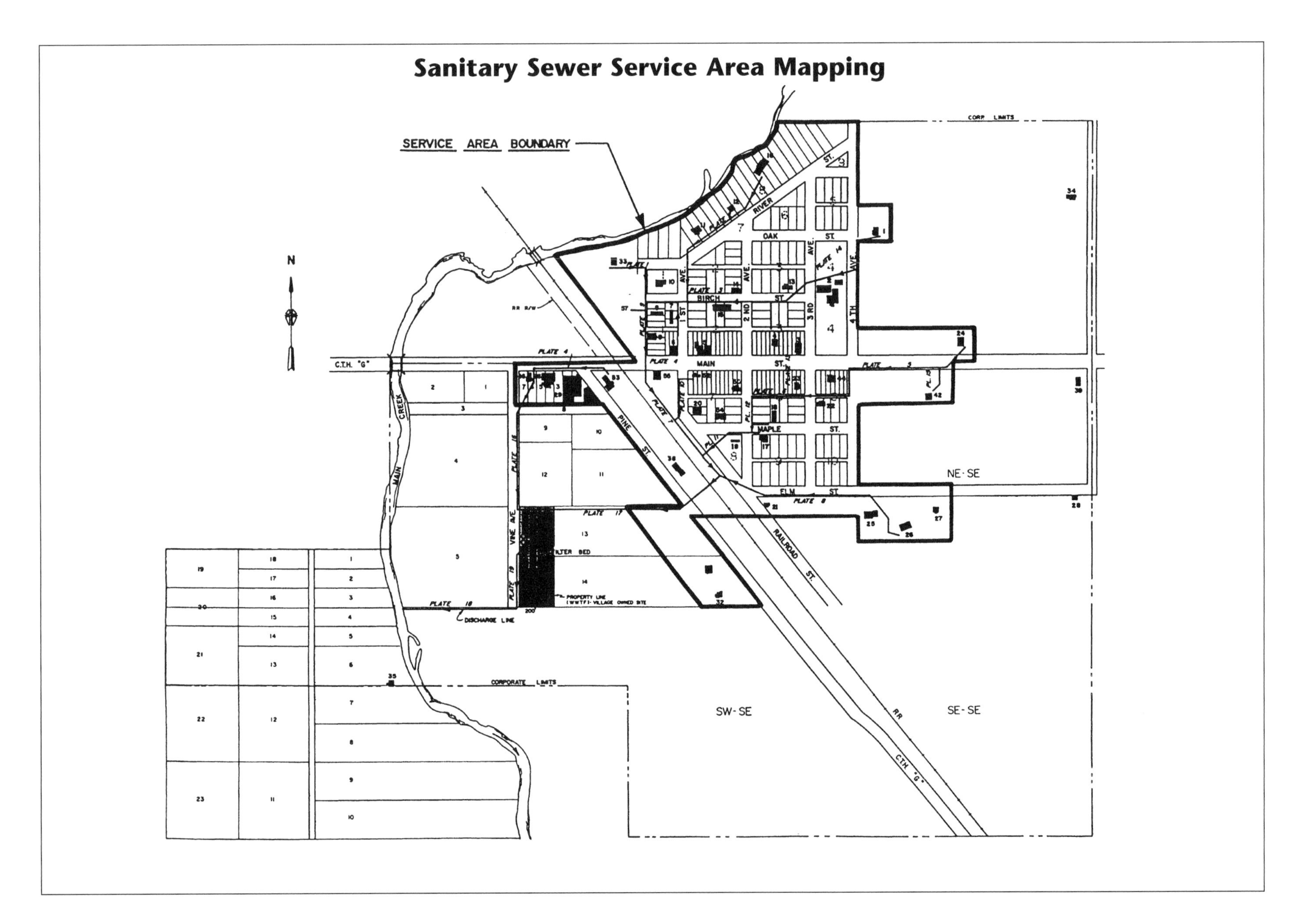

Source: Morgan & Parmley, Ltd.

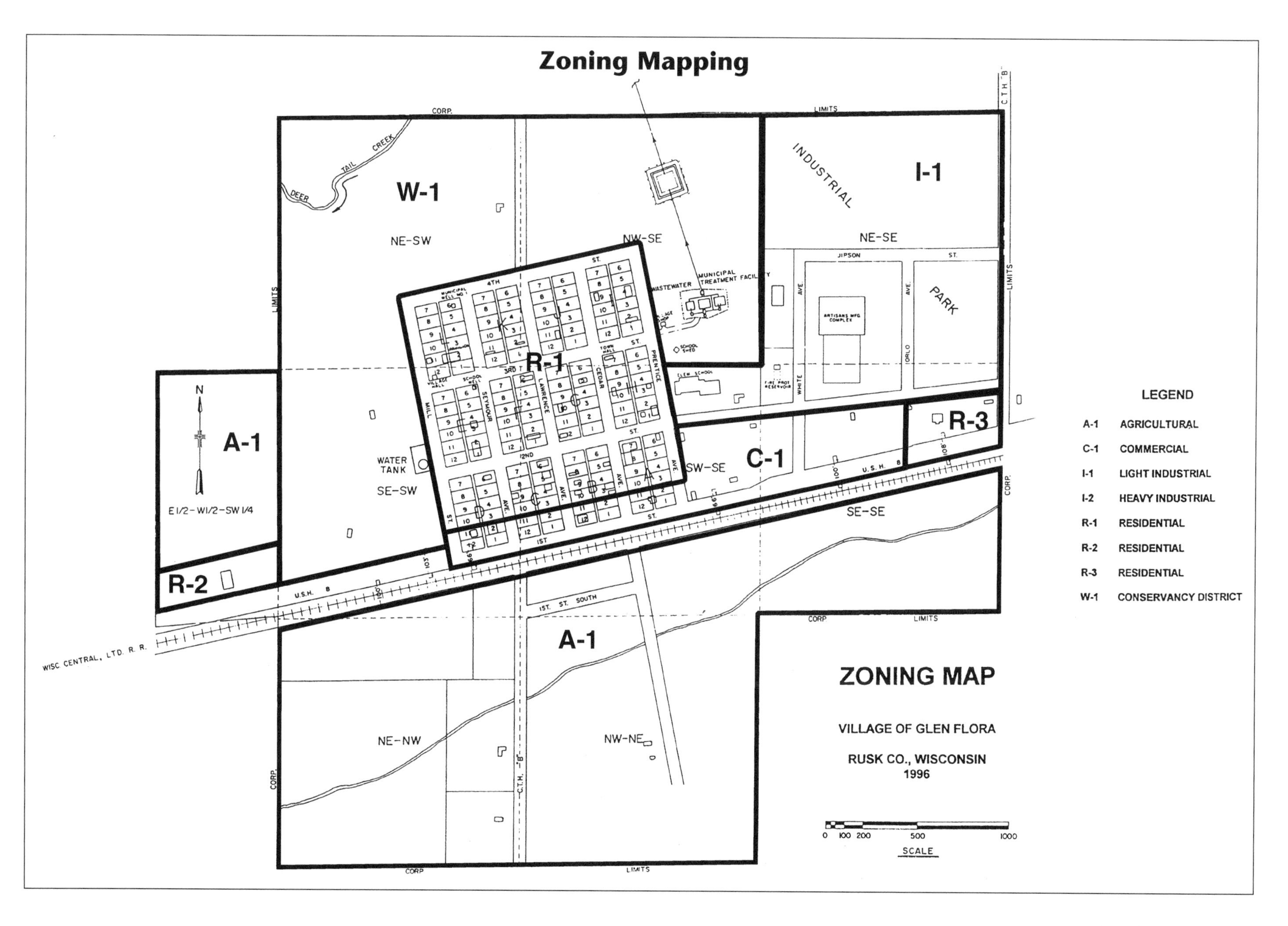

Source: Morgan & Parmley, Ltd.

Planning

PUBLIC MEETINGS

Section 4

Consulting engineers are continuously involved in a wide variety of public meetings during the performance of their professional duties. Most projects are exposed to at least one public meeting, and the consulting engineer is generally a key participate. Therefore, it is mandatory that the project engineer be well versed on the mechanics and types of public meetings.

TABLE OF CONTENTS

Regular Municipal Meetings

Elected officials of municipalities meet on a regular basis to conduct their governmental business in a publicly open format. These meetings are generally well publicized and follow a detailed agenda. Usually, the news media is present, and various citizens' groups attend when controversial issues or major projects will be discussed.

Whether these meetings are city councils, village boards, sewer commissions, sanitary districts, county boards, water commissions, township boards, school districts or similar municipalities, they all are basically similar in their concept and operational format. Therefore, it is strongly recommended that one obtain a copy of the classic book entitled, *Robert's Rules of Order* and study its contents. Most meetings do follow these rules and all meetings should to avoid confusion and improper procedures.

When an engineer is requested or required to participate in a municipal meeting, it usually is in direct relationship to an active project. This focus on a specific project mandates that the engineer be well prepared to respond to a wide range of questions and concerns. It is advisable that adequate visual aids and handouts be available for distribution to the audience to assist in the presentation. The engineer must remember to keep the project description general in nature and avoid technical jargon that may confuse the public. However, if specific technical questions do arise, the engineer must be able to respond in detail and have files available to support the answer if judged critical. While this is rare, one should never assume that detailed technical questions will not be asked.

Additionally, the engineer should evaluate the audience and municipal officials to determine how detailed and long the presentation should run. Finally, the engineer must maintain a relaxed, composed and professional manner at all times to avoid jeopardizing the progress of the project.

Committee Meetings

Committees usually have less members than the municipal boards they are subservient to and meet to discuss and make recommendations on specific topics. These committees include: zoning, forestry, finance, parks & grounds, public works, police & fire, personnel, welfare, transportation, sewer & water, utilities, etc.

As previously stated, these committees generally are smaller in membership and thereby easier to communicate with due to a more relaxed atmosphere. These groups have a smaller agenda and can focus on the topic without conflicting distractions. Committees gather and evaluate information to provide recommendations to their parent group; i.e., their respective municipal board. The engineer must command the committee's professional respect and develop a good working relationship or the project will begin to develop problems that may jeopardize the project itself.

Informational Meetings

Information meetings are held for a variety of purposes. The main reason, of course, is to inform the public of a proposed project while it is still in its conceptual stage. Basically, the municipal government, utility company, commission or regulatory agency is seeking input from the citizens before proceeding too far with a project or procedure.

These special meetings have a broad latitude and can be quite informal. This atmosphere is more relaxed and often this is reflected in good audience participation. Grassroot feedback for a specific project certainly helps the engineer in developing a more viable plan.

This type of meeting does require an in-depth agenda. However, a good moderator is needed to properly set the stage and control the participants in an orderly fashion. Someone, whether a secretary or clerk, should keep notes, and the project engineer must keep a record.

Assessment Hearings

Assessment hearings can generate some of the most heated comments and audience resistance because of costs to be assessed to specific private property. These hearings are very formal and often videotaped, tape-recorded and/or transcribed by a certified court reporter. Usually, these hearings are well attended and the news media covers the event.

Prior to the assessment hearing, the project is described in an engineering report which details the need for the project, conceptual design and a detailed estimated cost. The municipality presents the assessment list which summarizes the cost to be assessed to each affected property. This assessment list is made available to the public and certified copies mailed to each affected property owner well in advance of the public assessment hearing. All citizens will have access to the engineering report and assessment list for review before the hearing. Notices of the hearing are published and an agenda prepared.

At the public hearing, the project engineer will be required to describe the project, review the cost estimate and answer questions. Prior to the engineer's presentation, the official moderator will state the purpose of the hearing and set the ground rules. Often, to avoid confusion, any individual wishing to make a statement or ask a question will be required to complete a brief form listing name, address and representation so the official record will be accurate.

The engineer should have adequate visual aids and handouts to assist in the presentation. Remember, the moderator is in charge and must control the hearing; it is not the responsibility of the engineer. All questions should be held until the engineer's presentation is completed, and the engineer should only answer the questions that are asked. In an extremely rare case, if you do not know the answer, state that you will find the answer and respond in writing to the individual (with copy to proper municipal official) as soon as possible. At all times, maintain your composure and alertness. Be attentive to all participants and make your responses brief and to the point. Never give a statement that you can not back up with solid, reliable information.

Facility Planning Hearings

Like assessment hearings, facility planning hearings can become extremely controversial. Proposed construction of wastewater treatment facilities, power transmission lines, nuclear power plants, mining operations and similar projects usually impact environmentally sensitive areas and summon response from various environmental protection groups.

If the facility planning process did an adequate job in evaluating the potential impacts on sensitive areas and assuring proper protection, the proposed project will normally be allowed to continue. However, there are usually issues that are contested. Thus, the need for a public hearing. In addition it is necessary to keeping the public informed.

The format is usually very similar to the previously described assessment hearing. The facility planning document describes the proposed project, evaluates the cultural, historical, environmental, archaeological, social and project alternatives and selects the most cost-effective/environmentally sound alternative. This is followed by a project time table. This document is made available for review by the public and all applicable regulatory agencies prior to the hearing.

At the hearing, the engineer and support staff are called upon to make a formal presentation. This presentation should be in detail and often is lengthy because the audience probably has experts prepared to give technical responses.

Often, these hearings do not allow a question and answer period. However, comments and statements are allowed for the record. Many of these hearings are conducted by a regulatory agency who will grant or deny the approval to construct the proposed project. Therefore, the main purpose is to gather additional data and input to assist them in rendering a final decision. These hearings are very formal and an official transcript will be prepared for a permanent record. Usually, the hearing is left open for a specified time to receive written comments and statements to be added.

Pre-Construction Conferences

Every project should have a pre-construction conference prior to commencing any physical activity. This conference is very detailed and specific. All parties concerned; i.e., owner, contractor, sub-contractors, utilities, regulatory agencies, engineer, inspectors, etc. must attend. This is a major milestone in the overall process. The reader is directed to Section 26 of this sourcebook for a detailed presentation.

Ceremonies

The project engineer, on occasion, will be expected to attend and perhaps participate in ceremonies connected with some construction projects.

Often, a municipality will have a groundbreaking ceremony to publicize a large or special project. Usually, the news media is invited so the project receives its due recognition. The format of these groundbreaking events ranges from a photo shoot of the assembled officials to a formal agenda which includes a history of the project and speeches by the dignitaries. On a rare occasion, the project engineer will be called upon to address those assembled. This should be considered an honor and the comments be kept to a minimum.

Seminars & Lectures

An engineer is often called upon to participate in technical seminars. This is good public relations and gives your consulting firm a heightened image. Remember, you would not have been asked if you were not perceived as an expert. Therefore, it is wise to devote some extra time on preparing a good presentation.

Occasionally, a consulting engineer will be asked to give a lecture, usually to a high school or college class. This can be quite rewarding as well as challenging. Depending on the topic, this affords an engineer the opportunity to impart specific technical information to an attentive audience. Perhaps one of these lectures may spark an interest in engineering for someone who might otherwise never be exposed to this career. In any event, it is this writer's opinion that the consulting engineer has the responsibility to represent the profession in the highest manner and be willing to promote its virtues and opportunities.

Interviews with the News Media

Upon occasion, the consulting engineer will be interviewed by the press, relative to a current project that has the public interest. Whether it is an issue of opposition or just normal curiosity, the engineer should respond to questions in a positive manner. The engineer must remember that his primary duty is to protect the health and safety of the public and serve the client in a complete and professional way.

It is advisable to give the reporter a written description of the project and be certain that items be accurately communicated. Also, it is advisable to have a good, reproducible drawing of the site plan and a typical illustration or schematic of the proposed project. If it is a building, a floor plan, elevated views or perspective rendering adds greatly to the published article.

If the interview's main thrust is in response to opposition, the engineer must be very careful when answering any question. Some answers may be misleading or improperly reported. Therefore, beware of being verbally trapped. When necessary, decline to be interviewed, especially if there is a possibility of potential litigation.

Planning

REGULATORY APPROVALS

Section 5

A key element of a project is obtaining applicable approvals and permits from appropriate regulatory agencies.

TABLE OF CONTENTS

General Discussion

It is the professional responsibility of the designer to secure the required approvals and permits from all appropriate regulatory authorities, prior to commencing construction.

It certainly is good policy for the design firm to obtain all applicable approvals and permits early in the process so copies can be included in the bidding documents. This will provide the bidders with a clear understanding of any specific conditions attached to the approvals or permits. Sometimes a regulatory agency's approval will contain added special items that may affect the construction cost. Therefore, it is prudent that these items be displayed up front, well in advance of bid opening so the prospective contractors can factor this expense into their bids. This practice can avoid some potential costly change orders later in the project by a contractor claiming a modification in the scope of work after contract award.

These approvals and permits are extremely important to the project. Their conditions must comply with and have no modifications made without prior written approval from the affected regulatory agency. These documents are interconnected with the plans and specifications and are part of the construction contract. They are never to be taken for granted; rather their conditions are mandatory. It is the project engineer's responsibility to see that all conditions are enforced.

Regulatory Agencies

Civil engineering is that discipline of engineering devoted to planning, design and construction of municipal, public and private facilities. These areas include: sanitary sewer systems, wells, dams, water distribution networks, fire protection, buildings, structures, environmental controls, highways, roads, bridges, tunnels, airports, marine facilities and related specialties. All of these areas are directly related to the health and safety of the public. Therefore, applicable guidelines and codes have evolved over time to establish minimum standards for each type of project.

Appropriate governmental legislative bodies have enacted specific laws for the purpose of monitoring, regulating and enforcing these standards and codes. In turn, various regulatory agencies were established to ensure that these laws were complied with by those involved in designing and constructing these types of structures and facilities.

It is not the intent of this section to list all of these regulatory agencies or note every type of approval or permit required. However, the following tabulation summarizes most of the major areas where regulatory approvals (or permits) are required; whether federal, state, county or local municipality. Any omissions is entirely unintentional.

Abandonment Plans (Landfills, Dams, Wells, etc.)
Air Emission Control Facilities
Air Terminals
Airports & Runways
Athletic Facilities
Bridges
Buildings (Structural, HVAC, Plumbing & Fire Protection)
Campgrounds
Canals
Chemical Processing Facilities
Communication Transmission Systems
Dams (Hydroelectric & Water Control)
Electrical Systems
Electrical Transmission Networks

Educational Complex & Facilities
Hazardous Waste Landfills
Highways
Hospitals
HVAC Systems
Industrial Complexes
Man-Made Lakes & Wetlands
Manufacturing Facilities
Marine Facilities
Mining Operations
Nuclear Power Plants
Nursing Homes
Parking Lots, Ramps & Structures
Pipelines
Power Generating Facilities
Private Land Development
Pumping Systems
Railroad Networks
Reservoirs
Roads
Sanitary Collection Systems
Sanitary Landfills
Sanitary Sewage Systems
School Facilities
Storm Sewer Systems
Streets
Structures
Telecommunication Systems & Facilities
Trailer Courts
Tunnels
Wastewater Collection Systems
Wastewater Treatment Facilities
Water Distribution Networks
Water Tanks & Towers
Wells & Water Supply Facilities
Wetlands (Constructed)

Planning

COST ESTIMATING

Section 6

A project can not be successfully completed without adequate funding. Therefore, no project should ever proceed too far before a cost estimate is prepared and a sound financial plan developed.

TABLE OF CONTENTS

Project Estimating

Estimating the cost of a proposed construction project is a very complex process containing many variable factors. This is not a skill that is easily acquired. Proper study, training and experience are needed to become proficient in this area of engineering.

This section does not have ample space, nor is it the intent, to present a complete discussion on cost estimating. The primary purpose is to note the importance of this basic element for all projects and list several insights to aid the reader in fine tuning their estimates.

There are several categories that can have significant impacts on project costs. The estimator should be aware of them and properly evaluate their effects, prior to finalizing the cost estimate. Refer to the following:

1) **Similar Projects:** The best references are similar projects. Refer to their final cost items and related expenses as a sound basis. Experience with similar projects is invaluable.

2) **Material Costs:** Obtain reliable costs for materials and supplies, plus shipping charges, prior to commencing tabulation.

3) **Wage Rates:** Determine if the project will mandate state or federal wage rates. Also, check if local wage rates are required. It is mandatory to factor this into the estimate.

4) **Site Conditions:** Project site conditions that can increase construction costs are: poor soil conditions, wetlands, contaminated materials, conflicting utilities (buried pipe, cables, overhead lines, etc.), environmentally sensitivity area, ground water, river or stream crossings, heavy traffic, buried storage tanks, archaeological sites, endangered species habitat and similar existing conditions.

5) **Inflation Factor:** The presence of inflation is always a factor that can be extremely variable. When utilizing previous, similar projects as a primary basis for estimating, consider the Construction Cost Index as published in the *Engineering News Record*. This nationwide tabulation of the construction industry has been continuously recorded for decades.

6) **Bid Timing:** The timing of the bid opening can have a significant impact on obtaining a low bid. Seasonal variations in construction activity and conflicts with other bid openings are critical factors.

7) **Project Schedule:** The construction schedule can certainly affect the cost. If the project requires too aggressive of a time frame, generally the price increases, especially if there is a significant liquidated damages condition for failure to complete within a specified deadline. Conversely, if the award notice is beyond a reasonable time and the notice to proceed is indefinite, the contractors fear inflation of material costs and may have other projects that have priority. Therefore, most bidders will inflate their bids to protect against these conditions. Any time beyond 60 days may result in higher bids.

8) **Quality of Plans & Specifications:** There is no substitute for well-prepared plans and specifications. It is extremely important that every detail and component of the design be properly executed and fully described. Any vague wording or poorly drawn plan not only causes confusion, but places doubt in the contractor's mind which generally results in a higher bid.

9) **Reputation of Engineer:** If the project engineer or engineering firm has a good sound professional reputation with contractors, it is reflected in reasonably priced bids. If a contractor is comfortable working with a particular engineer, or engineering firm, the project runs smoother and therefore is more cost-effective.

10) **Granting Agency:** If a granting agency is involved in funding a portion of the project, contractors will take this into consideration when preparing their bids. Some granting agencies have considerable additional paperwork that is not normally required in a non-funded project. Sometimes this expected extra paperwork elevates the bid.

11) **Regulatory Requirements:** Sometimes there are conditions in regulatory agency approvals that will be costly to perform. Therefore, to be completely aboveboard with potential bidders, it is strongly recommended that copies of all regulatory approvals be contained in all bidding documents.

12) **Insurance Requirements:** General insurance requirements, such as performance bond, payment bond and contractors general liability are normal costs of doing business. However, there are special projects that require additional coverage. Railroad crossings are a prime example. Insurance premiums for these supplemental policies add to the project cost and must be considered up front.

13) **Size of Project:** The size and complexity of a project determines if local contractors have the capacity to execute the work. The larger and more intricate the proposed project is, the more it will potentially attract the attention of a broader number of prospective bidders. This is good for competition, but may increase mobilization costs.

14) **Locale of Work Site:** The locale of the proposed work can be a significant component in developing a realistic cost estimate. A rural setting usually has a limited labor force skilled in the construction trades. Therefore, the contractor must import tradesmen and generally pay per diem expenses; i.e., out-of-town lodging and related costs. Additionally, remote settings increase the charges for material shipment.

15) **Value Engineering:** Some agencies mandate that multi-million dollar projects perform a value engineering review, prior to finalizing the design or commencing the bidding process. Therefore, the estimator should be aware of this factor early in the process.

16) **Contingency:** The rule-of-thumb has historically added a 10% contingency on the construction total to cover those unforeseen costs that crop up as a project evolves. During times of high inflation or the limited amount of key construction materials and supplies, it is wise to increase the contingency to 15% or 20% for a more realistic estimate and provide a safety factor.

17) **Supplemental Studies & Investigations:** As stated in Item 4, some project sites will require special studies and/or investigations. Costs for this special work should be included in the initial cost estimate to avoid future surprises.

18) **Judgement:** In the final analysis, the best component of a good cost estimate is the art of practicing sound technical judgement. This factor is acquired by experience and the mentoring of senior personnel.

Cost Estimate

Public Water Supply System

for Glen Flora, Wisconsin

DESCRIPTION	TOTAL
WATER SUPPLY:	
Well w/Concrete Pump Base	$ 31,580
Pumphouse Structure	28,500
Turbine Pump w/Controls	12,500
Interior Piping	5,000
Electrical Service W/Main Panel & Controls	6,000
Electrical System & Lighting	3,520
Telemetry System & Electrical Retrofit	5,000
Chemical Feed Equipment	3,000
Auxiliary Power Mechanism	5,500
Surge Eliminator & Pressure Switch	750
Yard Piping	3,500
Yard Hydrant	1,500
Site Restoration	1,000
STORAGE:	
Tower Foundation	$22,000
Elevated Water Tank w/Recirculation Pump	121,000
Painting	19,800
AUXILIARY WELL:	
Purchase Existing School Well	10,000
Upgrade Well & Pump	15,000
Upgrade Pumphouse	8,500
Chemical Feed Equipment	3,000
Right Angle Gear Drive Unit	5,500
Yard Piping	3,500
Controls (Interconnect to Telemetry)	2,500
Interior Piping Retrofit	4,000
Electrical Upgrading	3,500
DISTRIBUTION SYSTEM:	
Watermain	142,450
Main Valves	11,550
Hydrants	25,725
Water Services w/Curb Stops	20,000
Water Meters	11 825
Street Repair & Restoration	47,115
SUB-TOTAL (CONSTRUCTION)	$584,315
Construction Contingency @ 12%	$ 57,141
Basic Engineering Services	49,700
Construction Staking	6,000
Inspection and Related Services	25,500
Testing Fees & Soil Borings	6,000
PSC Coordination	7,500
Wellhead Protection Plan	6,500
Legal Services	2,500
Administrative Costs	2,500
Accounting Fees	1,500
TOTAL ESTIMATED PROJECT	$749,156

NOTE: Does not include cost of grant administration, publication fees, property surveys, tower site purchase, audit fees, bond council services and related tasks.

Source: Morgan & Parmley, Ltd.

Present Worth Analysis

of Aerated Lagoon Facility

SHELDON, WISCONSIN — DESIGN YEAR 2016

COMPONENT DESCRIPTION	Initial Cost ($)	Service Life (Years)	Future Cost 15th Yr. ($)	Cost Per Yr. ($)	Salvage Value 20th Yr. ($)
MOBILIZATION	5,000	N/A	-0-	-0-	-0-
INTERCEPTOR:					
INFLUENT PIPING	4,000	50	-0-	80	2,400
MANHOLE (FINAL)	3,000	30	-0-	100	1,000
INFLUENT FLOW METER & SAMPLER	6,000	15	6,000	400	4,000
RAW SEWAGE PUMPING:					
ELECTRICAL SERVICE & CONTROLS	8,000	40	-0-	200	4,000
STRUCTURAL	25,000	40	-0-	625	12,500
MECHANICAL	15,000	20	-0-	750	-0-
TELEMETRY	15,000	30	-0-	500	5,000
FORCE MAIN:					
PIPING	75,000	50	-0-	1,500	45,000
AIR RELIEF M.H. W/ VALVING	10,000	30	-0-	333	3,333
HIGHWAY CROSSING (BORED/CASED)	15,000	50	-0-	300	9,000
EROSION CONTROL	5,000	N/A	-0-	-0-	-0-
TRAFFIC FLAGGING & SIGNAGE	2,500	N/A	-0-	-0-	-0-
RESTORATION (SEED, FER. & MULCH)	5,000	N/A	-0-	-0-	-0-
AERATED LAGOON SYSTEM:					
CLEARING & GRUBBING	5,000	N/A	-0-	-0-	-0-
CELL CONSTRUCTION	71,500	50	-0-	1,430	42,900
PVC LINER	38,000	50	-0-	760	22,800
RIP RAP	12,000	40	-0-	300	6,000
AIR SYSTEM (PIPING)	17,000	40	-0-	425	8,500
CELL PIPING (WET)	40,000	40	-0-	1,000	20,000
SECURITY FENCING & SIGNAGE	4,200	30	-0-	140	1,400
EROSION CONTROL	7,500	N/A	-0-	-0-	-0-
RESTORATION (SEED, FERT. & MULCH)	10,000	N/A	-0-	-0-	-0-
OUTFALL (EFFLUENT DISCHARGE)	66,000	50	-0-	1,320	39,600
CONTROL BUILDING:					
GENERAL CONSTRUCTION	45,000	40	-0-	1,125	22,500
OFFICE & LAB EQUIPMENT	4,000	20	-0-	200	-0-
ELECTRICAL (SERVICE)	5,000	30	-0-	167	1,667
ELECTRICAL / H & V (BUILDING)	12,000	30	-0-	400	4,000
ELECTRICAL (BLOWER SYSTEM)	18,000	20	-0-	900	-0-
MECHANICAL (BLOWER SYSTEM)	21,500	15	21,500	1,433	14,333
EFF. FLOW METER & SAMPLER	6,000	15	6,000	400	4,000
WATER SERVICE (POTABLE)	17,000	40	-0-	425	8,500
PLUMBING SYSTEM	6,000	30	-0-	200	2,000
UV DISINFECTION SYSTEM	18,000	30	-0-	600	6,000
AUXILIARY POWER: (DIESEL) L.S.:					
PORTABLE TRASH PUMP W / TRAILER	12,000	20	-0-	600	-0-
SAFETY EQUIPMENT	4,500	20	-0-	225	-0-
MAINTENANCE EQUIPMENT:					
SHOP TOOLS & DUCK BOAT	4,250	20	-0-	212	-0-
UTILITY TRACTOR W / MOWER	18,125	15	18,125	1,208	12,083
ACCESS DRIVE PAVING	10,000	20	-0-	500	-0-
SEPTAGE RECEIVING STATION (BASIC)	9,500	30	-0-	317	3,167
CONSTRUCTION SIGN	750	N/A	-0-	-0-	-0-
DEMOLITION OF EXISTING STP	5,000	N/A	-0-	-0-	-0-
CONTINGENCY @ 10%	68,132	N/A	-0-	-0-	-0-
ESTIMATED CONSTRUCTION	**$749,457**	**TOTALS**	**$51,625**	**$19,075**	**▲$305,683**
ENGINEERING SERVICES	89,935	N/A	-0-	-0-	-0-
LEGAL	5,000	N/A	-0-	-0-	-0-
ADMINISTRATION	3,500	N/A	-0-	-0-	-0-
CONSTRUCTION INSPECTION & STAKING	45,000	N/A	-0-	-0-	-0-
TESTING & LAB FEES	4,000	N/A	-0-	-0-	-0-
LAND PURCHASE	50,000	N/A	-0-	-0-	-0-
INITIAL ESTIMATED COST (TOTAL)	**$946,892**	—	—	—	**▲$73,656**

▲ DOES NOT INCLUDE LAND VALUE

Source: Morgan & Parmley, Ltd.

Loan Payment Schedule

Monthly Payment Necessary on a $1,000 Loan

Years	5%	5½%	6%	6½%	7%	7½%	8%	8½%	9%	9½%
1	$85.61	$85.84	$86.07	$86.30	$86.53	$86.76	$86.99	$87.22	$87.45	$87.68
2	43.87	44.10	44.32	44.55	44.77	45.00	45.22	45.45	45.67	45.90
3	**29.97**	**30.20**	**30.42**	**30.65**	**30.88**	**31.11**	**31.34**	**31.57**	**31.80**	**32.02**
4	23.03	23.26	23.49	23.72	23.95	24.18	24.42	24.65	24.89	25.11
5	18.87	19.10	19.33	19.57	19.80	20.04	20.27	20.50	20.74	20.97
6	**16.11**	**16.34**	**16.57**	**16.81**	**17.05**	**17.29**	**17.52**	**17.76**	**18.00**	**18.23**
7	14.13	14.37	14.61	14.85	15.09	15.34	15.59	15.83	16.08	16.32
8	12.66	12.90	13.14	13.39	13.64	13.89	14.14	14.39	14.63	14.88
9	**11.52**	**11.76**	**12.01**	**12.26**	**12.51**	**12.76**	**13.01**	**13.27**	**13.52**	**13.77**
10	10.61	10.85	11.10	11.36	11.60	11.87	12.13	12.39	12.64	12.90
11	9.86	10.11	10.37	10.62	10.89	11.15	11.41	11.66	11.92	12.18
12	**9.25**	**9.50**	**9.76**	**10.02**	**10.28**	**10.55**	**10.80**	**11.06**	**11.32**	**11.59**
13	8.73	8.99	9.24	9.51	9.78	10.05	10.32	10.58	10.85	11.11
14	8.29	8.55	8.81	9.08	9.35	9.63	9.89	10.16	10.43	10.70
15	**7.91**	**8.17**	**8.44**	**8.71**	**8.99**	**9.27**	**9.56**	**9.84**	**10.10**	**10.38**
16	7.58	7.84	8.11	8.39	8.67	8.96	9.24	9.51	9.78	10.06
17	7.29	7.56	7.83	8.11	8.40	8.69	8.96	9.24	9.51	9.79
18	**7.03**	**7.30**	**7.58**	**7.87**	**8.16**	**8.45**	**8.74**	**9.02**	**9.30**	**9.49**
19	6.80	7.08	7.36	7.65	7.94	8.24	8.50	8.78	9.06	9.35
20	6.60	6.88	7.16	7.46	7.75	8.06	8.34	8.63	8.92	9.22
25	**5.85**	**6.14**	**6.44**	**6.75**	**7.07**	**7.39**	**7.69**	**8.00**	**8.30**	**8.61**
30	5.37	5.68	6.00	6.32	6.65	6.99	7.29	7.60	7.93	8.24
35	5.05	5.37	5.70	6.04	6.39	6.74	7.06	7.40	7.74	8.08
40	**4.82**	**5.16**	**5.50**	**5.85**	**6.21**	**6.58**	**6.90**	**7.25**	**7.60**	**7.96**

Example: $4,100 Loan @ 7% for 10 years

Multiply $\frac{4,100}{1,000}$ X 11.60 = $47.56 per month

Compound Interest Amount of a Given Principal

The amount A at the end of n years of a given principal P placed at compound interest to-day is A = P X x, the interest (at the rate of r percent per annum) is compounded annually, the factor x being taken from the following tables. Values of x.

Years	r = 4	5	6	7	8	9	10	11	12
2	1.082	1.102	1.124	1.145	1.166	1.188	1.210	1.232	1.254
3	1.125	1.158	1.191	1.225	1.260	1.295	1.331	1.368	1.405
4	**1.170**	**1.216**	**1.262**	**1.311**	**1.360**	**1.412**	**1.464**	**1.518**	**1.574**
5	1.217	1.276	1.338	1.403	1.469	1.539	1.611	1.685	1.762
6	1.265	1.340	1.419	1.501	1.587	1.677	1.772	1.870	1.974
7	**1.316**	**1.407**	**1.504**	**1.606**	**1.714**	**1.828**	**1.949**	**2.076**	**2.211**
8	1.369	1.477	1.594	1.718	1.851	1.993	2.144	2.305	2.476
9	1.423	1.551	1.689	1.838	1.999	2.172	2.358	2.558	2.773
10	**1.480**	**1.629**	**1.791**	**1.967**	**2.159**	**2.367**	**2.594**	**2.839**	**3.106**
11	1.539	1.710	1.898	2.105	2.332	2.580	2.853	3.152	3.479
12	1.601	1.796	2.012	2.252	2.518	2.813	3.138	3.498	3.896
13	1.665	1.886	2.133	2.410	2.720	3.066	3.452	3.883	4.363
14	1.732	1.980	2.261	2.579	2.937	3.342	3.797	4.310	4.887
15	**1.801**	**2.079**	**2.397**	**2.759**	**3.172**	**3.642**	**4.177**	**4.785**	**5.474**
16	1.873	2.183	2.540	2.952	3.426	3.970	4.595	5.311	6.130
17	1.948	2.292	2.693	3.159	3.700	4.328	5.054	5.895	6.866
18	**2.026**	**2.407**	**2.854**	**3.380**	**3.996**	**4.717**	**5.560**	**6.543**	**7.690**
19	2.107	2.527	3.026	3.616	4.316	5.142	6.116	7.263	8.613
20	2.191	2.653	3.207	3.870	4.661	5.604	6.727	8.062	9.646
25	**2.666**	**3.386**	**4.292**	**5.427**	**6.848**	**8.623**	**10.834**	**13.585**	**17.000**
30	3.243	4.322	5.743	7.612	10.062	13.267	17.449	22.892	29.960
40	4.801	7.040	10.285	14.974	21.724	31.408	45.258	64.999	93.049

This table computed from the formula

$$x = [1 + (r/100)]^n$$

Amount of an Annuity

The amount S accumulated at the end of n years by a given annual payment Y set aside at the end of each year is $S = Y \times v$, where the factor v is to be taken from the following table. (Interest at r percent per annum, compounded annually.) Values of v.

Years	r = 4	5	6	7	8	9	10	11	12
2	2.040	2.050	2.060	2.070	2.080	2.090	2.100	2.110	2.120
3	3.122	3.152	3.184	3.215	3.246	3.278	3.310	3.342	3.374
4	**4.246**	**4.310**	**4.375**	**4.440**	**4.506**	**4.573**	**4.641**	**4.710**	**4.779**
5	5.416	5.526	5.637	5.751	5.867	5.985	6.105	6.228	6.353
6	6.633	6.802	6.975	7.153	7.336	7.523	7.716	7.913	8.115
7	**7.898**	**8.142**	**8.394**	**8.654**	**8.923**	**9.200**	**9.487**	**9.783**	**10.089**
8	9.214	9.549	9.897	10.260	10.637	11.028	11.436	11.859	12.300
9	10.583	11.026	11.491	11.978	12.487	13.021	13.579	14.164	14.776
10	**12.006**	**12.578**	**13.181**	**13.816**	**14.486**	**15.193**	**15.937**	**16.722**	**17.549**
11	13.486	14.206	14.971	15.783	16.645	17.560	18.531	19.561	20.654
12	15.026	15.917	16.870	17.888	18.977	20.140	21.384	22.713	24.133
13	16.627	17.712	18.882	20.140	21.495	22.953	24.522	26.211	28.029
14	18.292	19.598	21.015	22.550	24.215	26.019	27.975	30.095	32.392
15	**20.023**	**21.578**	**23.275**	**25.129**	**27.152**	**29.360**	**31.772**	**34.405**	**37.279**
16	21.824	23.657	25.672	27.887	30.324	33.003	35.949	39.189	42.753
17	23.697	25.840	28.212	30.840	33.750	36.973	40.544	44.500	48.883
18	**25.645**	**28.132**	**30.905**	**33.998**	**37.450**	**41.300**	**45.598**	**50.395**	**55.749**
19	27.671	30.538	33.759	37.378	41.446	46.017	51.158	56.939	63.439
20	29.777	33.065	36.785	40.995	45.761	51.159	57.274	64.202	72.052
25	**41.645**	**47.725**	**54.863**	**63.247**	**73.105**	**84.699**	**98.345**	**114.411**	**133.332**
30	56.083	66 436	79.055	94.458	113.281	136.303	164.489	199.017	241.330
40	95.023	20.794	154.755	199.628	259.050	337.869	442.576	581.810	767.079

Formula:

$$v = [\{1 + (r/100)\}^n - 1] \div (r/100)$$
$$= (x - 1) \div (r/100)$$

Principal Which Will Amount to a Given Sum

The principal P, which, if placed at compound interest to-day, will amount to a given sum A at the end of n years is $P = A \times x'$, the interest (at the rate of r percent per annum) is compounded annually, the factor x′ being taken from the following table. Values of x′.

Years	r = 4	5	6	7	8	9	10	11	12
2	0.925	0.907	0.890	0.873	0.857	0.842	0.826	0.812	0.797
3	0.889	0.864	0.840	0.816	0.794	0.772	0.751	0.731	0.712
4	**0.855**	**0.823**	**0.792**	**0.763**	**0.735**	**0.708**	**0.683**	**0.659**	**0.636**
5	0.822	0.784	0.747	0.713	0.681	0.650	0.621	0.593	0.567
6	0.790	0.746	0.705	0.666	0.630	0.596	0.564	0.535	0.507
7	**0.760**	**0.711**	**0.665**	**0.623**	**0.583**	**0.547**	**0.513**	**0.482**	**0.452**
8	0.731	0.677	0.627	0.582	0.540	0.502	0.467	0.434	0.404
9	0.703	0.645	0.592	0.544	0.500	0.460	0.424	0.391	0.361
10	**0.676**	**0.614**	**0.558**	**0.508**	**0.463**	**0.422**	**0.386**	**0.352**	**0.322**
11	0.650	0.585	0.527	0.475	0.429	0.388	0.350	0.317	0.287
12	0.625	0.557	0.497	0.444	0.397	0.356	0.319	0.286	0.257
13	0.601	0.530	0.469	0.415	0.368	0.326	0.290	0.258	0.229
14	0.577	0.505	0.442	0.388	0.340	0.299	0.263	0.232	0.205
15	**0.555**	**0.481**	**0.417**	**0.362**	**0.315**	**0.275**	**0.239**	**0.209**	**0.183**
16	0.534	0.458	0.394	0.339	0.292	0.252	0.218	0.188	0.163
17	0.513	0.436	0.371	0.317	0.270	0.231	0.198	0.170	0.146
18	**0.494**	**0.416**	**0.350**	**0.296**	**0.250**	**0.212**	**0.180**	**0.153**	**0.130**
19	0.475	0.396	0.331	0.277	0.232	0.194	0.164	0.138	0.116
20	0.456	0.377	0.312	0.258	0.215	0.178	0.149	0.124	0.104
25	**0.375**	**0.295**	**0.233**	**0.184**	**0.146**	**0.116**	**0.092**	**0.074**	**0.059**
30	0.308	0.231	0.174	0.131	0.099	0.075	0.057	0.044	0.033
40	0.208	0.142	0.097	0.067	0.046	0.032	0.022	0.015	0.011

Formula:

$$x' = [1 + (r/100)]^{-n} = 1/x$$

Annuity Which Will Amount to a Given Sum (Sinking Fund)

The annual payment, Y, which, if set aside at the end of each year, will amount with accumulated interest to a given sum S at the end of n years is Y=S×v′, where the factor v′ is given below. (Interest at r percent per annum, compounded annually.) Values of v′.

Years	r = 4	5	6	7	8	9	10	11	12
2	0.490	0.488	0.485	0.483	0.481	0.478	0.476	0.474	0.472
3	0.320	0.317	0.314	0.311	0.308	0.305	0.302	0.299	0.296
4	0.235	0.232	0.229	0.225	0.222	0.219	0.215	0.212	0.209
5	0.185	0.181	0.177	0.174	0.170	0.167	0.164	0.161	0.157
6	0.151	0.147	0.143	0.140	0.136	0.133	0.130	0.126	0.123
7	0.127	0.123	0.119	0.116	0.112	0.109	0.105	0.102	0.099
8	0.109	0.105	0.101	0.097	0.094	0.091	0.087	0.084	0.081
9	0.094	0.091	0.087	0.083	0.080	0.077	0.074	0.071	0.068
10	0.083	0.080	0.076	0.072	0.069	0.066	0.063	0.060	0.057
11	0.074	0.070	0.067	0.063	0.060	0.057	0.054	0.051	0.048
12	0.067	0.063	0.059	0.056	0.053	0.050	0.047	0.044	0.041
13	0.060	0.056	0.053	0.050	0.047	0.044	0.041	0.038	0.036
14	0.055	0.051	0.048	0.044	0.041	0.038	0.036	0.033	0.031
15	0.050	0.046	0.043	0.040	0.037	0.034	0.031	0.029	0.027
16	0.046	0.042	0.039	0.036	0.033	0.030	0.028	0.026	0.023
17	0.042	0.039	0.035	0.032	0.030	0.027	0.025	0.022	0.020
18	0.039	0.036	0.032	0.029	0.027	0.024	0.022	0.020	0.018
19	0.036	0.033	0.030	0.027	0.024	0.022	0.020	0.018	0.016
20	0.034	0.030	0.027	0.024	0.022	0.020	0.017	0.016	0.014
25	0.024	0.021	0.018	0.016	0.014	0.012	0.010	0.009	0.008
30	0.018	0.015	0.013	0.011	0.009	0.007	0.006	0.005	0.004
40	0.011	0.008	0.006	0.005	0.004	0.003	0.002	0.002	0.001

Formula:

$$v' = (r/100) + [\{1 + (r/100)\}^n - 1] = 1/v$$

Present Worth of an Annuity

The Capital C, which, if placed at interest to-day, will provide for a given annual payment Y for a term of n years before it is exhausted is C=Y×w, where the factor w is given below. (Interest at r percent per annum, compounded annually.) Values of w.

Years	r = 4	5	6	7	8	9	10	11	12
2	1.886	1.859	1.833	1.808	1.783	1.759	1.736	1.713	1.690
3	2.775	2.723	2.673	2.624	2.577	2.531	2.487	2.444	2.402
4	3.630	3.546	3.465	3.387	3.312	3.240	3.170	3.102	3.037
5	4.452	4.329	4.212	4.100	3.993	3.890	3.791	3.696	3.605
6	5.242	5.076	4.917	4.766	4.623	4.486	4.355	4.231	4.111
7	6.002	5.786	5.582	5.389	5.206	5.033	4.868	4.712	4.564
8	6.733	6.463	6.210	5.971	5.747	5.535	5.335	5.146	4.968
9	7.435	7.108	6.802	6.515	6.247	5.995	5.759	5.537	5.328
10	8.111	7.722	7.360	7.024	6.710	6.418	6.145	5.889	5.650
11	8.760	8.306	7.887	7.499	7.139	6.805	6.495	6.206	5.938
12	9.385	8.863	8.384	7.943	7.536	7.161	6.814	6.492	6.194
13	9.986	9.393	8.853	8.358	7.904	7.487	7.103	6.750	6.424
14	10.563	9.899	9.295	8.745	8.244	7.786	7.367	6.982	6.628
15	11.118	10.380	9.712	9.108	8.559	8.061	7.606	7.191	6.811
16	11.652	10.838	10.106	9.447	8.851	8.313	7.824	7.379	6.974
17	12.166	11.274	10.477	9.763	9.122	8.544	8.022	7.549	7.120
18	12.659	11.689	10.828	10.059	9.372	8.756	8.201	7.702	7.250
19	13.134	12.085	11.158	10.336	9.604	8.950	8.365	7.839	7.366
20	13.590	12.462	11.470	10.594	9.818	9.129	8.514	7.963	7.469
25	15.622	14.094	12.783	11.654	10.675	9.823	9.077	8.422	7.843
30	17.292	15.372	13.765	12.409	11.258	10.274	9.427	8.694	8.055
40	19.793	17.159	15.046	13.332	11.925	10.757	9.779	8.951	8.244

Formula:

$$w = [1 - \{1 + (r/100)]^{-n} \div [r/100] = v/x$$

Annuity Provided for by a Given Capital

The annual payment Y provided for a term of n years by a given capital C placed at interest to-day is Y=C×w′. (Interest at r percent per annum, compounded annually; the fund supposed to be exhausted at the end of the term.) Values of w′.

Years	r = 4	5	6	7	8	9	10	11	12
2	0.530	0.538	0.545	0.553	0.561	0.568	0.576	0.584	0.592
3	0.360	0.367	0.374	0.381	0.388	0.395	0.402	0.409	0.416
4	**0.275**	**0.282**	**0.289**	**0.295**	**0.302**	**0.309**	**0.315**	**0.322**	**0.329**
5	0.225	0.231	0.237	0.244	0.250	0.257	0.264	0.271	0.277
6	0.191	0.197	0.203	0.210	0.216	0.223	0.230	0.236	0.243
7	**0.167**	**0.173**	**0.179**	**0.186**	**0.192**	**0.199**	**0.205**	**0.212**	**0.219**
8	0.149	0.155	0.161	0.167	0.174	0.181	0.187	0.194	0.201
9	0.134	0.141	0.147	0.153	0.160	0.167	0.174	0.181	0.188
10	**0.123**	**0.130**	**0.136**	**0.142**	**0.149**	**0.156**	**0.163**	**0.170**	**0.177**
11	0.114	0.120	0.127	0.133	0.140	0.147	0.154	0.161	0.168
12	0.107	0.113	0.119	0.126	0.133	0.140	0.147	0.154	0.161
13	0.100	0.106	0.113	0.120	0.127	0.134	0.141	0.148	0.156
14	0.095	0.101	0.108	0.114	0.121	0.128	0.136	0.143	0.151
15	**0.090**	**0.096**	**0.103**	**0.110**	**0.117**	**0.124**	**0.131**	**0.139**	**0.147**
16	0.086	0.092	0.099	0.106	0.113	0.120	0.128	0.136	0.143
17	0.082	0.089	0.095	0.102	0.110	0.117	0.125	0.132	0.140
18	**0.079**	**0.086**	**0.092**	**0.099**	**0.107**	**0.114**	**0.122**	**0.130**	**0.138**
19	0.076	0.083	0.090	0.097	0.104	0.112	0.120	0.128	0.136
20	0.074	0.080	0.087	0.094	0.102	0.110	0.117	0.126	0.134
25	**0.064**	**0.071**	**0.078**	**0.086**	**0.094**	**0.102**	**0.110**	**0.119**	**0.127**
30	0.058	0.065	0.073	0.081	0.089	0.097	0.106	0.115	0.124
40	0.051	0.058	0.066	0.075	0.084	0.093	0.102	0.112	0.121

Formula:

$$w' = [r/100] \div [1 - \{1 + (r/100)\}^{-n}] = 1/w = v' + (r/100)$$

Design

Title Sheet Organization

Section 7

A title sheet is the cover page of the project plans and should properly display key elements of the proposed work. While each design professional certainly has the latitude to arrange the data in a style of their choosing, there are specific components that should be displayed. With this in mind, the following pages are presented to give a generic overview. The reader must be aware of specific requirements from applicable regulatory agencies before finalizing a title page layout, in addition to the basics noted in this section.

As the readers continues through this sourcebook, they should take note of the variety of title page designs. The broad range of actual project plans contained in the following pages will give the reader a rich resource of title page styles.

A typical blank plan sheet and generic plan & profile sheet are provided for general reference.

Table of Contents

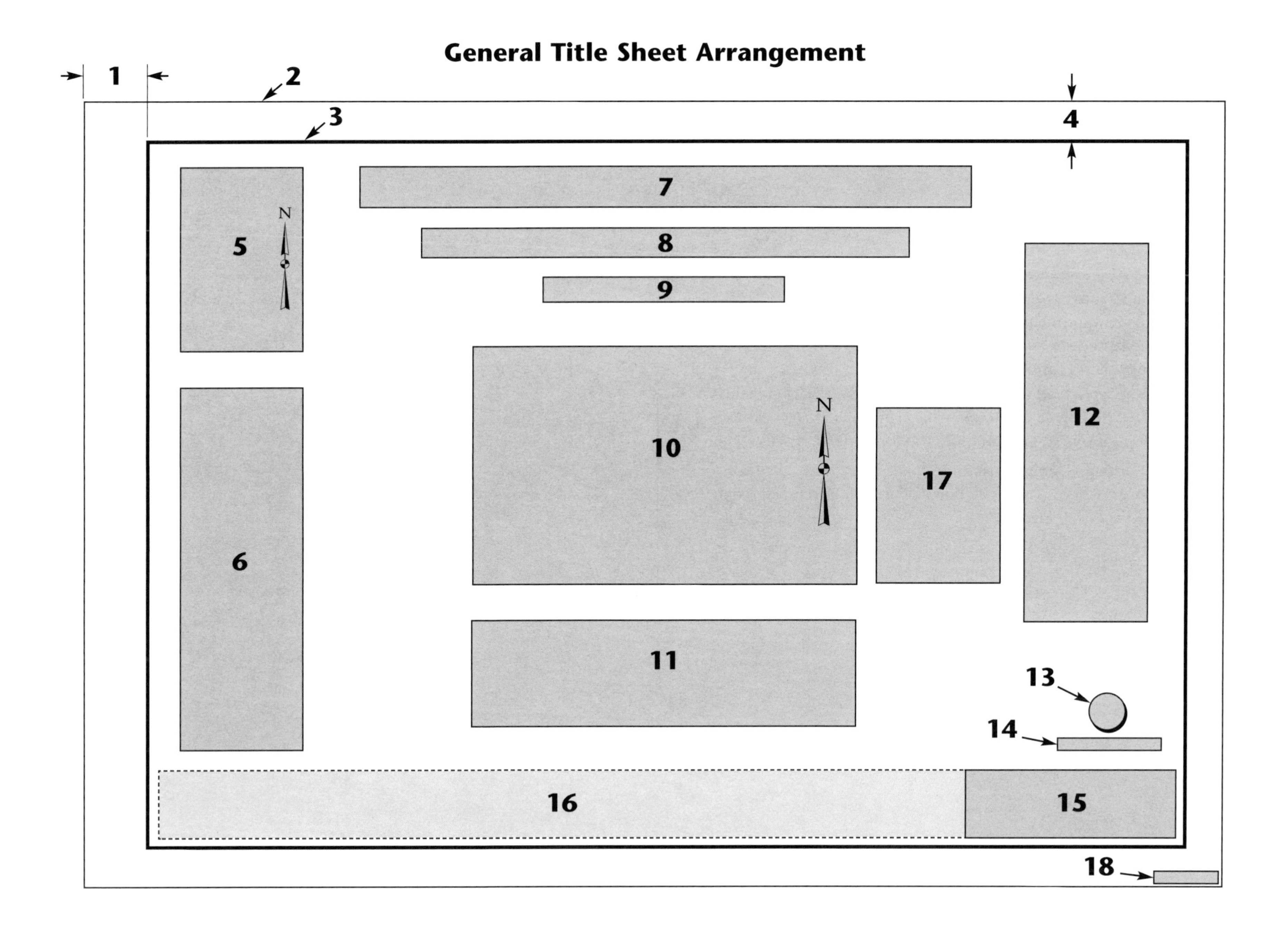
General Title Sheet Arrangement
1
2
3
4
5
N
7
8
9
6
10
N
17
12
11
13
14
16
15
18

Description of Title Sheet Components

1 – Room for Binding
2 – Edge of Sheet
3 – Border (usually heavy line)
4 – Sheet Trim Edge
5 – General Location Map with Project Site Highlighted
6 – General Specifications & Notes
7 – Project Title or Name
8 – Project Owner & Address
9 – Project Number & Date
10 – General Site Plan with Proposed Work Area Highlighted
11 – Special Notes & List of Affected Utilities
12 – Sheet Index of Plan Pages
13 – Area for Professional Seal
14 – Area for Signature of Design Professional
15 – Name & Address of Design Firm
16 – Extended Area for Listing Supplemental Information
17 – Graphic Legend
18 – In-House I.D. Number (if applicable)

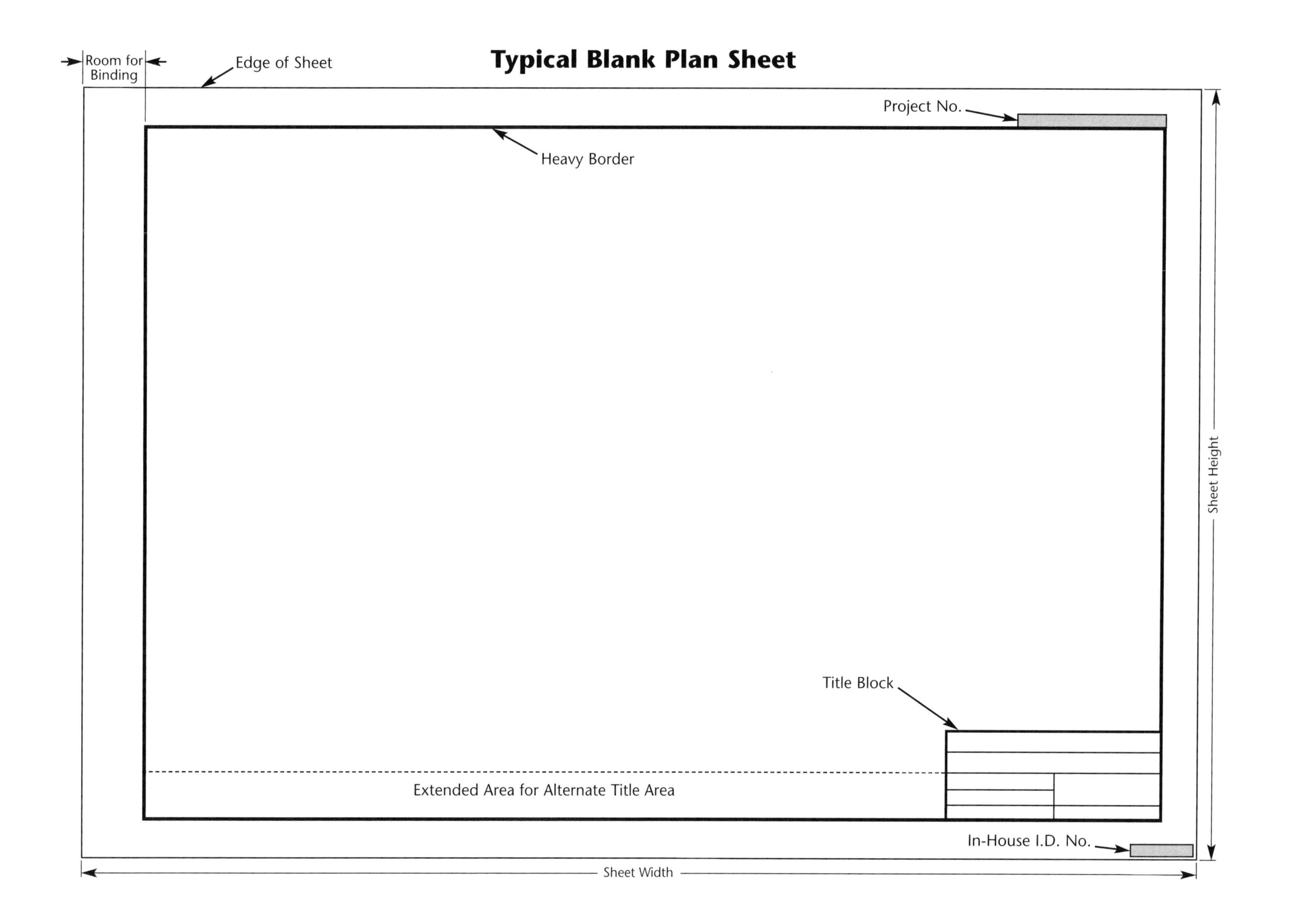
Typical Blank Plan Sheet
Room for Binding
Edge of Sheet
Project No.
Heavy Border
Title Block
Extended Area for Alternate Title Area
In-House I.D. No.
Sheet Height
Sheet Width

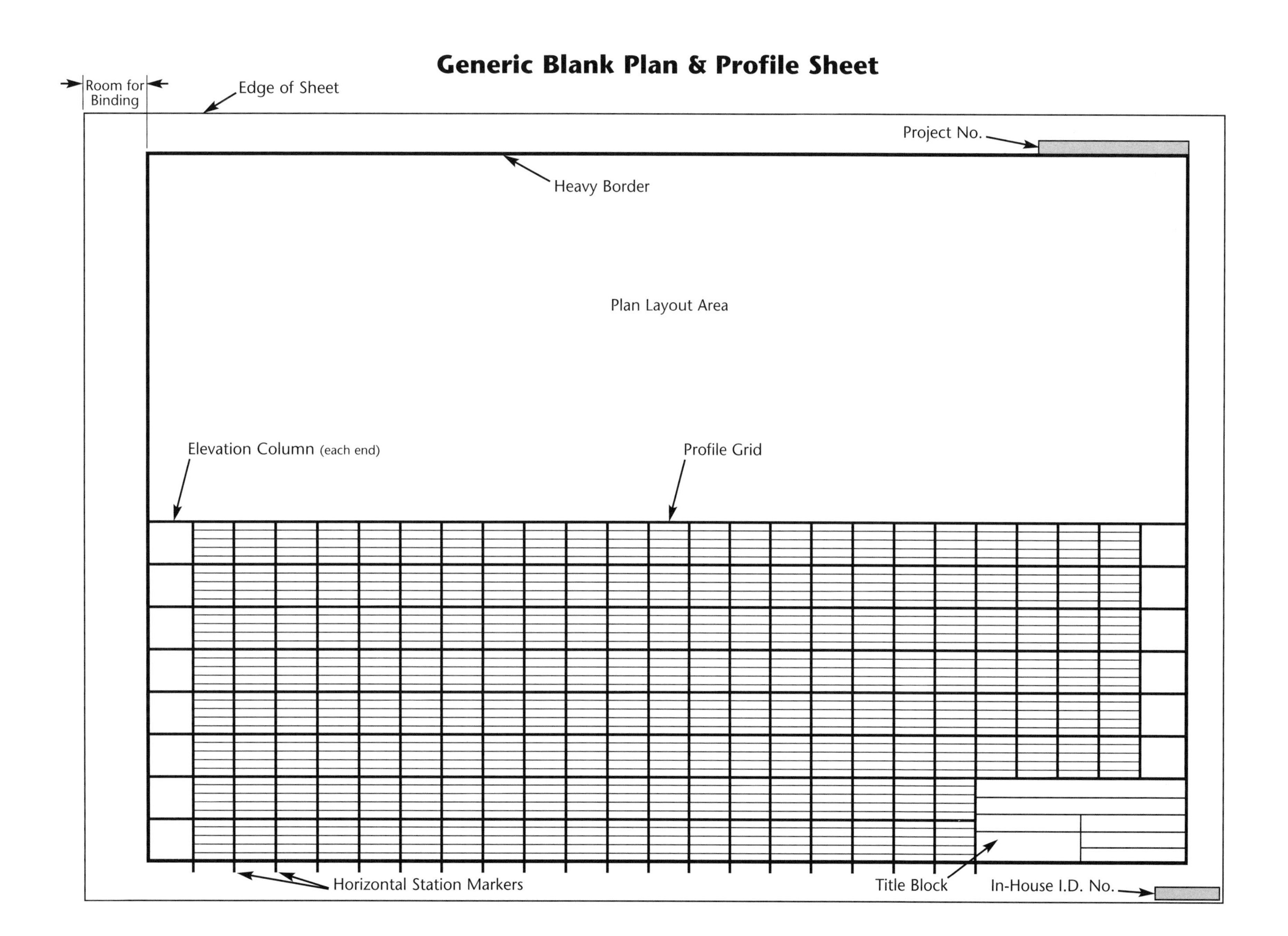
Generic Blank Plan & Profile Sheet
Room for Binding
Edge of Sheet
Project No.
Heavy Border
Plan Layout Area
Elevation Column (each end)
Profile Grid
Horizontal Station Markers
Title Block
In-House I.D. No.

Design

BUILDINGS

Section 8

Building design is a significant portion of a civil engineer's work. These structures include: city & village halls, libraries, service stations, warehouses, churches, schools, fire stations, parking structures, pumphouses, storage garages, maintenance buildings, manufacturing complexes, commercial buildings, and private structures.

Building plans include designs for support systems; i.e., HVAC, electrical, plumbing, telemetry and sprinkler installations. Usually these system layouts are incorporated into the total plan package.

The following plan examples represent a random selection of several types of buildings that are produced by civil engineers. The reader should note that buildings often are a small component of a much larger project, such as a service building for a wastewater treatment facility, and play a subordinate role.

TABLE OF CONTENTS

VILLAGE HALL & LIBRARY for the VILLAGE OF BRUCE BRUCE, WISCONSIN

BUILDING VOLUME= 109,200 CU. FT.
BUILDING AREA= 7,360 SQ. FT.

RUSK COUNTY

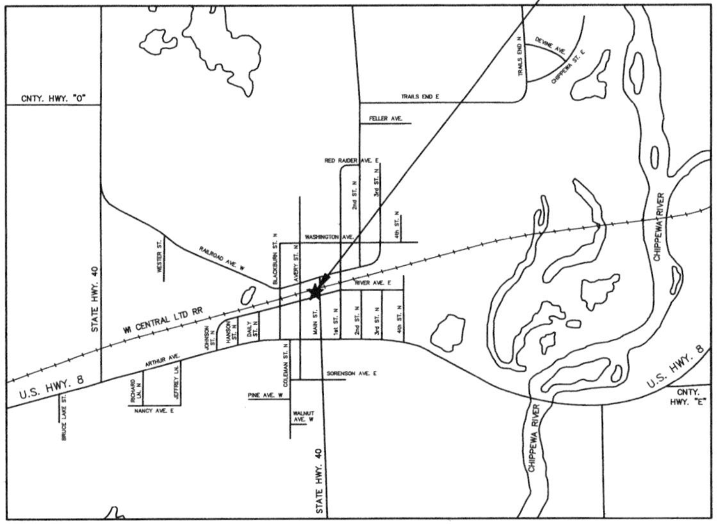

PROJECT LOCATION

OWNER:

VILLAGE OF BRUCE, RUSK COUNTY
VILLAGE HALL
BRUCE, WI 54819

INDEX OF SHEETS

SHEET NO.	DESCRIPTION
T1	TITLE PAGE
AP1	SITE PLAN
AP2	WEST SITE GRADING PLAN
AP3	EAST SITE GRADING PLAN
AP4	SITE MISCELLANEOUS DETAILS
B1	FLOOR PLAN
B2	BUILDING SECTIONS
B3	SECTIONS
B4	SECTIONS AND DETAILS
B5	RESTROOM DETAILS
B6	BUILDING ELEVATIONS
B7	ROOM FINISH & DOOR SCHEDULE AND DETAILS
B8	SECTIONS AND DETAILS
B9	CEILING PLAN
S1	FOUNDATION PLAN
S2	FOUNDATION DETAILS
S3	ROOF FRAMING PLAN
S4	ROOF FRAMING DETAILS & STRUCTURAL NOTES
P1	PLUMBING SITE PLAN
P2	UNDERGROUND PLUMBING PLAN
P3	PLUMBING FLOOR PLAN
P4	PLUMBING ISOMETRIC DIAGRAMS
M1	HVAC SCHEDULES
M2	HVAC FLOOR PLANS
M3	HVAC PARTIAL PLAN AND DETAILS
E1	ELECTRIC SYMBOLS
E2	ELECTRIC SITE PLAN
E3	LIGHTING PLAN
E4	POWER AND SYSTEMS PLAN
E5	ELECTRIC SCHEDULES

BRUCE VILLAGE HALL & LIBRARY
BRUCE, WISCONSIN
SCALE NONE
SHEET T1
DATE DEC. 31, 2001
MORGAN & PARMLEY, LTD.
CONSULTING ENGINEERS
LADYSMITH, WISCONSIN
DR. BY J.T.D.

Courtesy: Morgan & Parmley, Ltd.

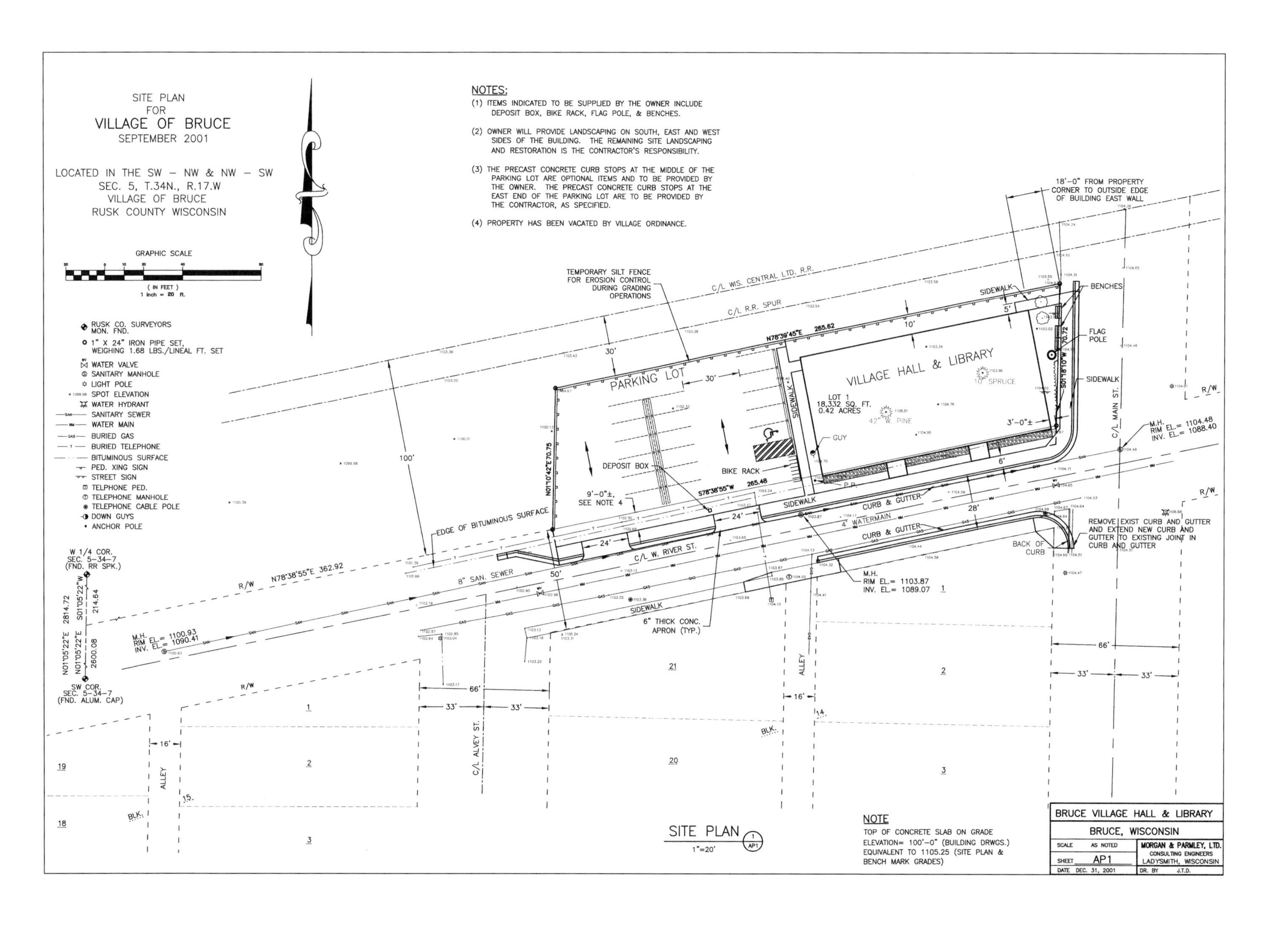
SITE PLAN
FOR
VILLAGE OF BRUCE
SEPTEMBER 2001
LOCATED IN THE SW – NW & NW – SW
SEC. 5, T.34N., R.17.W
VILLAGE OF BRUCE
RUSK COUNTY WISCONSIN
GRAPHIC SCALE
(IN FEET)
1 inch = 20 ft.
NOTES:
(1) ITEMS INDICATED TO BE SUPPLIED BY THE OWNER INCLUDE DEPOSIT BOX, BIKE RACK, FLAG POLE, & BENCHES.
(2) OWNER WILL PROVIDE LANDSCAPING ON SOUTH, EAST AND WEST SIDES OF THE BUILDING. THE REMAINING SITE LANDSCAPING AND RESTORATION IS THE CONTRACTOR'S RESPONSIBILITY.
(3) THE PRECAST CONCRETE CURB STOPS AT THE MIDDLE OF THE PARKING LOT ARE OPTIONAL ITEMS AND TO BE PROVIDED BY THE OWNER. THE PRECAST CONCRETE CURB STOPS AT THE EAST END OF THE PARKING LOT ARE TO BE PROVIDED BY THE CONTRACTOR, AS SPECIFIED.
(4) PROPERTY HAS BEEN VACATED BY VILLAGE ORDINANCE.
RUSK CO. SURVEYORS MON. FND.
1" X 24" IRON PIPE SET, WEIGHING 1.68 LBS./LINEAL FT. SET
WATER VALVE
SANITARY MANHOLE
LIGHT POLE
SPOT ELEVATION
WATER HYDRANT
SANITARY SEWER
WATER MAIN
BURIED GAS
BURIED TELEPHONE
BITUMINOUS SURFACE
PED. XING SIGN
STREET SIGN
TELPHONE PED.
TELEPHONE MANHOLE
TELEPHONE CABLE POLE
DOWN GUYS
ANCHOR POLE
TEMPORARY SILT FENCE FOR EROSION CONTROL DURING GRADING OPERATIONS
C/L WIS. CENTRAL LTD. R.R.
C/L R.R. SPUR
18'–0" FROM PROPERTY CORNER TO OUTSIDE EDGE OF BUILDING EAST WALL
SIDEWALK
BENCHES
FLAG POLE
SIDEWALK
PARKING LOT
VILLAGE HALL & LIBRARY
LOT 1
18,332 SQ. FT.
0.42 ACRES
42" W. PINE
10" SPRUCE
N78°39'45"E 265.62
S01°18'10"W 70.72
N01°10'42"E 70.75
S78°38'55"W 265.48
C/L MAIN ST.
M.H. RIM EL.= 1104.48 INV. EL.= 1088.40
R/W
DEPOSIT BOX
BIKE RACK
GUY
9'–0"±, SEE NOTE 4
EDGE OF BITUMINOUS SURFACE
SIDEWALK
CURB & GUTTER
4" WATERMAIN
CURB & GUTTER
BACK OF CURB
REMOVE EXIST CURB AND GUTTER AND EXTEND NEW CURB AND GUTTER TO EXISTING JOINT IN CURB AND GUTTER
C/L W. RIVER ST.
R/W
N78°38'55"E 362.92
8" SAN. SEWER
W 1/4 COR. SEC. 5–34–7 (FND. RR SPK.)
N01°05'22"E 2814.72
S01°05'22"W 214.64
N01°05'22"E 2600.08
SW COR. SEC. 5–34–7 (FND. ALUM. CAP)
M.H. RIM EL.= 1100.93 INV. EL.= 1090.41
SIDEWALK
6" THICK CONC. APRON (TYP.)
M.H. RIM EL.= 1103.87 INV. EL.= 1089.07
ALLEY
C/L ALLEY ST.
BLK.
SITE PLAN
1"=20'
NOTE
TOP OF CONCRETE SLAB ON GRADE ELEVATION= 100'–0" (BUILDING DRWGS.) EQUIVALENT TO 1105.25 (SITE PLAN & BENCH MARK GRADES)
BRUCE VILLAGE HALL & LIBRARY
BRUCE, WISCONSIN
SCALE AS NOTED
SHEET AP1
DATE DEC. 31, 2001
MORGAN & PARMLEY, LTD.
CONSULTING ENGINEERS
LADYSMITH, WISCONSIN
DR. BY J.T.D.

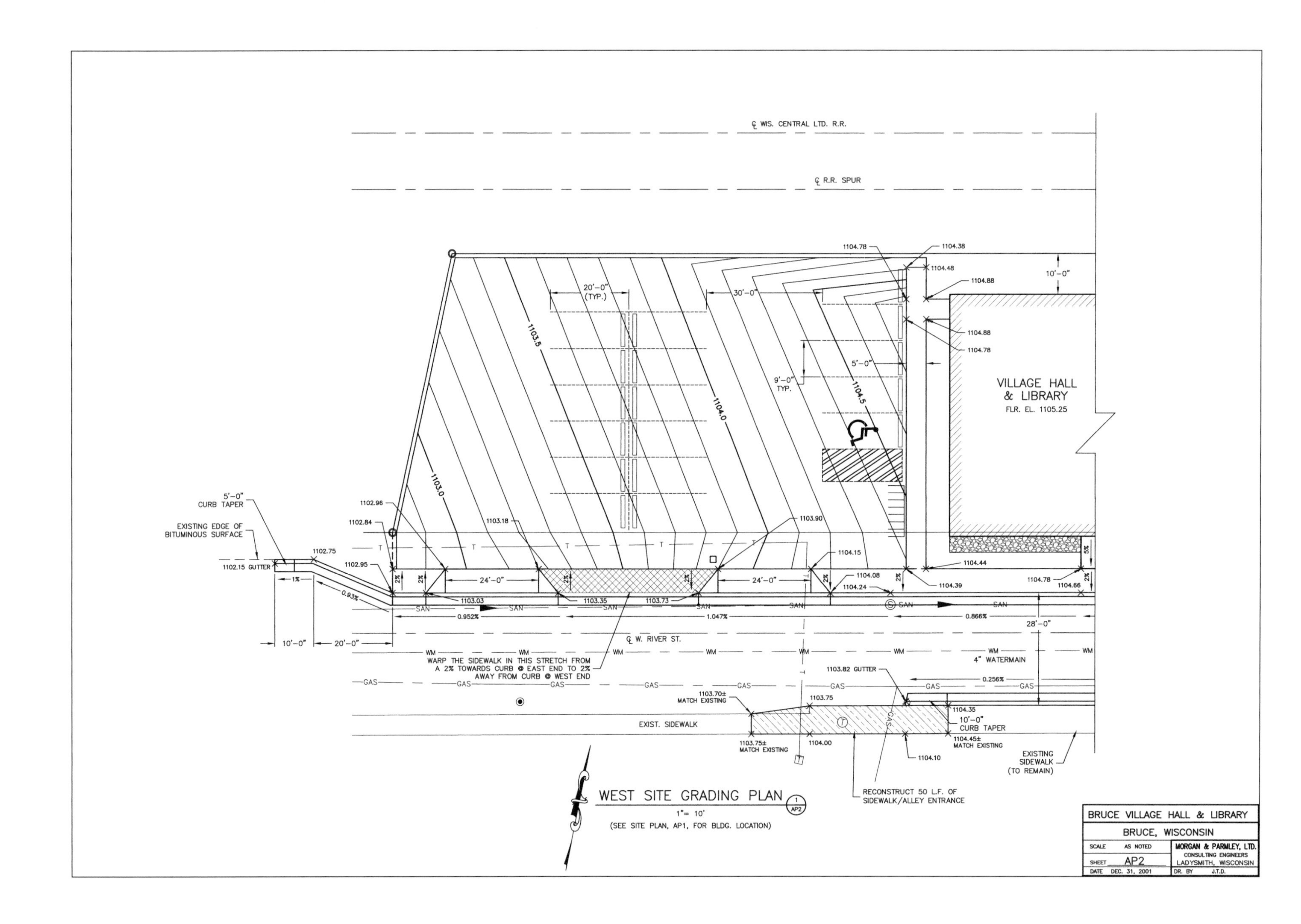

℄ WIS. CENTRAL LTD. R.R.
℄ R.R. SPUR
VILLAGE HALL & LIBRARY
FLR. EL. 1105.25
℄ W. RIVER ST.
WARP THE SIDEWALK IN THIS STRETCH FROM A 2% TOWARDS CURB @ EAST END TO 2% AWAY FROM CURB @ WEST END
4" WATERMAIN
EXIST. SIDEWALK
EXISTING SIDEWALK (TO REMAIN)
RECONSTRUCT 50 L.F. OF SIDEWALK/ALLEY ENTRANCE
5'-0" CURB TAPER
EXISTING EDGE OF BITUMINOUS SURFACE
10'-0" CURB TAPER
WEST SITE GRADING PLAN
1" = 10'
(SEE SITE PLAN, AP1, FOR BLDG. LOCATION)
BRUCE VILLAGE HALL & LIBRARY
BRUCE, WISCONSIN
SCALE AS NOTED
SHEET AP2
DATE DEC. 31, 2001
MORGAN & PARMLEY, LTD.
CONSULTING ENGINEERS
LADYSMITH, WISCONSIN
DR. BY J.T.D.

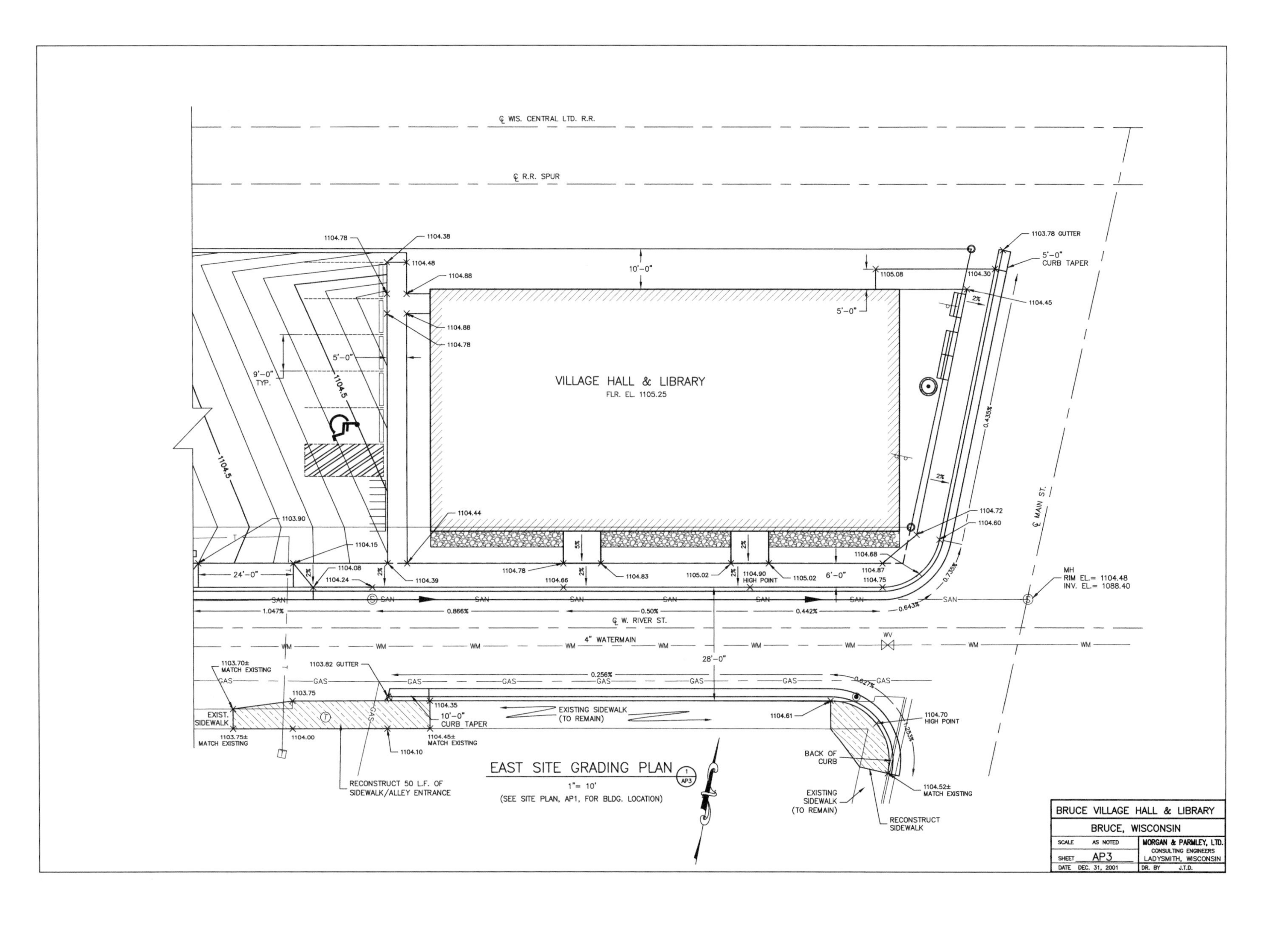
℄ WIS. CENTRAL LTD. R.R.
℄ R.R. SPUR
VILLAGE HALL & LIBRARY
FLR. EL. 1105.25
10'–0"
5'–0"
9'–0" TYP.
24'–0"
6'–0"
28'–0"
1103.78 GUTTER
5'–0" CURB TAPER
℄ MAIN ST.
MH
RIM EL.= 1104.48
INV. EL.= 1088.40
℄ W. RIVER ST.
4" WATERMAIN
1103.70± MATCH EXISTING
1103.82 GUTTER
EXIST. SIDEWALK
1103.75± MATCH EXISTING
10'–0" CURB TAPER
1104.45± MATCH EXISTING
EXISTING SIDEWALK (TO REMAIN)
1104.70 HIGH POINT
BACK OF CURB
1104.52± MATCH EXISTING
RECONSTRUCT 50 L.F. OF SIDEWALK/ALLEY ENTRANCE
EAST SITE GRADING PLAN
1"= 10'
(SEE SITE PLAN, AP1, FOR BLDG. LOCATION)
EXISTING SIDEWALK (TO REMAIN)
RECONSTRUCT SIDEWALK
BRUCE VILLAGE HALL & LIBRARY
BRUCE, WISCONSIN
SCALE AS NOTED
SHEET AP3
DATE DEC. 31, 2001
MORGAN & PARMLEY, LTD.
CONSULTING ENGINEERS
LADYSMITH, WISCONSIN
DR. BY J.T.D.

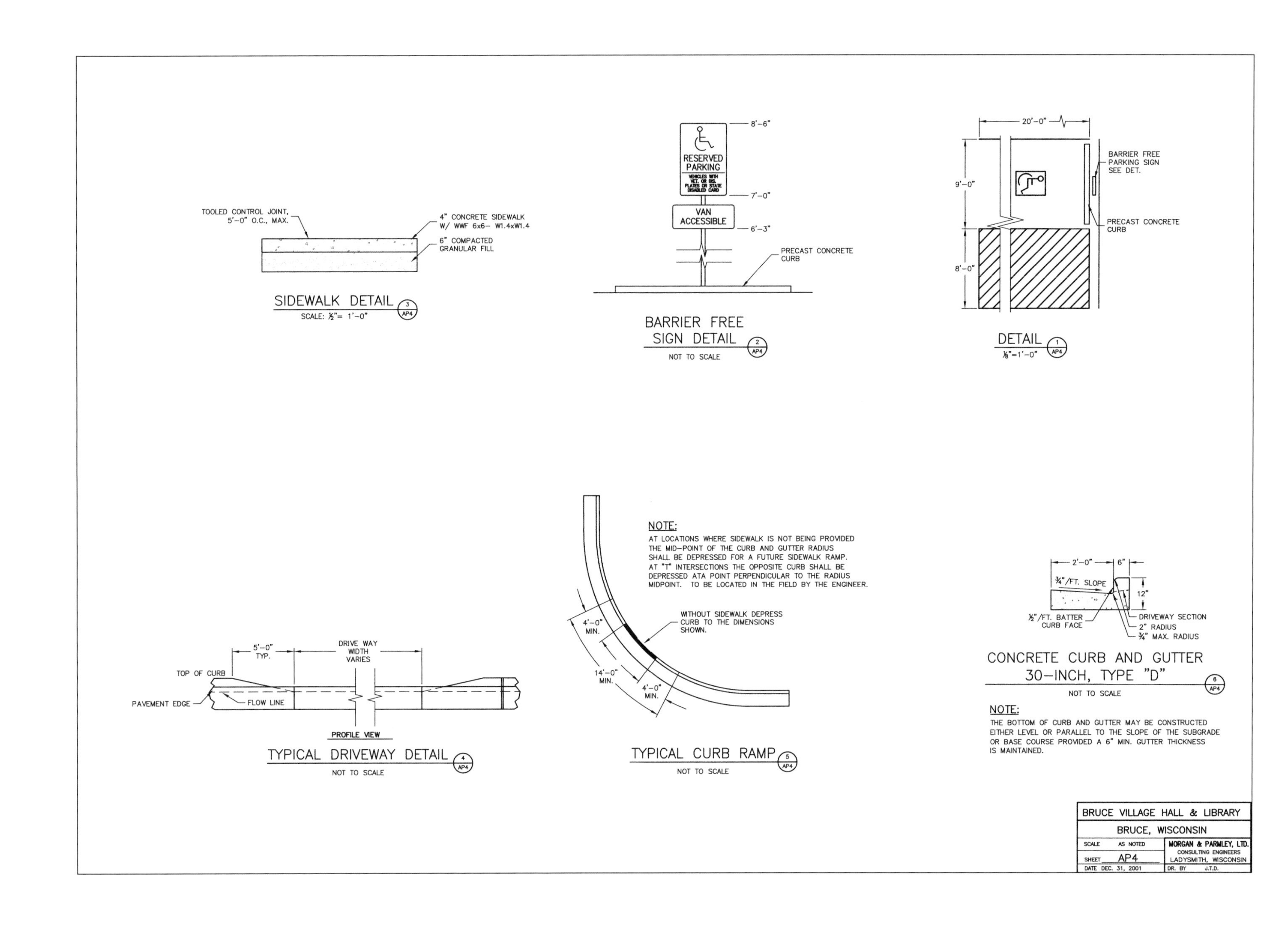
TOOLED CONTROL JOINT, 5'-0" O.C., MAX.
4" CONCRETE SIDEWALK W/ WWF 6x6- W1.4xW1.4
6" COMPACTED GRANULAR FILL
SIDEWALK DETAIL
SCALE: ½"= 1'-0"
3 AP4
8'-6"
RESERVED PARKING
7'-0"
VAN ACCESSIBLE
6'-3"
PRECAST CONCRETE CURB
BARRIER FREE SIGN DETAIL
NOT TO SCALE
2 AP4
20'-0"
BARRIER FREE PARKING SIGN SEE DET.
9'-0"
PRECAST CONCRETE CURB
8'-0"
DETAIL
⅛"=1'-0"
1 AP4
NOTE:
AT LOCATIONS WHERE SIDEWALK IS NOT BEING PROVIDED THE MID-POINT OF THE CURB AND GUTTER RADIUS SHALL BE DEPRESSED FOR A FUTURE SIDEWALK RAMP. AT "T" INTERSECTIONS THE OPPOSITE CURB SHALL BE DEPRESSED ATA POINT PERPENDICULAR TO THE RADIUS MIDPOINT. TO BE LOCATED IN THE FIELD BY THE ENGINEER.
WITHOUT SIDEWALK DEPRESS CURB TO THE DIMENSIONS SHOWN.
4'-0" MIN.
14'-0" MIN.
4'-0" MIN.
TYPICAL CURB RAMP
NOT TO SCALE
5 AP4
5'-0" TYP.
DRIVE WAY WIDTH VARIES
TOP OF CURB
PAVEMENT EDGE
FLOW LINE
PROFILE VIEW
TYPICAL DRIVEWAY DETAIL
NOT TO SCALE
4 AP4
2'-0"
6"
¾"/FT. SLOPE
12"
½"/FT. BATTER CURB FACE
DRIVEWAY SECTION
2" RADIUS
¾" MAX. RADIUS
CONCRETE CURB AND GUTTER 30-INCH, TYPE "D"
NOT TO SCALE
6 AP4
NOTE:
THE BOTTOM OF CURB AND GUTTER MAY BE CONSTRUCTED EITHER LEVEL OR PARALLEL TO THE SLOPE OF THE SUBGRADE OR BASE COURSE PROVIDED A 6" MIN. GUTTER THICKNESS IS MAINTAINED.
BRUCE VILLAGE HALL & LIBRARY
BRUCE, WISCONSIN
SCALE AS NOTED
SHEET AP4
DATE DEC. 31, 2001
MORGAN & PARMLEY, LTD.
CONSULTING ENGINEERS
LADYSMITH, WISCONSIN
DR. BY J.T.D.

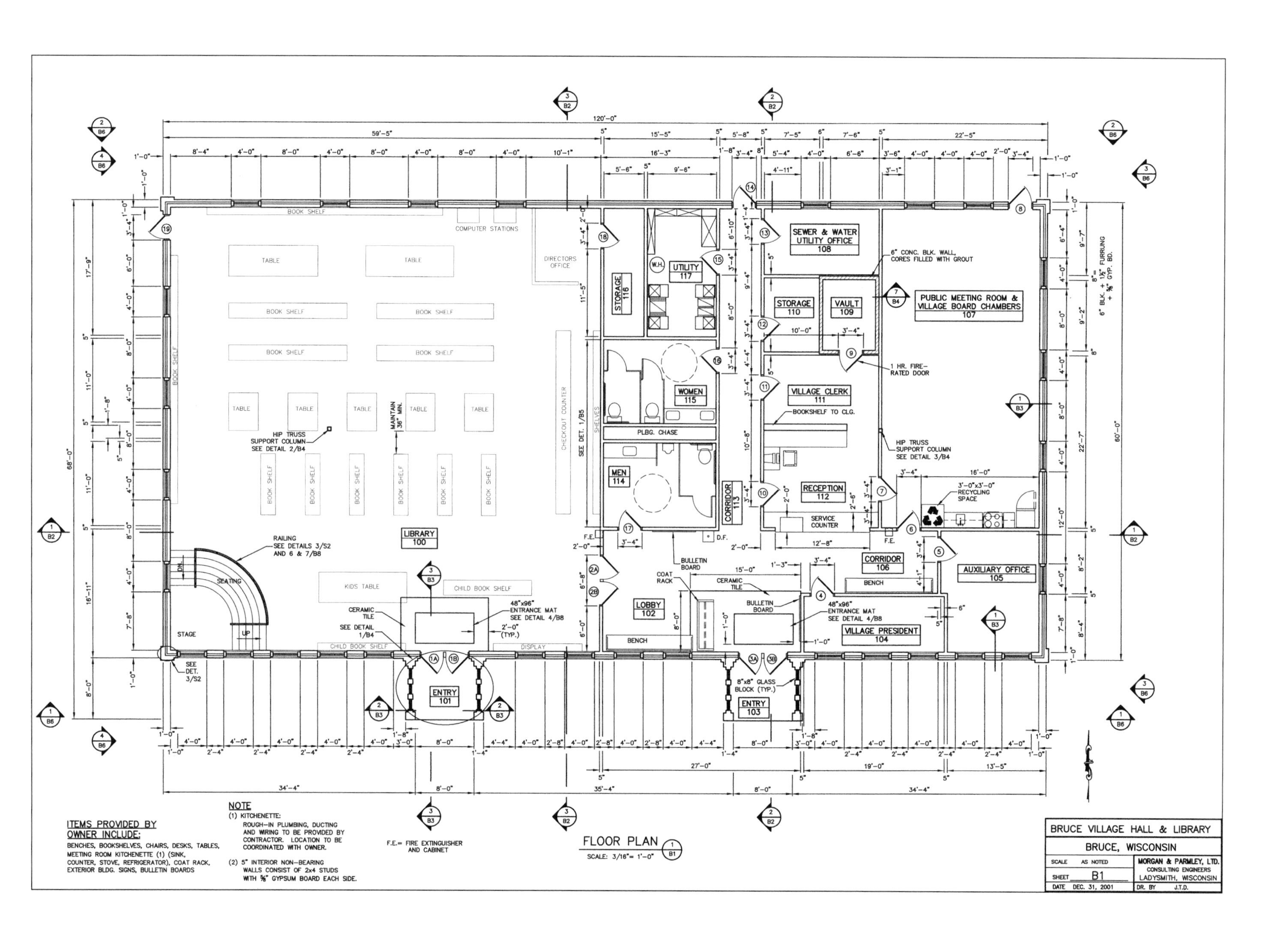

LIBRARY 100
STORAGE 116
UTILITY 117
WOMEN 115
MEN 114
CORRIDOR 113
SEWER & WATER UTILITY OFFICE 108
STORAGE 110
VAULT 109
PUBLIC MEETING ROOM & VILLAGE BOARD CHAMBERS 107
VILLAGE CLERK 111
RECEPTION 112
CORRIDOR 106
AUXILIARY OFFICE 105
LOBBY 102
VILLAGE PRESIDENT 104
ENTRY 101
ENTRY 103
ITEMS PROVIDED BY OWNER INCLUDE:
BENCHES, BOOKSHELVES, CHAIRS, DESKS, TABLES, MEETING ROOM KITCHENETTE (1) (SINK, COUNTER, STOVE, REFRIGERATOR), COAT RACK, EXTERIOR BLDG. SIGNS, BULLETIN BOARDS
NOTE
(1) KITCHENETTE: ROUGH-IN PLUMBING, DUCTING AND WIRING TO BE PROVIDED BY CONTRACTOR. LOCATION TO BE COORDINATED WITH OWNER.
(2) 5" INTERIOR NON-BEARING WALLS CONSIST OF 2x4 STUDS WITH ⅝" GYPSUM BOARD EACH SIDE.
F.E.= FIRE EXTINGUISHER AND CABINET
FLOOR PLAN
SCALE: 3/16"= 1'-0"
BRUCE VILLAGE HALL & LIBRARY
BRUCE, WISCONSIN
SCALE AS NOTED
SHEET B1
DATE DEC. 31, 2001
MORGAN & PARMLEY, LTD.
CONSULTING ENGINEERS
LADYSMITH, WISCONSIN
DR. BY J.T.D.

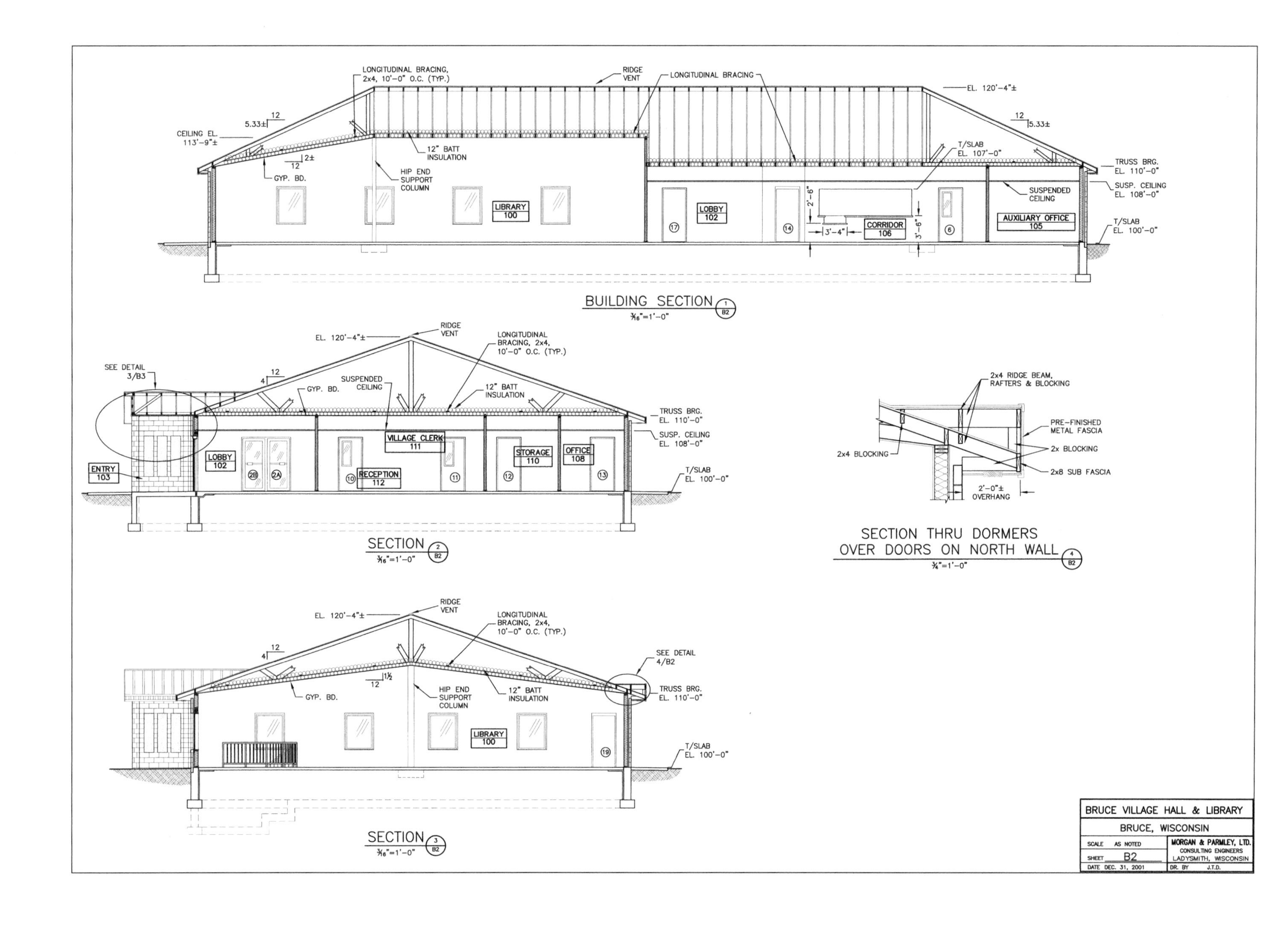
LONGITUDINAL BRACING, 2x4, 10'-0" O.C. (TYP.)
RIDGE VENT
LONGITUDINAL BRACING
EL. 120'-4"±
CEILING EL. 113'-9"±
12" BATT INSULATION
HIP END SUPPORT COLUMN
GYP. BD.
T/SLAB EL. 107'-0"
TRUSS BRG. EL. 110'-0"
SUSP. CEILING EL. 108'-0"
SUSPENDED CEILING
T/SLAB EL. 100'-0"
LIBRARY 100
LOBBY 102
CORRIDOR 106
AUXILIARY OFFICE 105
BUILDING SECTION
3/16"=1'-0"
SEE DETAIL 3/B3
SUSPENDED CEILING
VILLAGE CLERK 111
RECEPTION 112
STORAGE 110
OFFICE 108
ENTRY 103
SECTION
SECTION THRU DORMERS OVER DOORS ON NORTH WALL
3/4"=1'-0"
2x4 RIDGE BEAM, RAFTERS & BLOCKING
PRE-FINISHED METAL FASCIA
2x BLOCKING
2x4 BLOCKING
2x8 SUB FASCIA
2'-0"± OVERHANG
SEE DETAIL 4/B2
BRUCE VILLAGE HALL & LIBRARY
BRUCE, WISCONSIN
SCALE AS NOTED
SHEET B2
DATE DEC. 31, 2001
MORGAN & PARMLEY, LTD. CONSULTING ENGINEERS LADYSMITH, WISCONSIN
DR. BY J.T.D.

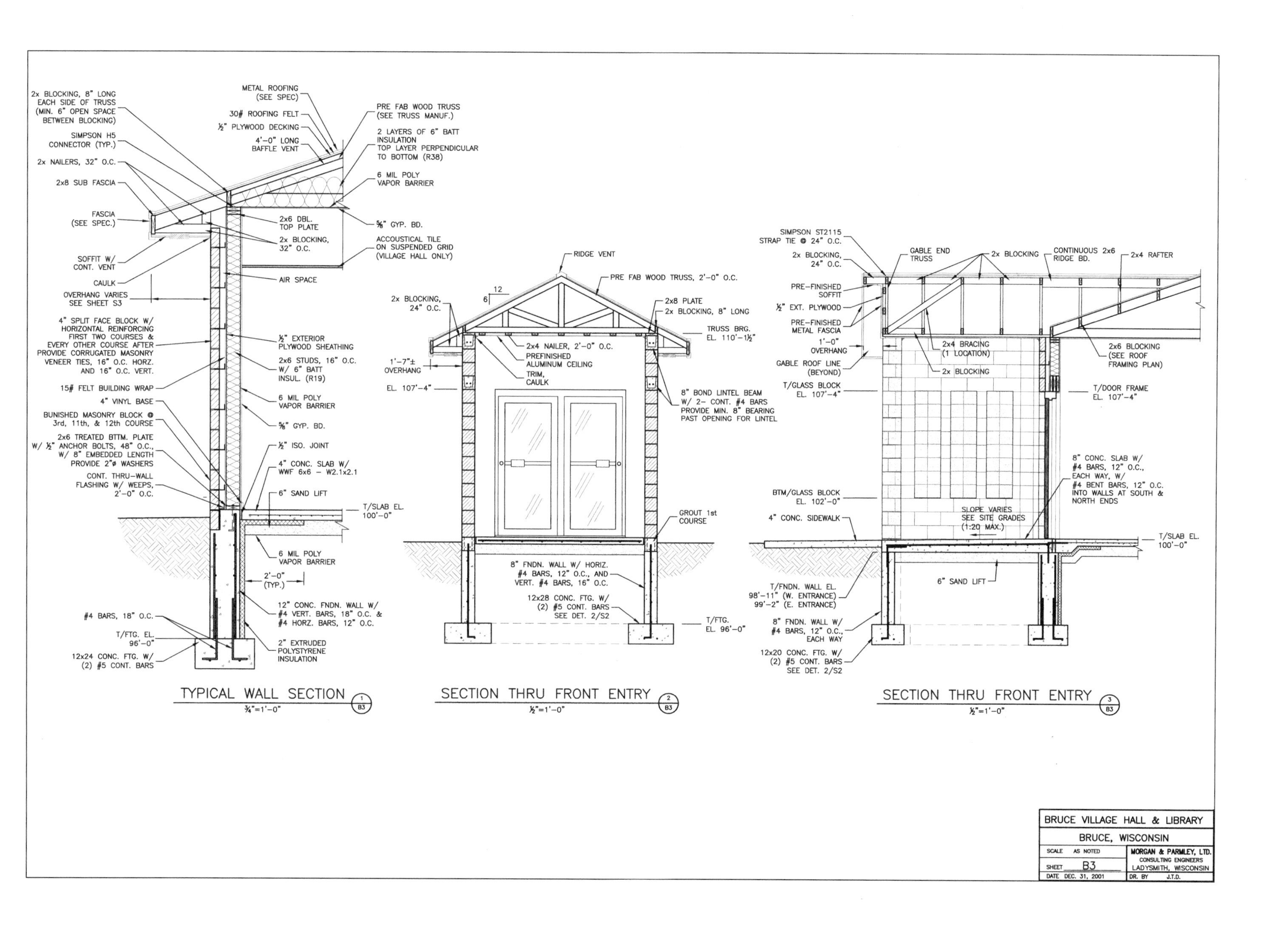
2x BLOCKING, 8" LONG EACH SIDE OF TRUSS (MIN. 6" OPEN SPACE BETWEEN BLOCKING)
SIMPSON H5 CONNECTOR (TYP.)
2x NAILERS, 32" O.C.
2x8 SUB FASCIA
FASCIA (SEE SPEC.)
SOFFIT W/ CONT. VENT
CAULK
OVERHANG VARIES SEE SHEET S3
4" SPLIT FACE BLOCK W/ HORIZONTAL REINFORCING FIRST TWO COURSES & EVERY OTHER COURSE AFTER PROVIDE CORRUGATED MASONRY VENEER TIES, 16" O.C. HORZ. AND 16" O.C. VERT.
15# FELT BUILDING WRAP
4" VINYL BASE
BUNISHED MASONRY BLOCK @ 3rd, 11th, & 12th COURSE
2x6 TREATED BTTM. PLATE W/ ½" ANCHOR BOLTS, 48" O.C., W/ 8" EMBEDDED LENGTH PROVIDE 2"ø WASHERS
CONT. THRU-WALL FLASHING W/ WEEPS, 2'-0" O.C.
#4 BARS, 18" O.C.
T/FTG. EL. 96'-0"
12x24 CONC. FTG. W/ (2) #5 CONT. BARS
METAL ROOFING (SEE SPEC)
30# ROOFING FELT
½" PLYWOOD DECKING
4'-0" LONG BAFFLE VENT
PRE FAB WOOD TRUSS (SEE TRUSS MANUF.)
2 LAYERS OF 6" BATT INSULATION TOP LAYER PERPENDICULAR TO BOTTOM (R38)
6 MIL POLY VAPOR BARRIER
2x6 DBL. TOP PLATE
2x BLOCKING, 32" O.C.
⅝" GYP. BD.
ACCOUSTICAL TILE ON SUSPENDED GRID (VILLAGE HALL ONLY)
AIR SPACE
½" EXTERIOR PLYWOOD SHEATHING
2x6 STUDS, 16" O.C. W/ 6" BATT INSUL. (R19)
6 MIL POLY VAPOR BARRIER
⅝" GYP. BD.
½" ISO. JOINT
4" CONC. SLAB W/ WWF 6x6 – W2.1x2.1
6" SAND LIFT
T/SLAB EL. 100'-0"
6 MIL POLY VAPOR BARRIER
2'-0" (TYP.)
12" CONC. FNDN. WALL W/ #4 VERT. BARS, 18" O.C. & #4 HORZ. BARS, 12" O.C.
2" EXTRUDED POLYSTYRENE INSULATION
TYPICAL WALL SECTION
¾"=1'-0"
1
B3
RIDGE VENT
PRE FAB WOOD TRUSS, 2'-0" O.C.
12
6
2x BLOCKING, 24" O.C.
2x8 PLATE
2x BLOCKING, 8" LONG
TRUSS BRG. EL. 110'-1½"
1'-7"± OVERHANG
EL. 107'-4"
2x4 NAILER, 2'-0" O.C.
PREFINISHED ALUMINUM CEILING
TRIM, CAULK
8" BOND LINTEL BEAM W/ 2- CONT. #4 BARS PROVIDE MIN. 8" BEARING PAST OPENING FOR LINTEL
GROUT 1st COURSE
8" FNDN. WALL W/ HORIZ. #4 BARS, 12" O.C., AND VERT. #4 BARS, 16" O.C.
12x28 CONC. FTG. W/ (2) #5 CONT. BARS SEE DET. 2/S2
T/FTG. EL. 96'-0"
SECTION THRU FRONT ENTRY
½"=1'-0"
2
B3
SIMPSON ST2115 STRAP TIE @ 24" O.C.
2x BLOCKING, 24" O.C.
PRE-FINISHED SOFFIT
½" EXT. PLYWOOD
PRE-FINISHED METAL FASCIA
1'-0" OVERHANG
GABLE ROOF LINE (BEYOND)
T/GLASS BLOCK EL. 107'-4"
GABLE END TRUSS
2x BLOCKING
CONTINUOUS 2x6 RIDGE BD.
2x4 RAFTER
2x4 BRACING (1 LOCATION)
2x BLOCKING
2x6 BLOCKING (SEE ROOF FRAMING PLAN)
T/DOOR FRAME EL. 107'-4"
8" CONC. SLAB W/ #4 BARS, 12" O.C., EACH WAY, W/ #4 BENT BARS, 12" O.C. INTO WALLS AT SOUTH & NORTH ENDS
BTM/GLASS BLOCK EL. 102'-0"
4" CONC. SIDEWALK
SLOPE VARIES SEE SITE GRADES (1:20 MAX.)
T/SLAB EL. 100'-0"
6" SAND LIFT
T/FNDN. WALL EL. 98'-11" (W. ENTRANCE) 99'-2" (E. ENTRANCE)
8" FNDN. WALL W/ #4 BARS, 12" O.C., EACH WAY
12x20 CONC. FTG. W/ (2) #5 CONT. BARS SEE DET. 2/S2
SECTION THRU FRONT ENTRY
½"=1'-0"
3
B3
BRUCE VILLAGE HALL & LIBRARY
BRUCE, WISCONSIN
SCALE AS NOTED
SHEET B3
DATE DEC. 31, 2001
MORGAN & PARMLEY, LTD.
CONSULTING ENGINEERS
LADYSMITH, WISCONSIN
DR. BY J.T.D.

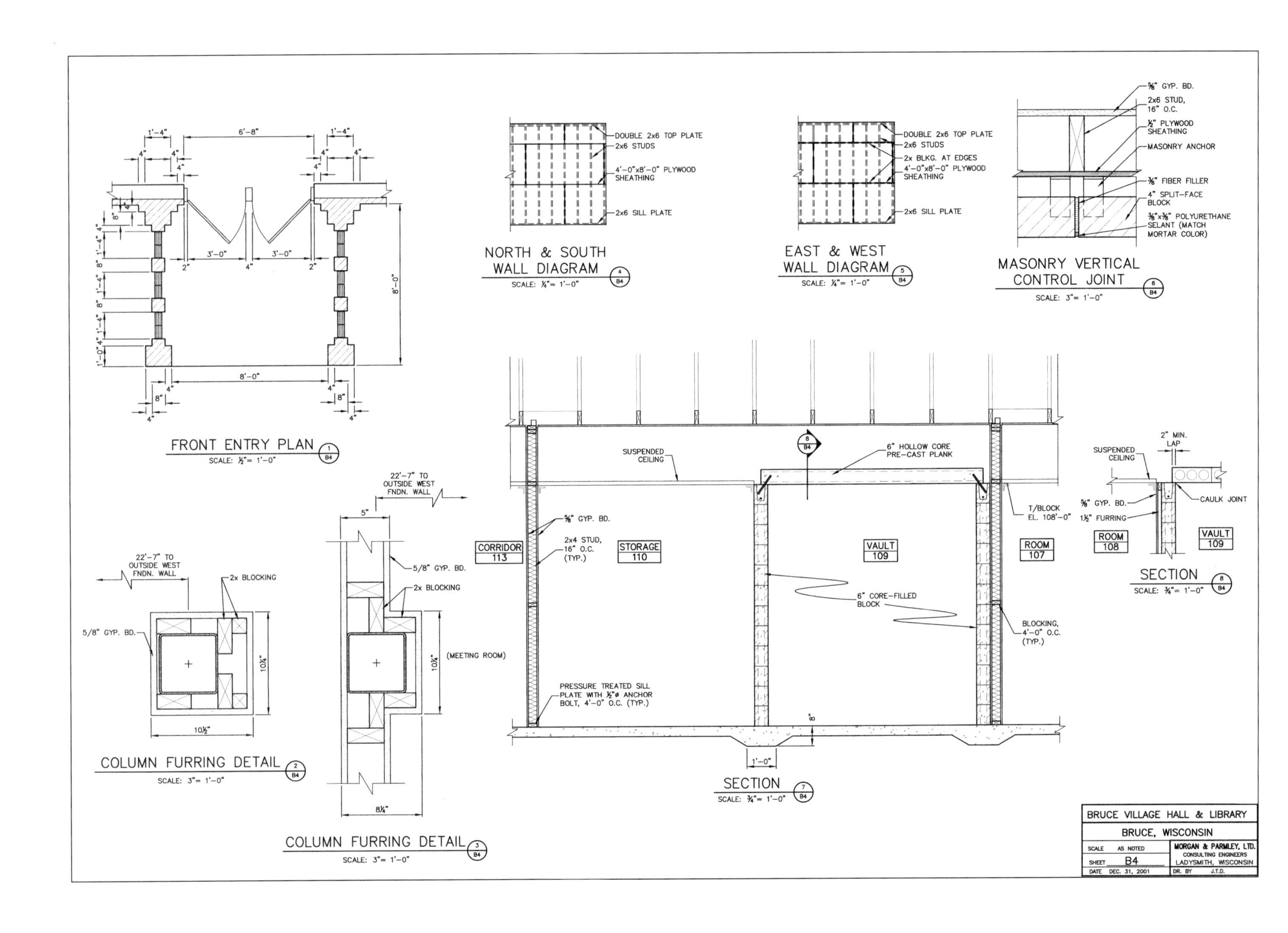

FRONT ENTRY PLAN
SCALE: ½"= 1'-0"
NORTH & SOUTH WALL DIAGRAM
SCALE: ¼"= 1'-0"
DOUBLE 2x6 TOP PLATE
2x6 STUDS
4'-0"x8'-0" PLYWOOD SHEATHING
2x6 SILL PLATE
EAST & WEST WALL DIAGRAM
SCALE: ¼"= 1'-0"
2x BLKG. AT EDGES
MASONRY VERTICAL CONTROL JOINT
SCALE: 3"= 1'-0"
⅝" GYP. BD.
2x6 STUD, 16" O.C.
½" PLYWOOD SHEATHING
MASONRY ANCHOR
⅜" FIBER FILLER
4" SPLIT-FACE BLOCK
⅜"x⅜" POLYURETHANE SELANT (MATCH MORTAR COLOR)
COLUMN FURRING DETAIL
SCALE: 3"= 1'-0"
22'-7" TO OUTSIDE WEST FNDN. WALL
5/8" GYP. BD.
2x BLOCKING
(MEETING ROOM)
SECTION
SCALE: ¾"= 1'-0"
SUSPENDED CEILING
6" HOLLOW CORE PRE-CAST PLANK
CORRIDOR 113
STORAGE 110
VAULT 109
ROOM 107
2x4 STUD, 16" O.C. (TYP.)
T/BLOCK EL. 108'-0"
6" CORE-FILLED BLOCK
BLOCKING, 4'-0" O.C. (TYP.)
PRESSURE TREATED SILL PLATE WITH ½"ø ANCHOR BOLT, 4'-0" O.C. (TYP.)
2" MIN. LAP
CAULK JOINT
1½" FURRING
ROOM 108
BRUCE VILLAGE HALL & LIBRARY
BRUCE, WISCONSIN
SCALE AS NOTED
SHEET B4
DATE DEC. 31, 2001
MORGAN & PARMLEY, LTD.
CONSULTING ENGINEERS
LADYSMITH, WISCONSIN
DR. BY J.T.D.

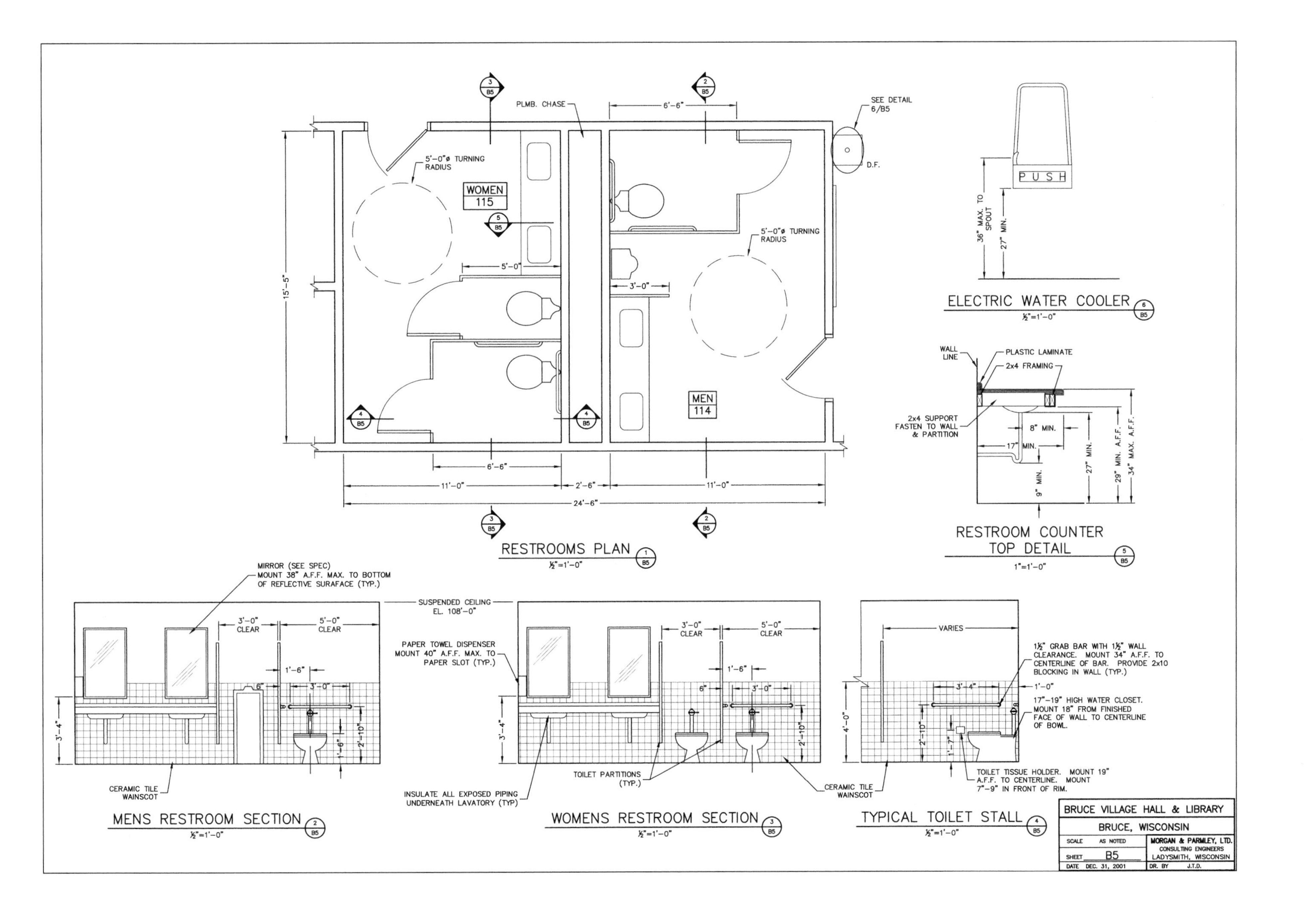
PLMB. CHASE
6'-6"
SEE DETAIL 6/B5
D.F.
5'-0"ø TURNING RADIUS
WOMEN 115
MEN 114
15'-5"
5'-0"
3'-0"
6'-6"
11'-0"
2'-6"
11'-0"
24'-6"
RESTROOMS PLAN
½"=1'-0"
PUSH
36" MAX. TO SPOUT
27" MIN.
ELECTRIC WATER COOLER
½"=1'-0"
WALL LINE
PLASTIC LAMINATE
2x4 FRAMING
2x4 SUPPORT FASTEN TO WALL & PARTITION
8" MIN.
17" MIN.
9" MIN.
27" MIN.
29" MIN. A.F.F.
34" MAX. A.F.F.
RESTROOM COUNTER TOP DETAIL
1"=1'-0"
MIRROR (SEE SPEC) MOUNT 38" A.F.F. MAX. TO BOTTOM OF REFLECTIVE SURAFACE (TYP.)
SUSPENDED CEILING EL. 108'-0"
3'-0" CLEAR
5'-0" CLEAR
1'-6"
3'-0"
6"
3'-4"
1'-6"
2'-10"
CERAMIC TILE WAINSCOT
MENS RESTROOM SECTION
½"=1'-0"
PAPER TOWEL DISPENSER MOUNT 40" A.F.F. MAX. TO PAPER SLOT (TYP.)
INSULATE ALL EXPOSED PIPING UNDERNEATH LAVATORY (TYP)
TOILET PARTITIONS (TYP.)
WOMENS RESTROOM SECTION
½"=1'-0"
VARIES
3'-4"
4'-0"
2'-10"
1'-7"
1½" GRAB BAR WITH 1½" WALL CLEARANCE. MOUNT 34" A.F.F. TO CENTERLINE OF BAR. PROVIDE 2x10 BLOCKING IN WALL (TYP.)
1'-0"
17"-19" HIGH WATER CLOSET. MOUNT 18" FROM FINISHED FACE OF WALL TO CENTERLINE OF BOWL.
TOILET TISSUE HOLDER. MOUNT 19" A.F.F. TO CENTERLINE. MOUNT 7"-9" IN FRONT OF RIM.
CERAMIC TILE WAINSCOT
TYPICAL TOILET STALL
½"=1'-0"
BRUCE VILLAGE HALL & LIBRARY
BRUCE, WISCONSIN
SCALE AS NOTED
SHEET B5
DATE DEC. 31, 2001
MORGAN & PARMLEY, LTD. CONSULTING ENGINEERS LADYSMITH, WISCONSIN
DR. BY J.T.D.

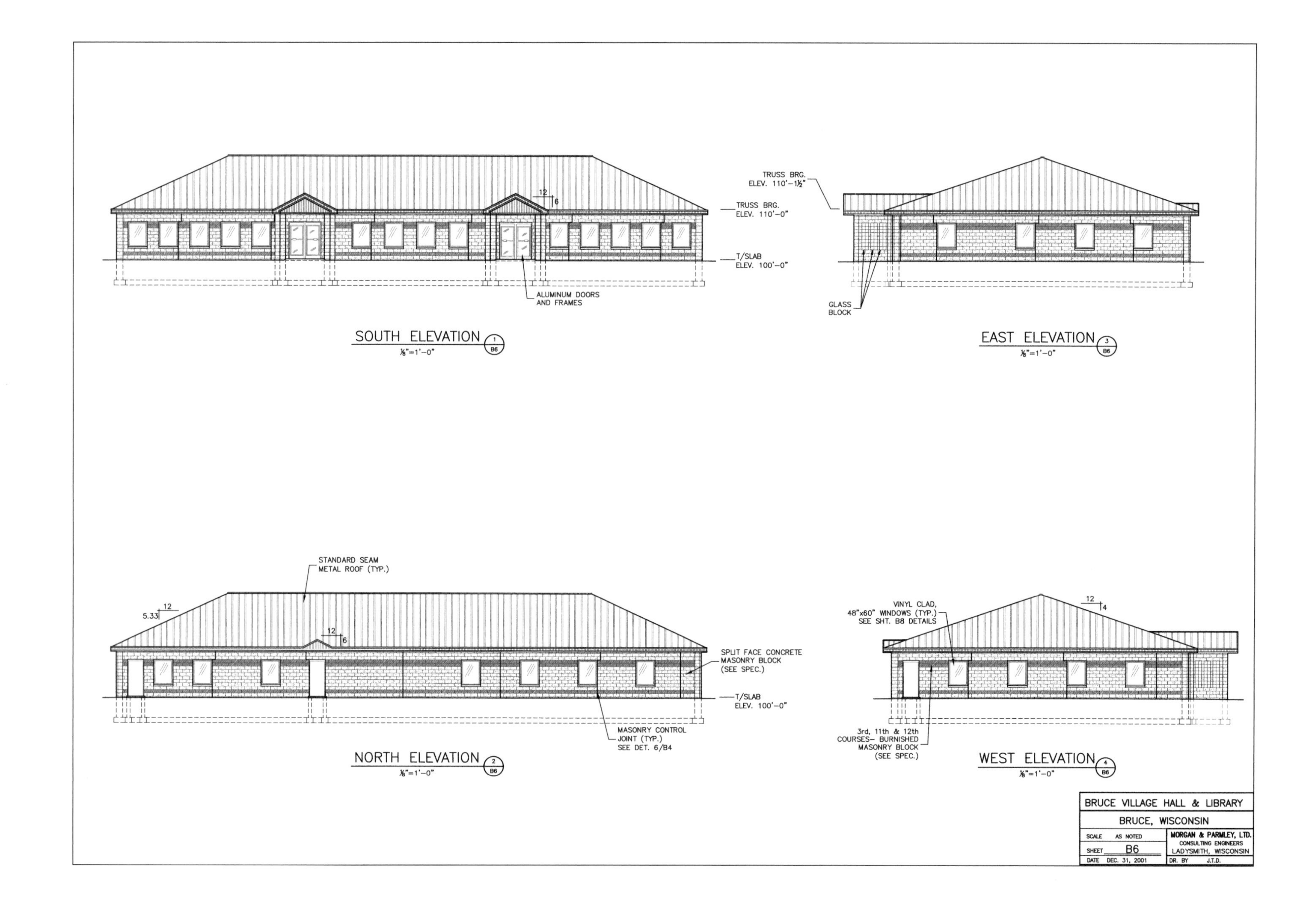
TRUSS BRG. ELEV. 110'–0"
T/SLAB ELEV. 100'–0"
ALUMINUM DOORS AND FRAMES
SOUTH ELEVATION
⅛"=1'–0"
TRUSS BRG. ELEV. 110'–1½"
GLASS BLOCK
EAST ELEVATION
⅛"=1'–0"
STANDARD SEAM METAL ROOF (TYP.)
SPLIT FACE CONCRETE MASONRY BLOCK (SEE SPEC.)
T/SLAB ELEV. 100'–0"
MASONRY CONTROL JOINT (TYP.) SEE DET. 6/B4
NORTH ELEVATION
⅛"=1'–0"
VINYL CLAD, 48"x60" WINDOWS (TYP.) SEE SHT. B8 DETAILS
3rd, 11th & 12th COURSES– BURNISHED MASONRY BLOCK (SEE SPEC.)
WEST ELEVATION
⅛"=1'–0"
BRUCE VILLAGE HALL & LIBRARY
BRUCE, WISCONSIN
SCALE AS NOTED
SHEET B6
DATE DEC. 31, 2001
MORGAN & PARMLEY, LTD. CONSULTING ENGINEERS LADYSMITH, WISCONSIN
DR. BY J.T.D.

DOOR SCHEDULE								
DOOR NO.	DOOR SIZE			DOOR		FRAME		NOTES
	WIDTH	HEIGHT	THICKNESS	MATERIAL	TYPE	MATERIAL	TYPE	
1A & 1B	3'–0"	7'–0"	THK.	ALUM./GLASS	C	ALUM.	B	FRONT ENTRANCE, EXT.
2A & 2B	3'–0"	7'–0"	THK.	ALUM./GLASS	C	ALUM.	B	LIBRARY, INT.
3A & 3B	3'–0"	7'–0"	THK.	ALUM./GLASS	C	ALUM.	B	FRONT ENTRANCE, EXT.
4	3'–0"	7'–0"	THK.	WOOD/GLASS	B	STEEL	A	VILLAGE PRESIDENT, INT.
5	3'–0"	7'–0"	THK.	WOOD/GLASS	B	STEEL	A	ASSESSOR & TOWN CLERK OFFICE, INT.
6	3'–0"	7'–0"	THK.	WOOD/GLASS	B	STEEL	A	PUBLIC MEETING ROOM & VILLAGE BOARD CHAMBERS, INT.
7	3'–0"	7'–0"	THK.	WOOD/GLASS	B	STEEL	A	PUBLIC MEETING ROOM & VILLAGE BOARD CHAMBERS, INT.
8	3'–0"	7'–0"	THK.	STEEL	A	STEEL	C	PUBLIC MEETING ROOM & VILLAGE BOARD CHAMBERS, EXT.
9	3'–0"	7'–0"	THK.	STEEL	A	STEEL	C	VAULT, INT., 1 HR. FIRE RATED
10	3'–0"	7'–0"	THK.	WOOD/GLASS	B	STEEL	A	RECEPTION, INT.
11	3'–0"	7'–0"	THK.	WOOD/GLASS	B	STEEL	A	VILLAGE CLERK, INT.
12	3'–0"	7'–0"	THK.	STEEL	A	STEEL	A	STORAGE, INT.
13	3'–0"	7'–0"	THK.	STEEL	B	STEEL	A	SEWER & WATER UTILITY OFFICE, INT.
14	3'–0"	7'–0"	THK.	STEEL	B	STEEL	C	CORRIDOR, EXT.
15	3'–0"	7'–0"	THK.	STEEL	A	STEEL	A	UTILITY, INT.
16	3'–0"	7'–0"	THK.	WOOD	A	STEEL	A	WOMEN'S BATHROOM, INT., PROVIDE 12"x12" LOUVER
17	3'–0"	7'–0"	THK.	WOOD	A	STEEL	A	MEN'S BATHROOM, INT., PROVIDE 12"x12" LOUVER
18	3'–0"	7'–0"	THK.	STEEL	A	STEEL	A	STORAGE, INT.
19	3'–0"	7'–0"	THK.	STEEL	A	STEEL	C	LIBRARY, EXT.

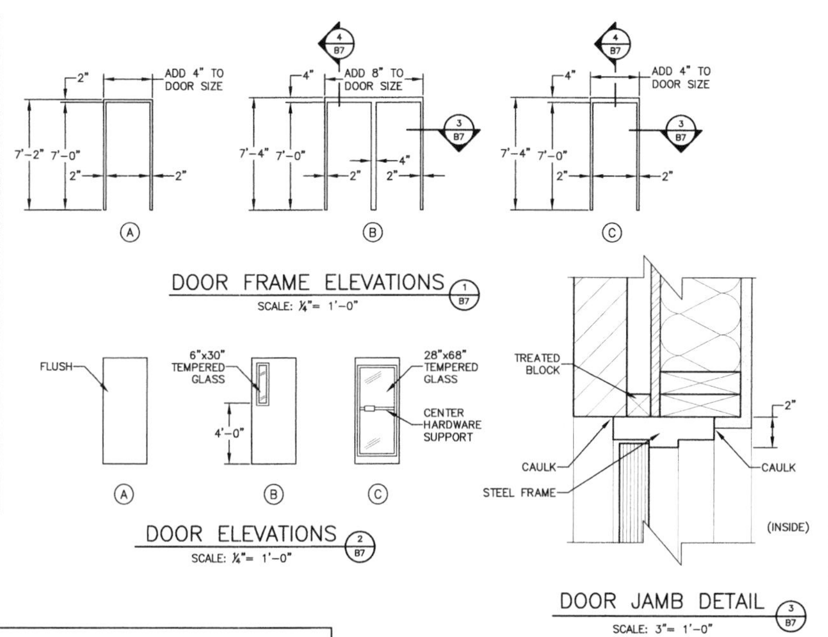

ROOM FINISH SCHEDULE					
ROOM NAME	FLOOR/BASE	WALLS	CEILING	CEILING HEIGHT	NOTES
LIBRARY (100)	CARPET/VINYL COVING	PAINTED GYP. BD.	PAINTED GYP. BD.	10'–0" – 13'–9"	–
ENTRY (101)	SEALED CONC.	CONC. BLK.	METAL SOFFIT	10'–0 1/2"	–
LOBBY (102)	CARPET & CERAMIC TILE/VINYL COVING	PAINTED GYP. BD.	SUSPENDED TILE	8'–0"	–
ENTRY (103)	SEALED CONC.	CONC. BLK.	METAL SOFFIT	10'–0 1/2"	–
VILLAGE PRESIDENT (104)	CARPET/VINYL COVING	PAINTED GYP. BD.	SUSPENDED TILE	8'–0"	–
AUXILIARY OFFICE (105)	CARPET/VINYL COVING	PAINTED GYP. BD.	SUSPENDED TILE	8'–0"	–
CORRIDOR (106)	CARPET/VINYL COVING	PAINTED GYP. BD.	SUSPENDED TILE	8'–0"	–
PUBLIC MEETING RM. & VILLAGE BD. CHAMBERS (107)	CARPET/VINYL COVING	PAINTED GYP. BD.	SUSPENDED TILE	8'–0"	–
SEWER & WATER UTILITY OFFICE (108)	SEALED CONC./VINYL COVING	PAINTED GYP. BD.	SUSPENDED TILE	8'–0"	–
VAULT (109)	SEALED CONC./VINYL COVING	CONC. BLK.	CONCRETE	8'–0"	–
STORAGE (110)	SEALED CONC./VINYL COVING	CONC. BLK./PAINTED GYP. BD.	SUSPENDED TILE	8'–0"	–
VILLAGE CLERK (111)	CARPET/VINYL COVING	PAINTED GYP. BD.	SUSPENDED TILE	8'–0"	–
RECEPTION (112)	CARPET/VINYL COVING	PAINTED GYP. BD.	SUSPENDED TILE	8'–0"	–
CORRIDOR (113)	CARPET/VINYL COVING	PAINTED GYP. BD.	SUSPENDED TILE	8'–0"	–
MEN'S BATHROOM (114)	CERAMIC TILE & COVING	PAINTED GYP. BD.	SUSPENDED TILE	8'–0"	4'–0" CERAMIC WAINSCOAT
WOMEN'S BATHROOM (115)	CERAMIC TILE & COVING	PAINTED GYP. BD.	SUSPENDED TILE	8'–0"	4'–0" CERAMIC WAINSCOAT
STORAGE (116)	SEALED CONC./VINYL COVING	PAINTED GYP. BD.	SUSPENDED TILE	8'–0"	–
UTILITY (117)	SEALED CONC./VINYL COVING	PAINTED GYP. BD.	PAINTED GYP. BD.	10'–0"	–

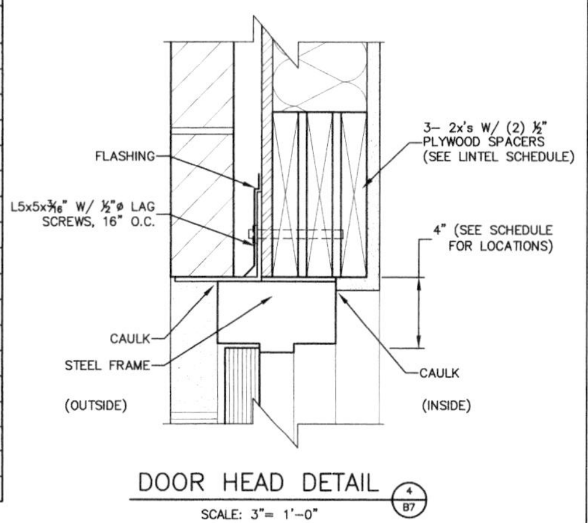

NOTES

CARPET WILL BE PROVIDED BY THE OWNER.
WALLS TO BE PAINTED IN ACCORDANCE WITH SPECIFICATIONS.

BRUCE VILLAGE HALL & LIBRARY
BRUCE, WISCONSIN

SCALE AS NOTED
SHEET B7
DATE DEC. 31, 2001

MORGAN & PARMLEY, LTD.
CONSULTING ENGINEERS
LADYSMITH, WISCONSIN
DR. BY J.T.D.

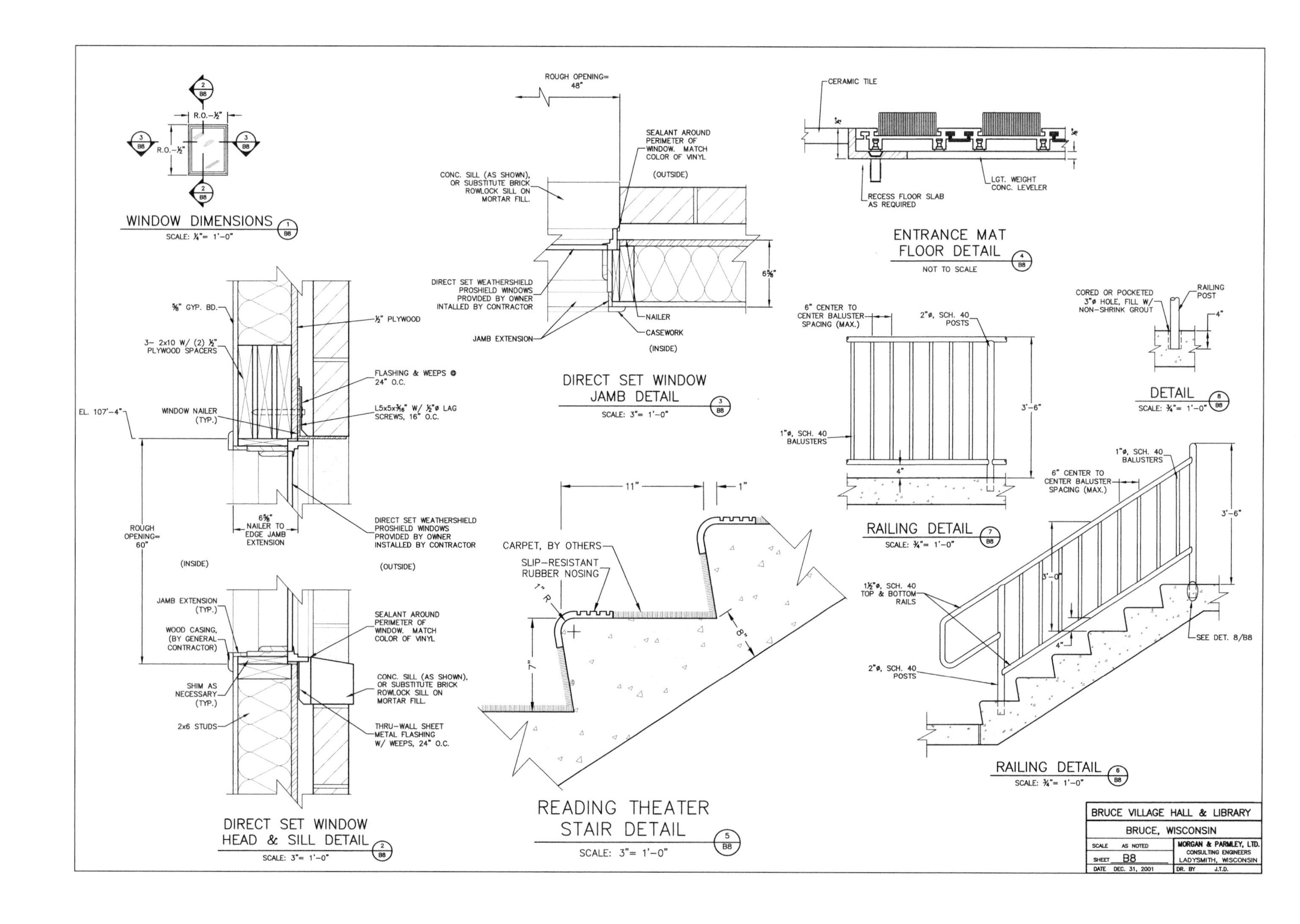

WINDOW DIMENSIONS
SCALE: ¼"= 1'-0"
R.O.-½"
DIRECT SET WINDOW HEAD & SILL DETAIL
SCALE: 3"= 1'-0"
⅝" GYP. BD.
3- 2x10 W/ (2) ½" PLYWOOD SPACERS
½" PLYWOOD
FLASHING & WEEPS @ 24" O.C.
L5x5x⁵⁄₁₆" W/ ½"ø LAG SCREWS, 16" O.C.
WINDOW NAILER (TYP.)
EL. 107'-4"
ROUGH OPENING= 60"
6⅝" NAILER TO EDGE JAMB EXTENSION
DIRECT SET WEATHERSHIELD PROSHIELD WINDOWS PROVIDED BY OWNER INSTALLED BY CONTRACTOR
(INSIDE)
(OUTSIDE)
JAMB EXTENSION (TYP.)
WOOD CASING, (BY GENERAL CONTRACTOR)
SHIM AS NECESSARY (TYP.)
2x6 STUDS
SEALANT AROUND PERIMETER OF WINDOW. MATCH COLOR OF VINYL
CONC. SILL (AS SHOWN), OR SUBSTITUTE BRICK ROWLOCK SILL ON MORTAR FILL.
THRU-WALL SHEET METAL FLASHING W/ WEEPS, 24" O.C.
ROUGH OPENING= 48"
DIRECT SET WEATHERSHIELD PROSHIELD WINDOWS PROVIDED BY OWNER INTALLED BY CONTRACTOR
JAMB EXTENSION
NAILER
CASEWORK
6⅝"
DIRECT SET WINDOW JAMB DETAIL
SCALE: 3"= 1'-0"
READING THEATER STAIR DETAIL
SCALE: 3"= 1'-0"
11"
1"
CARPET, BY OTHERS
SLIP-RESISTANT RUBBER NOSING
⅞" R
7"
8"
CERAMIC TILE
LGT. WEIGHT CONC. LEVELER
RECESS FLOOR SLAB AS REQUIRED
ENTRANCE MAT FLOOR DETAIL
NOT TO SCALE
6" CENTER TO CENTER BALUSTER SPACING (MAX.)
2"ø, SCH. 40 POSTS
1"ø, SCH. 40 BALUSTERS
3'-6"
4"
RAILING DETAIL
SCALE: ¾"= 1'-0"
CORED OR POCKETED 3"ø HOLE, FILL W/ NON-SHRINK GROUT
RAILING POST
DETAIL
SCALE: ¾"= 1'-0"
1½"ø, SCH. 40 TOP & BOTTOM RAILS
3'-0"
SEE DET. 8/B8
BRUCE VILLAGE HALL & LIBRARY
BRUCE, WISCONSIN
SCALE AS NOTED
SHEET B8
DATE DEC. 31, 2001
MORGAN & PARMLEY, LTD. CONSULTING ENGINEERS LADYSMITH, WISCONSIN
DR. BY J.T.D.

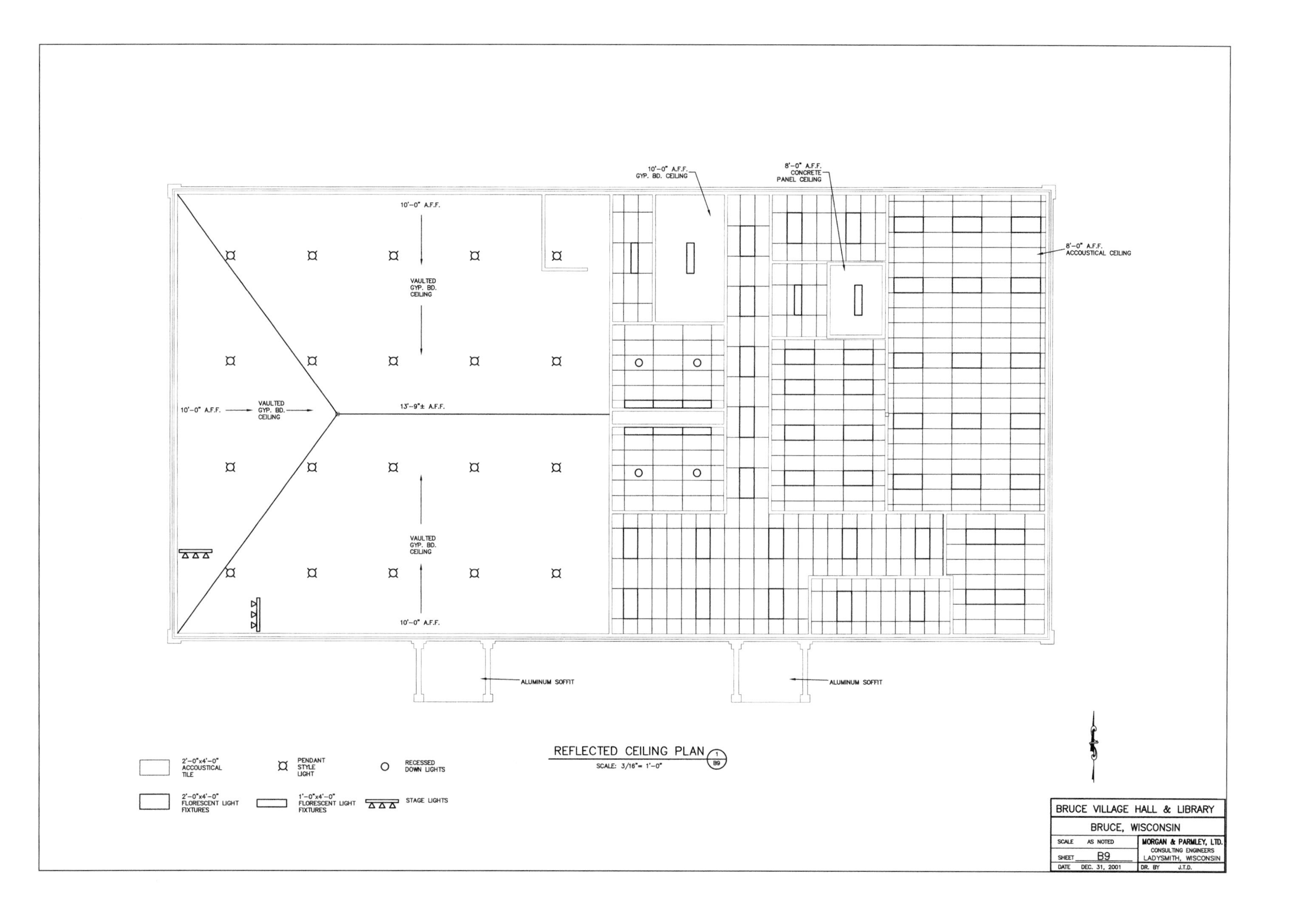

10'-0" A.F.F.
GYP. BD. CEILING
8'-0" A.F.F.
CONCRETE
PANEL CEILING
10'-0" A.F.F.
VAULTED
GYP. BD.
CEILING
8'-0" A.F.F.
ACCOUSTICAL CEILING
10'-0" A.F.F.
VAULTED
GYP. BD.
CEILING
13'-9"± A.F.F.
VAULTED
GYP. BD.
CEILING
10'-0" A.F.F.
ALUMINUM SOFFIT
ALUMINUM SOFFIT
REFLECTED CEILING PLAN
SCALE: 3/16"= 1'-0"
1
B9
2'-0"x4'-0"
ACCOUSTICAL
TILE
PENDANT
STYLE
LIGHT
RECESSED
DOWN LIGHTS
2'-0"x4'-0"
FLORESCENT LIGHT
FIXTURES
1'-0"x4'-0"
FLORESCENT LIGHT
FIXTURES
STAGE LIGHTS
BRUCE VILLAGE HALL & LIBRARY
BRUCE, WISCONSIN
SCALE AS NOTED
SHEET B9
DATE DEC. 31, 2001
MORGAN & PARMLEY, LTD.
CONSULTING ENGINEERS
LADYSMITH, WISCONSIN
DR. BY J.T.D.

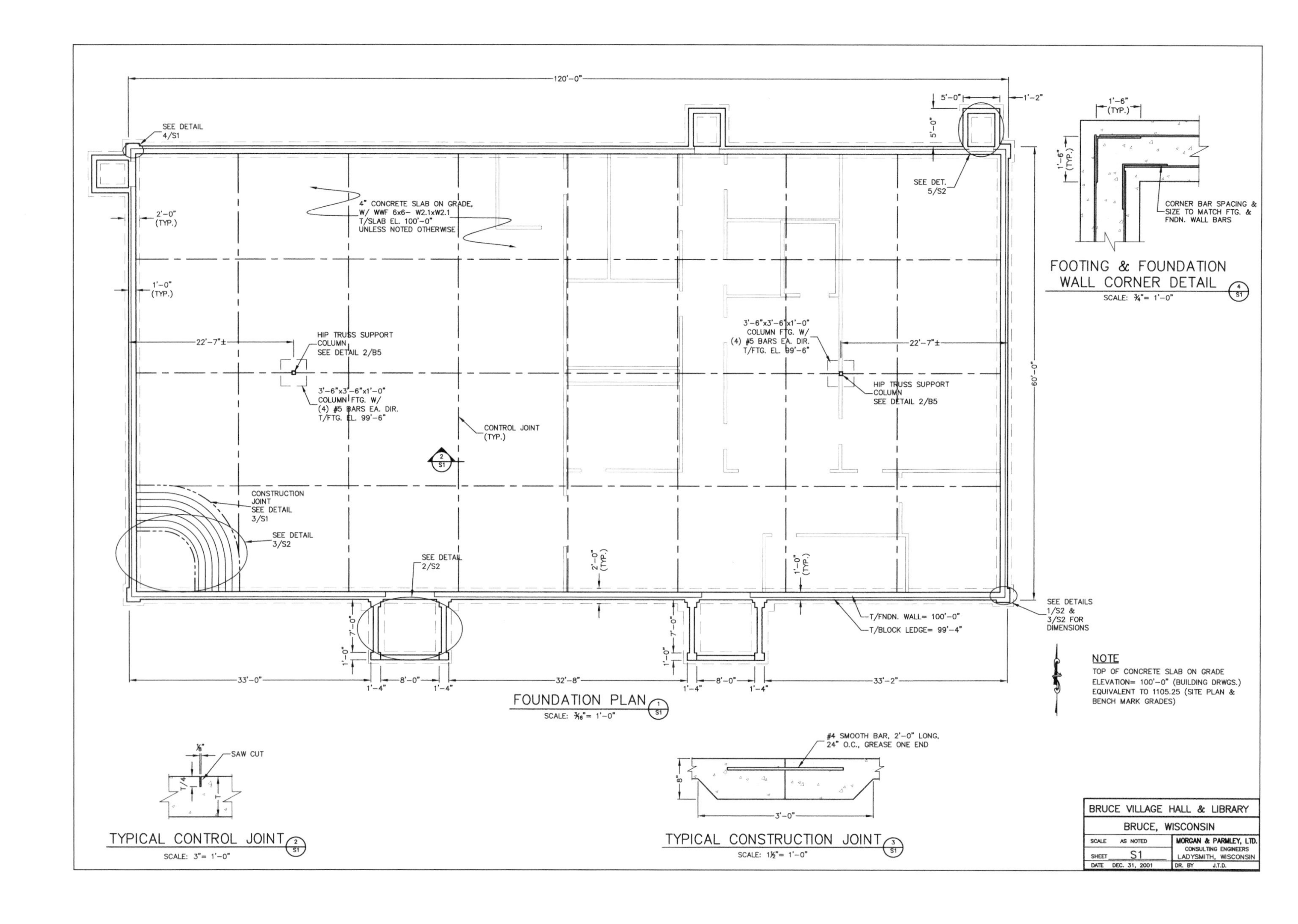

120'-0"
SEE DETAIL 4/S1
4" CONCRETE SLAB ON GRADE, W/ WWF 6x6- W2.1xW2.1 T/SLAB EL. 100'-0" UNLESS NOTED OTHERWISE
2'-0" (TYP.)
1'-0" (TYP.)
22'-7"±
HIP TRUSS SUPPORT COLUMN SEE DETAIL 2/B5
3'-6"x3'-6"x1'-0" COLUMN FTG. W/ (4) #5 BARS EA. DIR. T/FTG. EL. 99'-6"
CONTROL JOINT (TYP.)
CONSTRUCTION JOINT SEE DETAIL 3/S1
SEE DETAIL 3/S2
SEE DETAIL 2/S2
SEE DET. 5/S2
5'-0"
1'-2"
60'-0"
SEE DETAILS 1/S2 & 3/S2 FOR DIMENSIONS
T/FNDN. WALL= 100'-0"
T/BLOCK LEDGE= 99'-4"
7'-0"
1'-0"
33'-0"
8'-0"
1'-4"
32'-8"
33'-2"
FOUNDATION PLAN
SCALE: 3/16"= 1'-0"
1'-6" (TYP.)
CORNER BAR SPACING & SIZE TO MATCH FTG. & FNDN. WALL BARS
FOOTING & FOUNDATION WALL CORNER DETAIL
SCALE: 3/4"= 1'-0"
NOTE
TOP OF CONCRETE SLAB ON GRADE ELEVATION= 100'-0" (BUILDING DRWGS.) EQUIVALENT TO 1105.25 (SITE PLAN & BENCH MARK GRADES)
1/8"
SAW CUT
T/4
T
TYPICAL CONTROL JOINT
SCALE: 3"= 1'-0"
#4 SMOOTH BAR, 2'-0" LONG, 24" O.C., GREASE ONE END
8"
3'-0"
TYPICAL CONSTRUCTION JOINT
SCALE: 1½"= 1'-0"
BRUCE VILLAGE HALL & LIBRARY
BRUCE, WISCONSIN
SCALE AS NOTED
SHEET S1
DATE DEC. 31, 2001
MORGAN & PARMLEY, LTD.
CONSULTING ENGINEERS
LADYSMITH, WISCONSIN
DR. BY J.T.D.

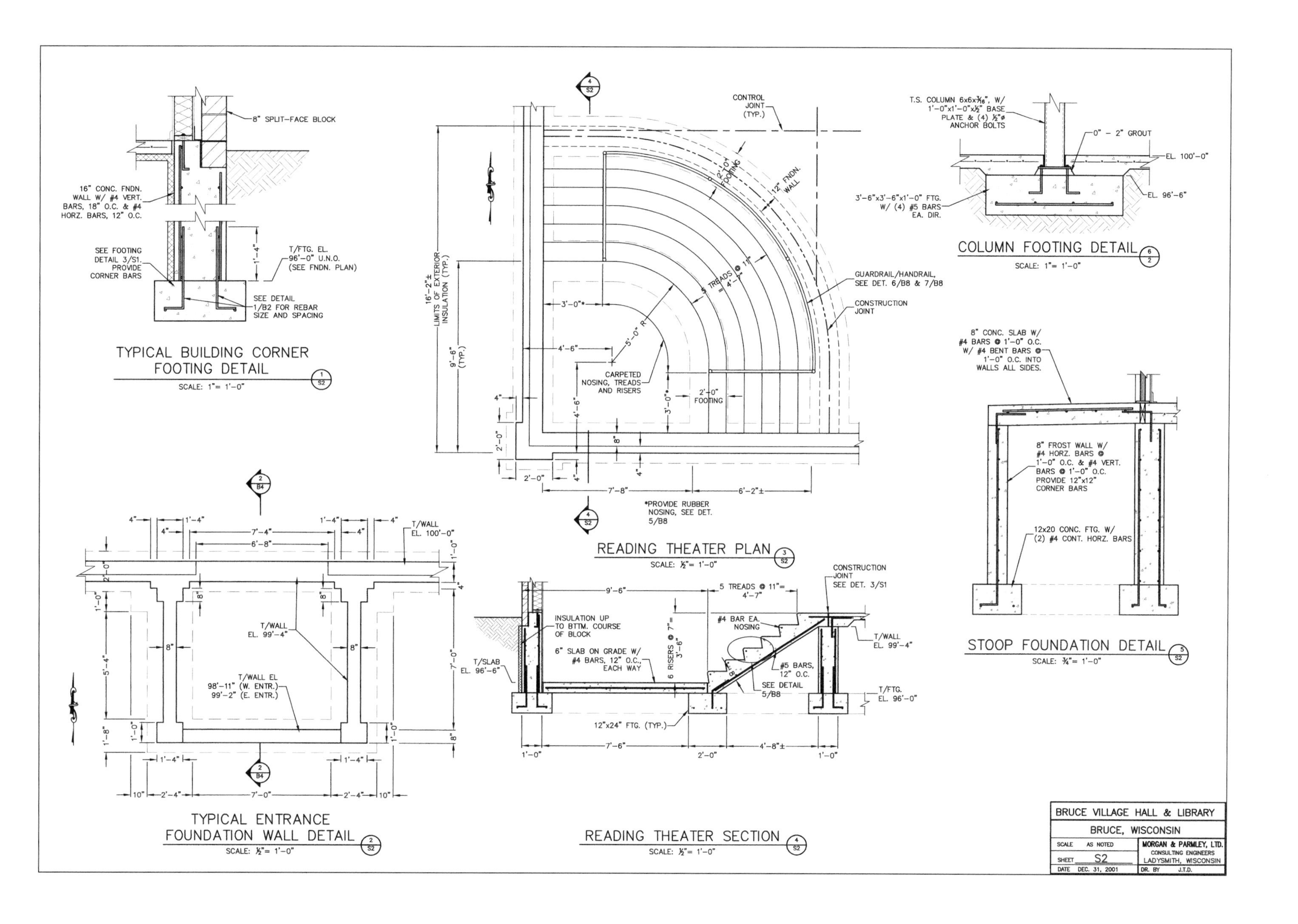

8" SPLIT-FACE BLOCK
16" CONC. FNDN. WALL W/ #4 VERT. BARS, 18" O.C. & #4 HORZ. BARS, 12" O.C.
SEE FOOTING DETAIL 3/S1. PROVIDE CORNER BARS
T/FTG. EL. 96'-0" U.N.O. (SEE FNDN. PLAN)
1'-4"
SEE DETAIL 1/B2 FOR REBAR SIZE AND SPACING
TYPICAL BUILDING CORNER FOOTING DETAIL
SCALE: 1"= 1'-0"
CONTROL JOINT (TYP.)
2'-0" FOOTING
12" FNDN. WALL
16'-2"± LIMITS OF EXTERIOR INSULATION (TYP.)
5 TREADS @ 11" = 4'-7"
GUARDRAIL/HANDRAIL, SEE DET. 6/B8 & 7/B8
CONSTRUCTION JOINT
3'-0"*
4'-6"
5'-0" R
9'-6" (TYP.)
CARPETED NOSING, TREADS AND RISERS
2'-0" FOOTING
4'-6"
3'-0"*
2'-0"
7'-8"
6'-2"±
*PROVIDE RUBBER NOSING, SEE DET. 5/B8
READING THEATER PLAN
SCALE: ½"= 1'-0"
T.S. COLUMN 6x6x3/16", W/ 1'-0"x1'-0"x½" BASE PLATE & (4) ½"ø ANCHOR BOLTS
0" – 2" GROUT
EL. 100'-0"
3'-6"x3'-6"x1'-0" FTG. W/ (4) #5 BARS EA. DIR.
EL. 96'-6"
COLUMN FOOTING DETAIL
SCALE: 1"= 1'-0"
8" CONC. SLAB W/ #4 BARS @ 1'-0" O.C. W/ #4 BENT BARS @ 1'-0" O.C. INTO WALLS ALL SIDES.
8" FROST WALL W/ #4 HORZ. BARS @ 1'-0" O.C. & #4 VERT. BARS @ 1'-0" O.C. PROVIDE 12"x12" CORNER BARS
12x20 CONC. FTG. W/ (2) #4 CONT. HORZ. BARS
STOOP FOUNDATION DETAIL
SCALE: ¾"= 1'-0"
T/WALL EL. 100'-0"
7'-4"
6'-8"
T/WALL EL. 99'-4"
T/WALL EL 98'-11" (W. ENTR.) 99'-2" (E. ENTR.)
5'-4"
7'-0"
10" 2'-4" 7'-0" 2'-4" 10"
TYPICAL ENTRANCE FOUNDATION WALL DETAIL
SCALE: ½"= 1'-0"
CONSTRUCTION JOINT SEE DET. 3/S1
9'-6"
5 TREADS @ 11"= 4'-7"
INSULATION UP TO BTTM. COURSE OF BLOCK
6" SLAB ON GRADE W/ #4 BARS, 12" O.C., EACH WAY
6 RISERS @ 7" = 3'-6"
#4 BAR EA. NOSING
T/WALL EL. 99'-4"
T/SLAB EL. 96'-6"
#5 BARS, 12" O.C.
SEE DETAIL 5/B8
T/FTG. EL. 96'-0"
12"x24" FTG. (TYP.)
1'-0" 7'-6" 2'-0" 4'-8"± 1'-0"
READING THEATER SECTION
SCALE: ½"= 1'-0"
BRUCE VILLAGE HALL & LIBRARY
BRUCE, WISCONSIN
SCALE AS NOTED
SHEET S2
DATE DEC. 31, 2001
MORGAN & PARMLEY, LTD. CONSULTING ENGINEERS LADYSMITH, WISCONSIN
DR. BY J.T.D.

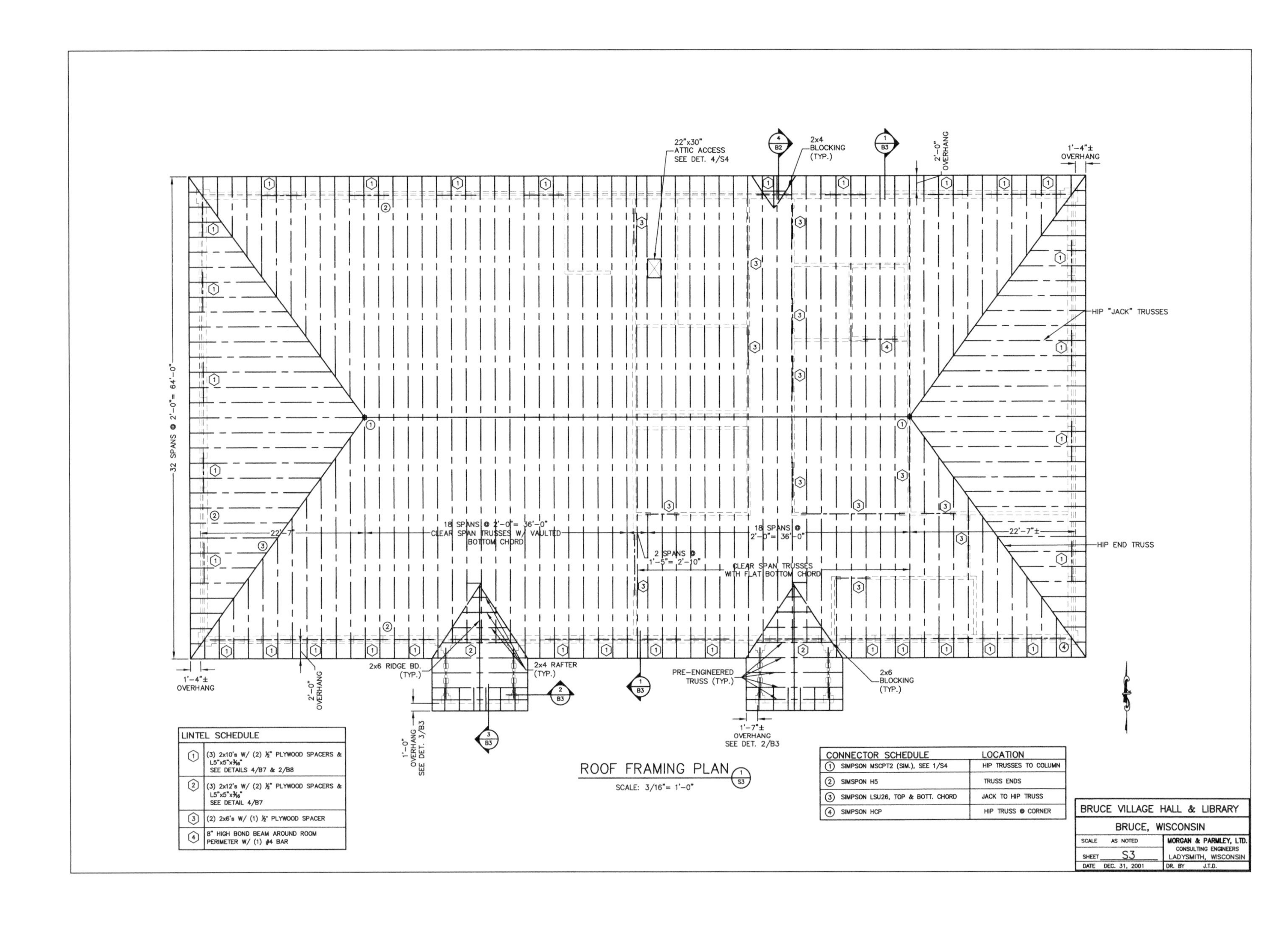
22"x30"
ATTIC ACCESS
SEE DET. 4/S4
2x4
BLOCKING
(TYP.)
2'-0"
OVERHANG
1'-4"±
OVERHANG
HIP "JACK" TRUSSES
HIP END TRUSS
32 SPANS @ 2'-0"= 64'-0"
22'-7"
18 SPANS @ 2'-0"= 36'-0"
CLEAR SPAN TRUSSES W/ VAULTED
BOTTOM CHORD
2 SPANS @
1'-5"= 2'-10"
18 SPANS @
2'-0"= 36'-0"
CLEAR SPAN TRUSSES
WITH FLAT BOTTOM CHORD
22'-7"±
1'-4"±
OVERHANG
2'-0"
OVERHANG
2x6 RIDGE BD.
(TYP.)
2x4 RAFTER
(TYP.)
PRE-ENGINEERED
TRUSS (TYP.)
2x6
BLOCKING
(TYP.)
1'-0"
OVERHANG
SEE DET. 3/B3
1'-7"±
OVERHANG
SEE DET. 2/B3
LINTEL SCHEDULE
1 (3) 2x10's W/ (2) ½" PLYWOOD SPACERS & L5"x5"x5/16" SEE DETAILS 4/B7 & 2/B8
2 (3) 2x12's W/ (2) ½" PLYWOOD SPACERS & L5"x5"x5/16" SEE DETAIL 4/B7
3 (2) 2x6's W/ (1) ½" PLYWOOD SPACER
4 8" HIGH BOND BEAM AROUND ROOM PERIMETER W/ (1) #4 BAR
ROOF FRAMING PLAN
SCALE: 3/16"= 1'-0"
CONNECTOR SCHEDULE
LOCATION
1 SIMPSON MSCPT2 (SIM.), SEE 1/S4 — HIP TRUSSES TO COLUMN
2 SIMSPON H5 — TRUSS ENDS
3 SIMPSON LSU26, TOP & BOTT. CHORD — JACK TO HIP TRUSS
4 SIMPSON HCP — HIP TRUSS @ CORNER
BRUCE VILLAGE HALL & LIBRARY
BRUCE, WISCONSIN
SCALE AS NOTED
SHEET S3
DATE DEC. 31, 2001
MORGAN & PARMLEY, LTD.
CONSULTING ENGINEERS
LADYSMITH, WISCONSIN
DR. BY J.T.D.

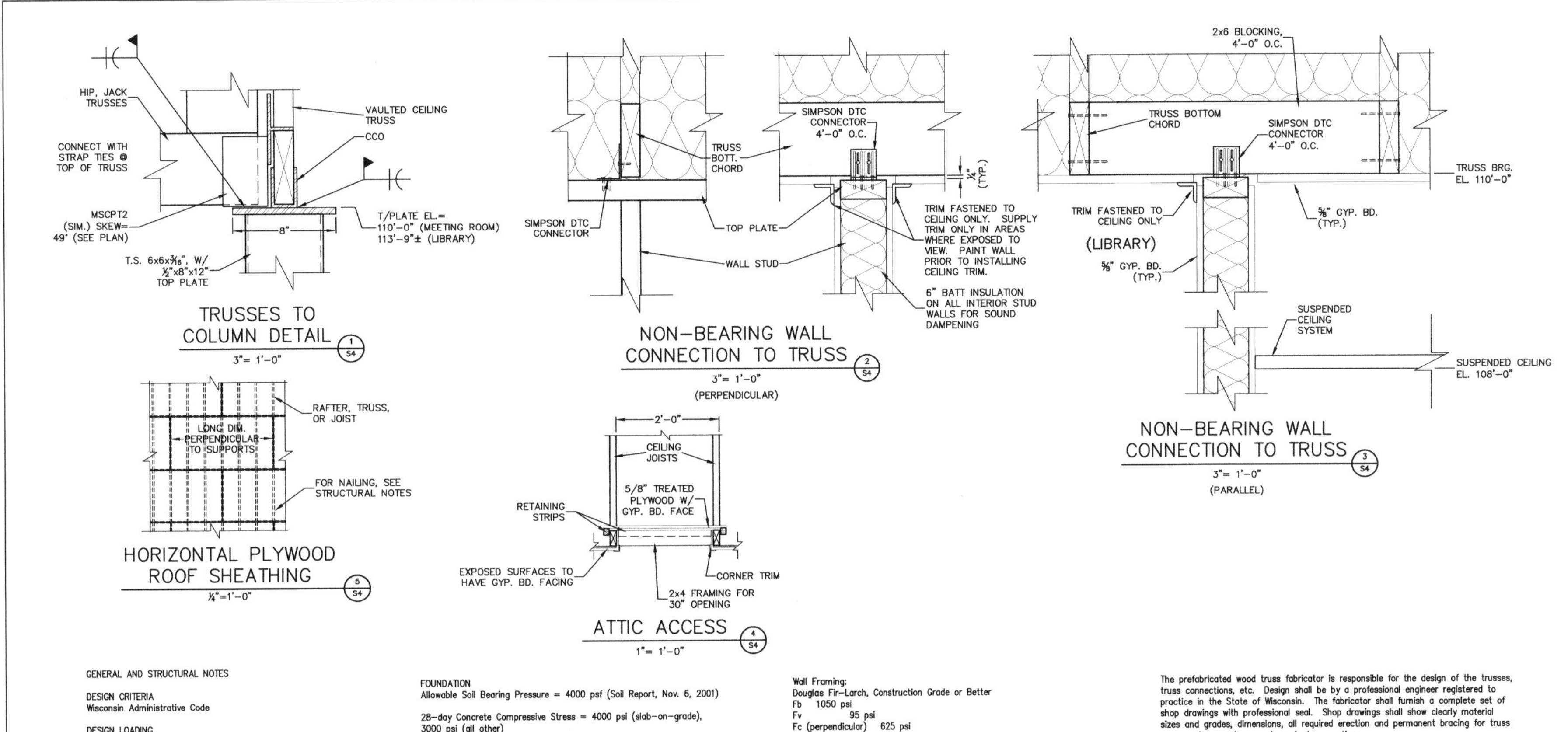

GENERAL AND STRUCTURAL NOTES

DESIGN CRITERIA
Wisconsin Administrative Code

DESIGN LOADING
Roof Gravity Loads:
Snow = 40 psf
Truss weight, plywood sheathing, metal roof deck = 8 psf
Collateral LL (truss bottom chord) = 7 psf
Wind (Projected Area Method) = 20 psf

OCCUPANCY AND SANITARY FIXTURE REQUIREMENTS
Village Offices Net Occupied Area = 2750 sq ft
No. Employees = 2750/75 sq ft/person = 37 people (19 males, 19 females)

Library Net Occupied Space = 2616 sq ft
No. Public = 2616/20 sq ft/person = 131 people (66 males, 66 females)

Total = 85 males, 85 females

Table 54.12-A:
Males – 1 WC, 1 LAV
Females – 1 WC, 1 LAV
1 DF
Table 54.12-B:
Males – 1 WC, 1 Urinal, 1 LAV
Females – 2 WC, 1 LAV
1 DF

RECYCLING SPACE (Comm 52.24, Appendix C)
Library – 2.2 cu ft / 1000 sq ft
Office (Assembly Occupancy) – 2.2 cu ft / 1000 sq ft
Required Recycling Space = 120 x 60 x 2.2 = 15.8 cubic feet
Provide 3' x 3' floor area x 2 ft high = 18 cubic feet OK

FOUNDATION
Allowable Soil Bearing Pressure = 4000 psf (Soil Report, Nov. 6, 2001)

28-day Concrete Compressive Stress = 4000 psi (slab-on-grade), 3000 psi (all other)

Maximum Slump = 4"
Maximum water/cement ratio – 0.53
All concrete exposed to weather shall be air-entrained 5 to 7

MASONRY
Units shall be standard weight, Grade N-1, 4" concrete block, set in running bond. Mortar to be Type M or S. See specifications for unit finishes and coloring, horizontal reinforcing and veneer tie requirements.

STRUCTURAL WOOD
Dimension Lumber: General construction requirements and provisions per Wisconsin Administrative Code Chapter 53.

Reference: American Forest and Paper Association; National Design Specification for Wood Construction; plus Supplement, "Design Values for Wood Construction"

Unless noted otherwise, dimension lumber shall have the following base design values:

Structural Framing, Lintels:
Douglas Fir-Larch, No.2 or Better
Fb 1450 psi
Fv 95 psi
Fc (perpendicular) 625 psi
Fc (parallel) 1000 psi
Ft 850 psi
E 1,700,000 psi

Wall Framing:
Douglas Fir-Larch, Construction Grade or Better
Fb 1050 psi
Fv 95 psi
Fc (perpendicular) 625 psi
Fc (parallel) 1150 psi
Ft 625 psi
E 1,500,000 psi

Top and Bottom Plates:
Douglas Fir-Larch, No.2 or Better
Fb 1450 psi
Fv 95 psi
Fc (perpendicular) 625 psi
Fc (parallel) 1000 psi
Ft 850 psi
E 1,700,000 psi

All member sizes indicated on the plans are nominal dimensions.

All nailing shall be in accordance with Table 53-XIX of the Wisconsin Administrative Code, unless noted otherwise.

All wood bearing on concrete or masonry shall be preservative treated.

PREFABRICATED ROOF TRUSSES
Trusses shall be designed to meet all loading and spans as indicated on the plans. Design trusses for concentrated loads and cantilevers as indicated on the architectural and structural plans.

Trusses shall comply with all provisions of the latest edition of the design specification for light metal plate wood trusses of the Truss Plate Institute.

The prefabricated wood truss fabricator is responsible for the design of the trusses, truss connections, etc. Design shall be by a professional engineer registered to practice in the State of Wisconsin. The fabricator shall furnish a complete set of shop drawings with professional seal. Shop drawings shall show clearly material sizes and grades, dimensions, all required erection and permanent bracing for truss compression members, and required connections.

The Contractor shall furnish all necessary blocking, bracing and connection material to provide a complete installation.

The Contractor shall be responsible for permanent bracing and bridging required as shown by the plans.

ROOF AND WALL FRAMING NOTES
Roof sheathing shall be as follows: 1/2 inch APA rated sheathing 32/16 EXP 1 plywood.

Plywood Roof and Wall Diaphragms: Shall be fastened with 10d (roof) and 8d (wall) nails, spaced 6" o/c along panel edges and 12" o/c at intermediate supports.

All nailing shall be accomplished with common nails and shall be driven flush but not fracture the sheathing surface. Box or sinker nails must be approved prior to use. to use.

As an alternate to hand nailing, the contractor shall submit for approval the size and type of nail used for automatic nailing with approved technical data for its use in nailing horizontal diaphragms.

BRUCE VILLAGE HALL & LIBRARY	
BRUCE, WISCONSIN	
SCALE AS NOTED	MORGAN & PARMLEY, LTD. CONSULTING ENGINEERS
SHEET S4	LADYSMITH, WISCONSIN
DATE DEC. 31, 2001	DR. BY J.T.D.

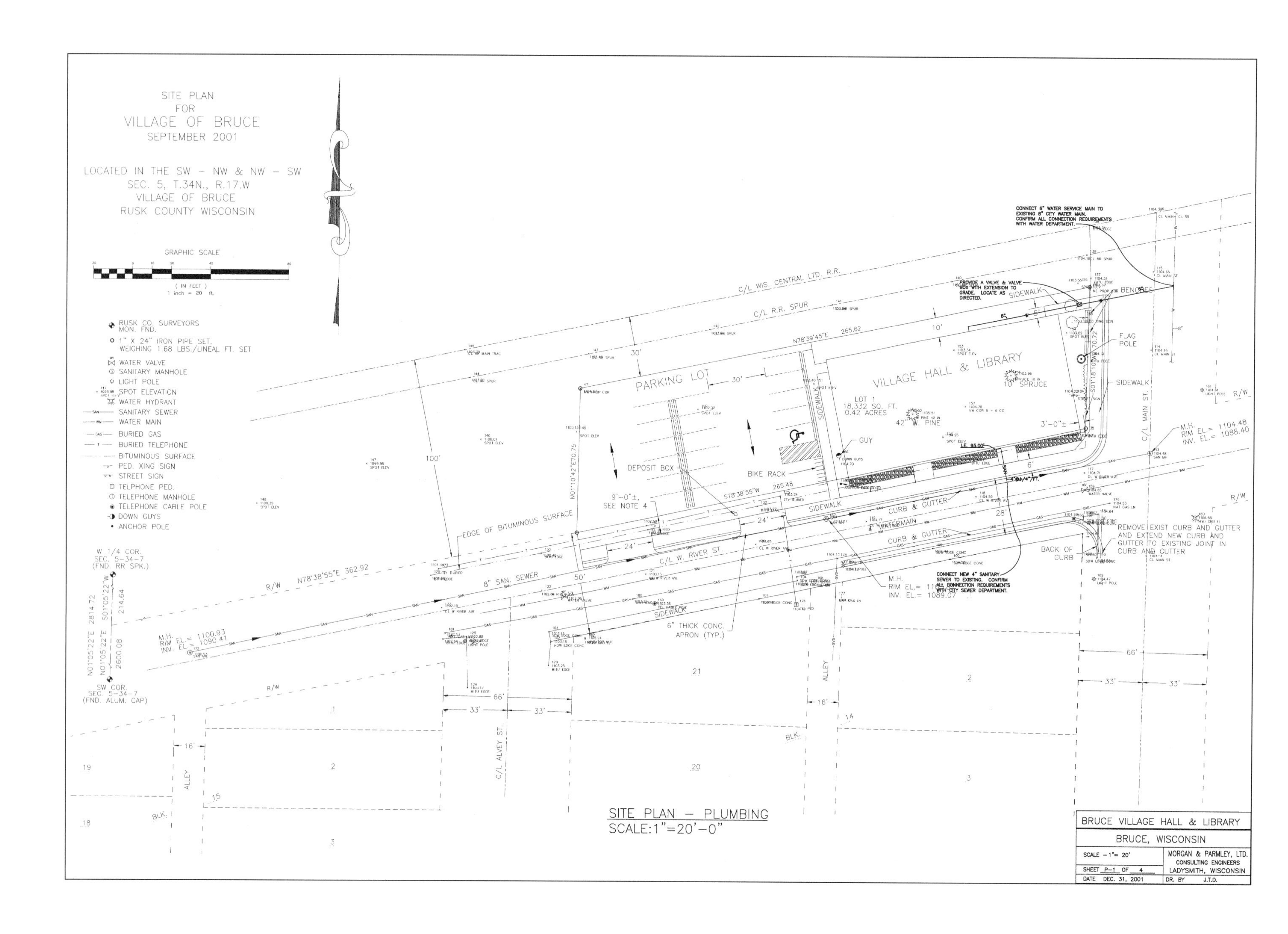
SITE PLAN
FOR
VILLAGE OF BRUCE
SEPTEMBER 2001
LOCATED IN THE SW – NW & NW – SW
SEC. 5, T.34N., R.17.W
VILLAGE OF BRUCE
RUSK COUNTY WISCONSIN
GRAPHIC SCALE
(IN FEET)
1 inch = 20 ft.
RUSK CO. SURVEYORS MON. FND.
1" X 24" IRON PIPE SET, WEIGHING 1.68 LBS./LINEAL FT. SET
WATER VALVE
SANITARY MANHOLE
LIGHT POLE
SPOT ELEVATION
WATER HYDRANT
SANITARY SEWER
WATER MAIN
BURIED GAS
BURIED TELEPHONE
BITUMINOUS SURFACE
PED. XING SIGN
STREET SIGN
TELPHONE PED.
TELEPHONE MANHOLE
TELEPHONE CABLE POLE
DOWN GUYS
ANCHOR POLE
W 1/4 COR. SEC. 5–34–7 (FND. RR SPK.)
SW COR. SEC. 5–34–7 (FND. ALUM. CAP)
C/L WIS. CENTRAL LTD. R.R.
C/L R.R. SPUR
PARKING LOT
VILLAGE HALL & LIBRARY
LOT 1
18,332 SQ. FT.
0.42 ACRES
42" W. PINE
10" SPRUCE
FLAG POLE
SIDEWALK
BENCHES
DEPOSIT BOX
BIKE RACK
GUY
9'-0"±, SEE NOTE 4
EDGE OF BITUMINOUS SURFACE
CURB & GUTTER
4" WATERMAIN
C/L W. RIVER ST.
8" SAN. SEWER
6" THICK CONC. APRON (TYP.)
M.H. RIM EL.= 1100.93 INV. EL.= 1090.41
M.H. RIM EL.= 1104.48 INV. EL.= 1088.40
INV. EL.= 1089.07
CONNECT 6" WATER SERVICE MAIN TO EXISTING 8" CITY WATER MAIN. CONFIRM ALL CONNECTION REQUIREMENTS WITH WATER DEPARTMENT.
PROVIDE A VALVE & VALVE BOX WITH EXTENSION TO GRADE. LOCATE AS DIRECTED.
CONNECT NEW 4" SANITARY SEWER TO EXISTING. CONFIRM ALL CONNECTION REQUIREMENTS WITH CITY SEWER DEPARTMENT.
REMOVE EXIST CURB AND GUTTER AND EXTEND NEW CURB AND GUTTER TO EXISTING JOINT IN CURB AND GUTTER
BACK OF CURB
C/L MAIN ST.
C/L ALVEY ST.
ALLEY
BLK.
R/W
N78°38'55"E 362.92
N78°39'45"E 265.62
S78°38'55"W 265.48
N01°10'42"E 70.75
S01°18'10"W 70.72
N01°05'22"E 2814.72
N01°05'22"E 2600.08
S01°05'22"W 214.64
SITE PLAN – PLUMBING
SCALE:1"=20'-0"
BRUCE VILLAGE HALL & LIBRARY
BRUCE, WISCONSIN
SCALE – 1"= 20'
SHEET P-1 OF 4
DATE DEC. 31, 2001
MORGAN & PARMLEY, LTD.
CONSULTING ENGINEERS
LADYSMITH, WISCONSIN
DR. BY J.T.D.

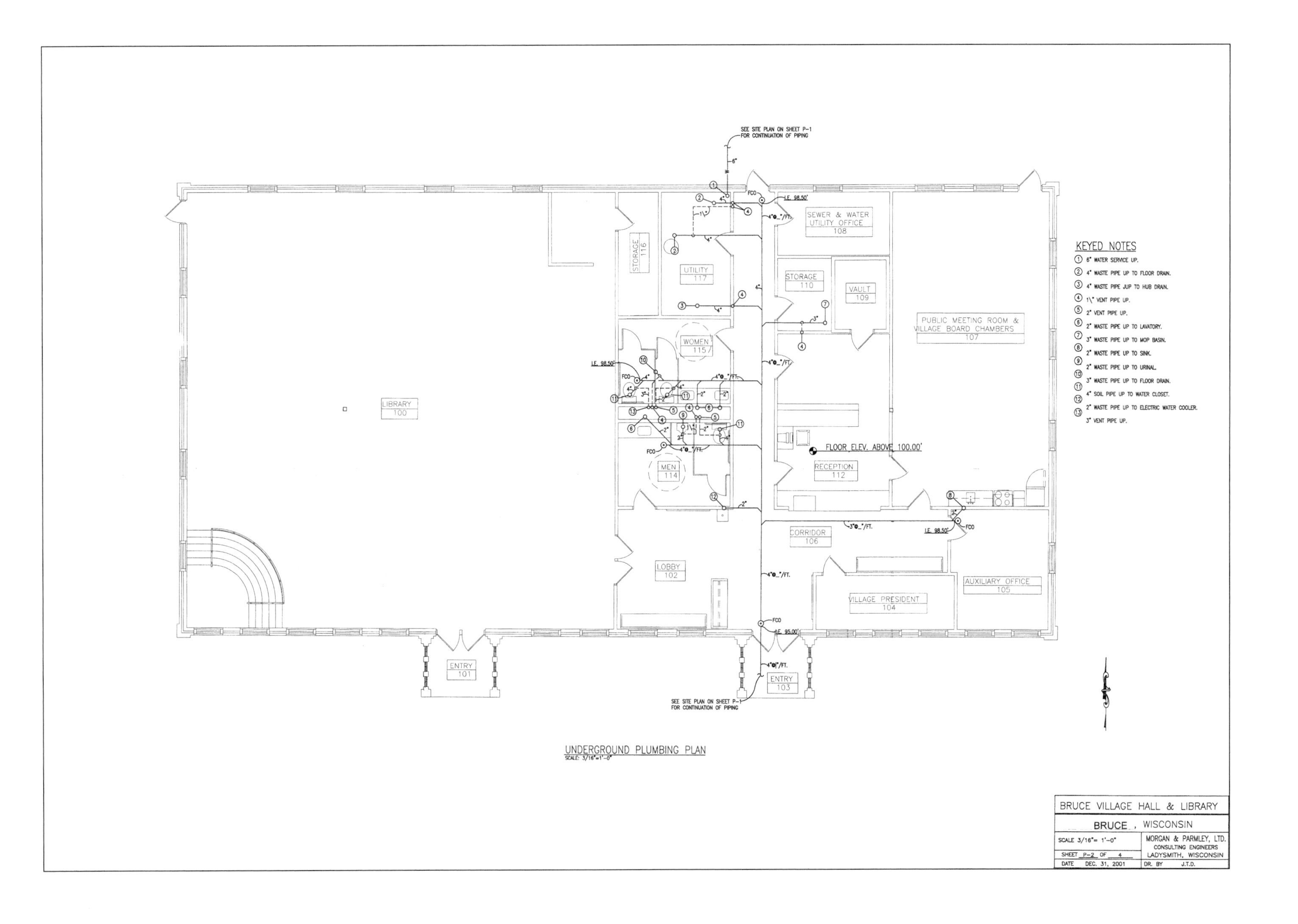
SEE SITE PLAN ON SHEET P-1
FOR CONTINUATION OF PIPING
SEWER & WATER UTILITY OFFICE 108
STORAGE 116
UTILITY 117
STORAGE 110
VAULT 109
PUBLIC MEETING ROOM & VILLAGE BOARD CHAMBERS 107
WOMEN 115
LIBRARY 100
MEN 114
RECEPTION 112
FLOOR ELEV. ABOVE 100.00'
CORRIDOR 106
LOBBY 102
AUXILIARY OFFICE 105
VILLAGE PRESIDENT 104
ENTRY 101
ENTRY 103
SEE SITE PLAN ON SHEET P-1
FOR CONTINUATION OF PIPING
KEYED NOTES
① 6" WATER SERVICE UP.
② 4" WASTE PIPE UP TO FLOOR DRAIN.
③ 4" WASTE PIPE JUP TO HUB DRAIN.
④ 1\" VENT PIPE UP.
⑤ 2" VENT PIPE UP.
⑥ 2" WASTE PIPE UP TO LAVATORY.
⑦ 3" WASTE PIPE UP TO MOP BASIN.
⑧ 2" WASTE PIPE UP TO SINK.
⑨ 2" WASTE PIPE UP TO URINAL.
⑩ 3" WASTE PIPE UP TO FLOOR DRAIN.
⑪ 4" SOIL PIPE UP TO WATER CLOSET.
⑫ 2" WASTE PIPE UP TO ELECTRIC WATER COOLER.
⑬ 3" VENT PIPE UP.
UNDERGROUND PLUMBING PLAN
SCALE: 3/16"=1'-0"
BRUCE VILLAGE HALL & LIBRARY
BRUCE, WISCONSIN
SCALE 3/16"= 1'-0"
MORGAN & PARMLEY, LTD.
CONSULTING ENGINEERS
LADYSMITH, WISCONSIN
SHEET P-2 OF 4
DATE DEC. 31, 2001
DR. BY J.T.D.

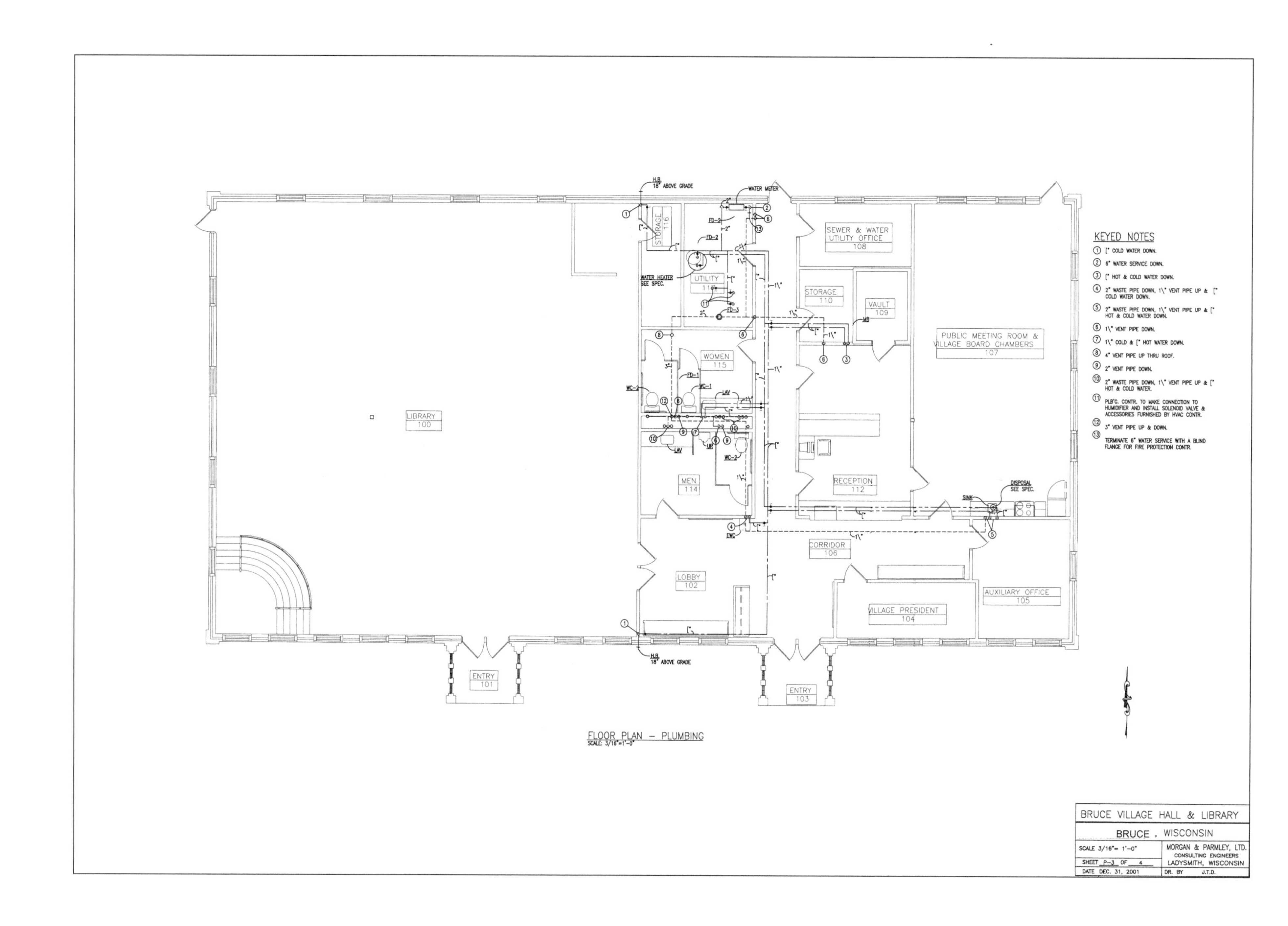
KEYED NOTES
① [" COLD WATER DOWN.
② 6" WATER SERVICE DOWN.
③ [" HOT & COLD WATER DOWN.
④ 2" WASTE PIPE DOWN, 1\" VENT PIPE UP & [" COLD WATER DOWN.
⑤ 2" WASTE PIPE DOWN, 1\" VENT PIPE UP & [" HOT & COLD WATER DOWN.
⑥ 1\" VENT PIPE DOWN.
⑦ 1\" COLD & [" HOT WATER DOWN.
⑧ 4" VENT PIPE UP THRU ROOF.
⑨ 2" VENT PIPE DOWN.
⑩ 2" WASTE PIPE DOWN, 1\" VENT PIPE UP & [" HOT & COLD WATER.
⑪ PLB'G. CONTR. TO MAKE CONNECTION TO HUMIDIFIER AND INSTALL SOLENOID VALVE & ACCESSORIES FURNISHED BY HVAC CONTR.
⑫ 3" VENT PIPE UP & DOWN.
⑬ TERMINATE 6" WATER SERVICE WITH A BLIND FLANGE FOR FIRE PROTECTION CONTR.
H.B. 18" ABOVE GRADE
WATER METER
WATER HEATER SEE SPEC.
STORAGE 116
UTILITY 111
WOMEN 115
MEN 114
LIBRARY 100
LOBBY 102
CORRIDOR 106
SEWER & WATER UTILITY OFFICE 108
STORAGE 110
VAULT 109
PUBLIC MEETING ROOM & VILLAGE BOARD CHAMBERS 107
RECEPTION 112
VILLAGE PRESIDENT 104
AUXILIARY OFFICE 105
ENTRY 101
ENTRY 103
DISPOSAL SEE SPEC.
SINK
EWC
FD-1
FD-2
FD-3
WC-1
WC-2
LAV
UR
FLOOR PLAN – PLUMBING
SCALE: 3/16"=1'-0"
BRUCE VILLAGE HALL & LIBRARY
BRUCE, WISCONSIN
SCALE 3/16"= 1'-0"
SHEET P-3 OF 4
DATE DEC. 31, 2001
MORGAN & PARMLEY, LTD.
CONSULTING ENGINEERS
LADYSMITH, WISCONSIN
DR. BY J.T.D.

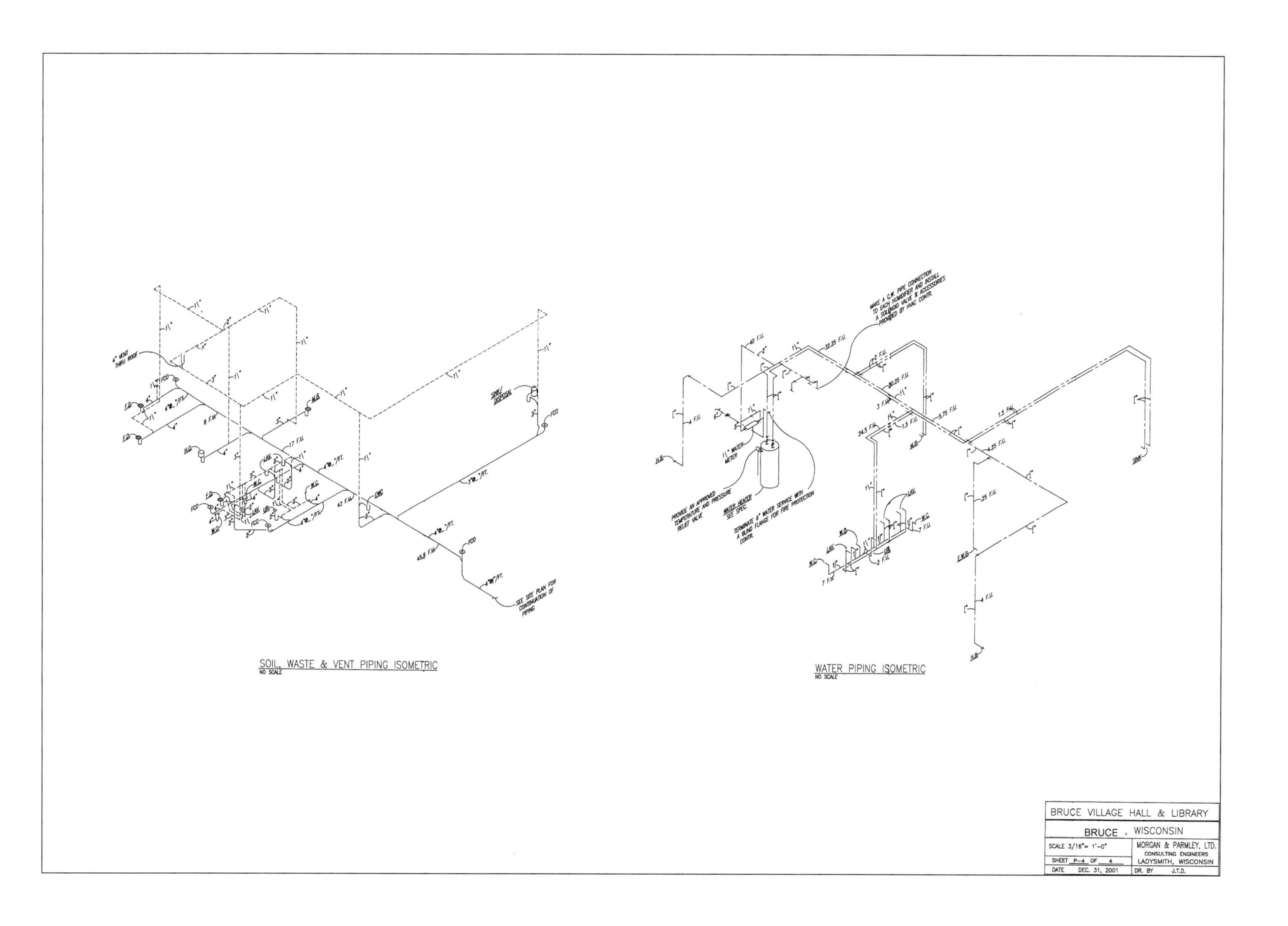
SOIL, WASTE & VENT PIPING ISOMETRIC
NO SCALE
WATER PIPING ISOMETRIC
NO SCALE
SEE SITE PLAN FOR CONTINUATION OF PIPING
PROVIDE AN APPROVED TEMPERATURE AND PRESSURE RELIEF VALVE
WATER HEATER SEE SPEC.
TERMINATE 6" WATER SERVICE WITH A BLIND FLANGE FOR FIRE PROTECTION CONTR.
MAKE A C.W. PIPE CONNECTION TO EACH HUMIDIFIER AND INSTALL A SOLENOID VALVE & ACCESSORIES PROVIDED BY HVAC CONTR.
BRUCE VILLAGE HALL & LIBRARY
BRUCE , WISCONSIN
SCALE 3/16"= 1'-0"
SHEET P-4 OF 4
DATE DEC. 31, 2001
MORGAN & PARMLEY, LTD.
CONSULTING ENGINEERS
LADYSMITH, WISCONSIN
DR. BY J.T.D.

GAS FIRED FURNACE SCHEDULE

UNIT NO.	AREA SERVED	MFR.	MODEL NO.	AIR HANDLING SECTION: SUP. CFM	OA CFM	E.S.P.	FAN RPM	FAN HP	HEATING SECTION: MBH INPUT	MBH OUTPUT	% EFF.	GAS PIPE SIZE	INTAKE/VENT SIZE	ELECT. RQMTS.: VOLTS	PHASE	REMARKS
F-1	LIBRARY-NORTH	CARRIER	58MCA080-20	1600	160	.8	HIGH	¾	80	74	91	¾"	2"	120V	1φ	①
F-2	LIBRARY-SOUTH	CARRIER	58MCA080-20	1600	160	.8	HIGH	¾	80	74	91	¾"	2"	120V	1φ	①
F-3	MEETING ROOM	CARRIER	58MCA080-16	1350	300	.8	HIGH	½	80	74	91	¾"	2"	120V	1φ	①
F-4	VILLAGE OFFICES	CARRIER	58MCA080-20	1600	160	.8	HIGH	¾	80	74	91	¾"	2"	120V	1φ	①

① CONCENTRIC COMB. AIR INTAKE/VENT KIT.
SIDE FILTER RACK W/ 1" THICK WASHABLE FILTER.
7 DAY ELECTRONIC PROGRAMMABLE THERMOSTAT.

FURNACE COOLING COIL SCHEDULE

UNIT NO.	UNIT SERVED	MFR.	MODEL NO.	TYPE	CFM	CLG. CAP. (MBH): TC	SC	EAT (°F): DB/WB	APD (IN H_2O)	SST COIL (°F)	REFRIG. PIPE SIZES: SUCT.	LIQ.
CC-1	F-1	CARRIER	CK5A048	CASED COIL	1600	46.0	32.7	80/67	.31	45	1⅛" OD.	½" OD.
CC-2	F-2	CARRIER	CK5A048	CASED COIL	1600	46.0	32.7	80/67	.31	45	1⅛" OD.	½" OD.
CC-3	F-3	CARRIER	CK5A036	CASED COIL	1350	34.0	25.7	80/67	.30	45	¾" OD.	½" OD.
CC-4	F-4	CARRIER	CK5A048	CASED COIL	1600	46.0	32.7	80/67	.31	45	1⅛" OD.	½" OD.

AIR COOLED CONDENSING UNIT SCHEDULE

(CU) — EQUIPMENT NAME / — UNIT NUMBER

UNIT NO.	UNIT SERVED	UNIT LOCATION	MFR.	MODEL NO.	NOM. CAP. AT 95°F (TONS)	ELECTRICAL REQUIREMENTS: VOLTS/PHASE	MIN. CIRC. AMPS	MAX FUSE SIZE	EER	UNIT WEIGHT	REMARKS
CU-1	F-1	ON GRADE	CARRIER	38 TUA-048	4	240V/1φ	30.5 A	50 A	11.0	200	①
CU-2	F-2	ON GRADE	CARRIER	38 TUA-048	4	240V/1φ	30.5 A	50 A	11.0	200	①
CU-3	F-3	ON GRADE	CARRIER	38 TUA-036	3	240V/1φ	23.4 A	40 A	11.0	140	①
CU-4	F-4	ON GRADE	CARRIER	38 TUA-048	4	240V/1φ	30.5 A	50 A	11.0	200	①

① BALL BEARING COND. FAN MOTOR.
COMPRESSOR START ASSIST CAPAC[illegible] & RELAY.
CRANKCASE HEATER.
CYCLE PROTECTOR.
EVAPORATOR FREEZE STAT.
FILTER DRIERS.
HIGH PRESSURE SWITCH.
LOW PRESSURE SWITCH.
CONDENSER FAN SPEED LOW AMBIENT CONTROL.
WINTER START CONTROL.

FAN SCHEDULE

(EF) — EQUIPMENT NAME / — UNIT NUMBER

UNIT NO.	AREA SERVED	MFR.	MODEL NO.	CFM	S.P.	RPM	H.P.	VOLTS	PHASE	SONES	NO. REQ'D	NOTES
EF-1	WOMENS TOILET	CARNES	VCDD030C	296	.375	905	2.7 A	120V	1φ	3.5	1	①
EF-2	MENS TOILET	CARNES	VCDD030C	296	.375	905	2.7 A	120V	1φ	3.5	1	①
EF-3	JANITORS CLOSET	CARNES	VCDD020C	150	.375	740	1.8 A	120V	1φ	2.9	1	①

① SPEED SWITCH.
SLOPING ROOF CAP.

SCHEDULE OF GRILLES, REGISTERS AND DIFFUSERS

TYPE	DESCRIPTION	MANUF.	MODEL NO.	MATERIAL	DAMPER	FINISH	REMARKS
SR-1	SUPPLY REGISTER	CARNES	RNDAH	ALUMINUM	YES	PRIME COAT	DOUBLE DEFLECTION, FLANGE FRAME
CD-1	CEILING DIFFUSER	CARNES	SETA	ALUMINUM	YES	OFF WHITE	LOUVER FACE, LAY-IN T-BAR FRAME
CD-2	CEILING DIFFUSER	CARNES	SEFA	ALUMINUM	YES	OFF WHITE	LOUVER FACE, FLANGE FRAME
CD-3	CEILING DIFFUSER	ACUTHERM THERMAUSER	TF-HC	STEEL	VAV DIFFUSER	OFF WHITE	VAV DIFFUSER, LAY-IN T-BAR FRAME
RG/EG/TG-1	RETURN/TRANSFER/ EXHAUST GRILLE	CARNES	R[illegible]AF	ALUMINUM	NO	OFF WHITE	EGG CRATE, FLANGE FRAME
RG/EG/TG-2	RETURN/TRANSFER/ EXHAUST GRILLE	CARNES	RA[illegible]AH	ALUMINUM	NO	OFF WHITE	EGG CRATE, LAY-IN T-BAR FRAME
RG/EG/TG-3	RETURN/TRANSFER/ EXHAUST GRILLE	CARNES	RALA[illegible]	ALUMINUM	NO	PRIME COAT	40° DEFLECT., LVR. FACE, FLANGE FRAME

NOTES: 1. SEE DRAWINGS FOR AIR QUANTITIES, NECK SIZES AND THROW PATTERN.
2. COORDINATE EXACT LOCATION OF CEILING OUTLETS [illegible] REFLECTED CEILING PLAN AND LIGHTING LAYOUT.

ELECTRIC HEATER SCHEDULE

(EW) — EQUIPMENT NAME / — UNIT NUMBER

UNIT NO.	AREA SERVED	MFR.	MODEL NO.	HEAT. CAP. (MBH)	HEAT. CAP. (KW)	AMPS	VOLTS/PHASE	MOUNT. HEIGHT (FT.)	REMARKS
EW-1	SEE DRAWINGS	Q-MARK	CWH-2101	3.4	1.0	8.4 A	120V/1φ	12'	①
EW-2	SEE DRAWINGS	Q-MARK	CWH-2151	5.1	1.5	12.6 A	120V/1φ	12'	①
EB-1	SEE DRAWINGS	Q-MARK	HBB-1257	3.4	1.0	8.3 A	120V/1φ	-	②

① RECESS MOUNTING BOX.
INTEGRAL TAMPERPROOF THERMOSTAT.
POWER DISCONNECT SWITCH.

② INTEGRAL THERMOSTAT SECTION.

SYMBOL SCHEDULE

SYMBOL	DESCRIPTION	SYMBOL	DESCRIPTION
—— RS ——	REFRIGERANT SUCTION PIPING	F	FURNACE
—— RL ——	REFRIGERANT LIQUID PIPING	H	HUMIDIFIER
—— G ——	GAS PIPING	CU	AIR COOLED CONDENSING UNIT
—— D ——	DRAIN PIPING	EF	EXHAUST FAN
—— V ——	COMBUSTION AIR VENT PIPING	RH	RANGE HOOD
—— I ——	COMBUSTION AIR INTAKE PIPING	EW	ELECTRIC WALL HEATER
(T)	THERMOSTAT/TEMPERATURE SENSOR	SR	SUPPLY REGISTER
(S)	STARTER/SPEED SWITCH	CD	CEILING DIFFUSER
[V]	VOLUME DAMPER	TG	TRANSFER GRILLE
[F]	FIRE DAMPER	EG	EXHAUST GRILLE
[A]	AUTOMATIC DAMPER	RG	RETURN GRILLE
[B]	BACKDRAFT DAMPER	DG	DOOR GRILLE

MOTOR AND MOTOR STARTER SCHEDULE

ITEM NO.	EQUIP. DESCRIPTION	EQUIP. LOCATION	HP	VOLTS	PHASE	STARTER MOD. NO (ALLEN BRADLEY)	NO. REQ'D	STARTER LOCATION	ATC INTRLK	EMERG. POWER	REMARKS
1	F-1	UTILITY ROOM	¾	120V	1φ	BY UNIT MFR.	1	IN UNIT	-	-	①
2	F-2	UTILITY ROOM	¾	120V	1φ	BY UNIT MFR.	1	IN UNIT	-	-	①
3	F-3	UTILITY ROOM	½	120V	1φ	BY UNIT MFR.	1	IN UNIT	-	-	①
4	F-4	UTILITY ROOM	¾	120V	1φ	BY UNIT MFR.	1	IN UNIT	-	-	①
5	H-1	UTILITY ROOM	.6 A	120V	1φ	BY UNIT MFR.	1	IN UNIT	1 & 2	-	①
6	H-2	UTILITY ROOM	.6 A	120V	1φ	BY UNIT MFR.	1	IN UNIT	3	-	①
7	H-3	UTILITY ROOM	.6 A	120V	1φ	BY UNIT MFR.	1	IN UNIT	4	-	①
8	CU-1	OUTSIDE ON GRADE	30.5 A	240V	1φ	BY UNIT MFR.	1	IN UNIT	1	-	②
9	CU-2	OUTSIDE ON GRADE	30.5 A	240V	1φ	BY UNIT MFR.	1	IN UNIT	2	-	②
10	CU-3	OUTSIDE ON GRADE	23.4 A	240V	1φ	BY UNIT MFR.	1	IN UNIT	3	-	②
11	CU-4	OUTSIDE ON GRADE	30.5 A	240V	1φ	BY UNIT MFR.	1	IN UNIT	4	-	②
12	EF-1	WOMEN'S TOILET	2.7 A	120V	1φ	600 TAX 9	1	UTILITY ROOM	1,2,3 & 4	-	③
13	EF-2	MEN'S TOILET	2.7 A	120V	1φ	600 TAX 9	1	UTILITY ROOM	1,2,3 & 4	-	③
14	EF-3	STORAGE	1.8 A	120V	1φ	600 TAX 9	1	UTILITY ROOM	1,2,3 & 4	-	③
15	EB-1	LIBRARY-STAGE	8.3 A	120V	1φ	BY UNIT MFR.	1	IN UNIT	-	-	⑤
16	EW-1	SEE DRAWING	8.4 A	120V	1φ	BY UNIT MFR.	2	IN UNIT	-	-	④⑤
17	EW-2	SEE DRAWING	12.6 A	120V	1φ	BY UNIT MFR.	2	IN UNIT	-	-	④⑤

NOTES: ① E.C. TO FURNISH AND INSTALL DISCONNECT SWITCH.
② E.C. TO FURNISH AND INSTALL WEATHERPROOF SWITCH.
③ E.C. TO INSTALL AND WIRE MFR. SUPPLIED SPEED SWITCH ON FAN HOUSING.
④ E.C. TO INSTALL AND WIRE MFR. SUPPLIED DISCONNECT SWITCH.
⑤ E.C. TO INSTALL AND WIRE ELECTRIC HEATER.

HUMIDIFIER SCHEDULE

UNIT NO.	UNIT SERVED	MFR.	MODEL NO.	EVAP. RATE AT 120°F PLENUM	ELECT. REQRMTS.	SUPPLY PLENUM CONNECTION	BYPASS DUCT SIZE	REMARKS
H-1	F-1/2	APRILAIRE	448	.75 GPH	.6 A/120V/1φ	13¾"W×11¼"H	6"φ	①
H-2	F-3	APRILAIRE	448	.75 GPH	.6 A/120V/1φ	13¾"W×11¼"H	6"φ	①
H-3	F-4	APRILAIRE	448	.75 GPH	.6 A/120V/1φ	13¾"W×11¼"H	6"φ	①

① LOW VOLTAGE WALL MOUNTED HUMIDISTAT.
MAKE-UP WATER CONTROL VALVE.

BRUCE VILLAGE HALL & LIBRARY	
BRUCE, WISCONSIN	
SCALE: NONE	MORGAN & PARMLEY, LTD. CONSULTING ENGINEERS LADYSMITH, WISCONSIN
SHEET: M-1 OF 3	
DATE: DEC. 31, 2001	DR. BY: FLM, INC.

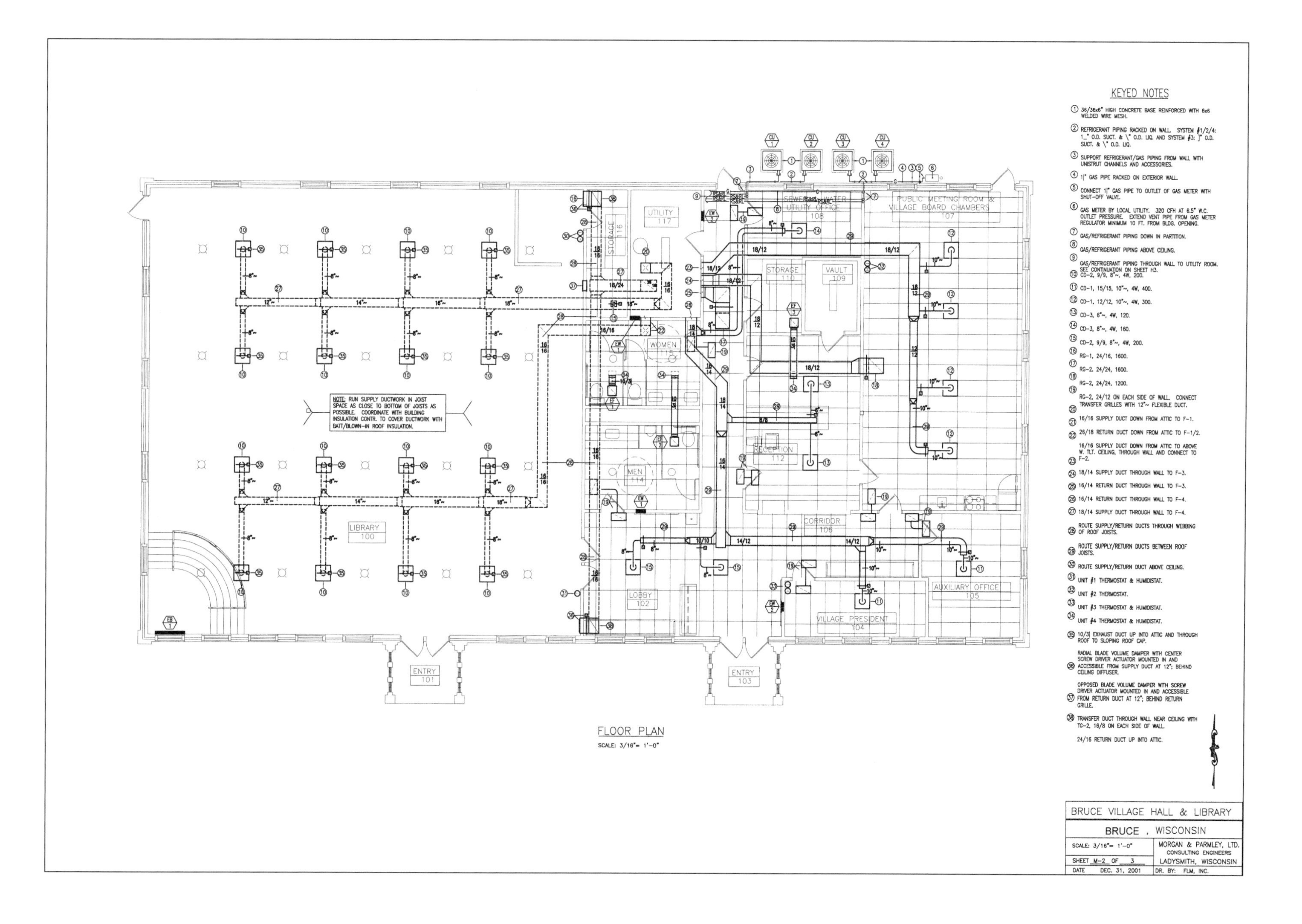
KEYED NOTES
① 36/36x6" HIGH CONCRETE BASE REINFORCED WITH 6x6 WELDED WIRE MESH.
② REFRIGERANT PIPING RACKED ON WALL. SYSTEM #1/2/4: 1_" O.D. SUCT. & \" O.D. LIQ. AND SYSTEM #3:]" O.D. SUCT. & \" O.D. LIQ.
③ SUPPORT REFRIGERANT/GAS PIPING FROM WALL WITH UNISTRUT CHANNELS AND ACCESSORIES.
④ 1|" GAS PIPE RACKED ON EXTERIOR WALL.
⑤ CONNECT 1|" GAS PIPE TO OUTLET OF GAS METER WITH SHUT-OFF VALVE.
⑥ GAS METER BY LOCAL UTILITY. 320 CFH AT 6.5" W.C. OUTLET PRESSURE. EXTEND VENT PIPE FROM GAS METER REGULATOR MINIMUM 10 FT. FROM BLDG. OPENING.
⑦ GAS/REFRIGERANT PIPING DOWN IN PARTITION.
⑧ GAS/REFRIGERANT PIPING ABOVE CEILING.
⑨ GAS/REFRIGERANT PIPING THROUGH WALL TO UTILITY ROOM. SEE CONTINUATION ON SHEET H3.
⑩ CD-2, 9/9, 8"~, 4W, 200.
⑪ CD-1, 15/15, 10"~, 4W, 400.
⑫ CD-1, 12/12, 10"~, 4W, 300.
⑬ CD-3, 6"~, 4W, 120.
⑭ CD-3, 8"~, 4W, 160.
⑮ CD-2, 9/9, 8"~, 4W, 200.
⑯ RG-1, 24/16, 1600.
⑰ RG-2. 24/24, 1600.
⑱ RG-2, 24/24, 1200.
⑲ RG-2, 24/12 ON EACH SIDE OF WALL. CONNECT TRANSFER GRILLES WITH 12"~ FLEXIBLE DUCT.
⑳ 16/16 SUPPLY DUCT DOWN FROM ATTIC TO F-1.
㉑ 26/18 RETURN DUCT DOWN FROM ATTIC TO F-1/2.
㉒ 16/16 SUPPLY DUCT DOWN FROM ATTIC TO ABOVE W. TLT. CEILING, THROUGH WALL AND CONNECT TO F-2.
㉓ 18/14 SUPPLY DUCT THROUGH WALL TO F-3.
㉔ 16/14 RETURN DUCT THROUGH WALL TO F-3.
㉕ 16/14 RETURN DUCT THROUGH WALL TO F-4.
㉖ 18/14 SUPPLY DUCT THROUGH WALL TO F-4.
㉗ ROUTE SUPPLY/RETURN DUCTS THROUGH WEBBING OF ROOF JOISTS.
㉘ ROUTE SUPPLY/RETURN DUCTS BETWEEN ROOF JOISTS.
㉙ ROUTE SUPPLY/RETURN DUCT ABOVE CEILING.
㉚ UNIT #1 THERMOSTAT & HUMIDISTAT.
㉛ UNIT #2 THERMOSTAT.
㉜ UNIT #3 THERMOSTAT & HUMIDISTAT.
㉝ UNIT #4 THERMOSTAT & HUMIDISTAT.
㉞ 10/3| EXHAUST DUCT UP INTO ATTIC AND THROUGH ROOF TO SLOPING ROOF CAP.
㉟ RADIAL BLADE VOLUME DAMPER WITH CENTER SCREW DRIVER ACTUATOR MOUNTED IN AND ACCESSIBLE FROM SUPPLY DUCT AT 12"; BEHIND CEILING DIFFUSER.
㊱ OPPOSED BLADE VOLUME DAMPER WITH SCREW DRIVER ACTUATOR MOUNTED IN AND ACCESSIBLE FROM RETURN DUCT AT 12"; BEHIND RETURN GRILLE.
㊲ TRANSFER DUCT THROUGH WALL NEAR CEILING WITH TG-2, 16/8 ON EACH SIDE OF WALL.
㊳ 24/16 RETURN DUCT UP INTO ATTIC.
NOTE: RUN SUPPLY DUCTWORK IN JOIST SPACE AS CLOSE TO BOTTOM OF JOISTS AS POSSIBLE. COORDINATE WITH BUILDING INSULATION CONTR. TO COVER DUCTWORK WITH BATT/BLOWN-IN ROOF INSULATION.
LIBRARY 100
ENTRY 101
LOBBY 102
ENTRY 103
VILLAGE PRESIDENT 104
AUXILIARY OFFICE 105
CORRIDOR 106
PUBLIC MEETING ROOM & VILLAGE BOARD CHAMBERS 107
SEWER & WATER UTILITY OFFICE 108
VAULT 109
STORAGE 110
RECEPTION 112
MEN 114
WOMEN 115
STORAGE 116
UTILITY 117
FLOOR PLAN
SCALE: 3/16"= 1'-0"
BRUCE VILLAGE HALL & LIBRARY
BRUCE, WISCONSIN
SCALE: 3/16"= 1'-0"
MORGAN & PARMLEY, LTD.
CONSULTING ENGINEERS
LADYSMITH, WISCONSIN
SHEET M-2 OF 3
DATE DEC. 31, 2001
DR. BY: FLM, INC.

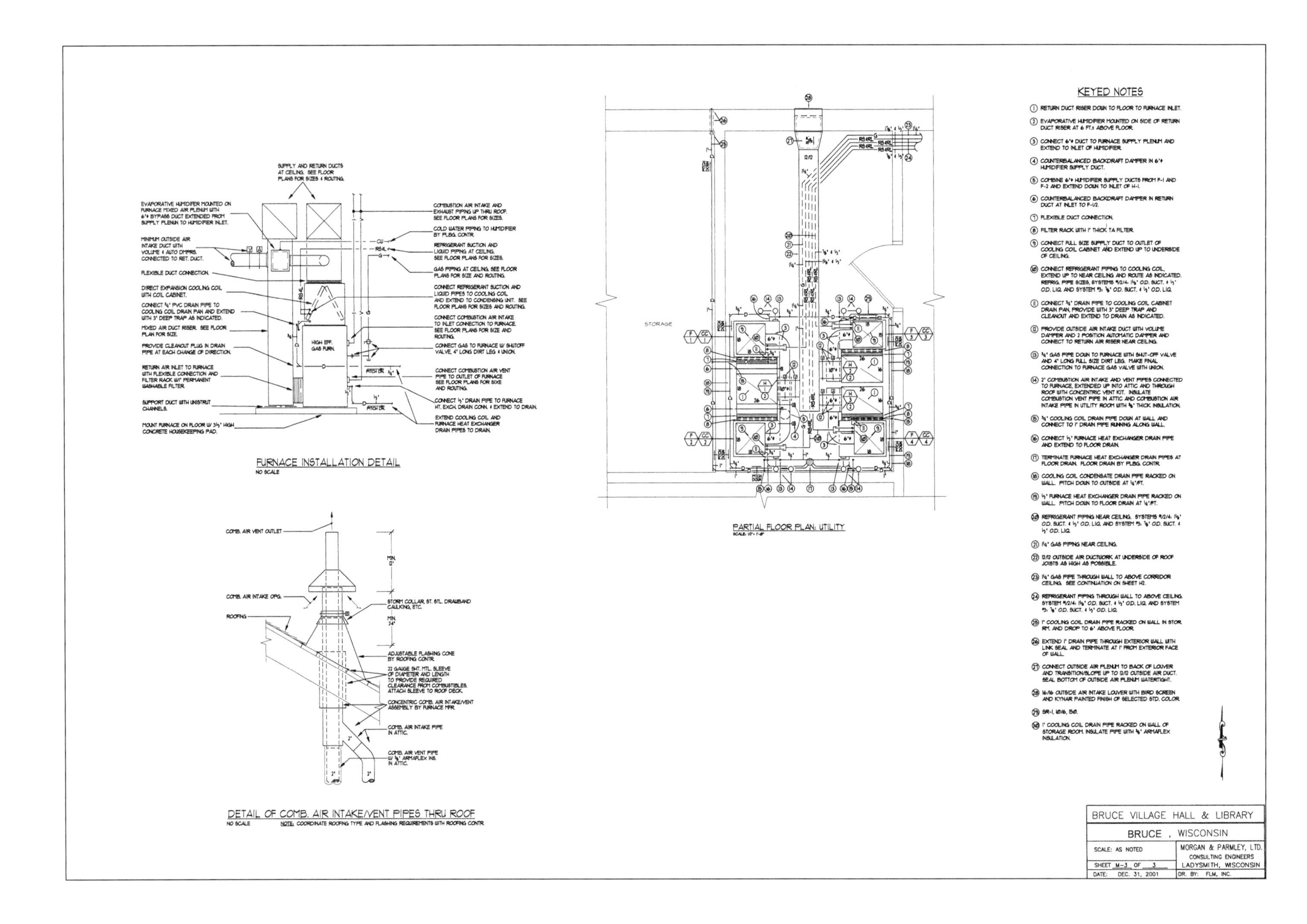
SUPPLY AND RETURN DUCTS AT CEILING. SEE FLOOR PLANS FOR SIZES & ROUTING.
EVAPORATIVE HUMIDIFER MOUNTED ON FURNACE MIXED AIR PLENUM WITH 6"ϕ BYPASS DUCT EXTENDED FROM SUPPLY PLENUM TO HUMIDIFIER INLET.
MINIMUM OUTSIDE AIR INTAKE DUCT WITH VOLUME & AUTO DMPRS. CONNECTED TO RET. DUCT.
FLEXIBLE DUCT CONNECTION.
DIRECT EXPANSION COOLING COIL WITH COIL CABINET.
CONNECT 3/4" PVC DRAIN PIPE TO COOLING COIL DRAIN PAN AND EXTEND WITH 3" DEEP TRAP AS INDICATED.
MIXED AIR DUCT RISER. SEE FLOOR PLAN FOR SIZE.
PROVIDE CLEANOUT PLUG IN DRAIN PIPE AT EACH CHANGE OF DIRECTION.
RETURN AIR INLET TO FURNACE WITH FLEXIBLE CONNECTION AND FILTER RACK W/1" PERMANENT WASHABLE FILTER.
SUPPORT DUCT WITH UNISTRUT CHANNELS.
MOUNT FURNACE ON FLOOR W/ 3 1/2" HIGH CONCRETE HOUSEKEEPING PAD.
HIGH EFF. GAS FURN.
CW
RS4L
G
PITCH DN. 3/4"
1/2"
PITCH DN.
COMBUSTION AIR INTAKE AND EXHAUST PIPING UP THRU ROOF. SEE FLOOR PLANS FOR SIZES.
COLD WATER PIPING TO HUMIDIFIER BY PLBG. CONTR.
REFRIGERANT SUCTION AND LIQUID PIPING AT CEILING. SEE FLOOR PLANS FOR SIZES.
GAS PIPING AT CEILING. SEE FLOOR PLANS FOR SIZE AND ROUTING.
CONNECT REFRIGERANT SUCTION AND LIQUID PIPES TO COOLING COIL AND EXTEND TO CONDENSING UNIT. SEE FLOOR PLANS FOR SIZES AND ROUTING.
CONNECT COMBUSTION AIR INTAKE TO INLET CONNECTION TO FURNACE. SEE FLOOR PLANS FOR SIZE AND ROUTING.
CONNECT GAS TO FURNACE W/ SHUTOFF VALVE, 4" LONG DIRT LEG & UNION.
CONNECT COMBUSTION AIR VENT PIPE TO OUTLET OF FURNACE SEE FLOOR PLANS FOR SIXE AND ROUTING.
CONNECT 1/2" DRAIN PIPE TO FURNACE HT. EXCH. DRAIN CONN. & EXTEND TO DRAIN.
EXTEND COOLING COIL AND FURNACE HEAT EXCHANGER DRAIN PIPES TO DRAIN.
FURNACE INSTALLATION DETAIL
NO SCALE
COMB. AIR VENT OUTLET
COMB. AIR INTAKE OPG.
ROOFING
MIN. 12"
MIN. 24"
STORM COLLAR, ST. STL. DRAWBAND CAULKING, ETC.
ADJUSTABLE FLASHING CONE BY ROOFING CONTR.
22 GAUGE SHT. MTL. SLEEVE OF DIAMETER AND LENGTH TO PROVIDE REQUIRED CLEARANCE FROM COMBUSTIBLES. ATTACH SLEEVE TO ROOF DECK.
CONCENTRIC COMB. AIR INTAKE/VENT ASSEMBLY BY FURNACE MFR.
COMB. AIR INTAKE PIPE IN ATTIC.
COMB. AIR VENT PIPE W/ 3/4" ARMAFLEX INS. IN ATTIC.
2"
DETAIL OF COMB. AIR INTAKE/VENT PIPES THRU ROOF
NO SCALE
NOTE: COORDINATE ROOFING TYPE AND FLASHING REQUIREMENTS WITH ROOFING CONTR.
STORAGE
PARTIAL FLOOR PLAN: UTILITY
KEYED NOTES
1 RETURN DUCT RISER DOWN TO FLOOR TO FURNACE INLET.
2 EVAPORATIVE HUMIDIFIER MOUNTED ON SIDE OF RETURN DUCT RISER AT 6 FT.± ABOVE FLOOR.
3 CONNECT 6"ϕ DUCT TO FURNACE SUPPLY PLENUM AND EXTEND TO INLET OF HUMIDIFIER.
4 COUNTERBALANCED BACKDRAFT DAMPER IN 6"ϕ HUMIDIFIER SUPPLY DUCT.
5 COMBINE 6"ϕ HUMIDIFIER SUPPLY DUCTS FROM F-1 AND F-2 AND EXTEND DOWN TO INLET OF H-1.
6 COUNTERBALANCED BACKDRAFT DAMPER IN RETURN DUCT AT INLET TO F-1/2.
7 FLEXIBLE DUCT CONNECTION.
8 FILTER RACK WITH 1" THICK TA FILTER.
9 CONNECT FULL SIZE SUPPLY DUCT TO OUTLET OF COOLING COIL CABINET AND EXTEND UP TO UNDERSIDE OF CEILING.
10 CONNECT REFRIGERANT PIPING TO COOLING COIL, EXTEND UP TO NEAR CEILING AND ROUTE AS INDICATED. REFRIG. PIPE SIZES, SYSTEMS #1/2/4: 1 1/8" O.D. SUCT. & 1/2" O.D. LIQ. AND SYSTEM #3: 7/8" O.D. SUCT. & 1/2" O.D. LIQ.
11 CONNECT 3/4" DRAIN PIPE TO COOLING COIL CABINET DRAIN PAN, PROVIDE WITH 3" DEEP TRAP AND CLEANOUT AND EXTEND TO DRAIN AS INDICATED.
12 PROVIDE OUTSIDE AIR INTAKE DUCT WITH VOLUME DAMPER AND 2 POSITION AUTOMATIC DAMPER AND CONNECT TO RETURN AIR RISER NEAR CEILING.
13 3/4" GAS PIPE DOWN TO FURNACE WITH SHUT-OFF VALVE AND 4" LONG FULL SIZE DIRT LEG. MAKE FINAL CONNECTION TO FURNACE GAS VALVE WITH UNION.
14 2" COMBUSTION AIR INTAKE AND VENT PIPES CONNECTED TO FURNACE, EXTENDED UP INTO ATTIC AND THROUGH ROOF WITH CONCENTRIC VENT KIT. INSULATE COMBUSTION VENT PIPE IN ATTIC AND COMBUSTION AIR INTAKE PIPE IN UTILITY ROOM WITH 3/8" THICK INSULATION.
15 3/4" COOLING COIL DRAIN PIPE DOWN AT WALL AND CONNECT TO 1" DRAIN PIPE RUNNING ALONG WALL.
16 CONNECT 1/2" FURNACE HEAT EXCHANGER DRAIN PIPE AND EXTEND TO FLOOR DRAIN.
17 TERMINATE FURNACE HEAT EXCHANGER DRAIN PIPES AT FLOOR DRAIN. FLOOR DRAIN BY PLBG. CONTR.
18 COOLING COIL CONDENSATE DRAIN PIPE RACKED ON WALL. PITCH DOWN TO OUTSIDE AT 1/4"/FT.
19 1/2" FURNACE HEAT EXCHANGER DRAIN PIPE RACKED ON WALL. PITCH DOWN TO FLOOR DRAIN AT 1/4"/FT.
20 REFRIGERANT PIPING NEAR CEILING. SYSTEMS #1/2/4: 1 1/8" O.D. SUCT. & 1/2" O.D. LIQ. AND SYSTEM #3: 7/8" O.D. SUCT. & 1/2" O.D. LIQ.
21 1 1/4" GAS PIPING NEAR CEILING.
22 12/12 OUTSIDE AIR DUCTWORK AT UNDERSIDE OF ROOF JOISTS AS HIGH AS POSSIBLE.
23 1 1/4" GAS PIPE THROUGH WALL TO ABOVE CORRIDOR CEILING. SEE CONTINUATION ON SHEET H2.
24 REFRIGERANT PIPING THROUGH WALL TO ABOVE CEILING. SYSTEM #1/2/4: 1 1/8" O.D. SUCT. & 1/2" O.D. LIQ. AND SYSTEM #3: 7/8" O.D. SUCT. & 1/2" O.D. LIQ.
25 1" COOLING COIL DRAIN PIPE RACKED ON WALL IN STOR. RM. AND DROP TO 6" ABOVE FLOOR.
26 EXTEND 1" DRAIN PIPE THROUGH EXTERIOR WALL WITH LINK SEAL AND TERMINATE AT 1" FROM EXTERIOR FACE OF WALL.
27 CONNECT OUTSIDE AIR PLENUM TO BACK OF LOUVER AND TRANSITION/SLOPE UP TO 12/12 OUTSIDE AIR DUCT. SEAL BOTTOM OF OUTSIDE AIR PLENUM WATERTIGHT.
28 16/16 OUTSIDE AIR INTAKE LOUVER WITH BIRD SCREEN AND KYNAR PAINTED FINISH OF SELECTED STD. COLOR.
29 SR-1, 10/6, 150.
30 1" COOLING COIL DRAIN PIPE RACKED ON WALL OF STORAGE ROOM. INSULATE PIPE WITH 3/8" ARMAFLEX INSULATION.
BRUCE VILLAGE HALL & LIBRARY
BRUCE , WISCONSIN
SCALE: AS NOTED
MORGAN & PARMLEY, LTD.
CONSULTING ENGINEERS
LADYSMITH, WISCONSIN
SHEET M-3 OF 3
DATE: DEC. 31, 2001
DR. BY: FLM, INC.

LIGHTING

FLUORESCENT LIGHTING FIXTURE
UNDERCABINET LIGHTING FIXTURE
WALL MOUNTED FLUORESCENT FIXTURE
EMERGENCY BATTERY UNIT
LIGHT TRACK ASSEMBLY
POLE MOUNTED LIGHT FIXTURE
POST TOP MOUNTED LIGHT FIXTURE OR BOLLARD
CEILING RECESSED LIGHTING FIXTURE
WALL RECESSED LIGHTING FIXTURE
CEILING MOUNTED LIGHTING FIXTURE
WALL MOUNTED LIGHTING FIXTURE
EXIT LIGHTING FIXTURE – WALL OR CEILING MOUNTED WITH DIRECTION ARROWS AS REQUIRED
LOCAL LINE VOLTAGE SWITCH
SWITCH DESIGNATION
SWITCH TYPE
- C – MOUNTED 6" ABOVE COUNTER
- DLS – DUAL LEVEL SWITCHING (INNER/OUTER LAMPS)
- PL – PILOT LIGHT
- E – EXPLOSION PROOF
- SSX – SWITCH STATION
- K – KEY OPERATED
- WP – WEATHER PROOF
- 2 – DOUBLE POLE
- 3 – 3 WAY
- 4 – 4 WAY

OS xx OCCUPANCY SENSOR
DENOTES THE FOLLOWING:
- W – WALL MOUNTED
- CD – CEILING DUAL TECHNOLOGY (PIR/ULTRASONIC)
- CU – CEILING ULTRASONIC

MOMENTARY CONTACT SWITCH
-SUBSCRIPT DESIGNATION INDICATES CONTACTOR CONTROLLED
DIMMING SWITCH – MOUNTED 48" AFF
LOCAL SWITCH, SINGLE POLE, WITH DUPLEX RECEPTACLE -MOUNTED 48" AFF
LOCAL SWITCH, SINGLE POLE, WITH DUPLEX RECEPTACLE -MOUNTED 6" ABOVE COUNTER
LOW VOLTAGE SWITCH STATION
X – DENOTE ZONE(S) OR STATION NUMBER
LIGHTING CONTROL PANEL

GENERAL

LIGHT LINE ———— EXISTING
DARK LINE ———— NEW

POWER & DIAGRAMS

DUPLEX RECEPTACLE – MOUNTED 18" AFF
DENOTES THE FOLLOWING:
- C – MOUNTED 6" ABOVE COUNTER
- GFI – GROUND FAULT CIRCUIT INTERRUPTER TYPE
- WP – WEATHER PROOF WITH GFI
- TP – TAMPER PROOF "CHILDSAFE" TYPE
- EQ – MOUNTED ON EQUIPMENT FURNISHED BY OTHERS
- H – MOUNTED HORIZONTALLY
- S – UPPER HALF SWITCH
- TS – MOUNTED HORIZONTALLY IN TOE SPACE
- IG – ISOLATED GROUND
- DD – FOURPLEX RECEPTACLE – TWO DUPLEX RECEPTACLES UNDER A COMMON COVERPLATE
- TVSS TRANSIENT VOLTAGE SURGE SUPPRESSOR WITH ISOLATED GROUND

FLUSH FLOOR BOX WITH DUPLEX RECEPTACLE(S) AND COMMUNICATIONS OUTLET(S)
FLUSH FLOOR OUTLET
DENOTES THE FOLLOWING:
- R – DUPLEX RECEPTACLE(S)
- FP – FURNITURE PARTITION POWER FEED
- FC – FURNITURE PARTITION COMMUNICATION FEED

SPECIAL PURPOSE OUTLET – SEE SCHEDULE
JUNCTION BOX
DENOTES THE FOLLOWING:
- FP – FURNITURE PARTITION POWER FEED
- FC – FURNITURE PARTITION COMMUNICATION FEED

EMERGENCY GENERATOR REMOTE ANNUNCIATOR
PUSH BUTTON
MOTOR – SEE SCHEDULE
DISCONNECT SWITCH
DISCONNECT SWITCH – FUSIBLE
PHOTOCELL
TIME CLOCK
CONTACTOR
METER
GENERATOR
ELECTRICAL DISTRIBUTION PANEL – EXISTING
ELECTRICAL DISTRIBUTION PANEL – NEW
TRANSFORMER
GROUND
TRANSOCKET/METER
CABLE TRAY
TYPE
- L – LADDER
- S – SOLID BOTTOM
- M – MESH

DEPTH
WIDTH
X–3 CIRCUITING
CIRCUIT NUMBER
PANELBOARD DESIGNATION

FIRE ALARM SYSTEM

FA xx FIRE ALARM DEVICE
DENOTES THE FOLLOWING:
- B – BELL
- CM – CONTROL MODULE
- DH – DOOR HOLDER
- DD – SMOKE DUCT DETECTOR
- FS – FLOW SWITCH
- HD – HEAT DETECTOR (RATE OF RISE/FIXED)
- HV1 – HORN W/ 15/75 CANDELA STROBE
- HV2 – HORN W/ 30 CANDELA STROBE
- HV3 – HORN W/ 75 CANDELA STROBE
- HV4 – HORN W/ 110 CANDELA STROBE
- MM – MONITOR MODULE
- PS – PULL STATION
- SO – SPEAKER ONLY
- SD – PHOTOELECTRIC SMOKE DETECTOR
- SV1 – SPEAKER W/ 15/75 CANDELA STROBE
- SV2 – SPEAKER W/ 30 CANDELA STROBE
- SV3 – SPEAKER W/ 75 CANDELA STROBE
- SV4 – SPEAKER W/ 110 CANDELA STROBE
- TS – TAMPER SWITCH
- VO – VISUAL UNIT ONLY
- VO1 – VISUAL UNIT ONLY, 15/75 CANDELA STROBE
- VO2 – VISUAL UNIT ONLY, 30 CANDELA STROBE
- VO3 – VISUAL UNIT ONLY, 75 CANDELA STROBE
- VO4 – VISUAL UNIT ONLY, 110 CANDELA STROBE
- WP – WEATHER PROOF

FIRE ALARM CONTROL PANEL
FIRE ALARM REMOTE ANNUNCIATOR
FIRE ALARM NOTIFICATION APPLIANCE CIRCUIT EXTENDER

SOUND SYSTEM

LOUDSPEAKER – FLUSH MOUNTED ON CEILING
AMP INTERCOM AMPLIFIER

NOTE:

ALL SYMBOLS SHOWN MAY NOT APPEAR ON DRAWINGS

ABBREVIATIONS

A/E	ARCHITECT/ENGINEER
ABV	ABOVE
AFF	ABOVE FINISHED FLOOR
ARCH	ARCHITECT
AFG	ABOVE FINAL GRADE
ALT SW	ALTERNATOR SWITCH
B	JUNCTION BOX
BKR	BREAKER
BFG	BELOW FINAL GRADE
BOL	BUILT-IN OVERLOAD PROTECTION
C	CONDUIT
CB	CIRCUIT BREAKER
CKT	CIRCUIT
CS	COMBINATION STARTER
DISC	DISCONNECT SWITCH
DN	DOWN
EWC	ELECTRIC WATER COOLER
E.C	ELECTRICAL CONTRACTOR
ELEV	ELEVATION
EMT	ELECTRICAL METALLIC TUBING
ENT	ELECTRICAL NON-METALLIC TUBING
F	FURNISHED
FIXT	FIXTURE
FLUOR	FLUORESCENT
FS	FREEZE-STAT
G.C.	GENERAL CONTRACTOR
GRC	GALVANIZED RIGID CONDUIT
GRDG	GROUNDING
HOA	HAND-OFF-AUTO SELECTOR SWITCH
HP	HORSEPOWER
HVAC	HEATING, VENTILATING, AND AIR CONDITIONING CONTRACTOR
HW	HEAVYWALL
I	INSTALLED
IL	INTERLOCK
IMC	INTERMEDIATE METAL CONDUIT
IU	IN UNIT
JC	JUNCTION BOX
KW	KILOWATT
LOC	LOCATION
LTG	LIGHTING
LVT	LINE VOLTAGE THERMOSTAT
MAG	MAGNETIC STARTER
MAN	MANUAL STARTER
MLO	MAIN LUGS ONLY
MNT	MOUNT
MSB	MAIN SWITCHBOARD
MTD	MOUNTED
NU	NEAR UNIT
NIC	NOT IN CONTRACT
OC	ON CENTER
OU	ON UNIT
P	POLE
PEND	PENDANT
PE SW	PNEUMATIC ELECTRIC SWITCH
PLBG	PLUMBING CONTRACTOR
PNL	PANEL
PB	PUSHBUTTON
R	RECEPTACLE
SURF	SURFACE
SW	SWITCH
SP SW	SPEED SWITCH
TC	TIME CLOCK
TCC	TEMPERATURE CONTROL CONTRACTOR
TO	TYPICAL OUTLET
TYP	TYPICAL
UO	UNIQUE OUTLET
VER	VERIFY
W	WIRED
WP	WEATHERPROOF
XFMR	TRANSFORMER

BRUCE VILLAGE HALL & LIBRARY	
BRUCE, WISCONSIN	
SCALE: NONE	MORGAN & PARMLEY, LTD. CONSULTING ENGINEERS
SHEET E-1 OF 5	LADYSMITH, WISCONSIN
DATE: DEC. 31, 2001	DR. BY TRP

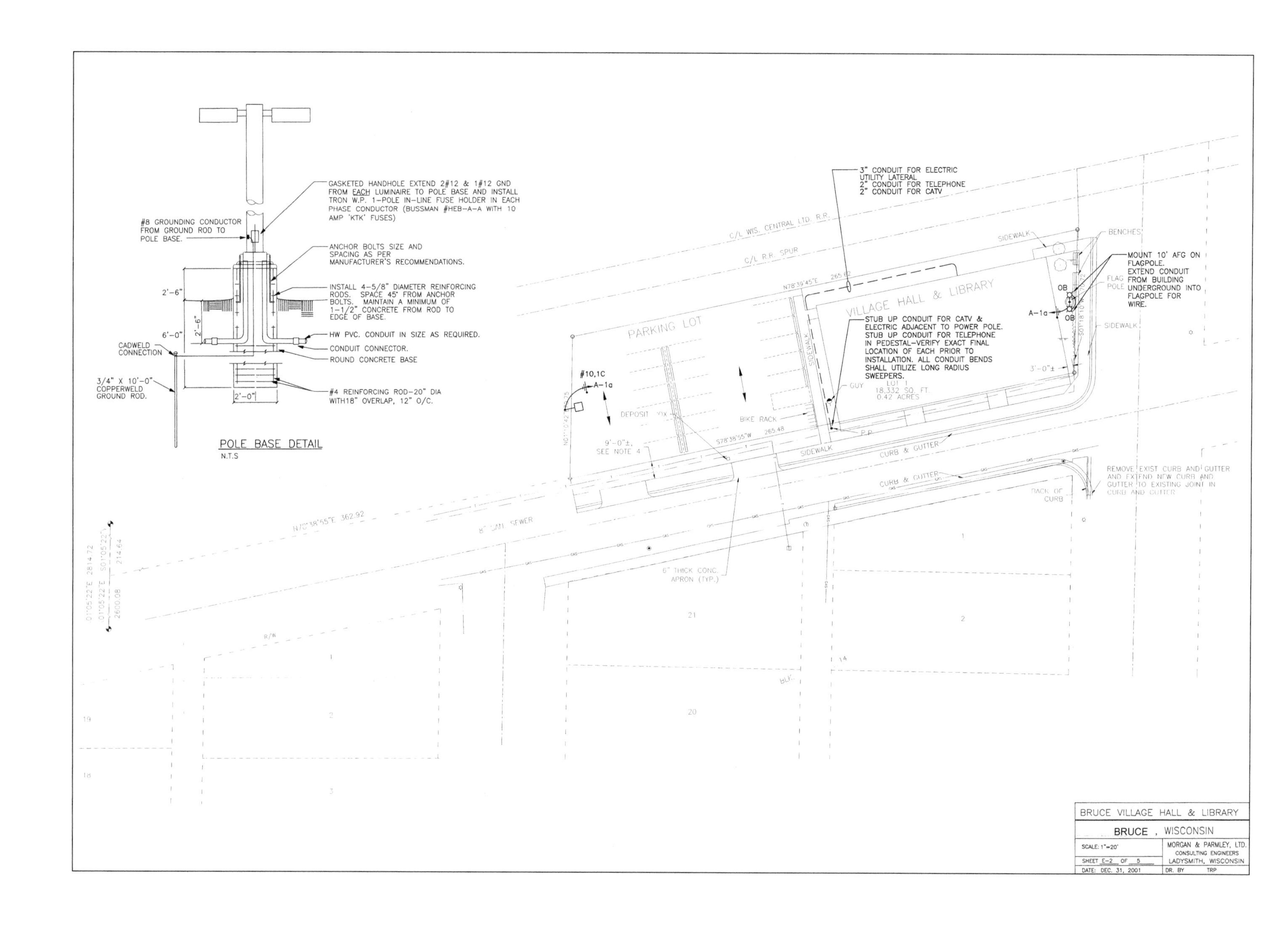
GASKETED HANDHOLE EXTEND 2#12 & 1#12 GND FROM EACH LUMINAIRE TO POLE BASE AND INSTALL TRON W.P. 1-POLE IN-LINE FUSE HOLDER IN EACH PHASE CONDUCTOR (BUSSMAN #HEB-A-A WITH 10 AMP 'KTK' FUSES)
#8 GROUNDING CONDUCTOR FROM GROUND ROD TO POLE BASE.
ANCHOR BOLTS SIZE AND SPACING AS PER MANUFACTURER'S RECOMMENDATIONS.
INSTALL 4-5/8" DIAMETER REINFORCING RODS. SPACE 45° FROM ANCHOR BOLTS. MAINTAIN A MINIMUM OF 1-1/2" CONCRETE FROM ROD TO EDGE OF BASE.
2'-6"
2'-6"
6'-0"
HW PVC. CONDUIT IN SIZE AS REQUIRED.
CADWELD CONNECTION
CONDUIT CONNECTOR.
ROUND CONCRETE BASE
3/4" X 10'-0" COPPERWELD GROUND ROD.
2'-0"
#4 REINFORCING ROD-20" DIA WITH18" OVERLAP, 12" O/C.
POLE BASE DETAIL
N.T.S
3" CONDUIT FOR ELECTRIC UTILITY LATERAL
2" CONDUIT FOR TELEPHONE
2" CONDUIT FOR CATV
C/L WIS. CENTRAL LTD. R.R.
C/L R.R. SPUR
SIDEWALK
BENCHES
MOUNT 10' AFG ON FLAGPOLE.
EXTEND CONDUIT FROM BUILDING UNDERGROUND INTO FLAGPOLE FOR WIRE.
FLAG POLE
N78°39'45"E 265.62
VILLAGE HALL & LIBRARY
OB
A-1a
OB
PARKING LOT
STUB UP CONDUIT FOR CATV & ELECTRIC ADJACENT TO POWER POLE. STUB UP CONDUIT FOR TELEPHONE IN PEDESTAL-VERIFY EXACT FINAL LOCATION OF EACH PRIOR TO INSTALLATION. ALL CONDUIT BENDS SHALL UTILIZE LONG RADIUS SWEEPERS.
SIDEWALK
3'-0"±
#10,1C
A-1a
GUY
LOT 1
18,332 SQ. FT.
0.42 ACRES
BIKE RACK
P.P.
9'-0"±, SEE NOTE 4
S78°38'55"W 265.48
SIDEWALK
CURB & GUTTER
CURB & GUTTER
REMOVE EXIST CURB AND GUTTER AND EXTEND NEW CURB AND GUTTER TO EXISTING JOINT IN CURB AND GUTTER
BACK OF CURB
N78°38'55"E 362.92
8" SAN. SEWER
6" THICK CONC. APRON (TYP.)
GAS
R/W
1
2
3
19
18
21
20
14
BLK
BRUCE VILLAGE HALL & LIBRARY
BRUCE, WISCONSIN
SCALE: 1"=20'
SHEET E-2 OF 5
DATE: DEC. 31, 2001
MORGAN & PARMLEY, LTD.
CONSULTING ENGINEERS
LADYSMITH, WISCONSIN
DR. BY TRP

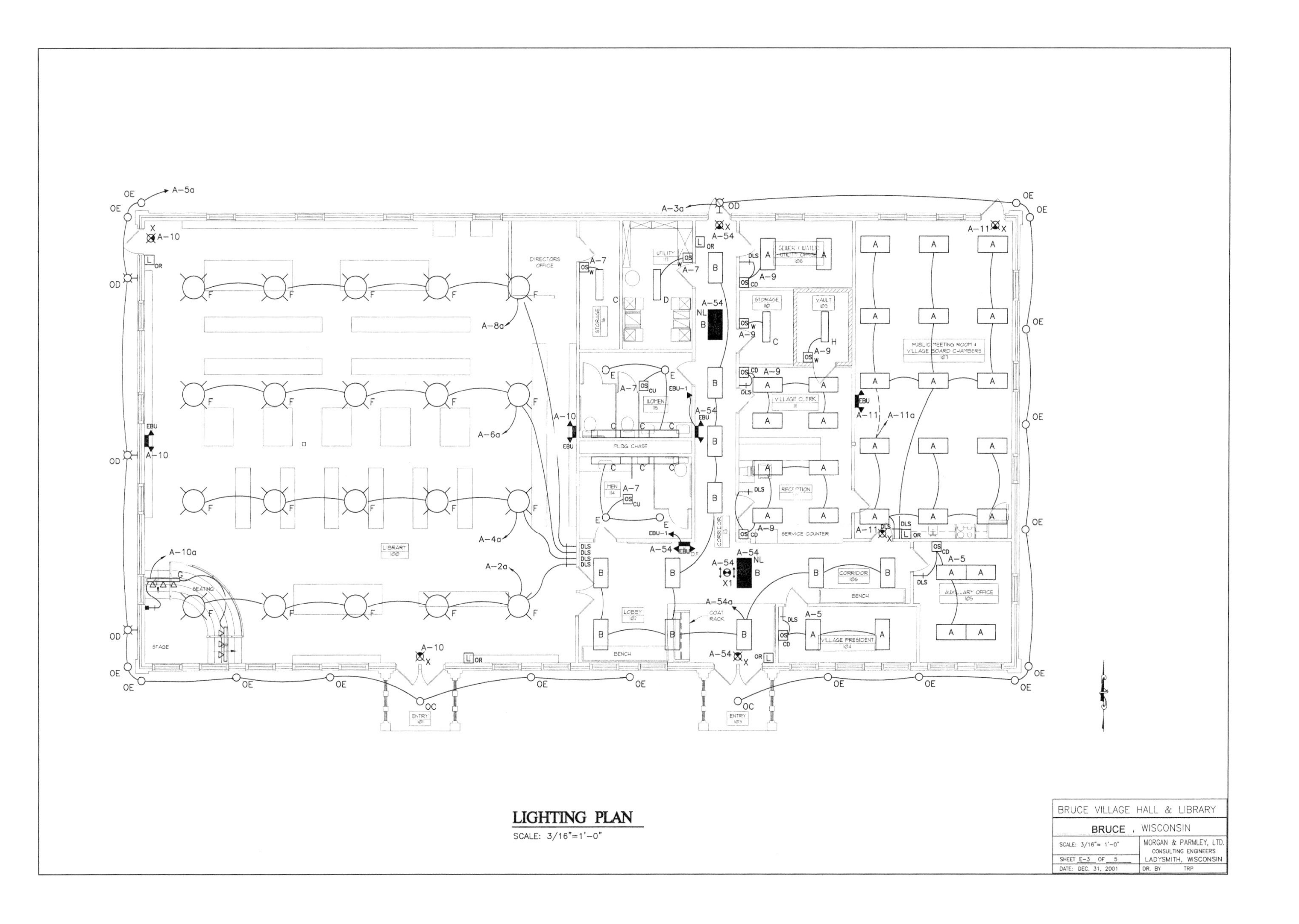
LIGHTING PLAN
SCALE: 3/16"=1'-0"
BRUCE VILLAGE HALL & LIBRARY
BRUCE , WISCONSIN
SCALE: 3/16"= 1'-0"
MORGAN & PARMLEY, LTD.
CONSULTING ENGINEERS
LADYSMITH, WISCONSIN
SHEET E-3 OF 5
DATE: DEC. 31, 2001
DR. BY TRP
LIBRARY 100
PUBLIC MEETING ROOM & VILLAGE BOARD CHAMBERS
VILLAGE CLERK
RECEPTION
SERVICE COUNTER
CORRIDOR 106
AUXILLARY OFFICE 105
VILLAGE PRESIDENT 104
LOBBY 102
COAT RACK
BENCH
ENTRY 101
ENTRY 103
STORAGE
VAULT
UTILITY
DIRECTORS OFFICE
WOMEN
MEN
PLBG CHASE
STAGE
SEATING

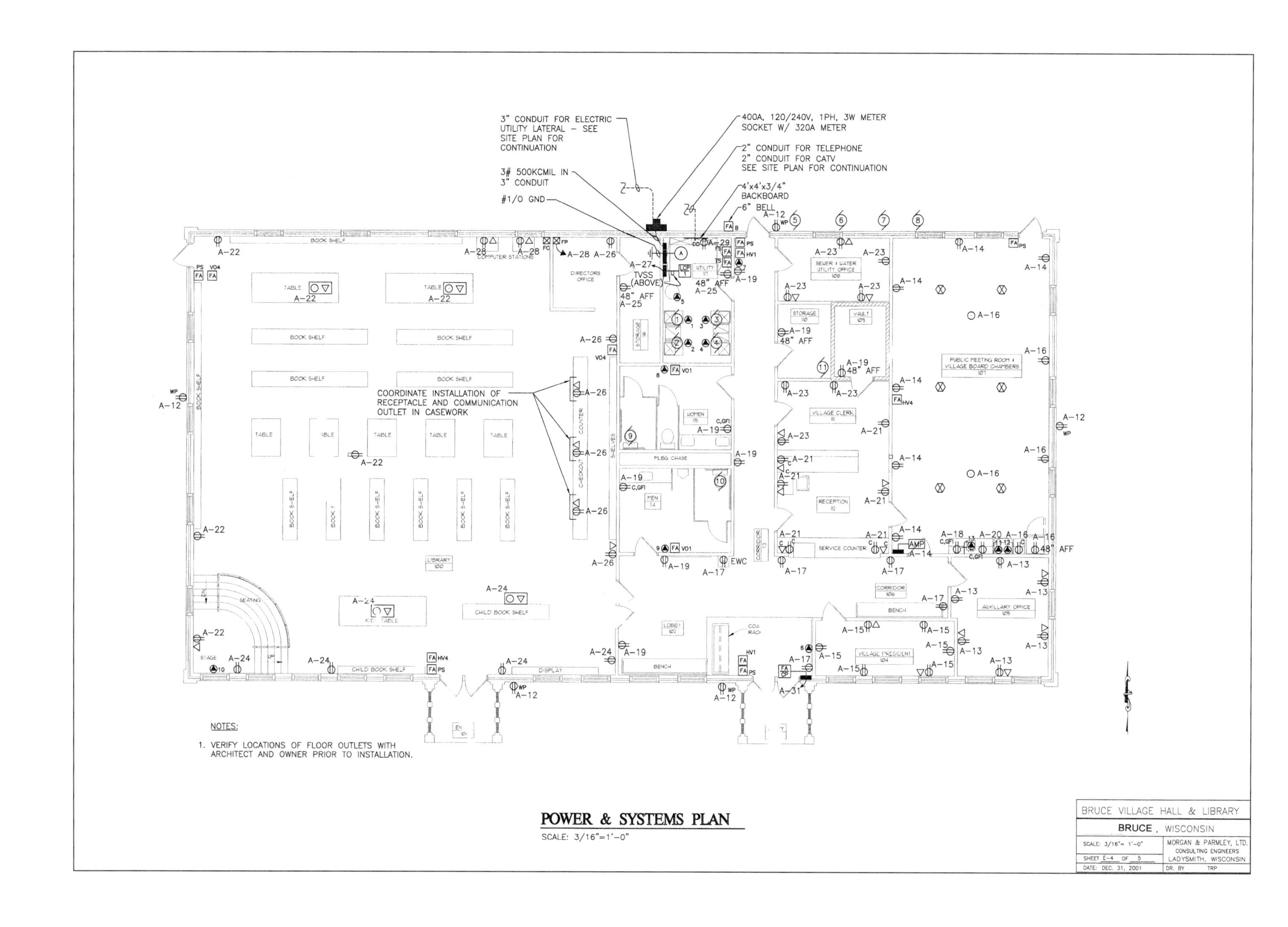
3" CONDUIT FOR ELECTRIC UTILITY LATERAL – SEE SITE PLAN FOR CONTINUATION
3# 500KCMIL IN 3" CONDUIT
#1/0 GND
400A, 120/240V, 1PH, 3W METER SOCKET W/ 320A METER
2" CONDUIT FOR TELEPHONE
2" CONDUIT FOR CATV
SEE SITE PLAN FOR CONTINUATION
4'x4'x3/4" BACKBOARD
6" BELL
TVSS (ABOVE)
48" AFF
COORDINATE INSTALLATION OF RECEPTACLE AND COMMUNICATION OUTLET IN CASEWORK
NOTES:
1. VERIFY LOCATIONS OF FLOOR OUTLETS WITH ARCHITECT AND OWNER PRIOR TO INSTALLATION.
POWER & SYSTEMS PLAN
SCALE: 3/16"=1'-0"
BRUCE VILLAGE HALL & LIBRARY
BRUCE, WISCONSIN
SCALE: 3/16"= 1'-0"
MORGAN & PARMLEY, LTD.
CONSULTING ENGINEERS
LADYSMITH, WISCONSIN
SHEET E-4 OF 5
DATE: DEC. 31, 2001
DR. BY TRP

SPECIAL OUTLET SCHEDULE ⊛

TERMINAL: R – RECEPTACLE D – DISCONNECT SWITCH CONNECTION B – JUNCTION BOX CONNECTION

NO.	TO FEED	LOC.	FEED FROM		BREAKER		WIRING				TERMINAL			VOLT	Ø	LOAD	SEE NOTE
			PANEL	CKT.	SIZE	POLE	NO.	SIZE	GND	COND.	R	D	B				
1	HUMIDIFIER	UTILITY	A	33	20	1	2	12	12	1/2		X		120	1	.6A	
2	HUMIDIFIER	UTILITY	A	35	20	1	2	12	12	1/2		X		120	1	.6A	
3	HUMIDIFIER	UTILITY	A	37	20	1	2	12	12	1/2		X		120	1	.6A	
4	HUMIDIFIER	UTILITY	A	39	20	1	2	12	12	1/2		X		120	1	.6A	
5	WATER HEATER	UTILITY	A	57,59	25	2	2	10	10	1/2		X		240	1	4500W MAX	
6	EW-2	CORRIDOR	A	44	20	1	2	12	12	1/2		X		120	1	12.6A	1,2
7	EW-2	CORRIDOR	A	46	20	1	2	12	12	1/2		X		120	1	12.6A	1,2
8	EW-1	WOMEN	A	48	20	1	2	12	12	1/2		X		120	1	8.4A	1,2
9	EW-1	MEN	A	50	20	1	2	12	12	1/2		X		120	1	8.4A	1,2
10	EB-1	LIBRARY	A	47	20	1	2	12	12	1/2			X	120	1	8.3A	2
11	STOVE	MTG	A	53,55	50	2	3	8	10	3/4	X			120/240	1	8KW	
12	HOOD	MTG	A	42	20	1	2	12	12	1/2			X	120	1	5A	
13	DISPOSER	MTG	A	52	20	1	2	12	12	1/2		X		120	1	6A	

NOTES:
1. INSTALL AND WIRE MFR. SUPPLIED DISCONNECT SWITCH.
2. INSTALL AND WIRE HEATER.

LIGHTING FIXTURE SCHEDULE

ABBREVIATIONS

ES – EXPOSED STRUCTURE
G – GYP BOARD
LG – LAYIN GRID
S – SURFACE
V – VARIES
W – WALL
F – FLUSH
U – UNIVERSAL
C – CONCRETE
R – RECESSED

DES.	DESCRIPTION	LAMP DATA		VOLT	DEPTH	LIGHTING FIXTURE		MTG.	MTG. SURF	SEE NOTE
		NO.	TYPE			MFR.	CAT. NO.			
OA	PARKING LOT LIGHTING UNIT	2	250W MH	120	–	LITHONIA	(2) KAD250MR4120SPD04 (1) SSA255G	POLE	CONC BASE	
OB	FLAGPOLE LIGHT	1	100W MH/PAR 38	120	–	DALTOR	FP/818–100MH–RAD	POLE	METAL	2
OC	CANOPY LIGHT	1	70W MH	120	7 3/8"	LITHONIA	LP6H70M120/6L4	R	V	
OD	WALL LIGHT	1	70W MH	120	–	LITHONIA	TWA70M120	S	CONC	3
OE	SOFFIT DOWNLIGHT	1	50W MH	120	7 3/8"	LITHONIA	LP6H50M120/6L4	R	V	4
A	2x4 PARABOLIC	3	F32T8/SP35	120	5 15/16"	LITHONIA	2PM3GB33218LD120GEB10	R	LG	
B	2x4 PARABOLIC	2	F32T8/SP35	120	5 15/16"	LITHONIA	2PM3GB23212LD120GEB10	R	LG	
C	1x4 LENSED	2	F32T8/SP35	120	4 7/8"	LITHONIA	SPG232A12125120GEB10	R	LG	
D	INDUSTRIAL	2	F32T8/SP35	120	–	LITHONIA	AF10232120GEB10	S	ES	
E	DOWNLIGHT	2	PL18/SP35	120	5 1/2"	LITHONIA	AF218DTT6WR120	R	LG	
F	PENDENT	6	BX50/SP35	120	–	VISA	CP36696F50120BA	P	GYP	1
G	TRACK	3	75W PAR30/FL	120	–	LITHONIA	(3) LTCBLSPPAR30WH (1) LT4WH/LTAIWH/LTA11	S	GYP	
H	SURFACE	2	F32T8/SP35	120	–	LITHONIA	LB232120 GEB10	S	C	
X	EXIT SIGN – 1 FACE	–	LED	120	–	LITHONIA	LES1G120 ELN SD	UNIV	V	
X1	EXIT SIGN – 2 FACES	–	LED	120	–	LITHONIA	LES2G120 ELN SD	UNIV	V	
EBU	BATTERY UNIT	2	5W MR16(INCL)	120	–	DUAL–LITE	LZ15PI	S	V	
EBU–1	REMOTE HEAD	1	5W MR16(INCL)	6	–	DUAL–LITE	LZR6V5W	S	V	

NOTES:
1. MOUNTING HEIGHT IS 9' AFF – PROVIDE STEM LENGTH AS REQUIRED.
2. FIXTURE DESIGNED TO BE INSTALLED FROM OUTSIDE OF FLAGPOLE. DRILL 1/2" HOLE AND PROVIDE GROMMET FOR WIRING IN POLE. FIXTURE TO HAVE FINISH TO MATCH POLE.
3. MOUNT 1'–0" BELOW SOFFIT.
4. LOCATE CENTERED IN SOFFIT. ADD AND/OR ADJUST 2x4's IN SOFFIT TO ACCOMMODATE FIXTURES AS NEEDED.

LIGHTING RELAY CABINET SCHEDULE

DESIGNATION: LCP			LOCATION: UTILITY				
RELAY NO(S)	CIRCUIT NO(S)	LOAD DESCRIPTION	PROG. HOURS OF OPERATION		LOW VOLTAGE SWITCH LOCATION		NOTES
			OPEN	DUSK/DAWN	ROOM	QUANTITY	
1	A–1a	EXTERIOR LIGHTING (SITE)		x			
2,3	A–3a,5a	EXTERIOR LIGHTING (BLDG)		x			
4	A–54a	CORRIDOR	x		CORRIDOR	2	
5,6,7,8,9	A–2a,4a,6a,8a,10a	LIBRARY	x		LIBRARY	2	
10	A–11a	MEETING ROOM	x		MEETING ROOM	1	
11–16	–	SPARE					

MOTOR WIRING SCHEDULE ◊

ABBREVIATIONS

IU – IN UNIT
OU – ON UNIT
NU – NEAR UNIT
WP – WEATHER PROOF
MCC – MOTOR CONTROL CENTER
MFR – MANUFACTURER
EC – ELECTRICAL CONTRACTOR
MC – MECHANICAL CONTRACTOR
MAN – MANUAL STARTER
PC – PLUMBING CONTRACTOR
ES – EQUIPMENT SUPPLIER
TS – TOGGLE SWITCH
CS – COMBINATION STARTER
MAG – MAGNETIC STARTER
OS – OCCUPANCY SENSOR
NF – NON FUSED
F – FUSED

NO	HP	VOLTS	Ø	LOC	DRIVING	FED FROM		STARTER					DISCONNECT					BKR		BR. WIRING				SEE NOTE
						PNL.	NO.	TYPE	F	I	W	LOC	TYPE	F	I	W	LOC	SIZE	POLE	NO	SIZE	GND	COND	
1	3/4	120	1	UTILITY	FURNACE	A	33	MAG	MFR	MFR	EC	IU	NF	EC	EC	EC	NU	20	1	2	12	12	1/2	
2	3/4	120	1	UTILITY	FURNACE	A	35	MAG	MFR	MFR	EC	IU	NF	EC	EC	EC	NU	20	1	2	12	12	1/2	
3	1/2	120	1	UTILITY	FURNACE	A	37	MAG	MFR	MFR	EC	IU	NF	EC	EC	EC	NU	20	1	2	12	12	1/2	
4	3/4	120	1	UTILITY	FURNACE	A	39	MAG	MFR	MFR	EC	IU	NF	EC	EC	EC	NU	20	1	2	12	12	1/2	
5	30.5A	240	1	EXT	CONDENSER	A	30,32	MAG	MFR	MFR	EC	IU	F/WP	EC	EC	EC	NU	50	2	2	8	10	3/4	
6	30.5A	240	1	EXT	CONDENSER	A	34,36	MAG	MFR	MFR	EC	IU	F/WP	EC	EC	EC	NU	50	2	2	8	10	3/4	
7	23.4A	240	1	EXT	CONDENSER	A	38,40	MAG	MFR	MFR	EC	IU	F/WP	EC	EC	EC	NU	40	2	2	8	10	3/4	
8	30.5A	240	1	EXT	CONDENSER	A	43,45	MAG	MFR	MFR	EC	IU	F/WP	EC	EC	EC	NU	50	2	2	8	10	3/4	
9	2.7A	120	1	WOMEN	EF–1	A	36	MAN	EC	EC	EC	117	–	–	–	–	–	20	1	2	12	12	1/2	1,2
10	2.7A	120	1	MEN	EF–2	A	36	MAN	EC	EC	EC	117	–	–	–	–	–	20	1	2	12	12	1/2	1,2
11	1.8A	120	1	STORAGE	EF–3	A	36	MAN	EC	EC	EC	117	–	–	–	–	–	20	1	2	12	12	1/2	1,2

NOTES:
1. WIRE STARTER THROUGH RELAY SUPPLIED BY CONTROL CONTRACTOR.
2. WIRE THROUGH MFR. SUPPLIED SPEED SWITCH INSTALLED ON SIDE OF FAN HOUSING (SPEED SWITCH HAS OFF POSITION).

PANEL A

PANEL: A	BUS AMPS: 400 VOLTAGE: 120/240 PHASE: 1 WIRE: 3	MAIN: CB 400A MCB	MOUNTING: SURFACE X / FLUSH	REMARKS: SERVICE ENTRANCE RATED, 22K AIC; (2) 42CKT PANELS

CIRCUIT DESCRIPTION	AMPS A	AMPS B	CB	CCT	CCT	CB	AMPS A	AMPS B	CIRCUIT DESCRIPTION
EXT LTG	5		20	1	2	20	12.5		LIBRARY LTG
EXT LTG		7.6	20	3	4	20		12.5	LIBRARY LTG
HALL/VP/OFFICE EXT. LTG	8.6		20	5	6	20	12.5		LIBRARY LTG
BATHRM/UTIL LTG		11.2	20	7	8	20		12.5	LIBRARY LTG
OFFICE LTG	8.5		20	9	10	20	8		LIBRARY LTG
MTG RM LTG		11.3	20	11	12	20		7.5	EXT REC
REC	9		20	13	14	20	7.5		REC
REC		9	20	15	16	20		10	REC
REC	6		20	17	18	20	10		REC
REC		7.5	20	19	20	20		10	REC
REC	7.5		20	21	22	20	6		REC
REC		7.5	20	23	24	20		6	REC
REC	6		20	25	26	20	6		REC
LCP–1		5	20	27	28	20		6	REC
BACKBOARD REC	6		20	29	30	50	30.5		MTR#5
FACP		10	20	31	32	2		30.5	MTR#5
MTR#1,SO#1	10		20	33	34	50	30.5		MTR#6
MTR#2,SO#2		10	20	35	36	2		30.5	MTR#6
MTR#3,SO#3	10		20	37	38	40	23.4		MTR#7
MTR#4,SO#4		10	20	39	40	2		23.4	MTR#7
SPARE	–		20	41	42	20	5		SO#12
MTR#8		30.5	50	43	44	20		8.4	SO#6
MTR#8	30.5		2	45	46	20	8.4		SO#7
SO#10		7.9	20	47	48	20		12.5	SO#8
TVSS	–		60	49	50	20	12.5		SO#9
TVSS		–	2	51	52	20		6	SO#13
SO#11	30		50	53	54	20	11		HALL/VP/OFFICE EXT. LTG
SO#11		30	2	55	56	20		7.2	MTR#9,#10,#11
SO#5	19		25	57	58	20	–		SPARE
SO#5		19	2	59	60	20		–	SPARE
	–			61	62		–		
		–		63	64			–	
	–			65	66		–		
		–		67	68			–	
	–			69	70		–		
		–		71	72			–	
	–			73	74		–		
		–		75	76			–	
	–			77	78		–		
		–		79	80			–	
	–			81	82		–		
		–		83	84			–	

BRUCE VILLAGE HALL & LIBRARY
BRUCE, WISCONSIN

SCALE: NONE	MORGAN & PARMLEY, LTD. CONSULTING ENGINEERS LADYSMITH, WISCONSIN
SHEET E–5 OF 5	
DATE: DEC. 31, 2001	DR. BY TRP

SPOKEN WORD CHURCH

LADYSMITH, WISCONSIN

PROJECT NO. 84-110

MAY, 1984

INDEX OF SHEETS

SHEET NO.	DESCRIPTION
1	TITLE SHEET W/ GENERAL LOCATION MAP
2	SITE PLAN
3	FLOOR PLAN
4	ELEVATION VIEWS
5	BUILDING CROSS-SECTION
6	ROOF FRAMING PLAN W/ CONSTRUCTION DETAILS
7	ELECTRICAL, HEATING & VENTILATING
8	PLUMBING

PREPARED BY:

MORGAN & PARMLEY, LTD.
CONSULTING ENGINEERS
LADYSMITH, WISCONSIN

Courtesy: Morgan & Parmley, Ltd.

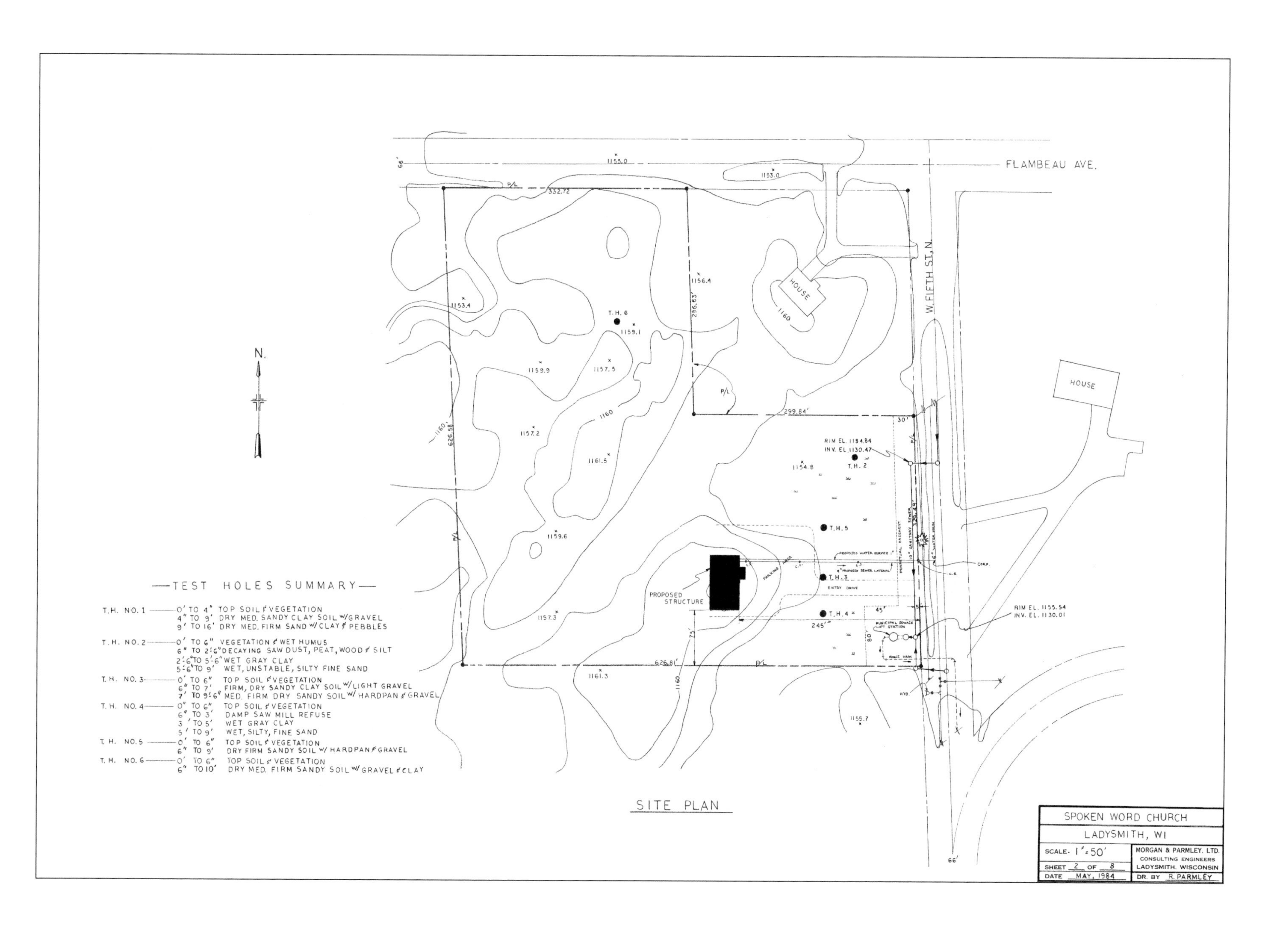
FLAMBEAU AVE.
W. FIFTH ST. N.
HOUSE
HOUSE
N.
T.H. 1
T.H. 2
T.H. 3
T.H. 4
T.H. 5
T.H. 6
PROPOSED STRUCTURE
ENTRY DRIVE
RIM EL. 1154.84
INV. EL. 1130.47
RIM EL. 1155.54
INV. EL. 1130.01
332.72
296.63'
299.84'
626.58'
626.81'
—TEST HOLES SUMMARY—
T.H. NO. 1 — 0' TO 4" TOP SOIL & VEGETATION
4" TO 9' DRY MED. SANDY CLAY SOIL W/GRAVEL
9' TO 16' DRY MED. FIRM SAND W/ CLAY & PEBBLES
T.H. NO. 2 — 0' TO 6" VEGETATION & WET HUMUS
6" TO 2'-6" DECAYING SAW DUST, PEAT, WOOD & SILT
2'-6" TO 5'-6" WET GRAY CLAY
5'-6" TO 9' WET, UNSTABLE, SILTY FINE SAND
T.H. NO. 3 — 0' TO 6" TOP SOIL & VEGETATION
6" TO 7' FIRM, DRY SANDY CLAY SOIL W/LIGHT GRAVEL
7' TO 9'-6" MED. FIRM DRY SANDY SOIL W/ HARDPAN & GRAVEL
T.H. NO. 4 — 0" TO 6" TOP SOIL & VEGETATION
6" TO 3' DAMP SAW MILL REFUSE
3' TO 5' WET GRAY CLAY
5' TO 9' WET, SILTY, FINE SAND
T.H. NO. 5 — 0' TO 6" TOP SOIL & VEGETATION
6" TO 9' DRY FIRM SANDY SOIL W/ HARDPAN & GRAVEL
T.H. NO. 6 — 0' TO 6" TOP SOIL & VEGETATION
6" TO 10' DRY MED. FIRM SANDY SOIL W/ GRAVEL & CLAY
SITE PLAN
SPOKEN WORD CHURCH
LADYSMITH, WI
SCALE: 1" = 50'
SHEET 2 OF 8
DATE MAY, 1984
MORGAN & PARMLEY, LTD.
CONSULTING ENGINEERS
LADYSMITH, WISCONSIN
DR. BY R. PARMLEY

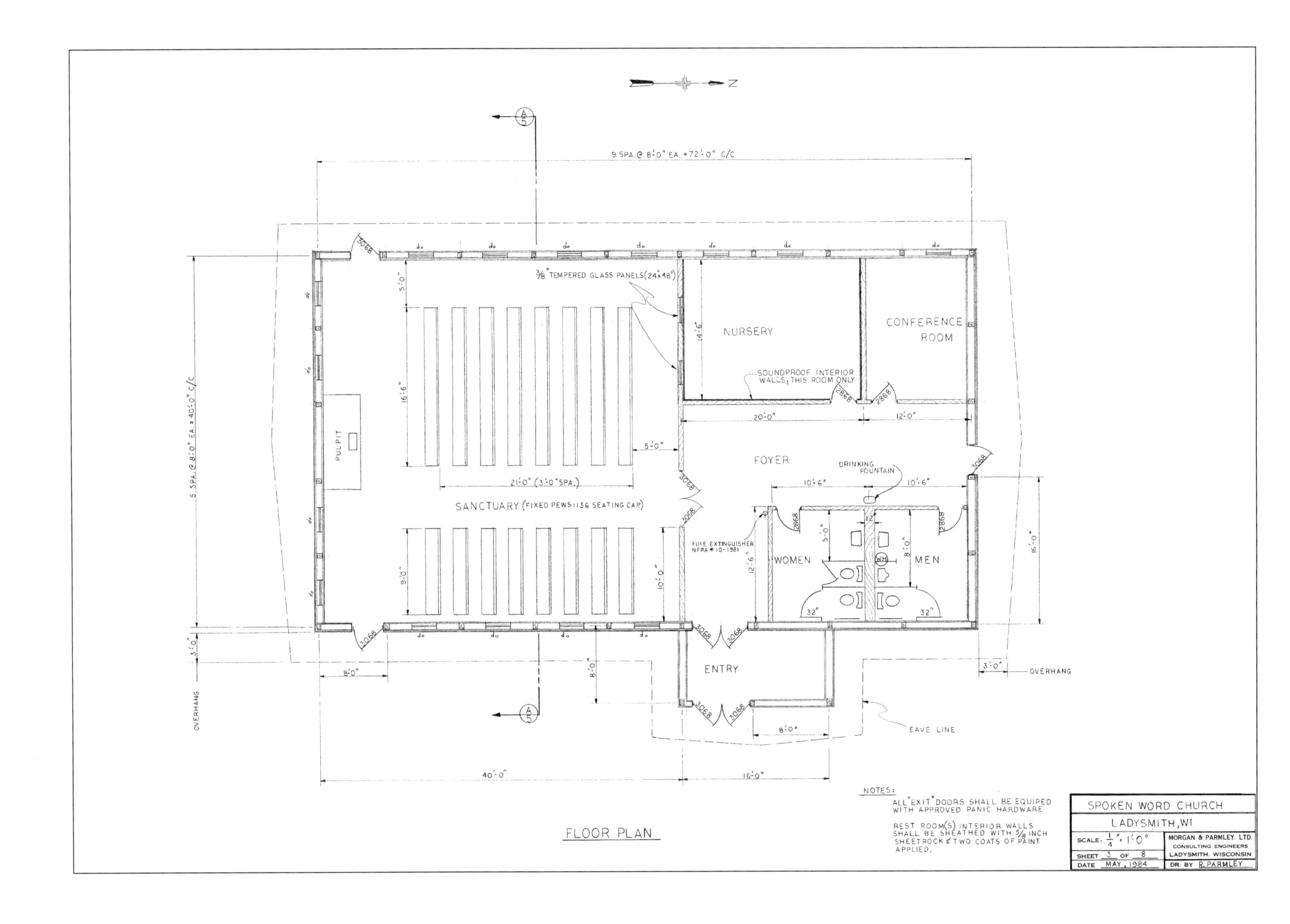
9 SPA. @ 8'-0" EA. = 72'-0" C/C
5 SPA. @ 8'-0" EA. = 40'-0" C/C
3/8" TEMPERED GLASS PANELS (24"x48")
NURSERY
CONFERENCE ROOM
SOUNDPROOF INTERIOR WALLS, THIS ROOM ONLY
PULPIT
SANCTUARY (FIXED PEWS: 136 SEATING CAP.)
21'-0" (3'-0" SPA.)
FOYER
DRINKING FOUNTAIN
FIRE EXTINGUISHER NFPA # 10-1981
WOMEN
MEN
ENTRY
OVERHANG
EAVE LINE
40'-0"
16'-0"
FLOOR PLAN
NOTES:
ALL "EXIT" DOORS SHALL BE EQUIPED WITH APPROVED PANIC HARDWARE
REST ROOM(S) INTERIOR WALLS SHALL BE SHEATHED WITH 5/8 INCH SHEETROCK & TWO COATS OF PAINT APPLIED.
SPOKEN WORD CHURCH
LADYSMITH, WI
SCALE - 1/4" = 1'-0"
SHEET 3 OF 8
DATE MAY, 1984
MORGAN & PARMLEY, LTD.
CONSULTING ENGINEERS
LADYSMITH, WISCONSIN
DR. BY R. PARMLEY

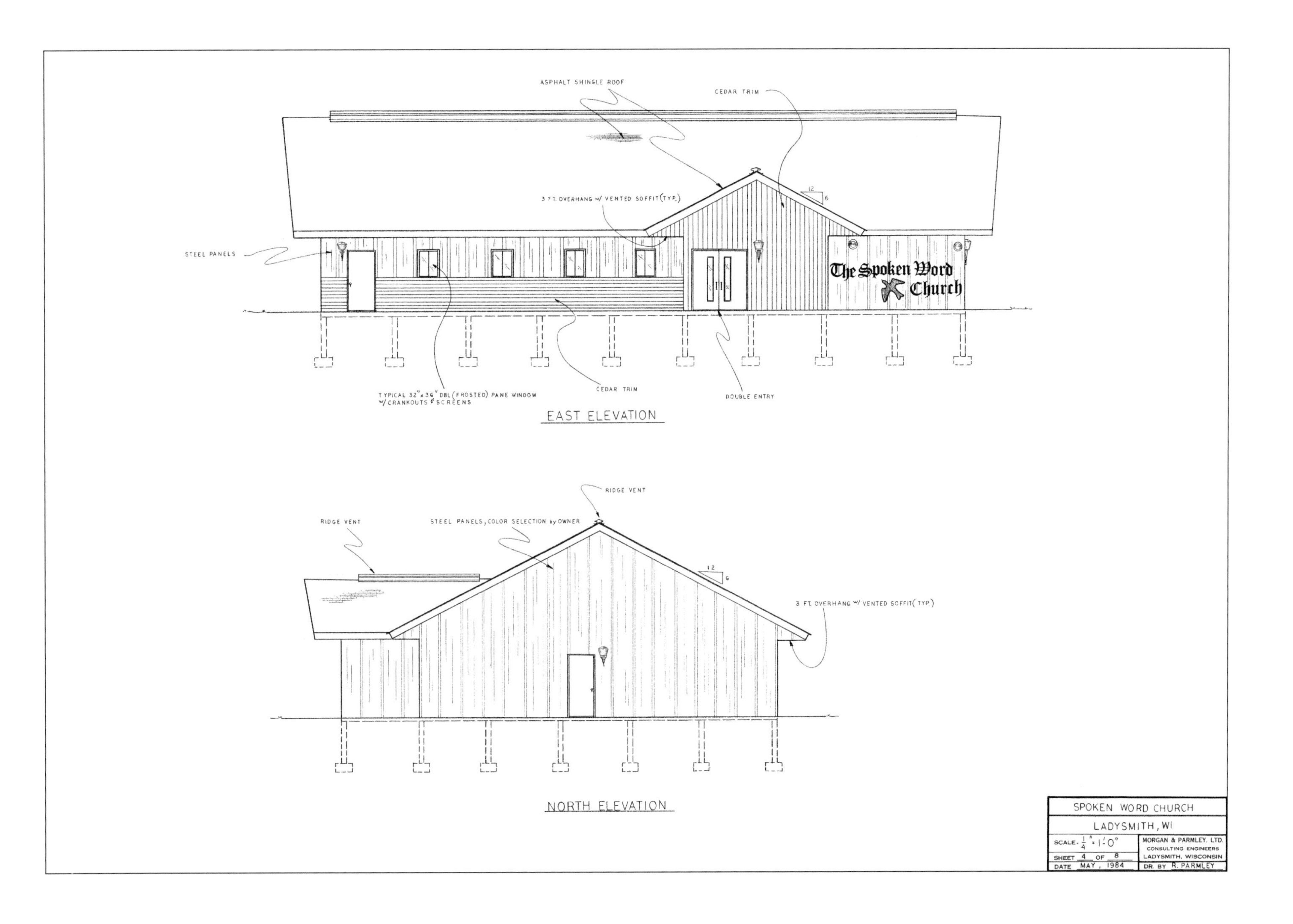
ASPHALT SHINGLE ROOF
CEDAR TRIM
3 FT. OVERHANG w/ VENTED SOFFIT (TYP.)
STEEL PANELS
The Spoken Word Church
TYPICAL 32"x36" DBL (FROSTED) PANE WINDOW w/ CRANKOUTS & SCREENS
CEDAR TRIM
DOUBLE ENTRY
EAST ELEVATION
RIDGE VENT
RIDGE VENT
STEEL PANELS, COLOR SELECTION by OWNER
3 FT. OVERHANG w/ VENTED SOFFIT (TYP.)
NORTH ELEVATION
SPOKEN WORD CHURCH
LADYSMITH, WI
SCALE - 1/4" = 1'-0"
SHEET 4 OF 8
DATE MAY, 1984
MORGAN & PARMLEY, LTD.
CONSULTING ENGINEERS
LADYSMITH, WISCONSIN
DR. BY R. PARMLEY

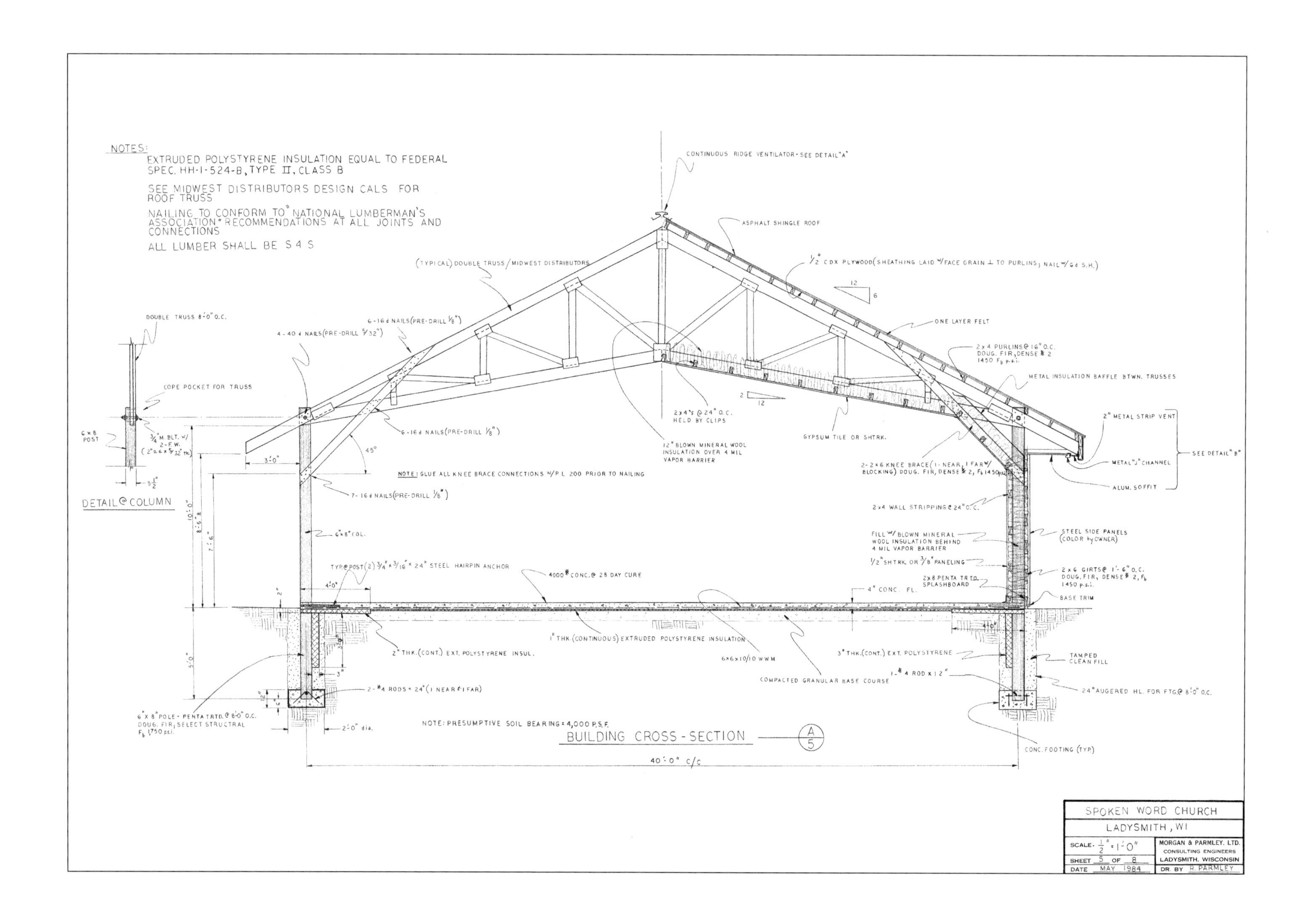

NOTES:
EXTRUDED POLYSTYRENE INSULATION EQUAL TO FEDERAL SPEC. HH-I-524-B, TYPE II, CLASS B
SEE MIDWEST DISTRIBUTORS DESIGN CALS FOR ROOF TRUSS
NAILING TO CONFORM TO "NATIONAL LUMBERMAN'S ASSOCIATION" RECOMMENDATIONS AT ALL JOINTS AND CONNECTIONS
ALL LUMBER SHALL BE S4S
CONTINUOUS RIDGE VENTILATOR - SEE DETAIL "A"
ASPHALT SHINGLE ROOF
(TYPICAL) DOUBLE TRUSS / MIDWEST DISTRIBUTORS
ONE LAYER FELT
METAL INSULATION BAFFLE BTWN. TRUSSES
GYPSUM TILE OR SHTRK.
2" METAL STRIP VENT
METAL "J" CHANNEL
ALUM. SOFFIT
SEE DETAIL "B"
DOUBLE TRUSS 8'-0" O.C.
COPE POCKET FOR TRUSS
6×8 POST
DETAIL @ COLUMN
NOTE: GLUE ALL KNEE BRACE CONNECTIONS W/ P.L. 200 PRIOR TO NAILING
2x4 WALL STRIPPING @ 24" O.C.
STEEL SIDE PANELS (COLOR by OWNER)
4000# CONC. @ 28 DAY CURE
4" CONC. FL.
BASE TRIM
6x6x10/10 WWM
COMPACTED GRANULAR BASE COURSE
TAMPED CLEAN FILL
NOTE: PRESUMPTIVE SOIL BEARING = 4,000 P.S.F.
BUILDING CROSS-SECTION
40'-0" c/c
CONC. FOOTING (TYP)
SPOKEN WORD CHURCH
LADYSMITH, WI
MORGAN & PARMLEY, LTD.
CONSULTING ENGINEERS
LADYSMITH, WISCONSIN
SHEET 5 OF 8
DATE MAY 1984
DR BY R. PARMLEY

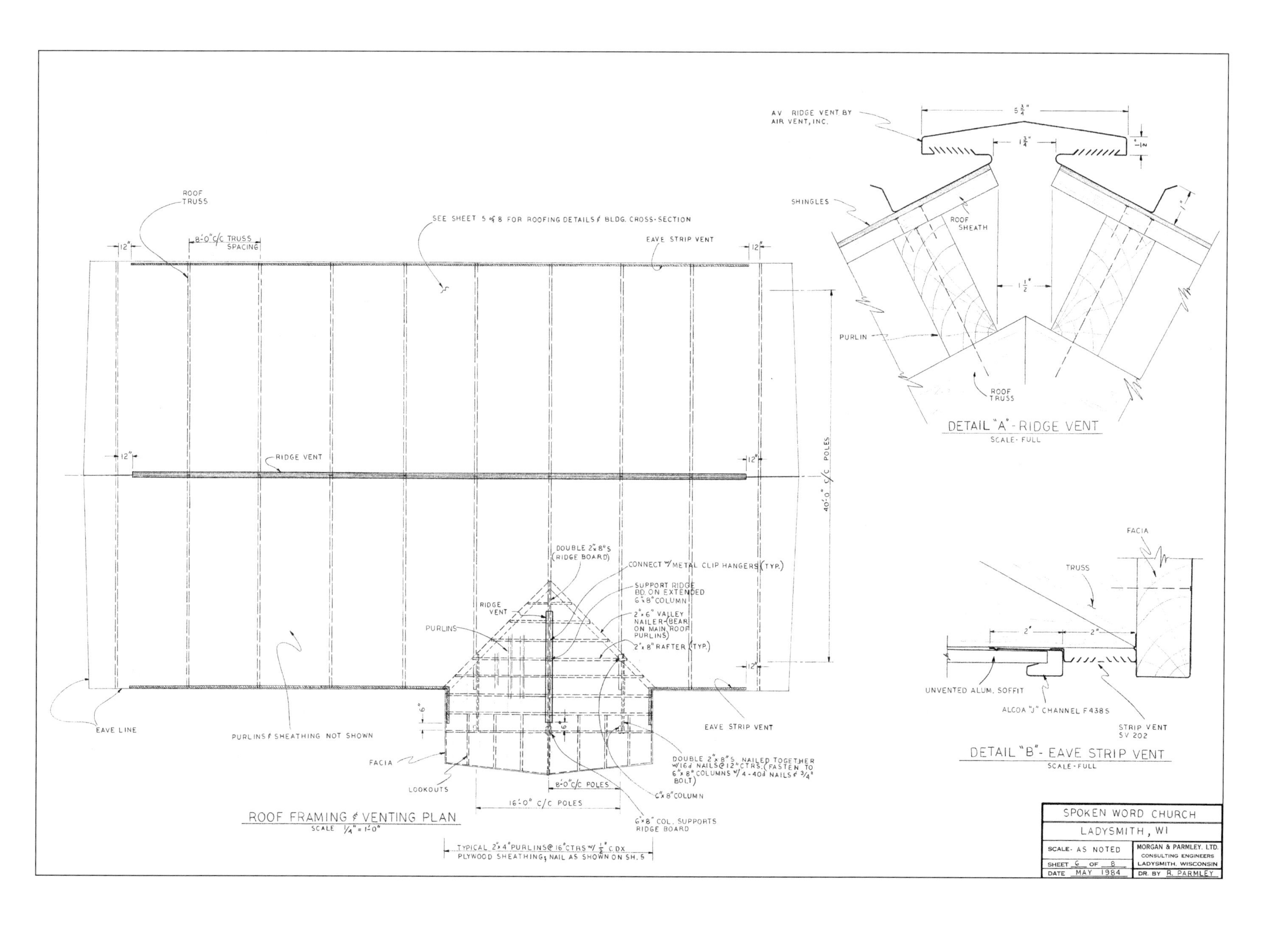

AV RIDGE VENT BY AIR VENT, INC.
SHINGLES
ROOF SHEATH
PURLIN
ROOF TRUSS
DETAIL "A" - RIDGE VENT
SCALE - FULL
ROOF TRUSS
SEE SHEET 5 of 8 FOR ROOFING DETAILS & BLDG. CROSS-SECTION
8'-0" C/C TRUSS SPACING
EAVE STRIP VENT
RIDGE VENT
40'-0" C/C POLES
DOUBLE 2"x 8"S (RIDGE BOARD)
CONNECT W/ METAL CLIP HANGERS (TYP.)
SUPPORT RIDGE BD. ON EXTENDED 6"x 8" COLUMN
2"x 6" VALLEY NAILER (BEAR ON MAIN ROOF PURLINS)
2"x 8" RAFTER (TYP.)
RIDGE VENT
PURLINS
EAVE LINE
PURLINS & SHEATHING NOT SHOWN
FACIA
LOOKOUTS
EAVE STRIP VENT
DOUBLE 2"x 8"S, NAILED TOGETHER W/ 16d NAILS @ 12" CTRS. (FASTEN TO 6"x 8" COLUMNS W/ 4-40d NAILS & 3/4" BOLT)
6"x 8" COLUMN
8'-0" C/C POLES
16'-0" C/C POLES
6"x 8" COL. SUPPORTS RIDGE BOARD
ROOF FRAMING & VENTING PLAN
SCALE 1/4" = 1'-0"
TYPICAL 2"x 4" PURLINS @ 16" CTRS W/ 1/2" C.DX PLYWOOD SHEATHING; NAIL AS SHOWN ON SH. 5
FACIA
TRUSS
UNVENTED ALUM. SOFFIT
ALCOA "J" CHANNEL F4385
STRIP VENT SV 202
DETAIL "B" - EAVE STRIP VENT
SCALE - FULL
SPOKEN WORD CHURCH
LADYSMITH, WI
SCALE - AS NOTED
SHEET 6 OF 8
DATE MAY 1984
MORGAN & PARMLEY, LTD.
CONSULTING ENGINEERS
LADYSMITH, WISCONSIN
DR. BY R. PARMLEY

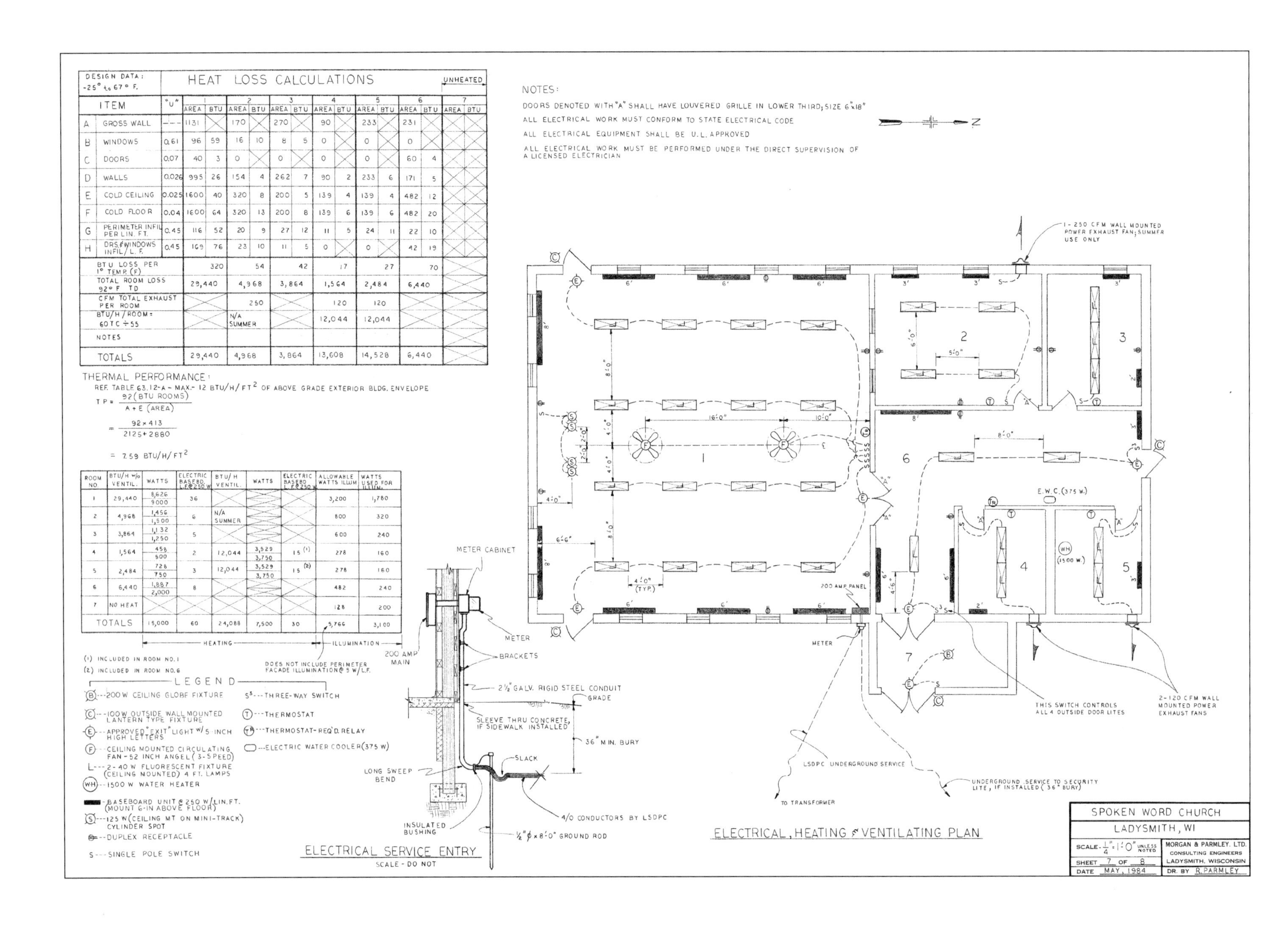

DESIGN DATA: -25° to 67° F.

HEAT LOSS CALCULATIONS

	ITEM	"U"	1 AREA	1 BTU	2 AREA	2 BTU	3 AREA	3 BTU	4 AREA	4 BTU	5 AREA	5 BTU	6 AREA	6 BTU	7 (UNHEATED) AREA	7 (UNHEATED) BTU
A	GROSS WALL	---	1131		170		270		90		233		231			
B	WINDOWS	0.61	96	59	16	10	8	5	0		0		0			
C	DOORS	0.07	40	3	0		0		0		0		60	4		
D	WALLS	0.026	995	26	154	4	262	7	90	2	233	6	171	5		
E	COLD CEILING	0.025	1600	40	320	8	200	5	139	4	139	4	482	12		
F	COLD FLOOR	0.04	1600	64	320	13	200	8	139	6	139	6	482	20		
G	PERIMETER INFIL PER LIN. FT.	0.45	116	52	20	9	27	12	11	5	24	11	22	10		
H	DRS.&WINDOWS INFIL/L.F.	0.45	169	76	23	10	11	5	0		0		42	19		
	BTU LOSS PER 1° TEMP. (F)			320		54		42		17		27		70		
	TOTAL ROOM LOSS 92° F TD		29,440		4,968		3,864		1,564		2,484		6,440			
	CFM TOTAL EXHAUST PER ROOM					250				120	120					
	BTU/H/ROOM = 60 TC ÷ 55				N/A SUMMER				12,044		12,044					
	NOTES															
	TOTALS		29,440		4,968		3,864		13,608		14,528		6,440			

THERMAL PERFORMANCE:

REF. TABLE 63.12-A – MAX.– 12 BTU/H/FT² OF ABOVE GRADE EXTERIOR BLDG. ENVELOPE

$$TP = \frac{92\,(\text{BTU ROOMS})}{A + E\,(\text{AREA})} = \frac{92 \times 413}{2125 + 2880} = 7.59\ \text{BTU/H/FT}^2$$

ROOM NO.	BTU/H w/o VENTIL.	WATTS	ELECTRIC BASEBD. L.F.@250 W	BTU/H VENTIL.	WATTS	ELECTRIC BASEBD L.F.@250 W	ALLOWABLE WATTS ILLUM	WATTS USED FOR ILLUM.
1	29,440	8,626 / 9000	36				3,200	1,780
2	4,968	1,456 / 1,500	6	N/A SUMMER			800	320
3	3,864	1,132 / 1,250	5				600	240
4	1,564	458 / 500	2	12,044	3,529 / 3,750	15 (1)	278	160
5	2,484	728 / 750	3	12,044	3,529 / 3,750	15 (2)	278	160
6	6,440	1,887 / 2,000	8				482	240
7	NO HEAT						128	200
TOTALS		15,000	60	24,088	7,500	30	5,766	3,100

HEATING — ILLUMINATION

(1) INCLUDED IN ROOM NO. 1
(2) INCLUDED IN ROOM NO. 6

DOES NOT INCLUDE PERIMETER FACADE ILLUMINATION @ 5 W/L.F.

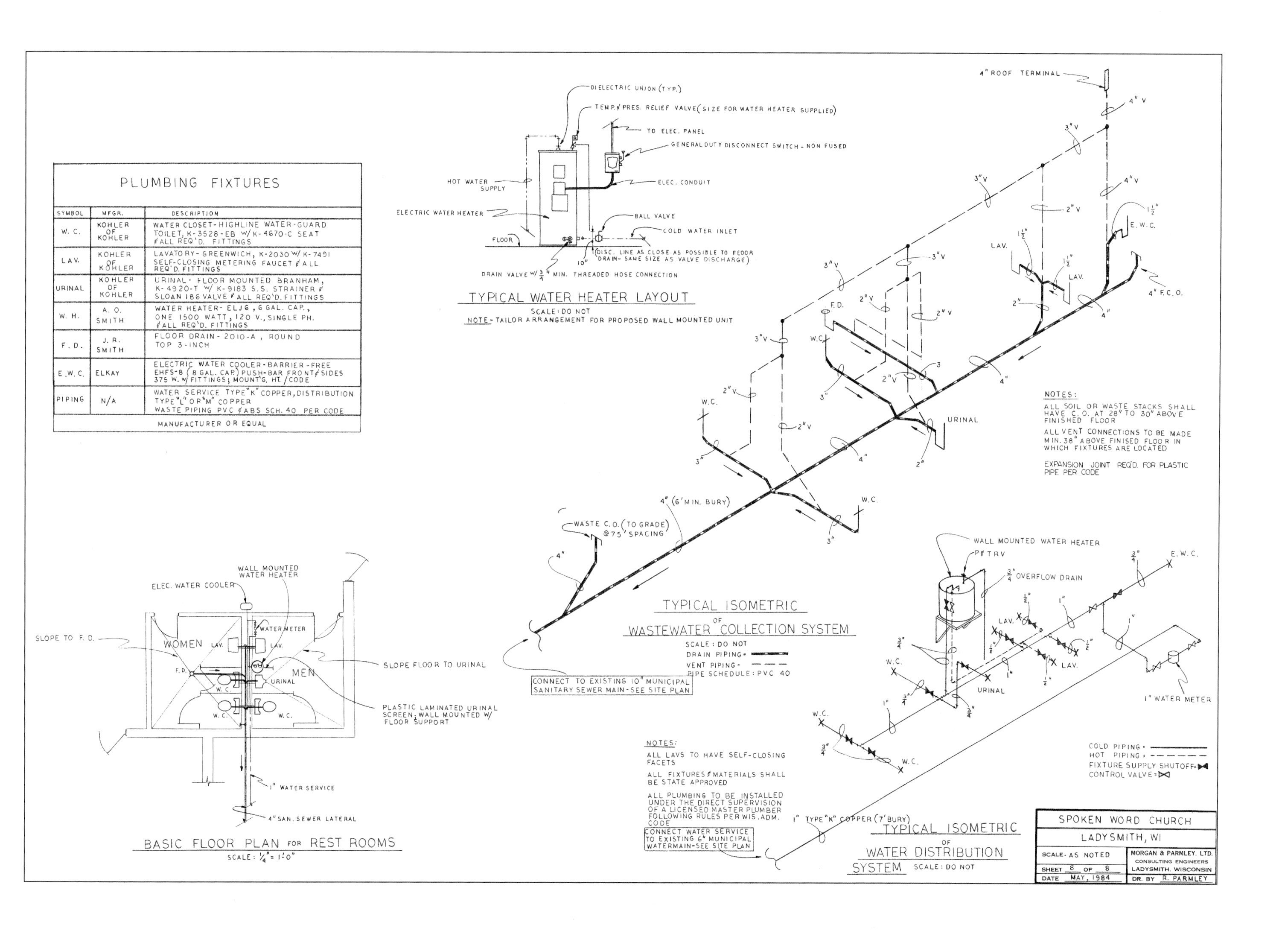

PLUMBING FIXTURES

SYMBOL	MFGR.	DESCRIPTION
W. C.	KOHLER OF KOHLER	WATER CLOSET-HIGHLINE WATER-GUARD TOILET, K-3528-EB W/ K-4670-C SEAT & ALL REQ'D. FITTINGS
LAV.	KOHLER OF KOHLER	LAVATORY-GREENWICH, K-2030 W/ K-7491 SELF-CLOSING METERING FAUCET & ALL REQ'D. FITTINGS
URINAL	KOHLER OF KOHLER	URINAL- FLOOR MOUNTED BRANHAM, K-4920-T W/ K-9183 S.S. STRAINER & SLOAN 186 VALVE & ALL REQ'D. FITTINGS
W. H.	A. O. SMITH	WATER HEATER- ELJ6, 6 GAL. CAP., ONE 1500 WATT, 120 V., SINGLE PH. & ALL REQ'D. FITTINGS
F. D.	J. R. SMITH	FLOOR DRAIN- 2010-A, ROUND TOP 3-INCH
E.W.C.	ELKAY	ELECTRIC WATER COOLER-BARRIER-FREE EHFS-8 (8 GAL. CAP.) PUSH-BAR FRONT & SIDES 375 W. W/ FITTINGS; MOUNT'G. HT. /CODE
PIPING	N/A	WATER SERVICE TYPE "K" COPPER, DISTRIBUTION TYPE "L" OR "M" COPPER WASTE PIPING PVC & ABS SCH. 40 PER CODE
MANUFACTURER OR EQUAL		

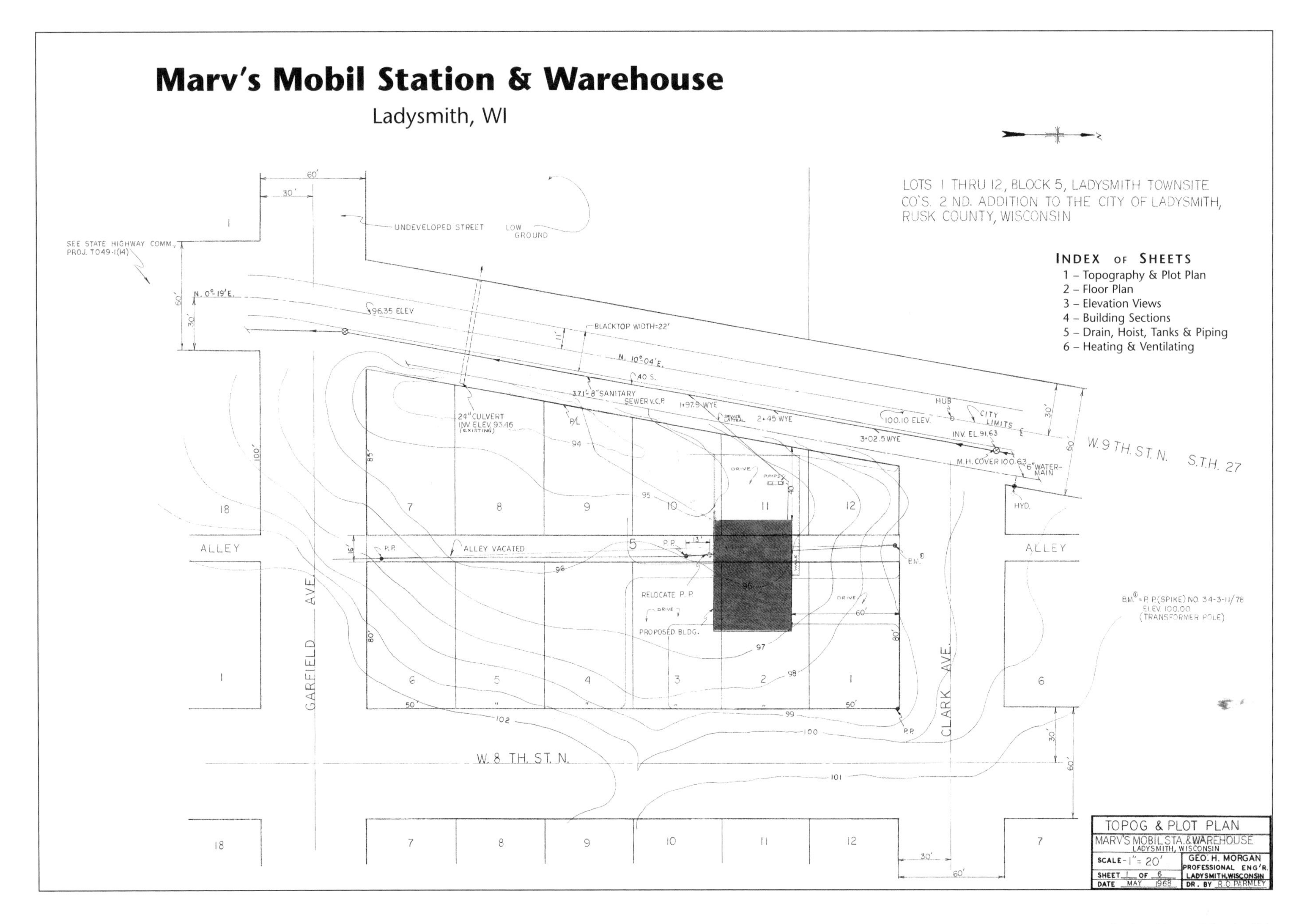

Courtesy: Morgan & Parmley, Ltd.

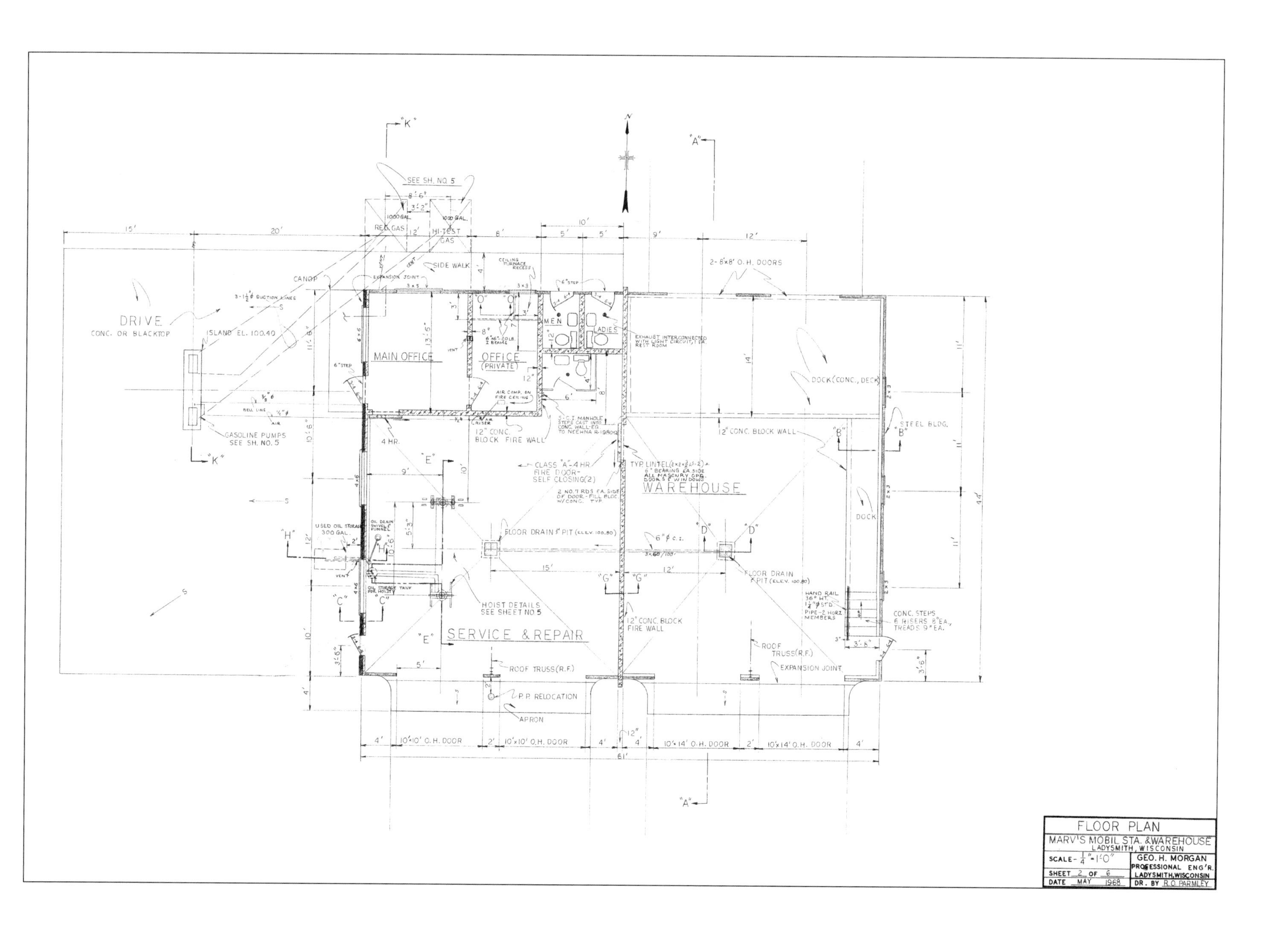
FLOOR PLAN
MARV'S MOBIL STA. & WAREHOUSE
LADYSMITH, WISCONSIN
SCALE- 1/4"=1'-0"
GEO. H. MORGAN
PROFESSIONAL ENG'R.
LADYSMITH, WISCONSIN
SHEET 2 OF 6
DATE MAY 1968
DR. BY R.O. PARMLEY
MAIN OFFICE
OFFICE (PRIVATE)
MEN
LADIES
WAREHOUSE
SERVICE & REPAIR
DRIVE
CONC. OR BLACKTOP
DOCK (CONC., DECK)
STEEL BLDG.
12" CONC. BLOCK WALL
12" CONC. BLOCK FIRE WALL
GASOLINE PUMPS
SEE SH. NO. 5
HOIST DETAILS
SEE SHEET NO 5
10'x10' O.H. DOOR
10'x14' O.H. DOOR
2-8'x8' O.H. DOORS
APRON
P.P. RELOCATION
ROOF TRUSS (R.F.)
EXPANSION JOINT
CONC. STEPS
6 RISERS 8" EA.
TREADS 9" EA.
61'
44'

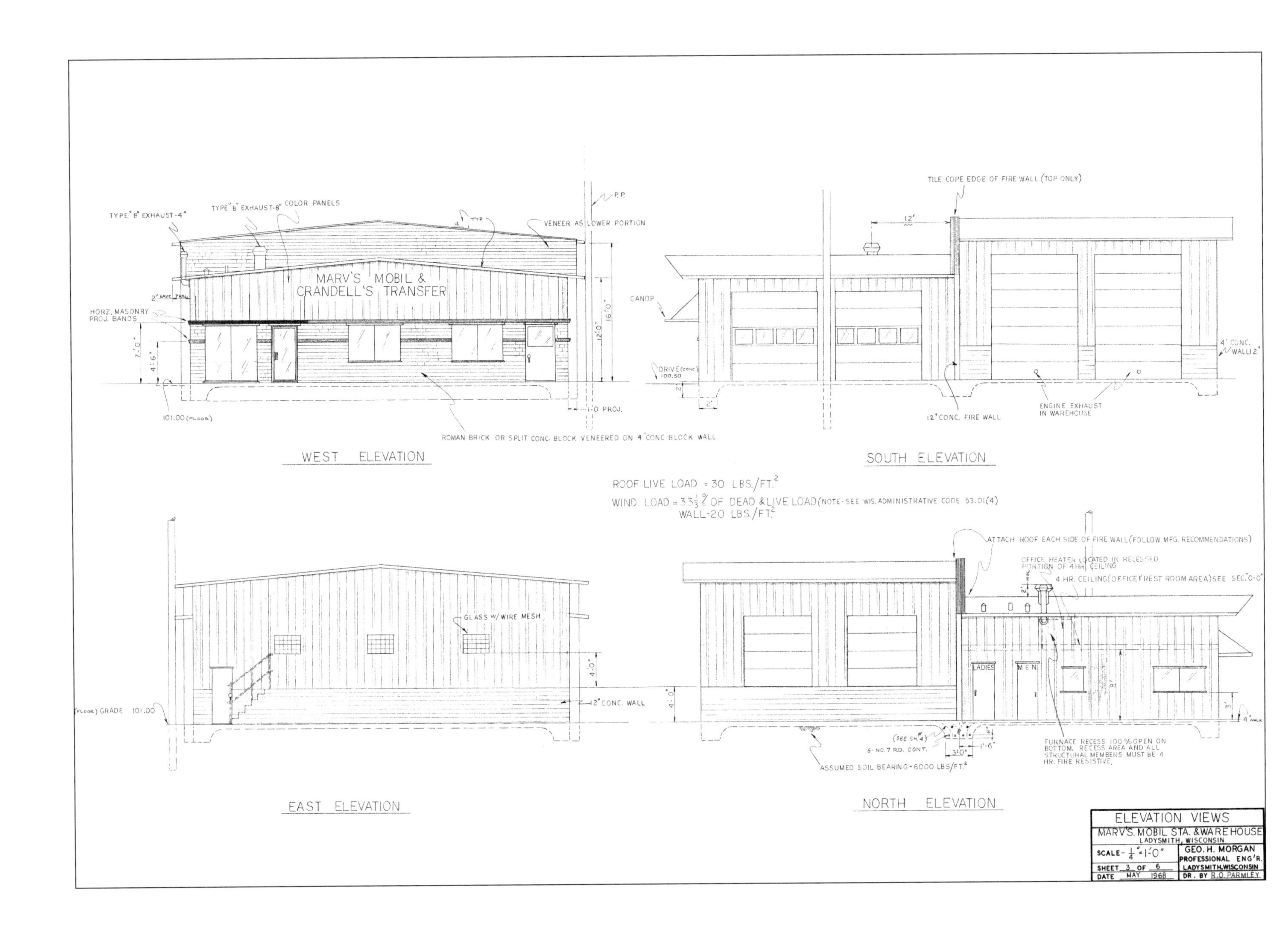

WEST ELEVATION
MARV'S MOBIL & CRANDELL'S TRANSFER
TYPE "B" EXHAUST-4"
TYPE "B" EXHAUST-8"
COLOR PANELS
VENEER AS LOWER PORTION
P.P.
HORZ. MASONRY PROJ. BANDS
2' EAVE PROJ.
101.00 (FLOOR)
1'-0 PROJ.
ROMAN BRICK OR SPLIT CONC. BLOCK VENEERED ON 4" CONC BLOCK WALL
SOUTH ELEVATION
TILE COPE EDGE OF FIRE WALL (TOP ONLY)
CANOPY
DRIVE (CONC) 100.50
12" CONC. FIRE WALL
ENGINE EXHAUST IN WAREHOUSE
4' CONC. WALL 12"
ROOF LIVE LOAD = 30 LBS./FT.²
WIND LOAD = 33⅓% OF DEAD & LIVE LOAD (NOTE-SEE WIS. ADMINISTRATIVE CODE 53.01(4)
WALL-20 LBS./FT.²
EAST ELEVATION
GLASS W/WIRE MESH
(FLOOR) GRADE 101.00
12" CONC. WALL
NORTH ELEVATION
ATTACH ROOF EACH SIDE OF FIRE WALL (FOLLOW MFG. RECOMMENDATIONS)
OFFICE HEATER LOCATED IN RECESSED PORTION OF 4 HR. CEILING
4 HR. CEILING (OFFICE & REST ROOM AREA) SEE SEC. O-O
LADIES
MEN
(SEE SH. 4)
6-NO. 7 RD. CONT.
ASSUMED SOIL BEARING-6000 LBS/FT.²
FURNACE RECESS 100% OPEN ON BOTTOM. RECESS AREA AND ALL STRUCTURAL MEMBERS MUST BE 4 HR. FIRE RESISTIVE.
ELEVATION VIEWS
MARV'S MOBIL STA. & WAREHOUSE
LADYSMITH, WISCONSIN
SCALE - 1/4" = 1'-0"
SHEET 3 OF 6
DATE MAY 1968
GEO. H. MORGAN
PROFESSIONAL ENG'R.
LADYSMITH, WISCONSIN
DR. BY R.O. PARMLEY

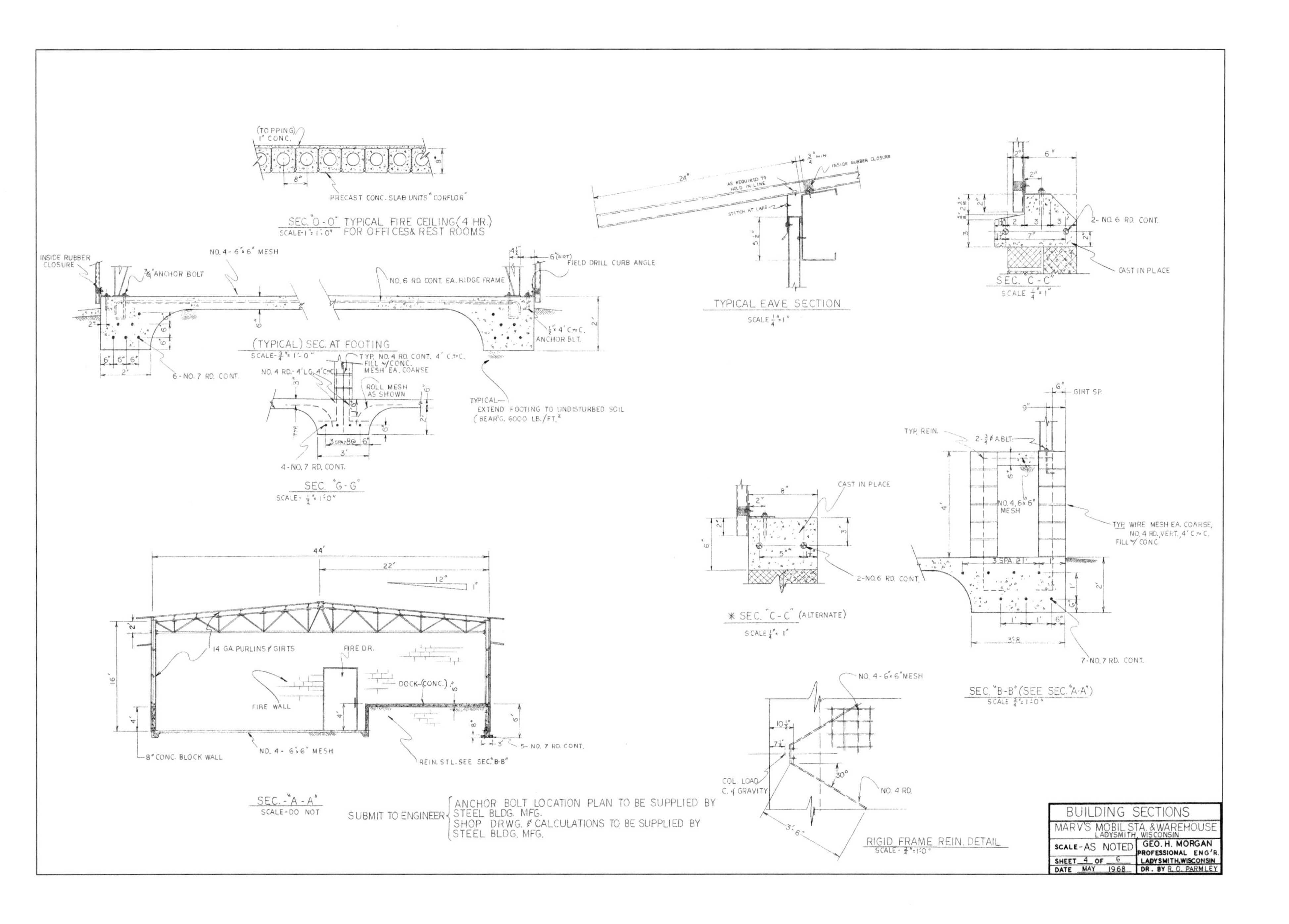
SEC. "O-O" TYPICAL FIRE CEILING (4 HR.) FOR OFFICES & REST ROOMS
SCALE-1"=1'-0"
PRECAST CONC. SLAB UNITS "CORFLOR"
(TOPPING) 1" CONC.
TYPICAL EAVE SECTION
SCALE 1/4"=1"
INSIDE RUBBER CLOSURE
SEC. "C-C"
SCALE 1/4"=1"
2-NO. 6 RD. CONT.
CAST IN PLACE
NO. 4-6"x6" MESH
3/4" ANCHOR BOLT
NO. 6 RD. CONT. EA. RIDGE FRAME
FIELD DRILL CURB ANGLE
(TYPICAL) SEC. AT FOOTING
6-NO. 7 RD. CONT.
ANCHOR BLT.
TYPICAL—EXTEND FOOTING TO UNDISTURBED SOIL (BEAR'G. 6000 LB./FT.²)
TYP. NO. 4 RD. CONT. 4' C. to C. FILL w/CONC. MESH EA. COARSE
ROLL MESH AS SHOWN
4-NO. 7 RD. CONT.
SEC. "G-G"
SCALE-1/2"=1'-0"
GIRT SP.
TYP. REIN.
NO. 4, 6x6" MESH
TYP. WIRE MESH EA. COARSE, NO. 4 RD., VERT., 4' C. to C. FILL w/ CONC.
7-NO. 7 RD. CONT.
SEC. "B-B" (SEE SEC. "A-A")
* SEC. "C-C" (ALTERNATE)
SCALE 1/4"=1"
2-NO. 6 RD. CONT.
14 GA. PURLINS & GIRTS
FIRE DR.
DOCK-(CONC.)
FIRE WALL
8" CONC. BLOCK WALL
NO. 4-6"x6" MESH
REIN. STL. SEE SEC. "B-B"
5-NO. 7 RD. CONT.
SEC. "A-A"
SCALE-DO NOT
SUBMIT TO ENGINEER: ANCHOR BOLT LOCATION PLAN TO BE SUPPLIED BY STEEL BLDG. MFG. SHOP DRWG. & CALCULATIONS TO BE SUPPLIED BY STEEL BLDG. MFG.
NO. 4-6"x6" MESH
COL. LOAD C. of GRAVITY
NO. 4 RD.
RIGID FRAME REIN. DETAIL
SCALE-1/2"=1'-0"
BUILDING SECTIONS
MARV'S MOBIL STA. & WAREHOUSE
LADYSMITH, WISCONSIN
SCALE-AS NOTED
GEO. H. MORGAN PROFESSIONAL ENG'R. LADYSMITH, WISCONSIN
SHEET 4 OF 6
DATE MAY 1968
DR. BY R.O. PARMLEY

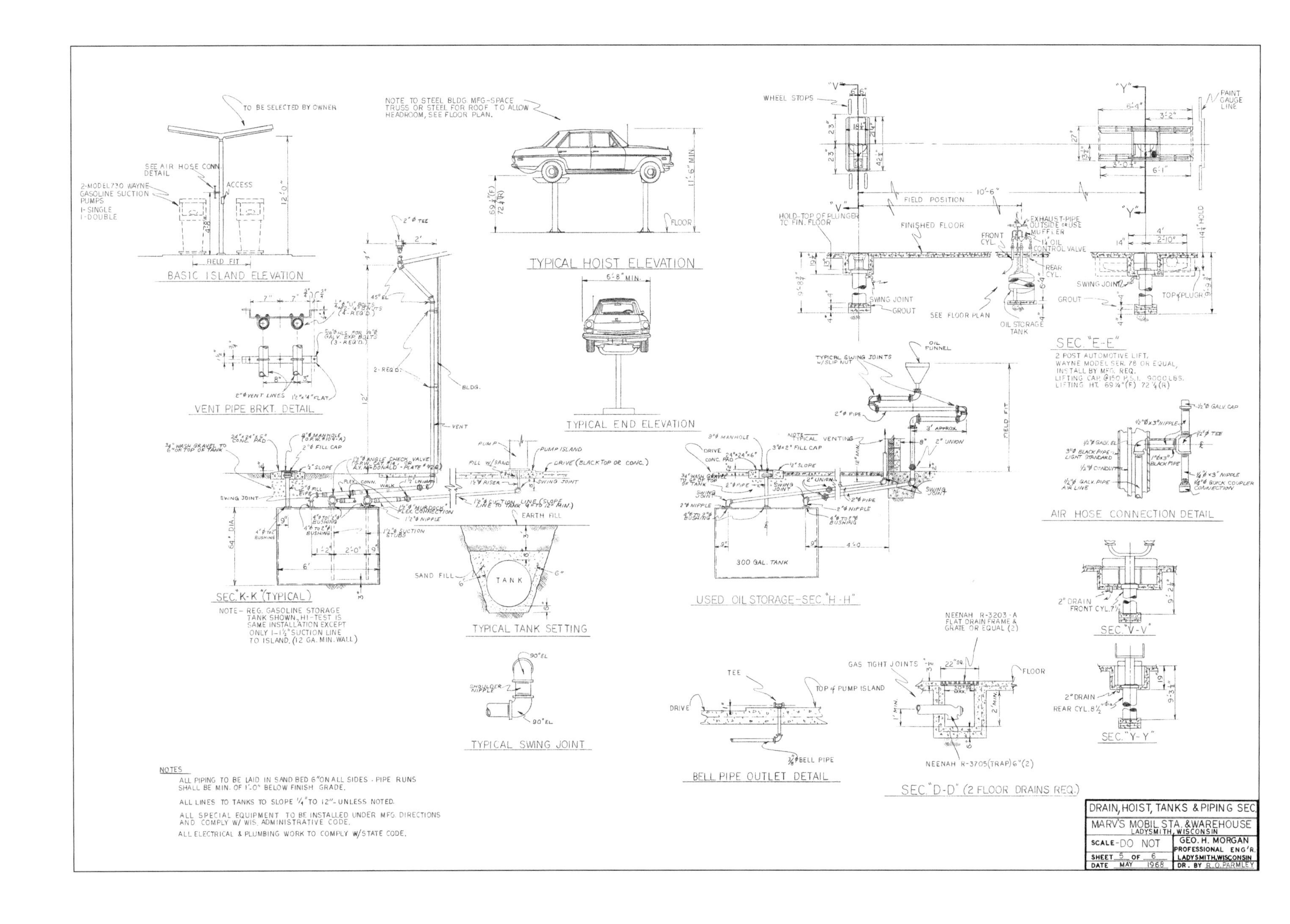

BASIC ISLAND ELEVATION
TO BE SELECTED BY OWNER
SEE AIR HOSE CONN. DETAIL
ACCESS
2-MODEL 730 WAYNE GASOLINE SUCTION PUMPS 1-SINGLE 1-DOUBLE
FIELD FIT
VENT PIPE BRKT. DETAIL
NOTE TO STEEL BLDG MFG-SPACE TRUSS OR STEEL FOR ROOF TO ALLOW HEADROOM, SEE FLOOR PLAN.
TYPICAL HOIST ELEVATION
FLOOR
TYPICAL END ELEVATION
WHEEL STOPS
FIELD POSITION
HOLD-TOP OF PLUNGER TO FIN. FLOOR
FINISHED FLOOR
SWING JOINT
GROUT
SEE FLOOR PLAN
OIL STORAGE TANK
SEC. "E-E"
2 POST AUTOMOTIVE LIFT, WAYNE MODEL SER. 78 OR EQUAL, INSTALL BY MFG. REQ.
PAINT GAUGE LINE
SEC.K-K (TYPICAL)
NOTE- REG. GASOLINE STORAGE TANK SHOWN, HI-TEST IS SAME INSTALLATION EXCEPT ONLY 1-1½" SUCTION LINE TO ISLAND. (12 GA. MIN. WALL)
TYPICAL TANK SETTING
EARTH FILL
SAND FILL
TANK
USED OIL STORAGE-SEC."H-H"
300 GAL. TANK
AIR HOSE CONNECTION DETAIL
SEC."V-V"
SEC."Y-Y"
TYPICAL SWING JOINT
BELL PIPE OUTLET DETAIL
SEC."D-D" (2 FLOOR DRAINS REQ.)
NEENAH R-3203-A FLAT DRAIN FRAME & GRATE OR EQUAL (2)
NEENAH R-3705 (TRAP) 6" (2)
NOTES
ALL PIPING TO BE LAID IN SAND BED 6" ON ALL SIDES - PIPE RUNS SHALL BE MIN. OF 1'-0" BELOW FINISH GRADE.
ALL LINES TO TANKS TO SLOPE ¼" TO 12" - UNLESS NOTED.
ALL SPECIAL EQUIPMENT TO BE INSTALLED UNDER MFG. DIRECTIONS AND COMPLY W/ WIS. ADMINISTRATIVE CODE.
ALL ELECTRICAL & PLUMBING WORK TO COMPLY W/STATE CODE.
DRAIN, HOIST, TANKS & PIPING SEC.
MARV'S MOBIL STA. & WAREHOUSE LADYSMITH, WISCONSIN
SCALE-DO NOT
GEO. H. MORGAN PROFESSIONAL ENG'R. LADYSMITH, WISCONSIN
SHEET 5 OF 6
DATE MAY 1968
DR. BY R.O. PARMLEY

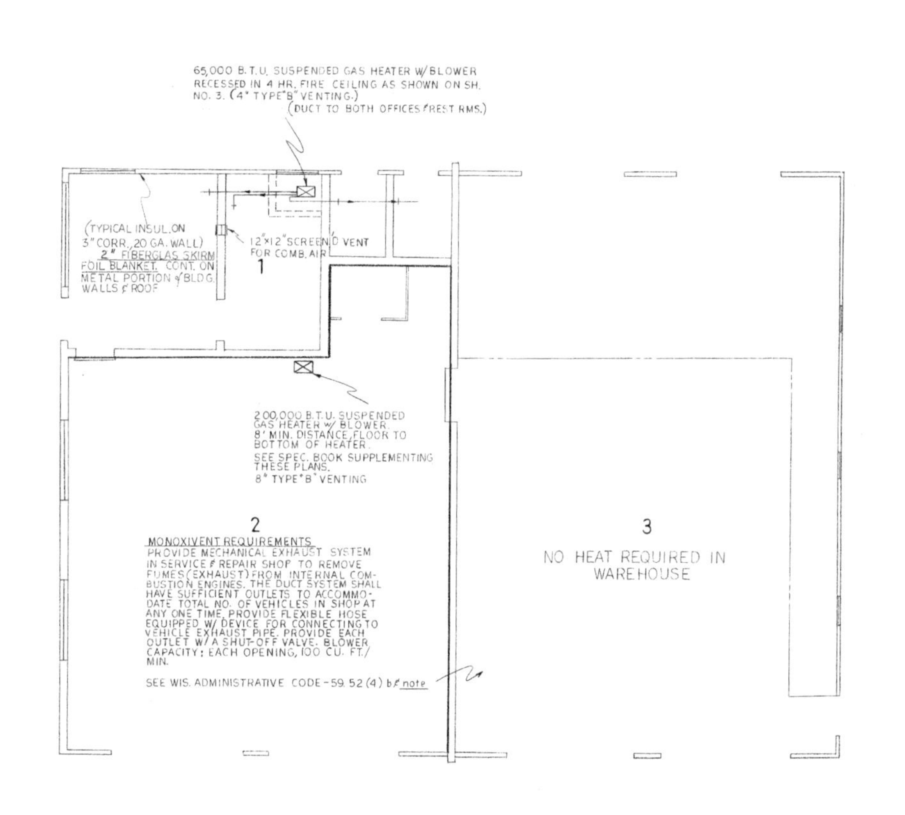
65,000 B.T.U. SUSPENDED GAS HEATER W/BLOWER RECESSED IN 4 HR. FIRE CEILING AS SHOWN ON SH. NO. 3. (4" TYPE "B" VENTING.)
(DUCT TO BOTH OFFICES & REST RMS.)
(TYPICAL INSUL. ON 3" CORR., 20 GA. WALL)
2" FIBERGLAS SKIRM FOIL BLANKET. CONT. ON METAL PORTION of BLDG. WALLS & ROOF
12"×12" SCREEN'D VENT FOR COMB. AIR
1
200,000 B.T.U. SUSPENDED GAS HEATER w/ BLOWER.
8' MIN. DISTANCE, FLOOR TO BOTTOM OF HEATER.
SEE SPEC. BOOK SUPPLEMENTING THESE PLANS.
8" TYPE "B" VENTING
2
MONOXIVENT REQUIREMENTS
PROVIDE MECHANICAL EXHAUST SYSTEM IN SERVICE & REPAIR SHOP TO REMOVE FUMES (EXHAUST) FROM INTERNAL COMBUSTION ENGINES. THE DUCT SYSTEM SHALL HAVE SUFFICIENT OUTLETS TO ACCOMMODATE TOTAL NO. OF VEHICLES IN SHOP AT ANY ONE TIME. PROVIDE FLEXIBLE HOSE EQUIPPED W/ DEVICE FOR CONNECTING TO VEHICLE EXHAUST PIPE. PROVIDE EACH OUTLET W/ A SHUT-OFF VALVE. BLOWER CAPACITY: EACH OPENING, 100 CU. FT./MIN.
SEE WIS. ADMINISTRATIVE CODE - 59.52 (4) b & note
3
NO HEAT REQUIRED IN WAREHOUSE

NOTE – FOR OTHER INFORMATION SEE SPEC. BOOK & SHEETS 1 THRU 5
ALL HEATING, VENTILATING, & MONOXIVENT REQUIREMENTS SHALL COMPLY WITH WIS. ADMINISTRATIVE CODE.

HEATING & VENTILATING	
MARV'S MOBIL STA. & WAREHOUSE LADYSMITH, WISCONSIN	
SCALE - DO NOT	GEO. H. MORGAN PROFESSIONAL ENG'R. LADYSMITH, WISCONSIN
SHEET 6 OF 6	
DATE MAY 1968	DR. BY R.O. PARMLEY

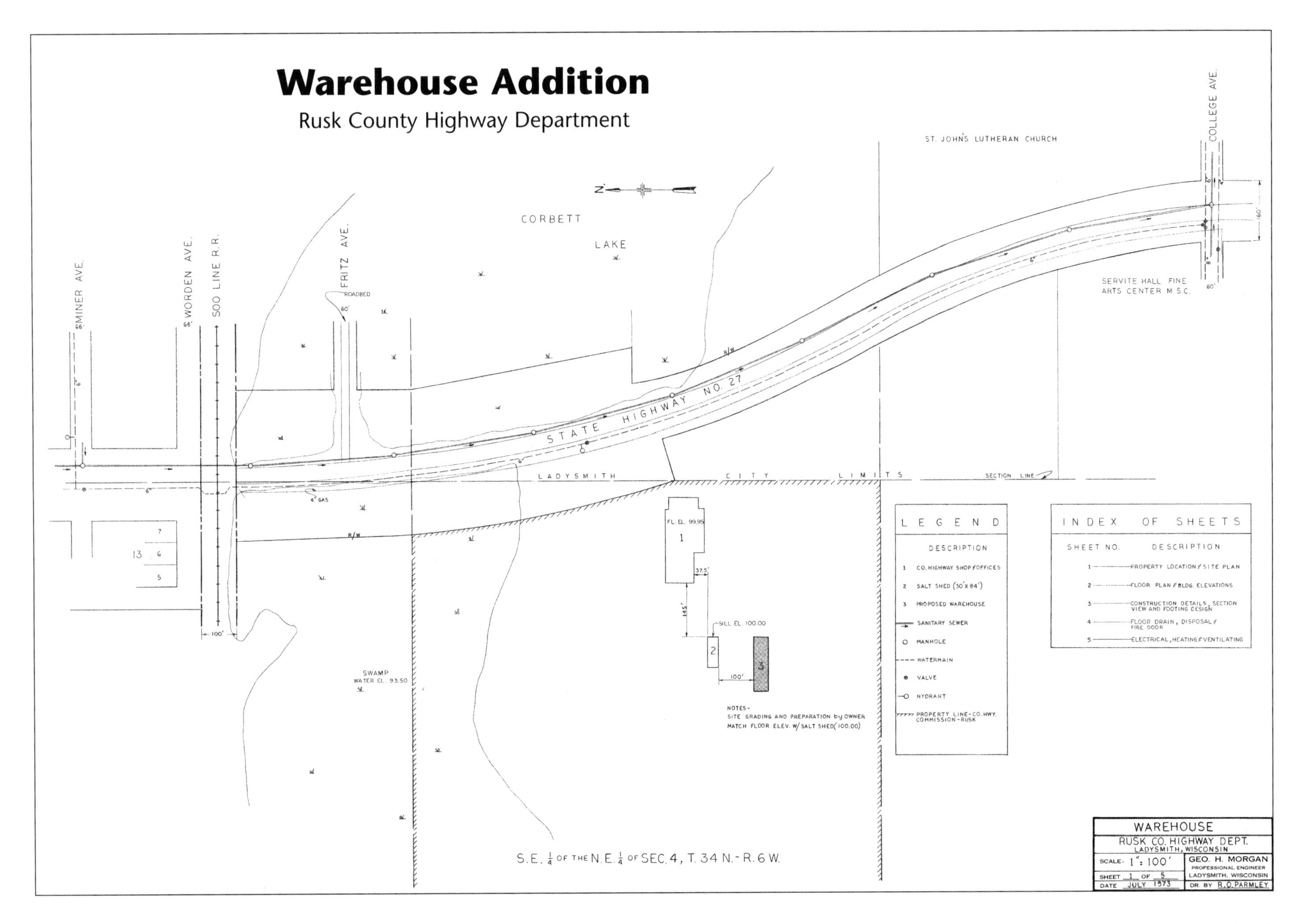

Courtesy: Morgan & Parmley, Ltd.

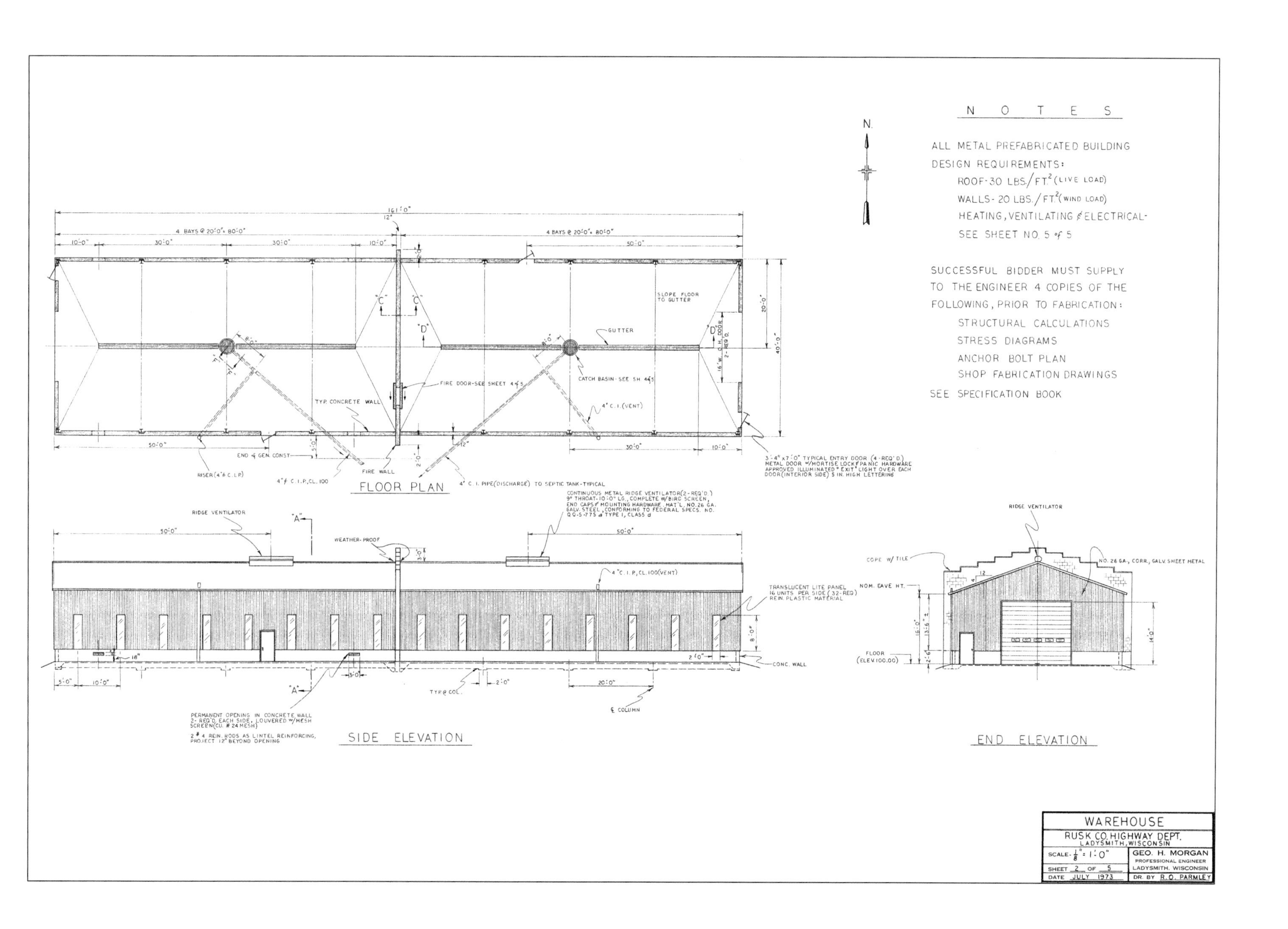
NOTES
ALL METAL PREFABRICATED BUILDING
DESIGN REQUIREMENTS:
ROOF-30 LBS/FT.² (LIVE LOAD)
WALLS-20 LBS./FT.² (WIND LOAD)
HEATING, VENTILATING & ELECTRICAL-
SEE SHEET NO. 5 of 5
SUCCESSFUL BIDDER MUST SUPPLY
TO THE ENGINEER 4 COPIES OF THE
FOLLOWING, PRIOR TO FABRICATION:
STRUCTURAL CALCULATIONS
STRESS DIAGRAMS
ANCHOR BOLT PLAN
SHOP FABRICATION DRAWINGS
SEE SPECIFICATION BOOK
FLOOR PLAN
SIDE ELEVATION
END ELEVATION
WAREHOUSE
RUSK CO. HIGHWAY DEPT.
LADYSMITH, WISCONSIN
SCALE: 1/8" = 1'-0"
SHEET 2 OF 5
DATE JULY 1973
GEO. H. MORGAN
PROFESSIONAL ENGINEER
LADYSMITH, WISCONSIN
DR. BY R.O. PARMLEY

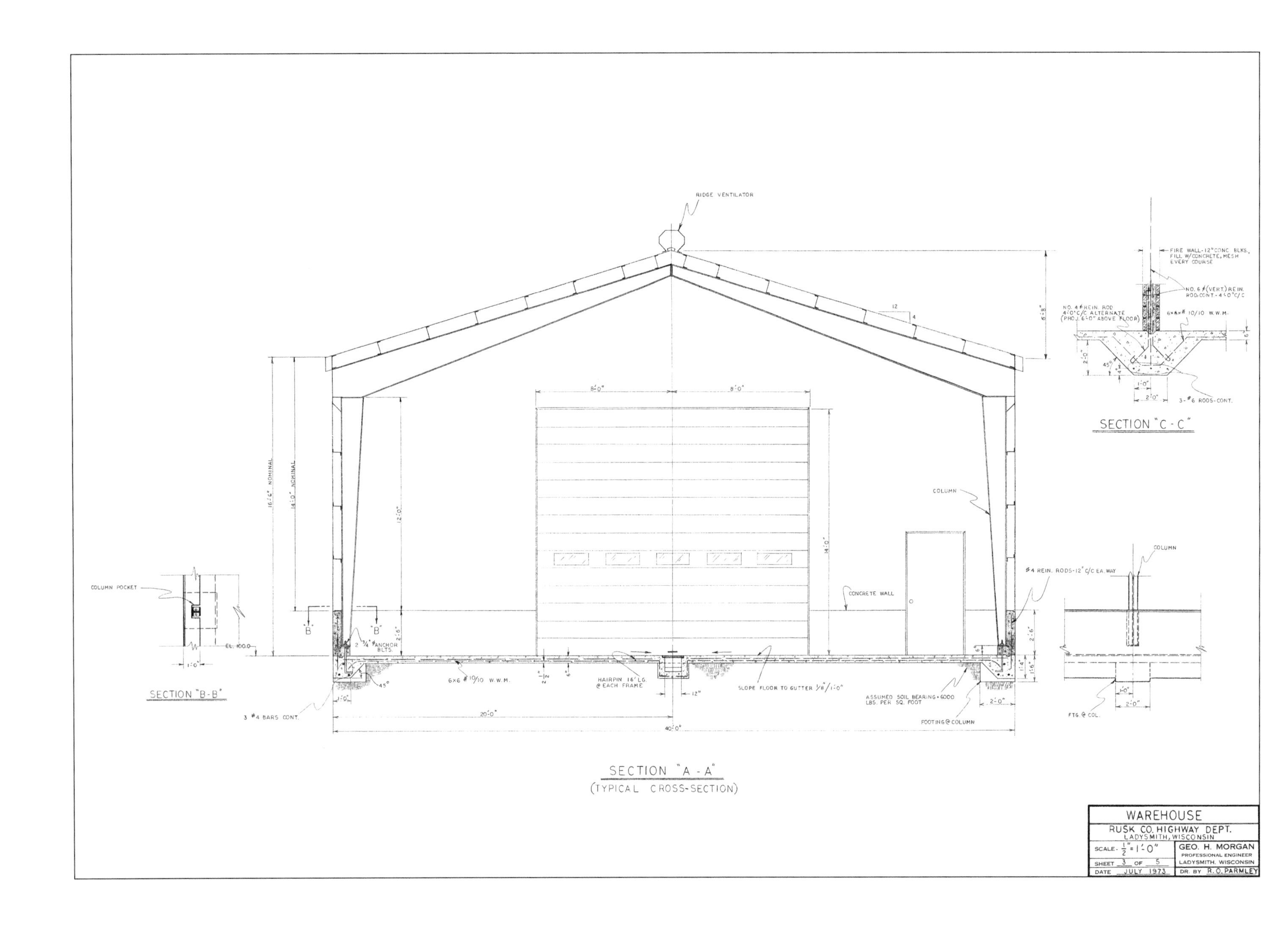
RIDGE VENTILATOR
12
4
8'-0"
8'-0"
16'-6" NOMINAL
14'-0" NOMINAL
12'-0"
14'-0"
COLUMN
CONCRETE WALL
#4 REIN. RODS-12" C/C EA. WAY
2 3/4" ANCHOR BLTS.
6x6 #10/10 W.W.M.
HAIRPIN 16' LG. @ EACH FRAME
SLOPE FLOOR TO GUTTER 1/8"/1'-0"
ASSUMED SOIL BEARING = 6000 LBS. PER SQ. FOOT
FOOTING @ COLUMN
3 #4 BARS CONT.
20'-0"
40'-0"
SECTION "A - A"
(TYPICAL CROSS-SECTION)
COLUMN POCKET
EL. 100.0
SECTION "B-B"
FIRE WALL-12" CONC. BLKS. FILL W/CONCRETE, MESH EVERY COURSE
NO. 6 # (VERT.) REIN. ROD CONT.-4'-0" C/C
NO. 4 # REIN. ROD 4'-0" C/C ALTERNATE (PROJ. 6'-0" ABOVE FLOOR)
6x6x# 10/10 W.W.M.
3-#6 RODS-CONT.
SECTION "C - C"
COLUMN
FTG. @ COL.
WAREHOUSE
RUSK CO. HIGHWAY DEPT.
LADYSMITH, WISCONSIN
SCALE- 1/2" = 1'-0"
SHEET 3 OF 5
DATE JULY 1973
GEO. H. MORGAN
PROFESSIONAL ENGINEER
LADYSMITH, WISCONSIN
DR. BY R. O. PARMLEY

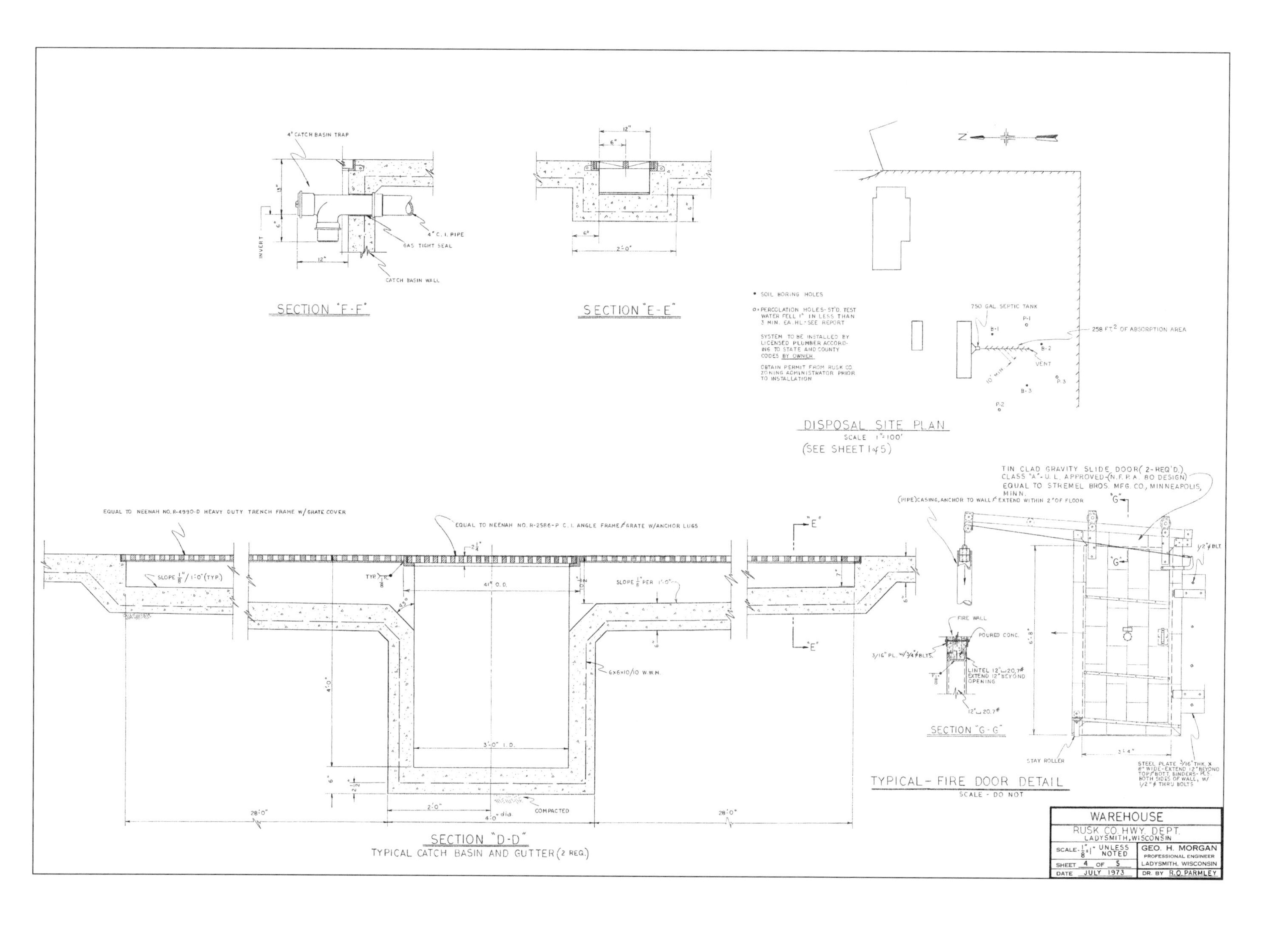
4" CATCH BASIN TRAP
4" C. I. PIPE
GAS TIGHT SEAL
CATCH BASIN WALL
INVERT
SECTION "F-F"
SECTION "E-E"
SOIL BORING HOLES
PERCOLATION HOLES- 5T'D. TEST WATER FELL 1" IN LESS THAN 3 MIN. EA. HL.-SEE REPORT
SYSTEM TO BE INSTALLED BY LICENSED PLUMBER ACCORDING TO STATE AND COUNTY CODES BY OWNER
OBTAIN PERMIT FROM RUSK CO. ZONING ADMINISTRATOR PRIOR TO INSTALLATION
750 GAL. SEPTIC TANK
258 FT.2 OF ABSORPTION AREA
VENT
DISPOSAL SITE PLAN
SCALE 1"=100'
(SEE SHEET 1 of 5)
TIN CLAD GRAVITY SLIDE DOOR (2-REQ'D.)
CLASS "A"- U. L. APPROVED-(N.F.P.A. 80 DESIGN)
EQUAL TO STREMEL BROS. MFG. CO., MINNEAPOLIS, MINN.
(PIPE)CASING, ANCHOR TO WALL & EXTEND WITHIN 2" OF FLOOR
EQUAL TO NEENAH NO. R-4990-D HEAVY DUTY TRENCH FRAME W/ GRATE COVER
EQUAL TO NEENAH NO. R-2586-P C. I. ANGLE FRAME & GRATE W/ANCHOR LUGS
SLOPE 1/8"/1'-0" (TYP)
41" O. D.
6x6-10/10 W.W.M.
3'-0" I. D.
COMPACTED
28'-0"
SECTION "D-D"
TYPICAL CATCH BASIN AND GUTTER (2 REQ.)
FIRE WALL
POURED CONC.
3/16" PL. w/ 3/4"φ BLTS.
LINTEL 12" ⊔ 20.7# EXTEND 12" BEYOND OPENING
SECTION "G-G"
STAY ROLLER
STEEL PLATE 3/16" THK. X 8" WIDE-EXTEND 12" BEYOND TOP & BOTT. BINDERS-PLS. BOTH SIDES OF WALL, W/ 1/2" φ THRU BOLTS
TYPICAL - FIRE DOOR DETAIL
SCALE - DO NOT
WAREHOUSE
RUSK CO. HWY. DEPT.
LADYSMITH, WISCONSIN
SCALE: 1/8" = 1' UNLESS NOTED
SHEET 4 OF 5
DATE JULY 1973
GEO. H. MORGAN
PROFESSIONAL ENGINEER
LADYSMITH, WISCONSIN
DR. BY R.O. PARMLEY

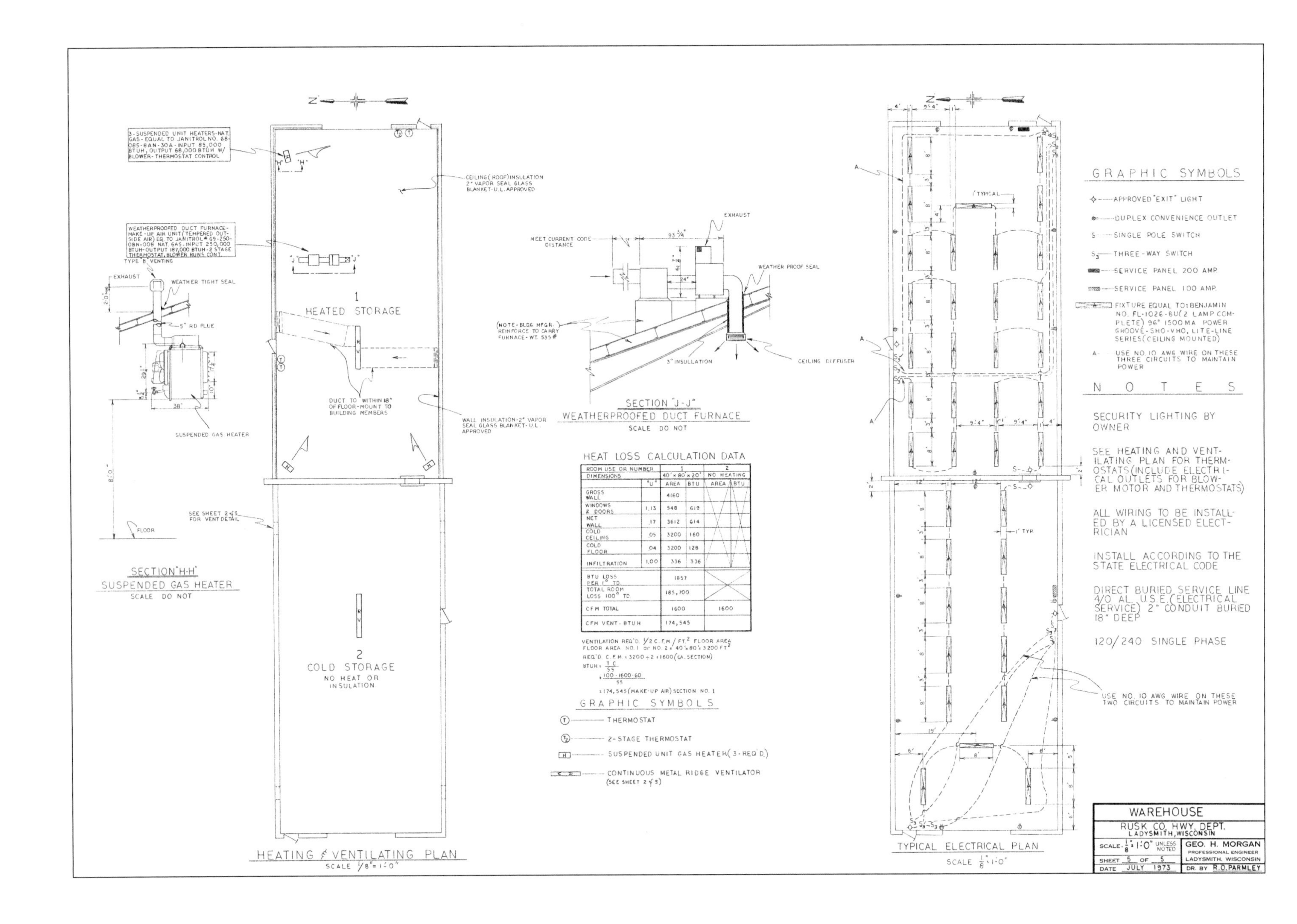

ROOM USE OR NUMBER		1		2	
DIMENSIONS		40'×80'×20'		NO HEATING	
	"U"	AREA	BTU	AREA	BTU
GROSS WALL		4160			
WINDOWS & DOORS	1.13	548	619		
NET WALL	.17	3612	614		
COLD CEILING	.05	3200	160		
COLD FLOOR	.04	3200	128		
INFILTRATION	1.00	336	336		
BTU LOSS PER 1° TD		1857			
TOTAL ROOM LOSS 100° TD		185,700			
CFM TOTAL		1600		1600	
CFM VENT-BTUH		174,545			

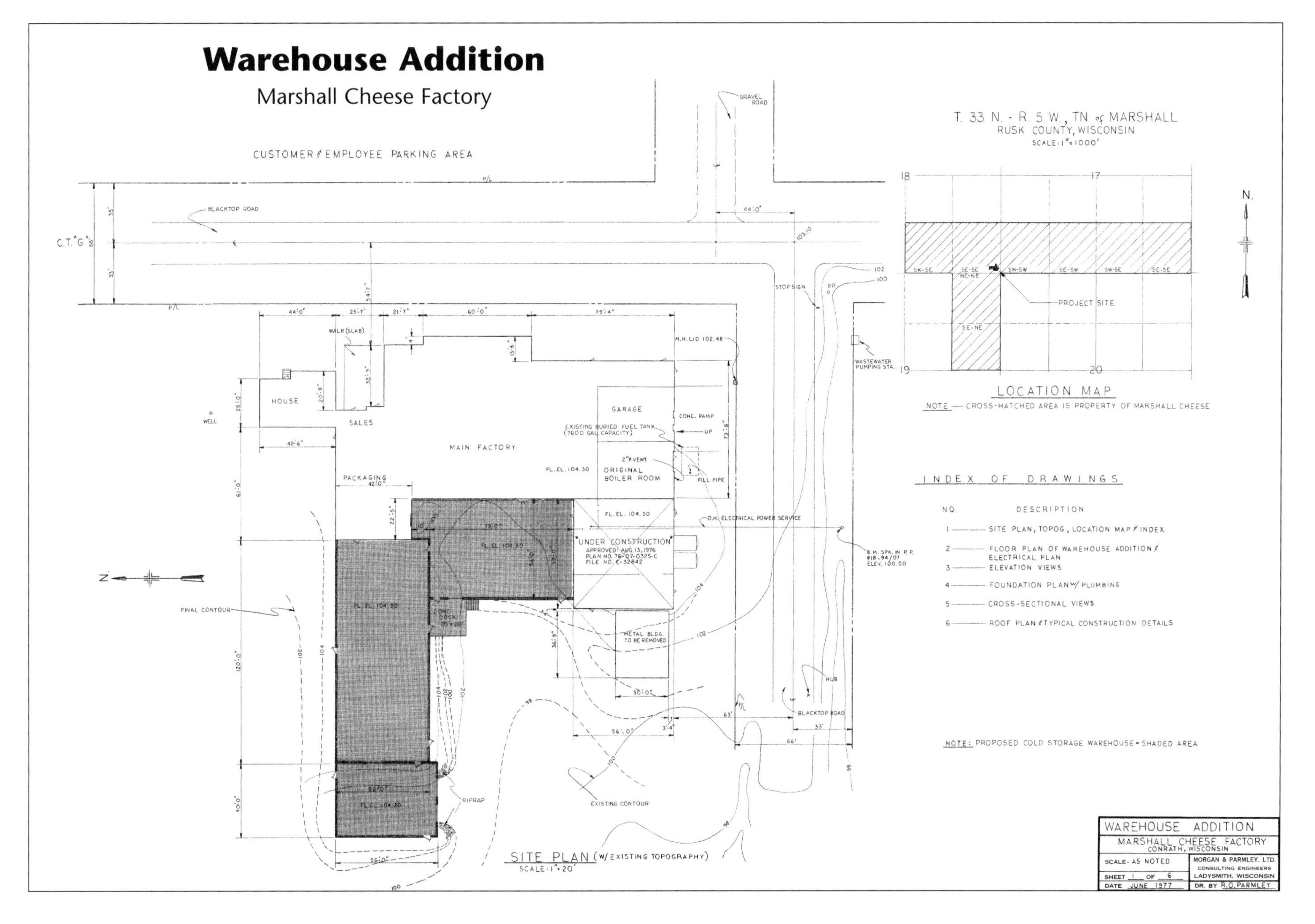

Courtesy: Morgan & Parmley, Ltd.

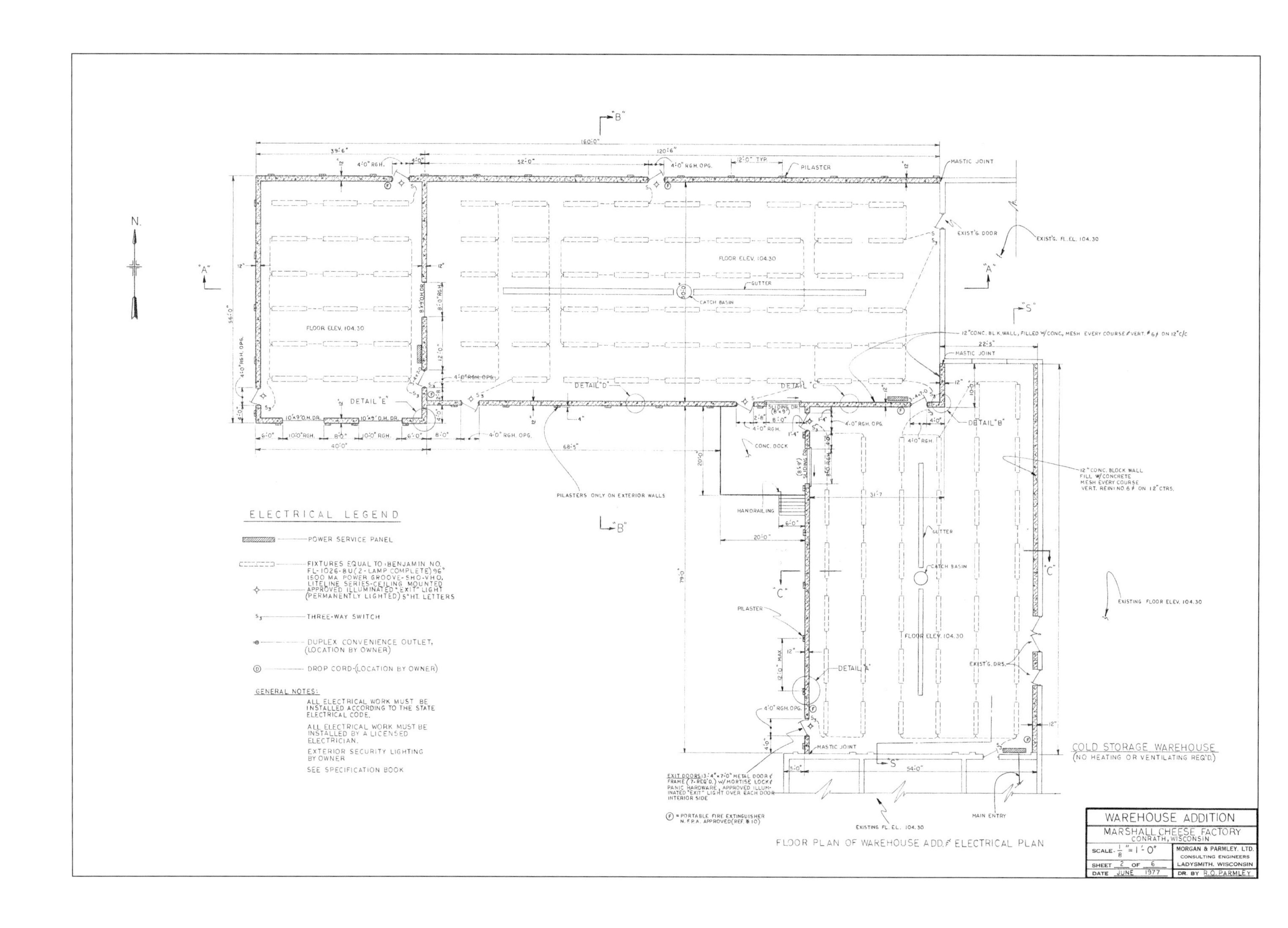
ELECTRICAL LEGEND
POWER SERVICE PANEL
FIXTURES EQUAL TO BENJAMIN NO. FL-1026-8U (2-LAMP COMPLETE) 96" 1500 MA. POWER GROOVE-SHO-VHO. LITELINE SERIES-CEILING MOUNTED
APPROVED ILLUMINATED "EXIT" LIGHT (PERMANENTLY LIGHTED) 5" HT. LETTERS
THREE-WAY SWITCH
DUPLEX CONVENIENCE OUTLET, (LOCATION BY OWNER)
DROP CORD-(LOCATION BY OWNER)
GENERAL NOTES:
ALL ELECTRICAL WORK MUST BE INSTALLED ACCORDING TO THE STATE ELECTRICAL CODE.
ALL ELECTRICAL WORK MUST BE INSTALLED BY A LICENSED ELECTRICIAN.
EXTERIOR SECURITY LIGHTING BY OWNER
SEE SPECIFICATION BOOK
COLD STORAGE WAREHOUSE
(NO HEATING OR VENTILATING REQ'D.)
FLOOR PLAN OF WAREHOUSE ADD.& ELECTRICAL PLAN
WAREHOUSE ADDITION
MARSHALL CHEESE FACTORY
CONRATH, WISCONSIN
SCALE-1/8"=1'-0"
SHEET 2 OF 6
DATE JUNE 1977
MORGAN & PARMLEY, LTD.
CONSULTING ENGINEERS
LADYSMITH, WISCONSIN
DR. BY R.O. PARMLEY
FLOOR ELEV. 104.30
CATCH BASIN
GUTTER
PILASTER
MASTIC JOINT
EXIST'G DOOR
EXIST'G. FL. EL. 104.30
DETAIL "A"
DETAIL "B"
DETAIL "C"
DETAIL "D"
DETAIL "E"
CONC. DOCK
HANDRAILING
PILASTERS ONLY ON EXTERIOR WALLS
EXISTING FLOOR ELEV. 104.30
EXIST'G. DRS.
MAIN ENTRY
EXISTING FL. EL. 104.30
12" CONC. BLOCK WALL FILL W/ CONCRETE MESH EVERY COURSE VERT. REIN. NO 6 # ON 12" CTRS.
EXIT DOORS: 3'-4" x 7'-0" METAL DOOR & FRAME (7-REQ'D.) W/ MORTISE LOCK & PANIC HARDWARE. APPROVED ILLUMINATED "EXIT" LIGHT OVER EACH DOOR-INTERIOR SIDE
(F) = PORTABLE FIRE EXTINGUISHER N.F.P.A. APPROVED (REF. #10)
N.

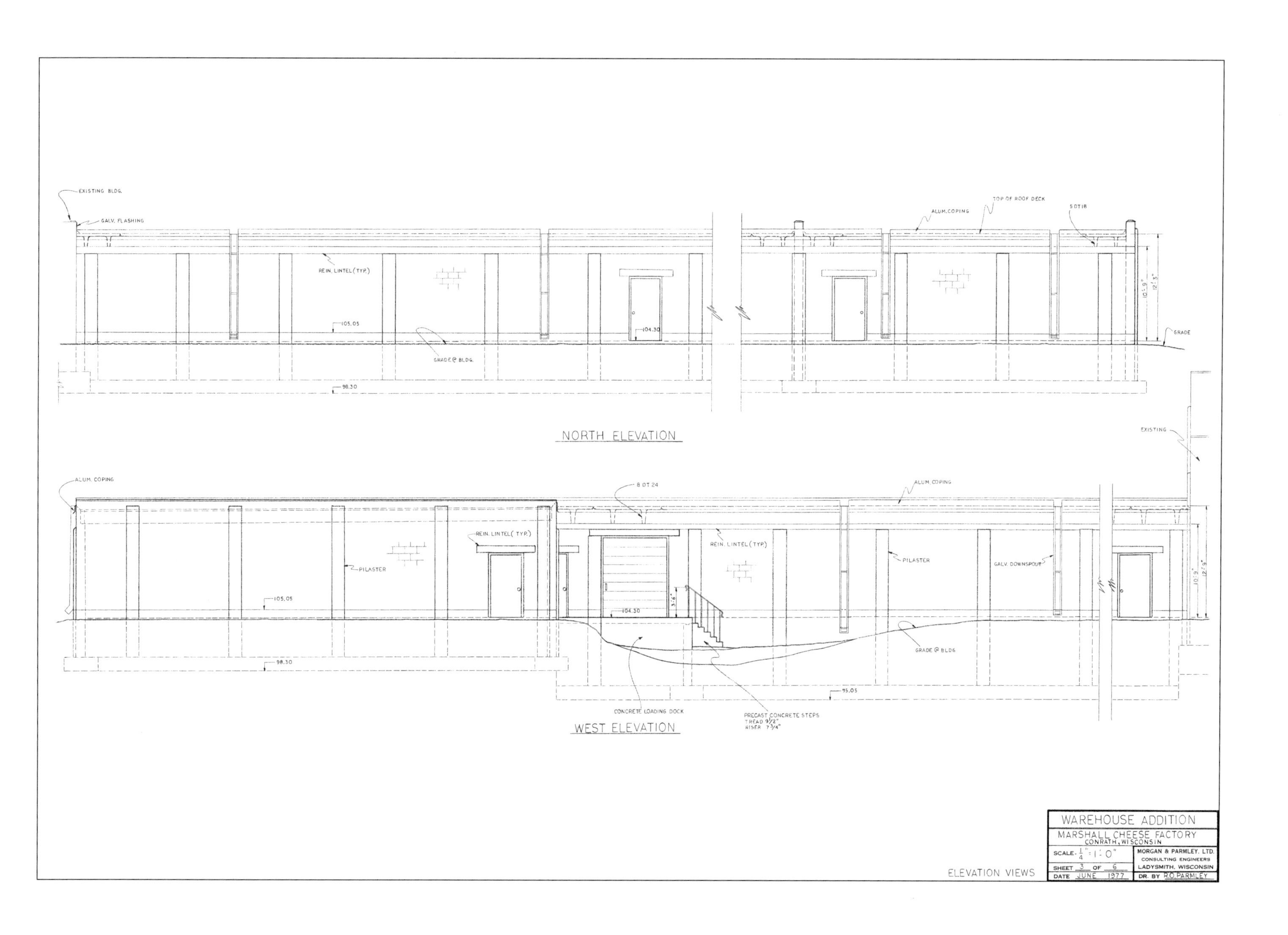
EXISTING BLDG.
GALV. FLASHING
REIN. LINTEL (TYP.)
105.05
104.30
GRADE @ BLDG.
98.30
ALUM. COPING
TOP OF ROOF DECK
GRADE
NORTH ELEVATION
EXISTING
ALUM. COPING
8 DT 24
REIN. LINTEL (TYP.)
PILASTER
105.05
104.30
GALV. DOWNSPOUT
GRADE @ BLDG.
98.30
95.05
CONCRETE LOADING DOCK
PRECAST CONCRETE STEPS
TREAD 9 1/2"
RISER 7 3/4"
WEST ELEVATION
ELEVATION VIEWS
WAREHOUSE ADDITION
MARSHALL CHEESE FACTORY
CONRATH, WISCONSIN
SCALE: 1/4" = 1'-0"
SHEET 3 OF 6
DATE JUNE 1977
MORGAN & PARMLEY, LTD.
CONSULTING ENGINEERS
LADYSMITH, WISCONSIN
DR. BY R.O. PARMLEY

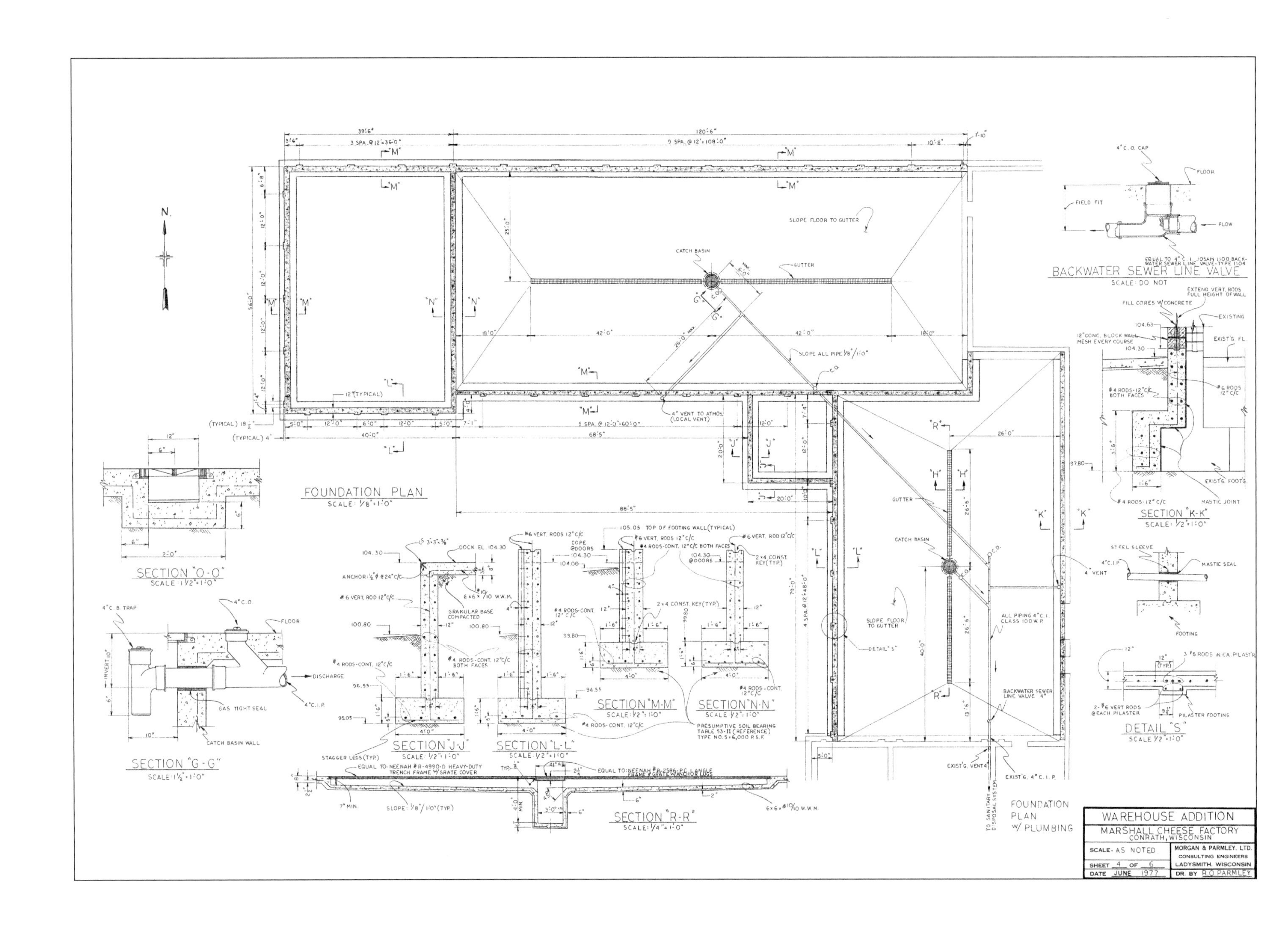
FOUNDATION PLAN
SCALE: 1/8"=1'-0"
SECTION "O-O"
SCALE 1 1/2"=1'-0"
SECTION "G-G"
SCALE 1 1/2"=1'-0"
SECTION "J-J"
SCALE 1/2"=1'-0"
SECTION "L-L"
SCALE 1/2"=1'-0"
SECTION "M-M"
SCALE 1/2"=1'-0"
SECTION "N-N"
SCALE 1/2"=1'-0"
SECTION "R-R"
SCALE 1/4"=1'-0"
BACKWATER SEWER LINE VALVE
SCALE: DO NOT
SECTION "K-K"
SCALE: 1/2"=1'-0"
DETAIL "S"
SCALE 1/2"=1'-0"
SLOPE FLOOR TO GUTTER
CATCH BASIN
GUTTER
SLOPE ALL PIPE 1/8"/1'-0"
4" VENT TO ATMOS. (LOCAL VENT)
ALL PIPING 4" C.I. CLASS 100 W.P.
BACKWATER SEWER LINE VALVE 4"
EXIST'G. VENT
EXIST'G. 4" C.I.P.
TO SANITARY DISPOSAL SYSTEM
PRESUMPTIVE SOIL BEARING TABLE 53-11 (REFERENCE) TYPE NO. 5 x 6,000 P.S.F.
FOUNDATION PLAN W/ PLUMBING
WAREHOUSE ADDITION
MARSHALL CHEESE FACTORY
CONRATH, WISCONSIN
SCALE- AS NOTED
SHEET 4 OF 6
DATE JUNE 1977
MORGAN & PARMLEY, LTD.
CONSULTING ENGINEERS
LADYSMITH, WISCONSIN
DR. BY R.O. PARMLEY

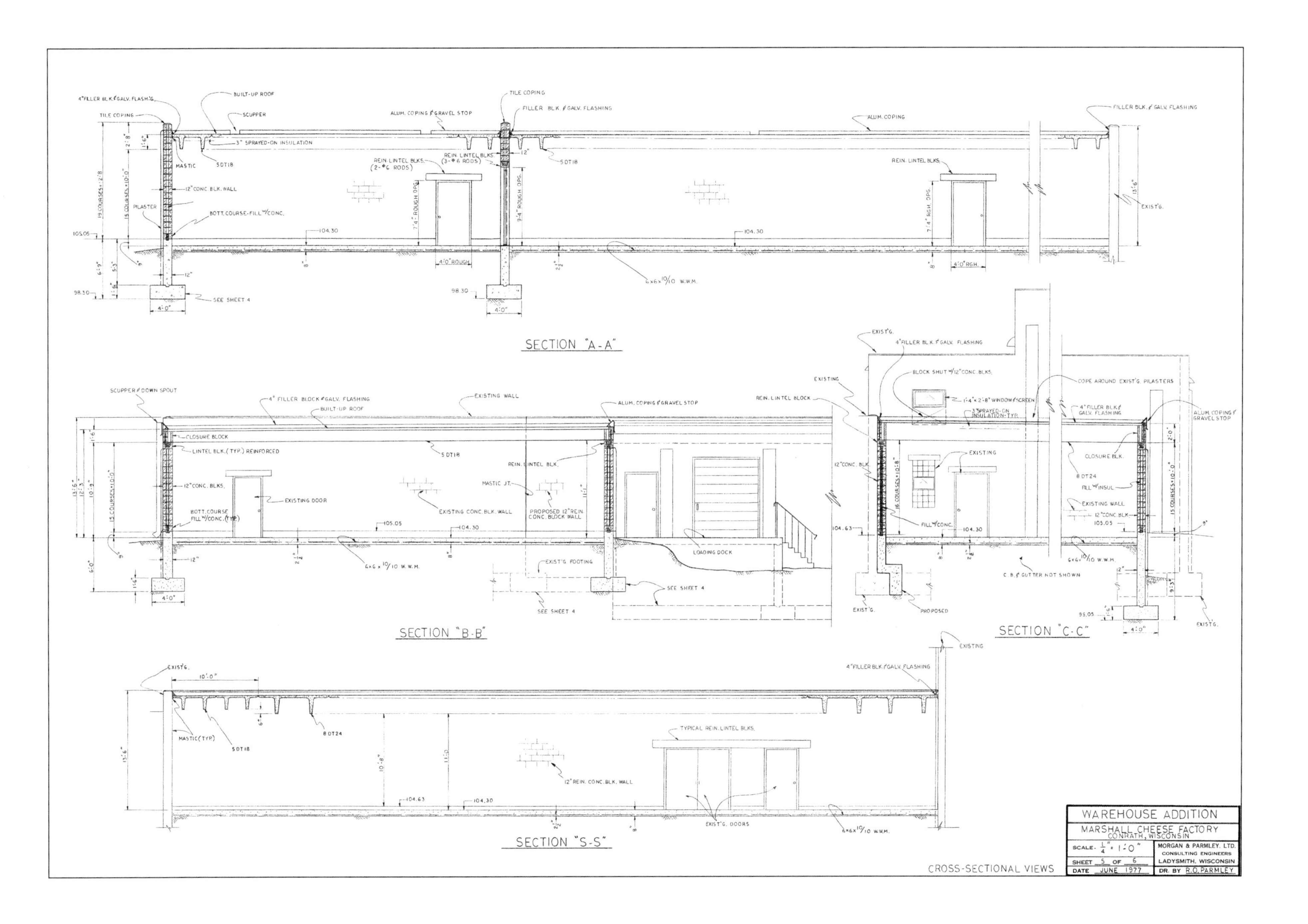
SECTION "A-A"
SECTION "B-B"
SECTION "C-C"
SECTION "S-S"
WAREHOUSE ADDITION
MARSHALL CHEESE FACTORY
CONRATH, WISCONSIN
SCALE - 1/4" = 1'-0"
SHEET 5 OF 6
DATE JUNE 1977
MORGAN & PARMLEY, LTD.
CONSULTING ENGINEERS
LADYSMITH, WISCONSIN
DR. BY R.O. PARMLEY
CROSS-SECTIONAL VIEWS

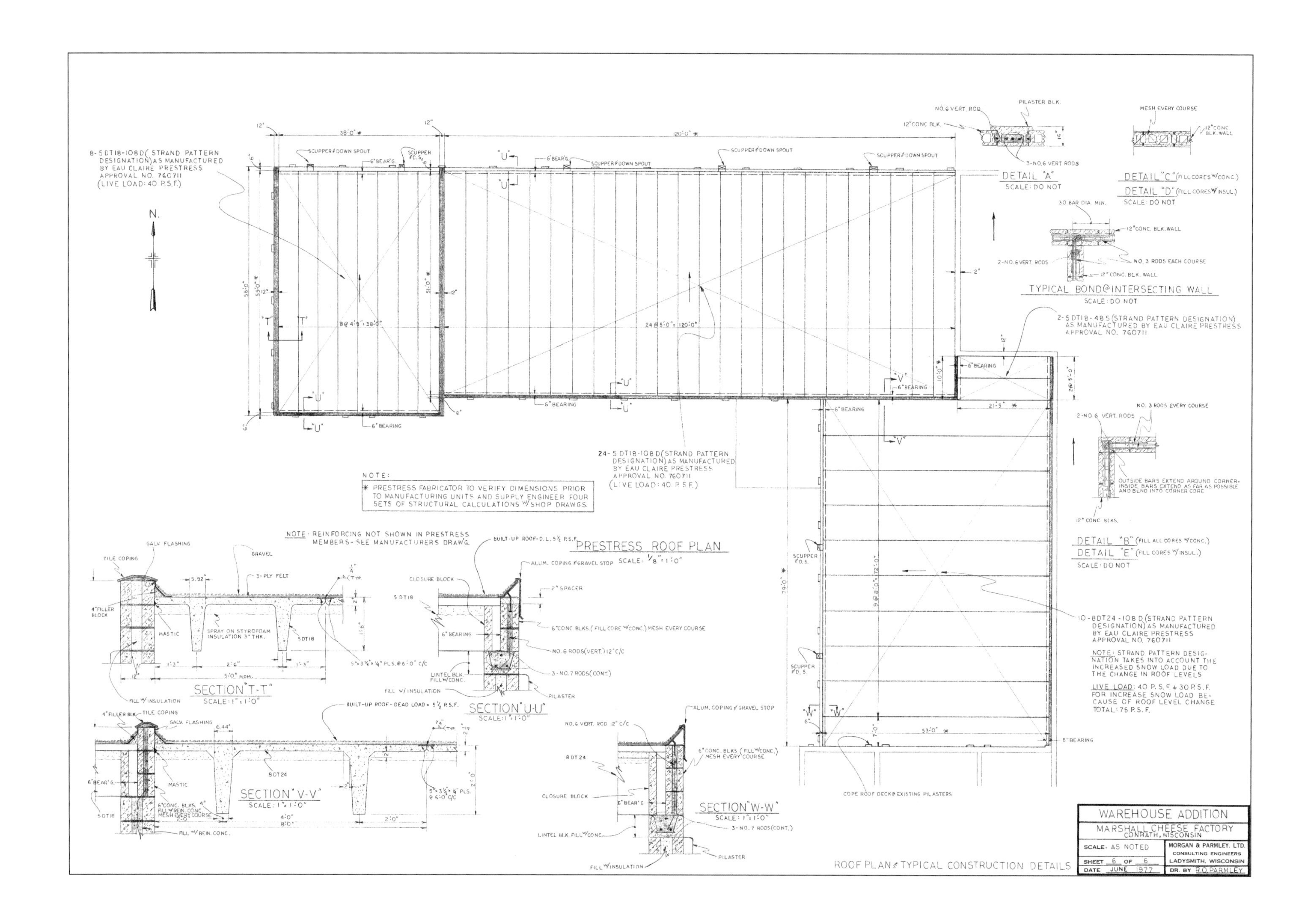
8-5DT18-108D (STRAND PATTERN DESIGNATION) AS MANUFACTURED BY EAU CLAIRE PRESTRESS APPROVAL NO. 760711 (LIVE LOAD: 40 P.S.F.)
N.
38'-0" *
120'-0" *
SCUPPER & DOWN SPOUT
6" BEAR'G.
8@4'-9"=38'-0"
24@5'-0"=120'-0"
6" BEARING
DETAIL "A"
SCALE: DO NOT
12" CONC. BLK.
PILASTER BLK.
3-NO. 6 VERT. RODS
MESH EVERY COURSE
DETAIL "C" (FILL CORES W/CONC.)
DETAIL "D" (FILL CORES W/INSUL.)
SCALE: DO NOT
12" CONC. BLK. WALL
2-NO. 6 VERT. RODS
NO. 3 RODS EACH COURSE
TYPICAL BOND @ INTERSECTING WALL
SCALE: DO NOT
2-5DT18-48S (STRAND PATTERN DESIGNATION) AS MANUFACTURED BY EAU CLAIRE PRESTRESS APPROVAL NO. 760711
21'-5" *
NO. 3 RODS EVERY COURSE
2-NO. 6 VERT. RODS
OUTSIDE BARS EXTEND AROUND CORNER-INSIDE BARS EXTEND AS FAR AS POSSIBLE AND BEND INTO CORNER CORE
12" CONC. BLKS.
DETAIL "B" (FILL ALL CORES W/CONC.)
DETAIL "E" (FILL CORES W/INSUL.)
SCALE: DO NOT
24-5DT18-108D (STRAND PATTERN DESIGNATION) AS MANUFACTURED BY EAU CLAIRE PRESTRESS APPROVAL NO. 760711 (LIVE LOAD: 40 P.S.F.)
NOTE:
* PRESTRESS FABRICATOR TO VERIFY DIMENSIONS PRIOR TO MANUFACTURING UNITS AND SUPPLY ENGINEER FOUR SETS OF STRUCTURAL CALCULATIONS W/SHOP DRAWGS.
NOTE: REINFORCING NOT SHOWN IN PRESTRESS MEMBERS- SEE MANUFACTURERS DRAW'G.
PRESTRESS ROOF PLAN
SCALE: 1/8"=1'-0"
10-8DT24-108D (STRAND PATTERN DESIGNATION) AS MANUFACTURED BY EAU CLAIRE PRESTRESS APPROVAL NO. 760711
NOTE: STRAND PATTERN DESIGNATION TAKES INTO ACCOUNT THE INCREASED SNOW LOAD DUE TO THE CHANGE IN ROOF LEVELS
LIVE LOAD: 40 P.S.F.+30 P.S.F. FOR INCREASE SNOW LOAD BECAUSE OF ROOF LEVEL CHANGE TOTAL: 75 P.S.F.
9@8'-0"=72'-0"
53'-0" *
COPE ROOF DECK @ EXISTING PILASTERS
GALV. FLASHING
TILE COPING
GRAVEL
3-PLY FELT
4" FILLER BLOCK
MASTIC
SPRAY ON STYROFOAM INSULATION 3" THK.
5DT18
5'-0" NOM.
FILL W/INSULATION
SECTION "T-T"
SCALE: 1"=1'-0"
BUILT-UP ROOF-D.L. 5 1/2 P.S.F.
ALUM. COPING & GRAVEL STOP
CLOSURE BLOCK
2" SPACER
6" BEARING
6" CONC. BLKS. (FILL CORE W/CONC.) MESH EVERY COURSE
NO. 6 RODS (VERT.) 12" C/C
LINTEL BLK. FILL W/CONC.
3-NO. 7 RODS (CONT.)
FILL W/INSULATION
PILASTER
SECTION "U-U"
SCALE: 1"=1'-0"
BUILT-UP ROOF-DEAD LOAD= 5 1/2 P.S.F.
8DT24
SECTION "V-V"
SCALE: 1"=1'-0"
FILL W/REIN. CONC.
NO. 6 VERT. ROD 12" C/C
6" CONC. BLKS. (FILL W/CONC.) MESH EVERY COURSE
CLOSURE BLOCK
LINTEL BLK. FILL W/CONC.
SECTION "W-W"
SCALE: 1"=1'-0"
3-NO. 7 RODS (CONT.)
PILASTER
FILL W/INSULATION
ROOF PLAN & TYPICAL CONSTRUCTION DETAILS
WAREHOUSE ADDITION
MARSHALL CHEESE FACTORY
CONRATH, WISCONSIN
SCALE- AS NOTED
SHEET 6 OF 6
DATE JUNE 1977
MORGAN & PARMLEY, LTD.
CONSULTING ENGINEERS
LADYSMITH, WISCONSIN
DR. BY R.O. PARMLEY

SOUTH SIDE ELEMENTARY SCHOOL ADDITION

LADYSMITH — HAWKINS SCHOOL DISTRICT

LADYSMITH, WISCONSIN

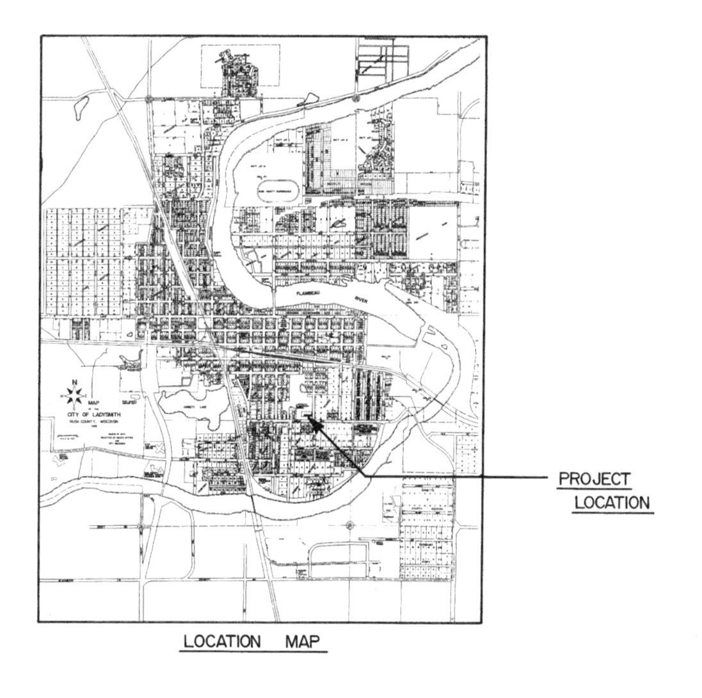

LOCATION MAP

INDEX OF SHEETS

SHEET NO.	*DESCRIPTION*
1	TITLE SHEET W/ MAP & INDEX.
2	SITE PLAN & EXISTING ELEVATIONS.
3	EXISTING & PROPOSED FLOOR PLANS, SECTIONS & CONSTRUCTION DETAILS.
4	SECTIONS & CONSTRUCTION DETAILS.
5	ELECTRICAL, HEATING & VENTILATING PLANS.

PREPARED BY: MORGAN & PARMLEY LTD. LADYSMITH, WISCONSIN

Courtesy: Morgan & Parmley, Ltd.

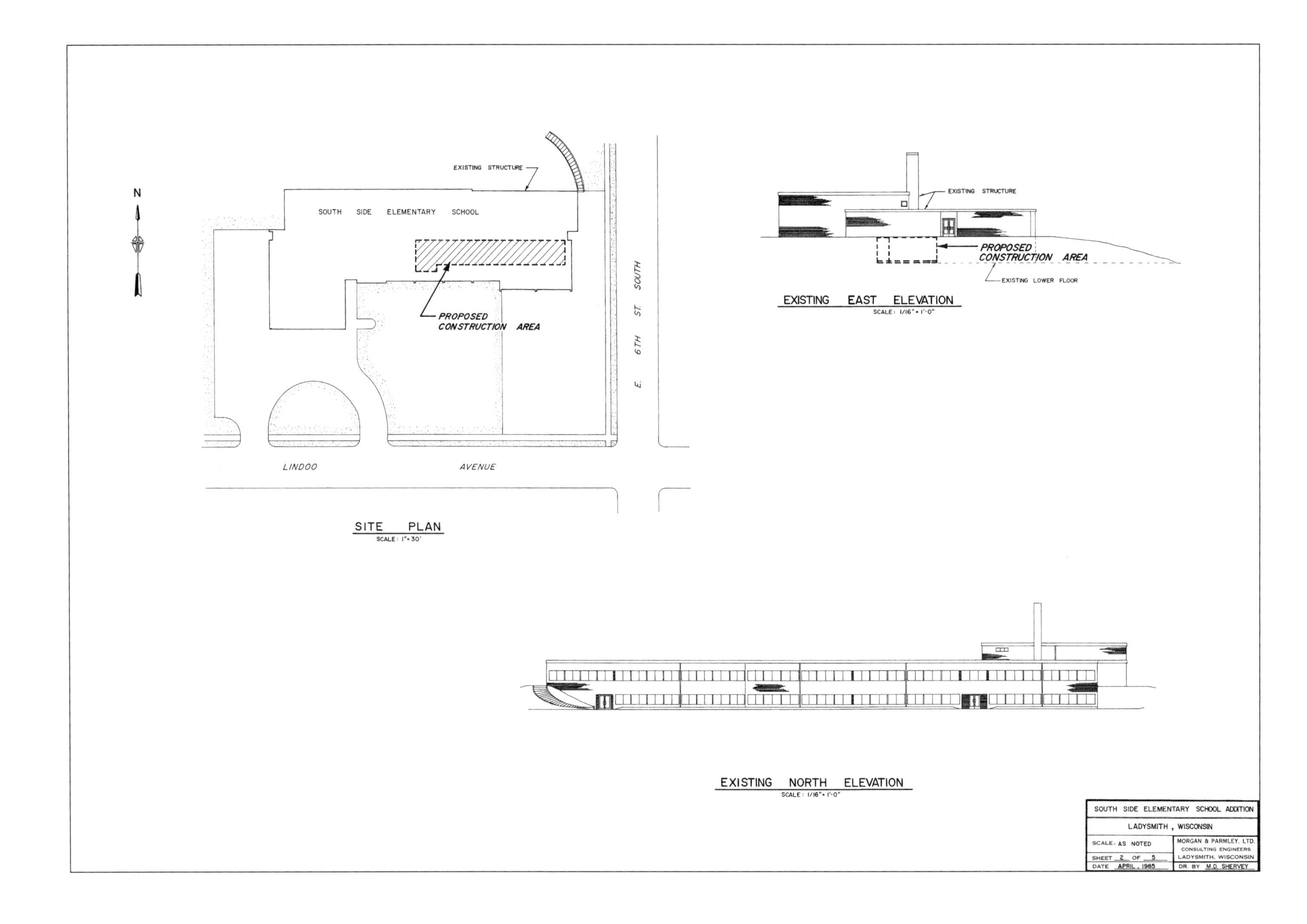
N
EXISTING STRUCTURE
SOUTH SIDE ELEMENTARY SCHOOL
PROPOSED CONSTRUCTION AREA
E. 6TH ST. SOUTH
LINDOO AVENUE
SITE PLAN
SCALE: 1" = 30'
EXISTING STRUCTURE
PROPOSED CONSTRUCTION AREA
EXISTING LOWER FLOOR
EXISTING EAST ELEVATION
SCALE: 1/16" = 1'-0"
EXISTING NORTH ELEVATION
SCALE: 1/16" = 1'-0"
SOUTH SIDE ELEMENTARY SCHOOL ADDITION
LADYSMITH, WISCONSIN
SCALE - AS NOTED
SHEET 2 OF 5
DATE APRIL, 1985
MORGAN & PARMLEY, LTD.
CONSULTING ENGINEERS
LADYSMITH, WISCONSIN
DR. BY M.D. SHERVEY

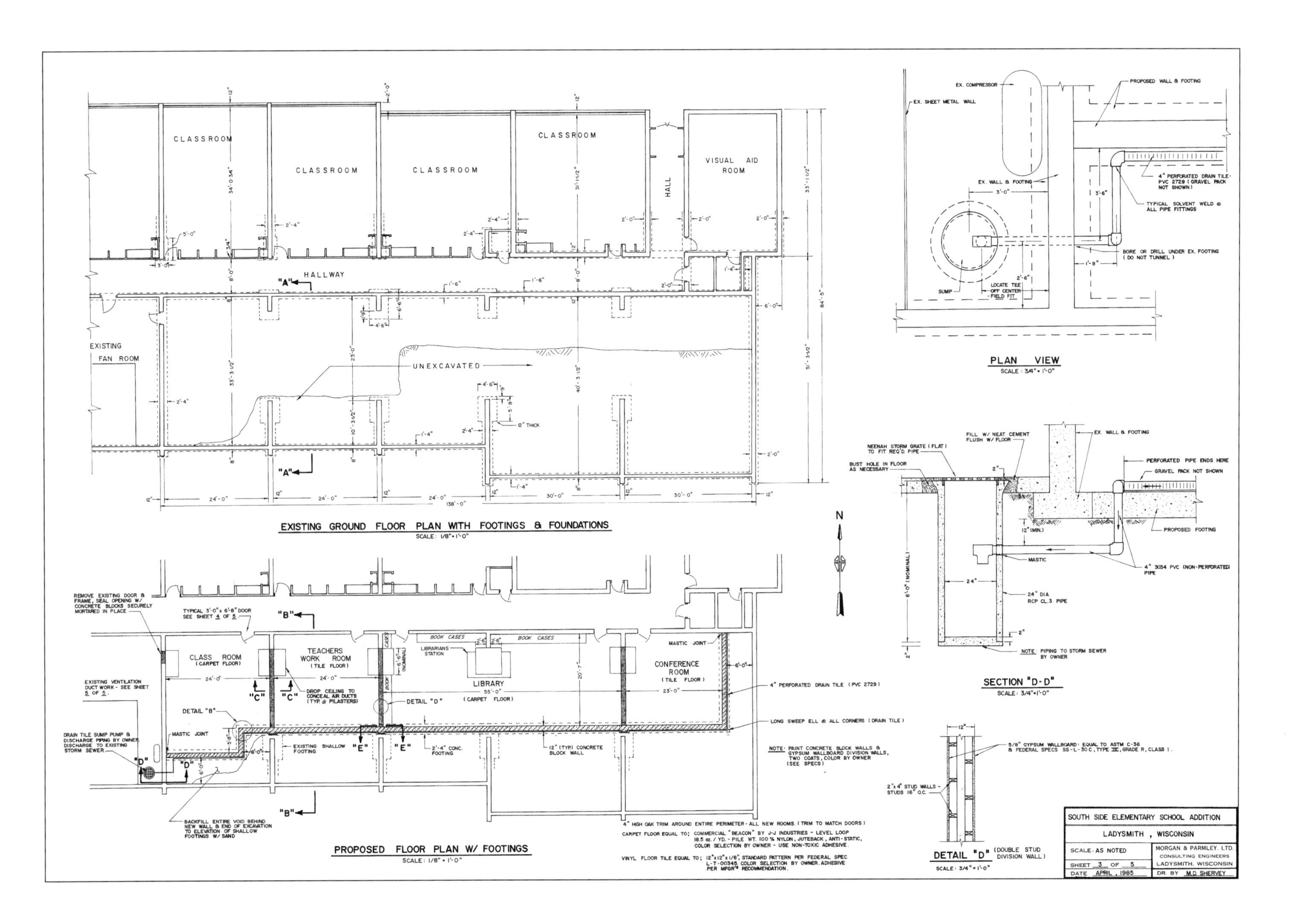
EXISTING GROUND FLOOR PLAN WITH FOOTINGS & FOUNDATIONS
SCALE: 1/8" = 1'-0"
PROPOSED FLOOR PLAN W/ FOOTINGS
SCALE: 1/8" = 1'-0"
PLAN VIEW
SCALE: 3/4" = 1'-0"
SECTION "D-D"
SCALE: 3/4" = 1'-0"
DETAIL "D" (DOUBLE STUD DIVISION WALL)
SCALE: 3/4" = 1'-0"
SOUTH SIDE ELEMENTARY SCHOOL ADDITION
LADYSMITH, WISCONSIN
MORGAN & PARMLEY, LTD. CONSULTING ENGINEERS LADYSMITH, WISCONSIN
SHEET 3 OF 5
DATE APRIL, 1985
DR. BY M.D. SHERVEY

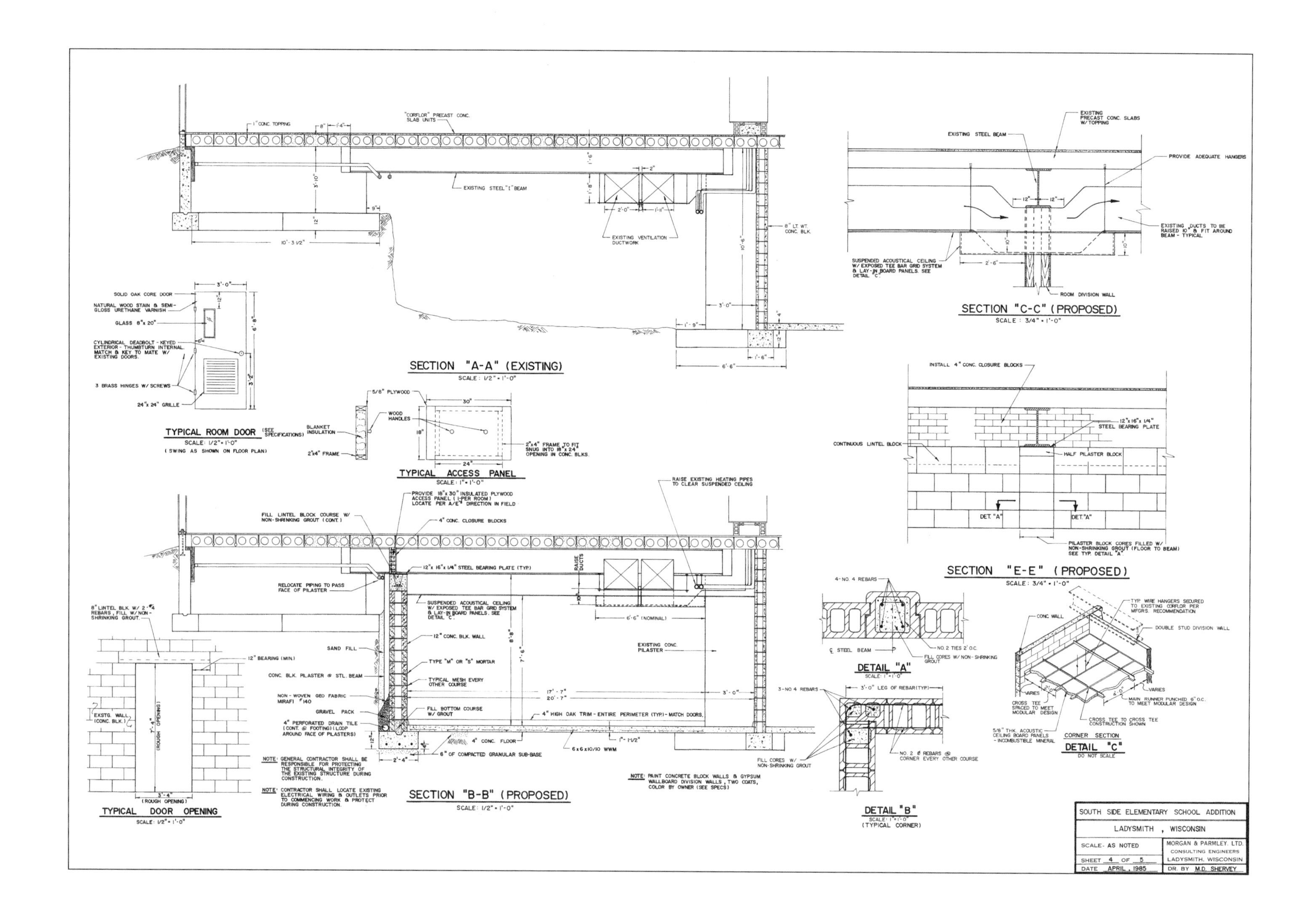
1" CONC. TOPPING
"CORFLOR" PRECAST CONC. SLAB UNITS
EXISTING STEEL "I" BEAM
EXISTING VENTILATION DUCTWORK
8" LT. WT. CONC. BLK.
10'-3 1/2"
SECTION "A-A" (EXISTING)
SCALE: 1/2" = 1'-0"
EXISTING STEEL BEAM
EXISTING PRECAST CONC. SLABS W/ TOPPING
PROVIDE ADEQUATE HANGERS
EXISTING DUCTS TO BE RAISED 10" & FIT AROUND BEAM - TYPICAL
SUSPENDED ACOUSTICAL CEILING W/ EXPOSED TEE BAR GRID SYSTEM & LAY-IN BOARD PANELS. SEE DETAIL "C".
ROOM DIVISION WALL
SECTION "C-C" (PROPOSED)
SCALE: 3/4" = 1'-0"
SOLID OAK CORE DOOR
NATURAL WOOD STAIN & SEMI-GLOSS URETHANE VARNISH
GLASS 8"x 20"
CYLINDRICAL DEADBOLT - KEYED EXTERIOR - THUMBTURN INTERNAL. MATCH & KEY TO MATE W/ EXISTING DOORS.
3 BRASS HINGES W/ SCREWS
24"x 24" GRILLE
TYPICAL ROOM DOOR (SEE SPECIFICATIONS)
SCALE: 1/2" = 1'-0"
(SWING AS SHOWN ON FLOOR PLAN)
5/8" PLYWOOD
WOOD HANDLES
BLANKET INSULATION
2"x4" FRAME
2"x4" FRAME TO FIT SNUG INTO 18"x 24" OPENING IN CONC. BLKS.
TYPICAL ACCESS PANEL
SCALE: 1" = 1'-0"
INSTALL 4" CONC. CLOSURE BLOCKS
12"x 16"x 1/4" STEEL BEARING PLATE
CONTINUOUS LINTEL BLOCK
HALF PILASTER BLOCK
DET. "A"
PILASTER BLOCK CORES FILLED W/ NON-SHRINKING GROUT (FLOOR TO BEAM) SEE TYP. DETAIL "A".
SECTION "E-E" (PROPOSED)
SCALE: 3/4" = 1'-0"
PROVIDE 18"x 30" INSULATED PLYWOOD ACCESS PANEL (1-PER ROOM) LOCATE PER A/E'S DIRECTION IN FIELD
RAISE EXISTING HEATING PIPES TO CLEAR SUSPENDED CEILING
FILL LINTEL BLOCK COURSE W/ NON-SHRINKING GROUT (CONT.)
4" CONC. CLOSURE BLOCKS
12"x 16"x 1/4" STEEL BEARING PLATE (TYP)
RELOCATE PIPING TO PASS FACE OF PILASTER
RAISE DUCTS
SUSPENDED ACOUSTICAL CEILING W/ EXPOSED TEE BAR GRID SYSTEM & LAY-IN BOARD PANELS. SEE DETAIL "C".
6'-6" (NOMINAL)
12" CONC. BLK. WALL
EXISTING CONC. PILASTER
SAND FILL
TYPE "M" OR "S" MORTAR
CONC. BLK. PILASTER @ STL. BEAM
TYPICAL MESH EVERY OTHER COURSE
NON-WOVEN GEO FABRIC MIRAFI #140
GRAVEL PACK
FILL BOTTOM COURSE W/ GROUT
4" PERFORATED DRAIN TILE (CONT. @ FOOTING) (LOOP AROUND FACE OF PILASTERS)
4" HIGH OAK TRIM - ENTIRE PERIMETER (TYP) - MATCH DOORS.
4" CONC. FLOOR
6" OF COMPACTED GRANULAR SUB-BASE
6x6x10/10 WWM
NOTE: GENERAL CONTRACTOR SHALL BE RESPONSIBLE FOR PROTECTING THE STRUCTURAL INTEGRITY OF THE EXISTING STRUCTURE DURING CONSTRUCTION.
NOTE: CONTRACTOR SHALL LOCATE EXISTING ELECTRICAL WIRING & OUTLETS PRIOR TO COMMENCING WORK & PROTECT DURING CONSTRUCTION.
SECTION "B-B" (PROPOSED)
SCALE: 1/2" = 1'-0"
NOTE: PAINT CONCRETE BLOCK WALLS & GYPSUM WALLBOARD DIVISION WALLS, TWO COATS, COLOR BY OWNER (SEE SPECS)
8" LINTEL BLK. W/ 2-#4 REBARS, FILL W/ NON-SHRINKING GROUT.
12" BEARING (MIN.)
EXSTG. WALL (CONC. BLK.)
(ROUGH OPENING)
TYPICAL DOOR OPENING
SCALE: 1/2" = 1'-0"
4-NO. 4 REBARS
NO. 2 TIES 2' O.C.
STEEL BEAM
FILL CORES W/ NON-SHRINKING GROUT
DETAIL "A"
SCALE: 1" = 1'-0"
3-NO. 4 REBARS
3'-0" LEG OF REBAR (TYP)
FILL CORES W/ NON-SHRINKING GROUT
NO. 2 @ REBARS @ CORNER EVERY OTHER COURSE
DETAIL "B"
SCALE: 1" = 1'-0"
(TYPICAL CORNER)
CONC. WALL
TYP. WIRE HANGERS SECURED TO EXISTING CORFLOR PER MFGRS. RECOMMENDATION
DOUBLE STUD DIVISION WALL
VARIES
CROSS TEE SPACED TO MEET MODULAR DESIGN
MAIN RUNNER PUNCHED 6" O.C. TO MEET MODULAR DESIGN
CROSS TEE TO CROSS TEE CONSTRUCTION SHOWN.
5/8" THK. ACOUSTIC CEILING BOARD PANELS - INCOMBUSTIBLE MINERAL
CORNER SECTION
DETAIL "C"
DO NOT SCALE
SOUTH SIDE ELEMENTARY SCHOOL ADDITION
LADYSMITH, WISCONSIN
SCALE - AS NOTED
SHEET 4 OF 5
DATE APRIL, 1985
MORGAN & PARMLEY LTD.
CONSULTING ENGINEERS
LADYSMITH, WISCONSIN
DR. BY M.D. SHERVEY

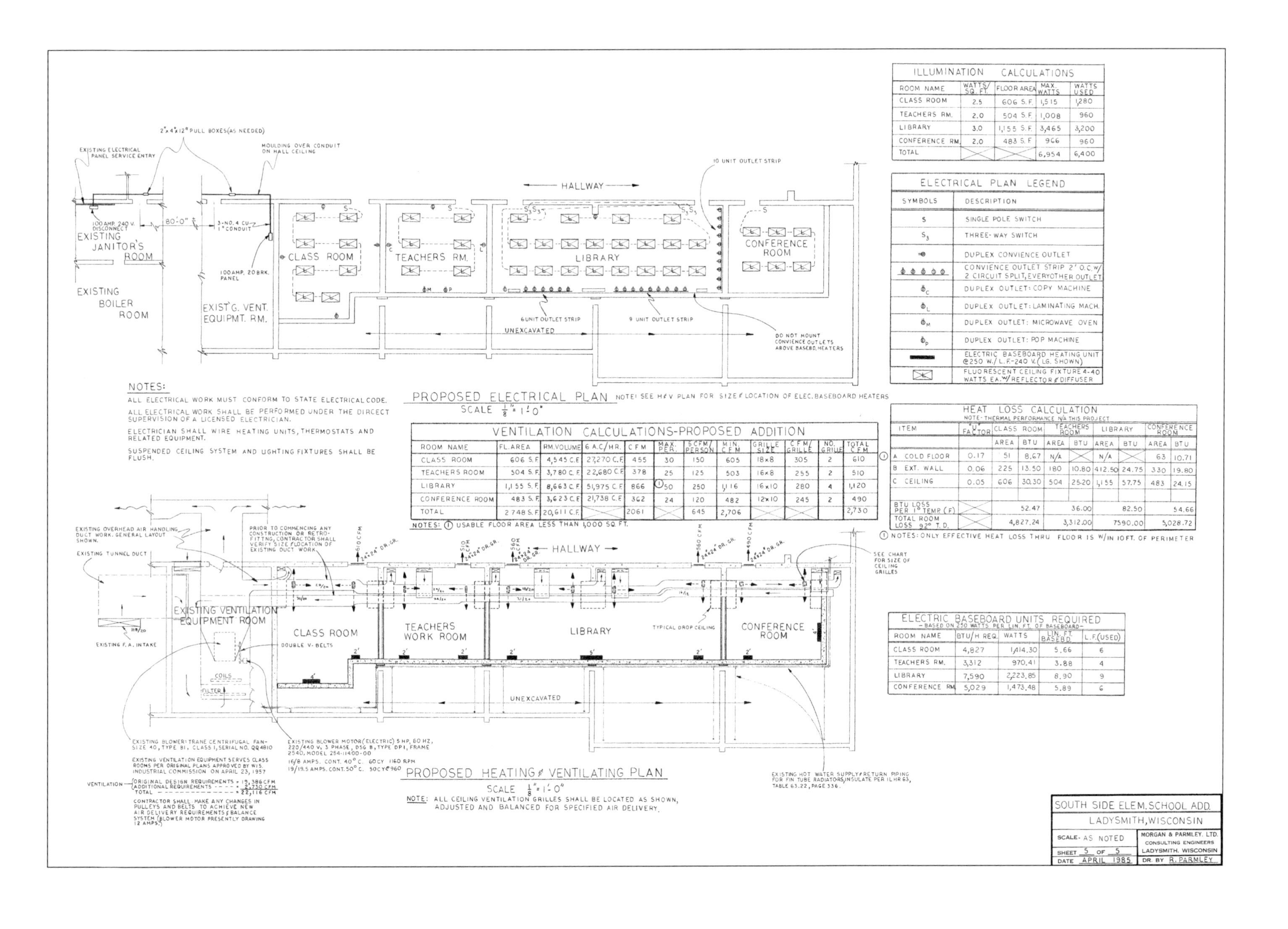

ILLUMINATION CALCULATIONS

ROOM NAME	WATTS/SQ.FT.	FLOOR AREA	MAX. WATTS	WATTS USED
CLASS ROOM	2.5	606 S.F.	1,515	1,280
TEACHERS RM.	2.0	504 S.F.	1,008	960
LIBRARY	3.0	1,155 S.F.	3,465	3,200
CONFERENCE RM.	2.0	483 S.F.	966	960
TOTAL			6,954	6,400

ELECTRICAL PLAN LEGEND

SYMBOLS	DESCRIPTION
S	SINGLE POLE SWITCH
S_3	THREE-WAY SWITCH
	DUPLEX CONVIENCE OUTLET
	CONVIENCE OUTLET STRIP 2' O.C. W/ 2 CIRCUIT SPLIT, EVERYOTHER OUTLET
ϕ_C	DUPLEX OUTLET: COPY MACHINE
ϕ_L	DUPLEX OUTLET: LAMINATING MACH.
ϕ_M	DUPLEX OUTLET: MICROWAVE OVEN
ϕ_P	DUPLEX OUTLET: POP MACHINE
	ELECTRIC BASEBOARD HEATING UNIT @250 W./L.F.-240 V. (LG. SHOWN)
	FLUORESCENT CEILING FIXTURE 4-40 WATTS EA. W/ REFLECTOR & DIFFUSER

VENTILATION CALCULATIONS-PROPOSED ADDITION

ROOM NAME	FL. AREA	RM. VOLUME	6 A.C/HR.	CFM	MAX. PER.	5 CFM/PERSON	MIN. CFM	GRILLE SIZE	CFM/GRILLE	NO. GRILLE	TOTAL CFM
CLASS ROOM	606 S.F.	4,545 C.F.	27,270 C.F.	455	30	150	605	18×8	305	2	610
TEACHERS ROOM	504 S.F.	3,780 C.F.	22,680 C.F.	378	25	125	503	16×8	255	2	510
LIBRARY	1,155 S.F.	8,663 C.F.	51,975 C.F.	866	(1) 50	250	1,116	16×10	280	4	1,120
CONFERENCE ROOM	483 S.F.	3,623 C.F.	21,738 C.F.	362	24	120	482	12×10	245	2	490
TOTAL	2748 S.F.	20,611 C.F.		2061		645	2,706				2,730

NOTES: (1) USABLE FLOOR AREA LESS THAN 1,000 SQ. FT.

HEAT LOSS CALCULATION
NOTE: THERMAL PERFORMANCE N/A THIS PROJECT

ITEM	"U" FACTOR	CLASS ROOM AREA	CLASS ROOM BTU	TEACHERS ROOM AREA	TEACHERS ROOM BTU	LIBRARY AREA	LIBRARY BTU	CONFERENCE ROOM AREA	CONFERENCE ROOM BTU
(1) A COLD FLOOR	0.17	51	8.67	N/A		N/A		63	10.71
B EXT. WALL	0.06	225	13.50	180	10.80	412.50	24.75	330	19.80
C CEILING	0.05	606	30.30	504	25.20	1,155	57.75	483	24.15
BTU LOSS PER 1° TEMP (F)			52.47		36.00		82.50		54.66
TOTAL ROOM LOSS 92° T.D.			4,827.24		3,312.00		7590.00		5,028.72

(1) NOTES: ONLY EFFECTIVE HEAT LOSS THRU FLOOR IS W/IN 10 FT. OF PERIMETER

ELECTRIC BASEBOARD UNITS REQUIRED
– BASED ON 250 WATTS PER LIN. FT. OF BASEBOARD –

ROOM NAME	BTU/H REQ.	WATTS	LIN. FT. BASEBD.	L.F. (USED)
CLASS ROOM	4,827	1,414.30	5.66	6
TEACHERS RM.	3,312	970.41	3.88	4
LIBRARY	7,590	2,223.85	8.90	9
CONFERENCE RM.	5,029	1,473.48	5.89	6

Design

WATER SUPPLY & DISTRIBUTION

Section 9

Design of municipal water supply and distribution systems is a major segment of the civil engineer's professional activity. A continuous ample supply of potable water is the life blood of the modern municipality. The source, storage and distribution of safe drinking water are regulated and monitored by applicable governmental agencies. All plans and specifications for these facilities must be officially approved by the controlling agencies before commencing construction.

TABLE OF CONTENTS

MUNICIPAL WELL NO. 4

VILLAGE OF
GILMAN, WISCONSIN

APRIL, 1987

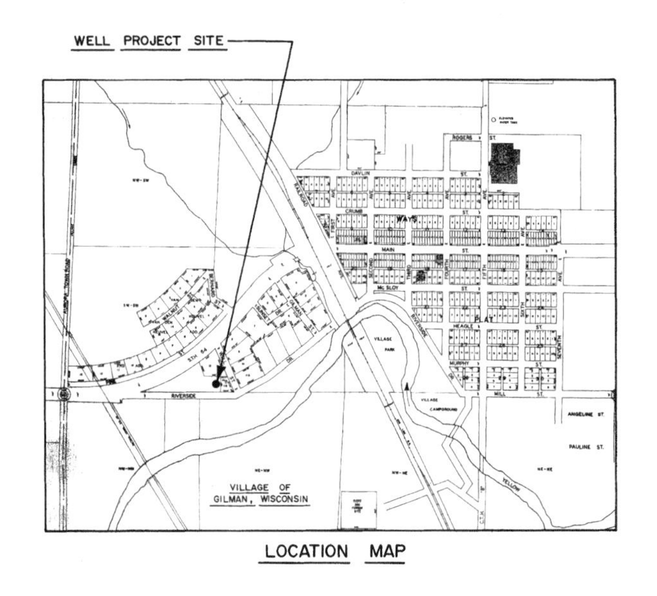

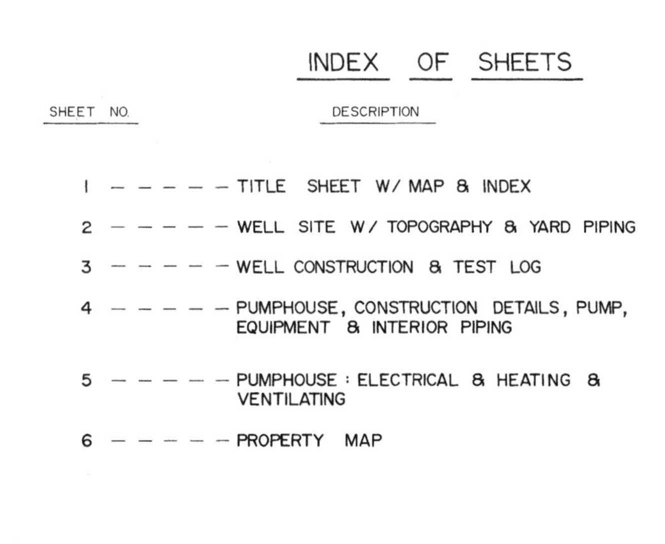

INDEX OF SHEETS

SHEET NO.	DESCRIPTION
1	TITLE SHEET W/ MAP & INDEX
2	WELL SITE W/ TOPOGRAPHY & YARD PIPING
3	WELL CONSTRUCTION & TEST LOG
4	PUMPHOUSE, CONSTRUCTION DETAILS, PUMP, EQUIPMENT & INTERIOR PIPING
5	PUMPHOUSE : ELECTRICAL & HEATING & VENTILATING
6	PROPERTY MAP

PREPARED BY: MORGAN & PARMLEY, LTD. LADYSMITH, WISCONSIN

Courtesy: Morgan & Parmley, Ltd.

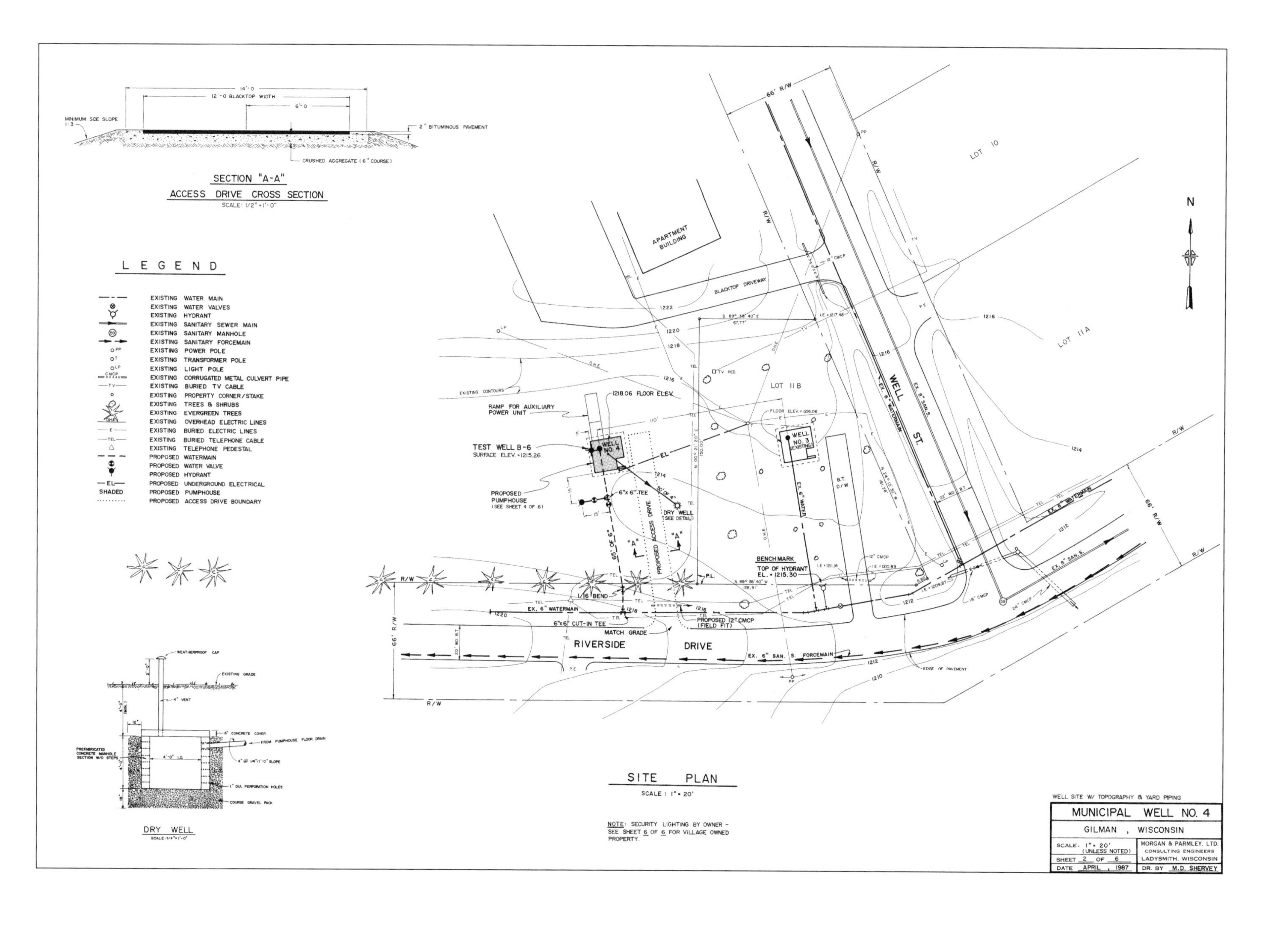
SECTION "A-A"
ACCESS DRIVE CROSS SECTION
SCALE: 1/2" = 1'-0"
14'-0
12'-0 BLACKTOP WIDTH
6'-0
MINIMUM SIDE SLOPE 1:3
2" BITUMINOUS PAVEMENT
CRUSHED AGGREGATE (6" COURSE)
LEGEND
EXISTING WATER MAIN
EXISTING WATER VALVES
EXISTING HYDRANT
EXISTING SANITARY SEWER MAIN
EXISTING SANITARY MANHOLE
EXISTING SANITARY FORCEMAIN
EXISTING POWER POLE
EXISTING TRANSFORMER POLE
EXISTING LIGHT POLE
EXISTING CORRUGATED METAL CULVERT PIPE
EXISTING BURIED TV CABLE
EXISTING PROPERTY CORNER/STAKE
EXISTING TREES & SHRUBS
EXISTING EVERGREEN TREES
EXISTING OVERHEAD ELECTRIC LINES
EXISTING BURIED ELECTRIC LINES
EXISTING BURIED TELEPHONE CABLE
EXISTING TELEPHONE PEDESTAL
PROPOSED WATERMAIN
PROPOSED WATER VALVE
PROPOSED HYDRANT
PROPOSED UNDERGROUND ELECTRICAL
SHADED
PROPOSED PUMPHOUSE
PROPOSED ACCESS DRIVE BOUNDARY
DRY WELL
SCALE: 3/4" = 1'-0"
WEATHERPROOF CAP
EXISTING GRADE
4" VENT
6" CONCRETE COVER
FROM PUMPHOUSE FLOOR DRAIN
PREFABRICATED CONCRETE MANHOLE SECTION W/O STEPS
4'-0" I.D.
1" DIA. PERFORATION HOLES
COURSE GRAVEL PACK
SITE PLAN
SCALE: 1" = 20'
NOTE: SECURITY LIGHTING BY OWNER - SEE SHEET 6 OF 6 FOR VILLAGE OWNED PROPERTY.
APARTMENT BUILDING
BLACKTOP DRIVEWAY
LOT 10
LOT 11A
LOT 11B
WELL ST.
EX. 6" WATERMAIN
EX. 8" SAN. S.
RIVERSIDE DRIVE
EX. 6" SAN. S. FORCEMAIN
EX. 6" WATERMAIN
6"x6" CUT-IN TEE
MATCH GRADE
PROPOSED 12" CMCP (FIELD FIT)
1/16 BEND
PROPOSED ACCESS DRIVE
DRY WELL (SEE DETAIL)
WELL NO. 4
WELL NO. 3 (EXISTING)
TEST WELL B-6
SURFACE ELEV. = 1215.26
1218.06 FLOOR ELEV.
RAMP FOR AUXILIARY POWER UNIT
PROPOSED PUMPHOUSE (SEE SHEET 4 OF 6)
6"x6" TEE
65' OF 6"
30' OF 4"
BENCH MARK
TOP OF HYDRANT
EL. = 1215.30
EXISTING CONTOURS
EDGE OF PAVEMENT
66' R/W
N
WELL SITE W/ TOPOGRAPHY & YARD PIPING
MUNICIPAL WELL NO. 4
GILMAN, WISCONSIN
SCALE - 1" = 20' (UNLESS NOTED)
SHEET 2 OF 6
DATE APRIL, 1987
MORGAN & PARMLEY, LTD.
CONSULTING ENGINEERS
LADYSMITH, WISCONSIN
DR. BY M.D. SHERVEY

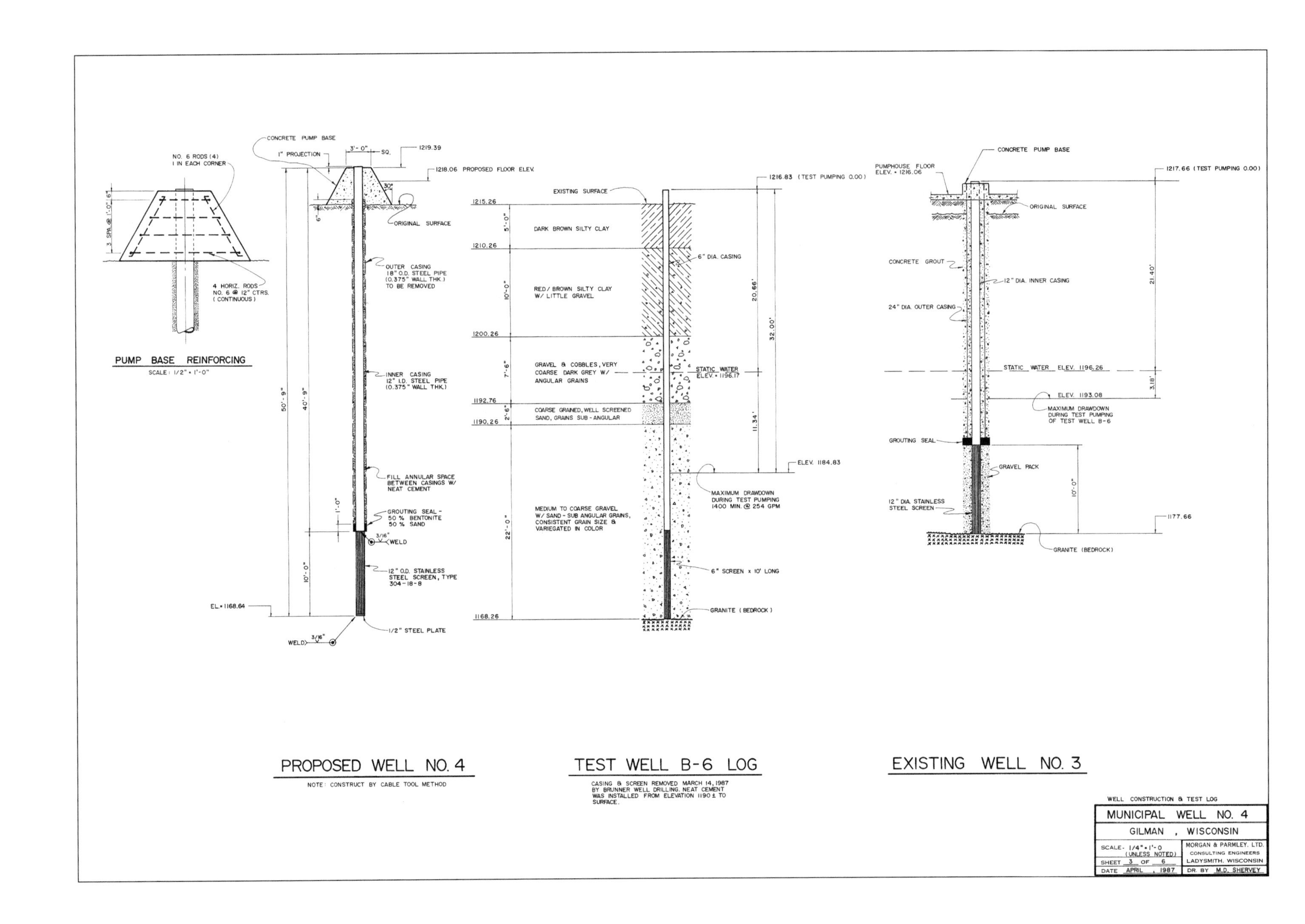

NO. 6 RODS (4)
1 IN EACH CORNER
3 SPA. @ 1'-0"
6"
4 HORIZ. RODS
NO. 6 @ 12" CTRS.
(CONTINUOUS)
PUMP BASE REINFORCING
SCALE: 1/2" = 1'-0"
CONCRETE PUMP BASE
1" PROJECTION
3'-0" SQ.
1219.39
1218.06 PROPOSED FLOOR ELEV.
30°
6"
ORIGINAL SURFACE
OUTER CASING
18" O.D. STEEL PIPE
(0.375" WALL THK.)
TO BE REMOVED
INNER CASING
12" I.D. STEEL PIPE
(0.375" WALL THK.)
50'-9"
40'-9"
FILL ANNULAR SPACE
BETWEEN CASINGS W/
NEAT CEMENT
1'-0"
GROUTING SEAL -
50 % BENTONITE
50 % SAND
3/16"
WELD
10'-0"
12" O.D. STAINLESS
STEEL SCREEN, TYPE
304-18-8
EL.= 1168.64
1/2" STEEL PLATE
WELD 3/16"
PROPOSED WELL NO. 4
NOTE: CONSTRUCT BY CABLE TOOL METHOD
EXISTING SURFACE
1216.83 (TEST PUMPING 0.00)
1215.26
5'-0"
DARK BROWN SILTY CLAY
1210.26
6" DIA. CASING
10'-0"
RED/BROWN SILTY CLAY
W/ LITTLE GRAVEL
20.66'
32.00'
1200.26
7'-6"
GRAVEL & COBBLES, VERY
COARSE DARK GREY W/
ANGULAR GRAINS
STATIC WATER
ELEV. = 1196.17
1192.76
2'-6"
COARSE GRAINED, WELL SCREENED
SAND, GRAINS SUB-ANGULAR
1190.26
11.34'
ELEV. 1184.83
MAXIMUM DRAWDOWN
DURING TEST PUMPING
1400 MIN. @ 254 GPM
22'-0"
MEDIUM TO COARSE GRAVEL
W/ SAND - SUB ANGULAR GRAINS,
CONSISTENT GRAIN SIZE &
VARIEGATED IN COLOR
6" SCREEN x 10' LONG
1168.26
GRANITE (BEDROCK)
TEST WELL B-6 LOG
CASING & SCREEN REMOVED MARCH 14, 1987
BY BRUNNER WELL DRILLING. NEAT CEMENT
WAS INSTALLED FROM ELEVATION 1190 ± TO
SURFACE.
CONCRETE PUMP BASE
PUMPHOUSE FLOOR
ELEV. = 1216.06
1217.66 (TEST PUMPING 0.00)
ORIGINAL SURFACE
CONCRETE GROUT
12" DIA. INNER CASING
24" DIA. OUTER CASING
21.40'
STATIC WATER ELEV. 1196.26
3.18'
ELEV. 1193.08
MAXIMUM DRAWDOWN
DURING TEST PUMPING
OF TEST WELL B-6
GROUTING SEAL
GRAVEL PACK
10'-0"
12" DIA. STAINLESS
STEEL SCREEN
1177.66
GRANITE (BEDROCK)
EXISTING WELL NO. 3
WELL CONSTRUCTION & TEST LOG
MUNICIPAL WELL NO. 4
GILMAN, WISCONSIN
SCALE- 1/4" = 1'-0
(UNLESS NOTED)
SHEET 3 OF 6
DATE APRIL, 1987
MORGAN & PARMLEY, LTD.
CONSULTING ENGINEERS
LADYSMITH, WISCONSIN
DR. BY M.D. SHERVEY

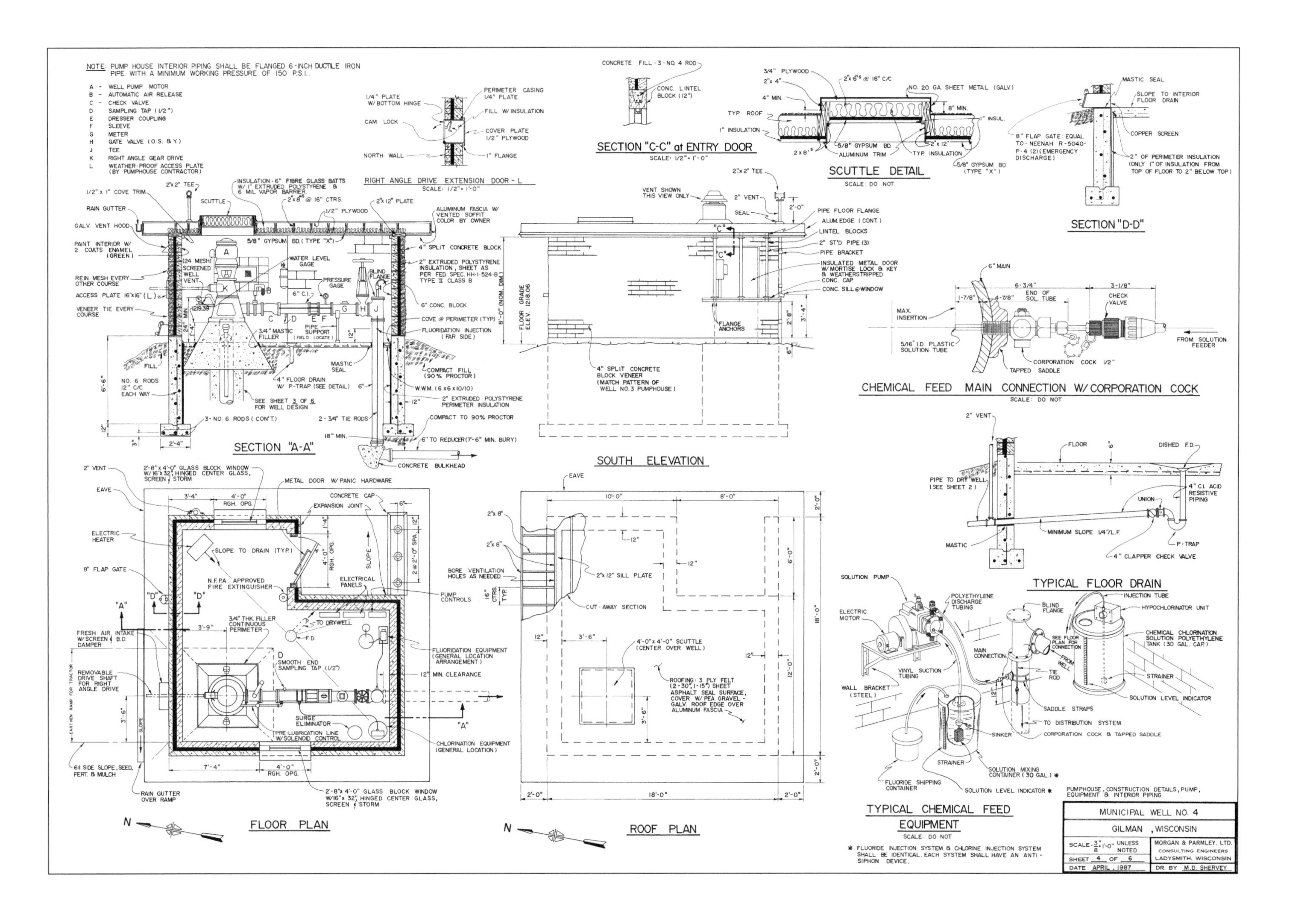
SECTION "A-A"
FLOOR PLAN
ROOF PLAN
SOUTH ELEVATION
SECTION "C-C" at ENTRY DOOR
SCUTTLE DETAIL
SECTION "D-D"
RIGHT ANGLE DRIVE EXTENSION DOOR - L
CHEMICAL FEED MAIN CONNECTION W/ CORPORATION COCK
TYPICAL FLOOR DRAIN
TYPICAL CHEMICAL FEED EQUIPMENT
PUMPHOUSE, CONSTRUCTION DETAILS, PUMP, EQUIPMENT & INTERIOR PIPING
MUNICIPAL WELL NO. 4
GILMAN, WISCONSIN
MORGAN & PARMLEY, LTD. CONSULTING ENGINEERS LADYSMITH, WISCONSIN
SHEET 4 OF 6
DATE APRIL, 1987
DR. BY M.D. SHERVEY

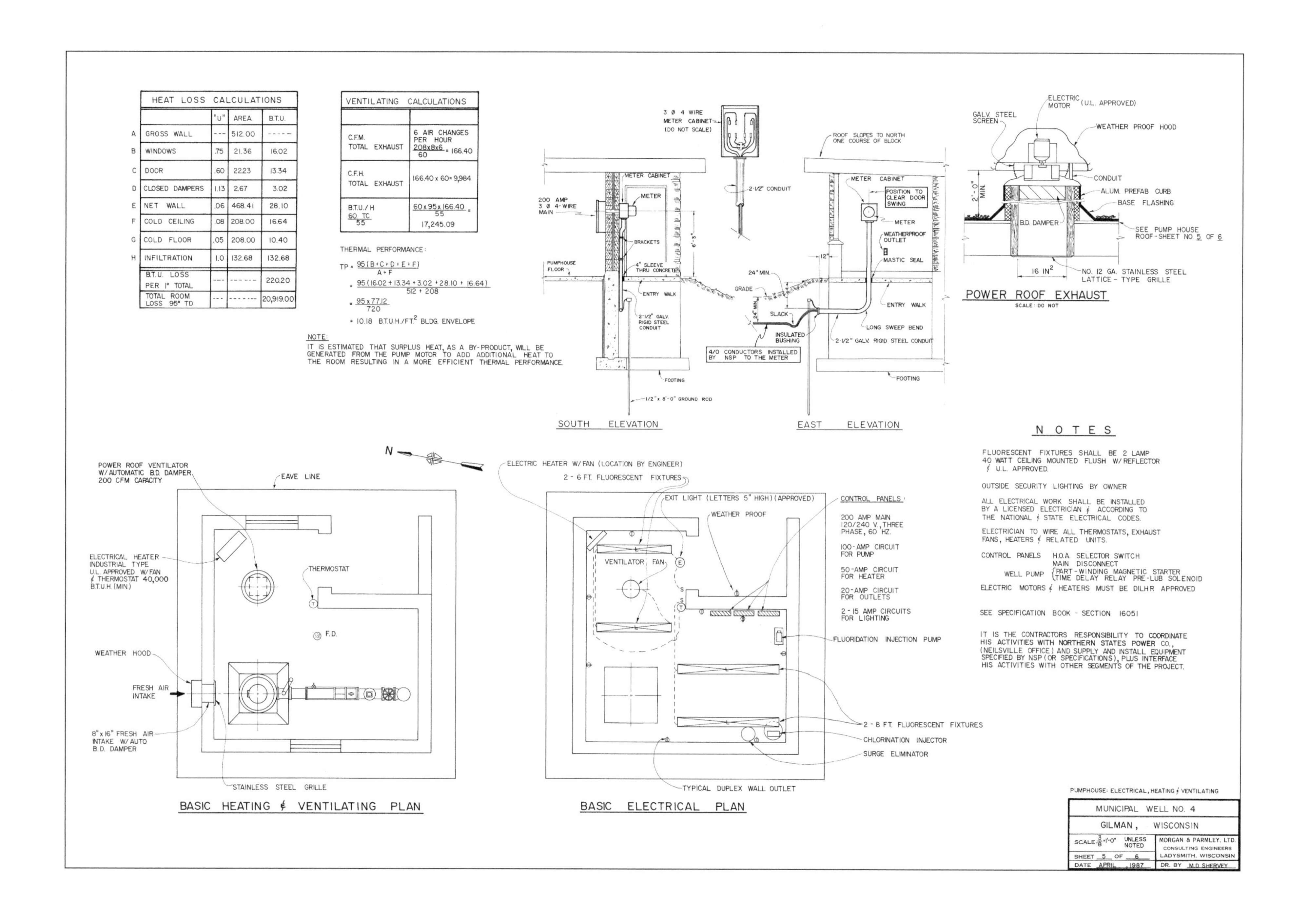
HEAT LOSS CALCULATIONS
"U" AREA B.T.U.
A GROSS WALL --- 512.00 -----
B WINDOWS .75 21.36 16.02
C DOOR .60 22.23 13.34
D CLOSED DAMPERS 1.13 2.67 3.02
E NET WALL .06 468.41 28.10
F COLD CEILING .08 208.00 16.64
G COLD FLOOR .05 208.00 10.40
H INFILTRATION 1.0 132.68 132.68
B.T.U. LOSS PER 1° TOTAL 220.20
TOTAL ROOM LOSS 95° TD 20,919.00
VENTILATING CALCULATIONS
C.F.M. TOTAL EXHAUST
6 AIR CHANGES PER HOUR 208x8x6 / 60 = 166.40
C.F.H. TOTAL EXHAUST
166.40 x 60 = 9,984
B.T.U./H 60 TC / 55
60 x 95 x 166.40 / 55 = 17,245.09
THERMAL PERFORMANCE:
TP = 95 (B+C+D+E+F) / A+F
= 95 (16.02 + 13.34 + 3.02 + 28.10 + 16.64) / 512 + 208
= 95 x 77.12 / 720
= 10.18 B.T.U.H./FT.² BLDG. ENVELOPE
NOTE:
IT IS ESTIMATED THAT SURPLUS HEAT, AS A BY-PRODUCT, WILL BE GENERATED FROM THE PUMP MOTOR TO ADD ADDITIONAL HEAT TO THE ROOM RESULTING IN A MORE EFFICIENT THERMAL PERFORMANCE.
3 Ø 4 WIRE METER CABINET (DO NOT SCALE)
2-1/2" CONDUIT
200 AMP 3 Ø 4-WIRE MAIN
METER CABINET
METER
BRACKETS
6'-3"
PUMPHOUSE FLOOR
4" SLEEVE THRU CONCRETE
ENTRY WALK
2-1/2" GALV. RIGID STEEL CONDUIT
FOOTING
1/2" x 8'-0" GROUND ROD
SOUTH ELEVATION
ROOF SLOPES TO NORTH ONE COURSE OF BLOCK
METER CABINET
POSITION TO CLEAR DOOR SWING
METER
WEATHERPROOF OUTLET
MASTIC SEAL
12"
24" MIN.
GRADE
24" MIN.
SLACK
INSULATED BUSHING
4/0 CONDUCTORS INSTALLED BY NSP TO THE METER
ENTRY WALK
LONG SWEEP BEND
2-1/2" GALV. RIGID STEEL CONDUIT
FOOTING
EAST ELEVATION
ELECTRIC MOTOR (U.L. APPROVED)
GALV. STEEL SCREEN
WEATHER PROOF HOOD
CONDUIT
ALUM. PREFAB CURB
BASE FLASHING
2'-0" MIN.
B.D. DAMPER
SEE PUMP HOUSE ROOF-SHEET NO. 5 OF 6
16 IN²
NO. 12 GA. STAINLESS STEEL LATTICE - TYPE GRILLE
POWER ROOF EXHAUST
SCALE: DO NOT
N O T E S
FLUORESCENT FIXTURES SHALL BE 2 LAMP 40 WATT CEILING MOUNTED FLUSH W/ REFLECTOR & U.L. APPROVED.
OUTSIDE SECURITY LIGHTING BY OWNER
ALL ELECTRICAL WORK SHALL BE INSTALLED BY A LICENSED ELECTRICIAN & ACCORDING TO THE NATIONAL & STATE ELECTRICAL CODES.
ELECTRICIAN TO WIRE ALL THERMOSTATS, EXHAUST FANS, HEATERS & RELATED UNITS.
CONTROL PANELS H.O.A. SELECTOR SWITCH MAIN DISCONNECT
WELL PUMP PART-WINDING MAGNETIC STARTER TIME DELAY RELAY PRE-LUB SOLENOID
ELECTRIC MOTORS & HEATERS MUST BE DILHR APPROVED
SEE SPECIFICATION BOOK - SECTION 16051
IT IS THE CONTRACTORS RESPONSIBILITY TO COORDINATE HIS ACTIVITIES WITH NORTHERN STATES POWER CO., (NEILSVILLE OFFICE) AND SUPPLY AND INSTALL EQUIPMENT SPECIFIED BY NSP (OR SPECIFICATIONS), PLUS INTERFACE HIS ACTIVITIES WITH OTHER SEGMENTS OF THE PROJECT.
N
POWER ROOF VENTILATOR W/ AUTOMATIC B.D. DAMPER, 200 CFM CAPACITY
EAVE LINE
ELECTRICAL HEATER INDUSTRIAL TYPE U.L. APPROVED W/ FAN & THERMOSTAT 40,000 B.T.U.H. (MIN.)
THERMOSTAT
F.D.
WEATHER HOOD
FRESH AIR INTAKE
8" x 16" FRESH AIR INTAKE W/ AUTO B.D. DAMPER
STAINLESS STEEL GRILLE
BASIC HEATING & VENTILATING PLAN
ELECTRIC HEATER W/ FAN (LOCATION BY ENGINEER)
2 - 6 FT. FLUORESCENT FIXTURES
EXIT LIGHT (LETTERS 5" HIGH) (APPROVED)
WEATHER PROOF
VENTILATOR FAN
CONTROL PANELS:
200 AMP MAIN 120/240 V., THREE PHASE, 60 HZ.
100-AMP CIRCUIT FOR PUMP
50-AMP CIRCUIT FOR HEATER
20-AMP CIRCUIT FOR OUTLETS
2 - 15 AMP CIRCUITS FOR LIGHTING
FLUORIDATION INJECTION PUMP
2 - 8 FT. FLUORESCENT FIXTURES
CHLORINATION INJECTOR
SURGE ELIMINATOR
TYPICAL DUPLEX WALL OUTLET
BASIC ELECTRICAL PLAN
PUMPHOUSE: ELECTRICAL, HEATING & VENTILATING
MUNICIPAL WELL NO. 4
GILMAN, WISCONSIN
SCALE: 3/8"=1'-0" UNLESS NOTED
MORGAN & PARMLEY, LTD. CONSULTING ENGINEERS LADYSMITH, WISCONSIN
SHEET 5 OF 6
DATE APRIL, 1987
DR. BY M.D. SHERVEY

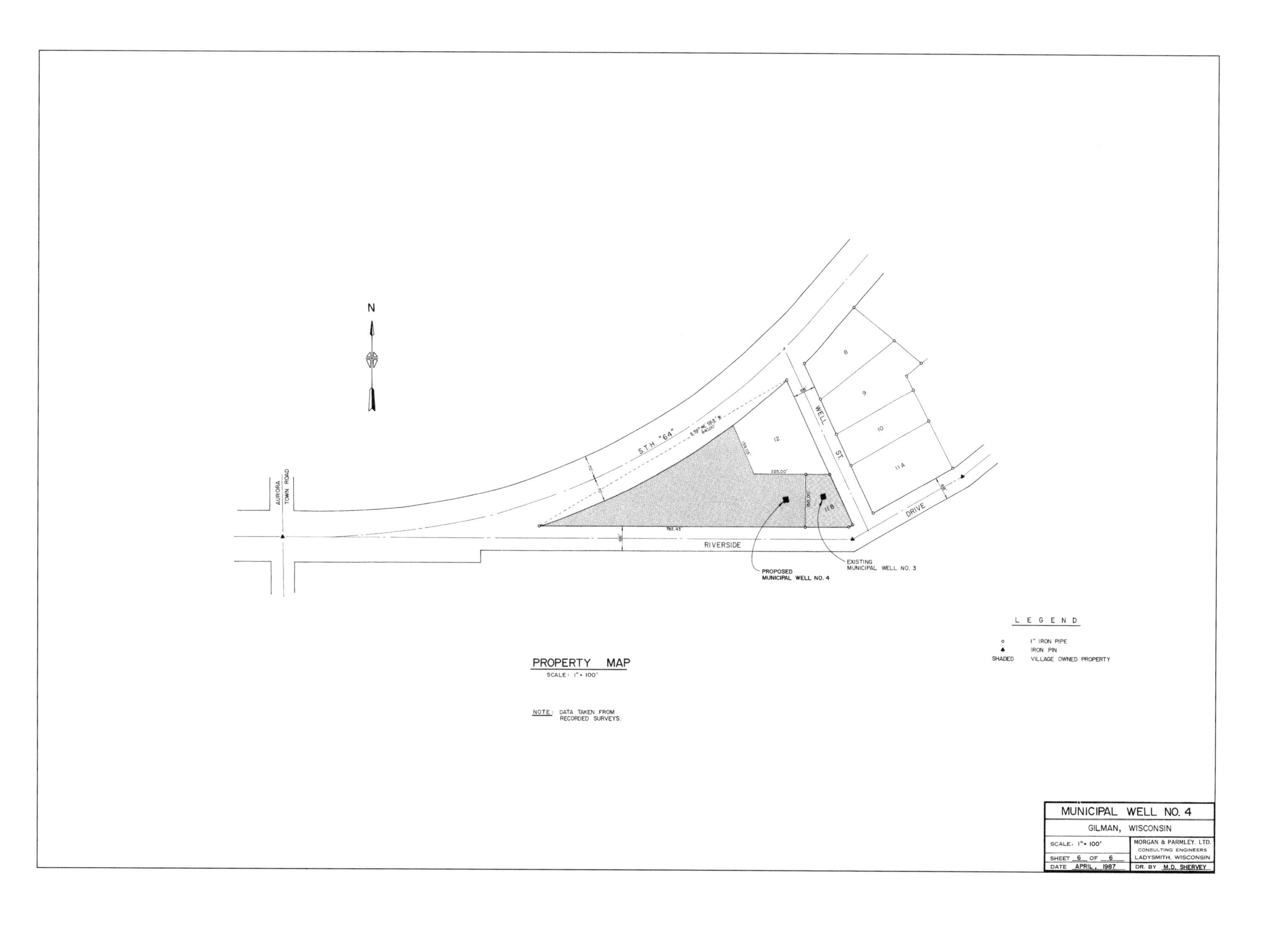
N
AURORA TOWN ROAD
STH "64"
WELL ST.
RIVERSIDE
DRIVE
12
8
9
10
11 A
11 B
225.00'
160.00'
782.45'
PROPOSED MUNICIPAL WELL NO. 4
EXISTING MUNICIPAL WELL NO. 3
PROPERTY MAP
SCALE: 1" = 100'
NOTE: DATA TAKEN FROM RECORDED SURVEYS.
LEGEND
1" IRON PIPE
IRON PIN
SHADED VILLAGE OWNED PROPERTY
MUNICIPAL WELL NO. 4
GILMAN, WISCONSIN
SCALE: 1" = 100'
SHEET 6 OF 6
DATE APRIL, 1987
MORGAN & PARMLEY, LTD.
CONSULTING ENGINEERS
LADYSMITH, WISCONSIN
DR. BY M.D. SHERVEY

WATER SUPPLY, PUMPHOUSE & TRANSMISSION MAIN

FOR

INDUSTRIAL PARK COMPLEX

E.D.A. PROJECT NO. 06-01-02571

VILLAGE OF GLEN FLORA, WISCONSIN

NOV. , 1992

UTILITIES

CONTACT - NORTHERN STATES POWER CO.
711 W. 9th Street, North
Ladysmith, WI. 54848

Attn: John Rymarkiewicz
715-532-6226

CONTACT _ UNIVERSAL TELEPHONE CO.
OF NORTHERN WISCONSIN
Highway "8"
Hawkins, WI 54530

Attn: Jim Arquette
715-585-7707

CONTACT - DIGGERS HOTLINE

1-800 242-8511

CONTACT - VILLAGE OF GLEN FLORA WWCTF

Les Evjen, Operator
P.O. Box 253
Glen Flora, WI. 54526

715-322-5511

CONTACT - GLEN FLORA ELEMENTARY SCHOOL
SCHOOL WELL & PIPELINE
Larry Johnson, Custodian

715-322-5271

VILLAGE OF GLEN FLORA, RUSK CO, WISCONSIN

GENERAL LOCATION MAP

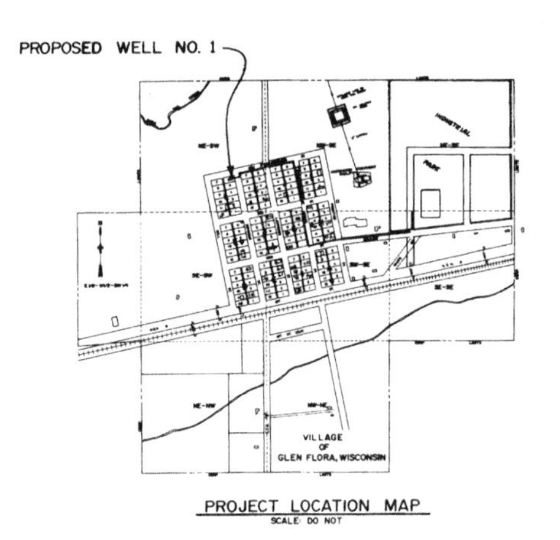

PROJECT LOCATION MAP
SCALE: DO NOT

INDEX OF SHEETS

NO.	DESCRIPTION
1	TITLE SHEET, GENERAL LOCATION & INDEX
2	WELL SITE W/ TOPOGRAPHY & YARD PIPING
3	WELL CONSTRUCTION & TEST LOG
4	PUMPHOUSE, CONSTRUCTION DETAILS, PUMP, EQUIPMENT & INTERIOR PIPING
5	PUMPHOUSE: ELECTRICAL, HEATING & VENTILATING
6	TRANSMISSION MAIN
7	TYPICAL EROSION CONTROL DETAILS — This Sheet Not Shown

THESE PLANS REPRESENT ONLY 1 SECTION OF THE TOTAL PROJECT

THE TOTAL PROJECT EDA NO. 06-01-02571 IS PARTIALLY FUNDED BY A 60% GRANT FROM THE UNITED STATES ECONOMIC DEVELOPMENT ADMINISTRATION IN THE TOTAL GRANT AMOUNT OF $590,520

PREPARED BY:
MORGAN & PARMLEY LTD.
CONSULTING ENGINEERS
LADYSMITH, WISCONSIN

Courtesy: Morgan & Parmley, Ltd.

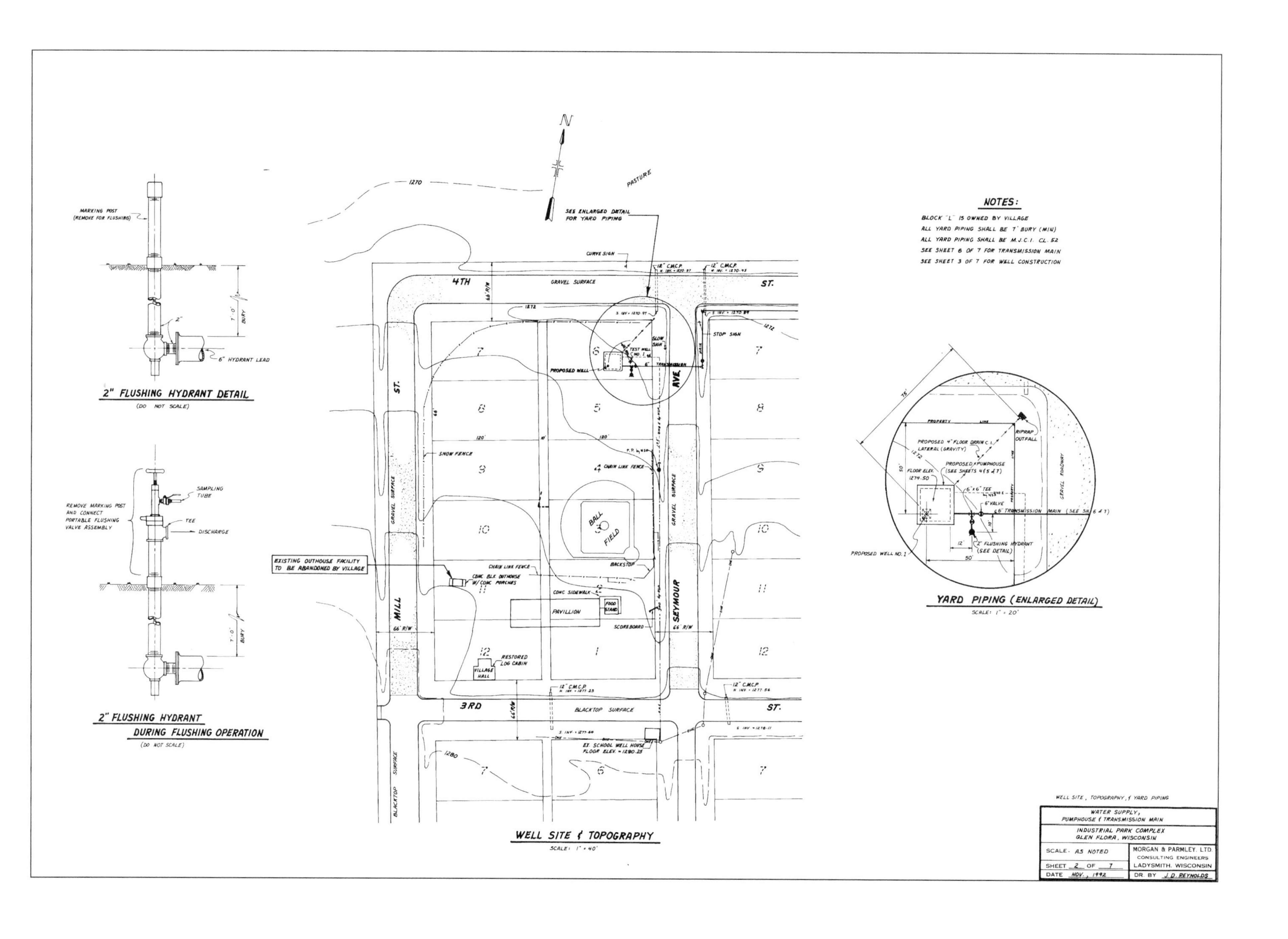
NOTES:
BLOCK "L" IS OWNED BY VILLAGE
ALL YARD PIPING SHALL BE 7' BURY (MIN)
ALL YARD PIPING SHALL BE M.J.C.I. CL. 52
SEE SHEET 6 OF 7 FOR TRANSMISSION MAIN
SEE SHEET 3 OF 7 FOR WELL CONSTRUCTION
2" FLUSHING HYDRANT DETAIL
(DO NOT SCALE)
2" FLUSHING HYDRANT DURING FLUSHING OPERATION
(DO NOT SCALE)
WELL SITE & TOPOGRAPHY
SCALE: 1" = 40'
YARD PIPING (ENLARGED DETAIL)
SCALE: 1" = 20'
EXISTING OUTHOUSE FACILITY TO BE ABANDONED BY VILLAGE
SEE ENLARGED DETAIL FOR YARD PIPING
WATER SUPPLY, PUMPHOUSE & TRANSMISSION MAIN
INDUSTRIAL PARK COMPLEX
GLEN FLORA, WISCONSIN
SCALE: AS NOTED
SHEET 2 OF 7
DATE NOV., 1992
MORGAN & PARMLEY, LTD.
CONSULTING ENGINEERS
LADYSMITH, WISCONSIN
DR. BY J.D. REYNOLDS

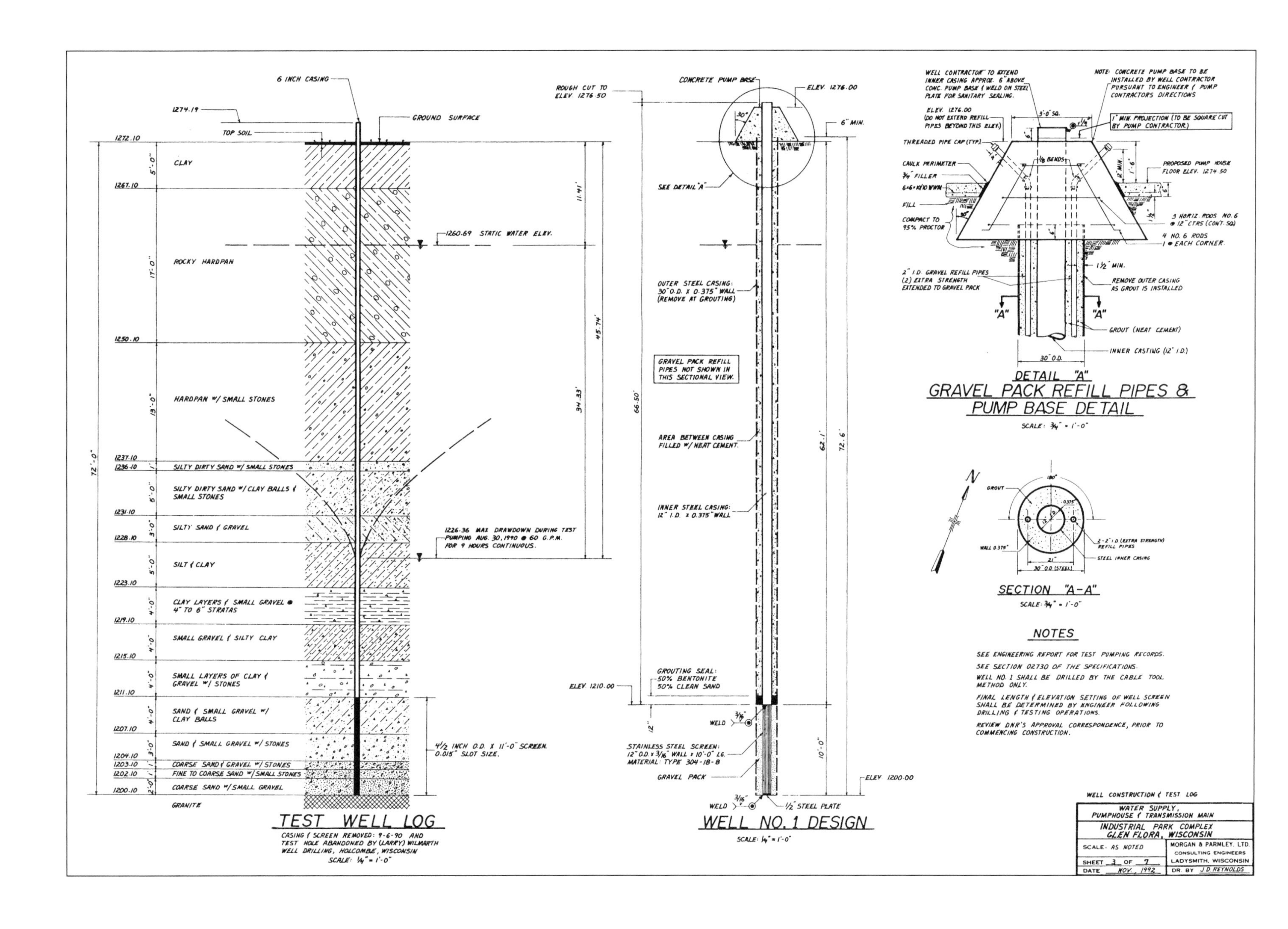
TEST WELL LOG
CASING & SCREEN REMOVED: 9-6-90 AND TEST HOLE ABANDONED BY (LARRY) WILMARTH WELL DRILLING, HOLCOMBE, WISCONSIN
SCALE: 1/4" = 1'-0"
WELL NO. 1 DESIGN
SCALE: 1/4" = 1'-0"
DETAIL "A"
GRAVEL PACK REFILL PIPES & PUMP BASE DETAIL
SCALE: 3/4" = 1'-0"
SECTION "A-A"
SCALE: 3/4" = 1'-0"
NOTES
SEE ENGINEERING REPORT FOR TEST PUMPING RECORDS.
SEE SECTION 02730 OF THE SPECIFICATIONS.
WELL NO. 1 SHALL BE DRILLED BY THE CABLE TOOL METHOD ONLY.
FINAL LENGTH & ELEVATION SETTING OF WELL SCREEN SHALL BE DETERMINED BY ENGINEER FOLLOWING DRILLING & TESTING OPERATIONS.
REVIEW DNR'S APPROVAL CORRESPONDENCE, PRIOR TO COMMENCING CONSTRUCTION.
WELL CONSTRUCTION & TEST LOG
WATER SUPPLY, PUMPHOUSE & TRANSMISSION MAIN
INDUSTRIAL PARK COMPLEX
GLEN FLORA, WISCONSIN
MORGAN & PARMLEY, LTD.
CONSULTING ENGINEERS
LADYSMITH, WISCONSIN
SCALE: AS NOTED
SHEET 3 OF 7
DATE NOV. 1992
DR. BY J.D. REYNOLDS

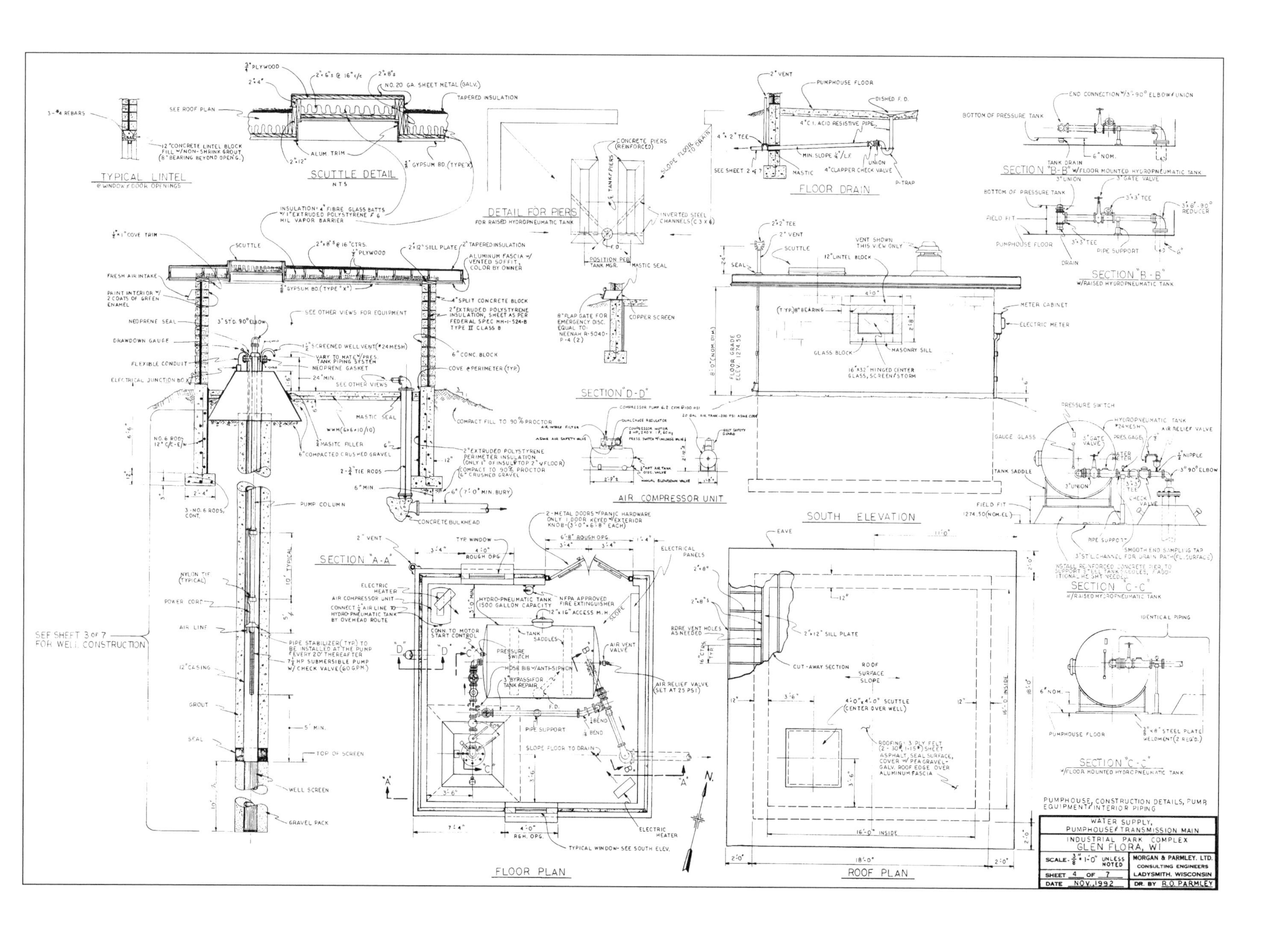
TYPICAL LINTEL
@ WINDOW & DOOR OPENINGS
3-#4 REBARS
12" CONCRETE LINTEL BLOCK FILL W/NON-SHRINK GROUT (8" BEARING BEYOND OPEN'G.)
SCUTTLE DETAIL
NTS
3/4" PLYWOOD
2"x4"
2"x6"s @ 16" c/c
2"x8"s
NO. 20 GA. SHEET METAL (GALV.)
TAPERED INSULATION
SEE ROOF PLAN
ALUM. TRIM
2"x12"
3/4" GYPSUM BD. (TYPE "X")
INSULATION: 4" FIBRE GLASS BATTS W/1" EXTRUDED POLYSTYRENE & 6 MIL VAPOR BARRIER
DETAIL FOR PIERS
FOR RAISED HYDROPNEUMATIC TANK
CONCRETE PIERS (REINFORCED)
INVERTED STEEL CHANNELS (C 3 X 6)
POSITION PER TANK MGR.
SLOPE FLOOR TO DRAIN
SECTION "D-D"
8" FLAP GATE FOR EMERGENCY DISC. EQUAL TO: NEENAH R-5040-P-4 (2)
COPPER SCREEN
MASTIC SEAL
FLOOR DRAIN
2" VENT
PUMPHOUSE FLOOR
DISHED F.D.
4" C.I. ACID RESISTIVE PIPE
4"x2" TEE
MIN. SLOPE 1/4"/L.F.
UNION
SEE SHEET 2 of 7
MASTIC
4" CLAPPER CHECK VALVE
P-TRAP
SECTION "B-B" W/FLOOR MOUNTED HYDROPNEUMATIC TANK
END CONNECTION W/3"-90° ELBOW & UNION
BOTTOM OF PRESSURE TANK
TANK DRAIN
6" NOM.
3" UNION
3" GATE VALVE
SECTION "B-B"
W/RAISED HYDROPNEUMATIC TANK
3"x3" TEE
3"x6"-90° REDUCER
FIELD FIT
PUMPHOUSE FLOOR
PIPE SUPPORT
DRAIN
3/4"x1" COVE TRIM
SCUTTLE
2"x8"s @ 16" CTRS.
1/2" PLYWOOD
2"x12" SILL PLATE
2" TAPERED INSULATION
ALUMINUM FASCIA W/ VENTED SOFFIT, COLOR BY OWNER
FRESH AIR INTAKE
PAINT INTERIOR W/ 2 COATS OF GREEN ENAMEL
5/8" GYPSUM BD. (TYPE "X")
4" SPLIT CONCRETE BLOCK
2" EXTRUDED POLYSTYRENE INSULATION, SHEET AS PER FEDERAL SPEC HH-I-524-B TYPE II CLASS B
SEE OTHER VIEWS FOR EQUIPMENT
NEOPRENE SEAL
3" STD. 90° ELBOW
DRAWDOWN GAUGE
1 1/2" SCREENED WELL VENT (#24 MESH)
VARY TO MATE W/PRES. TANK PIPING SYSTEM
NEOPRENE GASKET
FLEXIBLE CONDUIT
ELECTRICAL JUNCTION BOX
24" MIN.
SEE OTHER VIEWS
6" CONC. BLOCK
COVE @ PERIMETER (TYP.)
MASTIC SEAL
COMPACT FILL TO 90% PROCTOR
WWM (6x6x10/10)
MASITC FILLER
6" COMPACTED CRUSHED GRAVEL
NO. 6 RODS 12" C/C-E/W
2" EXTRUDED POLYSTYRENE PERIMETER INSULATION (ONLY 1" OF INSUL. @ TOP 2" OF FLOOR)
COMPACT TO 90% PROCTOR
6" CRUSHED GRAVEL
2-3/4" TIE RODS
6" MIN.
6" (7'-0" MIN. BURY)
3-NO. 6 RODS, CONT.
CONCRETE BULKHEAD
PUMP COLUMN
SECTION "A-A"
2" VENT
NYLON TIE (TYPICAL)
POWER CORD
AIR LINE
SEE SHEET 3 OF 7 FOR WELL CONSTRUCTION
PIPE STABILIZER (TYP.) TO BE INSTALLED AT THE PUMP & EVERY 20' THEREAFTER
7 1/2 HP SUBMERSIBLE PUMP W/ CHECK VALVE (60 GPM)
12" CASING
GROUT
SEAL
5' MIN.
TOP OF SCREEN
WELL SCREEN
GRAVEL PACK
10' TYPICAL
AIR COMPRESSOR UNIT
SOUTH ELEVATION
2"x2" TEE
2" VENT
SCUTTLE
VENT SHOWN THIS VIEW ONLY
12" LINTEL BLOCK
SEAL
(TYP.) 8" BEARING
GLASS BLOCK
MASONRY SILL
16"x32" HINGED CENTER GLASS, SCREEN & STORM
METER CABINET
ELECTRIC METER
FLOOR GRADE ELEV. 1274.50
8'-0" (NOM. DIM.)
PRESSURE SWITCH
HYDROPNEUMATIC TANK
AIR RELIEF VALVE
GAUGE GLASS
3" GATE VALVE
PRES. GAGE
WATER METER
1/2" NIPPLE
3" 90° ELBOW
TANK SADDLE
3" UNION
CHECK VALVE
FIELD FIT
1274.50 (NOM. EL.)
PIPE SUPPORT
SMOOTH END SAMPLING TAP
3" ST'L CHANNEL FOR DRAIN PATH (FL. SURFACE)
SECTION "C-C"
W/RAISED HYDROPNEUMATIC TANK
IDENTICAL PIPING
6" NOM.
PUMPHOUSE FLOOR
3/8"x8" STEEL PLATE WELDMENT (2 REQ'D.)
SECTION "C-C"
W/FLOOR MOUNTED HYDROPNEUMATIC TANK
2-METAL DOORS W/PANIC HARDWARE ONLY 1 DOOR KEYED W/EXTERIOR KNOB-(3'-0"x6'-8" EACH)
6'-8" ROUGH OPG
TYP. WINDOW
4'-0" ROUGH OPG
ELECTRIC HEATER
AIR COMPRESSOR UNIT
CONNECT 1/2" AIR LINE TO HYDRO-PNEUMATIC TANK BY OVEHEAD ROUTE
HYDRO-PNEUMATIC TANK (1500 GALLON CAPACITY)
NFPA APPROVED FIRE EXTINGUISHER
12"x16" ACCESS M.H.
ELECTRICAL PANELS
CONN. TO MOTOR START CONTROL
TANK SADDLES
PRESSURE SWITCH
AIR VENT VALVE
HOSE BIB W/ANTI-SIPHON
3" BYPASS FOR TANK REPAIR
AIR RELIEF VALVE (SET AT 25 PSI)
F.D.
PIPE SUPPORT
SLOPE FLOOR TO DRAIN
BEND
ELECTRIC HEATER
TYPICAL WINDOW- SEE SOUTH ELEV.
FLOOR PLAN
EAVE
2"x8"
2"x8"s
BORE VENT HOLES AS NEEDED
2"x12" SILL PLATE
CUT-AWAY SECTION
ROOF SURFACE SLOPE
4'-0"x4'-0" SCUTTLE (CENTER OVER WELL)
ROOFING: 3 PLY FELT (2-30#, 1-15#) SHEET ASPHALT, SEAL SURFACE, COVER W/ PEA GRAVEL, GALV. ROOF EDGE OVER ALUMINUM FASCIA
16'-0" INSIDE
18'-0"
ROOF PLAN
PUMPHOUSE, CONSTRUCTION DETAILS, PUMP EQUIPMENT & INTERIOR PIPING
WATER SUPPLY, PUMPHOUSE & TRANSMISSION MAIN
INDUSTRIAL PARK COMPLEX
GLEN FLORA, WI
SCALE: 3/8"=1'-0" UNLESS NOTED
SHEET 4 OF 7
DATE NOV. 1992
MORGAN & PARMLEY, LTD.
CONSULTING ENGINEERS
LADYSMITH, WISCONSIN
DR. BY R.O. PARMLEY

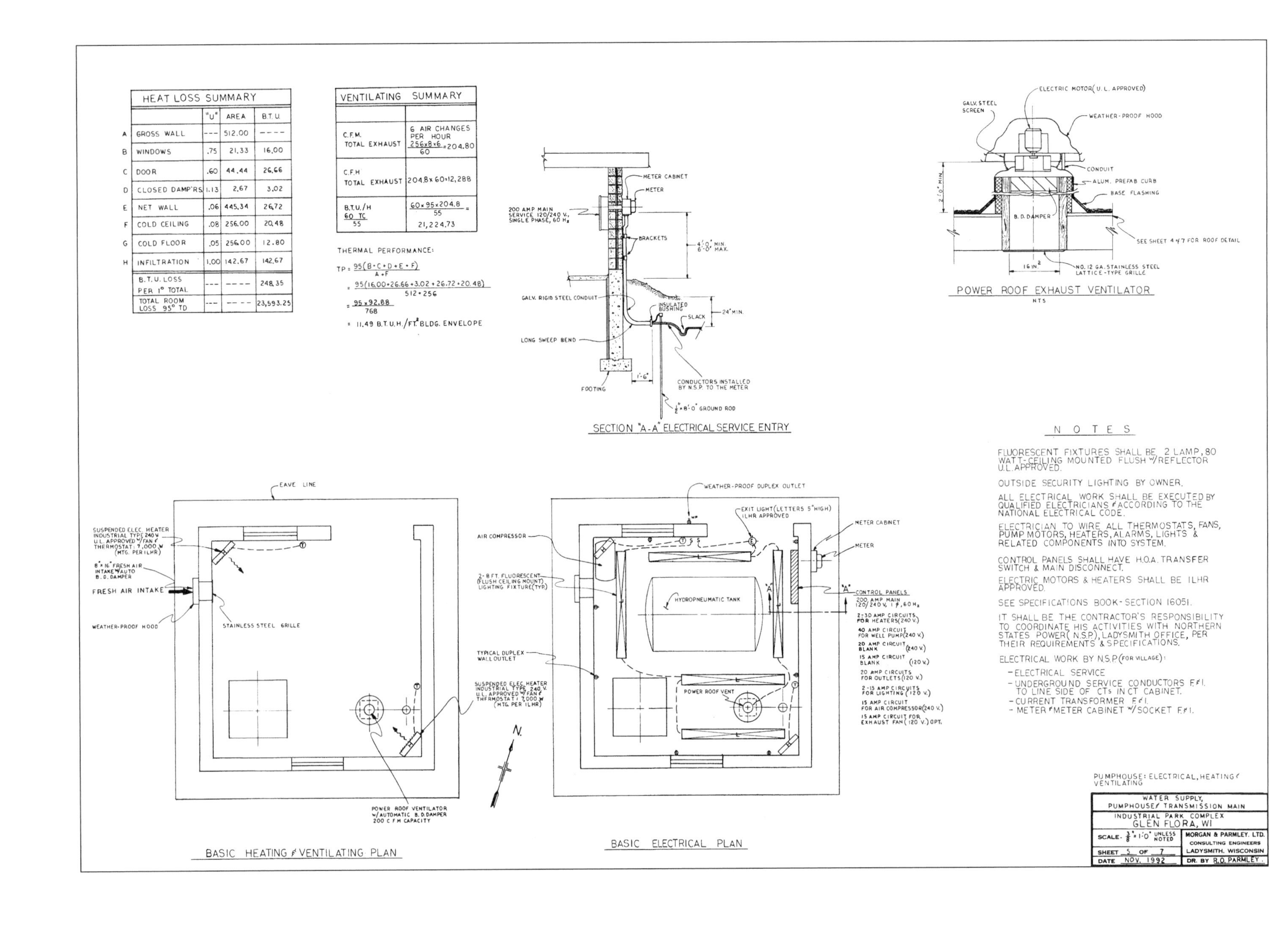
HEAT LOSS SUMMARY
"U" AREA B.T.U.
A GROSS WALL --- 512.00 ----
B WINDOWS .75 21.33 16.00
C DOOR .60 44.44 26.66
D CLOSED DAMP'RS 1.13 2.67 3.02
E NET WALL .06 445.34 26.72
F COLD CEILING .08 256.00 20.48
G COLD FLOOR .05 256.00 12.80
H INFILTRATION 1.00 142.67 142.67
B.T.U. LOSS PER 1° TOTAL --- ---- 248.35
TOTAL ROOM LOSS 95° TD --- ---- 23,593.25
VENTILATING SUMMARY
C.F.M. TOTAL EXHAUST 6 AIR CHANGES PER HOUR 256x8x6/60 = 204.80
C.F.H TOTAL EXHAUST 204.8x60=12,288
B.T.U./H 60 TC 55 60x95x204.8/55 = 21,224.73
THERMAL PERFORMANCE:
TP = 95(B+C+D+E+F)/A+F
= 95(16.00+26.66+3.02+26.72+20.48)/512+256
= 95x92.88/768
= 11.49 B.T.U.H./FT² BLDG. ENVELOPE
METER CABINET
METER
200 AMP MAIN SERVICE 120/240 V., SINGLE PHASE, 60 Hz
BRACKETS
4'-0" MIN. 6'-0" MAX.
GALV. RIGID STEEL CONDUIT
INSULATED BUSHING
SLACK
24" MIN.
LONG SWEEP BEND
FOOTING
1'-6"
CONDUCTORS INSTALLED BY N.S.P. TO THE METER
1/2"x8'-0" GROUND ROD
SECTION "A-A" ELECTRICAL SERVICE ENTRY
ELECTRIC MOTOR (U.L. APPROVED)
GALV. STEEL SCREEN
WEATHER-PROOF HOOD
CONDUIT
ALUM. PREFAB CURB
BASE FLASHING
B.D. DAMPER
2'-0" MIN.
SEE SHEET 4 of 7 FOR ROOF DETAIL
16 IN.²
NO. 12 GA. STAINLESS STEEL LATTICE-TYPE GRILLE
POWER ROOF EXHAUST VENTILATOR
NTS
NOTES
FLUORESCENT FIXTURES SHALL BE 2 LAMP, 80 WATT-CEILING MOUNTED FLUSH W/REFLECTOR U.L. APPROVED.
OUTSIDE SECURITY LIGHTING BY OWNER.
ALL ELECTRICAL WORK SHALL BE EXECUTED BY QUALIFIED ELECTRICIANS & ACCORDING TO THE NATIONAL ELECTRICAL CODE.
ELECTRICIAN TO WIRE ALL THERMOSTATS, FANS, PUMP MOTORS, HEATERS, ALARMS, LIGHTS & RELATED COMPONENTS INTO SYSTEM.
CONTROL PANELS SHALL HAVE H.O.A. TRANSFER SWITCH & MAIN DISCONNECT.
ELECTRIC MOTORS & HEATERS SHALL BE ILHR APPROVED.
SEE SPECIFICATIONS BOOK-SECTION 16051.
IT SHALL BE THE CONTRACTOR'S RESPONSIBILITY TO COORDINATE HIS ACTIVITIES WITH NORTHERN STATES POWER (N.S.P.), LADYSMITH OFFICE, PER THEIR REQUIREMENTS & SPECIFICATIONS.
ELECTRICAL WORK BY N.S.P. (FOR VILLAGE):
- ELECTRICAL SERVICE
- UNDERGROUND SERVICE CONDUCTORS F&I. TO LINE SIDE OF CTs IN CT CABINET.
- CURRENT TRANSFORMER F&I.
- METER & METER CABINET W/SOCKET F&I.
EAVE LINE
SUSPENDED ELEC. HEATER INDUSTRIAL TYPE 240 V. U.L. APPROVED W/FAN & THERMOSTAT: 7,000 W (MTG. PER ILHR)
8"x16" FRESH AIR INTAKE W/AUTO B.D. DAMPER
FRESH AIR INTAKE
WEATHER-PROOF HOOD
STAINLESS STEEL GRILLE
POWER ROOF VENTILATOR W/AUTOMATIC B.D. DAMPER 200 C.F.M. CAPACITY
BASIC HEATING & VENTILATING PLAN
N.
WEATHER-PROOF DUPLEX OUTLET
EXIT LIGHT (LETTERS 5" HIGH) ILHR APPROVED
METER CABINET
METER
AIR COMPRESSOR
2-8 FT. FLUORESCENT (FLUSH CEILING MOUNT) LIGHTING FIXTURE (TYP)
HYDROPNEUMATIC TANK
CONTROL PANELS
200 AMP MAIN 120/240 V. 1 Φ, 60 Hz
2-30 AMP CIRCUITS FOR HEATERS (240 V.)
40 AMP CIRCUIT FOR WELL PUMP (240 V.)
20 AMP CIRCUIT BLANK (240 V.)
15 AMP CIRCUIT BLANK (120 V.)
20 AMP CIRCUITS FOR OUTLETS (120 V.)
2-15 AMP CIRCUITS FOR LIGHTING (120 V.)
15 AMP CIRCUIT FOR AIR COMPRESSOR (240 V.)
15 AMP CIRCUIT FOR EXHAUST FAN (120 V.) OPT.
TYPICAL DUPLEX WALL OUTLET
SUSPENDED ELEC. HEATER INDUSTRIAL TYPE 240 V. U.L. APPROVED W/FAN & THERMOSTAT: 7,000 W (MTG. PER ILHR)
POWER ROOF VENT
BASIC ELECTRICAL PLAN
PUMPHOUSE: ELECTRICAL, HEATING & VENTILATING
WATER SUPPLY, PUMPHOUSE & TRANSMISSION MAIN
INDUSTRIAL PARK COMPLEX
GLEN FLORA, WI
SCALE- 3/8"=1'-0" UNLESS NOTED
MORGAN & PARMLEY, LTD. CONSULTING ENGINEERS LADYSMITH, WISCONSIN
SHEET 5 OF 7
DATE NOV. 1992
DR. BY R.O. PARMLEY

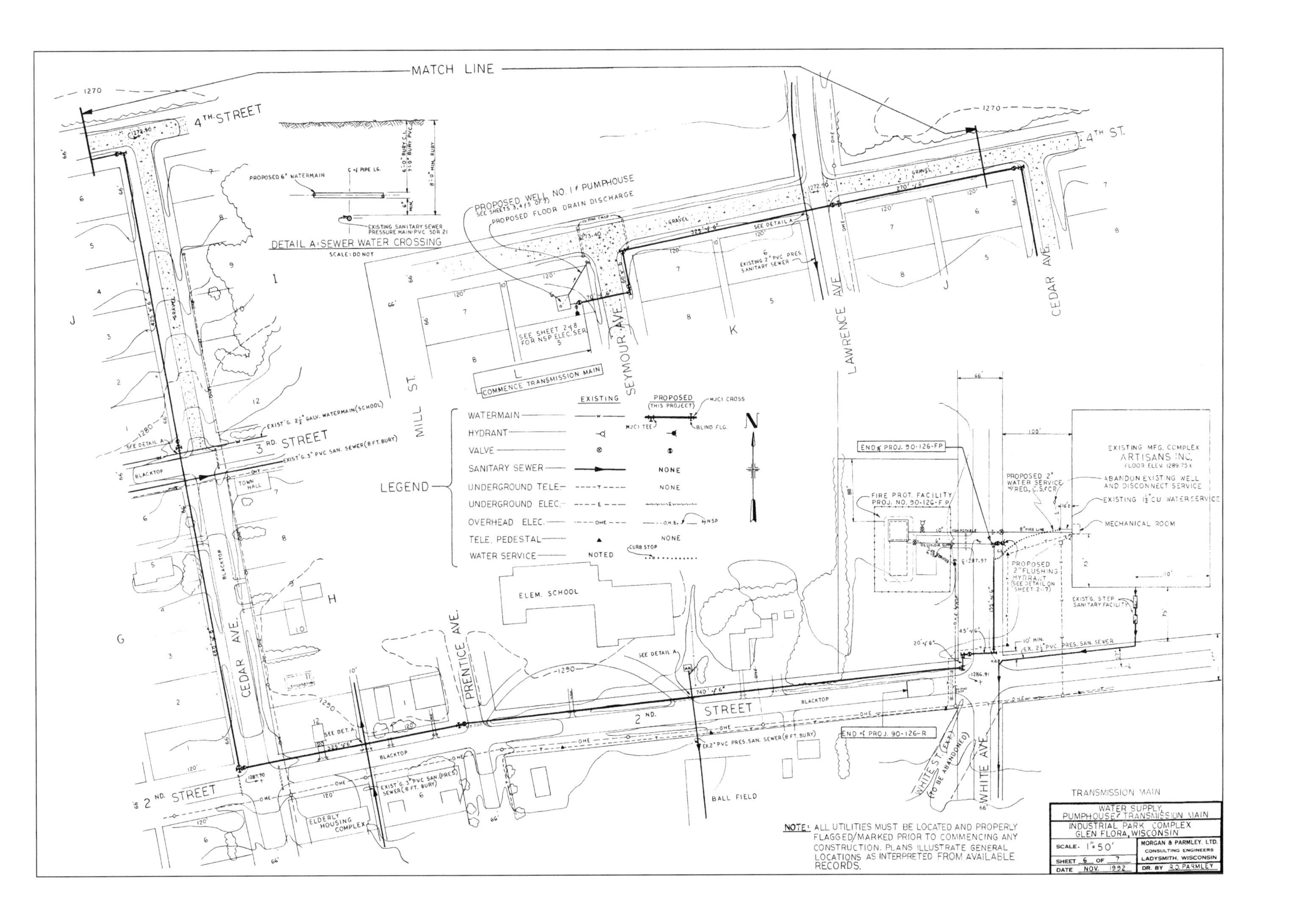
MATCH LINE
4TH STREET
4TH ST.
3RD STREET
2ND STREET
CEDAR AVE.
MILL ST.
SEYMOUR AVE.
LAWRENCE AVE.
PRENTICE AVE.
WHITE AVE.
WHITE ST. (EXT.) (TO BE ABANDONED)
DETAIL A: SEWER WATER CROSSING
SCALE: DO NOT
PROPOSED 6" WATERMAIN
EXISTING SANITARY SEWER PRESSURE MAIN PVC SDR 21
PROPOSED WELL NO. 1 & PUMPHOUSE
SEE SHEETS 3, 4 & 5 OF 7
PROPOSED FLOOR DRAIN DISCHARGE
COMMENCE TRANSMISSION MAIN
SEE SHEET 2 & 8 FOR NSP ELEC. SER.
EXISTING 2" PVC PRES. SANITARY SEWER
EXIST'G. 2½" GALV. WATERMAIN (SCHOOL)
EXIST'G. 3" PVC SAN. SEWER (8 FT. BURY)
LEGEND
EXISTING
PROPOSED (THIS PROJECT)
WATERMAIN
HYDRANT
VALVE
SANITARY SEWER
UNDERGROUND TELE.
UNDERGROUND ELEC.
OVERHEAD ELEC.
TELE. PEDESTAL
WATER SERVICE
NONE
NOTED
MJCI TEE
MJCI CROSS
BLIND FLG.
CURB STOP
NSP
N
TOWN HALL
BLACKTOP
GRAVEL
ELEM. SCHOOL
ELDERLY HOUSING COMPLEX
BALL FIELD
SEE DETAIL A
END ℄ PROJ. 90-126-FP
END ℄ PROJ. 90-126-R
FIRE PROT. FACILITY PROJ. NO. 90-126-F.P.
EXISTING MFG. COMPLEX ARTISANS INC.
PROPOSED 2" WATER SERVICE
ABANDON EXISTING WELL AND DISCONNECT SERVICE
MECHANICAL ROOM
PROPOSED 2" FLUSHING HYDRANT (SEE DETAIL ON SHEET 2-7)
EXIST'G. STEP SANITARY FACILITY
10' MIN.
EX. 2½" PVC PRES. SAN. SEWER
EX. 2" PVC PRES. SAN. SEWER (8 FT. BURY)
TRANSMISSION MAIN
NOTE: ALL UTILITIES MUST BE LOCATED AND PROPERLY FLAGGED/MARKED PRIOR TO COMMENCING ANY CONSTRUCTION. PLANS ILLUSTRATE GENERAL LOCATIONS AS INTERPRETED FROM AVAILABLE RECORDS.
WATER SUPPLY, PUMPHOUSE & TRANSMISSION MAIN
INDUSTRIAL PARK COMPLEX
GLEN FLORA, WISCONSIN
SCALE: 1"=50'
SHEET 6 OF 7
DATE NOV. 1992
MORGAN & PARMLEY, LTD.
CONSULTING ENGINEERS
LADYSMITH, WISCONSIN
DR. BY R.O. PARMLEY

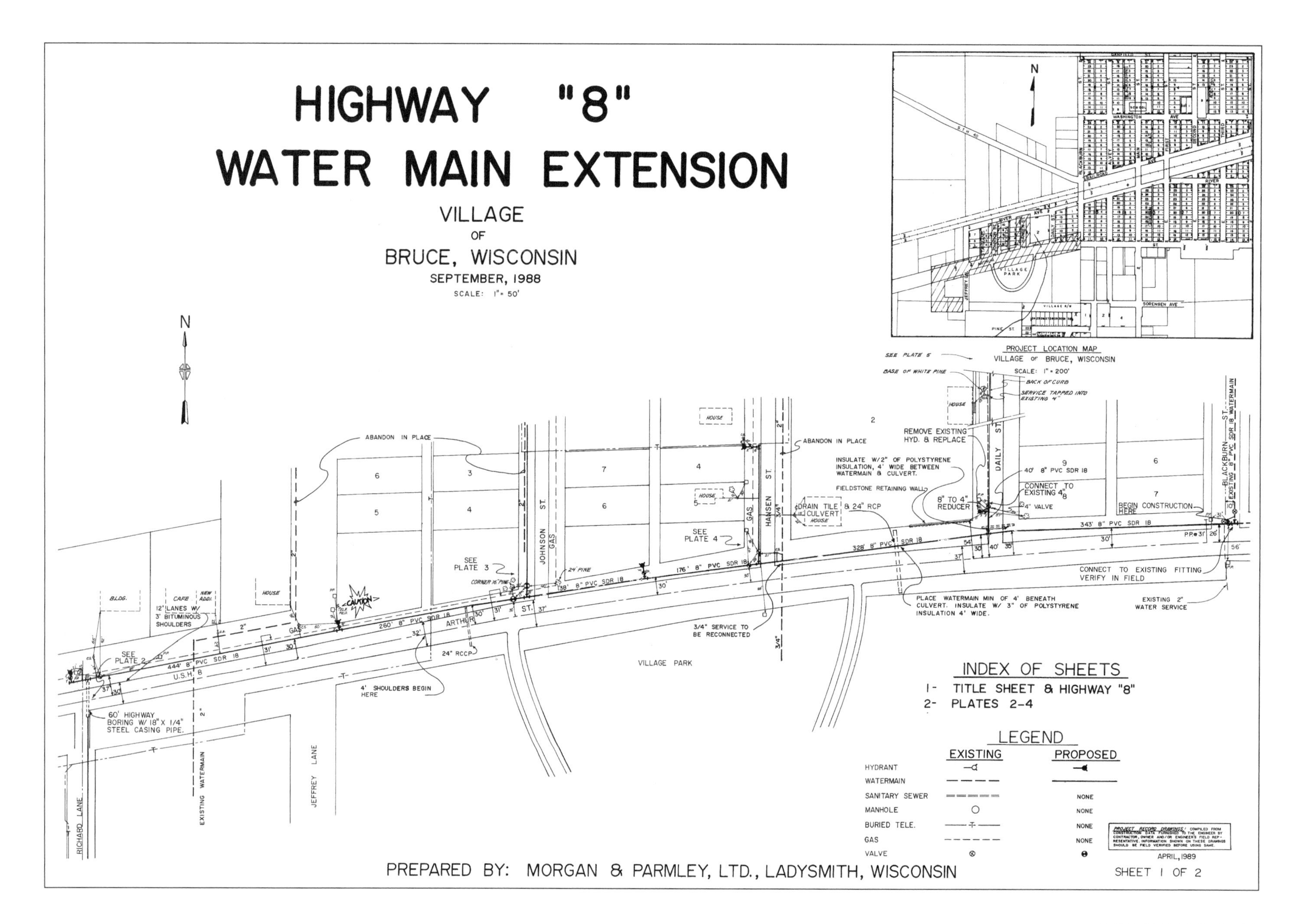

Courtesy: Morgan & Parmley, Ltd.

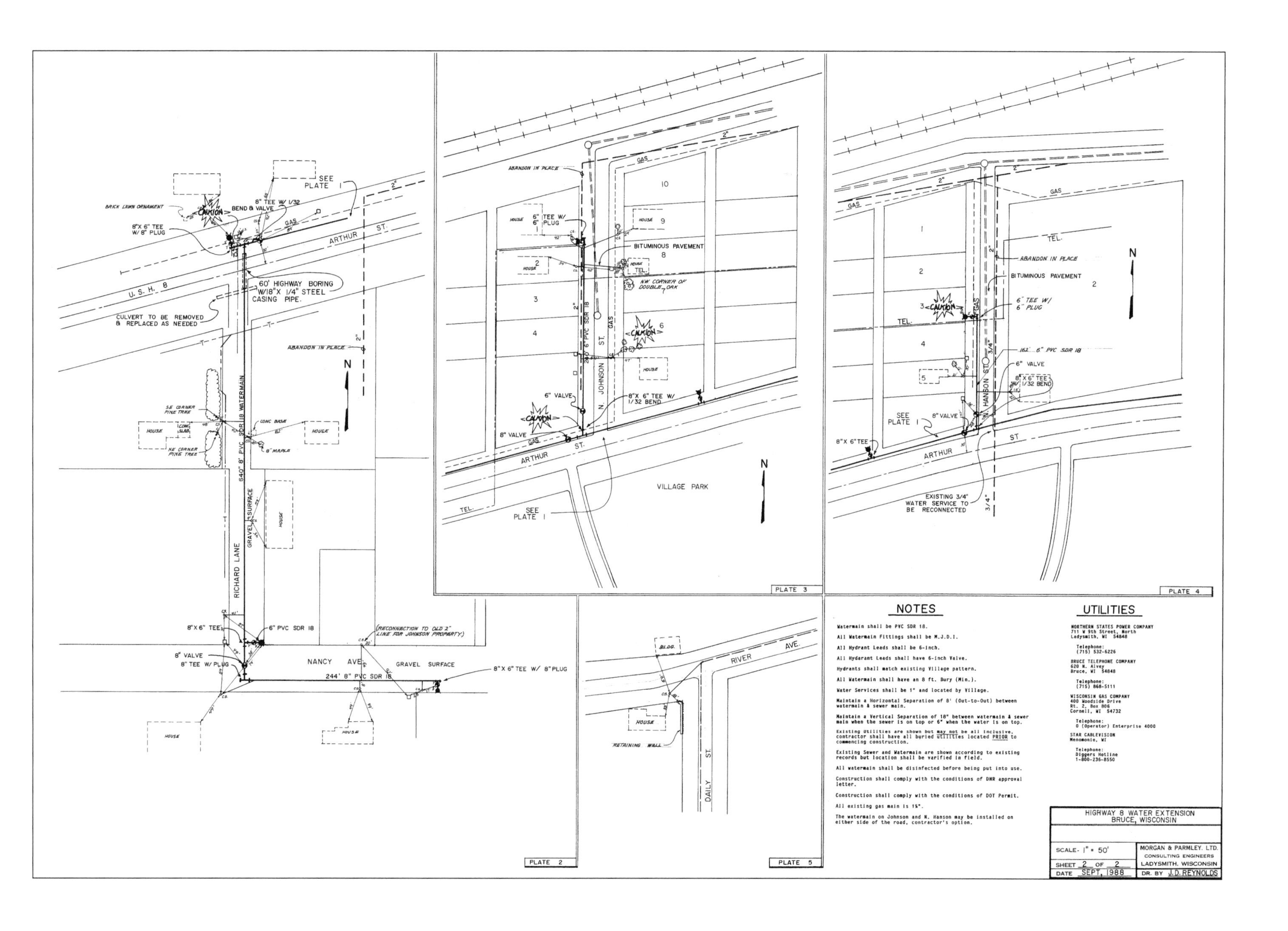
SEE PLATE 1
BRICK LAWN ORNAMENT
CAUTION
BEND & VALVE
8" TEE W/ 1/32
8"X 6" TEE W/ 8" PLUG
GAS
ARTHUR ST.
U. S. H. 8
60' HIGHWAY BORING W/18" X 1/4" STEEL CASING PIPE.
CULVERT TO BE REMOVED & REPLACED AS NEEDED
ABANDON IN PLACE
N
SE CORNER PINE TREE
HOUSE
CONC. SLAB
NE CORNER PINE TREE
CONC BASE
8" MAPLE
540' 8" PVC SDR 18 WATERMAIN
GRAVEL SURFACE
RICHARD LANE
8"X 6" TEE
6" PVC SDR 18
(RECONNECTION TO OLD 2" LINE FOR JOHNSON PROPERTY)
8" VALVE
8" TEE W/ PLUG
NANCY AVE.
GRAVEL SURFACE
244' 8" PVC SDR 18
8"X 6" TEE W/ 8" PLUG
PLATE 2
ABANDON IN PLACE
6" TEE W/ 6" PLUG
BITUMINOUS PAVEMENT
TEL.
NW CORNER OF DOUBLE OAK
N. JOHNSON ST.
6" VALVE
8"X 6" TEE W/ 1/32 BEND
8" VALVE
VILLAGE PARK
SEE PLATE 1
PLATE 3
ABANDON IN PLACE
BITUMINOUS PAVEMENT
6" TEE W/ 6" PLUG
162' 6" PVC SDR 18
6" VALVE
8"X 6" TEE W/ 1/32 BEND
N. HANSON ST.
SEE PLATE 1
8" VALVE
8"X 6" TEE
EXISTING 3/4" WATER SERVICE TO BE RECONNECTED
PLATE 4
BLDG.
RIVER AVE.
HOUSE
RETAINING WALL
DAILY ST.
PLATE 5
NOTES
Watermain shall be PVC SDR 18.
All Watermain Fittings shall be M.J.D.I.
All Hydrant Leads shall be 6-inch.
All Hydarant Leads shall have 6-inch Valve.
Hydrants shall match existing Village pattern.
All Watermain shall have an 8 ft. Bury (Min.).
Water Services shall be 1" and located by Village.
Maintain a Horizontal Separation of 8' (Out-to-Out) between watermain & sewer main.
Maintain a Vertical Separation of 18" between watermain & sewer main when the sewer is on top or 6" when the water is on top.
Existing Utilities are shown but may not be all inclusive, contractor shall have all buried utilities located PRIOR to commencing construction.
Existing Sewer and Watermain are shown according to existing records but location shall be varified in field.
All watermain shall be disinfected before being put into use.
Construction shall comply with the conditions of DNR approval letter.
Construction shall comply with the conditions of DOT Permit.
All existing gas main is 1¼".
The watermain on Johnson and N. Hanson may be installed on either side of the road, contractor's option.
UTILITIES
NORTHERN STATES POWER COMPANY
711 W 9th Street, North
Ladysmith, WI 54848
Telephone:
(715) 532-6226
BRUCE TELEPHONE COMPANY
620 N. Alvey
Bruce, WI 54848
Telephone:
(715) 868-5111
WISCONSIN GAS COMPANY
400 Woodside Drive
Rt. 2, Box 806
Cornell, WI 54732
Telephone:
0 (Operator) Enterprise 4000
STAR CABLEVISION
Menomonie, WI
Telephone:
Diggers Hotline
1-800-236-8550
HIGHWAY 8 WATER EXTENSION
BRUCE, WISCONSIN
SCALE- 1" = 50'
SHEET 2 OF 2
DATE SEPT, 1988
MORGAN & PARMLEY, LTD.
CONSULTING ENGINEERS
LADYSMITH, WISCONSIN
DR. BY J.D. REYNOLDS

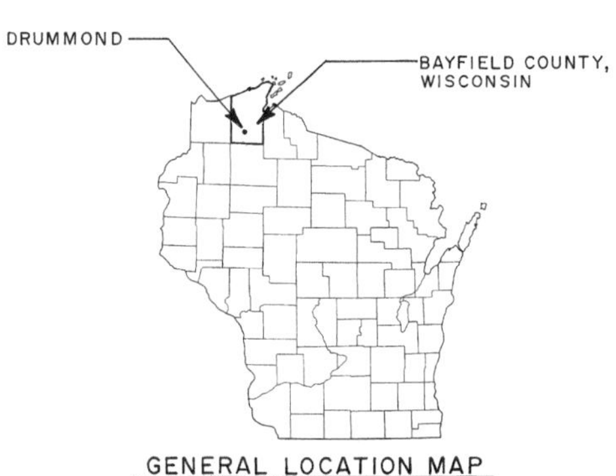

GENERAL LOCATION MAP

WATERMAIN EXTENSION

DRUMMOND SANITARY DISTRICT

DRUMMOND, WISCONSIN

M. & P. PROJECT NO. 99-157-S

JULY, 2002

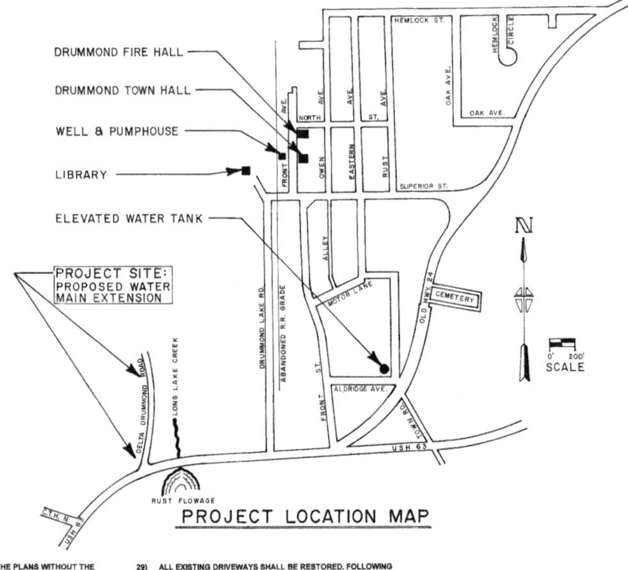

PROJECT LOCATION MAP

-NOTES-

1) ALL ELEVATIONS BASED ON DATUM OBTAINED FROM PRIOR PLANS AND SANITARY DISTRICT UTILITY RECORDS.

2) CONSTRUCTION SHALL COMPLY WITH THE CONDITIONS OF DNR APPROVAL LETTER, AND OTHER REGULATORY APPROVALS AS LISTED IN THE SPECIFICATION BOOK.

3) ALL WATERMAIN SHALL BE PVC DR18 C900, SIZED AS SHOWN ON PLANS.

4) ALL WATERMAIN FITTINGS SHALL BE M.J.D.I.

5) HYDRANT LEAD SHALL BE 6-INCH D.I., CL. 52.

6) HYDRANT LEAD SHALL HAVE A 6-INCH GATE VALVE W/RISER.

7) HYDRANT SHALL MATCH THE VILLAGE OF DRUMMOND'S PATTERN.

8) ALL WATERMAIN SHALL HAVE A 8 FT. BURY (MIN.)

9) 8 FT. (MINIMUM) HORIZONTAL DISTANCE (℄ TO ℄) BETWEEN WATERMAIN AND SANITARY SEWER OR STORM SEWER.

10) 18" (MINIMUM) VERTICAL DISTANCE (OUT TO OUT) BETWEEN WATERMAIN UNDER SANITARY SEWER OR STORM SEWER @ THEIR INTERSECTION.

11) 6" (MINIMUM) VERTICAL DISTANCE (OUT TO OUT) BETWEEN WATERMAIN OVER SANITARY SEWER OR STORM SEWER @ THEIR INTERSECTION.

12) VALVES SHALL BE SET TO GRADE SHOWN ON PLANS, UNLESS DIRECTED OTHERWISE.

13) WATERMAIN SHALL BE INSULATED AT ALL STORM SEWER, CULVERT, STREET AND DITCH CROSSINGS WITH 2" THICK EXTRUDED POLYSTYRENE INSULATION BOARD AND JOINTS LAPPED A MINIMUM OF 2 FT.

14) ALL WATERMAIN SHALL BE DISINFECTED AND RECEIVE A "SAFE" LAB CERTIFICATION BEFORE BEING PUT INTO SERVICE.

15) EXISTING UTILITIES ARE SHOWN BUT MAY NOT BE ALL INCLUSIVE, CONTRACTOR SHALL HAVE ALL BURIED UTILITIES LOCATED PRIOR TO COMMENCING CONSTRUCTION.

16) EXISTING SANITARY SEWER IS SHOWN ACCORDING TO EXISTING RECORDS, BUT LOCATION SHALL BE VERIFIED IN FIELD.

17) EROSION CONTROL MEASURES SHALL BE IN PLACE BEFORE CONSTRUCTION BEGINS.

18) THE CONTRACTOR SHALL MINIMIZE EROSION DURING CONSTRUCTION USING GOOD CONSTRUCTION TECHNIQUES AND UTILIZING SILT FENCES AND BALE DITCH CHECKS. THE CONTRACTOR SHALL RELEASE RUNOFF FROM THE SITE IN A NUISANCE FREE MANNER.

19) THE CONTRACTOR IS RESPONSIBLE TO MAINTAIN THE SITE IN A SAFE CONDITION. THE SOLE RESPONSIBILITY FOR WARNING SIGNS, BARRICADES, FLAGGING PERSONNEL AND ALL ASPECTS OF SAFETY LIE WITH THE CONTRACTOR.

20) ALL DISTRUBED YARD & ROADWAY DITCHES SHALL BE RESTORED W /SEEDING AS FOLLOWS:

A) REPLACE ALL SALVAGED TOPSOIL
B) SEED MIXUTRE: PERMANENT SEEDING – (LAWN TYPE TURF)
35% KENTUCKY BLUEGRASS
25% IMPROVED FINE PERENNIAL RYEGRASS
15% CREEPING RED FESCUE
15% IMPROVED HARD FESCUE
10% WHITE CLOVER
C) SEED RATE: 2# / 1000 SQ. FT.
D) MULCH RATE: 3 TON / ACRE
E) PERMANENT SEEDING: ALLOWED TO SEPT. 7
TEMPORARY SEEDING: SEPT. 8 THROUGH Nov. 10 – (4 BU. OATS / ACRE)
F) SEEDING MAY BE BROADCAST BUT MUST BE COVERED WITH A MAXIMUM OF ½" SOIL. MULCH MUST BE ANCHORED WITH A MULCH TILLER OR BLOWN W / 75 – 100 GAL. OF ASPHALT / TON OF MULCH.
G) FERTILIZER: 500# / ACRE 20-10-10

21) THERE SHALL BE NO DEVIATION FROM THE PLANS WITHOUT THE DESIGN ENGINEER'S APPROVAL.

22) CONSTRUCTION STAKES DESTROYED BY THE CONTRACTOR THAT NEED TO BE REPLACED, WILL BE CHARGED TO THE CONTRACTOR.

23) PROPERTY CORNERS KNOWN TO EXIST SHALL NOT BE DISTURBED. DAMAGED CORNERS WILL BE REPLACED AT THE CONTRACTOR'S EXPENSE. PRESERVE ALL SURVEY MONUMENTS.

24) SQUARE CUT ALL ASPHALT PAVEMENT PRIOR TO TRENCH EXCAVATION. THOROUGHLY COMPACT BACKFILL TO 90% PROCTOR. INSTALL A MINIMUM OF 9 INCHES OF GRANULAR SUB-BASE AND 6 INCHES OF BASE COURSE. PAVE WITH 2 – 1½ INCH LIFTS OF ASPHALT.

25) SEE SPECIFICATION BOOK AND FOLLOWING PLAN SHEETS FOR ADDITIONAL DETAILS AND NOTES.

26) IT SHALL BE THE CONTRACTOR'S RESPONSIBILITY TO PROPERLY NOTIFY ALL UTILITIES & DIGGERS HOTLINE PRIOR TO COMMENCING ANY CONSTRUCTION ON THIS PROJECT.

27) DO NOT CAST, DEPOSIT OR STOCKPILE ANY MATERIAL ON EXISTING ASPHALT PAVEMENT.

28) NO HEAVY CONSTRUCTION EQUIPMENT ALLOWED ON ASPHALT PAVEMENT.

29) ALL EXISTING DRIVEWAYS SHALL BE RESTORED, FOLLOWING WATERMAIN INSTALLATION AND NEW CULVERTS (12") INSTALLED WITH ENDWALLS.

30) AN (APPROVED) TRENCH BOX SHALL BE USED FOR WATERMAIN INSTALLATION.

UTILITIES

MUNICIPAL WATER SYSTEM:

DRUMMOND SANITARY DISTRICT
P. O. BOX 43 – FRONT AVE.
DRUMMOND, WISCONSIN 54832
MARK JEROME: WATERWORKS OPERATOR
TELE. NO. 715/ 739-6741

TELEPHONE COMPANY:

CHEQUAMEGON TELEPHONE CO.
1st AVE. – P. O. BOX 67
CABLE, WISCONSIN 54821
TELE. NO. 800/ 250-8927

ELECTRICAL SERVICE:

XCEL ENERGY
(A.K.A. NSP)
16048 ELECTRIC AVE.
HAYWARD, WISCONSIN 54843
KEN DISHER: FIELD ENGINEER
TELE. NO. 800 / 895 – 4999

FIRE DEPARTMENT

DRUMMOND FIRE & RESCUE
MARK JEROME: CHIEF
TELE. NO. 715 / 739 - 6696

DIGGERS HOTLINE

TELE. NO. 800 / 242 - 8511

-INDEX OF SHEETS-

SHEET NO.	DESCRIPTION
1	TITLE SHEET W/LOCATION MAPS AND NOTES
2	WATERMAIN PLAN & PROFILE
3 & 4	TYPICAL EROSION CONTROL DETAILS & NOTES This Sheet Not Shown

GENERAL SPECIFICATIONS: STANDARD SPECIFICATIONS FOR SEWER AND WATER CONSTRUCTION IN WISCONSIN (5TH EDITION)

WISCONSIN CONSTRUCTION SITE BEST MANAGEMENT PRACTIC HANDBOOK

SPECIFIC SPECIFICATIONS: MORGAN & PARMLEY, LTD. SPEC. BOOK

BIDDING DOCUMENTS: MORGAN & PARMLEY, LTD. SPEC. BOOK

PREPARED BY:
MORGAN & PARMLEY, Ltd.
PROFESSIONAL CONSULTING ENGINEERS
115 W. 2nd STREET SOUTH
LADYSMITH, WISCONSIN 54848

Courtesy: Morgan & Parmley, Ltd.

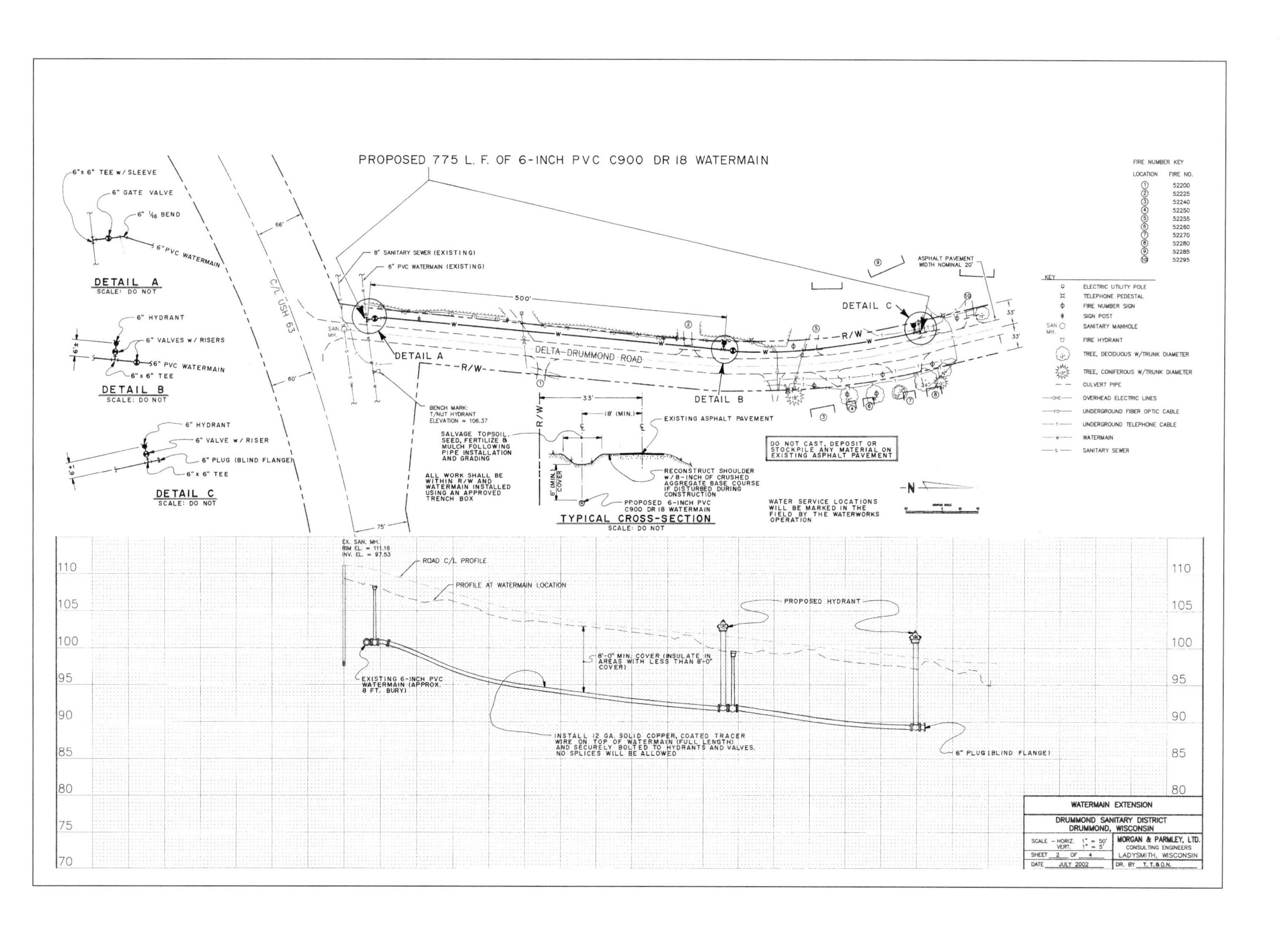
PROPOSED 775 L. F. OF 6-INCH PVC C900 DR 18 WATERMAIN
6"x 6" TEE w/ SLEEVE
6" GATE VALVE
6" 1/16 BEND
6" PVC WATERMAIN
DETAIL A
SCALE: DO NOT
6" HYDRANT
6" VALVES w/ RISERS
6" PVC WATERMAIN
6" x 6" TEE
DETAIL B
SCALE: DO NOT
6" HYDRANT
6" VALVE w/ RISER
6" PLUG (BLIND FLANGE)
6" x 6" TEE
DETAIL C
SCALE: DO NOT
C/L USH 63
8" SANITARY SEWER (EXISTING)
6" PVC WATERMAIN (EXISTING)
DELTA-DRUMMOND ROAD
BENCH MARK: T/NUT HYDRANT ELEVATION = 106.37
SALVAGE TOPSOIL, SEED, FERTILIZE & MULCH FOLLOWING PIPE INSTALLATION AND GRADING
ALL WORK SHALL BE WITHIN R/W AND WATERMAIN INSTALLED USING AN APPROVED TRENCH BOX
EXISTING ASPHALT PAVEMENT
RECONSTRUCT SHOULDER w/ 8-INCH OF CRUSHED AGGREGATE BASE COURSE IF DISTURBED DURING CONSTRUCTION
PPOPOSED 6-INCH PVC C900 DR 18 WATERMAIN
TYPICAL CROSS-SECTION
SCALE: DO NOT
DO NOT CAST, DEPOSIT OR STOCKPILE ANY MATERIAL ON EXISTING ASPHALT PAVEMENT
WATER SERVICE LOCATIONS WILL BE MARKED IN THE FIELD BY THE WATERWORKS OPERATION
ASPHALT PAVEMENT WIDTH NOMINAL 20'
FIRE NUMBER KEY
LOCATION FIRE NO.
1 52200
2 52225
3 52240
4 52250
5 52255
6 52260
7 52270
8 52280
9 52285
10 52295
KEY
ELECTRIC UTILITY POLE
TELEPHONE PEDESTAL
FIRE NUMBER SIGN
SIGN POST
SANITARY MANHOLE
FIRE HYDRANT
TREE, DECIDUOUS W/TRUNK DIAMETER
TREE, CONIFEROUS W/TRUNK DIAMETER
CULVERT PIPE
OVERHEAD ELECTRIC LINES
UNDERGROUND FIBER OPTIC CABLE
UNDERGROUND TELEPHONE CABLE
WATERMAIN
SANITARY SEWER
EX. SAN. MH. RIM EL. = 111.16 INV. EL. = 97.53
ROAD C/L PROFILE
PROFILE AT WATERMAIN LOCATION
PROPOSED HYDRANT
8'-0" MIN. COVER (INSULATE IN AREAS WITH LESS THAN 8'-0" COVER)
EXISTING 6-INCH PVC WATERMAIN (APPROX. 8 FT. BURY)
INSTALL 12 GA. SOLID COPPER, COATED TRACER WIRE ON TOP OF WATERMAIN (FULL LENGTH) AND SECURELY BOLTED TO HYDRANTS AND VALVES. NO SPLICES WILL BE ALLOWED
6" PLUG (BLIND FLANGE)
WATERMAIN EXTENSION
DRUMMOND SANITARY DISTRICT
DRUMMOND, WISCONSIN
SCALE - HORIZ. 1" = 50' VERT. 1" = 5'
SHEET 2 OF 4
DATE JULY 2002
MORGAN & PARMLEY, LTD.
CONSULTING ENGINEERS
LADYSMITH, WISCONSIN
DR. BY T. T. & D.N.

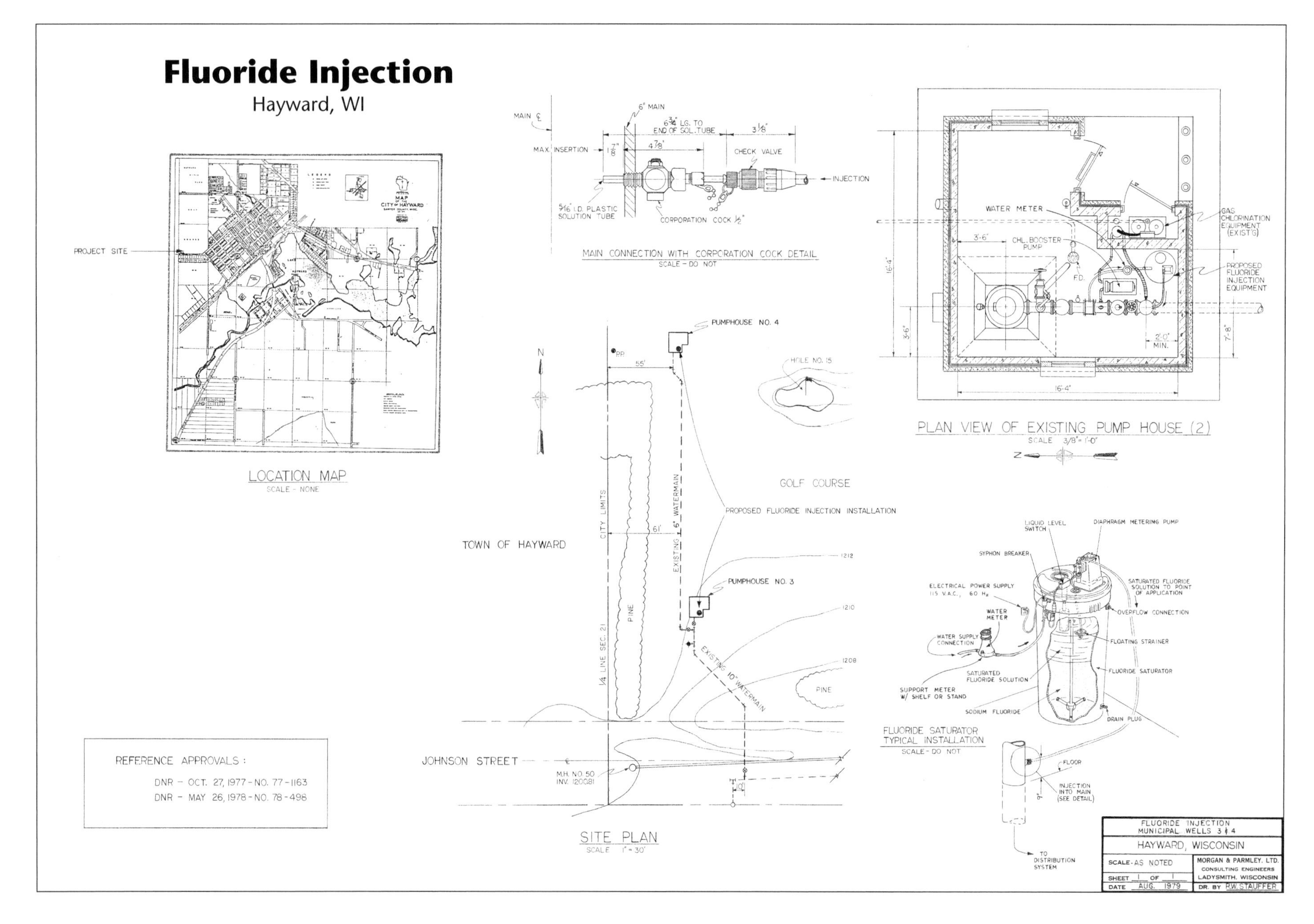

Courtesy: Morgan & Parmley, Ltd.

Design

FIRE PROTECTION

Section 10

Fire protection, whether provided by public water supply systems or non-potable facilities, is a key element in safeguarding our society from the ravages of fire and its costly consequences.

This section begins with a set of plans for an elevated water reservoir for storage and to maintain adequate head pressure on the potable water distribution system. Access to water for fire fighting is available from hydrants dispersed throughout the piping network. The reader is referred to the previous section for water distribution systems.

Non-potable water may be used for fire protection serving facilities where the supply is limited. With this parameter in mind, plans have been included for such a facility that was designed for a rural industrial complex. The manufacturing structure has a complete "wet" sprinkler system which is served by emergency pumps. The lead pump is electric but the standby pump operates on diesel fuel and automatically activates when electrical power fails.

TABLE OF CONTENTS

-PROPOSED-

ELEVATED WATER RESERVOIR

VILLAGE of GILMAN, WISCONSIN

JUNE, 1998

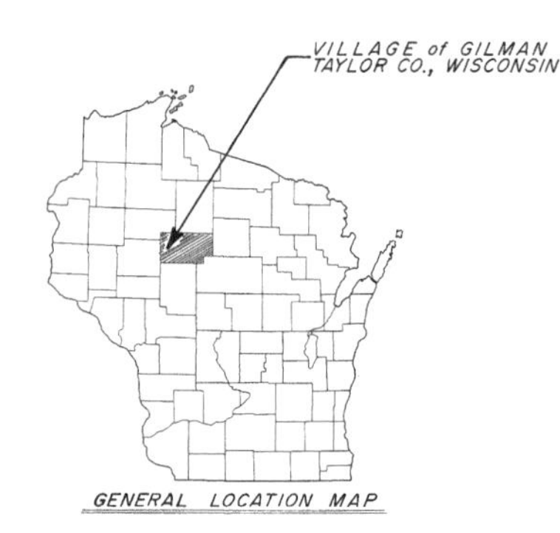

GENERAL LOCATION MAP

-NOTES-

1 ...... ALL ELEVATIONS BASED ON USGS DATUM.
2 ...... CONSTRUCTION SHALL COMPLY WITH THE CONDITIONS OF DNR APPROVAL LETTER, PSC AUTHORIZATION TO CONSTRUCT AND OTHER REGULATORY APPROVALS AS LISTED IN THE SPECIFICATION BOOK.
3 ...... ALL WATERMAIN SHALL BE M.J.D.I., CL. 52; SIZED AS SHOWN ON PLANS.
4 ...... ALL WATERMAIN FITTINGS SHALL BE M.J.D.I.
5 ...... ALL HYDRANT LEADS SHALL BE 6-INCH M.J.D.I., CL. 52.
6 ...... ALL HYDRANT LEADS SHALL HAVE A 6-INCH GATE VALVE.
7 ...... HYDRANTS SHALL MATCH THE VILLAGE OF GILMAN'S PATTERN.
8 ...... ALL WATER MAIN SHALL HAVE AN 8 FT. BURY (MIN.).
9 ...... 8 FT. (MINIMUM) HORIZONTAL DISTANCE (℄ TO ℄) BETWEEN WATERMAIN AND SANITARY SEWER OR STORM SEWER.
10 ...... 18" (MINIMUM) VERTICAL DISTANCE (OUT TO OUT) BETWEEN WATERMAIN UNDER SANITARY SEWER OR STORM SEWER @ THEIR INTERSECTION.
11 ...... 6" (MINIMUM) VERTICAL DISTANCE (OUT TO OUT) BETWEEN WATERMAIN OVER SANITARY SEWER OR STORM SEWER @ THEIR INTERSECTION.
12 ...... VALVES SHALL BE SET TO GRADE SHOWN ON PLANS, UNLESS DIRECTED OTHERWISE.
13 ...... WATERMAIN SHALL BE INSULATED AT ALL STORM SEWER, CULVERT AND DITCH CROSSINGS WITH 2" THICK EXTRUDED POLYSTYRENE INSULATION BOARD.
14 ...... ALL WATERMAIN SHALL BE DISINFECTED BEFORE BEING PUT INTO SERVICE.
15 ...... EXISTING UTILITIES ARE SHOWN BUT MAY NOT BE ALL INCLUSIVE, CONTRACTOR SHALL HAVE ALL BURIED UTILITIES LOCATED PRIOR TO COMMENCING CONSTRUCTION.
16 ...... EXISTING SANITARY SEWER IS SHOWN ACCORDING TO EXISTING RECORDS, BUT LOCATION SHALL BE VERIFIED IN FIELD.
17 ...... EROSION CONTROL MEASURES SHALL BE IN PLACE BEFORE CONSTRUCTION BEGINS.
18 ...... THE CONTRACTORS SHALL MINIMIZE EROSION DURING CONSTRUCTION USING GOOD CONSTRUCTION TECHNIQUES AND UTILIZING SILT FENCES AND BALE DITCH CHECKS. THE CONTRACTOR SHALL RELEASE RUNOFF FROM THE SITE IN A NUISANCE FREE MANNER.
19 ...... THE CONTRACTOR IS RESPONSIBLE TO MAINTAIN THE SITE IN A SAFE CONDITION. THE SOLE RESPONSIBILITY FOR WARNING SIGNS, BARRICADES, FLAGGING PERSONNEL AND ALL ASPECTS OF SAFETY LIE WITH THE CONTRACTOR.
20 ...... ALL DISTURBED AREAS SHALL BE SEEDED, FERTILIZED AND MULCHED, OR COVERED W/BASE COURSE AFTER CONSTRUCTION, OR RESTORED WITH BITUMINOUS PAVING WHERE APPLICABLE.
21 ...... THERE SHALL BE NO DEVIATION FROM THE PLANS WITHOUT THE DESIGN ENGINEER'S APPROVAL.
22 ...... CONSTRUCTION STAKES DESTROYED BY THE CONTRACTOR THAT NEED TO BE REPLACED, WILL BE CHARGED TO THE CONTRACTOR.
23 ...... PROPERTY CORNERS KNOWN TO EXIST SHALL NOT BE DISTURBED. DAMAGED CORNERS WILL BE REPLACED AT THE CONTRACTOR'S EXPENSE. PRESERVE ALL SURVEY MONUMENTS.
24 ...... SQUARE CUT ALL BLACKTOP PAVEMENT PRIOR TO TRENCH EXCAVATION.
25 ...... SEE SPECIFICATION BOOK AND FOLLOWING PLAN SHEETS FOR ADDITIONAL DETAILS AND NOTES.
26 ...... SEE SPECIFICATION BOOK AND PLANS FOR ELEVATED TANK & TOWER.

PROJECT LOCATION SEE PROJECT SITE LOCATION

EXISTING ELEVATED TANK

VILLAGE OF GILMAN, WISCONSIN

LOCATION MAP

GENERAL SPECIFICATIONS: STANDARD SPECIFICATIONS FOR SEWER AND WATER CONSTRUCTION IN WISCONSIN (5TH EDITION)

WISCONSIN CONSTRUCTION SITE BEST MANAGEMENT PRACTICE HANDBOOK

SPECIFIC SPECIFICATIONS: MORGAN & PARMLEY, LTD. SPEC. BOOK

BIDDING DOCUMENTS: MORGAN & PARMLEY, LTD. SPEC. BOOK

PROJECT SITE LOCATION

NOTE - SEE TAYLOR CO. REGISTER OF DEEDS' RECORDS; VOL. 35 PAGES 92 & 93 FOR COMPLETE DESCRIPTION OF THIS C.S.M. DOCUMENT.

TAYLOR COUNTY CERTIFIED SURVEY MAP NO. 693

LOT 1 · LOT 2 · OUTLOT 1 · PROJECT SITE

CURVE DATA TABLE

UTILITIES

S & K TV SYSTEMS
508 W. MINER
LADYSMITH, WI 54848
ATTN: RANDY SCOTT

LAKEHEAD PIPELINE
803 HIGHLAND AVENUE
P.O. BOX 308
FT. ATKINSON, WI 53538
ATTN: MARK KINBLOM

NORTHERN STATES POWER CO.
500 N. 5TH STREET
ABBOTSFORD, WI 54405
ATTN: DAVE PEPPER

DIGGER'S HOTLINE

CENTURY TELEPHONE OF NORTHERN WISCONSIN, INC.
425 ELLINGSON AVE.
P.O. BOX 78
HAWKINS, WISCONSIN 54530
ATTN: JAMES ARQUETTE

INDEX OF DRAWINGS

SHEET NO.	DESCRIPTION
1	TITLE SHEET, LOCATION MAP, INDEX & NOTES
2	PROJECT SITE PLAN
3	BASIC GEOMETRY of ELEVATED WATER RESERVOIR
4	DETAILS of ELEVATED WATER RESERVOIR
5	TYPICAL EROSION CONTROL DETAILS — This Sheet Not Shown

PREPARED BY:
MORGAN & PARMLEY, Ltd.
PROFESSIONAL CONSULTING ENGINEERS
LADYSMITH, WISCONSIN

Courtesy: Morgan & Parmley, Ltd.

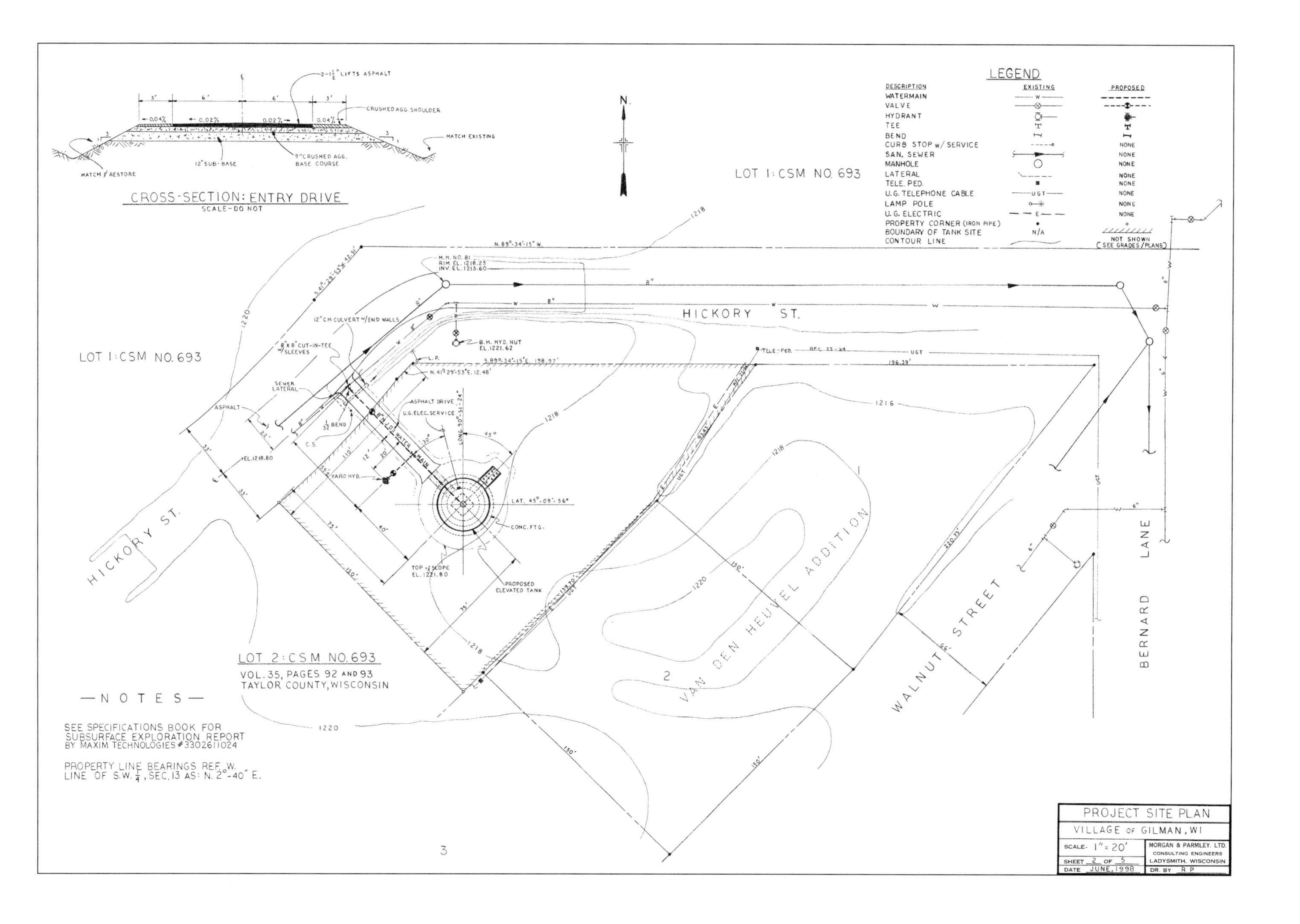
CROSS-SECTION: ENTRY DRIVE
SCALE-DO NOT
2-1½" LIFTS ASPHALT
CRUSHED AGG. SHOULDER
MATCH EXISTING
12" SUB-BASE
9" CRUSHED AGG. BASE COURSE
MATCH & RESTORE
LEGEND
DESCRIPTION
EXISTING
PROPOSED
WATERMAIN
VALVE
HYDRANT
TEE
BEND
CURB STOP w/ SERVICE
SAN. SEWER
MANHOLE
LATERAL
TELE. PED.
U.G. TELEPHONE CABLE
LAMP POLE
U.G. ELECTRIC
PROPERTY CORNER (IRON PIPE)
BOUNDARY OF TANK SITE
CONTOUR LINE
NONE
N/A
NOT SHOWN (SEE GRADES/PLANS)
N.
LOT 1: CSM NO. 693
HICKORY ST.
LOT 1: CSM NO. 693
HICKORY ST.
LOT 2: CSM NO. 693
VOL. 35, PAGES 92 AND 93
TAYLOR COUNTY, WISCONSIN
VAN DEN HEUVEL ADDITION
WALNUT STREET
BERNARD LANE
PROPOSED ELEVATED TANK
CONC. FTG.
ASPHALT DRIVE
U.G. ELEC. SERVICE
YARD HYD.
B.M. HYD. NUT EL. 1221.62
M.H. NO. 81
RIM EL. 1218.25
INV. EL. 1213.60
12" CM CULVERT w/ END WALLS
8"x8" CUT-IN-TEE w/ SLEEVES
SEWER LATERAL
ASPHALT
TOP of SLOPE EL. 1221.80
LAT. 45°-09'-56"
LONG. 90°-51'-24"
N. 89°-34'-15" W.
S. 89°-34'-15" E. 198.97'
N. 41°-29'-53" E. 12.48'
196.39'
TELE-PED.
—NOTES—
SEE SPECIFICATIONS BOOK FOR SUBSURFACE EXPLORATION REPORT BY MAXIM TECHNOLOGIES #3302611024
PROPERTY LINE BEARINGS REF. W. LINE OF S.W. ¼, SEC. 13 AS: N. 2°-40" E.
PROJECT SITE PLAN
VILLAGE OF GILMAN, WI
SCALE- 1" = 20'
SHEET 2 OF 5
DATE JUNE, 1998
MORGAN & PARMLEY, LTD.
CONSULTING ENGINEERS
LADYSMITH, WISCONSIN
DR. BY R.P.

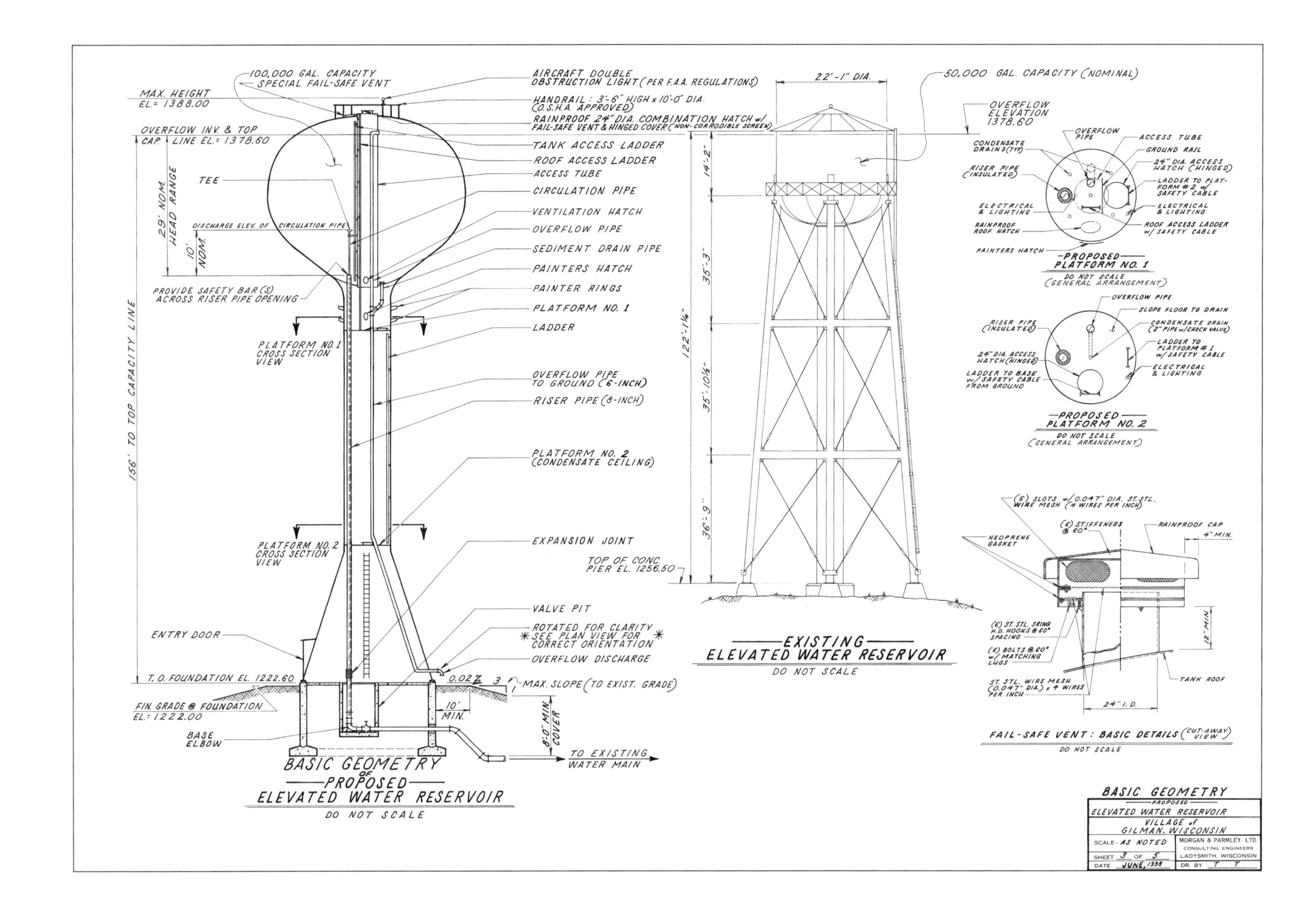
100,000 GAL. CAPACITY
SPECIAL FAIL-SAFE VENT
MAX. HEIGHT EL.= 1388.00
OVERFLOW INV. & TOP CAP. LINE EL.= 1378.60
AIRCRAFT DOUBLE OBSTRUCTION LIGHT (PER F.A.A. REGULATIONS)
HANDRAIL: 3'-6" HIGH x 10'-0" DIA. (O.S.H.A. APPROVED)
RAINPROOF 24" DIA. COMBINATION HATCH w/ FAIL-SAFE VENT & HINGED COVER (NON-CORRODIBLE SCREEN)
TANK ACCESS LADDER
ROOF ACCESS LADDER
ACCESS TUBE
CIRCULATION PIPE
VENTILATION HATCH
OVERFLOW PIPE
SEDIMENT DRAIN PIPE
PAINTERS HATCH
PAINTER RINGS
PLATFORM NO. 1
LADDER
TEE
29' NOM. HEAD RANGE
DISCHARGE ELEV. OF CIRCULATION PIPE
10' NOM.
PROVIDE SAFETY BAR(S) ACROSS RISER PIPE OPENING
PLATFORM NO. 1 CROSS SECTION VIEW
156' TO TOP CAPACITY LINE
OVERFLOW PIPE TO GROUND (6-INCH)
RISER PIPE (8-INCH)
PLATFORM NO. 2 (CONDENSATE CEILING)
PLATFORM NO. 2 CROSS SECTION VIEW
EXPANSION JOINT
VALVE PIT
ROTATED FOR CLARITY
* SEE PLAN VIEW FOR CORRECT ORIENTATION *
OVERFLOW DISCHARGE
ENTRY DOOR
T. O. FOUNDATION EL. 1222.60
0.02 Z
3
1
MAX. SLOPE (TO EXIST. GRADE)
FIN. GRADE @ FOUNDATION EL.= 1222.00
10' MIN.
8'-0" MIN. COVER
BASE ELBOW
TO EXISTING WATER MAIN
BASIC GEOMETRY of PROPOSED ELEVATED WATER RESERVOIR
DO NOT SCALE
22'-1" DIA.
50,000 GAL. CAPACITY (NOMINAL)
OVERFLOW ELEVATION 1378.60
14'-2"
35'-3"
122'-1¼"
35'-10½"
36'-9"
TOP OF CONC. PIER EL. 1256.50
EXISTING ELEVATED WATER RESERVOIR
DO NOT SCALE
OVERFLOW PIPE
ACCESS TUBE
CONDENSATE DRAINS (TYP.)
GROUND RAIL
RISER PIPE (INSULATED)
24" DIA. ACCESS HATCH (HINGED)
LADDER TO PLATFORM #2 w/ SAFETY CABLE
ELECTRICAL & LIGHTING
RAINPROOF ROOF HATCH
ROOF ACCESS LADDER w/ SAFETY CABLE
PAINTERS HATCH
PROPOSED PLATFORM NO. 1
DO NOT SCALE (GENERAL ARRANGEMENT)
OVERFLOW PIPE
SLOPE FLOOR TO DRAIN
RISER PIPE (INSULATED)
CONDENSATE DRAIN (2" PIPE w/CHECK VALVE)
LADDER TO PLATFORM #1 w/ SAFETY CABLE
24" DIA. ACCESS HATCH (HINGED)
ELECTRICAL & LIGHTING
LADDER TO BASE w/ SAFETY CABLE FROM GROUND
PROPOSED PLATFORM NO. 2
DO NOT SCALE (GENERAL ARRANGEMENT)
(6) SLOTS w/0.047" DIA. ST. STL. WIRE MESH (4 WIRES PER INCH)
(6) STIFFENERS @ 60°
RAINPROOF CAP
4" MIN.
NEOPRENE GASKET
(6) ST. STL. SPRING H.D. HOOKS @ 60° SPACING
(6) BOLTS @ 60° w/ MATCHING LUGS
12" MIN.
ST. STL. WIRE MESH (0.047" DIA.) x 4 WIRES PER INCH
TANK ROOF
24" I.D.
FAIL-SAFE VENT: BASIC DETAILS (CUT-AWAY VIEW)
DO NOT SCALE
BASIC GEOMETRY
PROPOSED
ELEVATED WATER RESERVOIR
VILLAGE of GILMAN, WISCONSIN
SCALE- AS NOTED
MORGAN & PARMLEY, LTD.
CONSULTING ENGINEERS
LADYSMITH, WISCONSIN
SHEET 3 OF 5
DATE JUNE, 1998
DR. BY T T

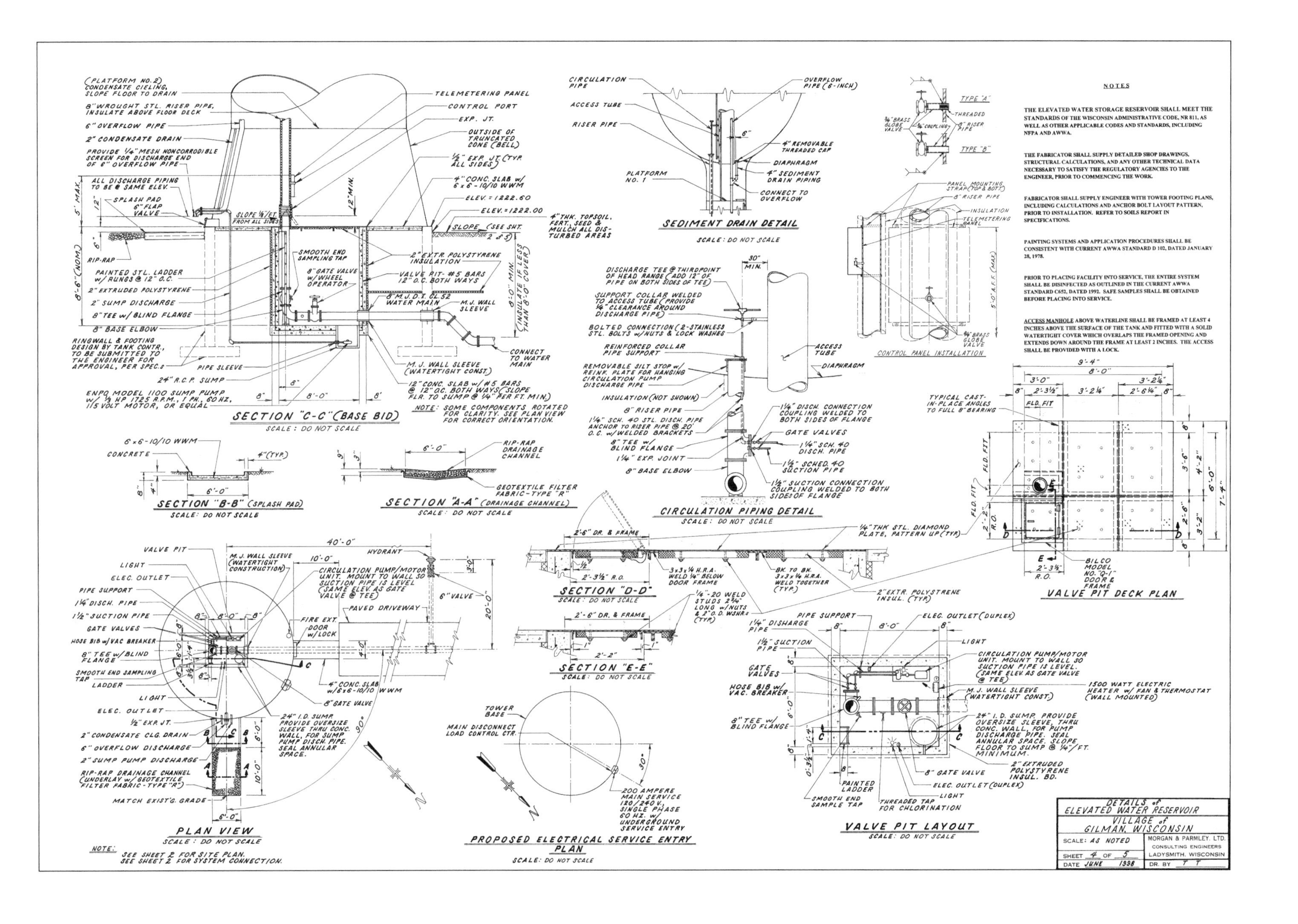
(PLATFORM NO. 2) CONDENSATE CIELING, SLOPE FLOOR TO DRAIN
8" WROUGHT STL. RISER PIPE, INSULATE ABOVE FLOOR DECK
6" OVERFLOW PIPE
2" CONDENSATE DRAIN
PROVIDE 1/4" MESH NONCORRODIBLE SCREEN FOR DISCHARGE END OF 6" OVERFLOW PIPE
ALL DISCHARGE PIPING TO BE @ SAME ELEV.
SPLASH PAD
6" FLAP VALVE
RIP-RAP
PAINTED STL. LADDER w/ RUNGS @ 12" O.C.
2" EXTRUDED POLYSTYRENE
2" SUMP DISCHARGE
8" TEE w/ BLIND FLANGE
8" BASE ELBOW
RINGWALL & FOOTING DESIGN BY TANK CONTR., TO BE SUBMITTED TO THE ENGINEER FOR APPROVAL, PER SPEC.s
PIPE SLEEVE
24" R.C.P. SUMP
ENPO MODEL 1100 SUMP PUMP w/ 1/3 HP 1725 R.P.M., 1 PH, 60 HZ, 115 VOLT MOTOR, OR EQUAL
TELEMETERING PANEL
CONTROL PORT
EXP. JT.
OUTSIDE OF TRUNCATED CONE (BELL)
1/2" EXP. JT. (TYP. ALL SIDES)
4" CONC. SLAB w/ 6 x 6 - 10/10 WWM
ELEV. = 1222.60
ELEV. = 1222.00
SLOPE 1/8"/FT. FROM ALL SIDES
SLOPE (SEE SHT. 2 & 5)
4" THK. TOPSOIL, FERT., SEED & MULCH ALL DISTURBED AREAS
SMOOTH END SAMPLING TAP
8" GATE VALVE w/ WHEEL OPERATOR
2" EXTR. POLYSTYRENE INSULATION
VALVE PIT - #5 BARS 12" O.C. BOTH WAYS
8" M.J. D.I. CL.52 WATER MAIN
M.J. WALL SLEEVE
8'-0" MIN. (INSULATE IF LESS THAN 8'-0" COVER)
CONNECT TO WATER MAIN
M. J. WALL SLEEVE (WATERTIGHT CONST.)
12" CONC. SLAB w/ #5 BARS @ 12" O.C. BOTH WAYS (SLOPE FLR. TO SUMP @ 1/4" PER FT. MIN.)
NOTE: SOME COMPONENTS ROTATED FOR CLARITY. SEE PLAN VIEW FOR CORRECT ORIENTATION.
SECTION "C-C" (BASE BID)
SCALE: DO NOT SCALE
6 x 6 - 10/10 WWM
CONCRETE
4" (TYP.)
SECTION "B-B" (SPLASH PAD)
SCALE: DO NOT SCALE
RIP-RAP DRAINAGE CHANNEL
GEOTEXTILE FILTER FABRIC-TYPE "R"
SECTION "A-A" (DRAINAGE CHANNEL)
SCALE: DO NOT SCALE
CIRCULATION PIPE
ACCESS TUBE
RISER PIPE
PLATFORM NO. 1
OVERFLOW PIPE (6-INCH)
4" REMOVABLE THREADED CAP
DIAPHRAGM
4" SEDIMENT DRAIN PIPING
CONNECT TO OVERFLOW
SEDIMENT DRAIN DETAIL
SCALE: DO NOT SCALE
DISCHARGE TEE @ THIRDPOINT OF HEAD RANGE (ADD 12" OF PIPE ON BOTH SIDES OF TEE)
SUPPORT COLLAR WELDED TO ACCESS TUBE (PROVIDE 1/4" CLEARANCE AROUND DISCHARGE PIPE)
BOLTED CONNECTION (2-STAINLESS STL. BOLTS w/ NUTS & LOCK WASHES)
REINFORCED COLLAR PIPE SUPPORT
REMOVABLE SILT STOP w/ REINF. PLATE FOR HANGING CIRCULATION PUMP DISCHARGE PIPE
INSULATION (NOT SHOWN)
8" RISER PIPE
1 1/4" SCH. 40 STL. DISCH. PIPE ANCHOR TO RISER PIPE @ 20' O.C. w/ WELDED BRACKETS
8" TEE w/ BLIND FLANGE
1 1/4" EXP. JOINT
8" BASE ELBOW
ACCESS TUBE
DIAPHRAGM
1 1/4" DISCH. CONNECTION COUPLING WELDED TO BOTH SIDES OF FLANGE
GATE VALVES
1 1/4" SCH. 40 DISCH. PIPE
1 1/2" SCHED. 40 SUCTION PIPE
1 1/2" SUCTION CONNECTION COUPLING WELDED TO BOTH SIDES OF FLANGE
CIRCULATION PIPING DETAIL
SCALE: DO NOT SCALE
TYPE "A"
TYPE "B"
THREADED
3/4" BRASS GLOBE VALVE
3/4" COUPLING
8" RISER PIPE
PANEL MOUNTING STRAP (TOP & BOTT.)
INSULATION
TELEMETERING PANEL
5'-0" A.F.F. (MAX)
CONTROL PANEL INSTALLATION
NOTES
THE ELEVATED WATER STORAGE RESERVOIR SHALL MEET THE STANDARDS OF THE WISCONSIN ADMINISTRATIVE CODE, NR 811, AS WELL AS OTHER APPLICABLE CODES AND STANDARDS, INCLUDING NFPA AND AWWA.
THE FABRICATOR SHALL SUPPLY DETAILED SHOP DRAWINGS, STRUCTURAL CALCULATIONS, AND ANY OTHER TECHNICAL DATA NECESSARY TO SATISFY THE REGULATORY AGENCIES TO THE ENGINEER, PRIOR TO COMMENCING THE WORK.
FABRICATOR SHALL SUPPLY ENGINEER WITH TOWER FOOTING PLANS, INCLUDING CALCULATIONS AND ANCHOR BOLT LAYOUT PATTERN, PRIOR TO INSTALLATION. REFER TO SOILS REPORT IN SPECIFICATIONS.
PAINTING SYSTEMS AND APPLICATION PROCEDURES SHALL BE CONSISTENT WITH CURRENT AWWA STANDARD D 102, DATED JANUARY 28, 1978.
PRIOR TO PLACING FACILITY INTO SERVICE, THE ENTIRE SYSTEM SHALL BE DISINFECTED AS OUTLINED IN THE CURRENT AWWA STANDARD C652, DATED 1992. SAFE SAMPLES SHALL BE OBTAINED BEFORE PLACING INTO SERVICE.
ACCESS MANHOLE ABOVE WATERLINE SHALL BE FRAMED AT LEAST 4 INCHES ABOVE THE SURFACE OF THE TANK AND FITTED WITH A SOLID WATERTIGHT COVER WHICH OVERLAPS THE FRAMED OPENING AND EXTENDS DOWN AROUND THE FRAME AT LEAST 2 INCHES. THE ACCESS SHALL BE PROVIDED WITH A LOCK.
TYPICAL CAST-IN-PLACE ANGLES TO FULL 8" BEARING
FLD. FIT
BILCO MODEL NO. "Q-1" DOOR & FRAME
VALVE PIT DECK PLAN
1/4" THK STL. DIAMOND PLATE, PATTERN UP (TYP.)
2'-6" DR. & FRAME
3 x 3 x 1/4 H.R.A. WELD 1/4" BELOW DOOR FRAME
BK. TO BK. 3 x 3 x 1/4 H.R.A. WELD TOGETHER (TYP.)
2" EXTR. POLYSTRENE INSUL. (TYP.)
SECTION "D-D"
SCALE: DO NOT SCALE
1/4"-20 WELD STUDS 2 3/4" LONG w/ NUTS & 2" O.D. WSHR.s (TYP.)
SECTION "E-E"
SCALE: DO NOT SCALE
VALVE PIT
LIGHT
ELEC. OUTLET
PIPE SUPPORT
1 1/4" DISCH. PIPE
1 1/2" SUCTION PIPE
GATE VALVES
HOSE BIB w/ VAC BREAKER
8" TEE w/ BLIND FLANGE
SMOOTH END SAMPLING TAP
LADDER
1/2" EXP. JT.
2" CONDENSATE CLG. DRAIN
6" OVERFLOW DISCHARGE
2" SUMP PUMP DISCHARGE
RIP-RAP DRAINAGE CHANNEL (UNDERLAY w/ GEOTEXTILE FILTER FABRIC-TYPE "R")
MATCH EXIST'G. GRADE
M.J. WALL SLEEVE (WATERTIGHT CONSTRUCTION)
HYDRANT
CIRCULATION PUMP/MOTOR UNIT. MOUNT TO WALL SO SUCTION PIPE IS LEVEL (SAME ELEV. AS GATE VALVE @ TEE)
6" VALVE
PAVED DRIVEWAY
FIRE EXT.
DOOR w/ LOCK
4" CONC. SLAB w/ 6 x 6 - 10/10 WWM
8" GATE VALVE
24" I.D. SUMP. PROVIDE OVERSIZE SLEEVE THRU CONC. WALL, FOR SUMP PUMP DISCH. PIPE. SEAL ANNULAR SPACE.
PLAN VIEW
SCALE: DO NOT SCALE
NOTE: SEE SHEET 2 FOR SITE PLAN. SEE SHEET 2 FOR SYSTEM CONNECTION.
TOWER BASE
MAIN DISCONNECT LOAD CONTROL CTR.
200 AMPERE MAIN SERVICE 120/240 V., SINGLE PHASE 60 HZ. w/ UNDERGROUND SERVICE ENTRY
PROPOSED ELECTRICAL SERVICE ENTRY PLAN
SCALE: DO NOT SCALE
PIPE SUPPORT
1 1/4" DISHARGE PIPE
1 1/2" SUCTION PIPE
GATE VALVES
HOSE BIB w/ VAC. BREAKER
8" TEE w/ BLIND FLANGE
ELEC. OUTLET (DUPLEX)
LIGHT
CIRCULATION PUMP/MOTOR UNIT. MOUNT TO WALL SO SUCTION PIPE IS LEVEL. (SAME ELEV. AS GATE VALVE @ TEE)
M. J. WALL SLEEVE (WATERTIGHT CONST.)
1500 WATT ELECTRIC HEATER w/ FAN & THERMOSTAT (WALL MOUNTED)
24" I. D. SUMP. PROVIDE OVERSIZE SLEEVE, THRU CONC. WALL, FOR PUMP DISCHARGE PIPE. SEAL ANNULAR SPACE. SLOPE FLOOR TO SUMP @ 1/4"/FT. MINIMUM.
2" EXTRUDED POLYSTYRENE INSUL. BD.
8" GATE VALVE
ELEC. OUTLET (DUPLEX)
LIGHT
PAINTED LADDER
SMOOTH END SAMPLE TAP
THREADED TAP FOR CHLORINATION
VALVE PIT LAYOUT
SCALE: DO NOT SCALE
DETAILS of ELEVATED WATER RESERVOIR
VILLAGE of GILMAN, WISCONSIN
SCALE: AS NOTED
SHEET 4 OF 5
DATE JUNE 1998
MORGAN & PARMLEY, LTD.
CONSULTING ENGINEERS
LADYSMITH, WISCONSIN
DR. BY T T

FIRE PROTECTION FACILITY
FOR
INDUSTRIAL PARK COMPLEX

E.D.A. PROJECT NO. 06-01-02571

VILLAGE
OF
GLEN FLORA, WISCONSIN
SEPT., 1992

UTILITIES

CONTACT - NORTHERN STATES POWER CO.
711 W. 9th Street, North
Ladysmith, WI 54848

Attn: John Rymarkiewicz
715-532-6226

CONTACT - UNIVERSAL TELEPHONE CO.
OF NORTHERN WISCONSIN
Highway "8"
Hawkins, WI 54530

Attn: Jim Arquette
715-585-7707

CONTACT - DIGGER'S HOTLINE

1-800-242-8511

Ticket's Nos.: 1598953
1598962

CONTACT - VILLAGE OF GLEN FLORA WWCTF

Les Evjen, Operator
P.O. Box 253
Glen Flora, WI 54526

715-322-5511

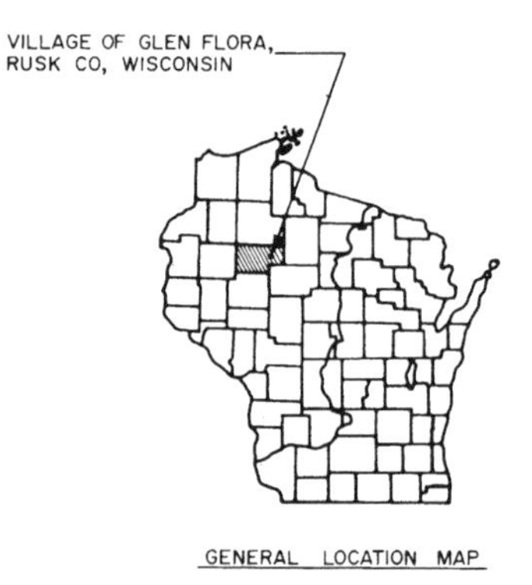

GENERAL LOCATION MAP

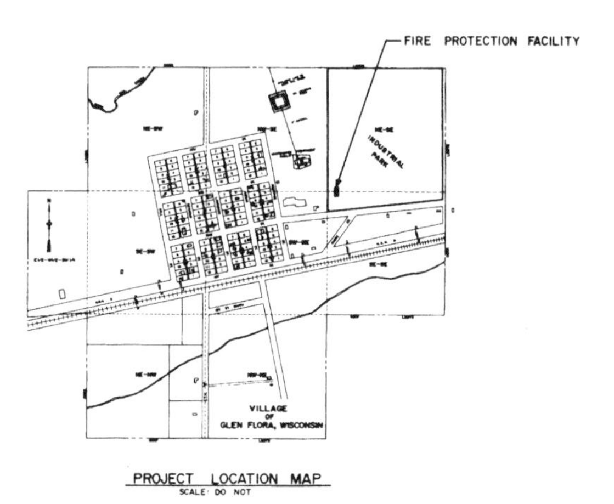

PROJECT LOCATION MAP
SCALE: DO NOT

INDEX

NO.	DESCRIPTION
1	TITLE SHEET, GENERAL LOCATION & INDEX
2	GENERAL SITE PLAN W/ SOIL BORINGS & TOPOG.
3	RESERVOIR PLAN, LONGITUDINAL-SECTION & YARD PIPING
4	RESERVOIR ROOF PLAN & PUMPHOUSE FLOOR
5	RESERVOIR DETAILS
6	PUMPHOUSE EQUIPMENT ARRANGEMENT
7	EQUIPMENT DETAILS & CROSS-SECTIONS
8	PUMPHOUSE LAYOUT & DETAILS
9	PUMPHOUSE HEATING, VENTILATING & ELEC.
10	MISC. DIAGRAMS, LAYOUTS, FUEL TANK, PUMPHOUSE SPRINKLER SYSTEM & DETAILS
11	TYPICAL EROSION CONTROL DETAILS This Detail Sheet Not Shown

PREPARED BY:
MORGAN & PARMLEY LTD.
CONSULTING ENGINEERS
LADYSMITH, WISCONSIN

Courtesy: Morgan & Parmley, Ltd.

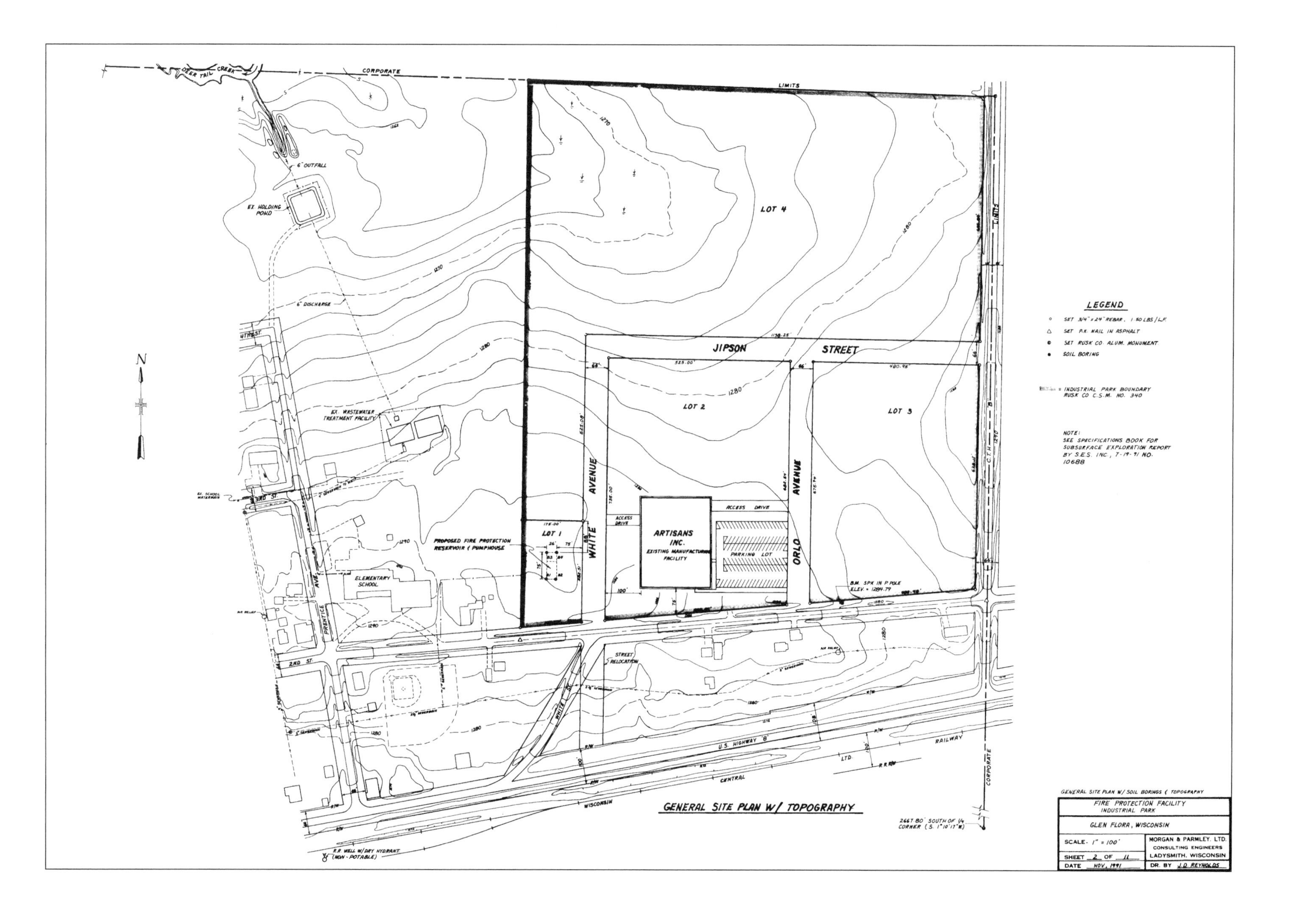
CORPORATE
LIMITS
DEER TAIL CREEK
6" OUTFALL
EX. HOLDING POND
6" DISCHARGE
EX. WASTEWATER TREATMENT FACILITY
LOT 4
JIPSON
STREET
LOT 2
LOT 3
LOT 1
WHITE
AVENUE
ORLO
AVENUE
ARTISANS INC.
EXISTING MANUFACTURING FACILITY
ACCESS DRIVE
PARKING LOT
PROPOSED FIRE PROTECTION RESERVOIR & PUMPHOUSE
ELEMENTARY SCHOOL
PRENTICE AVE.
3RD ST.
2ND ST.
4TH ST.
STREET RELOCATION
U.S. HIGHWAY "8"
RAILWAY
WISCONSIN
CENTRAL
LTD.
C.T.H. "B"
B.M. SPK IN P. POLE ELEV. = 1284.79
R.R. WELL W/ DRY HYDRANT (NON-POTABLE)
2667.80' SOUTH OF 1/4 CORNER (S. 1°10'17"W)
GENERAL SITE PLAN W/ TOPOGRAPHY
LEGEND
SET 3/4" x 24" REBAR, 1.50 LBS/L.F.
SET P.K. NAIL IN ASPHALT
SET RUSK CO. ALUM. MONUMENT
SOIL BORING
= INDUSTRIAL PARK BOUNDARY RUSK CO C.S.M. NO. 340
NOTE:
SEE SPECIFICATIONS BOOK FOR SUBSURFACE EXPLORATION REPORT BY S.E.S. INC., 7-19-91 NO. 10688
GENERAL SITE PLAN W/ SOIL BORINGS & TOPOGRAPHY
FIRE PROTECTION FACILITY
INDUSTRIAL PARK
GLEN FLORA, WISCONSIN
SCALE- 1" = 100'
SHEET 2 OF 11
DATE NOV, 1991
MORGAN & PARMLEY, LTD.
CONSULTING ENGINEERS
LADYSMITH, WISCONSIN
DR. BY J.D. REYNOLDS

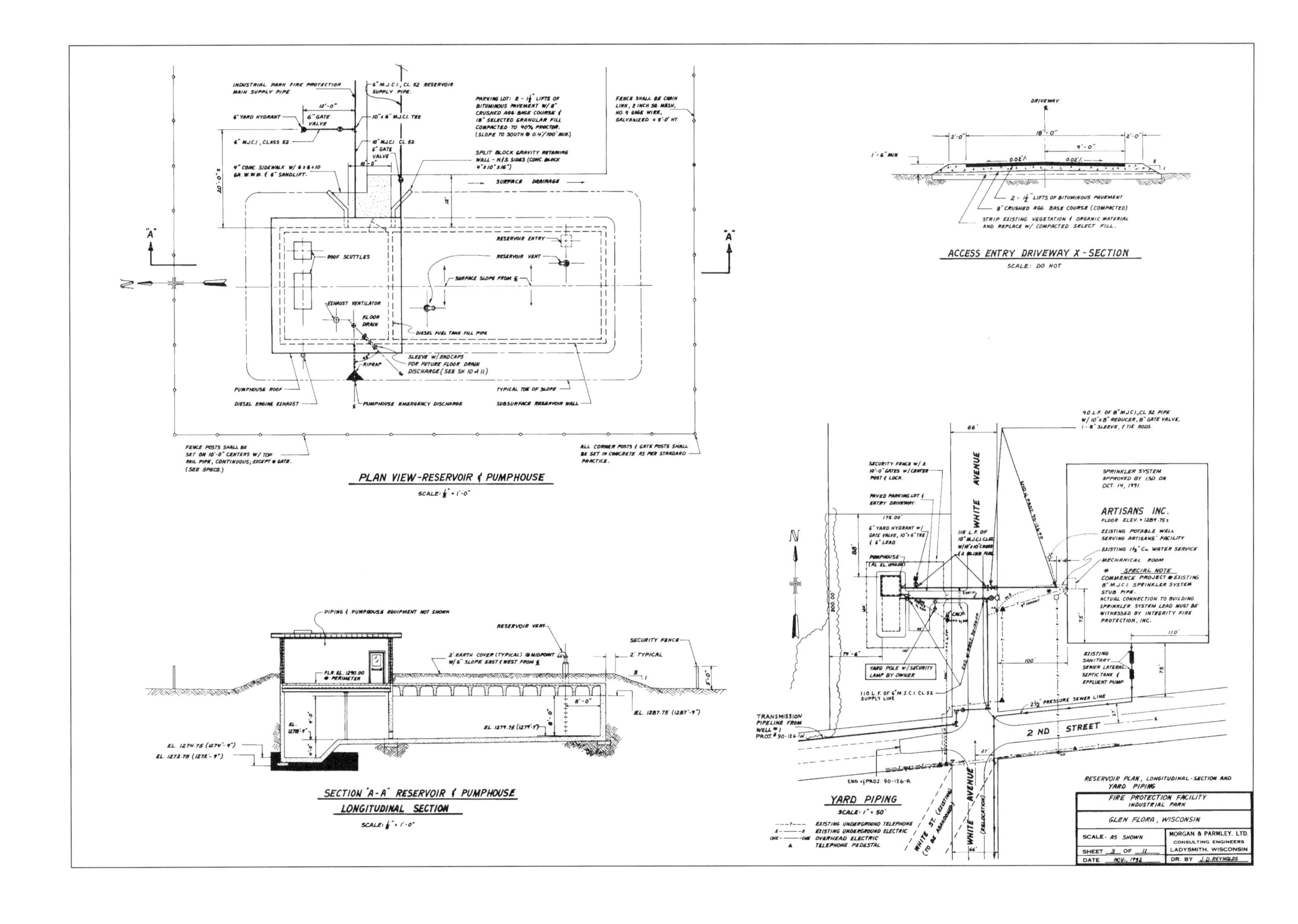
INDUSTRIAL PARK FIRE PROTECTION MAIN SUPPLY PIPE
6" M.J.C.I., CL. 52 RESERVOIR SUPPLY PIPE.
12'-0"
6" YARD HYDRANT
6" GATE VALVE
10" x 6" M.J.C.I. TEE
6" M.J.C.I., CLASS 52
10" M.J.C.I. CL. 52
6" GATE VALVE
10'-0"
4" CONC. SIDEWALK W/ 6 x 6 x 10 GA. W.W.M. & 6" SANDLIFT.
20'-0" ±
PARKING LOT: 2 - 1½" LIFTS OF BITUMINOUS PAVEMENT W/ 8" CRUSHED AGG. BASE COURSE & 18" SELECTED GRANULAR FILL COMPACTED TO 90% PROCTOR. (SLOPE TO SOUTH @ 0.4'/100' MIN.)
SPLIT BLOCK GRAVITY RETAINING WALL - N/S SIDES (CONC. BLOCK 4" x 10" x 16")
SURFACE DRAINAGE
FENCE SHALL BE CHAIN LINK, 2 INCH SQ. MESH, NO. 9 GAGE WIRE, GALVANIZED x 8'-0" HT.
12'
"A"
RESERVOIR ENTRY
RESERVOIR VENT
ROOF SCUTTLES
SURFACE SLOPE FROM ℄
EXHAUST VENTILATOR
FLOOR DRAIN
DIESEL FUEL TANK FILL PIPE
SLEEVE W/ ENDCAPS FOR FUTURE FLOOR DRAIN DISCHARGE (SEE SH. 10 & 11)
RIPRAP
PUMPHOUSE ROOF
DIESEL ENGINE EXHAUST
PUMPHOUSE EMERGENCY DISCHARGE
TYPICAL TOE OF SLOPE
SUBSURFACE RESERVOIR WALL
FENCE POSTS SHALL BE SET ON 10'-0" CENTERS W/ TOP RAIL PIPE, CONTINUOUS; EXCEPT @ GATE. (SEE SPECS.)
ALL CORNER POSTS & GATE POSTS SHALL BE SET IN CONCRETE AS PER STANDARD PRACTICE.
PLAN VIEW-RESERVOIR & PUMPHOUSE
SCALE: ⅛" = 1'-0"
DRIVEWAY
2'-0"
18'-0"
9'-0"
1'-6" MIN.
0.02'/.
2 - 1½" LIFTS OF BITUMINOUS PAVEMENT
8" CRUSHED AGG. BASE COURSE (COMPACTED)
STRIP EXISTING VEGETATION & ORGANIC MATERIAL AND REPLACE W/ COMPACTED SELECT FILL.
ACCESS ENTRY DRIVEWAY X-SECTION
SCALE: DO NOT
PIPING & PUMPHOUSE EQUIPMENT NOT SHOWN
RESERVOIR VENT
SECURITY FENCE
2' EARTH COVER (TYPICAL) @ MIDPOINT W/ 6" SLOPE EAST & WEST FROM ℄
2' TYPICAL
FLR. EL. 1290.00 @ PERIMETER
8'-0"
EL. 1287.75 (1287'-9")
EL. 1279.75 (1279'-9")
EL. 1278'-9"
EL. 1274.75 (1274'-9")
EL. 1272.75 (1272'-9")
SECTION "A-A" RESERVOIR & PUMPHOUSE
LONGITUDINAL SECTION
SCALE: ⅛" = 1'-0"
90 L.F. OF 8" M.J.C.I., CL 52 PIPE W/ 10" x 8" REDUCER, 8" GATE VALVE, 1 - 8" SLEEVE, & TIE RODS.
66'
SECURITY FENCE W/ 2 10'-0" GATES W/ CENTER POST & LOCK.
PAVED PARKING LOT & ENTRY DRIVEWAY.
WHITE AVENUE
SPRINKLER SYSTEM APPROVED BY I.S.O. ON OCT. 14, 1991.
ARTISANS INC.
FLOOR ELEV. = 1289.75 ±
EXISTING POTABLE WELL SERVING ARTISANS' FACILITY
EXISTING 1½" Cu. WATER SERVICE
MECHANICAL ROOM
* SPECIAL NOTE
COMMENCE PROJECT @ EXISTING 8" M.J.C.I. SPRINKLER SYSTEM STUB PIPE.
ACTUAL CONNECTION TO BUILDING SPRINKLER SYSTEM LEAD MUST BE WITNESSED BY INTEGRITY FIRE PROTECTION, INC.
175.00'
88'
6" YARD HYDRANT W/ GATE VALVE, 10" x 6" TEE & 6" LEAD
115 L.F. OF 10" M.J.C.I. CL. 52
PUMPHOUSE
300.00'
110'
75'
100'
YARD POLE W/ SECURITY LAMP BY OWNER
EXISTING SANITARY SEWER LATERAL, SEPTIC TANK & EFFLUENT PUMP
110 L.F. OF 6" M.J.C.I. CL 52 SUPPLY LINE
2½" PRESSURE SEWER LINE
TRANSMISSION PIPELINE FROM WELL # 1 PROJ. # 90-126-W
2 ND STREET
27'
END # PROJ. 90-126-R
WHITE ST. (EXISTING) (TO BE ABANDONED)
WHITE AVENUE (RELOCATION)
YARD PIPING
SCALE: 1" = 50'
EXISTING UNDERGROUND TELEPHONE
EXISTING UNDERGROUND ELECTRIC
OVERHEAD ELECTRIC
TELEPHONE PEDESTAL
RESERVOIR PLAN, LONGITUDINAL-SECTION AND YARD PIPING
FIRE PROTECTION FACILITY
INDUSTRIAL PARK
GLEN FLORA, WISCONSIN
SCALE- AS SHOWN
SHEET 3 OF 11
DATE NOV. 1992
MORGAN & PARMLEY, LTD.
CONSULTING ENGINEERS
LADYSMITH, WISCONSIN
DR. BY J.D. REYNOLDS

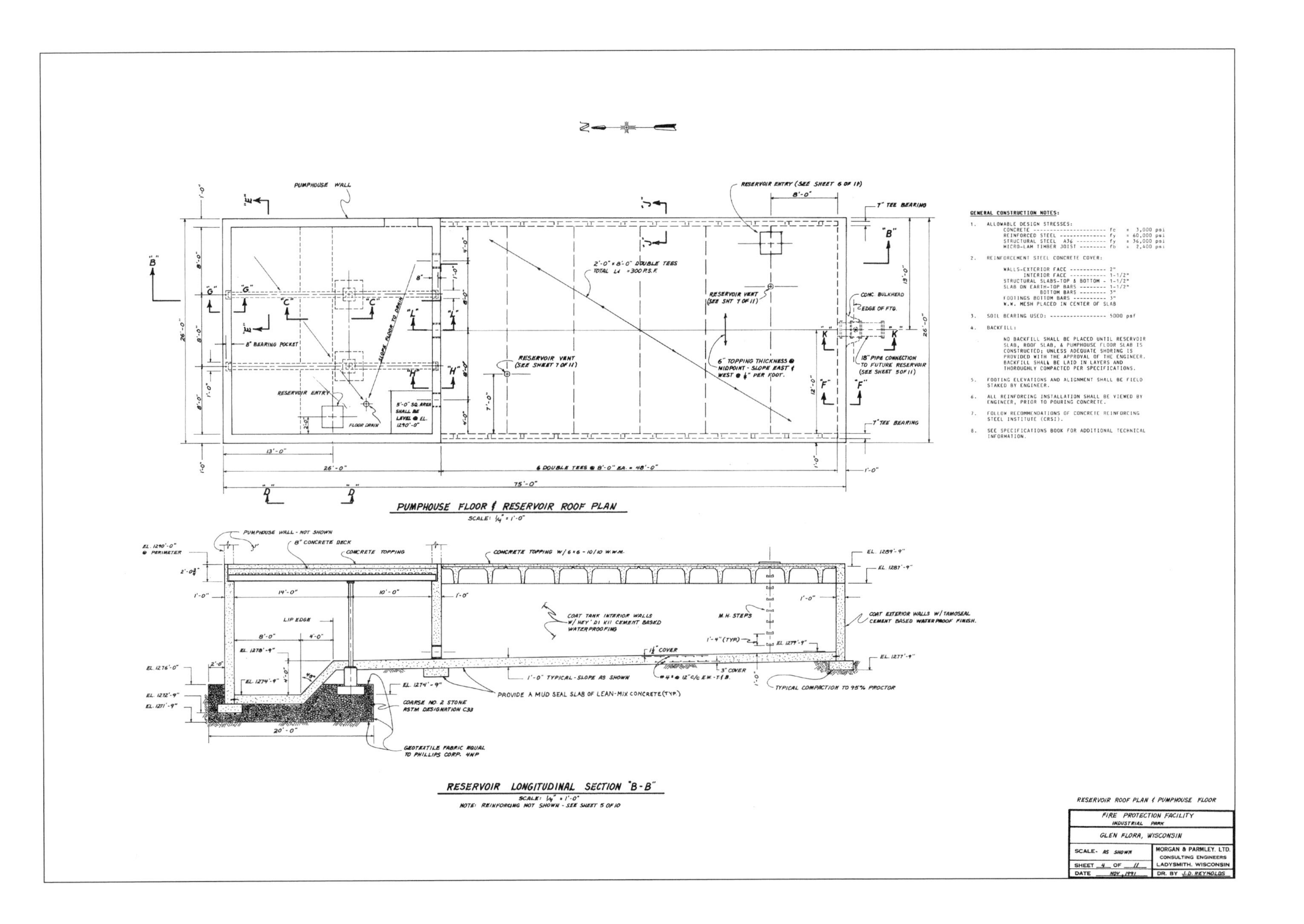
PUMPHOUSE WALL
RESERVOIR ENTRY (SEE SHEET 6 OF 11)
7" TEE BEARING
2'-0" x 8'-0" DOUBLE TEES
TOTAL LL = 300 P.S.F.
RESERVOIR VENT (SEE SHT 7 OF 11)
CONC. BULKHEAD
EDGE OF FTG.
8" BEARING POCKET
SLOPE FLOOR TO DRAIN
RESERVOIR VENT (SEE SHEET 7 OF 11)
6" TOPPING THICKNESS @ MIDPOINT - SLOPE EAST & WEST @ 1/8" PER FOOT.
18" PIPE CONNECTION TO FUTURE RESERVOIR (SEE SHEET 5 OF 11)
RESERVOIR ENTRY
5'-0" SQ. AREA SHALL BE LEVEL @ EL. 1290'-0"
FLOOR DRAIN
6 DOUBLE TEES @ 8'-0" EA. = 48'-0"
75'-0"
PUMPHOUSE FLOOR & RESERVOIR ROOF PLAN
SCALE: 1/4" = 1'-0"
GENERAL CONSTRUCTION NOTES:
1. ALLOWABLE DESIGN STRESSES:
CONCRETE ---------------------- fc = 3,000 psi
REINFORCED STEEL -------------- fy = 60,000 psi
STRUCTURAL STEEL A36 --------- fy = 36,000 psi
MICRO-LAM TIMBER JOIST -------- fb = 2,400 psi
2. REINFORCEMENT STEEL CONCRETE COVER:
WALLS-EXTERIOR FACE ----------- 2"
INTERIOR FACE ----------- 1-1/2"
STRUCTURAL SLABS-TOP & BOTTOM - 1-1/2"
SLAB ON EARTH-TOP BARS -------- 1-1/2"
BOTTOM BARS -------- 3"
FOOTINGS BOTTOM BARS ---------- 3"
W.W. MESH PLACED IN CENTER OF SLAB
3. SOIL BEARING USED: ---------------- 5000 psf
4. BACKFILL:
NO BACKFILL SHALL BE PLACED UNTIL RESERVOIR SLAB, ROOF SLAB, & PUMPHOUSE FLOOR SLAB IS CONSTRUCTED; UNLESS ADEQUATE SHORING IS PROVIDED WITH THE APPROVAL OF THE ENGINEER. BACKFILL SHALL BE LAID IN LAYERS AND THOROUGHLY COMPACTED PER SPECIFICATIONS.
5. FOOTING ELEVATIONS AND ALIGNMENT SHALL BE FIELD STAKED BY ENGINEER.
6. ALL REINFORCING INSTALLATION SHALL BE VIEWED BY ENGINEER, PRIOR TO POURING CONCRETE.
7. FOLLOW RECOMMENDATIONS OF CONCRETE REINFORCING STEEL INSTITUTE (CRSI).
8. SEE SPECIFICATIONS BOOK FOR ADDITIONAL TECHNICAL INFORMATION.
PUMPHOUSE WALL - NOT SHOWN
8" CONCRETE DECK
CONCRETE TOPPING
CONCRETE TOPPING W/ 6x6 - 10/10 W.W.M.
EL. 1290'-0" @ PERIMETER
EL. 1289'-9"
EL. 1287'-9"
LIP EDGE
COAT TANK INTERIOR WALLS W/ HEY' DI KII CEMENT BASED WATERPROOFING
M.H. STEPS
COAT EXTERIOR WALLS W/ TAMOSEAL CEMENT BASED WATERPROOF FINISH.
1 1/2" COVER
3" COVER
1'-0" TYPICAL - SLOPE AS SHOWN
#4 @ 12" C/C E.W. - T & B.
EL. 1277'-9"
PROVIDE A MUD SEAL SLAB OF LEAN-MIX CONCRETE (TYP.)
TYPICAL COMPACTION TO 95% PROCTOR
COARSE NO. 2 STONE ASTM DESIGNATION C33
GEOTEXTILE FABRIC EQUAL TO PHILLIPS CORP. 4NP
RESERVOIR LONGITUDINAL SECTION "B-B"
SCALE: 1/4" = 1'-0"
NOTE: REINFORCING NOT SHOWN - SEE SHEET 5 OF 10
RESERVOIR ROOF PLAN & PUMPHOUSE FLOOR
FIRE PROTECTION FACILITY
INDUSTRIAL PARK
GLEN FLORA, WISCONSIN
SCALE- AS SHOWN
SHEET 4 OF 11
DATE NOV. 1991
MORGAN & PARMLEY, LTD.
CONSULTING ENGINEERS
LADYSMITH, WISCONSIN
DR. BY J.D. REYNOLDS

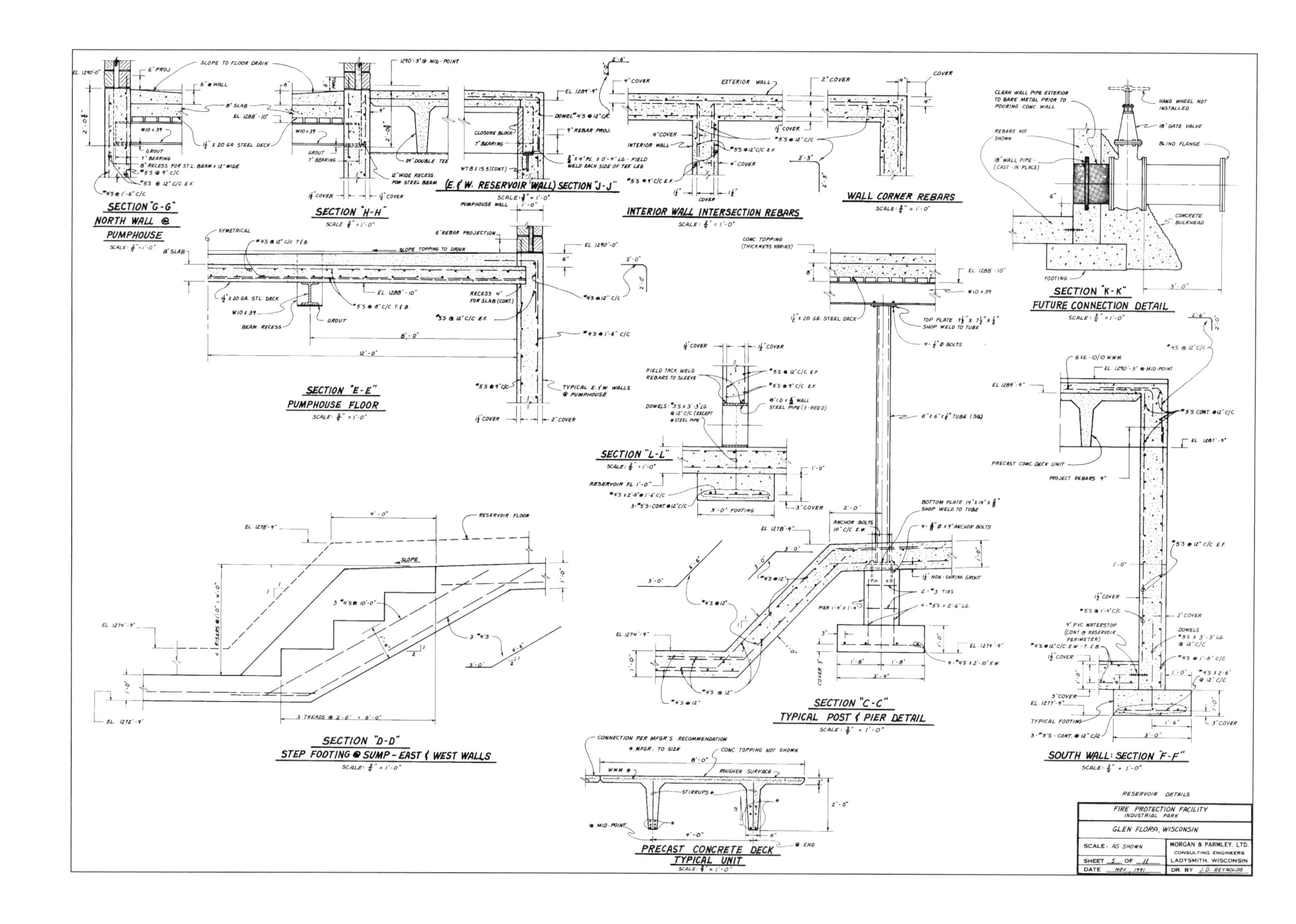
SECTION "G-G"
NORTH WALL @
PUMPHOUSE
SCALE: 3/4" = 1'-0"
SECTION "H-H"
SCALE: 3/4" = 1'-0"
(E. & W. RESERVOIR WALL) SECTION "J-J"
SCALE: 3/4" = 1'-0"
INTERIOR WALL INTERSECTION REBARS
SCALE: 3/4" = 1'-0"
WALL CORNER REBARS
SCALE: 3/4" = 1'-0"
SECTION "K-K"
FUTURE CONNECTION DETAIL
SCALE: 3/4" = 1'-0"
SECTION "E-E"
PUMPHOUSE FLOOR
SCALE: 3/4" = 1'-0"
SECTION "L-L"
SCALE: 3/4" = 1'-0"
SECTION "C-C"
TYPICAL POST & PIER DETAIL
SCALE: 3/4" = 1'-0"
SECTION "D-D"
STEP FOOTING @ SUMP - EAST & WEST WALLS
SCALE: 3/4" = 1'-0"
SOUTH WALL: SECTION "F-F"
SCALE: 3/4" = 1'-0"
PRECAST CONCRETE DECK
TYPICAL UNIT
SCALE: 3/4" = 1'-0"
SLOPE TO FLOOR DRAIN
6" PROJ.
6" @ WALL
8" SLAB
EL. 1288'-10"
1 1/2" X 20 GA STEEL DECK
GROUT
7" BEARING
8" RECESS FOR ST'L. BEAM x 12" WIDE
1290'-3" @ MID-POINT
24" DOUBLE TEE
12" WIDE RECESS FOR STEEL BEAM
CLOSURE BLOCK
WT 8 x 15.5 (CONT.)
EXTERIOR WALL
INTERIOR WALL
CLEAN WALL PIPE EXTERIOR TO BARE METAL PRIOR TO POURING CONC. WALL
HAND WHEEL NOT INSTALLED
18" GATE VALVE
BLIND FLANGE
18" WALL PIPE - (CAST - IN - PLACE)
CONCRETE BULKHEAD
FOOTING
SYMETRICAL
6" REBAR PROJECTION
SLOPE TOPPING TO DRAIN
EL. 1290'-0"
BEAM RECESS
TYPICAL E. & W. WALLS @ PUMPHOUSE
CONC. TOPPING (THICKNESS VARIES)
6" x 6" x 1/4" TUBE (SQ)
TOP PLATE 7 1/2" x 7 1/2" x 1/2" SHOP WELD TO TUBE
BOTTOM PLATE 14" x 14" x 5/8" SHOP WELD TO TUBE
1 1/2" NON-SHRINK GROUT
FIELD TACK WELD REBARS TO SLEEVE
3'-0" FOOTING
RESERVOIR FLOOR
SLOPE
3 TREADS @ 2'-0" = 6'-0"
PRECAST CONC. DECK UNIT
PROJECT REBARS 9"
9" PVC WATERSTOP (CONT. @ RESERVOIR PERIMETER)
TYPICAL FOOTING
CONNECTION PER MFGR'S RECOMMENDATION
* MFGR. TO SIZE
CONC. TOPPING NOT SHOWN
ROUGHEN SURFACE
STIRRUPS
@ MID-POINT
@ END
RESERVOIR DETAILS
FIRE PROTECTION FACILITY
INDUSTRIAL PARK
GLEN FLORA, WISCONSIN
SCALE - AS SHOWN
SHEET 5 OF 11
DATE NOV. 1991
MORGAN & PARMLEY, LTD.
CONSULTING ENGINEERS
LADYSMITH, WISCONSIN
DR. BY J.D. REYNOLDS

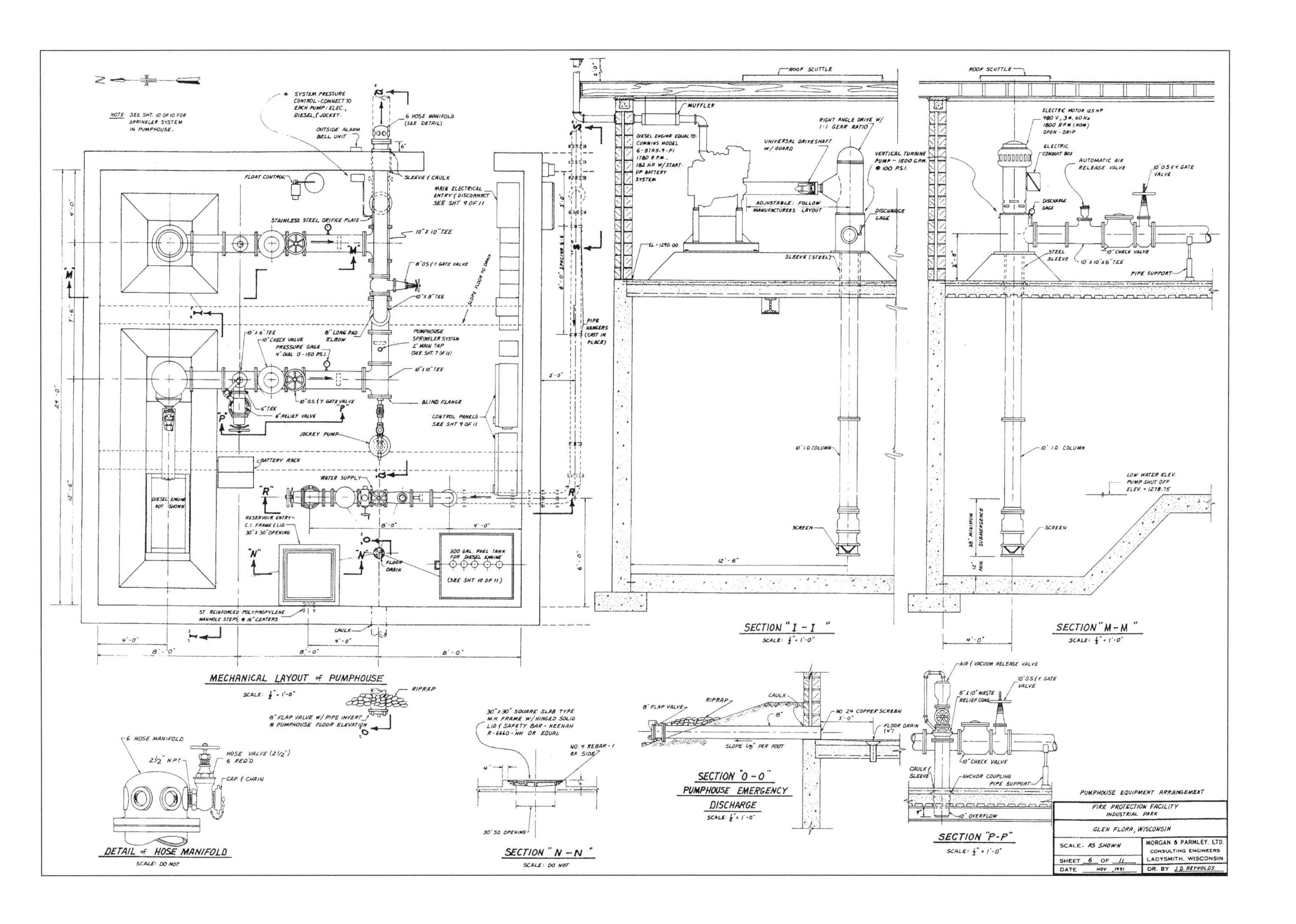
MECHANICAL LAYOUT of PUMPHOUSE
SCALE: ½" = 1'-0"
SECTION "I - I"
SCALE: ½" = 1'-0"
SECTION "M-M"
SCALE: ½" = 1'-0"
DETAIL of HOSE MANIFOLD
SCALE: DO NOT
SECTION "N - N"
SCALE: DO NOT
SECTION "O - O"
PUMPHOUSE EMERGENCY
DISCHARGE
SCALE: ½" = 1'-0"
SECTION "P-P"
SCALE: ½" = 1'-0"
PUMPHOUSE EQUIPMENT ARRANGEMENT
FIRE PROTECTION FACILITY
INDUSTRIAL PARK
GLEN FLORA, WISCONSIN
SCALE- AS SHOWN
SHEET 6 OF 11
DATE NOV 1991
MORGAN & PARMLEY, LTD.
CONSULTING ENGINEERS
LADYSMITH, WISCONSIN
DR. BY J.D. REYNOLDS
NOTE: SEE SHT. 10 OF 10 FOR SPRINKLER SYSTEM IN PUMPHOUSE.
ROOF SCUTTLE
MUFFLER
DIESEL ENGINE EQUAL TO: CUMMINS MODEL 6-BTA5.9-F1 1780 R.P.M., 182 HP W/ START-UP BATTERY SYSTEM
RIGHT ANGLE DRIVE W/ 1:1 GEAR RATIO
UNIVERSAL DRIVESHAFT W/ GUARD
VERTICAL TURBINE PUMP - 1500 G.P.M. @ 100 P.S.I.
ELECTRIC MOTOR 125 HP 480 V, 3Φ, 60 Hz 1800 RPM (NOM) OPEN - DRIP
AUTOMATIC AIR RELEASE VALVE
10" O.S.& Y. GATE VALVE
10" I.D. COLUMN
SCREEN
LOW WATER ELEV. PUMP SHUT OFF ELEV. = 1278.75'
300 GAL. FUEL TANK FOR DIESEL ENGINE (SEE SHT 10 OF 11)
JOCKEY PUMP
BATTERY RACK
WATER SUPPLY
CONTROL PANELS SEE SHT 9 OF 11
6 HOSE MANIFOLD
HOSE VALVE (2½") 6 REQ'D
CAP & CHAIN
2½" NPT
RIPRAP
8" FLAP VALVE
NO. 24 COPPER SCREEN
FLOOR DRAIN (4")
SLOPE ⅛" PER FOOT
AIR & VACUUM RELEASE VALVE
10" CHECK VALVE
ANCHOR COUPLING
PIPE SUPPORT
10" OVERFLOW
30" SQ. OPENING
NO. 4 REBAR - 1 EA. SIDE

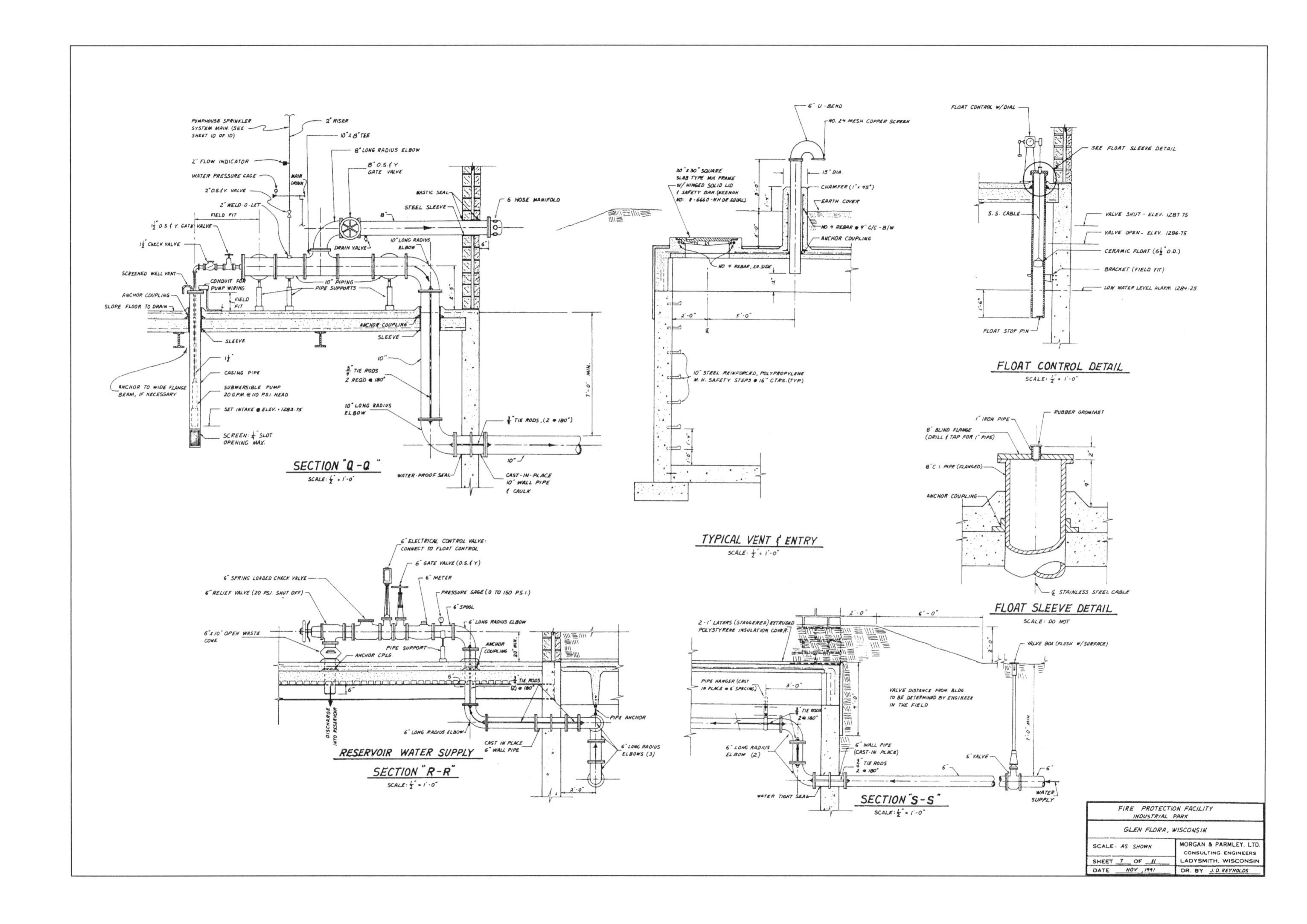

PUMPHOUSE SPRINKLER SYSTEM MAIN. (SEE SHEET 10 OF 10)
2" RISER
10" X 8" TEE
8" LONG RADIUS ELBOW
2" FLOW INDICATOR
WATER PRESSURE GAGE
MAIN DRAIN
8" O.S. & Y GATE VALVE
2" O.S. & Y. VALVE
2" WELD-O-LET
FIELD FIT
MASTIC SEAL
STEEL SLEEVE
6 HOSE MANIFOLD
1½" O.S. & Y. GATE VALVE
1½" CHECK VALVE
DRAIN VALVE
10" LONG RADIUS ELBOW
SCREENED WELL VENT
CONDUIT FOR PUMP WIRING
10" PIPING
PIPE SUPPORTS
ANCHOR COUPLING
SLOPE FLOOR TO DRAIN
FIELD FIT
SLEEVE
1½"
CASING PIPE
SUBMERSIBLE PUMP 20 G.P.M. @ 110 P.S.I. HEAD
SET INTAKE @ ELEV. = 1283.75
SCREEN: ⅛" SLOT OPENING MAX.
ANCHOR TO WIDE FLANGE BEAM, IF NECESSARY
¾" TIE RODS 2 REQD @ 180°
10" LONG RADIUS ELBOW
¾" TIE RODS, (2 @ 180°)
7'-0" MIN.
WATER-PROOF SEAL
CAST-IN-PLACE 10" WALL PIPE & CAULK
SECTION "Q-Q"
SCALE: ½" = 1'-0"
6" U-BEND
NO. 24 MESH COPPER SCREEN
30" X 30" SQUARE SLAB TYPE MH FRAME W/ HINGED SOLID LID & SAFETY BAR (NEENAH NO: R-6660-NH OR EQUAL)
15" DIA.
CHAMFER (1" × 45°)
EARTH COVER
NO. 4 REBAR @ 9" C/C - B/W
ANCHOR COUPLING
NO. 4 REBAR, EA. SIDE
2'-0"
5'-0"
10" STEEL REINFORCED, POLYPROPYLENE M.H. SAFETY STEPS @ 16" CTRS. (TYP.)
TYPICAL VENT & ENTRY
SCALE: ½" = 1'-0"
FLOAT CONTROL W/ DIAL
SEE FLOAT SLEEVE DETAIL
S.S. CABLE
VALVE SHUT - ELEV. 1287.75
VALVE OPEN - ELEV. 1286.75
CERAMIC FLOAT (6½" O.D.)
BRACKET (FIELD FIT)
LOW WATER LEVEL ALARM 1284.25
FLOAT STOP PIN
FLOAT CONTROL DETAIL
SCALE: ½" = 1'-0"
RUBBER GROMMET
1" IRON PIPE
8" BLIND FLANGE (DRILL & TAP FOR 1" PIPE)
8" C.I. PIPE (FLANGED)
ANCHOR COUPLING
STAINLESS STEEL CABLE
FLOAT SLEEVE DETAIL
SCALE: DO NOT
6" ELECTRICAL CONTROL VALVE: CONNECT TO FLOAT CONTROL
6" GATE VALVE (O.S. & Y.)
6" SPRING LOADED CHECK VALVE
6" METER
6" RELIEF VALVE (20 PSI. SHUT OFF)
PRESSURE GAGE (0 TO 150 P.S.I.)
6" SPOOL
6" LONG RADIUS ELBOW
6" X 10" OPEN WASTE CONE
PIPE SUPPORT
ANCHOR CPLG
ANCHOR COUPLING
20" MIN.
¾" TIE RODS (2) @ 180°
DISCHARGE INTO RESERVOIR
PIPE ANCHOR
6" LONG RADIUS ELBOW
CAST IN PLACE 6" WALL PIPE
6" LONG RADIUS ELBOWS (3)
RESERVOIR WATER SUPPLY
SECTION "R-R"
SCALE: ½" = 1'-0"
2-1" LAYERS (STAGGERED) EXTRUDED POLYSTYRENE INSULATION COVER.
2'-0"
6'-0"
VALVE BOX (FLUSH W/ SURFACE)
PIPE HANGER (CAST IN PLACE @ 6' SPACING)
3'-0"
VALVE DISTANCE FROM BLDG. TO BE DETERMINED BY ENGINEER IN THE FIELD
¾" TIE RODS 2 @ 180°
6" LONG RADIUS ELBOW (2)
6" WALL PIPE (CAST-IN-PLACE)
¾" TIE RODS 2 @ 180°
7'-0" MIN.
6" VALVE
WATER TIGHT SEAL
WATER SUPPLY
SECTION "S-S"
SCALE: ½" = 1'-0"
FIRE PROTECTION FACILITY INDUSTRIAL PARK
GLEN FLORA, WISCONSIN
SCALE - AS SHOWN
SHEET 7 OF 11
DATE NOV., 1991
MORGAN & PARMLEY, LTD. CONSULTING ENGINEERS LADYSMITH, WISCONSIN
DR. BY J.D. REYNOLDS

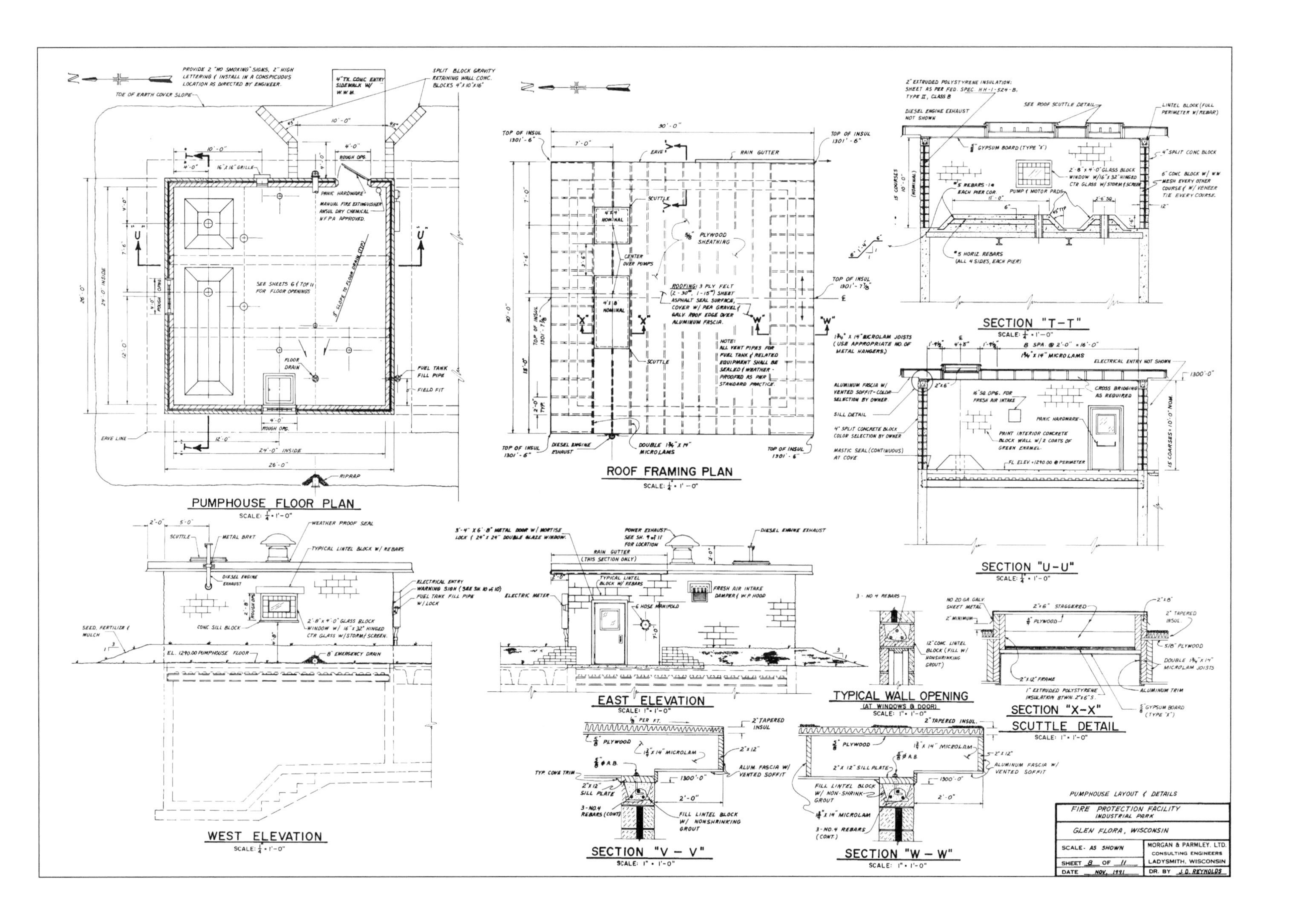
PUMPHOUSE FLOOR PLAN
SCALE: 1/4" = 1'-0"
ROOF FRAMING PLAN
SCALE: 1/4" = 1'-0"
SECTION "T-T"
SCALE: 1/4" = 1'-0"
SECTION "U-U"
SCALE: 1/4" = 1'-0"
WEST ELEVATION
SCALE: 1/4" = 1'-0"
EAST ELEVATION
SCALE: 1" = 1'-0"
TYPICAL WALL OPENING
(AT WINDOWS & DOOR)
SCALE: 1" = 1'-0"
SECTION "X-X"
SCUTTLE DETAIL
SCALE: 1" = 1'-0"
SECTION "V - V"
SCALE: 1" = 1'-0"
SECTION "W - W"
SCALE: 1" = 1'-0"
PUMPHOUSE LAYOUT & DETAILS
FIRE PROTECTION FACILITY
INDUSTRIAL PARK
GLEN FLORA, WISCONSIN
SCALE- AS SHOWN
SHEET 8 OF 11
DATE NOV. 1991
MORGAN & PARMLEY, LTD.
CONSULTING ENGINEERS
LADYSMITH, WISCONSIN
DR. BY J.D. REYNOLDS

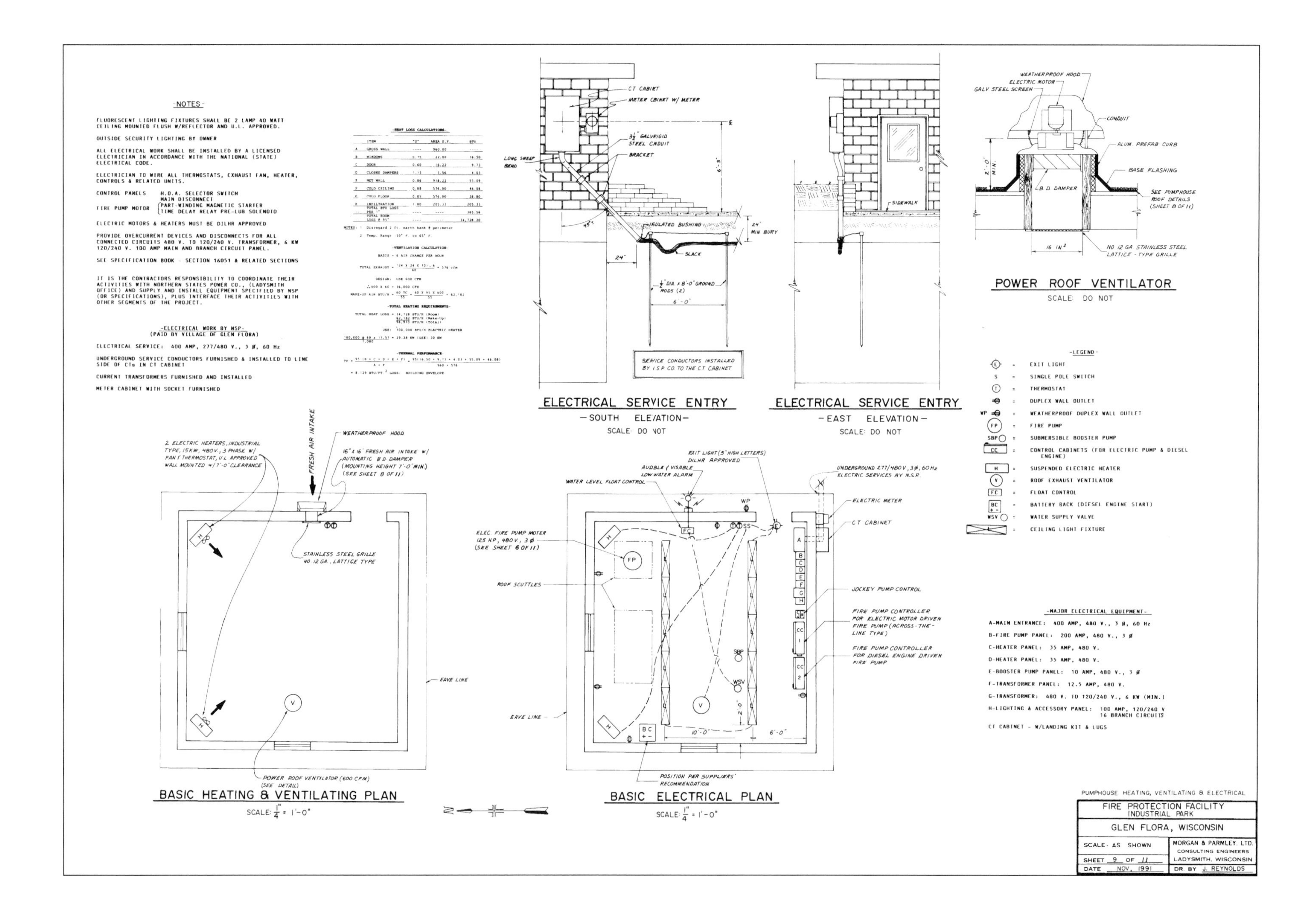
-NOTES-
FLUORESCENT LIGHTING FIXTURES SHALL BE 2 LAMP 40 WATT CEILING MOUNTED FLUSH W/REFLECTOR AND U.L. APPROVED.
OUTSIDE SECURITY LIGHTING BY OWNER
ALL ELECTRICAL WORK SHALL BE INSTALLED BY A LICENSED ELECTRICIAN IN ACCORDANCE WITH THE NATIONAL (STATE) ELECTRICAL CODE.
ELECTRICIAN TO WIRE ALL THERMOSTATS, EXHAUST FAN, HEATER, CONTROLS & RELATED UNITS.
CONTROL PANELS H.O.A. SELECTOR SWITCH
MAIN DISCONNECT
FIRE PUMP MOTOR PART-WINDING MAGNETIC STARTER
TIME DELAY RELAY PRE-LUB SOLENOID
ELECTRIC MOTORS & HEATERS MUST BE DILHR APPROVED
PROVIDE OVERCURRENT DEVICES AND DISCONNECTS FOR ALL CONNECTED CIRCUITS 480 V. TO 120/240 V. TRANSFORMER, 6 KW 120/240 V. 100 AMP MAIN AND BRANCH CIRCUIT PANEL.
SEE SPECIFICATION BOOK - SECTION 16051 & RELATED SECTIONS
IT IS THE CONTRACTORS RESPONSIBILITY TO COORDINATE THEIR ACTIVITIES WITH NORTHERN STATES POWER CO., (LADYSMITH OFFICE) AND SUPPLY AND INSTALL EQUIPMENT SPECIFIED BY NSP (OR SPECIFICATIONS), PLUS INTERFACE THEIR ACTIVITIES WITH OTHER SEGMENTS OF THE PROJECT.
-ELECTRICAL WORK BY NSP-
(PAID BY VILLAGE OF GLEN FLORA)
ELECTRICAL SERVICE: 400 AMP, 277/480 V., 3 Ø, 60 Hz
UNDERGROUND SERVICE CONDUCTORS FURNISHED & INSTALLED TO LINE SIDE OF CTs IN CT CABINET
CURRENT TRANSFORMERS FURNISHED AND INSTALLED
METER CABINET WITH SOCKET FURNISHED
CT CABINET
METER CBINET W/ METER
3½" GALV/RIGID STEEL CNDUIT
BRACKET
LONG SWEEP BEND
INSULATED BUSHING
24" MIN BURY
SLACK
½" DIA x 8'-0" GROUND RODS (2)
SERVICE CONDUCTORS INSTALLED BY NSP CO. TO THE CT CABINET
ELECTRICAL SERVICE ENTRY
-SOUTH ELEVATION-
SCALE: DO NOT
SIDEWALK
ELECTRICAL SERVICE ENTRY
-EAST ELEVATION-
SCALE: DO NOT
WEATHERPROOF HOOD
ELECTRIC MOTOR
GALV STEEL SCREEN
CONDUIT
ALUM. PREFAB CURB
BASE FLASHING
B.D. DAMPER
SEE PUMPHOUSE ROOF DETAILS (SHEET 8 OF 11)
NO 12 GA STAINLESS STEEL LATTICE - TYPE GRILLE
POWER ROOF VENTILATOR
SCALE: DO NOT
-LEGEND-
EXIT LIGHT
SINGLE POLE SWITCH
THERMOSTAT
DUPLEX WALL OUTLET
WEATHERPROOF DUPLEX WALL OUTLET
FIRE PUMP
SUBMERSIBLE BOOSTER PUMP
CONTROL CABINETS (FOR ELECTRIC PUMP & DIESEL ENGINE)
SUSPENDED ELECTRIC HEATER
ROOF EXHAUST VENTILATOR
FLOAT CONTROL
BATTERY BACK (DIESEL ENGINE START)
WATER SUPPLY VALVE
CEILING LIGHT FIXTURE
2 ELECTRIC HEATERS, INDUSTRIAL TYPE, 15KW, 480V, 3 PHASE W/ FAN (THERMOSTAT, U.L APPROVED WALL MOUNTED W/ 7'-0" CLEARANCE
FRESH AIR INTAKE
WEATHERPROOF HOOD
16"x16" FRESH AIR INTAKE W/ AUTOMATIC B.D DAMPER (MOUNTING HEIGHT 7'-0" MIN) (SEE SHEET 8 OF 11)
STAINLESS STEEL GRILLE NO 12 GA., LATTICE TYPE
EAVE LINE
POWER ROOF VENTILATOR (600 CFM) (SEE DETAIL)
BASIC HEATING & VENTILATING PLAN
SCALE: ¼" = 1'-0"
EXIT LIGHT (5" HIGH LETTERS) DILHR APPROVED
AUDBLE (VISABLE LOW WATER ALARM
WATER LEVEL FLOAT CONTROL
UNDERGROUND 277/480V, 3Ø, 60Hz ELECTRIC SERVICES BY N.S.P.
ELECTRIC METER
CT CABINET
ELEC FIRE PUMP MOTER 125 HP, 480V, 3Ø (SEE SHEET 6 OF 11)
ROOF SCUTTLES
JOCKEY PUMP CONTROL
FIRE PUMP CONTROLLER FOR ELECTRIC MOTOR DRIVEN FIRE PUMP (ACROSS-THE-LINE TYPE)
FIRE PUMP CONTROLLER FOR DIESEL ENGINE DRIVEN FIRE PUMP
EAVE LINE
POSITION PER SUPPLIERS' RECOMMENDATION
BASIC ELECTRICAL PLAN
SCALE: ¼" = 1'-0"
-MAJOR ELECTRICAL EQUIPMENT-
A-MAIN ENTRANCE: 400 AMP, 480 V., 3 Ø, 60 Hz
B-FIRE PUMP PANEL: 200 AMP, 480 V., 3 Ø
C-HEATER PANEL: 35 AMP, 480 V.
D-HEATER PANEL: 35 AMP, 480 V.
E-BOOSTER PUMP PANEL: 10 AMP, 480 V., 3 Ø
F-TRANSFORMER PANEL: 12.5 AMP, 480 V.
G-TRANSFORMER: 480 V. TO 120/240 V., 6 KW (MIN.)
H-LIGHTING & ACCESSORY PANEL: 100 AMP, 120/240 V 16 BRANCH CIRCUITS
CT CABINET - W/LANDING KIT & LUGS
PUMPHOUSE HEATING, VENTILATING & ELECTRICAL
FIRE PROTECTION FACILITY INDUSTRIAL PARK
GLEN FLORA, WISCONSIN
SCALE: AS SHOWN
SHEET 9 OF 11
DATE NOV. 1991
MORGAN & PARMLEY, LTD. CONSULTING ENGINEERS LADYSMITH, WISCONSIN
DR. BY J. REYNOLDS

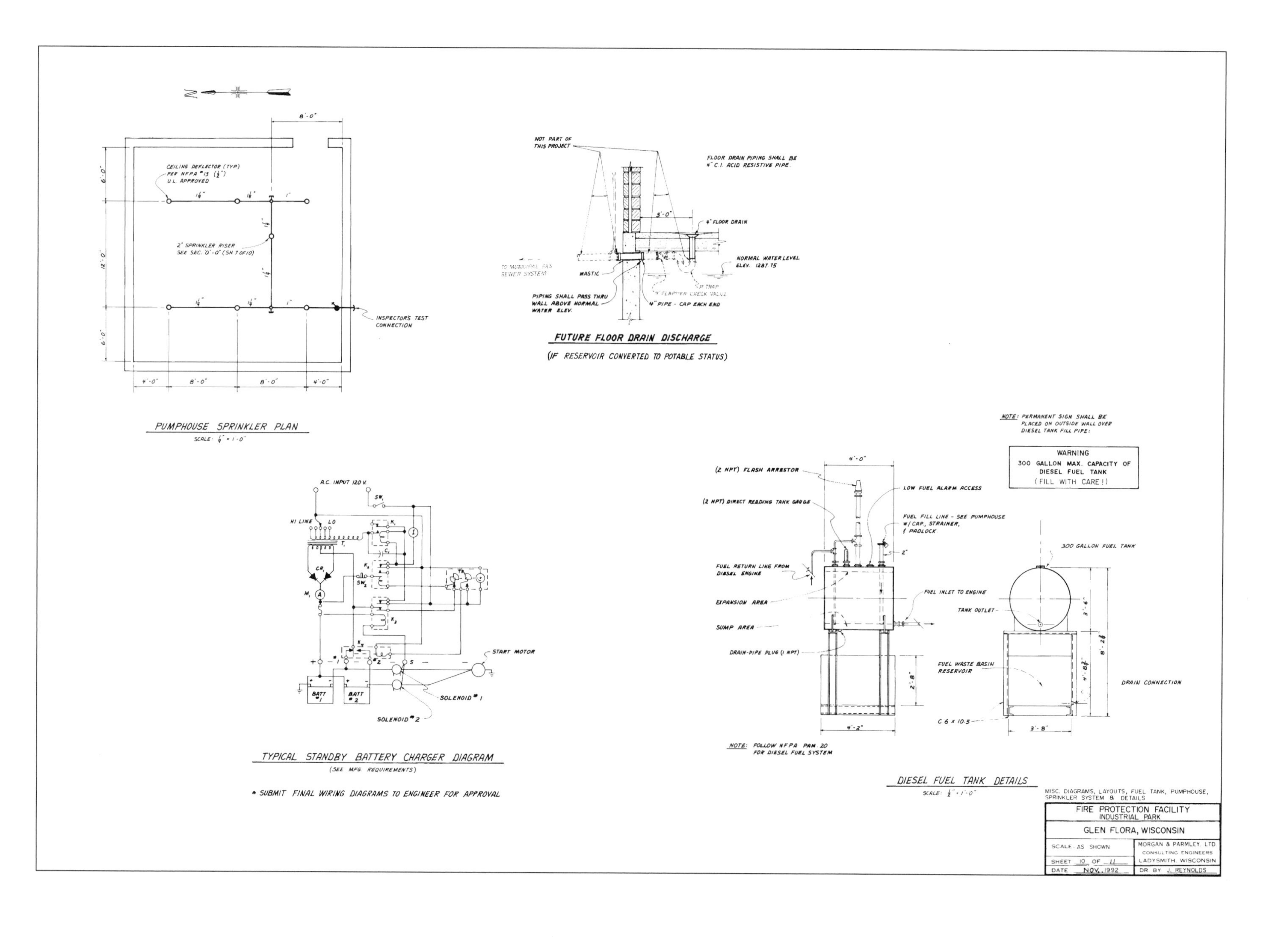
PUMPHOUSE SPRINKLER PLAN
SCALE: 1/4" = 1'-0"
CEILING DEFLECTOR (TYP) PER NFPA #13 (1/2") U.L. APPROVED
2" SPRINKLER RISER SEE SEC. 'Q'-'Q' (SH 7 OF 10)
INSPECTORS TEST CONNECTION
FUTURE FLOOR DRAIN DISCHARGE
(IF RESERVOIR CONVERTED TO POTABLE STATUS)
NOT PART OF THIS PROJECT
FLOOR DRAIN PIPING SHALL BE 4" C.I. ACID RESISTIVE PIPE.
4" FLOOR DRAIN
NORMAL WATER LEVEL ELEV. 1287.75
TO MUNICIPAL SAN. SEWER SYSTEM
MASTIC
PIPING SHALL PASS THRU WALL ABOVE NORMAL WATER ELEV.
4" PIPE - CAP EACH END
TYPICAL STANDBY BATTERY CHARGER DIAGRAM
(SEE MFG. REQUIREMENTS)
* SUBMIT FINAL WIRING DIAGRAMS TO ENGINEER FOR APPROVAL
A.C. INPUT 120 V.
HI LINE
LO
START MOTOR
SOLENOID #1
SOLENOID #2
BATT #1
BATT #2
NOTE: PERMANENT SIGN SHALL BE PLACED ON OUTSIDE WALL OVER DIESEL TANK FILL PIPE:
WARNING
300 GALLON MAX. CAPACITY OF DIESEL FUEL TANK
(FILL WITH CARE!)
(2 NPT) FLASH ARRESTOR
(2 NPT) DIRECT READING TANK GAUGE
LOW FUEL ALARM ACCESS
FUEL FILL LINE - SEE PUMPHOUSE W/ CAP, STRAINER, & PADLOCK
300 GALLON FUEL TANK
FUEL RETURN LINE FROM DIESEL ENGINE
EXPANSION AREA
FUEL INLET TO ENGINE
TANK OUTLET
SUMP AREA
DRAIN-PIPE PLUG (1 NPT)
FUEL WASTE BASIN RESERVOIR
DRAIN CONNECTION
C 6 x 10.5
NOTE: FOLLOW NFPA PAM. 30 FOR DIESEL FUEL SYSTEM
DIESEL FUEL TANK DETAILS
SCALE: 1/2" = 1'-0"
MISC. DIAGRAMS, LAYOUTS, FUEL TANK, PUMPHOUSE, SPRINKLER SYSTEM & DETAILS
FIRE PROTECTION FACILITY
INDUSTRIAL PARK
GLEN FLORA, WISCONSIN
SCALE: AS SHOWN
SHEET 10 OF 11
DATE NOV. 1992
MORGAN & PARMLEY, LTD.
CONSULTING ENGINEERS
LADYSMITH, WISCONSIN
DR. BY J. REYNOLDS

Design

WASTEWATER COLLECTION & TREATMENT

Section 11

In order to maintain public health, it is mandatory to safely collect and adequately treat municipally generated sewage. The final effluent must be treated to levels that can be discharged to surface waters or the groundwater without endangering the environment.

All new wastewater treatment facilities and modifications to existing sewage plants must conform to strict regulatory rules and regulations. Initial planning is generally known as the Step 1 Phase or Facility Planning Process. The purpose of this multi-faceted study is to explore all feasible treatment alternatives and select the most cost-effective and environmentally sound design prior to commencing the preparation of construction plans.

The first set of plans in this section details a three-cell aerated lagoon with ultraviolet disinfection designed to serve a community of 5,000. A master lift station and forcemain to transport raw sewage from the municipality to the remote treatment facility is also contained in the plans. The second example is an alternative/innovative wastewater collection and treatment facility for a small rural village with less than sixty buildings to serve. This facility commences with individual septic tanks whose septage is transported to a sub-surface rapid sand filter via a small diameter combination gravity/pressure collection system. The third set of plans show a proposed modification to an existing industrial treatment plan. The last two plan examples are for sanitary sewer collection extensions to an existing system. The reader should note that these plans also incorporate water main extensions which is a common practice for municipal service expansions.

TABLE OF CONTENTS

M & P PROJECT NO.
(DESIGN) 87-110
(CONSTRUCTION) 88-161

WASTEWATER TREATMENT FACILITY, LIFT STATION & FORCEMAIN

LADYSMITH, WISCONSIN
APRIL, 1988

PROJECT NO. 0551098-02 (DESIGN)
0551098-03 (CONSTRUCTION)

GENERAL LOCATION MAP
SCALE– DO NOT

INDEX OF SHEETS

SHEET NO.	DESCRIPTION
1	TITLE SHEET w/ GENERAL LOCATION MAP & INDEX
2	LIFT STATION: SITE PLAN w/ ELEVATION VIEWS & EXISTING STRUCTURE
3	LIFT STATION: FLOOR PLAN w/ CROSS-SECTIONS & DETAILS
4	LIFT STATION: CROSS-SECTION VIEWS
5	LIFT STATION: FRAMING PLANS w/ CROSS-SECTIONS & DETAILS
6	LIFT STATION: PLUMBING, ELECTRICAL, HEATING & VENTILATING LAYOUTS
7	FORCEMAIN: ROUTE - PLAN & PROFILE
8	FORCEMAIN: ROUTE - PLAN & PROFILE
9	FORCEMAIN: ROUTE - PLAN & PROFILE
10	FORCEMAIN: ROUTE - PLAN & PROFILE
11	LAGOON SITE LAYOUT w/ CONTOURS & SOIL BORINGS
12	AERATION CELLS: TYPICAL CONSTRUCTION DETAILS
13	AERATION CELLS: CROSS-SECTIONS
14	AERATION CELLS: CROSS-SECTIONS
15	AERATION CELLS: CROSS-SECTIONS
16	AERATION CELLS: CROSS-SECTIONS
17	AERATION CELLS: CROSS-SECTIONS
18	AERATION CELLS: CROSS-SECTIONS
19	AERATION CELL & WET PIPING LAYOUTS
20	SERVICE BUILDING: FLOOR PLAN w/ CROSS-SECTIONS
21	SERVICE BUILDING: PLUMBING, ELECTRICAL, HEATING & VENTILATING LAYOUTS
22	SERVICE BUILDING: ELEVATION VIEWS
23	SERVICE BUILDING: BLOWER ROOM FLOOR PLAN w/ CROSS-SECTIONS
24	SERVICE BUILDING: UV DISINFECTION & VALVE ROOM FLOOR PLAN w/ CROSS-SECTIONS
25	SERVICE BUILDING: VALVE VAULT PIPING, REINFORCING & PERSPECTIVE VIEWS
26	AIR DISTRIBUTION SYSTEM - STATIC TUBE DESIGN
27	AERATION CELLS: APPURTENANCE DETAILS
28	TYPICAL DETAILS

PREPARED BY: MORGAN & PARMLEY LTD. LADYSMITH, WISCONSIN

Source: Morgan & Parmley, Ltd.

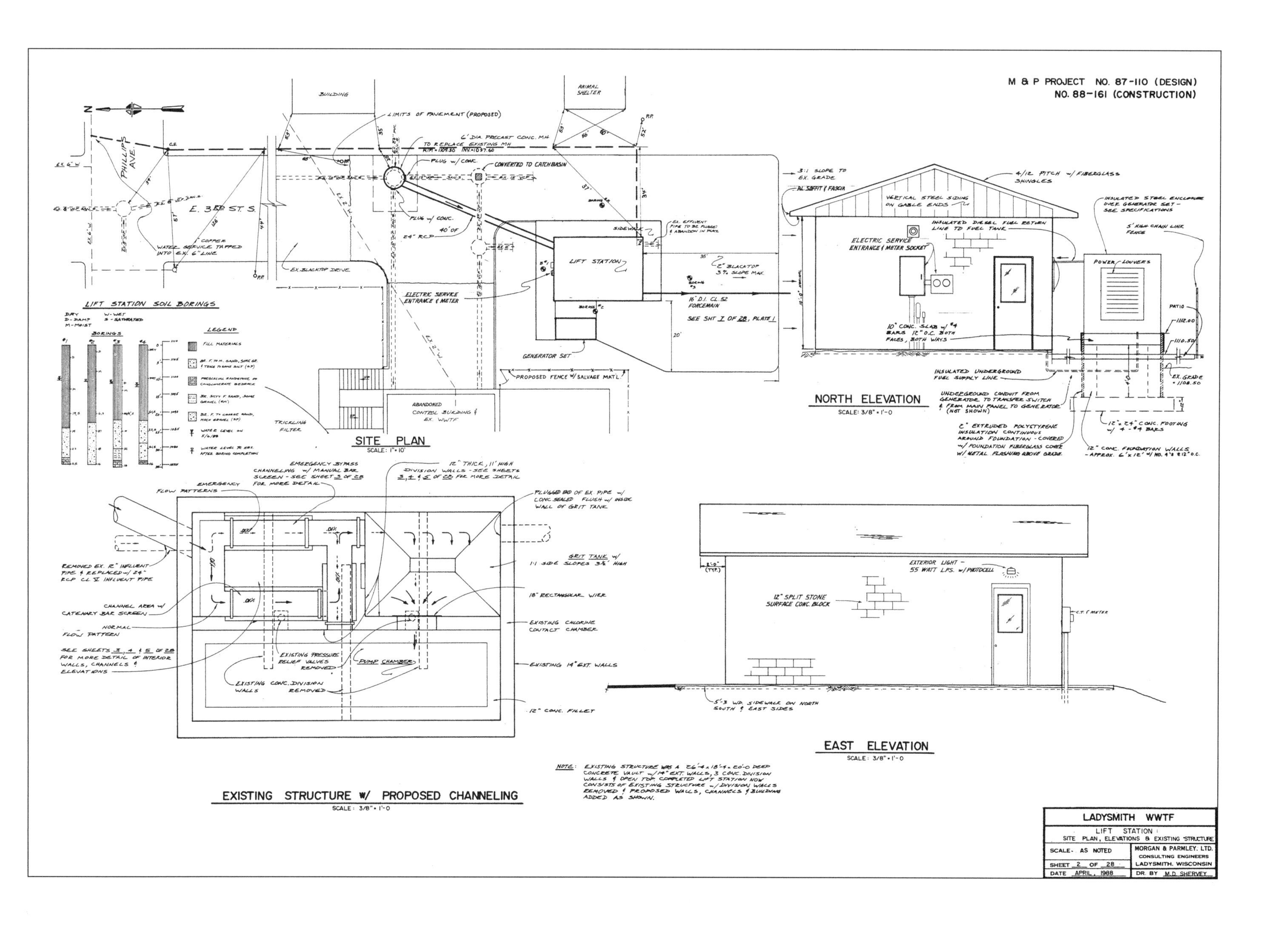
M & P PROJECT NO. 87-110 (DESIGN)
NO. 88-161 (CONSTRUCTION)
BUILDING
ANIMAL SHELTER
LIMITS OF PAVEMENT (PROPOSED)
6' DIA. PRECAST CONC. MH TO REPLACE EXISTING MH
PLUG w/ CONC.
CONVERTED TO CATCH BASIN
PHILLIPS AVE.
E. 3RD ST. S.
COPPER WATER SERVICE TAPPED INTO EX. 6" LINE
24" RCP
EX. BLACKTOP DRIVE
LIFT STATION
ELECTRIC SERVICE ENTRANCE & METER
GENERATOR SET
PROPOSED FENCE w/ SALVAGE MAT'L.
2" BLACKTOP 3% SLOPE MAX.
16" D.I. CL.52 FORCEMAIN
ABANDONED CONTROL BUILDING & EX. WWTF
TRICKLING FILTER
LIFT STATION SOIL BORINGS
LEGEND
SITE PLAN
SCALE: 1" = 10'
3:1 SLOPE TO EX. GRADE
AL. SOFFIT & FASCIA
VERTICAL STEEL SIDING ON GABLE ENDS
4/12 PITCH w/ FIBERGLASS SHINGLES
INSULATED STEEL ENCLOSURE OVER GENERATOR SET - SEE SPECIFICATIONS
5' HIGH CHAIN LINK FENCE
ELECTRIC SERVICE ENTRANCE & METER SOCKET
INSULATED DIESEL FUEL RETURN LINE TO FUEL TANK
POWER LOUVERS
PATIO
10" CONC. SLAB w/ #4 BARS 12" O.C. BOTH FACES, BOTH WAYS
INSULATED UNDERGROUND FUEL SUPPLY LINE
UNDERGROUND CONDUIT FROM GENERATOR TO TRANSFER SWITCH & FROM MAIN PANEL TO GENERATOR (NOT SHOWN)
2" EXTRUDED POLYSTYRENE INSULATION CONTINUOUS AROUND FOUNDATION - COVERED w/ FOUNDATION FIBERGLASS COVER w/ METAL FLASHING ABOVE GRADE
12" x 24" CONC. FOOTING w/ 4 - #4 BARS
12" CONC. FOUNDATION WALLS
NORTH ELEVATION
SCALE: 3/8" = 1'-0
EMERGENCY BYPASS CHANNELING w/ MANUAL BAR SCREEN - SEE SHEET 3 OF 28 FOR MORE DETAIL
12" THICK, 11' HIGH DIVISION WALLS - SEE SHEETS 3, 4 & 5 OF 28 FOR MORE DETAIL
EMERGENCY FLOW PATTERNS
PLUGGED END OF EX. PIPE w/ CONC. SEALED FLUSH w/ INSIDE WALL OF GRIT TANK
GRIT TANK w/ 1:1 SIDE SLOPES 3 1/2' HIGH
18" RECTANGULAR WIER
EXISTING CHLORINE CONTACT CHAMBER
EXISTING 14" EXT. WALLS
12" CONC. FILLET
CHANNEL AREA w/ CATENARY BAR SCREEN
NORMAL FLOW PATTERN
SEE SHEETS 3, 4 & 5 OF 28 FOR MORE DETAIL OF INTERIOR WALLS, CHANNELS & ELEVATIONS
EXISTING PRESSURE RELIEF VALVES REMOVED
PUMP CHAMBER
EXISTING CONC. DIVISION WALLS REMOVED
EXISTING STRUCTURE w/ PROPOSED CHANNELING
SCALE: 3/8" = 1'-0
EXTERIOR LIGHT - 55 WATT LPS. w/ PHOTOCELL
12" SPLIT STONE SURFACE CONC. BLOCK
C.T. & METER
5'-3 WD. SIDEWALK ON NORTH SOUTH & EAST SIDES
EAST ELEVATION
SCALE: 3/8" = 1'-0
NOTE: EXISTING STRUCTURE WAS A 26'-4 x 18'-4 x 20'-0 DEEP CONCRETE VAULT w/ 14" EXT. WALLS, 3 CONC. DIVISION WALLS & OPEN TOP. COMPLETED LIFT STATION NOW CONSISTS OF EXISTING STRUCTURE w/ DIVISION WALLS REMOVED & PROPOSED WALLS, CHANNELS & BUILDING ADDED AS SHOWN.
LADYSMITH WWTF
LIFT STATION:
SITE PLAN, ELEVATIONS & EXISTING STRUCTURE
SCALE: AS NOTED
MORGAN & PARMLEY, LTD.
CONSULTING ENGINEERS
LADYSMITH, WISCONSIN
SHEET 2 OF 28
DATE APRIL, 1988
DR. BY M.D. SHERVEY

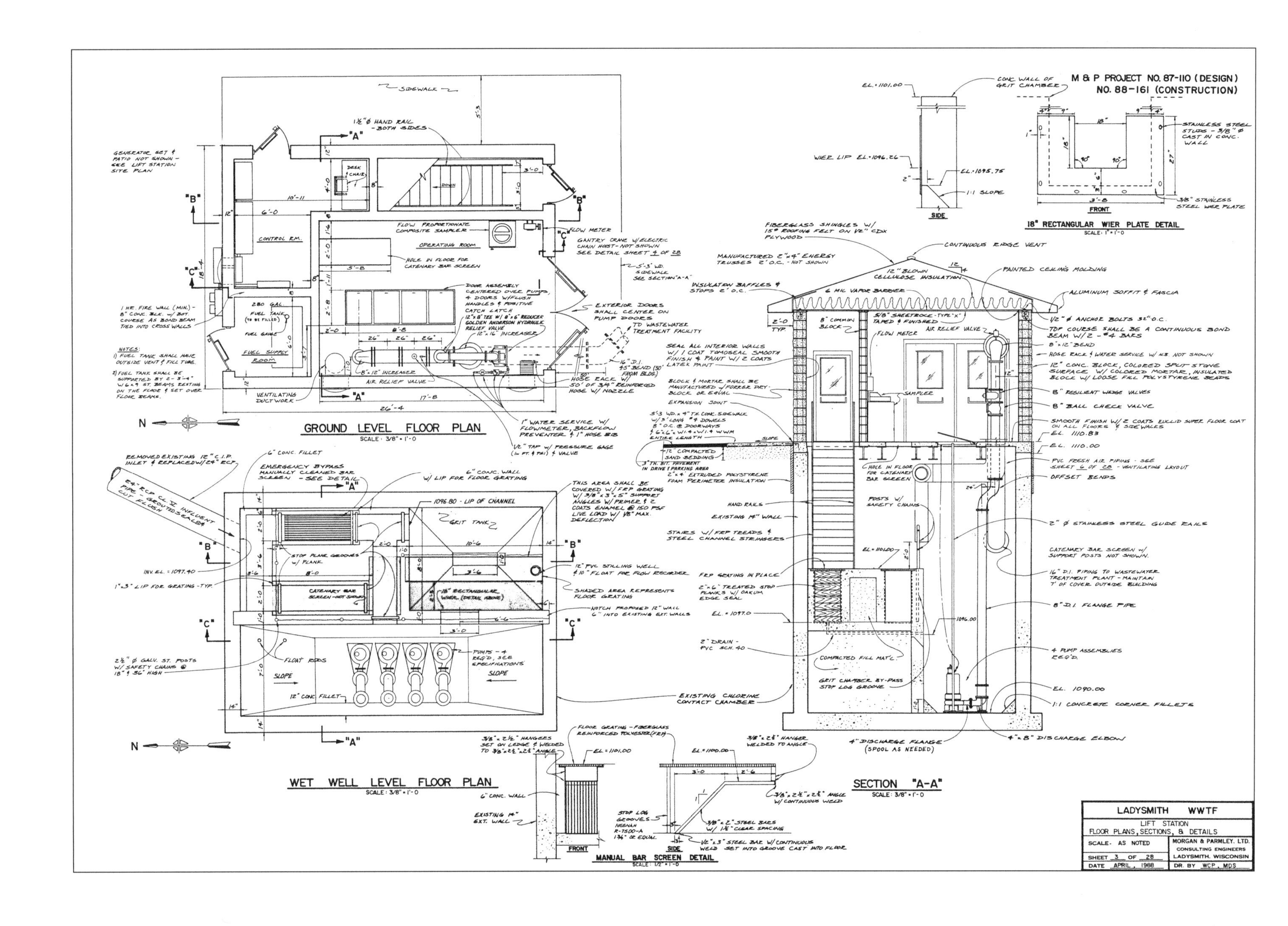
M & P PROJECT NO. 87-110 (DESIGN)
NO. 88-161 (CONSTRUCTION)
GROUND LEVEL FLOOR PLAN
SCALE: 3/8" = 1'-0
WET WELL LEVEL FLOOR PLAN
SCALE: 3/8" = 1'-0
SECTION "A-A"
SCALE: 3/8" = 1'-0
18" RECTANGULAR WIER PLATE DETAIL
SCALE: 1" = 1'-0
MANUAL BAR SCREEN DETAIL
SCALE: 1/2" = 1'-0
SIDE
FRONT
CONTROL RM.
OPERATING ROOM
FUEL SUPPLY ROOM
GRIT TANK
SLOPE
N
LADYSMITH WWTF
LIFT STATION
FLOOR PLANS, SECTIONS, & DETAILS
SCALE- AS NOTED
SHEET 3 OF 28
DATE APRIL, 1988
MORGAN & PARMLEY, LTD.
CONSULTING ENGINEERS
LADYSMITH, WISCONSIN
DR. BY WCP, MDS

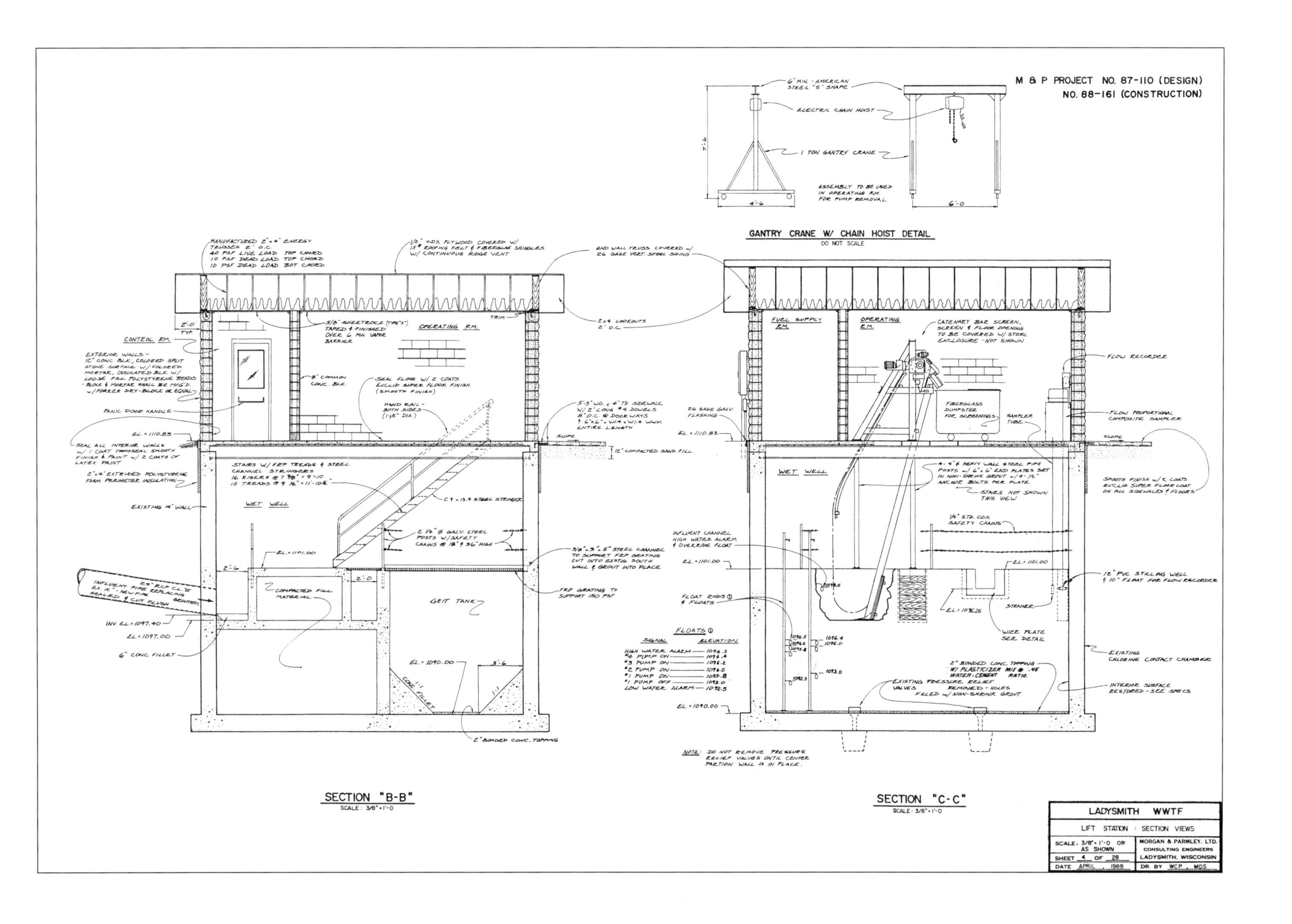
M & P PROJECT NO. 87-110 (DESIGN)
NO. 88-161 (CONSTRUCTION)
GANTRY CRANE W/ CHAIN HOIST DETAIL
DO NOT SCALE
SECTION "B-B"
SCALE: 3/8"=1'-0
SECTION "C-C"
SCALE: 3/8"=1'-0
LADYSMITH WWTF
LIFT STATION : SECTION VIEWS
SCALE- 3/8"=1'-0 OR AS SHOWN
SHEET 4 OF 28
DATE APRIL , 1988
MORGAN & PARMLEY, LTD.
CONSULTING ENGINEERS
LADYSMITH, WISCONSIN
DR. BY WCP , MDS

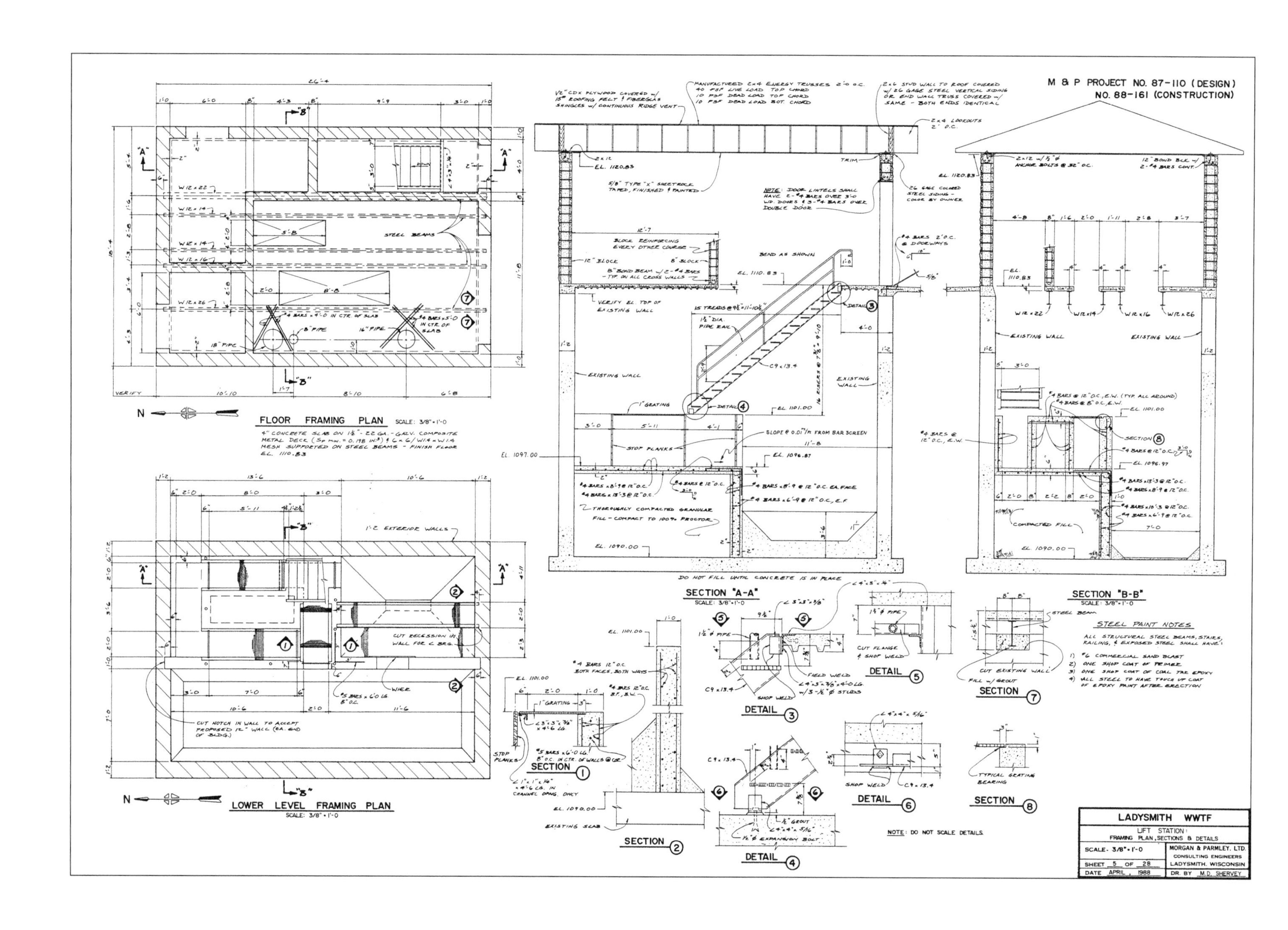
M & P PROJECT NO. 87-110 (DESIGN)
NO. 88-161 (CONSTRUCTION)
FLOOR FRAMING PLAN SCALE: 3/8" = 1'-0
LOWER LEVEL FRAMING PLAN SCALE: 3/8" = 1'-0
SECTION "A-A" SCALE: 3/8" = 1'-0
SECTION "B-B" SCALE: 3/8" = 1'-0
SECTION 1
SECTION 2
DETAIL 3
DETAIL 4
DETAIL 5
DETAIL 6
SECTION 7
SECTION 8
STEEL PAINT NOTES
NOTE: DO NOT SCALE DETAILS.
LADYSMITH WWTF
LIFT STATION
FRAMING PLAN, SECTIONS & DETAILS
SCALE: 3/8" = 1'-0
SHEET 5 OF 28
DATE APRIL, 1988
MORGAN & PARMLEY, LTD.
CONSULTING ENGINEERS
LADYSMITH, WISCONSIN
DR. BY M.D. SHERVEY

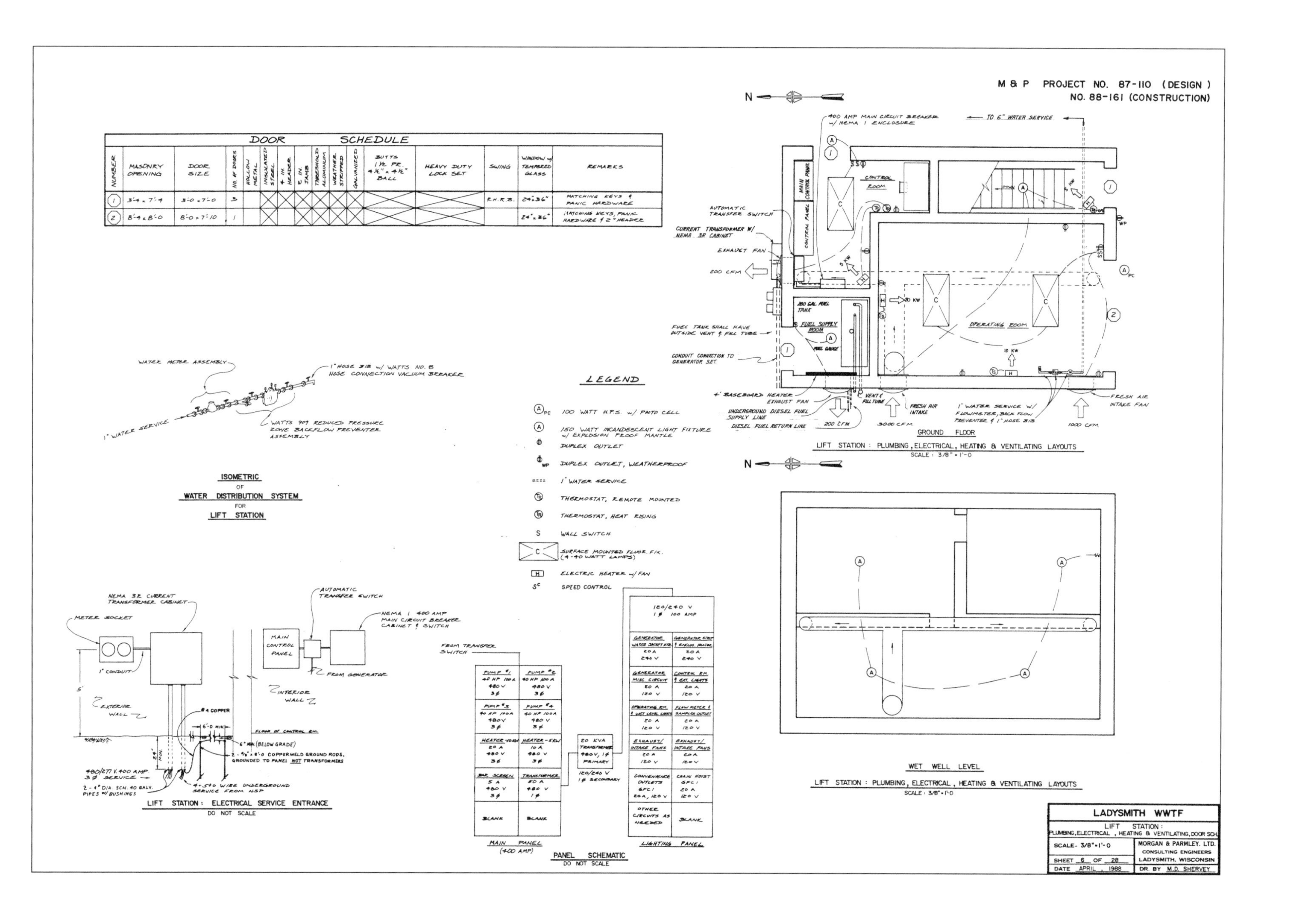
M & P PROJECT NO. 87-110 (DESIGN)
NO. 88-161 (CONSTRUCTION)
DOOR SCHEDULE
ISOMETRIC OF WATER DISTRIBUTION SYSTEM FOR LIFT STATION
LEGEND
GROUND FLOOR
LIFT STATION : PLUMBING, ELECTRICAL, HEATING & VENTILATING LAYOUTS
WET WELL LEVEL
LIFT STATION : ELECTRICAL SERVICE ENTRANCE
DO NOT SCALE
PANEL SCHEMATIC
MAIN PANEL (400 AMP)
LIGHTING PANEL
LADYSMITH WWTF
MORGAN & PARMLEY, LTD.
CONSULTING ENGINEERS
LADYSMITH, WISCONSIN

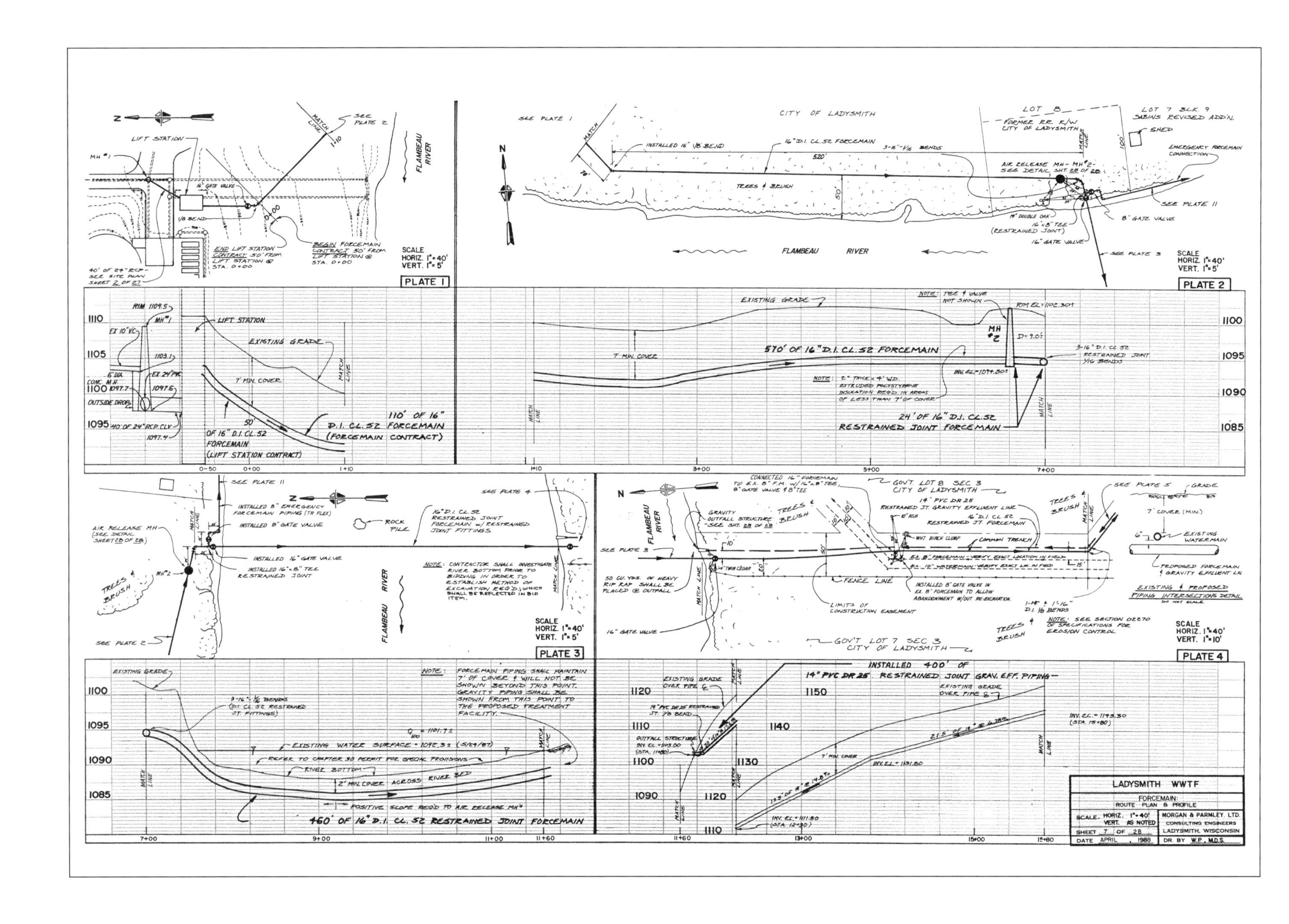
PLATE 1
LIFT STATION
MH #1
1/8 BEND
END LIFT STATION CONTRACT 50' FROM LIFT STATION @ STA. 0+00
BEGIN FORCEMAIN CONTRACT 50' FROM LIFT STATION @ STA. 0+00
FLAMBEAU RIVER
SCALE HORIZ. 1"=40' VERT. 1"=5'
EXISTING GRADE
7' MIN. COVER
110' OF 16" D.I. CL.52 FORCEMAIN (FORCEMAIN CONTRACT)
OF 16" D.I. CL.52 FORCEMAIN (LIFT STATION CONTRACT)
PLATE 2
CITY OF LADYSMITH
16" D.I. CL.52 FORCEMAIN
TREES & BRUSH
FLAMBEAU RIVER
SEE PLATE 1
SEE PLATE 11
SEE PLATE 3
8" GATE VALVE
16" GATE VALVE
EMERGENCY FORCEMAIN CONNECTION
SCALE HORIZ. 1"=40' VERT. 1"=5'
EXISTING GRADE
570' OF 16" D.I. CL.52 FORCEMAIN
24' OF 16" D.I. CL.52 RESTRAINED JOINT FORCEMAIN
7' MIN. COVER
PLATE 3
SEE PLATE 11
SEE PLATE 4
SEE PLATE 2
AIR RELEASE MH
INSTALLED 8" GATE VALVE
INSTALLED 16" GATE VALVE
ROCK PILE
FLAMBEAU RIVER
TREES & BRUSH
SCALE HORIZ. 1"=40' VERT. 1"=5'
EXISTING GRADE
EXISTING WATER SURFACE
RIVER BOTTOM
2' MIN. COVER ACROSS RIVER BED
POSITIVE SLOPE REQ'D TO AIR RELEASE MH
460' OF 16" D.I. CL.52 RESTRAINED JOINT FORCEMAIN
PLATE 4
GOV'T LOT 8 SEC 3 CITY OF LADYSMITH
GOV'T LOT 7 SEC 3 CITY OF LADYSMITH
FLAMBEAU RIVER
SEE PLATE 3
SEE PLATE 5
FENCE LINE
LIMITS OF CONSTRUCTION EASEMENT
16" GATE VALVE
TREES & BRUSH
COMMON TRENCH
EXISTING & PROPOSED PIPING INTERSECTIONS DETAIL
SCALE HORIZ. 1"=40' VERT. 1"=10'
INSTALLED 400' OF 14" PVC DR 25 RESTRAINED JOINT GRAV. EFF. PIPING
LADYSMITH WWTF
FORCEMAIN: ROUTE PLAN & PROFILE
MORGAN & PARMLEY, LTD. CONSULTING ENGINEERS LADYSMITH, WISCONSIN
SHEET 7 OF 28
DATE APRIL, 1988

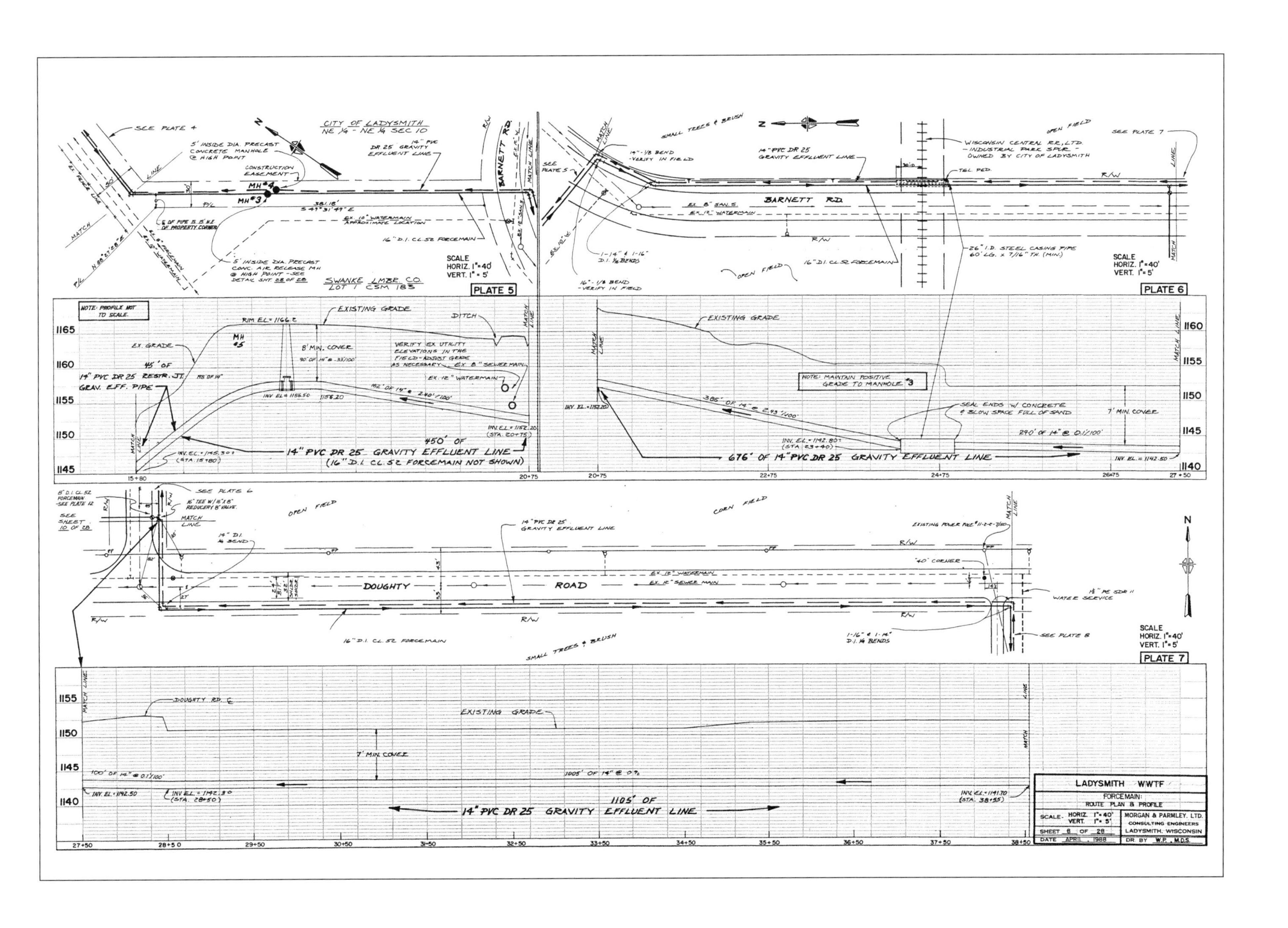
PLATE 5
PLATE 6
PLATE 7
CITY OF LADYSMITH
NE 1/4 - NE 1/4 SEC 10
BARNETT RD.
DOUGHTY ROAD
OPEN FIELD
CORN FIELD
SMALL TREES & BRUSH
SCALE
HORIZ. 1"=40'
VERT. 1"=5'
EXISTING GRADE
7' MIN. COVER
14" PVC DR 25 GRAVITY EFFLUENT LINE
(16" D.I. CL.52 FORCEMAIN NOT SHOWN)
676' OF 14" PVC DR 25 GRAVITY EFFLUENT LINE
1105' OF 14" PVC DR 25 GRAVITY EFFLUENT LINE
NOTE: MAINTAIN POSITIVE GRADE TO MANHOLE #3
NOTE: PROFILE NOT TO SCALE.
LADYSMITH WWTF
FORCEMAIN: ROUTE PLAN & PROFILE
MORGAN & PARMLEY, LTD.
CONSULTING ENGINEERS
LADYSMITH, WISCONSIN
SHEET 6 OF 28
DATE APRIL, 1988
DR BY W.P., M.D.S.

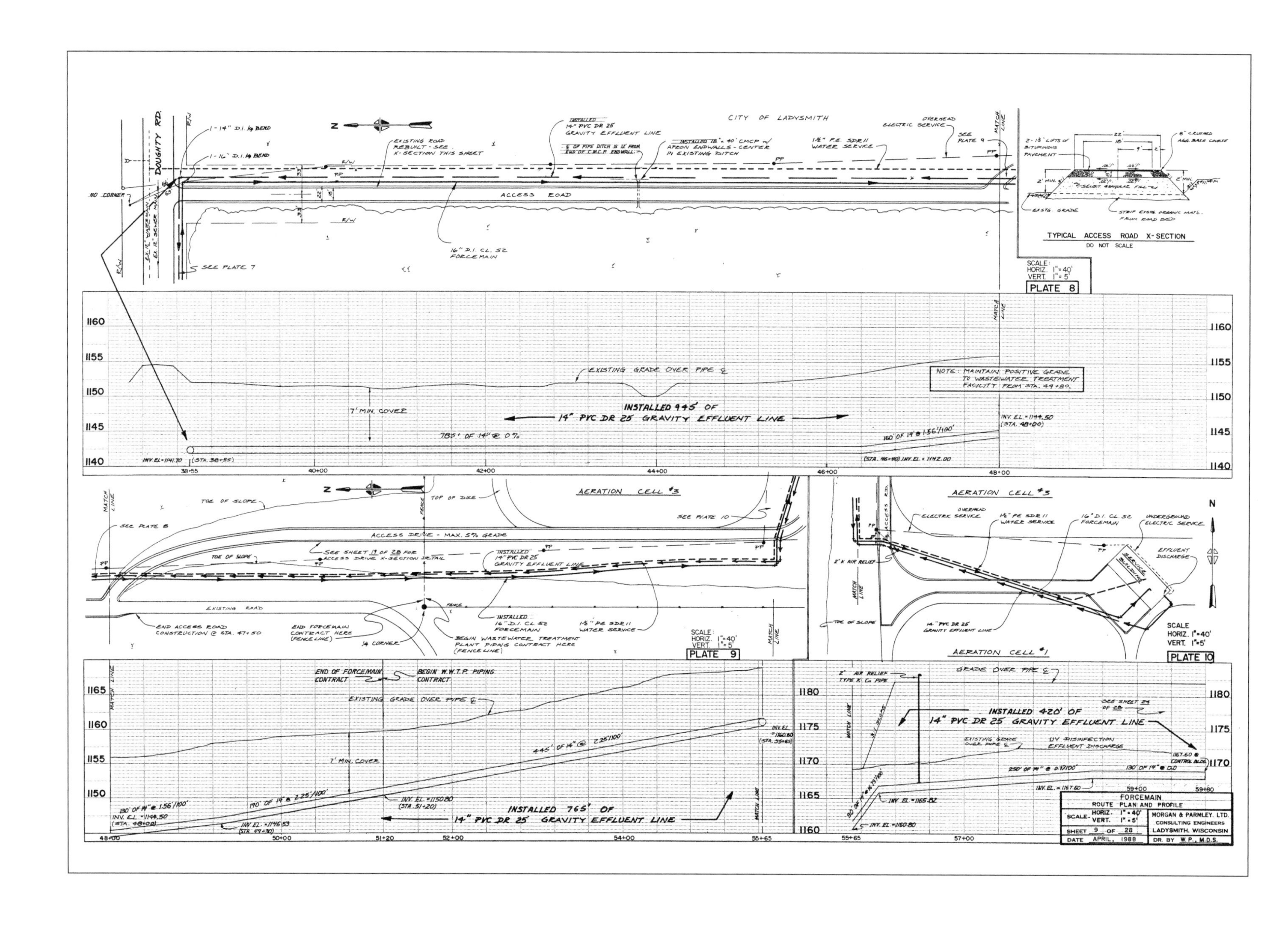
CITY OF LADYSMITH
DOUGHTY RD.
ACCESS ROAD
TYPICAL ACCESS ROAD X-SECTION
DO NOT SCALE
SCALE:
HORIZ. 1"=40'
VERT. 1"=5'
PLATE 8
NOTE: MAINTAIN POSITIVE GRADE TO WASTEWATER TREATMENT FACILITY FROM STA. 49+80.
INSTALLED 945' OF 14" PVC DR 25 GRAVITY EFFLUENT LINE
7' MIN. COVER
785' OF 14" @ 0%
EXISTING GRADE OVER PIPE ℄
AERATION CELL #3
ACCESS DRIVE - MAX. 5% GRADE
EXISTING ROAD
PLATE 9
PLATE 10
AERATION CELL #1
INSTALLED 765' OF 14" PVC DR 25 GRAVITY EFFLUENT LINE
INSTALLED 420' OF 14" PVC DR 25 GRAVITY EFFLUENT LINE
END OF FORCEMAIN CONTRACT
BEGIN W.W.T.P. PIPING CONTRACT
FORCEMAIN
ROUTE PLAN AND PROFILE
MORGAN & PARMLEY, LTD.
CONSULTING ENGINEERS
LADYSMITH, WISCONSIN
SHEET 9 OF 28
DATE APRIL, 1988
DR. BY W.P., M.D.S.

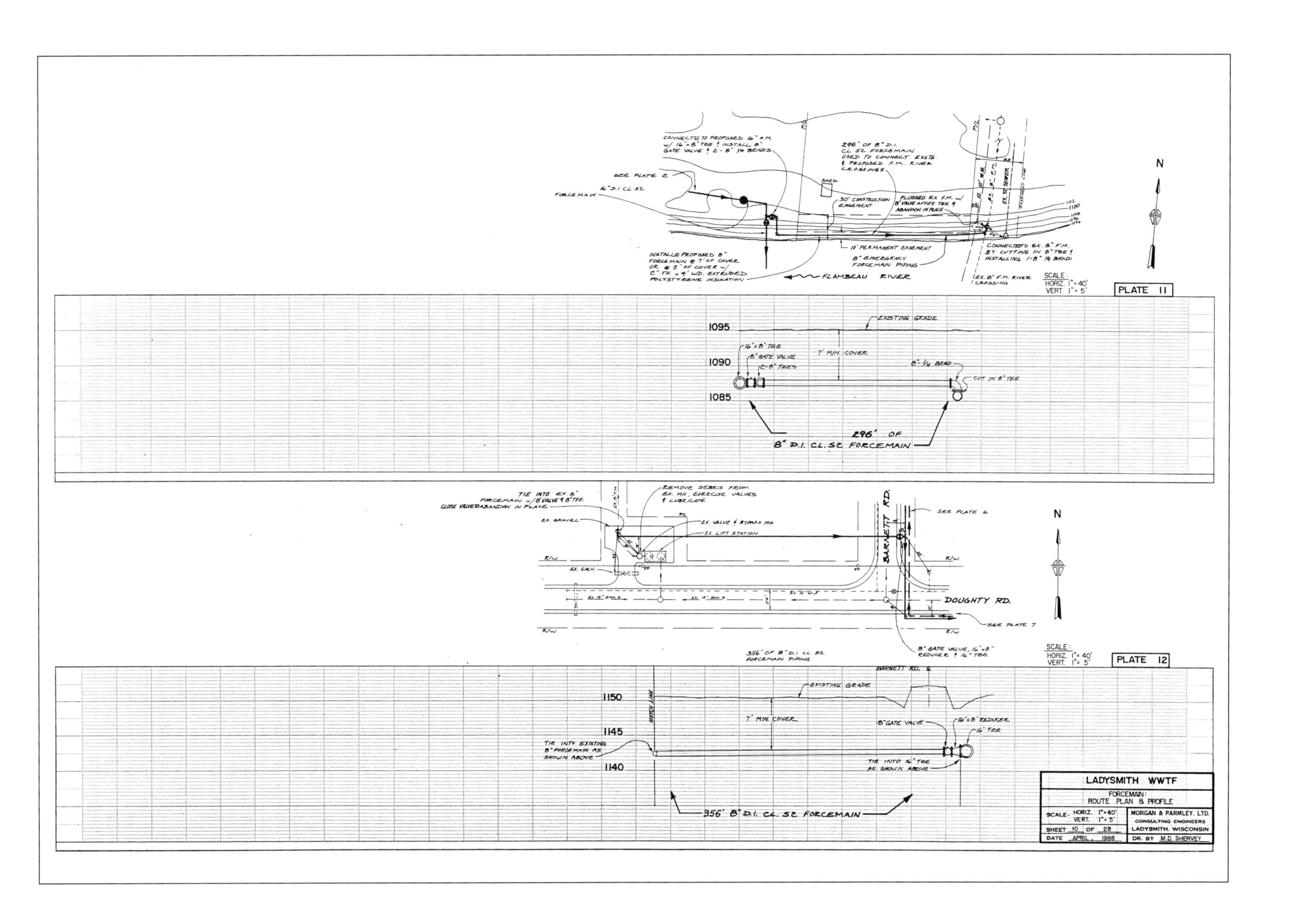
N
SEE PLATE 2
16" D.I. CL 52
FORCEMAIN
FLAMBEAU RIVER
10' PERMANENT EASEMENT
8" EMERGENCY
FORCEMAIN PIPING
SCALE:
HORIZ. 1" = 40'
VERT. 1" = 5'
PLATE 11
1095
1090
1085
EXISTING GRADE
7' MIN. COVER
16" x 8" TEE
8" GATE VALVE
8" - 1/4 BEND
CUT IN 8" TEE
296' OF
8" D.I. CL. 52 FORCEMAIN
BARNETT RD.
SEE PLATE 6
DOUGHTY RD.
SEE PLATE 7
R/W
N
SCALE:
HORIZ. 1" = 40'
VERT. 1" = 5'
PLATE 12
1150
1145
1140
EXISTING GRADE
7' MIN. COVER
8" GATE VALVE
16" x 8" REDUCER
16" TEE
TIE INTO EXISTING
8" FORCEMAIN AS
SHOWN ABOVE
356' 8" D.I. CL. 52 FORCEMAIN
LADYSMITH WWTF
FORCEMAIN:
ROUTE PLAN & PROFILE
SCALE: HORIZ. 1" = 40'
VERT. 1" = 5'
SHEET 10 OF 28
DATE APRIL, 1988
MORGAN & PARMLEY, LTD.
CONSULTING ENGINEERS
LADYSMITH, WISCONSIN
DR. BY M.D. SHERVEY

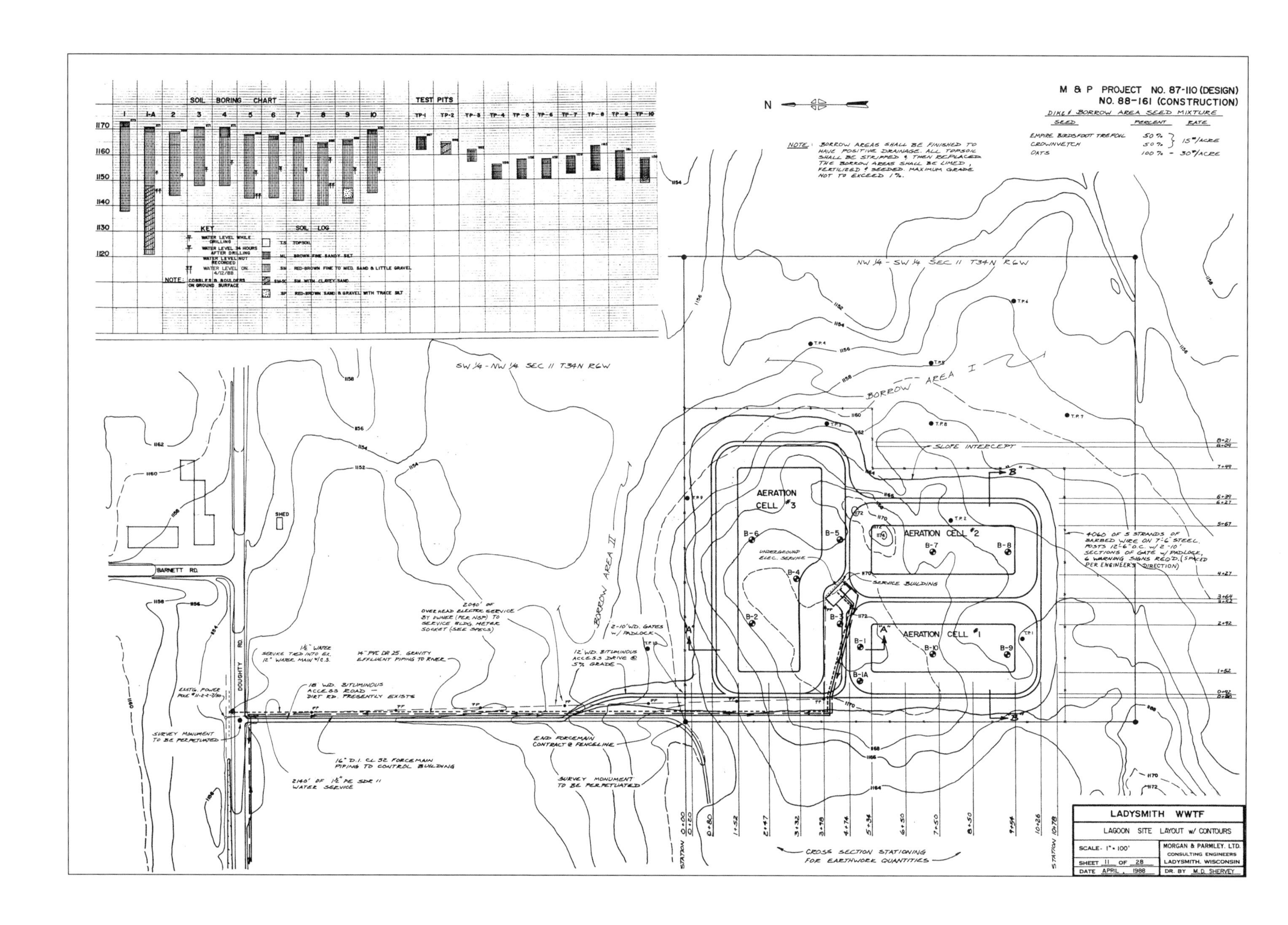

SOIL BORING CHART
TEST PITS
1 1-A 2 3 4 5 6 7 8 9 10
TP-1 TP-2 TP-3 TP-4 TP-5 TP-6 TP-7 TP-8 TP-9 TP-10
1170
1160
1150
1140
1130
1120
KEY
SOIL LOG
WATER LEVEL WHILE DRILLING
WATER LEVEL 24 HOURS AFTER DRILLING
WATER LEVEL NOT RECORDED
WATER LEVEL ON 4/12/88
NOTE: COBBLES & BOULDERS ON GROUND SURFACE
T.S. TOPSOIL
ML BROWN FINE SANDY SILT
SM RED-BROWN FINE TO MED. SAND & LITTLE GRAVEL
SM-SC SM WITH CLAYEY SAND
SP RED-BROWN SAND & GRAVEL WITH TRACE SILT
N
M & P PROJECT NO. 87-110 (DESIGN)
NO. 88-161 (CONSTRUCTION)
DIKE & BORROW AREA SEED MIXTURE
SEED PERCENT RATE
EMPIRE BIRDSFOOT TREFOIL 50 %
CROWNVETCH 50 % 15#/ACRE
OATS 100 % - 30#/ACRE
NOTE: BORROW AREAS SHALL BE FINISHED TO HAVE POSITIVE DRAINAGE. ALL TOPSOIL SHALL BE STRIPPED & THEN REPLACED. THE BORROW AREAS SHALL BE LIMED, FERTILIZED & SEEDED. MAXIMUM GRADE NOT TO EXCEED 1%.
NW 1/4 - SW 1/4 SEC 11 T34N R6W
SW 1/4 - NW 1/4 SEC 11 T34N R6W
BORROW AREA I
BORROW AREA II
SLOPE INTERCEPT
AERATION CELL #3
AERATION CELL #2
AERATION CELL #1
SERVICE BUILDING
UNDERGROUND ELEC. SERVICE
4060' OF 5 STRANDS OF BARBED WIRE ON 7'-6" STEEL POSTS 12'-6" O.C. w/ 2-10' SECTIONS OF GATE w/ PADLOCK, 6 WARNING SIGNS REQ'D. (SPACED PER ENGINEER'S DIRECTION)
SHED
BARNETT RD.
DOUGHTY RD.
2040' OF OVERHEAD ELECTRIC SERVICE BY OWNER (PER NSP) TO SERVICE BLDG. METER SOCKET (SEE SPECS)
2-10' WD. GATES w/ PADLOCK
12' WD. BITUMINOUS ACCESS DRIVE @ 5% GRADE
1½" WATER SERVICE TAP INTO EX. 12" WATER MAIN w/C.S.
14" PVC DR 25. GRAVITY EFFLUENT PIPING TO RIVER
18' WD. BITUMINOUS ACCESS ROAD - DIRT RD. PRESENTLY EXISTS
EXIST'G. POWER POLE
SURVEY MONUMENT TO BE PERPETUATED
END FORCEMAIN CONTRACT @ FENCELINE
16" D.I. CL 52 FORCEMAIN PIPING TO CONTROL BUILDING
2140' OF 1½" PE SDR 11 WATER SERVICE
SURVEY MONUMENT TO BE PERPETUATED
STATION 0+00
STATION 10+78
CROSS SECTION STATIONING FOR EARTHWORK QUANTITIES
LADYSMITH WWTF
LAGOON SITE LAYOUT w/ CONTOURS
SCALE- 1" = 100'
SHEET 11 OF 28
DATE APRIL, 1988
MORGAN & PARMLEY, LTD.
CONSULTING ENGINEERS
LADYSMITH, WISCONSIN
DR. BY M.D. SHERVEY

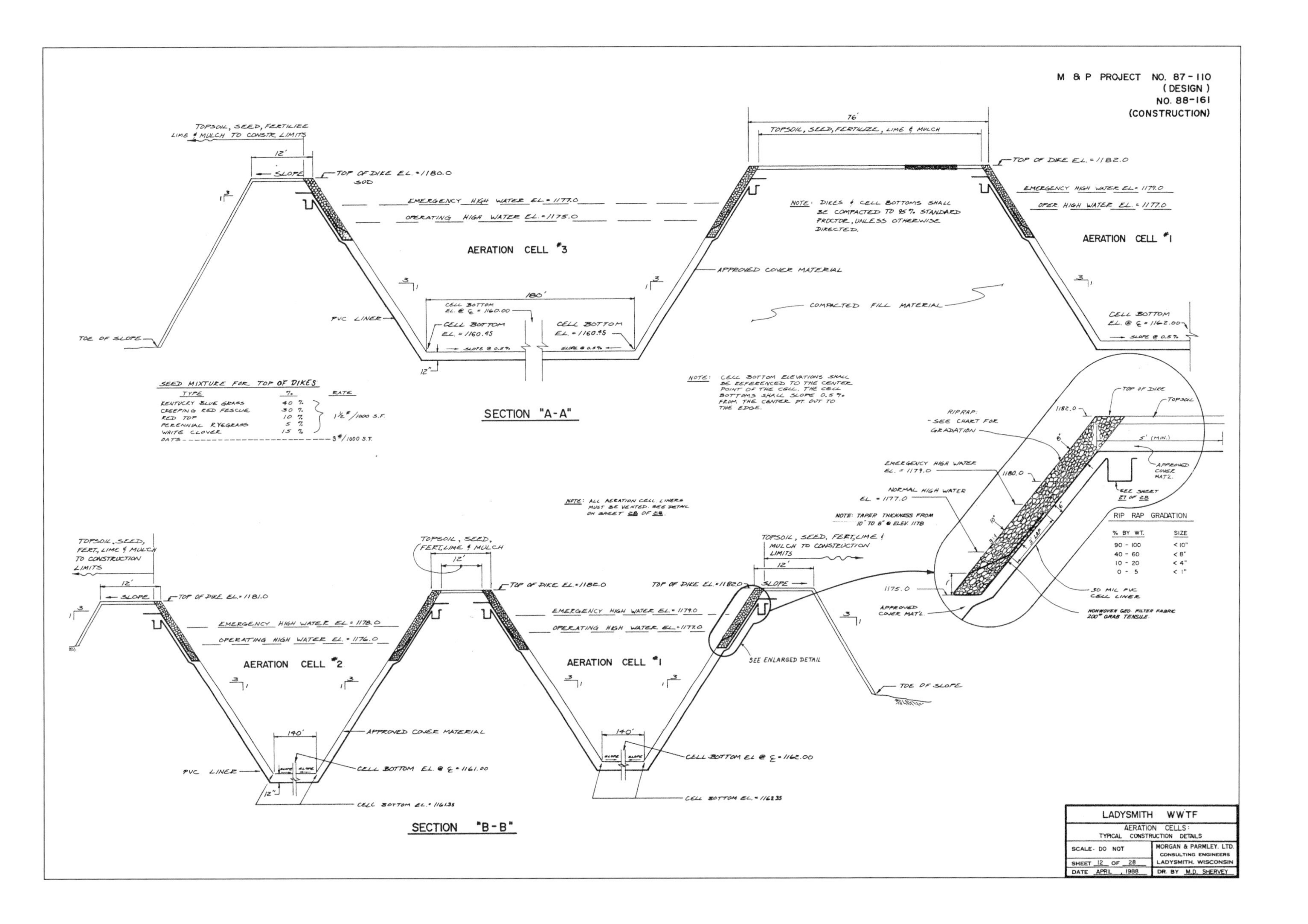
M & P PROJECT NO. 87-110 (DESIGN) NO. 88-161 (CONSTRUCTION)
TOPSOIL, SEED, FERTILIZE LIME & MULCH TO CONSTR. LIMITS
TOP OF DIKE EL.=1180.0 SOD
EMERGENCY HIGH WATER EL.=1177.0
OPERATING HIGH WATER EL.=1175.0
AERATION CELL #3
TOPSOIL, SEED, FERTILIZE, LIME & MULCH
TOP OF DIKE EL.=1182.0
EMERGENCY HIGH WATER EL.=1179.0
OPER. HIGH WATER EL.=1177.0
NOTE: DIKES & CELL BOTTOMS SHALL BE COMPACTED TO 95% STANDARD PROCTOR, UNLESS OTHERWISE DIRECTED.
AERATION CELL #1
APPROVED COVER MATERIAL
COMPACTED FILL MATERIAL
PVC LINER
CELL BOTTOM EL. @ ℄ = 1160.00
CELL BOTTOM EL.=1160.45
CELL BOTTOM EL. @ ℄ = 1162.00
TOE OF SLOPE
SEED MIXTURE FOR TOP OF DIKES
TYPE % RATE
KENTUCKY BLUE GRASS 40%
CREEPING RED FESCUE 30%
RED TOP 10%
PERENNIAL RYEGRASS 5%
WHITE CLOVER 15%
1½#/1000 S.F.
OATS 3#/1000 S.F.
SECTION "A-A"
NOTE: CELL BOTTOM ELEVATIONS SHALL BE REFERENCED TO THE CENTER POINT OF THE CELL. THE CELL BOTTOMS SHALL SLOPE 0.5% FROM THE CENTER PT. OUT TO THE EDGE.
RIPRAP: - SEE CHART FOR GRADATION
TOP OF DIKE
TOPSOIL
5' (MIN.)
APPROVED COVER MAT'L.
SEE SHEET 27 OF 28
EMERGENCY HIGH WATER EL.=1179.0
NORMAL HIGH WATER EL=1177.0
NOTE: TAPER THICKNESS FROM 10" TO 8" @ ELEV. 1178
RIP RAP GRADATION
% BY WT. SIZE
90 - 100 < 10"
40 - 60 < 8"
10 - 20 < 4"
0 - 5 < 1"
30 MIL PVC CELL LINER
NONWOVEN GEO FILTER FABRIC 200# GRAB TENSILE
NOTE: ALL AERATION CELL LINERS MUST BE VENTED. SEE DETAIL ON SHEET 28 OF 28.
TOPSOIL, SEED, FERT, LIME & MULCH TO CONSTRUCTION LIMITS
TOP OF DIKE EL.=1181.0
EMERGENCY HIGH WATER EL.=1178.0
OPERATING HIGH WATER EL.=1176.0
AERATION CELL #2
TOPSOIL, SEED, FERT, LIME & MULCH
TOP OF DIKE EL.=1182.0
EMERGENCY HIGH WATER EL.=1179.0
OPERATING HIGH WATER EL.=1177.0
AERATION CELL #1
SEE ENLARGED DETAIL
TOE OF SLOPE
CELL BOTTOM EL. @ ℄ = 1161.00
CELL BOTTOM EL.=1161.35
CELL BOTTOM EL @ ℄ = 1162.00
CELL BOTTOM EL.=1162.35
SECTION "B-B"
LADYSMITH WWTF
AERATION CELLS: TYPICAL CONSTRUCTION DETAILS
SCALE- DO NOT
MORGAN & PARMLEY, LTD. CONSULTING ENGINEERS LADYSMITH, WISCONSIN
SHEET 12 OF 28
DATE APRIL, 1988
DR. BY M.D. SHERVEY

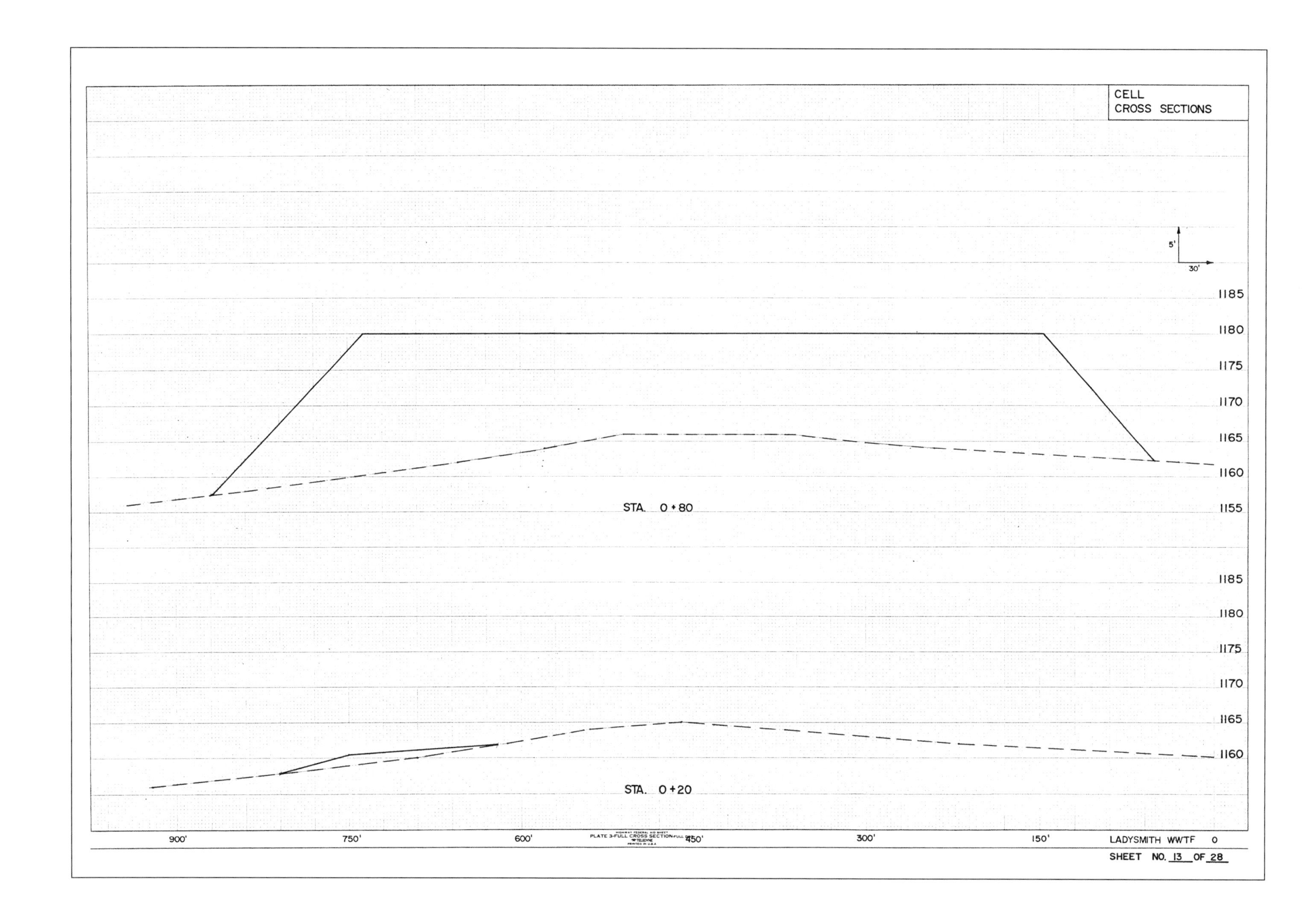
CELL
CROSS SECTIONS
5'
30'
1185
1180
1175
1170
1165
1160
1155
STA. 0 + 80
1185
1180
1175
1170
1165
1160
STA. 0 + 20
900'
750'
600'
450'
300'
150'
LADYSMITH WWTF 0
SHEET NO. 13 OF 28

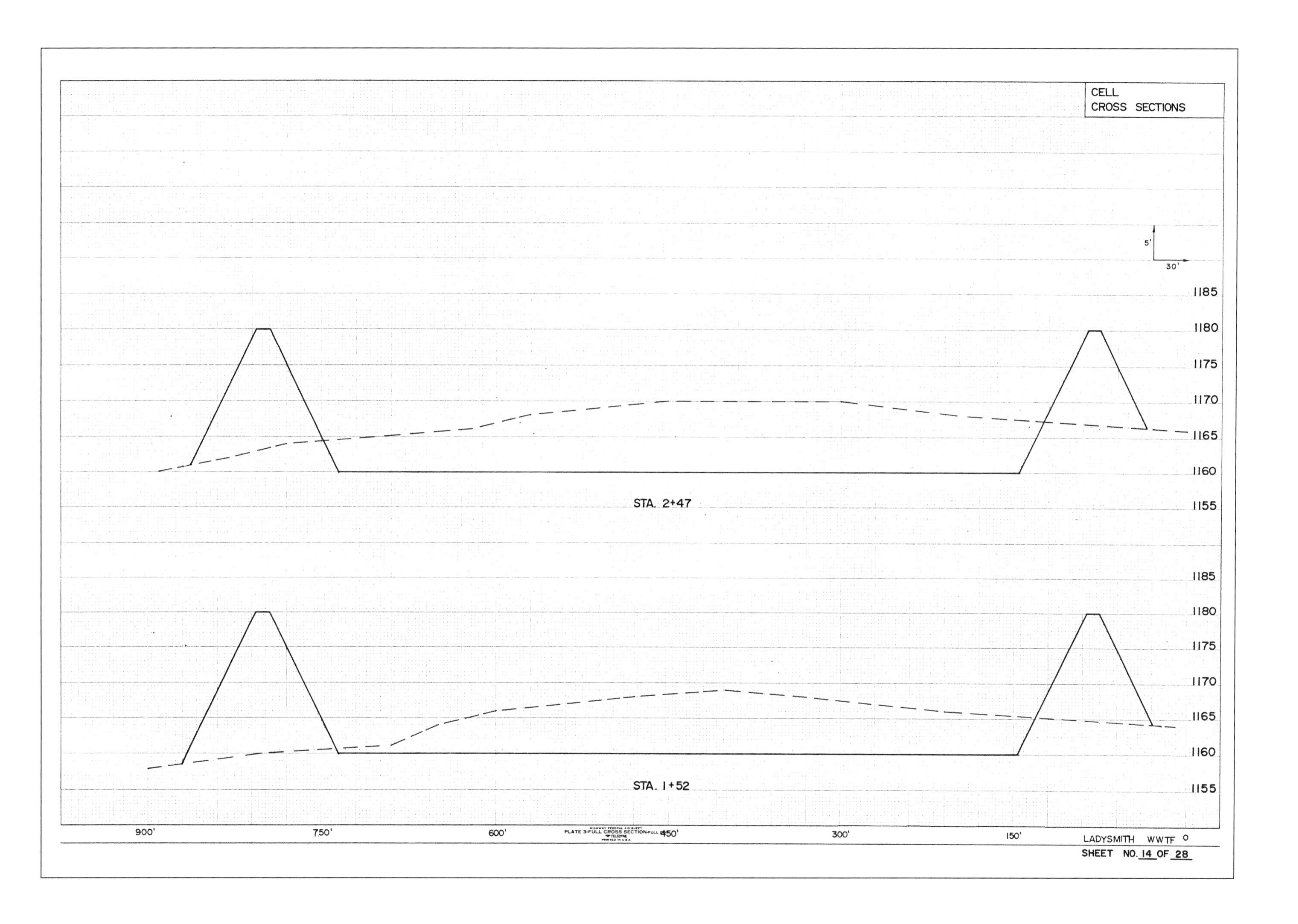
CELL
CROSS SECTIONS
5'
30'
1185
1180
1175
1170
1165
1160
1155
STA. 2+47
1185
1180
1175
1170
1165
1160
1155
STA. 1+52
900'
750'
600'
450'
300'
150'
0
LADYSMITH WWTF
SHEET NO. 14 OF 28

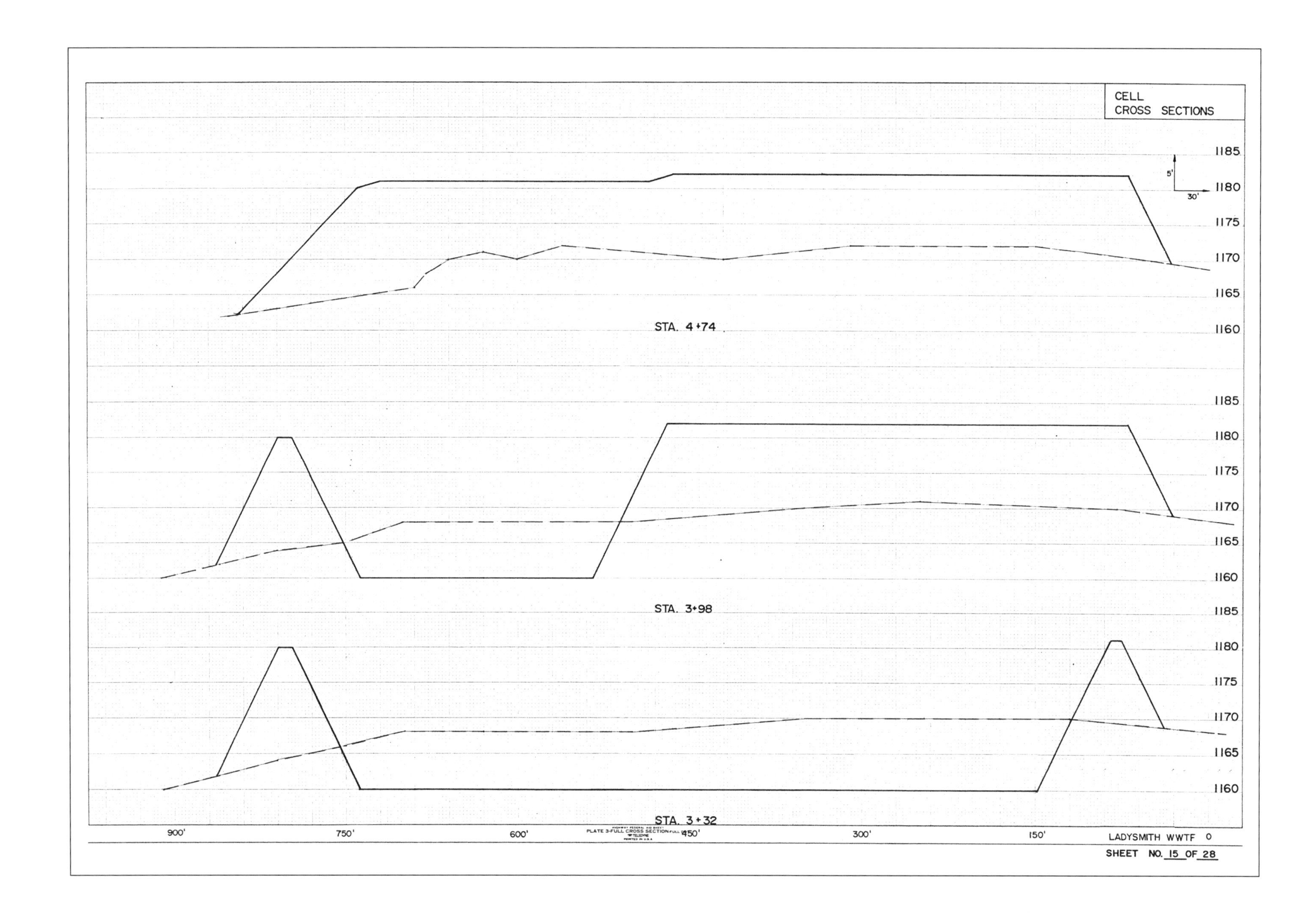
CELL
CROSS SECTIONS
1185
1180
1175
1170
1165
1160
5'
30'
STA. 4+74
STA. 3+98
STA. 3+32
900'
750'
600'
450'
300'
150'
0
LADYSMITH WWTF
SHEET NO. 15 OF 28

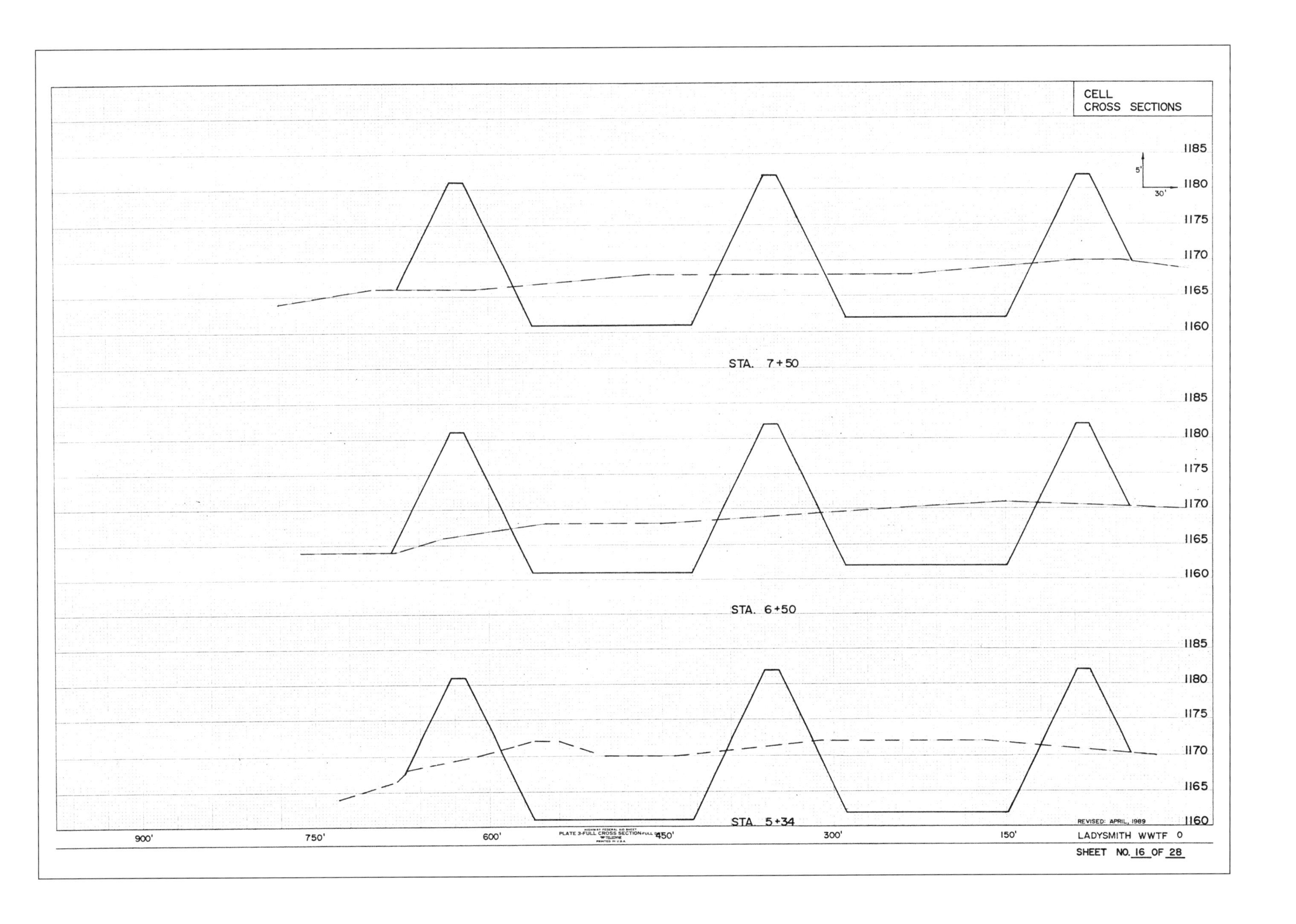
CELL
CROSS SECTIONS
1185
1180
1175
1170
1165
1160
5'
30'
STA. 7+50
STA. 6+50
STA. 5+34
900'
750'
600'
450'
300'
150'
REVISED: APRIL, 1989
LADYSMITH WWTF
SHEET NO. 16 OF 28

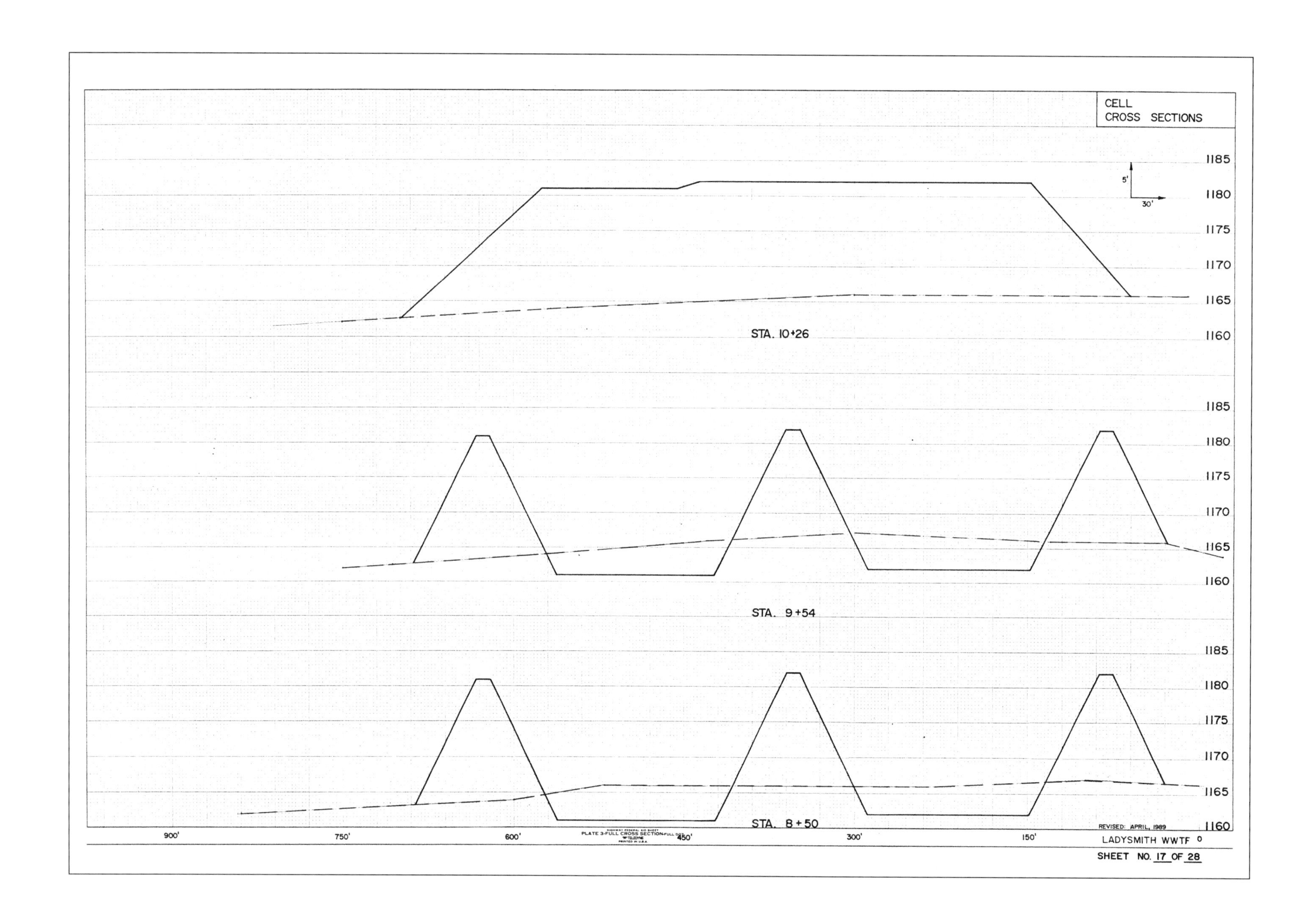
CELL
CROSS SECTIONS
1185
1180
1175
1170
1165
1160
5'
30'
STA. 10+26
STA. 9+54
STA. 8+50
900'
750'
600'
450'
300'
150'
0
REVISED: APRIL, 1989
LADYSMITH WWTF
SHEET NO. 17 OF 28

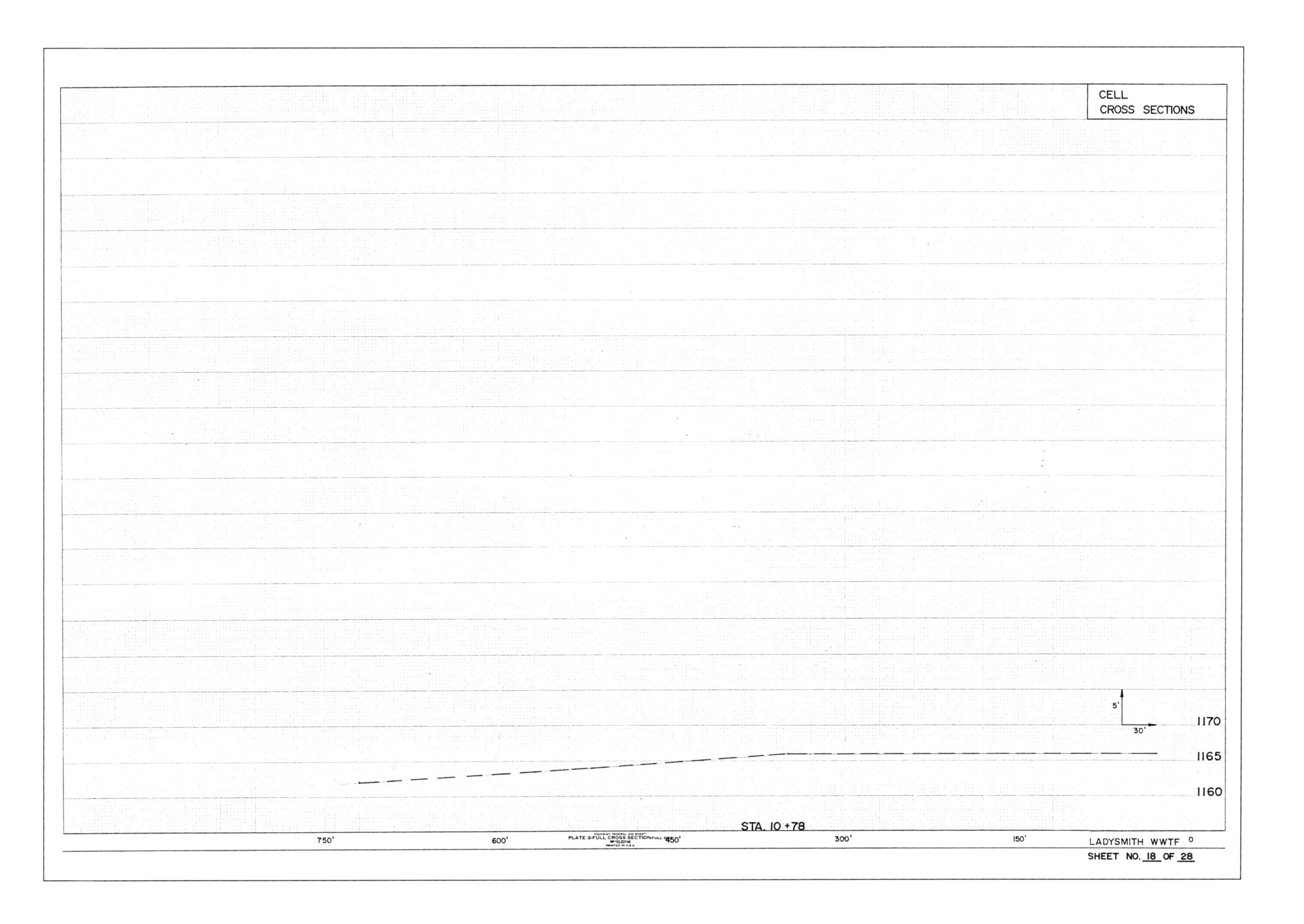
CELL
CROSS SECTIONS
5'
30'
1170
1165
1160
STA. 10 +78
750'
600'
PLATE 3-FULL CROSS SECTION
450'
300'
150'
LADYSMITH WWTF
0
SHEET NO. 18 OF 28

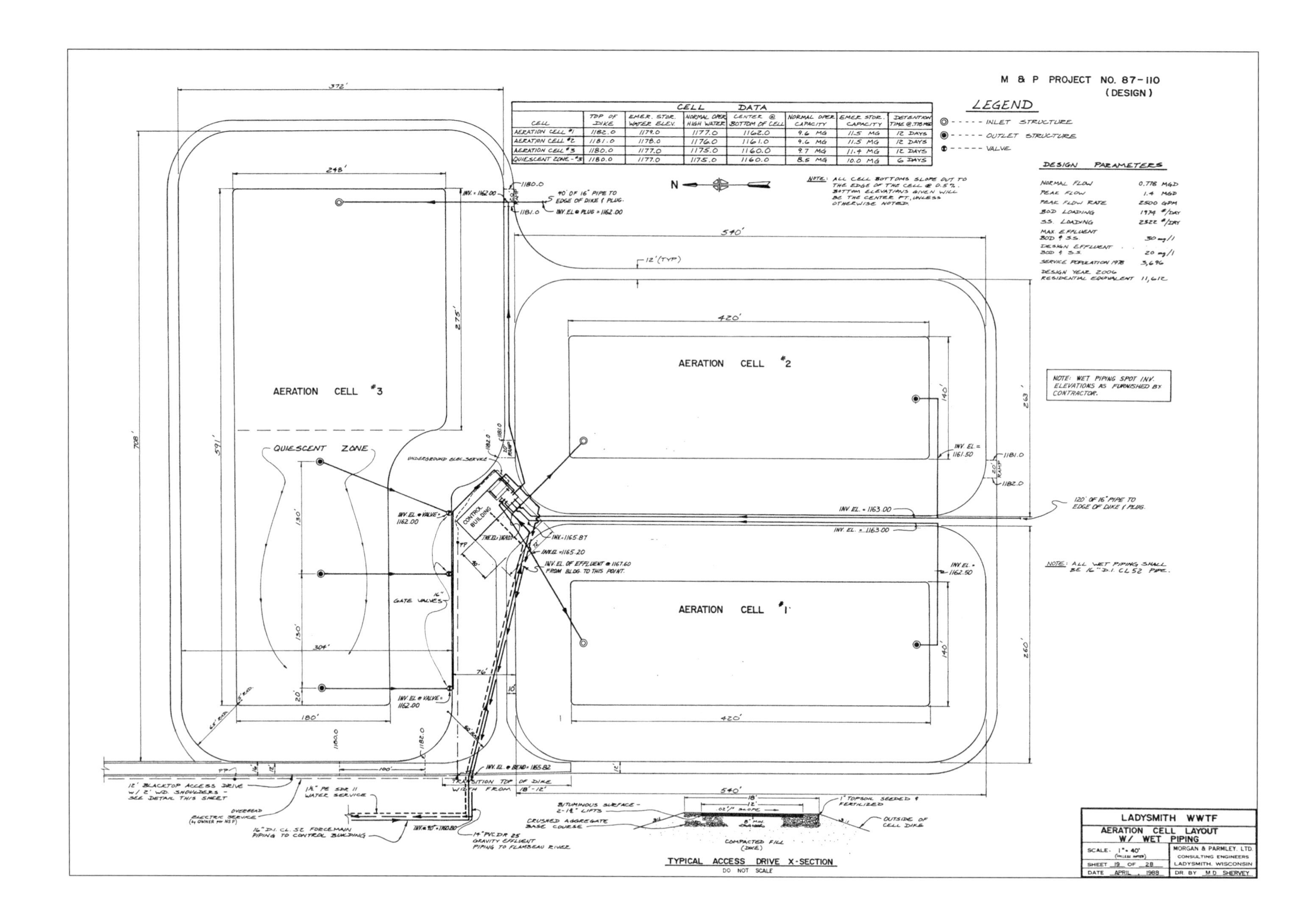

CELL DATA							
CELL	TOP OF DIKE	EMER. STOR. WATER ELEV.	NORMAL OPER HIGH WATER	CENTER @ BOTTOM OF CELL	NORMAL OPER CAPACITY	EMER. STOR. CAPACITY	DETENTION TIME @ .778 MGD
AERATION CELL #1	1182.0	1179.0	1177.0	1162.0	9.6 MG	11.5 MG	12 DAYS
AERATION CELL #2	1181.0	1178.0	1176.0	1161.0	9.6 MG	11.5 MG	12 DAYS
AERATION CELL #3	1180.0	1177.0	1175.0	1160.0	9.7 MG	11.4 MG	12 DAYS
QUIESCENT ZONE - #3	1180.0	1177.0	1175.0	1160.0	8.5 MG	10.0 MG	6 DAYS

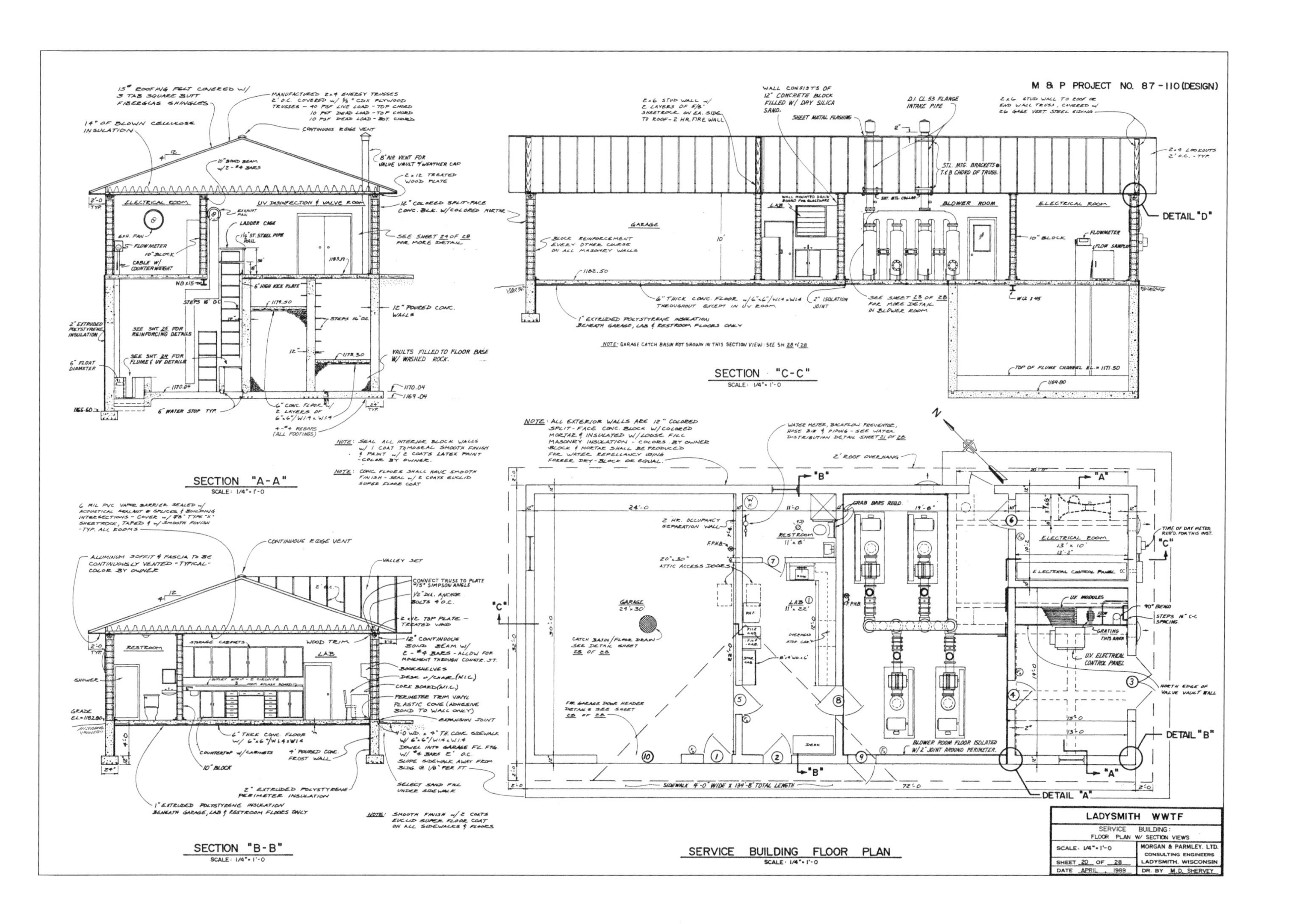
M & P PROJECT NO. 87-110 (DESIGN)
SECTION "A-A"
SCALE: 1/4" = 1'-0
SECTION "C-C"
SCALE: 1/4" = 1'-0
SECTION "B-B"
SCALE: 1/4" = 1'-0
SERVICE BUILDING FLOOR PLAN
SCALE: 1/4" = 1'-0
DETAIL "D"
DETAIL "B"
DETAIL "A"
LADYSMITH WWTF
SERVICE BUILDING: FLOOR PLAN W/ SECTION VIEWS
SCALE- 1/4" = 1'-0
SHEET 20 OF 28
DATE APRIL, 1988
MORGAN & PARMLEY, LTD. CONSULTING ENGINEERS LADYSMITH, WISCONSIN
DR. BY M.D. SHERVEY

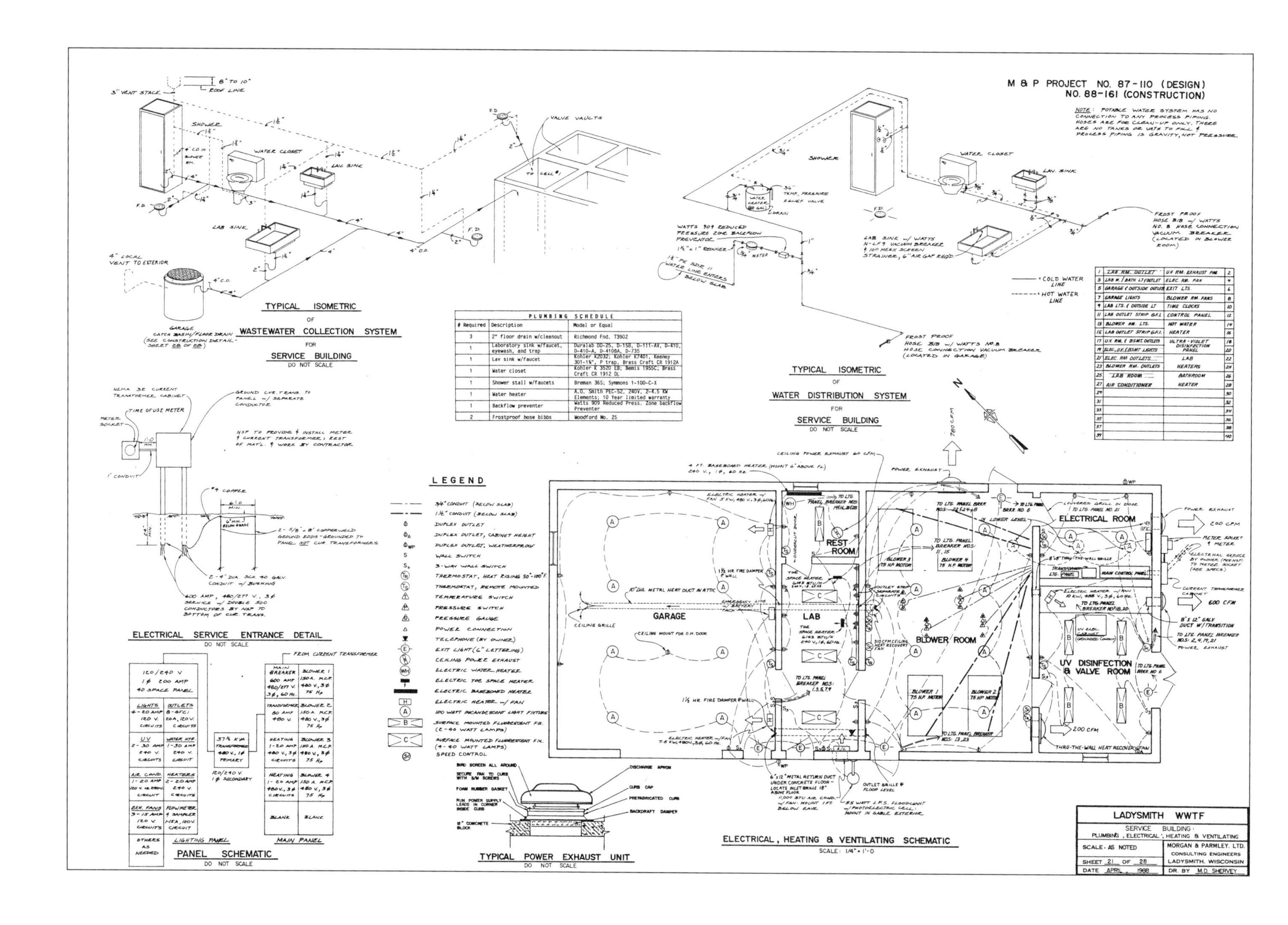

PLUMBING SCHEDULE

# Required	Description	Model or Equal
3	2" floor drain w/cleanout	Richmond Fnd. T3902
1	Laboratory sink w/faucet, eyewash, and trap	Duralab DD-25, D-158, D-111-AV, D-410, D-410-A, D-410BA, D-735
1	Lav sink w/faucet	Kohler K2032; Kohler K7401, Keeney 301-1¼", P trap, Brass Craft CR 1912A
1	Water closet	Kohler K 3520 EB; Bemis 1955C; Brass Craft CR 1912 DL
1	Shower stall w/faucets	Breman 36S; Symmons 1-100-C-X
1	Water heater	A.O. Smith PEC-52, 240V, 2-4.5 KW Elements; 10 Year limited warranty
1	Backflow preventer	Watts 909 Reduced Press. Zone backflow Preventer
2	Frostproof hose bibbs	Woodford No. 25

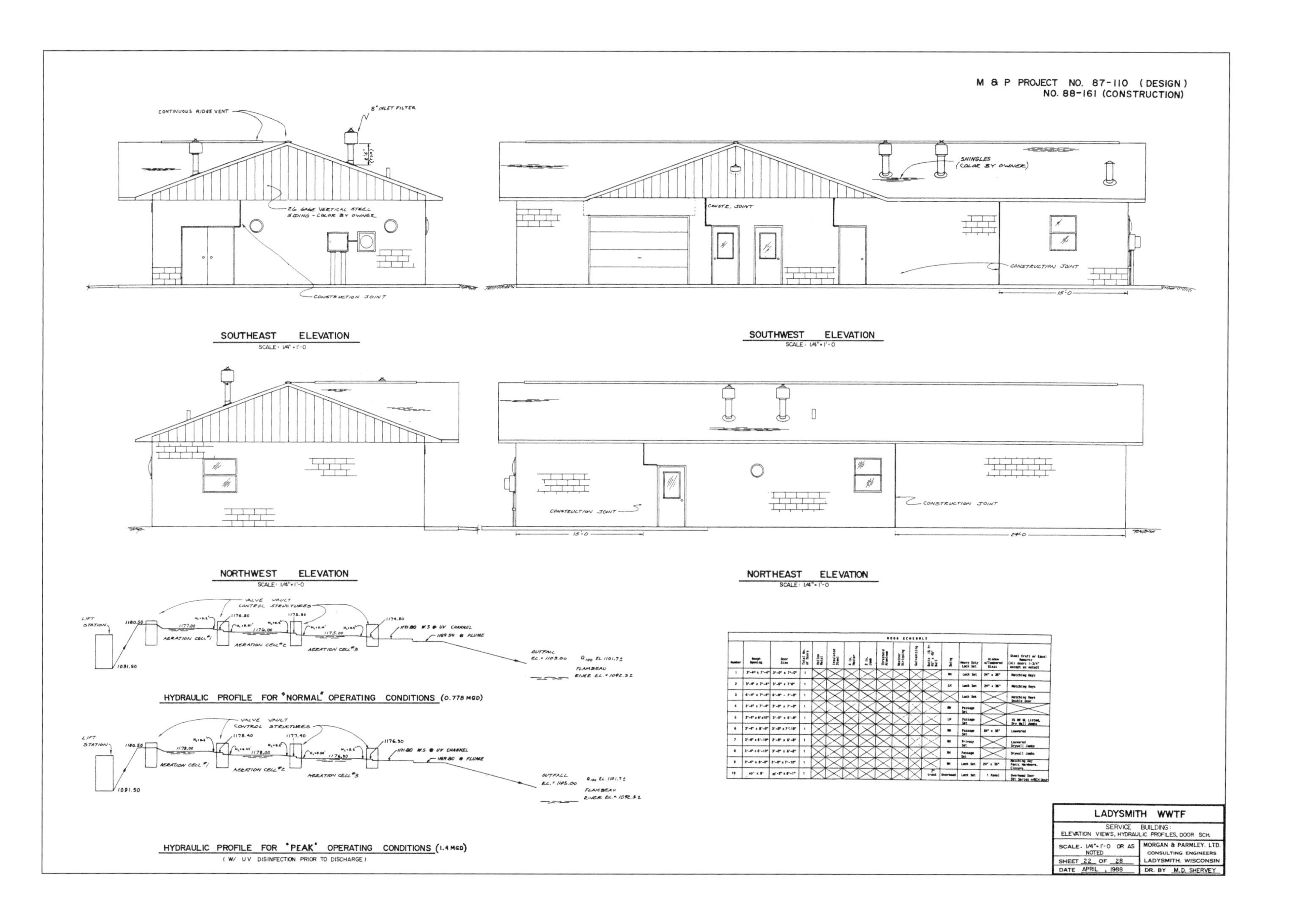
M & P PROJECT NO. 87-110 (DESIGN)
NO. 88-161 (CONSTRUCTION)
CONTINUOUS RIDGE VENT
8" INLET FILTER
26 GAGE VERTICAL STEEL SIDING - COLOR BY OWNER
CONSTRUCTION JOINT
SHINGLES (COLOR BY OWNER)
CONSTR. JOINT
15'-0
SOUTHEAST ELEVATION
SCALE: 1/4" = 1'-0
SOUTHWEST ELEVATION
SCALE: 1/4" = 1'-0
24'-0
NORTHWEST ELEVATION
SCALE: 1/4" = 1'-0
NORTHEAST ELEVATION
SCALE: 1/4" = 1'-0
VALVE VAULT CONTROL STRUCTURES
LIFT STATION
1180.50
1177.00
1176.80
1176.00
1175.80
1175.00
1174.80
1171.80 W.S. @ UV CHANNEL
1169.54 @ FLUME
AERATION CELL #1
AERATION CELL #2
AERATION CELL #3
1091.50
OUTFALL EL. = 1103.00
Q_{100} EL. 1101.7±
FLAMBEAU RIVER EL. = 1092.3±
HYDRAULIC PROFILE FOR "NORMAL" OPERATING CONDITIONS (0.778 MGD)
1178.40
1178.00
1177.40
1176.50
1176.30
1171.80 W.S. @ UV CHANNEL
1169.80 @ FLUME
HYDRAULIC PROFILE FOR "PEAK" OPERATING CONDITIONS (1.4 MGD)
(W/ UV DISINFECTION PRIOR TO DISCHARGE)
DOOR SCHEDULE
LADYSMITH WWTF
SERVICE BUILDING:
ELEVATION VIEWS, HYDRAULIC PROFILES, DOOR SCH.
SCALE: 1/4" = 1'-0 OR AS NOTED
MORGAN & PARMLEY, LTD.
CONSULTING ENGINEERS
LADYSMITH, WISCONSIN
SHEET 22 OF 28
DATE APRIL, 1988
DR. BY M.D. SHERVEY

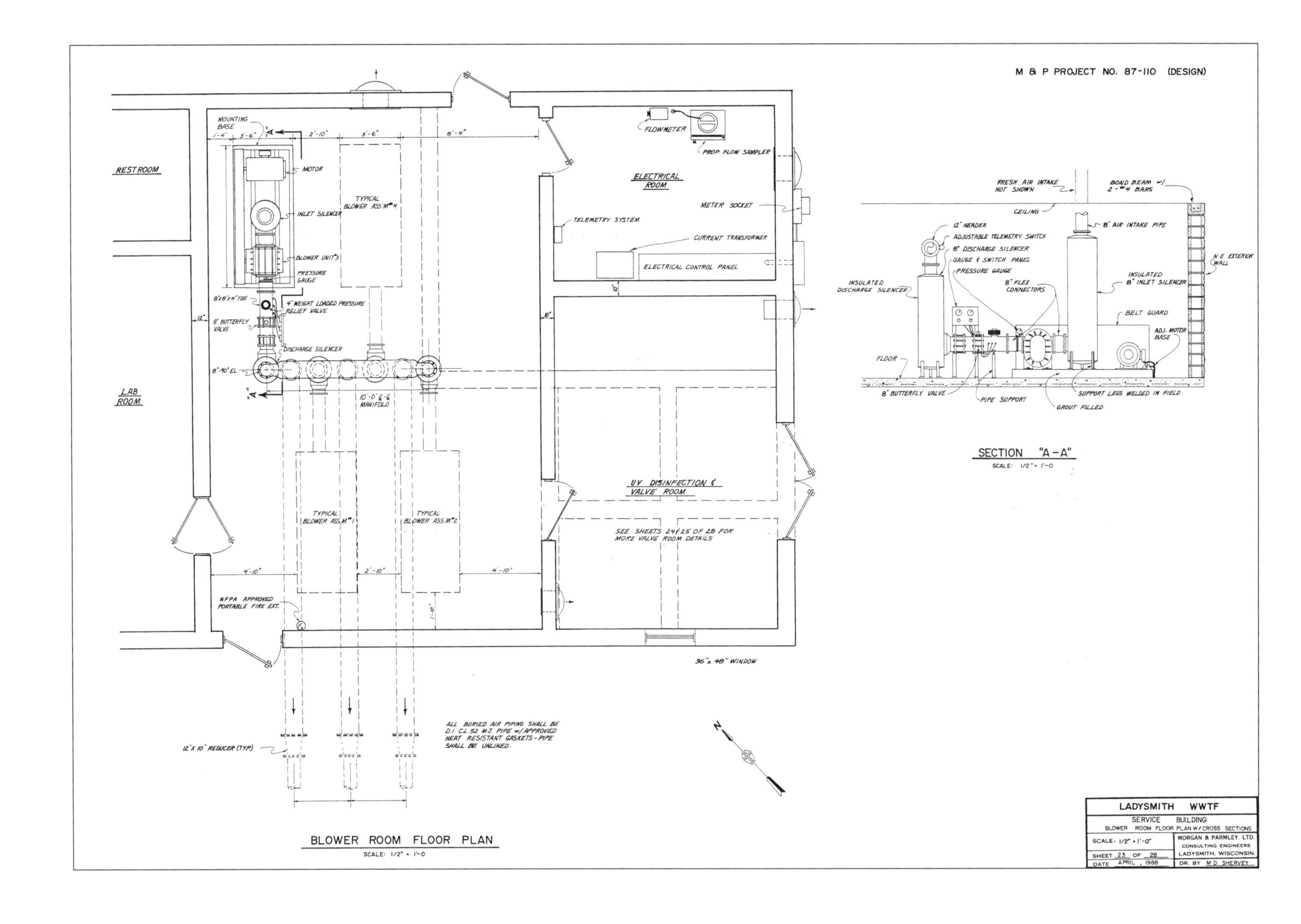
M & P PROJECT NO. 87-110 (DESIGN)
RESTROOM
LAB ROOM
MOUNTING BASE
MOTOR
INLET SILENCER
BLOWER UNIT #3
PRESSURE GAUGE
8"x8"x4" TEE
4" WEIGHT LOADED PRESSURE RELIEF VALVE
8" BUTTERFLY VALVE
DISCHARGE SILENCER
8"-90° EL
10'-0" ℄-℄ MANIFOLD
TYPICAL BLOWER ASS'M #4
TYPICAL BLOWER ASS'M #1
TYPICAL BLOWER ASS'M #2
NFPA APPROVED PORTABLE FIRE EXT.
12"x 10" REDUCER (TYP.)
ALL BURIED AIR PIPING SHALL BE D.I. CL.52 MJ PIPE w/ APPROVED HEAT RESISTANT GASKETS - PIPE SHALL BE UNLINED.
FLOWMETER
PROP. FLOW SAMPLER
ELECTRICAL ROOM
METER SOCKET
TELEMETRY SYSTEM
CURRENT TRANSFORMER
ELECTRICAL CONTROL PANEL
UV DISINFECTION & VALVE ROOM
SEE SHEETS 24 & 25 OF 28 FOR MORE VALVE ROOM DETAILS
36" x 48" WINDOW
N
BLOWER ROOM FLOOR PLAN
SCALE: 1/2" = 1'-0
FRESH AIR INTAKE NOT SHOWN
BOND BEAM w/ 2-#4 BARS
CEILING
12" HEADER
ADJUSTABLE TELEMETRY SWITCH
8" DISCHARGE SILENCER
GAUGE & SWITCH PANEL
PRESSURE GAUGE
8" FLEX CONNECTORS
8" AIR INTAKE PIPE
N-E EXTERIOR WALL
INSULATED DISCHARGE SILENCER
INSULATED 8" INLET SILENCER
BELT GUARD
ADJ. MOTOR BASE
FLOOR
8" BUTTERFLY VALVE
PIPE SUPPORT
GROUT FILLED
SUPPORT LEGS WELDED IN FIELD
SECTION "A-A"
SCALE: 1/2" = 1'-0
LADYSMITH WWTF
SERVICE BUILDING:
BLOWER ROOM FLOOR PLAN W/CROSS SECTIONS
SCALE - 1/2" = 1'-0"
MORGAN & PARMLEY, LTD. CONSULTING ENGINEERS LADYSMITH, WISCONSIN
SHEET 23 OF 28
DATE APRIL, 1988
DR. BY M.D. SHERVEY

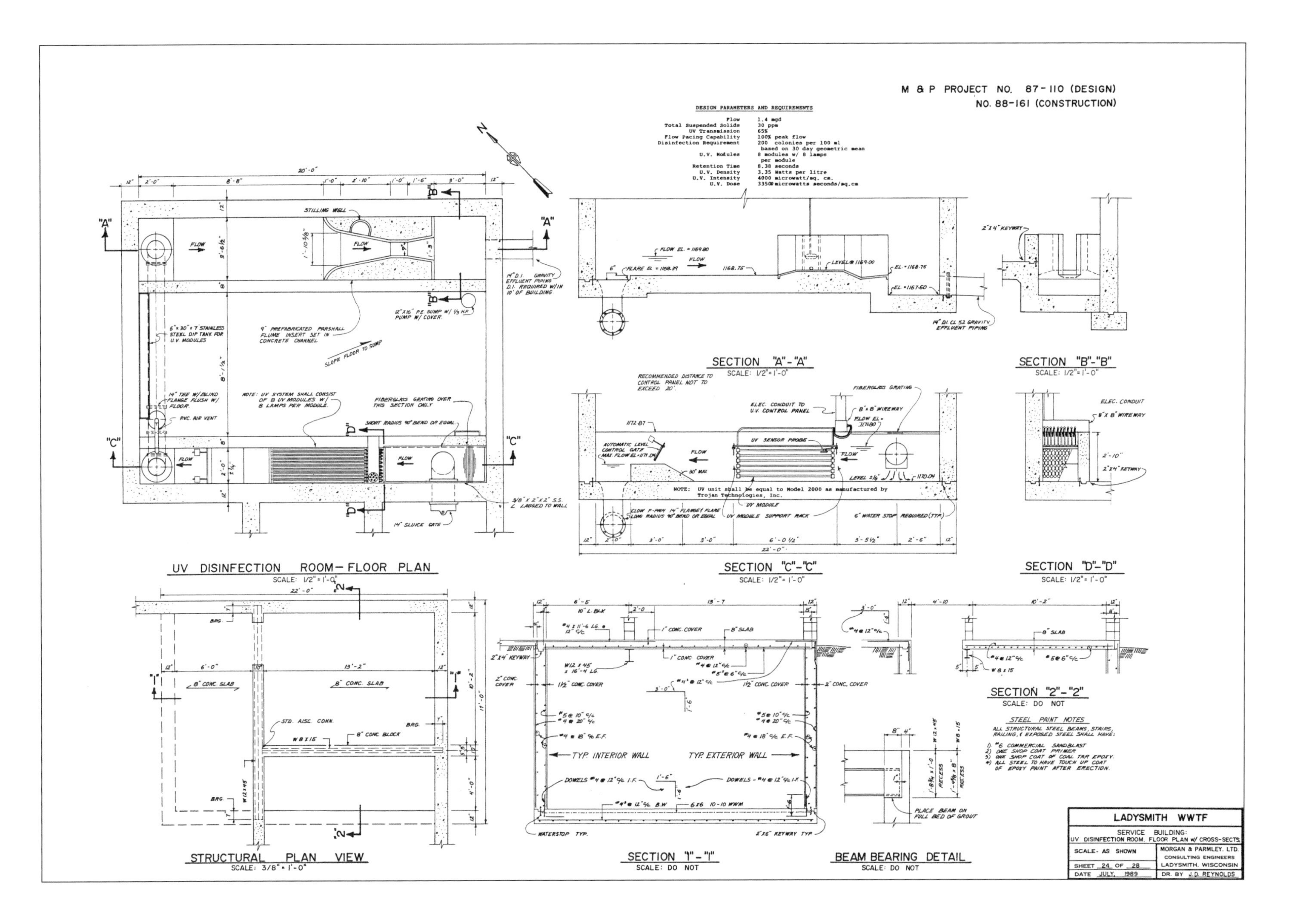
M & P PROJECT NO. 87-110 (DESIGN)
NO. 88-161 (CONSTRUCTION)
DESIGN PARAMETERS AND REQUIREMENTS
Flow 1.4 mgd
Total Suspended Solids 30 ppm
UV Transmission 65%
Flow Pacing Capability 100% peak flow
Disinfection Requirement 200 colonies per 100 ml based on 30 day geometric mean
U.V. Modules 8 modules w/ 8 lamps per module
Retention Time 8.38 seconds
U.V. Density 3.35 Watts per litre
U.V. Intensity 4000 microwatt/sq. cm.
U.V. Dose 33500 microwatts seconds/sq.cm
UV DISINFECTION ROOM—FLOOR PLAN
SCALE: 1/2" = 1'-0"
SECTION "A"-"A"
SCALE: 1/2" = 1'-0"
SECTION "B"-"B"
SCALE: 1/2" = 1'-0"
SECTION "C"-"C"
SCALE: 1/2" = 1'-0"
SECTION "D"-"D"
SCALE: 1/2" = 1'-0"
NOTE: UV unit shall be equal to Model 2000 as manufactured by Trojan Technologies, Inc.
STRUCTURAL PLAN VIEW
SCALE: 3/8" = 1'-0"
SECTION "1"-"1"
SCALE: DO NOT
BEAM BEARING DETAIL
SCALE: DO NOT
SECTION "2"-"2"
SCALE: DO NOT
STEEL PAINT NOTES
ALL STRUCTURAL STEEL BEAMS, STAIRS, RAILING, & EXPOSED STEEL SHALL HAVE:
1) #6 COMMERCIAL SANDBLAST
2) ONE SHOP COAT PRIMER
3) ONE SHOP COAT OF COAL TAR EPOXY.
4) ALL STEEL TO HAVE TOUCH UP COAT OF EPOXY PAINT AFTER ERECTION.
LADYSMITH WWTF
SERVICE BUILDING:
UV DISINFECTION ROOM, FLOOR PLAN w/ CROSS-SECTS.
SCALE- AS SHOWN
SHEET 24 OF 28
DATE JULY, 1989
MORGAN & PARMLEY, LTD.
CONSULTING ENGINEERS
LADYSMITH, WISCONSIN
DR. BY J.D. REYNOLDS

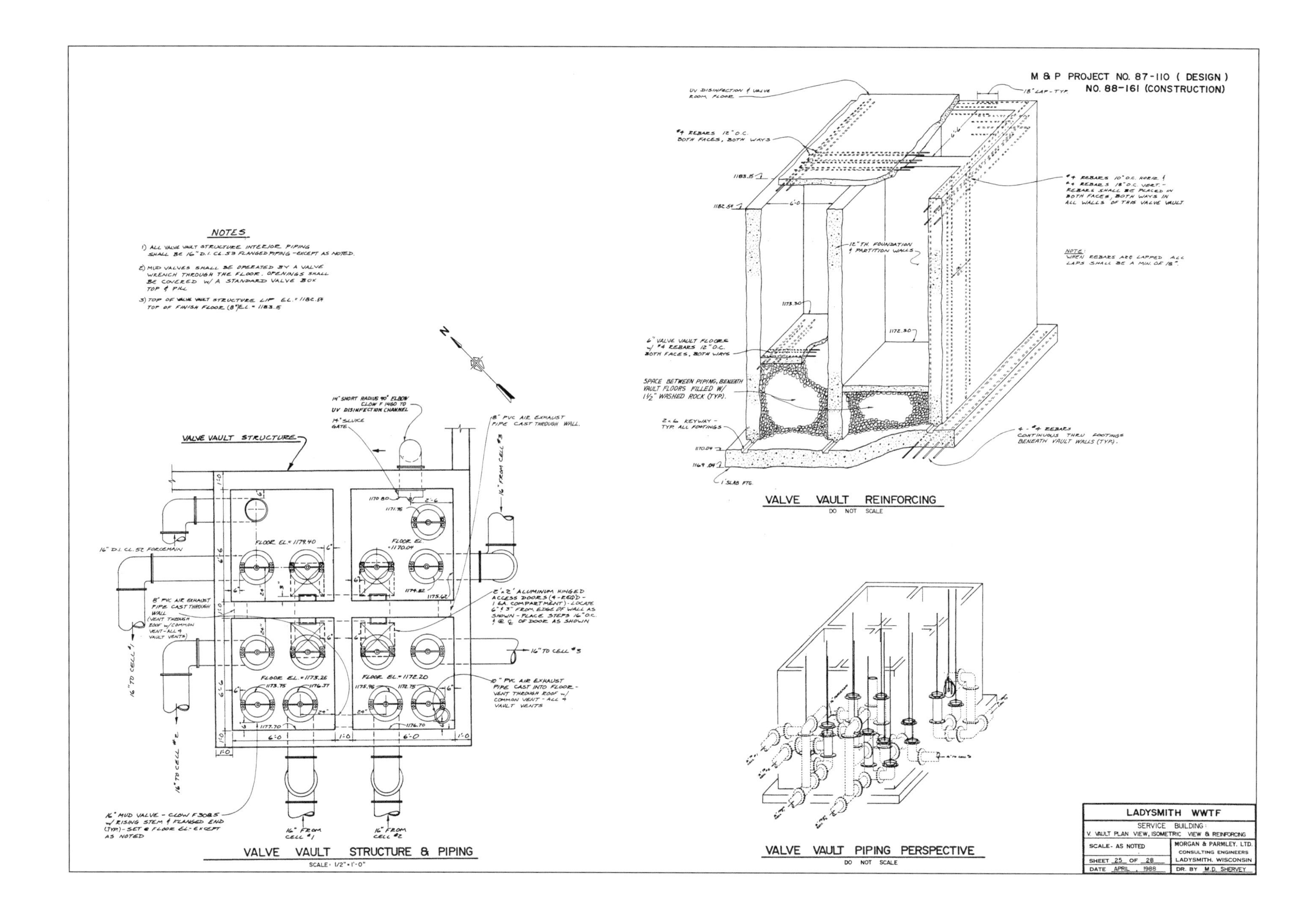

M & P PROJECT NO. 87-110 (DESIGN)
NO. 88-161 (CONSTRUCTION)
NOTES
1) ALL VALVE VAULT STRUCTURE INTERIOR PIPING SHALL BE 16" D.I. CL.53 FLANGED PIPING - EXCEPT AS NOTED.
2) MUD VALVES SHALL BE OPERATED BY A VALVE WRENCH THROUGH THE FLOOR. OPENINGS SHALL BE COVERED W/ A STANDARD VALVE BOX TOP & FILL
3) TOP OF VALVE VAULT STRUCTURE LIP EL.= 1182.54
TOP OF FINISH FLOOR (8")EL.= 1183.15
VALVE VAULT STRUCTURE
16" D.I. CL.52 FORCEMAIN
FLOOR EL.= 1179.40
FLOOR EL. = 1170.09
FLOOR EL.= 1173.26
FLOOR EL.= 1172.20
14" SHORT RADIUS 90° ELBOW
UV DISINFECTION CHANNEL
14" SLUICE GATE
8" PVC AIR EXHAUST PIPE CAST THROUGH WALL.
16" FROM CELL #3
16" TO CELL #5
16" TO CELL #4
16" TO CELL #2
16" FROM CELL #1
16" FROM CELL #2
2'x2' ALUMINUM HINGED ACCESS DOORS
10" PVC AIR EXHAUST PIPE CAST INTO FLOOR - VENT THROUGH ROOF w/ COMMON VENT - ALL 4 VAULT VENTS
16" MUD VALVE - CLOW F3085 w/ RISING STEM & FLANGED END (TYP.) - SET @ FLOOR EL. - EXCEPT AS NOTED
VALVE VAULT STRUCTURE & PIPING
SCALE: 1/2" = 1'-0"
#4 REBARS 12" O.C. BOTH FACES, BOTH WAYS
12" TH. FOUNDATION & PARTITION WALLS
6" VALVE VAULT FLOORS w/ #4 REBARS 12" O.C. BOTH FACES, BOTH WAYS
SPACE BETWEEN PIPING, BENEATH VAULT FLOORS FILLED W/ 1½" WASHED ROCK (TYP).
2x6 KEYWAY - TYP. ALL FOOTINGS
1' SLAB FTG.
NOTE: WHEN REBARS ARE LAPPED ALL LAPS SHALL BE A MIN. OF 18".
4 - #4 REBARS CONTINUOUS THRU FOOTINGS BENEATH VAULT WALLS (TYP.).
VALVE VAULT REINFORCING
DO NOT SCALE
VALVE VAULT PIPING PERSPECTIVE
DO NOT SCALE
LADYSMITH WWTF
SERVICE BUILDING:
V. VAULT PLAN VIEW, ISOMETRIC VIEW & REINFORCING
SCALE- AS NOTED
SHEET 25 OF 28
DATE APRIL, 1988
MORGAN & PARMLEY, LTD.
CONSULTING ENGINEERS
LADYSMITH, WISCONSIN
DR. BY M.D. SHERVEY

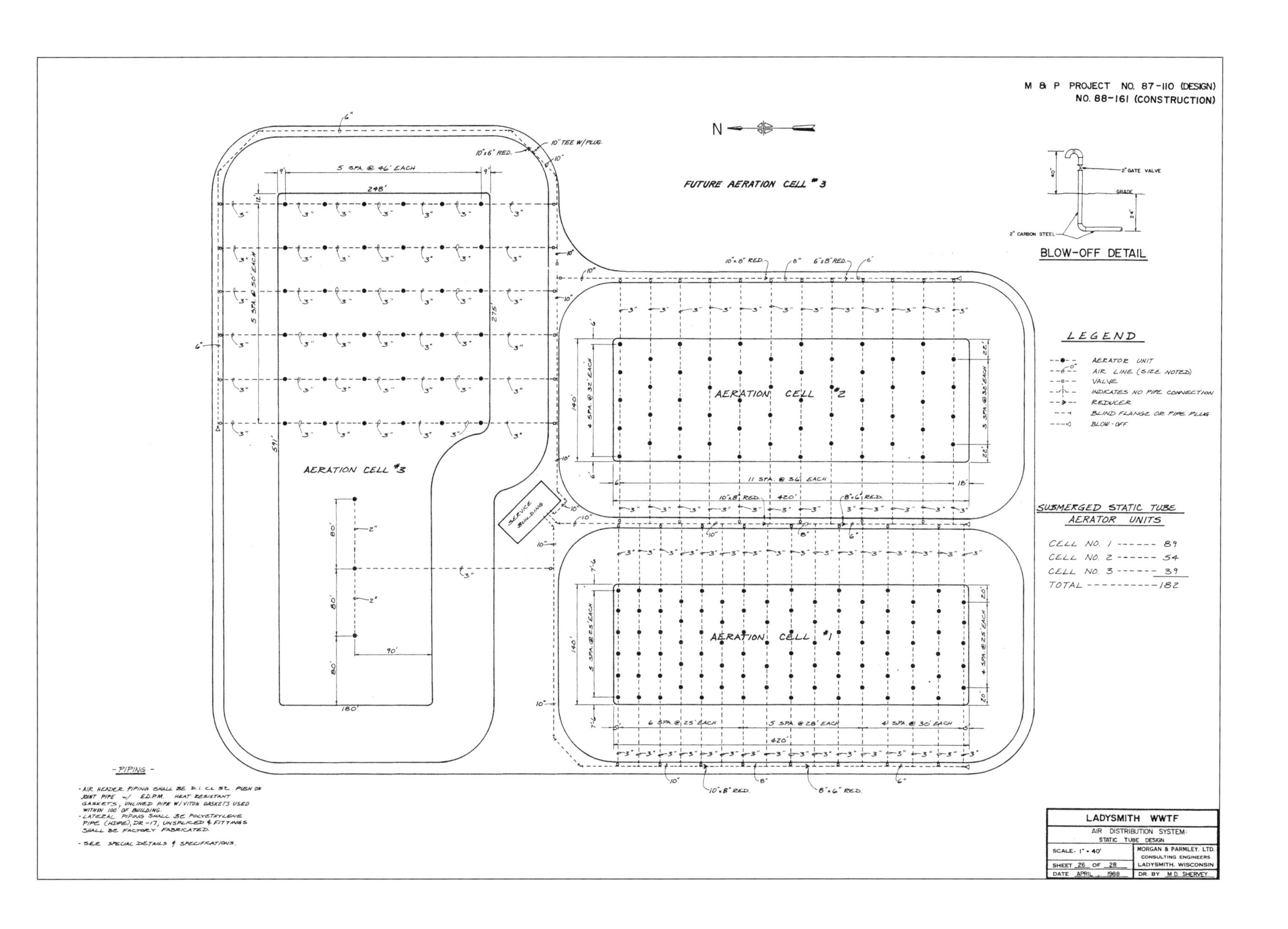
M & P PROJECT NO. 87-110 (DESIGN)
NO. 88-161 (CONSTRUCTION)
N
FUTURE AERATION CELL # 3
10" TEE W/PLUG
10"x6" RED.
5 SPA. @ 46' EACH
248'
5 SPA. @ 50' EACH
275'
591'
AERATION CELL #3
SERVICE BUILDING
90'
180'
AERATION CELL #2
4 SPA. @ 32' EACH
3 SPA. @ 32' EACH
11 SPA. @ 36' EACH
420'
10"x8" RED.
8"x6" RED.
6"x8" RED.
AERATION CELL #1
5 SPA. @ 25' EACH
4 SPA. @ 25' EACH
6 SPA. @ 25' EACH
5 SPA. @ 28' EACH
4 SPA. @ 30' EACH
2" GATE VALVE
GRADE
2" CARBON STEEL
BLOW-OFF DETAIL
LEGEND
AERATOR UNIT
AIR LINE (SIZE NOTED)
VALVE
INDICATES NO PIPE CONNECTION
REDUCER
BLIND FLANGE OR PIPE PLUG
BLOW-OFF
SUBMERGED STATIC TUBE AERATOR UNITS
CELL NO. 1 ------ 89
CELL NO. 2 ------ 54
CELL NO. 3 ------ 39
TOTAL ---------- 182
- PIPING -
- AIR HEADER PIPING SHALL BE D.I. CL 52 PUSH ON JOINT PIPE W/ E.D.P.M. HEAT RESISTANT GASKETS, UNLINED PIPE W/ VITON GASKETS USED WITHIN 100' OF BUILDING.
- LATERAL PIPING SHALL BE POLYETHYLENE PIPE (HDPE), DR-17, UNSPLICED & FITTINGS SHALL BE FACTORY FABRICATED.
- SEE SPECIAL DETAILS & SPECIFICATIONS.
LADYSMITH WWTF
AIR DISTRIBUTION SYSTEM:
STATIC TUBE DESIGN
SCALE- 1" = 40'
SHEET 26 OF 28
DATE APRIL 1988
MORGAN & PARMLEY, LTD.
CONSULTING ENGINEERS
LADYSMITH, WISCONSIN
DR. BY M.D. SHERVEY

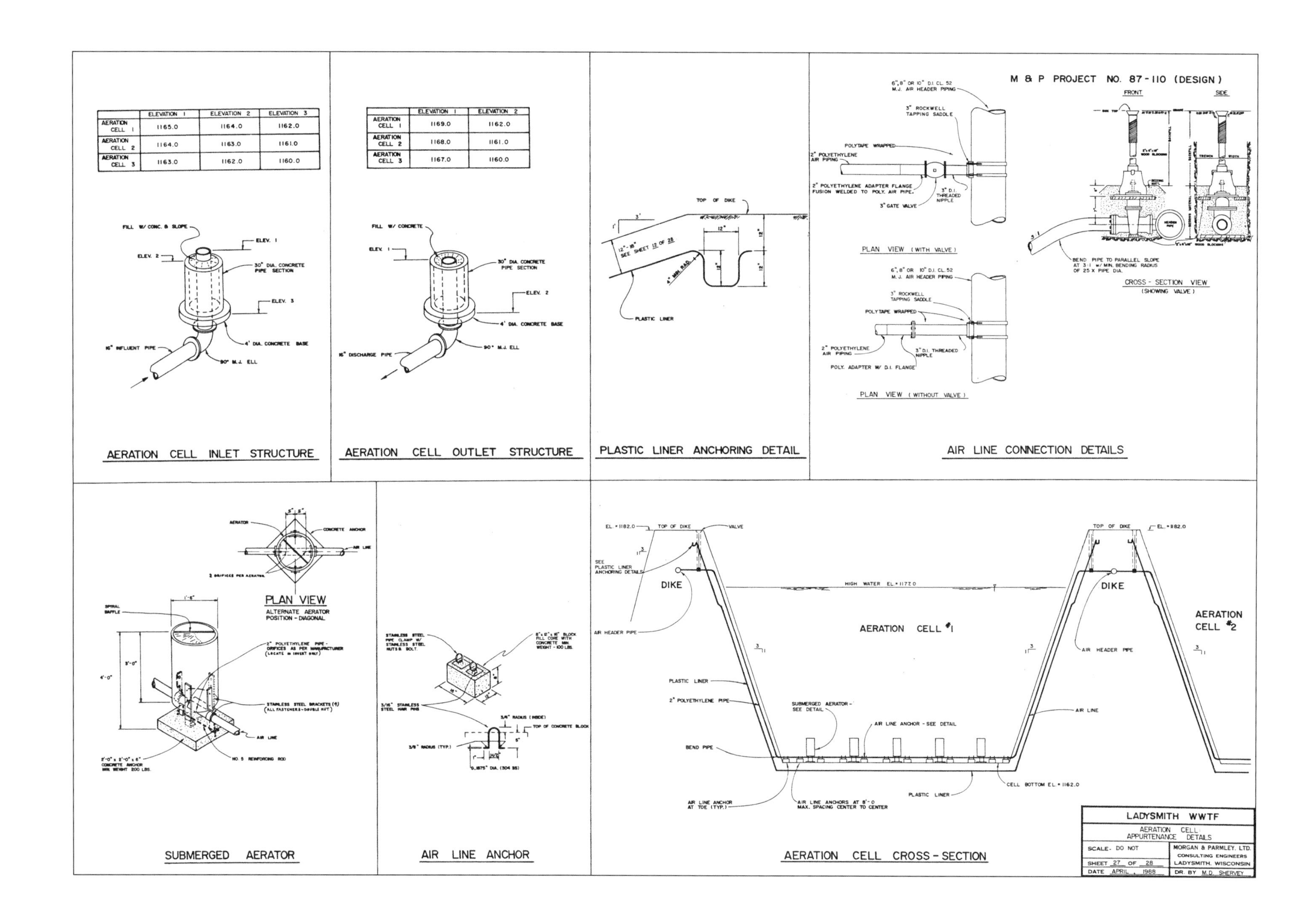

	ELEVATION 1	ELEVATION 2	ELEVATION 3
AERATION CELL 1	1165.0	1164.0	1162.0
AERATION CELL 2	1164.0	1163.0	1161.0
AERATION CELL 3	1163.0	1162.0	1160.0

	ELEVATION 1	ELEVATION 2
AERATION CELL 1	1169.0	1162.0
AERATION CELL 2	1168.0	1161.0
AERATION CELL 3	1167.0	1160.0

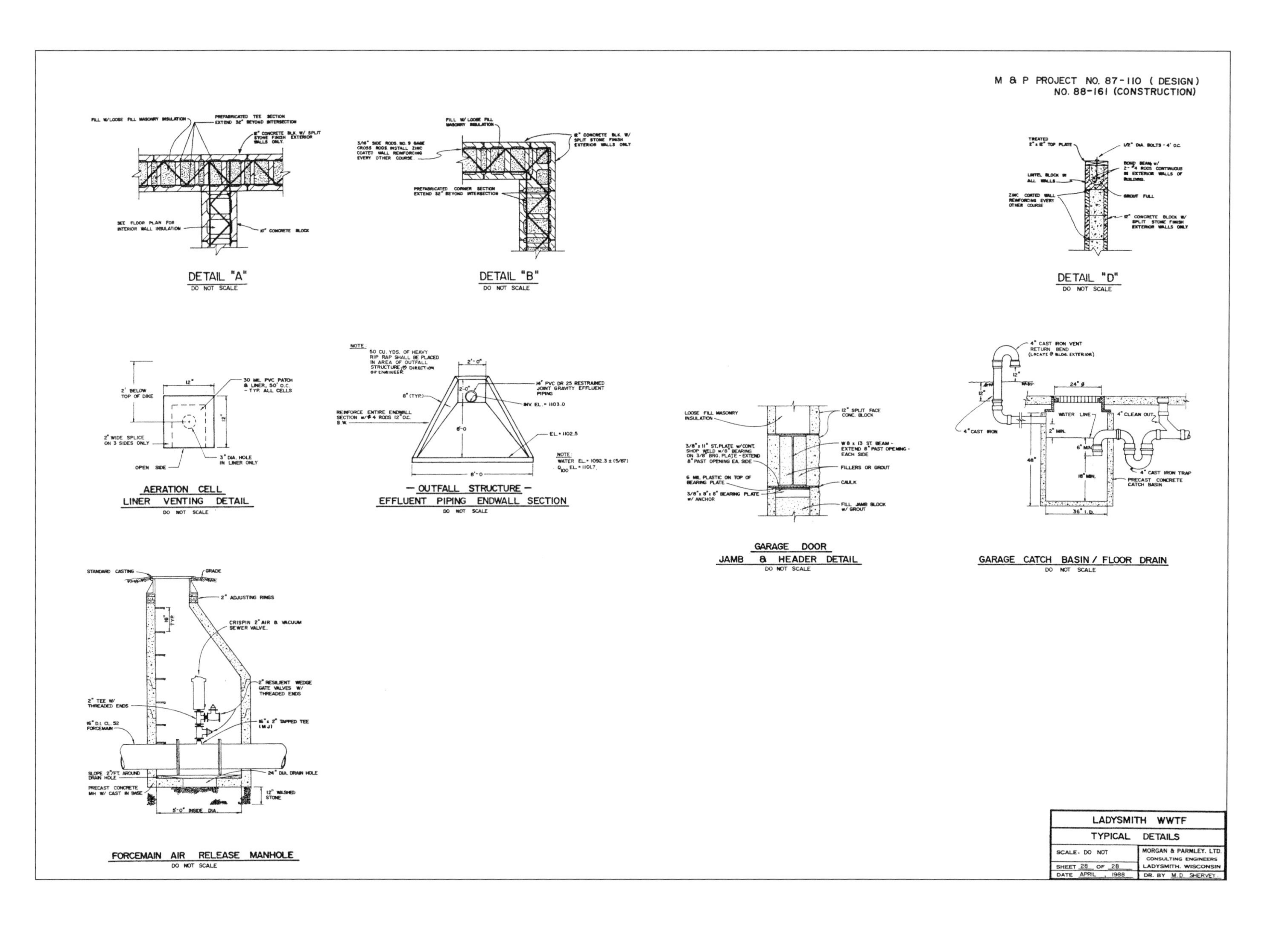
M & P PROJECT NO. 87-110 (DESIGN)
NO. 88-161 (CONSTRUCTION)
DETAIL "A"
DO NOT SCALE
DETAIL "B"
DO NOT SCALE
DETAIL "D"
DO NOT SCALE
AERATION CELL
LINER VENTING DETAIL
DO NOT SCALE
— OUTFALL STRUCTURE —
EFFLUENT PIPING ENDWALL SECTION
DO NOT SCALE
GARAGE DOOR
JAMB & HEADER DETAIL
DO NOT SCALE
GARAGE CATCH BASIN / FLOOR DRAIN
DO NOT SCALE
FORCEMAIN AIR RELEASE MANHOLE
DO NOT SCALE
LADYSMITH WWTF
TYPICAL DETAILS
SCALE- DO NOT
SHEET 28 OF 28
DATE APRIL, 1988
MORGAN & PARMLEY, LTD.
CONSULTING ENGINEERS
LADYSMITH, WISCONSIN
DR. BY M.D. SHERVEY

M & P PROJECT NO. 85-182
NO. 88-116

WASTEWATER COLLECTION & TREATMENT FACILITY

VILLAGE
OF
CONRATH, WISCONSIN

PROJECT NO. C551225-02 (DESIGN)
C551225-03 (CONSTRUCTION)

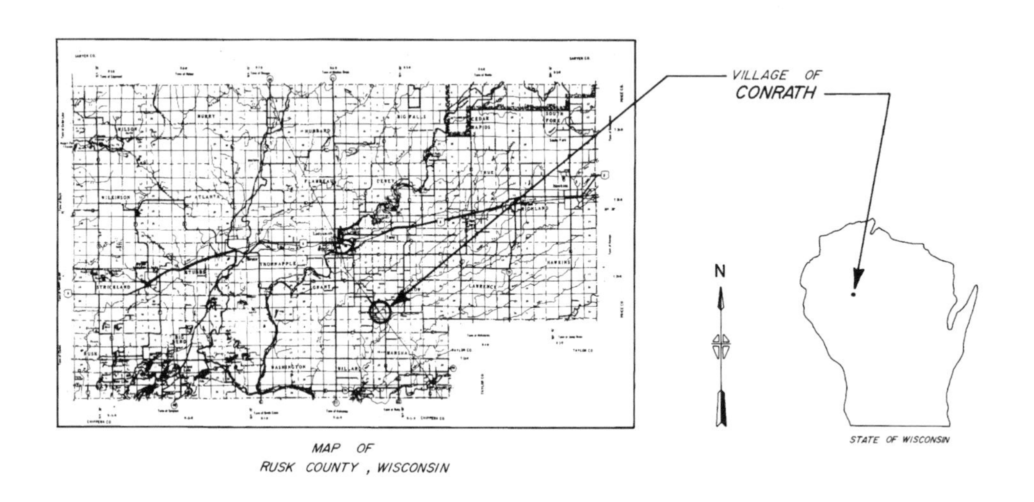

MAP OF
RUSK COUNTY, WISCONSIN

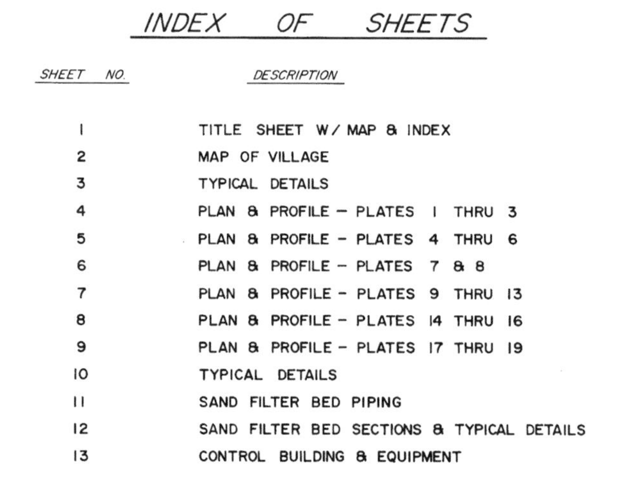

INDEX OF SHEETS

SHEET NO.	DESCRIPTION
1	TITLE SHEET W/ MAP & INDEX
2	MAP OF VILLAGE
3	TYPICAL DETAILS
4	PLAN & PROFILE – PLATES 1 THRU 3
5	PLAN & PROFILE – PLATES 4 THRU 6
6	PLAN & PROFILE – PLATES 7 & 8
7	PLAN & PROFILE – PLATES 9 THRU 13
8	PLAN & PROFILE – PLATES 14 THRU 16
9	PLAN & PROFILE – PLATES 17 THRU 19
10	TYPICAL DETAILS
11	SAND FILTER BED PIPING
12	SAND FILTER BED SECTIONS & TYPICAL DETAILS
13	CONTROL BUILDING & EQUIPMENT

PREPARED BY: MORGAN & PARMLEY, LTD. LADYSMITH, WISCONSIN

Courtesy: Morgan & Parmley, Ltd.

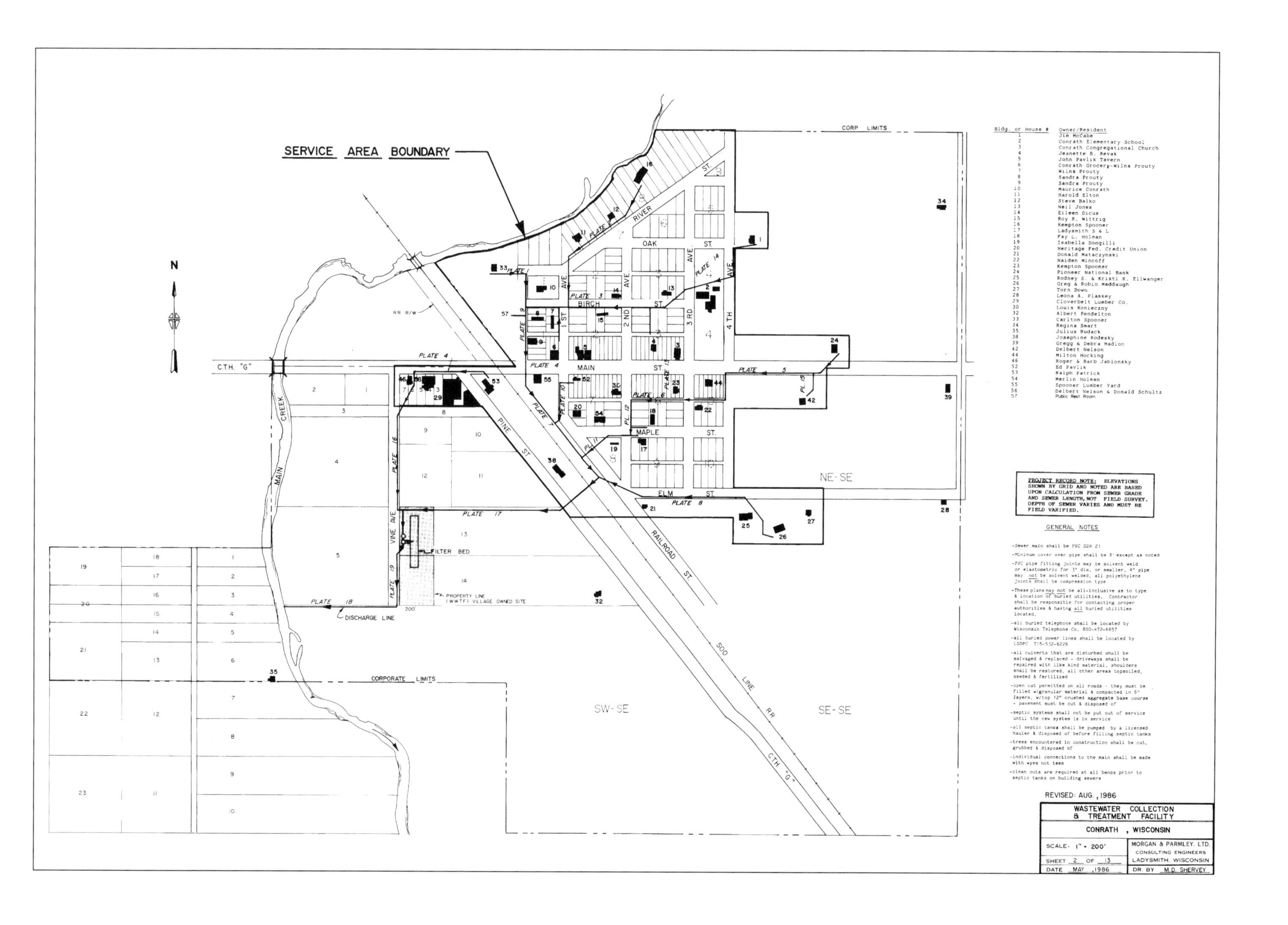
SERVICE AREA BOUNDARY
N
CORP LIMITS
CORPORATE LIMITS
C.T.H. "G"
MAIN CREEK
RIVER ST.
OAK ST.
BIRCH ST.
MAIN ST.
MAPLE ST.
ELM ST.
1 ST
2 ND
3 RD
4 TH AVE
PINE ST
VINE AVE
RAILROAD ST
SOO LINE RR
CTH "G"
RR R/W
FILTER BED
PROPERTY LINE (WWTF) VILLAGE OWNED SITE
DISCHARGE LINE
NE-SE
SW-SE
SE-SE
Bldg. or House # Owner/Resident
1 Jim McCabe
2 Conrath Elementary School
3 Conrath Congregational Church
4 Jeanette B. Revak
5 John Pavlik Tavern
6 Conrath Grocery-Wilna Prouty
7 Wilna Prouty
8 Sandra Prouty
9 Sandra Prouty
10 Maurice Conrath
11 Harold Elton
12 Steve Balko
13 Neil Jones
14 Eileen Dicus
15 Roy R. Wittrig
16 Kempton Spooner
17 Ladysmith S & L
18 Fay L. Holman
19 Isabella Dongilli
20 Heritage Fed. Credit Union
21 Donald Mataczynski
22 Naiden Mincoff
23 Kempton Spooner
24 Pioneer National Bank
25 Rodney S. & Kristi K. Ellwanger
26 Greg & Robin Meddaugh
27 Torn Down
28 Leona A. Plaskey
29 Cloverbelt Lumber Co.
30 Louis Konieczny
32 Albert Pendelton
33 Carlton Spooner
34 Regina Smart
35 Julius Rudack
38 Josephine Rodesky
39 Gregg & Debra Madlon
42 Delbert Nelson
44 Milton Hocking
46 Roger & Barb Jablonsky
52 Ed Pavlik
53 Ralph Patrick
54 Merlin Holman
55 Spooner Lumber Yard
56 Delbert Nelson & Donald Schultz
57 Public Rest Room
PROJECT RECORD NOTE: ELEVATIONS SHOWN BY GRID AND NOTED ARE BASED UPON CALCULATION FROM SEWER GRADE AND SEWER LENGTH, NOT FIELD SURVEY. DEPTH OF SEWER VARIES AND MUST BE FIELD VARIFIED.
GENERAL NOTES
-Sewer main shall be PVC SDR 21
-Minimum cover over pipe shall be 8' except as noted
-PVC pipe fitting joints may be solvent weld or elastometric for 3" dia. or smaller, 4" pipe may not be solvent welded, all polyethylene joints shall be compression type
-These plans may not be all-inclusive as to type & location of buried utilities. Contractor shall be responsible for contacting proper authorities & having all buried utilities located.
-all buried telephone shall be located by Wisconsin Telephone Co. 800-472-6657
-all buried power lines shall be located by LSDPC 715-532-6226
-all culverts that are disturbed shall be salvaged & replaced - driveways shall be repaired with like kind material, shoulders shall be restored, all other areas topsoiled, seeded & fertilized
-open cut permitted on all roads - they must be filled w/granular material & compacted in 6" layers, w/top 12" crushed aggregate base course - pavement must be cut & disposed of
-septic systems shall not be put out of service until the new system is in service
-all septic tanks shall be pumped by a licensed hauler & disposed of before filling septic tanks
-trees encountered in construction shall be cut, grubbed & disposed of
-individual connections to the main shall be made with wyes not tees
-clean outs are required at all bends prior to septic tanks on building sewers
REVISED: AUG., 1986
WASTEWATER COLLECTION & TREATMENT FACILITY
CONRATH, WISCONSIN
SCALE- 1" = 200'
SHEET 2 OF 13
DATE MAY, 1986
MORGAN & PARMLEY, LTD.
CONSULTING ENGINEERS
LADYSMITH, WISCONSIN
DR. BY M.D. SHERVEY

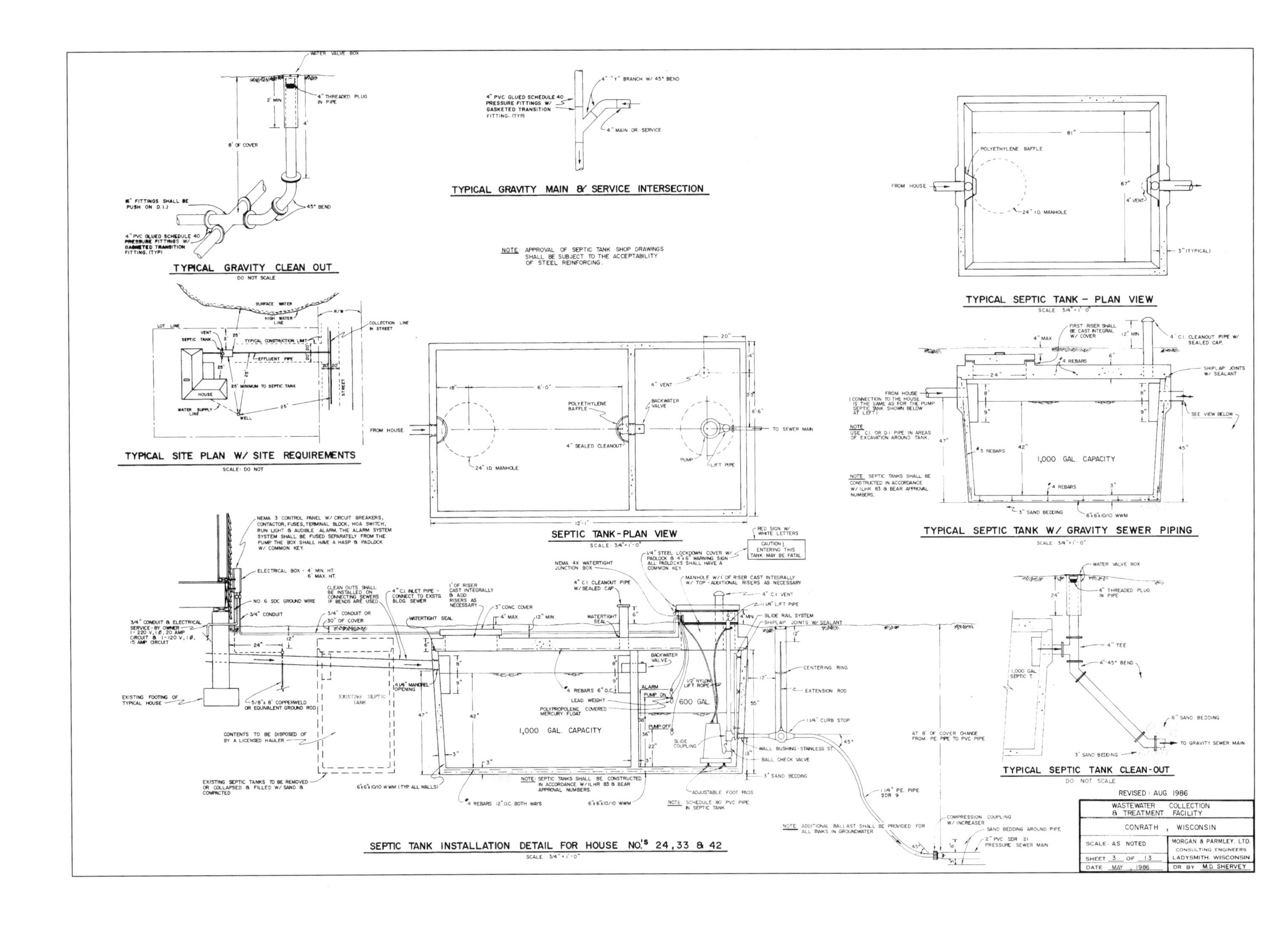
TYPICAL GRAVITY CLEAN OUT
DO NOT SCALE
TYPICAL GRAVITY MAIN & SERVICE INTERSECTION
NOTE: APPROVAL OF SEPTIC TANK SHOP DRAWINGS SHALL BE SUBJECT TO THE ACCEPTABILITY OF STEEL REINFORCING.
TYPICAL SEPTIC TANK - PLAN VIEW
SCALE 3/4" = 1'-0"
TYPICAL SITE PLAN W/ SITE REQUIREMENTS
SCALE: DO NOT
SEPTIC TANK-PLAN VIEW
SCALE: 3/4" = 1'-0"
TYPICAL SEPTIC TANK W/ GRAVITY SEWER PIPING
SCALE 3/4" = 1'-0"
1,000 GAL. CAPACITY
TYPICAL SEPTIC TANK CLEAN-OUT
DO NOT SCALE
SEPTIC TANK INSTALLATION DETAIL FOR HOUSE NO.'S 24, 33 & 42
SCALE 3/4" = 1'-0"
REVISED: AUG. 1986
WASTEWATER COLLECTION & TREATMENT FACILITY
CONRATH, WISCONSIN
SCALE: AS NOTED
SHEET 3 OF 13
DATE MAY, 1986
MORGAN & PARMLEY, LTD.
CONSULTING ENGINEERS
LADYSMITH, WISCONSIN
DR BY M.D. SHERVEY

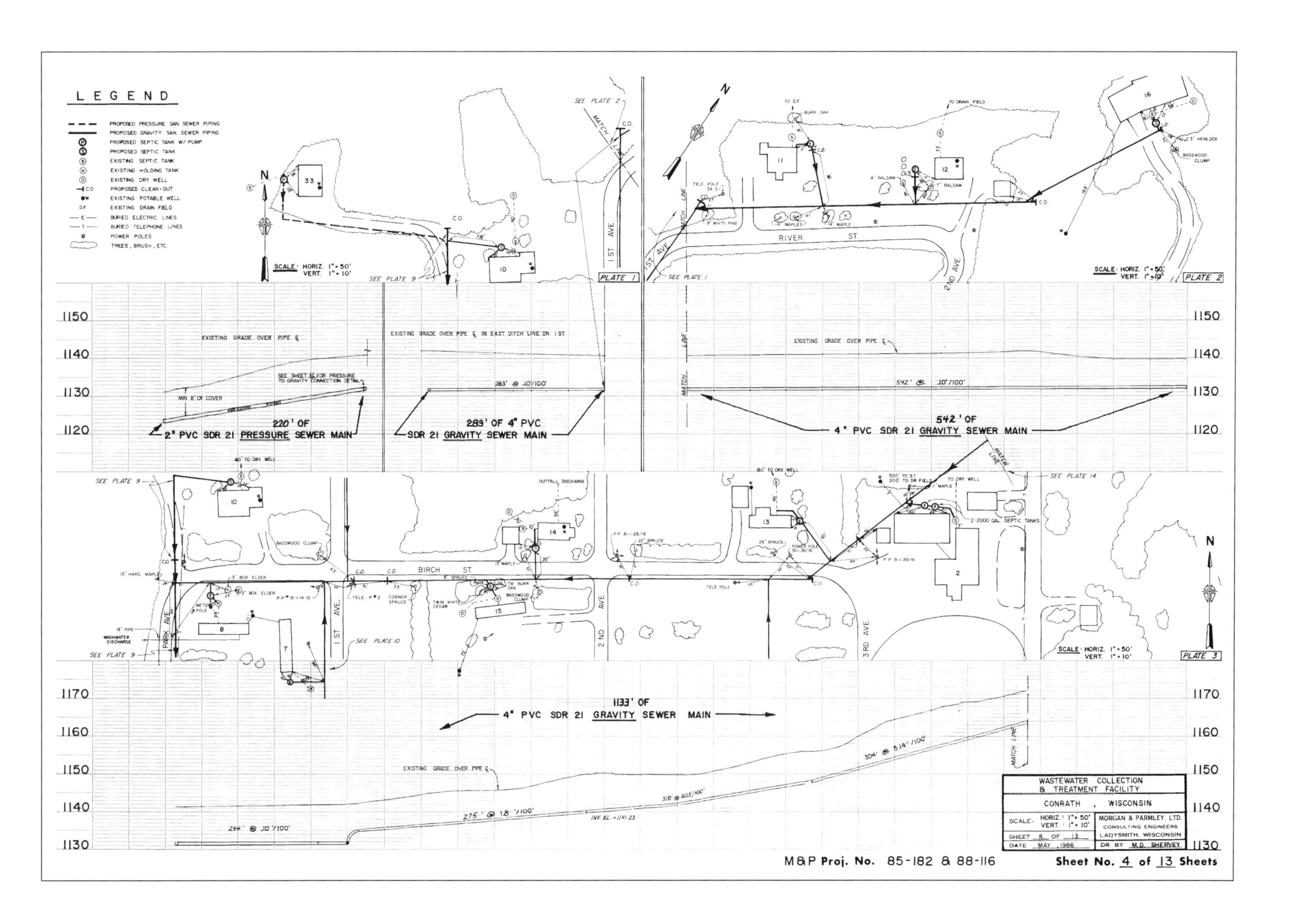

LEGEND
PROPOSED PRESSURE SAN. SEWER PIPING
PROPOSED GRAVITY SAN. SEWER PIPING
PROPOSED SEPTIC TANK W/ PUMP
PROPOSED SEPTIC TANK
EXISTING SEPTIC TANK
EXISTING HOLDING TANK
EXISTING DRY WELL
PROPOSED CLEAN-OUT
EXISTING POTABLE WELL
EXISTING DRAIN FIELD
BURIED ELECTRIC LINES
BURIED TELEPHONE LINES
POWER POLES
TREES, BRUSH, ETC.
SCALE: HORIZ. 1" = 50'
VERT. 1" = 10'
PLATE 1
PLATE 2
PLATE 3
SEE PLATE 2
SEE PLATE 9
SEE PLATE 1
SEE PLATE 10
SEE PLATE 14
MATCH LINE
1ST AVE.
2ND AVE.
3RD AVE.
PARK AVE.
RIVER ST.
BIRCH ST.
EXISTING GRADE OVER PIPE ℄
EXISTING GRADE OVER PIPE ℄ IN EAST DITCH LINE ON 1ST
MIN. 8' OF COVER
SEE SHEET FOR PRESSURE TO GRAVITY CONNECTION DETAIL
220' OF
2" PVC SDR 21 PRESSURE SEWER MAIN
283' @ .10'/100'
283' OF 4" PVC
SDR 21 GRAVITY SEWER MAIN
542' @ .10'/100'
542' OF
4" PVC SDR 21 GRAVITY SEWER MAIN
1133' OF
4" PVC SDR 21 GRAVITY SEWER MAIN
244' @ .10'/100'
275' @ 1.8'/100'
310' @ 2.03'/100'
304' @ 5.14'/100'
INV. EL. = 1141.23
1150
1140
1130
1120
1170
1160
WASTEWATER COLLECTION & TREATMENT FACILITY
CONRATH, WISCONSIN
SCALE - HORIZ.: 1" = 50' VERT.: 1" = 10'
MORGAN & PARMLEY, LTD.
CONSULTING ENGINEERS
LADYSMITH, WISCONSIN
SHEET 4 OF 13
DATE MAY, 1986
DR. BY M.D. SHERVEY
M&P Proj. No. 85-182 & 88-116
Sheet No. 4 of 13 Sheets

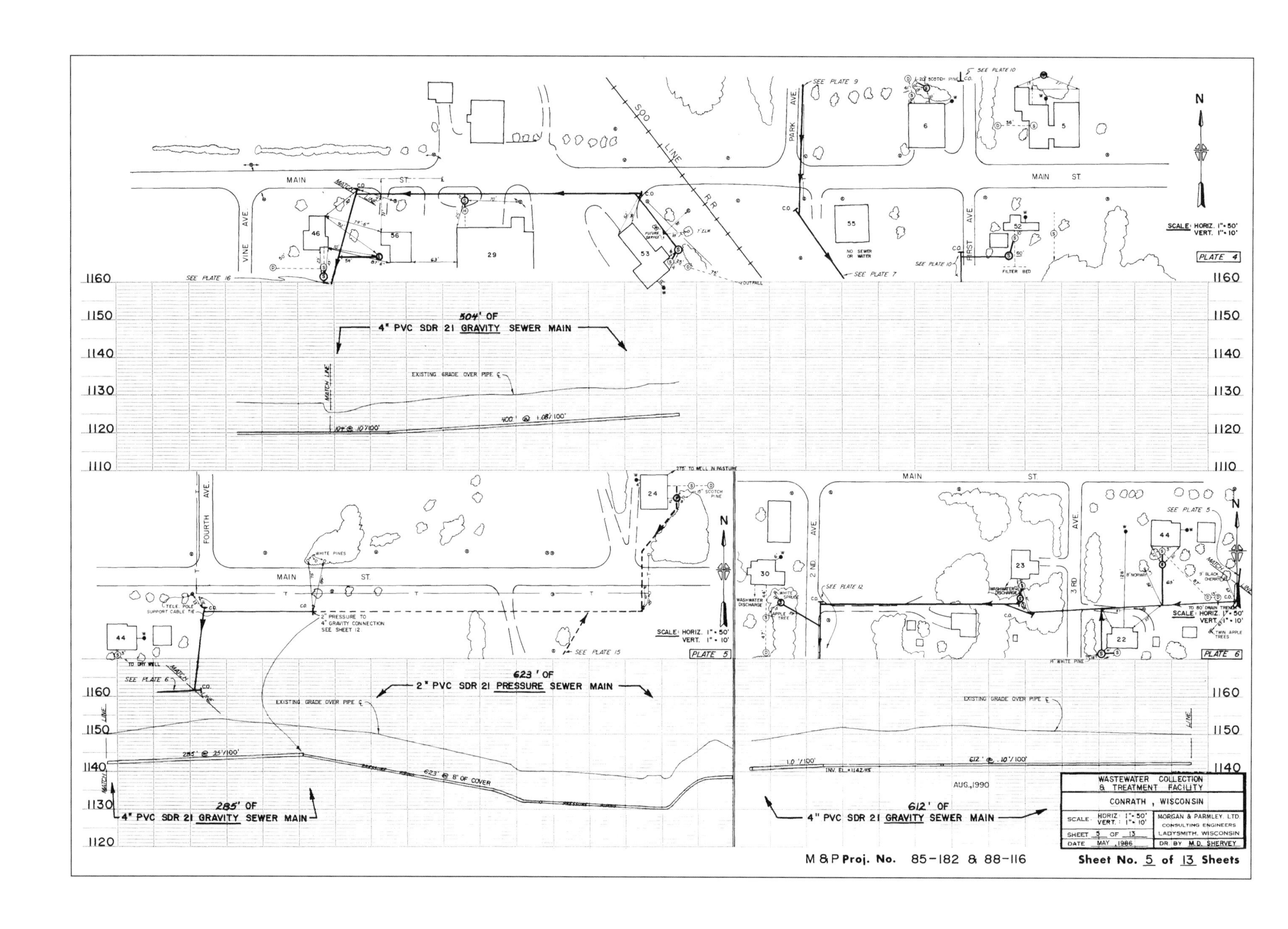

504' OF
4" PVC SDR 21 GRAVITY SEWER MAIN
EXISTING GRADE OVER PIPE ℄
400' @ +.08'/100'
104' @ .10'/100'
MAIN ST.
VINE AVE.
PARK AVE.
FIRST AVE.
SOO LINE R.R.
SEE PLATE 16
SEE PLATE 9
SEE PLATE 7
SEE PLATE 10
NO SEWER OR WATER
FILTER BED
SCALE: HORIZ. 1" = 50' VERT. 1" = 10'
PLATE 4
623' OF
2" PVC SDR 21 PRESSURE SEWER MAIN
285' @ .25'/100'
623' @ 8' OF COVER
285' OF
4" PVC SDR 21 GRAVITY SEWER MAIN
FOURTH AVE.
2" PRESSURE TO 4" GRAVITY CONNECTION SEE SHEET 12
SEE PLATE 6
SEE PLATE 15
275' TO WELL IN PASTURE
TO DRY WELL
PLATE 5
612' OF
4" PVC SDR 21 GRAVITY SEWER MAIN
612' @ .10'/100'
1.0'/100'
INV. EL. = 1142.45
AUG., 1990
2 ND. AVE.
3 RD. AVE.
SEE PLATE 12
SEE PLATE 5
WASHWATER DISCHARGE
TO 80' DRAIN TRENCH
PLATE 6
1160
1150
1140
1130
1120
1110
WASTEWATER COLLECTION & TREATMENT FACILITY
CONRATH, WISCONSIN
SCALE: HORIZ: 1" = 50' VERT.: 1" = 10'
SHEET 5 OF 13
DATE MAY, 1986
MORGAN & PARMLEY LTD. CONSULTING ENGINEERS LADYSMITH, WISCONSIN
DR BY M.D. SHERVEY
M & P Proj. No. 85-182 & 88-116
Sheet No. 5 of 13 Sheets

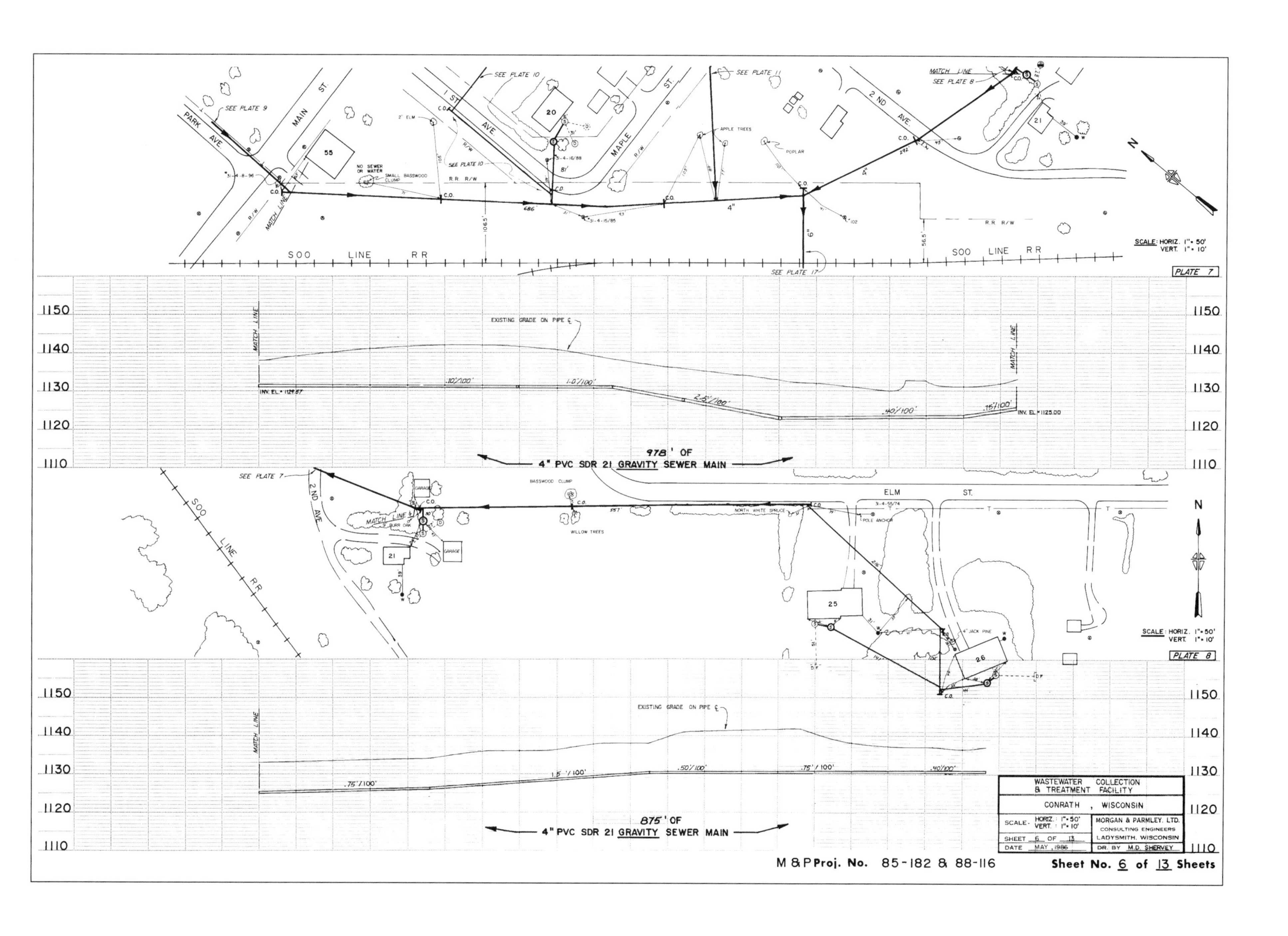
SEE PLATE 9
PARK AVE.
MAIN ST.
1 ST. AVE
MAPLE ST.
2 ND AVE.
SEE PLATE 10
SEE PLATE 11
SEE PLATE 8
MATCH LINE
NO SEWER OR WATER
SMALL BASSWOOD CLUMP
APPLE TREES
POPLAR
SOO LINE RR
SEE PLATE 17
SCALE: HORIZ. 1" = 50' VERT. 1" = 10'
PLATE 7
EXISTING GRADE ON PIPE ℄
INV. EL. = 1125.00
978' OF
4" PVC SDR 21 GRAVITY SEWER MAIN
SEE PLATE 7
BASSWOOD CLUMP
WILLOW TREES
NORTH WHITE SPRUCE
POLE ANCHOR
ELM ST.
GARAGE
BURR OAK
JACK PINE
PLATE 8
875' OF
4" PVC SDR 21 GRAVITY SEWER MAIN
WASTEWATER COLLECTION & TREATMENT FACILITY
CONRATH, WISCONSIN
MORGAN & PARMLEY, LTD.
CONSULTING ENGINEERS
LADYSMITH, WISCONSIN
SHEET 6 OF 13
DATE MAY, 1986
DR. BY M.D. SHERVEY
M & P Proj. No. 85-182 & 88-116
Sheet No. 6 of 13 Sheets

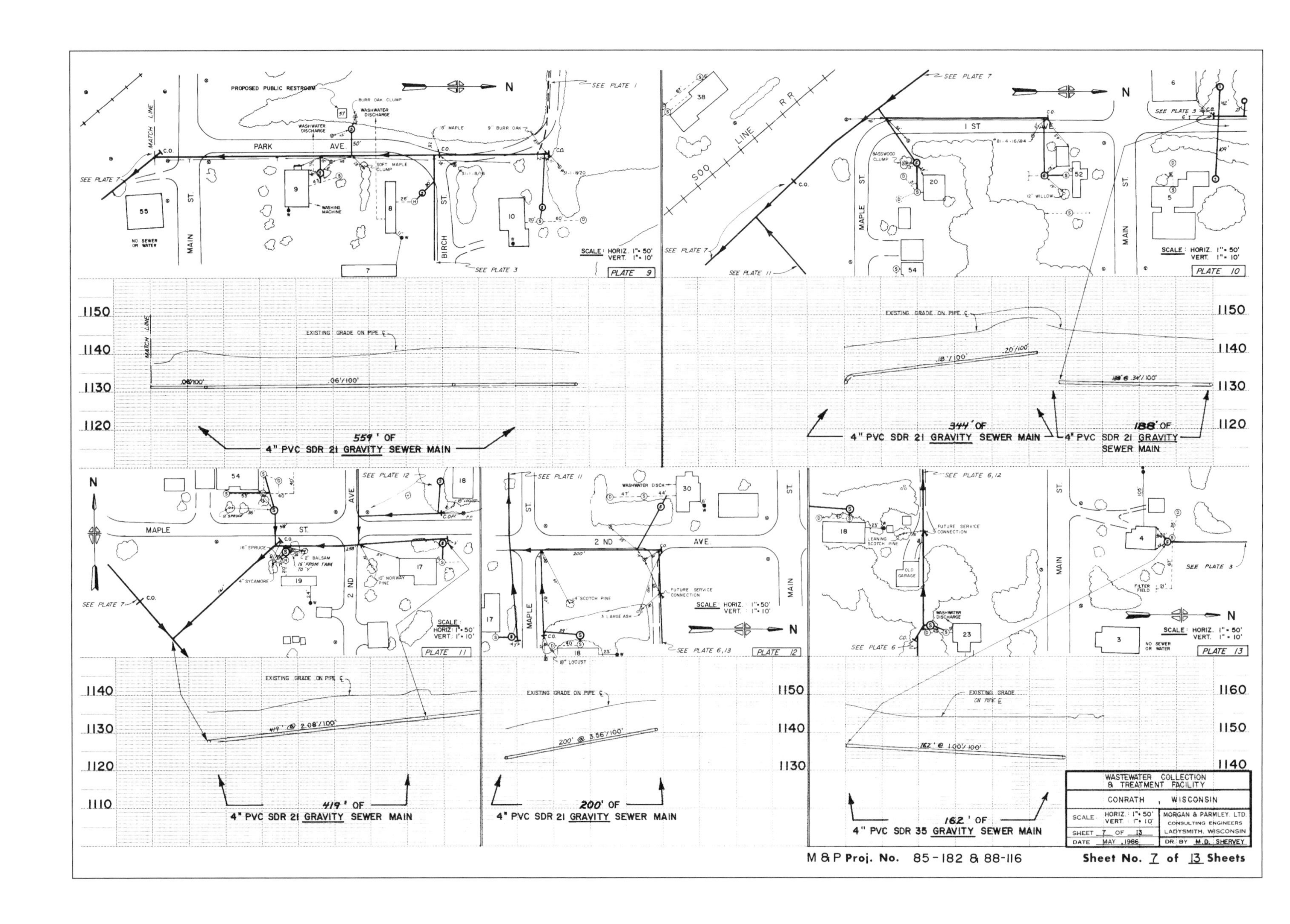
PROPOSED PUBLIC RESTROOM
BURR OAK CLUMP
WASHWATER DISCHARGE
PARK AVE.
MAIN ST.
BIRCH ST.
SOFT MAPLE CLUMP
WASHING MACHINE
NO SEWER OR WATER
SEE PLATE 7
SEE PLATE 1
SEE PLATE 3
MATCH LINE
SCALE: HORIZ. 1" = 50' VERT. 1" = 10'
PLATE 9
EXISTING GRADE ON PIPE ℄
.06'/100'
.06'/100'
559' OF
4" PVC SDR 21 GRAVITY SEWER MAIN
SOO LINE R.R.
I ST
MAPLE ST.
MAIN ST.
BASSWOOD CLUMP
12" WILLOW
SEE PLATE 11
PLATE 10
.18'/100'
.20'/100'
.34'/100'
344' OF
4" PVC SDR 21 GRAVITY SEWER MAIN
188' OF
4" PVC SDR 21 GRAVITY SEWER MAIN
1150
1140
1130
1120
MAPLE ST.
2 ND AVE.
SEE PLATE 12
16" SPRUCE
BALSAM
15' FROM TANK TO "Y"
4" SYCAMORE
10" NORWAY PINE
PLATE 11
419' @ 2.08'/100'
419' OF
4" PVC SDR 21 GRAVITY SEWER MAIN
WASHWATER DISCH.
4" SCOTCH PINE
3 LARGE ASH
FUTURE SERVICE CONNECTION
18" LOCUST
SEE PLATE 6,13
PLATE 12
200' @ 3.56'/100'
200' OF
4" PVC SDR 21 GRAVITY SEWER MAIN
SEE PLATE 6,12
LEANING SCOTCH PINE
OLD GARAGE
FILTER FIELD
SEE PLATE 6
PLATE 13
162' @ 1.00'/100'
162' OF
4" PVC SDR 35 GRAVITY SEWER MAIN
1160
1110
N
WASTEWATER COLLECTION & TREATMENT FACILITY
CONRATH, WISCONSIN
SCALE HORIZ. 1" = 50' VERT. 1" = 10'
MORGAN & PARMLEY, LTD. CONSULTING ENGINEERS LADYSMITH, WISCONSIN
SHEET 7 OF 13
DATE MAY 1986
DR. BY M.D. SHERVEY
M & P Proj. No. 85-182 & 88-116
Sheet No. 7 of 13 Sheets

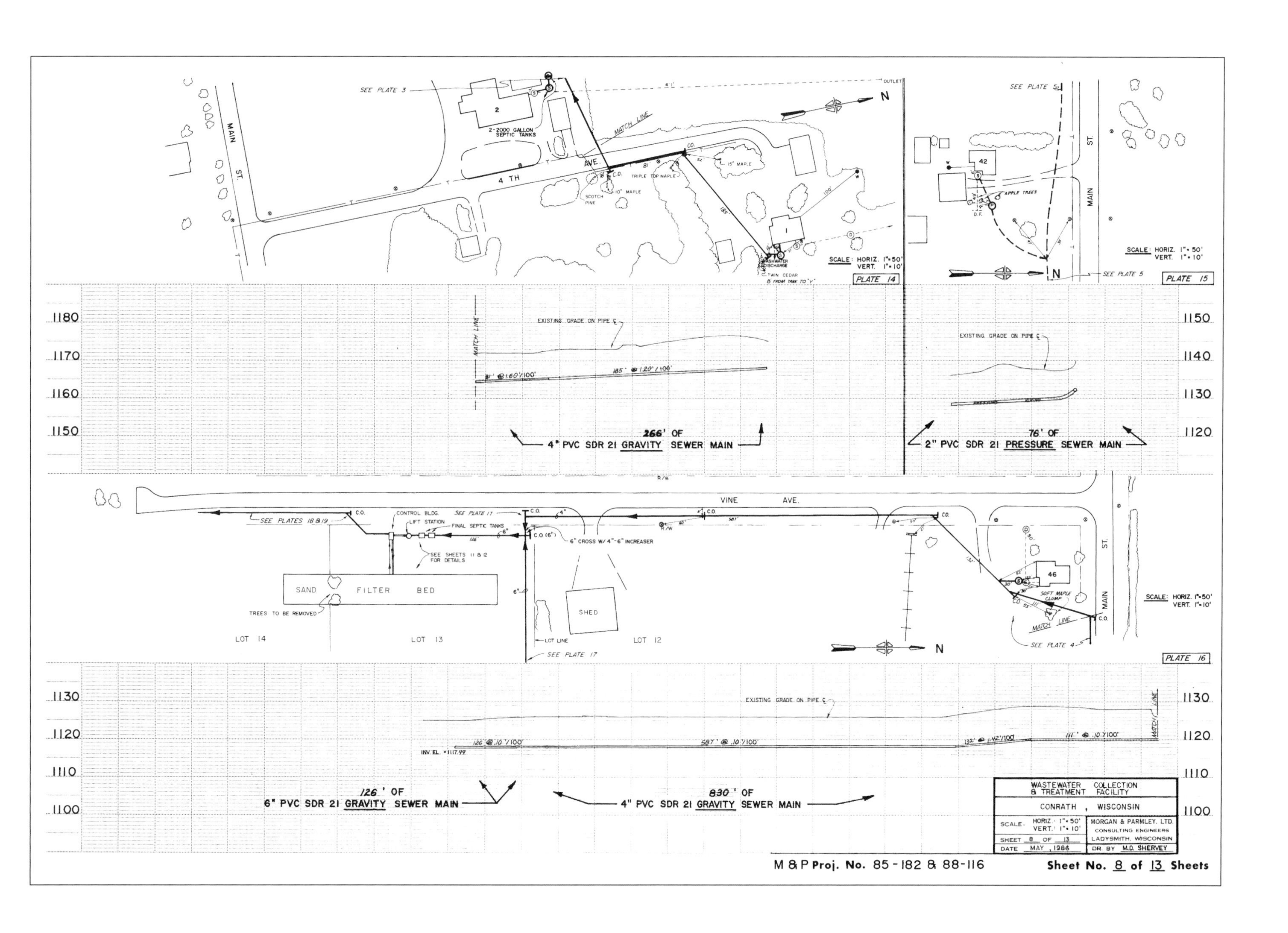
SEE PLATE 3
2-2000 GALLON SEPTIC TANKS
MATCH LINE
MAIN ST.
4 TH AVE.
SCALE: HORIZ. 1"=50' VERT. 1"=10'
PLATE 14
PLATE 15
SEE PLATE 5
MAIN ST.
EXISTING GRADE ON PIPE ℄
266' OF 4" PVC SDR 21 GRAVITY SEWER MAIN
76' OF 2" PVC SDR 21 PRESSURE SEWER MAIN
VINE AVE.
SEE PLATES 18 & 19
CONTROL BLDG.
LIFT STATION
FINAL SEPTIC TANKS
SEE SHEETS 11 & 12 FOR DETAILS
SAND FILTER BED
TREES TO BE REMOVED
SHED
LOT 14
LOT 13
LOT 12
SEE PLATE 17
SEE PLATE 4
PLATE 16
126' OF 6" PVC SDR 21 GRAVITY SEWER MAIN
830' OF 4" PVC SDR 21 GRAVITY SEWER MAIN
WASTEWATER COLLECTION & TREATMENT FACILITY
CONRATH, WISCONSIN
SCALE- HORIZ.: 1"=50' VERT.: 1"=10'
SHEET 8 OF 13
DATE MAY, 1986
MORGAN & PARMLEY, LTD. CONSULTING ENGINEERS LADYSMITH, WISCONSIN
DR. BY M.D. SHERVEY
M & P Proj. No. 85-182 & 88-116
Sheet No. 8 of 13 Sheets

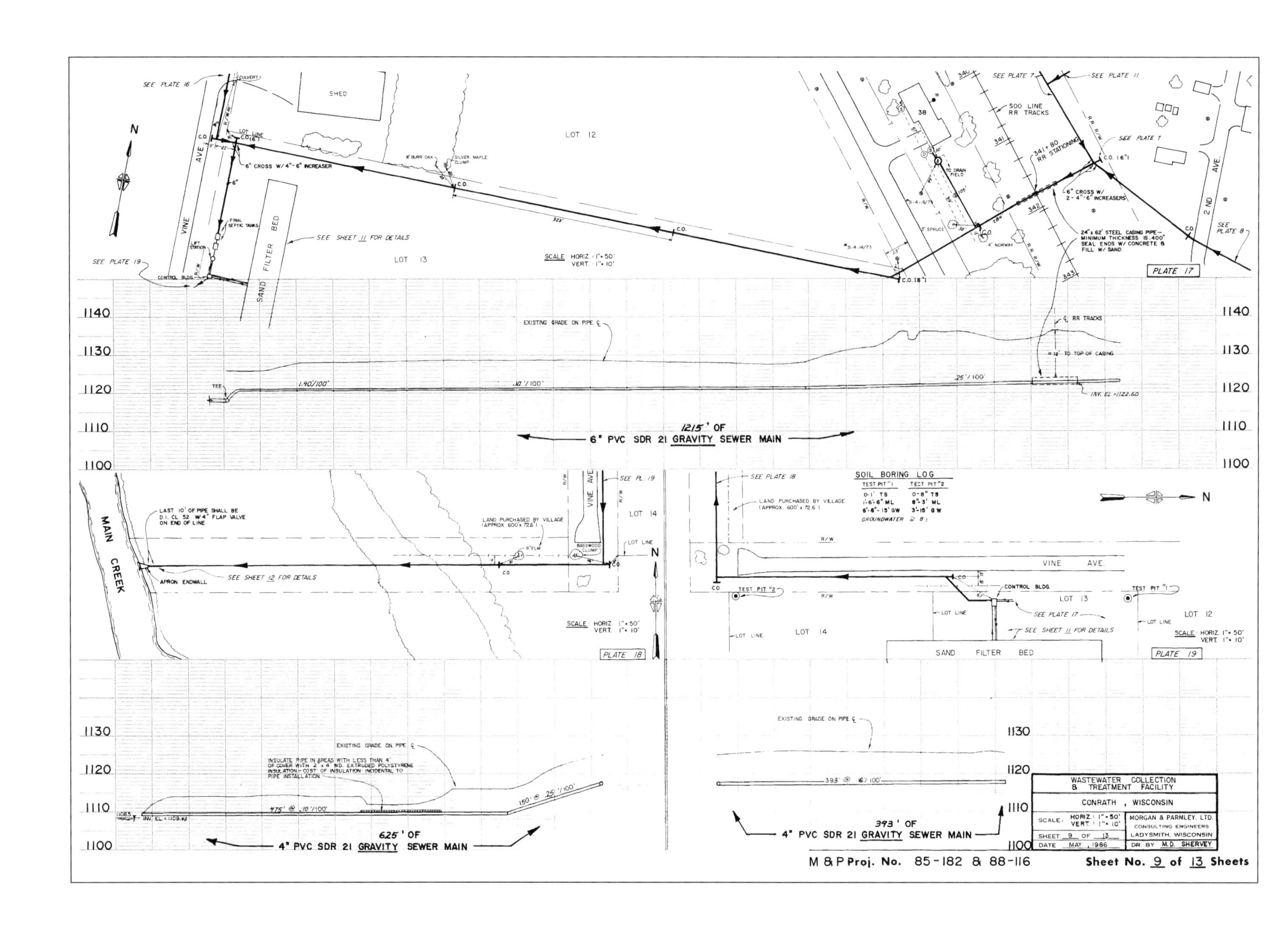
SEE PLATE 16
SHED
LOT 12
LOT 13
VINE AVE.
SAND FILTER BED
6" CROSS W/ 4"-6" INCREASER
SEE SHEET 11 FOR DETAILS
SEE PLATE 19
CONTROL BLDG.
SCALE: HORIZ.: 1"=50' VERT.: 1"=10'
SEE PLATE 7
SEE PLATE 11
SOO LINE RR TRACKS
341+80 RR STATIONING
6" CROSS W/ 2-4"-6" INCREASERS
24" x 62' STEEL CASING PIPE—MINIMUM THICKNESS IS .400" SEAL ENDS W/ CONCRETE & FILL W/ SAND
2 ND AVE.
SEE PLATE 8
PLATE 17
EXISTING GRADE ON PIPE ℄
RR TRACKS
12' TO TOP OF CASING
INV. EL.=1122.60
1215' OF
6" PVC SDR 21 GRAVITY SEWER MAIN
MAIN CREEK
LAST 10' OF PIPE SHALL BE D.I. CL. 52 W/ 4" FLAP VALVE ON END OF LINE
APRON ENDWALL
SEE SHEET 12 FOR DETAILS
LAND PURCHASED BY VILLAGE (APPROX. 600' x 72.6')
LOT 14
PLATE 18
SEE PLATE 18
SOIL BORING LOG
TEST PIT "1: 0-1' TS, 1'-6'-6" ML, 6'-6"-15' GW
TEST PIT "2: 0-8" TS, 8"-3' ML, 3'-15' GW
GROUNDWATER @ 8'±
VINE AVE.
CONTROL BLDG.
SEE PLATE 17
SEE SHEET 11 FOR DETAILS
SAND FILTER BED
PLATE 19
INSULATE PIPE IN AREAS WITH LESS THAN 4' OF COVER WITH 2" x 4' WD. EXTRUDED POLYSTYRENE INSULATION - COST OF INSULATION INCIDENTAL TO PIPE INSTALLATION
475' @ .10'/100'
150' @ .25'/100'
625' OF
4" PVC SDR 21 GRAVITY SEWER MAIN
393' @ .16'/100'
393' OF
4" PVC SDR 21 GRAVITY SEWER MAIN
WASTEWATER COLLECTION & TREATMENT FACILITY
CONRATH, WISCONSIN
MORGAN & PARMLEY, LTD. CONSULTING ENGINEERS LADYSMITH, WISCONSIN
SHEET 9 OF 13
DATE MAY, 1986
DR. BY M.D. SHERVEY
M & P Proj. No. 85-182 & 88-116
Sheet No. 9 of 13 Sheets

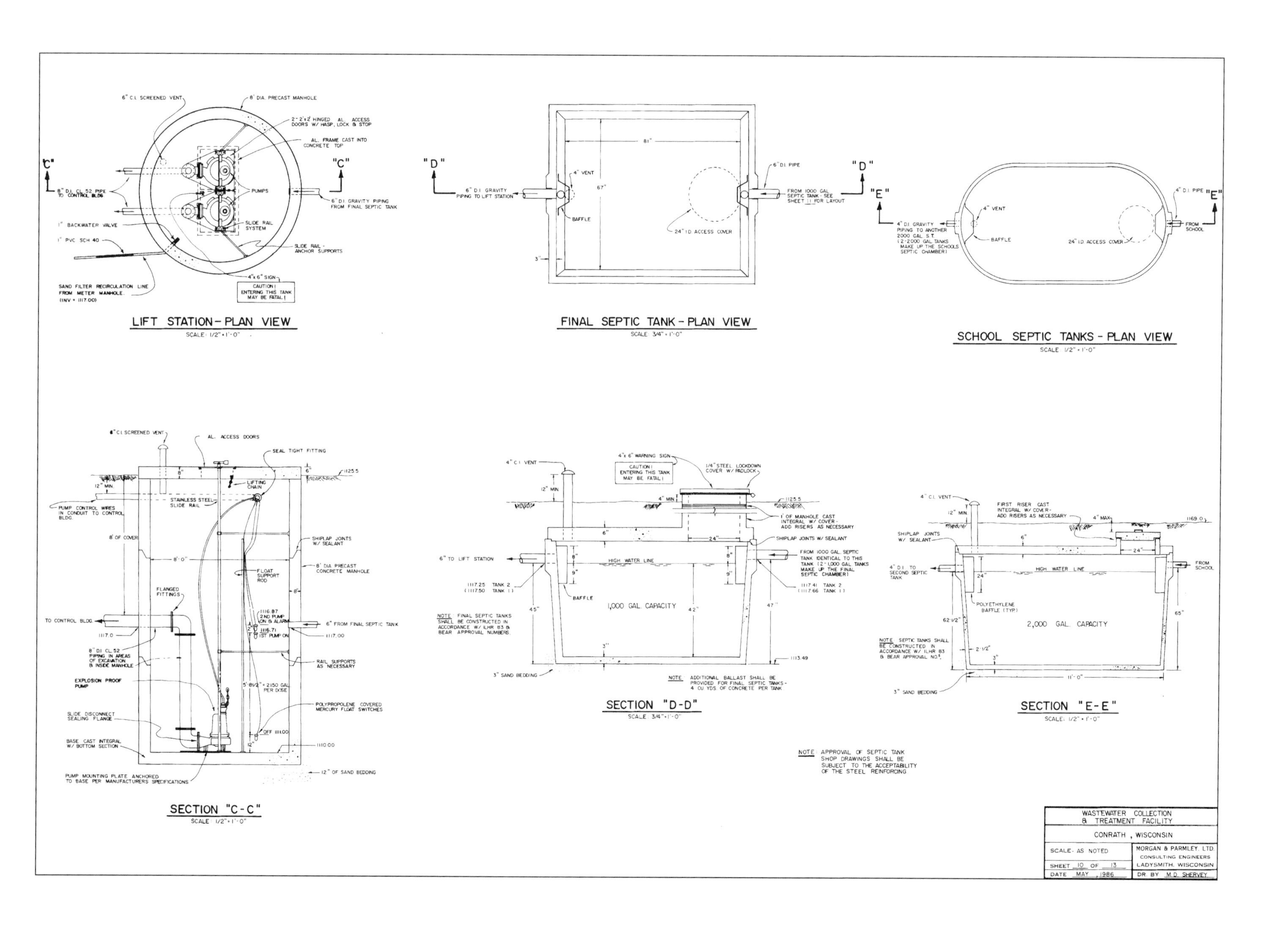

LIFT STATION - PLAN VIEW
SCALE: 1/2" = 1'-0"
FINAL SEPTIC TANK - PLAN VIEW
SCALE: 3/4" = 1'-0"
SCHOOL SEPTIC TANKS - PLAN VIEW
SCALE: 1/2" = 1'-0"
SECTION "C-C"
SCALE: 1/2" = 1'-0"
SECTION "D-D"
SCALE: 3/4" = 1'-0"
SECTION "E-E"
SCALE: 1/2" = 1'-0"
1,000 GAL. CAPACITY
2,000 GAL. CAPACITY
HIGH WATER LINE
CAUTION! ENTERING THIS TANK MAY BE FATAL!
NOTE: APPROVAL OF SEPTIC TANK SHOP DRAWINGS SHALL BE SUBJECT TO THE ACCEPTABILITY OF THE STEEL REINFORCING
WASTEWATER COLLECTION & TREATMENT FACILITY
CONRATH, WISCONSIN
SCALE - AS NOTED
SHEET 10 OF 13
DATE MAY, 1986
MORGAN & PARMLEY, LTD.
CONSULTING ENGINEERS
LADYSMITH, WISCONSIN
DR. BY M.D. SHERVEY

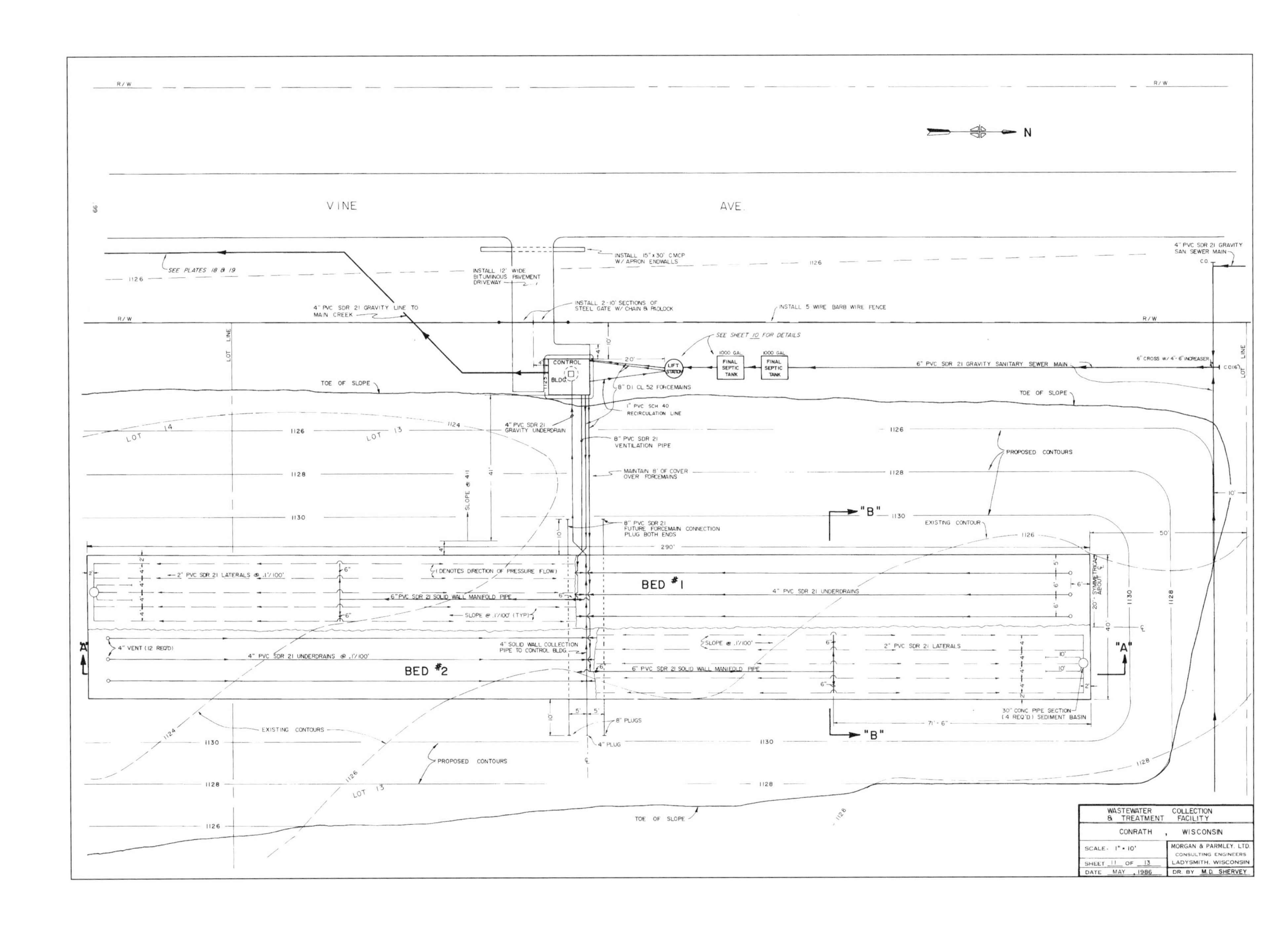

VINE AVE.
R/W
N
INSTALL 15" x 30' CMCP W/ APRON ENDWALLS
INSTALL 12' WIDE BITUMINOUS PAVEMENT DRIVEWAY
SEE PLATES 18 & 19
INSTALL 2-10' SECTIONS OF STEEL GATE W/ CHAIN & PADLOCK
INSTALL 5 WIRE BARB WIRE FENCE
4" PVC SDR 21 GRAVITY LINE TO MAIN CREEK
4" PVC SDR 21 GRAVITY SAN. SEWER MAIN
SEE SHEET 10 FOR DETAILS
1000 GAL. FINAL SEPTIC TANK
1000 GAL. FINAL SEPTIC TANK
6" PVC SDR 21 GRAVITY SANITARY SEWER MAIN
6" CROSS W/ 4"-6" INCREASER
CONTROL BLDG.
LIFT STATION
LOT LINE
TOE OF SLOPE
8" D.I. CL 52 FORCEMAINS
1" PVC SCH 40 RECIRCULATION LINE
4" PVC SDR 21 GRAVITY UNDERDRAIN
8" PVC SDR 21 VENTILATION PIPE
MAINTAIN 8' OF COVER OVER FORCEMAINS
PROPOSED CONTOURS
EXISTING CONTOUR
8" PVC SDR 21 FUTURE FORCEMAIN CONNECTION PLUG BOTH ENDS
SLOPE @ 4:1
290'
50'
BED #1
BED #2
2" PVC SDR 21 LATERALS @ .1'/100'
(DENOTES DIRECTION OF PRESSURE FLOW)
6" PVC SDR 21 SOLID WALL MANIFOLD PIPE
SLOPE @ .1'/100' (TYP)
4" PVC SDR 21 UNDERDRAINS
20' SYMMETRICAL ABOUT ℄
4" VENT (12 REQ'D)
4" SOLID WALL COLLECTION PIPE TO CONTROL BLDG.
4" PVC SDR 21 UNDERDRAINS @ .1'/100'
2" PVC SDR 21 LATERALS
30" CONC. PIPE SECTION (4 REQ'D) SEDIMENT BASIN
8" PLUGS
4" PLUG
EXISTING CONTOURS
PROPOSED CONTOURS
"A"
"B"
LOT 13
LOT 14
WASTEWATER COLLECTION & TREATMENT FACILITY
CONRATH, WISCONSIN
SCALE - 1" = 10'
SHEET 11 OF 13
DATE MAY, 1986
MORGAN & PARMLEY, LTD.
CONSULTING ENGINEERS
LADYSMITH, WISCONSIN
DR. BY M.D. SHERVEY

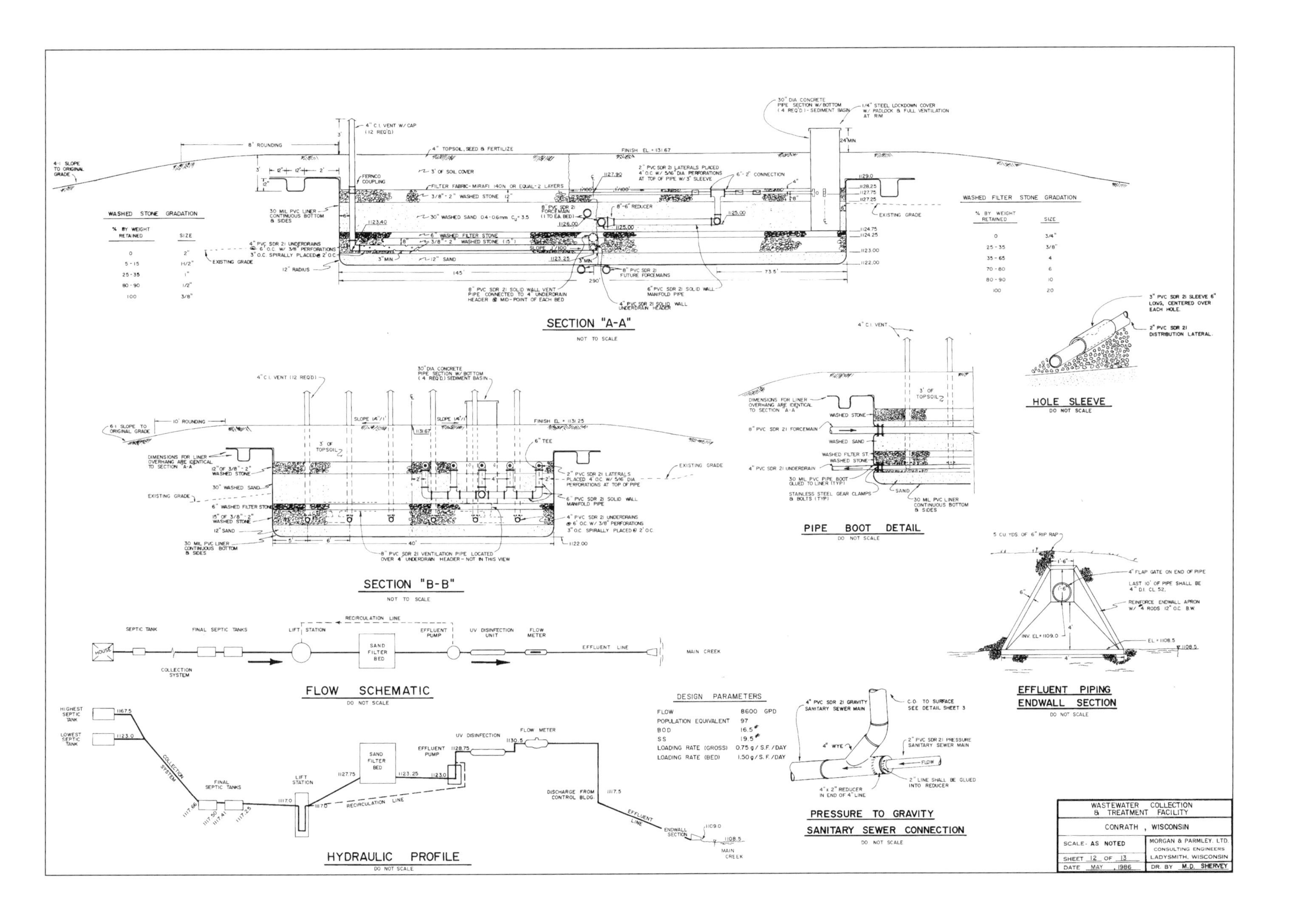

SECTION "A-A"
NOT TO SCALE
SECTION "B-B"
NOT TO SCALE
FLOW SCHEMATIC
DO NOT SCALE
HYDRAULIC PROFILE
DO NOT SCALE
PIPE BOOT DETAIL
DO NOT SCALE
HOLE SLEEVE
DO NOT SCALE
EFFLUENT PIPING
ENDWALL SECTION
DO NOT SCALE
PRESSURE TO GRAVITY
SANITARY SEWER CONNECTION
DO NOT SCALE
WASHED STONE GRADATION
WASHED FILTER STONE GRADATION
DESIGN PARAMETERS
FLOW 8600 GPD
POPULATION EQUIVALENT 97
BOD 16.5#
SS 19.5#
LOADING RATE (GROSS) 0.75 g / S.F. / DAY
LOADING RATE (BED) 1.50 g / S.F. / DAY
WASTEWATER COLLECTION & TREATMENT FACILITY
CONRATH, WISCONSIN
SCALE- AS NOTED
MORGAN & PARMLEY, LTD.
CONSULTING ENGINEERS
LADYSMITH, WISCONSIN
SHEET 12 OF 13
DATE MAY, 1986
DR. BY M.D. SHERVEY

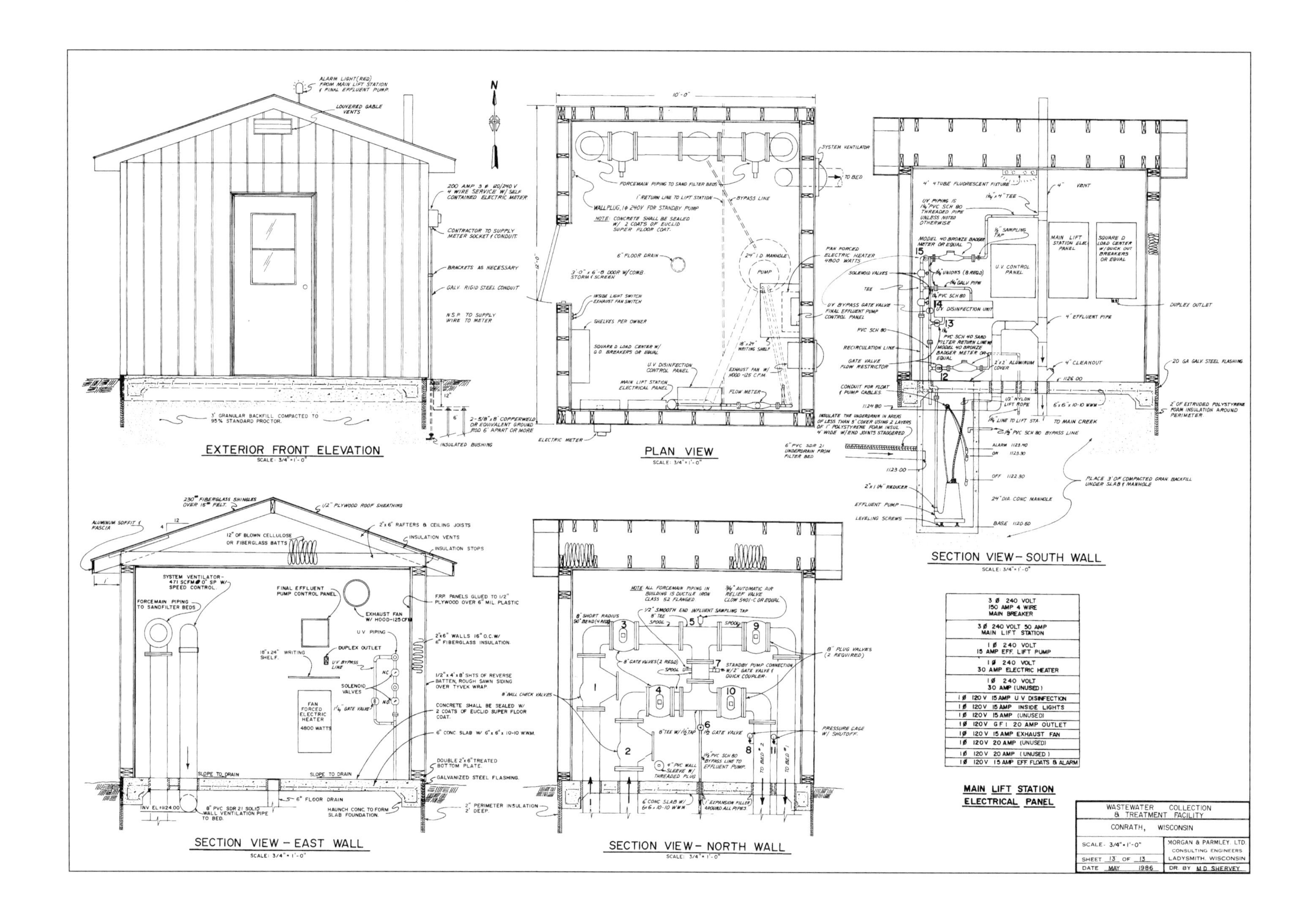

EXTERIOR FRONT ELEVATION
SCALE: 3/4"=1'-0"
ALARM LIGHT (RED) FROM MAIN LIFT STATION & FINAL EFFLUENT PUMP
LOUVERED GABLE VENTS
200 AMP 3 Ø 120/240 V 4 WIRE SERVICE W/ SELF CONTAINED ELECTRIC METER
CONTRACTOR TO SUPPLY METER SOCKET & CONDUIT
BRACKETS AS NECESSARY
GALV. RIGID STEEL CONDUIT
N.S.P. TO SUPPLY WIRE TO METER
3' GRANULAR BACKFILL COMPACTED TO 95% STANDARD PROCTOR.
2-5/8" x 8' COPPERWELD OR EQUIVALENT GROUND ROD 6' APART OR MORE
INSULATED BUSHING
PLAN VIEW
SCALE: 3/4"=1'-0"
10'-0"
12'-0"
SYSTEM VENTILATOR
TO BED
FORCEMAIN PIPING TO SAND FILTER BEDS
1" RETURN LINE TO LIFT STATION
BYPASS LINE
WALLPLUG, 1Ø 240V FOR STANDBY PUMP
NOTE: CONCRETE SHALL BE SEALED W/ 2 COATS OF EUCLID SUPER FLOOR COAT.
6" FLOOR DRAIN
24" I.D. MANHOLE
PUMP
FAN FORCED ELECTRIC HEATER 4800 WATTS
3'-0" x 6'-8 DOOR W/COMB STORM & SCREEN
INSIDE LIGHT SWITCH EXHAUST FAN SWITCH
SHELVES PER OWNER
SQUARE D LOAD CENTER W/ Q.O. BREAKERS OR EQUAL
18" x 24" WRITING SHELF
U.V. DISINFECTION CONTROL PANEL
MAIN LIFT STATION ELECTRICAL PANEL
EXHAUST FAN W/ HOOD-125 C.F.M.
FLOW METER
ELECTRIC METER
6" PVC SDR 21 UNDERDRAIN FROM FILTER BED
INSULATE THE UNDERDRAIN IN AREAS OF LESS THAN 5' COVER USING 2 LAYERS OF 1" POLYSTYRENE FOAM INSUL. 4' WIDE W/END JOINTS STAGGERED
SECTION VIEW - SOUTH WALL
SCALE: 3/4"=1'-0"
4' 4 TUBE FLUORESCENT FIXTURE
4" VENT
UV PIPING IS 1¼" PVC SCH 80 THREADED PIPE UNLESS NOTED OTHERWISE
1¼" x 4" TEE
½" SAMPLING TAP
MODEL 40 BRONZE BADGER METER OR EQUAL
MAIN LIFT STATION ELEC. PANEL
SQUARE D LOAD CENTER W/ QUICK OUT BREAKERS OR EQUAL
U.V. CONTROL PANEL
SOLENOID VALVES
1¼" UNIONS (8 REQ'D)
1¼" GALV PIPE
TEE
1¼" PVC SCH 80
U.V. DISINFECTION UNIT
DUPLEX OUTLET
UV BYPASS GATE VALVE
FINAL EFFLUENT PUMP CONTROL PANEL
4" EFFLUENT PIPE
PVC SCH 80
PVC SCH 40 SAND FILTER RETURN LINE W/ MODEL 40 BRONZE BADGER METER OR EQUAL
RECIRCULATION LINE
2" x 2" ALUMINUM COVER
GATE VALVE FLOW RESTRICTOR
4" CLEANOUT
20 GA GALV. STEEL FLASHING
CONDUIT FOR FLOAT & PUMP CABLES
1126.00
1124.80
1/2" NYLON LIFT ROPE
6" x 6" x 10-10 WWM
1¼" LINE TO LIFT STA.
TO MAIN CREEK
1½" PVC SCH 80 BYPASS LINE
2" OF EXTRUDED POLYSTYRENE FOAM INSULATION AROUND PERIMETER
ALARM 1123.40
ON 1123.30
1123.00
OFF 1122.30
2" x 1¼" REDUCER
PLACE 3' OF COMPACTED GRAN. BACKFILL UNDER SLAB & MANHOLE
24" DIA. CONC MANHOLE
EFFLUENT PUMP
LEVELING SCREWS
BASE 1120.60
SECTION VIEW - EAST WALL
SCALE: 3/4"=1'-0"
230# FIBERGLASS SHINGLES OVER 15# FELT
1/2" PLYWOOD ROOF SHEATHING
ALUMINUM SOFFIT & FASCIA
2" x 6" RAFTERS & CEILING JOISTS
12" OF BLOWN CELLULOSE OR FIBERGLASS BATTS
INSULATION VENTS
INSULATION STOPS
SYSTEM VENTILATOR - 471 SCFM @ 0" SP W/ SPEED CONTROL.
FINAL EFFLUENT PUMP CONTROL PANEL
FORCEMAIN PIPING TO SANDFILTER BEDS
EXHAUST FAN W/ HOOD-125 CFM
F.R.P. PANELS GLUED TO 1/2" PLYWOOD OVER 6 MIL PLASTIC
U.V. PIPING
DUPLEX OUTLET
2" x 6" WALLS 16" O.C. W/ 6" FIBERGLASS INSULATION.
18" x 24" WRITING SHELF.
UV BYPASS LINE
SOLENOID VALVES
1/2" x 4' x 8' SHTS OF REVERSE BATTEN, ROUGH SAWN SIDING OVER TYVEK WRAP.
FAN FORCED ELECTRIC HEATER 4800 WATTS
1¼" GATE VALVE
CONCRETE SHALL BE SEALED W/ 2 COATS OF EUCLID SUPER FLOOR COAT.
6" CONC SLAB W/ 6" x 6" x 10-10 W.W.M.
SLOPE TO DRAIN
DOUBLE 2" x 6" TREATED BOTTOM PLATE.
GALVANIZED STEEL FLASHING.
6" FLOOR DRAIN
INV EL=1124.00
8" PVC SDR 21 SOLID WALL VENTILATION PIPE TO BED.
HAUNCH CONC. TO FORM SLAB FOUNDATION.
2" PERIMETER INSULATION 2' DEEP.
SECTION VIEW - NORTH WALL
SCALE: 3/4"=1'-0"
NOTE: ALL FORCEMAIN PIPING IN BUILDING IS DUCTILE IRON CLASS 52 FLANGED.
3/4" AUTOMATIC AIR RELIEF VALVE CLOW 5401-C OR EQUAL
1/2" SMOOTH END INFLUENT SAMPLING TAP
8" SHORT RADIUS 90° BEND (4 REQ'D)
8" TEE
SPOOL
8" GATE VALVES (2 REQ'D)
STANDBY PUMP CONNECTION W/ 2" GATE VALVE & QUICK COUPLER.
8" PLUG VALVES (2 REQUIRED)
8" BALL CHECK VALVES
8" TEE W/ 1½" TAP
1½" GATE VALVE
PRESSURE GAGE W/ SHUTOFF.
4" PVC WALL SLEEVE W/ THREADED PLUG
1½" PVC SCH 80 BYPASS LINE TO EFFLUENT PUMP.
TO BED #2
TO BED #1
6" CONC SLAB W/ 6 x 6 x 10-10 W.W.M.
1" EXPANSION FILLER AROUND ALL PIPES
3 Ø 240 VOLT 150 AMP 4 WIRE MAIN BREAKER
3 Ø 240 VOLT 50 AMP MAIN LIFT STATION
1 Ø 240 VOLT 15 AMP EFF. LIFT PUMP
1 Ø 240 VOLT 30 AMP ELECTRIC HEATER
1 Ø 240 VOLT 30 AMP (UNUSED)
1 Ø 120 V 15 AMP U V DISINFECTION
1 Ø 120 V 15 AMP INSIDE LIGHTS
1 Ø 120 V 15 AMP (UNUSED)
1 Ø 120 V G F I 20 AMP OUTLET
1 Ø 120 V 15 AMP EXHAUST FAN
1 Ø 120 V 20 AMP (UNUSED)
1 Ø 120 V 20 AMP (UNUSED)
1 Ø 120 V 15 AMP EFF. FLOATS & ALARM
MAIN LIFT STATION ELECTRICAL PANEL
WASTEWATER COLLECTION & TREATMENT FACILITY
CONRATH, WISCONSIN
SCALE: 3/4"=1'-0"
SHEET 13 OF 13
DATE MAY 1986
MORGAN & PARMLEY LTD. CONSULTING ENGINEERS LADYSMITH, WISCONSIN
DR. BY M.D. SHERVEY

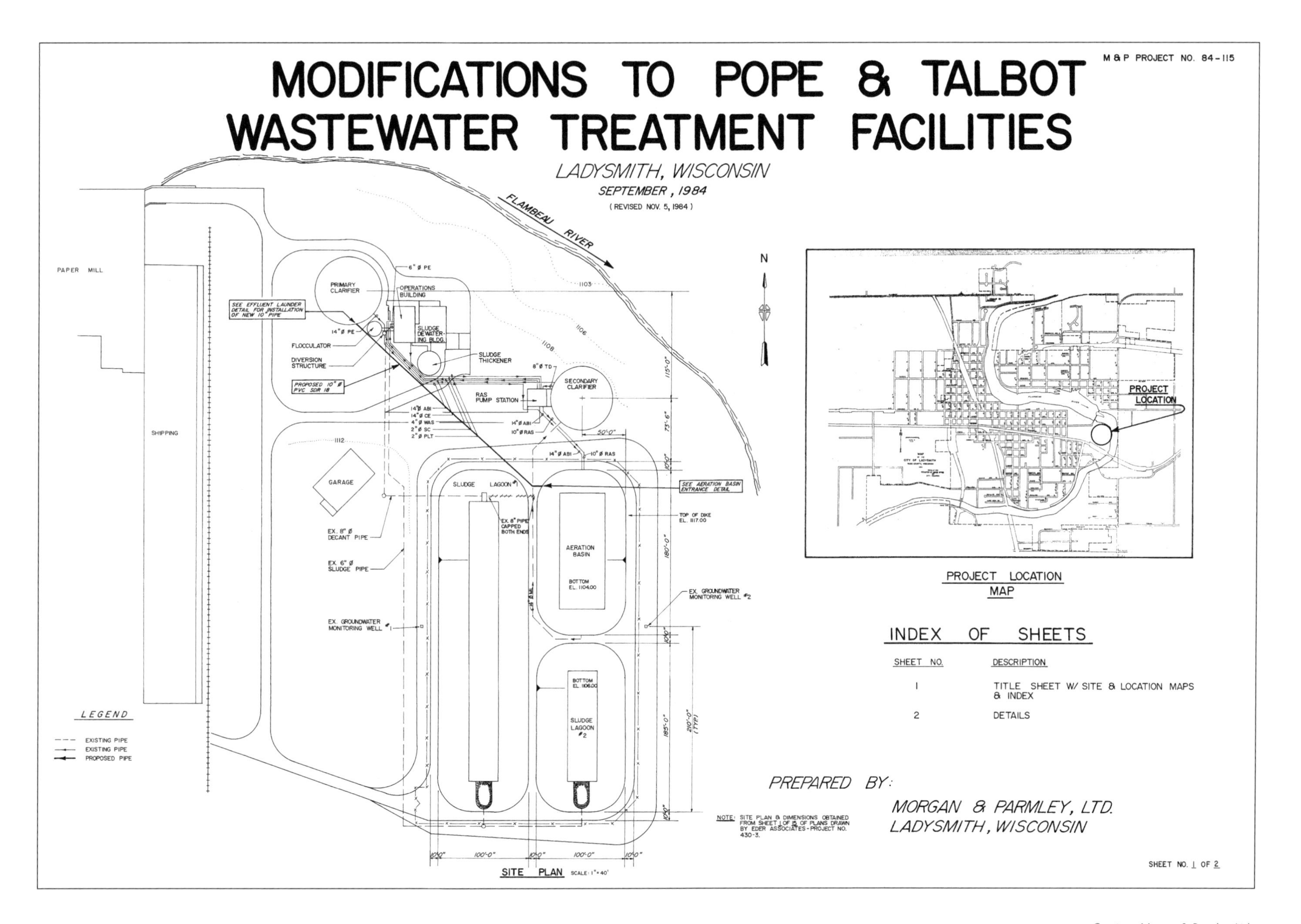

Courtesy: Morgan & Parmley, Ltd.

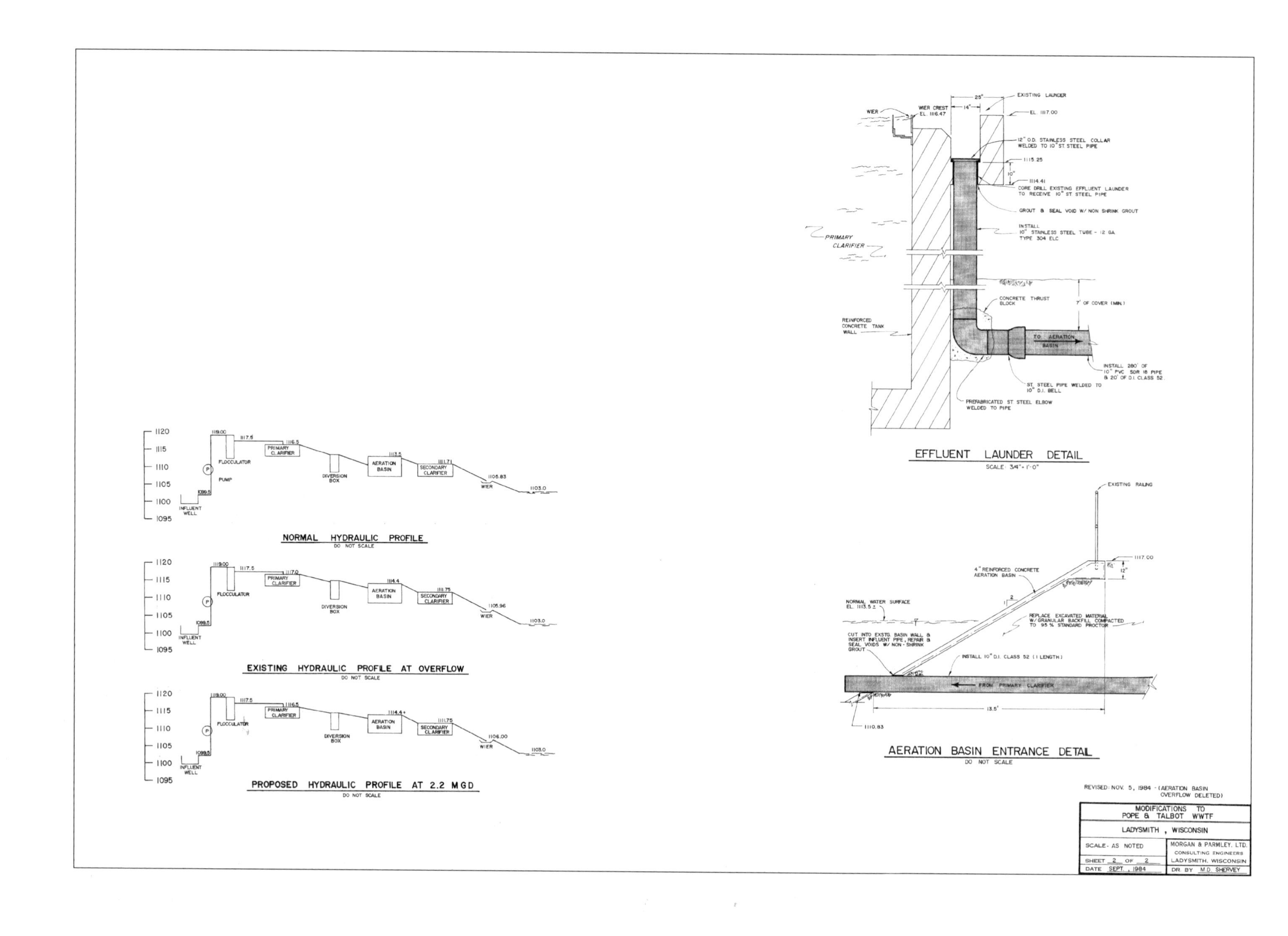
EXISTING LAUNDER
WIER CREST EL. 1116.47
WIER
EL. 1117.00
12" O.D. STAINLESS STEEL COLLAR WELDED TO 10" ST. STEEL PIPE
1115.25
1114.41
CORE DRILL EXISTING EFFLUENT LAUNDER TO RECEIVE 10" ST STEEL PIPE
GROUT & SEAL VOID W/ NON SHRINK GROUT
INSTALL 10" STAINLESS STEEL TUBE - 12 GA TYPE 304 ELC
PRIMARY CLARIFIER
CONCRETE THRUST BLOCK
7' OF COVER (MIN.)
REINFORCED CONCRETE TANK WALL
TO AERATION BASIN
INSTALL 280' OF 10" PVC SDR 18 PIPE & 20' OF D.I. CLASS 52
ST. STEEL PIPE WELDED TO 10" D.I. BELL
PREFABRICATED ST. STEEL ELBOW WELDED TO PIPE
EFFLUENT LAUNDER DETAIL
SCALE: 3/4" = 1'-0"
NORMAL HYDRAULIC PROFILE
DO NOT SCALE
EXISTING HYDRAULIC PROFILE AT OVERFLOW
DO NOT SCALE
PROPOSED HYDRAULIC PROFILE AT 2.2 MGD
DO NOT SCALE
INFLUENT WELL
PUMP
FLOCCULATOR
PRIMARY CLARIFIER
DIVERSION BOX
AERATION BASIN
SECONDARY CLARIFIER
WIER
EXISTING RAILING
4" REINFORCED CONCRETE AERATION BASIN
NORMAL WATER SURFACE EL. 1113.5 ±
REPLACE EXCAVATED MATERIAL W/ GRANULAR BACKFILL COMPACTED TO 95 % STANDARD PROCTOR
CUT INTO EXSTG. BASIN WALL & INSERT INFLUENT PIPE, REPAIR & SEAL VOIDS W/ NON - SHRINK GROUT
INSTALL 10" D.I. CLASS 52 (1 LENGTH)
FROM PRIMARY CLARIFIER
1110.83
AERATION BASIN ENTRANCE DETAIL
DO NOT SCALE
REVISED: NOV. 5, 1984 - (AERATION BASIN OVERFLOW DELETED)
MODIFICATIONS TO POPE & TALBOT WWTF
LADYSMITH, WISCONSIN
SCALE - AS NOTED
SHEET 2 OF 2
DATE SEPT., 1984
MORGAN & PARMLEY, LTD.
CONSULTING ENGINEERS
LADYSMITH, WISCONSIN
DR. BY M.D. SHERVEY

-UTILITY EXTENSIONS-

SANITARY SEWER, LIFT STATION & TRANSMISSION WATERMAIN

for the

VILLAGE

of

TONY, WISCONSIN

M. & P. PROJECT NO. 97-127

FEBRUARY, 1998

R5W
TONY VILLAGE LIMITS
ELEV. WATER TANK 50,000 GAL.
EXISTING SEWAGE LIFT STATION NO. 1
TOWNSHIP OF DEWEY
EXISTING 4" FORCE MAIN
TONY VILLAGE LIMITS
SCHULTZ ROAD
STA. 96+56
STA. 70+00
SANITARY SEWER FORCEMAIN (4-INCH)
T35N
T34N
TOWNLINE ROAD
8 – INCH WATERMAIN EXTENSION
TOWNSHIP OF GROW
STA. 0+00
STA. 0+00
GENERAL LOCATION of NEW FLAMBEAU SCHOOL COMPLEX
NO. 2 LIFT STATION SITE
FLOOR ELEV. 1204.33
SOUTH 1/2 OF THE SE 1/4 OF SEC. 4, T34N, R5W, TN. of GROW, RUSK CO., WI.

PROJECT LOCATION MAP

SCALE IN FEET

STATE OF WISCONSIN
TONY
N.
TOWNSHIP OF DEWEY
VILLAGE OF TONY
TOWNSHIP OF GROW
RUSK COUNTY

VICINITY MAP

DO NOT SCALE

REFERENCE SPECIFICATIONS

- * WISCONSIN CONSTRUCTION SITE BEST MANAGEMENT PRACTICE HANDBOOK
- * STANDARD SPECIFICATIONS FOR ROAD AND BRIDGE CONSTRUCTION 1989 EDITION WIS. D.O.T.
- * STANDARD SPECIFICATIONS FOR SEWER & WATER CONSTRUCTION IN WISCONSIN, FIFTH EDITION
- * MORGAN & PARMLEY, LTD., ACCOMPANYING SPECIFICATIONS

PROJECT RECORD DRAWINGS: COMPILED FROM CONSTRUCTION DATA FURNISHED TO THE ENGINEER BY CONTRACTOR, OWNER AND/OR ENGINEER'S FIELD REPRESENTATIVE. INFORMATION SHOWN ON THESE DRAWINGS SHOULD BE FIELD VERIFIED BEFORE USING SAME.

GENERAL NOTES

1. ALL WATER MAIN SHALL BE PVC DR 18 C900, EXCEPT AS NOTED.
2. ALL WATERMAIN FITTINGS SHALL BE M.J.D.I., WITH TEES AND BENDS BLOCKED AS SHOWN IN THE DETAILS.
3. ALL HYDRANT LEADS SHALL HAVE A 6-INCH VALVE.
4. HYDRANTS SHALL BE WATEROUS PACER, HAVE A 22" BREAKOFF & MEET CITY OF LADYSMITH'S PATTERN FOR THREADS, ROTATION AND NUT.
5. ALL WATERMAIN SHALL HAVE AN 8 FT. COVER (MINIMUM).
6. SANITARY SEWER FORCEMAIN SHALL BE 4" PVC DR 18 C900.
7. SEE SHEET 9 FOR TYPICAL (MINIMUM) HORIZONTAL SEPARATION BETWEEN SANITARY SEWER FORCEMAIN & WATERMAIN.
8. 18" (MINIMUM) VERTICAL DISTANCE (OUT TO OUT) BETWEEN WATERMAIN UNDER SANITARY SEWER FORCEMAIN @ THEIR INTERSECTION.
9. 6" (MINIMUM) VERTICAL DISTANCE (OUT TO OUT) BETWEEN WATERMAIN OVER SANITARY SEWER FORCEMAIN @ THEIR INTERSECTION.
10. ALL ELEVATIONS BASED ON USGS DATUM.
11. EROSION CONTROL MEASURES SHALL BE IN PLACE BEFORE CONSTRUCTION BEGINS. (SEE SHEET 12 & SPECIFICATIONS)
12. THE CONTRACTORS SHALL MINIMIZE EROSION DURING CONSTRUCTION USING GOOD CONSTRUCTION TECHNIQUES AND UTILIZING SILT FENCES AND BALE DITCH CHECKS. THE CONTRACTORS SHALL RELEASE RUNOFF FROM THE SITE IN A NUISANCE FREE MANNER.
13. THE CONTRACTOR IS RESPONSIBLE TO MAINTAIN THE SITE IN A SAFE CONDITION. THE SOLE RESPONSIBILITY FOR WARNING SIGNS, BARRICADES AND ALL ASPECTS OF SAFETY LIE WITH THE CONTRACTOR.
14. EXISTING UTILITIES ARE SHOWN BUT MAY NOT BE ALL INCLUSIVE, CONTRACTOR SHALL HAVE ALL BURIED UTILITIES LOCATED PRIOR TO COMMENCING CONSTRUCTION.
15. EXISTING SEWER FORCEMAIN AND WATERMAIN ARE SHOWN ACCORDING TO EXISTING RECORDS BUT LOCATION SHALL BE VERIFIED IN FIELD.
16. WATERMAIN SHALL BE INSULATED AT ALL CULVERT AND DITCH CROSSINGS WITH 2" THICK EXTRUDED POLYSTYRENE INSULATION BOARD; LAPPED & STAGGERED AT JOINTS.
17. PROPERTY CORNERS KNOWN TO EXIST SHALL NOT BE DISTURBED. DAMAGED CORNERS WILL BE REPLACED AT THE CONTRACTOR'S EXPENSE.
18. ALL WATERMAIN SHALL BE DISINFECTED AND A "SAFE' SAMPLE OBTAINED BEFORE PLACING IN SERVICE.
19. VALVES SHALL BE SET TO GRADE SHOWN ON PLANS, UNLESS DIRECTED OTHERWISE.
20. CONSTRUCTION SHALL COMPLY WITH THE CONDITIONS OF DNR'S APPROVAL LETTER.
21. ALL DISTURBED AREAS SHALL BE SEEDED, FERTILIZED AND MULCHED, OR COVERED W / BASE COURSE AFTER CONSTRUCTION.
22. ALL DISTURBED AREAS TO BE SEEDED SHALL BE RESTORED AS FOLLOWS:
    - A) SEED MIXTURE:

BIRDSFOOT TREFOIL	15#/AC.
SMOOTH BROMEGRASS	40#/AC.
TALL FESCUE	25#/AC.

    - B) SEED BED - PREPARE A THOROUGHLY TILLED BUT FIRM SEED BED WHEREVER CONDITIONS WILL PERMIT.
    - C) FERTILIZER: 500 #/AC. 20-10-10
    - D) SEED TO BE SOWN WITH A HYDROSEEDER.
    - E) MULCH WITH CLEAN HAY OR STRAW AT THE RATE OF 2 TONS/ACRE. ANCHOR MULCH WITH 75 - 100 GALLONS OF EMUILSIFIED ASPHALT PER TON OF MULCH.
    - F) SEED PRIOR TO SEPT. 1ST.
23. CONTRACTOR SHALL DETERMINE CONSTRUCTION SEQUENCE IN ORDER TO MINIMIZE INSTALLATION CONFLICTS.
24. THERE SHALL BE NO DEVIATION FROM THE PLANS WITHOUT THE DESIGN ENGINEER'S APPROVAL.
25. CONSTRUCTION STAKES DESTROYED BY THE CONTRACTOR THAT NEED TO BE REPLACED, WILL BE CHARGED TO THE CONTRACTOR.
26. NO TREES ARE TO BE REMOVED WITHOUT THE APPROVAL OF THE ENGINEER.
27. THE CONTRACTOR MUST REMOVE AND REPLACE ALL EFFECTED STREET SIGNS AFTER PROPERLY SIGNING AND BARRICADING THE CONSTRUCTION AREA. ALL SIGNS MUST BE REPLACED PRIOR TO OPENING THE STREET TO TRAFFIC.
28. CONFLICTING POWER POLES WILL BE MOVED OR STABILIZED BY NSP, IF WITHIN ROAD R/W. HOWEVER, IF POWER POLES ARE BEYOND ROAD R/W, NSP'S CHARGES SHALL BE PAID BY CONTRACTOR.
29. SEE ACCOMPANYING TECHNICAL SPECIFICATIONS AND SECTION 01010 "SPECIAL PROVISIONS".
30. SEE SHEET 9 FOR TYPICAL X-SECTIONS & LEGEND.

DRAWING INDEX

SHEET NO.	DESCRIPTION
1	TITLE SHEET, LOCATION MAPS & NOTES
2	PLAN & PROFILE: STA. 0+00 to 14+00
3	PLAN & PROFILE STA. 14+00 to 28+00
4	PLAN & PROFILE STA. 28+00 to 42+00
5	PLAN & PROFILE STA. 42+00 to 56+00
6	PLAN & PROFILE STA. 56+00 to 70+00
7	PLAN & PROFILE STA. 70+00 to 84+00
8	PLAN & PROFILE STA. 84+00 to 96+37
9	TYPICAL CONSTRUCTION X-SECTIONS, ACCESS DRIVEWAY & DETAILS w/ NOTES & LEGEND
10	SITE PLANS: "A" INTERSECTION SOUTH AVE. & C.T.H. "I"; "B" LIFT STATION
11	LIFT STATION PLAN, X-SECTION & DETAILS
12	EROSION CONTROL DETAILS w/ TYPICAL NOTES This Sheet Not Shown

-UTILITIES-

SEWER & WATER: VILLAGE OF TONY
BILLY MECHELKE
N5297 LITTLE X RD.
TONY, WI 54563
1-715-532-3183 (W)
1-715-532-7046 (H)

ELECTRIC:
NORTHERN STATES POWER COMPANY
133 NORTH LAKE AVENUE
PHILLIPS, WI 54555
ATTN: JOE PERKINS
1-800-445-1275

TELEPHONE:
AMERITECH
304 S. DEWEY STREET
EAU CLAIRE, WI 54701
ATTN: PAM PARKER
1-800-305-5815

WISCONSIN GAS COMPANY:
104 WEST SOUTH STREET
RICE LAKE, WI 54868
ATTN: DON WEDIN, MGR.
1-715-387-0189

DIGGER'S HOTLINE:
1-800-242-8511

PREPARED BY:
MORGAN & PARMLEY, Ltd.
PROFESSIONAL CONSULTING ENGINEERS
LADYSMITH, WISCONSIN

Courtesy: Morgan & Parmley, Ltd.

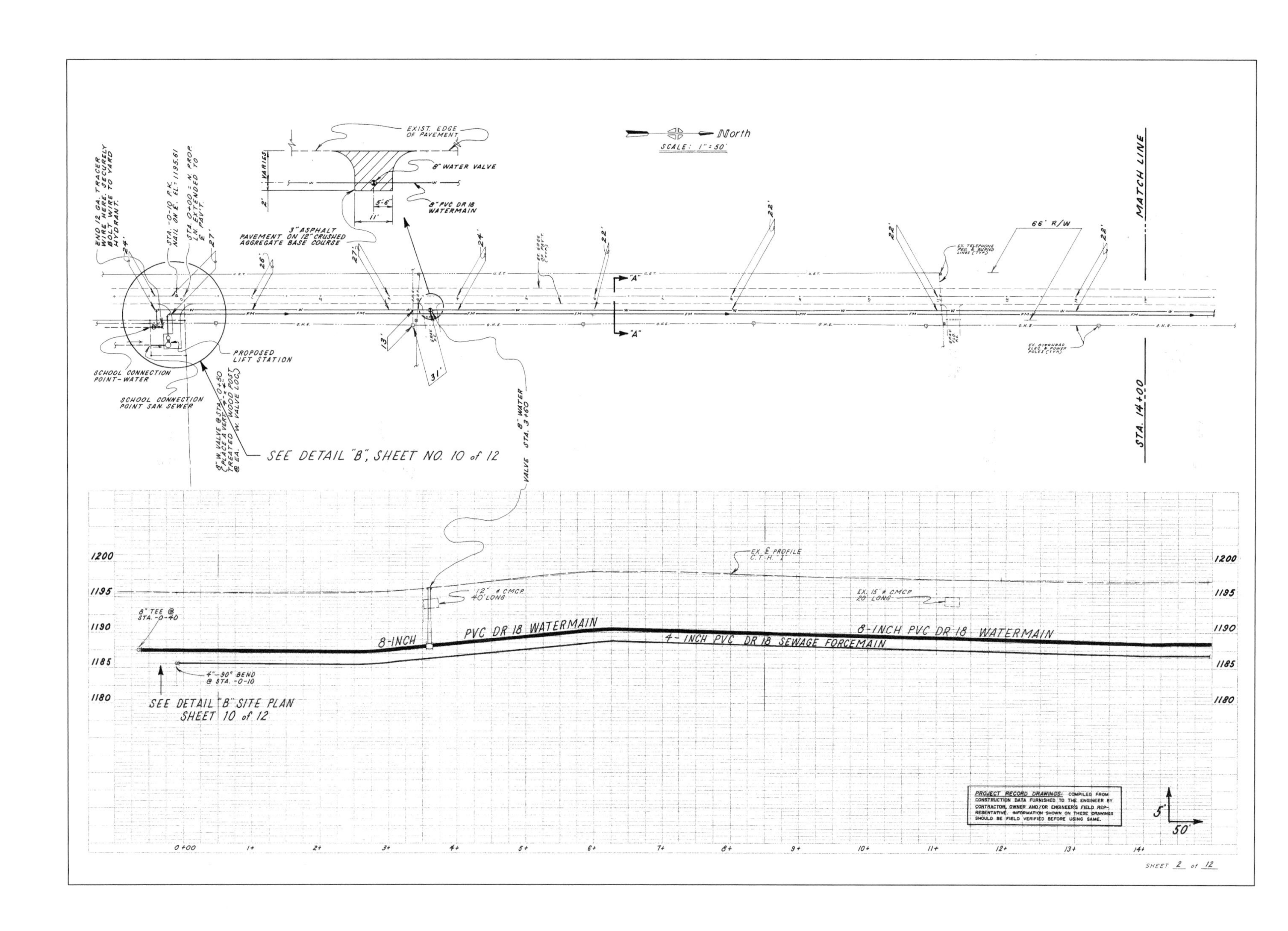

EXIST. EDGE OF PAVEMENT
8" WATER VALVE
8" PVC DR 18 WATERMAIN
3" ASPHALT PAVEMENT ON 12" CRUSHED AGGREGATE BASE COURSE
North
SCALE: 1" = 50'
MATCH LINE
STA. 14+00
66' R/W
PROPOSED LIFT STATION
SCHOOL CONNECTION POINT-WATER
SCHOOL CONNECTION POINT SAN. SEWER
SEE DETAIL "B", SHEET NO. 10 of 12
8" WATER VALVE STA. 3+80
EX. ℄ PROFILE C.T.H. "I"
12" ⌀ CMCP 40' LONG
EX. 15" ⌀ CMCP 20' LONG
8" TEE @ STA. -0-40
4"-90° BEND @ STA. -0-10
8-INCH PVC DR 18 WATERMAIN
4-INCH PVC DR 18 SEWAGE FORCEMAIN
8-INCH PVC DR 18 WATERMAIN
SEE DETAIL "B" SITE PLAN SHEET 10 of 12
PROJECT RECORD DRAWINGS: COMPILED FROM CONSTRUCTION DATA FURNISHED TO THE ENGINEER BY CONTRACTOR, OWNER AND/OR ENGINEER'S FIELD REPRESENTATIVE. INFORMATION SHOWN ON THESE DRAWINGS SHOULD BE FIELD VERIFIED BEFORE USING SAME.
SHEET 2 of 12

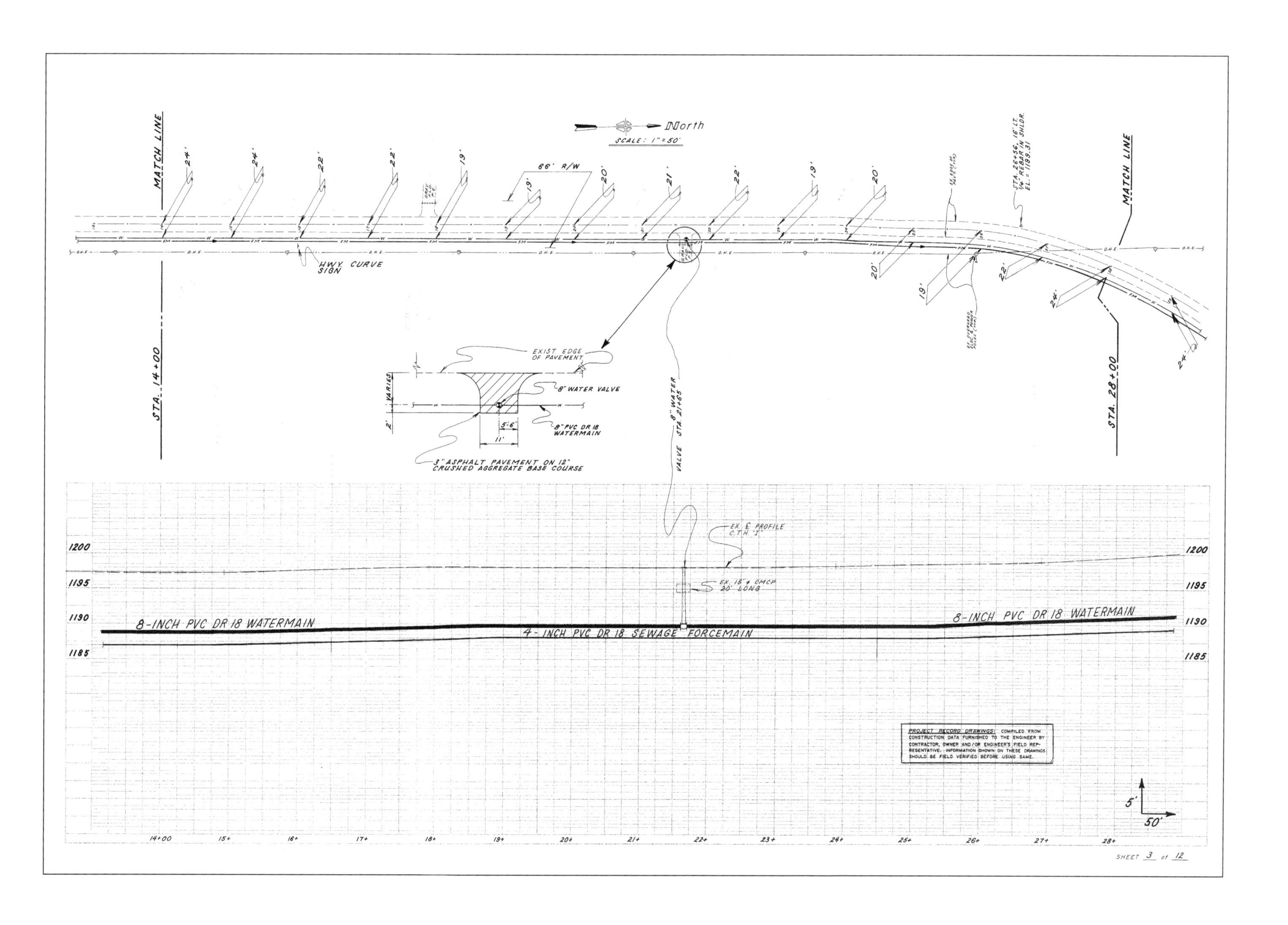

North
SCALE: 1" = 50'
MATCH LINE
STA. 14+00
MATCH LINE
STA. 28+00
66' R/W
HWY. CURVE SIGN
STA. 26+56, 16' LT.
1/4" REBAR IN SHLDR.
EL. = 1199.31
EXIST. EDGE OF PAVEMENT
8" WATER VALVE
8" PVC DR 18 WATERMAIN
VARIES
2'
5'-6"
11'
3" ASPHALT PAVEMENT ON 12" CRUSHED AGGREGATE BASE COURSE
8" WATER VALVE STA. 21+65
EX. ℄ PROFILE C.T.H. "I"
EX. 15" ø CMCP 20' LONG
1200
1195
1190
1185
8-INCH PVC DR 18 WATERMAIN
4-INCH PVC DR 18 SEWAGE FORCEMAIN
8-INCH PVC DR 18 WATERMAIN
PROJECT RECORD DRAWINGS: COMPILED FROM CONSTRUCTION DATA FURNISHED TO THE ENGINEER BY CONTRACTOR, OWNER AND/OR ENGINEER'S FIELD REPRESENTATIVE. INFORMATION SHOWN ON THESE DRAWINGS SHOULD BE FIELD VERIFIED BEFORE USING SAME.
5'
50'
14+00
15+
16+
17+
18+
19+
20+
21+
22+
23+
24+
25+
26+
27+
28+
SHEET 3 of 12

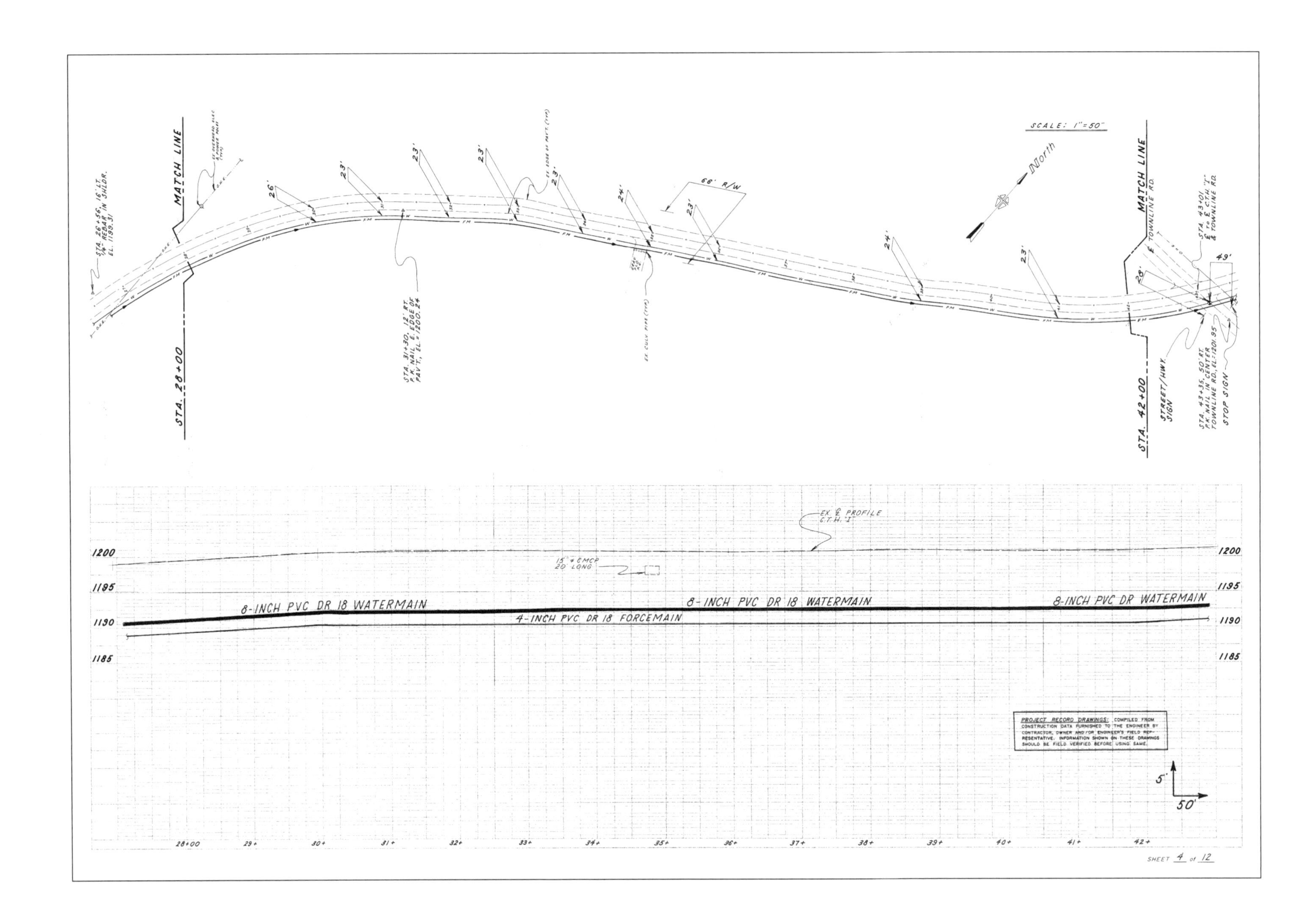
SCALE: 1"=50"
North
MATCH LINE
MATCH LINE
STA. 28+00
STA. 42+00
66' R/W
STA. 31+30, 12' RT.
P.K. NAIL E. EDGE OF
PAV'T., EL.=1200.24
STREET/HWY. SIGN
STOP SIGN
EX. ℄ PROFILE
C.T.H. "I"
15" ⌀ CMCP
20' LONG
8-INCH PVC DR 18 WATERMAIN
8-INCH PVC DR 18 WATERMAIN
8-INCH PVC DR WATERMAIN
4-INCH PVC DR 18 FORCEMAIN
1200
1195
1190
1185
28+00
29+
30+
31+
32+
33+
34+
35+
36+
37+
38+
39+
40+
41+
42+
PROJECT RECORD DRAWINGS: COMPILED FROM CONSTRUCTION DATA FURNISHED TO THE ENGINEER BY CONTRACTOR, OWNER AND/OR ENGINEER'S FIELD REPRESENTATIVE. INFORMATION SHOWN ON THESE DRAWINGS SHOULD BE FIELD VERIFIED BEFORE USING SAME.
5'
50'
SHEET 4 of 12

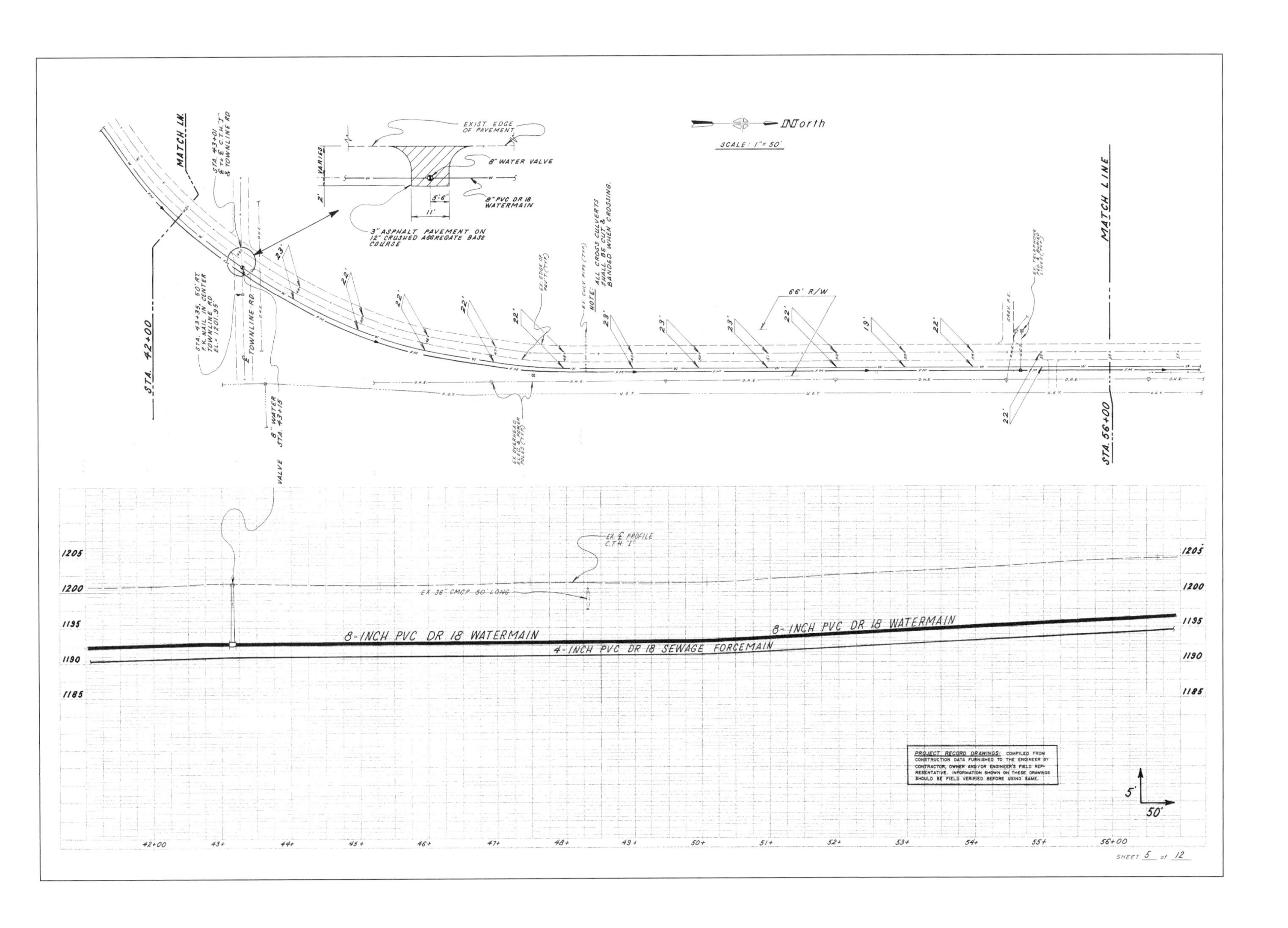

North
SCALE: 1" = 50'
MATCH LN.
MATCH LINE
STA. 42+00
STA. 56+00
EXIST EDGE OF PAVEMENT
8" WATER VALVE
8" PVC DR 18 WATERMAIN
3" ASPHALT PAVEMENT ON 12" CRUSHED AGGREGATE BASE COURSE
TOWNLINE RD.
66' R/W
NOTE: ALL CROSS CULVERTS SHALL BE CUT & BANDED WHEN CROSSING.
8" WATER VALVE STA. 43+15
EX. ℄ PROFILE C.T.H. "I"
EX. 36" CMCP 50' LONG
8-INCH PVC DR 18 WATERMAIN
4-INCH PVC DR 18 SEWAGE FORCEMAIN
PROJECT RECORD DRAWINGS: COMPILED FROM CONSTRUCTION DATA FURNISHED TO THE ENGINEER BY CONTRACTOR, OWNER AND/OR ENGINEER'S FIELD REPRESENTATIVE. INFORMATION SHOWN ON THESE DRAWINGS SHOULD BE FIELD VERIFIED BEFORE USING SAME.
SHEET 5 of 12

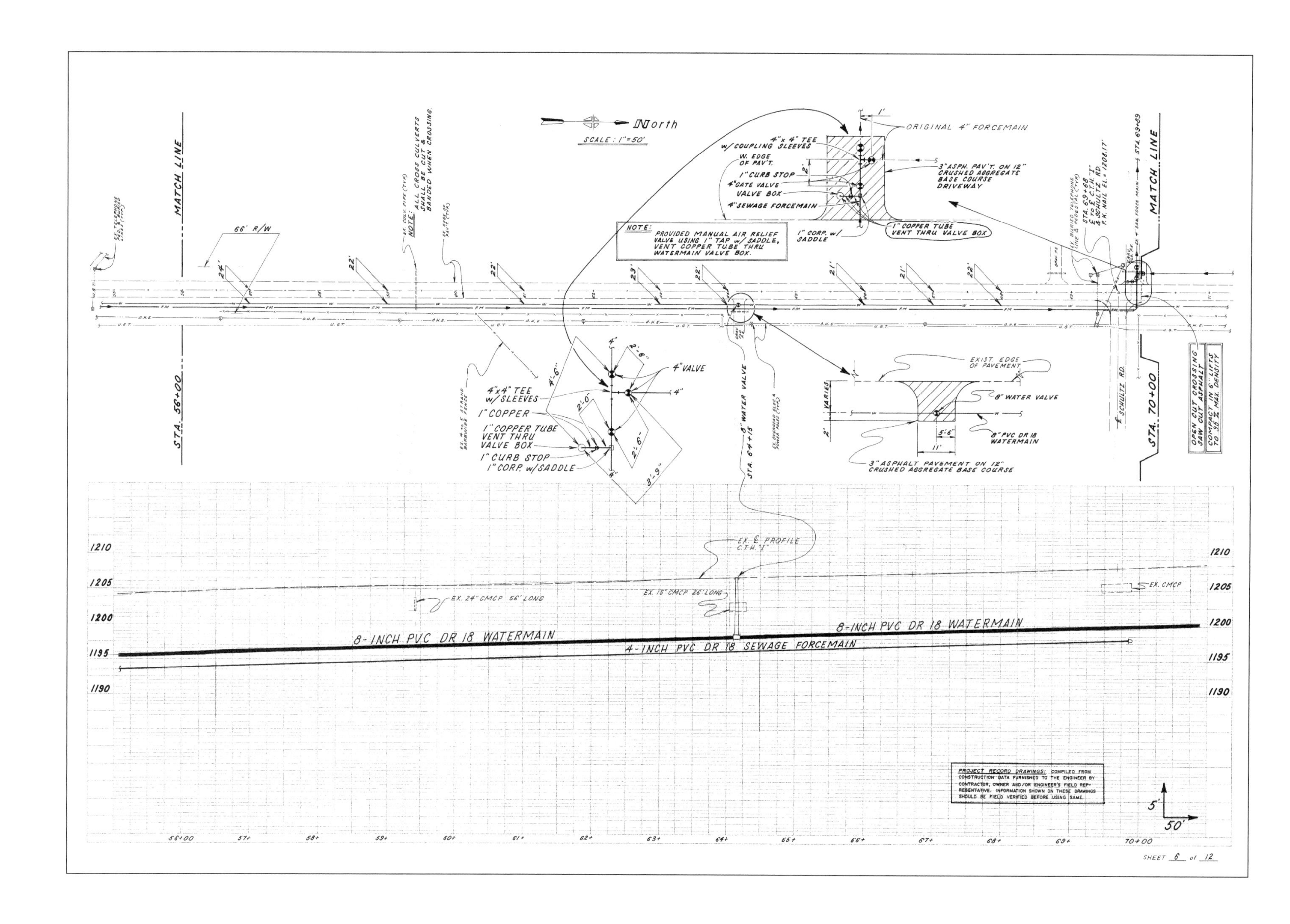
North
SCALE: 1"=50'
MATCH LINE
STA. 56+00
66' R/W
NOTE: ALL CROSS CULVERTS SHALL BE CUT & BANDED WHEN CROSSING.
NOTE: PROVIDED MANUAL AIR RELIEF VALVE USING 1" TAP w/ SADDLE, VENT COPPER TUBE THRU WATERMAIN VALVE BOX.
ORIGINAL 4" FORCEMAIN
4"x 4" TEE w/ COUPLING SLEEVES
W. EDGE OF PAV'T.
1" CURB STOP
4" GATE VALVE
VALVE BOX
4" SEWAGE FORCEMAIN
1" CORP. w/ SADDLE
3" ASPH. PAV'T. ON 12" CRUSHED AGGREGATE BASE COURSE DRIVEWAY
1" COPPER TUBE VENT THRU VALVE BOX
4" VALVE
4"x4" TEE w/ SLEEVES
1" COPPER
1" COPPER TUBE VENT THRU VALVE BOX
1" CURB STOP
1" CORP. w/SADDLE
8" WATER VALVE
STA. 64+15
EXIST. EDGE OF PAVEMENT
VARIES
8" WATER VALVE
8" PVC DR 18 WATERMAIN
3" ASPHALT PAVEMENT ON 12" CRUSHED AGGREGATE BASE COURSE
℄ SCHULTZ RD.
STA. 69+68 ℄ TO ℄ C.T.H. "I" & SCHULTZ RD. P.K. NAIL EL. = 1208.17'
STA. 70+00
MATCH LINE
OPEN CUT CROSSING SAW CUT ASPHALT
COMPACT IN 6" LIFTS TO 95 % MAX. DENSITY
EX. ℄ PROFILE C.T.H. "I"
EX. 24" CMCP 56' LONG
EX. 18" CMCP 26' LONG
EX. CMCP
8-INCH PVC DR 18 WATERMAIN
4-INCH PVC DR 18 SEWAGE FORCEMAIN
1210
1205
1200
1195
1190
PROJECT RECORD DRAWINGS: COMPILED FROM CONSTRUCTION DATA FURNISHED TO THE ENGINEER BY CONTRACTOR, OWNER AND/OR ENGINEER'S FIELD REPRESENTATIVE. INFORMATION SHOWN ON THESE DRAWINGS SHOULD BE FIELD VERIFIED BEFORE USING SAME.
5'
50'
56+00 57+ 58+ 59+ 60+ 61+ 62+ 63+ 64+ 65+ 66+ 67+ 68+ 69+ 70+00
SHEET 6 of 12

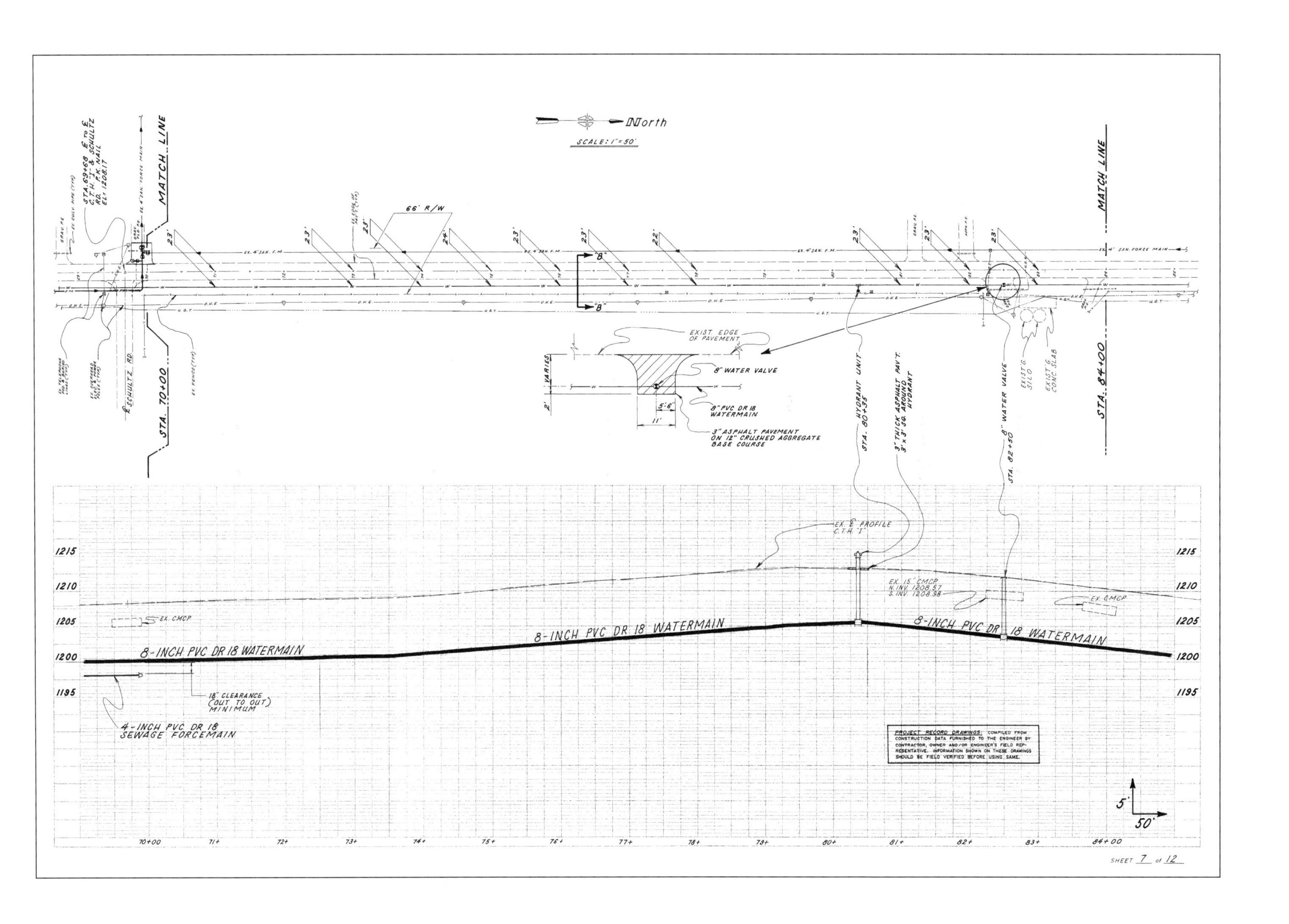
North
SCALE: 1" = 50'
MATCH LINE
STA. 70+00
STA. 84+00
66' R/W
EXIST EDGE OF PAVEMENT
8" WATER VALVE
8" PVC DR 18 WATERMAIN
3" ASPHALT PAVEMENT ON 12" CRUSHED AGGREGATE BASE COURSE
HYDRANT UNIT STA. 80+35
3" THICK ASPHALT PAV'T. 3' x 3' SQ. AROUND HYDRANT
8" WATER VALVE STA. 82+50
EX. ℄ PROFILE C.T.H. "I"
EX. 15" CMCP
EX. CMCP
8-INCH PVC DR 18 WATERMAIN
18" CLEARANCE (OUT TO OUT) MINIMUM
4-INCH PVC DR 18 SEWAGE FORCEMAIN
PROJECT RECORD DRAWINGS: COMPILED FROM CONSTRUCTION DATA FURNISHED TO THE ENGINEER BY CONTRACTOR, OWNER AND/OR ENGINEER'S FIELD REPRESENTATIVE. INFORMATION SHOWN ON THESE DRAWINGS SHOULD BE FIELD VERIFIED BEFORE USING SAME.
SHEET 7 of 12

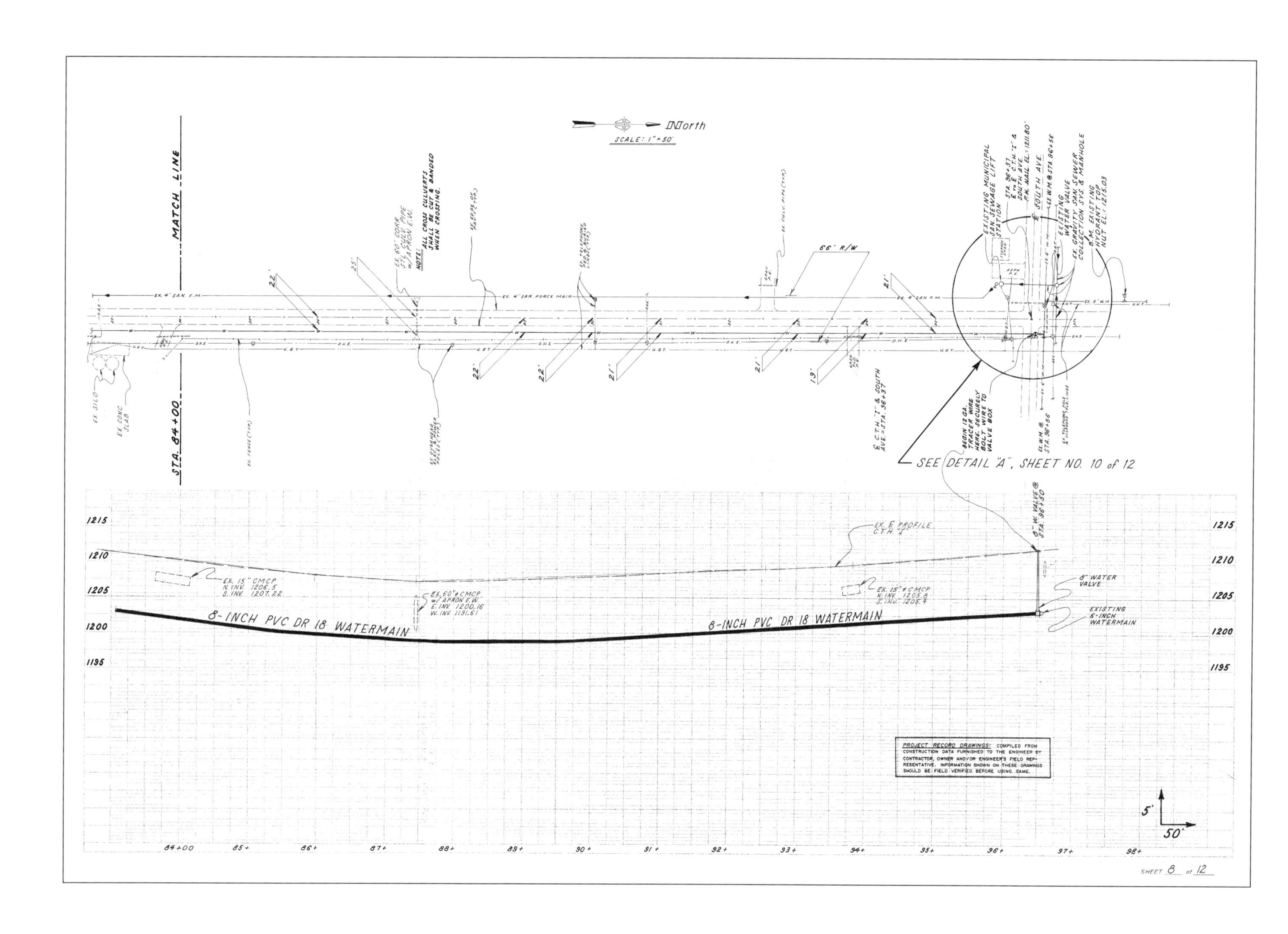
North
SCALE: 1" = 50'
MATCH LINE
STA. 84+00
NOTE: ALL CROSS CULVERTS SHALL BE CUT & BANDED WHEN CROSSING.
66' R/W
SEE DETAIL "A", SHEET NO. 10 of 12
8-INCH PVC DR 18 WATERMAIN
8-INCH PVC DR 18 WATERMAIN
8" WATER VALVE
EXISTING 6-INCH WATERMAIN
PROJECT RECORD DRAWINGS: COMPILED FROM CONSTRUCTION DATA FURNISHED TO THE ENGINEER BY CONTRACTOR, OWNER AND/OR ENGINEER'S FIELD REPRESENTATIVE. INFORMATION SHOWN ON THESE DRAWINGS SHOULD BE FIELD VERIFIED BEFORE USING SAME.
SHEET 8 of 12

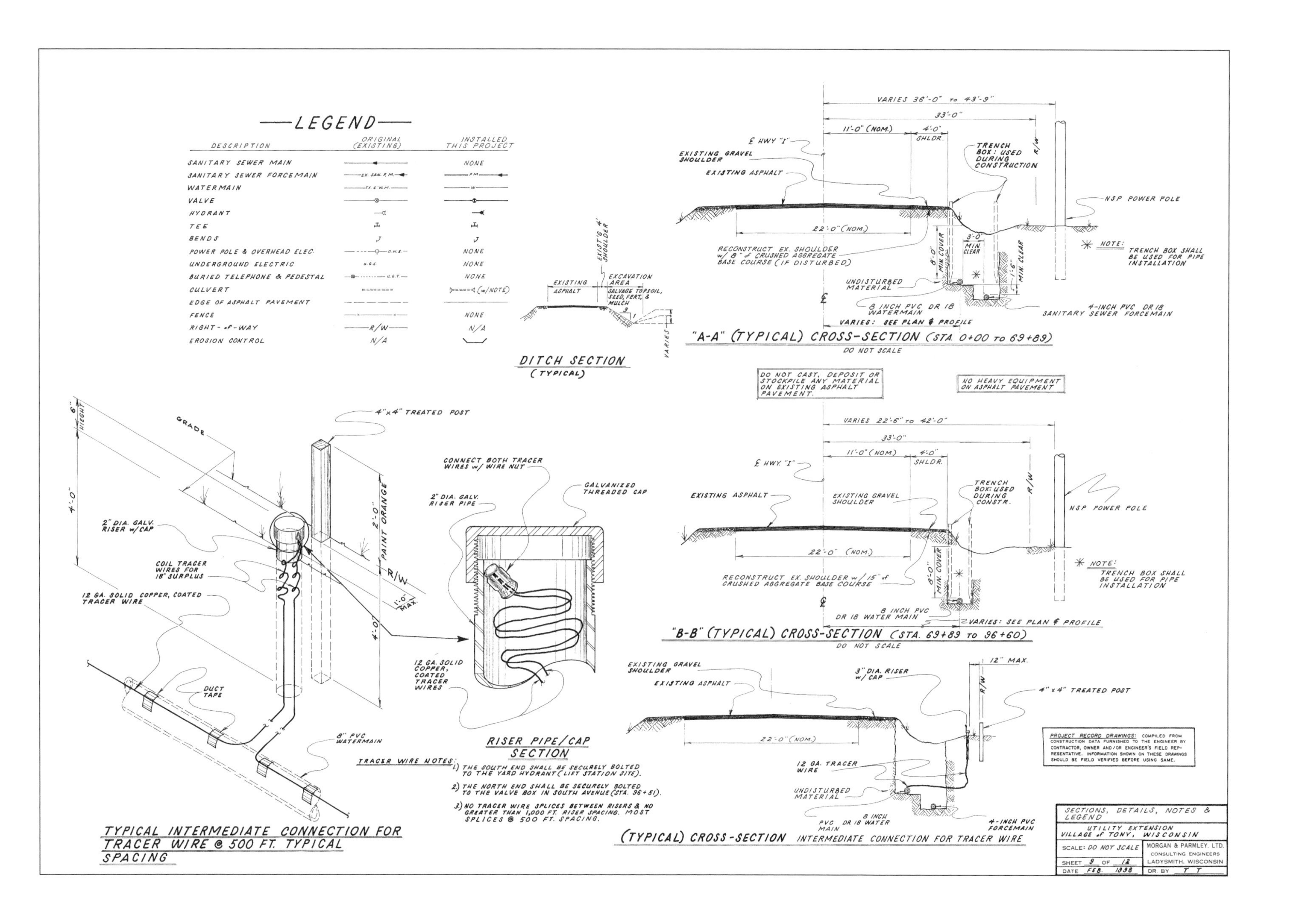
LEGEND
DESCRIPTION
ORIGINAL (EXISTING)
INSTALLED THIS PROJECT
SANITARY SEWER MAIN
NONE
SANITARY SEWER FORCEMAIN
WATERMAIN
VALVE
HYDRANT
TEE
BENDS
POWER POLE & OVERHEAD ELEC.
NONE
UNDERGROUND ELECTRIC
NONE
BURIED TELEPHONE & PEDESTAL
NONE
CULVERT
(w/NOTE)
EDGE OF ASPHALT PAVEMENT
FENCE
NONE
RIGHT-of-WAY
R/W
N/A
EROSION CONTROL
N/A
EXIST'G 4' SHOULDER
EXISTING ASPHALT
EXCAVATION AREA
SALVAGE TOPSOIL, SEED, FERT, & MULCH
VARIES
DITCH SECTION
(TYPICAL)
VARIES 36'-0" TO 43'-9"
33'-0"
11'-0" (NOM.)
4'-0" SHLDR.
℄ HWY "I"
EXISTING GRAVEL SHOULDER
EXISTING ASPHALT
TRENCH BOX: USED DURING CONSTRUCTION
R/W
NSP POWER POLE
22'-0" (NOM.)
RECONSTRUCT EX. SHOULDER w/ 8" of CRUSHED AGGREGATE BASE COURSE (IF DISTURBED)
8'-0" MIN. COVER
3'-0" MIN CLEAR
1'-6" MIN. CLEAR
* NOTE: TRENCH BOX SHALL BE USED FOR PIPE INSTALLATION
UNDISTURBED MATERIAL
8 INCH PVC DR 18 WATERMAIN
4-INCH PVC DR 18 SANITARY SEWER FORCEMAIN
VARIES: SEE PLAN & PROFILE
"A-A" (TYPICAL) CROSS-SECTION (STA. 0+00 TO 69+89)
DO NOT SCALE
DO NOT CAST, DEPOSIT OR STOCKPILE ANY MATERIAL ON EXISTING ASPHALT PAVEMENT.
NO HEAVY EQUIPMENT ON ASPHALT PAVEMENT
VARIES 22'-6" TO 42'-0"
33'-0"
11'-0" (NOM.)
4'-0" SHLDR.
℄ HWY "I"
EXISTING ASPHALT
EXISTING GRAVEL SHOULDER
TRENCH BOX: USED DURING CONSTR.
R/W
NSP POWER POLE
22'-0" (NOM.)
RECONSTRUCT EX. SHOULDER w/ 15" of CRUSHED AGGREGATE BASE COURSE
8'-0" MIN. COVER
* NOTE: TRENCH BOX SHALL BE USED FOR PIPE INSTALLATION
8 INCH PVC DR 18 WATER MAIN
VARIES: SEE PLAN & PROFILE
"B-B" (TYPICAL) CROSS-SECTION (STA. 69+89 TO 96+60)
DO NOT SCALE
6" HIEGHT
GRADE
4'-0"
4"x4" TREATED POST
2'-0" PAINT ORANGE
2" DIA. GALV. RISER w/CAP
COIL TRACER WIRES FOR 18" SURPLUS
R/W
1'-0" MAX.
12 GA. SOLID COPPER, COATED TRACER WIRE
4'-0"
DUCT TAPE
8" PVC WATERMAIN
CONNECT BOTH TRACER WIRES w/ WIRE NUT
2" DIA. GALV. RISER PIPE
GALVANIZED THREADED CAP
12 GA. SOLID COPPER, COATED TRACER WIRES
RISER PIPE/CAP SECTION
TRACER WIRE NOTES:
1) THE SOUTH END SHALL BE SECURELY BOLTED TO THE YARD HYDRANT (LIFT STATION SITE).
2) THE NORTH END SHALL BE SECURELY BOLTED TO THE VALVE BOX IN SOUTH AVENUE (STA. 96+51).
3) NO TRACER WIRE SPLICES BETWEEN RISERS & NO GREATER THAN 1,000 FT. RISER SPACING. MOST SPLICES @ 500 FT. SPACING.
TYPICAL INTERMEDIATE CONNECTION FOR TRACER WIRE @ 500 FT. TYPICAL SPACING
EXISTING GRAVEL SHOULDER
EXISTING ASPHALT
3" DIA. RISER w/ CAP
12" MAX.
R/W
4" x 4" TREATED POST
22'-0" (NOM.)
12 GA. TRACER WIRE
UNDISTURBED MATERIAL
8 INCH PVC DR 18 WATER MAIN
4-INCH PVC FORCEMAIN
(TYPICAL) CROSS-SECTION INTERMEDIATE CONNECTION FOR TRACER WIRE
PROJECT RECORD DRAWINGS: COMPILED FROM CONSTRUCTION DATA FURNISHED TO THE ENGINEER BY CONTRACTOR, OWNER AND/OR ENGINEER'S FIELD REPRESENTATIVE. INFORMATION SHOWN ON THESE DRAWINGS SHOULD BE FIELD VERIFIED BEFORE USING SAME.
SECTIONS, DETAILS, NOTES & LEGEND
UTILITY EXTENSION
VILLAGE of TONY, WISCONSIN
SCALE: DO NOT SCALE
MORGAN & PARMLEY, LTD. CONSULTING ENGINEERS LADYSMITH, WISCONSIN
SHEET 9 OF 12
DATE FEB. 1998
DR BY T T

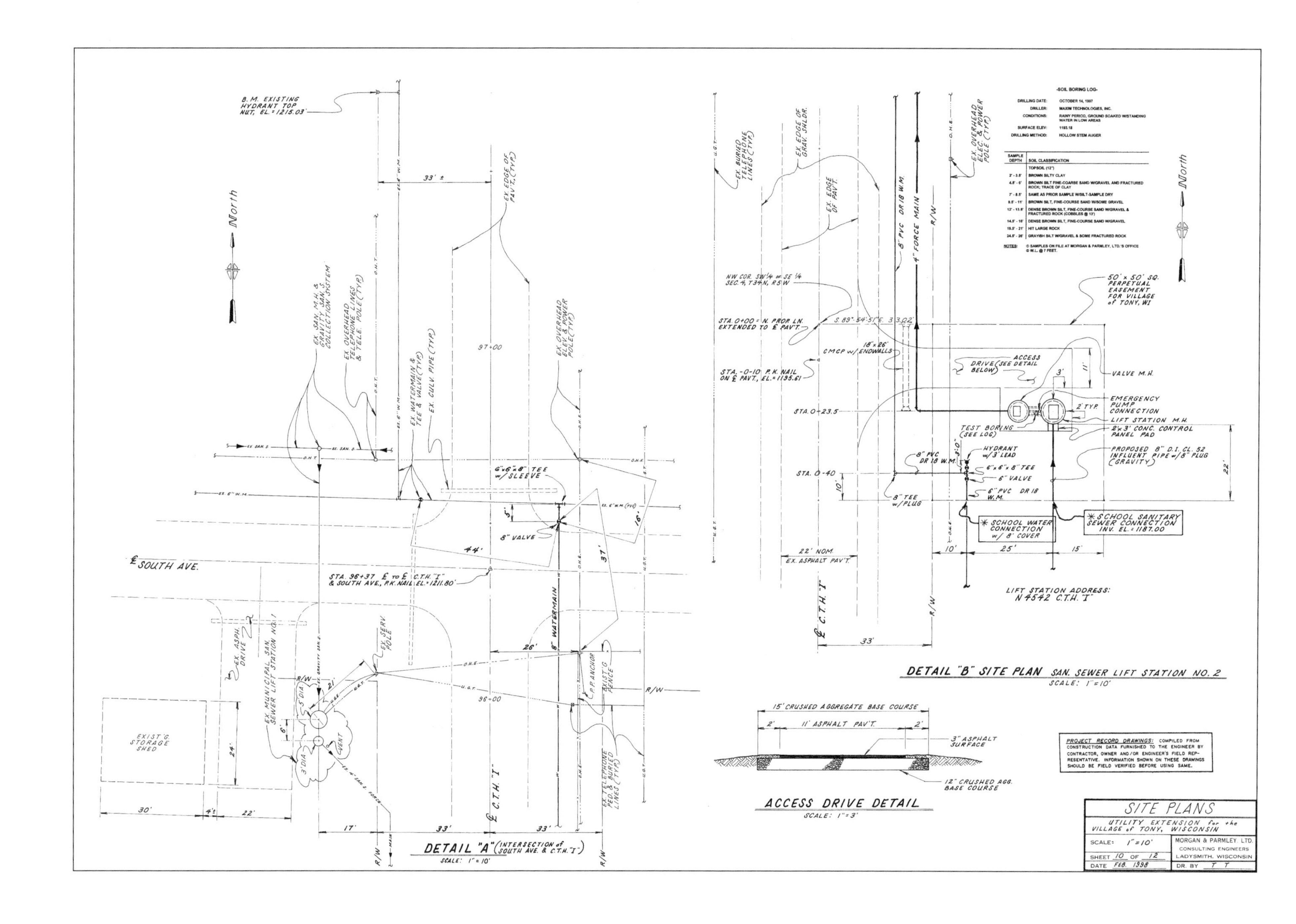
B.M. EXISTING HYDRANT TOP NUT, EL.=1215.03'
North
EX. SAN. M.H. & GRAVITY SAN. S. COLLECTION SYSTEM
EX. OVERHEAD TELEPHONE LINES & TELE. POLE (TYP.)
EX. WATERMAIN & TEE & VALVE (TYP.)
EX. CULV. PIPE (TYP.)
EX. EDGE OF PAV'T. (TYP.)
EX. OVERHEAD ELEC. & POWER POLE (TYP.)
33' ±
97+00
6"x6"x8" TEE w/ SLEEVE
8" VALVE
44'
16'
37'
℄ SOUTH AVE.
STA. 96+37 ℄ TO ℄ C.T.H."I" & SOUTH AVE., P.K. NAIL EL.=1211.80'
8" WATERMAIN
EX. ASPH. DRIVE
EX. MUNICIPAL SAN. SEWER LIFT STATION NO. 1
EX. SERV. POLE
26'
P.P. ANCHOR
EXIST'G FENCE
R/W
96+00
EXIST'G STORAGE SHED
5' DIA.
3' DIA.
VENT
EX. TELEPHONE PED. & BURIED LINES (TYP.)
℄ C.T.H. "I"
30'
4'±
22'
24'
17'
33'
33'
DETAIL "A" (INTERSECTION of SOUTH AVE. & C.T.H. "I")
SCALE: 1"=10'
EX. BURIED TELEPHONE LINES (TYP.)
EX. EDGE OF GRAV. SHLDR.
EX. EDGE OF PAV'T.
8" PVC DR 18 W.M.
4" FORCE MAIN
R/W
EX. OVERHEAD ELEC. & POWER POLE (TYP.)
-SOIL BORING LOG-
DRILLING DATE: OCTOBER 14, 1997
DRILLER: MAXIM TECHNOLOGIES, INC.
CONDITIONS: RAINY PERIOD, GROUND SOAKED W/STANDING WATER IN LOW AREAS
SURFACE ELEV: 1193.18
DRILLING METHOD: HOLLOW STEM AUGER
SAMPLE DEPTH | SOIL CLASSIFICATION
TOPSOIL (12")
2' - 3.5' | BROWN SILTY CLAY
4.5' - 6' | BROWN SILT FINE-COARSE SAND W/GRAVEL AND FRACTURED ROCK; TRACE OF CLAY
7' - 8.5' | SAME AS PRIOR SAMPLE W/SILT-SAMPLE DRY
9.5' - 11' | BROWN SILT, FINE-COURSE SAND W/SOME GRAVEL
12' - 13.5' | DENSE BROWN SILT, FINE-COURSE SAND W/GRAVEL & FRACTURED ROCK (COBBLES @ 13')
14.5' - 16' | DENSE BROWN SILT, FINE-COURSE SAND W/GRAVEL
19.5' - 21' | HIT LARGE ROCK
24.5' - 26' | GRAYISH SILT W/GRAVEL & SOME FRACTURED ROCK
NOTES: ① SAMPLES ON FILE AT MORGAN & PARMLEY, LTD.'S OFFICE ② W.L. @ 7 FEET.
North
50' x 50' SQ. PERPETUAL EASEMENT FOR VILLAGE of TONY, WI
NW COR. SW¼ of SE¼ SEC. 4, T34N, R5W
STA. 0+00 = N. PROP. LN. EXTENDED TO ℄ PAV'T.
S. 89°-54'-51" E. 33.02'
18"x26" CMCP w/ ENDWALLS
ACCESS DRIVE (SEE DETAIL BELOW)
STA. -0-10 P.K. NAIL ON ℄ PAV'T., EL.=1135.61
VALVE M.H.
STA. 0+23.5
EMERGENCY PUMP CONNECTION
LIFT STATION M.H.
2'x3' CONC. CONTROL PANEL PAD
TEST BORING (SEE LOG)
2' TYP.
HYDRANT w/ 3' LEAD
8" PVC DR 18 W.M.
6"x6"x8" TEE
6" VALVE
6" PVC DR 18 W.M.
STA. 0-40
8" TEE w/ PLUG
PROPOSED 8" D.I. CL. 52 INFLUENT PIPE w/ 8" PLUG (GRAVITY)
✱ SCHOOL WATER CONNECTION w/ 8' COVER
✱ SCHOOL SANITARY SEWER CONNECTION INV. EL.=1187.00
22'
10'
25'
15'
22' NOM.
EX. ASPHALT PAV'T.
℄ C.T.H. "I"
33'
LIFT STATION ADDRESS: N4542 C.T.H. "I"
DETAIL "B" SITE PLAN SAN. SEWER LIFT STATION NO. 2
SCALE: 1"=10'
15' CRUSHED AGGREGATE BASE COURSE
2'
11' ASPHALT PAV'T.
2'
3" ASPHALT SURFACE
12" CRUSHED AGG. BASE COURSE
ACCESS DRIVE DETAIL
SCALE: 1"=3'
PROJECT RECORD DRAWINGS: COMPILED FROM CONSTRUCTION DATA FURNISHED TO THE ENGINEER BY CONTRACTOR, OWNER AND/OR ENGINEER'S FIELD REPRESENTATIVE. INFORMATION SHOWN ON THESE DRAWINGS SHOULD BE FIELD VERIFIED BEFORE USING SAME.
SITE PLANS
UTILITY EXTENSION for the VILLAGE of TONY, WISCONSIN
SCALE: 1"=10'
SHEET 10 OF 12
DATE FEB. 1998
MORGAN & PARMLEY LTD.
CONSULTING ENGINEERS
LADYSMITH, WISCONSIN
DR. BY T T

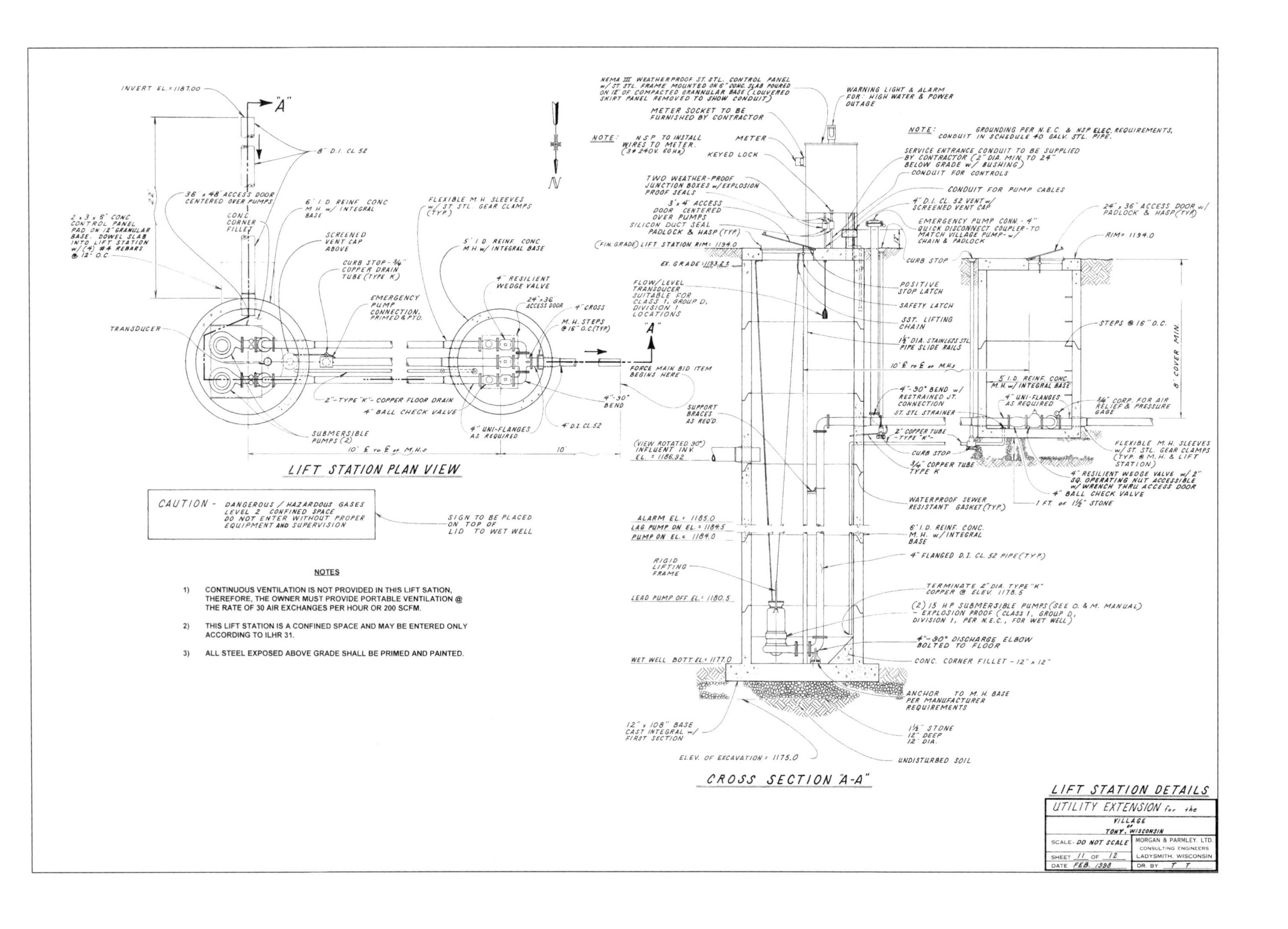
INVERT EL.=1187.00
"A"
8" D.I. CL 52
22
36" x 48" ACCESS DOOR CENTERED OVER PUMPS
2 x 3 x 6" CONC CONTROL PANEL PAD ON 12" GRANULAR BASE. DOWEL SLAB INTO LIFT STATION w/(4) #4 REBARS @ 12" O.C.
CONC CORNER FILLET
6' I.D. REINF CONC M.H. w/ INTEGRAL BASE
SCREENED VENT CAP ABOVE
FLEXIBLE M.H. SLEEVES w/ ST. STL. GEAR CLAMPS (TYP.)
5' I.D. REINF CONC M.H. w/ INTEGRAL BASE
CURB STOP - 3/4" COPPER DRAIN TUBE (TYPE K)
4" RESILIENT WEDGE VALVE
EMERGENCY PUMP CONNECTION, PRIMED & PTD.
24" x 36" ACCESS DOOR
4" CROSS
M.H. STEPS @ 16" O.C.(TYP)
N
TRANSDUCER
FORCE MAIN BID ITEM BEGINS HERE
2"-TYPE "K"- COPPER FLOOR DRAIN
4" BALL CHECK VALVE
4"-30° BEND
4" UNI-FLANGES AS REQUIRED
4" D.I. CL.52
SUBMERSIBLE PUMPS (2)
10' ℄ TO ℄ OF M.H.S
10'
LIFT STATION PLAN VIEW
CAUTION - DANGEROUS / HAZARDOUS GASES LEVEL 2 CONFINED SPACE DO NOT ENTER WITHOUT PROPER EQUIPMENT AND SUPERVISION
SIGN TO BE PLACED ON TOP OF LID TO WET WELL
NOTES
1) CONTINUOUS VENTILATION IS NOT PROVIDED IN THIS LIFT SATION, THEREFORE, THE OWNER MUST PROVIDE PORTABLE VENTILATION @ THE RATE OF 30 AIR EXCHANGES PER HOUR OR 200 SCFM.
2) THIS LIFT STATION IS A CONFINED SPACE AND MAY BE ENTERED ONLY ACCORDING TO ILHR 31.
3) ALL STEEL EXPOSED ABOVE GRADE SHALL BE PRIMED AND PAINTED.
NEMA III WEATHERPROOF ST. STL. CONTROL PANEL w/ ST. STL. FRAME MOUNTED ON 6" CONC. SLAB POURED ON 12" OF COMPACTED GRANNULAR BASE (LOUVERED SKIRT PANEL REMOVED TO SHOW CONDUIT)
METER SOCKET TO BE FURNISHED BY CONTRACTOR
NOTE: NSP TO INSTALL WIRES TO METER. (3# 240V. 60Hz)
METER
KEYED LOCK
WARNING LIGHT & ALARM FOR: HIGH WATER & POWER OUTAGE
NOTE: GROUNDING PER N.E.C. & NSP ELEC. REQUIREMENTS. CONDUIT IN SCHEDULE 40 GALV. STL. PIPE.
SERVICE ENTRANCE CONDUIT TO BE SUPPLIED BY CONTRACTOR (2" DIA. MIN. TO 24" BELOW GRADE w/ BUSHING)
CONDUIT FOR CONTROLS
CONDUIT FOR PUMP CABLES
TWO WEATHER-PROOF JUNCTION BOXES w/ EXPLOSION PROOF SEALS
3' x 4' ACCESS DOOR CENTERED OVER PUMPS
SILICON DUCT SEAL
PADLOCK & HASP (TYP.)
4" D.I. CL. 52 VENT w/ SCREENED VENT CAP
EMERGENCY PUMP CONN. - 4" QUICK DISCONNECT COUPLER-TO MATCH VILLAGE PUMP- w/ CHAIN & PADLOCK
24" x 36" ACCESS DOOR w/ PADLOCK & HASP (TYP)
RIM= 1194.0
(FIN. GRADE) LIFT STATION RIM= 1194.0
EX. GRADE = 1193.23
CURB STOP
POSITIVE STOP LATCH
SAFETY LATCH
SST. LIFTING CHAIN
1/2" DIA. STAINLESS STL PIPE SLIDE RAILS
FLOW/LEVEL TRANSDUCER SUITABLE FOR CLASS 1, GROUP D, DIVISION 1 LOCATIONS
STEPS @ 16" O.C.
8' COVER MIN.
10' ℄ TO ℄ OF M.H.S
5' I.D. REINF. CONC. M.H. w/ INTEGRAL BASE
4"-90° BEND w/ RESTRAINED JT. CONNECTION
4" UNI-FLANGES AS REQUIRED
3/4" CORP. FOR AIR RELIEF & PRESSURE GAGE
ST. STL. STRAINER
SUPPORT BRACES AS REQ'D.
2" COPPER TUBE -TYPE "K"-
CURB STOP
(VIEW ROTATED 90°) INFLUENT INV. EL. = 1186.92
3/4" COPPER TUBE TYPE K
FLEXIBLE M.H. SLEEVES w/ ST. STL. GEAR CLAMPS (TYP. @ M.H. & LIFT STATION)
4" RESILIENT WEDGE VALVE w/ 2" SQ. OPERATING NUT ACCESSIBLE w/ WRENCH THRU ACCESS DOOR
4" BALL CHECK VALVE
WATERPROOF SEWER RESISTANT GASKET (TYP.)
1 FT. OF 1 1/2" STONE
ALARM EL.= 1185.0
LAG PUMP ON EL.= 1184.5
PUMP ON EL.= 1184.0
6' I.D. REINF. CONC. M.H. w/ INTEGRAL BASE
4" FLANGED D.I. CL. 52 PIPE (TYP.)
RIGID LIFTING FRAME
LEAD PUMP OFF EL.= 1180.5
TERMINATE 2" DIA. TYPE "K" COPPER @ ELEV. 1178.5
(2) 15 HP SUBMERSIBLE PUMPS (SEE O. & M. MANUAL) - EXPLOSION PROOF (CLASS 1, GROUP D, DIVISION 1, PER N.E.C., FOR WET WELL)
4"-90° DISCHARGE ELBOW BOLTED TO FLOOR
WET WELL BOTT. EL.= 1177.0
CONC. CORNER FILLET - 12" x 12"
ANCHOR TO M.H. BASE PER MANUFACTURER REQUIREMENTS
12" x 108" BASE CAST INTEGRAL w/ FIRST SECTION
1 1/2" STONE 12" DEEP 12' DIA.
ELEV. OF EXCAVATION = 1175.0
UNDISTURBED SOIL
CROSS SECTION "A-A"
LIFT STATION DETAILS
UTILITY EXTENSION for the VILLAGE of TONY, WISCONSIN
SCALE: DO NOT SCALE
MORGAN & PARMLEY, LTD. CONSULTING ENGINEERS LADYSMITH, WISCONSIN
SHEET 11 OF 12
DATE FEB. 1998
DR. BY T T

SANITARY SEWER & WATER MAIN EXTENSIONS
(W. 11th SREET. N., EASEMENTS, BAKER AVENUE & W. 13th STREET. N.)

CITY
of
LADYSMITH, WISCONSIN
FEBRUARY, 2000

PROJECT LOCATION

GENERAL LOCATION MAP

UTILITIES

WATER & SEWER:
LADYSMITH DEPARTMENT OF PUBLIC WORKS
ATTN: BILL CHRISTIANSON, DIRECTOR
120 WEST MINER AVENUE
P.O. BOX 431
LADYSMITH, WISCONSIN 54848
TELEPHONE: (715) 532-2601

ELECTRIC:
NORTHERN STATES POWER COMPANY
310 HICKORY HILL LANE
PHILLIPS, WISOCNSIN 54555
ATTN: JOE PERKINS
TELEPHONE: (715) 836-1198

NATURAL GAS:
WISCONSIN GAS COMPANY
104 WEST SOUTH STREET
RICE LAKE, WISCONSIN 54868
ATTN: DON WEDIN
TELEPHONE: (715) 236-2104

CABLE TELEVISION:
CHARTER COMMUNICATIONS
1725 S. MAIN ST.
RICE LAKE, WISCONSIN 54868
TELEPHONE: (800) 262-2578

TELEPHONE:
CENTURYTEL
425 ELLINGSON AVENUE
P.O. BOX 78
HAWKINS, WISCONSIN 54530
TELEPHONE: (715) 585-7707
ATTN: JAMES ARQUETTE

DIGGER'S HOTLINE:
1-800-242-8511

GENERAL NOTES

1. ALL WATER MAIN SHALL BE 8" DI CL 52.
2. ALL WATERMAIN FITTINGS SHALL BE M.J.D.I., W / RESTRAINED JOINT LUGS.
3. ALL HYDRANT LEADS SHALL BE 6-INCH DI CL 52.
4. ALL HYDRANT LEADS SHALL HAVE A 6-INCH VALVE.
5. HYDRANTS SHALL BE WATEROUS WB 67-250 (4" PUMPER NOZZLE), 22" BREAKOFF WITH THREADS MATCHING CITY PATTERN.
6. ALL WATER MAIN SHALL HAVE AN 8 FT. BURY (MIN.).
7. WATER SERVICES SHALL BE 1" TYPE K COPPER W / MUELLER OR FORD BRASS AND LOCATED BY OWNER.
8. SANITARY SERVICES SHALL BE 4" PVC SCH. 40 W / WYES LAID TO PROVIDE FOR MAXIMUM DEPTH OR 10' @ R / W, LOCATION BY OWNER.
9. SEWER MAIN SHALL BE 8" PVC SDR 35.
10. MANHOLES SHALL BE SET WITH A MINIMUM OF 6" OF RINGS W / 9" CASTINGS.
11. MANHOLES SHALL BE 4' DIA. PRECASE CONCRETE W / INTEGRAL BASES AND BOOTS.
12. 8' (MINIMUM) HORIZONTAL DISTANCE (℄ TO ℄) BETWEEN WATERMAIN AND SANITARY SEWER OR STORM SEWER.
13. 18" (MINIMUM) VERTICAL DISTANCE (OUT TO OUT) BETWEEN WATERMAIN UNDER SANITARY SEWER OR STORM SEWER @ THEIR INTERSECTION.
14. 6" (MINIMUM) VERTICAL DISTANCE (OUT TO OUT) BETWEEN WATERMAIN OVER SANITARY SEWER OR STORM SEWER @ THEIR INTERSECTION.
15. ALL ELEVATIONS BASED ON CITY DATUM FROM B.M. SHOWN.
16. EROSION CONTROL MEASURES SHALL BE IN PLACE BEFORE CONSTRUCTION BEGINS.
17. THE CONTRACTORS SHALL MINIMIZE EROSION DURING CONSTRUCTION USING GOOD CONSTRUCTION TECHNIQUES AND UTILIZING SILT FENCES AND BALE DITCH CHECKS. THE CONTRACTOR SHALL RELEASE RUNOFF FROM THE SITE IN A NUISANCE FREE MANNER.
18. THE CONTRACTOR IS RESPONSIBLE TO MAINTAIN THE SITE IN A SAFE CONDITION. THE SOLE RESPONSIBILITY FOR WARNING SIGNS, BARRICADES AND ALL ASPECTS OF SAFETY LIE WITH THE CONTRACTOR.
19. EXISTING UTILITIES ARE SHOWN BUT MAY NOT BE ALL INCLUSIVE. CONTRACTOR SHALL HAVE ALL BURIED UTILITIES LOCATED PRIOR TO COMMENCING CONSTRUCTION.
20. EXISTING SEWER AND WATERMAIN ARE SHOWN ACCORDING TO EXISTING RECORDS BUT LOCATION SHALL BE VERIFIED IN FIELD.
21. WATERMAIN & SERVICES SHALL BE INSULATED AT ALL STORM SEWER, CULVERT AND DITCH CROSSINGS WITH 2" THICK EXTRUDED POLYSTYRENE INSULATION BOARD.
22. PROPERTY CORNERS KNOWN TO EXIST SHALL NOT BE DISTURBED. DAMAGED CORNERS WILL BE REPLACED AT THE CONTRACTOR'S EXPENSE. PRESERVE ALL SURVEY MONUMENTS.
23. ALL WATERMAIN SHALL BE DISINFECTED BEFORE BEING PUT INTO USE.
24. VALVES AND MANHOLES SHALL BE SET TO GRADE SHOWN ON PLANS, UNLESS DIRECTED OTHERWISE.
25. CONSTRUCTION SHALL COMPLY WITH THE CONDITIONS OF DNR APPROVAL LETTER, AND OTHER REGULATORY APPROVALS AS LISTED IN THE SPECIFICATION BOOK.
26. ALL DISTURBED AREAS SHALL BE SEEDED, FERTILIZED AND MULCHED, OR COVERED W / BASE COURSE AFTER CONSTRUCTION, OR RESTORED WITH BITUMINOUS PAVEMENT WHERE APPLICABLE.
27. ALL DISTURBED ARES IN EASEMENT, ALLEY & ROAD DITCH SHALL BE SEEDED, FERTILIZED AND MULCHED.
28. ALL DISTURBED AREAS SHALL BE RESTORED W / SEEDING AS FOLLOWS:

    A) REPLACED ALL SALVAGED TOPSOIL
    B) SEED MIXUTRE: PERMANENT SEEDING – (LAWN TYPE TURF)

        35% KENTUCKY BLUEGRASS
        25% IMPROVED FINE PERENNIAL RYEGRASS
        15% CREEEPING RED FESCUE
        15% IMPROVED HARD FESCUE
        10% WHITE CLOVER

    C) SEED RATE: 2# / 1000 SQ. FT.
    D) MULCH RATE: 3 TON / ACRE
    E) PERMANENT SEEDING: ALLOWWED TO SEPT. 7
    TEMPORARY SEEDING: SEPT. 8 THROUGH Nov. 10
    (4 BU. OATS / ACRE)
    DORMANT SEEDING; SEED PERM. SEEDING INTO TEMP. SEEDING AFTER NOV. 10
    F) SEEDING MAY BE BROADCAST BUT MUST BE COVERED WITH A MAXIMUM OF ½" SOIL. MULCH MUST BE ANCHORED WITH A MULCH TILLER OR BLOWN W / 75-100 GAL. OF ASPHALT / TON OF MULCH.
    G) FERTILIZER: 500# / ACRE 20-10-10
29. CONTRACTOR SHALL DETERMINE CONSTRUCTION SEQUENCE IN ORDER TO MINIMIZE INSTALLATION CONFLICTS WITH PIPE CROSSINGS.
30. THERE SHALL BE NO DEVIATION FROM THE PLANS WITHOUT THE DESIGN ENGINEER'S APPROVAL.
31. CONSTRUCTION STAKES DESTROYED BY THE CONTRACTOR THAT NEED TO BE REPLACED, WILL BE CHARGED TO THE CONTRACTOR.
32. CONTRACTOR SHALL COORDINATOR WORK TO AVOID UNDUE CONFLICT WITH ADJACENT PROPERTY.
33. NO TREES ARE TO BE REMOVED WITHOUT THE APPROVAL OF THE ENGINEER.
34. THE CONTRACTOR MUST REMOVE AND REPLACE ALL EFFECTED STREET SIGNS AFTER PROPERLY SIGNING AND BARRICADING THE CONSTRUCTION AREA. ALL SIGNS MUST BE REPLACED PRIOR TO OPENING THE STREET TO TRAFFIC.
35. CONFLICTING POWER POLES WILL BE MOVED BY NSP.
36. SEE SPECIFICATION BOOK, SECTION 01010 AND FOLLOWING PLAN SHEETS FOR ADDITIONAL DETAILS.

GENERAL SPECIFICATIONS: STANDARD SPECIFICATIONS FOR SEWER AND WATER CONSTRUCTION IN WISCONSIN (5TH EDITION)

WISCONSIN CONSTRUCTION SITE BEST MANAGEMENT PRACTICE HANDBOOK

SPECIFIC SPECIFICATIONS: MORGAN & PARMLEY, LTD. SPEC. BOOK

BIDDING DOCUMENTS: MORGAN & PARMLEY, LTD. SPEC. BOOK

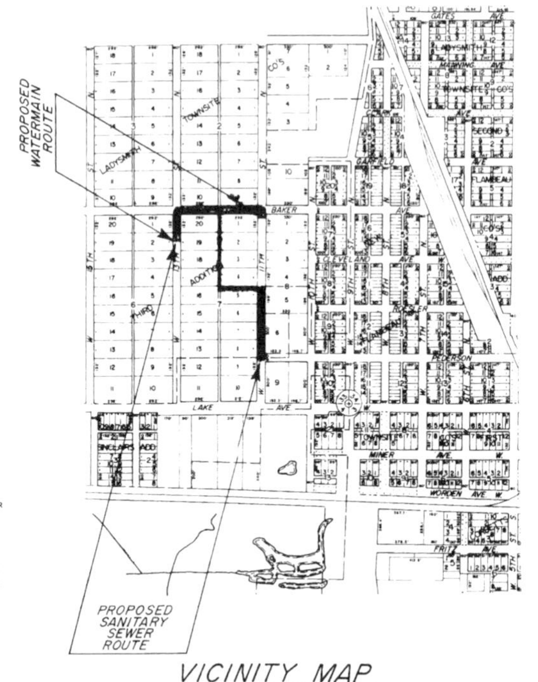

DRAWING INDEX

SHEET NO.	DESCRIPTION
1	TITLE SHEET, LOCATION MAPS, INDEX & NOTES
2	PLAN & PROFILE: SANITARY SEWER
3	PLAN & PROFILE: SANITARY SEWER & WATERMAIN
4	PLAN & PROFILE: SANITARY SEWER & WATERMAIN
5	TYPICAL EROSION CONTROL DETAILS This Sheet Not Shown

PREPARED BY:
MORGAN & PARMLEY, LTD.
PROFESSIONAL CONSULTING ENGINEERS
LADYSMITH, WISCONSIN

Courtesy: Morgan & Parmley, Ltd.

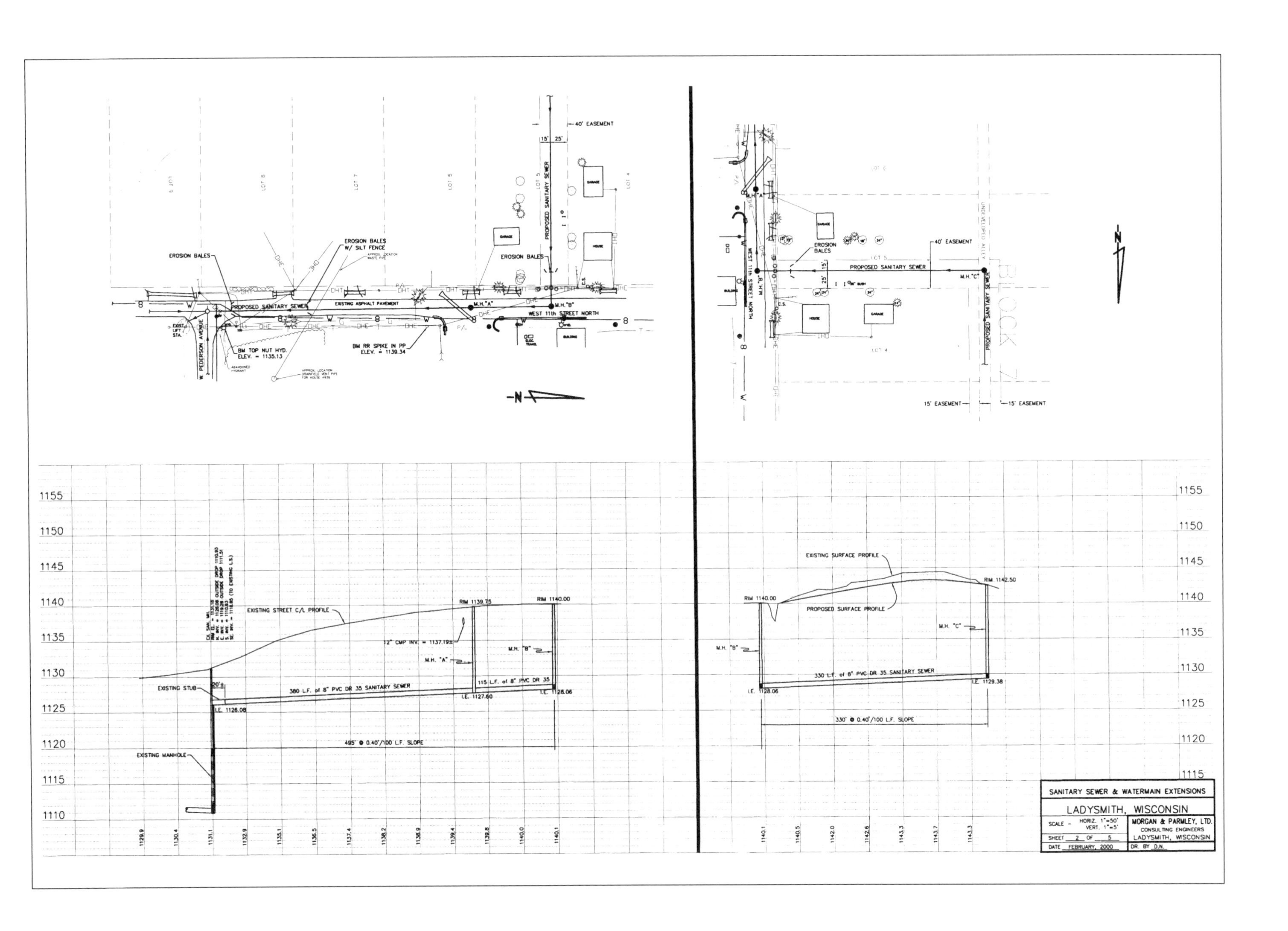
40' EASEMENT
15'
25'
PROPOSED SANITARY SEWER
EROSION BALES
EROSION BALES W/ SILT FENCE
EXISTING ASPHALT PAVEMENT
M.H. "A"
M.H. "B"
WEST 11th STREET NORTH
BM RR SPIKE IN PP ELEV. = 1139.34
BM TOP NUT HYD. ELEV. = 1135.13
W. PEDERSON AVENUE
GARAGE
HOUSE
M.H. "C"
WEST 11th STREET NORTH
15' EASEMENT
15' EASEMENT
1155
1150
1145
1140
1135
1130
1125
1120
1115
1110
EXISTING STREET C/L PROFILE
RIM 1139.75
RIM 1140.00
12" CMP INV. = 1137.19±
M.H. "A"
M.H. "B"
EXISTING STUB
20'±
380 L.F. of 8" PVC DR 35 SANITARY SEWER
115 L.F. of 8" PVC DR 35
I.E. 1126.06
I.E. 1127.60
I.E. 1128.06
495' @ 0.40'/100 L.F. SLOPE
EXISTING MANHOLE
1129.9
1130.4
1131.1
1132.9
1135.1
1136.5
1137.4
1138.2
1138.9
1139.4
1139.8
1140.0
1140.1
EXISTING SURFACE PROFILE
PROPOSED SURFACE PROFILE
RIM 1140.00
RIM 1142.50
M.H. "B"
M.H. "C"
330 L.F. of 8" PVC DR 35 SANITARY SEWER
I.E. 1128.06
I.E. 1129.38
330' @ 0.40'/100 L.F. SLOPE
1140.1
1140.5
1142.0
1142.6
1143.3
1143.7
1143.3
SANITARY SEWER & WATERMAIN EXTENSIONS
LADYSMITH, WISCONSIN
SCALE – HORIZ. 1"=50' VERT. 1"=5'
SHEET 2 OF 5
DATE FEBRUARY, 2000
MORGAN & PARMLEY, LTD.
CONSULTING ENGINEERS
LADYSMITH, WISCONSIN
DR. BY D.N.

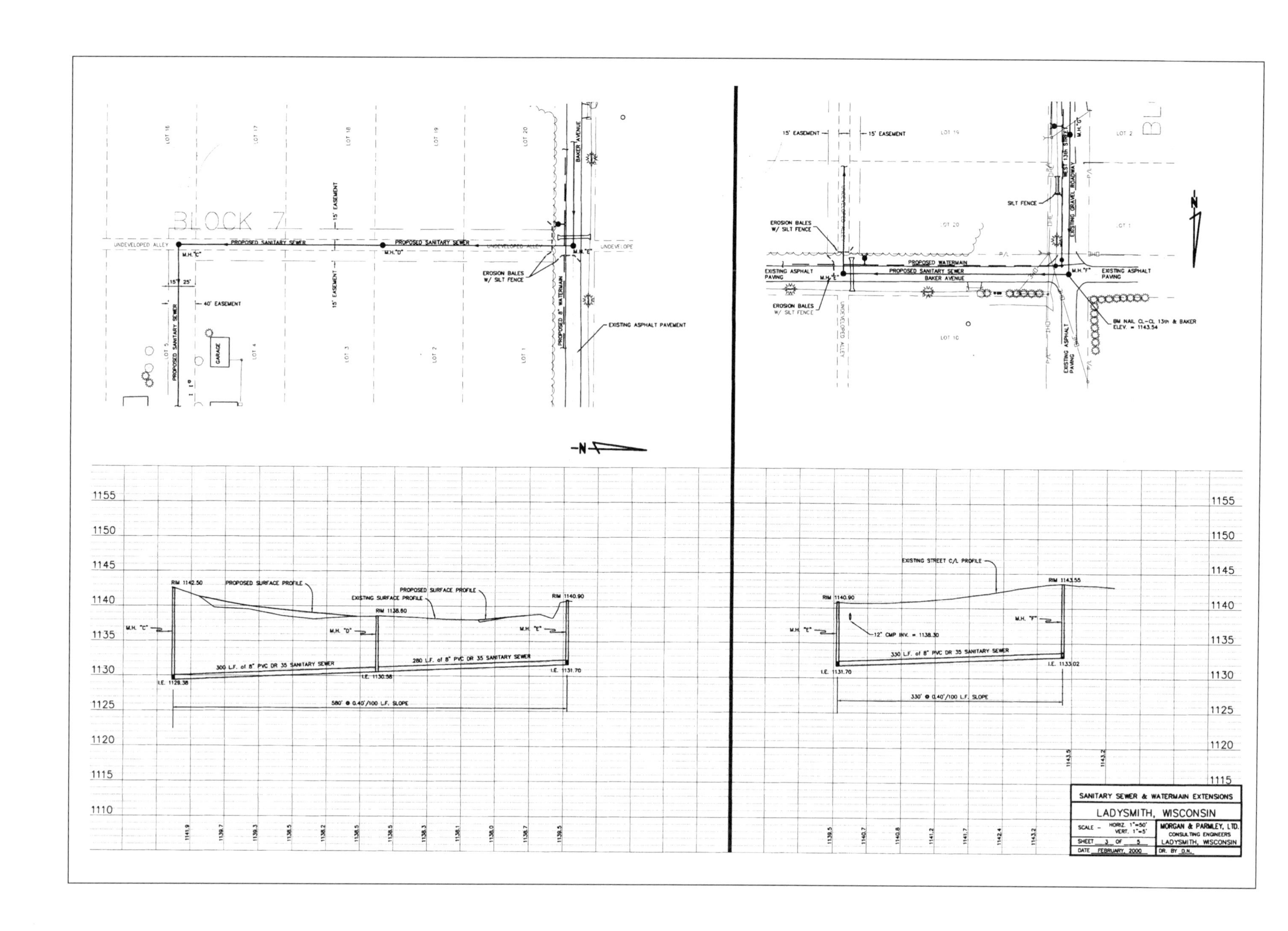
BLOCK 7
UNDEVELOPED ALLEY
PROPOSED SANITARY SEWER
M.H. "C"
M.H. "D"
M.H. "E"
15' EASEMENT
40' EASEMENT
GARAGE
BAKER AVENUE
EROSION BALES W/ SILT FENCE
PROPOSED 8" WATERMAIN
EXISTING ASPHALT PAVEMENT
LOT 16
LOT 17
LOT 18
LOT 19
LOT 20
LOT 5
LOT 4
LOT 3
LOT 2
LOT 1
15' EASEMENT
SILT FENCE
WEST 13th STREET
EXISTING GRAVEL ROADWAY
PROPOSED WATERMAIN
PROPOSED SANITARY SEWER
EXISTING ASPHALT PAVING
M.H. "F"
BM NAIL CL-CL 13th & BAKER ELEV. = 1143.54
UNDEVELOPED ALLEY
LOT 10
RIM 1142.50
PROPOSED SURFACE PROFILE
EXISTING SURFACE PROFILE
RIM 1138.60
RIM 1140.90
300 L.F. of 8" PVC DR 35 SANITARY SEWER
280 L.F. of 8" PVC DR 35 SANITARY SEWER
I.E. 1129.38
I.E. 1130.58
I.E. 1131.70
580' @ 0.40'/100 L.F. SLOPE
EXISTING STREET C/L PROFILE
RIM 1143.55
12" CMP INV. = 1138.30
330 L.F. of 8" PVC DR 35 SANITARY SEWER
I.E. 1133.02
330' @ 0.40'/100 L.F. SLOPE
1155
1150
1145
1140
1135
1130
1125
1120
1115
1110
1141.9
1139.7
1139.3
1138.5
1138.2
1138.5
1138.5
1138.3
1138.1
1138.0
1138.7
1139.5
1139.5
1140.7
1140.8
1141.2
1141.7
1142.4
1143.2
1143.5
1143.2
SANITARY SEWER & WATERMAIN EXTENSIONS
LADYSMITH, WISCONSIN
SCALE - HORIZ. 1"=50' VERT. 1"=5'
MORGAN & PARMLEY, LTD.
CONSULTING ENGINEERS
LADYSMITH, WISCONSIN
SHEET 3 OF 5
DATE FEBRUARY, 2000
DR. BY D.N.

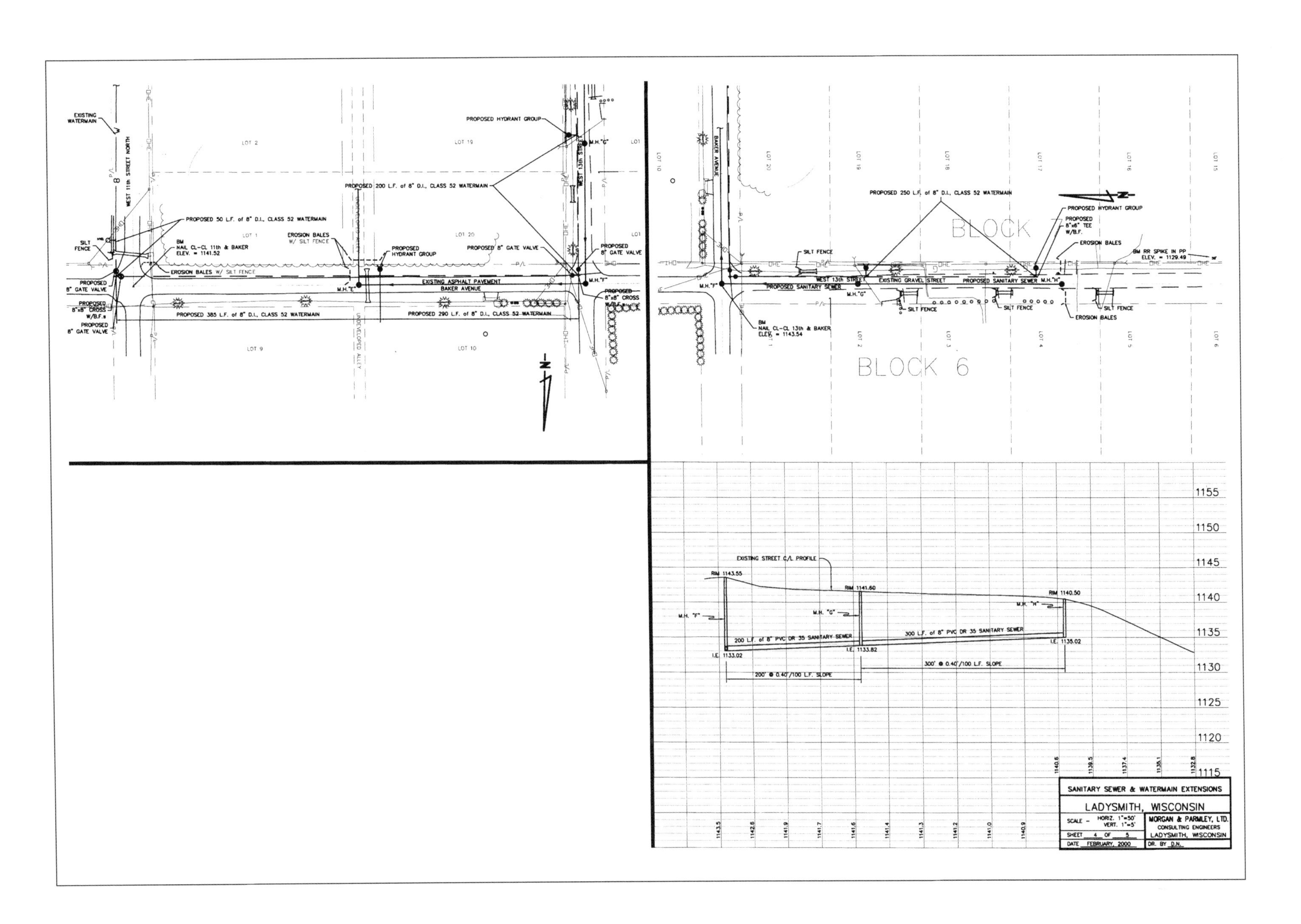
EXISTING WATERMAIN
WEST 11th STREET NORTH
LOT 2
LOT 19
LOT 1
LOT 20
LOT 9
LOT 10
PROPOSED HYDRANT GROUP
M.H."G"
WEST 13th STREET
PROPOSED 200 L.F. of 8" D.I., CLASS 52 WATERMAIN
PROPOSED 50 L.F. of 8" D.I., CLASS 52 WATERMAIN
SILT FENCE
BM NAIL CL-CL 11th & BAKER ELEV. = 1141.52
EROSION BALES W/ SILT FENCE
PROPOSED HYDRANT GROUP
PROPOSED 8" GATE VALVE
PROPOSED 8" GATE VALVE
M.H."F"
PROPOSED 8"x8" CROSS
EXISTING ASPHALT PAVEMENT
BAKER AVENUE
M.H."E"
PROPOSED 8" GATE VALVE
PROPOSED 8"x8" CROSS W/B.F.s
PROPOSED 8" GATE VALVE
PROPOSED 385 L.F. of 8" D.I., CLASS 52 WATERMAIN
PROPOSED 290 L.F. of 8" D.I., CLASS 52 WATERMAIN
UNDEVELOPED ALLEY
BAKER AVENUE
PROPOSED 250 L.F. of 8" D.I., CLASS 52 WATERMAIN
PROPOSED HYDRANT GROUP
PROPOSED 8"x6" TEE W/B.F.
BLOCK 7
EROSION BALES
BM RR SPIKE IN PP ELEV. = 1129.49
SILT FENCE
WEST 13th STREET
EXISTING GRAVEL STREET
PROPOSED SANITARY SEWER
M.H."H"
M.H."F"
PROPOSED SANITARY SEWER
M.H."G"
SILT FENCE
SILT FENCE
SILT FENCE
EROSION BALES
BM NAIL CL-CL 13th & BAKER ELEV. = 1143.54
BLOCK 6
EXISTING STREET C/L PROFILE
RIM 1143.55
RIM 1141.60
RIM 1140.50
M.H. "F"
M.H. "G"
M.H. "H"
200 L.F. of 8" PVC DR 35 SANITARY SEWER
300 L.F. of 8" PVC DR 35 SANITARY SEWER
I.E. 1133.02
I.E. 1133.82
I.E. 1135.02
200' @ 0.40'/100 L.F. SLOPE
300' @ 0.40'/100 L.F. SLOPE
1155
1150
1145
1140
1135
1130
1125
1120
1115
1143.5
1142.6
1141.9
1141.7
1141.6
1141.4
1141.3
1141.2
1141.0
1140.9
1140.6
1139.5
1137.4
1135.1
1132.8
SANITARY SEWER & WATERMAIN EXTENSIONS
LADYSMITH, WISCONSIN
SCALE – HORIZ. 1"=50' VERT. 1"=5'
MORGAN & PARMLEY, LTD. CONSULTING ENGINEERS LADYSMITH, WISCONSIN
SHEET 4 OF 5
DATE FEBRUARY, 2000
DR. BY D.N.

Design

STORM SEWER SYSTEMS

Section 12

All storm water, runoff, snow-melt and related precipitation that falls on municipal areas, roadways, parking lots and similar improved sites must be properly dealt with or serious problems may result.

Design of storm sewer systems can be very complex and incorporate additional components in the total design, such as replacement or sewer and water utility piping.

This section contains several plans for small storm sewer systems. The presentation commences with a one page preliminary layout for an existing elementary school site. The second plan is a parking lot for a new commercial building complex. The third plan details the design for a small development in a rural setting. The next two sets of plans are for reconstruction of municipal streets to upgrade their storm water drainage. The final set of plans detail a basic storm sewer extension.

TABLE OF CONTENTS

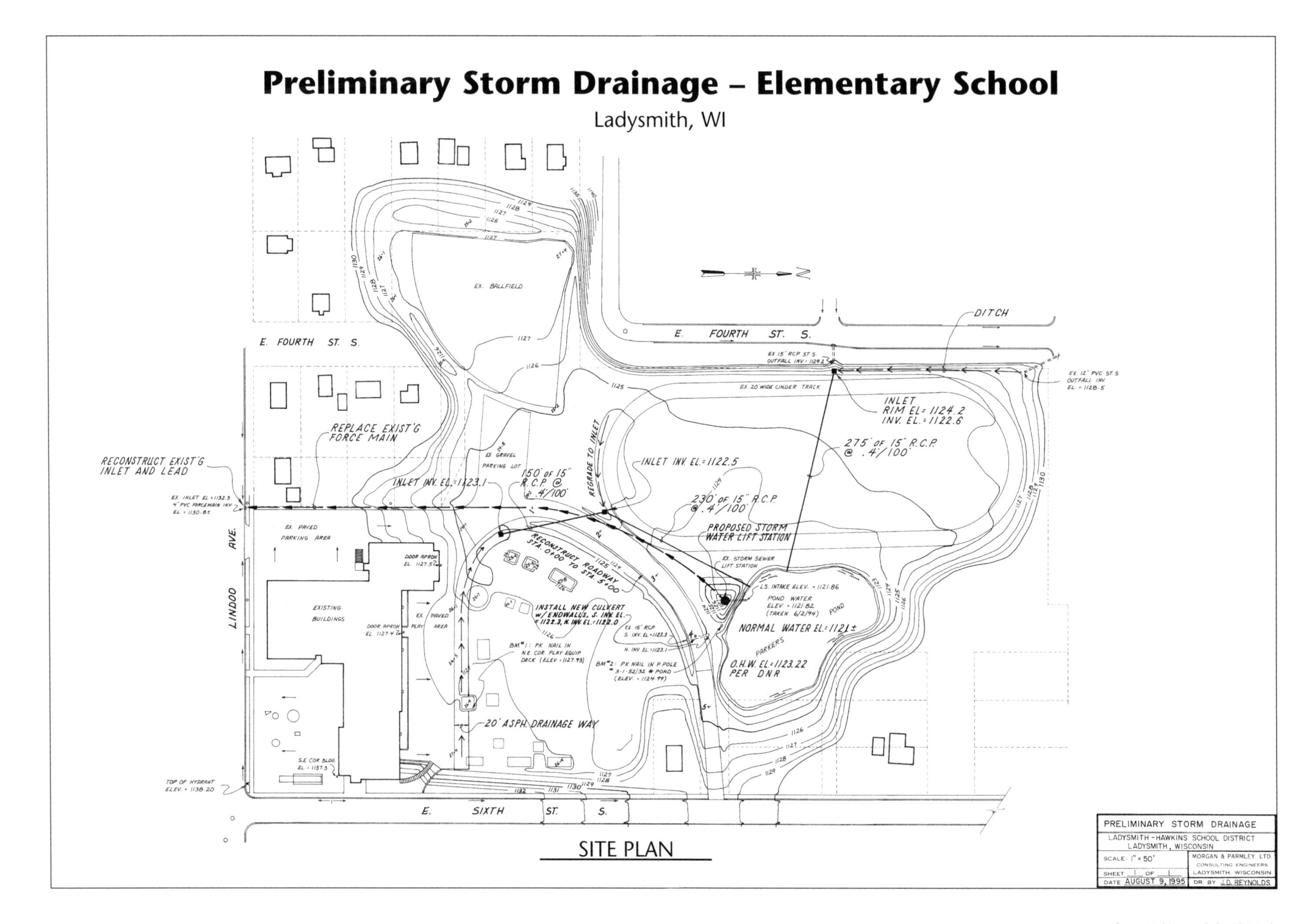

Courtesy: Morgan & Parmley, Ltd.

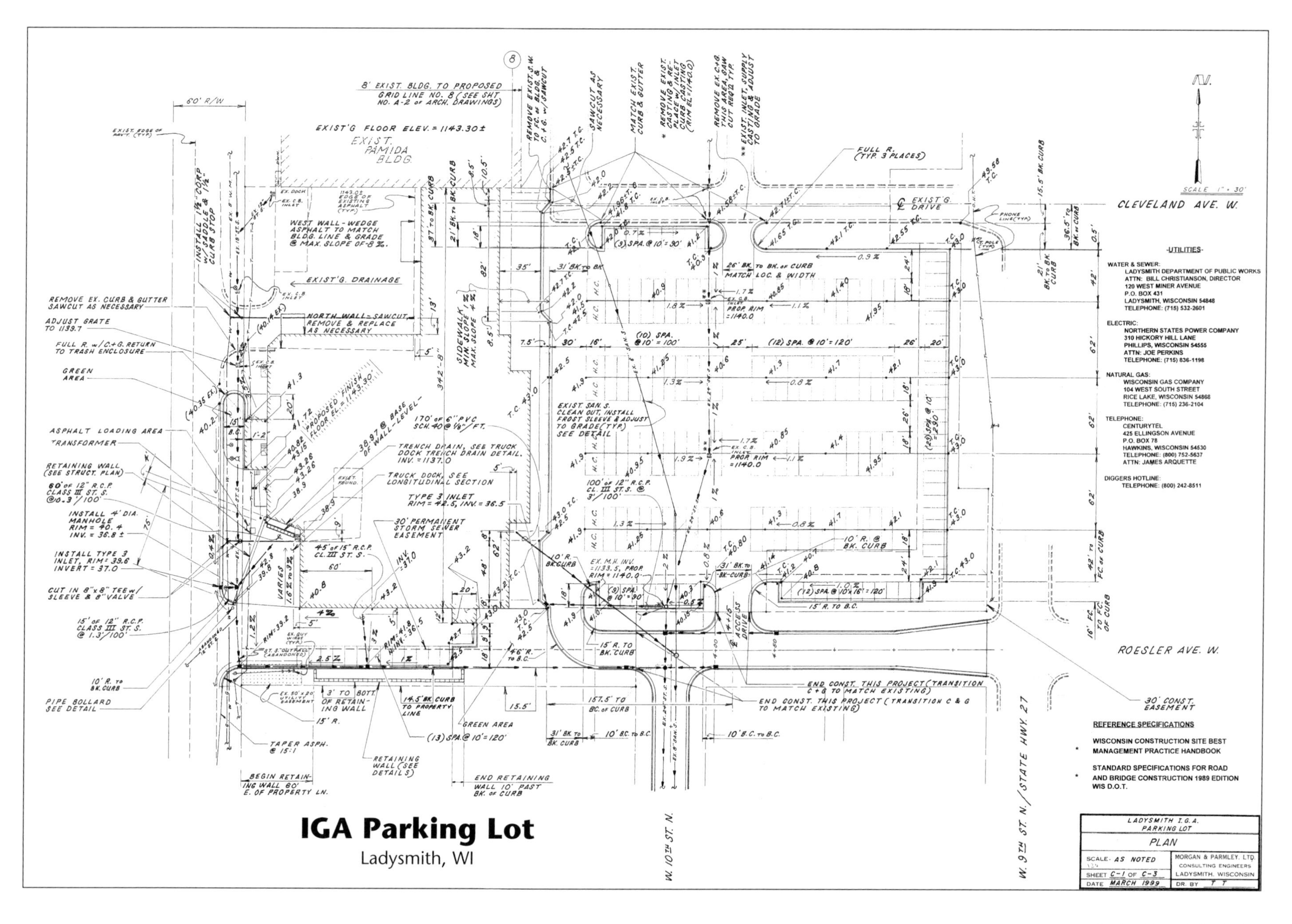

IGA Parking Lot

Ladysmith, WI

Courtesy: Morgan & Parmley, Ltd.

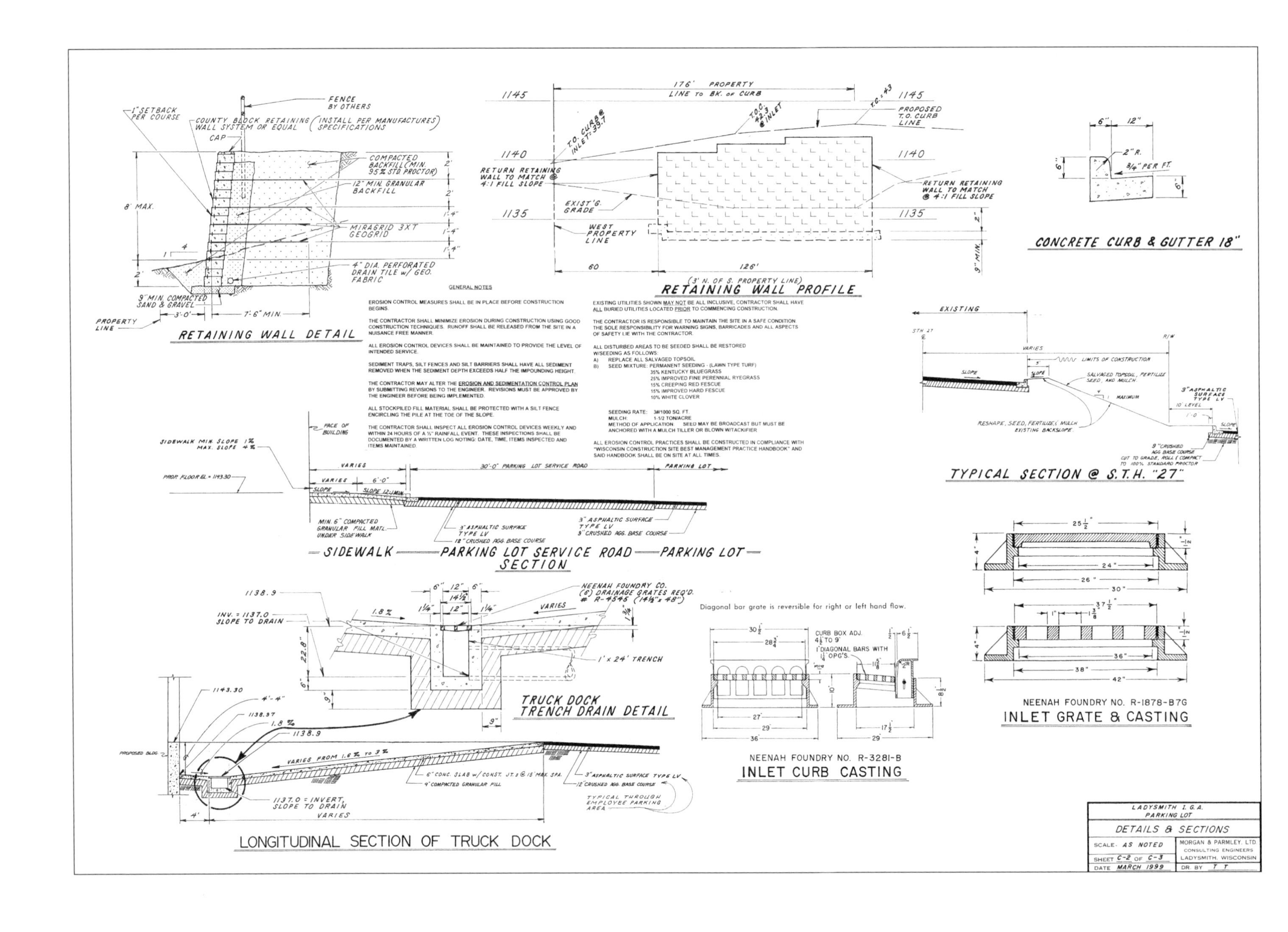

FENCE BY OTHERS
1" SETBACK PER COURSE
COUNTY BLOCK RETAINING WALL SYSTEM OR EQUAL (INSTALL PER MANUFACTURES SPECIFICATIONS)
CAP
COMPACTED BACKFILL (MIN. 95% STD PROCTOR)
12" MIN. GRANULAR BACKFILL
8' MAX.
MIRAGRID 3XT GEOGRID
4" DIA. PERFORATED DRAIN TILE w/ GEO. FABRIC
9" MIN. COMPACTED SAND & GRAVEL
PROPERTY LINE
3'-0"
7'-6" MIN.
RETAINING WALL DETAIL
176' PROPERTY LINE TO BK. OF CURB
PROPOSED T.O. CURB LINE
RETURN RETAINING WALL TO MATCH @ 4:1 FILL SLOPE
EXIST'G. GRADE
WEST PROPERTY LINE
60
126'
(3' N. OF S. PROPERTY LINE)
RETAINING WALL PROFILE
CONCRETE CURB & GUTTER 18"
GENERAL NOTES
EROSION CONTROL MEASURES SHALL BE IN PLACE BEFORE CONSTRUCTION BEGINS.
THE CONTRACTOR SHALL MINIMIZE EROSION DURING CONSTRUCTION USING GOOD CONSTRUCTION TECHNIQUES. RUNOFF SHALL BE RELEASED FROM THE SITE IN A NUISANCE FREE MANNER.
ALL EROSION CONTROL DEVICES SHALL BE MAINTAINED TO PROVIDE THE LEVEL OF INTENDED SERVICE.
SEDIMENT TRAPS, SILT FENCES AND SILT BARRIERS SHALL HAVE ALL SEDIMENT REMOVED WHEN THE SEDIMENT DEPTH EXCEEDS HALF THE IMPOUNDING HEIGHT.
THE CONTRACTOR MAY ALTER THE EROSION AND SEDIMENTATION CONTROL PLAN BY SUBMITTING REVISIONS TO THE ENGINEER. REVISIONS MUST BE APPROVED BY THE ENGINEER BEFORE BEING IMPLEMENTED.
ALL STOCKPILED FILL MATERIAL SHALL BE PROTECTED WITH A SILT FENCE ENCIRCLING THE PILE AT THE TOE OF THE SLOPE.
THE CONTRACTOR SHALL INSPECT ALL EROSION CONTROL DEVICES WEEKLY AND WITHIN 24 HOURS OF A ½" RAINFALL EVENT. THESE INSPECTIONS SHALL BE DOCUMENTED BY A WRITTEN LOG NOTING: DATE, TIME, ITEMS INSPECTED AND ITEMS MAINTAINED.
EXISTING UTILITIES SHOWN MAY NOT BE ALL INCLUSIVE, CONTRACTOR SHALL HAVE ALL BURIED UTILITIES LOCATED PRIOR TO COMMENCING CONSTRUCTION.
THE CONTRACTOR IS RESPONSIBLE TO MAINTAIN THE SITE IN A SAFE CONDITION. THE SOLE RESPONSIBILITY FOR WARNING SIGNS, BARRICADES AND ALL ASPECTS OF SAFETY LIE WITH THE CONTRACTOR.
ALL DISTURBED AREAS TO BE SEEDED SHALL BE RESTORED W/SEEDING AS FOLLOWS:
A) REPLACE ALL SALVAGED TOPSOIL
B) SEED MIXTURE: PERMANENT SEEDING - (LAWN TYPE TURF)
35% KENTUCKY BLUEGRASS
25% IMPROVED FINE PERENNIAL RYEGRASS
15% CREEPING RED FESCUE
15% IMPROVED HARD FESCUE
10% WHITE CLOVER
SEEDING RATE: 3#/1000 SQ. FT.
MULCH: 1-1/2 TON/ACRE
METHOD OF APPLICATION: SEED MAY BE BROADCAST BUT MUST BE ANCHORED WITH A MULCH TILLER OR BLOWN W/TACKIFIER
ALL EROSION CONTROL PRACTICES SHALL BE CONSTRUCTED IN COMPLIANCE WITH "WISCONSIN CONSTRUCTION SITE BEST MANAGEMENT PRACTICE HANDBOOK" AND SAID HANDBOOK SHALL BE ON SITE AT ALL TIMES.
EXISTING
LIMITS OF CONSTRUCTION
SALVAGED TOPSOIL, FERTILIZE SEED, AND MULCH
3" ASPHALTIC SURFACE TYPE LV
RESHAPE, SEED, FERTILIZE, MULCH EXISTING BACKSLOPE
9" CRUSHED AGG. BASE COURSE
CUT TO GRADE, ROLL & COMPACT TO 100% STANDARD PROCTOR
TYPICAL SECTION @ S.T.H. "27"
SIDEWALK MIN. SLOPE 1% MAX. SLOPE 4%
FACE OF BUILDING
PROP. FLOOR EL. = 1143.30
VARIES
6'-0"
30'-0" PARKING LOT SERVICE ROAD
PARKING LOT
SLOPE 12:1 MIN.
MIN. 6" COMPACTED GRANULAR FILL MATL. UNDER SIDEWALK
3" ASPHALTIC SURFACE TYPE LV
12" CRUSHED AGG. BASE COURSE
3" ASPHALTIC SURFACE TYPE LV
8" CRUSHED AGG. BASE COURSE
SIDEWALK — PARKING LOT SERVICE ROAD — PARKING LOT SECTION
NEENAH FOUNDRY CO. (6) DRAINAGE GRATES REQ'D. # R-4545 (14½" x 48")
INV. = 1137.0 SLOPE TO DRAIN
1' x 24' TRENCH
TRUCK DOCK TRENCH DRAIN DETAIL
Diagonal bar grate is reversible for right or left hand flow.
CURB BOX ADJ. 4½ TO 9
1" DIAGONAL BARS WITH 1¼" OPG'S.
NEENAH FOUNDRY NO. R-3281-B
INLET CURB CASTING
NEENAH FOUNDRY NO. R-1878-B7G
INLET GRATE & CASTING
PROPOSED BLDG.
VARIES FROM 1.6% TO 3%
6" CONC. SLAB w/ CONST. JT.s @ 15' MAX. SPA.
4" COMPACTED GRANULAR FILL
3" ASPHALTIC SURFACE TYPE LV
12" CRUSHED AGG. BASE COURSE
TYPICAL THROUGH EMPLOYEE PARKING AREA
1137.0 = INVERT, SLOPE TO DRAIN
LONGITUDINAL SECTION OF TRUCK DOCK
LADYSMITH I.G.A. PARKING LOT
DETAILS & SECTIONS
SCALE: AS NOTED
SHEET C-2 OF C-3
DATE MARCH 1999
MORGAN & PARMLEY, LTD. CONSULTING ENGINEERS LADYSMITH, WISCONSIN
DR. BY T T

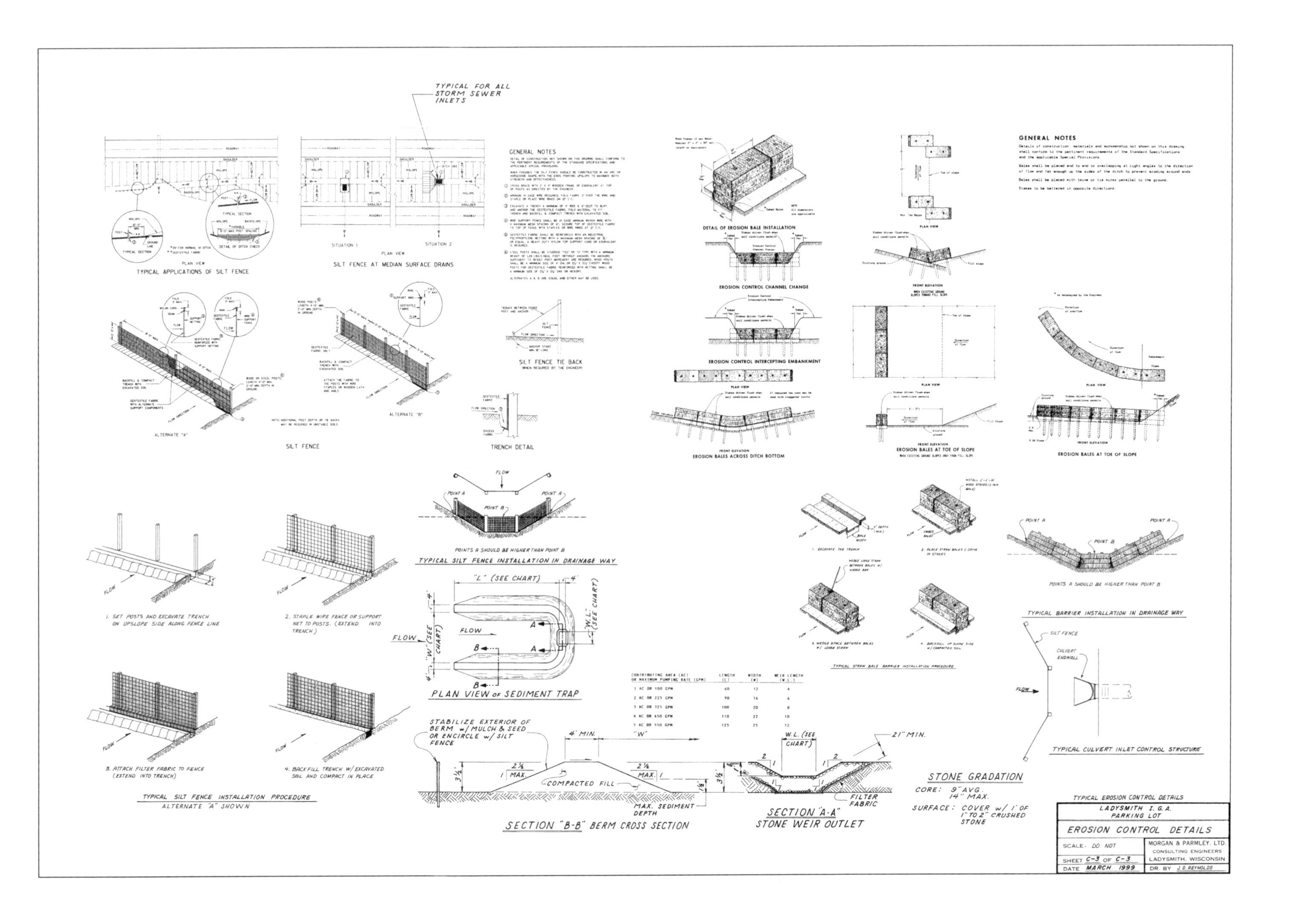

CONTRIBUTING AREA (AC) OR MAXIMUM PUMPING RATE (GPM)	LENGTH (L)	WIDTH (W)	WEIR LENGTH (W.L.)
1 AC OR 100 GPM	60	12	4
2 AC OR 225 GPM	90	16	6
3 AC OR 325 GPM	100	20	8
4 AC OR 450 GPM	110	22	10
5 AC OR 550 GPM	125	25	12

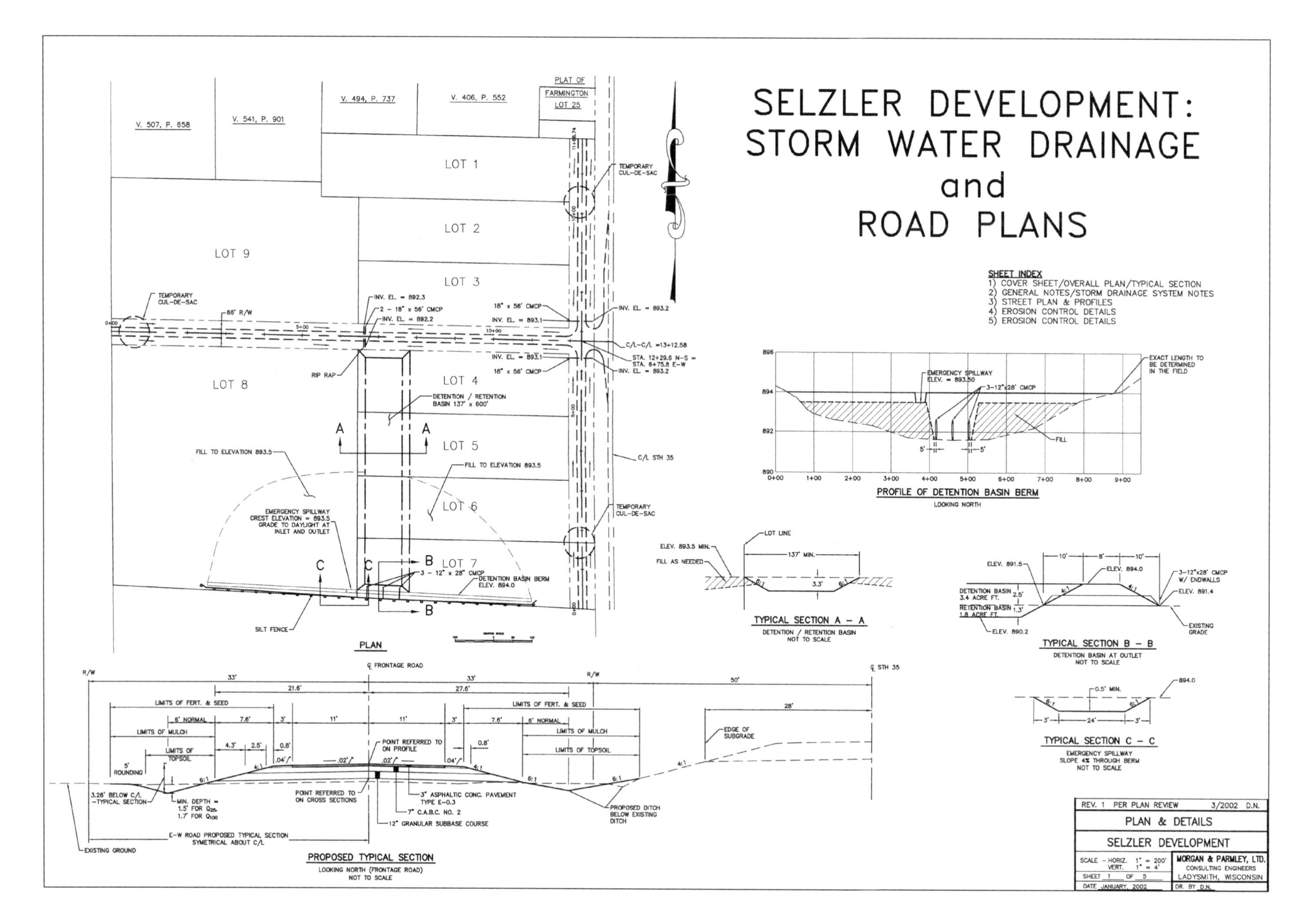

Courtesy: Morgan & Parmley, Ltd.

GENERAL NOTES

1. EROSION CONTROL MEASURES SHALL BE IN PLACE BEFORE CONSTRUCTION BEGINS.
2. THE CONTRACTORS SHALL MINIMIZE EROSION DURING CONSTRUCTION USING GOOD CONSTRUCTION TECHNIQUES AND UTILIZING SILT FENCES AND BALE DITCH CHECKS. THE CONTRACTOR SHALL RELEASE RUNOFF FROM THE SITE IN A NUISANCE FREE MANNER.
3. THE CONTRACTOR IS RESPONSIBLE TO MAINTAIN THE SITE IN A SAFE CONDITION. THE SOLE RESPONSIBILITY FOR WARNING SIGNS, BARRICADES, FLAGGING PERSONNEL AND ALL ASPECTS OF SAFETY LIE WITH THE CONTRACTOR.
4. EXISTING UTILITIES ARE NOT SHOWN. CONTRACTOR SHALL HAVE ALL BURIED UTILITIES LOCATED PRIOR TO COMMENCING CONSTRUCTION.
5. PROPERTY CORNERS KNOWN TO EXIST SHALL NOT BE DISTURBED. DAMAGED CORNERS WILL BE REPLACED AT THE CONTRACTOR'S EXPENSE. PRESERVE ALL SURVEY MONUMENTS.
6. CONSTRUCTION SHALL COMPLY WITH THE CONDITIONS OF DOT, COUNTY AND TOWNSHIP APPROVALS.
7. ALL DISTURBED AREAS SHALL BE SEEDED, FERTILIZED AND MULCHED, OR COVERED W/BASE COURSE AFTER CONSTRUCTION, OR RESTORED WITH BITUMINOUS PAVEMENT WHERE APPLICABLE.
8. THERE SHALL BE NO DEVIATION FROM THE PLANS WITHOUT THE DESIGN ENGINEER'S APPROVAL.
9. CONSTRUCTION STAKES DESTROYED BY THE CONTRACTOR THAT NEED TO BE REPLACED, WILL BE CHARGED TO THE CONTRACTOR.
10. ALL DISTURBED AREAS TO BE SEEDED SHALL BE RESTORED AS FOLLOWS:

    A) REPLACE ALL SALVAGED AND IMPORTED TOPSOIL
    B) SEED MIXTURE: PERMANENT SEEDING–(LAWN TYPE TURF)

        35% KENTUCKY BLUE GRASS
        25% IMPROVED FINE PERENNIAL RYEGRASS
        20% CREEPING RED FESCUE
        20% IMPROVED HARD FESCUE

    C) SEED RATE: 2#/100 SQ. FT.
    D) MULCH RATE: 2 TON/ACRE
    E) PERMANENT SEEDING: ALLOWED TO SEPT. 7
    TEMPORARY SEEDING: SEPT. 8 THROUGH NOV. 10 (4 BU. OATS/ACRE)
    DORMANT SEEDING: SEED PERM. SEEDING INTO TEMP. SEEDING AFTER NOV. 10
    F) SEEDING MAY BE BROADCAST BUT MUST BE COVERED WITH A MAXIMUM OF ½" SOIL. MULCH MUST BE ANCHORED WITH A MULCH TILLER, TACKIFIER OR NETTING.
    G) FERTILIZER: 500#/ACRE 20–10–10

11. IF MORE THAN 5 ACRES ARE DISTURBED THE CONTRACTOR SHALL OBTAIN A STORM WATER DISCHARGE PERMIT FROM THE DNR.
12. IF IN FIELD CONDITIONS ARE ENCOUNTERED THAT DIFFER FROM THE PLANS, NOTIFY THE DESIGN ENGINEER.

SPECIFICATIONS:

WISCONSIN CONSTRUCTION SITE BEST MANAGEMENT PRACTICE HANDBOOK

STANDARD SPECIFICATIONS FOR ROAD AND BRIDGE CONSTRUCTION 1996 EDITION WIS. D.O.T.

STORM DRAINAGE SYSTEM NOTES.

1. ALL CULVERT PIPES SHALL BE CMCP WITH ENDWALLS.
2. CULVERT PIPE LENGTHS SHALL BE ADJUSTED IN THE FIELD TO MATCH INSLOPES.
3. DITCHES SHALL HAVE A MINIMUM DEPTH OF 1.7'.
4. THE DETENTION BASIN SHALL CONTAIN A MINIMUM OF 3.4 ACRE FEET OF STORAGE ABOVE THE INVERT OF THE OUTLET STRUCTURE AND BELOW THE CREST OF THE EMERGENCY SPILLWAY. THE RETENTION BASIN SHALL CONTAIN A MINIMUM OF 1.8 ACRE FEET BELOW THE INVERT OF THE OUTLET STRUCTURE. THE OWNER MAY SLOPE THE BOTTOM OF THE BASIN TO ACCOMMODATE DEVELOPMENT AND TO FACILITATE DRAINAGE, STORAGE VOLUME MUST BE MAINTAINED AS STATED.
5. THE LENGTH OF THE DETENTION BASIN BERM SHALL BE DETERMINED IN THE FIELD BY INTERSECTING THE EXISTING OR PROPOSED GRADE AT ELEVATION 894.0 ALONG THE SOUTH PROPERTY LINE.
6. STORMWATER FROM THE DETENTION BASIN SHALL LEAVE THE SITE AT A SAFE VELOCITY. SOD, EROSION MAT, TURF REINFORCEMENT OR RIP RAP SHALL PROTECT THE DETENTION BASIN OUTLET AND EMERGENCY SPILLWAY.

DESIGN SUMMARY (TR 55 GR.)

Q_{25} EXISTING	20.6 CFS
Q_{25} PROPOSED	25.5 CFS
MAXIMUM CAPACITY OF OUTLET STRUCTURE	9.5 CFS
DETENTION BASIN STORAGE	3.4 AC. FT.
RETENTION BASIN STORAGE	1.8 AC. FT.
TOTAL STORAGE	5.2 AC. FT.
MAXIMUM CAPACITY OF EMERGENCY SPILLWAY (EXCEEDS Q_{100})	47 CFS
MAXIMUM CAPACITY OF TWIN 18" CMCP Q_{25}	13 CFS

REV. 1 PER PLAN REVIEW	3/2002 D.N.
GENERAL NOTES	
SELZLER DEVELOPMENT	
SCALE – HORIZ. 1" = NONE VERT. 1" = NONE	MORGAN & PARMLEY, LTD. CONSULTING ENGINEERS
SHEET 2 OF 5	LADYSMITH, WISCONSIN
DATE JANUARY, 2002	DR. BY D.N.

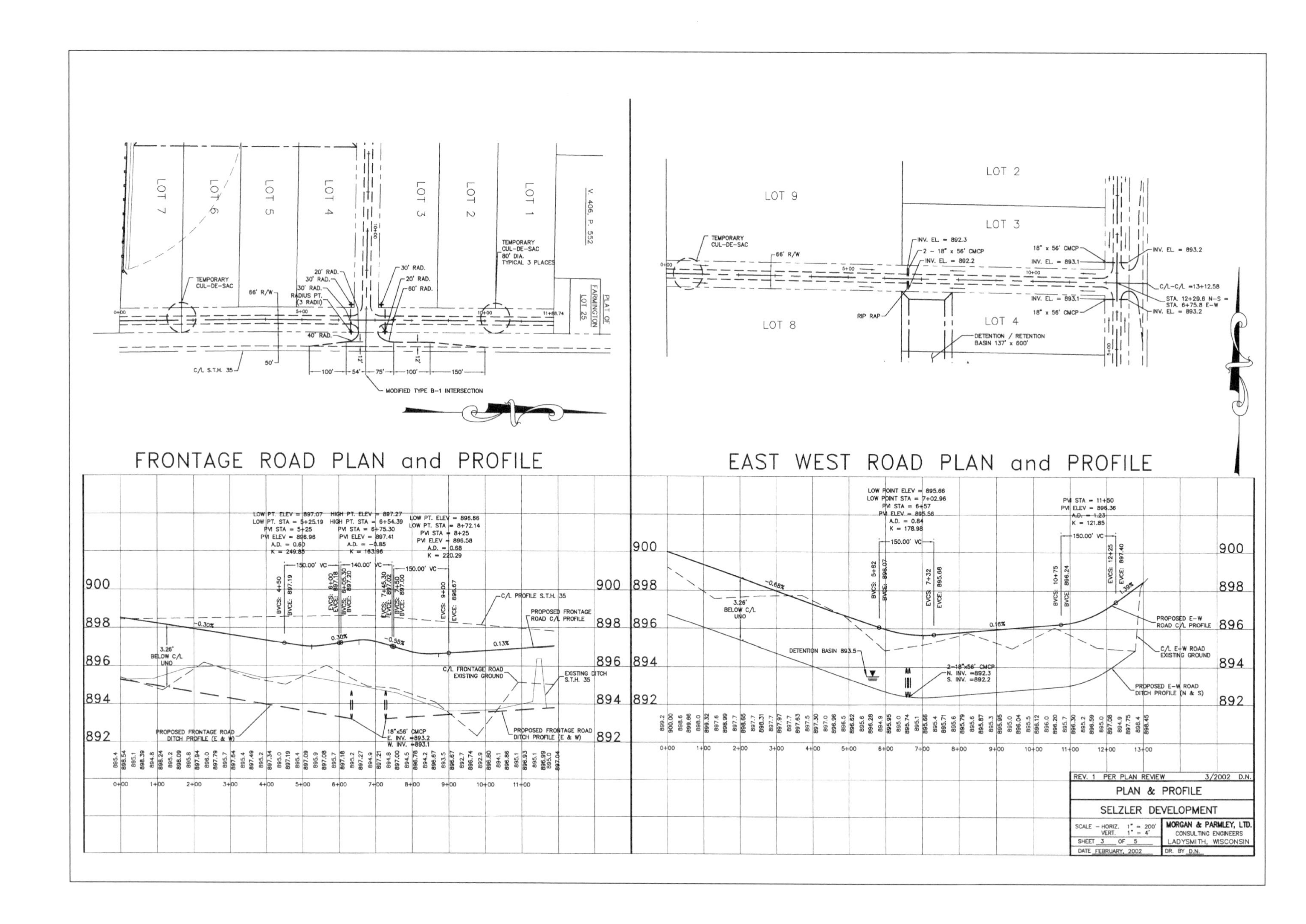

FRONTAGE ROAD PLAN and PROFILE
EAST WEST ROAD PLAN and PROFILE
LOT 7
LOT 6
LOT 5
LOT 4
LOT 3
LOT 2
LOT 1
V. 406, P. 552
PLAT OF FARMINGTON LOT 25
TEMPORARY CUL-DE-SAC
TEMPORARY CUL-DE-SAC 80' DIA. TYPICAL 3 PLACES
66' R/W
C/L S.T.H. 35
MODIFIED TYPE B-1 INTERSECTION
LOT 9
LOT 8
DETENTION / RETENTION BASIN 137' x 600'
RIP RAP
REV. 1 PER PLAN REVIEW 3/2002 D.N.
PLAN & PROFILE
SELZLER DEVELOPMENT
SCALE - HORIZ. 1" = 200' VERT. 1" = 4'
SHEET 3 OF 5
DATE FEBRUARY, 2002
MORGAN & PARMLEY, LTD.
CONSULTING ENGINEERS
LADYSMITH, WISCONSIN
DR. BY D.N.

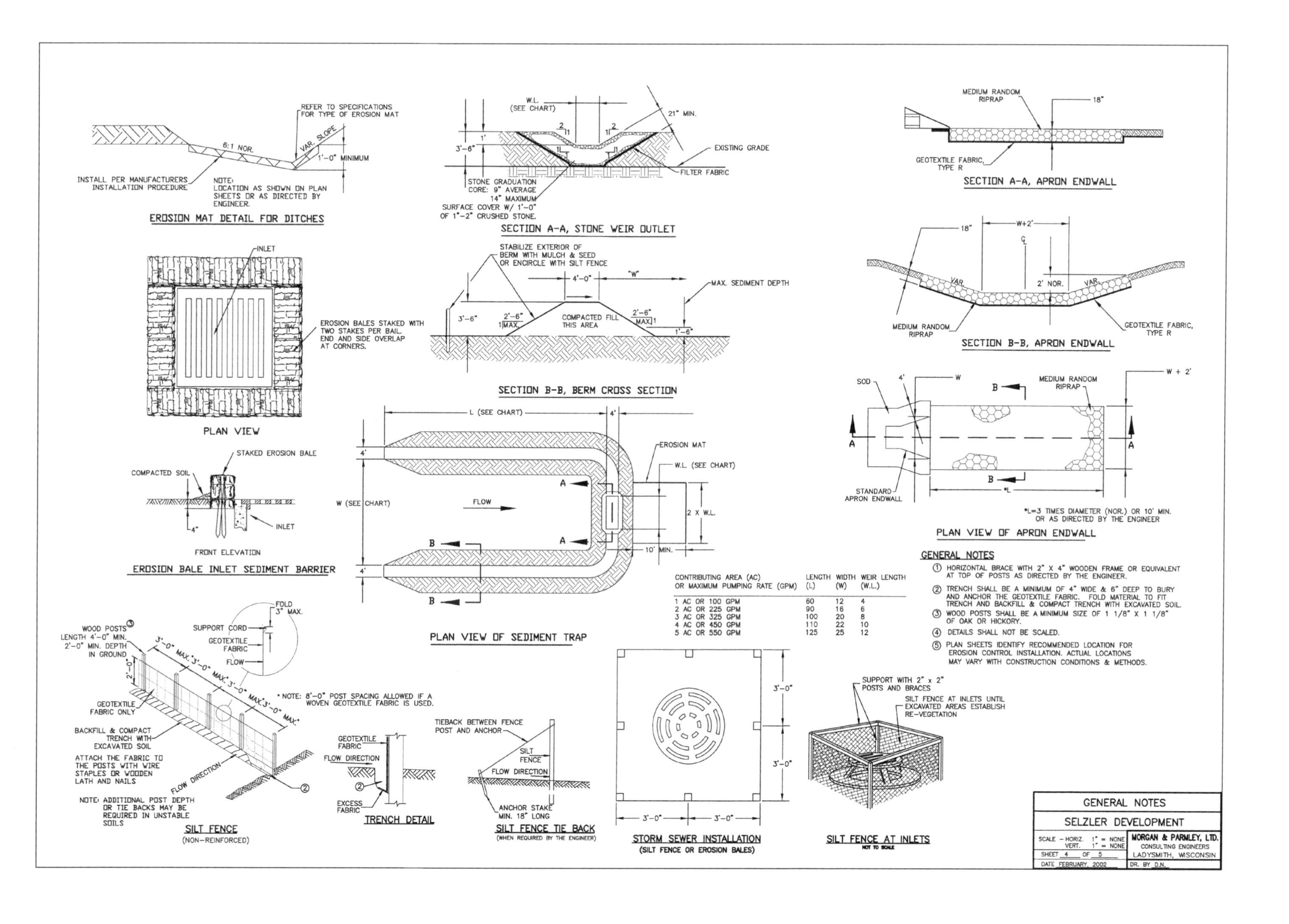
REFER TO SPECIFICATIONS FOR TYPE OF EROSION MAT
6:1 NOR.
VAR. SLOPE
1'-0" MINIMUM
INSTALL PER MANUFACTURERS INSTALLATION PROCEDURE
NOTE: LOCATION AS SHOWN ON PLAN SHEETS OR AS DIRECTED BY ENGINEER.
EROSION MAT DETAIL FOR DITCHES
W.L. (SEE CHART)
21" MIN.
3'-6"
1'
EXISTING GRADE
FILTER FABRIC
STONE GRADUATION CORE: 9" AVERAGE 14" MAXIMUM
SURFACE COVER W/ 1'-0" OF 1"-2" CRUSHED STONE.
SECTION A-A, STONE WEIR OUTLET
MEDIUM RANDOM RIPRAP
18"
GEOTEXTILE FABRIC, TYPE R
SECTION A-A, APRON ENDWALL
INLET
EROSION BALES STAKED WITH TWO STAKES PER BAIL. END AND SIDE OVERLAP AT CORNERS.
PLAN VIEW
STABILIZE EXTERIOR OF BERM WITH MULCH & SEED OR ENCIRCLE WITH SILT FENCE
4'-0"
"W"
MAX. SEDIMENT DEPTH
3'-6"
2'-6" 1 MAX.
COMPACTED FILL THIS AREA
2'-6" MAX. 1
1'-6"
SECTION B-B, BERM CROSS SECTION
18"
W+2'
2' NOR.
VAR.
MEDIUM RANDOM RIPRAP
GEOTEXTILE FABRIC, TYPE R
SECTION B-B, APRON ENDWALL
STAKED EROSION BALE
COMPACTED SOIL
4"
INLET
FRONT ELEVATION
EROSION BALE INLET SEDIMENT BARRIER
L (SEE CHART)
4'
EROSION MAT
W.L. (SEE CHART)
W (SEE CHART)
FLOW
2 X W.L.
10' MIN.
PLAN VIEW OF SEDIMENT TRAP
SOD
W
MEDIUM RANDOM RIPRAP
W + 2'
STANDARD APRON ENDWALL
*L
*L=3 TIMES DIAMETER (NOR.) OR 10' MIN. OR AS DIRECTED BY THE ENGINEER
PLAN VIEW OF APRON ENDWALL
CONTRIBUTING AREA (AC) OR MAXIMUM PUMPING RATE (GPM) | LENGTH (L) | WIDTH (W) | WEIR LENGTH (W.L.)
1 AC OR 100 GPM | 60 | 12 | 4
2 AC OR 225 GPM | 90 | 16 | 6
3 AC OR 325 GPM | 100 | 20 | 8
4 AC OR 450 GPM | 110 | 22 | 10
5 AC OR 550 GPM | 125 | 25 | 12
GENERAL NOTES
1 HORIZONTAL BRACE WITH 2" X 4" WOODEN FRAME OR EQUIVALENT AT TOP OF POSTS AS DIRECTED BY THE ENGINEER.
2 TRENCH SHALL BE A MINIMUM OF 4" WIDE & 6" DEEP TO BURY AND ANCHOR THE GEOTEXTILE FABRIC. FOLD MATERIAL TO FIT TRENCH AND BACKFILL & COMPACT TRENCH WITH EXCAVATED SOIL.
3 WOOD POSTS SHALL BE A MINIMUM SIZE OF 1 1/8" X 1 1/8" OF OAK OR HICKORY.
4 DETAILS SHALL NOT BE SCALED.
5 PLAN SHEETS IDENTIFY RECOMMENDED LOCATION FOR EROSION CONTROL INSTALLATION. ACTUAL LOCATIONS MAY VARY WITH CONSTRUCTION CONDITIONS & METHODS.
WOOD POSTS LENGTH 4'-0" MIN. 2'-0" MIN. DEPTH IN GROUND
SUPPORT CORD
FOLD 3" MAX.
GEOTEXTILE FABRIC
FLOW
3'-0" MAX.
2'-0"
* NOTE: 8'-0" POST SPACING ALLOWED IF A WOVEN GEOTEXTILE FABRIC IS USED.
GEOTEXTILE FABRIC ONLY
BACKFILL & COMPACT TRENCH WITH EXCAVATED SOIL
ATTACH THE FABRIC TO THE POSTS WITH WIRE STAPLES OR WOODEN LATH AND NAILS
FLOW DIRECTION
NOTE: ADDITIONAL POST DEPTH OR TIE BACKS MAY BE REQUIRED IN UNSTABLE SOILS
SILT FENCE (NON-REINFORCED)
GEOTEXTILE FABRIC
FLOW DIRECTION
EXCESS FABRIC
TRENCH DETAIL
TIEBACK BETWEEN FENCE POST AND ANCHOR
SILT FENCE
FLOW DIRECTION
ANCHOR STAKE MIN. 18" LONG
SILT FENCE TIE BACK (WHEN REQUIRED BY THE ENGINEER)
3'-0"
STORM SEWER INSTALLATION (SILT FENCE OR EROSION BALES)
SUPPORT WITH 2" x 2" POSTS AND BRACES
SILT FENCE AT INLETS UNTIL EXCAVATED AREAS ESTABLISH RE-VEGETATION
SILT FENCE AT INLETS NOT TO SCALE
GENERAL NOTES
SELZLER DEVELOPMENT
SCALE - HORIZ. 1" = NONE VERT. 1" = NONE
MORGAN & PARMLEY, LTD. CONSULTING ENGINEERS LADYSMITH, WISCONSIN
SHEET 4 OF 5
DATE FEBRUARY, 2002
DR. BY D.N.

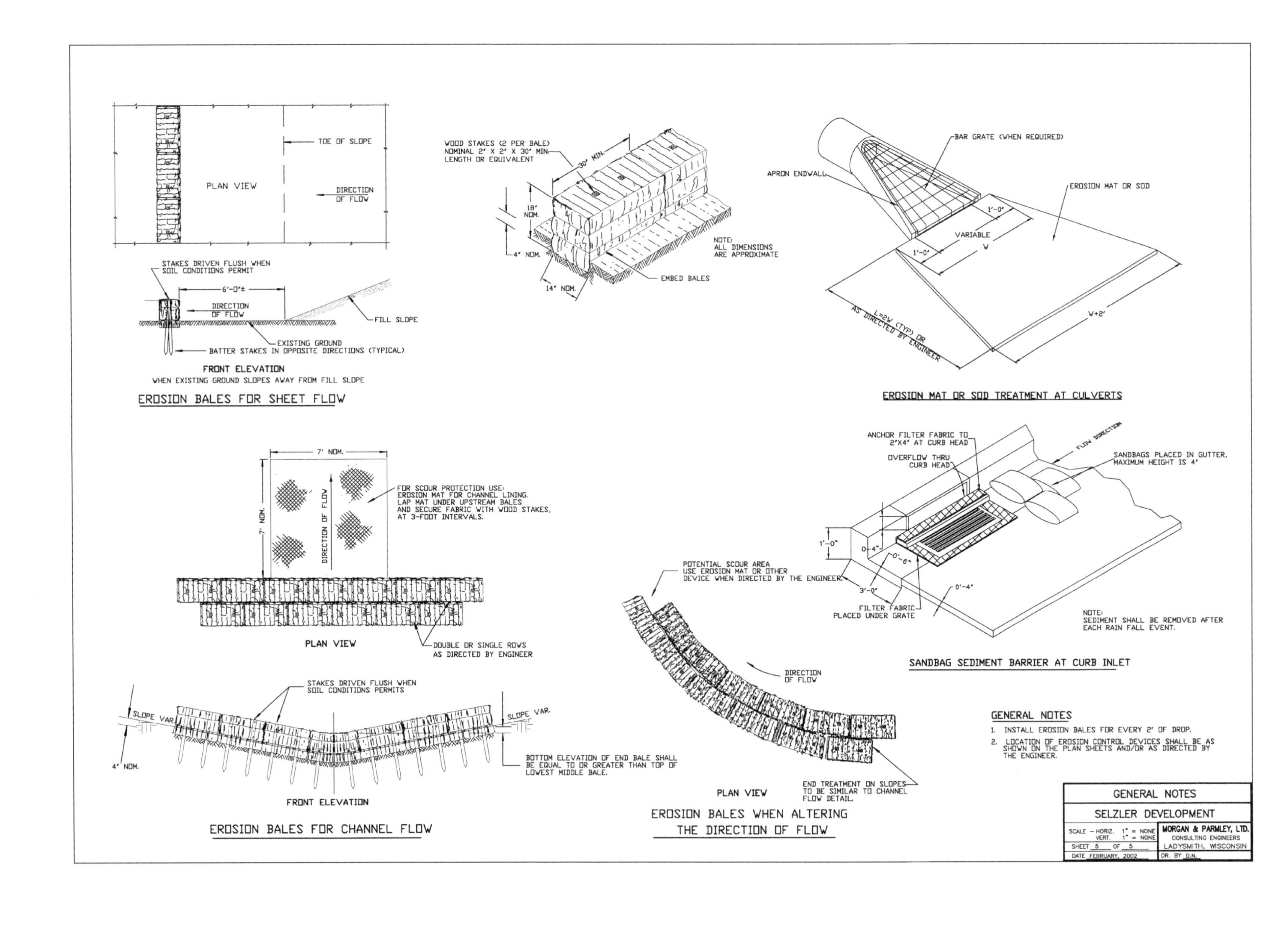
TOE OF SLOPE
PLAN VIEW
DIRECTION OF FLOW
STAKES DRIVEN FLUSH WHEN SOIL CONDITIONS PERMIT
6'-0"±
DIRECTION OF FLOW
FILL SLOPE
EXISTING GROUND
BATTER STAKES IN OPPOSITE DIRECTIONS (TYPICAL)
FRONT ELEVATION
WHEN EXISTING GROUND SLOPES AWAY FROM FILL SLOPE
EROSION BALES FOR SHEET FLOW
WOOD STAKES (2 PER BALE) NOMINAL 2" X 2" X 30" MIN. LENGTH OR EQUIVALENT
30" MIN.
18" NOM.
4" NOM.
14" NOM.
NOTE: ALL DIMENSIONS ARE APPROXIMATE
EMBED BALES
BAR GRATE (WHEN REQUIRED)
APRON ENDWALL
EROSION MAT OR SOD
1'-0"
VARIABLE
W
1'-0"
L=2W (TYP) OR AS DIRECTED BY ENGINEER
W+2'
EROSION MAT OR SOD TREATMENT AT CULVERTS
7' NOM.
7' NOM.
DIRECTION OF FLOW
FOR SCOUR PROTECTION USE: EROSION MAT FOR CHANNEL LINING. LAP MAT UNDER UPSTREAM BALES AND SECURE FABRIC WITH WOOD STAKES, AT 3-FOOT INTERVALS.
PLAN VIEW
DOUBLE OR SINGLE ROWS AS DIRECTED BY ENGINEER
STAKES DRIVEN FLUSH WHEN SOIL CONDITIONS PERMITS
SLOPE VAR.
SLOPE VAR.
4' NOM.
BOTTOM ELEVATION OF END BALE SHALL BE EQUAL TO OR GREATER THAN TOP OF LOWEST MIDDLE BALE.
FRONT ELEVATION
EROSION BALES FOR CHANNEL FLOW
ANCHOR FILTER FABRIC TO 2"X4" AT CURB HEAD
OVERFLOW THRU CURB HEAD
FLOW DIRECTION
SANDBAGS PLACED IN GUTTER, MAXIMUM HEIGHT IS 4'
1'-0"
0'-4"
0'-6"
3'-0"
0'-4"
POTENTIAL SCOUR AREA USE EROSION MAT OR OTHER DEVICE WHEN DIRECTED BY THE ENGINEER.
FILTER FABRIC PLACED UNDER GRATE
NOTE: SEDIMENT SHALL BE REMOVED AFTER EACH RAIN FALL EVENT.
SANDBAG SEDIMENT BARRIER AT CURB INLET
DIRECTION OF FLOW
END TREATMENT ON SLOPES TO BE SIMILAR TO CHANNEL FLOW DETAIL.
PLAN VIEW
EROSION BALES WHEN ALTERING THE DIRECTION OF FLOW
GENERAL NOTES
1. INSTALL EROSION BALES FOR EVERY 2' OF DROP.
2. LOCATION OF EROSION CONTROL DEVICES SHALL BE AS SHOWN ON THE PLAN SHEETS AND/OR AS DIRECTED BY THE ENGINEER.
GENERAL NOTES
SELZLER DEVELOPMENT
SCALE – HORIZ. 1" = NONE VERT. 1" = NONE
SHEET 5 OF 5
DATE FEBRUARY, 2002
MORGAN & PARMLEY, LTD.
CONSULTING ENGINEERS
LADYSMITH, WISCONSIN
DR. BY D.N.

CDBG PROJECT NO. PF FY 99-0124
Mc SLOY STREET AREA RECONSTRUCTION

FOR THE VILLAGE OF

GILMAN, WISCONSIN

M. & P. PROJECT NO. 99-120

MARCH 2000

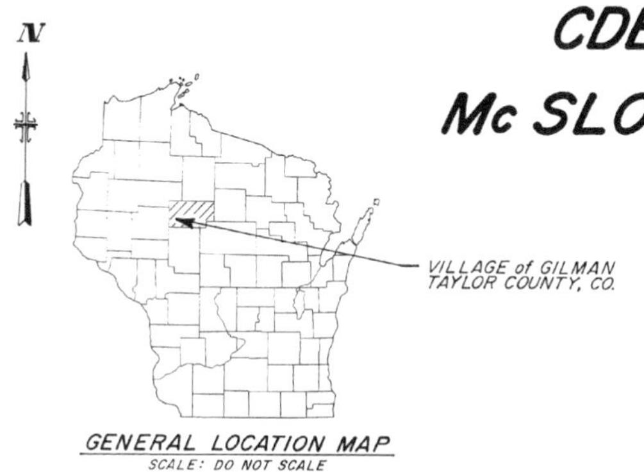

GENERAL LOCATION MAP

SCALE: DO NOT SCALE

SPECIFICATIONS:

WISCONSIN CONSTRUCTION SITE BEST MANAGEMENT PRACTICE HANDBOOK

STANDARD SPECIFICATIONS FOR SEWER AND WATER CONSTRUCTION IN WISCONSIN 5TH EDITION WITH ADDENDUM NO. 1

STANDARD SPECIFICATIONS FOR ROAD AND BRIDGE CONSTRUCTION 1996 EDITION WIS D.O.T.

GENERAL NOTES

1. ALL WATER MAIN SHALL BE 6" PVC DR 18.
2. ALL WATERMAIN FITTINGS SHALL BE M.J.D.I., W / RESTRAINED JOINT LUGS.
3. ALL HYDRANT LEADS SHALL BE 6-INCH VALVE.
4. HYDRANTS SHALL BE WATEROUS WB 67-250 (4" PUMPER NOZZLE), 22" BREAKOFF WITH THREADS MATCHING VILLAGE'S PATTERN.
5. ALL WATER MAIN SHALL HAVE AN 8 FT. BURY (MIN.).
6. WATER SERVICES SHALL BE 1" TYPE K COPPER W / MUELLER OR FORD BRASS.
7. 8' (MINIMUM) HORIZONTAL DISTANCE (℄ TO ℄) BETWEEN WATERMAIN AND SANITARY SEWER OR STORM SEWER.
8. 18" (MINIMUM) VERTICAL DISTANCE (OUT TO OUT) BETWEEN WATERMAIN UNDER SANITARY SEWER OR STORM SEWER @ THEIR INTERSECTION.
9. 6" (MINIMUM) VERTICAL DISTANCE (OUT TO OUT) BETWEEN WATERMAIN OVER SANITARY SEWER OR STORM SEWER @ THEIR INTERSECTION.
10. SANITARY SERVICES SHALL BE 4" SCH. 40 PVC W / WYES LAID TO PROVIDE FOR MAXIMUM DEPTH OR 10' AT THE R / W.
11. SEWERMAIN SHALL BE 8" PVC SDR 35.
12. MANHOLES SHALL BE SET WITH A MINIMUM OF 6" OF RINGS AND 9" CASTINGS.
13. MANHOLES SHALL BE 4' DIAMETER PRECAST CONCRETE W / INTEGRAL BASES AND BOOTS.
14. ALL ELEVATIONS BASED ON USGS DATUM.
15. EROSION CONTROL MEASURES SHALL BE IN PLACE BEFORE CONSTRUCTION BEGINS.
16. THE CONTRACTORS SHALL MINIMIZE EROSION DURING CONSTRUCTION USING GOOD CONSTRUCTION TECHNIQUES AND UTILIZING SILT FENCES AND BALE DITCH CHECKS. THE CONTRACTOR SHALL RELEASE RUNOFF FROM THE SITE IN A NUISANCE FREE MANNER.
17. THE CONTRACTOR IS RESPONSIBLE TO MAINTAIN THE SITE IN A SAFE CONDITION. THE SOLE RESPONSIBILITY FOR WARNING SIGNS, BARRICADES, FLAGGING PERSONNEL AND ALL ASPECTS OF SAFETY LIE WITH THE CONTRACTOR.
18. EXISTING UTILITIES ARE SHOWN BUT MAY NOT BE ALL INCLUSIVE; CONTRACTOR SHALL HAVE ALL BURIED UTILITIES LOCATED PRIOR TO COMMENCING CONSTRUCTION.
19. EXISTING SEWER AND WATERMAIN ARE SHOWN ACCORDING TO EXISTING RECORDS BUT LOCATION SHALL BE VERIFIED IN FIELD.
20. WATERMAIN & SERVICES SHALL BE INSULATED AT ALL STORM SEWER, CULVERT AND DITCH CROSSINGS WITH 2" THICK EXTRUDED POLYSTYRENE INSULATION BOARD.
21. PROPERTY CORNERS KNOWN TO EXIST SHALL NOT BE DISTURBED. DAMAGED CORNERS WILL BE REPLACED AT THE CONTRACTOR'S EXPENSE. PRESERVE ALL SURVEY MONUMENTS.
22. ALL WATERMAIN SHALL BE DISINFECTED BEFORE BEING PUT INTO USE.
23. VALVES AND MANHOLES SHALL BE ADJUSTED TO FINAL GRADE, UNLESS DIRECTED OTHERWISE.
24. CONSTRUCTION SHALL COMPLY WITH THE CONDITIONS OF DNR APPROVAL ELTTER, AND OTHER REGULATORY APPROVALS AS LISTED IN THE SPECIFICATION BOOK.
25. ALL DISTURBED AREAS SHALL BE SEEDED, FERTILIZED AND MULCHED, OR COVERED W / BASE COURSE AFTER CONSTRUCTION, OR RESTORED WITH BITUMINOUS PAVEMENT WHERE APPLICABLE.
26. THERE SHALL BE NO DEVIATION FROM THE PLANS WITHOUT THE DESIGN ENGINEER'S APPROVAL.
27. CONSTRUCTION STAKES DESTROYED BY THE CONTRACTOR THAT NEED TO BE REPLACED, WILL BE CHARGED TO THE CONTRACTOR.
28. ALL DISTURBED AREAS TO BE SEEDED SHALL BE RESTORED AS FOLLOWS:

    A) REPLACED ALL SALVAGED AND IMPORTED TOPSOIL
    B) SEED MIXUTRE: PERMANENT SEEDING – (LAWN TYPE TURF)

    35% KENTUCKY BLUEGRASS
    25% IMPROVED FINE PERENNIAL RYEGRASS
    20% CREEPING RED FESCUE
    20% IMPROVED HARD FESCUE

    C) SEED RATE: 2# / 100 SQ. FT.
    D) MULCH RATE: 2 TON / ACRE
    E) PERMANENT SEEDING: ALLOWED TO SEPT. 7
    TEMPORARY SEEDING: SEPT. 8 THROUGH NOV. 10
    (4 BU. OATS / ACRE)
    DORMANT SEEDING; SEED PERM. SEEDING INTO TEMP. SEEDING AFTER NOV. 10
    F) SEEDING MAY BE BRAODCAST BUT MUST BE COVERED WITH A MAXIMUM OF ½" SOIL. MULCH MUST BE ANCHORED WITH A MULCH TILLER, TACKIFIER OR NETTING.
    G) FERTILIZER: 500# / ACRE 20-10-10
29. CONTRACTOR MUST REMOVE AND REPLACE ALL AFFECTED STREET SIGNS AFTER PROPERLY SIGNING AND BARRICADING THE CONSTRUCTION AREA. ALL SIGNS MUST BE REPLACED PRIOR TO OPENING THE STREET TO TRAFFIC.
30. CONTRACTOR SHALL DETERMINE CONSTRUCTION SEQUENCE IN ORDER TO MINIMIZE INSTALLATION CONFLICTS WITH PIPE CROSSINGS.
31. STORM SEWER PIPE SHALL BE CL III RCP OR SMOOTH WALL PE AS NOTED ON THE PLANS.
32. NO TREES ARE TO BE REMOVED WITHOUT THE APPROVAL OF THE ENGINEER.

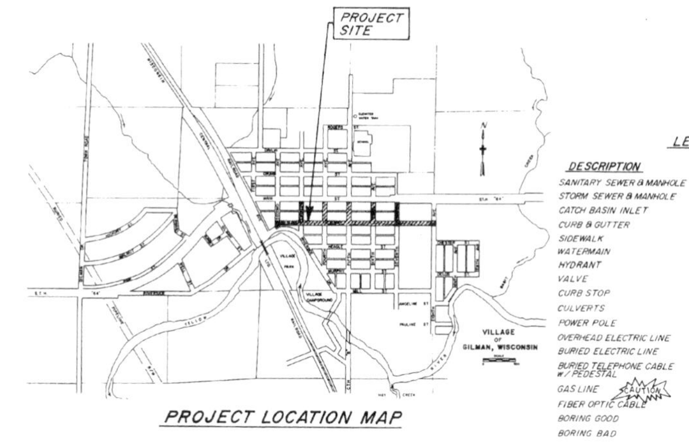

PROJECT LOCATION MAP

INDEX OF SHEETS

SHEET NO.	DESCRIPTION
1	TITLE SHEET, LOCATION MAP, INDEX, NOTES & TYPICAL CROSS SECTION
2	PLAN & PROFILE – McSLOY STREET STA. 1+00 TO STA. 12+50
3	PLAN & PROFILE – McSLOY STREET STA. 12+50 TO STA. 25+14
4	PLAN & PROFILE – THIRD AVENUE & FOURTH AVENUE
5	PLAN & PROFILE – FIFTH AVENUE & STORM SEWER SCHEDULE & CROSS SECTION
6	PLAN & PROFILE – SIXTH AVENUE & SEVENTH AVENUE
7	TYPICAL EROSION CONTROL DETAILS
8 – 21	CROSS SECTIONS – McSLOY STREET
22 – 23	CROSS SECTIONS – THIRD AVENUE
24 – 25	CROSS SECTIONS – FOURTH AVENUE
26 – 29	CROSS SECTIONS – FIFTH AVENUE
30 – 32	CROSS SECTIONS – SIXTH AVENUE
33 – 35	CROSS SECTIONS – SEVENTH AVENUE

NOTE: Sheets 7 & 9 thru 35 Not Shown

LEGEND

DESCRIPTION	EXISTING	PROPOSED
SANITARY SEWER & MANHOLE		
STORM SEWER & MANHOLE		
CATCH BASIN INLET		
CURB & GUTTER		
SIDEWALK		NONE
WATERMAIN		
HYDRANT		
VALVE		
CURB STOP		
CULVERTS		NONE
POWER POLE		NONE
OVERHEAD ELECTRIC LINE		NONE
BURIED ELECTRIC LINE		NONE
BURIED TELEPHONE CABLE w/ PEDESTAL		NONE
GAS LINE		NONE
FIBER OPTIC CABLE		NONE
BORING GOOD		NONE
BORING BAD		NONE

CAUTION

—UTILITIES—

TELEPHONE:
CENTURYTEL OF NORTHERN WISCONSIN, INC.
428 ELLINGSON AVE.
P. O. BOX 78
HAWKINS, WISCONSIN 54530
ATTN: JAMES ARQUETTE
715-585-7707

CABLE:
S & K TV SYSTEMS
508 W. MINER AVE.
LADYSMITH, WISCONSIN 54848
ATTN: RANDY SCOTT
715-532-7321

ELECTRIC:
NORTHERN STATES POWER CO.
500 N. 5th STREET
ABBOTSFORD, WISCONSIN 54405
ATTN: DAVE PEPPER
715-839-2678

DIGGERS HOTLINE:
800-242-8511

R/W RIGHT-OF-WAY 66' R/W

TYPICAL CROSS SECTION

EARTHWORK QUANTITIES

EXCAVATION:		14,550	CU. YD.s
FILL:	x 1.3 =	1,300	CU. YD.s
EXCESS WASTE:		13,250	CU. YD.s

PREPARED BY:
MORGAN & PARMLEY, Ltd.
PROFESSIONAL CONSULTING ENGINEERS
LADYSMITH, WISCONSIN

Courtesy: Morgan & Parmley, Ltd.

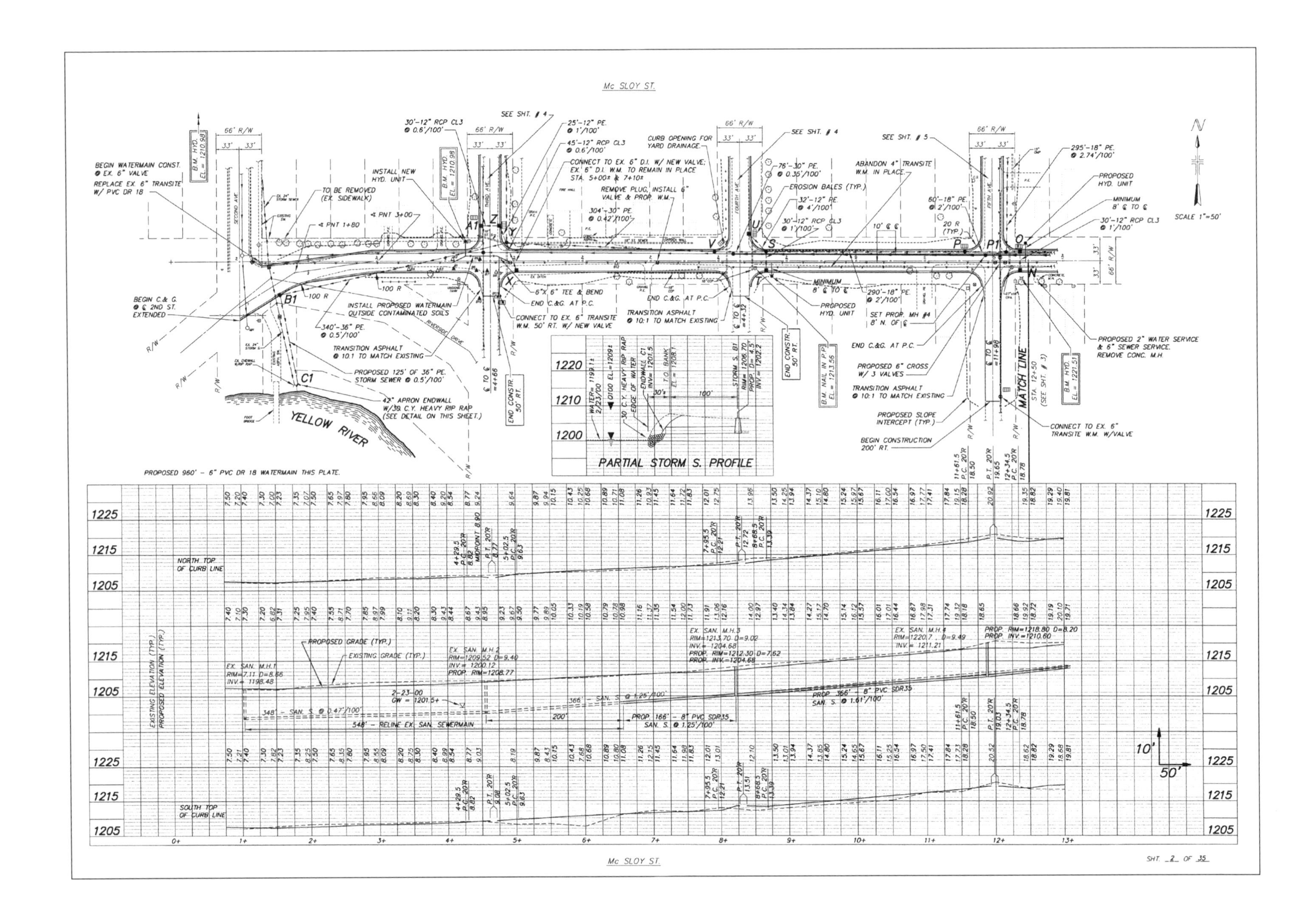
Mc SLOY ST.
BEGIN WATERMAIN CONST. @ EX. 6" VALVE
REPLACE EX. 6" TRANSITE W/ PVC DR 18
BEGIN C.& G. @ ℄ 2ND ST. EXTENDED
TO BE REMOVED (EX. SIDEWALK)
INSTALL NEW HYD. UNIT
SEE SHT. # 4
30'-12" RCP CL3 @ 0.6'/100'
25'-12" PE. @ 1'/100'
45'-12" RCP CL3 @ 0.6'/100'
CONNECT TO EX. 6" D.I. W/ NEW VALVE; EX. 6" D.I. W.M. TO REMAIN IN PLACE STA. 5+00± & 7+10±
REMOVE PLUG, INSTALL 6" VALVE & PROP. W.M.
304'-30" PE. @ 0.42'/100'
CURB OPENING FOR YARD DRAINAGE.
76'-30" PE. @ 0.35'/100'
EROSION BALES (TYP.)
32'-12" PE. @ 4'/100'
30'-12" RCP CL3 @ 1'/100'
ABANDON 4" TRANSITE W.M. IN PLACE
SEE SHT. # 5
60'-18" PE. @ 2'/100'
295'-18" PE. @ 2.74'/100'
PROPOSED HYD. UNIT
MINIMUM 8' ℄ TO ℄
30'-12" RCP CL3 @ 1'/100'
SCALE 1"=50'
100 R
340'-36" PE. @ 0.5'/100'
INSTALL PROPOSED WATERMAIN OUTSIDE CONTAMINATED SOILS
TRANSITION ASPHALT @ 10:1 TO MATCH EXISTING
PROPOSED 125' OF 36" PE. STORM SEWER @ 0.5'/100'
42" APRON ENDWALL W/30 C.Y. HEAVY RIP RAP (SEE DETAIL ON THIS SHEET.)
YELLOW RIVER
6"X 6" TEE & BEND
END C.&G. AT P.C.
CONNECT TO EX. 6" TRANSITE W.M. 50' RT. W/ NEW VALVE
TRANSITION ASPHALT @ 10:1 TO MATCH EXISTING
MINIMUM 8' ℄ TO ℄
290'-18" PE. @ 2'/100'
PROPOSED HYD. UNIT
SET PROP. MH #4 8' N. OF ℄
END C.&G. AT P.C.
PROPOSED 6" CROSS W/ 3 VALVES
TRANSITION ASPHALT @ 10:1 TO MATCH EXISTING
PROPOSED SLOPE INTERCEPT (TYP.)
BEGIN CONSTRUCTION 200' RT.
PROPOSED 2" WATER SERVICE & 6" SEWER SERVICE. REMOVE CONC. M.H.
CONNECT TO EX. 6" TRANSITE W.M. W/VALVE
MATCH LINE
STA. 12+50 (SEE SHT. # 3)
B.M. HYD. EL = 1210.98
B.M. HYD. EL = 1221.51
B.M. NAIL IN P.P. EL = 1213.56
END CONSTR. 50' RT.
PARTIAL STORM S. PROFILE
WATER= 1199.1± 2/23/00
Q100 EL=1209±
30 C.Y. HEAVY RIP RAP
EDGE OF WATER
ENDWALL C1 INV= 1201.5
T.O. BANK EL = 1208.1
STORM S. B1 RIM = 1208.70 PROP. D= 4.5 INV = 1202.2
PROPOSED 960' - 6" PVC DR 18 WATERMAIN THIS PLATE.
NORTH TOP OF CURB LINE
SOUTH TOP OF CURB LINE
EXISTING ELEVATION (TYP.)
PROPOSED ELEVATION (TYP.)
PROPOSED GRADE (TYP.)
EXISTING GRADE (TYP.)
EX. SAN. M.H.1 RIM=7.11 D=8.66 INV.= 1198.48
EX. SAN. M.H.2 RIM=1209.52 D=9.40 INV.= 1200.12 PROP. RIM=1208.77
EX. SAN. M.H.3 RIM=1213.70 D=9.02 INV.= 1204.68 PROP. RIM=1212.30 D=7.62 PROP. INV.=1204.68
EX. SAN. M.H.4 RIM=1220.7 , D=9.49 INV.= 1211.21
PROP. RIM=1218.80 D=8.20 PROP. INV.=1210.60
2-23-00 GW = 1201.5+
348' - SAN. S. @ 0.47'/100'
548' - RELINE EX. SAN. SEWERMAIN
366' - SAN. S. @ 1.25'/100'
PROP. 166' - 8" PVC SDR35 SAN. S. @ 1.25'/100'
PROP. 366' - 8" PVC SDR35 SAN. S. @ 1.61'/100'
Mc SLOY ST.
SHT. 2 OF 35

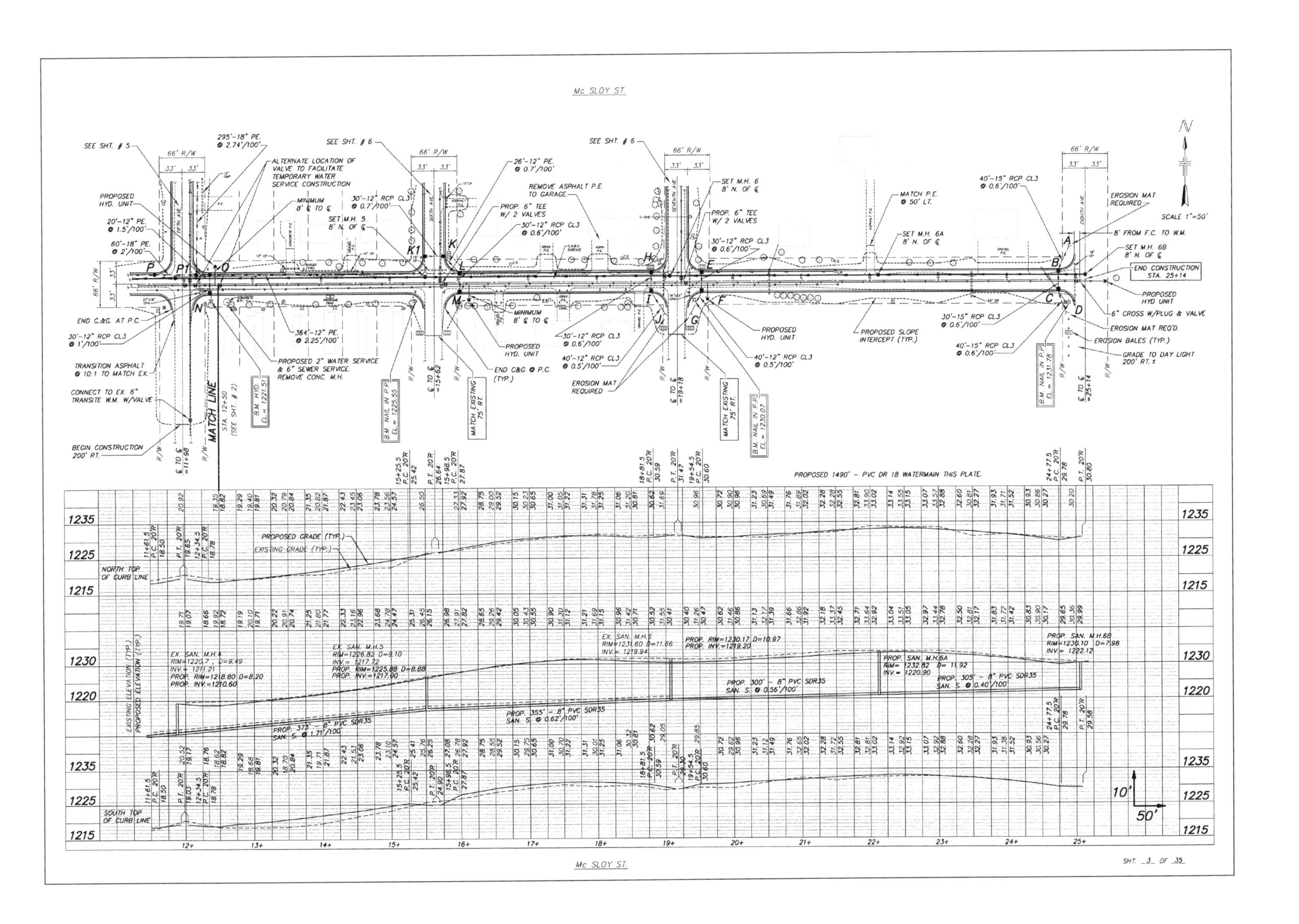
Mc SLOY ST.
PROPOSED 1490' - PVC DR 18 WATERMAIN THIS PLATE.
MATCH LINE
PROPOSED GRADE (TYP.)
EXISTING GRADE (TYP.)
NORTH TOP OF CURB LINE
SOUTH TOP OF CURB LINE
EXISTING ELEVATION (TYP.)
PROPOSED ELEVATION (TYP.)
SCALE 1"=50'
END CONSTRUCTION STA. 25+14
BEGIN CONSTRUCTION 200' RT.
SHT. 3 OF 35

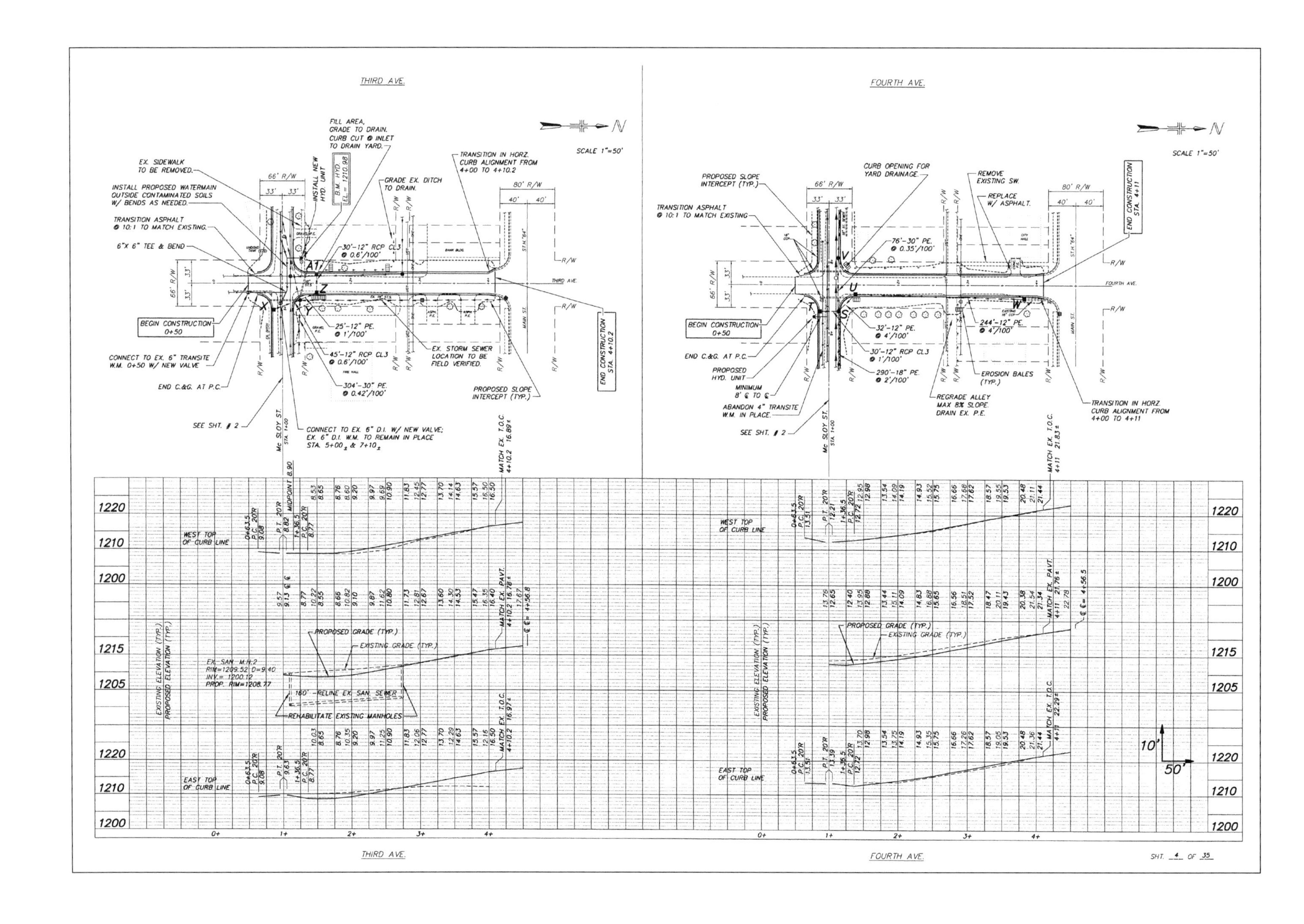
THIRD AVE.
FOURTH AVE.
SCALE 1"=50'
EX. SIDEWALK TO BE REMOVED.
INSTALL PROPOSED WATERMAIN OUTSIDE CONTAMINATED SOILS W/ BENDS AS NEEDED.
TRANSITION ASPHALT @ 10:1 TO MATCH EXISTING.
6"X 6" TEE & BEND
FILL AREA, GRADE TO DRAIN. CURB CUT @ INLET TO DRAIN YARD.
INSTALL NEW HYD. UNIT
B.M. HYD. EL.= 1210.98
GRADE EX. DITCH TO DRAIN.
TRANSITION IN HORZ. CURB ALIGNMENT FROM 4+00 TO 4+10.2
30'-12" RCP CL3 @ 0.6'/100'
25'-12" PE. @ 1'/100'
45'-12" RCP CL3 @ 0.6'/100'
304'-30" PE. @ 0.42'/100'
EX. STORM SEWER LOCATION TO BE FIELD VERIFIED.
PROPOSED SLOPE INTERCEPT (TYP.)
BEGIN CONSTRUCTION 0+50
CONNECT TO EX. 6" TRANSITE W.M. 0+50 W/ NEW VALVE
END C.&G. AT P.C.
SEE SHT. # 2
CONNECT TO EX. 6" D.I. W/ NEW VALVE; EX. 6" D.I. W.M. TO REMAIN IN PLACE STA. 5+00± & 7+10±
END CONSTRUCTION STA. 4+10.2
Mc SLOY ST.
MAIN ST.
PROPOSED SLOPE INTERCEPT (TYP.)
TRANSITION ASPHALT @ 10:1 TO MATCH EXISTING
CURB OPENING FOR YARD DRAINAGE.
REMOVE EXISTING SW.
REPLACE W/ ASPHALT.
76'-30" PE. @ 0.35'/100'
32'-12" PE. @ 4'/100'
30'-12" RCP CL3 @ 1'/100'
290'-18" PE. @ 2'/100'
244'-12" PE. @ 4'/100'
EROSION BALES (TYP.)
REGRADE ALLEY MAX 8% SLOPE. DRAIN EX. P.E.
END C.&G. AT P.C.
PROPOSED HYD. UNIT
MINIMUM 8' ℄ TO ℄
ABANDON 4" TRANSITE W.M. IN PLACE.
TRANSITION IN HORZ. CURB ALIGNMENT FROM 4+00 TO 4+11
END CONSTRUCTION STA. 4+11
WEST TOP OF CURB LINE
EAST TOP OF CURB LINE
EXISTING ELEVATION (TYP.)
PROPOSED ELEVATION (TYP.)
PROPOSED GRADE (TYP.)
EXISTING GRADE (TYP.)
EX. SAN. M.H.2 RIM=1209.52 D=9.40 INV.= 1200.12 PROP. RIM=1208.77
160' RELINE EX. SAN. SEWER
REHABILITATE EXISTING MANHOLES
MATCH EX. T.O.C. 4+10.2 16.89±
MATCH EX. PAVT. 4+10.2 16.78±
MATCH EX. T.O.C. 4+11 21.83±
MATCH EX. PAVT. 4+11 21.76±
10'
50'
SHT. 4 OF 35

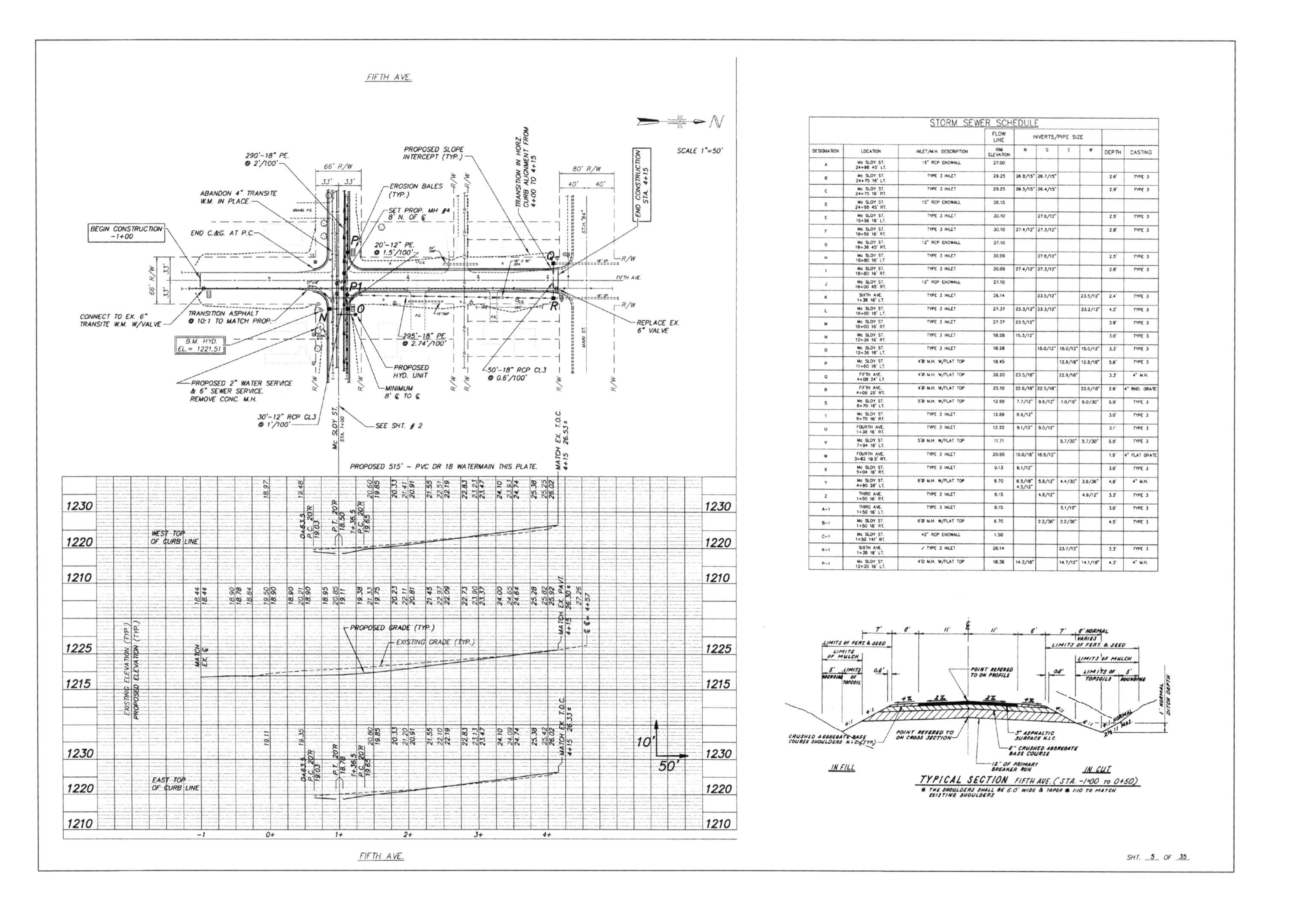

STORM SEWER SCHEDULE

DESIGNATION	LOCATION	INLET/M.H. DESCRIPTION	FLOW LINE RIM ELEVATION	INVERTS/PIPE SIZE N	S	E	W	DEPTH	CASTING
A	Mc SLOY ST. 24+98 45' LT.	15" RCP ENDWALL	27.00						
B	Mc SLOY ST. 24+75 16' LT.	TYPE 3 INLET	29.25	26.8/15"	26.7/15"			2.6'	TYPE 3
C	Mc SLOY ST. 24+75 16' RT.	TYPE 3 INLET	29.25	26.5/15"	26.4/15"			2.9'	TYPE 3
D	Mc SLOY ST. 24+98 45' RT.	15" RCP ENDWALL	28.15						
E	Mc SLOY ST. 19+56 16' LT.	TYPE 3 INLET	30.10		27.6/12"			2.5'	TYPE 3
F	Mc SLOY ST. 19+56 16' RT.	TYPE 3 INLET	30.10	27.4/12"	27.3/12"			2.8'	TYPE 3
G	Mc SLOY ST. 19+38 45' RT.	12" RCP ENDWALL	27.10						
H	Mc SLOY ST. 18+80 16' LT.	TYPE 3 INLET	30.09		27.6/12"			2.5'	TYPE 3
I	Mc SLOY ST. 18+80 16' RT.	TYPE 3 INLET	30.09	27.4/12"	27.3/12"			2.8'	TYPE 3
J	Mc SLOY ST. 19+00 45' RT.	12" RCP ENDWALL	27.10						
K	SIXTH AVE. 1+38 16' LT.	TYPE 3 INLET	26.14		23.5/12"		23.5/12"	2.4'	TYPE 3
L	Mc SLOY ST. 16+00 16' LT.	TYPE 3 INLET	27.37	23.3/12"	23.3/12"		23.2/12"	4.2'	TYPE 3
M	Mc SLOY ST. 16+00 16' RT.	TYPE 3 INLET	27.37	23.5/12"				3.9'	TYPE 3
N	Mc SLOY ST. 12+36 16' RT.	TYPE 3 INLET	18.28	15.3/12"				3.0'	TYPE 3
O	Mc SLOY ST. 12+36 16' LT.	TYPE 3 INLET	18.28		15.0/12"	15.0/12"	15.0/12"	3.3'	TYPE 3
P	Mc SLOY ST. 11+60 16' LT.	4'Ø M.H. W/FLAT TOP	18.45			12.9/18"	12.8/18"	5.6'	TYPE 3
Q	FIFTH AVE. 4+08 24' LT	4'Ø M.H. W/FLAT TOP	26.20	23.5/18"		22.9/18"		3.3'	4" M.H.
R	FIFTH AVE. 4+08 25' RT.	4'Ø M.H. W/FLAT TOP	25.10	22.6/18"	22.5/18"		22.6/18"	2.6'	4" RND. GRATE
S	Mc SLOY ST. 8+70 16' LT.	5'Ø M.H. W/FLAT TOP	12.89	7.7/12"	9.6/12"	7.0/18"	6.0/30"	6.9'	TYPE 3
T	Mc SLOY ST. 8+70 16' RT.	TYPE 3 INLET	12.89	9.9/12"				3.0'	TYPE 3
U	FOURTH AVE. 1+38 16' RT.	TYPE 3 INLET	12.22	9.1/12"	9.0/12"			3.1'	TYPE 3
V	Mc SLOY ST. 7+94 16' LT.	5'Ø M.H. W/FLAT TOP	11.71			5.7/30"	5.7/30"	6.0'	TYPE 3
W	FOURTH AVE. 3+82 19.5' RT.	TYPE 3 INLET	20.90	19.0/18"	18.9/12"			1.9'	4" FLAT GRATE
X	Mc SLOY ST. 5+04 16' RT.	TYPE 3 INLET	9.13	6.1/12"				3.0'	TYPE 3
Y	Mc SLOY ST. 4+90 26' LT.	6'Ø M.H. W/FLAT TOP	8.70	6.5/18" 4.5/12"	5.8/12"	4.4/30"	3.9/36"	4.8'	4" M.H.
Z	THIRD AVE. 1+50 16' RT.	TYPE 3 INLET	8.15		4.8/12"		4.9/12"	3.3'	TYPE 3
A-1	THIRD AVE. 1+50 16' LT.	TYPE 3 INLET	8.15			5.1/12"		3.0'	TYPE 3
B-1	Mc SLOY ST. 1+50 16' RT.	6'Ø M.H. W/FLAT TOP	6.70		2.2/36"	2.2/36"		4.5'	TYPE 3
C-1	Mc SLOY ST. 1+50 141' RT.	42" RCP ENDWALL	1.50						
K-1	SIXTH AVE. 1+38 16' LT.	/ TYPE 3 INLET	26.14			23.7/12"		3.3'	TYPE 3
P-1	Mc SLOY ST. 12+20 16' LT.	4'O M.H. W/FLAT TOP	18.36	14.2/18"		14.7/12"	14.1/18"	4.3'	4" M.H.

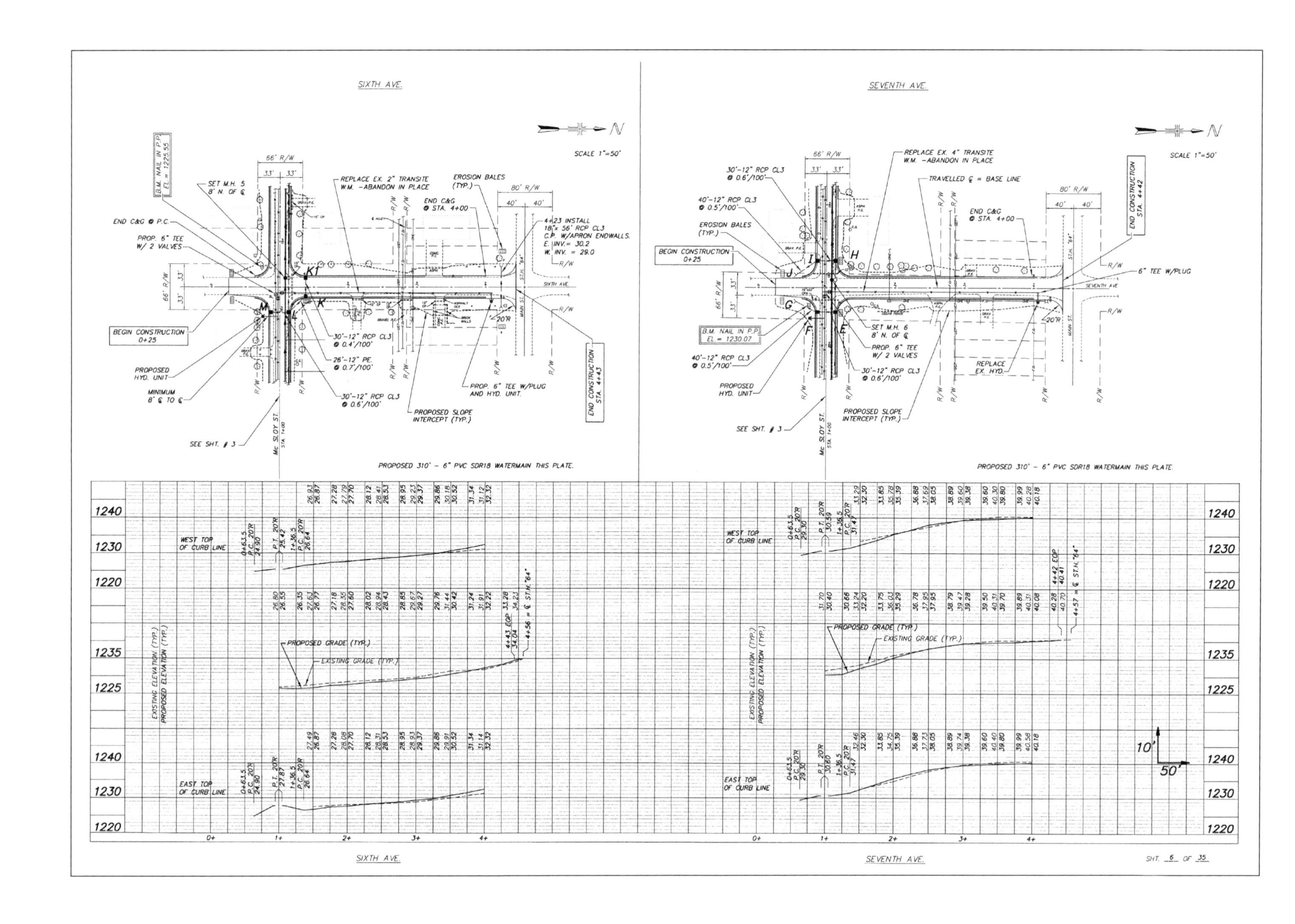
SIXTH AVE.
SEVENTH AVE.
SCALE 1"=50'
B.M. NAIL IN P.P. EL.= 1225.55
66' R/W
SET M.H. 5 8' N. OF ℄
END C&G @ P.C.
PROP. 6" TEE W/ 2 VALVES
REPLACE EX. 2" TRANSITE W.M. -ABANDON IN PLACE
EROSION BALES (TYP.)
END C&G @ STA. 4+00
80' R/W
4+23 INSTALL 18"x 56' RCP CL3 C.P. W/APRON ENDWALLS. E. INV.= 30.2 W. INV. = 29.0
BEGIN CONSTRUCTION 0+25
30'-12" RCP CL3 @ 0.4'/100'
26'-12" PE. @ 0.7'/100'
30'-12" RCP CL3 @ 0.6'/100'
PROPOSED HYD. UNIT
MINIMUM 8' ℄ TO ℄
PROP. 6" TEE W/PLUG AND HYD. UNIT.
PROPOSED SLOPE INTERCEPT (TYP.)
END CONSTRUCTION STA. 4+43
SEE SHT. # 3
Mc SLOY ST. STA. 1+00
PROPOSED 310' - 6" PVC SDR18 WATERMAIN THIS PLATE.
REPLACE EX. 4" TRANSITE W.M. -ABANDON IN PLACE
30'-12" RCP CL3 @ 0.6'/100'
40'-12" RCP CL3 @ 0.5'/100'
TRAVELLED ℄ = BASE LINE
END CONSTRUCTION STA. 4+42
6" TEE W/PLUG
B.M. NAIL IN P.P. EL.= 1230.07
SET M.H. 6 8' N. OF ℄
PROP. 6" TEE W/ 2 VALVES
REPLACE EX. HYD.
WEST TOP OF CURB LINE
EAST TOP OF CURB LINE
PROPOSED GRADE (TYP.)
EXISTING GRADE (TYP.)
EXISTING ELEVATION (TYP.)
PROPOSED ELEVATION (TYP.)
10'
50'
SHT. 6 OF 35

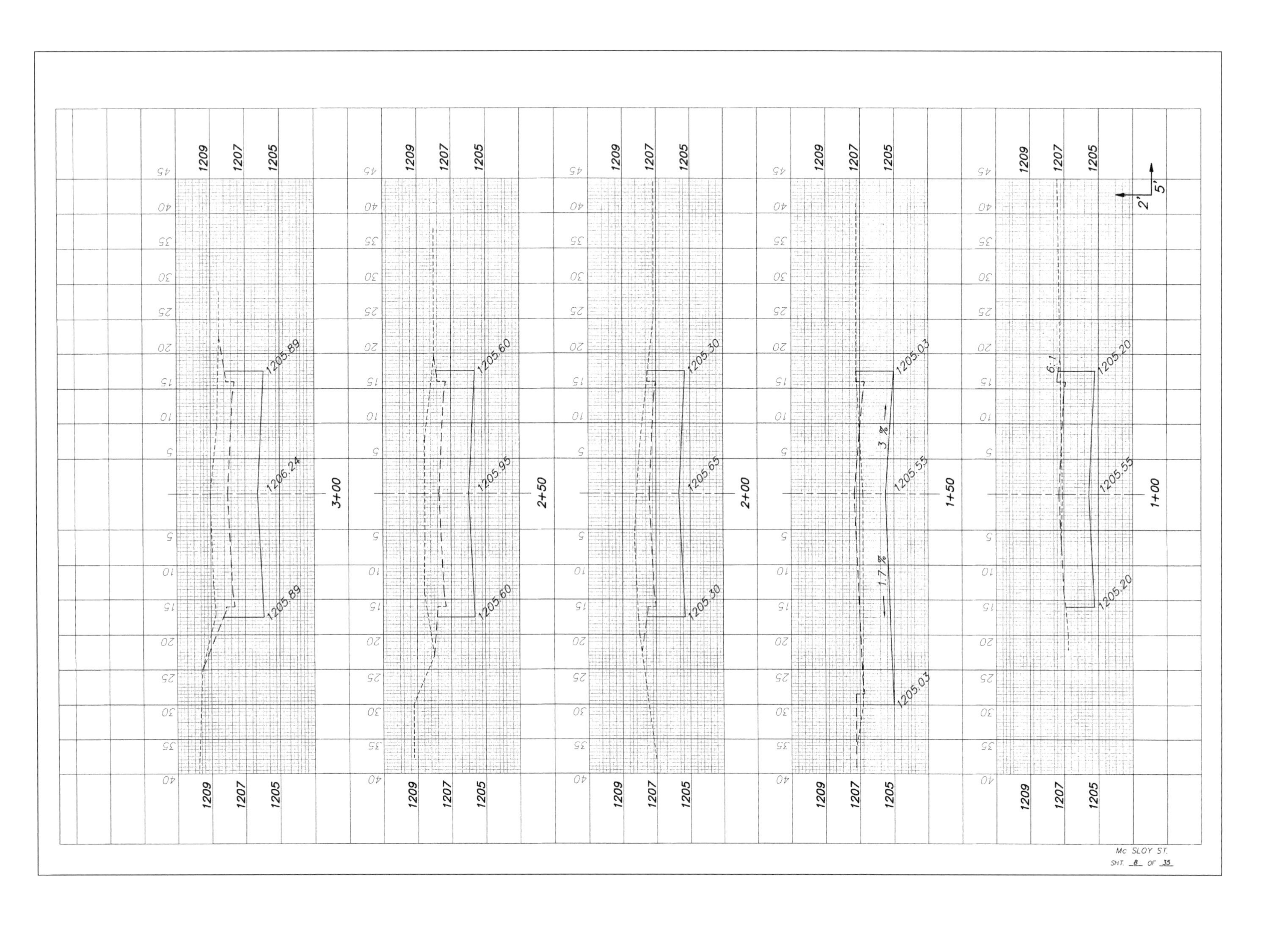
1209
1207
1205
3+00
1205.89
1206.24
1205.89
2+50
1205.60
1205.95
1205.60
2+00
1205.30
1205.65
1205.30
1+50
1205.03
3 %
1205.55
1.7 %
1205.03
1+00
6:1
1205.20
1205.55
1205.20
5'
2'
Mc SLOY ST.
SHT. 8 OF 35

STREET RECONSTRUCTION for the VILLAGE of HAWKINS, WISCONSIN SCANDINAVIA AVENUE

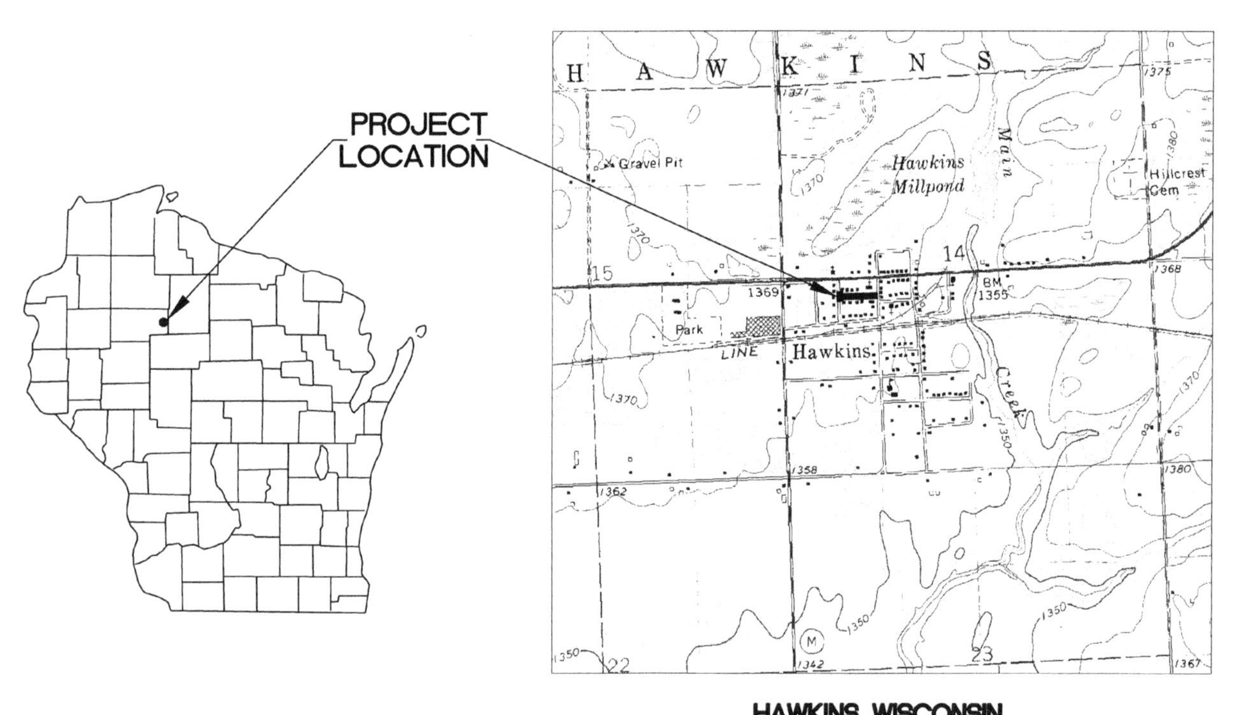

INDEX OF SHEETS

SHEET NO.	DESCRIPTION
1	TITLE SHEET
2	EXISTING CONDITIONS
3	SANITARY SEWER & WATER
4	CURB & GUTTER
5	TYPICAL CROSS SECTIONS & GENERAL NOTES
6-13	CROSS SECTIONS
14	EROSION CONTROL

NOTE: Sheets 7 thru 14 Not Shown

TITLE SHEET	
HAWKINS, WISCONSIN	
SCALE – NONE	MORGAN & PARMLEY, LTD. CONSULTING ENGINEERS
SHEET 1 OF 14	LADYSMITH, WISCONSIN
DATE MARCH, 2000	DR. BY D.A.N.

Courtesy: Morgan & Parmley, Ltd.

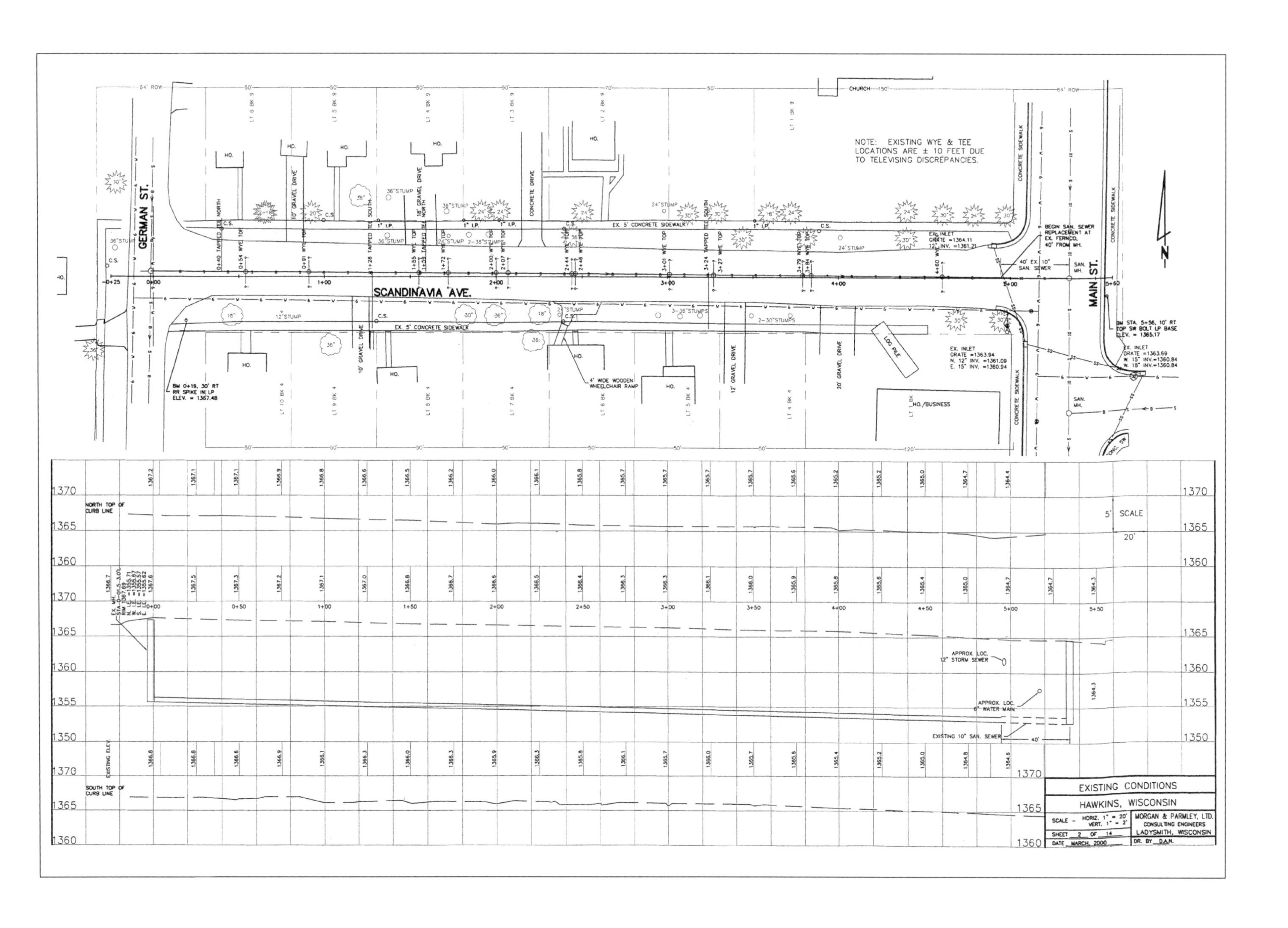

NOTE: EXISTING WYE & TEE LOCATIONS ARE ± 10 FEET DUE TO TELEVISING DISCREPANCIES.
GERMAN ST.
SCANDINAVIA AVE.
MAIN ST.
CHURCH
EX. 5' CONCRETE SIDEWALK
CONCRETE SIDEWALK
BEGIN SAN. SEWER REPLACEMENT AT EX. FERNCO, 40' FROM MH.
40' EX. 10" SAN. SEWER
SAN. MH.
BM STA. 5+56, 10' RT TOP SW BOLT LP BASE ELEV. = 1365.17
EX. INLET GRATE =1363.69 W. 15" INV.=1360.84 W. 18" INV.=1360.84
EX. INLET GRATE =1363.94 N. 12" INV. =1361.09 E. 15" INV. =1360.94
EX. INLET GRATE =1364.11 12" INV. =1361.21
BM 0+19, 30' RT RR SPIKE IN LP ELEV. = 1367.48
4' WIDE WOODEN WHEELCHAIR RAMP
LOG PILE
HO./BUSINESS
NORTH TOP OF CURB LINE
SOUTH TOP OF CURB LINE
EXISTING ELEV.
APPROX. LOC. 12" STORM SEWER
APPROX. LOC. 6" WATER MAIN
EXISTING 10" SAN. SEWER
5' SCALE 20'
EXISTING CONDITIONS
HAWKINS, WISCONSIN
SCALE - HORIZ. 1" = 20' VERT. 1" = 2'
MORGAN & PARMLEY, LTD. CONSULTING ENGINEERS LADYSMITH, WISCONSIN
SHEET 2 OF 14
DATE MARCH, 2000
DR. BY D.A.N.

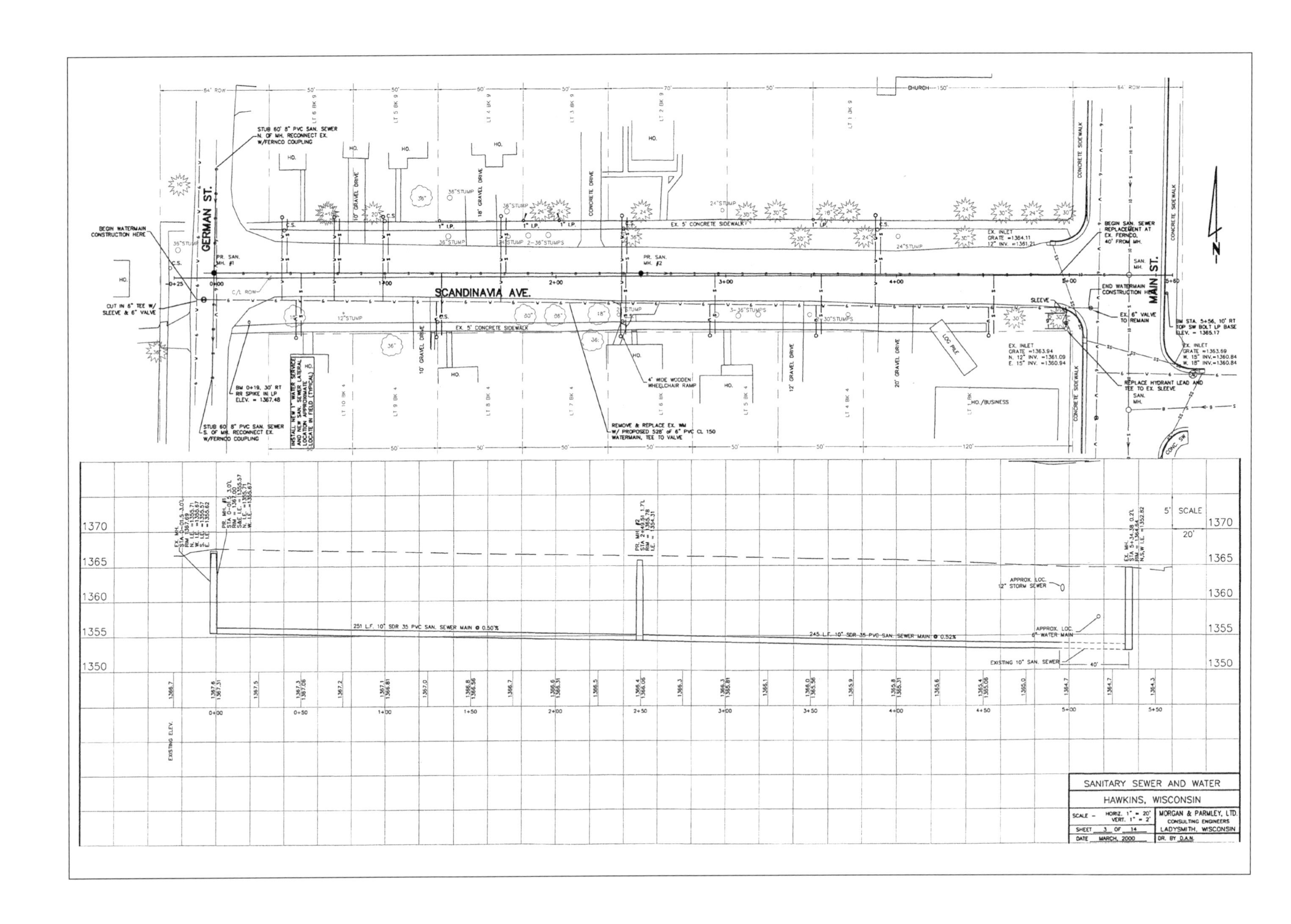
SANITARY SEWER AND WATER
HAWKINS, WISCONSIN
SCALE – HORIZ. 1" = 20' VERT. 1" = 2'
MORGAN & PARMLEY, LTD.
CONSULTING ENGINEERS
LADYSMITH, WISCONSIN
SHEET 3 OF 14
DATE MARCH, 2000
DR. BY D.A.N.
SCANDINAVIA AVE.
GERMAN ST.
MAIN ST.

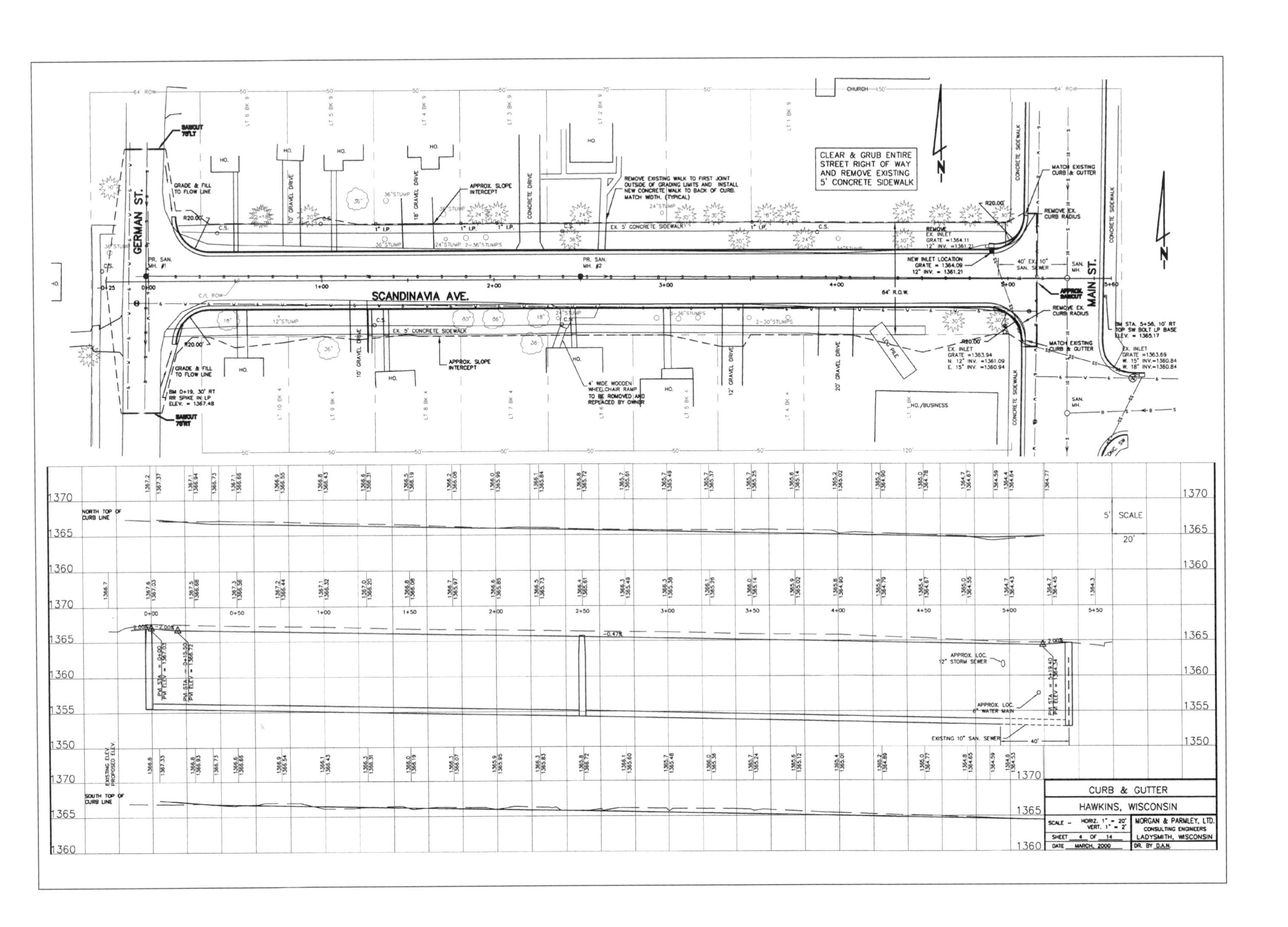
CHURCH
CLEAR & GRUB ENTIRE STREET RIGHT OF WAY AND REMOVE EXISTING 5' CONCRETE SIDEWALK
GERMAN ST.
SCANDINAVIA AVE.
MAIN ST.
CONCRETE SIDEWALK
CONCRETE DRIVE
10' GRAVEL DRIVE
18' GRAVEL DRIVE
12' GRAVEL DRIVE
20' GRAVEL DRIVE
EX. 5' CONCRETE SIDEWALK
GRADE & FILL TO FLOW LINE
APPROX. SLOPE INTERCEPT
REMOVE EXISTING WALK TO FIRST JOINT OUTSIDE OF GRADING LIMITS AND INSTALL NEW CONCRETE WALK TO BACK OF CURB. MATCH WIDTH. (TYPICAL)
PR. SAN. MH. #1
PR. SAN. MH. #2
NEW INLET LOCATION GRATE = 1364.09 12" INV. = 1361.21
MATCH EXISTING CURB & GUTTER
REMOVE EX. CURB RADIUS
4' WIDE WOODEN WHEELCHAIR RAMP TO BE REMOVED AND REPLACED BY OWNER
BM 0+19, 30' RT RR SPIKE IN LP ELEV. = 1367.48
BM STA. 5+56, 10' RT TOP SW BOLT LP BASE ELEV. = 1365.17
64' R.O.W.
HO./BUSINESS
NORTH TOP OF CURB LINE
SOUTH TOP OF CURB LINE
EXISTING ELEV. PROPOSED ELEV.
APPROX. LOC. 12" STORM SEWER
APPROX. LOC. 6" WATER MAIN
EXISTING 10" SAN. SEWER
SCALE
CURB & GUTTER
HAWKINS, WISCONSIN
SCALE – HORIZ. 1" = 20' VERT. 1" = 2'
SHEET 4 OF 14
DATE MARCH, 2000
MORGAN & PARMLEY, LTD. CONSULTING ENGINEERS LADYSMITH, WISCONSIN
DR. BY D.A.N.

GENERAL NOTES

1. ALL WATER MAIN SHALL BE 6" PVC DR 18.
2. ALL WATERMAIN FITTINGS SHALL BE M.J.D.I., W / RESTRAINED JOINT LUGS.
3. ALL HYDRANT LEADS SHALL BE 6-INCH VALVE.
4. HYDRANTS SHALL BE WATEROUS WB 67-250 (4" PUMPER NOZZLE), 22" BREAKOFF WITH THREADS MATCHING VILLAGE'S PATTERN.
5. ALL WATER MAIN SHALL HAVE AN 8 FT. BURY (MIN.).
6. WATER SERVICES SHALL BE 1" TYPE K COPPER W / MUELLER OR FORD BRASS.
7. 8' (MINIMUM) HORIZONTAL DISTANCE (℄ TO ℄) BETWEEN WATERMAIN AND SANITARY SEWER OR STORM SEWER.
8. 18" (MINIMUM) VERTICAL DISTANCE (OUT TO OUT) BETWEEN WATERMAIN UNDER SANITARY SEWER OR STORM SEWER @ THEIR INTERSECTION.
9. 6" (MINIMUM) VERTICAL DISTANCE (OUT TO OUT) BETWEEN WATERMAIN OVER SANITARY SEWER OR STORM SEWER @ THEIR INTERSECTION.
10. SANITARY SERVICES SHALL BE 4" SCH. 40 PVC W / WYES LAID TO PROVIDE FOR MAXIMUM DEPTH OR 10' AT THE R / W.
11. SEWERMAIN SHALL BE PVC SDR 35.
12. MANHOLES SHALL BE SET WITH A MINIMUM OF 6" OF RINGS AND 9" CASTINGS.
13. MANHOLES SHALL BE 4' DIAMETER PRECAST CONCRETE W / INTEGRAL BASES AND BOOTS.
14. ALL ELEVATIONS BASED ON USGS DATUM.
15. EROSION CONTROL MEASURES SHALL BE IN PLACE BEFORE CONSTRUCTION BEGINS.
16. THE CONTRACTORS SHALL MINIMIZE EROSION DURING CONSTRUCTION USING GOOD CONSTRUCTION TECHNIQUES AND UTILIZING SILT FENCES AND BALE DITCH CHECKS. THE CONTRACTOR SHALL RELEASE RUNOFF FROM THE SITE IN A NUISANCE FREE MANNER.
17. THE CONTRACTOR IS RESPONSIBLE TO MAINTAIN THE SITE IN A SAFE CONDITION. THE SOLE RESPONSIBILITY FOR WARNING SIGNS, BARRICADES, FLAGGING PERSONNEL AND ALL ASPECTS OF SAFETY LIE WITH THE CONTRACTOR.
18. EXISTING UTILITIES ARE SHOWN BUT MAY NOT BE ALL INCLUSIVE; CONTRACTOR SHALL HAVE ALL BURIED UTILITIES LOCATED PRIOR TO COMMENCING CONSTRUCTION.
19. EXISTING SEWER AND WATERMAIN ARE SHOWN ACCORDING TO EXISTING RECORDS BUT LOCATION SHALL BE VERIFIED IN FIELD.
20. WATERMAIN & SERVICES SHALL BE INSULATED AT ALL STORM SEWER, CULVERT AND DITCH CROSSINGS WITH 2" THICK EXTRUDED POLYSTYRENE INSULATION BOARD.
21. PROPERTY CORNERS KNOWN TO EXIST SHALL NOT BE DISTURBED. DAMAGED CORNERS WILL BE REPLACED AT THE CONTRACTOR'S EXPENSE. PRESERVE ALL SURVEY MONUMENTS.
22. ALL WATERMAIN SHALL BE DISINFECTED BEFORE BEING PUT INTO USE.
23. VALVES AND MANHOLES SHALL BE ADJUSTED TO FINAL GRADE, UNLESS DIRECTED OTHERWISE.
24. CONSTRUCTION SHALL COMPLY WITH THE CONDITIONS OF DNR APPROVAL ELTTER, AND OTHER REGULATORY APPROVALS AS LISTED IN THE SPECIFICATION BOOK.
25. ALL DISTURBED AREAS SHALL BE SEEDED, FERTILIZED AND MULCHED, OR COVERED W / BASE COURSE AFTER CONSTRUCTION, OR RESTORED WITH BITUMINOUS PAVEMENT WHERE APPLICABLE.
26. THERE SHALL BE NO DEVIATION FROM THE PLANS WITHOUT THE DESIGN ENGINEER'S APPROVAL.

28. ALL DISTURBED AREAS TO BE SEEDED SHALL BE RESTORED AS FOLLOWS:

A) REPLACED ALL SALVAGED AND IMPORTED TOPSOIL
B) SEED MIXUTRE: PERMANENT SEEDING – (LAWN TYPE TURF)

35% KENTUCKY BLUEGRASS
25% IMPROVED FINE PERENNIAL RYEGRASS
20% CREEPING RED FESCUE
20% IMPROVED HARD FESCUE

C) SEED RATE: 2# / 100 SQ. FT.
D) MULCH RATE: 2 TON / ACRE
E) PERMANENT SEEDING: ALLOWED TO SEPT. 7
TEMPORARY SEEDING: SEPT. 8 THROUGH NOV. 10
(4 BU. OATS / ACRE)
DORMANT SEEDING; SEED PERM. SEEDING INTO TEMP. SEEDING AFTER NOV. 10
F) SEEDING MAY BE BRAODCAST BUT MUST BE COVERED WITH A MAXIMUM OF ½" SOIL. MULCH MUST BE ANCHORED WITH A MULCH TILLER, TACKIFIER OR NETTING.
G) FERTILIZER: 500# / ACRE 20-10-10

29. CONTRACTOR MUST REMOVE AND REPLACE ALL AFFECTED STREET SIGNS AFTER PROPERLY SIGNING AND BARRICADING THE CONSTRUCTION AREA. ALL SIGNS MUST BE REPLACED PRIOR TO OPENING THE STREET TO TRAFFIC.
30. CONTRACTOR SHALL DETERMINE CONSTRUCTION SEQUENCE IN ORDER TO MINIMIZE INSTALLATION CONFLICTS WITH PIPE CROSSINGS.
31. NO TREES ARE TO BE REMOVED WITHOUT THE APPROVAL OF THE ENGINEER.

SPECIFICATIONS:

WISCONSIN CONSTRUCTION SITE BEST MANAGEMENT PRACTICE HANDBOOK

STANDARD SPECIFICATIONS FOR SEWER AND WATER CONSTRUCTION IN WISCONSIN (5TH EDITION)

STANDARD SPECIFICATIONS FOR ROAD AND BRIDGE CONSTRUCITON 1996 EDITION WIS D.O.T.

-UTILIITES-

WATER & SEWER:
VILLAGE OF HAWKINS
DAVE HETTINGER
HAWKINS VILLAGE HALL
P.O. BOX 108
HAWKINS, WI 54530
1-715-585-6322

ELECTRIC:
NORTHERN STATES
POWER COMPANY
310 HICKORY HILL LANE
PHILLIPS, WI 54555
JOE PERKINS
1-800-445-1275

CABLE TELEVISION:
KRM CABLEVISION
BOX 77
MELLON, WI 54546
1-800-327-4330

TELEPHONE:
CENTURYTEL
425 ELLINGSON AVE.
P.O. BOX 78
HAWKINS, WI 54530
JIM ARQUETTE
1-715-585-7707

DIGGERS HOTLINE:
1-800-242-8511

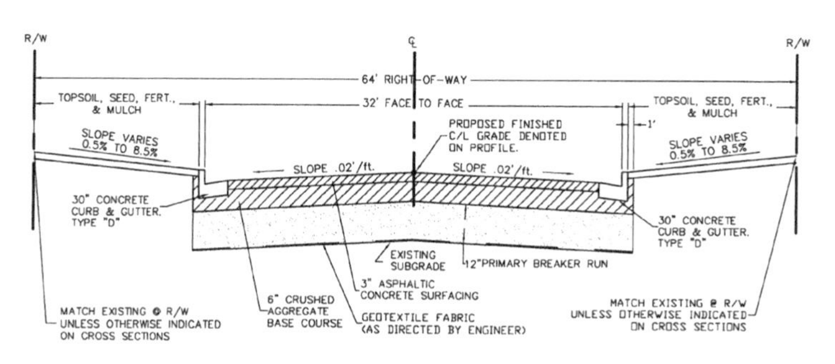

FINISHED STREET TYPICAL SECTION
NOT TO SCALE

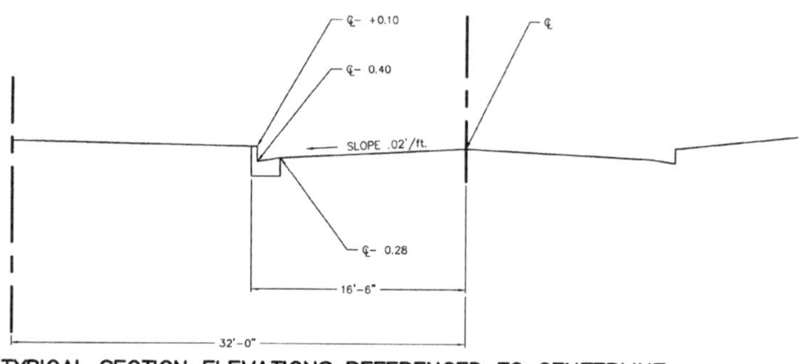

TYPICAL SECTION ELEVATIONS REFERENCED TO CENTERLINE
NOT TO SCALE

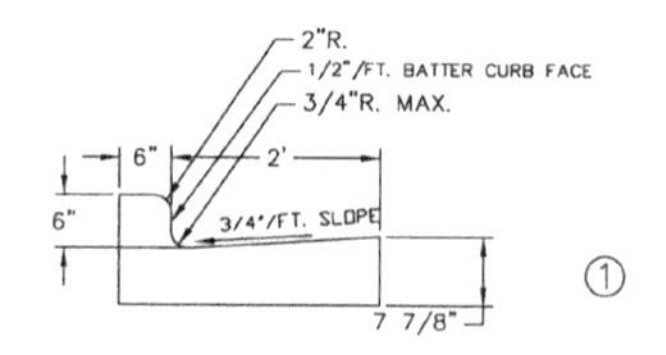

① THE BOTTOM OF CURB AND GUTTER MAY BE CONSTRUCTED EITHER LEVEL OR PARALLEL TO THE SLOPE OF THE SUBGRADE OR BASE COURSE PROVIDED A 6" MIN. GUTTER THICKNESS IS MAINTAINED.

TYPE "D" CURB

TYPICAL CROSS SECTIONS	
HAWKINS, WISCONSIN	
SCALE - NONE	MORGAN & PARMLEY, LTD. CONSULTING ENGINEERS LADYSMITH, WISCONSIN
SHEET 5 OF 14	
DATE MARCH, 2000	DR. BY D.A.N.

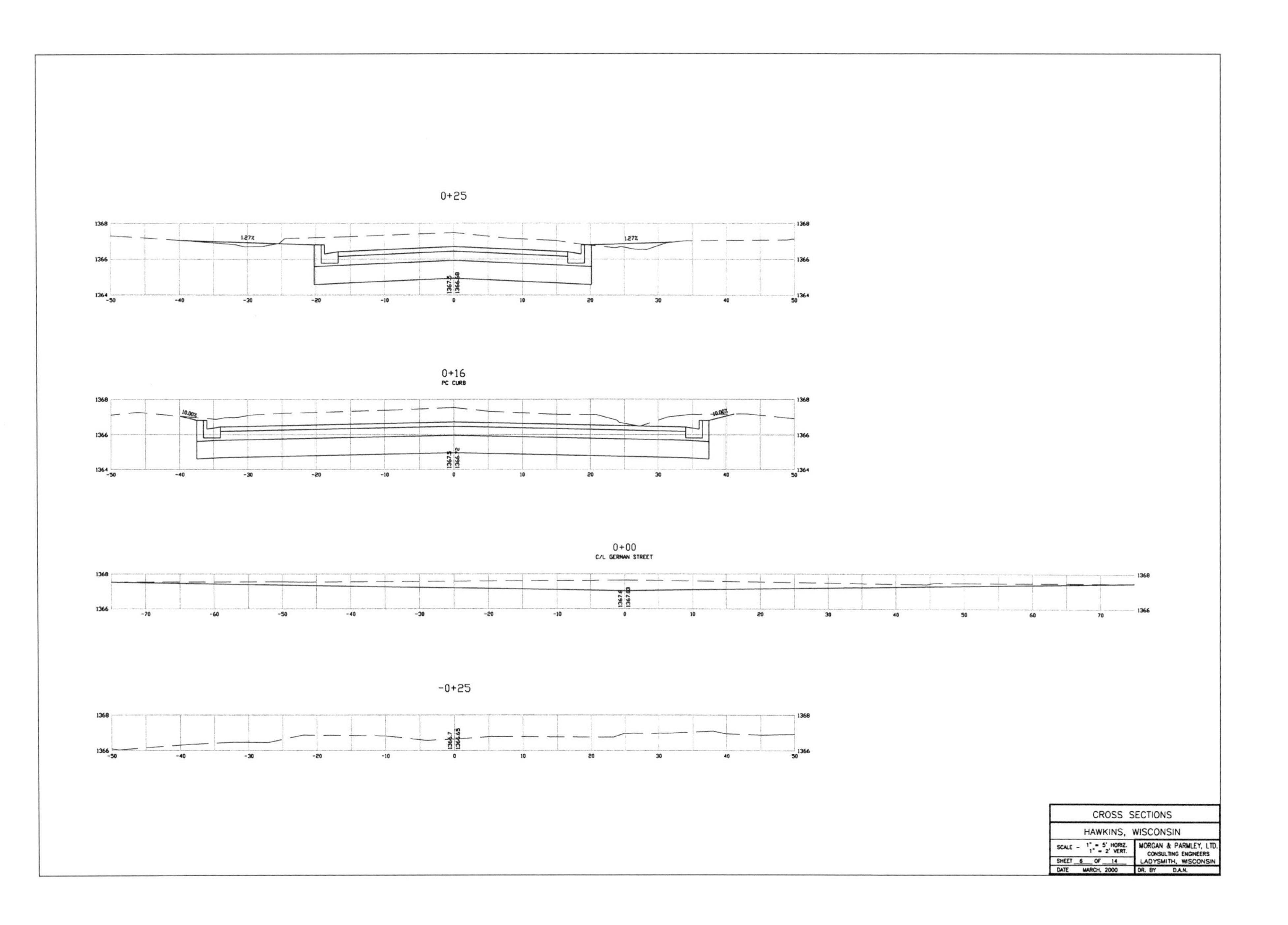

0+25
1.27%
1.27%
1367.5
1366.58
0+16
PC CURB
10.00%
10.00%
1367.5
1366.72
0+00
C/L GERMAN STREET
1367.6
1367.03
-0+25
1366.7
1366.65
CROSS SECTIONS
HAWKINS, WISCONSIN
SCALE - 1" = 5' HORIZ. 1" = 2' VERT.
MORGAN & PARMLEY, LTD.
CONSULTING ENGINEERS
LADYSMITH, WISCONSIN
SHEET 6 OF 14
DATE MARCH, 2000
DR. BY D.A.N.

STORM SEWER EXTENSION
HAYWARD, WISCONSIN

SECOND ST., WISCONSIN AVE. & THIRD ST.

APRIL, 1980

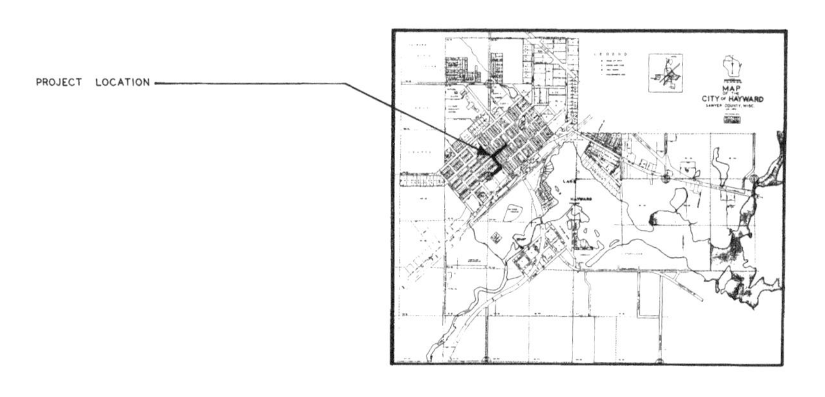

GENERAL MATERIAL LIST

DESCRIPTION	TOTALS
* 24" CONCRETE PIPE, CLASS IV, MORTAR JOINT W/ PERFORATION HOLES	370 L.F.
* 24" CONCRETE PIPE, CLASS III, MORTAR JOINT	575 L.F.
* 18" CONCRETE PIPE, CLASS IV, MORTAR JOINT	370 L.F.
* REINFORCED CONCRETE M.H. - 48" DIA.	52 V.F. (7 UNITS)
* REINFORCED CONCRETE M.H. BASES	7 UNITS
* REINFORCED CONCRETE M.H. COVERS - FLAT 6"	7 UNITS
* CATCH BASINS (24")	100 V.F. (17 UNITS)
* CATCH BASIN BASES	17 UNITS
* CATCH BASIN FRAME & COVER (C.I.)	17 UNITS
* 12" CONCRETE PIPE, CLASS IV (C.B. LEADS)	420 L.F.
* MANHOLE FRAME & COVER (C.I.)	7 UNITS

INDEX OF SHEETS

SHEET	DESCRIPTION
1	TITLE SHEET W/ PROJECT LOCATION
2	PLAN & PROFILE DETAILS
3	TYPICAL STORM SEWER DETAILS

PREPARED BY: MORGAN & PARMLEY, LTD. LADYSMITH, WISCONSIN

Courtesy: Morgan & Parmley, Ltd.

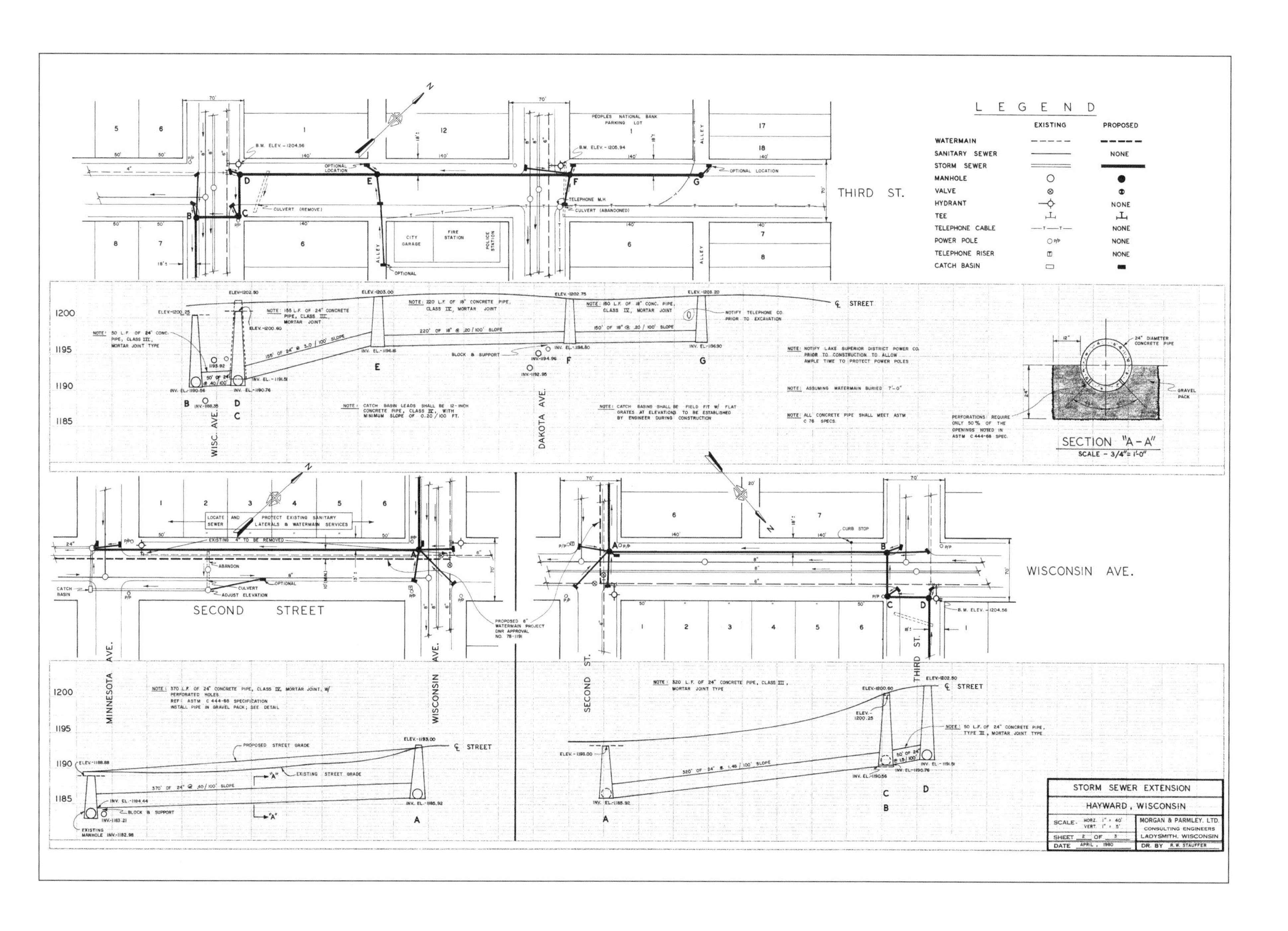
LEGEND
EXISTING
PROPOSED
WATERMAIN
SANITARY SEWER
STORM SEWER
MANHOLE
VALVE
HYDRANT
TEE
TELEPHONE CABLE
POWER POLE
TELEPHONE RISER
CATCH BASIN
NONE
THIRD ST.
WISCONSIN AVE.
SECOND STREET
MINNESOTA AVE.
WISC. AVE.
DAKOTA AVE.
SECOND ST.
PEOPLES NATIONAL BANK PARKING LOT
CITY GARAGE
FIRE STATION
POLICE STATION
ALLEY
OPTIONAL LOCATION
CULVERT (REMOVE)
CULVERT (ABANDONED)
TELEPHONE M.H.
B.M. ELEV. - 1204.56
B.M. ELEV. - 1205.94
NOTE: 50 L.F. OF 24" CONC. PIPE, CLASS III, MORTAR JOINT TYPE
NOTE: 155 L.F. OF 24" CONCRETE PIPE, CLASS III, MORTAR JOINT
NOTE: 220 L.F. OF 18" CONCRETE PIPE, CLASS IV, MORTAR JOINT
NOTE: 150 L.F. OF 18" CONC. PIPE, CLASS IV, MORTAR JOINT
NOTIFY TELEPHONE CO. PRIOR TO EXCAVATION
NOTE: NOTIFY LAKE SUPERIOR DISTRICT POWER CO. PRIOR TO CONSTRUCTION TO ALLOW AMPLE TIME TO PROTECT POWER POLES
NOTE: ASSUMING WATERMAIN BURIED 7'-0"
NOTE: ALL CONCRETE PIPE SHALL MEET ASTM C 76 SPECS.
NOTE: CATCH BASIN LEADS SHALL BE 12-INCH CONCRETE PIPE, CLASS IV, WITH MINIMUM SLOPE OF 0.20/100 FT.
NOTE: CATCH BASINS SHALL BE FIELD FIT W/ FLAT GRATES AT ELEVATIONS TO BE ESTABLISHED BY ENGINEER DURING CONSTRUCTION
BLOCK & SUPPORT
PERFORATIONS REQUIRE ONLY 50% OF THE OPENINGS NOTED IN ASTM C 444-68 SPEC.
24" DIAMETER CONCRETE PIPE
GRAVEL PACK
SECTION "A-A"
SCALE - 3/4" = 1'-0"
LOCATE AND PROTECT EXISTING SANITARY SEWER LATERALS & WATERMAIN SERVICES
EXISTING 4" TO BE REMOVED
ABANDON
ADJUST ELEVATION
CATCH BASIN
PROPOSED 8" WATERMAIN PROJECT DNR APPROVAL NO. 78-1191
CURB STOP
NOTE: 370 L.F. OF 24" CONCRETE PIPE, CLASS IV, MORTAR JOINT, W/ PERFORATED HOLES. REF: ASTM C 444-68 SPECIFICATION INSTALL PIPE IN GRAVEL PACK; SEE DETAIL
PROPOSED STREET GRADE
EXISTING STREET GRADE
EXISTING MANHOLE INV.-1182.98
NOTE: 320 L.F. OF 24" CONCRETE PIPE, CLASS III, MORTAR JOINT TYPE
NOTE: 50 L.F. OF 24" CONCRETE PIPE, TYPE III, MORTAR JOINT TYPE
℄ STREET
STORM SEWER EXTENSION
HAYWARD, WISCONSIN
SCALE - HORZ. 1" = 40' VERT. 1" = 5'
SHEET 2 OF 3
DATE APRIL, 1980
MORGAN & PARMLEY, LTD. CONSULTING ENGINEERS LADYSMITH, WISCONSIN
DR. BY R.W. STAUFFER

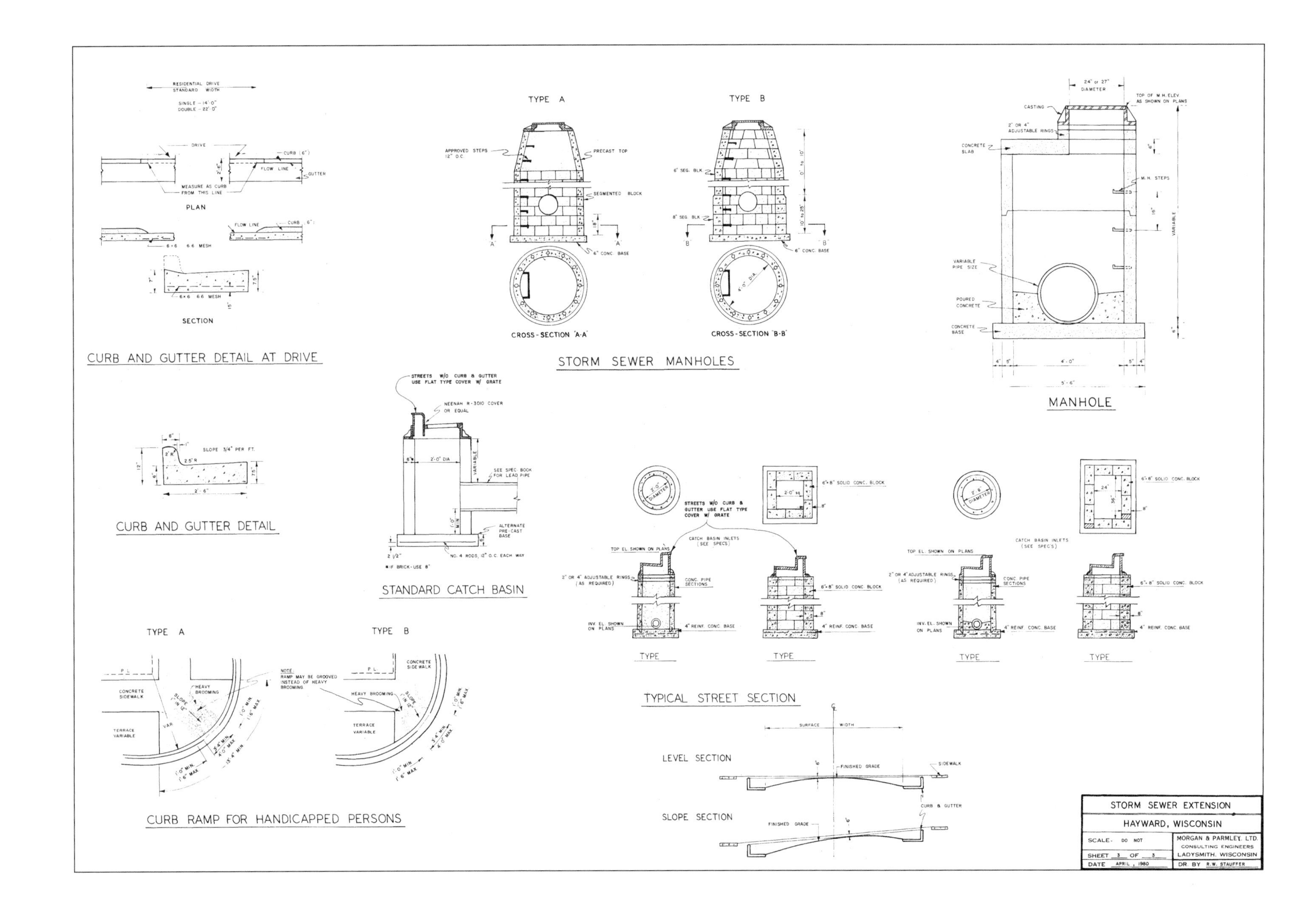
CURB AND GUTTER DETAIL AT DRIVE
PLAN
SECTION
RESIDENTIAL DRIVE STANDARD WIDTH
SINGLE – 14'-0"
DOUBLE – 22'-0"
DRIVE
CURB (6")
FLOW LINE
GUTTER
MEASURE AS CURB FROM THIS LINE
6×6 6 6 MESH
STORM SEWER MANHOLES
TYPE A
TYPE B
APPROVED STEPS 12" O.C.
PRECAST TOP
SEGMENTED BLOCK
6" SEG. BLK
8" SEG. BLK
6" CONC. BASE
CROSS-SECTION 'A-A'
CROSS-SECTION 'B-B'
MANHOLE
24" or 27" DIAMETER
TOP OF M.H. ELEV AS SHOWN ON PLANS
CASTING
2" OR 4" ADJUSTABLE RINGS
CONCRETE SLAB
M.H. STEPS
VARIABLE PIPE SIZE
POURED CONCRETE
CONCRETE BASE
VARIABLE
CURB AND GUTTER DETAIL
SLOPE 3/4" PER FT.
STANDARD CATCH BASIN
STREETS W/O CURB & GUTTER USE FLAT TYPE COVER W/ GRATE
NEENAH R-3010 COVER OR EQUAL
2'-0" DIA
SEE SPEC. BOOK FOR LEAD PIPE
ALTERNATE PRE-CAST BASE
NO. 4 RODS, 12" O.C. EACH WAY
*IF BRICK-USE 8"
CATCH BASIN INLETS (SEE SPECS)
TOP EL. SHOWN ON PLANS
2" OR 4" ADJUSTABLE RINGS (AS REQUIRED)
CONC. PIPE SECTIONS
6"×8" SOLID CONC. BLOCK
INV. EL. SHOWN ON PLANS
4" REINF. CONC. BASE
TYPE
TYPICAL STREET SECTION
SURFACE WIDTH
LEVEL SECTION
SLOPE SECTION
FINISHED GRADE
SIDEWALK
CURB & GUTTER
CURB RAMP FOR HANDICAPPED PERSONS
TYPE A
TYPE B
P.L.
CONCRETE SIDEWALK
TERRACE VARIABLE
HEAVY BROOMING
NOTE: RAMP MAY BE GROOVED INSTEAD OF HEAVY BROOMING
STORM SEWER EXTENSION
HAYWARD, WISCONSIN
SCALE: DO NOT
SHEET 3 OF 3
DATE APRIL, 1980
MORGAN & PARMLEY, LTD.
CONSULTING ENGINEERS
LADYSMITH, WISCONSIN
DR. BY R.W. STAUFFER

Design

DAMS & RESERVOIRS

Section 13

Impoundments of water by constructing dams to form lakes, ponds and reservoirs is a time-tested method of controlling nature's most common liquid. These structures and facilities are built for power generation, flood control, irrigation, storage, recreational uses, navigational locks, land development, erosion control and conservation.

This section presents three examples of plans for small dams. The first, Murphy Dam Construction, is a totally new structure designed to replace a dam built in the 1930s by the WPA. This structure failed during a major storm event in the early 1970s. Two decades later, officials decided to reestablish the facility.

The second example is a plan for a private, low head dam that creates a shallow pond at a recreational retreat.

The last presentation is a set of plans for retrofitting an existing dam, constructed in the early 1930s, with radial control gates and related components added.

TABLE OF CONTENTS

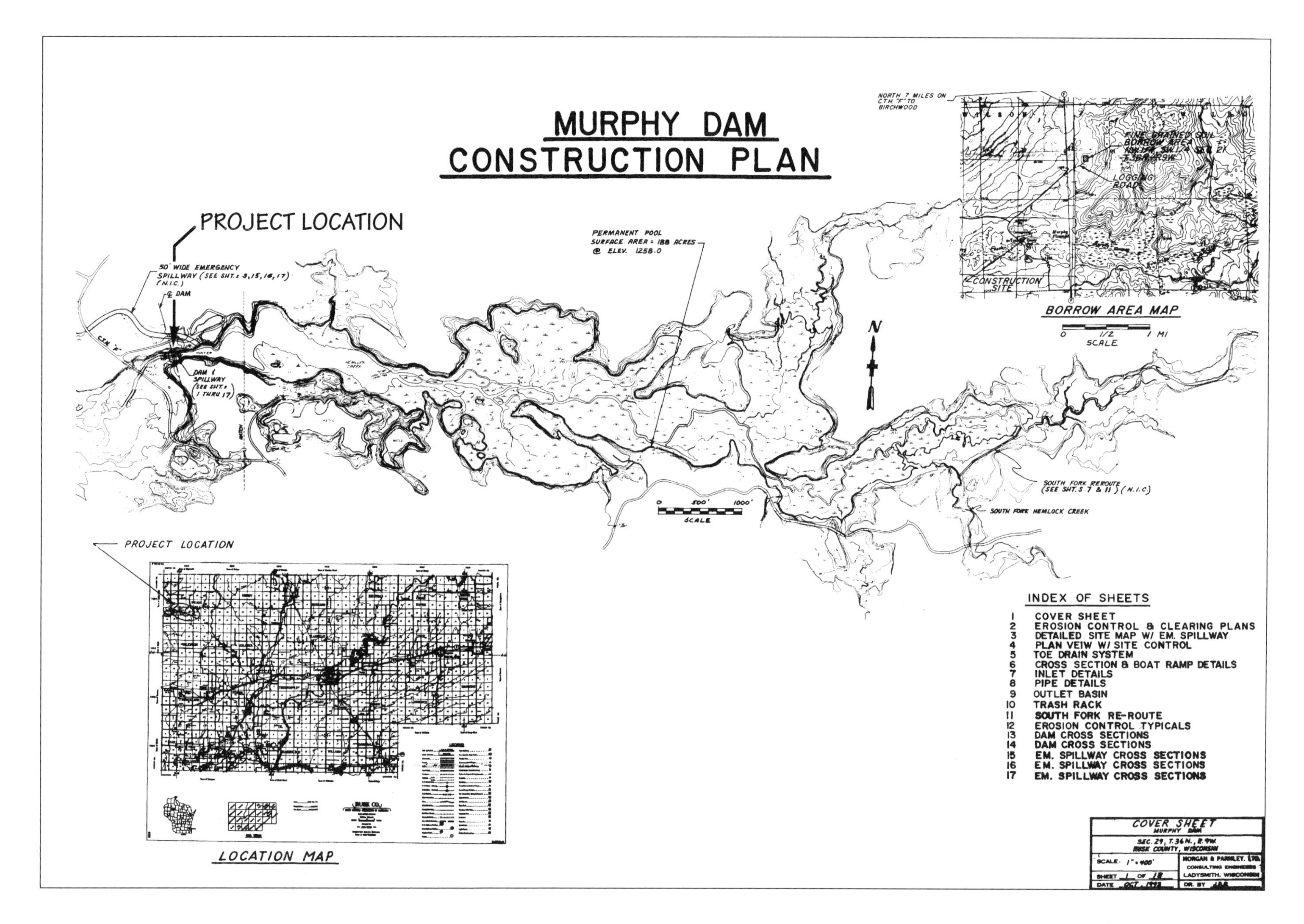

MURPHY DAM
CONSTRUCTION PLAN
PROJECT LOCATION
50' WIDE EMERGENCY SPILLWAY (SEE SHT.s 3,15,16,17) (N.I.C.)
℄ DAM
DAM & SPILLWAY (SEE SHT.s 1 THRU 17)
PERMANENT POOL SURFACE AREA = 188 ACRES @ ELEV. 1258.0
NORTH 7 MILES ON CTH "F" TO BIRCHWOOD
FINE GRAINED SOIL BORROW AREA
LOGGING ROAD
CONSTRUCTION SITE
BORROW AREA MAP
0 1/2 1 MI
SCALE
N
0 500' 1000'
SCALE
SOUTH FORK REROUTE (SEE SHT.S 7 & 11) (N.I.C.)
SOUTH FORK HEMLOCK CREEK
PROJECT LOCATION
LOCATION MAP
INDEX OF SHEETS
1 COVER SHEET
2 EROSION CONTROL & CLEARING PLANS
3 DETAILED SITE MAP W/ EM. SPILLWAY
4 PLAN VEIW W/ SITE CONTROL
5 TOE DRAIN SYSTEM
6 CROSS SECTION & BOAT RAMP DETAILS
7 INLET DETAILS
8 PIPE DETAILS
9 OUTLET BASIN
10 TRASH RACK
11 SOUTH FORK RE-ROUTE
12 EROSION CONTROL TYPICALS
13 DAM CROSS SECTIONS
14 DAM CROSS SECTIONS
15 EM. SPILLWAY CROSS SECTIONS
16 EM. SPILLWAY CROSS SECTIONS
17 EM. SPILLWAY CROSS SECTIONS
COVER SHEET
MURPHY DAM
SEC. 29, T.36N., R.9W
RUSK COUNTY, WISCONSIN
SCALE 1" = 400'
SHEET 1 OF 18
DATE OCT. 1998
MORGAN & PARMLEY, LTD.
CONSULTING ENGINEERS
LADYSMITH, WISCONSIN
DR. BY

Source: Morgan & Parmley, Ltd.

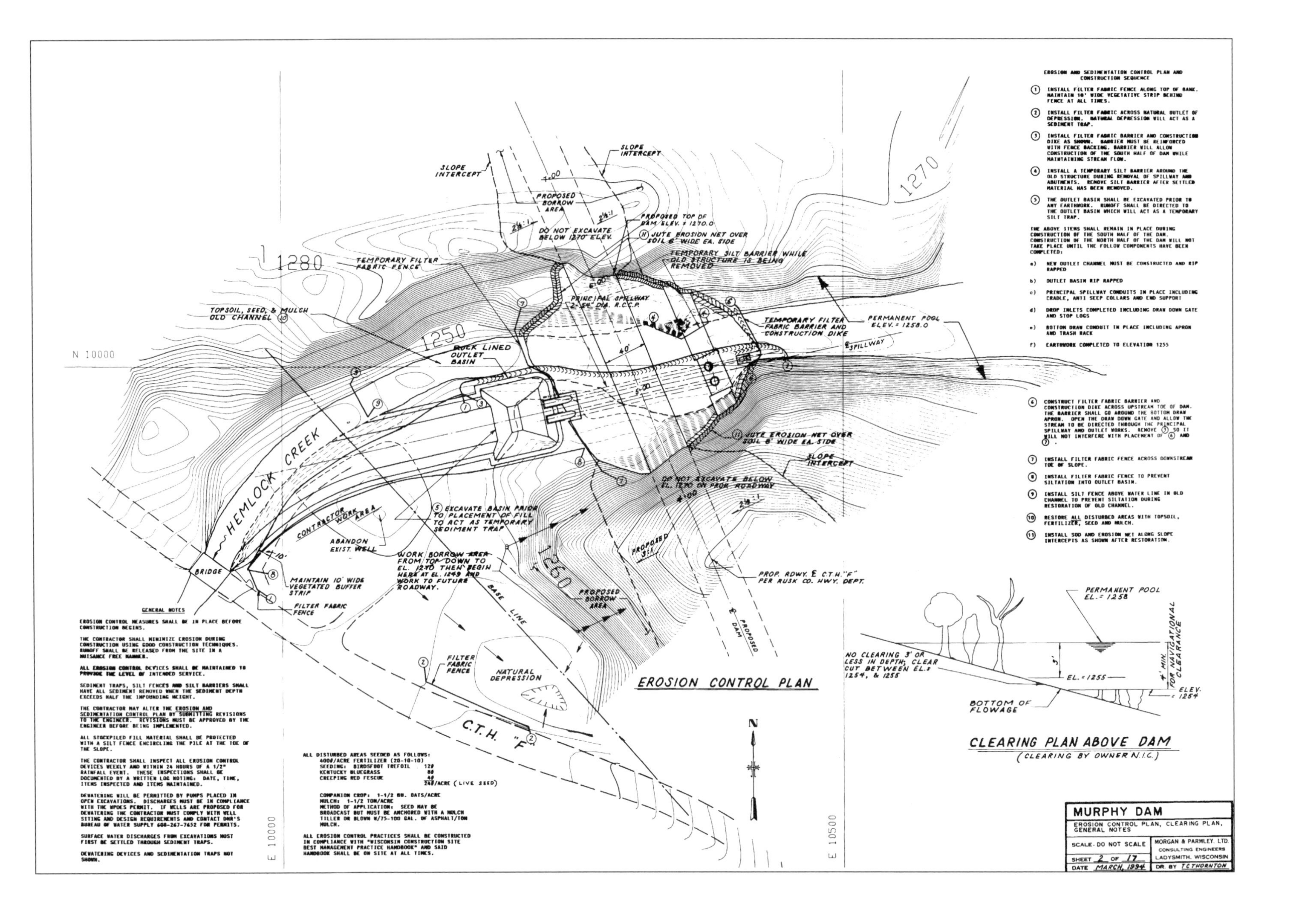
EROSION AND SEDIMENTATION CONTROL PLAN AND CONSTRUCTION SEQUENCE
① INSTALL FILTER FABRIC FENCE ALONG TOP OF BANK. MAINTAIN 10' WIDE VEGETATIVE STRIP BEHIND FENCE AT ALL TIMES.
② INSTALL FILTER FABRIC ACROSS NATURAL OUTLET OF DEPRESSION. NATURAL DEPRESSION WILL ACT AS A SEDIMENT TRAP.
③ INSTALL FILTER FABRIC BARRIER AND CONSTRUCTION DIKE AS SHOWN. BARRIER MUST BE REINFORCED WITH FENCE BACKING. BARRIER WILL ALLOW CONSTRUCTION OF THE SOUTH HALF OF DAM WHILE MAINTAINING STREAM FLOW.
④ INSTALL A TEMPORARY SILT BARRIER AROUND THE OLD STRUCTURE DURING REMOVAL OF SPILLWAY AND ABUTMENTS. REMOVE SILT BARRIER AFTER SETTLED MATERIAL HAS BEEN REMOVED.
⑤ THE OUTLET BASIN SHALL BE EXCAVATED PRIOR TO ANY EARTHWORK. RUNOFF SHALL BE DIRECTED TO THE OUTLET BASIN WHICH WILL ACT AS A TEMPORARY SILT TRAP.
THE ABOVE ITEMS SHALL REMAIN IN PLACE DURING CONSTRUCTION OF THE SOUTH HALF OF THE DAM. CONSTRUCTION OF THE NORTH HALF OF THE DAM WILL NOT TAKE PLACE UNTIL THE FOLLOW COMPONENTS HAVE BEEN COMPLETED:
a) NEW OUTLET CHANNEL MUST BE CONSTRUCTED AND RIP RAPPED
b) OUTLET BASIN RIP RAPPED
c) PRINCIPAL SPILLWAY CONDUITS IN PLACE INCLUDING CRADLE, ANTI SEEP COLLARS AND END SUPPORT
d) DROP INLETS COMPLETED INCLUDING DRAW DOWN GATE AND STOP LOGS
e) BOTTOM DRAW CONDUIT IN PLACE INCLUDING APRON AND TRASH RACK
f) EARTHWORK COMPLETED TO ELEVATION 1255
⑥ CONSTRUCT FILTER FABRIC BARRIER AND CONSTRUCTION DIKE ACROSS UPSTREAM TOE OF DAM. THE BARRIER SHALL GO AROUND THE BOTTOM DRAW APRON. OPEN THE DRAW DOWN GATE AND ALLOW THE STREAM TO BE DIRECTED THROUGH THE PRINCIPAL SPILLWAY AND OUTLET WORKS. REMOVE ③ SO IT WILL NOT INTERFERE WITH PLACEMENT OF ⑥ AND ⑦.
⑦ INSTALL FILTER FABRIC FENCE ACROSS DOWNSTREAM TOE OF SLOPE.
⑧ INSTALL FILTER FABRIC FENCE TO PREVENT SILTATION INTO OUTLET BASIN.
⑨ INSTALL SILT FENCE ABOVE WATER LINE IN OLD CHANNEL TO PREVENT SILTATION DURING RESTORATION OF OLD CHANNEL.
⑩ RESTORE ALL DISTURBED AREAS WITH TOPSOIL, FERTILIZER, SEED AND MULCH.
⑪ INSTALL SOD AND EROSION NET ALONG SLOPE INTERCEPTS AS SHOWN AFTER RESTORATION.
SLOPE INTERCEPT
PROPOSED BORROW AREA
DO NOT EXCAVATE BELOW 1270 ELEV.
PROPOSED TOP OF DAM ELEV. = 1270.0
⑪ JUTE EROSION NET OVER SOIL 6' WIDE EA. SIDE
TEMPORARY SILT BARRIER WHILE OLD STRUCTURE IS BEING REMOVED
TEMPORARY FILTER FABRIC FENCE
1280
1270
1250
1260
PRINCIPAL SPILLWAY 2-54" DIA. R.C.C.P.
TOPSOIL, SEED, & MULCH OLD CHANNEL ⑩
ROCK LINED OUTLET BASIN
TEMPORARY FILTER FABRIC BARRIER AND CONSTRUCTION DIKE
PERMANENT POOL ELEV. = 1258.0
℄ SPILLWAY
N 10000
40'
HEMLOCK CREEK
DO NOT EXCAVATE BELOW EL. 1270 ON PROP. ROADWAY
⑤ EXCAVATE BASIN PRIOR TO PLACEMENT OF FILL TO ACT AS TEMPORARY SEDIMENT TRAP
CONTRACTOR WORK AREA
ABANDON EXIST. WELL
WORK BORROW AREA FROM TOP DOWN TO EL. 1270 THEN BEGIN HERE AT EL. 1249 AND WORK TO FUTURE ROADWAY.
PROPOSED 3:1
PROP. RDWY. ℄ C.T.H. "F" PER RUSK CO. HWY. DEPT.
BRIDGE
MAINTAIN 10' WIDE VEGETATED BUFFER STRIP
FILTER FABRIC FENCE
BASE LINE
PROPOSED BORROW AREA
℄ PROPOSED DAM
NATURAL DEPRESSION
C.T.H. "F"
EROSION CONTROL PLAN
N
E 10000
E 10500
GENERAL NOTES
EROSION CONTROL MEASURES SHALL BE IN PLACE BEFORE CONSTRUCTION BEGINS.
THE CONTRACTOR SHALL MINIMIZE EROSION DURING CONSTRUCTION USING GOOD CONSTRUCTION TECHNIQUES. RUNOFF SHALL BE RELEASED FROM THE SITE IN A NUISANCE FREE MANNER.
ALL EROSION CONTROL DEVICES SHALL BE MAINTAINED TO PROVIDE THE LEVEL OF INTENDED SERVICE.
SEDIMENT TRAPS, SILT FENCES AND SILT BARRIERS SHALL HAVE ALL SEDIMENT REMOVED WHEN THE SEDIMENT DEPTH EXCEEDS HALF THE IMPOUNDING HEIGHT.
THE CONTRACTOR MAY ALTER THE EROSION AND SEDIMENTATION CONTROL PLAN BY SUBMITTING REVISIONS TO THE ENGINEER. REVISIONS MUST BE APPROVED BY THE ENGINEER BEFORE BEING IMPLEMENTED.
ALL STOCKPILED FILL MATERIAL SHALL BE PROTECTED WITH A SILT FENCE ENCIRCLING THE PILE AT THE TOE OF THE SLOPE.
THE CONTRACTOR SHALL INSPECT ALL EROSION CONTROL DEVICES WEEKLY AND WITHIN 24 HOURS OF A 1/2" RAINFALL EVENT. THESE INSPECTIONS SHALL BE DOCUMENTED BY A WRITTEN LOG NOTING: DATE, TIME, ITEMS INSPECTED AND ITEMS MAINTAINED.
DEWATERING WILL BE PERMITTED BY PUMPS PLACED IN OPEN EXCAVATIONS. DISCHARGES MUST BE IN COMPLIANCE WITH THE WPDES PERMIT. IF WELLS ARE PROPOSED FOR DEWATERING THE CONTRACTOR MUST COMPLY WITH WELL SITING AND DESIGN REQUIREMENTS AND CONTACT DNR'S BUREAU OF WATER SUPPLY 608-267-7652 FOR PERMITS.
SURFACE WATER DISCHARGES FROM EXCAVATIONS MUST FIRST BE SETTLED THROUGH SEDIMENT TRAPS.
DEWATERING DEVICES AND SEDIMENTATION TRAPS NOT SHOWN.
ALL DISTURBED AREAS SEEDED AS FOLLOWS:
400#/ACRE FERTILIZER (20-10-10)
SEEDING: BIRDSFOOT TREFOIL 12#
KENTUCKY BLUEGRASS 8#
CREEPING RED FESCUE 4#
24#/ACRE (LIVE SEED)
COMPANION CROP: 1-1/2 BU. OATS/ACRE
MULCH: 1-1/2 TON/ACRE
METHOD OF APPLICATION: SEED MAY BE BROADCAST BUT MUST BE ANCHORED WITH A MULCH TILLER OR BLOWN W/75-100 GAL. OF ASPHALT/TON MULCH.
ALL EROSION CONTROL PRACTICES SHALL BE CONSTRUCTED IN COMPLIANCE WITH "WISCONSIN CONSTRUCTION SITE BEST MANAGEMENT PRACTICE HANDBOOK" AND SAID HANDBOOK SHALL BE ON SITE AT ALL TIMES.
PERMANENT POOL EL. = 1258
NO CLEARING 3' OR LESS IN DEPTH; CLEAR CUT BETWEEN EL.s 1254, & 1255
3'
EL. = 1255
4' MIN. FOR NAVIGATIONAL CLEARANCE
ELEV. = 1254
BOTTOM OF FLOWAGE
CLEARING PLAN ABOVE DAM
(CLEARING BY OWNER N.I.C.)
MURPHY DAM
EROSION CONTROL PLAN, CLEARING PLAN, GENERAL NOTES
SCALE- DO NOT SCALE
MORGAN & PARMLEY, LTD.
CONSULTING ENGINEERS
LADYSMITH, WISCONSIN
SHEET 2 OF 17
DATE MARCH, 1994
DR. BY T.C. THORNTON

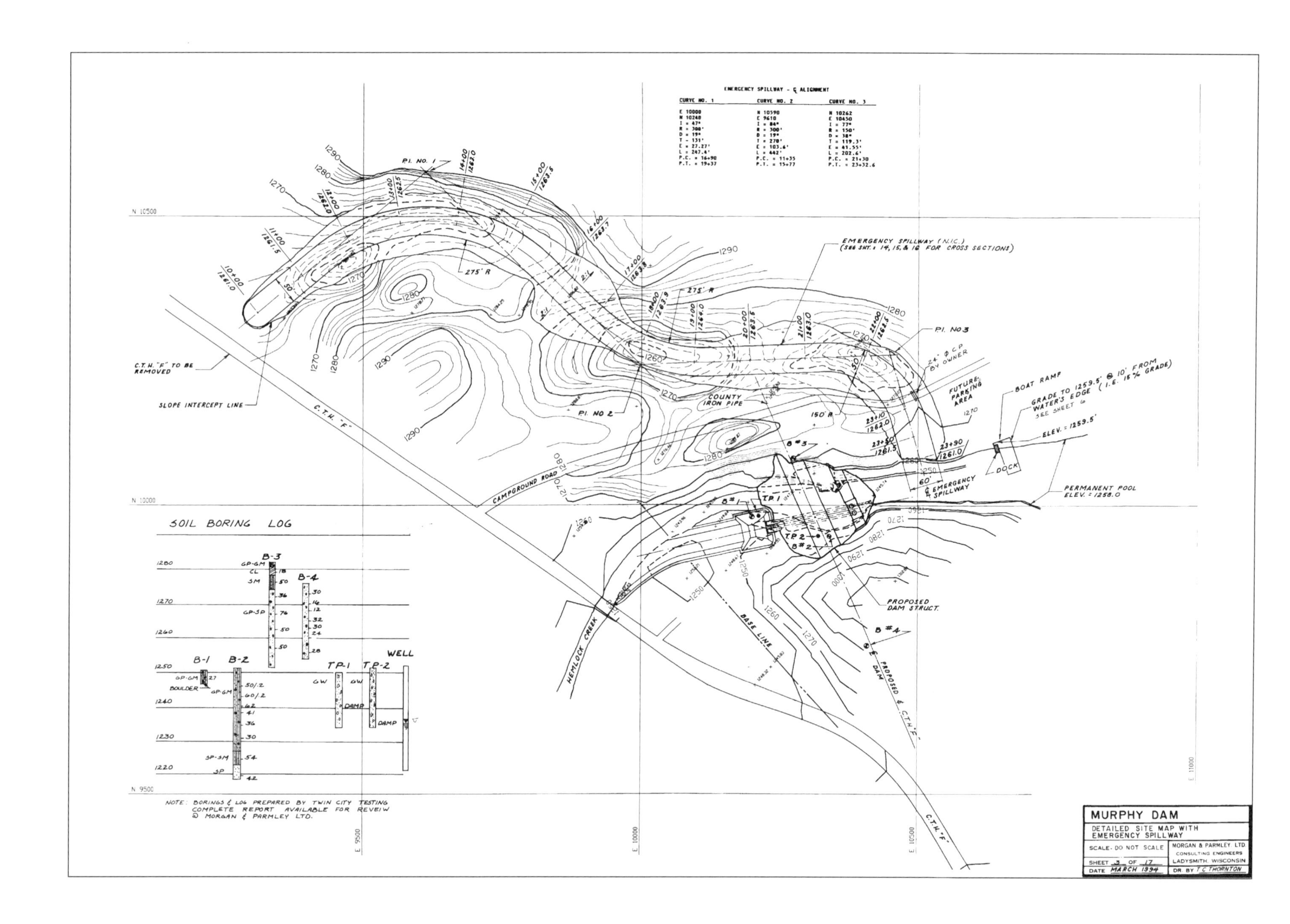
EMERGENCY SPILLWAY - ℄ ALIGNMENT
CURVE NO. 1
E 10000
N 10240
I = 47°
R = 300'
D = 19°
T - 131'
E = 27.27'
L = 247.4'
P.C. = 16+90
P.T. = 19+37
CURVE NO. 2
N 10590
E 9610
I = 84°
R = 300'
D = 19°
T = 270'
E = 103.6'
L = 442'
P.C. = 11+35
P.T. = 15+77
CURVE NO. 3
N 10262
E 10450
I = 77°
R = 150'
D = 38°
T = 119.3'
E = 41.55'
L = 202.6'
P.C. = 21+30
P.T. = 23+32.6
P.I. NO. 1
P.I. NO 2
P.I. NO3
275' R
150' R
EMERGENCY SPILLWAY (N.I.C.)
(SEE SHT.s 14, 15, & 16 FOR CROSS SECTIONS)
C.T.H. "F" TO BE REMOVED
SLOPE INTERCEPT LINE
C.T.H. "F"
COUNTY IRON PIPE
CAMPGROUND ROAD
24" Ø CP BY OWNER
FUTURE PARKING AREA
BOAT RAMP
GRADE TO 1259.5' @ 10' FROM WATER'S EDGE (I.E. 15% GRADE)
SEE SHEET 6
ELEV. = 1259.5'
DOCK
60'
℄ EMERGENCY SPILLWAY
PERMANENT POOL ELEV. = 1258.0
TP 1
TP 2
B #1
B #2
B #3
B #4
PROPOSED DAM STRUCT.
BASE LINE
℄ DAM
PROPOSED ℄ C.T.H. "F"
HEMLOCK CREEK
N 10500
N 10000
N 9500
E 9500
E 10000
E 10500
E 11000
SOIL BORING LOG
B-1
B-2
B-3
B-4
TP-1
TP-2
WELL
GP-GM
CL
SM
GP-SP
BOULDER
SP-SM
SP
GW
DAMP
NOTE: BORINGS & LOG PREPARED BY TWIN CITY TESTING
COMPLETE REPORT AVAILABLE FOR REVIEW
@ MORGAN & PARMLEY LTD.
MURPHY DAM
DETAILED SITE MAP WITH
EMERGENCY SPILLWAY
SCALE- DO NOT SCALE
MORGAN & PARMLEY LTD.
CONSULTING ENGINEERS
LADYSMITH, WISCONSIN
SHEET 3 OF 17
DATE MARCH 1994
DR. BY T C THORNTON

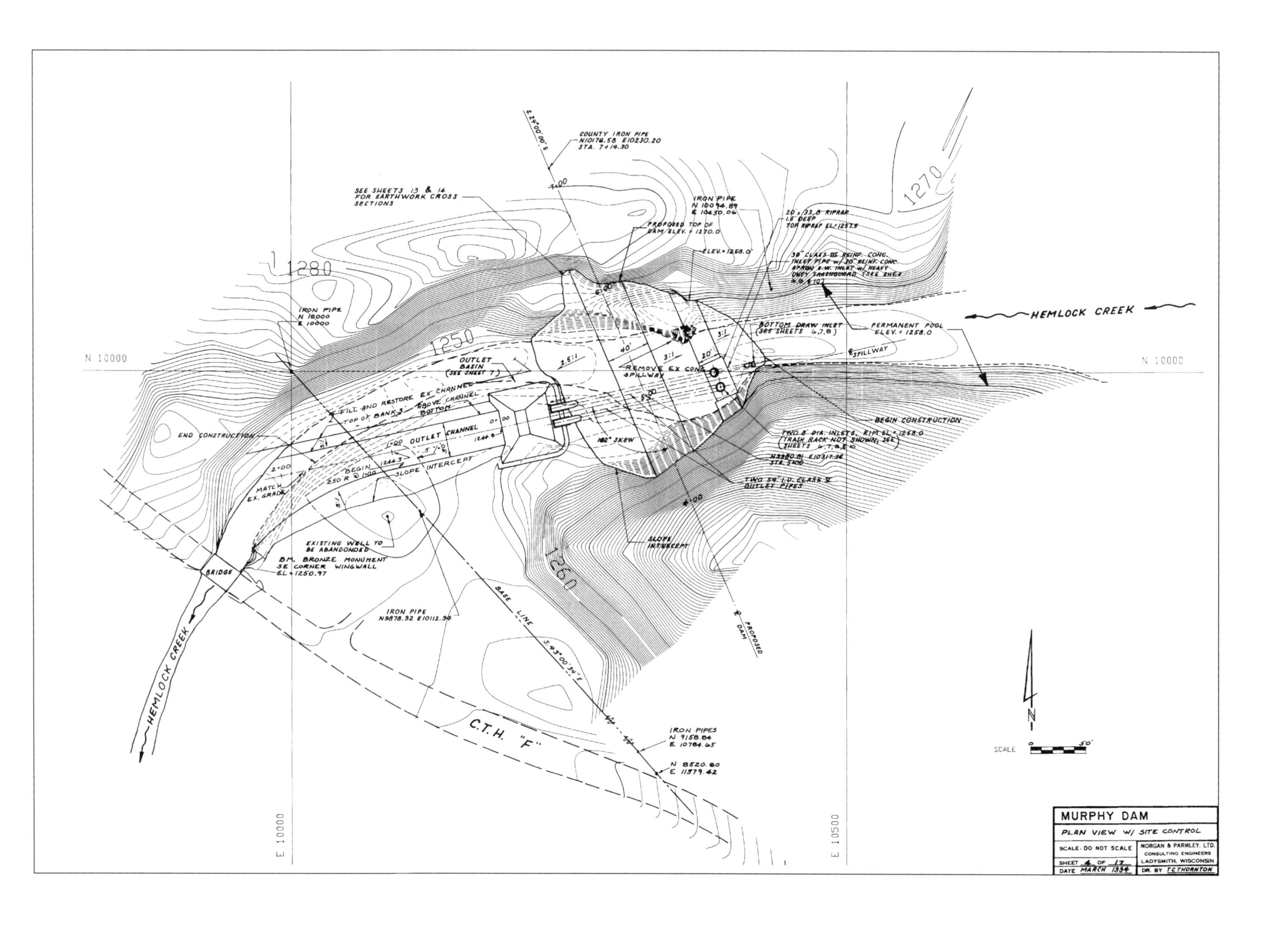
MURPHY DAM
PLAN VIEW W/ SITE CONTROL
SCALE- DO NOT SCALE
MORGAN & PARMLEY, LTD.
CONSULTING ENGINEERS
LADYSMITH, WISCONSIN
SHEET 4 OF 17
DATE MARCH 1994
DR. BY T.C.THORNTON
HEMLOCK CREEK
C.T.H. "F"
BASE LINE
COUNTY IRON PIPE
N10176.58 E10230.20
STA. 7+14.30
SEE SHEETS 13 & 14 FOR EARTHWORK CROSS SECTIONS
PROPOSED TOP OF DAM ELEV. = 1270.0
IRON PIPE N 10094.89 E 10430.06
PERMANENT POOL ELEV. = 1258.0
BOTTOM DRAW INLET (SEE SHEETS 6,7,8)
REMOVE EX CONC SPILLWAY
OUTLET BASIN (SEE SHEET 7)
FILL AND RESTORE EX CHANNEL
OUTLET CHANNEL
SLOPE INTERCEPT
END CONSTRUCTION
BEGIN CONSTRUCTION
EXISTING WELL TO BE ABANDONDED
BM. BRONZE MONUMENT SE CORNER WINGWALL EL = 1250.97
BRIDGE
IRON PIPE N9878.92 E10112.94
IRON PIPES N 9158.84 E 10784.65
N 8520.80 E 11379.42
TWO 54" I.D. CLASS V OUTLET PIPES
N 10000
E 10000
E 10500
SCALE

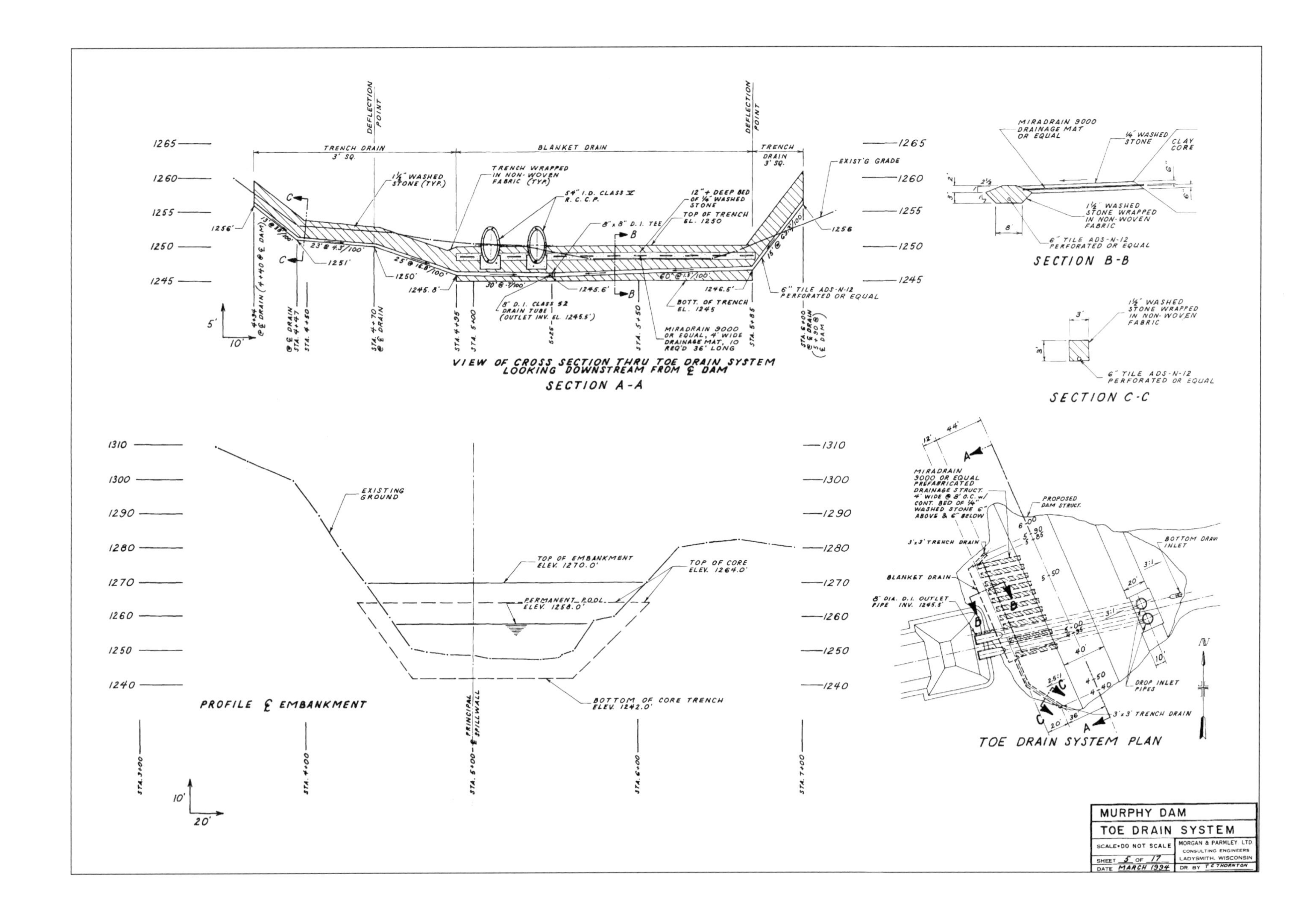
VIEW OF CROSS SECTION THRU TOE DRAIN SYSTEM
LOOKING DOWNSTREAM FROM ℄ DAM
SECTION A-A
SECTION B-B
SECTION C-C
PROFILE ℄ EMBANKMENT
TOE DRAIN SYSTEM PLAN
TRENCH DRAIN 3' SQ.
BLANKET DRAIN
DEFLECTION POINT
EXIST'G GRADE
1½" WASHED STONE (TYP.)
TRENCH WRAPPED IN NON-WOVEN FABRIC (TYP.)
54" I.D. CLASS V R.C.C.P.
8" x 8" D.I. TEE
TOP OF TRENCH EL. 1250
BOTT. OF TRENCH EL. 1245
6" TILE ADS-N-12 PERFORATED OR EQUAL
MIRADRAIN 9000 DRAINAGE MAT OR EQUAL
CLAY CORE
EXISTING GROUND
TOP OF EMBANKMENT ELEV. 1270.0'
TOP OF CORE ELEV. 1264.0'
PERMANENT POOL ELEV. 1258.0'
BOTTOM OF CORE TRENCH ELEV. 1242.0'
PRINCIPAL ℄ SPILLWALL
3'x3' TRENCH DRAIN
BLANKET DRAIN
PROPOSED DAM STRUCT.
DROP INLET PIPES
MURPHY DAM
TOE DRAIN SYSTEM
SCALE-DO NOT SCALE
MORGAN & PARMLEY LTD
CONSULTING ENGINEERS
LADYSMITH, WISCONSIN
SHEET 5 OF 17
DATE MARCH 1994

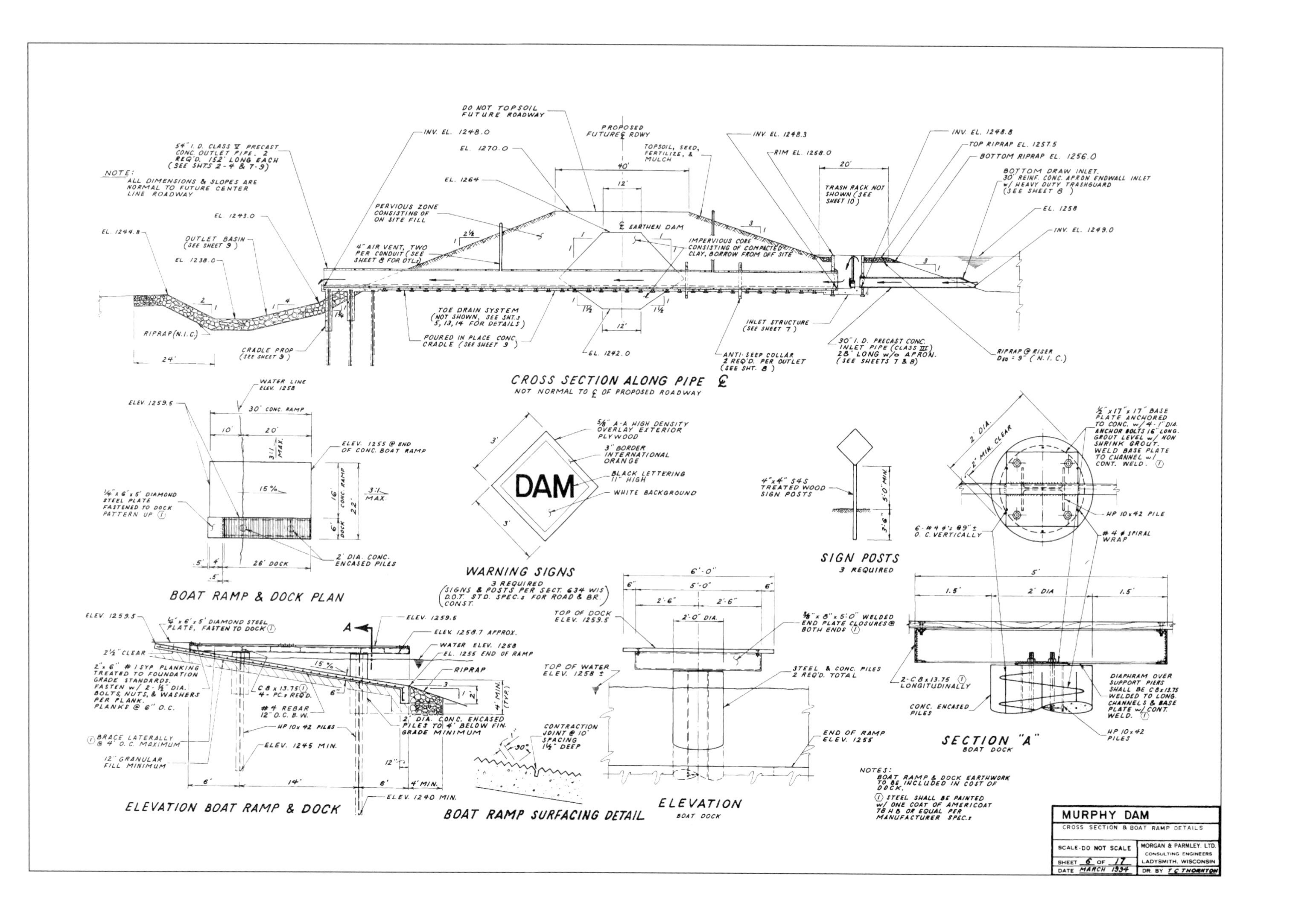
DO NOT TOPSOIL FUTURE ROADWAY
PROPOSED FUTURE ℄ RDWY
INV. EL. 1248.0
EL. 1270.0
TOPSOIL, SEED, FERTILIZE, & MULCH
INV. EL. 1248.3
INV. EL. 1248.8
TOP RIPRAP EL. 1257.5
BOTTOM RIPRAP EL. 1256.0
54" I.D. CLASS V PRECAST CONC. OUTLET PIPE. 2 REQ'D. 152' LONG EACH (SEE SHTS 2-4 & 7-9)
NOTE: ALL DIMENSIONS & SLOPES ARE NORMAL TO FUTURE CENTER LINE ROADWAY
RIM EL. 1268.0
40'
12'
20'
EL. 1264
TRASH RACK NOT SHOWN (SEE SHEET 10)
BOTTOM DRAW INLET. 30" REINF. CONC. APRON ENDWALL INLET w/ HEAVY DUTY TRASHGUARD (SEE SHEET 8)
PERVIOUS ZONE CONSISTING OF ON SITE FILL
EL. 1243.0
EL. 1258
EL. 1244.8
℄ EARTHEN DAM
INV. EL. 1249.0
OUTLET BASIN (SEE SHEET 9)
IMPERVIOUS CORE CONSISTING OF COMPACTED CLAY, BORROW FROM OFF SITE
EL. 1238.0
4" AIR VENT, TWO PER CONDUIT (SEE SHEET 8 FOR DTL)
RIPRAP (N.I.C.)
TOE DRAIN SYSTEM (NOT SHOWN, SEE SHTS 5, 13, 14 FOR DETAILS)
POURED IN PLACE CONC. CRADLE (SEE SHEET 9)
INLET STRUCTURE (SEE SHEET 7)
CRADLE PROP (SEE SHEET 9)
24'
EL. 1242.0
ANTI-SEEP COLLAR 2 REQ'D. PER OUTLET (SEE SHT. 8)
30" I.D. PRECAST CONC. INLET PIPE (CLASS III) 28' LONG w/o APRON. (SEE SHEETS 7 & 8)
RIPRAP @ RISER D50 = 9" (N.I.C.)
CROSS SECTION ALONG PIPE ℄
NOT NORMAL TO ℄ OF PROPOSED ROADWAY
WATER LINE ELEV. 1258
ELEV. 1259.5
30' CONC. RAMP
ELEV. 1255 @ END OF CONC. BOAT RAMP
1/4" x 6' x 5' DIAMOND STEEL PLATE FASTENED TO DOCK PATTERN UP ①
15%
3:1 MAX.
2' DIA. CONC. ENCASED PILES
26' DOCK
BOAT RAMP & DOCK PLAN
5/8" A-A HIGH DENSITY OVERLAY EXTERIOR PLYWOOD
3" BORDER INTERNATIONAL ORANGE
BLACK LETTERING 11" HIGH
WHITE BACKGROUND
DAM
WARNING SIGNS
3 REQUIRED
(SIGNS & POSTS PER SECT. 634 WIS D.O.T. STD. SPEC.s FOR ROAD & BR. CONST.
4"x4" S4S TREATED WOOD SIGN POSTS
SIGN POSTS
3 REQUIRED
1/2"x17"x17" BASE PLATE ANCHORED TO CONC. w/ 4-1" DIA. ANCHOR BOLTS 16" LONG. GROUT LEVEL w/ NON SHRINK GROUT. WELD BASE PLATE TO CHANNEL w/ CONT. WELD. ①
2' DIA.
2" MIN. CLEAR
HP 10x42 PILE
6-#4 ϕ's @ 9" ± O.C. VERTICALLY
#4 ϕ SPIRAL WRAP
ELEV. 1259.5
1/4" x 6' x 5' DIAMOND STEEL PLATE, FASTEN TO DOCK ①
ELEV. 1259.5
ELEV. 1258.7 APPROX.
WATER ELEV. 1258
EL. 1255 END OF RAMP
RIPRAP
2 1/2" CLEAR
2"x6" #1 SYP PLANKING TREATED TO FOUNDATION GRADE STANDARDS. FASTEN w/ 2-1/2" DIA. BOLTS, NUTS, & WASHERS PER PLANK. PLANKS @ 6" O.C.
C 8 x 13.75 ① 4-PC.s REQ'D.
#4 REBAR 12" O.C. B.W.
2' DIA. CONC. ENCASED PILES TO 4' BELOW FIN. GRADE MINIMUM
HP 10x42 PILES
① BRACE LATERALLY @ 4' O.C. MAXIMUM
ELEV. 1245 MIN.
12" GRANULAR FILL MINIMUM
ELEV. 1240 MIN.
ELEVATION BOAT RAMP & DOCK
CONTRACTION JOINT @ 10' SPACING 1 1/4" DEEP
BOAT RAMP SURFACING DETAIL
TOP OF DOCK ELEV. 1259.5
TOP OF WATER ELEV. 1258 ±
6'-0"
5'-0"
2'-0" DIA.
3/8" x 8" x 5'0" WELDED END PLATE CLOSURES @ BOTH ENDS ①
STEEL & CONC. PILES 2 REQ'D. TOTAL
END OF RAMP ELEV. 1255
ELEVATION
BOAT DOCK
5'
1.5'
2' DIA
2-C 8 x 13.75 ① LONGITUDINALLY
CONC. ENCASED PILES
DIAPHRAM OVER SUPPORT PIERS SHALL BE C8x13.75 WELDED TO LONG. CHANNELS & BASE PLATE w/ CONT. WELD. ①
HP 10x42 PILES
SECTION "A"
BOAT DOCK
NOTES:
BOAT RAMP & DOCK EARTHWORK TO BE INCLUDED IN COST OF DOCK.
① STEEL SHALL BE PAINTED w/ ONE COAT OF AMERICOAT 78 H.B. OR EQUAL PER MANUFACTURER SPEC.s
MURPHY DAM
CROSS SECTION & BOAT RAMP DETAILS
SCALE-DO NOT SCALE
MORGAN & PARMLEY, LTD. CONSULTING ENGINEERS LADYSMITH, WISCONSIN
SHEET 6 OF 17
DATE MARCH 1994
DR. BY T.C. THORNTON

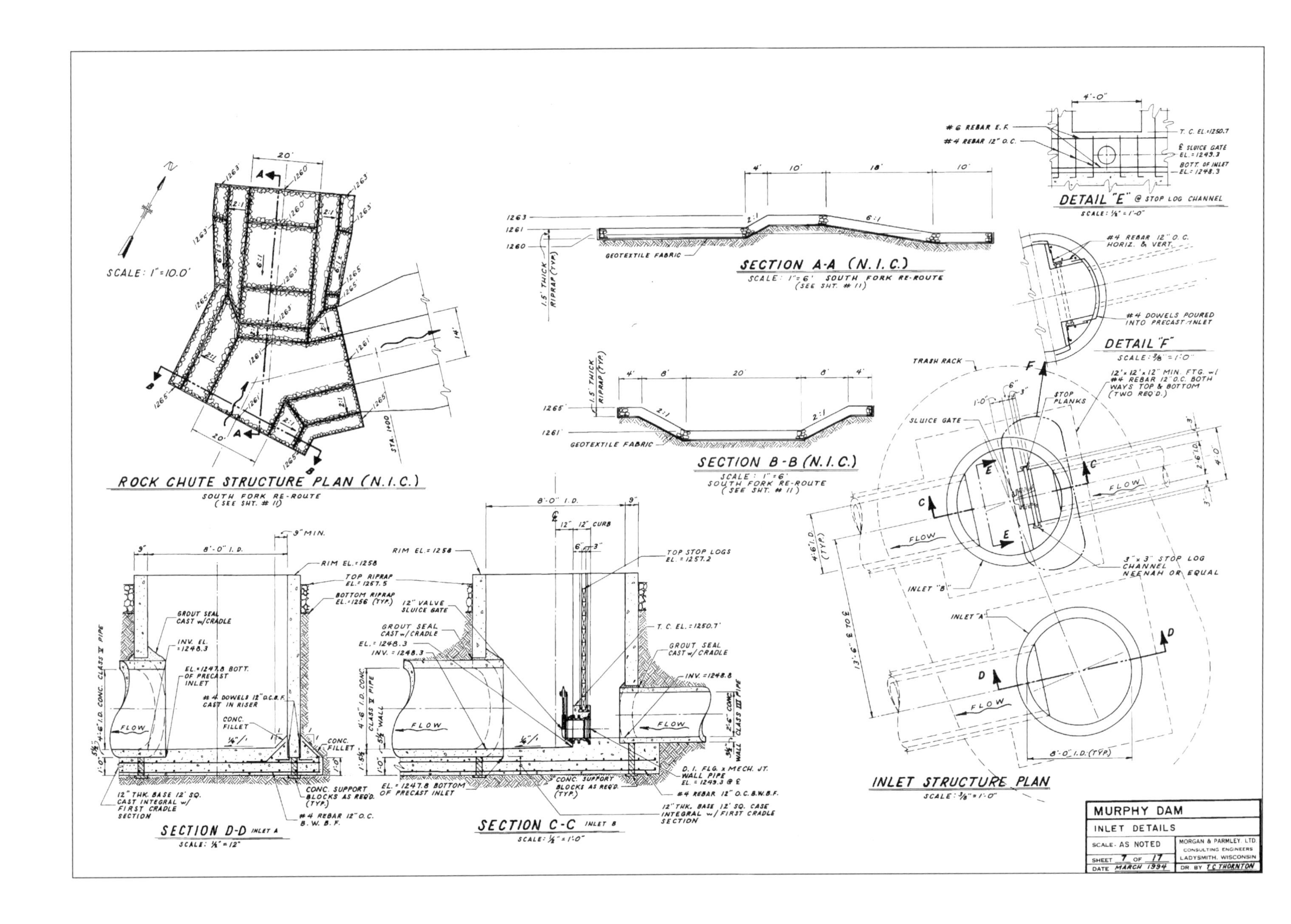
ROCK CHUTE STRUCTURE PLAN (N.I.C.)
SECTION A-A (N.I.C.)
SECTION B-B (N.I.C.)
DETAIL "E" @ STOP LOG CHANNEL
DETAIL "F"
INLET STRUCTURE PLAN
SECTION D-D INLET A
SECTION C-C INLET B
MURPHY DAM
INLET DETAILS
SCALE- AS NOTED
SHEET 7 OF 17
DATE MARCH 1994
MORGAN & PARMLEY, LTD
CONSULTING ENGINEERS
LADYSMITH, WISCONSIN
DR. BY T.C. THORNTON

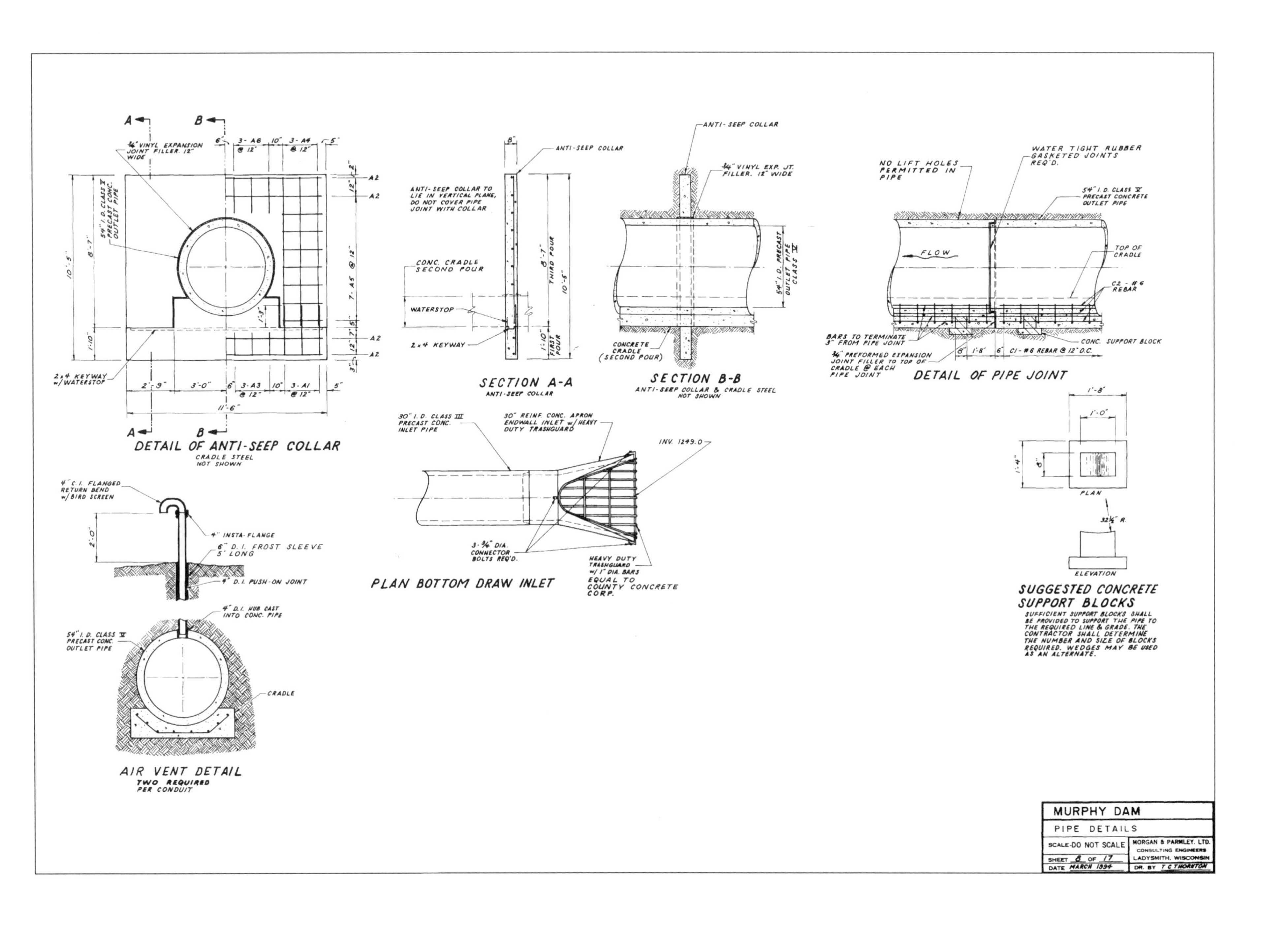
DETAIL OF ANTI-SEEP COLLAR
CRADLE STEEL NOT SHOWN
¾" VINYL EXPANSION JOINT FILLER. 12" WIDE
54" I.D. CLASS V PRECAST CONC. OUTLET PIPE
2×4 KEYWAY w/WATERSTOP
SECTION A-A
ANTI-SEEP COLLAR
ANTI-SEEP COLLAR TO LIE IN VERTICAL PLANE, DO NOT COVER PIPE JOINT WITH COLLAR
CONC. CRADLE SECOND POUR
WATERSTOP
2×4 KEYWAY
THIRD POUR
FIRST POUR
SECTION B-B
ANTI-SEEP COLLAR & CRADLE STEEL NOT SHOWN
ANTI-SEEP COLLAR
¾" VINYL EXP. JT. FILLER, 12" WIDE
54" I.D. PRECAST OUTLET PIPE CLASS V
CONCRETE CRADLE (SECOND POUR)
DETAIL OF PIPE JOINT
NO LIFT HOLES PERMITTED IN PIPE
WATER TIGHT RUBBER GASKETED JOINTS REQ'D.
54" I.D. CLASS V PRECAST CONCRETE OUTLET PIPE
FLOW
TOP OF CRADLE
C2 - #6 REBAR
CONC. SUPPORT BLOCK
BARS TO TERMINATE 3" FROM PIPE JOINT
¾" PREFORMED EXPANSION JOINT FILLER TO TOP OF CRADLE @ EACH PIPE JOINT
C1 - #6 REBAR @ 12" O.C.
PLAN BOTTOM DRAW INLET
30" I.D. CLASS III PRECAST CONC. INLET PIPE
30" REINF. CONC. APRON ENDWALL INLET w/HEAVY DUTY TRASHGUARD
INV. 1249.0
3-¾" DIA. CONNECTOR BOLTS REQ'D.
HEAVY DUTY TRASHGUARD w/1" DIA. BARS EQUAL TO COUNTY CONCRETE CORP.
AIR VENT DETAIL
TWO REQUIRED PER CONDUIT
4" C.I. FLANGED RETURN BEND w/BIRD SCREEN
4" INSTA-FLANGE
6" D.I. FROST SLEEVE 5' LONG
4" D.I. PUSH-ON JOINT
4" D.I. HUB CAST INTO CONC. PIPE
54" I.D. CLASS V PRECAST CONC. OUTLET PIPE
CRADLE
SUGGESTED CONCRETE SUPPORT BLOCKS
SUFFICIENT SUPPORT BLOCKS SHALL BE PROVIDED TO SUPPORT THE PIPE TO THE REQUIRED LINE & GRADE. THE CONTRACTOR SHALL DETERMINE THE NUMBER AND SIZE OF BLOCKS REQUIRED. WEDGES MAY BE USED AS AN ALTERNATE.
PLAN
ELEVATION
MURPHY DAM
PIPE DETAILS
SCALE-DO NOT SCALE
SHEET 8 OF 17
DATE MARCH 1994
MORGAN & PARMLEY, LTD. CONSULTING ENGINEERS LADYSMITH, WISCONSIN
DR. BY T.C. THORNTON

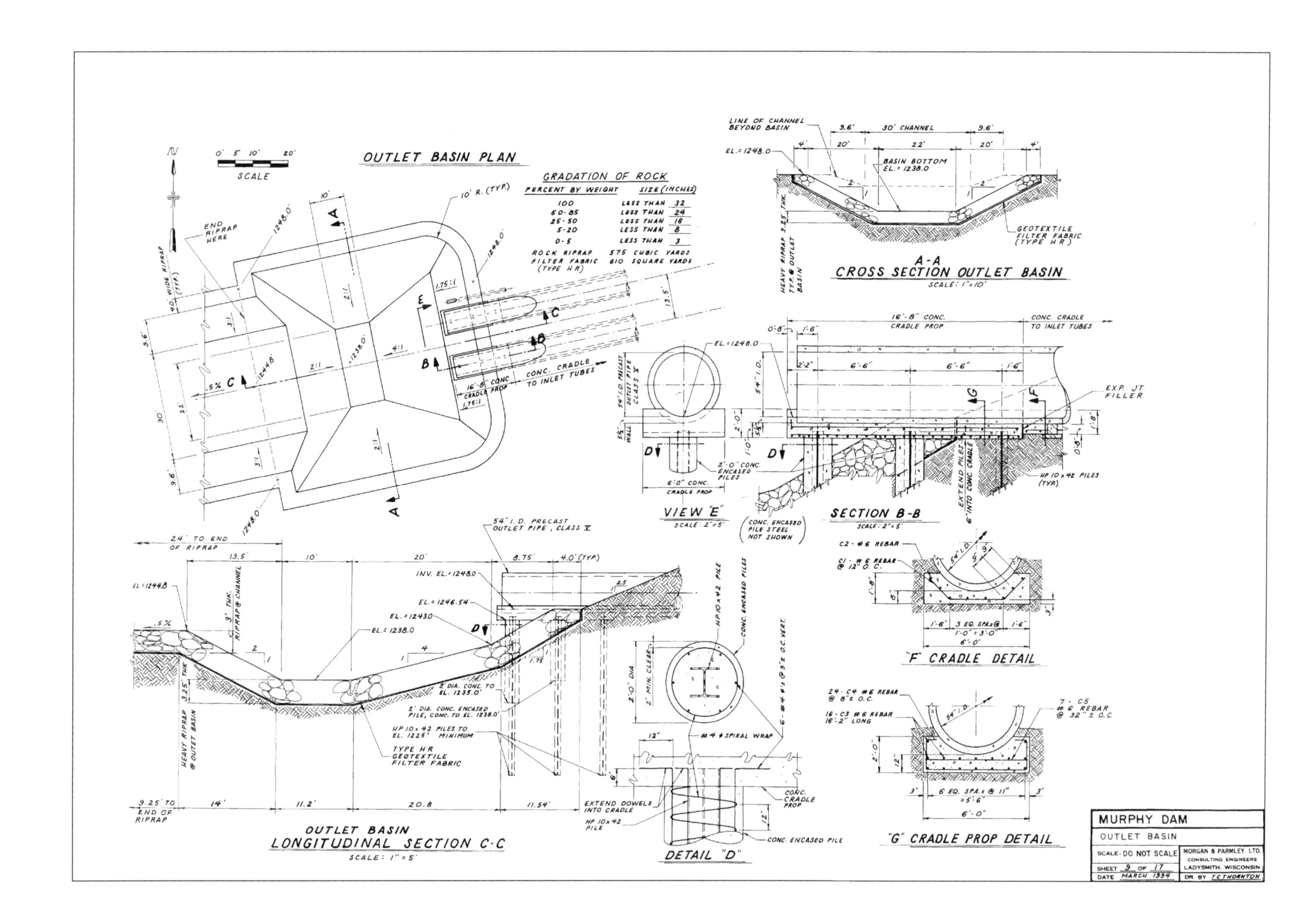
OUTLET BASIN PLAN
SCALE
GRADATION OF ROCK
PERCENT BY WEIGHT SIZE (INCHES)
100 LESS THAN 32
60-85 LESS THAN 24
25-50 LESS THAN 16
5-20 LESS THAN 8
0-5 LESS THAN 3
ROCK RIPRAP 575 CUBIC YARDS
FILTER FABRIC 610 SQUARE YARDS (TYPE HR)
END RIPRAP HERE
4.0' WIDE RIPRAP (TYP.)
10' R. (TYP.)
CONC. CRADLE TO INLET TUBES
A-A CROSS SECTION OUTLET BASIN
SCALE: 1"=10'
LINE OF CHANNEL BEYOND BASIN
30' CHANNEL
BASIN BOTTOM EL.= 1238.0
GEOTEXTILE FILTER FABRIC (TYPE HR)
HEAVY RIPRAP 3.25' THK. TYP. @ OUTLET BASIN
VIEW "E"
SCALE: 2"=5'
SECTION B-B
SCALE: 2"=5'
16'-8" CONC. CRADLE PROP
CONC. CRADLE TO INLET TUBES
EXP. JT FILLER
HP 10 x 42 PILES (TYP.)
EXTEND PILES 6" INTO CONC. CRADLE
CONC. ENCASED PILE STEEL NOT SHOWN
2'-0" CONC. ENCASED PILES
54" I.D. PRECAST OUTLET PIPE CLASS V
"F" CRADLE DETAIL
C2 - #6 REBAR
C1 - #6 REBAR @ 12" O.C.
"G" CRADLE PROP DETAIL
24 - C4 #6 REBAR @ 8"± O.C.
16 - C3 #6 REBAR 16'-2" LONG
7 - C5 #6 REBAR @ 32"± O.C.
6 EQ. SPA.s @ 11" = 5'-6"
OUTLET BASIN LONGITUDINAL SECTION C-C
SCALE: 1"=5'
24' TO END OF RIPRAP
54" I.D. PRECAST OUTLET PIPE, CLASS V
INV. EL.=1248.0
EL.=1246.54
EL.=1243.0
EL.=1238.0
3' THK. RIPRAP @ CHANNEL
HEAVY RIPRAP @ OUTLET BASIN
2' DIA. CONC. TO EL. 1235.0'
2' DIA. CONC. ENCASED PILE, CONC. TO EL. 1238.0'
HP 10 x 42 PILES TO EL. 1225' MINIMUM
TYPE HR GEOTEXTILE FILTER FABRIC
9.25' TO END OF RIPRAP
DETAIL "D"
HP 10 x 42 PILE
CONC. ENCASED PILES
#4 SPIRAL WRAP
6-#4's @ 3"± O.C. VERT.
EXTEND DOWELS INTO CRADLE
CONC. CRADLE PROP
CONC. ENCASED PILE
MURPHY DAM
OUTLET BASIN
SCALE - DO NOT SCALE
SHEET 9 OF 17
DATE MARCH 1994
MORGAN & PARMLEY LTD. CONSULTING ENGINEERS LADYSMITH, WISCONSIN
DR. BY T.C. THORNTON

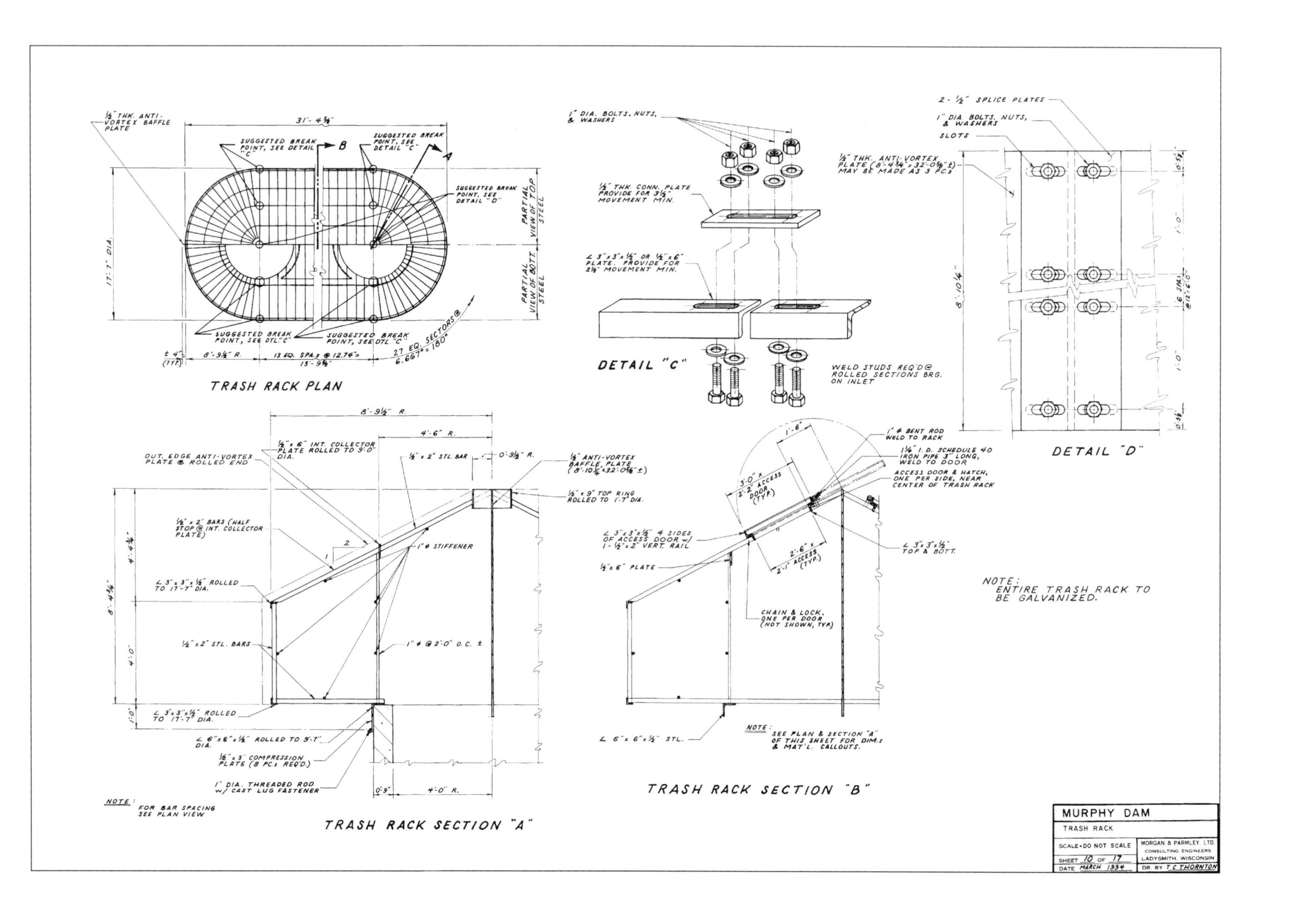
TRASH RACK PLAN
DETAIL "C"
DETAIL "D"
TRASH RACK SECTION "A"
TRASH RACK SECTION "B"
NOTE: ENTIRE TRASH RACK TO BE GALVANIZED.
MURPHY DAM
TRASH RACK
SCALE-DO NOT SCALE
MORGAN & PARMLEY LTD.
CONSULTING ENGINEERS
LADYSMITH, WISCONSIN
SHEET 10 OF 17
DATE MARCH 1994
DR. BY T.C. THORNTON

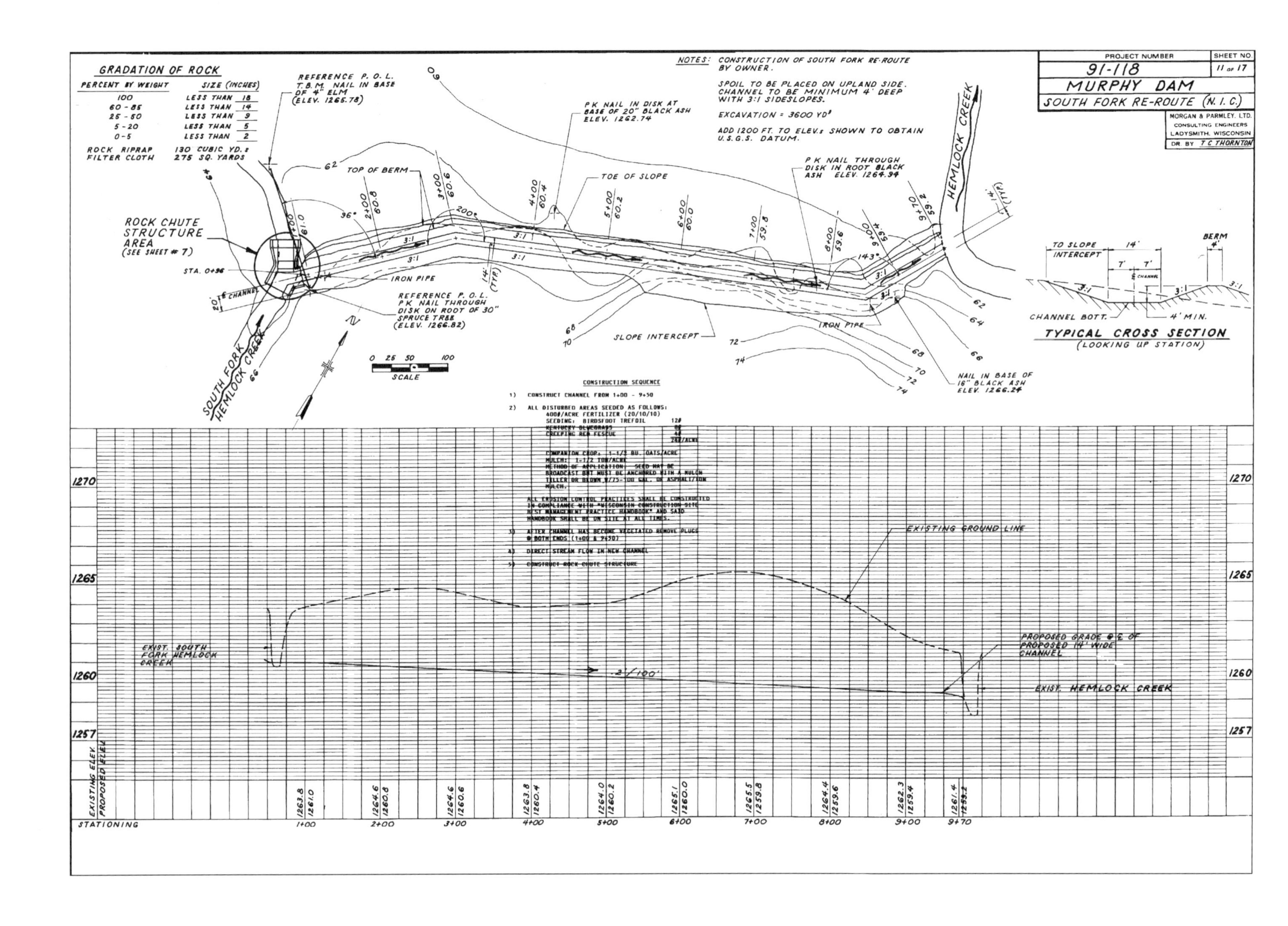
PROJECT NUMBER
91-118
SHEET NO.
11 OF 17
MURPHY DAM
SOUTH FORK RE-ROUTE (N. I. C.)
MORGAN & PARMLEY, LTD.
CONSULTING ENGINEERS
LADYSMITH, WISCONSIN
DR. BY T C THORNTON
GRADATION OF ROCK
PERCENT BY WEIGHT
SIZE (INCHES)
100 LESS THAN 18
60 - 85 LESS THAN 14
25 - 50 LESS THAN 9
5 - 20 LESS THAN 5
0 - 5 LESS THAN 2
ROCK RIPRAP 130 CUBIC YD.±
FILTER CLOTH 275 SQ. YARDS
REFERENCE P. O. L. T.B.M. NAIL IN BASE OF 4" ELM (ELEV. 1265.78)
NOTES: CONSTRUCTION OF SOUTH FORK RE-ROUTE BY OWNER.
SPOIL TO BE PLACED ON UPLAND SIDE. CHANNEL TO BE MINIMUM 4' DEEP WITH 3:1 SIDESLOPES.
EXCAVATION = 3600 YD³
ADD 1200 FT. TO ELEV.s SHOWN TO OBTAIN U.S.G.S. DATUM.
PK NAIL IN DISK AT BASE OF 20" BLACK ASH ELEV. 1262.74
P K NAIL THROUGH DISK IN ROOT BLACK ASH ELEV. 1264.94
HEMLOCK CREEK
TOP OF BERM
TOE OF SLOPE
ROCK CHUTE STRUCTURE AREA (SEE SHEET # 7)
STA. 0+96
℄ CHANNEL
IRON PIPE
REFERENCE P. O. L. PK NAIL THROUGH DISK ON ROOT OF 30" SPRUCE TREE (ELEV. 1266.82)
SLOPE INTERCEPT
NAIL IN BASE OF 16" BLACK ASH ELEV. 1266.24
SOUTH FORK HEMLOCK CREEK
SCALE
TYPICAL CROSS SECTION
(LOOKING UP STATION)
TO SLOPE INTERCEPT
BERM
CHANNEL BOTT.
4' MIN.
CONSTRUCTION SEQUENCE
1) CONSTRUCT CHANNEL FROM 1+00 - 9+50
2) ALL DISTURBED AREAS SEEDED AS FOLLOWS:
400#/ACRE FERTILIZER (20/10/10)
SEEDING: BIRDSFOOT TREFOIL 12#
KENTUCKY BLUEGRASS 8#
CREEPING RED FESCUE 4#
24#/ACRE
COMPANION CROP: 1-1/2 BU. OATS/ACRE
MULCH: 1-1/2 TON/ACRE
METHOD OF APPLICATION: SEED MAY BE BROADCAST BUT MUST BE ANCHORED WITH A MULCH TILLER OR BLOWN W/75-100 GAL. OF ASPHALT/TON MULCH.
ALL EROSION CONTROL PRACTICES SHALL BE CONSTRUCTED IN COMPLIANCE WITH "WISCONSIN CONSTRUCTION SITE BEST MANAGEMENT PRACTICE HANDBOOK" AND SAID HANDBOOK SHALL BE ON SITE AT ALL TIMES.
3) AFTER CHANNEL HAS BECOME VEGETATED REMOVE PLUGS @ BOTH ENDS (1+00 & 9+50)
4) DIRECT STREAM FLOW IN NEW CHANNEL
5) CONSTRUCT ROCK CHUTE STRUCTURE
EXISTING GROUND LINE
EXIST. SOUTH FORK HEMLOCK CREEK
.2'/100'
PROPOSED GRADE @ ℄ OF PROPOSED 14' WIDE CHANNEL
EXIST. HEMLOCK CREEK
1270
1265
1260
1257
EXISTING ELEV.
PROPOSED ELEV.
STATIONING
1+00 1263.8 1261.0
2+00 1264.6 1260.8
3+00 1264.6 1260.6
4+00 1263.8 1260.4
5+00 1264.0 1260.2
6+00 1265.1 1260.0
7+00 1265.5 1259.8
8+00 1264.4 1259.6
9+00 1262.3 1259.4
9+70 1261.4 1259.2

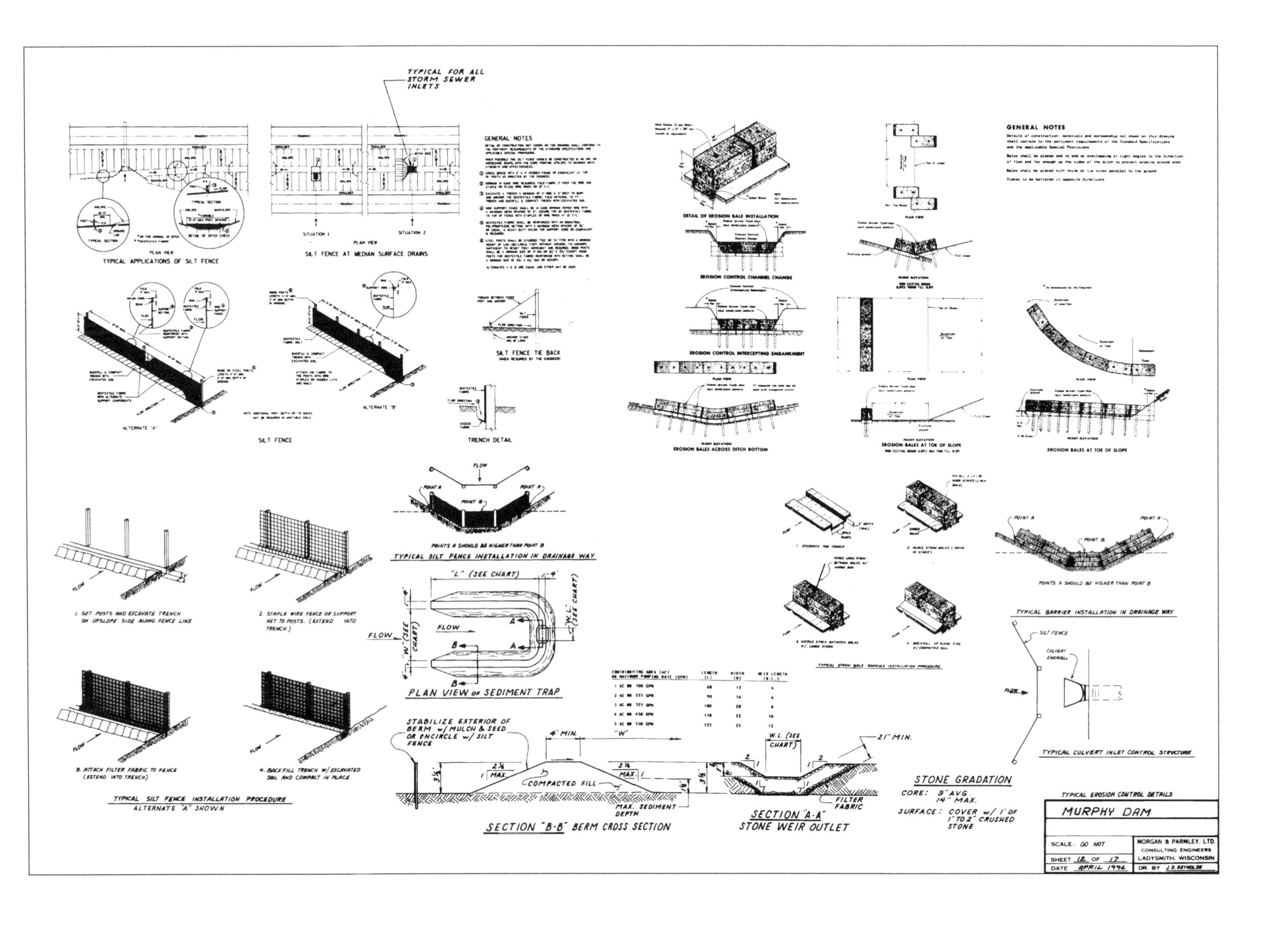
TYPICAL FOR ALL STORM SEWER INLETS
PLAN VIEW
TYPICAL APPLICATIONS OF SILT FENCE
SITUATION 1
SITUATION 2
PLAN VIEW
SILT FENCE AT MEDIAN SURFACE DRAINS
GENERAL NOTES
ALTERNATE "A"
ALTERNATE "B"
SILT FENCE
SILT FENCE TIE BACK
TRENCH DETAIL
DETAIL OF EROSION BALE INSTALLATION
EROSION CONTROL CHANNEL CHANGE
EROSION CONTROL INTERCEPTING EMBANKMENT
PLAN VIEW
FRONT ELEVATION
EROSION BALES ACROSS DITCH BOTTOM
EROSION BALES AT TOE OF SLOPE
GENERAL NOTES
EROSION BALES AT TOE OF SLOPE
FLOW
POINT A
POINT B
POINTS A SHOULD BE HIGHER THAN POINT B
TYPICAL SILT FENCE INSTALLATION IN DRAINAGE WAY
1. SET POSTS AND EXCAVATE TRENCH ON UPSLOPE SIDE ALONG FENCE LINE
2. STAPLE WIRE FENCE OR SUPPORT NET TO POSTS. (EXTEND INTO TRENCH)
3. ATTACH FILTER FABRIC TO FENCE (EXTEND INTO TRENCH)
4. BACKFILL TRENCH W/ EXCAVATED SOIL AND COMPACT IN PLACE
TYPICAL SILT FENCE INSTALLATION PROCEDURE
ALTERNATE "A" SHOWN
"L" (SEE CHART)
4'
"W" (SEE CHART)
W.L. (SEE CHART)
FLOW
A
B
PLAN VIEW OF SEDIMENT TRAP
CONTRIBUTING AREA (AC) OR MAXIMUM PUMPING RATE (GPM) | LENGTH (L) | WIDTH (W) | WEIR LENGTH (W.L.)
1 AC OR 100 GPM | 68 | 12 | 4
2 AC OR 225 GPM | 90 | 16 | 6
3 AC OR 325 GPM | 100 | 20 | 8
4 AC OR 450 GPM | 118 | 22 | 10
5 AC OR 550 GPM | 125 | 25 | 12
STABILIZE EXTERIOR OF BERM w/ MULCH & SEED OR ENCIRCLE w/ SILT FENCE
4' MIN.
"W"
2½ MAX.
COMPACTED FILL
MAX. SEDIMENT DEPTH
SECTION "B-B" BERM CROSS SECTION
W.L. (SEE CHART)
21" MIN.
FILTER FABRIC
SECTION "A-A" STONE WEIR OUTLET
STONE GRADATION
CORE: 9" AVG. 14" MAX.
SURFACE: COVER w/ 1' OF 1" TO 2" CRUSHED STONE
POINT A
POINT B
POINTS A SHOULD BE HIGHER THAN POINT B
TYPICAL BARRIER INSTALLATION IN DRAINAGE WAY
SILT FENCE
CULVERT ENDWALL
FLOW
TYPICAL CULVERT INLET CONTROL STRUCTURE
TYPICAL STRAW BALE BARRIER INSTALLATION PROCEDURE
TYPICAL EROSION CONTROL DETAILS
MURPHY DAM
SCALE: DO NOT
SHEET 12 OF 17
DATE APRIL 1994
MORGAN & PARMLEY, LTD.
CONSULTING ENGINEERS
LADYSMITH, WISCONSIN
DR. BY J.D. REYNOLDS

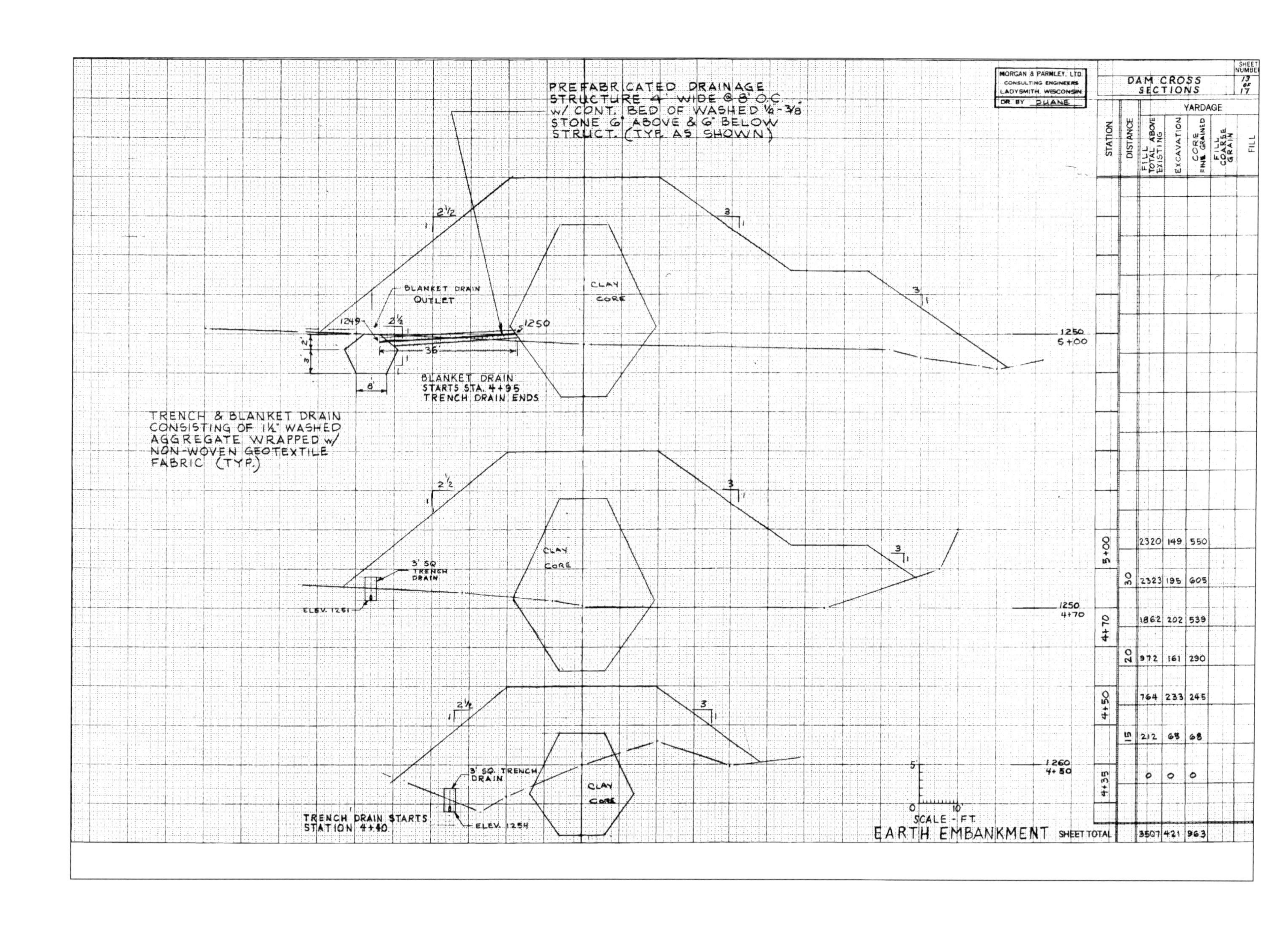
MORGAN & PARMLEY, LTD.
CONSULTING ENGINEERS
LADYSMITH, WISCONSIN
DR. BY DUANE
DAM CROSS SECTIONS
SHEET NUMBER 13 OF 17
PREFABRICATED DRAINAGE STRUCTURE 4' WIDE @ 8' O.C. w/ CONT. BED OF WASHED ½-⅜ STONE 6" ABOVE & 6" BELOW STRUCT. (TYP. AS SHOWN)
BLANKET DRAIN OUTLET
1249
1250
36'
8'
BLANKET DRAIN STARTS STA. 4+95 TRENCH DRAIN ENDS
CLAY CORE
1250 5+00
TRENCH & BLANKET DRAIN CONSISTING OF 1½" WASHED AGGREGATE WRAPPED w/ NON-WOVEN GEOTEXTILE FABRIC (TYP.)
3' SQ. TRENCH DRAIN
ELEV. 1251
1250 4+70
3' SQ. TRENCH DRAIN
ELEV. 1254
TRENCH DRAIN STARTS STATION 4+40
1260 4+50
SCALE - FT.
EARTH EMBANKMENT
YARDAGE
STATION
DISTANCE
FILL TOTAL ABOVE EXISTING
EXCAVATION
CORE FINE GRAINED
FILL COARSE GRAIN
FILL
5+00
2320 149 550
30
2323 195 605
4+70
1862 202 539
20
972 161 290
4+50
764 233 245
15
212 65 68
4+35
0 0 0
SHEET TOTAL
3507 421 963

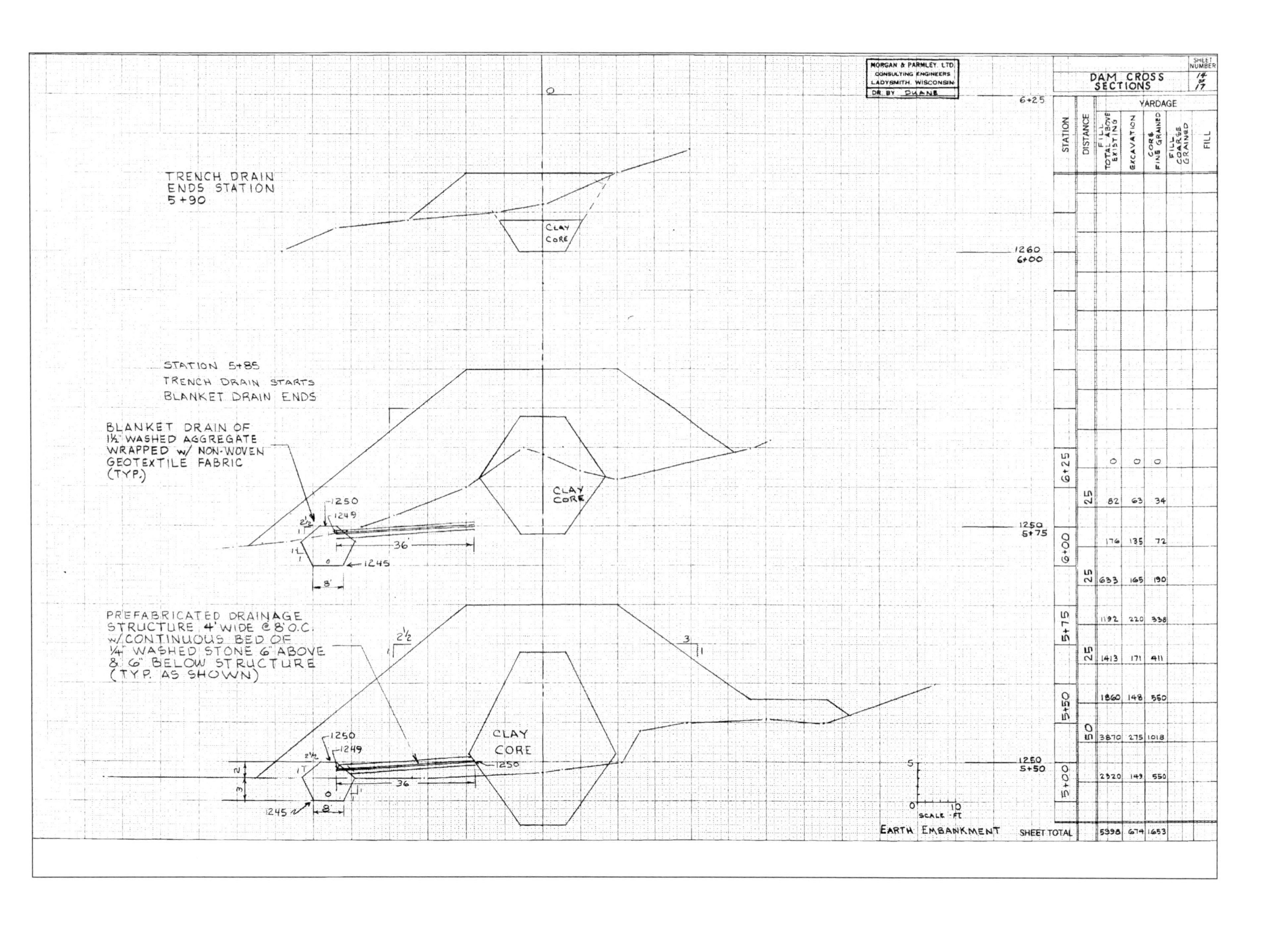

STATION	DISTANCE	FILL TOTAL ABOVE EXISTING	EXCAVATION	CORE FINE GRAINED	FILL COARSE GRAINED	FILL
		YARDAGE				
6+25		0	0	0		
	25	82	63	34		
6+00		176	135	72		
	25	633	165	190		
5+75		1192	220	338		
	25	1413	171	411		
5+50		1860	148	550		
	50	3870	275	1018		
5+00		2320	149	550		
SHEET TOTAL		5998	674	1653		

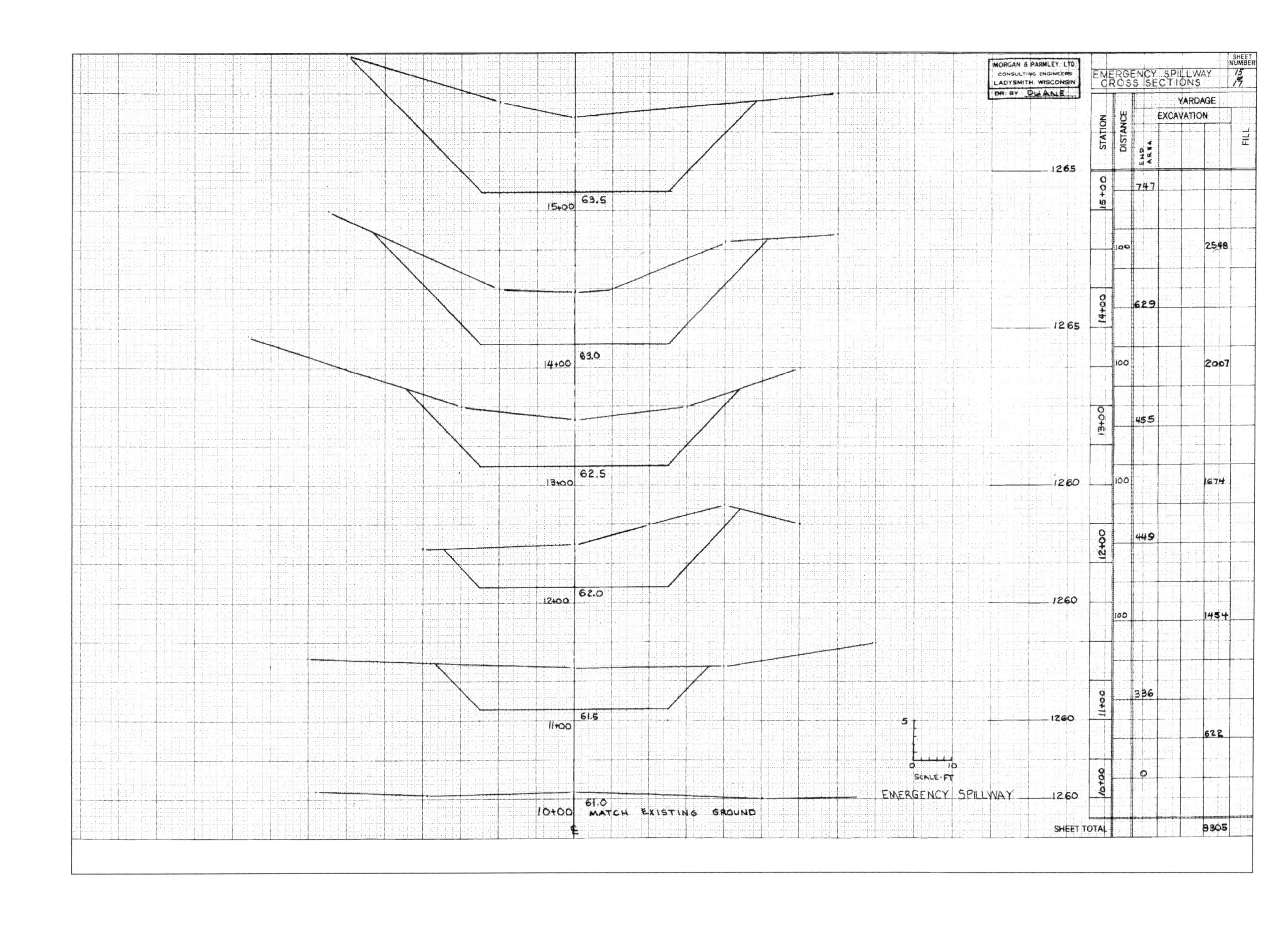

MORGAN & PARMLEY, LTD.
CONSULTING ENGINEERS
LADYSMITH, WISCONSIN
DR. BY DUANE
EMERGENCY SPILLWAY
CROSS SECTIONS
SHEET NUMBER
YARDAGE
EXCAVATION
STATION
DISTANCE
END AREA
FILL
1265
15+00
63.5
747
100
2548
14+00
63.0
629
1265
100
2007
13+00
62.5
455
1260
100
1674
12+00
62.0
449
1260
100
1454
11+00
61.5
336
1260
622
10+00
61.0
0
1260
10+00 MATCH EXISTING GROUND
EMERGENCY SPILLWAY
SCALE-FT
SHEET TOTAL
8305

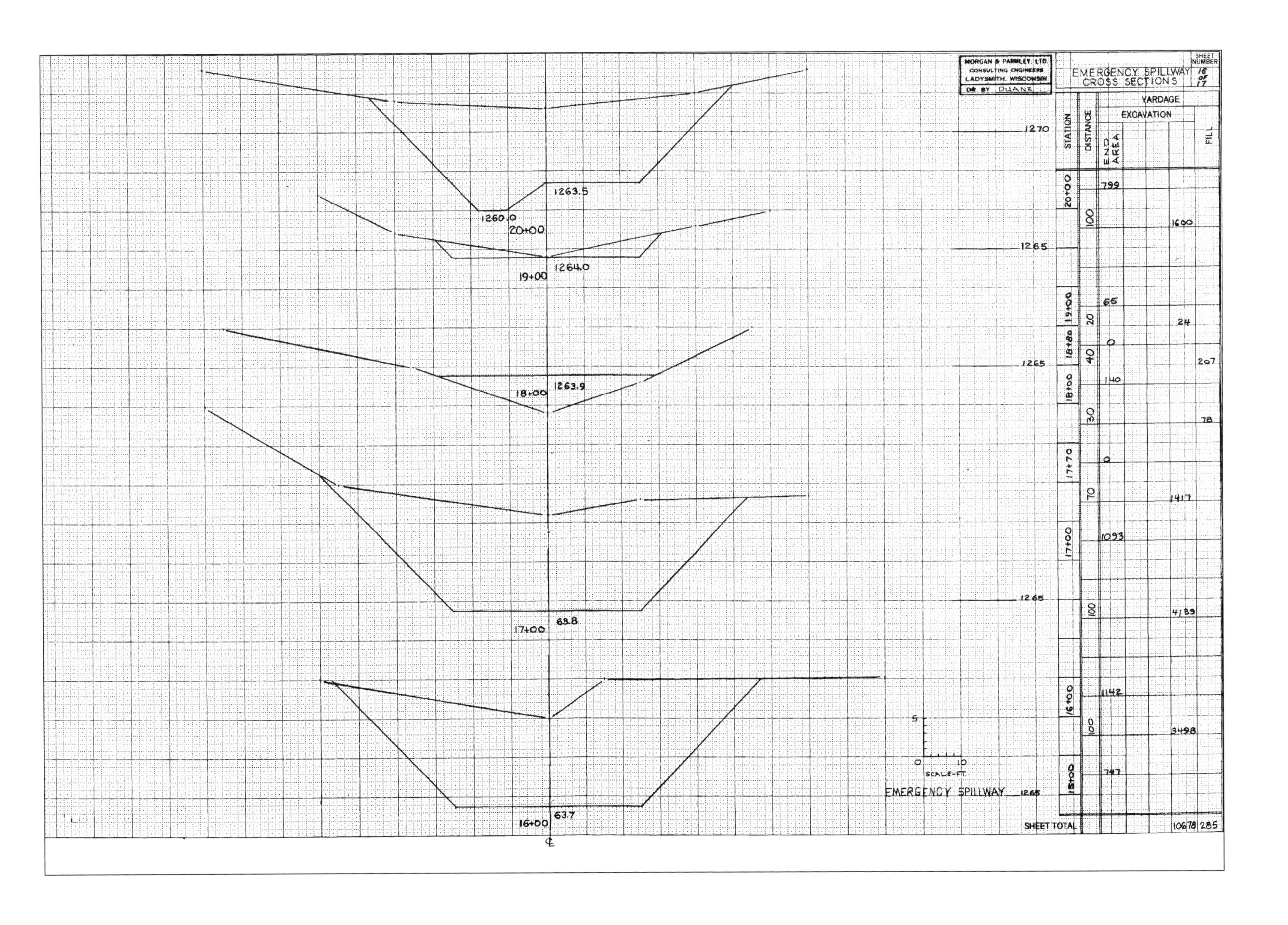
MORGAN & PARMLEY LTD.
CONSULTING ENGINEERS
LADYSMITH, WISCONSIN
DR BY DUANE
EMERGENCY SPILLWAY
CROSS SECTIONS
SHEET NUMBER
16 of 17
YARDAGE
EXCAVATION
FILL
STATION
DISTANCE
END AREA
1270
1263.5
1260.0
20+00
1265
1264.0
19+00
1263.9
18+00
17+00
63.8
16+00
63.7
20+00
100
799
1600
19+00
20
65
24
18+80
40
0
207
18+00
140
30
78
17+70
0
70
1417
17+00
1093
100
4189
16+00
1142
100
3498
15+00
747
SCALE-FT.
EMERGENCY SPILLWAY
SHEET TOTAL
10678
285

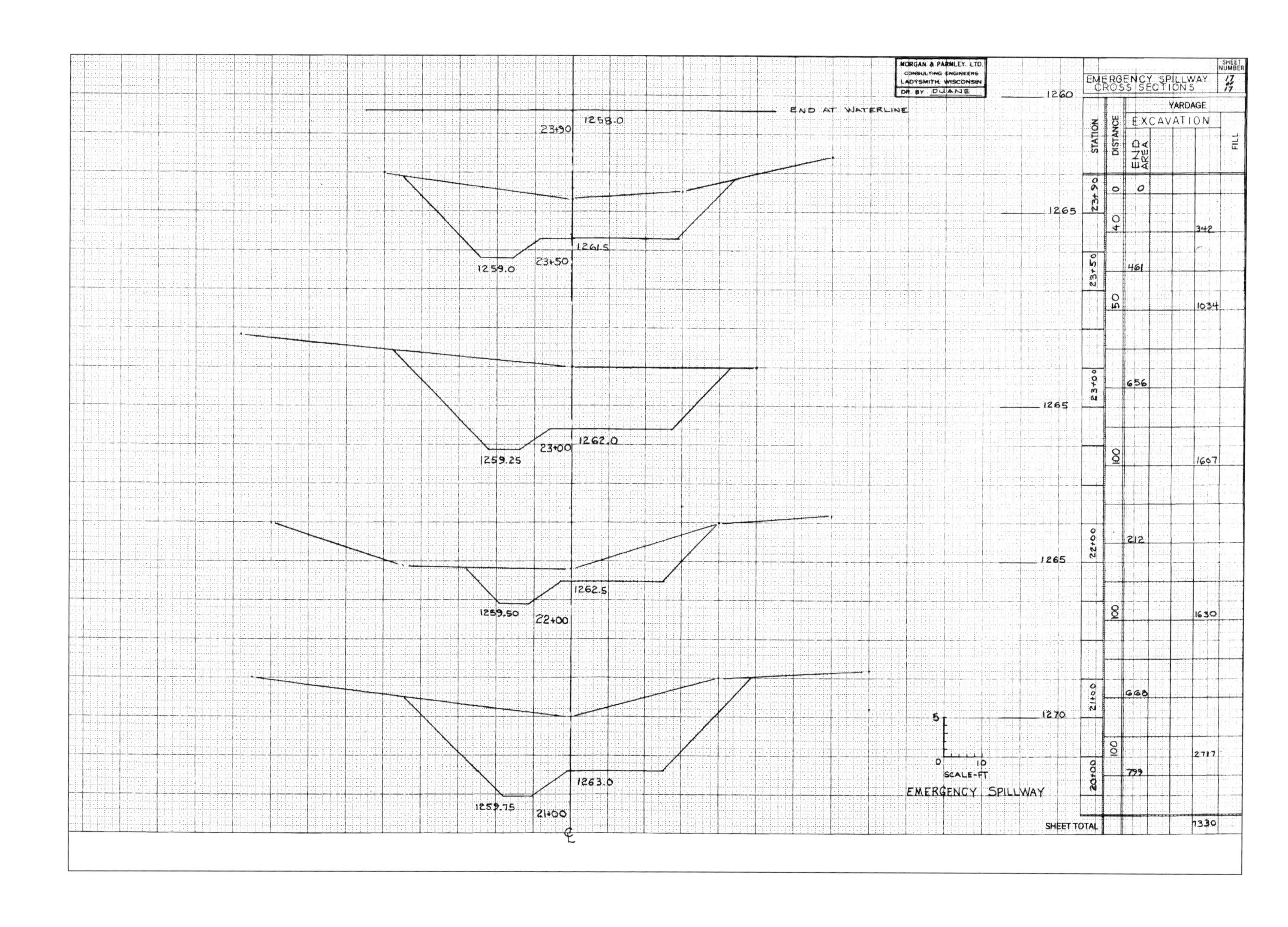
MORGAN & PARMLEY, LTD.
CONSULTING ENGINEERS
LADYSMITH, WISCONSIN
DR. BY DUANE
EMERGENCY SPILLWAY CROSS SECTIONS
SHEET NUMBER 17 of 17
YARDAGE
STATION
DISTANCE
EXCAVATION
END AREA
FILL
END AT WATERLINE
1260
1265
1265
1265
1270
23+90
1258.0
1261.5
1259.0
23+50
1262.0
1259.25
23+00
1262.5
1259.50
22+00
1263.0
1259.75
21+00
SCALE-FT
EMERGENCY SPILLWAY
SHEET TOTAL
7330

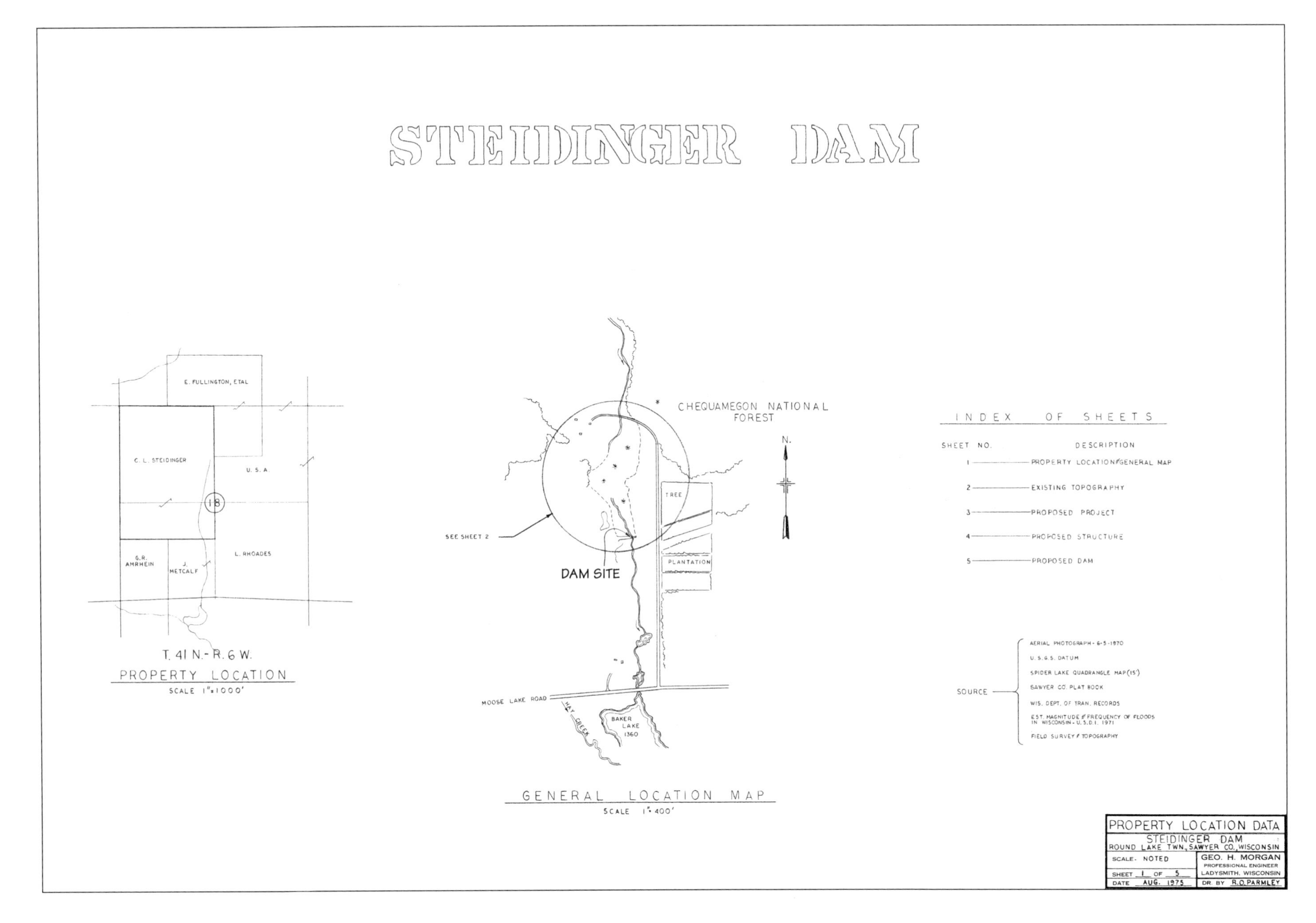

Source: Morgan & Parmley, Ltd.

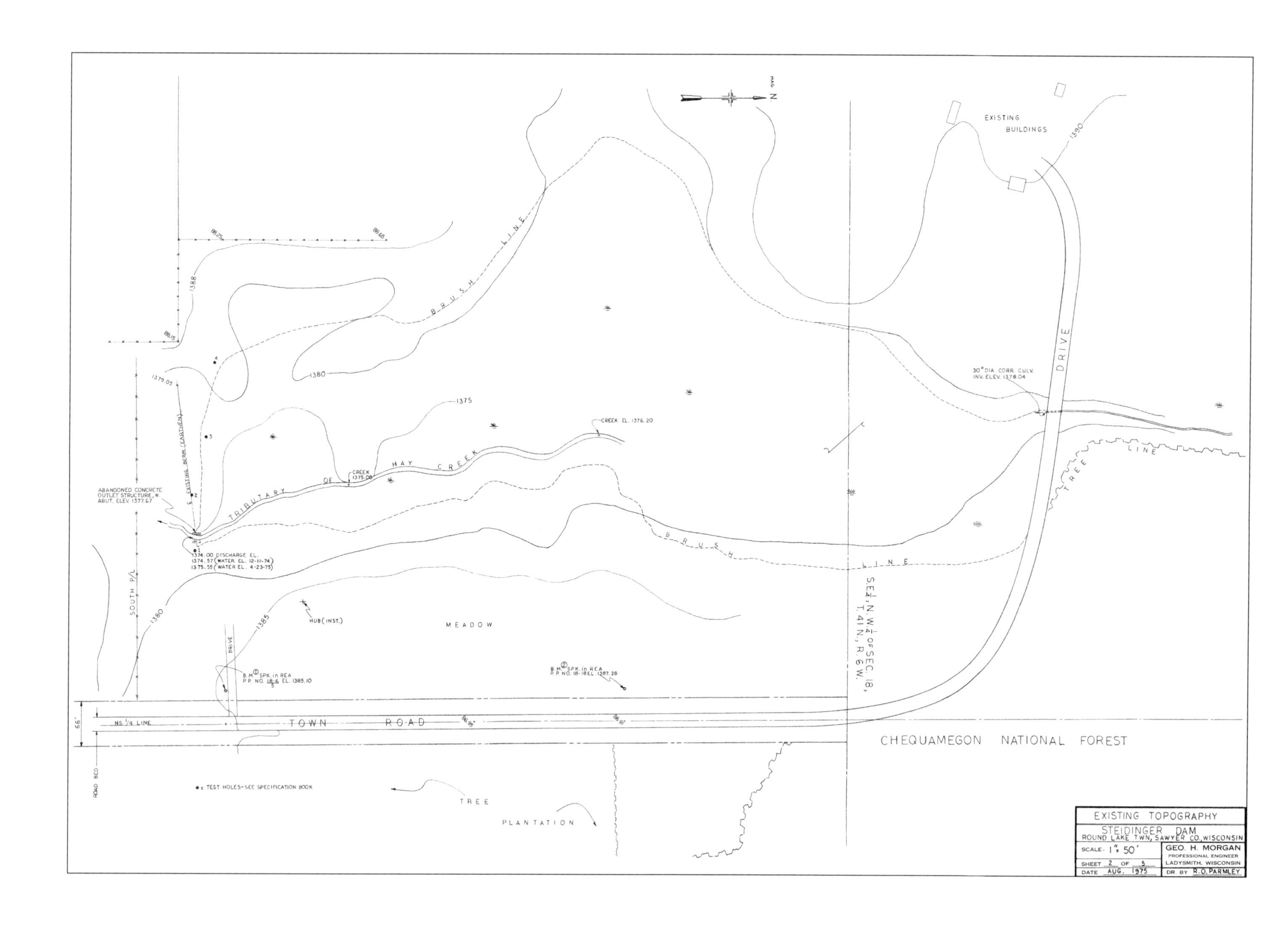
MAG. N.
EXISTING BUILDINGS
1390
1388
BRUSH LINE
1380
1375
1379.05
EXISTING BERM (EARTHEN)
CREEK EL. 1376.20
CREEK 1375.00
TRIBUTARY OF HAY CREEK
ABANDONED CONCRETE OUTLET STRUCTURE, W. ABUT. ELEV. 1377.67
1374.00 DISCHARGE EL.
1374.57 (WATER EL. 12-11-74)
1375.55 (WATER EL. 4-23-75)
30" DIA. CORR. CULV. INV. ELEV. 1378.04
DRIVE
TREE LINE
SOUTH P/L
1385
HUB (INST.)
MEADOW
S.E.¼, N.W.¼ OF SEC. 18, T. 41 N., R. 6 W.
B.M.① SPK. in REA P.P. NO. 18-6 EL. 1385.10
B.M.② SPK. in REA P.P. NO. 18-18 EL. 1387.28
66'
NS ¼ LINE
TOWN ROAD
CHEQUAMEGON NATIONAL FOREST
ROAD BED
● TEST HOLES-SEE SPECIFICATION BOOK
TREE PLANTATION
EXISTING TOPOGRAPHY
STEIDINGER DAM
ROUND LAKE TWN, SAWYER CO., WISCONSIN
SCALE- 1" = 50'
SHEET 2 OF 5
DATE AUG. 1975
GEO. H. MORGAN
PROFESSIONAL ENGINEER
LADYSMITH, WISCONSIN
DR. BY R.O. PARMLEY

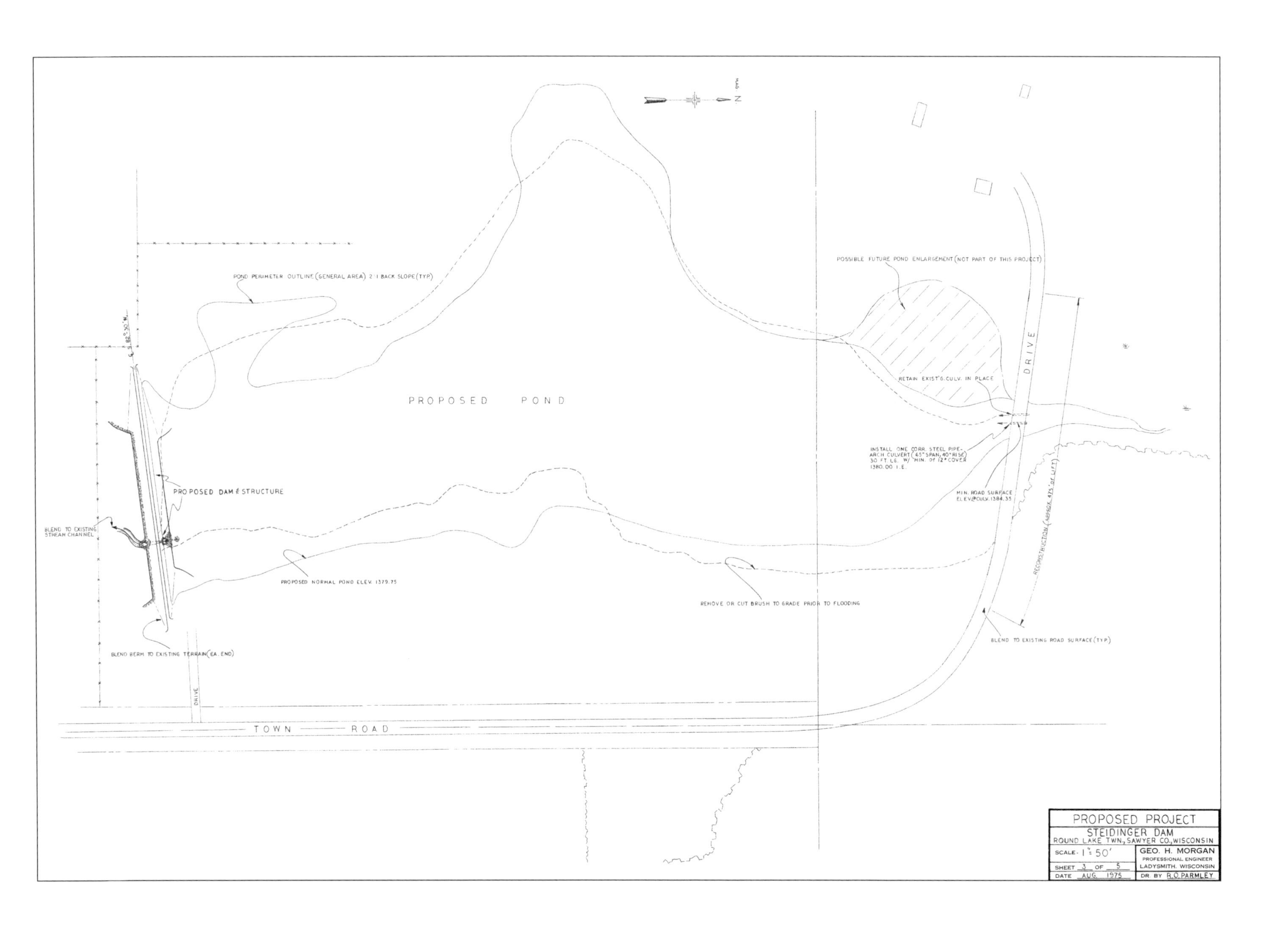
N
PROPOSED POND
POND PERIMETER OUTLINE (GENERAL AREA) 2:1 BACK SLOPE (TYP.)
S 82°-30' W
PROPOSED DAM & STRUCTURE
BLEND TO EXISTING STREAM CHANNEL
PROPOSED NORMAL POND ELEV. 1379.75
BLEND BERM TO EXISTING TERRAIN (EA. END)
DRIVE
TOWN ROAD
POSSIBLE FUTURE POND ENLARGEMENT (NOT PART OF THIS PROJECT)
RETAIN EXIST'G. CULV. IN PLACE
DRIVE
INSTALL ONE CORR. STEEL PIPE-ARCH CULVERT (65" SPAN, 40" RISE) 30 FT. LG. W/ MIN. OF 12" COVER 1380.00 I.E.
MIN. ROAD SURFACE ELEV @ CULV. 1384.35
RECONSTRUCTION (APPROX. 475' OF LIFT)
REMOVE OR CUT BRUSH TO GRADE PRIOR TO FLOODING
BLEND TO EXISTING ROAD SURFACE (TYP.)
PROPOSED PROJECT
STEIDINGER DAM
ROUND LAKE TWN., SAWYER CO., WISCONSIN
SCALE - 1" = 50'
SHEET 3 OF 5
DATE AUG. 1975
GEO. H. MORGAN
PROFESSIONAL ENGINEER
LADYSMITH, WISCONSIN
DR. BY R.O. PARMLEY

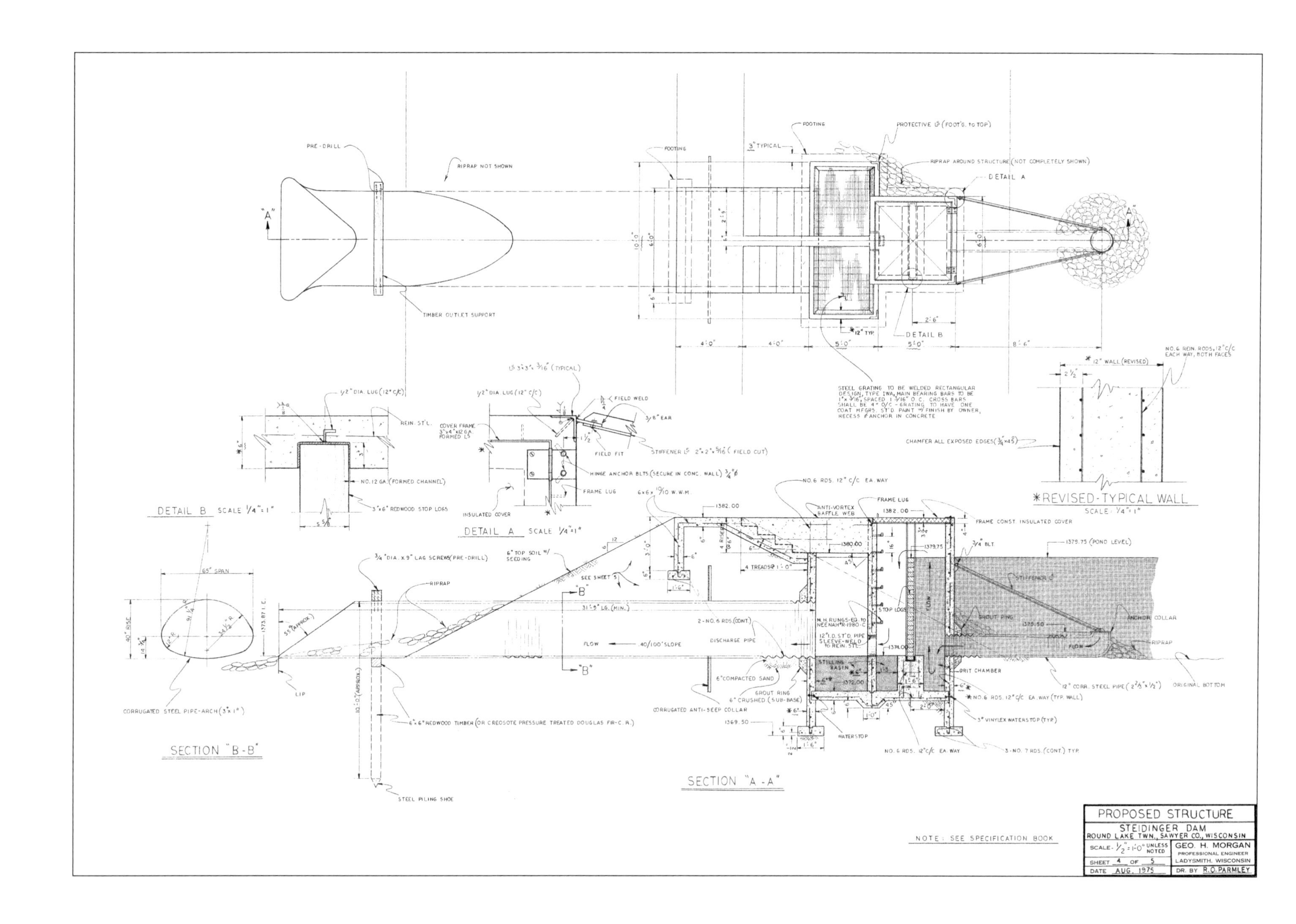
PRE-DRILL
RIPRAP NOT SHOWN
TIMBER OUTLET SUPPORT
FOOTING
3" TYPICAL
PROTECTIVE L5 (FOOT'G. TO TOP)
RIPRAP AROUND STRUCTURE (NOT COMPLETELY SHOWN)
DETAIL A
DETAIL B
*12" TYP
STEEL GRATING TO BE WELDED RECTANGULAR DESIGN, TYPE IWA, MAIN BEARING BARS TO BE 1"x3/16", SPACED 1-3/16" O.C. CROSS BARS SHALL BE 4" O/C - GRATING TO HAVE ONE COAT MFGRS. ST'D PAINT - FINISH BY OWNER, RECESS & ANCHOR IN CONCRETE
CHAMFER ALL EXPOSED EDGES (3/4"x45°)
*12" WALL (REVISED)
NO. 6 REIN. RODS, 12" C/C EACH WAY, BOTH FACES
*REVISED-TYPICAL WALL
SCALE: 1/4"=1"
1/2" DIA. LUG (12" C/C)
REIN. ST'L.
NO. 12 GA. (FORMED CHANNEL)
3"x6" REDWOOD STOP LOGS
DETAIL B SCALE 1/4"=1"
COVER FRAME 3"x4"x12 GA. FORMED L5
FIELD WELD
3/8" EAR
FIELD FIT
STIFFENER L5 2"x2"x3/16" (FIELD CUT)
HINGE ANCHOR BLTS. (SECURE IN CONC. WALL) 3/4"ø
FRAME LUG
INSULATED COVER
DETAIL A SCALE 1/4"=1"
6x6x10/10 W.W.M.
NO. 6 RDS. 12" C/C EA. WAY
ANTI-VORTEX BAFFLE WEB
FRAME LUG
FRAME CONST. INSULATED COVER
1382.00
1380.00
1379.75
1375.75 (POND LEVEL)
4 TREADS @ 1'-0"
STOP LOGS
FLOW
STIFFENER
GROUT RING
ANCHOR COLLAR
RIPRAP
M.H. RUNGS-EQ. TO NEENAH R-1980-C
12" I.D. ST'D. PIPE SLEEVE-WELD TO REIN. ST'L.
STILLING BASIN
GRIT CHAMBER
12" CORR. STEEL PIPE (2 2/3"x1/2")
ORIGINAL BOTTOM
*NO. 6 RDS. 12" C/C EA. WAY (TYP. WALL)
3" VINYLEX WATERSTOP (TYP.)
3-NO. 7 RDS. (CONT.) TYP.
NO. 6 RDS. 12" C/C EA. WAY
WATERSTOP
1369.50
CORRUGATED ANTI-SEEP COLLAR
6" CRUSHED (SUB-BASE)
GROUT RING
6" COMPACTED SAND
DISCHARGE PIPE
2-NO. 6 RDS. (CONT.)
FLOW 40/100' SLOPE
31'-9" LG. (MIN.)
SEE SHEET 5
6" TOP SOIL W/ SEEDING
3/4" DIA. x 9" LAG SCREWS (PRE-DRILL)
RIPRAP
65" SPAN
40" RISE
55° (APPROX.)
10'-0" (APPROX.)
LIP
CORRUGATED STEEL PIPE-ARCH (3"x1")
6"x6" REDWOOD TIMBER (OR CREOSOTE PRESSURE TREATED DOUGLAS FIR-C.R.)
STEEL PILING SHOE
SECTION "B-B"
SECTION "A-A"
NOTE: SEE SPECIFICATION BOOK
PROPOSED STRUCTURE
STEIDINGER DAM
ROUND LAKE TWN., SAWYER CO., WISCONSIN
SCALE- 1/2"=1'-0" UNLESS NOTED
SHEET 4 OF 5
DATE AUG. 1975
GEO. H. MORGAN
PROFESSIONAL ENGINEER
LADYSMITH, WISCONSIN
DR. BY R.O. PARMLEY

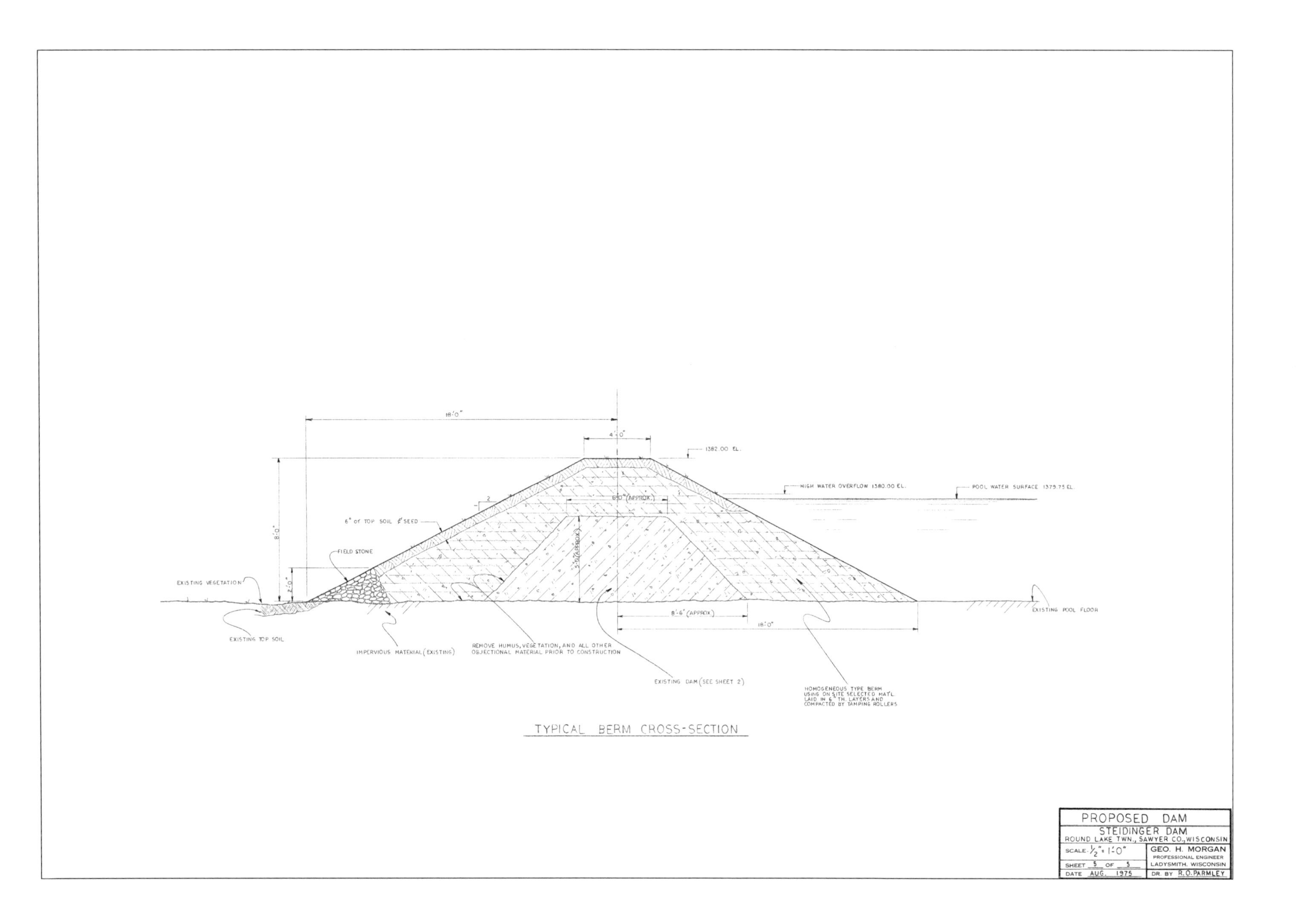
18'-0"
4'-0"
1382.00 EL.
HIGH WATER OVERFLOW 1380.00 EL.
POOL WATER SURFACE 1379.75 EL.
2
1
6'-0" (APPROX.)
6" OF TOP SOIL & SEED
8'-0"
5'-0" (APPROX.)
FIELD STONE
2'-0"
EXISTING VEGETATION
EXISTING POOL FLOOR
8'-6" (APPROX)
18'-0"
EXISTING TOP SOIL
IMPERVIOUS MATERIAL (EXISTING)
REMOVE HUMUS, VEGETATION, AND ALL OTHER OBJECTIONAL MATERIAL PRIOR TO CONSTRUCTION
EXISTING DAM (SEE SHEET 2)
HOMOGENEOUS TYPE BERM USING ON SITE SELECTED MAT'L. LAID IN 6" TH. LAYERS AND COMPACTED BY TAMPING ROLLERS
TYPICAL BERM CROSS-SECTION
PROPOSED DAM
STEIDINGER DAM
ROUND LAKE TWN., SAWYER CO., WISCONSIN
SCALE 1/2" = 1'-0"
GEO. H. MORGAN
PROFESSIONAL ENGINEER
LADYSMITH, WISCONSIN
SHEET 5 OF 5
DATE AUG. 1975
DR. BY R.O. PARMLEY

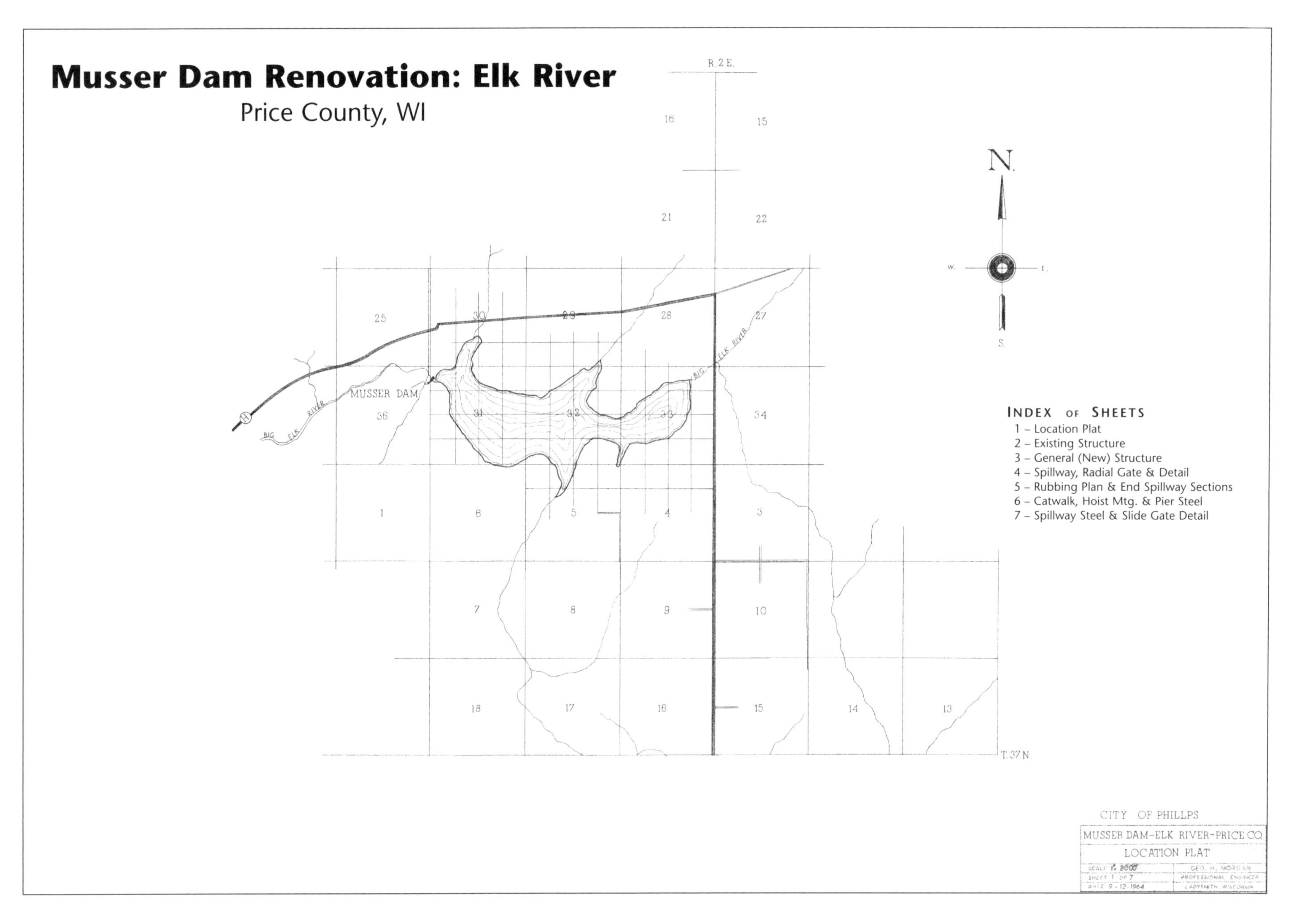

Source: Morgan & Parmley, Ltd.

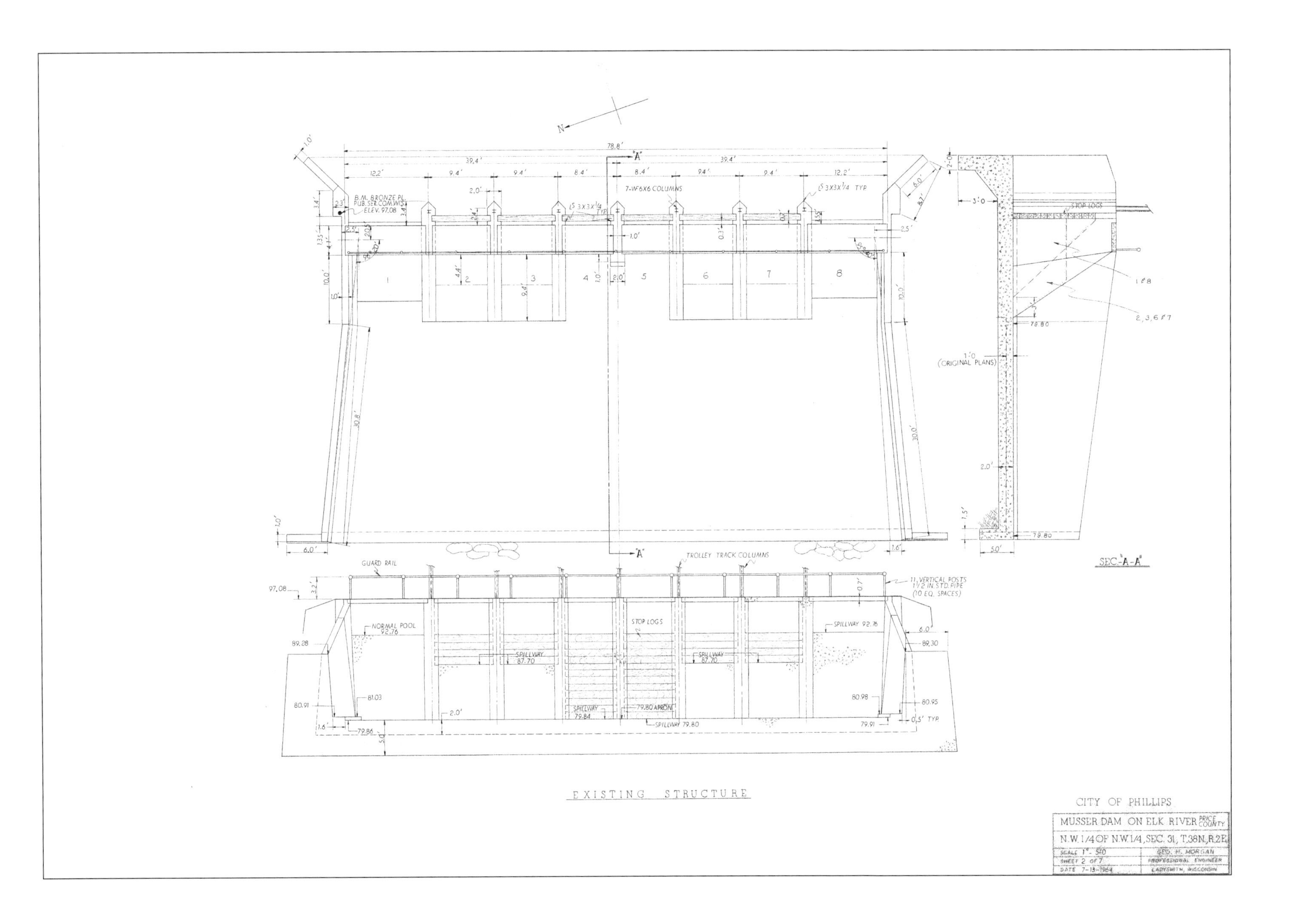
N
78.8'
39.4'
12.2'
9.4'
8.4'
7-WF6X6 COLUMNS
L5 3X3X1/4 TYP
B.M. BRONZE PL.
PUB.SER.COM.WIS.
ELEV. 97.08
1
2
3
4
5
6
7
8
"A"
STOP LOGS
1 & 8
2, 3, 6 & 7
79.80
1'-0
(ORIGINAL PLANS)
SEC. "A-A"
GUARD RAIL
TROLLEY TRACK COLUMNS
11 VERTICAL POSTS
1 1/2 IN. STD. PIPE
(10 EQ. SPACES)
97.08
NORMAL POOL
92.76
SPILLWAY 92.76
SPILLWAY
87.70
89.28
89.30
81.03
80.91
80.98
80.95
SPILLWAY
79.84
79.80 APRON
SPILLWAY 79.80
79.86
79.91
0.5' TYP
EXISTING STRUCTURE
CITY OF PHILLIPS
MUSSER DAM ON ELK RIVER PRICE COUNTY
N.W. 1/4 OF N.W. 1/4, SEC. 31, T.38N, R.2E.
SCALE 1" - 5'-0
SHEET 2 OF 7
DATE 7-13-1964
GEO. H. MORGAN
PROFESSIONAL ENGINEER
LADYSMITH, WISCONSIN

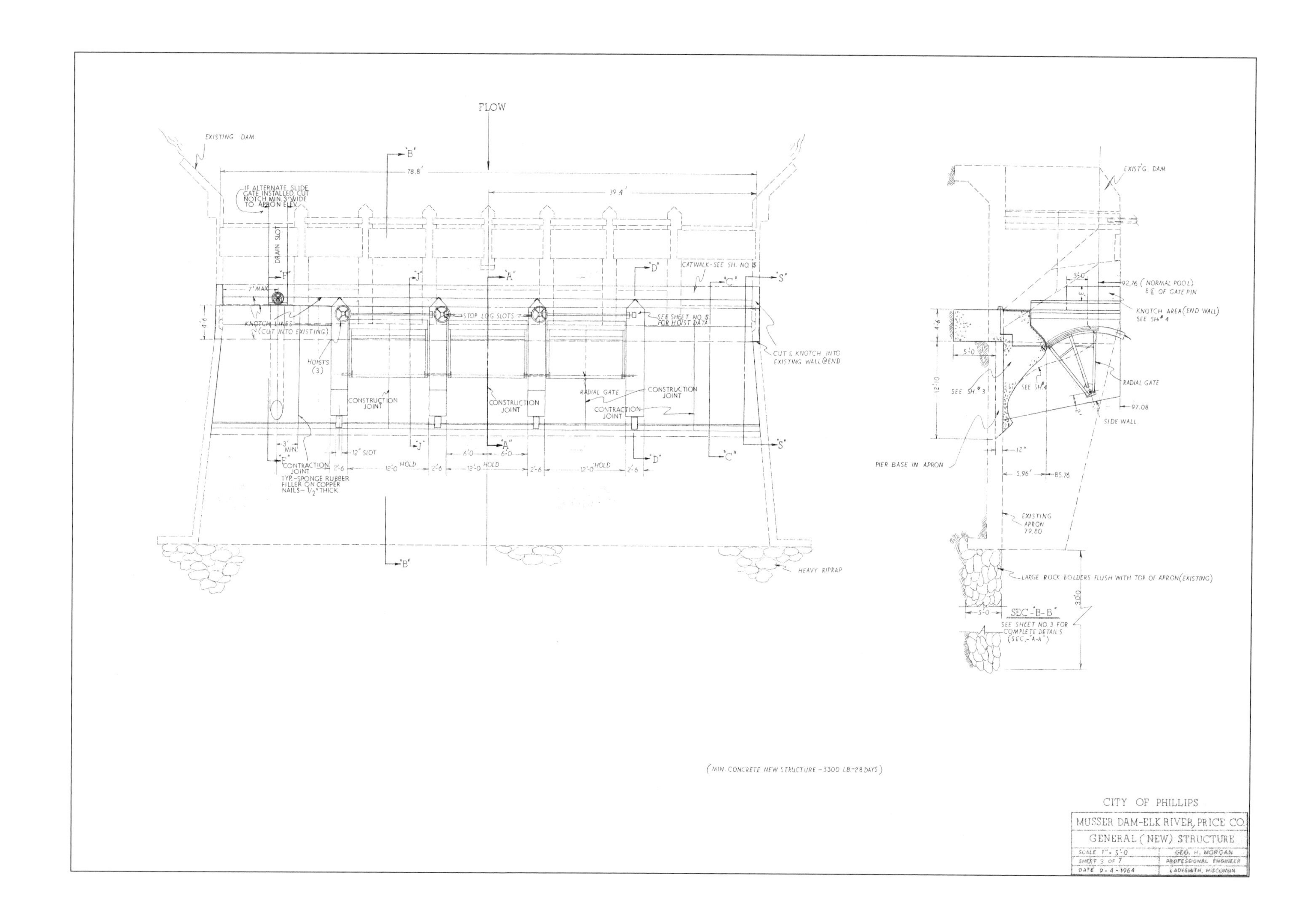

FLOW
EXISTING DAM
IF ALTERNATE SLIDE GATE INSTALLED CUT NOTCH MIN. 3' WIDE TO APRON ELEV.
DRAIN SLOT
CATWALK-SEE SH. NO. 5
STOP LOG SLOTS
SEE SHEET NO. 5 FOR HOIST DATA
KNOTCH LINES (CUT INTO EXISTING)
CUT & KNOTCH INTO EXISTING WALL @ END
HOISTS (3)
CONSTRUCTION JOINT
RADIAL GATE
CONTRACTION JOINT
CONTRACTION JOINT TYP.-SPONGE RUBBER FILLER ON COPPER NAILS-1/2" THICK
12" SLOT
HOLD
HEAVY RIPRAP
EXIST'G. DAM
92.76 (NORMAL POOL) & ℄ OF GATE PIN
KNOTCH AREA (END WALL) SEE SH. # 4
RADIAL GATE
SIDE WALL
PIER BASE IN APRON
EXISTING APRON 79.80
LARGE ROCK BOLDERS FLUSH WITH TOP OF APRON (EXISTING)
SEC-"B-B"
SEE SHEET NO. 3 FOR COMPLETE DETAILS (SEC.-"A-A")
(MIN. CONCRETE NEW STRUCTURE -3300 LB.-28 DAYS)
CITY OF PHILLIPS
MUSSER DAM-ELK RIVER, PRICE CO.
GENERAL (NEW) STRUCTURE
SCALE 1" = 5'-0
SHEET 3 OF 7
DATE 9-4-1964
GEO. H. MORGAN
PROFESSIONAL ENGINEER
LADYSMITH, WISCONSIN

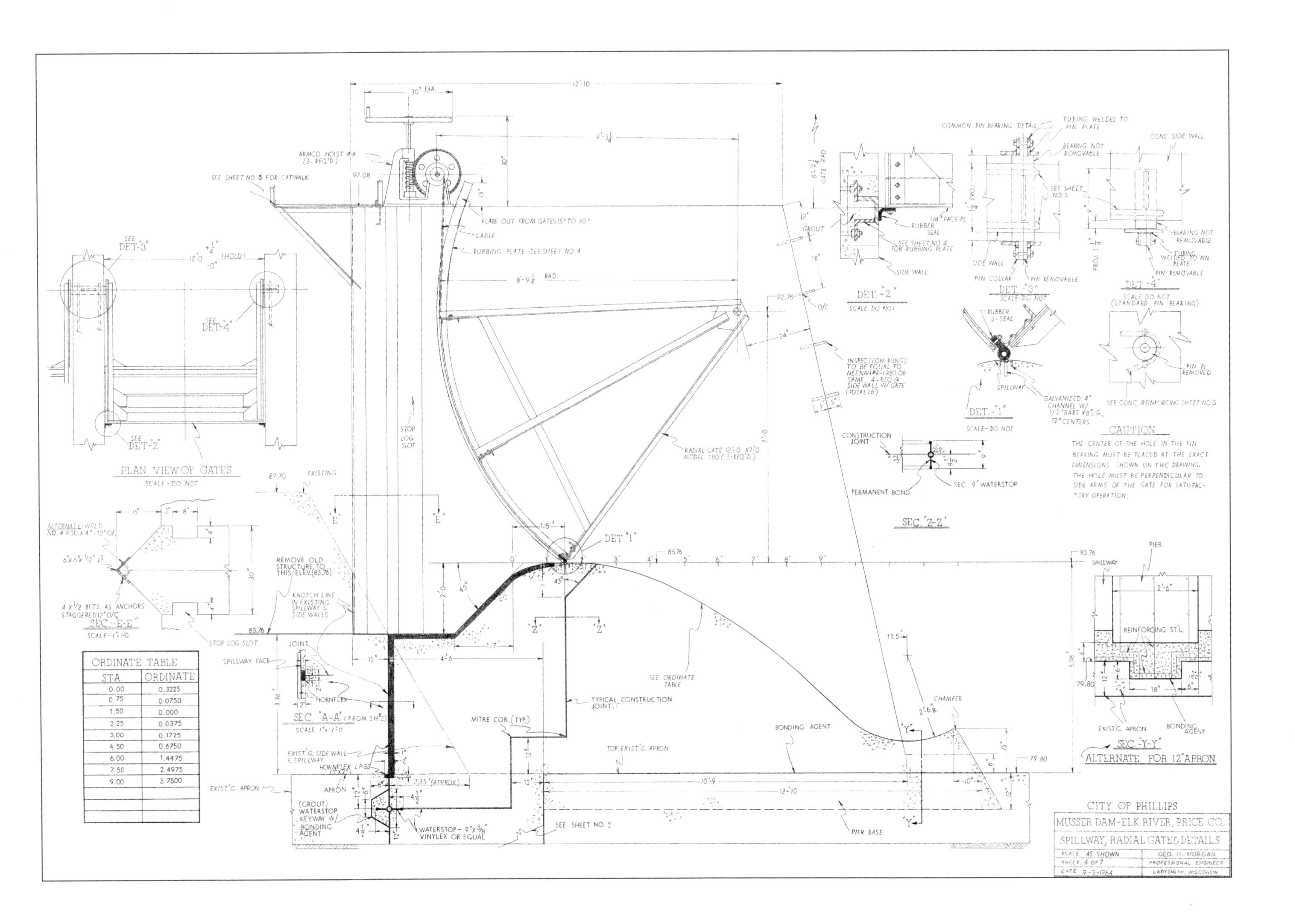
30" DIA.
12'-10
9'-3 1/4
ARMCO HOIST 4-A (3-REQ'D.)
SEE SHEET NO. 5 FOR CATWALK
97.08
FLARE OUT FROM GATES 15° TO 30°
CABLE
RUBBING PLATE - SEE SHEET NO. 4
8'-9 1/4 RAD.
8'-9 1/4 GATE RAD.
92.76
O/C
24"
INSPECTION RUNGS TO BE EQUAL TO NEENAH R-1980 OR SAME 4-REQ @ SIDE WALL W/ GATE (TOTAL 16)
STOP LOG SLOT
RADIAL GATE 12'-0 X 7'-0 MODEL 180 (3-REQ'D.)
DET. "1"
PLAN VIEW OF GATES
SCALE - DO NOT
SEE DET-"3"
SEE DET-"4"
SEE DET-"2"
12'-0 +1/4" -0" (HOLD)
87.70
EXISTING
ALTERNATE WELD NO. 4 ROD X 4" - 12" O/C
6"x6"x1/2" LS
4 X 1/2 BLTS. AS ANCHORS STAGGERED 12" O/C
SEC.-"E-E"
SCALE- 1"=1'-0
REMOVE OLD STRUCTURE TO THIS ELEV. (83.76)
KNOTCH LINE IN EXISTING SPILLWAY & SIDE WALLS
83.76
STOP LOG SLOT
JOINT
SPILLWAY FACE
HORNFLEX
SEC. "A-A" (FROM SH. 2)
SCALE 1" = 1'-0
EXIST'G. SIDE WALL & SPILLWAY
HORNFLEX LP-32
EXIST'G. APRON
APRON
(GROUT) WATERSTOP KEYWAY W/ BONDING AGENT
WATERSTOP - 9"x3/8" VINYLEX OR EQUAL
2.35' (APPROX.)
4'-6
1.7'
1.5'
85.76
"Z"
SEE ORDINATE TABLE
TYPICAL CONSTRUCTION JOINT
MITRE COR. (TYP.)
TOP EXIST'G. APRON
BONDING AGENT
10'-9
12'-10
SEE SHEET NO. 2
PIER BASE
11.5'
CHAMFER
2'-6" R.
"Y"
79.80
ORDINATE TABLE
STA. | ORDINATE
0.00 | 0.3225
0.75 | 0.0750
1.50 | 0.000
2.25 | 0.0375
3.00 | 0.1725
4.50 | 0.6750
6.00 | 1.4475
7.50 | 2.4975
9.00 | 3.7500
GROUT
SEE SHEET NO. 4 FOR RUBBING PLATE
RUBBER SEAL
1/4" FACE PL.
SIDE WALL
DET.-"2"
SCALE DO NOT
COMMON PIN BEARING DETAIL
TUBING WELDED TO PIN PLATE
BEARING NOT REMOVABLE
SEE SHEET NO. 5
PIN COLLAR
PIN REMOVABLE
DET.-"3"
SCALE-DO NOT
CONC. SIDE WALL
BEARING NOT REMOVABLE
TUBING WELDED TO PIN PLATE
PIN REMOVABLE
DET.-"4"
SCALE DO NOT (STANDARD PIN BEARING)
RUBBER J-SEAL
SPILLWAY
GALVANIZED 4" CHANNEL W/ 1/2" BARS X 8" LG. 12" CENTERS
DET.-"1"
SCALE- DO NOT
PIN PL. REMOVED
SEE CONC. REINFORCING SHEET NO. 5
CAUTION
THE CENTER OF THE HOLE IN THE PIN BEARING MUST BE PLACED AT THE EXACT DIMENSIONS SHOWN ON THIS DRAWING. THE HOLE MUST BE PERPENDICULAR TO SIDE ARMS OF THE GATE FOR SATISFACTORY OPERATION.
CONSTRUCTION JOINT
PERMANENT BOND
SEC. 9" WATERSTOP
SEC. "Z-Z"
PIER
SPILLWAY
REINFORCING ST'L.
EXIST'G. APRON
BONDING AGENT
SEC. "Y-Y"
ALTERNATE FOR 12" APRON
CITY OF PHILLIPS
MUSSER DAM-ELK RIVER, PRICE CO.
SPILLWAY, RADIAL GATE & DETAILS
SCALE AS SHOWN
SHEET 4 OF 7
DATE 9-3-1964
GEO. H. MORGAN
PROFESSIONAL ENGINEER
LADYSMITH, WISCONSIN

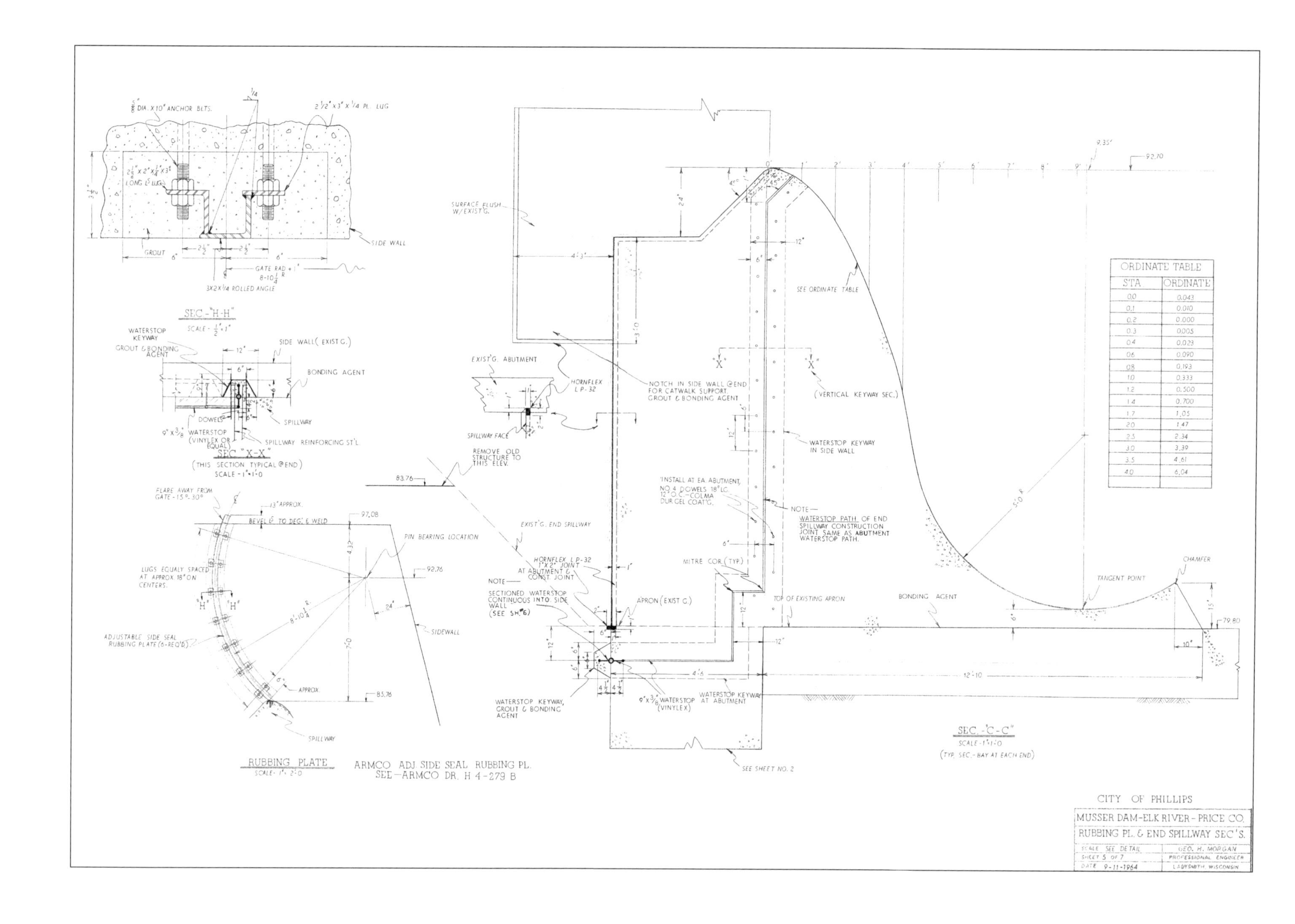

ORDINATE TABLE

STA	ORDINATE
0.0	0.043
0.1	0.010
0.2	0.000
0.3	0.005
0.4	0.023
0.6	0.090
0.8	0.193
1.0	0.333
1.2	0.500
1.4	0.700
1.7	1.05
2.0	1.47
2.5	2.34
3.0	3.39
3.5	4.61
4.0	6.04

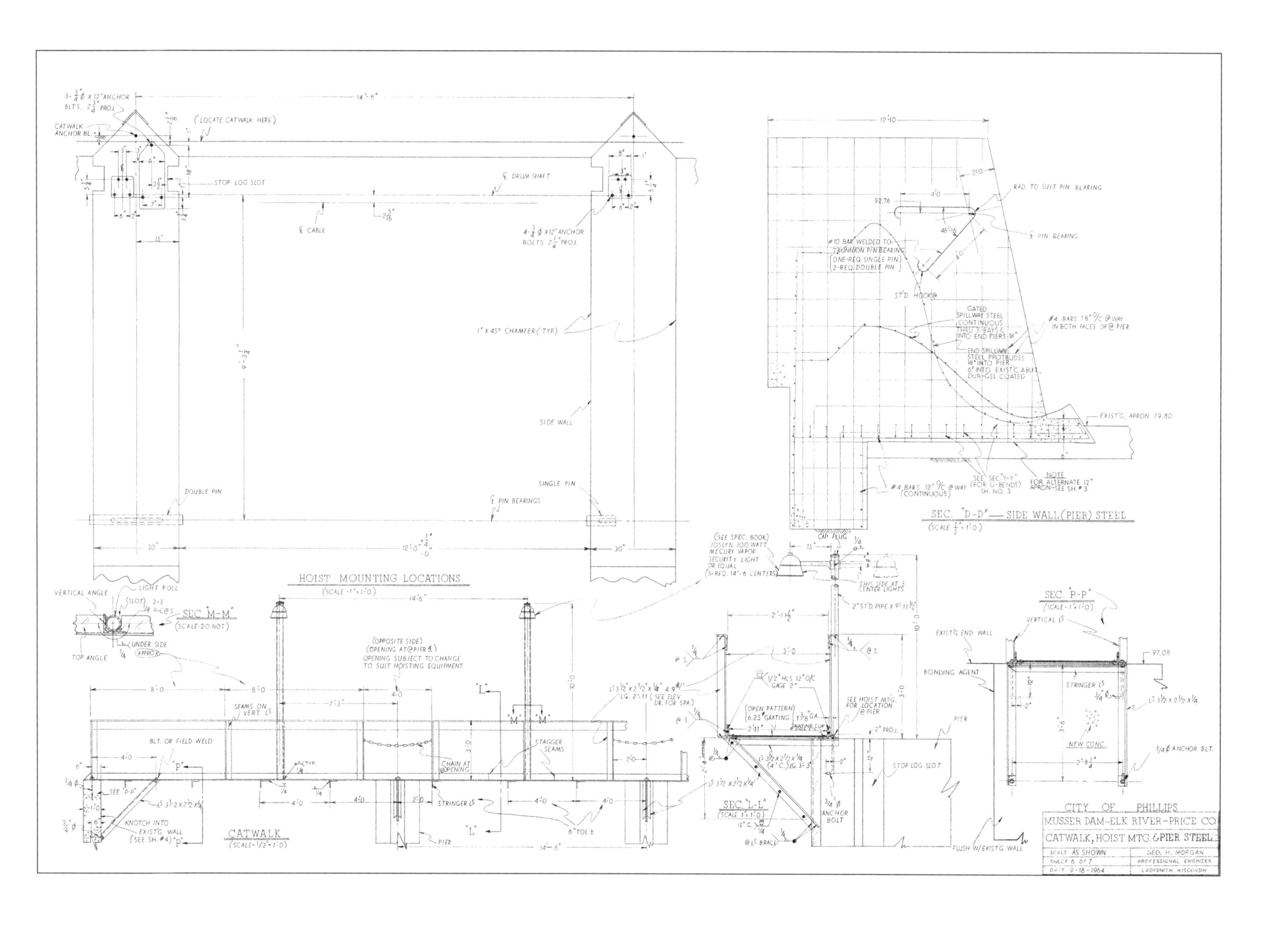
HOIST MOUNTING LOCATIONS
SEC. "D-D" — SIDE WALL (PIER) STEEL
SEC "M-M"
SEC "P-P"
SEC "L-L"
CATWALK
CITY OF PHILLIPS
MUSSER DAM-ELK RIVER-PRICE CO.
CATWALK, HOIST MTG. & PIER STEEL
SCALE AS SHOWN
SHEET 6 OF 7
DATE 9-18-1964
GEO. H. MORGAN
PROFESSIONAL ENGINEER
LADYSMITH WISCONSIN

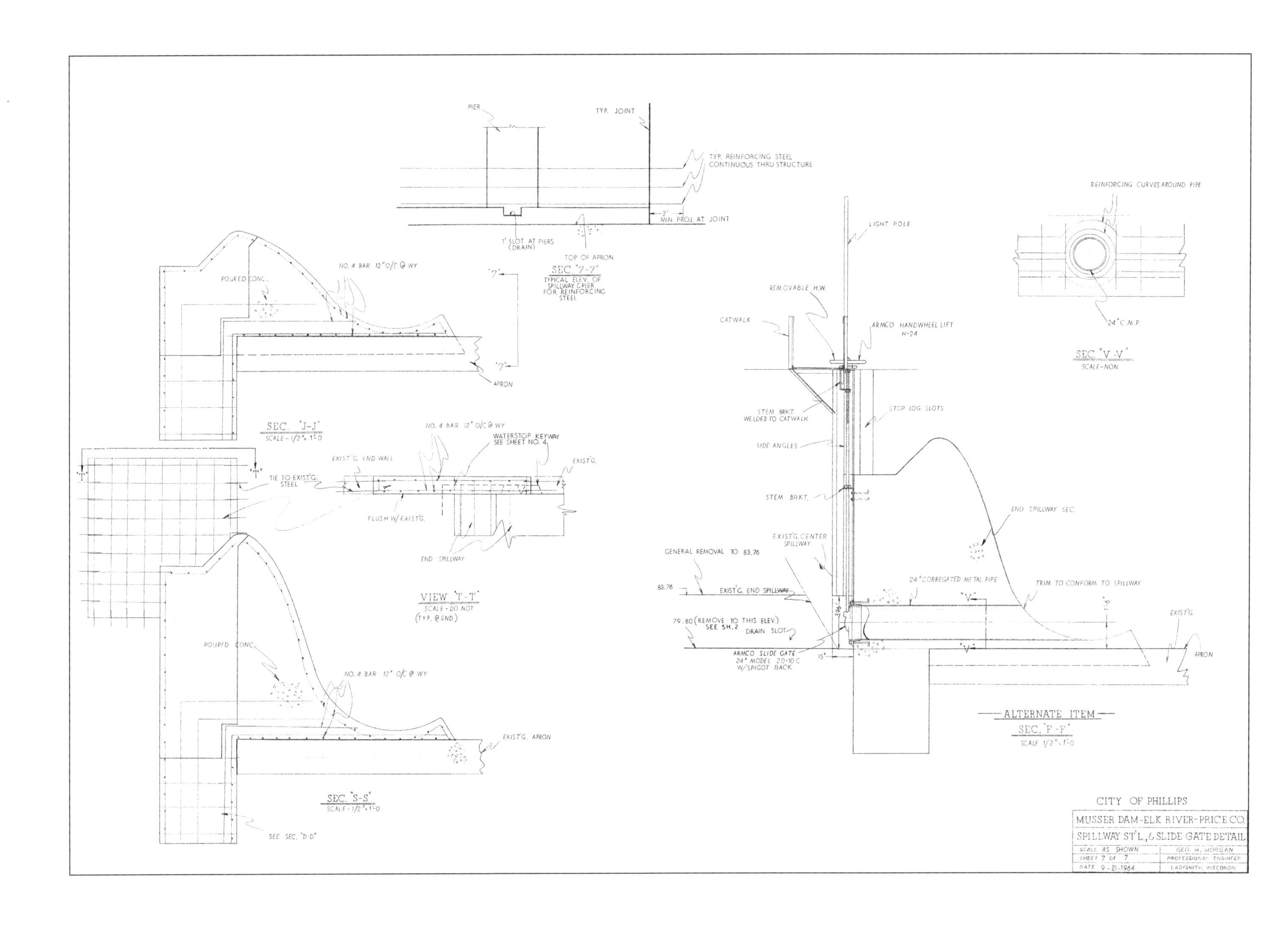
PIER
TYP. JOINT
TYP. REINFORCING STEEL CONTINUOUS THRU STRUCTURE
2' MIN. PROJ. AT JOINT
1' SLOT AT PIERS (DRAIN)
TOP OF APRON
SEC. "7-7"
TYPICAL ELEV. OF SPILLWAY & PIER FOR REINFORCING STEEL
NO. 4 BAR 12" O/C @ WY
POURED CONC.
APRON
SEC. "J-J"
SCALE - 1/2" = 1'-0
NO. 4 BAR 12" O/C @ WY
WATERSTOP KEYWAY SEE SHEET NO. 4
EXIST'G. END WALL
EXIST'G.
TIE TO EXIST'G. STEEL
FLUSH W/EXIST'G.
END SPILLWAY
VIEW "T-T"
SCALE - DO NOT
(TYP. @ END)
POURED CONC.
NO. 4 BAR 12" O/C @ WY
EXIST'G. APRON
SEC. "S-S"
SCALE - 1/2" = 1'-0
SEE SEC. "D-D"
LIGHT POLE
REMOVABLE H.W.
CATWALK
ARMCO HANDWHEEL LIFT H-24
STEM BRKT. WELDED TO CATWALK
STOP LOG SLOTS
SIDE ANGLES
STEM BRKT.
END SPILLWAY SEC.
EXIST'G. CENTER SPILLWAY
GENERAL REMOVAL TO 83.76
83.76
EXIST'G. END SPILLWAY
24" CORREGATED METAL PIPE
TRIM TO CONFORM TO SPILLWAY
79.80 (REMOVE TO THIS ELEV.) SEE SH. 2
DRAIN SLOT
ARMCO SLIDE GATE 24" MODEL 20-10 C W/SPIGOT BACK
EXIST'G.
APRON
ALTERNATE ITEM
SEC. "F-F"
SCALE 1/2" = 1'-0
REINFORCING CURVES AROUND PIPE
24" C.M.P.
SEC. "V-V"
SCALE - NON
CITY OF PHILLIPS
MUSSER DAM-ELK RIVER-PRICE CO.
SPILLWAY ST'L., & SLIDE GATE DETAIL
SCALE AS SHOWN
SHEET 7 OF 7
DATE 9-21-1964
GEO. H. MORGAN
PROFESSIONAL ENGINEER
LADYSMITH, WISCONSIN

Design

STREETS, ROADS & HIGHWAYS

Section 14

A large area of a civil engineer's involvement is in the public transportation arena; i.e., streets, roads and highways.

The majority of this work is the reconstruction of existing transportation pathways. Samples of actual project plans are reproduced on the following pages.

TABLE OF CONTENTS

INDUSTRIAL ROADS & STREET RECONSTRUCTION

INDUSTRIAL PARK COMPLEX

FOR

VILLAGE OF GLEN FLORA, WISCONSIN

PROJECT NO.S:
M. & P. NO: 90-126-R
DOT NO: 8441-02-00,71
EDA NO: 06-01-02571
MAY, 1992

UTILITIES

CONTACT - NORTHERN STATES POWER CO.
711 W. 9th Street, North
Ladysmith, WI 54848

Attn: John Rymarkiewicz
715-532-6226

CONTACT - UNIVERSAL TELEPHONE CO.
OF NORTHERN WISCONSIN
Highway "8"
Hawkins, WI 54530

Attn: Jim Arquette
715-585-7707

CONTACT - DIGGER'S HOTLINE

1-800-242-8511

Ticket's Nos.: 1598953
1598962

CONTACT - VILLAGE OF GLEN FLORA WWCTF

Les Evjen, Operator
P.O. Box 253
Glen Flora, WI 54526

715-322-5511

VILLAGE OF GLEN FLORA,
RUSK CO, WISCONSIN

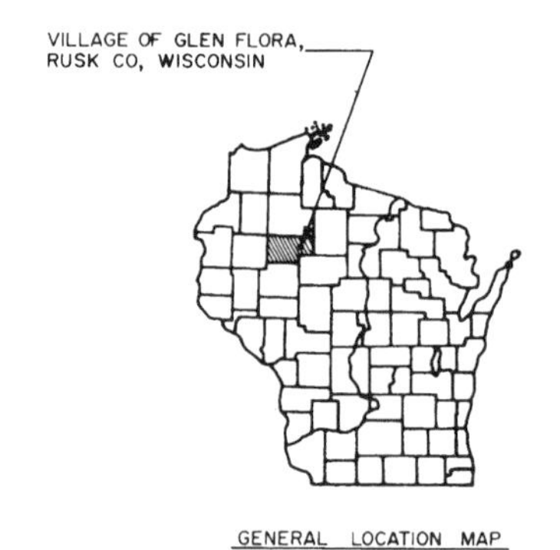

GENERAL LOCATION MAP

PROJECT LOCATION

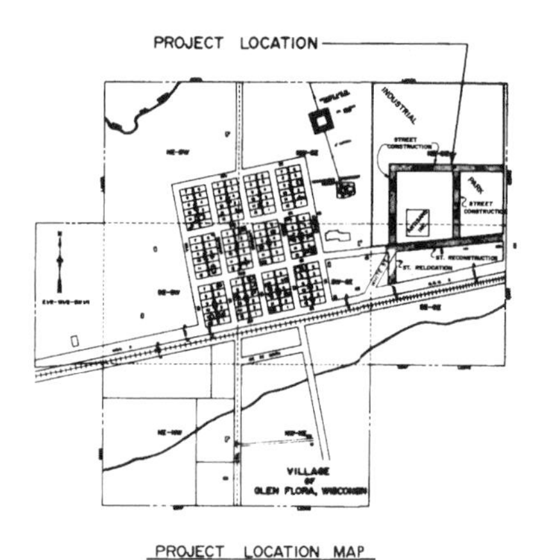

PROJECT LOCATION MAP

INDEX

NO.	DESCRIPTION
1	TITLE SHEET W/ INDEX
2	WHITE AVE. RELOCATION & CONST. 2ND ST. RECONSTRUCTION
3	ORLO AVE. CONSTRUCTION JIPSON ST. CONSTRUCTION
4	TYPICAL DETAILS
5-10	CROSS SECTIONS (SECOND STREET)
11-13	CROSS SECTIONS (ORLO AVENUE)
14-18	CROSS SECTIONS (WHITE AVENUE)
19-23	CROSS SECTIONS (JIPSON STREET)
24	TYPICAL EROSION CONTROL

NOTE: Sheets 6 thru 24
Have Been Omited

PREPARED BY:
MORGAN & PARMLEY LTD.
CONSULTING ENGINEERS
LADYSMITH, WISCONSIN

Courtesy: Morgan & Parmley, Ltd.

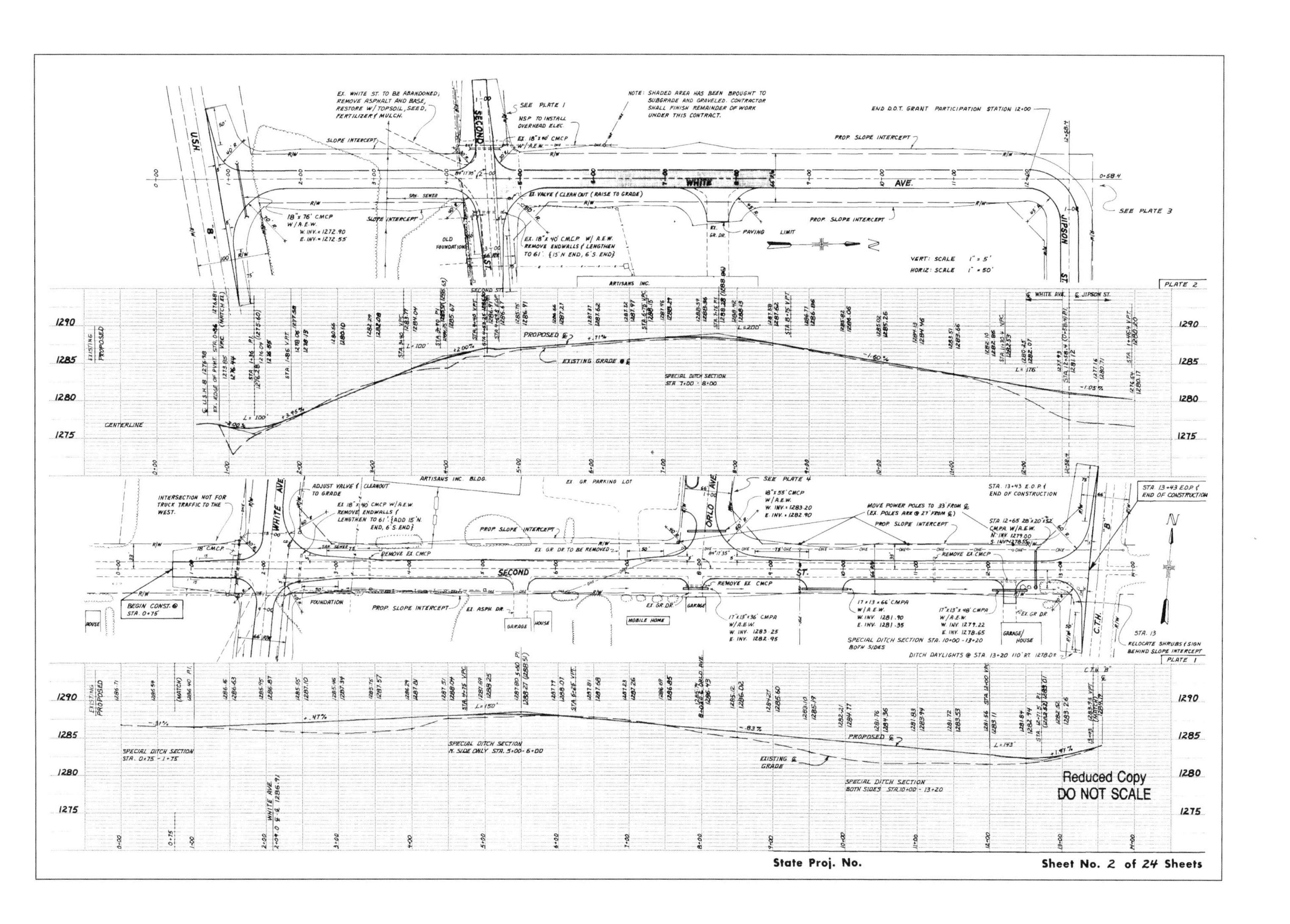
NOTE: SHADED AREA HAS BEEN BROUGHT TO SUBGRADE AND GRAVELED. CONTRACTOR SHALL FINISH REMAINDER OF WORK UNDER THIS CONTRACT.
END D.O.T. GRANT PARTICIPATION STATION 12+00
SEE PLATE 1
SEE PLATE 3
SEE PLATE 4
WHITE AVE.
SECOND ST.
JIPSON ST.
ORLO AVE
C.T.H. "B"
USH "B"
ARTISANS INC.
ARTISANS INC. BLDG.
PROP SLOPE INTERCEPT
SLOPE INTERCEPT
PAVING LIMIT
VERT. SCALE 1" = 5'
HORIZ. SCALE 1" = 50'
PLATE 2
PLATE 1
PROPOSED ℄
EXISTING GRADE ℄
CENTERLINE
SPECIAL DITCH SECTION STA. 7+00 - 8+00
INTERSECTION NOT FOR TRUCK TRAFFIC TO THE WEST.
BEGIN CONST. @ STA. 0+75
STA. 13+43 E.O.P. & END OF CONSTRUCTION
MOVE POWER POLES TO 33' FROM ℄ (EX. POLES ARE @ 27' FROM ℄)
REMOVE EX. CMCP
FOUNDATION
GARAGE
HOUSE
MOBILE HOME
SPECIAL DITCH SECTION STA. 10+00 - 13+20 BOTH SIDES
SPECIAL DITCH SECTION STA. 0+75 - 1+75
SPECIAL DITCH SECTION N. SIDE ONLY STA. 5+00 - 6+00
EXISTING ℄ GRADE
1290
1285
1280
1275
Reduced Copy
DO NOT SCALE
State Proj. No.
Sheet No. 2 of 24 Sheets

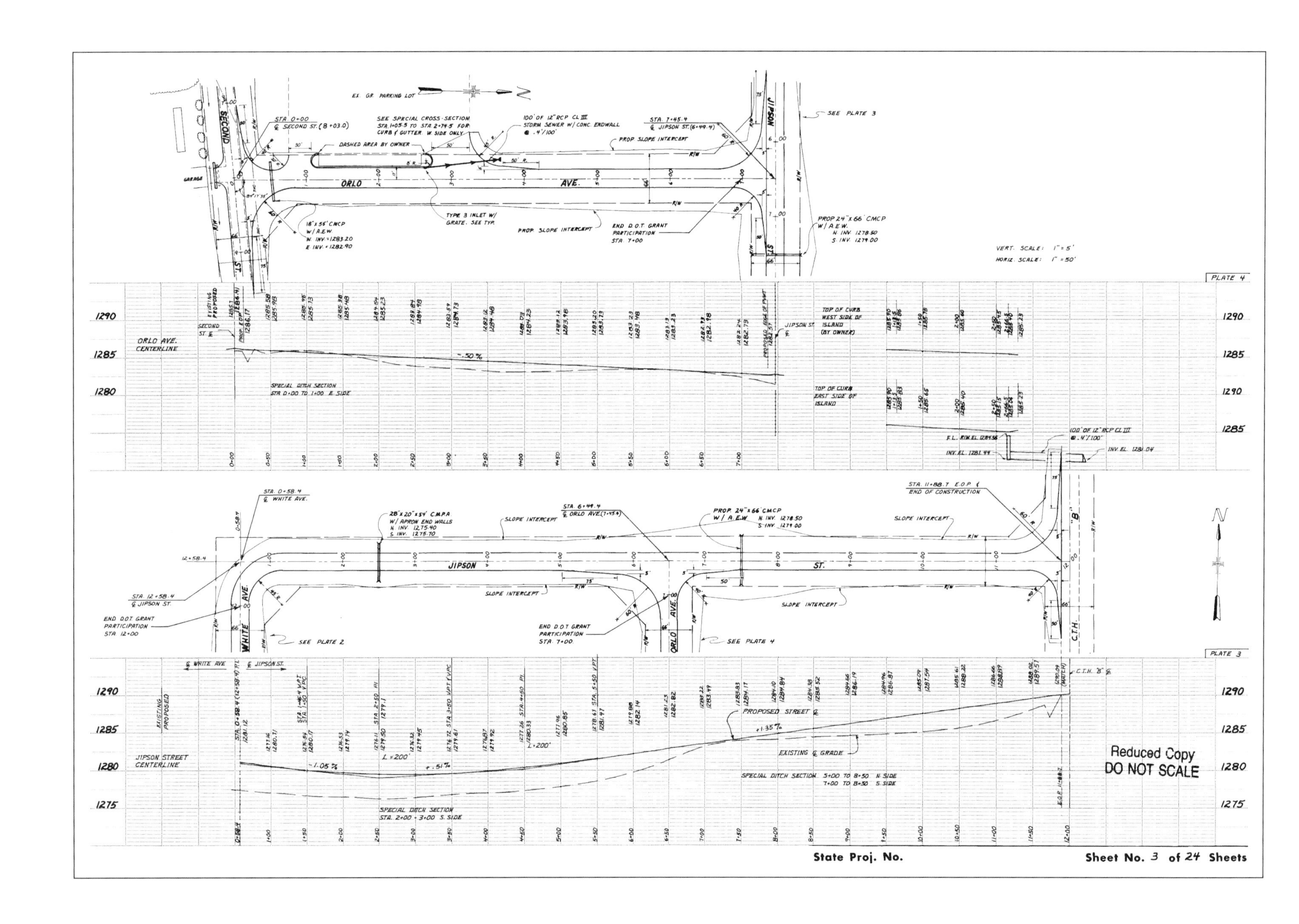
ORLO AVE.
SECOND ST.
JIPSON ST.
SEE PLATE 3
VERT. SCALE: 1" = 5'
HORIZ. SCALE: 1" = 50'
PLATE 4
ORLO AVE. CENTERLINE
SPECIAL DITCH SECTION STA 0+00 TO 1+00 E. SIDE
TOP OF CURB WEST SIDE OF ISLAND (BY OWNER)
TOP OF CURB EAST SIDE OF ISLAND
JIPSON ST.
WHITE AVE.
C.T.H. "B"
SEE PLATE 2
SEE PLATE 4
END D.O.T. GRANT PARTICIPATION STA 12+00
END D.O.T. GRANT PARTICIPATION STA. 7+00
STA. 11+88.7 E.O.P. END OF CONSTRUCTION
PLATE 3
JIPSON STREET CENTERLINE
PROPOSED STREET
EXISTING GRADE
SPECIAL DITCH SECTION 5+00 TO 8+50 N. SIDE 7+00 TO 8+50 S. SIDE
SPECIAL DITCH SECTION STA. 2+00 - 3+00 S. SIDE
Reduced Copy
DO NOT SCALE
State Proj. No.
Sheet No. 3 of 24 Sheets

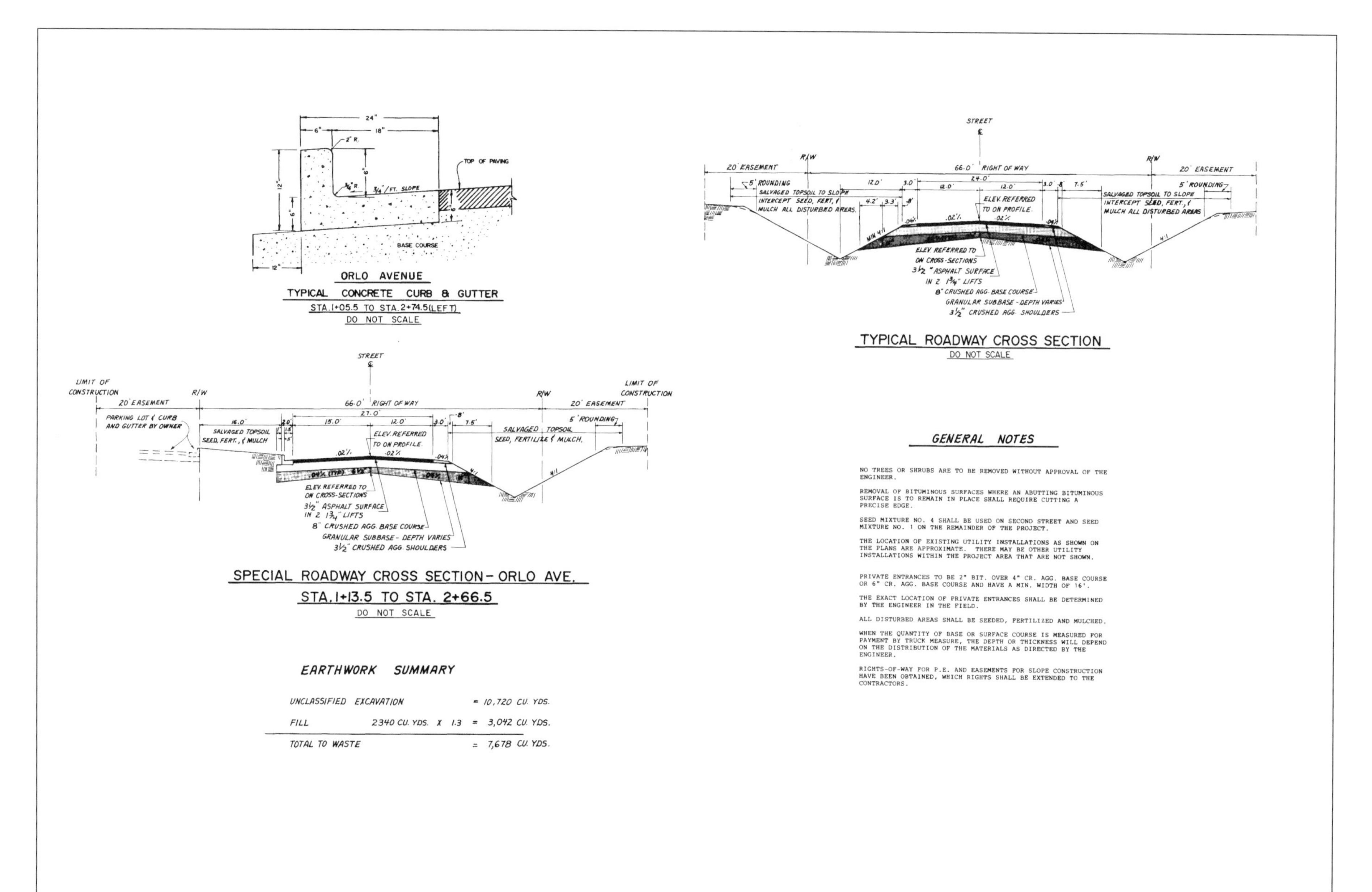
24"
6"
18"
2" R.
TOP OF PAVING
3/4"/FT. SLOPE
BASE COURSE
12"
ORLO AVENUE
TYPICAL CONCRETE CURB & GUTTER
STA. 1+05.5 TO STA. 2+74.5 (LEFT)
DO NOT SCALE
STREET
R/W
20' EASEMENT
66.0' RIGHT OF WAY
24.0'
12.0'
3.0'
7.5'
5' ROUNDING
SALVAGED TOPSOIL TO SLOPE INTERCEPT SEED, FERT, & MULCH ALL DISTURBED AREAS.
ELEV. REFERRED TO ON PROFILE.
.02%
.04%
4:1
ELEV. REFERRED TO ON CROSS-SECTIONS
3 1/2" ASPHALT SURFACE IN 2 1 3/4" LIFTS
8" CRUSHED AGG. BASE COURSE
GRANULAR SUBBASE - DEPTH VARIES
3 1/2" CRUSHED AGG. SHOULDERS
TYPICAL ROADWAY CROSS SECTION
DO NOT SCALE
LIMIT OF CONSTRUCTION
PARKING LOT & CURB AND GUTTER BY OWNER
16.0'
15.0'
27.0'
SALVAGED TOPSOIL SEED, FERT., & MULCH
SALVAGED TOPSOIL SEED, FERTILIZE & MULCH.
SPECIAL ROADWAY CROSS SECTION - ORLO AVE.
STA. 1+13.5 TO STA. 2+66.5
DO NOT SCALE
EARTHWORK SUMMARY
UNCLASSIFIED EXCAVATION = 10,720 CU. YDS.
FILL 2340 CU. YDS. X 1.3 = 3,042 CU. YDS.
TOTAL TO WASTE = 7,678 CU. YDS.
GENERAL NOTES
NO TREES OR SHRUBS ARE TO BE REMOVED WITHOUT APPROVAL OF THE ENGINEER.
REMOVAL OF BITUMINOUS SURFACES WHERE AN ABUTTING BITUMINOUS SURFACE IS TO REMAIN IN PLACE SHALL REQUIRE CUTTING A PRECISE EDGE.
SEED MIXTURE NO. 4 SHALL BE USED ON SECOND STREET AND SEED MIXTURE NO. 1 ON THE REMAINDER OF THE PROJECT.
THE LOCATION OF EXISTING UTILITY INSTALLATIONS AS SHOWN ON THE PLANS ARE APPROXIMATE. THERE MAY BE OTHER UTILITY INSTALLATIONS WITHIN THE PROJECT AREA THAT ARE NOT SHOWN.
PRIVATE ENTRANCES TO BE 2" BIT. OVER 4" CR. AGG. BASE COURSE OR 6" CR. AGG. BASE COURSE AND HAVE A MIN. WIDTH OF 16'.
THE EXACT LOCATION OF PRIVATE ENTRANCES SHALL BE DETERMINED BY THE ENGINEER IN THE FIELD.
ALL DISTURBED AREAS SHALL BE SEEDED, FERTILIZED AND MULCHED.
WHEN THE QUANTITY OF BASE OR SURFACE COURSE IS MEASURED FOR PAYMENT BY TRUCK MEASURE, THE DEPTH OR THICKNESS WILL DEPEND ON THE DISTRIBUTION OF THE MATERIALS AS DIRECTED BY THE ENGINEER.
RIGHTS-OF-WAY FOR P.E. AND EASEMENTS FOR SLOPE CONSTRUCTION HAVE BEEN OBTAINED, WHICH RIGHTS SHALL BE EXTENDED TO THE CONTRACTORS.
SHEET NO. 4 OF 24 SHEETS

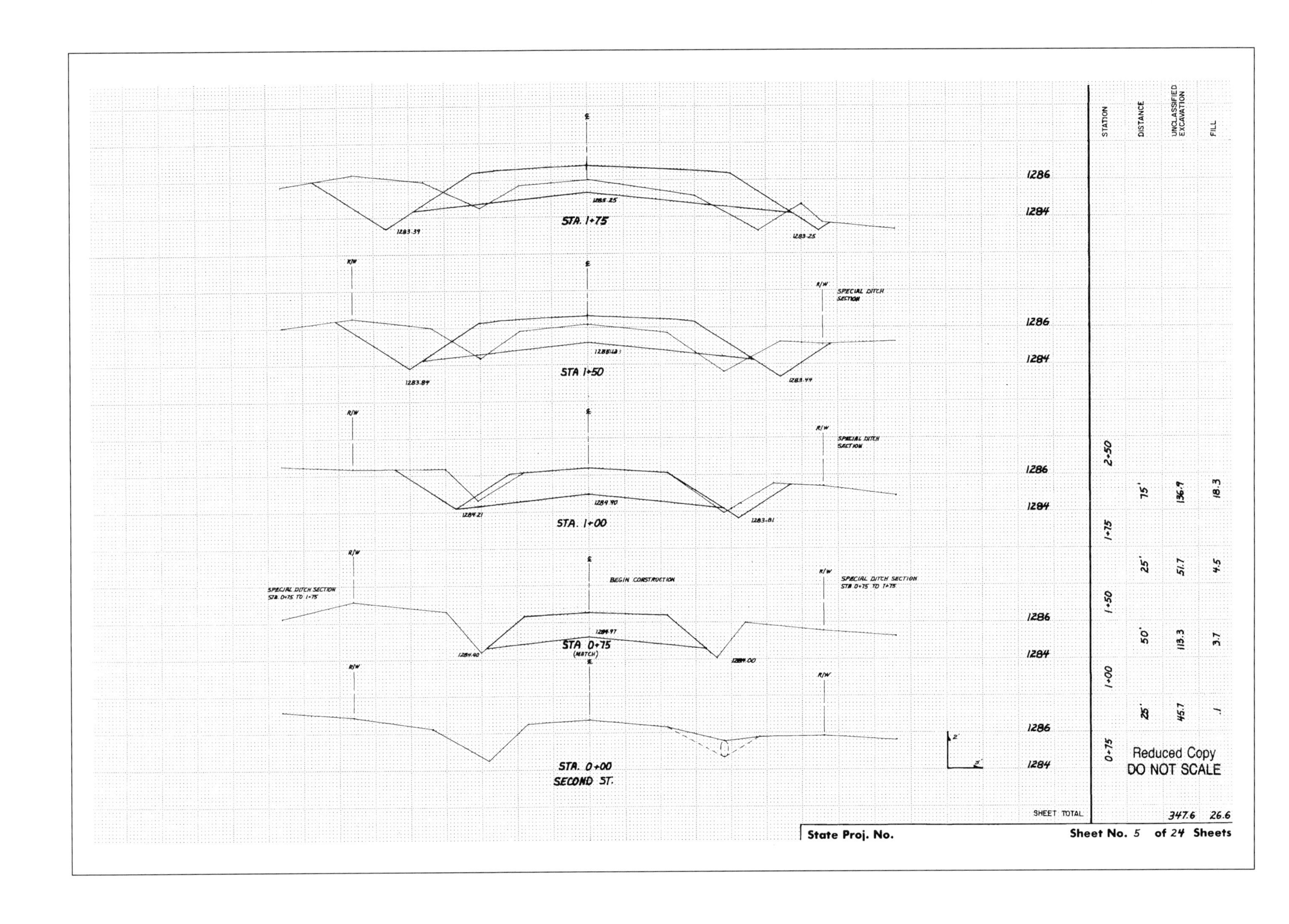
STA. 1+75
STA 1+50
STA. 1+00
STA 0+75
(MATCH)
STA. 0+00
SECOND ST.
R/W
SPECIAL DITCH SECTION
SPECIAL DITCH SECTION STA. 0+75 TO 1+75
BEGIN CONSTRUCTION
1286
1284
STATION
DISTANCE
UNCLASSIFIED EXCAVATION
FILL
2+50
1+75
1+50
1+00
0+75
75'
25'
50'
25'
136.9
51.7
113.3
45.7
18.3
4.5
3.7
.1
Reduced Copy
DO NOT SCALE
SHEET TOTAL
347.6
26.6
State Proj. No.
Sheet No. 5 of 24 Sheets

DOUGHTY ROAD – ROAD RECONSTRUCTION
for the
CITY OF LADYSMITH
LADYSMITH, WI 54848
T.E.A. GRANT, I.D. 8955-02-71

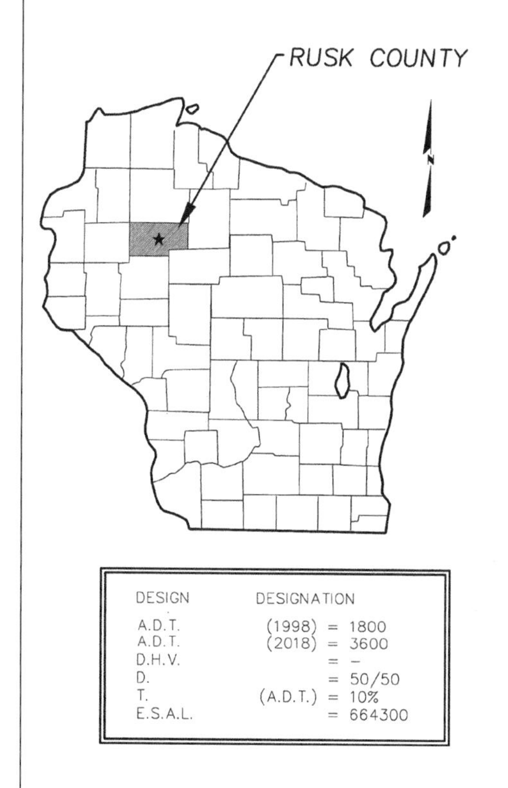

DESIGN	DESIGNATION	
A.D.T.	(1998)	= 1800
A.D.T.	(2018)	= 3600
D.H.V.		= –
D.		= 50/50
T.	(A.D.T.)	= 10%
E.S.A.L.		= 664300

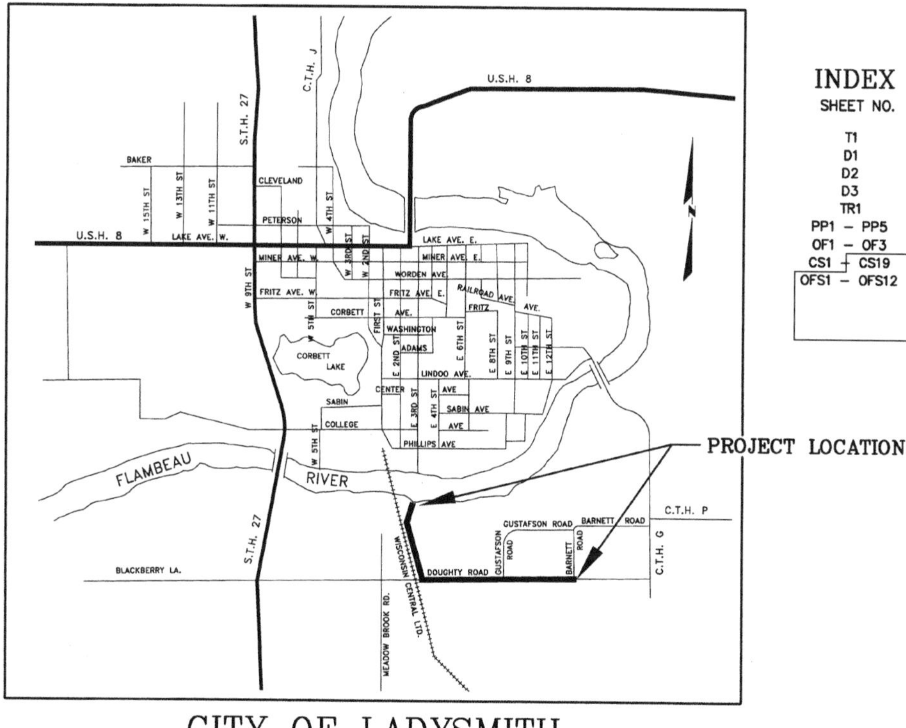

CITY OF LADYSMITH

PROJECT LOCATION

INDEX OF SHEETS

SHEET NO.	DESCRIPTION
T1	TITLE SHEET
D1	TYPICAL SECTIONS
D2	RAILROAD DETAILS
D3	STORM SEWER DETAILS
TR1	TRAFFIC CONTROL
PP1 – PP5	DOUGHTY RD. PLAN & PROFILES
OF1 – OF3	STORM SEWER OUTFALL PLAN & PROFILES
CS1 – CS19	DOUGHTY ROAD CROSS SECTIONS
OFS1 – OFS12	OUTFALL CROSS SECTIONS

NOTE: Sheets Beyond CS4 Have Been Omitted

REVISION	DATE	DESCRIPTION
2	1/4/99	D1 ✱
1	7/9/98	MISC. CHANGES

DOUGHTY ROAD ROAD RECONSTRUCTION

SCALE NONE | PROJECT ENGINEER Robert O. Parmley | DRAWN BY M.R.E.

DATE 5-12-98

MORGAN & PARMLEY, LTD.
Professional Consulting Engineers
115 West 2nd Street, Ladysmith, Wisconsin 54848

TELEPHONE 715-532-3721 | SHEET DESCRIPTION TITLE SHEET | PROJECT NO. 97-144 | SHEET NO. T1

Courtesy: Morgan & Parmley, Ltd.

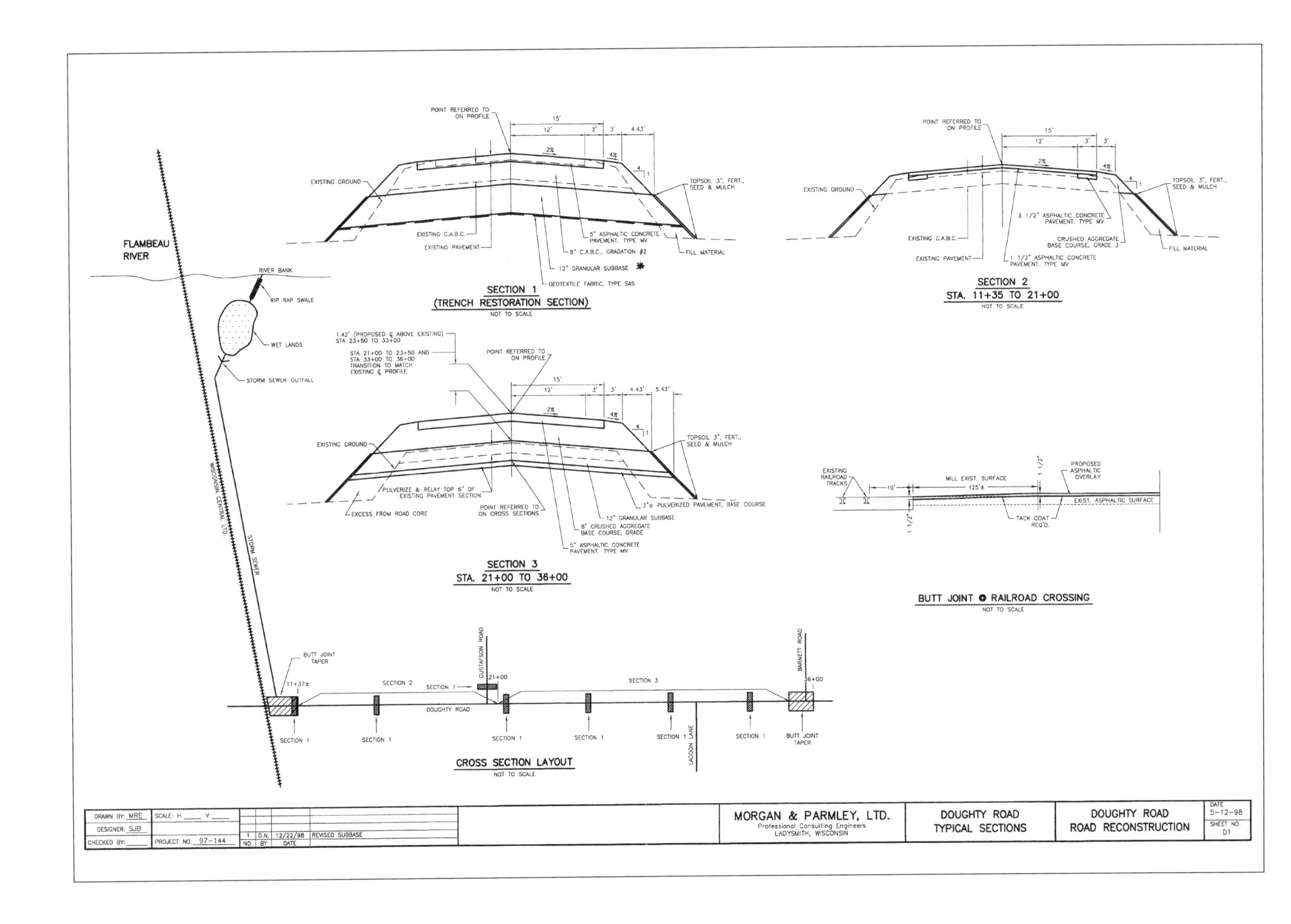
SECTION 1
(TRENCH RESTORATION SECTION)
NOT TO SCALE
SECTION 2
STA. 11+35 TO 21+00
NOT TO SCALE
SECTION 3
STA. 21+00 TO 36+00
NOT TO SCALE
BUTT JOINT @ RAILROAD CROSSING
NOT TO SCALE
CROSS SECTION LAYOUT
NOT TO SCALE
POINT REFERRED TO ON PROFILE
EXISTING GROUND
TOPSOIL 3", FERT., SEED & MULCH
EXISTING C.A.B.C.
EXISTING PAVEMENT
5" ASPHALTIC CONCRETE PAVEMENT, TYPE MV
8" C.A.B.C., GRADATION #2
12" GRANULAR SUBBASE
GEOTEXTILE FABRIC, TYPE SAS
FILL MATERIAL
3 1/2" ASPHALTIC CONCRETE PAVEMENT, TYPE MV
CRUSHED AGGREGATE BASE COURSE, GRADE 3
1 1/2" ASPHALTIC CONCRETE PAVEMENT, TYPE MV
1.42' (PROPOSED ℄ ABOVE EXISTING) STA 23+50 TO 33+00
STA 21+00 TO 23+50 AND STA 33+00 TO 36+00 TRANSITION TO MATCH EXISTING ℄ PROFILE
PULVERIZE & RELAY TOP 6" OF EXISTING PAVEMENT SECTION
EXCESS FROM ROAD CORE
POINT REFERRED TO ON CROSS SECTIONS
3"± PULVERIZED PAVEMENT, BASE COURSE
8" CRUSHED AGGREGATE BASE COURSE, GRADE
EXISTING RAILROAD TRACKS
MILL EXIST. SURFACE
PROPOSED ASPHALTIC OVERLAY
EXIST. ASPHALTIC SURFACE
TACK COAT REQ'D.
FLAMBEAU RIVER
RIVER BANK
RIP RAP SWALE
WET LANDS
STORM SEWER OUTFALL
WISCONSIN CENTRAL LTD.
STORM SEWER
BUTT JOINT TAPER
GUSTAFSON ROAD
BARNETT ROAD
LAGOON LANE
DOUGHTY ROAD
DRAWN BY: MRE
DESIGNER: SJB
CHECKED BY:
PROJECT NO. 97-144
1 D.N. 12/22/98 REVISED SUBBASE
MORGAN & PARMLEY, LTD.
Professional Consulting Engineers
LADYSMITH, WISCONSIN
DOUGHTY ROAD TYPICAL SECTIONS
DOUGHTY ROAD ROAD RECONSTRUCTION
DATE 5-12-98
SHEET NO. D1

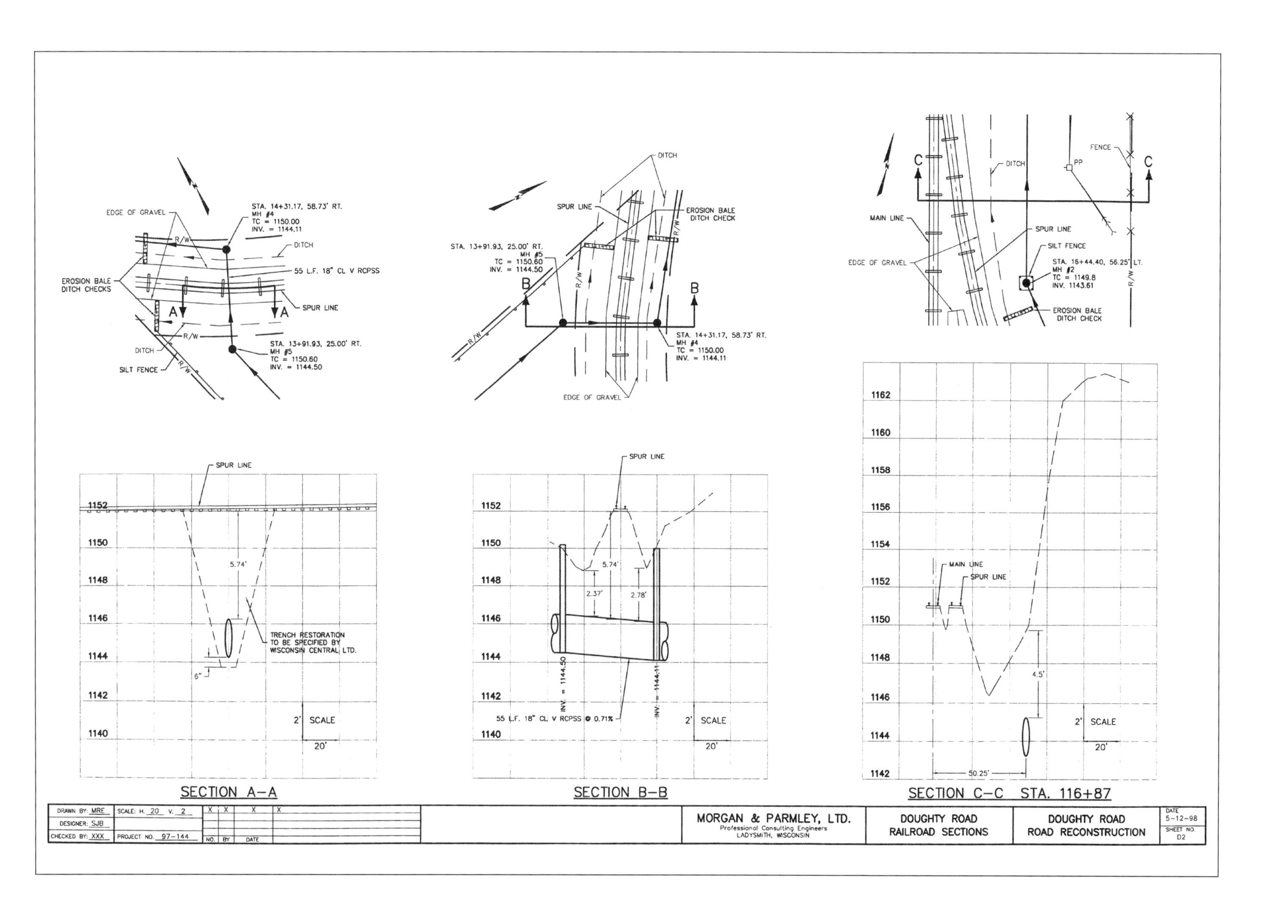
EDGE OF GRAVEL
STA. 14+31.17, 58.73' RT.
MH #4
TC = 1150.00
INV. = 1144.11
DITCH
55 L.F. 18" CL V RCPSS
EROSION BALE DITCH CHECKS
SPUR LINE
STA. 13+91.93, 25.00' RT.
MH #5
TC = 1150.60
INV. = 1144.50
DITCH
SILT FENCE
R/W
SPUR LINE
EROSION BALE DITCH CHECK
EDGE OF GRAVEL
MAIN LINE
SILT FENCE
STA. 16+44.40, 56.25' LT.
MH #2
TC = 1149.8
INV. 1143.61
FENCE
PP
TRENCH RESTORATION TO BE SPECIFIED BY WISCONSIN CENTRAL LTD.
5.74'
2.37'
2.78'
4.5'
50.25'
INV. = 1144.50
INV. = 1144.11
55 L.F. 18" CL V RCPSS @ 0.71%
2' SCALE
20'
SECTION A-A
SECTION B-B
SECTION C-C STA. 116+87
DRAWN BY: MRE
DESIGNER: SJB
CHECKED BY: XXX
SCALE: H. 20 V. 2
PROJECT NO. 97-144
MORGAN & PARMLEY, LTD.
Professional Consulting Engineers
LADYSMITH, WISCONSIN
DOUGHTY ROAD
RAILROAD SECTIONS
DOUGHTY ROAD
ROAD RECONSTRUCTION
DATE 5-12-98
SHEET NO. D2

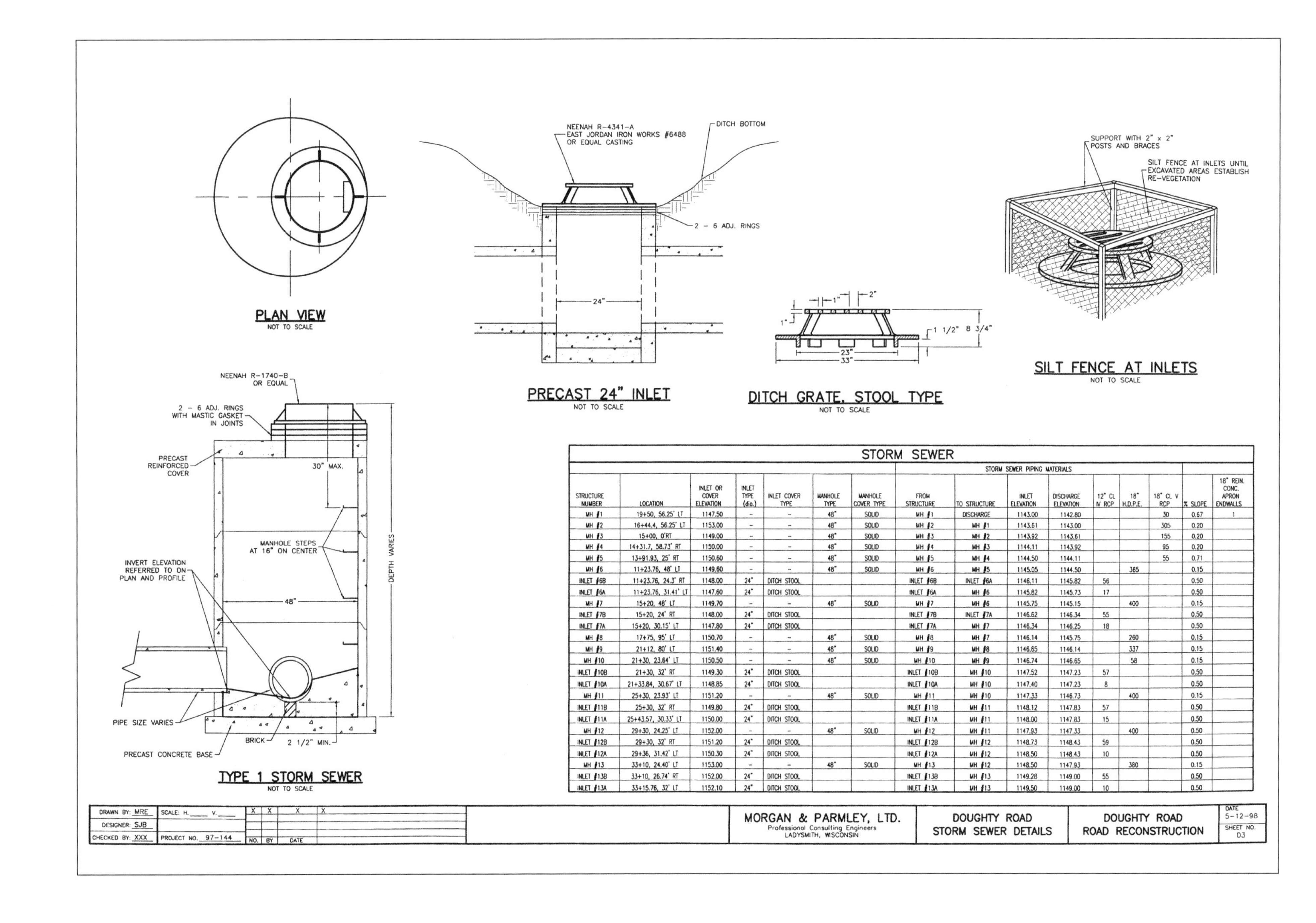

STORM SEWER

STRUCTURE NUMBER	LOCATION	INLET OR COVER ELEVATION	INLET TYPE (dia.)	INLET COVER TYPE	MANHOLE TYPE	MANHOLE COVER TYPE	STORM SEWER PIPING MATERIALS: FROM STRUCTURE	TO STRUCTURE	INLET ELEVATION	DISCHARGE ELEVATION	12" CL IV RCP	18" H.D.P.E.	18" CL V RCP	% SLOPE	18" REIN. CONC. APRON ENDWALLS
MH #1	19+50, 56.25' LT	1147.50	-	-	48"	SOLID	MH #1	DISCHARGE	1143.00	1142.80			30	0.67	1
MH #2	16+44.4, 56.25' LT	1153.00	-	-	48"	SOLID	MH #2	MH #1	1143.61	1143.00			305	0.20	
MH #3	15+00, 0'RT	1149.00	-	-	48"	SOLID	MH #3	MH #2	1143.92	1143.61			155	0.20	
MH #4	14+31.7, 58.73' RT	1150.00	-	-	48"	SOLID	MH #4	MH #3	1144.11	1143.92			95	0.20	
MH #5	13+91.93, 25' RT	1150.60	-	-	48"	SOLID	MH #5	MH #4	1144.50	1144.11			55	0.71	
MH #6	11+23.76, 48' LT	1149.60	-	-	48"	SOLID	MH #6	MH #5	1145.05	1144.50		365		0.15	
INLET #6B	11+23.76, 24.3' RT	1148.00	24"	DITCH STOOL			INLET #6B	INLET #6A	1146.11	1145.82	56			0.50	
INLET #6A	11+23.76, 31.41' LT	1147.60	24"	DITCH STOOL			INLET #6A	MH #6	1145.82	1145.73	17			0.50	
MH #7	15+20, 48' LT	1149.70	-	-	48"	SOLID	MH #7	MH #6	1145.75	1145.15		400		0.15	
INLET #7B	15+20, 24' RT	1148.00	24"	DITCH STOOL			INLET #7B	INLET #7A	1146.62	1146.34	55			0.50	
INLET #7A	15+20, 30.15' LT	1147.80	24"	DITCH STOOL			INLET #7A	MH #7	1146.34	1146.25	18			0.50	
MH #8	17+75, 95' LT	1150.70	-	-	48"	SOLID	MH #8	MH #7	1146.14	1145.75		260		0.15	
MH #9	21+12, 80' LT	1151.40	-	-	48"	SOLID	MH #9	MH #8	1146.65	1146.14		337		0.15	
MH #10	21+30, 23.64' LT	1150.50	-	-	48"	SOLID	MH #10	MH #9	1146.74	1146.65		58		0.15	
INLET #10B	21+30, 32' RT	1149.30	24"	DITCH STOOL			INLET #10B	MH #10	1147.52	1147.23	57			0.50	
INLET #10A	21+33.84, 30.67' LT	1148.85	24"	DITCH STOOL			INLET #10A	MH #10	1147.40	1147.23	8			0.50	
MH #11	25+30, 23.93' LT	1151.20	-	-	48"	SOLID	MH #11	MH #10	1147.33	1146.73		400		0.15	
INLET #11B	25+30, 32' RT	1149.80	24"	DITCH STOOL			INLET #11B	MH #11	1148.12	1147.83	57			0.50	
INLET #11A	25+43.57, 30.33' LT	1150.00	24"	DITCH STOOL			INLET #11A	MH #11	1148.00	1147.83	15			0.50	
MH #12	29+30, 24.25' LT	1152.00	-	-	48"	SOLID	MH #12	MH #11	1147.93	1147.33		400		0.50	
INLET #12B	29+30, 32' RT	1151.20	24"	DITCH STOOL			INLET #12B	MH #12	1148.73	1148.43	59			0.50	
INLET #12A	29+36, 31.42' LT	1150.30	24"	DITCH STOOL			INLET #12A	MH #12	1148.50	1148.43	10			0.50	
MH #13	33+10, 24.40' LT	1153.00	-	-	48"	SOLID	MH #13	MH #12	1148.50	1147.93		380		0.15	
INLET #13B	33+10, 26.74' RT	1152.00	24"	DITCH STOOL			INLET #13B	MH #13	1149.28	1149.00	55			0.50	
INLET #13A	33+15.76, 32' LT	1152.10	24"	DITCH STOOL			INLET #13A	MH #13	1149.50	1149.00	10			0.50	

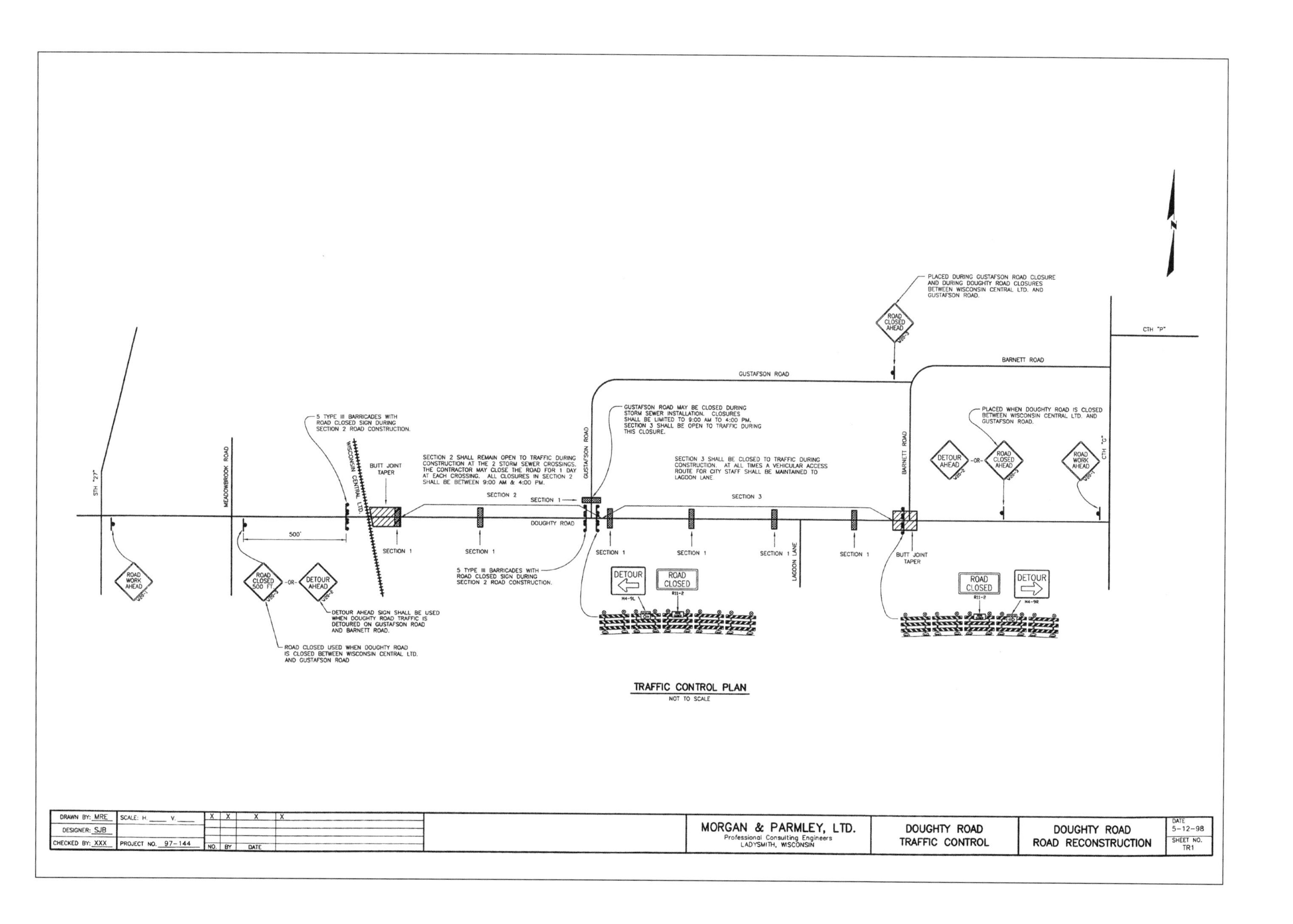
PLACED DURING GUSTAFSON ROAD CLOSURE AND DURING DOUGHTY ROAD CLOSURES BETWEEN WISCONSIN CENTRAL LTD. AND GUSTAFSON ROAD.
ROAD CLOSED AHEAD
CTH "P"
BARNETT ROAD
GUSTAFSON ROAD
GUSTAFSON ROAD MAY BE CLOSED DURING STORM SEWER INSTALLATION. CLOSURES SHALL BE LIMITED TO 9:00 AM TO 4:00 PM. SECTION 3 SHALL BE OPEN TO TRAFFIC DURING THIS CLOSURE.
PLACED WHEN DOUGHTY ROAD IS CLOSED BETWEEN WISCONSIN CENTRAL LTD. AND GUSTAFSON ROAD.
5 TYPE III BARRICADES WITH ROAD CLOSED SIGN DURING SECTION 2 ROAD CONSTRUCTION.
SECTION 2 SHALL REMAIN OPEN TO TRAFFIC DURING CONSTRUCTION AT THE 2 STORM SEWER CROSSINGS. THE CONTRACTOR MAY CLOSE THE ROAD FOR 1 DAY AT EACH CROSSING. ALL CLOSURES IN SECTION 2 SHALL BE BETWEEN 9:00 AM & 4:00 PM.
SECTION 3 SHALL BE CLOSED TO TRAFFIC DURING CONSTRUCTION. AT ALL TIMES A VEHICULAR ACCESS ROUTE FOR CITY STAFF SHALL BE MAINTAINED TO LAGOON LANE.
STH "27"
MEADOWBROOK ROAD
WISCONSIN CENTRAL LTD.
BUTT JOINT TAPER
GUSTAFSON ROAD
BARNETT ROAD
DETOUR AHEAD
-OR-
ROAD WORK AHEAD
CTH "G"
SECTION 2
SECTION 1
SECTION 3
DOUGHTY ROAD
500'
SECTION 1
LAGOON LANE
BUTT JOINT TAPER
5 TYPE III BARRICADES WITH ROAD CLOSED SIGN DURING SECTION 2 ROAD CONSTRUCTION.
DETOUR
ROAD CLOSED
ROAD CLOSED 500 FT.
DETOUR AHEAD SIGN SHALL BE USED WHEN DOUGHTY ROAD TRAFFIC IS DETOURED ON GUSTAFSON ROAD AND BARNETT ROAD.
ROAD CLOSED USED WHEN DOUGHTY ROAD IS CLOSED BETWEEN WISCONSIN CENTRAL LTD. AND GUSTAFSON ROAD
TRAFFIC CONTROL PLAN
NOT TO SCALE
DRAWN BY: MRE
DESIGNER: SJB
CHECKED BY: XXX
SCALE: H. V.
PROJECT NO. 97-144
NO. BY DATE
MORGAN & PARMLEY, LTD.
Professional Consulting Engineers
LADYSMITH, WISCONSIN
DOUGHTY ROAD TRAFFIC CONTROL
DOUGHTY ROAD ROAD RECONSTRUCTION
DATE 5-12-98
SHEET NO. TR1

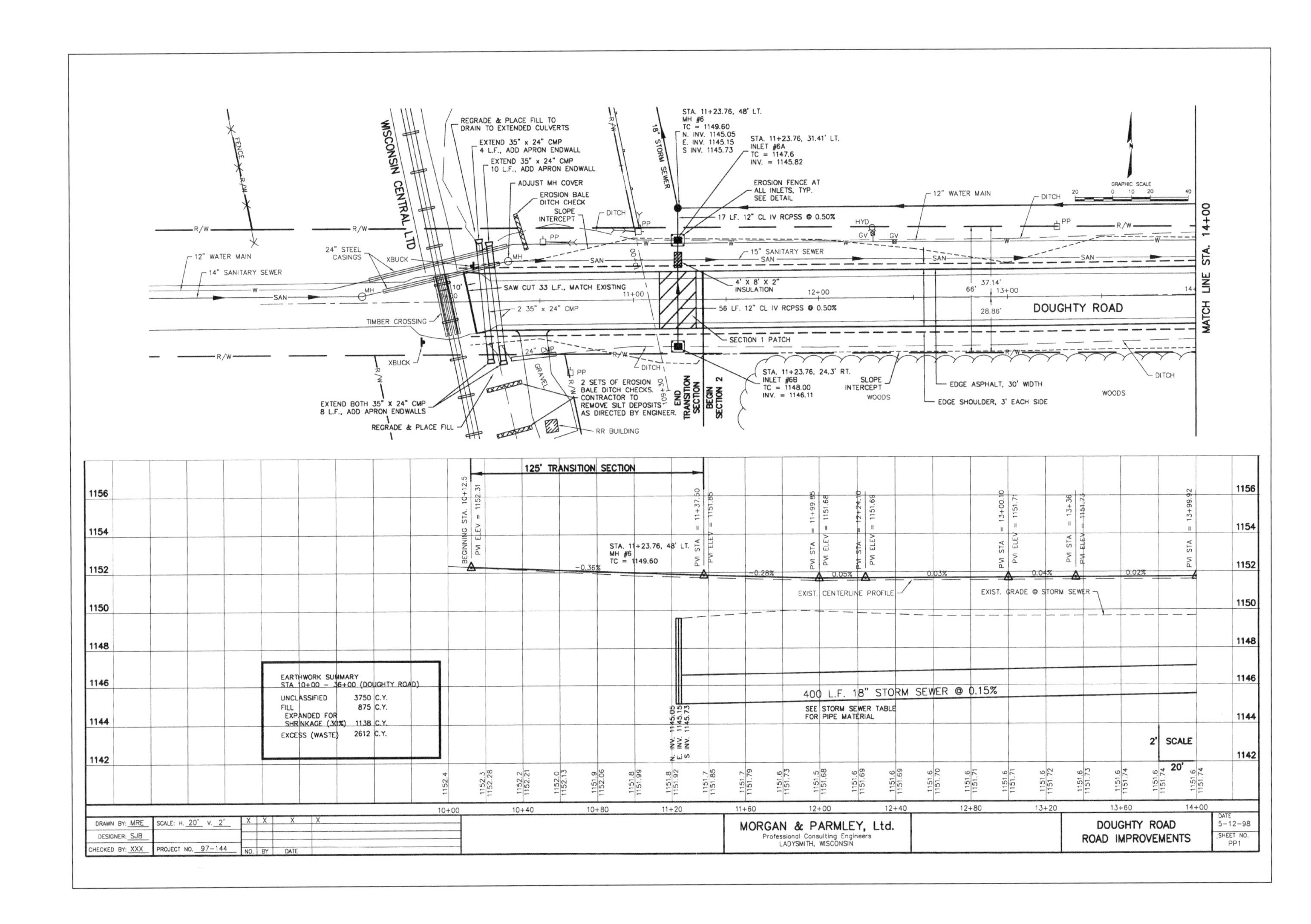

WISCONSIN CENTRAL LTD
DOUGHTY ROAD
MATCH LINE STA. 14+00
125' TRANSITION SECTION
400 L.F. 18" STORM SEWER @ 0.15%
SEE STORM SEWER TABLE FOR PIPE MATERIAL
EARTHWORK SUMMARY
STA 10+00 - 36+00 (DOUGHTY ROAD)
UNCLASSIFIED 3750 C.Y.
FILL 875 C.Y.
EXPANDED FOR SHRINKAGE (30%) 1138 C.Y.
EXCESS (WASTE) 2612 C.Y.
MORGAN & PARMLEY, Ltd.
Professional Consulting Engineers
LADYSMITH, WISCONSIN
DOUGHTY ROAD ROAD IMPROVEMENTS
DATE 5-12-98
SHEET NO. PP1
DRAWN BY: MRE
DESIGNER: SJB
CHECKED BY: XXX
SCALE: H. 20' V. 2'
PROJECT NO. 97-144

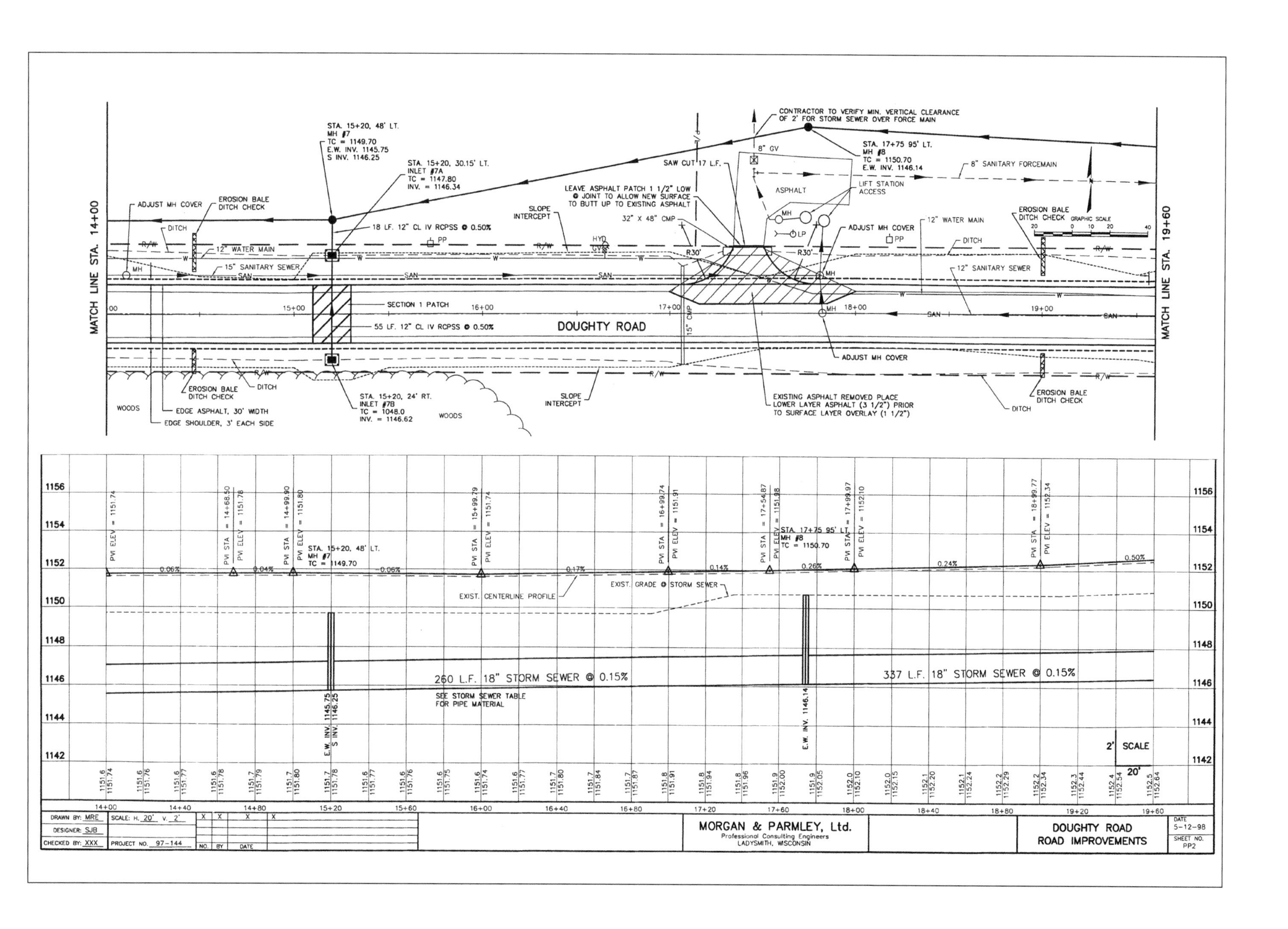

MATCH LINE STA. 14+00
MATCH LINE STA. 19+60
STA. 15+20, 48' LT.
MH #7
TC = 1149.70
E.W. INV. 1145.75
S INV. 1146.25
STA. 15+20, 30.15' LT.
INLET #7A
TC = 1147.80
INV. = 1146.34
ADJUST MH COVER
EROSION BALE DITCH CHECK
DITCH
12" WATER MAIN
15" SANITARY SEWER
18 LF. 12" CL IV RCPSS @ 0.50%
SECTION 1 PATCH
55 LF. 12" CL IV RCPSS @ 0.50%
DOUGHTY ROAD
SLOPE INTERCEPT
LEAVE ASPHALT PATCH 1 1/2" LOW @ JOINT TO ALLOW NEW SURFACE TO BUTT UP TO EXISTING ASPHALT
32" X 48" CMP
SAW CUT 17 L.F.
CONTRACTOR TO VERIFY MIN. VERTICAL CLEARANCE OF 2' FOR STORM SEWER OVER FORCE MAIN
8" GV
ASPHALT
LIFT STATION ACCESS
STA. 17+75 95' LT.
MH #8
TC = 1150.70
E.W. INV. 1146.14
8" SANITARY FORCEMAIN
12" WATER MAIN
ADJUST MH COVER
12" SANITARY SEWER
GRAPHIC SCALE
EROSION BALE DITCH CHECK
WOODS
EDGE ASPHALT, 30' WIDTH
EDGE SHOULDER, 3' EACH SIDE
STA. 15+20, 24' RT.
INLET #7B
TC = 1048.0
INV. = 1146.62
EXISTING ASPHALT REMOVED PLACE LOWER LAYER ASPHALT (3 1/2") PRIOR TO SURFACE LAYER OVERLAY (1 1/2")
EXIST. CENTERLINE PROFILE
EXIST. GRADE @ STORM SEWER
260 L.F. 18" STORM SEWER @ 0.15%
337 L.F. 18" STORM SEWER @ 0.15%
SEE STORM SEWER TABLE FOR PIPE MATERIAL
E.W. INV. 1145.75
S INV. 1146.25
E.W. INV. 1146.14
MORGAN & PARMLEY, Ltd.
Professional Consulting Engineers
LADYSMITH, WISCONSIN
DOUGHTY ROAD
ROAD IMPROVEMENTS
DRAWN BY: MRE
DESIGNER: SJB
CHECKED BY: XXX
SCALE: H. 20' V. 2'
PROJECT NO. 97-144
DATE 5-12-98
SHEET NO. PP2

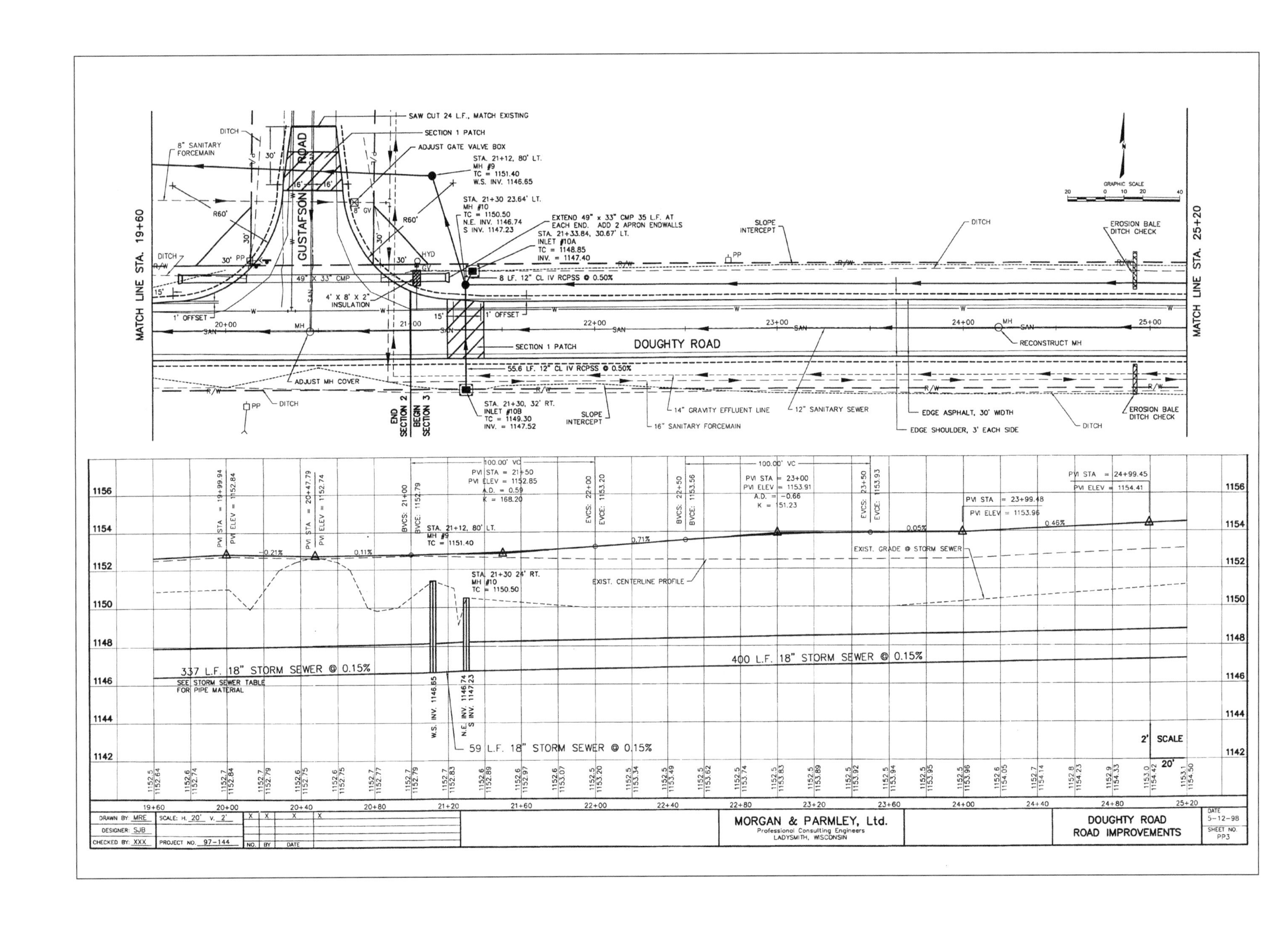
SAW CUT 24 L.F., MATCH EXISTING
SECTION 1 PATCH
ADJUST GATE VALVE BOX
STA. 21+12, 80' LT.
MH #9
TC = 1151.40
W.S. INV. 1146.65
STA. 21+30 23.64' LT.
MH #10
TC = 1150.50
N.E. INV. 1146.74
S INV. 1147.23
EXTEND 49" x 33" CMP 35 L.F. AT EACH END. ADD 2 APRON ENDWALLS
STA. 21+33.84, 30.67' LT.
INLET #10A
TC = 1148.85
INV. = 1147.40
8 LF. 12" CL IV RCPSS @ 0.50%
55.6 LF. 12" CL IV RCPSS @ 0.50%
STA. 21+30, 32' RT.
INLET #10B
TC = 1149.30
INV. = 1147.52
GUSTAFSON ROAD
DOUGHTY ROAD
MATCH LINE STA. 19+60
MATCH LINE STA. 25+20
8" SANITARY FORCEMAIN
DITCH
SLOPE INTERCEPT
EROSION BALE DITCH CHECK
RECONSTRUCT MH
ADJUST MH COVER
14" GRAVITY EFFLUENT LINE
12" SANITARY SEWER
16" SANITARY FORCEMAIN
EDGE ASPHALT, 30' WIDTH
EDGE SHOULDER, 3' EACH SIDE
4' X 8' X 2" INSULATION
END SECTION 2
BEGIN SECTION 3
GRAPHIC SCALE
337 L.F. 18" STORM SEWER @ 0.15%
SEE STORM SEWER TABLE FOR PIPE MATERIAL
400 L.F. 18" STORM SEWER @ 0.15%
59 L.F. 18" STORM SEWER @ 0.15%
EXIST. CENTERLINE PROFILE
EXIST. GRADE @ STORM SEWER
PVI STA = 21+50
PVI ELEV = 1152.85
A.D. = 0.59
K = 168.20
PVI STA = 23+00
PVI ELEV = 1153.91
A.D. = -0.66
K = 151.23
PVI STA = 23+99.48
PVI ELEV = 1153.96
PVI STA = 24+99.45
PVI ELEV = 1154.41
DRAWN BY: MRE
DESIGNER: SJB
CHECKED BY: XXX
SCALE: H. 20' V. 2'
PROJECT NO. 97-144
MORGAN & PARMLEY, Ltd.
Professional Consulting Engineers
LADYSMITH, WISCONSIN
DOUGHTY ROAD
ROAD IMPROVEMENTS
DATE 5-12-98
SHEET NO. PP3

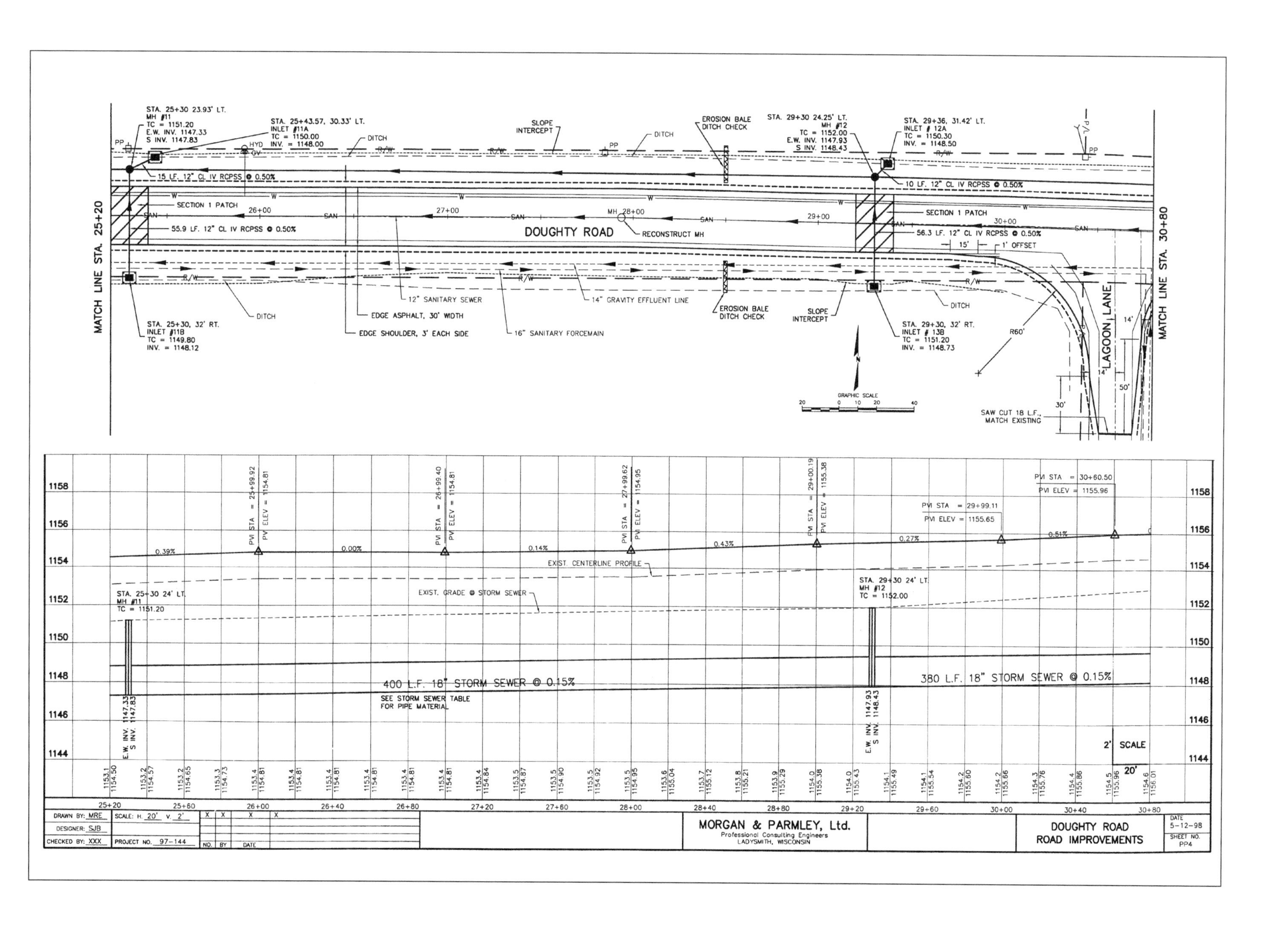
MATCH LINE STA. 25+20
MATCH LINE STA. 30+80
DOUGHTY ROAD
LAGOON LANE
STA. 25+30 23.93' LT.
MH #11
TC = 1151.20
E.W. INV. 1147.33
S INV. 1147.83
STA. 25+43.57, 30.33' LT.
INLET #11A
TC = 1150.00
INV. = 1148.00
15 LF. 12" CL IV RCPSS @ 0.50%
SECTION 1 PATCH
55.9 LF. 12" CL IV RCPSS @ 0.50%
STA. 25+30, 32' RT.
INLET #11B
TC = 1149.80
INV. = 1148.12
12" SANITARY SEWER
EDGE ASPHALT, 30' WIDTH
EDGE SHOULDER, 3' EACH SIDE
16" SANITARY FORCEMAIN
14" GRAVITY EFFLUENT LINE
RECONSTRUCT MH
EROSION BALE DITCH CHECK
SLOPE INTERCEPT
STA. 29+30 24.25' LT.
MH #12
TC = 1152.00
E.W. INV. 1147.93
S INV. 1148.43
STA. 29+36, 31.42' LT.
INLET # 12A
TC = 1150.30
INV. = 1148.50
10 LF. 12" CL IV RCPSS @ 0.50%
56.3 LF. 12" CL IV RCPSS @ 0.50%
1' OFFSET
STA. 29+30, 32' RT.
INLET # 13B
TC = 1151.20
INV. = 1148.73
SAW CUT 18 L.F., MATCH EXISTING
GRAPHIC SCALE
EXIST. CENTERLINE PROFILE
EXIST. GRADE @ STORM SEWER
400 L.F. 18" STORM SEWER @ 0.15%
SEE STORM SEWER TABLE FOR PIPE MATERIAL
380 L.F. 18" STORM SEWER @ 0.15%
PVI STA = 30+60.50
PVI ELEV = 1155.96
PVI STA = 29+99.11
PVI ELEV = 1155.65
DRAWN BY: MRE
DESIGNER: SJB
CHECKED BY: XXX
SCALE: H. 20' V. 2'
PROJECT NO. 97-144
MORGAN & PARMLEY, Ltd.
Professional Consulting Engineers
LADYSMITH, WISCONSIN
DOUGHTY ROAD
ROAD IMPROVEMENTS
DATE 5-12-98
SHEET NO. PP4

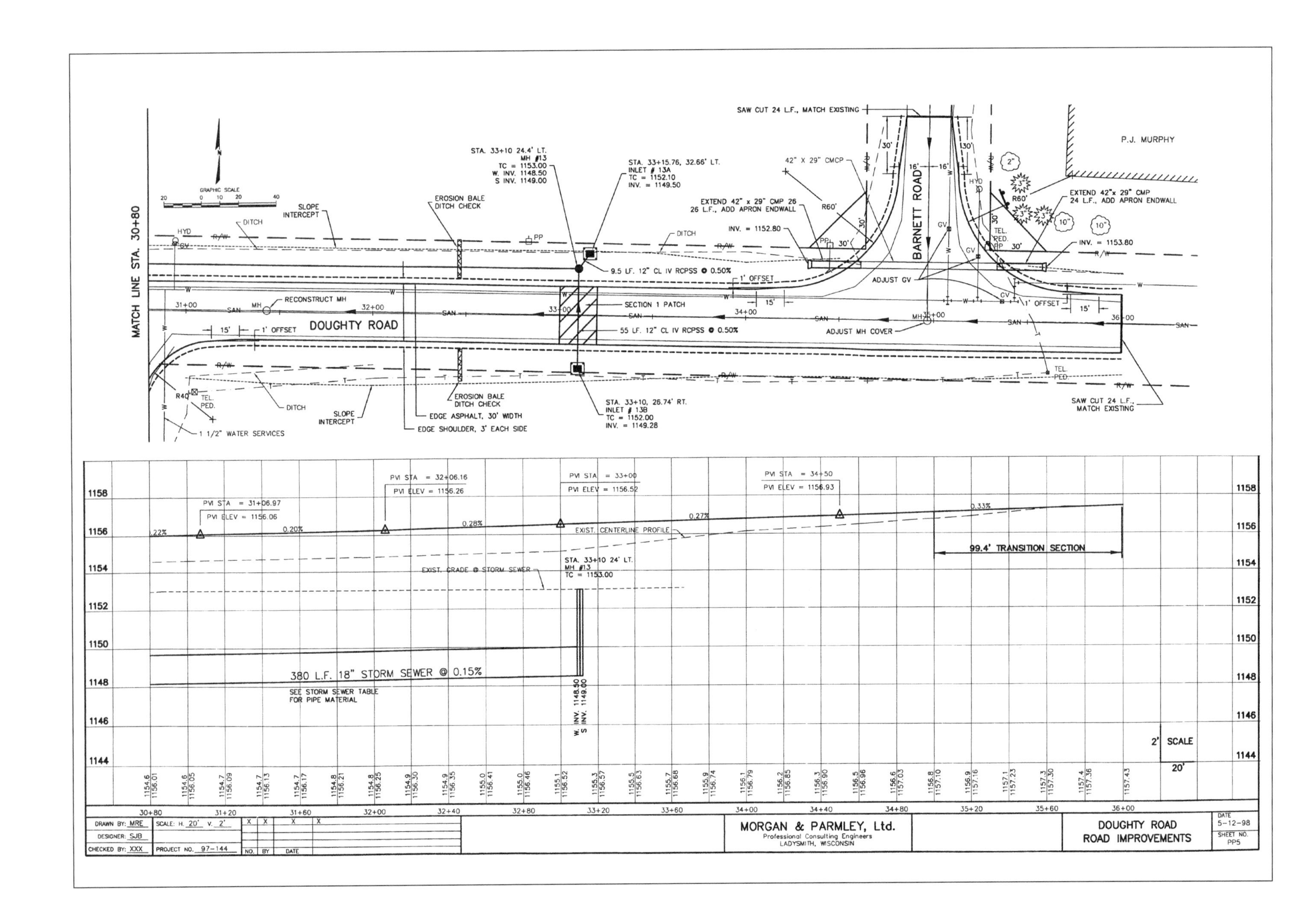
MATCH LINE STA. 30+80
DOUGHTY ROAD
BARNETT ROAD
P.J. MURPHY
SAW CUT 24 L.F., MATCH EXISTING
STA. 33+10 24.4' LT. MH #13 TC = 1153.00 W. INV. 1148.50 S INV. 1149.00
STA. 33+15.76, 32.66' LT. INLET # 13A TC = 1152.10 INV. = 1149.50
EROSION BALE DITCH CHECK
SLOPE INTERCEPT
EXTEND 42" x 29" CMP 26 L.F., ADD APRON ENDWALL
EXTEND 42" x 29" CMP 24 L.F., ADD APRON ENDWALL
INV. = 1152.80
INV. = 1153.80
9.5 LF. 12" CL IV RCPSS @ 0.50%
55 LF. 12" CL IV RCPSS @ 0.50%
SECTION 1 PATCH
RECONSTRUCT MH
ADJUST GV
ADJUST MH COVER
STA. 33+10, 26.74' RT. INLET # 13B TC = 1152.00 INV. = 1149.28
EDGE ASPHALT, 30' WIDTH
EDGE SHOULDER, 3' EACH SIDE
1 1/2" WATER SERVICES
SAW CUT 24 L.F., MATCH EXISTING
PVI STA = 31+06.97 PVI ELEV = 1156.06
PVI STA = 32+06.16 PVI ELEV = 1156.26
PVI STA = 33+00 PVI ELEV = 1156.52
PVI STA = 34+50 PVI ELEV = 1156.93
EXIST. CENTERLINE PROFILE
EXIST. GRADE @ STORM SEWER
99.4' TRANSITION SECTION
380 L.F. 18" STORM SEWER @ 0.15%
SEE STORM SEWER TABLE FOR PIPE MATERIAL
W. INV. 1148.50 S INV. 1149.00
MORGAN & PARMLEY, Ltd.
Professional Consulting Engineers
LADYSMITH, WISCONSIN
DOUGHTY ROAD ROAD IMPROVEMENTS
DATE 5-12-98
SHEET NO. PP5
DRAWN BY: MRE
DESIGNER: SJB
CHECKED BY: XXX
SCALE: H. 20' V. 2'
PROJECT NO. 97-144

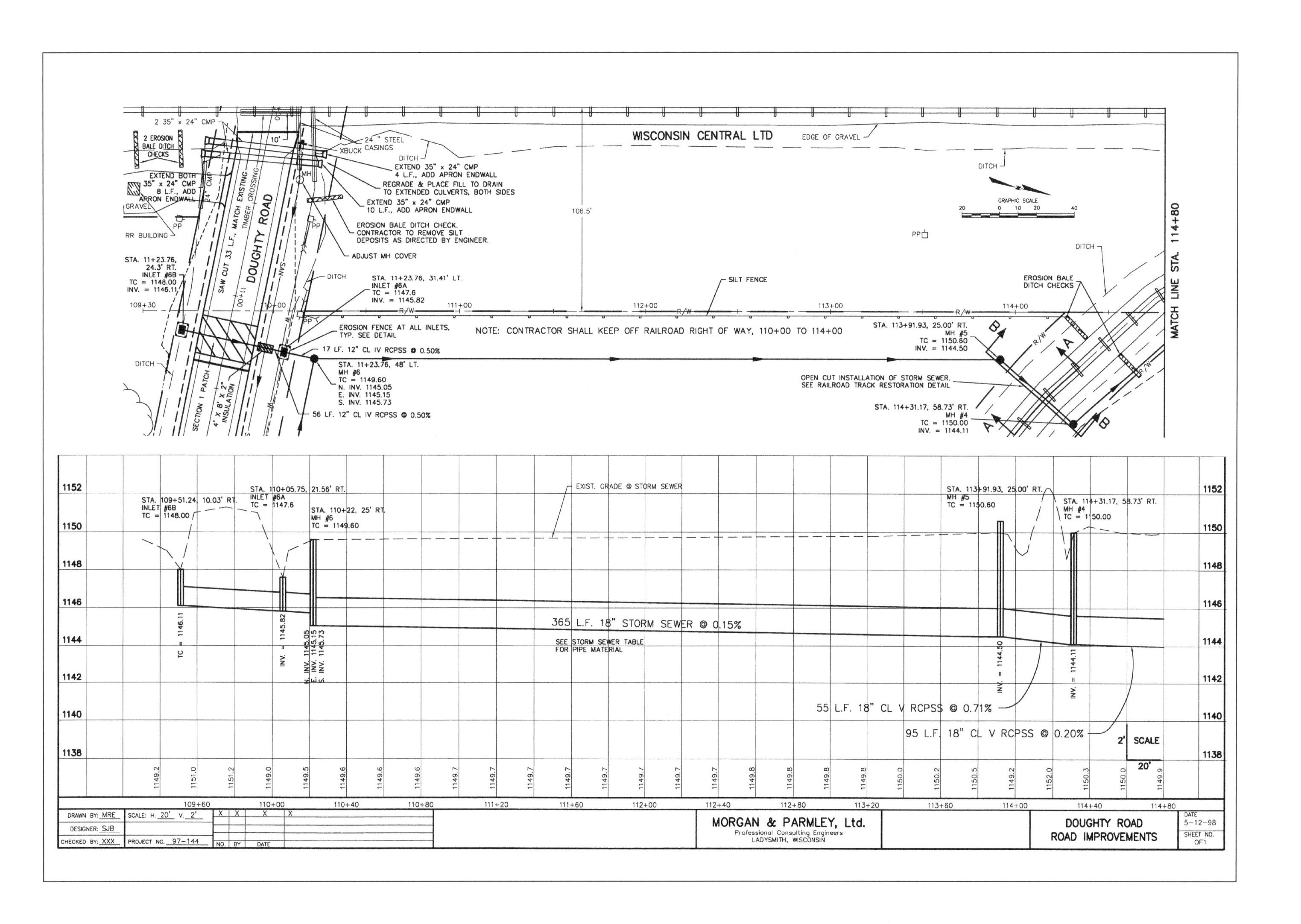
WISCONSIN CENTRAL LTD
DOUGHTY ROAD
NOTE: CONTRACTOR SHALL KEEP OFF RAILROAD RIGHT OF WAY, 110+00 TO 114+00
MATCH LINE STA. 114+80
365 L.F. 18" STORM SEWER @ 0.15%
55 L.F. 18" CL V RCPSS @ 0.71%
95 L.F. 18" CL V RCPSS @ 0.20%
MORGAN & PARMLEY, Ltd.
Professional Consulting Engineers
LADYSMITH, WISCONSIN
DOUGHTY ROAD
ROAD IMPROVEMENTS

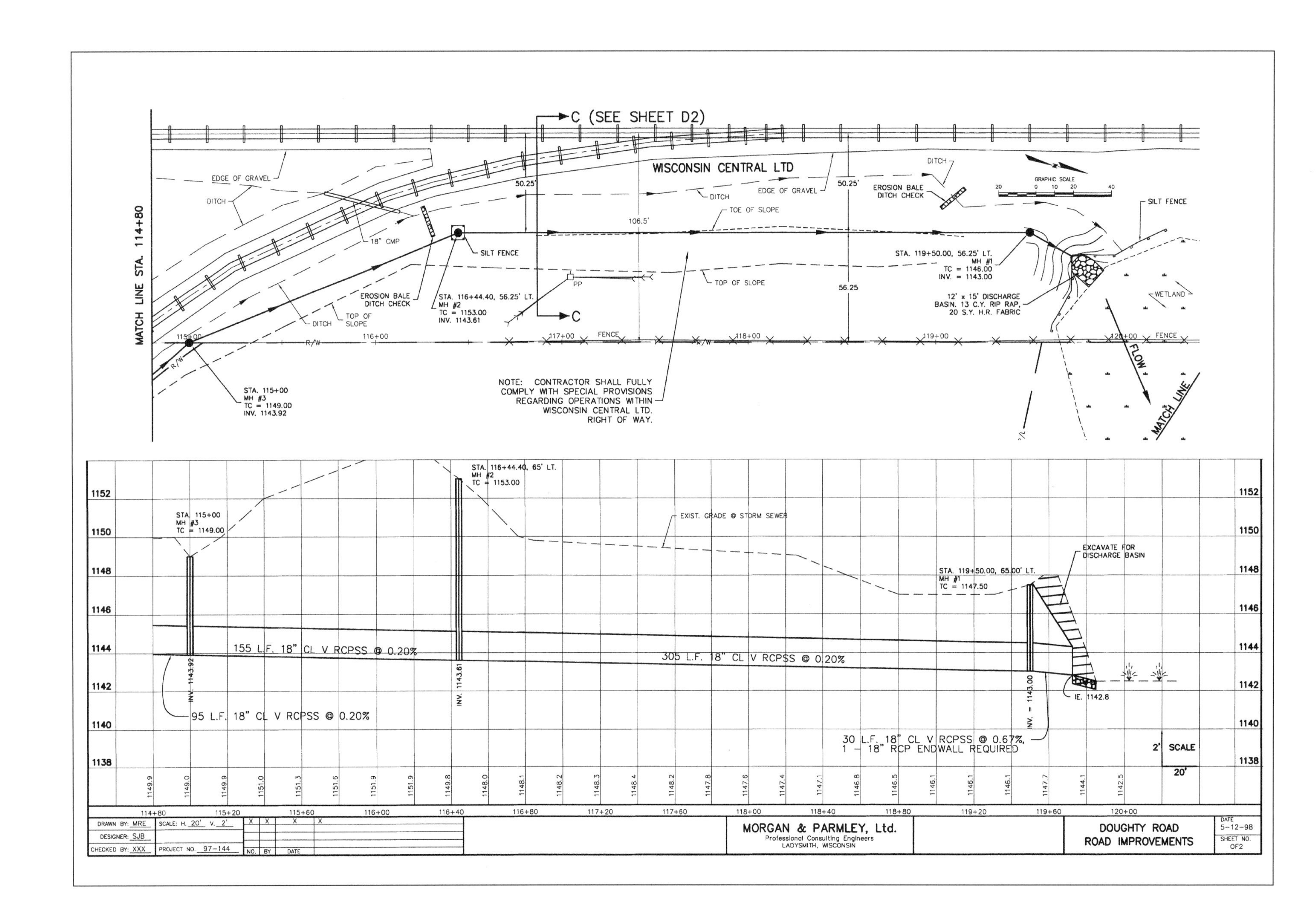

C (SEE SHEET D2)
WISCONSIN CENTRAL LTD
MATCH LINE STA. 114+80
EDGE OF GRAVEL
DITCH
18" CMP
SILT FENCE
EROSION BALE DITCH CHECK
TOP OF SLOPE
TOE OF SLOPE
STA. 115+00 MH #3 TC = 1149.00 INV. 1143.92
STA. 116+44.40, 56.25' LT. MH #2 TC = 1153.00 INV. 1143.61
STA. 119+50.00, 56.25' LT. MH #1 TC = 1146.00 INV. = 1143.00
12' x 15' DISCHARGE BASIN. 13 C.Y. RIP RAP, 20 S.Y. H.R. FABRIC
WETLAND
FLOW
MATCH LINE
GRAPHIC SCALE
NOTE: CONTRACTOR SHALL FULLY COMPLY WITH SPECIAL PROVISIONS REGARDING OPERATIONS WITHIN WISCONSIN CENTRAL LTD. RIGHT OF WAY.
155 L.F. 18" CL V RCPSS @ 0.20%
305 L.F. 18" CL V RCPSS @ 0.20%
95 L.F. 18" CL V RCPSS @ 0.20%
30 L.F. 18" CL V RCPSS @ 0.67%, 1 – 18" RCP ENDWALL REQUIRED
EXIST. GRADE @ STORM SEWER
EXCAVATE FOR DISCHARGE BASIN
STA. 116+44.40, 65' LT. MH #2 TC = 1153.00
STA. 119+50.00, 65.00' LT. MH #1 TC = 1147.50
IE. 1142.8
MORGAN & PARMLEY, Ltd.
Professional Consulting Engineers
LADYSMITH, WISCONSIN
DOUGHTY ROAD ROAD IMPROVEMENTS
DATE 5-12-98
SHEET NO. OF2
DRAWN BY: MRE
DESIGNER: SJB
CHECKED BY: XXX
SCALE: H. 20' V. 2'
PROJECT NO. 97-144

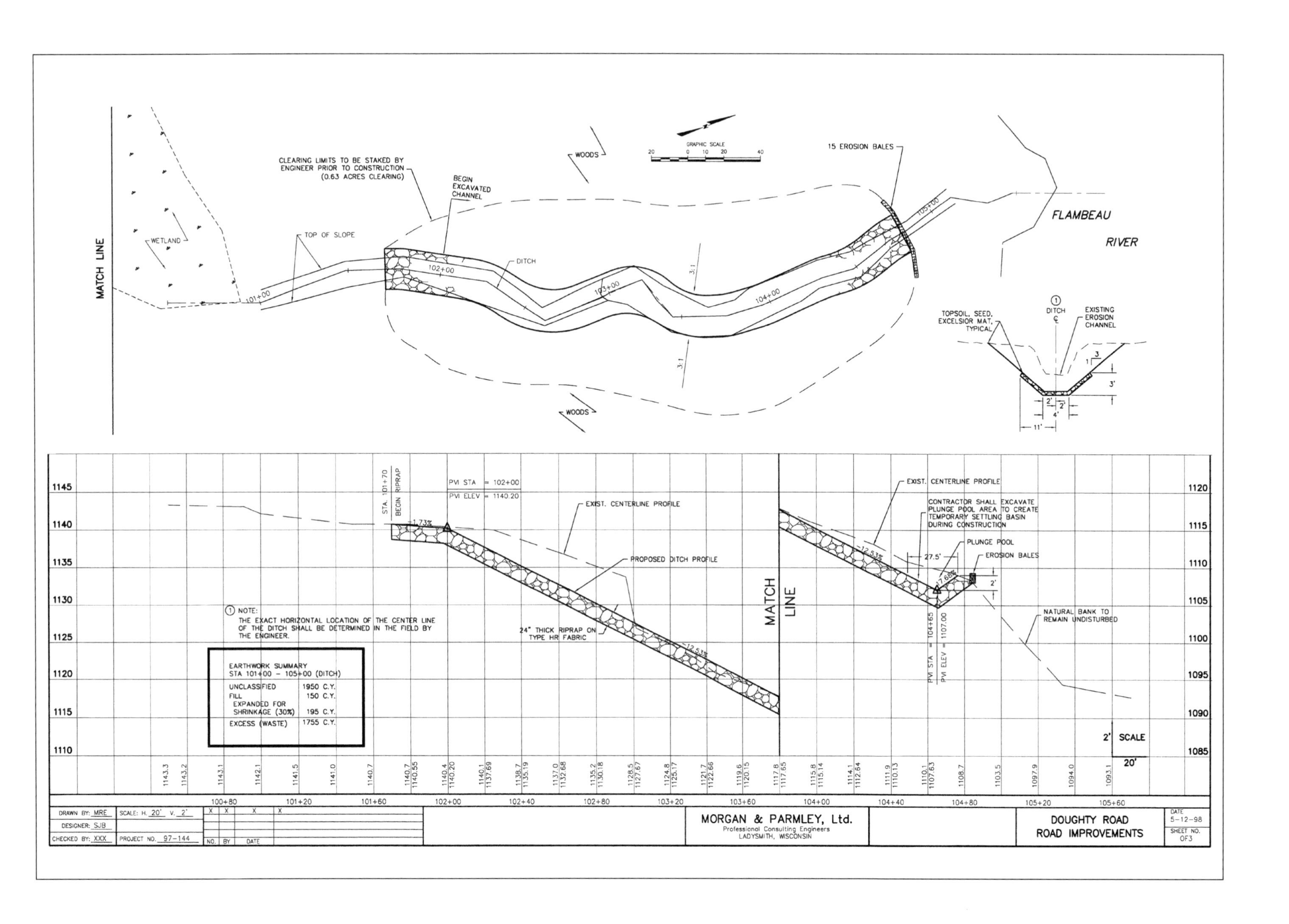
MATCH LINE
WETLAND
CLEARING LIMITS TO BE STAKED BY ENGINEER PRIOR TO CONSTRUCTION (0.63 ACRES CLEARING)
TOP OF SLOPE
BEGIN EXCAVATED CHANNEL
WOODS
DITCH
GRAPHIC SCALE
15 EROSION BALES
FLAMBEAU
RIVER
TOPSOIL, SEED, EXCELSIOR MAT, TYPICAL
DITCH
EXISTING EROSION CHANNEL
PVI STA = 102+00
PVI ELEV = 1140.20
STA. 101+70 BEGIN RIPRAP
EXIST. CENTERLINE PROFILE
PROPOSED DITCH PROFILE
24" THICK RIPRAP ON TYPE HR FABRIC
NOTE:
THE EXACT HORIZONTAL LOCATION OF THE CENTER LINE OF THE DITCH SHALL BE DETERMINED IN THE FIELD BY THE ENGINEER.
EARTHWORK SUMMARY
STA 101+00 – 105+00 (DITCH)
UNCLASSIFIED 1950 C.Y.
FILL 150 C.Y.
EXPANDED FOR SHRINKAGE (30%) 195 C.Y.
EXCESS (WASTE) 1755 C.Y.
MATCH LINE
CONTRACTOR SHALL EXCAVATE PLUNGE POOL AREA TO CREATE TEMPORARY SETTLING BASIN DURING CONSTRUCTION
PLUNGE POOL
EROSION BALES
NATURAL BANK TO REMAIN UNDISTURBED
PVI STA = 104+65
PVI ELEV = 1107.00
SCALE
DRAWN BY: MRE
DESIGNER: SJB
CHECKED BY: XXX
SCALE: H. 20' V. 2'
PROJECT NO. 97-144
MORGAN & PARMLEY, Ltd.
Professional Consulting Engineers
LADYSMITH, WISCONSIN
DOUGHTY ROAD
ROAD IMPROVEMENTS
DATE 5-12-98
SHEET NO. OF3

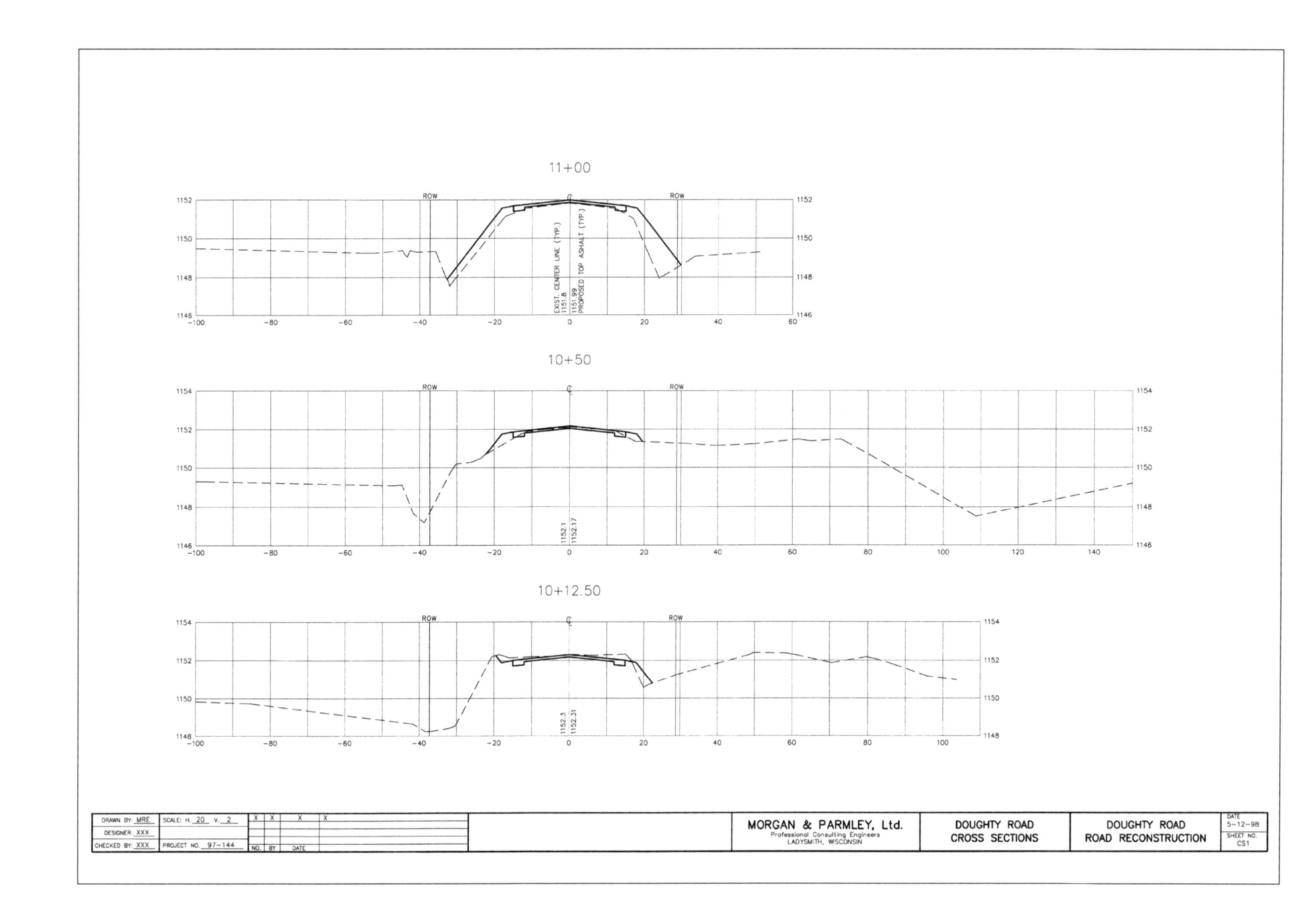
11+00
ROW
ROW
EXIST. CENTER LINE (TYP.)
PROPOSED TOP ASHALT (TYP.)
1151.8
1151.99
10+50
1152.1
1152.17
10+12.50
1152.3
1152.31
DRAWN BY: MRE
DESIGNER: XXX
CHECKED BY: XXX
SCALE: H. 20 V. 2
PROJECT NO. 97-144
NO. BY DATE
MORGAN & PARMLEY, Ltd.
Professional Consulting Engineers
LADYSMITH, WISCONSIN
DOUGHTY ROAD
CROSS SECTIONS
DOUGHTY ROAD
ROAD RECONSTRUCTION
DATE
5-12-98
SHEET NO.
CS1

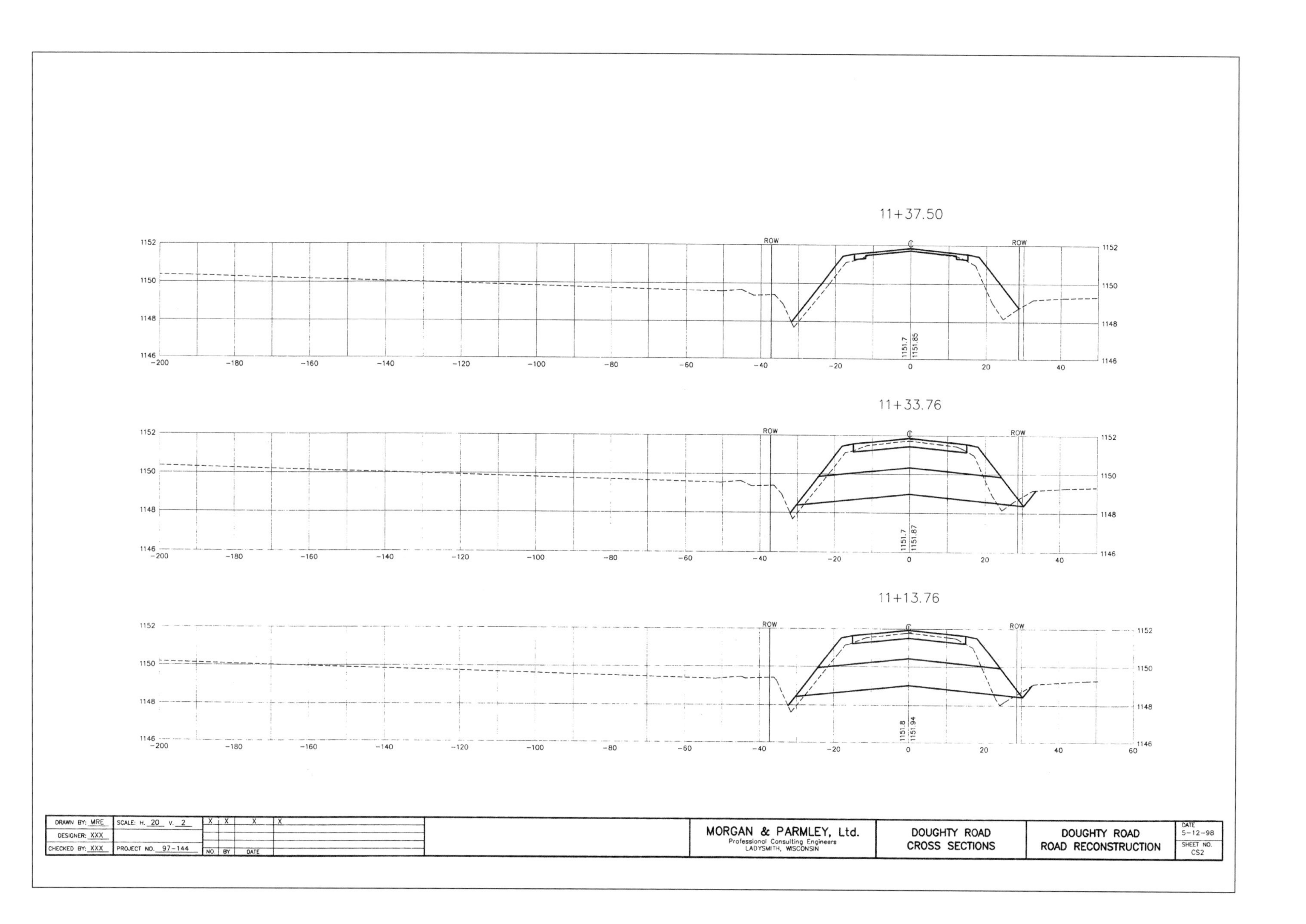
11+37.50
11+33.76
11+13.76
ROW
1152
1150
1148
1146
-200
-180
-160
-140
-120
-100
-80
-60
-40
-20
0
20
40
60
1151.7
1151.85
1151.87
1151.8
1151.94
DRAWN BY: MRE
DESIGNER: XXX
CHECKED BY: XXX
SCALE: H. 20 V. 2
PROJECT NO. 97-144
NO. BY DATE
MORGAN & PARMLEY, Ltd.
Professional Consulting Engineers
LADYSMITH, WISCONSIN
DOUGHTY ROAD
CROSS SECTIONS
DOUGHTY ROAD
ROAD RECONSTRUCTION
DATE
5-12-98
SHEET NO.
CS2

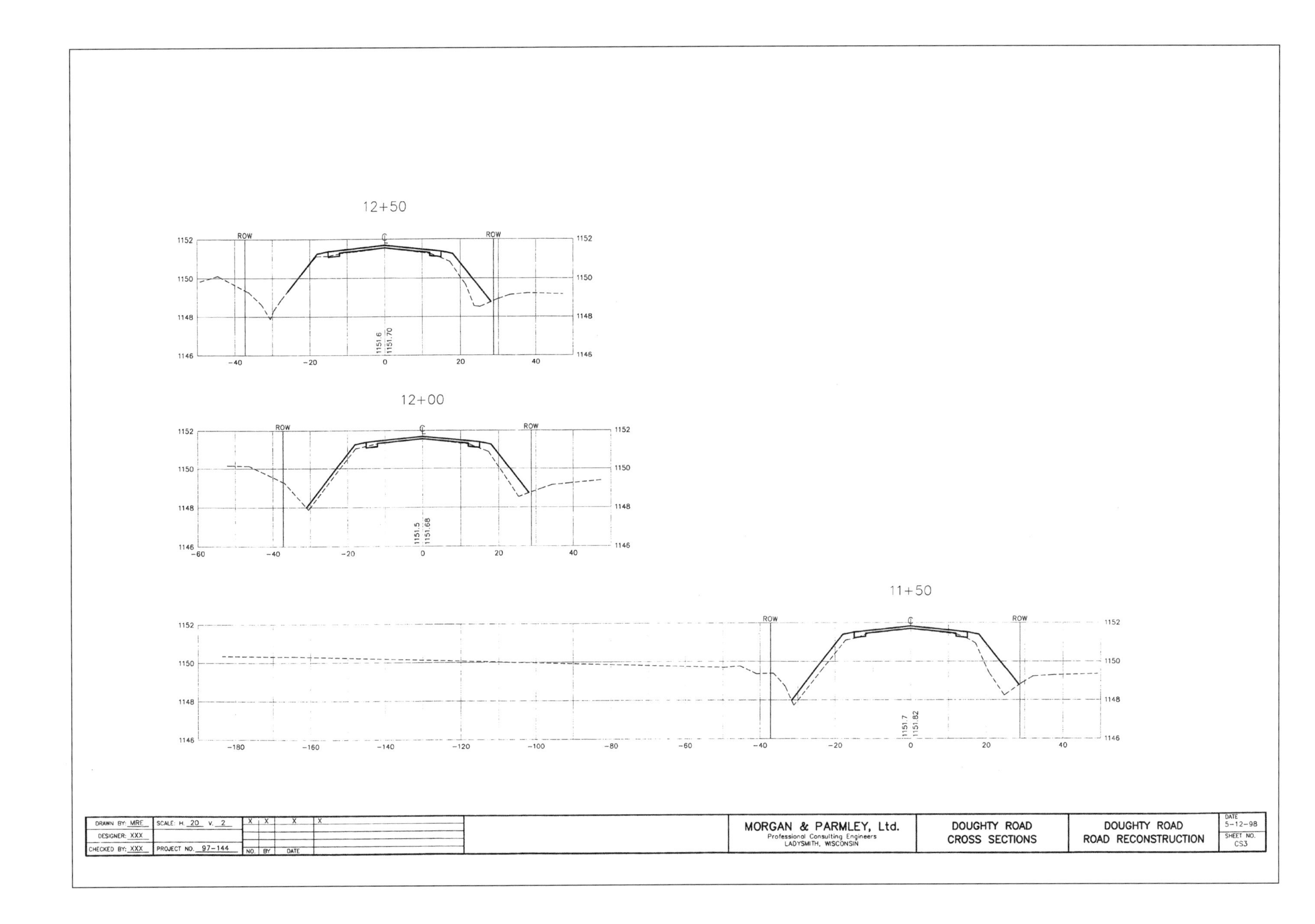
12+50
ROW
ROW
1152
1150
1148
1146
1151.6
1151.70
-40
-20
0
20
40
12+00
ROW
ROW
1151.5
1151.68
-60
11+50
1151.7
1151.82
-180
-160
-140
-120
-100
-80
-60
DRAWN BY: MRE
DESIGNER: XXX
CHECKED BY: XXX
SCALE: H. 20 V. 2
PROJECT NO. 97-144
NO. BY DATE
MORGAN & PARMLEY, Ltd.
Professional Consulting Engineers
LADYSMITH, WISCONSIN
DOUGHTY ROAD
CROSS SECTIONS
DOUGHTY ROAD
ROAD RECONSTRUCTION
DATE
5-12-98
SHEET NO.
CS3

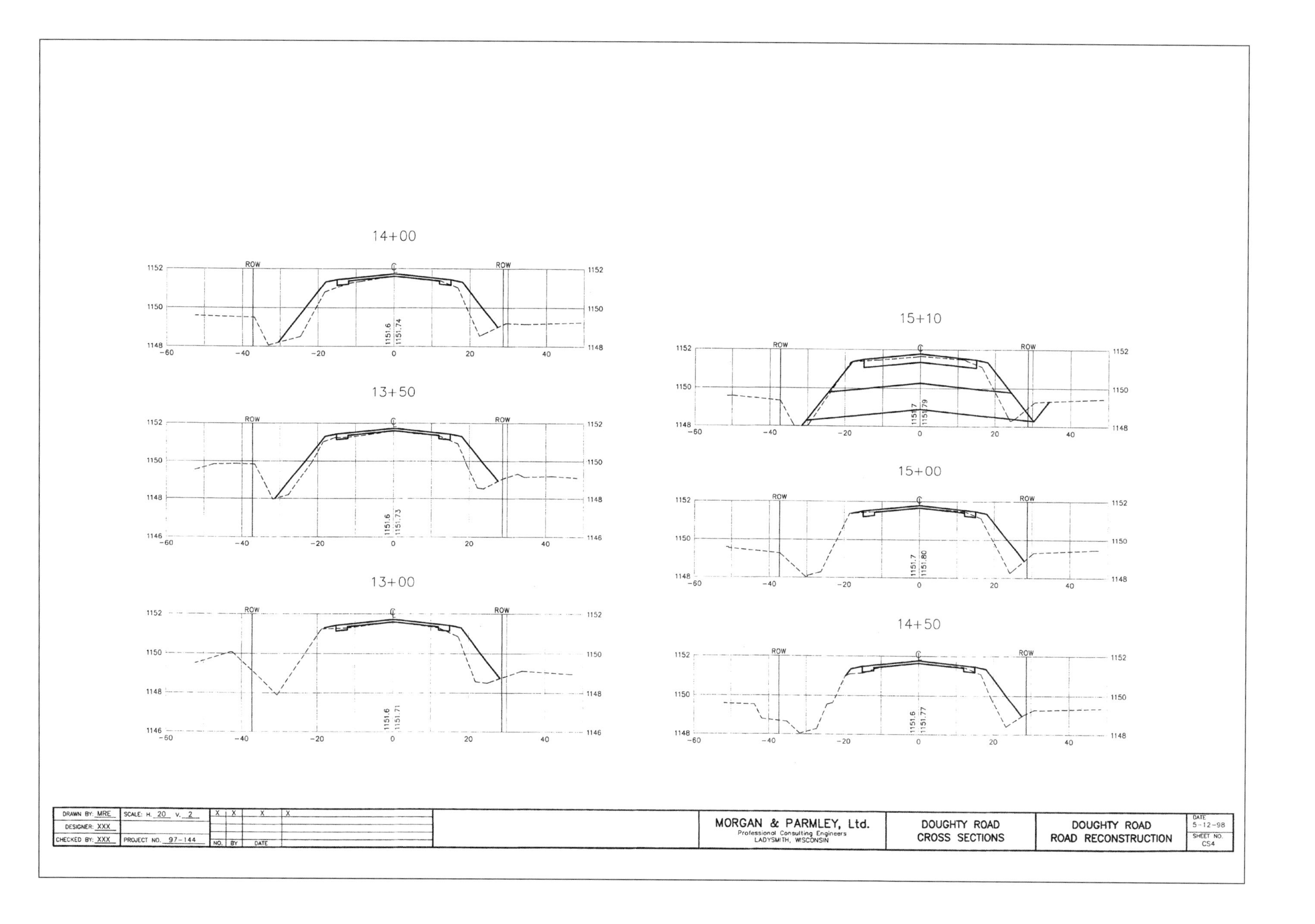
14+00
13+50
13+00
15+10
15+00
14+50
ROW
DRAWN BY: MRE
DESIGNER: XXX
CHECKED BY: XXX
SCALE: H. 20 V. 2
PROJECT NO. 97-144
MORGAN & PARMLEY, Ltd.
Professional Consulting Engineers
LADYSMITH, WISCONSIN
DOUGHTY ROAD
CROSS SECTIONS
DOUGHTY ROAD
ROAD RECONSTRUCTION
DATE
5-12-98
SHEET NO.
CS4

INDEX OF SHEETS

STATE OF WISCONSIN
DEPARTMENT OF TRANSPORTATION
PLAN OF PROPOSED IMPROVEMENT

RAMES ROAD – NORTH 1.163 MILES

C.T.H. A
RUSK COUNTY

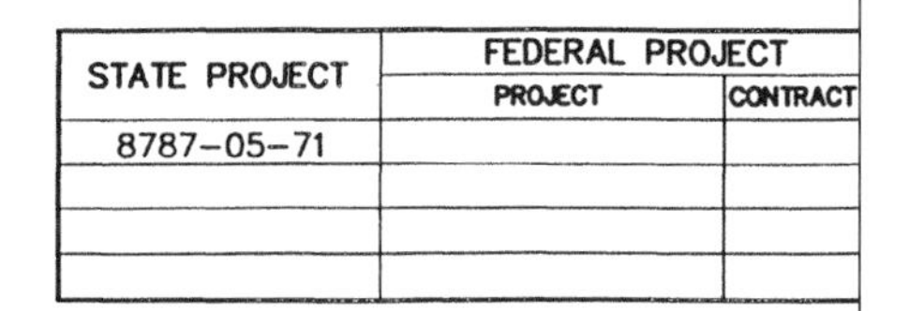

STATE PROJECT	FEDERAL PROJECT	
	PROJECT	CONTRACT
8787-05-71		

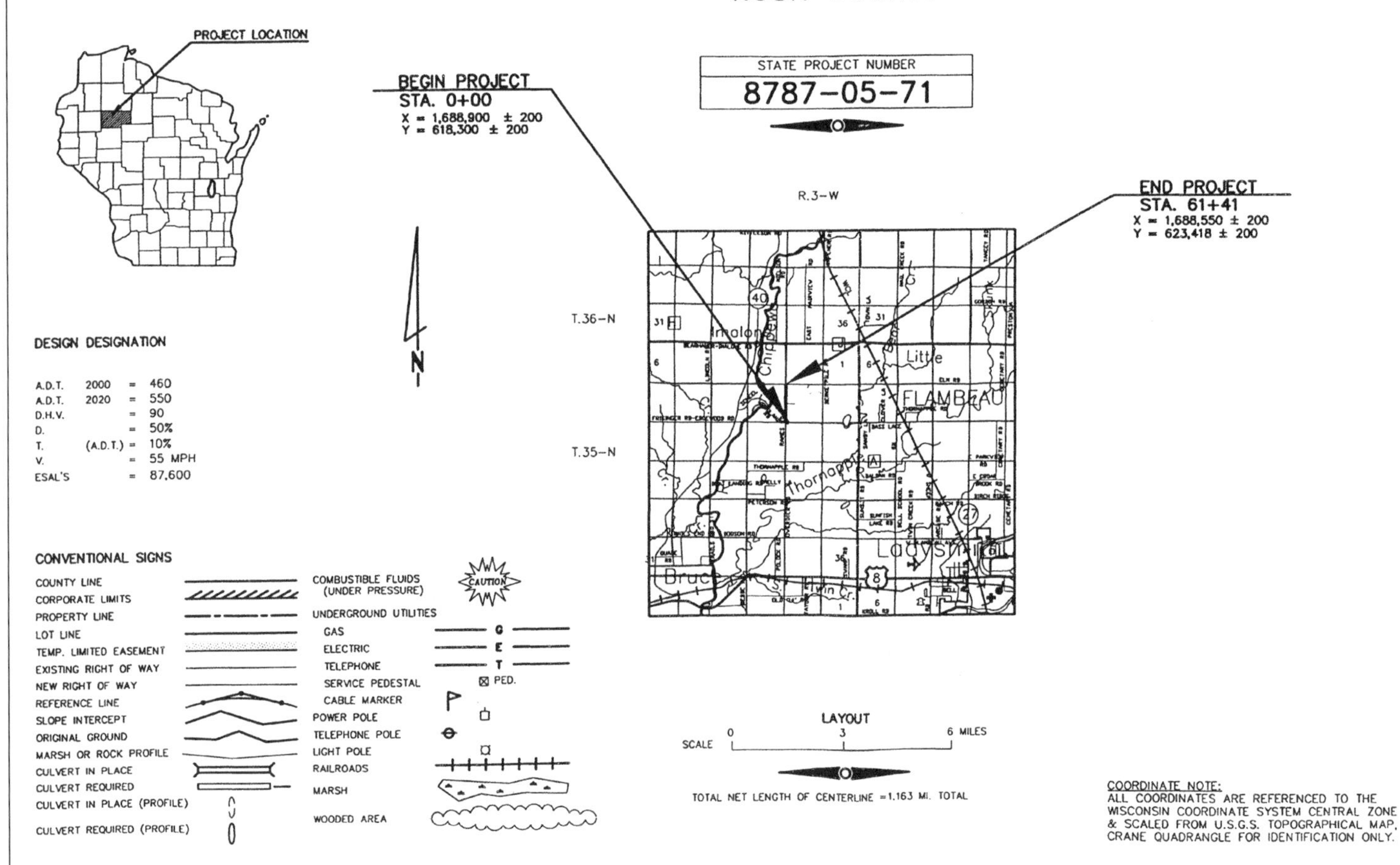

Courtesy: Morgan & Parmley, Ltd.

STATE PROJECT NUMBER	SHEET NO.
8787-05-71	2
TYPICAL SECTIONS–DETAILED SUMMARY OF MISCELLANEOUS QUANTITIES	

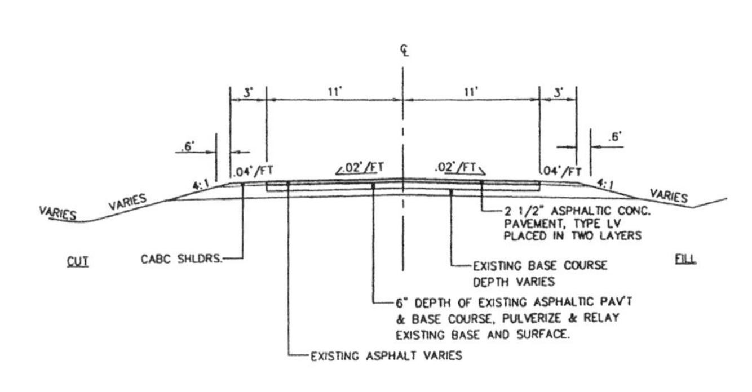

TYPICAL SECTION

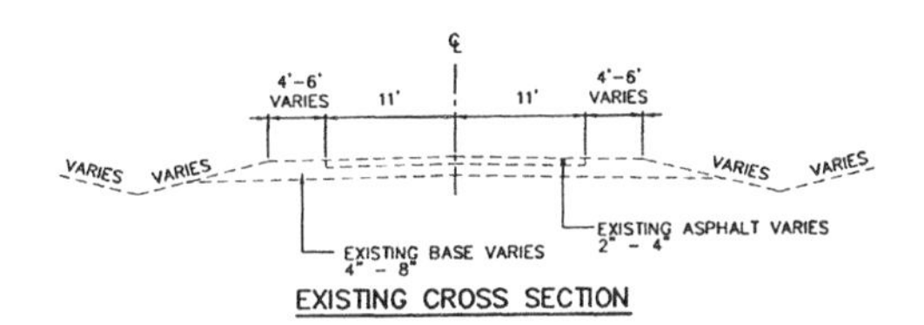

EXISTING CROSS SECTION

CALL DIGGERS' HOTLINE

1-800-242-8511

TOLL FREE

2040 W. WISCONSIN AVE.
SUITE 10
MILWAUKEE, WI 53233

RUSK COUNTY SURVEYOR
DAVE KAISER, RLS
(715) 532-2165

OTHER CONTACTS

DNR LIASON

DAN MICHELS
D.N.R. NORTHWESTERN DISTRICT H.Q.
P.O. BOX 309
SPOONER, WI. 54801 (715) 635-4228

DESIGN CONSULTANT

LARRY GOTHAM
MORGAN & PARMLEY LTD.
115 W. 2ND STREET SOUTH
LADYSMITH WI. 54848 (715) 532-3721

UTILITIES

JUMP RIVER ELECTRIC CO-OP

ATTN: HANK LEW
1102 W. 9TH STREET NORTH
LADYSMITH, WI. 54848
(715) 532-5524

UTC OF NORTHERN WISCONSIN CENTURY TELEPHONE

ATTN: JAMES ARQUETTE
425 ELLINGSON AVE
P.O. BOX 78
HAWKINS, WI. 54530
(800) 752-5637

DETAILED SUMMARY OF MISCELLANEOUS QUANTITIES

ASPHALTIC MATERIAL FOR PLANT MIXES

STATION	TO	STATION	LOCATION	TON
0+00		61+41	MAINLINE	129
P.E.'S & SIDEROADS		------	UNDISTRIBUTED	3
TOTAL				132

TEMPORARY PAVEMENT MARKING, 4"

STATION	TO	STATION	LOCATION	L.F.
0+00		61+41	CENTERLINE	9,800
TOTAL				9,800

PAVEMENT MARKING, 4", PAINT

STATION	TO	STATION	LOCATION	L.F.
0+00		61+41	CENTERLINE	9,800
0+00		61+41	RT. & LT. EDGE LINES	12,300
TOTAL				22,100

PULVERIZE AND RELAY EXISTING BASE & SURFACE

STATION	TO	STATION	LOCATION	S.Y.
0+00		61+41	MAINLINE	15,080
		------	UNDISTRIBUTED	20
TOTAL				15,100

CRUSHED AGGREGATE BASE COURSE

STATION	TO	STATION	LOCATION	C.Y.
0+00		61+41	MAINLINE	750
0+00		61+41	RT. & LT. SHLDR.	450
P.E.'S & SIDEROADS			UNDISTRIBUTED	100
TOTAL				1,300

ASPHALTIC MATERIAL FOR TACK COAT

STATION	TO	STATION	APPLICATIONS	RATE	GAL
0+00		61+41	1	0.025 GAL/SQ. YD.	375
P.E.'S & SIDEROADS		------	1	0.025 GAL/SQ. YD.	10
TOTAL					385

ASPHALTIC CONCRETE PAVEMENT, TYPE LV

STATION	TO	STATION	LOCATION	TON
0+00		61+41	MAINLINE	2,150
P.E.'S & SIDEROADS		------	UNDISTRIBUTED	50
TOTAL				2,200

GENERAL NOTES:

PRIOR TO PULVERIZING AND RELAYING, CRUSHED AGGREGATE BASE COURSE SHALL BE EVENLY PLACED ACROSS THE ROADWAY IN A VOLUME SUFFICIENT TO FILL EXISTING RUTS AND CREATE A 0.02'/FT CROWN, OR TO CONFORM TO SUPERELEVATION RATE.

THRU ALL PRIVATE ENTRANCES AND ON THE INSIDE OF ALL HORIZONTAL CURVES, THE EDGE OF ASPHALT SHALL BE EXTENDED 18".

THERE ARE UTILITY FACILITIES IN THE PROJECT AREA WHICH ARE NOT SHOWN ON THE PLANS. THERE ARE NO KNOWN CONFLICTS.

THE 2 1/2" ASPHALTIC CONCRETE PAVEMENT SHALL BE PLACED IN ONE 1 1/4" LOWER LAYER AND ONE 1 1/4" UPPER LAYER.

TACK COAT SHALL BE APPLIED BETWEEN THE ASPHALT LAYERS AT A RATE OF 0.025 GAL/SQ. YD.

STANDARD DETAIL DRAWINGS

15C8-8a	PAVEMENT MARKINGS (MAINLINE)
15C8-8b	PAVEMENT MARKINGS (INTERSECTIONS)
15C12-2	TRAFFIC CONTROL FOR LANE CLOSURES (SUITABLE FOR MOVING OPERATIONS)

LIST OF STANDARD ABBREVIATIONS

ABUT	ABUTMENT
AC	ACRES
AGG	AGGREGATE
AH	AHEAD
ADT	AVERAGE DAILY TRAFFIC
AVE	AVENUE
AVG	AVERAGE
ASPH	ASPHALTIC
BK	BACK
BM	BENCHMARK
△	CENTRAL ANGLE
℄, C/L	CENTERLINE
C & G	CURB AND GUTTER
CABC	CRUSHED AGGREGATE BASE COURSE
CONC	CONCRETE
CONST	CONSTRUCTION
COR	CORNER
CORR	CORRUGATED
CSCP	CORRUGATED STEEL CULVERT PIPE
CSPA	CORRUGATED STEEL PIPE ARCH
CTH	COUNTY TRUNK HIGHWAY
CP	CULVERT PIPE
CY	CUBIC YARD
CWT	HUNDRED WEIGHT
DIA	DIAMETER
D	DEGREE OF CURVE
DHV	DESIGN HOURLY VOLUME
DWY	DRIVEWAY
EBS	EXC. BELOW SUBGRADE
ELEV, EL	ELEVATION
ELEC	ELECTRICAL
EXC	EXCAVATION
EXIST	EXISTING
E	EAST
FE	FIELD ENTRANCE

FF	FACE TO FACE
FL, F/L	FLOW LINE
G	GARAGE
GN	GRID NORTH
H	HOUSE
HOR	HORIZONTAL
HWY	HIGHWAY
HYD	HYDRANDT
I	INTERSECTION ANGLE
INTERS	INTERSECTION
INV	INVERT
IP	IRON PIN OR PIPE
LC	LONG CHORD OF CURVE
LEN	LENGTH
LF	LINEAR FOOT
LHF	LEFT HAND FORWARD
L	LENGTH OF CURVE
LHE	LIMITED HIGHWAY EASEMENT
LT	LEFT
LS	LUMP SUM
m	METER
mm	MILLIMETERS
MH	MANHOLE
N	NORTH
PAVT	PAVEMENT
PC	POINT OF CURVATURE
PE	PRIVATE ENTRANCE
PI	POINT OF INTERSECTION
PL	PROPERTY LINE
PP	POWER POLE
PROP	PROPOSED
PT	POINT OF TANGENCY
R	RANGE , RADIUS
RCCP	REINFORCED CONCRETE CULVERT PIPE
RD	ROAD
REBAR	REINFORCEMENT BAR

REM	REMOVE
REQD	REQUIRED
RDWY	ROADWAY
RHF	RIGHT HAND FORWARD
RL, R/L	REFERENCE LINE
RR	RAILROAD
RT	RIGHT
R/W	RIGHT-OF-WAY
S	SOUTH
SAN S	SANITARY SEWER
SDD	STANDARD DETAIL DRAWING
SE	SOUTHEAST , SUPERELEVATION
SHLDR	SHOULDER
SPECS	SPECIFICATIONS
SQ	SQUARE
SS	STORM SEWER
STH	STATE TRUNK HIGHWAY
ST	STREET
STA	STATION
SW	SIDEWALK
TAN	TANGENT
T	TANGENT LENGTH OF CURVE, TRUCKS
TC	TOP OF CURB
TL , T/L	TRANSIT LINE
TEL	TELEPHONE
TEMP	TEMPORARY
TLE	TEMPORARY LIMITED EASEMENT
TYP	TYPICAL
UNCL	UNCLASSIFIED
USH	UNITED STATES HIGHWAY
UG	UNDERGROUND
V	DESIGN SPEED
VAR	VARIABLE
VERT	VERTICAL
W	WEST
YD	YARD

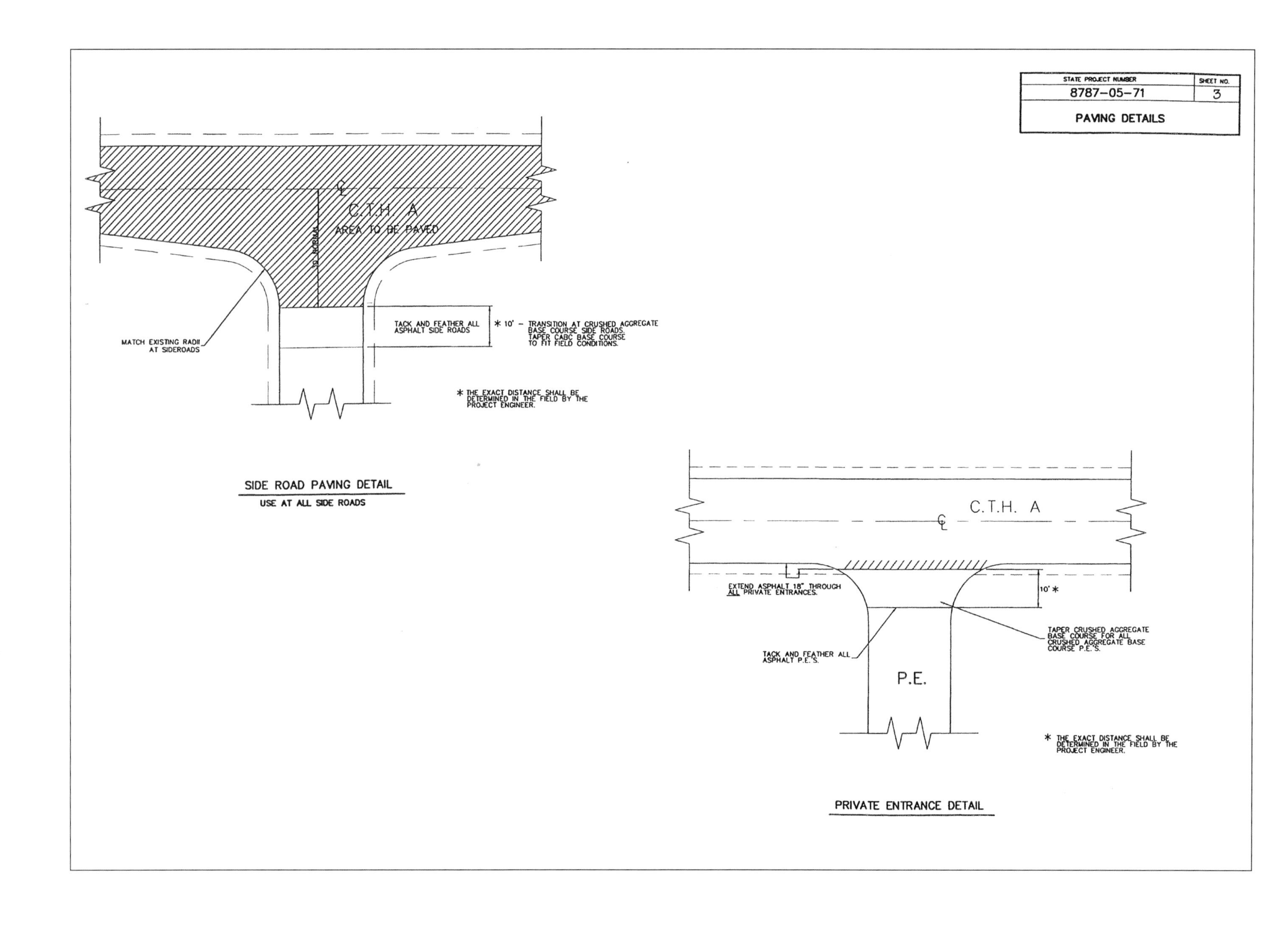
STATE PROJECT NUMBER
SHEET NO.
8787-05-71
3
PAVING DETAILS
C.T.H. A
AREA TO BE PAVED
30' NORMAL
MATCH EXISTING RADII AT SIDEROADS
TACK AND FEATHER ALL ASPHALT SIDE ROADS
* 10' - TRANSITION AT CRUSHED AGGREGATE BASE COURSE SIDE ROADS. TAPER CABC BASE COURSE TO FIT FIELD CONDITIONS.
* THE EXACT DISTANCE SHALL BE DETERMINED IN THE FIELD BY THE PROJECT ENGINEER.
SIDE ROAD PAVING DETAIL
USE AT ALL SIDE ROADS
C.T.H. A
EXTEND ASPHALT 18" THROUGH ALL PRIVATE ENTRANCES.
10' *
TAPER CRUSHED AGGREGATE BASE COURSE FOR ALL CRUSHED AGGREGATE BASE COURSE P.E.'S.
TACK AND FEATHER ALL ASPHALT P.E.'S.
P.E.
* THE EXACT DISTANCE SHALL BE DETERMINED IN THE FIELD BY THE PROJECT ENGINEER.
PRIVATE ENTRANCE DETAIL

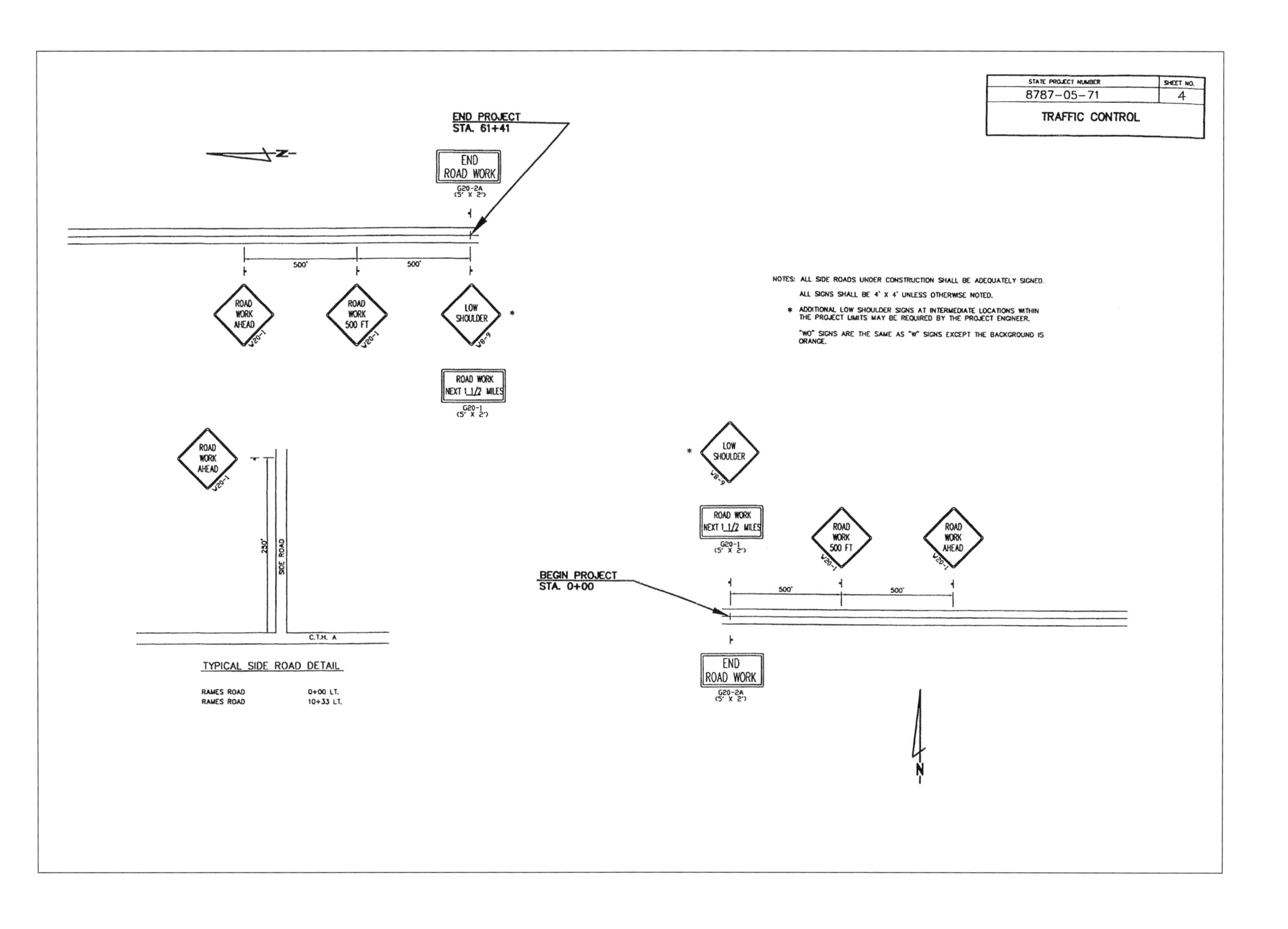
STATE PROJECT NUMBER
SHEET NO.
8787-05-71
4
TRAFFIC CONTROL
END PROJECT
STA. 61+41
END
ROAD WORK
G20-2A
(5' X 2')
500'
500'
ROAD
WORK
AHEAD
W20-1
ROAD
WORK
500 FT
W20-1
LOW
SHOULDER
W8-9
*
ROAD WORK
NEXT 1 1/2 MILES
G20-1
(5' X 2')
NOTES: ALL SIDE ROADS UNDER CONSTRUCTION SHALL BE ADEQUATELY SIGNED.
ALL SIGNS SHALL BE 4' X 4' UNLESS OTHERWISE NOTED.
* ADDITIONAL LOW SHOULDER SIGNS AT INTERMEDIATE LOCATIONS WITHIN THE PROJECT LIMITS MAY BE REQUIRED BY THE PROJECT ENGINEER.
"WO" SIGNS ARE THE SAME AS "W" SIGNS EXCEPT THE BACKGROUND IS ORANGE.
250'
SIDE ROAD
C.T.H. A
TYPICAL SIDE ROAD DETAIL
RAMES ROAD 0+00 LT.
RAMES ROAD 10+33 LT.
BEGIN PROJECT
STA. 0+00
N

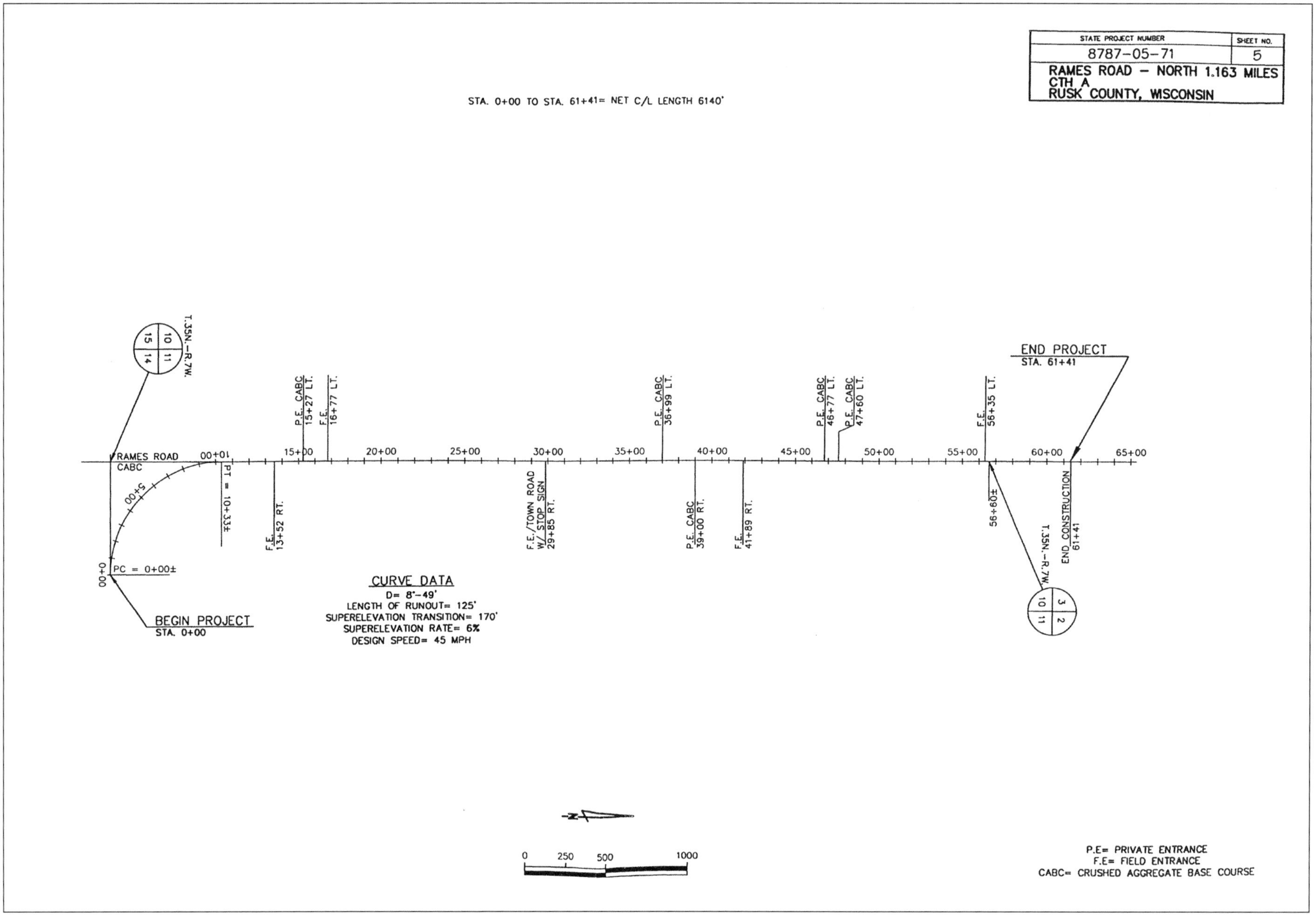

STATE PROJECT NUMBER	SHEET NO.
8787-05-71	5

RAMES ROAD – NORTH 1.163 MILES
CTH A
RUSK COUNTY, WISCONSIN

STA. 0+00 TO STA. 61+41= NET C/L LENGTH 6140'

T.35N.–R.7W.
15 10 14 11

RAMES ROAD
CABC

BEGIN PROJECT
STA. 0+00

PC = 0+00±
PT = 10+33±

P.E. CABC 15+27 LT.
F.E. 16+77 LT.
F.E. 13+52 RT.
F.E./TOWN ROAD W/ STOP SIGN 29+85 RT.
P.E. CABC 36+99 LT.
P.E. CABC 39+00 RT.
F.E. 41+89 RT.
P.E. CABC 46+77 LT.
P.E. CABC 47+60 LT.
F.E. 56+35 LT.
56+60±

END PROJECT
STA. 61+41

END CONSTRUCTION 61+41

T.35N.–R.7W.
3 10 2 11

CURVE DATA
D= 8°–49'
LENGTH OF RUNOUT= 125'
SUPERELEVATION TRANSITION= 170'
SUPERELEVATION RATE= 6%
DESIGN SPEED= 45 MPH

0 250 500 1000

P.E= PRIVATE ENTRANCE
F.E= FIELD ENTRANCE
CABC= CRUSHED AGGREGATE BASE COURSE

INDEX OF SHEETS

Sheet No. 1 – Title
Sheet No. 2 – Typical Sections - Detailed Summary of Misc. Qty.
Sheet No. 3 – Paving Details
Sheet No. 4 – Road Stationing
Sheet No. 5 – Road Stationing

STATE OF WISCONSIN
DEPARTMENT OF TRANSPORTATION

PLAN OF PROPOSED IMPROVEMENT

C.T.H. "I" – C.T.H. "P"

C.T.H. G
RUSK COUNTY

STATE PROJECT	FEDERAL PROJECT	
	PROJECT	CONTRACT
8794–08–71		

PROJECT LOCATION

STATE PROJECT NUMBER
8794–08–71

END PROJECT
STA. 206+87
X = 1,723,100 ± 200
Y = 581,400 ± 200

NET EXCEPTION TO C/L LENGTH
STA. 151+02 TO STA. 151+73
STR. P 54–47

BEGIN PROJECT
STA. 1+80
X = 1,732,450 ± 200
Y = 568,150 ± 200

R.6–W

T.34–N

N

Ladysmith
Dairyland Reservoir
Tony
GRANT
Conrath

NET EXCEPTION TO C/L LENGTH
STA. 6+25 TO STA. 7+23
STR. B 54–86

LAYOUT
SCALE 0 3 6 MILES

TOTAL NET LENGTH OF CENTERLINE =3.852 MI. TOTAL

DESIGN DESIGNATION

A.D.T.	2001	=	1100
A.D.T.	2021	=	1500
D.H.V.		=	200
D.		=	50%
T.	(A.D.T.)	=	10%
V.		=	55 MPH
ESAL'S		=	350,000

CONVENTIONAL SIGNS

COUNTY LINE
CORPORATE LIMITS
PROPERTY LINE
LOT LINE
TEMP. LIMITED EASEMENT
EXISTING RIGHT OF WAY
NEW RIGHT OF WAY
REFERENCE LINE
SLOPE INTERCEPT
ORIGINAL GROUND
MARSH OR ROCK PROFILE
CULVERT IN PLACE
CULVERT REQUIRED
CULVERT IN PLACE (PROFILE)
CULVERT REQUIRED (PROFILE)

COMBUSTIBLE FLUIDS (UNDER PRESSURE) — CAUTION
UNDERGROUND UTILITIES
GAS — G
ELECTRIC — E
TELEPHONE — T
SERVICE PEDESTAL — PED.
CABLE MARKER
POWER POLE
TELEPHONE POLE
LIGHT POLE
RAILROADS
MARSH
WOODED AREA

COORDINATE NOTE:
ALL COORDINATES ARE REFERENCED TO THE WISCONSIN COORDINATE SYSTEM CENTRAL ZONE & SCALED FROM U.S.G.S. TOPOGRAPHICAL MAP, LADYSMITH QUADRANGLE FOR IDENTIFICATION ONLY.

ACCEPTED FOR
COUNTY of RUSK

3-23-01 (Date) — Richard A. Brand (Highway Commissioner)

ORIGINAL PLAN PREPARED BY:
MORGAN & PARMLEY, LTD.
CONSULTING ENGINEERS, LADYSMITH, WISCONSIN

WISCONSIN PROFESSIONAL ENGINEER
LARRY F. GOTHAM
E-19463
NEW AUBURN, WI

3/23/1 (Date) — Larry Gotham (Signature)

STATE OF WISCONSIN
DEPARTMENT OF TRANSPORTATION

PREPARED BY

Surveyor	MORGAN & PARMLEY, LTD.
Designer	MORGAN & PARMLEY, LTD.
District Examiner	CHRISTINE KOSKI
District Supervisor	RICK WASHKUHN
Project Dev. Engineer	
C.O. Examiner	

Courtesy: Morgan & Parmley, Ltd.

STATE PROJECT NUMBER	SHEET NO.
8794-08-71	2

TYPICAL SECTIONS–DETAILED SUMMARY OF MISCELLANEOUS QUANTITIES

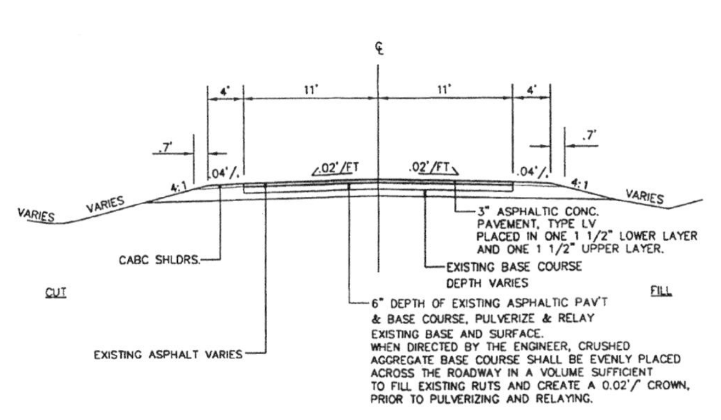

TYPICAL SECTION

STA. 1+80 TO STA. 206+87
(NOT TO SCALE)

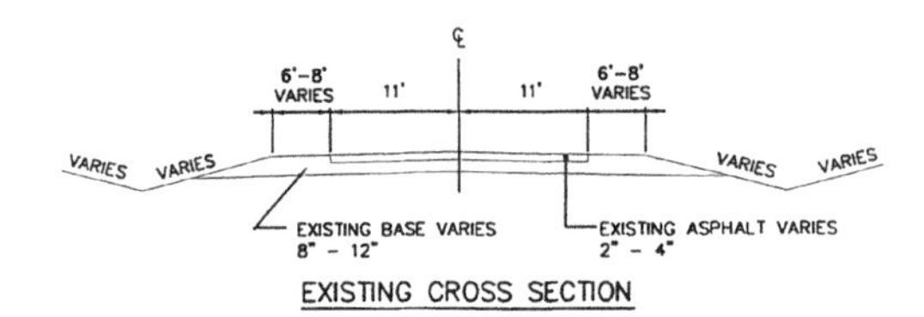

EXISTING CROSS SECTION

CALL DIGGERS' HOTLINE
1-800-242-8511
TOLL FREE
2040 W. WISCONSIN AVE.
SUITE 10
MILWAUKEE, WI 53233

RUSK COUNTY SURVEYOR
DAVE KAISER, RLS
(715) 532-2165

DETAILED SUMMARY OF MISCELLANEOUS QUANTITIES

ASPHALTIC MATERIAL FOR PLANT MIXES

STATION	TO	STATION	LOCATION	TON
1+80		206+87	MAINLINE	516
P.E.'S & SIDEROADS		——	UNDISTRIBUTED	12
TOTAL				528

ASPHALTIC MATERIAL FOR TACK COAT

STATION	TO	STATION	APPLICATIONS	RATE	GAL
1+80		206+87	1	0.025 GAL/SQ. YD.	1,300
P.E.'S & SIDEROADS		——	1	0.025 GAL/SQ. YD.	20
TOTAL					1,320

ASPHALTIC CONCRETE PAVEMENT, TYPE LV

STATION	TO	STATION	LOCATION	TON
1+80		206+87	MAINLINE	8,600
P.E.'S & SIDEROADS		——	UNDISTRIBUTED	200
TOTAL				8,800

CRUSHED AGGREGATE BASE COURSE

STATION	TO	STATION	LOCATION	TON
1+80		206+87	MAINLINE	3,500
1+80		206+87	RT. & LT. SHLDR.	3,000
P.E.'S & SIDEROADS			UNDISTRIBUTED	350
TOTAL				6,850

PULVERIZE AND RELAY EXISTING BASE & SURFACE

STATION	TO	STATION	LOCATION	S.Y.
1+80		206+87	MAINLINE	50,950
		——	UNDISTRIBUTED	50
TOTAL				51,000

PAVEMENT MARKING, 4-INCH, PAINT

STATION	TO	STATION	LOCATION	L.F.
1+80		206+87	RT. & LT. EDGE LINES	41,000
TOTAL				41,000

PAVEMENT MARKING, SAME DAY, 4-INCH, PAINT

STATION	TO	STATION	LOCATION	L.F.
1+80		206+87	CENTERLINE	30,000
TOTAL				30,000

SAWING EXISTING PAVEMENT

STATION	LOCATION	L.F.
1+80	RT. & LT.	22
206+87	RT. & LT.	22
	UNDISTRIBUTED	56
TOTAL		100

TEMPORARY PAVEMENT MARKING, 4-INCH

STATION	TO	STATION	LOCATION	L.F.
1+80		206+87	CENTERLINE	20,000
TOTAL				20,000

GENERAL NOTES:

THROUGH ALL PRIVATE ENTRANCES AND ON THE INSIDE OF ALL HORIZONTAL CURVES, THE EDGE OF ASPHALT SHALL BE EXTENDED 18".

THERE ARE UTILITY FACILITIES IN THE PROJECT AREA WHICH ARE NOT SHOWN ON THE PLANS. THERE ARE NO KNOWN CONFLICTS.

TACK COAT SHALL BE APPLIED BETWEEN THE ASPHALT LAYERS AT A RATE OF 0.025 GAL/SQ. YD.

OTHER CONTACTS

DNR LIASON

DAN MICHELS
D.N.R. NORTHWESTERN DISTRICT H.Q.
P.O. BOX 309
SPOONER, WI. 54801 (715) 635-4228

DESIGN CONSULTANT

LARRY GOTHAM
MORGAN & PARMLEY LTD.
115 W. 2ND STREET SOUTH
LADYSMITH WI. 54848 (715) 532-3721

UTILITIES

EXCEL

ATTN: JOE PERKINS
310 HICKORY HILL LANE
PHILLIPS, WI. 54555
(715) 836-1198

JUMP RIVER ELECTRIC CO-OP

ATTN: HANK LEW
1102 W. 9TH STREET
LADYSMITH, WI. 54848
(715) 532-5524

CENTURYTEL

ATTN: JAMES ARQUETTE
425 ELLINGTON ROAD
P.O. BOX 78
HAWKINS, WI. 54530
(715) 752-5637

STANDARD DETAIL DRAWINGS

15C4-1	TRAFFIC CONTROL ADVANCE WARNING
15C6-4	SIGNING AND MARKING AT 2-LANE BRIDGE
15C8-9a	PAVEMENT MARKINGS (MAINLINE)
15C8-9b	PAVEMENT MARKINGS (INTERSECTIONS)
15C12-2	TRAFFIC CONTROL FOR LANE CLOSURES (SUITABLE FOR MOVING OPERATIONS)

LIST OF STANDARD ABBREVIATIONS

ABUT	ABUTMENT
AC	ACRES
AGG	AGGREGATE
AH	AHEAD
ADT	AVERAGE DAILY TRAFFIC
AVE	AVENUE
AVG	AVERAGE
ASPH	ASPHALTIC
BK	BACK
BM	BENCHMARK
△	CENTRAL ANGLE
℄ , C/L	CENTERLINE
C & G	CURB AND GUTTER
CABC	CRUSHED AGGREGATE BASE COURSE
CONC	CONCRETE
CONST	CONSTRUCTION
COR	CORNER
CORR	CORRUGATED
CSCP	CORRUGATED STEEL CULVERT PIPE
CSPA	CORRUGATED STEEL PIPE ARCH
CTH	COUNTY TRUNK HIGHWAY
CP	CULVERT PIPE
CY	CUBIC YARD
CWT	HUNDRED WEIGHT
DIA	DIAMETER
D	DEGREE OF CURVE
DHV	DESIGN HOURLY VOLUME
DWY	DRIVEWAY
EBS	EXC. BELOW SUBGRADE
ELEV, EL	ELEVATION
ELEC	ELECTRICAL
EXC	EXCAVATION
EXIST	EXISTING
E	EAST
FE	FIELD ENTRANCE
FF	FACE TO FACE
FL, F/L	FLOW LINE
G	GARAGE
GN	GRID NORTH
H	HOUSE
HOR	HORIZONTAL
HWY	HIGHWAY
HYD	HYDRANDT
I	INTERSECTION ANGLE
INTERS	INTERSECTION
INV	INVERT
IP	IRON PIN OR PIPE
LC	LONG CHORD OF CURVE
LEN	LENGTH
LF	LINEAR FOOT
LHF	LEFT HAND FORWARD
L	LENGTH OF CURVE
LHE	LIMITED HIGHWAY EASEMENT
LT	LEFT
LS	LUMP SUM
m	METER
mm	MILLIMETERS
MH	MANHOLE
N	NORTH
PAVT	PAVEMENT
PC	POINT OF CURVATURE
PE	PRIVATE ENTRANCE
PI	POINT OF INTERSECTION
PL	PROPERTY LINE
PP	POWER POLE
PROP	PROPOSED
PT	POINT OF TANGENCY
R	RANGE , RADIUS
RCCP	REINFORCED CONCRETE CULVERT PIPE
RD	ROAD
REBAR	REINFORCEMENT BAR
REM	REMOVE
REQD	REQUIRED
RDWY	ROADWAY
RHF	RIGHT HAND FORWARD
RL, R/L	REFERENCE LINE
RR	RAILROAD
RT	RIGHT
R/W	RIGHT-OF-WAY
S	SOUTH
SAN S	SANITARY SEWER
SDD	STANDARD DETAIL DRAWING
SE	SOUTHEAST , SUPERELEVATION
SHLDR	SHOULDER
SPECS	SPECIFICATIONS
SQ	SQUARE
SS	STORM SEWER
STH	STATE TRUNK HIGHWAY
ST	STREET
STA	STATION
SW	SIDEWALK
TAN	TANGENT
T	TANGENT LENGTH OF CURVE, TRUCKS
TC	TOP OF CURB
TL , T/L	TRANSIT LINE
TEL	TELEPHONE
TEMP	TEMPORARY
TLE	TEMPORARY LIMITED EASEMENT
TYP	TYPICAL
UNCL	UNCLASSIFIED
USH	UNITED STATES HIGHWAY
UG	UNDERGROUND
V	DESIGN SPEED
VAR	VARIABLE
VERT	VERTICAL
W	WEST
YD	YARD

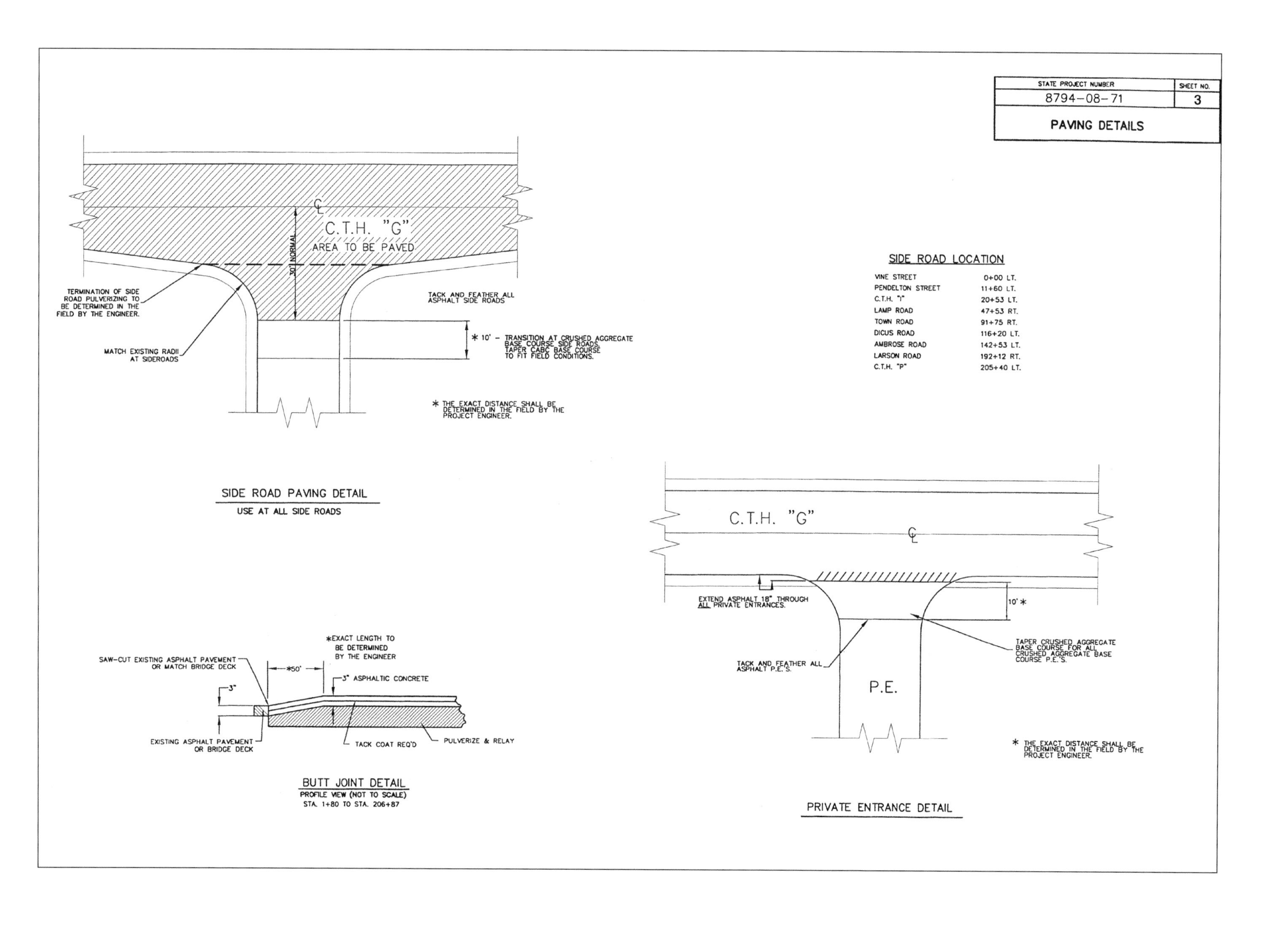
STATE PROJECT NUMBER
8794-08-71
SHEET NO.
3
PAVING DETAILS
C.T.H. "G"
AREA TO BE PAVED
30' NORMAL
TERMINATION OF SIDE ROAD PULVERIZING TO BE DETERMINED IN THE FIELD BY THE ENGINEER.
MATCH EXISTING RADII AT SIDEROADS
TACK AND FEATHER ALL ASPHALT SIDE ROADS
* 10' - TRANSITION AT CRUSHED AGGREGATE BASE COURSE SIDE ROADS. TAPER CABC BASE COURSE TO FIT FIELD CONDITIONS.
* THE EXACT DISTANCE SHALL BE DETERMINED IN THE FIELD BY THE PROJECT ENGINEER.
SIDE ROAD PAVING DETAIL
USE AT ALL SIDE ROADS
SIDE ROAD LOCATION
VINE STREET 0+00 LT.
PENDELTON STREET 11+60 LT.
C.T.H. "I" 20+53 LT.
LAMP ROAD 47+53 RT.
TOWN ROAD 91+75 RT.
DICUS ROAD 116+20 LT.
AMBROSE ROAD 142+53 LT.
LARSON ROAD 192+12 RT.
C.T.H. "P" 205+40 LT.
C.T.H. "G"
EXTEND ASPHALT 18" THROUGH ALL PRIVATE ENTRANCES.
10' *
TAPER CRUSHED AGGREGATE BASE COURSE FOR ALL CRUSHED AGGREGATE BASE COURSE P.E.'S.
TACK AND FEATHER ALL ASPHALT P.E.'S.
P.E.
* THE EXACT DISTANCE SHALL BE DETERMINED IN THE FIELD BY THE PROJECT ENGINEER.
PRIVATE ENTRANCE DETAIL
*EXACT LENGTH TO BE DETERMINED BY THE ENGINEER
SAW-CUT EXISTING ASPHALT PAVEMENT OR MATCH BRIDGE DECK
*50'
3" ASPHALTIC CONCRETE
3"
EXISTING ASPHALT PAVEMENT OR BRIDGE DECK
TACK COAT REQ'D
PULVERIZE & RELAY
BUTT JOINT DETAIL
PROFILE VIEW (NOT TO SCALE)
STA. 1+80 TO STA. 206+87

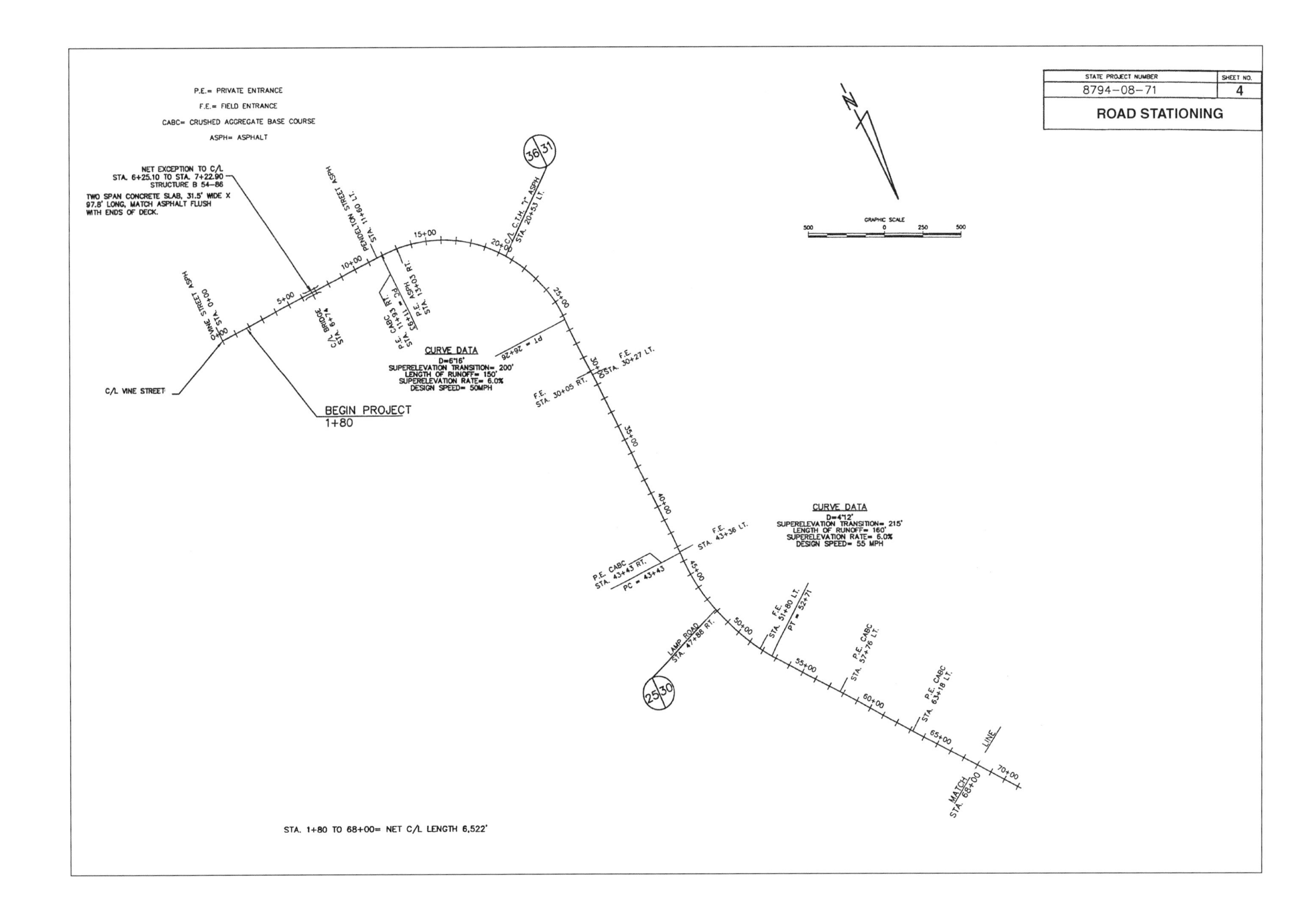
STATE PROJECT NUMBER
SHEET NO.
8794-08-71
4
ROAD STATIONING
P.E.= PRIVATE ENTRANCE
F.E.= FIELD ENTRANCE
CABC= CRUSHED AGGREGATE BASE COURSE
ASPH= ASPHALT
NET EXCEPTION TO C/L
STA. 6+25.10 TO STA. 7+22.90
STRUCTURE B 54-86
TWO SPAN CONCRETE SLAB, 31.5' WIDE X 97.8' LONG, MATCH ASPHALT FLUSH WITH ENDS OF DECK.
GRAPHIC SCALE
C/L VINE STREET
BEGIN PROJECT
1+80
CURVE DATA
D=6°16'
SUPERELEVATION TRANSITION= 200'
LENGTH OF RUNOFF= 150'
SUPERELEVATION RATE= 6.0%
DESIGN SPEED= 50MPH
CURVE DATA
D=4°12'
SUPERELEVATION TRANSITION= 215'
LENGTH OF RUNOFF= 160'
SUPERELEVATION RATE= 6.0%
DESIGN SPEED= 55 MPH
PT = 26+26
PC = 43+43
PT = 52+71
LAMP ROAD STA. 47+88 RT.
MATCH LINE STA. 68+00
STA. 1+80 TO 68+00= NET C/L LENGTH 6,522'

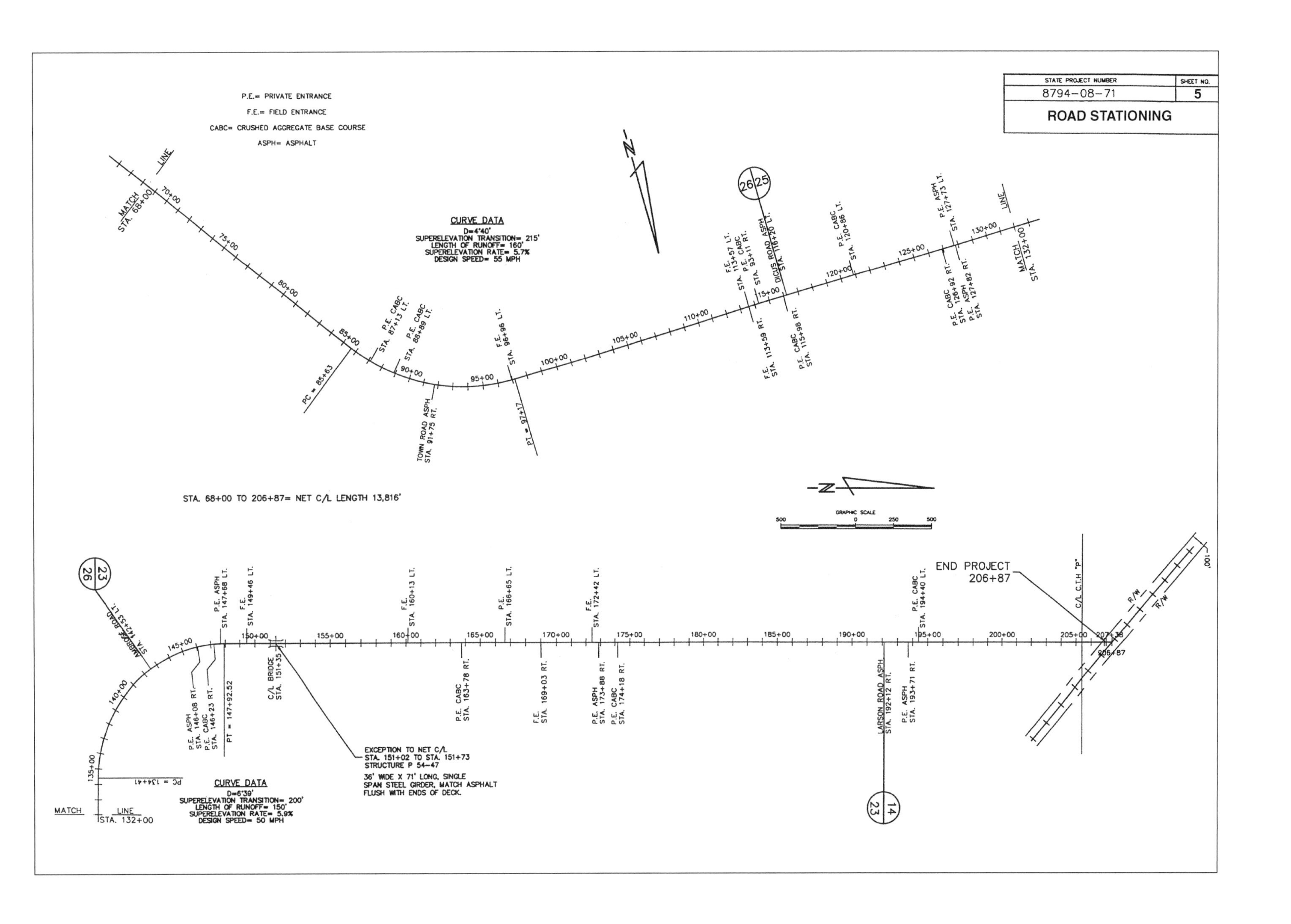
STATE PROJECT NUMBER
8794–08–71
SHEET NO.
5
ROAD STATIONING
P.E.= PRIVATE ENTRANCE
F.E.= FIELD ENTRANCE
CABC= CRUSHED AGGREGATE BASE COURSE
ASPH= ASPHALT
CURVE DATA
D=4°40'
SUPERELEVATION TRANSITION= 215'
LENGTH OF RUNOFF= 160'
SUPERELEVATION RATE= 5.7%
DESIGN SPEED= 55 MPH
MATCH LINE STA. 68+00
PC = 85+63
PT = 97+17
TOWN ROAD ASPH STA. 91+75 RT.
DICUS ROAD ASPH STA. 116+20 LT.
MATCH LINE STA. 132+00
STA. 68+00 TO 206+87= NET C/L LENGTH 13,816'
GRAPHIC SCALE
END PROJECT 206+87
CURVE DATA
D=6°39'
SUPERELEVATION TRANSITION= 200'
LENGTH OF RUNOFF= 150'
SUPERELEVATION RATE= 5.9%
DESIGN SPEED= 50 MPH
MATCH LINE STA. 132+00
PC = 134+41
PT = 147+92.52
C/L BRIDGE STA. 151+35
EXCEPTION TO NET C/L
STA. 151+02 TO STA. 151+73
STRUCTURE P 54–47
36' WIDE X 71' LONG, SINGLE SPAN STEEL GIRDER, MATCH ASPHALT FLUSH WITH ENDS OF DECK.
LARSON ROAD ASPH STA. 192+12 RT.
C/L C.T.H "P"
R/W

Design

BRIDGES

Section 15

Most bridge designs must meet state and federal standards. Prior to construction, the plans must be reviewed by the regulating Department of Transportation. The first set of plans reflect the current format, while the second plan represents the typical layout used 25 years ago. However, the design and details remain relatively similar. The third set of plans detail a timber abutment design. The last set of bridge plans are for a snowmobile bridge spanning a navigable waterway. Local rules usually apply on this type of structure, not highway standards.

TABLE OF CONTENTS

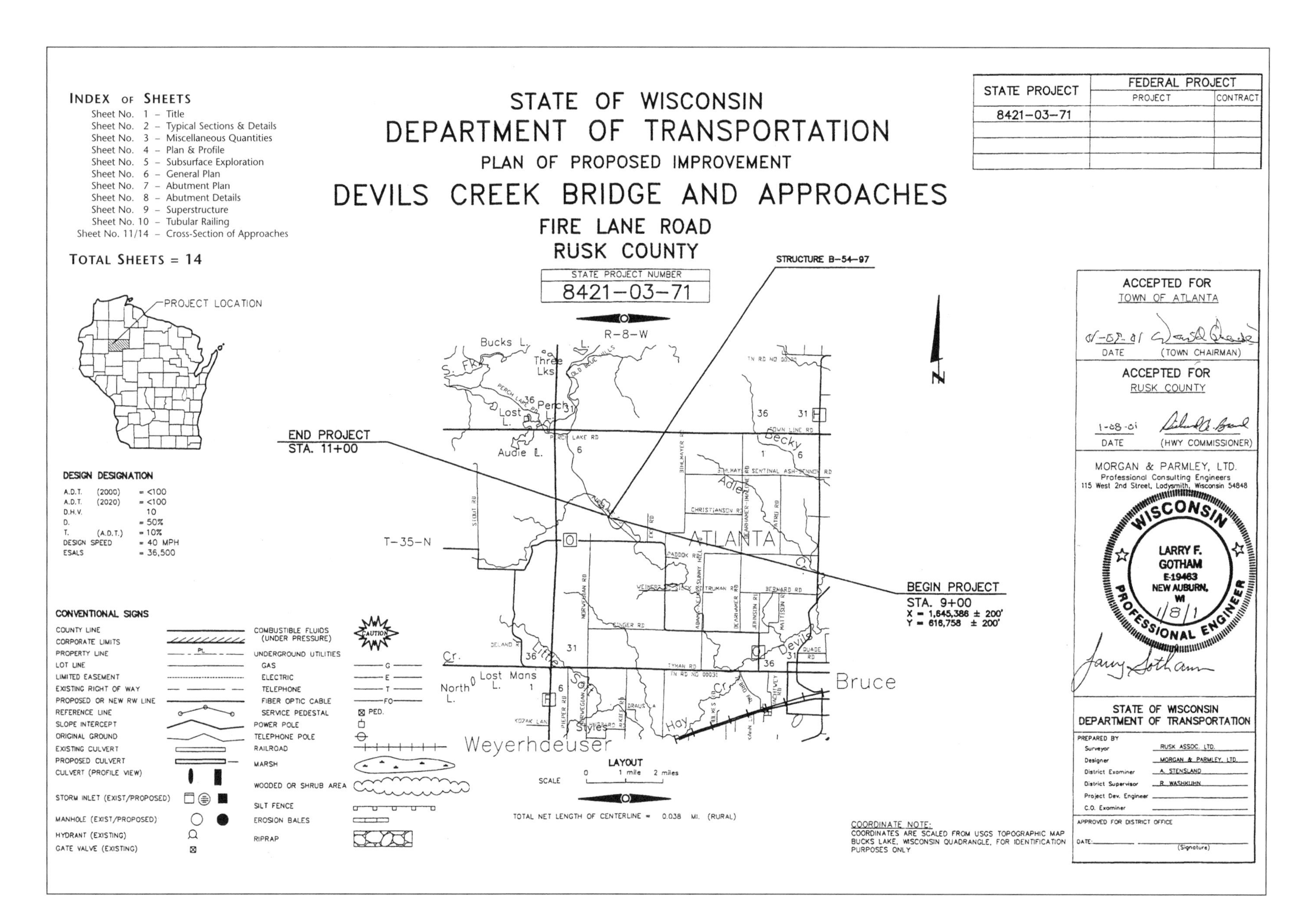

Courtesy: Morgan & Parmley, Ltd.

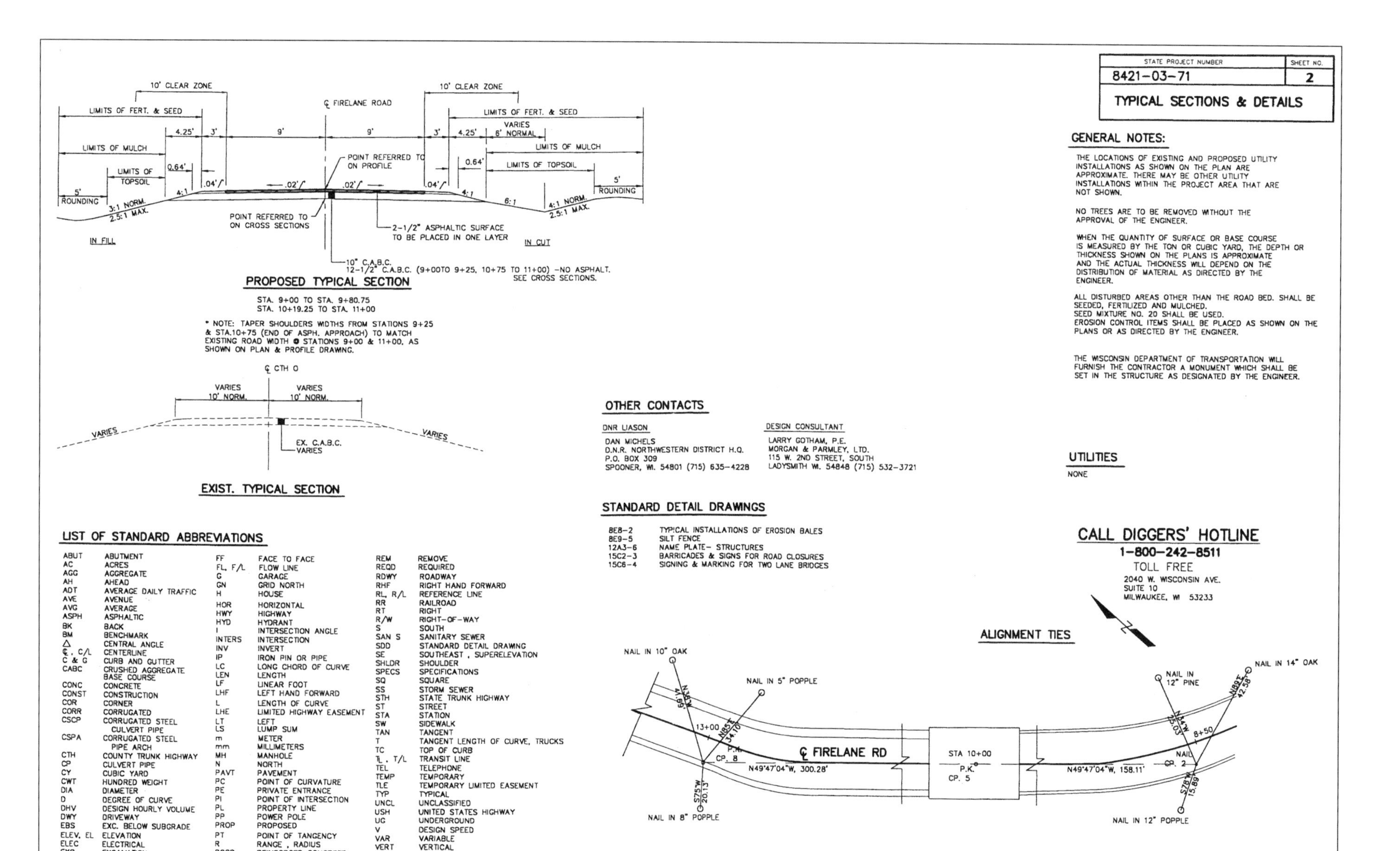
STATE PROJECT NUMBER
8421-03-71
SHEET NO.
2
TYPICAL SECTIONS & DETAILS
10' CLEAR ZONE
LIMITS OF FERT. & SEED
℄ FIRELANE ROAD
LIMITS OF MULCH
LIMITS OF TOPSOIL
4.25'
3'
9'
VARIES 6' NORMAL
0.64'
POINT REFERRED TO ON PROFILE
5' ROUNDING
3:1 NORM. 2.5:1 MAX.
4:1
.04'/'
.02'/'
6:1
4:1 NORM. 2.5:1 MAX.
POINT REFERRED TO ON CROSS SECTIONS
2-1/2" ASPHALTIC SURFACE TO BE PLACED IN ONE LAYER
IN FILL
IN CUT
10" C.A.B.C.
12-1/2" C.A.B.C. (9+00TO 9+25, 10+75 TO 11+00) -NO ASPHALT. SEE CROSS SECTIONS.
PROPOSED TYPICAL SECTION
STA. 9+00 TO STA. 9+80.75
STA. 10+19.25 TO STA. 11+00
* NOTE: TAPER SHOULDERS WIDTHS FROM STATIONS 9+25 & STA.10+75 (END OF ASPH. APPROACH) TO MATCH EXISTING ROAD WIDTH @ STATIONS 9+00 & 11+00, AS SHOWN ON PLAN & PROFILE DRAWING.
℄ CTH O
VARIES 10' NORM.
VARIES
EX. C.A.B.C. VARIES
EXIST. TYPICAL SECTION
GENERAL NOTES:
THE LOCATIONS OF EXISTING AND PROPOSED UTILITY INSTALLATIONS AS SHOWN ON THE PLAN ARE APPROXIMATE. THERE MAY BE OTHER UTILITY INSTALLATIONS WITHIN THE PROJECT AREA THAT ARE NOT SHOWN.
NO TREES ARE TO BE REMOVED WITHOUT THE APPROVAL OF THE ENGINEER.
WHEN THE QUANTITY OF SURFACE OR BASE COURSE IS MEASURED BY THE TON OR CUBIC YARD, THE DEPTH OR THICKNESS SHOWN ON THE PLANS IS APPROXIMATE AND THE ACTUAL THICKNESS WILL DEPEND ON THE DISTRIBUTION OF MATERIAL AS DIRECTED BY THE ENGINEER.
ALL DISTURBED AREAS OTHER THAN THE ROAD BED. SHALL BE SEEDED, FERTILIZED AND MULCHED.
SEED MIXTURE NO. 20 SHALL BE USED.
EROSION CONTROL ITEMS SHALL BE PLACED AS SHOWN ON THE PLANS OR AS DIRECTED BY THE ENGINEER.
THE WISCONSIN DEPARTMENT OF TRANSPORTATION WILL FURNISH THE CONTRACTOR A MONUMENT WHICH SHALL BE SET IN THE STRUCTURE AS DESIGNATED BY THE ENGINEER.
OTHER CONTACTS
DNR LIASON
DAN MICHELS
D.N.R. NORTHWESTERN DISTRICT H.Q.
P.O. BOX 309
SPOONER, WI. 54801 (715) 635-4228
DESIGN CONSULTANT
LARRY GOTHAM, P.E.
MORGAN & PARMLEY, LTD.
115 W. 2ND STREET, SOUTH
LADYSMITH WI. 54848 (715) 532-3721
UTILITIES
NONE
STANDARD DETAIL DRAWINGS
8E8-2 TYPICAL INSTALLATIONS OF EROSION BALES
8E9-5 SILT FENCE
12A3-6 NAME PLATE- STRUCTURES
15C2-3 BARRICADES & SIGNS FOR ROAD CLOSURES
15C6-4 SIGNING & MARKING FOR TWO LANE BRIDGES
CALL DIGGERS' HOTLINE
1-800-242-8511
TOLL FREE
2040 W. WISCONSIN AVE.
SUITE 10
MILWAUKEE, WI 53233
LIST OF STANDARD ABBREVIATIONS
ABUT ABUTMENT
AC ACRES
AGG AGGREGATE
AH AHEAD
ADT AVERAGE DAILY TRAFFIC
AVE AVENUE
AVG AVERAGE
ASPH ASPHALTIC
BK BACK
BM BENCHMARK
Δ CENTRAL ANGLE
℄, C/L CENTERLINE
C & G CURB AND GUTTER
CABC CRUSHED AGGREGATE BASE COURSE
CONC CONCRETE
CONST CONSTRUCTION
COR CORNER
CORR CORRUGATED
CSCP CORRUGATED STEEL CULVERT PIPE
CSPA CORRUGATED STEEL PIPE ARCH
CTH COUNTY TRUNK HIGHWAY
CP CULVERT PIPE
CY CUBIC YARD
CWT HUNDRED WEIGHT
DIA DIAMETER
D DEGREE OF CURVE
DHV DESIGN HOURLY VOLUME
DWY DRIVEWAY
EBS EXC. BELOW SUBGRADE
ELEV, EL ELEVATION
ELEC ELECTRICAL
EXC EXCAVATION
EXIST EXISTING
E EAST
FE FIELD ENTRANCE
FF FACE TO FACE
FL, F/L FLOW LINE
G GARAGE
GN GRID NORTH
H HOUSE
HOR HORIZONTAL
HWY HIGHWAY
HYD HYDRANT
I INTERSECTION ANGLE
INTERS INTERSECTION
INV INVERT
IP IRON PIN OR PIPE
LC LONG CHORD OF CURVE
LEN LENGTH
LF LINEAR FOOT
LHF LEFT HAND FORWARD
L LENGTH OF CURVE
LHE LIMITED HIGHWAY EASEMENT
LT LEFT
LS LUMP SUM
m METER
mm MILLIMETERS
MH MANHOLE
N NORTH
PAVT PAVEMENT
PC POINT OF CURVATURE
PE PRIVATE ENTRANCE
PI POINT OF INTERSECTION
PL PROPERTY LINE
PP POWER POLE
PROP PROPOSED
PT POINT OF TANGENCY
R RANGE , RADIUS
RCCP REINFORCED CONCRETE CULVERT PIPE
RD ROAD
REBAR REINFORCEMENT BAR
REM REMOVE
REQD REQUIRED
RDWY ROADWAY
RHF RIGHT HAND FORWARD
RL, R/L REFERENCE LINE
RR RAILROAD
RT RIGHT
R/W RIGHT-OF-WAY
S SOUTH
SAN S SANITARY SEWER
SDD STANDARD DETAIL DRAWING
SE SOUTHEAST , SUPERELEVATION
SHLDR SHOULDER
SPECS SPECIFICATIONS
SQ SQUARE
SS STORM SEWER
STH STATE TRUNK HIGHWAY
ST STREET
STA STATION
SW SIDEWALK
TAN TANGENT
T TANGENT LENGTH OF CURVE, TRUCKS
TC TOP OF CURB
TL, T/L TRANSIT LINE
TEL TELEPHONE
TEMP TEMPORARY
TLE TEMPORARY LIMITED EASEMENT
TYP TYPICAL
UNCL UNCLASSIFIED
USH UNITED STATES HIGHWAY
UG UNDERGROUND
V DESIGN SPEED
VAR VARIABLE
VERT VERTICAL
W WEST
YD YARD
ALIGNMENT TIES
NAIL IN 10" OAK
NAIL IN 5" POPPLE
NAIL IN 8" POPPLE
NAIL IN 12" PINE
NAIL IN 14" OAK
NAIL IN 12" POPPLE
13+00
CP. 8
℄ FIRELANE RD
N49°47'04"W, 300.28'
STA 10+00
P.K.
CP. 5
N49°47'04"W, 158.11'
8+50
CP. 2
S75°W 20.13'

STATE PROJECT NUMBER	SHEET NO.
8421-03-71	3
FIRELANE ROAD MISCELLANEOUS QUANTITIES	

CLEARING

STA.	TO	STA.	LOC.	STA.
9+00		9+75	LT. & RT.	.75
10+50		11+00	LT. & RT.	.5
			TOTAL =	1.25

GRUBBING

STA.	TO	STA.	LOC.	STA.
9+00		9+75	LT. & RT.	.75
10+50		11+00	LT. & RT.	.5
			TOTAL =	1.25

EARTHWORK SUMMARY

STA.	TO	STA.	COMMON EXCAV. C.Y.	FILL EXP. 25% C.Y.	SELECT BORROW C.Y.	WASTE C.Y.
9+00		9+61	30	248	218	0
10+39		11+00	42	102	60	0
		TOTALS	72	350	278	0

ASPHALTIC SURFACE

STA.	TO	STA.	TONS
9+25		9+80.75	16
10+19.25		10+75	16
		TOTALS =	32

CRUSHED AGGREGATE BASE COURSE

STA.	TO	STA.	CU. YDS.
9+00		9+80.75	100
10+19.25		11+00	100
PARKING LOT ENTRANCE			20
		TOTAL =	220

CRUSHED AGGREGATE BASE COURSE FOR SHOULDERS

STA.	TO	STA.	CU. YDS.
9+00		9+80.75	6
10+19.25		11+00	6
		TOTAL =	12

TOPSOIL

STA.	TO	STA.	LOC.	TOPSOIL S.Y.
9+00		9+80.75	LT.	156
9+00		9+80.75	RT.	138
10+19.25		11+00	LT.	91
10+19.25		11+00	RT.	142
			TOTAL =	527

MULCH

STA.	TO	STA.	LOC.	MULCH S.Y.
9+00		9+80.75	LT.	211
9+00		9+80.75	RT.	188
10+19.25		11+00	LT.	139
10+19.25		11+00	RT.	193
			TOTAL =	731

FERTILIZER

STA.	TO	STA.	LOC.	FERTILIZER TYPE B CWT
9+00		9+80.75	LT./RT.	20
10+19.25		11+00	LT./RT.	15
			TOTAL =	35

SEEDING, MIXTURE NO. 20

STA.	TO	STA.	LOC.	SEED #20 LB.
9+00		9+80.75	LT./RT.	9
10+19.25		11+00	LT./RT.	7
			TOTAL =	16

EROSION MAT CLASS II, TYPE B

STA.	TO	STA.	LOC.	DELIVERED S.Y.	INSTALLED S.Y.
9+60		9+80	LT.	26	26
9+60		9+80	RT.	29	29
10+19		10+39	LT.	19	19
10+19		10+39	RT.	17	17
UNDISTRIBUTED				9	9
			TOTALS	100	100

SILT FENCE

STA.	TO	STA.	LOC.	DELIVERED L.F.	INSTALLED L.F.	MAINTENANCE L.F.
9+00		9+80.75	LT./RT.	209	209	209
10+19.25		11+00	LT./RT.	183	183	183
UNDISTRIBUTED				58	58	58
			TOTALS	450	450	450

EROSION BALES

LOCATION	DELIVERED EACH	INSTALLED EACH
UNDISTRIBUTED	10	10

WOOD POSTS, 4 x 4 INCH x 10 FEET

LOCATION	EACH
END OF DECK (OBJECT MARKER)	4

SIGNS, TYPE II, REFLECTIVE (W5-52 LT./RT.)

LOCATION	S.F.
END OF DECK (OBJECT MARKER)	12

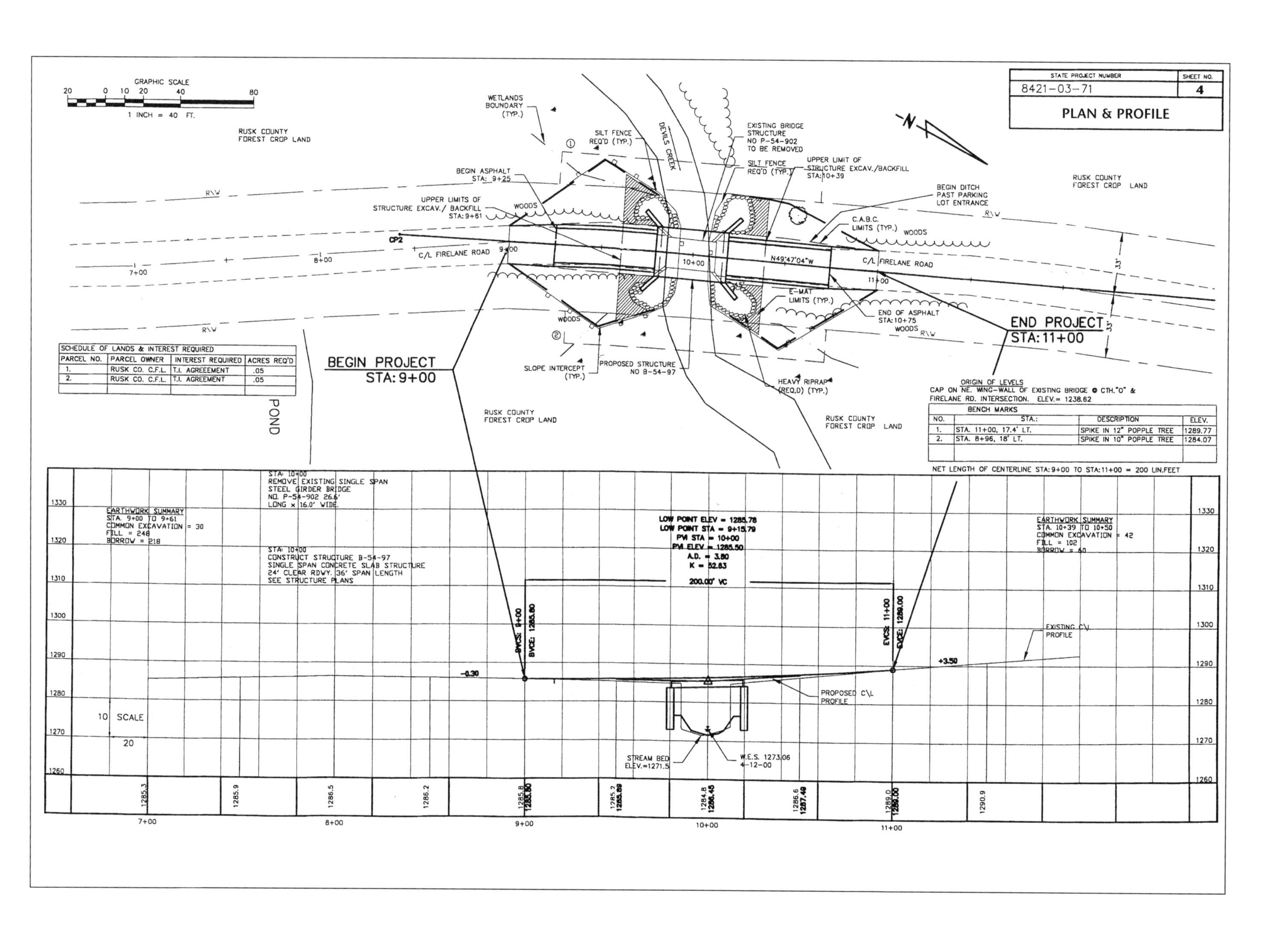
STATE PROJECT NUMBER
8421-03-71
SHEET NO.
4
PLAN & PROFILE
GRAPHIC SCALE
1 INCH = 40 FT.
RUSK COUNTY FOREST CROP LAND
WETLANDS BOUNDARY (TYP.)
SILT FENCE REQ'D (TYP.)
DEVILS CREEK
EXISTING BRIDGE STRUCTURE NO P-54-902 TO BE REMOVED
UPPER LIMIT OF STRUCTURE EXCAV./BACKFILL STA:10+39
BEGIN ASPHALT STA: 9+25
UPPER LIMITS OF STRUCTURE EXCAV./ BACKFILL STA: 9+61
BEGIN DITCH PAST PARKING LOT ENTRANCE
C.A.B.C. LIMITS (TYP.)
WOODS
C/L FIRELANE ROAD
N49°47'04"W
E-MAT LIMITS (TYP.)
END OF ASPHALT STA:10+75
END PROJECT STA: 11+00
BEGIN PROJECT STA: 9+00
SLOPE INTERCEPT (TYP.)
PROPOSED STRUCTURE NO B-54-97
HEAVY RIPRAP (REQ,D) (TYP.)
POND
SCHEDULE OF LANDS & INTEREST REQUIRED
PARCEL NO. | PARCEL OWNER | INTEREST REQUIRED | ACRES REQ'D
1. | RUSK CO. C.F.L. | T.I. AGREEMENT | .05
2. | RUSK CO. C.F.L. | T.I. AGREEMENT | .05
ORIGIN OF LEVELS
CAP ON NE. WING-WALL OF EXISTING BRIDGE @ CTH."O" & FIRELANE RD. INTERSECTION. ELEV.= 1238.62
BENCH MARKS
NO. | STA.: | DESCRIPTION | ELEV.
1. | STA. 11+00, 17.4' LT. | SPIKE IN 12" POPPLE TREE | 1289.77
2. | STA. 8+96, 18' LT. | SPIKE IN 10" POPPLE TREE | 1284.07
NET LENGTH OF CENTERLINE STA:9+00 TO STA:11+00 = 200 LIN.FEET
STA: 10+00 REMOVE EXISTING SINGLE SPAN STEEL GIRDER BRIDGE NO. P-54-902 26.6' LONG x 16.0' WIDE
EARTHWORK SUMMARY STA. 9+00 TO 9+61 COMMON EXCAVATION = 30 FILL = 248 BORROW = 218
STA: 10+00 CONSTRUCT STRUCTURE B-54-97 SINGLE SPAN CONCRETE SLAB STRUCTURE 24' CLEAR RDWY. 36' SPAN LENGTH SEE STRUCTURE PLANS
LOW POINT ELEV = 1285.78
LOW POINT STA = 9+15.79
PVI STA = 10+00
PVI ELEV = 1285.50
A.D. = 3.80
K = 52.63
200.00' VC
EARTHWORK SUMMARY STA. 10+39 TO 10+50 COMMON EXCAVATION = 42 FILL = 102 BORROW = 60
BVCS: 9+00
BVCE: 1285.80
EVCS: 11+00
EVCE: 1289.00
-0.30
+3.50
EXISTING C\L PROFILE
PROPOSED C\L PROFILE
10 SCALE 20
STREAM BED ELEV.=1271.5
W.E.S. 1273.06 4-12-00
1285.3
1285.9
1286.5
1286.2
1285.8 1285.80
1285.2 1285.69
1284.8 1285.45
1286.6 1287.49
1289.0 1289.00
1290.9
7+00
8+00
9+00
10+00
11+00

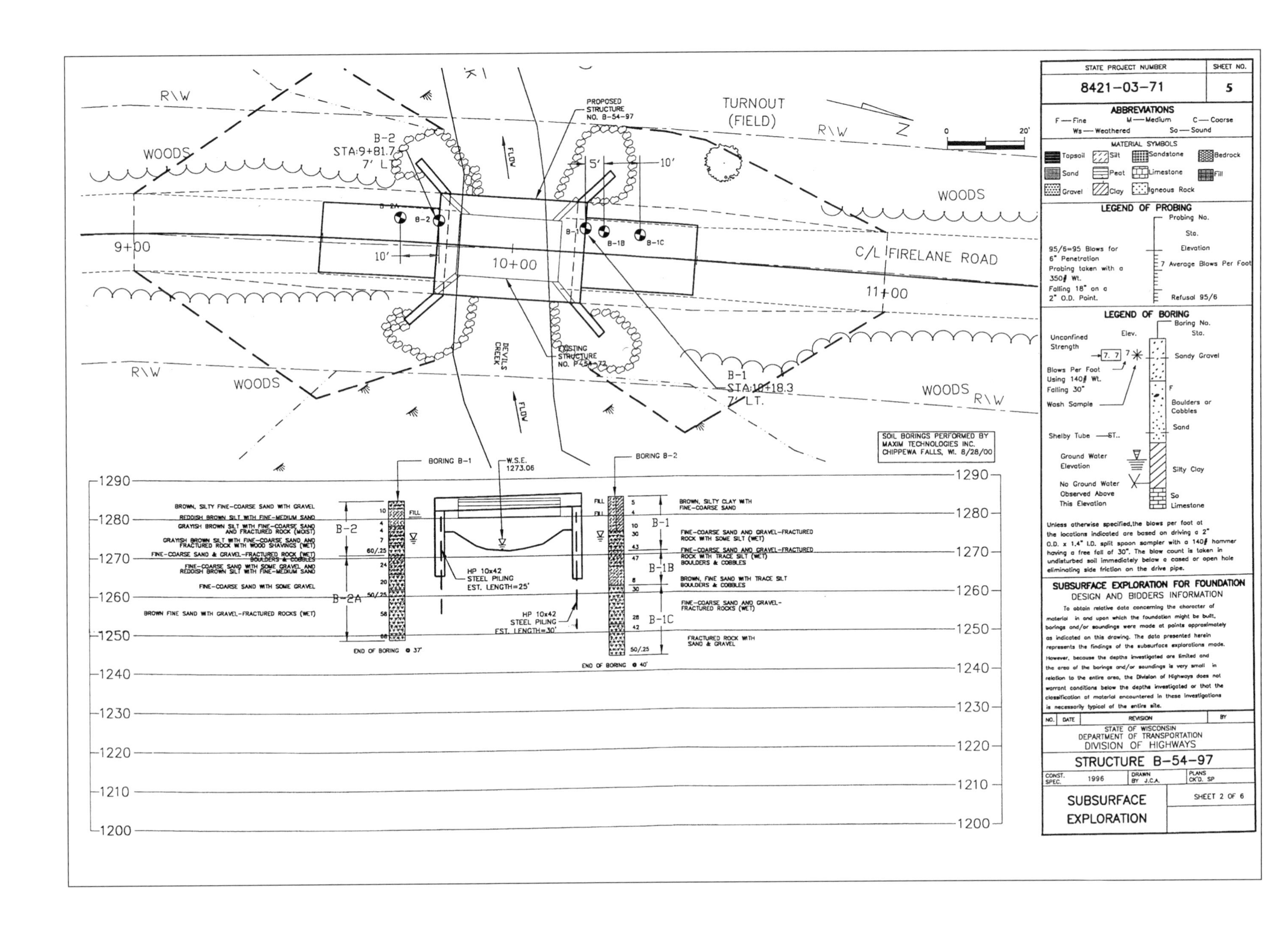

STATE PROJECT NUMBER
8421-03-71
SHEET NO. 5
ABBREVIATIONS
F — Fine M — Medium C — Coarse
Ws — Weathered So — Sound
MATERIAL SYMBOLS
Topsoil Silt Sandstone Bedrock
Sand Peat Limestone Fill
Gravel Clay Igneous Rock
LEGEND OF PROBING
95/6=95 Blows for 6" Penetration
Probing taken with a 350# Wt. Falling 18" on a 2" O.D. Point.
LEGEND OF BORING
SUBSURFACE EXPLORATION FOR FOUNDATION DESIGN AND BIDDERS INFORMATION
STATE OF WISCONSIN
DEPARTMENT OF TRANSPORTATION
DIVISION OF HIGHWAYS
STRUCTURE B-54-97
SUBSURFACE EXPLORATION
SHEET 2 OF 6
TURNOUT (FIELD)
WOODS
R\W
C/L FIRELANE ROAD
9+00
10+00
11+00
PROPOSED STRUCTURE NO. B-54-97
EXISTING STRUCTURE
DEVILS CREEK
FLOW
B-2 STA:9+81.7 7' LT.
B-1 STA:10+18.3 7' LT.
SOIL BORINGS PERFORMED BY MAXIM TECHNOLOGIES INC. CHIPPEWA FALLS, WI. 8/28/00
BORING B-1
BORING B-2
W.S.E. 1273.06
HP 10x42 STEEL PILING EST. LENGTH=25'
HP 10x42 STEEL PILING EST. LENGTH=30'
END OF BORING @ 37'
END OF BORING @ 40'

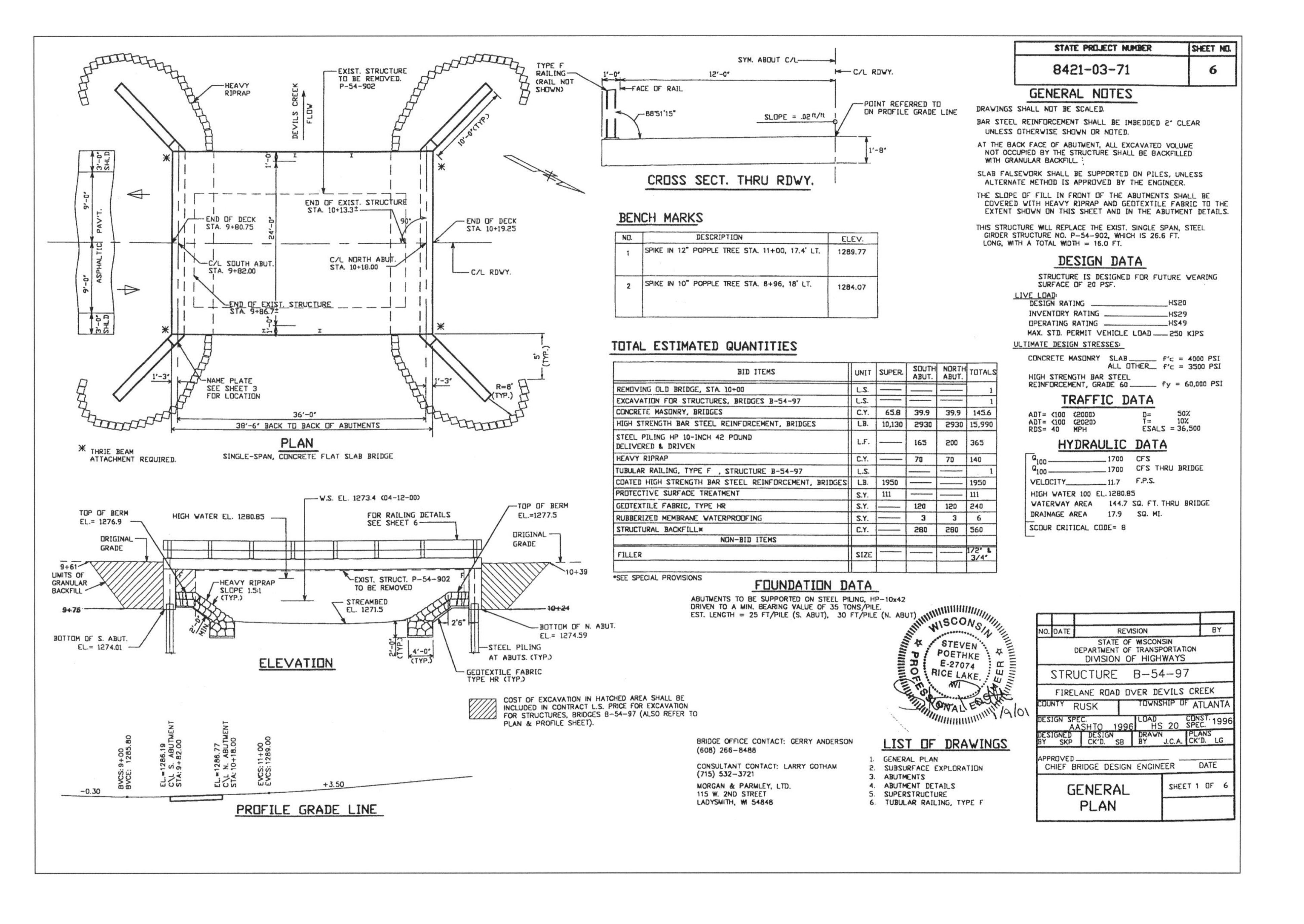
STATE PROJECT NUMBER
8421-03-71
SHEET NO.
6
GENERAL NOTES
DRAWINGS SHALL NOT BE SCALED.
BAR STEEL REINFORCEMENT SHALL BE IMBEDDED 2" CLEAR UNLESS OTHERWISE SHOWN OR NOTED.
AT THE BACK FACE OF ABUTMENT, ALL EXCAVATED VOLUME NOT OCCUPIED BY THE STRUCTURE SHALL BE BACKFILLED WITH GRANULAR BACKFILL.
SLAB FALSEWORK SHALL BE SUPPORTED ON PILES, UNLESS ALTERNATE METHOD IS APPROVED BY THE ENGINEER.
THE SLOPE OF FILL IN FRONT OF THE ABUTMENTS SHALL BE COVERED WITH HEAVY RIPRAP AND GEOTEXTILE FABRIC TO THE EXTENT SHOWN ON THIS SHEET AND IN THE ABUTMENT DETAILS.
THIS STRUCTURE WILL REPLACE THE EXIST. SINGLE SPAN, STEEL GIRDER STRUCTURE NO. P-54-902, WHICH IS 26.6 FT. LONG, WITH A TOTAL WIDTH = 16.0 FT.
DESIGN DATA
STRUCTURE IS DESIGNED FOR FUTURE WEARING SURFACE OF 20 PSF.
LIVE LOAD:
DESIGN RATING HS20
INVENTORY RATING HS29
OPERATING RATING HS49
MAX. STD. PERMIT VEHICLE LOAD 250 KIPS
ULTIMATE DESIGN STRESSES:
CONCRETE MASONRY SLAB f'c = 4000 PSI
ALL OTHER f'c = 3500 PSI
HIGH STRENGTH BAR STEEL REINFORCEMENT, GRADE 60 fy = 60,000 PSI
TRAFFIC DATA
ADT= <100 (2000)
ADT= <100 (2020)
RDS= 40 MPH
D= 50%
T= 10%
ESALS = 36,500
HYDRAULIC DATA
Q100 1700 CFS
Q100 1700 CFS THRU BRIDGE
VELOCITY 11.7 F.P.S.
HIGH WATER 100 EL. 1280.85
WATERWAY AREA 144.7 SQ. FT. THRU BRIDGE
DRAINAGE AREA 17.9 SQ. MI.
SCOUR CRITICAL CODE= 8
HEAVY RIPRAP
EXIST. STRUCTURE TO BE REMOVED. P-54-902
DEVILS CREEK
FLOW
TYPE F RAILING (RAIL NOT SHOWN)
10'-0" (TYP.)
END OF EXIST. STRUCTURE STA. 10+13.3±
END OF DECK STA. 9+80.75
END OF DECK STA. 10+19.25
C/L SOUTH ABUT. STA. 9+82.00
C/L NORTH ABUT. STA. 10+18.00
C/L RDWY.
END OF EXIST. STRUCTURE STA. 9+86.7±
3'-0" SHLD
9'-0"
ASPHALTIC PAV'T.
24'-0"
1'-0"
90°
5' (TYP.)
R=8' (TYP.)
1'-3"
NAME PLATE SEE SHEET 3 FOR LOCATION
36'-0"
38'-6" BACK TO BACK OF ABUTMENTS
PLAN
SINGLE-SPAN, CONCRETE FLAT SLAB BRIDGE
* THRIE BEAM ATTACHMENT REQUIRED.
SYM. ABOUT C/L
C/L RDWY.
12'-0"
FACE OF RAIL
88°51'15"
SLOPE = .02 ft/ft
POINT REFERRED TO ON PROFILE GRADE LINE
1'-8"
CROSS SECT. THRU RDWY.
BENCH MARKS
NO. | DESCRIPTION | ELEV.
1 | SPIKE IN 12" POPPLE TREE STA. 11+00, 17.4' LT. | 1289.77
2 | SPIKE IN 10" POPPLE TREE STA. 8+96, 18' LT. | 1284.07
TOTAL ESTIMATED QUANTITIES
BID ITEMS | UNIT | SUPER. | SOUTH ABUT. | NORTH ABUT. | TOTALS
REMOVING OLD BRIDGE, STA. 10+00 | L.S. | — | — | — | 1
EXCAVATION FOR STRUCTURES, BRIDGES B-54-97 | L.S. | — | — | — | 1
CONCRETE MASONRY, BRIDGES | C.Y. | 65.8 | 39.9 | 39.9 | 145.6
HIGH STRENGTH BAR STEEL REINFORCEMENT, BRIDGES | LB. | 10,130 | 2930 | 2930 | 15,990
STEEL PILING HP 10-INCH 42 POUND DELIVERED & DRIVEN | L.F. | — | 165 | 200 | 365
HEAVY RIPRAP | C.Y. | — | 70 | 70 | 140
TUBULAR RAILING, TYPE F , STRUCTURE B-54-97 | L.S. | — | — | — | 1
COATED HIGH STRENGTH BAR STEEL REINFORCEMENT, BRIDGES | LB. | 1950 | — | — | 1950
PROTECTIVE SURFACE TREATMENT | S.Y. | 111 | — | — | 111
GEOTEXTILE FABRIC, TYPE HR | S.Y. | — | 120 | 120 | 240
RUBBERIZED MEMBRANE WATERPROOFING | S.Y. | — | 3 | 3 | 6
STRUCTURAL BACKFILL* | C.Y. | — | 280 | 280 | 560
NON-BID ITEMS
FILLER | SIZE | — | — | — | 1/2" & 3/4"
*SEE SPECIAL PROVISIONS
FOUNDATION DATA
ABUTMENTS TO BE SUPPORTED ON STEEL PILING, HP-10x42 DRIVEN TO A MIN. BEARING VALUE OF 35 TONS/PILE.
EST. LENGTH = 25 FT/PILE (S. ABUT), 30 FT/PILE (N. ABUT)
W.S. EL. 1273.4 (04-12-00)
TOP OF BERM EL.= 1276.9
HIGH WATER EL. 1280.85
FOR RAILING DETAILS SEE SHEET 6
TOP OF BERM EL.=1277.5
ORIGINAL GRADE
9+61
LIMITS OF GRANULAR BACKFILL
HEAVY RIPRAP SLOPE 1.5:1 (TYP.)
EXIST. STRUCT. P-54-902 TO BE REMOVED
STREAMBED EL. 1271.5
10+39
10+24
9+76
2'-0" MIN.
2'6"
BOTTOM OF N. ABUT. EL.= 1274.59
BOTTOM OF S. ABUT. EL.= 1274.01
2'-0" (TYP.)
4'-0" (TYP.)
STEEL PILING AT ABUTS. (TYP.)
GEOTEXTILE FABRIC TYPE HR (TYP.)
ELEVATION
COST OF EXCAVATION IN HATCHED AREA SHALL BE INCLUDED IN CONTRACT L.S. PRICE FOR EXCAVATION FOR STRUCTURES, BRIDGES B-54-97 (ALSO REFER TO PLAN & PROFILE SHEET).
BVCS:9+00
BVCE: 1285.80
EL.=1286.19
C/L S. ABUTMENT
STA:9+82.00
EL.=1286.77
C/L N. ABUTMENT
STA:10+18.00
EVCS:11+00
EVCS:1289.00
-0.30
+3.50
PROFILE GRADE LINE
WISCONSIN
STEVEN POETHKE E-27074 RICE LAKE, WI
PROFESSIONAL ENGINEER
1/9/01
BRIDGE OFFICE CONTACT: GERRY ANDERSON (608) 266-8488
CONSULTANT CONTACT: LARRY GOTHAM (715) 532-3721
MORGAN & PARMLEY, LTD.
115 W. 2ND STREET
LADYSMITH, WI 54848
LIST OF DRAWINGS
1. GENERAL PLAN
2. SUBSURFACE EXPLORATION
3. ABUTMENTS
4. ABUTMENT DETAILS
5. SUPERSTRUCTURE
6. TUBULAR RAILING, TYPE F
NO. DATE REVISION BY
STATE OF WISCONSIN
DEPARTMENT OF TRANSPORTATION
DIVISION OF HIGHWAYS
STRUCTURE B-54-97
FIRELANE ROAD OVER DEVILS CREEK
COUNTY RUSK
TOWNSHIP OF ATLANTA
DESIGN SPEC. AASHTO 1996
LOAD HS 20
CONST. SPEC. 1996
DESIGNED BY SKP
DESIGN CK'D. SB
DRAWN BY J.C.A.
PLANS CK'D. LG
APPROVED
CHIEF BRIDGE DESIGN ENGINEER
DATE
GENERAL PLAN
SHEET 1 OF 6

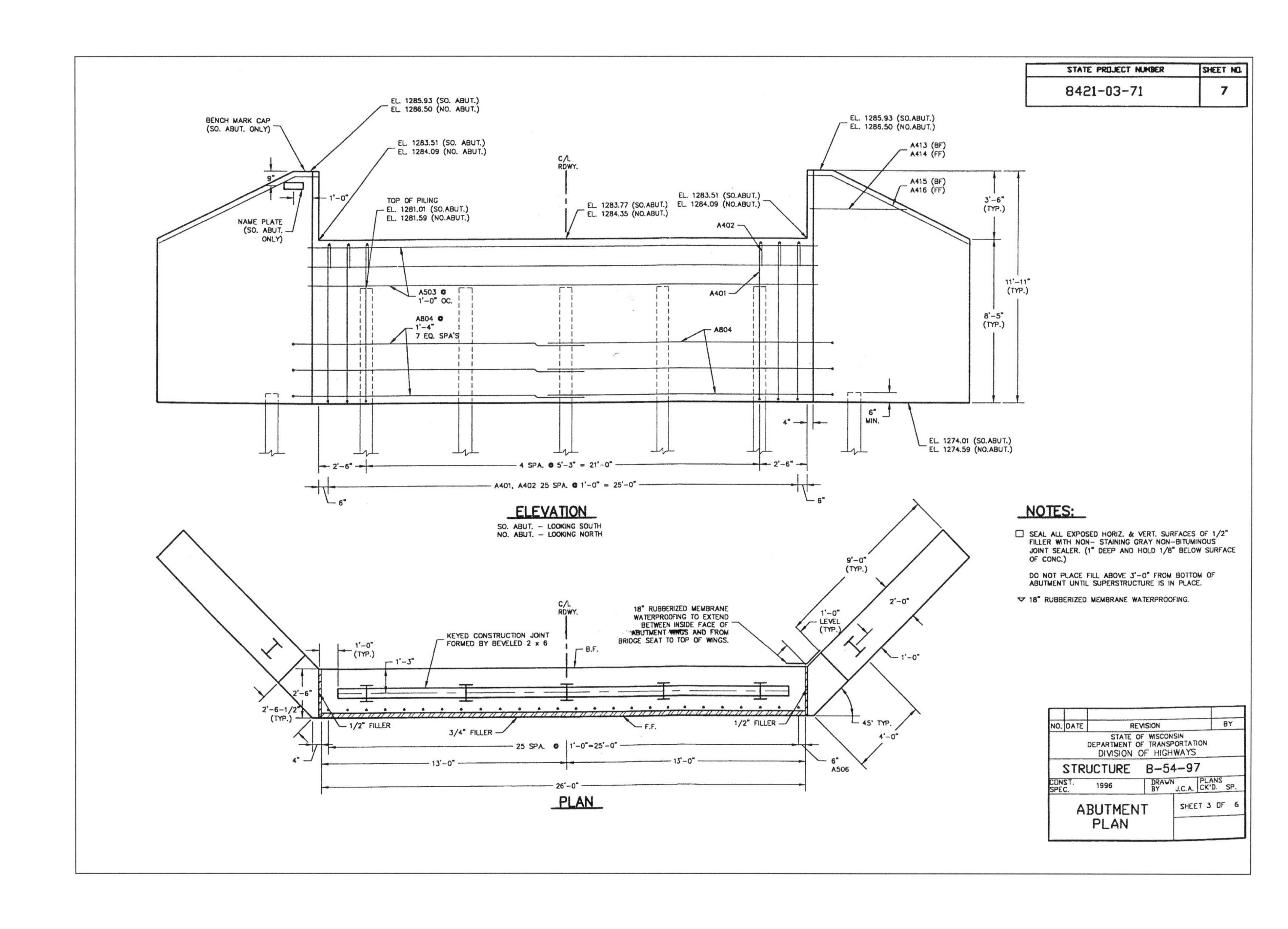
STATE PROJECT NUMBER
SHEET NO.
8421-03-71
7
ELEVATION
SO. ABUT. – LOOKING SOUTH
NO. ABUT. – LOOKING NORTH
PLAN
NOTES:
SEAL ALL EXPOSED HORIZ. & VERT. SURFACES OF 1/2" FILLER WITH NON- STAINING GRAY NON-BITUMINOUS JOINT SEALER. (1" DEEP AND HOLD 1/8" BELOW SURFACE OF CONC.)
DO NOT PLACE FILL ABOVE 3'-0" FROM BOTTOM OF ABUTMENT UNTIL SUPERSTRUCTURE IS IN PLACE.
18" RUBBERIZED MEMBRANE WATERPROOFING.
STATE OF WISCONSIN
DEPARTMENT OF TRANSPORTATION
DIVISION OF HIGHWAYS
STRUCTURE B-54-97
ABUTMENT PLAN
SHEET 3 OF 6

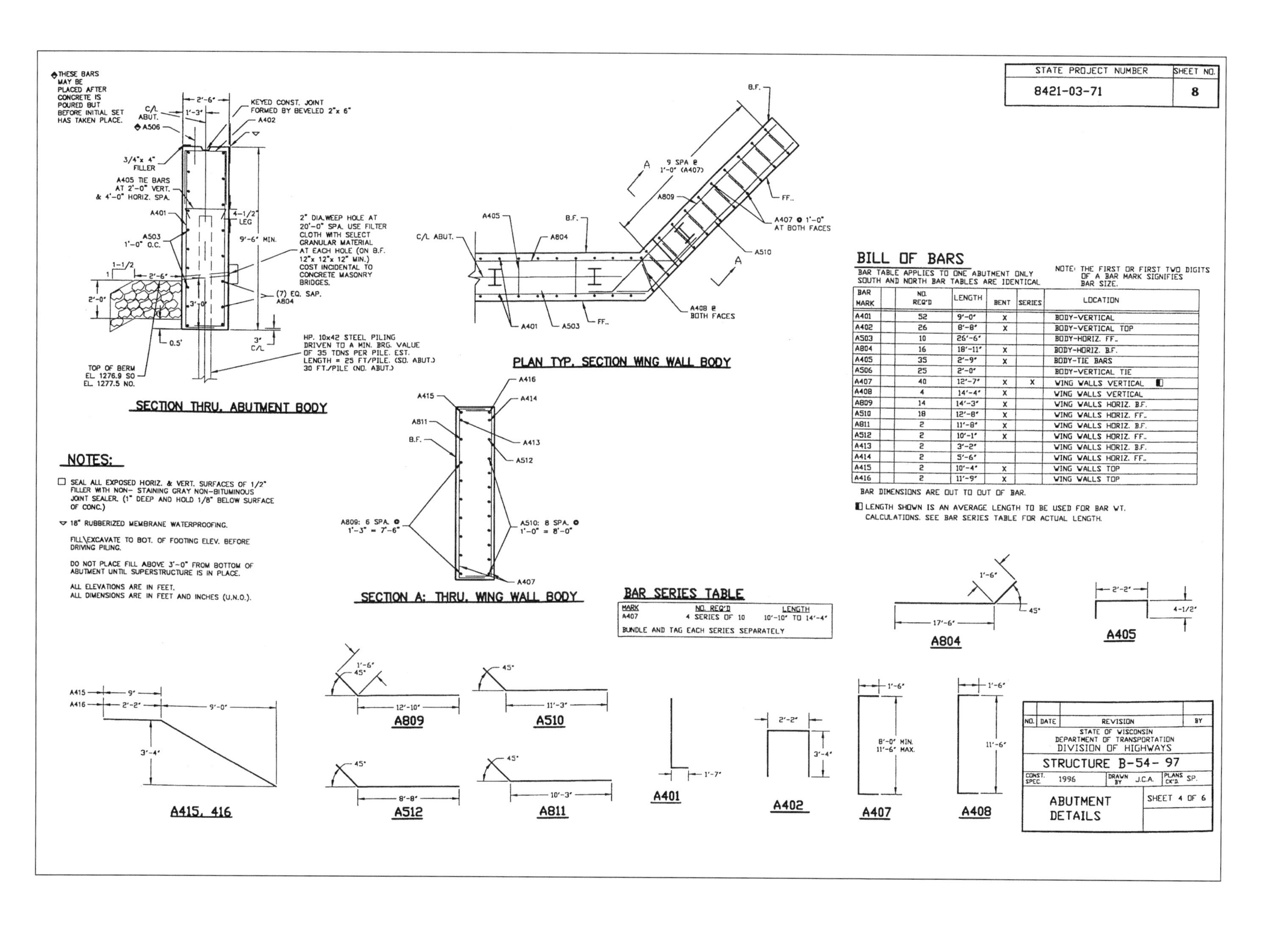

BAR MARK		NO. REQ'D	LENGTH	BENT	SERIES	LOCATION
A401		52	9'-0"	X		BODY-VERTICAL
A402		26	8'-8"	X		BODY-VERTICAL TOP
A503		10	26'-6"			BODY-HORIZ. FF..
A804		16	18'-11"	X		BODY-HORIZ. B.F.
A405		35	2'-9"	X		BODY-TIE BARS
A506		25	2'-0"			BODY-VERTICAL TIE
A407		40	12'-7"	X	X	WING WALLS VERTICAL ◧
A408		4	14'-4"	X		WING WALLS VERTICAL
A809		14	14'-3"	X		WING WALLS HORIZ. B.F.
A510		18	12'-8"	X		WING WALLS HORIZ. FF..
A811		2	11'-8"	X		WING WALLS HORIZ. B.F.
A512		2	10'-1"	X		WING WALLS HORIZ. FF..
A413		2	3'-2"			WING WALLS HORIZ. B.F.
A414		2	5'-6"			WING WALLS HORIZ. FF..
A415		2	10'-4"	X		WING WALLS TOP
A416		2	11'-9"	X		WING WALLS TOP

MARK	NO. REQ'D	LENGTH
A407	4 SERIES OF 10	10'-10" TO 14'-4"

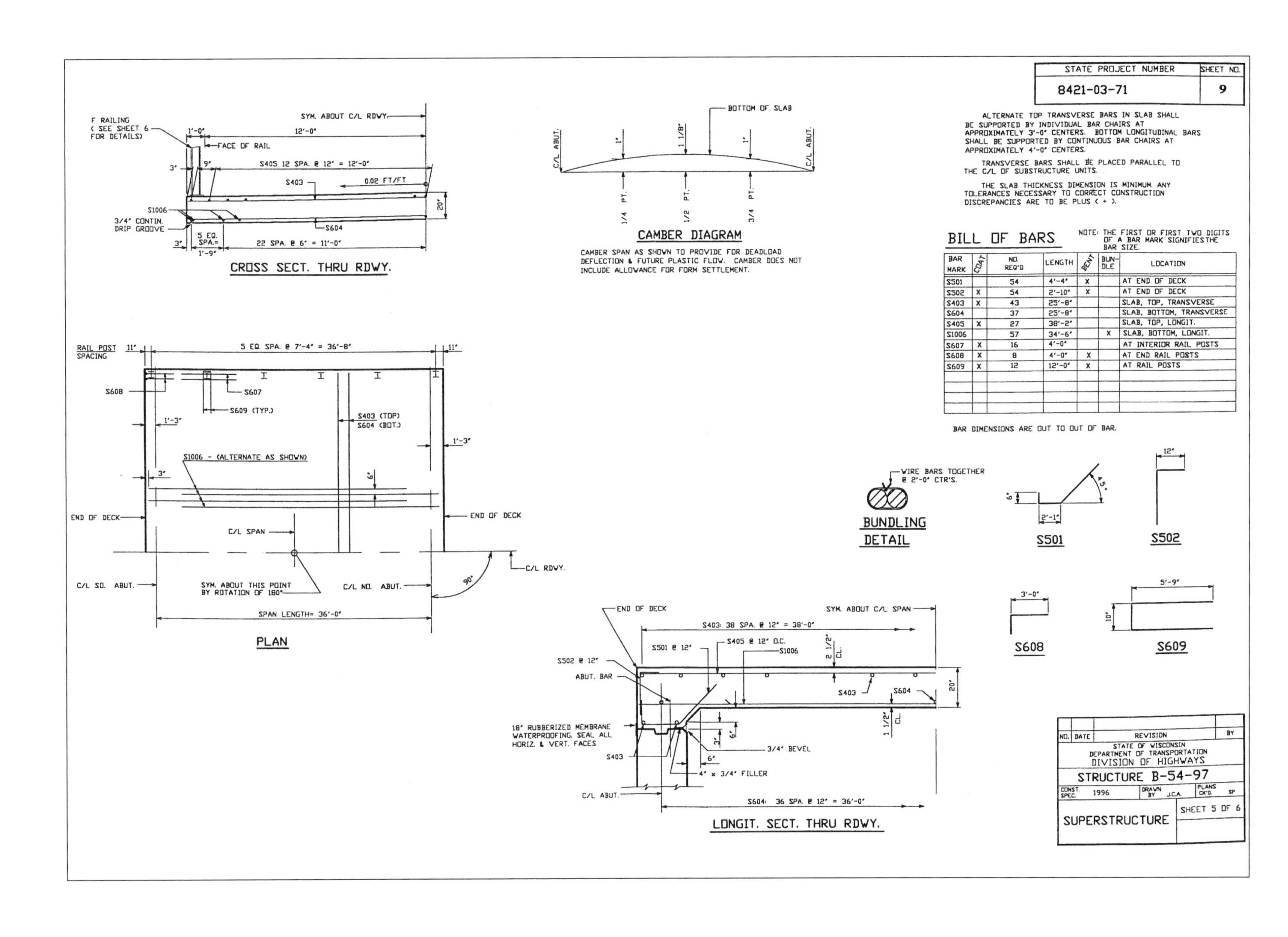

BAR MARK	COAT	NO. REQ'D	LENGTH	BENT	BUN-DLE	LOCATION
S501		54	4'-4"	X		AT END OF DECK
S502	X	54	2'-10"	X		AT END OF DECK
S403	X	43	25'-8"			SLAB, TOP, TRANSVERSE
S604		37	25'-8"			SLAB, BOTTOM, TRANSVERSE
S405	X	27	38'-2"			SLAB, TOP, LONGIT.
S1006		57	34'-6"		X	SLAB, BOTTOM, LONGIT.
S607	X	16	4'-0"			AT INTERIOR RAIL POSTS
S608	X	8	4'-0"	X		AT END RAIL POSTS
S609	X	12	12'-0"	X		AT RAIL POSTS

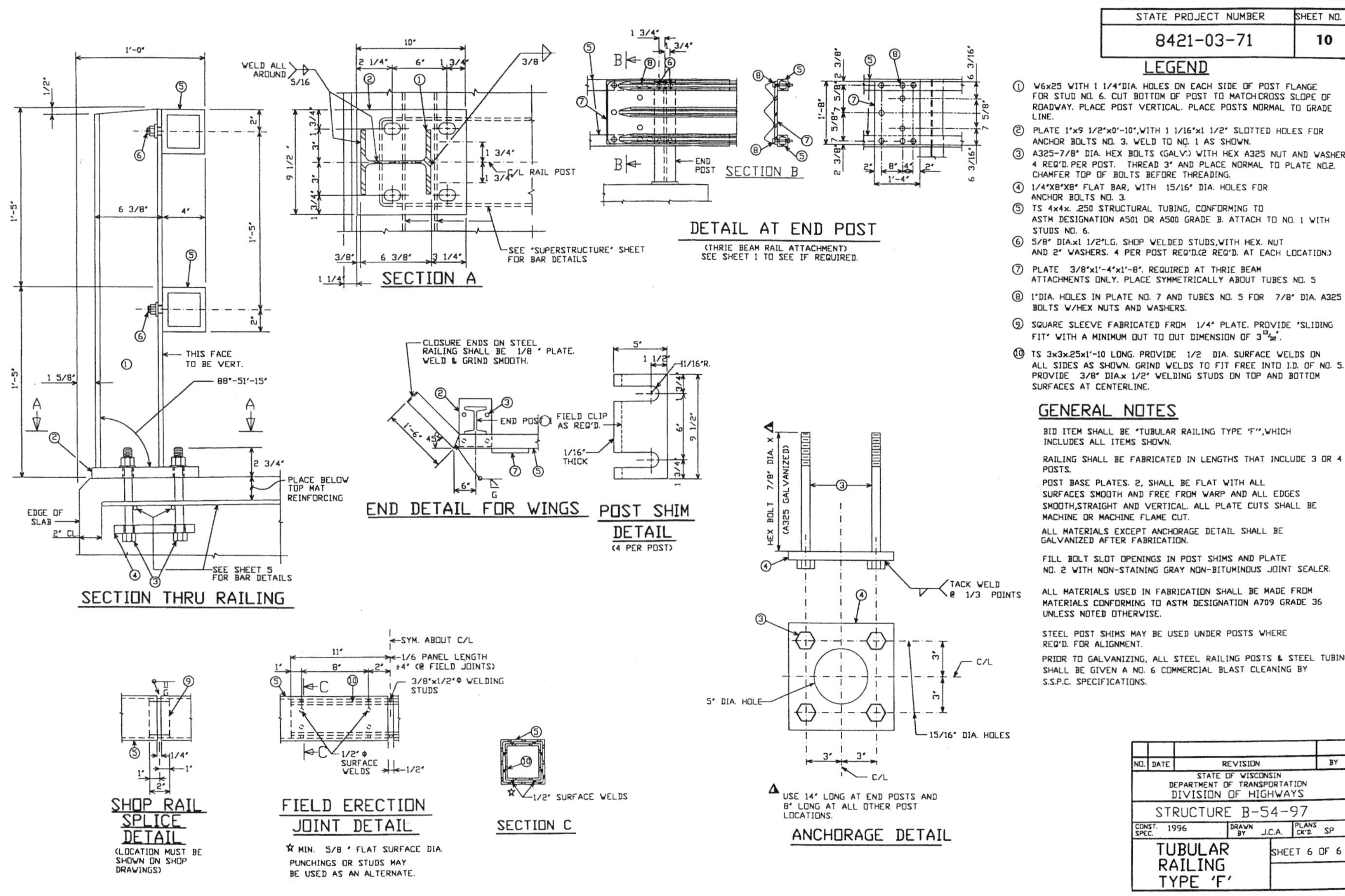
STATE PROJECT NUMBER
SHEET NO.
8421-03-71
10
LEGEND
① W6x25 WITH 1 1/4"DIA. HOLES ON EACH SIDE OF POST FLANGE FOR STUD NO. 6. CUT BOTTOM OF POST TO MATCH CROSS SLOPE OF ROADWAY. PLACE POST VERTICAL. PLACE POSTS NORMAL TO GRADE LINE.
② PLATE 1"x9 1/2"x0'-10",WITH 1 1/16"x1 1/2" SLOTTED HOLES FOR ANCHOR BOLTS NO. 3. WELD TO NO. 1 AS SHOWN.
③ A325-7/8" DIA. HEX BOLTS (GALV.) WITH HEX A325 NUT AND WASHER. 4 REQ'D. PER POST. THREAD 3" AND PLACE NORMAL TO PLATE NO.2. CHAMFER TOP OF BOLTS BEFORE THREADING.
④ 1/4"X8"X8" FLAT BAR, WITH 15/16" DIA. HOLES FOR ANCHOR BOLTS NO. 3.
⑤ TS 4x4x. .250 STRUCTURAL TUBING, CONFORMING TO ASTM DESIGNATION A501 OR A500 GRADE B. ATTACH TO NO. 1 WITH STUDS NO. 6.
⑥ 5/8" DIA.x1 1/2"LG. SHOP WELDED STUDS,WITH HEX. NUT AND 2" WASHERS. 4 PER POST REQ'D.(2 REQ'D. AT EACH LOCATION.)
⑦ PLATE 3/8"x1'-4"x1'-8". REQUIRED AT THRIE BEAM ATTACHMENTS ONLY. PLACE SYMMETRICALLY ABOUT TUBES NO. 5
⑧ 1"DIA. HOLES IN PLATE NO. 7 AND TUBES NO. 5 FOR 7/8" DIA. A325 BOLTS W/HEX NUTS AND WASHERS.
⑨ SQUARE SLEEVE FABRICATED FROM 1/4" PLATE. PROVIDE "SLIDING FIT" WITH A MINIMUM OUT TO OUT DIMENSION OF 3 13/32".
⑩ TS 3x3x.25x1'-10 LONG. PROVIDE 1/2 DIA. SURFACE WELDS ON ALL SIDES AS SHOWN. GRIND WELDS TO FIT FREE INTO I.D. OF NO. 5. PROVIDE 3/8" DIA.x 1/2" WELDING STUDS ON TOP AND BOTTOM SURFACES AT CENTERLINE.
GENERAL NOTES
BID ITEM SHALL BE "TUBULAR RAILING TYPE "F"",WHICH INCLUDES ALL ITEMS SHOWN.
RAILING SHALL BE FABRICATED IN LENGTHS THAT INCLUDE 3 OR 4 POSTS.
POST BASE PLATES. 2, SHALL BE FLAT WITH ALL SURFACES SMOOTH AND FREE FROM WARP AND ALL EDGES SMOOTH,STRAIGHT AND VERTICAL. ALL PLATE CUTS SHALL BE MACHINE OR MACHINE FLAME CUT.
ALL MATERIALS EXCEPT ANCHORAGE DETAIL SHALL BE GALVANIZED AFTER FABRICATION.
FILL BOLT SLOT OPENINGS IN POST SHIMS AND PLATE NO. 2 WITH NON-STAINING GRAY NON-BITUMINOUS JOINT SEALER.
ALL MATERIALS USED IN FABRICATION SHALL BE MADE FROM MATERIALS CONFORMING TO ASTM DESIGNATION A709 GRADE 36 UNLESS NOTED OTHERWISE.
STEEL POST SHIMS MAY BE USED UNDER POSTS WHERE REQ'D. FOR ALIGNMENT.
PRIOR TO GALVANIZING, ALL STEEL RAILING POSTS & STEEL TUBING SHALL BE GIVEN A NO. 6 COMMERCIAL BLAST CLEANING BY S.S.P.C. SPECIFICATIONS.
SECTION THRU RAILING
THIS FACE TO BE VERT.
88°-51'-15"
PLACE BELOW TOP MAT REINFORCING
EDGE OF SLAB
SEE SHEET 5 FOR BAR DETAILS
WELD ALL AROUND
SECTION A
SEE "SUPERSTRUCTURE" SHEET FOR BAR DETAILS
C/L RAIL POST
DETAIL AT END POST
(THRIE BEAM RAIL ATTACHMENT)
SEE SHEET 1 TO SEE IF REQUIRED.
END POST
SECTION B
CLOSURE ENDS ON STEEL RAILING SHALL BE 1/8" PLATE. WELD & GRIND SMOOTH.
END POST
FIELD CLIP AS REQ'D.
END DETAIL FOR WINGS
POST SHIM DETAIL
(4 PER POST)
HEX BOLT 7/8" DIA. X (A325 GALVANIZED)
TACK WELD @ 1/3 POINTS
5" DIA. HOLE
15/16" DIA. HOLES
USE 14" LONG AT END POSTS AND 8" LONG AT ALL OTHER POST LOCATIONS.
ANCHORAGE DETAIL
SHOP RAIL SPLICE DETAIL
(LOCATION MUST BE SHOWN ON SHOP DRAWINGS)
SYM. ABOUT C/L
1/6 PANEL LENGTH ±4" (@ FIELD JOINTS)
3/8"x1/2"Φ WELDING STUDS
1/2"Φ SURFACE WELDS
FIELD ERECTION JOINT DETAIL
☆ MIN. 5/8" FLAT SURFACE DIA. PUNCHINGS OR STUDS MAY BE USED AS AN ALTERNATE.
1/2" SURFACE WELDS
SECTION C
NO. DATE REVISION BY
STATE OF WISCONSIN
DEPARTMENT OF TRANSPORTATION
DIVISION OF HIGHWAYS
STRUCTURE B-54-97
CONST. SPEC. 1996
DRAWN BY J.C.A.
PLANS CK'D. SP
TUBULAR RAILING TYPE 'F'
SHEET 6 OF 6

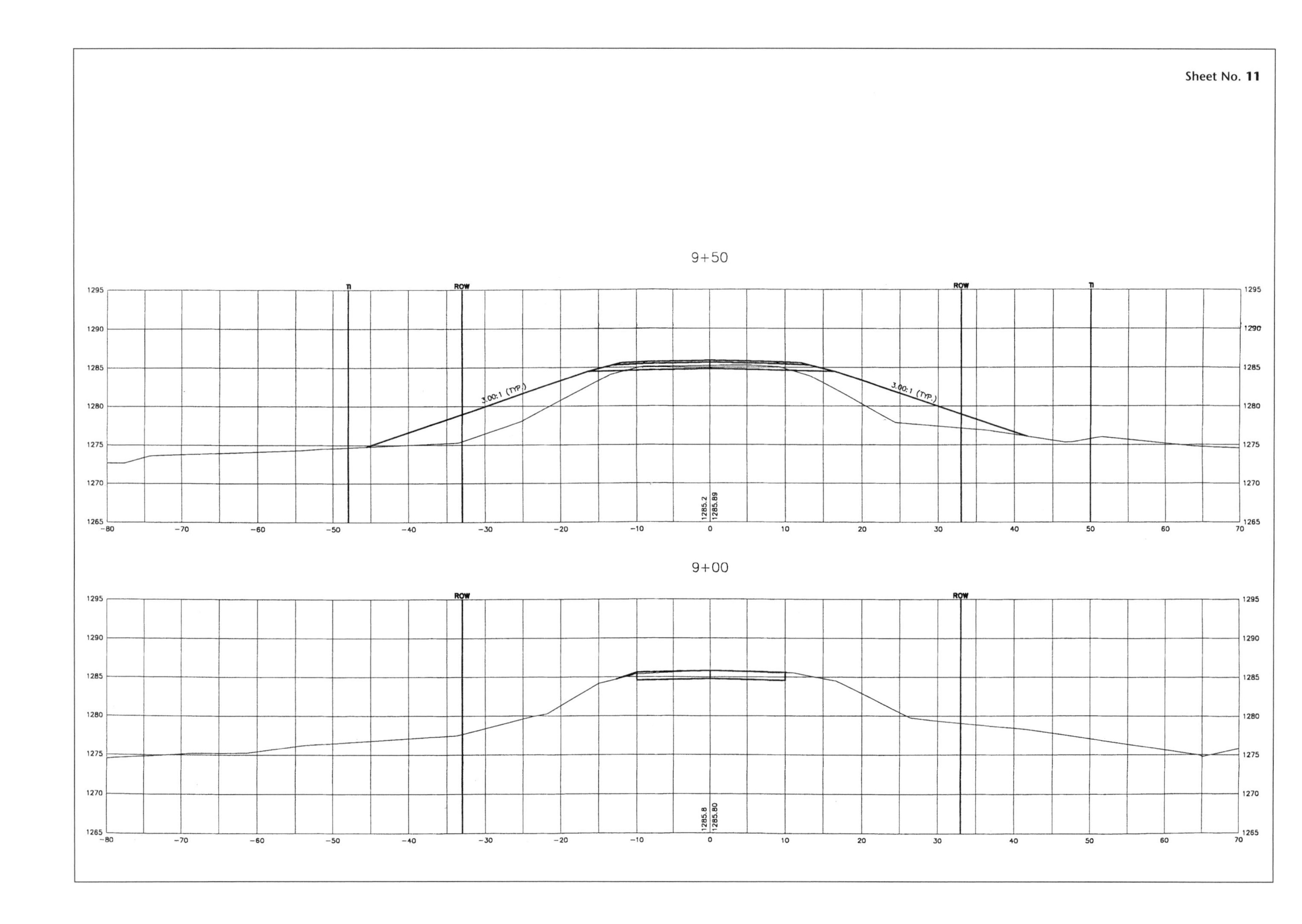
Sheet No. 11
9+50
TI
ROW
ROW
TI
3.00:1 (TYP.)
3.00:1 (TYP.)
1285.2
1285.89
1295
1290
1285
1280
1275
1270
1265
-80
-70
-60
-50
-40
-30
-20
-10
0
10
20
30
40
50
60
70
9+00
ROW
ROW
1285.8
1285.80

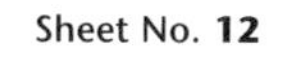

STA: 10+00 CENTERLINE OF STRUCTURE

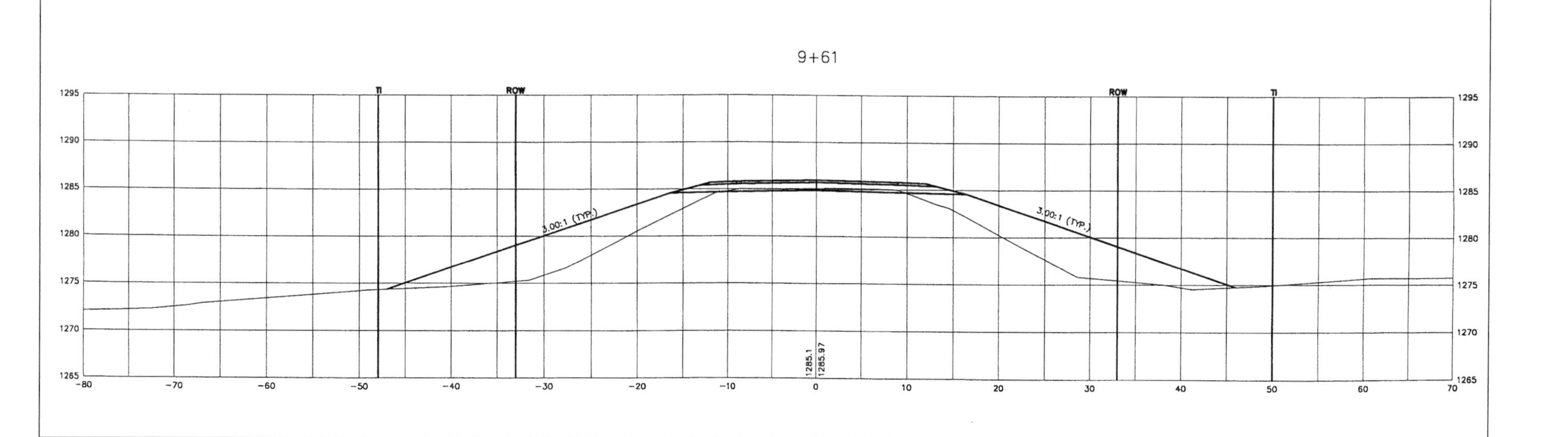

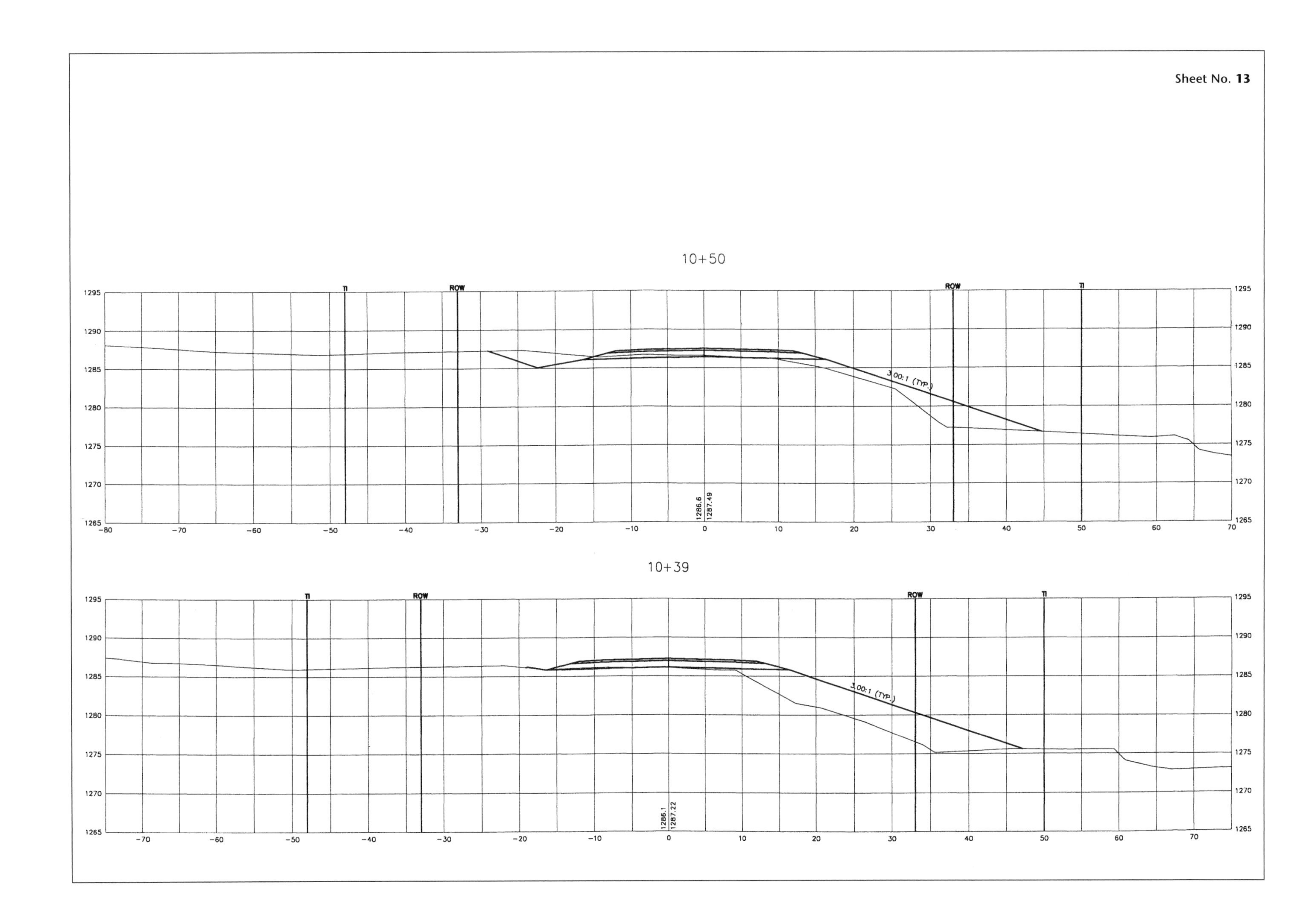
Sheet No. 13
10+50
TI
ROW
ROW
TI
3.00:1 (TYP.)
1286.6
1287.49
10+39
TI
ROW
ROW
TI
3.00:1 (TYP.)
1286.1
1287.22

Sheet No. 14

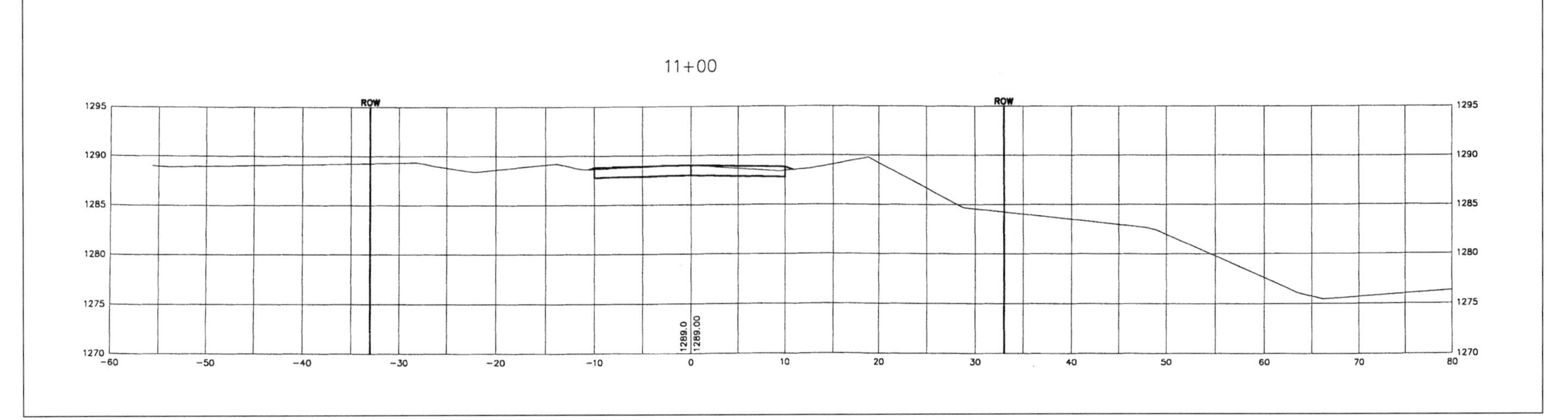
11+00
ROW
ROW
1295
1290
1285
1280
1275
1270
1289.0
1289.00
−60
−50
−40
−30
−20
−10
0
10
20
30
40
50
60
70
80

CLOVERLAND ROAD BRIDGE

— MAIN CREEK —

TOWN OF GROW
RUSK COUNTY, WI.

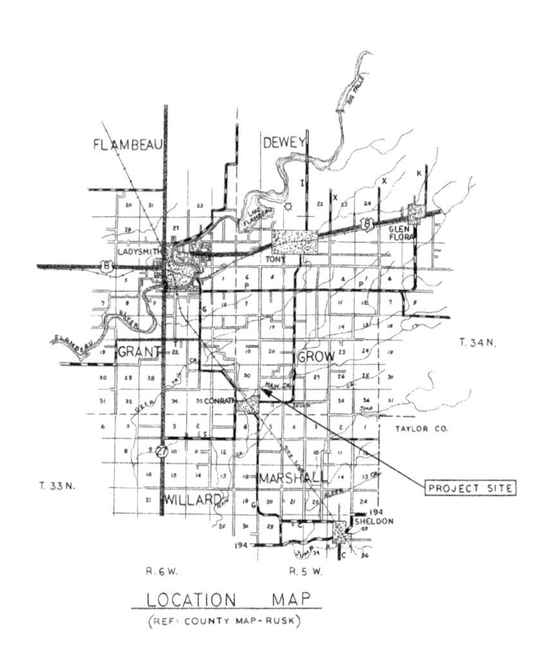

INDEX OF SHEETS

NO.	DESCRIPTION
1	TITLE SHEET & INDEX W/ LOCATION MAP
2	TOPOGRAPHY & GENERAL MAP
3	PLAN & ROAD PROFILE
4	GENERAL PLAN W/ DESIGN & HYDRAULIC DATA
5	SUBSURFACE EXPLORATION
6	FOUNDATIONS
7	ABUTMENTS
8	SUPERSTRUCTURE W/ DETAILS
9	PRESTRESSED GIRDER DETAILS
10	TUBULAR STEEL RAILING

CLOVERLAND ROAD BRIDGE	
TITLE SHEET AND INDEX	
SCALE- DO NOT	MORGAN & PARMLEY, LTD. CONSULTING ENGINEERS LADYSMITH, WISCONSIN
SHEET 1 OF 10	
DATE MAY 1976	DR. BY R.O. PARMLEY

Courtesy: Morgan & Parmley, Ltd.

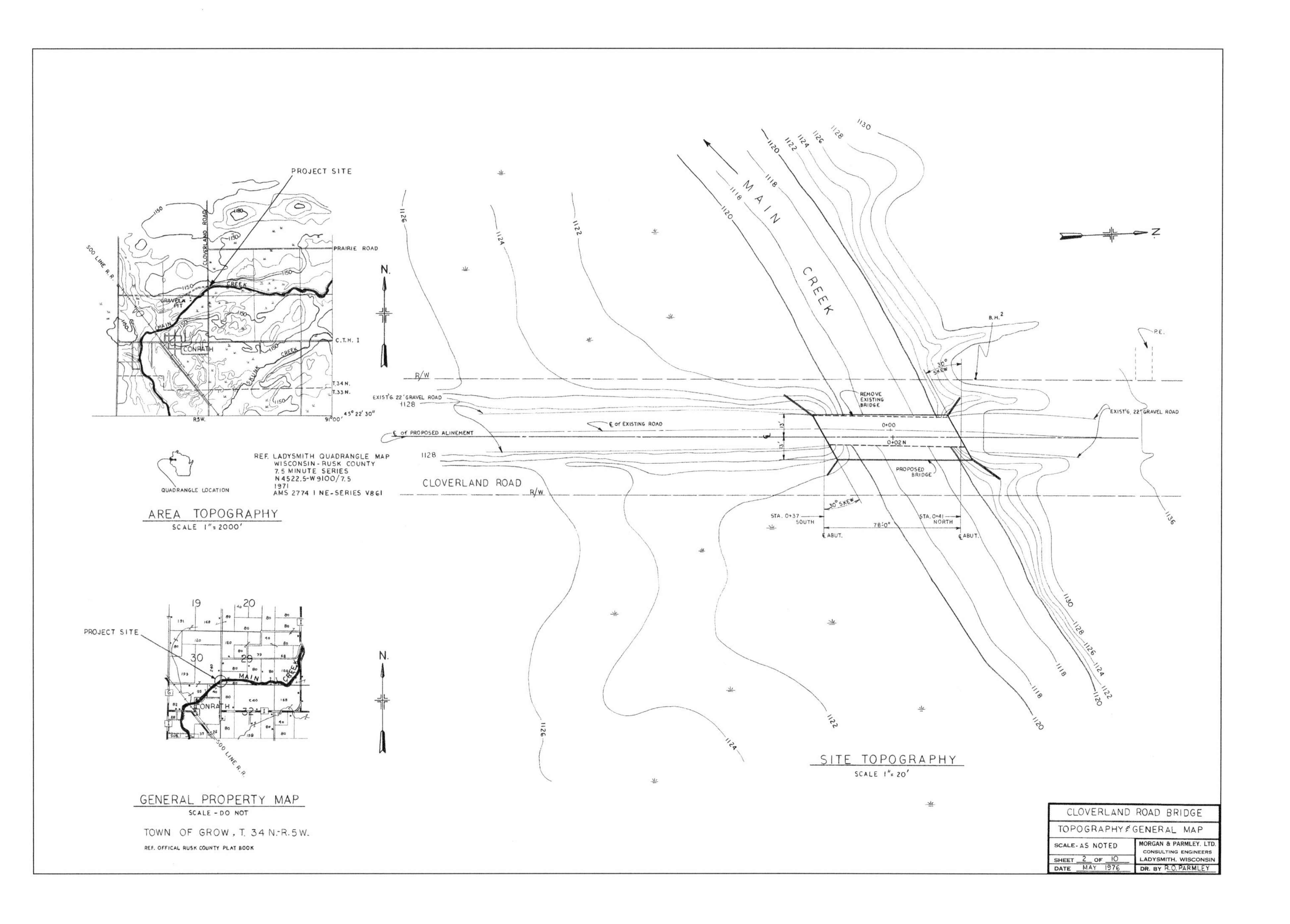

PROJECT SITE
PRAIRIE ROAD
C.T.H. I
500 LINE R.R.
CLOVERLAND ROAD
GRAVEL PIT
MAIN
CREEK
CONRATH
SKUNK CREEK
T.34 N.
T.33 N.
R5W.
91°00'
45° 22' 30"
N.
QUADRANGLE LOCATION
REF. LADYSMITH QUADRANGLE MAP
WISCONSIN - RUSK COUNTY
7.5 MINUTE SERIES
N4522.5-W9100/7.5
1971
AMS 2774 I NE-SERIES V861
AREA TOPOGRAPHY
SCALE 1" = 2000'
GENERAL PROPERTY MAP
SCALE - DO NOT
TOWN OF GROW, T. 34 N.-R. 5 W.
REF. OFFICAL RUSK COUNTY PLAT BOOK
MAIN CREEK
R/W
EXIST'G 22' GRAVEL ROAD
℄ of EXISTING ROAD
℄ of PROPOSED ALINEMENT
CLOVERLAND ROAD
REMOVE EXISTING BRIDGE
PROPOSED BRIDGE
0+00
0+02 N
30° SKEW
B.M. 2
P.E.
STA. 0+37 SOUTH
STA. 0+41 NORTH
78'-0"
℄ ABUT.
SITE TOPOGRAPHY
SCALE 1" = 20'
CLOVERLAND ROAD BRIDGE
TOPOGRAPHY & GENERAL MAP
SCALE - AS NOTED
SHEET 2 OF 10
DATE MAY 1976
MORGAN & PARMLEY, LTD.
CONSULTING ENGINEERS
LADYSMITH, WISCONSIN
DR. BY R.O. PARMLEY

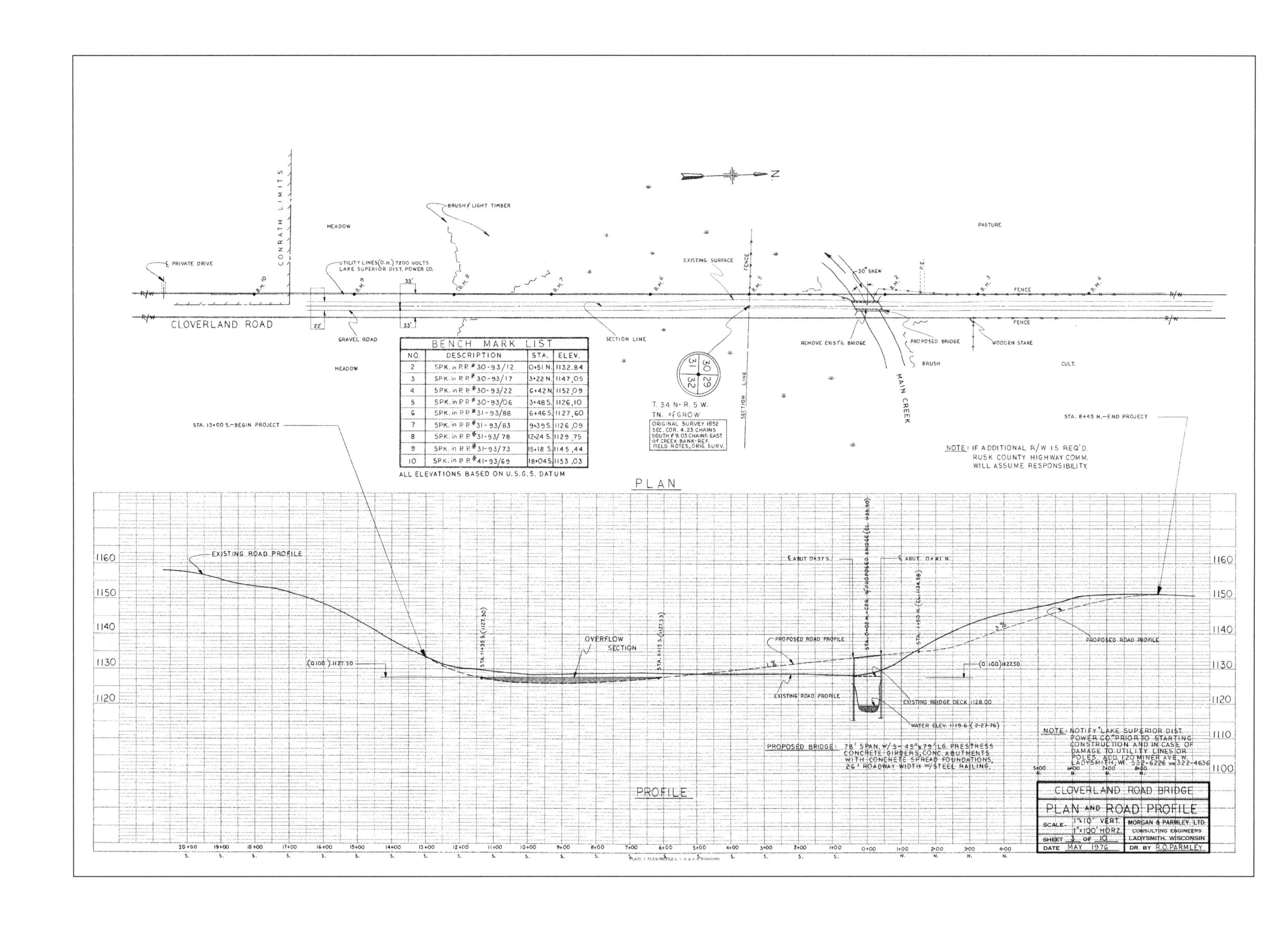

BENCH MARK LIST

NO.	DESCRIPTION	STA.	ELEV.
2	SPK. in P.P. #30-93/12	0+51 N.	1132.84
3	SPK. in P.P. #30-93/17	3+22 N.	1147.05
4	SPK. in P.P. #30-93/22	6+42 N.	1152.09
5	SPK. in P.P. #30-93/06	3+48 S.	1126.10
6	SPK. in P.P. #31-93/88	6+46 S.	1127.60
7	SPK. in P.P. #31-93/83	9+39 S.	1126.09
8	SPK. in P.P. #31-93/78	12+24 S.	1129.75
9	SPK. in P.P. #31-93/73	15+18 S.	1145.44
10	SPK. in P.P. #41-93/69	18+04 S.	1153.03

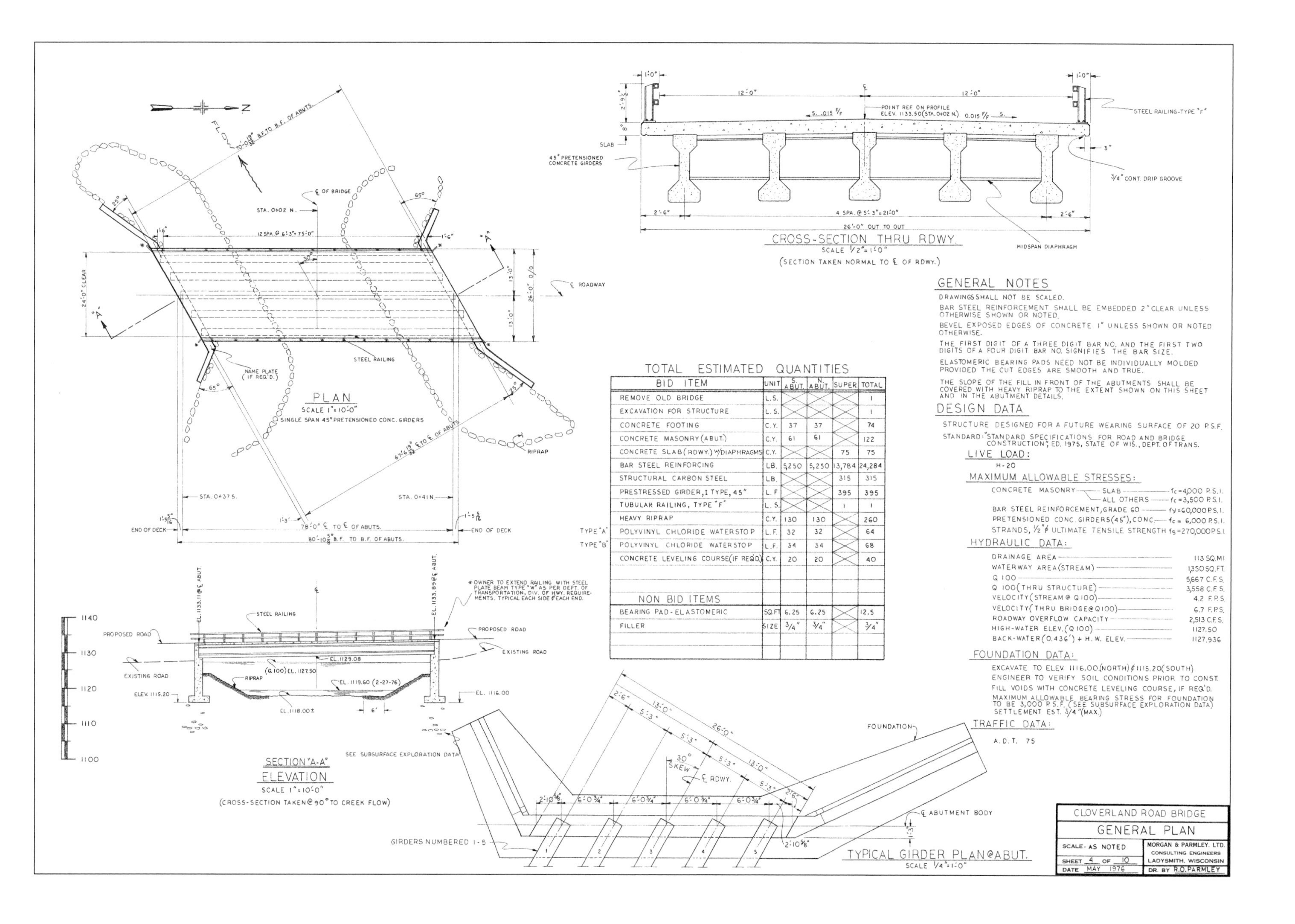

TOTAL ESTIMATED QUANTITIES

	BID ITEM	UNIT	S. ABUT.	N. ABUT.	SUPER.	TOTAL
	REMOVE OLD BRIDGE	L.S.				1
	EXCAVATION FOR STRUCTURE	L.S.				1
	CONCRETE FOOTING	C.Y.	37	37		74
	CONCRETE MASONRY (ABUT.)	C.Y.	61	61		122
	CONCRETE SLAB (RDWY.) W/DIAPHRAGMS	C.Y.			75	75
	BAR STEEL REINFORCING	LB.	5,250	5,250	13,784	24,284
	STRUCTURAL CARBON STEEL	LB.			315	315
	PRESTRESSED GIRDER, I TYPE, 45"	L.F.			395	395
	TUBULAR RAILING, TYPE "F"	L.S.			1	1
	HEAVY RIPRAP	C.Y.	130	130		260
TYPE "A"	POLYVINYL CHLORIDE WATERSTOP	L.F.	32	32		64
TYPE "B"	POLYVINYL CHLORIDE WATERSTOP	L.F.	34	34		68
	CONCRETE LEVELING COURSE (IF REQ'D.)	C.Y.	20	20		40
	NON BID ITEMS					
	BEARING PAD-ELASTOMERIC	SQ.FT.	6.25	6.25		12.5
	FILLER	SIZE	3/4"	3/4"		3/4"

GENERAL NOTES

DRAWINGS SHALL NOT BE SCALED.

BAR STEEL REINFORCEMENT SHALL BE EMBEDDED 2" CLEAR UNLESS OTHERWISE SHOWN OR NOTED.

BEVEL EXPOSED EDGES OF CONCRETE 1" UNLESS SHOWN OR NOTED OTHERWISE.

THE FIRST DIGIT OF A THREE DIGIT BAR NO. AND THE FIRST TWO DIGITS OF A FOUR DIGIT BAR NO. SIGNIFIES THE BAR SIZE.

ELASTOMERIC BEARING PADS NEED NOT BE INDIVIDUALLY MOLDED PROVIDED THE CUT EDGES ARE SMOOTH AND TRUE.

THE SLOPE OF THE FILL IN FRONT OF THE ABUTMENTS SHALL BE COVERED WITH HEAVY RIPRAP TO THE EXTENT SHOWN ON THIS SHEET AND IN THE ABUTMENT DETAILS.

DESIGN DATA

STRUCTURE DESIGNED FOR A FUTURE WEARING SURFACE OF 20 P.S.F.

STANDARD: "STANDARD SPECIFICATIONS FOR ROAD AND BRIDGE CONSTRUCTION", ED. 1975, STATE OF WIS., DEPT. OF TRANS.

LIVE LOAD:

H-20

MAXIMUM ALLOWABLE STRESSES:

CONCRETE MASONRY — SLAB — f_c = 4,000 P.S.I.
— ALL OTHERS — f_c = 3,500 P.S.I.
BAR STEEL REINFORCEMENT, GRADE 60 — f_y = 60,000 P.S.I.
PRETENSIONED CONC. GIRDERS (45"), CONC. — f_c = 6,000 P.S.I.
STRANDS, 1/2" ϕ ULTIMATE TENSILE STRENGTH f_s = 270,000 P.S.I.

HYDRAULIC DATA:

DRAINAGE AREA — 113 SQ.MI.
WATERWAY AREA (STREAM) — 1,350 SQ.FT.
Q 100 — 5,667 C.F.S.
Q 100 (THRU STRUCTURE) — 3,558 C.F.S.
VELOCITY (STREAM @ Q 100) — 4.2 F.P.S.
VELOCITY (THRU BRIDGE @ Q 100) — 6.7 F.P.S.
ROADWAY OVERFLOW CAPACITY — 2,513 C.F.S.
HIGH-WATER ELEV. (Q 100) — 1127.50
BACK-WATER (0.436') + H.W. ELEV. — 1127.936

FOUNDATION DATA:

EXCAVATE TO ELEV. 1116.00 (NORTH) & 1115.20 (SOUTH)
ENGINEER TO VERIFY SOIL CONDITIONS PRIOR TO CONST.
FILL VOIDS WITH CONCRETE LEVELING COURSE, IF REQ'D.
MAXIMUM ALLOWABLE BEARING STRESS FOR FOUNDATION TO BE 3,000 P.S.F. (SEE SUBSURFACE EXPLORATION DATA) SETTLEMENT EST. 3/4" (MAX.)

TRAFFIC DATA:

A.D.T. 75

CLOVERLAND ROAD BRIDGE
GENERAL PLAN
SCALE- AS NOTED
SHEET 4 OF 10
DATE MAY 1976
MORGAN & PARMLEY, LTD.
CONSULTING ENGINEERS
LADYSMITH, WISCONSIN
DR. BY R.O. PARMLEY

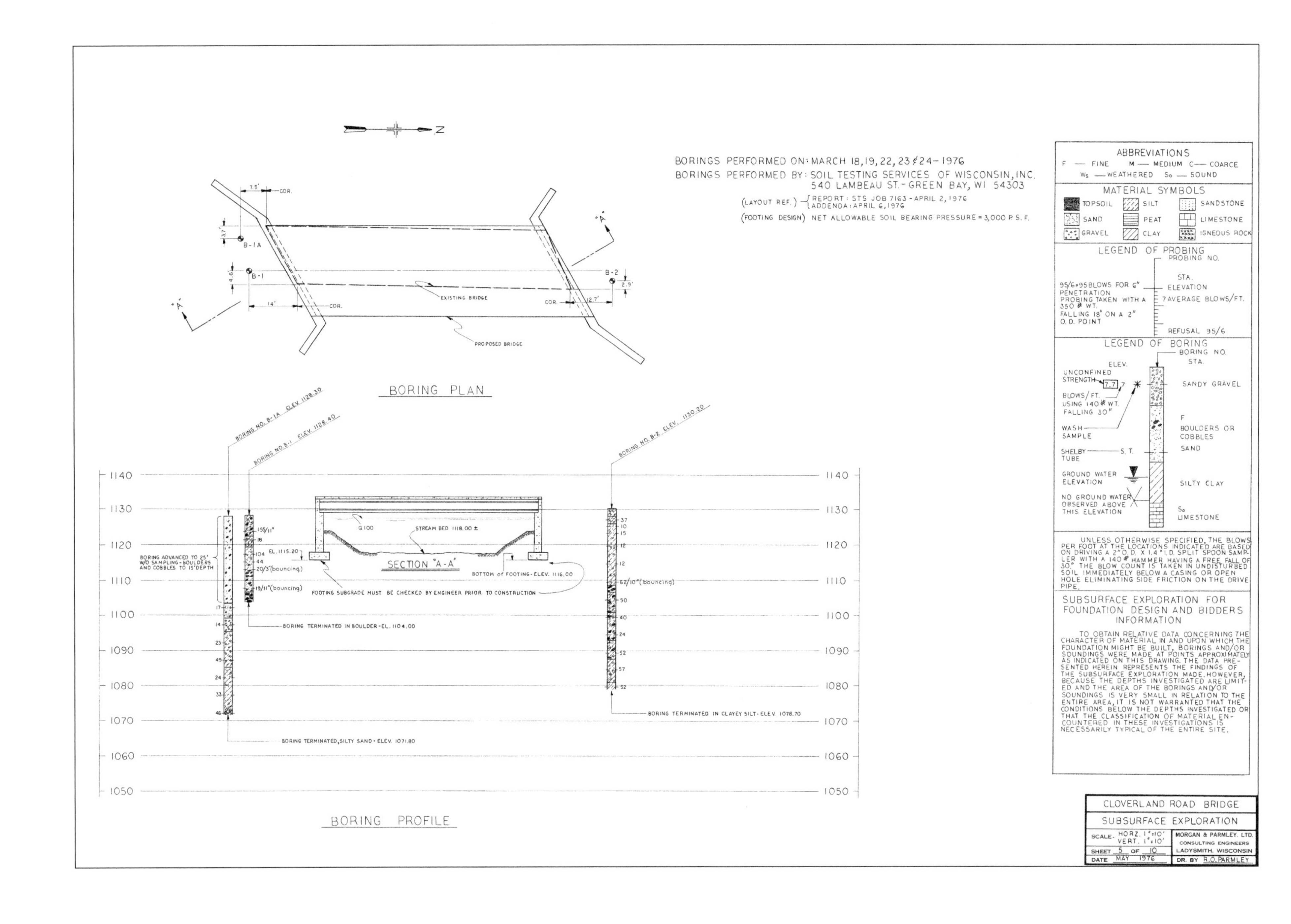

BORINGS PERFORMED ON: MARCH 18, 19, 22, 23 & 24 - 1976
BORINGS PERFORMED BY: SOIL TESTING SERVICES OF WISCONSIN, INC.
540 LAMBEAU ST. - GREEN BAY, WI 54303
(LAYOUT REF.) REPORT: STS JOB 7163 - APRIL 2, 1976
ADDENDA: APRIL 6, 1976
(FOOTING DESIGN) NET ALLOWABLE SOIL BEARING PRESSURE = 3,000 P.S.F.
BORING PLAN
EXISTING BRIDGE
PROPOSED BRIDGE
B-1A
B-1
B-2
COR.
BORING PROFILE
SECTION "A-A"
STREAM BED 1118.00 ±
BOTTOM of FOOTING - ELEV. 1116.00
FOOTING SUBGRADE MUST BE CHECKED BY ENGINEER PRIOR TO CONSTRUCTION
BORING NO. B-1A ELEV. 1128.30
BORING NO. B-1 ELEV. 1128.40
BORING NO. B-2 ELEV. 1130.20
BORING ADVANCED TO 25' W/O SAMPLING - BOULDERS AND COBBLES TO 15' DEPTH
BORING TERMINATED IN BOULDER - EL. 1104.00
BORING TERMINATED, SILTY SAND - ELEV. 1071.80
BORING TERMINATED IN CLAYEY SILT - ELEV. 1078.70
ABBREVIATIONS
F — FINE M — MEDIUM C — COARCE
Ws — WEATHERED So — SOUND
MATERIAL SYMBOLS
TOPSOIL SILT SANDSTONE
SAND PEAT LIMESTONE
GRAVEL CLAY IGNEOUS ROCK
LEGEND OF PROBING
PROBING NO.
STA.
ELEVATION
95/6 = 95 BLOWS FOR 6" PENETRATION
PROBING TAKEN WITH A 350 # WT. FALLING 18" ON A 2" O.D. POINT
7 AVERAGE BLOWS/FT.
REFUSAL 95/6
LEGEND OF BORING
BORING NO.
STA.
ELEV.
UNCONFINED STRENGTH
BLOWS/FT. USING 140 # WT. FALLING 30"
WASH SAMPLE
SHELBY TUBE
S.T.
GROUND WATER ELEVATION
NO GROUND WATER OBSERVED ABOVE THIS ELEVATION
SANDY GRAVEL
BOULDERS OR COBBLES
SAND
SILTY CLAY
So LIMESTONE
UNLESS OTHERWISE SPECIFIED, THE BLOWS PER FOOT AT THE LOCATIONS INDICATED ARE BASED ON DRIVING A 2" O.D. X 1.4" I.D. SPLIT SPOON SAMPLER WITH A 140 # HAMMER HAVING A FREE FALL OF 30." THE BLOW COUNT IS TAKEN IN UNDISTURBED SOIL IMMEDIATELY BELOW A CASING OR OPEN HOLE ELIMINATING SIDE FRICTION ON THE DRIVE PIPE.
SUBSURFACE EXPLORATION FOR FOUNDATION DESIGN AND BIDDERS INFORMATION
TO OBTAIN RELATIVE DATA CONCERNING THE CHARACTER OF MATERIAL IN AND UPON WHICH THE FOUNDATION MIGHT BE BUILT, BORINGS AND/OR SOUNDINGS WERE MADE AT POINTS APPROXIMATELY AS INDICATED ON THIS DRAWING. THE DATA PRESENTED HEREIN REPRESENTS THE FINDINGS OF THE SUBSURFACE EXPLORATION MADE. HOWEVER, BECAUSE THE DEPTHS INVESTIGATED ARE LIMITED AND THE AREA OF THE BORINGS AND/OR SOUNDINGS IS VERY SMALL IN RELATION TO THE ENTIRE AREA, IT IS NOT WARRANTED THAT THE CONDITIONS BELOW THE DEPTHS INVESTIGATED OR THAT THE CLASSIFICATION OF MATERIAL ENCOUNTERED IN THESE INVESTIGATIONS IS NECESSARILY TYPICAL OF THE ENTIRE SITE.
CLOVERLAND ROAD BRIDGE
SUBSURFACE EXPLORATION
SCALE - HORZ. 1" = 10' VERT. 1" = 10'
SHEET 5 OF 10
DATE MAY 1976
MORGAN & PARMLEY, LTD.
CONSULTING ENGINEERS
LADYSMITH, WISCONSIN
DR. BY R.O. PARMLEY

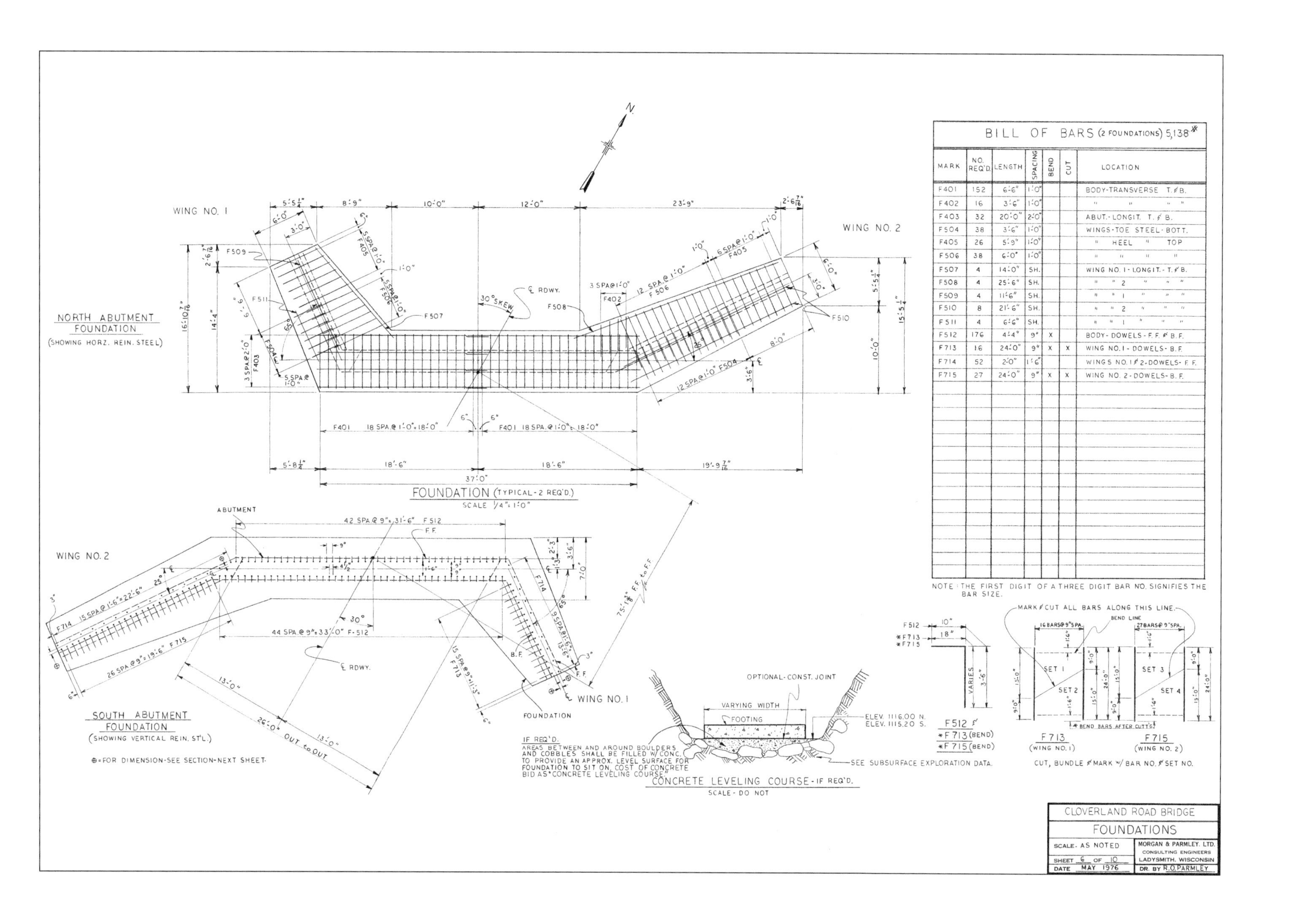

BILL OF BARS (2 FOUNDATIONS) 5,138#

MARK	NO. REQ'D.	LENGTH	SPACING	BEND	CUT	LOCATION
F401	152	6'-6"	1'-0"			BODY-TRANSVERSE T. & B.
F402	16	3'-6"	1'-0"			" " " "
F403	32	20'-0"	2'-0"			ABUT.-LONGIT. T. & B.
F504	38	3'-6"	1'-0"			WINGS-TOE STEEL-BOTT.
F405	26	5'-9"	1'-0"			" HEEL " TOP
F506	38	6'-0"	1'-0"			" " " "
F507	4	14'-0"	SH.			WING NO. 1-LONGIT.-T. & B.
F508	4	25'-6"	SH.			" " 2 " " "
F509	4	11'-6"	SH.			" " 1 " " "
F510	8	21'-6"	SH.			" " 2 " " "
F511	4	6'-6"	SH.			" " 1 " " "
F512	176	4'-4"	9"	X		BODY-DOWELS-F.F. & B.F.
F713	16	24'-0"	9"	X	X	WING NO. 1-DOWELS-B.F.
F714	52	2'-0"	1'-6"			WINGS NO. 1 & 2-DOWELS-F.F.
F715	27	24'-0"	9"	X	X	WING NO. 2-DOWELS-B.F.

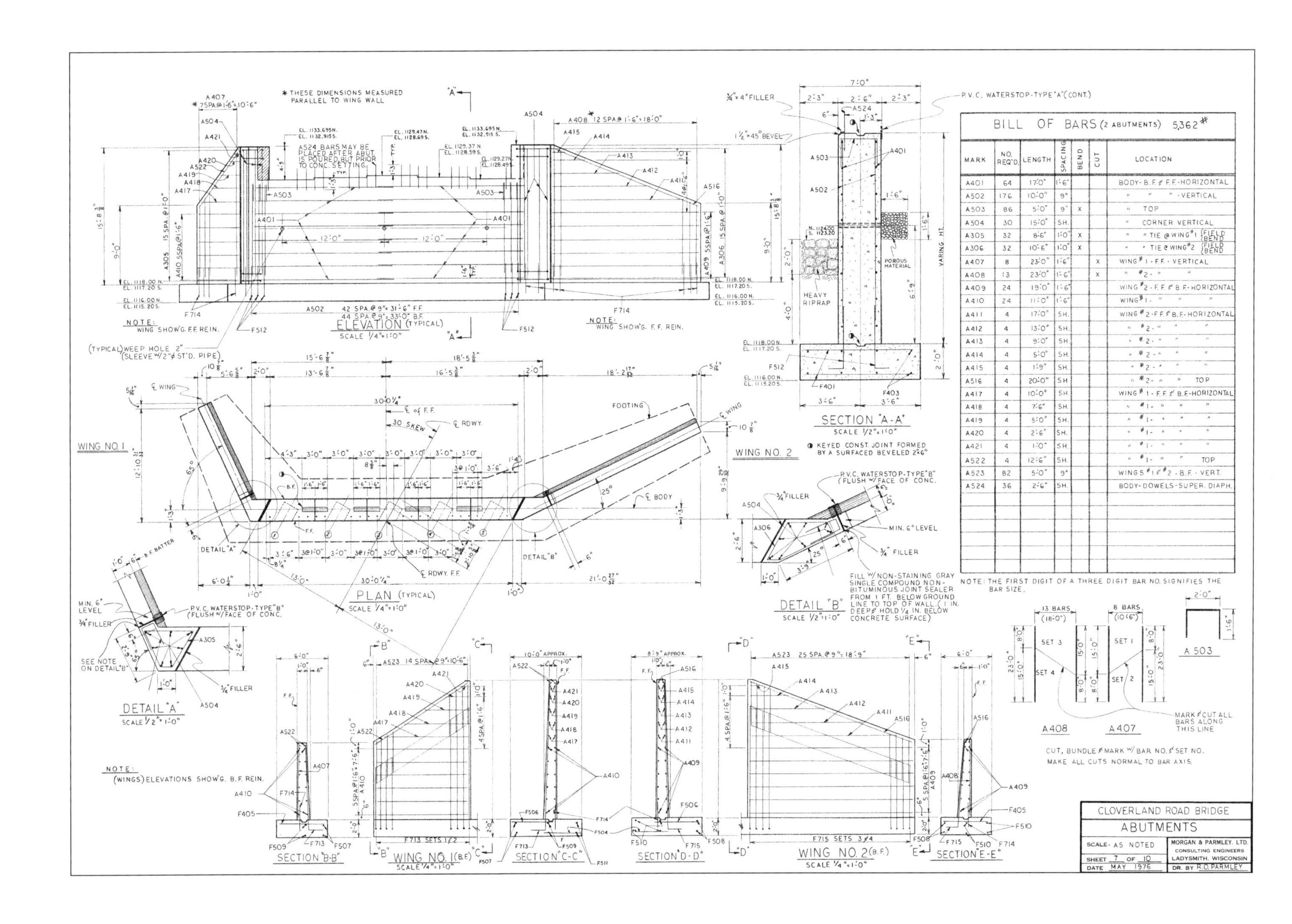

BILL OF BARS (2 ABUTMENTS) 5,362#

MARK	NO. REQ'D.	LENGTH	SPACING	BEND	CUT	LOCATION
A401	64	17'-0"	1'-6"			BODY- B.F. & F.F.-HORIZONTAL
A502	176	10'-0"	9"			" " " -VERTICAL
A503	86	5'-0"	9"	X		" TOP
A504	30	15'-0"	SH.			" CORNER VERTICAL
A305	32	8'-6"	1'-0"	X		" " TIE @ WING #1 {FIELD BEND
A306	32	10'-6"	1'-0"	X		" " TIE @ WING #2 {FIELD BEND
A407	8	23'-0"	1'-6"		X	WING #1 - F.F. - VERTICAL
A408	13	23'-0"	1'-6"		X	" #2 - " "
A409	24	19'-0"	1'-6"			WING #2 - F.F. & B.F.- HORIZONTAL
A410	24	11'-0"	1'-6"			WING #1 - " " "
A411	4	17'-0"	SH.			WING #2 - F.F. & B.F.- HORIZONTAL
A412	4	13'-0"	SH.			" #2 - " " "
A413	4	9'-0"	SH.			" #2 - " " "
A414	4	5'-0"	SH.			" #2 - " " "
A415	4	1'-9"	SH.			" #2 - " " "
A516	4	20'-0"	SH			" #2 - " " TOP
A417	4	10'-0"	SH.			WING #1 - F.F. & B.F.-HORIZONTAL
A418	4	7'-6"	SH.			" #1 - " " "
A419	4	5'-0"	SH.			" #1 - " " "
A420	4	2'-6"	SH.			" #1 - " " "
A421	4	1'-0"	SH.			" #1 - " " "
A522	4	12'-6"	SH.			" #1 - " " TOP
A523	82	5'-0"	9"			WINGS #1 & #2 - B.F. - VERT.
A524	36	2'-6"	SH.			BODY-DOWELS-SUPER. DIAPH.

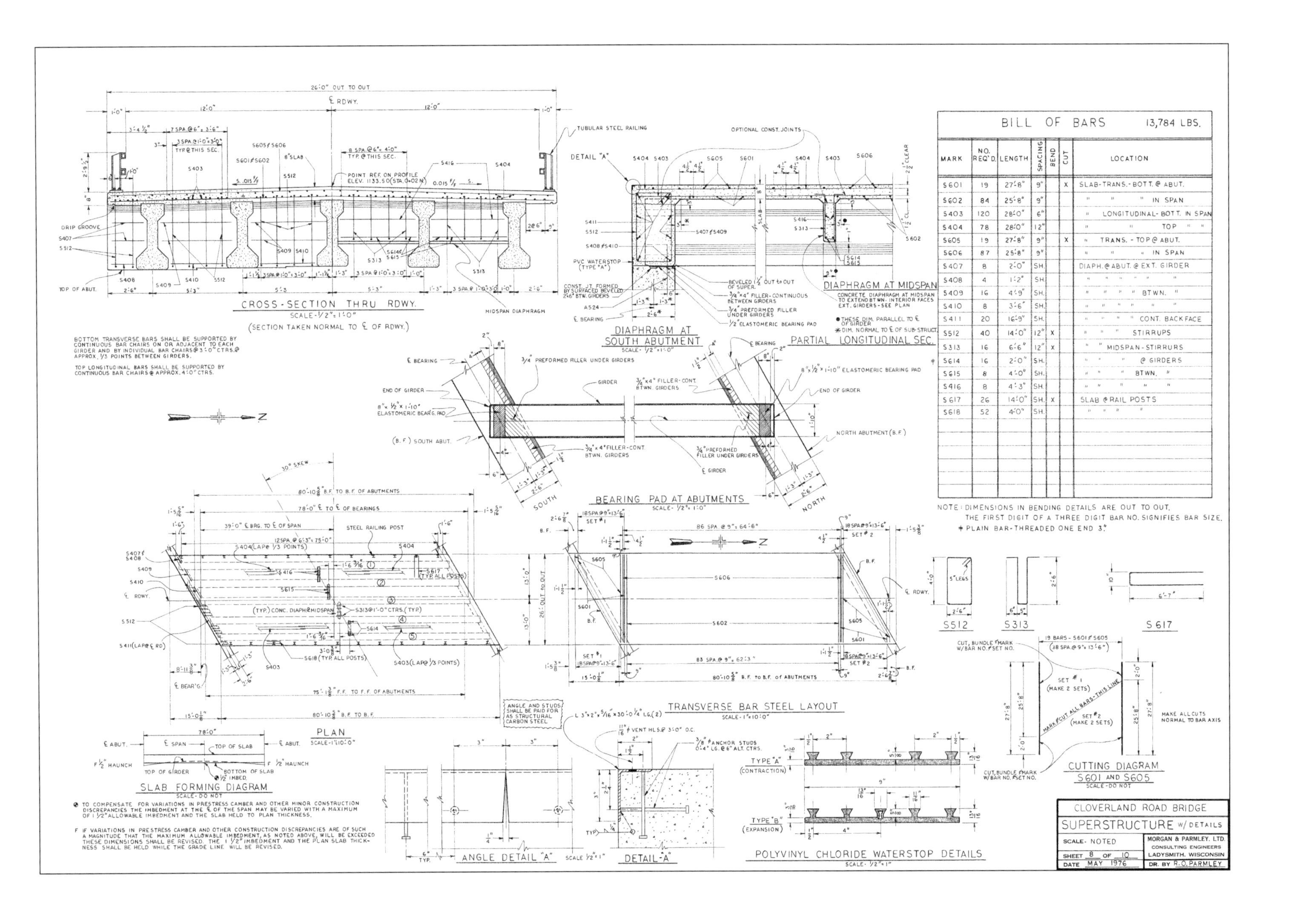

BILL OF BARS 13,784 LBS.

MARK	NO. REQ'D.	LENGTH	SPACING	BEND	CUT	LOCATION
S601	19	27'-8"	9"		X	SLAB-TRANS.-BOTT. @ ABUT.
S602	84	25'-8"	9"			" " " IN SPAN
S403	120	28'-0"	6"			" LONGITUDINAL-BOTT. IN SPAN
S404	78	28'-0"	12"			" " TOP " "
S605	19	27'-8"	9"		X	" TRANS. - TOP @ ABUT.
S606	87	25'-8"	9"			" " " IN SPAN
S407	8	2'-0"	SH.			DIAPH. @ ABUT. @ EXT. GIRDER
S408	4	1'-2"	SH.			" " " " " "
S409	16	4'-9"	SH.			" " " " BTWN. "
S410	8	3'-6"	SH.			" " " " " "
S411	20	16'-9"	SH.			" " " " CONT. BACK FACE
S512	40	14'-0"	12"	X		" " " STIRRUPS
S313	16	6'-6"	12"	X		" " MIDSPAN-STIRRURS
‡ S614	16	2'-0"	SH.			" " " @ GIRDERS
S615	8	4'-0"	SH.			" " " BTWN. "
S416	8	4'-3"	SH.			" " " " "
S617	26	14'-0"	SH.	X		SLAB @ RAIL POSTS
S618	52	4'-0"	SH.			" " " "

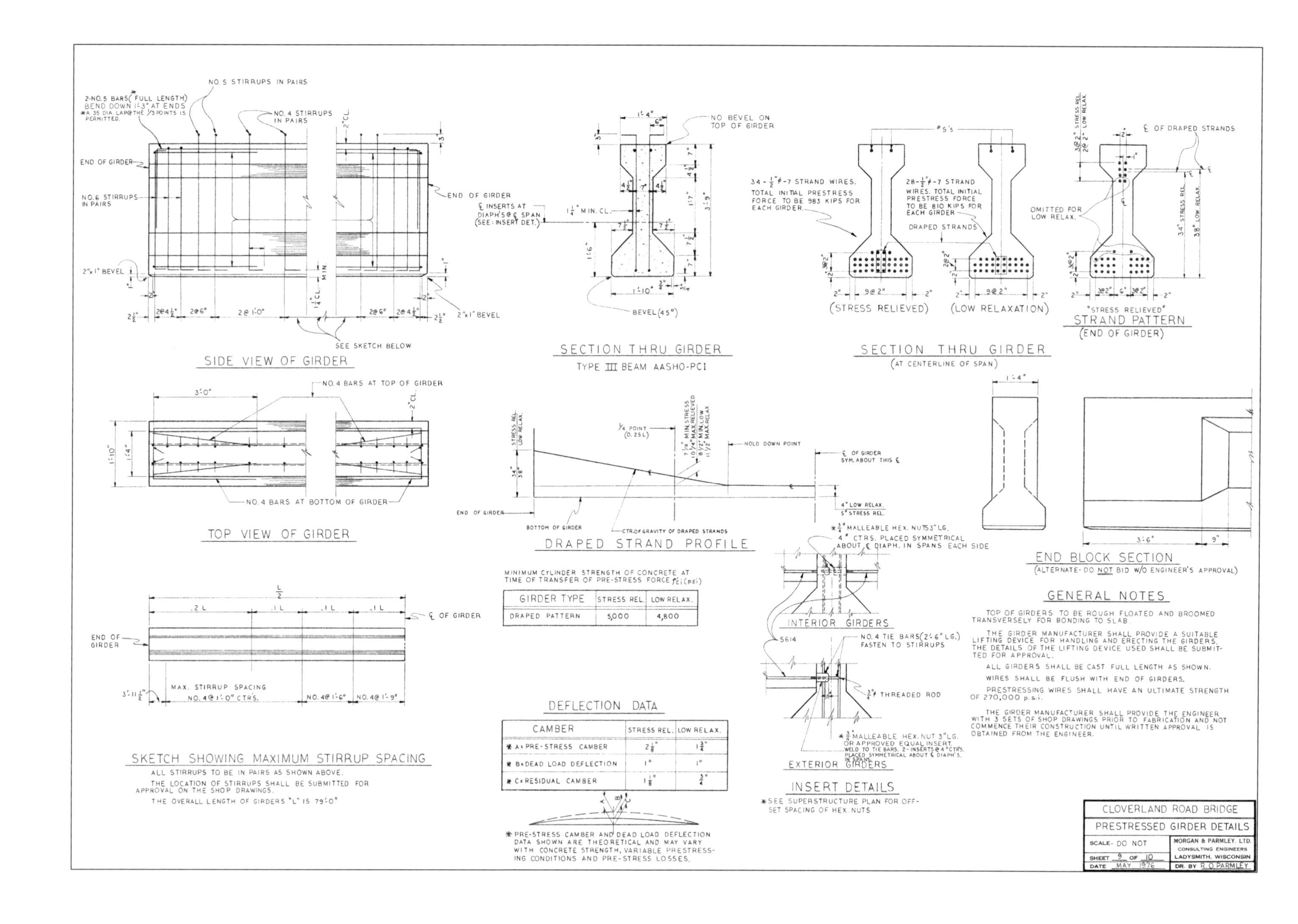

2-NO.5 BARS (FULL LENGTH) BEND DOWN 1'-3" AT ENDS
* A 35 DIA. LAP@THE 1/3 POINTS IS PERMITTED.
NO. 5 STIRRUPS IN PAIRS
NO. 4 STIRRUPS IN PAIRS
END OF GIRDER
NO. 6 STIRRUPS IN PAIRS
2"x1" BEVEL
SEE SKETCH BELOW
SIDE VIEW OF GIRDER
NO BEVEL ON TOP OF GIRDER
℄ INSERTS AT DIAPH'S@℄ SPAN (SEE: INSERT DET.)
1 1/4" MIN. CL.
BEVEL (45°)
SECTION THRU GIRDER
TYPE III BEAM AASHO-PCI
34 - 1/2"φ-7 STRAND WIRES. TOTAL INITIAL PRESTRESS FORCE TO BE 983 KIPS FOR EACH GIRDER.
28 - 1/2"φ-7 STRAND WIRES. TOTAL INITIAL PRESTRESS FORCE TO BE 810 KIPS FOR EACH GIRDER
DRAPED STRANDS
(STRESS RELIEVED)
(LOW RELAXATION)
SECTION THRU GIRDER
(AT CENTERLINE OF SPAN)
℄ OF DRAPED STRANDS
OMITTED FOR LOW RELAX.
"STRESS RELIEVED"
STRAND PATTERN
(END OF GIRDER)
NO. 4 BARS AT TOP OF GIRDER
NO. 4 BARS AT BOTTOM OF GIRDER
TOP VIEW OF GIRDER
1/4 POINT (0.25L)
HOLD DOWN POINT
℄ OF GIRDER SYM. ABOUT THIS ℄
END OF GIRDER
BOTTOM OF GIRDER
CTR. OF GRAVITY OF DRAPED STRANDS
4" LOW RELAX.
5" STRESS REL.
DRAPED STRAND PROFILE
END BLOCK SECTION
(ALTERNATE - DO NOT BID W/O ENGINEER'S APPROVAL)
℄ OF GIRDER
END OF GIRDER
MAX. STIRRUP SPACING
NO. 4@1'-0" CTR'S.
NO. 4@1'-6"
NO. 4@1'-9"
SKETCH SHOWING MAXIMUM STIRRUP SPACING
ALL STIRRUPS TO BE IN PAIRS AS SHOWN ABOVE.
THE LOCATION OF STIRRUPS SHALL BE SUBMITTED FOR APPROVAL ON THE SHOP DRAWINGS.
THE OVERALL LENGTH OF GIRDERS "L" IS 79'-0"
MINIMUM CYLINDER STRENGTH OF CONCRETE AT TIME OF TRANSFER OF PRE-STRESS FORCE f'ci (psi)
GIRDER TYPE | STRESS REL. | LOW RELAX.
DRAPED PATTERN | 5,000 | 4,800
DEFLECTION DATA
CAMBER | STRESS REL. | LOW RELAX.
* A = PRE-STRESS CAMBER | 2 1/8" | 1 3/4"
* B = DEAD LOAD DEFLECTION | 1" | 1"
* C = RESIDUAL CAMBER | 1 1/8" | 3/4"
* PRE-STRESS CAMBER AND DEAD LOAD DEFLECTION DATA SHOWN ARE THEORETICAL AND MAY VARY WITH CONCRETE STRENGTH, VARIABLE PRESTRESSING CONDITIONS AND PRE-STRESS LOSSES.
* 3/4" MALLEABLE HEX. NUTS 3" LG. 4" CTRS. PLACED SYMMETRICAL ABOUT ℄ DIAPH. IN SPANS EACH SIDE
INTERIOR GIRDERS
S614
NO. 4 TIE BARS (2'-6" LG.) FASTEN TO STIRRUPS
3/4" THREADED ROD
* 3/4" MALLEABLE HEX. NUT 3" LG. OR APPROVED EQUAL INSERT.
EXTERIOR GIRDERS
INSERT DETAILS
* SEE SUPERSTRUCTURE PLAN FOR OFF-SET SPACING OF HEX. NUTS
GENERAL NOTES
TOP OF GIRDERS TO BE ROUGH FLOATED AND BROOMED TRANSVERSELY FOR BONDING TO SLAB.
THE GIRDER MANUFACTURER SHALL PROVIDE A SUITABLE LIFTING DEVICE FOR HANDLING AND ERECTING THE GIRDERS. THE DETAILS OF THE LIFTING DEVICE USED SHALL BE SUBMITTED FOR APPROVAL.
ALL GIRDERS SHALL BE CAST FULL LENGTH AS SHOWN.
WIRES SHALL BE FLUSH WITH END OF GIRDERS.
PRESTRESSING WIRES SHALL HAVE AN ULTIMATE STRENGTH OF 270,000 p.s.i.
THE GIRDER MANUFACTURER SHALL PROVIDE THE ENGINEER WITH 3 SETS OF SHOP DRAWINGS PRIOR TO FABRICATION AND NOT COMMENCE THEIR CONSTRUCTION UNTIL WRITTEN APPROVAL IS OBTAINED FROM THE ENGINEER.
CLOVERLAND ROAD BRIDGE
PRESTRESSED GIRDER DETAILS
SCALE - DO NOT
SHEET 9 OF 10
DATE MAY 1976
MORGAN & PARMLEY, LTD.
CONSULTING ENGINEERS
LADYSMITH, WISCONSIN
DR. BY R.O. PARMLEY

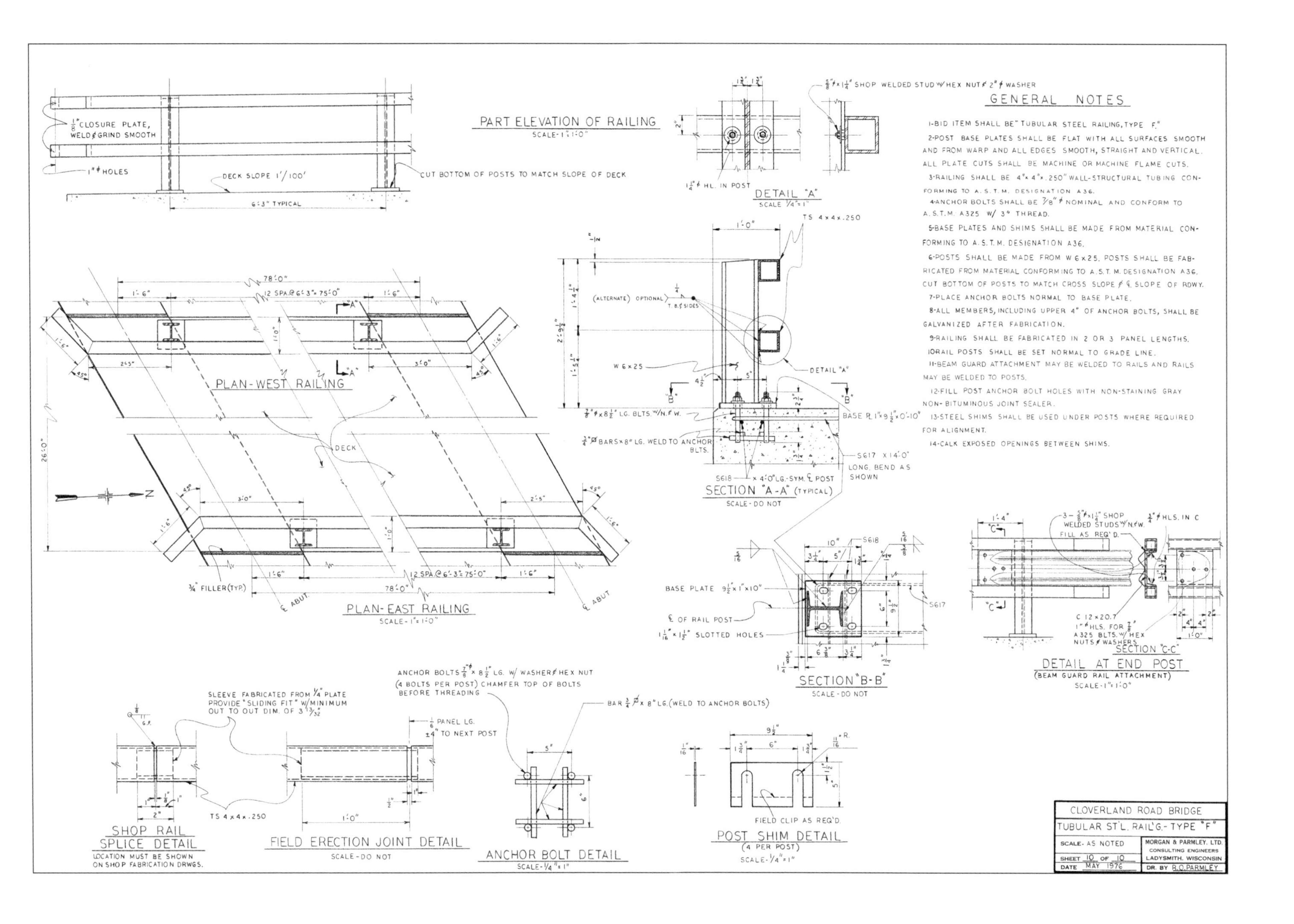
PART ELEVATION OF RAILING
SCALE-1"=1'-0"
1/8" CLOSURE PLATE, WELD & GRIND SMOOTH
1" ϕ HOLES
DECK SLOPE 1'/100'
CUT BOTTOM OF POSTS TO MATCH SLOPE OF DECK
6'-3" TYPICAL
DETAIL "A"
SCALE 1/4"=1"
5/8" ϕ x 1 1/4" SHOP WELDED STUD W/HEX NUT & 2" ϕ WASHER
1 1/4" ϕ HL. IN POST
GENERAL NOTES
1-BID ITEM SHALL BE "TUBULAR STEEL RAILING, TYPE F."
2-POST BASE PLATES SHALL BE FLAT WITH ALL SURFACES SMOOTH AND FROM WARP AND ALL EDGES SMOOTH, STRAIGHT AND VERTICAL. ALL PLATE CUTS SHALL BE MACHINE OR MACHINE FLAME CUTS.
3-RAILING SHALL BE 4"x 4"x .250" WALL-STRUCTURAL TUBING CONFORMING TO A.S.T.M. DESIGNATION A36.
4-ANCHOR BOLTS SHALL BE 7/8" ϕ NOMINAL AND CONFORM TO A.S.T.M. A325 W/ 3" THREAD.
5-BASE PLATES AND SHIMS SHALL BE MADE FROM MATERIAL CONFORMING TO A.S.T.M. DESIGNATION A36.
6-POSTS SHALL BE MADE FROM W 6x25. POSTS SHALL BE FABRICATED FROM MATERIAL CONFORMING TO A.S.T.M. DESIGNATION A36. CUT BOTTOM OF POSTS TO MATCH CROSS SLOPE & ℄ SLOPE OF RDWY.
7-PLACE ANCHOR BOLTS NORMAL TO BASE PLATE.
8-ALL MEMBERS, INCLUDING UPPER 4" OF ANCHOR BOLTS, SHALL BE GALVANIZED AFTER FABRICATION.
9-RAILING SHALL BE FABRICATED IN 2 OR 3 PANEL LENGTHS.
10-RAIL POSTS SHALL BE SET NORMAL TO GRADE LINE.
11-BEAM GUARD ATTACHMENT MAY BE WELDED TO RAILS AND RAILS MAY BE WELDED TO POSTS.
12-FILL POST ANCHOR BOLT HOLES WITH NON-STAINING GRAY NON-BITUMINOUS JOINT SEALER.
13-STEEL SHIMS SHALL BE USED UNDER POSTS WHERE REQUIRED FOR ALIGNMENT.
14-CALK EXPOSED OPENINGS BETWEEN SHIMS.
TS 4x4x.250
(ALTERNATE) OPTIONAL
T. B. & SIDES
W 6x25
DETAIL "A"
7/8" ϕ x 8 1/2" LG. BLTS. W/N. & W.
BASE PL 1"x 9 1/2"x 0'-10"
3/4" ϕ BARS x 8" LG. WELD TO ANCHOR BLTS.
S617 x 14'-0" LONG, BEND AS SHOWN
S618 x 4'-0" LG.-SYM. ℄ POST
SECTION "A-A" (TYPICAL)
SCALE-DO NOT
78'-0"
12 SPA.@ 6'-3" = 75'-0"
1'-6"
PLAN-WEST RAILING
DECK
26'-0"
N
3/4" FILLER (TYP.)
℄ ABUT.
PLAN-EAST RAILING
SCALE-1"=1'-0"
BASE PLATE 9 1/2"x 1"x 10"
℄ OF RAIL POST
1 1/16"x 1 1/2" SLOTTED HOLES
S618
S617
SECTION "B-B"
SCALE-DO NOT
3-5/8" ϕ x 1 1/4" SHOP WELDED STUDS W/N. & W.
FILL AS REQ'D.
1/4" ϕ HLS. IN C
C 12x20.7
1" ϕ HLS. FOR 7/8" A325 BLTS. W/ HEX NUTS & WASHERS
SECTION "C-C"
DETAIL AT END POST
(BEAM GUARD RAIL ATTACHMENT)
SCALE-1"=1'-0"
ANCHOR BOLTS 7/8" ϕ x 8 1/2" LG. W/ WASHER & HEX NUT (4 BOLTS PER POST) CHAMFER TOP OF BOLTS BEFORE THREADING
SLEEVE FABRICATED FROM 1/4" PLATE PROVIDE "SLIDING FIT" W/MINIMUM OUT TO OUT DIM. OF 3 13/32"
BAR 3/4" ϕ x 8" LG. (WELD TO ANCHOR BOLTS)
1/6 PANEL LG. ±4" TO NEXT POST
TS 4x4x.250
SHOP RAIL SPLICE DETAIL
LOCATION MUST BE SHOWN ON SHOP FABRICATION DRWGS.
FIELD ERECTION JOINT DETAIL
SCALE-DO NOT
ANCHOR BOLT DETAIL
SCALE-1/4"=1"
FIELD CLIP AS REQ'D.
POST SHIM DETAIL
(4 PER POST)
SCALE-1/4"=1"
CLOVERLAND ROAD BRIDGE
TUBULAR ST'L. RAIL'G.- TYPE "F"
SCALE- AS NOTED
SHEET 10 OF 10
DATE MAY 1976
MORGAN & PARMLEY, LTD.
CONSULTING ENGINEERS
LADYSMITH, WISCONSIN
DR. BY R.O.PARMLEY

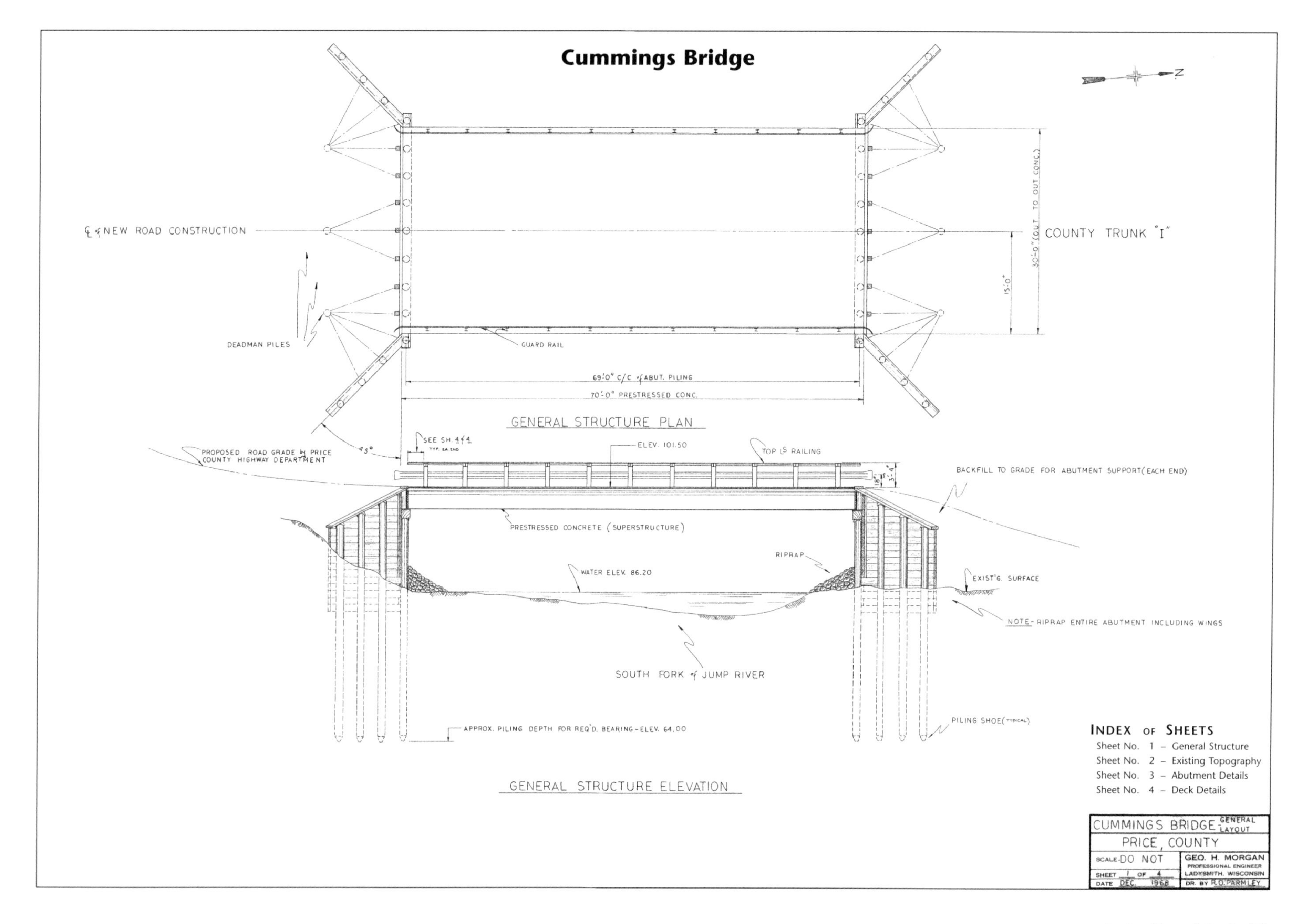

Courtesy: Morgan & Parmley, Ltd.

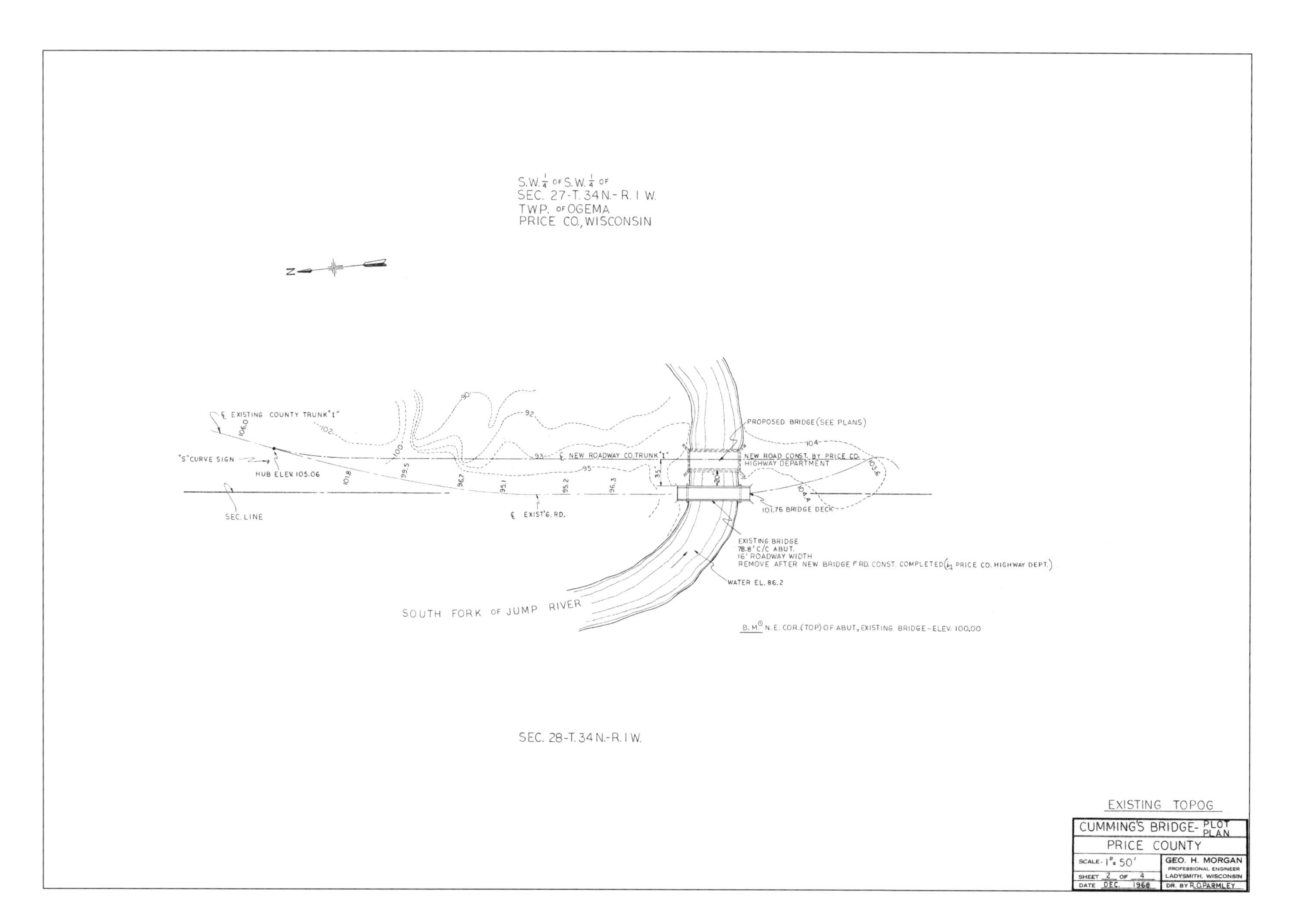
S.W.¼ OF S.W.¼ OF
SEC. 27-T. 34N.-R. I W.
TWP. OF OGEMA
PRICE CO., WISCONSIN
N
℄ EXISTING COUNTY TRUNK "I"
"S" CURVE SIGN
HUB ELEV. 105.06
SEC. LINE
℄ NEW ROADWAY CO. TRUNK "I"
℄ EXIST'G. RD.
PROPOSED BRIDGE (SEE PLANS)
NEW ROAD CONST. BY PRICE CO. HIGHWAY DEPARTMENT
101.76 BRIDGE DECK
EXISTING BRIDGE
78.8' C/C ABUT.
16' ROADWAY WIDTH
REMOVE AFTER NEW BRIDGE & RD. CONST. COMPLETED (by PRICE CO. HIGHWAY DEPT.)
WATER EL. 86.2
SOUTH FORK OF JUMP RIVER
B.M. N. E. COR. (TOP) OF ABUT., EXISTING BRIDGE - ELEV. 100.00
SEC. 28-T. 34 N.-R. I W.
EXISTING TOPOG
CUMMING'S BRIDGE- PLOT PLAN
PRICE COUNTY
SCALE- 1" = 50'
SHEET 2 OF 4
DATE DEC. 1968
GEO. H. MORGAN
PROFESSIONAL ENGINEER
LADYSMITH, WISCONSIN
DR. BY R.G. PARMLEY

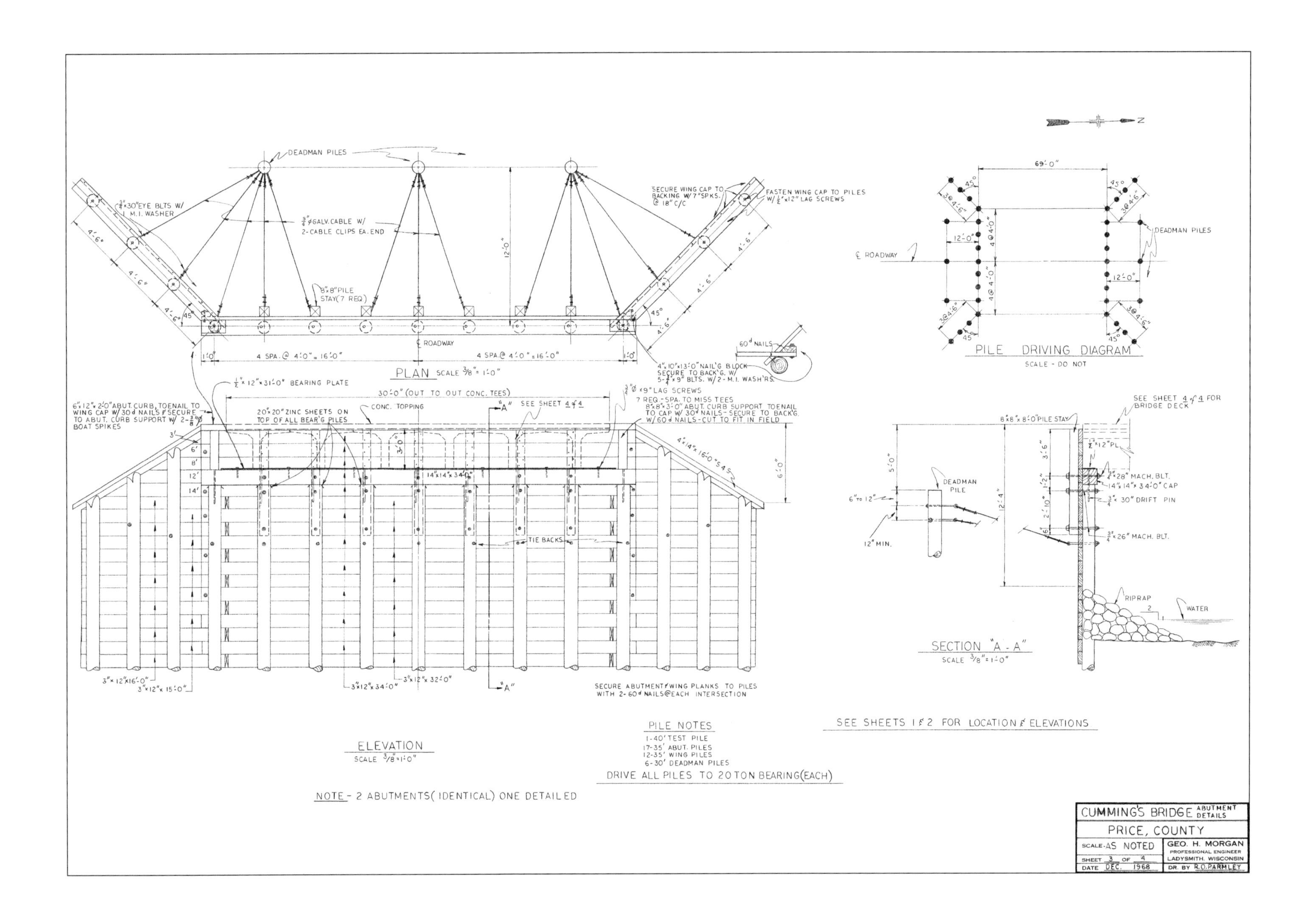

DEADMAN PILES
PLAN SCALE 3/8" = 1'-0"
ELEVATION SCALE 3/8" = 1'-0"
PILE DRIVING DIAGRAM SCALE - DO NOT
SECTION "A - A" SCALE 3/8" = 1'-0"
PILE NOTES
1-40' TEST PILE
17-35' ABUT. PILES
12-35' WING PILES
6-30' DEADMAN PILES
DRIVE ALL PILES TO 20 TON BEARING(EACH)
SEE SHEETS 1 & 2 FOR LOCATION & ELEVATIONS
NOTE - 2 ABUTMENTS(IDENTICAL) ONE DETAILED
SECURE ABUTMENT & WING PLANKS TO PILES WITH 2-60d NAILS@EACH INTERSECTION
CUMMING'S BRIDGE ABUTMENT DETAILS
PRICE, COUNTY
SCALE-AS NOTED
GEO. H. MORGAN
PROFESSIONAL ENGINEER
LADYSMITH, WISCONSIN
SHEET 3 OF 4
DATE DEC. 1968
DR. BY R.O.PARMLEY

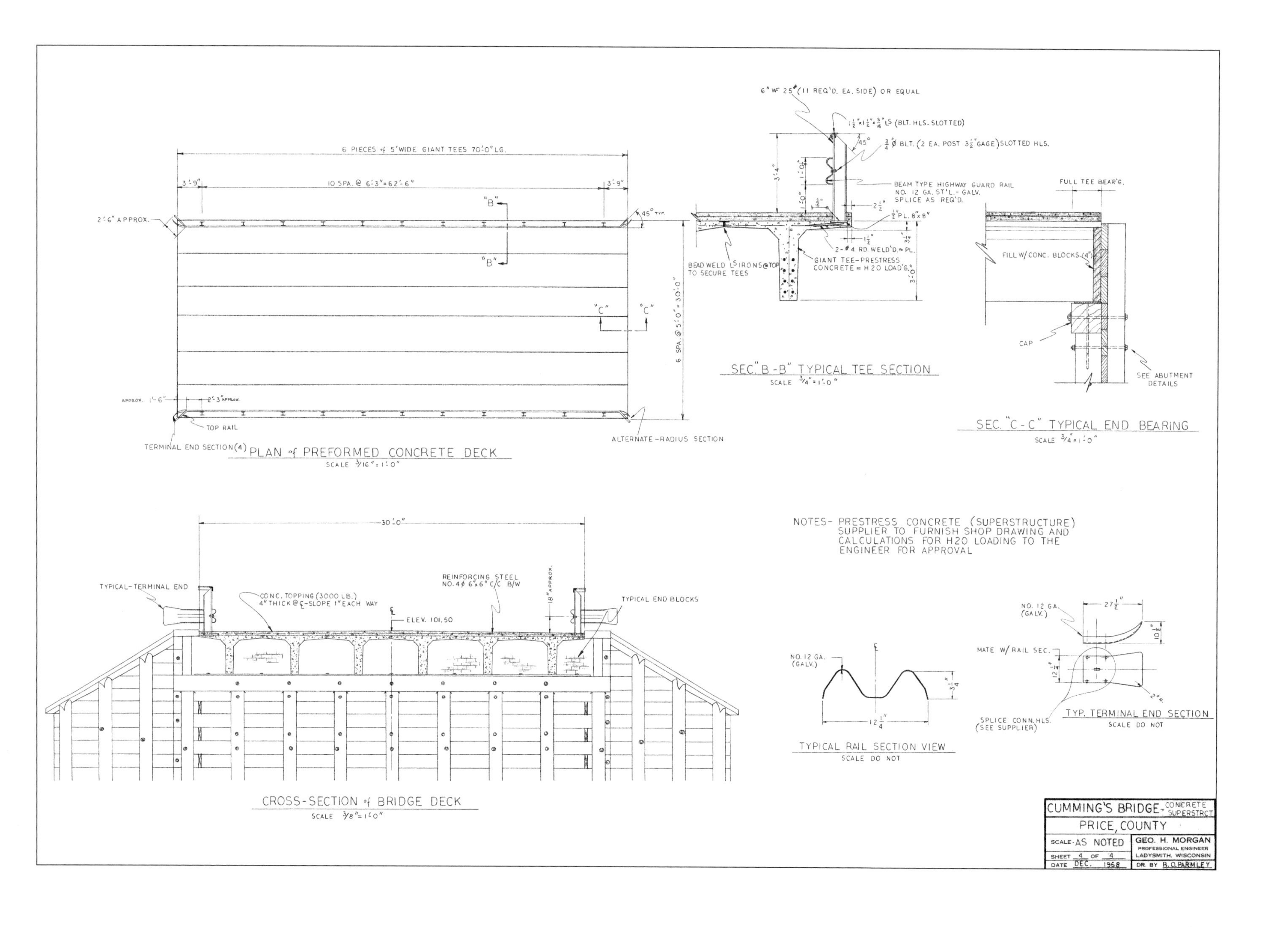
6 PIECES of 5' WIDE GIANT TEES 70'-0" LG.
10 SPA. @ 6'-3" = 62'-6"
3'-9"
2'-6" APPROX.
45° TYP.
"B"
"C"
6 SPA. @ 5'-0" = 30'-0"
APPROX. 1'-6"
2'-3" APPROX.
TOP RAIL
TERMINAL END SECTION(4)
ALTERNATE-RADIUS SECTION
PLAN of PREFORMED CONCRETE DECK
SCALE 3/16" = 1'-0"
6" WF 25# (11 REQ'D. EA. SIDE) OR EQUAL
BEAM TYPE HIGHWAY GUARD RAIL NO. 12 GA. ST'L. - GALV. SPLICE AS REQ'D.
2-#4 RD. WELD'D. = PL.
BEAD WELD L'S IRONS @ TOP TO SECURE TEES
GIANT TEE-PRESTRESS CONCRETE = H20 LOAD'G.
SEC. "B-B" TYPICAL TEE SECTION
SCALE 3/4" = 1'-0"
FULL TEE BEAR'G.
FILL W/ CONC. BLOCKS (4")
CAP
SEE ABUTMENT DETAILS
SEC. "C-C" TYPICAL END BEARING
SCALE 3/4" = 1'-0"
30'-0"
TYPICAL-TERMINAL END
CONC. TOPPING (3000 LB.) 4" THICK @ ℄-SLOPE 1" EACH WAY
REINFORCING STEEL NO. 4 ϕ 6"x6" C/C B/W
18" APPROX.
ELEV. 101.50
TYPICAL END BLOCKS
CROSS-SECTION of BRIDGE DECK
SCALE 3/8" = 1'-0"
NOTES- PRESTRESS CONCRETE (SUPERSTRUCTURE) SUPPLIER TO FURNISH SHOP DRAWING AND CALCULATIONS FOR H20 LOADING TO THE ENGINEER FOR APPROVAL
NO. 12 GA. (GALV.)
12 1/4"
3 1/4"
TYPICAL RAIL SECTION VIEW
SCALE DO NOT
27 1/2"
MATE W/ RAIL SEC.
SPLICE CONN. HLS. (SEE SUPPLIER)
TYP. TERMINAL END SECTION
SCALE DO NOT
CUMMING'S BRIDGE - CONCRETE SUPERSTRCT
PRICE, COUNTY
SCALE - AS NOTED
GEO. H. MORGAN
PROFESSIONAL ENGINEER
LADYSMITH, WISCONSIN
SHEET 4 OF 4
DATE DEC. 1968
DR. BY R. O. PARMLEY

DEVILS CREEK SNOWMOBILE BRIDGE

RUSK COUNTY FORESTRY DEPARTMENT

RUSK COUNTY, WISCONSIN

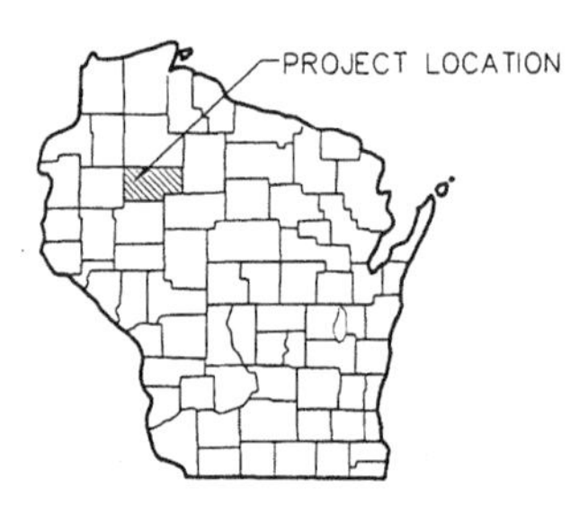

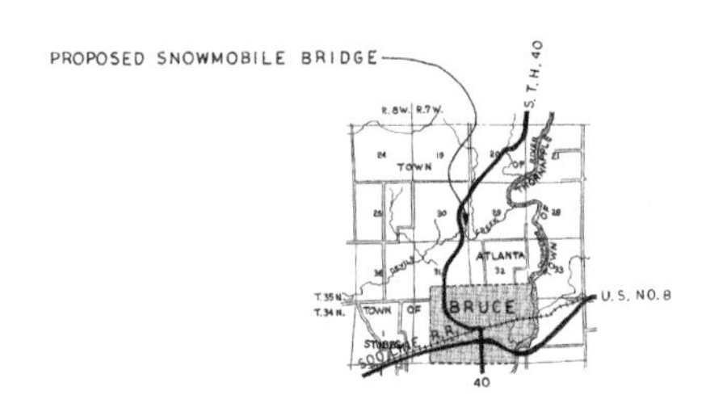

PROJECT LOCATION

SCALE 1 MILE =

I N D E X

1 — PROJECT LOCATION & INDEX

2 — EXISTING SITE PLAN, TOPOGRAPHY, CROSS-SECTIONS & GENERAL MAP

3 — HYDRAULIC CROSS-SECTION, PLAN VIEW & RELATED DATA

4 — BRIDGE PLAN, ELEVATION & SECTION

5 — BRIDGE ABUTMENT, DETAILS, PILING NOTES & RAIL POSTS

NOTE: SEE SPECIFICATION BOOK

SNOWMOBILE BRIDGE	
DEVILS CREEK CROSSING RUSK COUNTY, WISCONSIN	
SCALE: NOTED	GEO. H. MORGAN PROFESSIONAL ENGINEER LADYSMITH, WISCONSIN
SHEET 1 OF 5	
DATE JAN. 1976	DR. BY R.O. PARMLEY

Courtesy: Morgan & Parmley, Ltd.

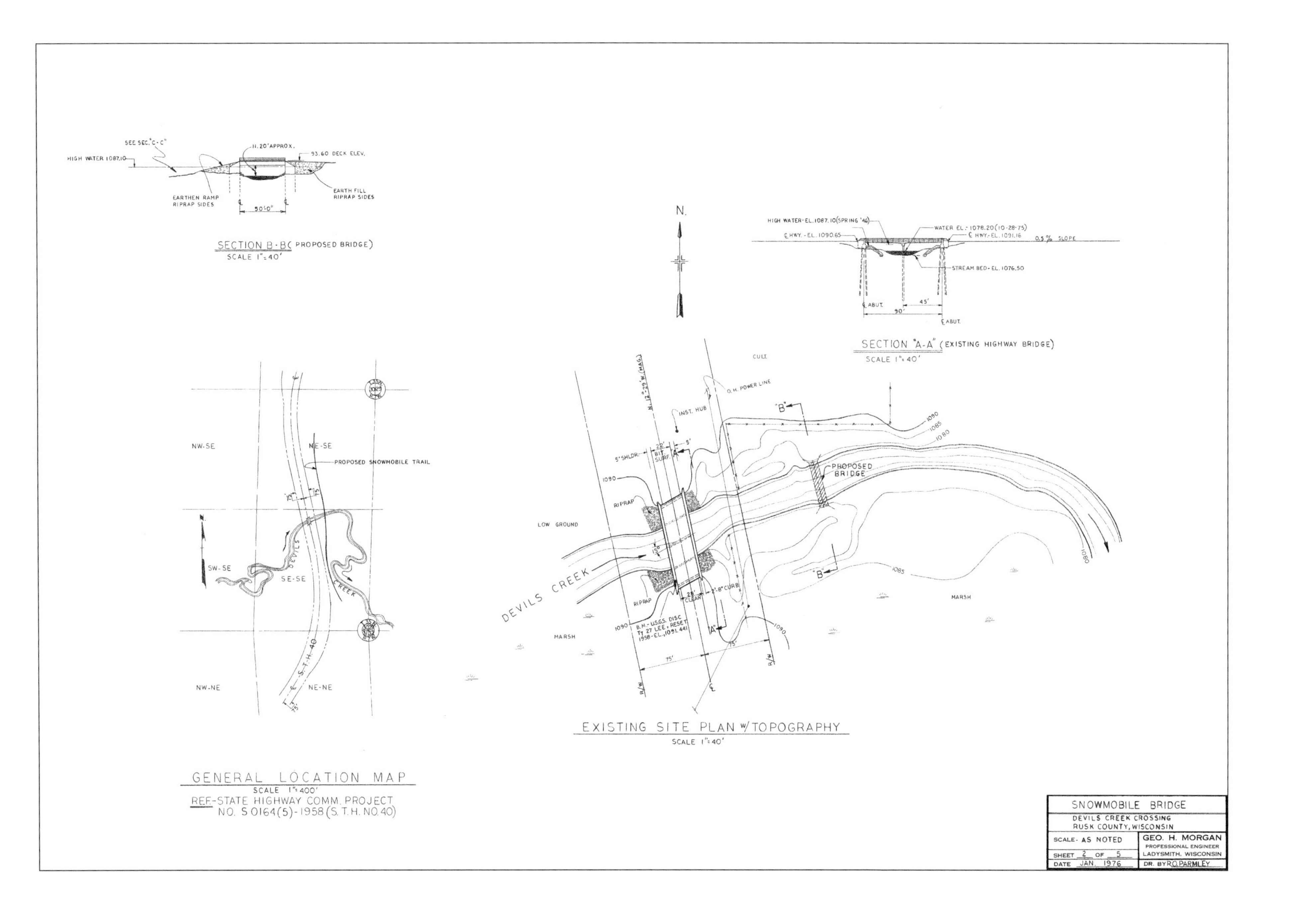
SECTION B-B (PROPOSED BRIDGE)
SCALE 1"=40'
SECTION "A-A" (EXISTING HIGHWAY BRIDGE)
SCALE 1"=40'
GENERAL LOCATION MAP
SCALE 1"=400'
REF-STATE HIGHWAY COMM. PROJECT
NO. S 0164(5)-1958 (S.T.H. NO. 40)
EXISTING SITE PLAN w/TOPOGRAPHY
SCALE 1"=40'
SNOWMOBILE BRIDGE
DEVILS CREEK CROSSING
RUSK COUNTY, WISCONSIN
SCALE- AS NOTED
SHEET 2 OF 5
DATE JAN. 1976
GEO. H. MORGAN
PROFESSIONAL ENGINEER
LADYSMITH, WISCONSIN
DR. BY R.O. PARMLEY

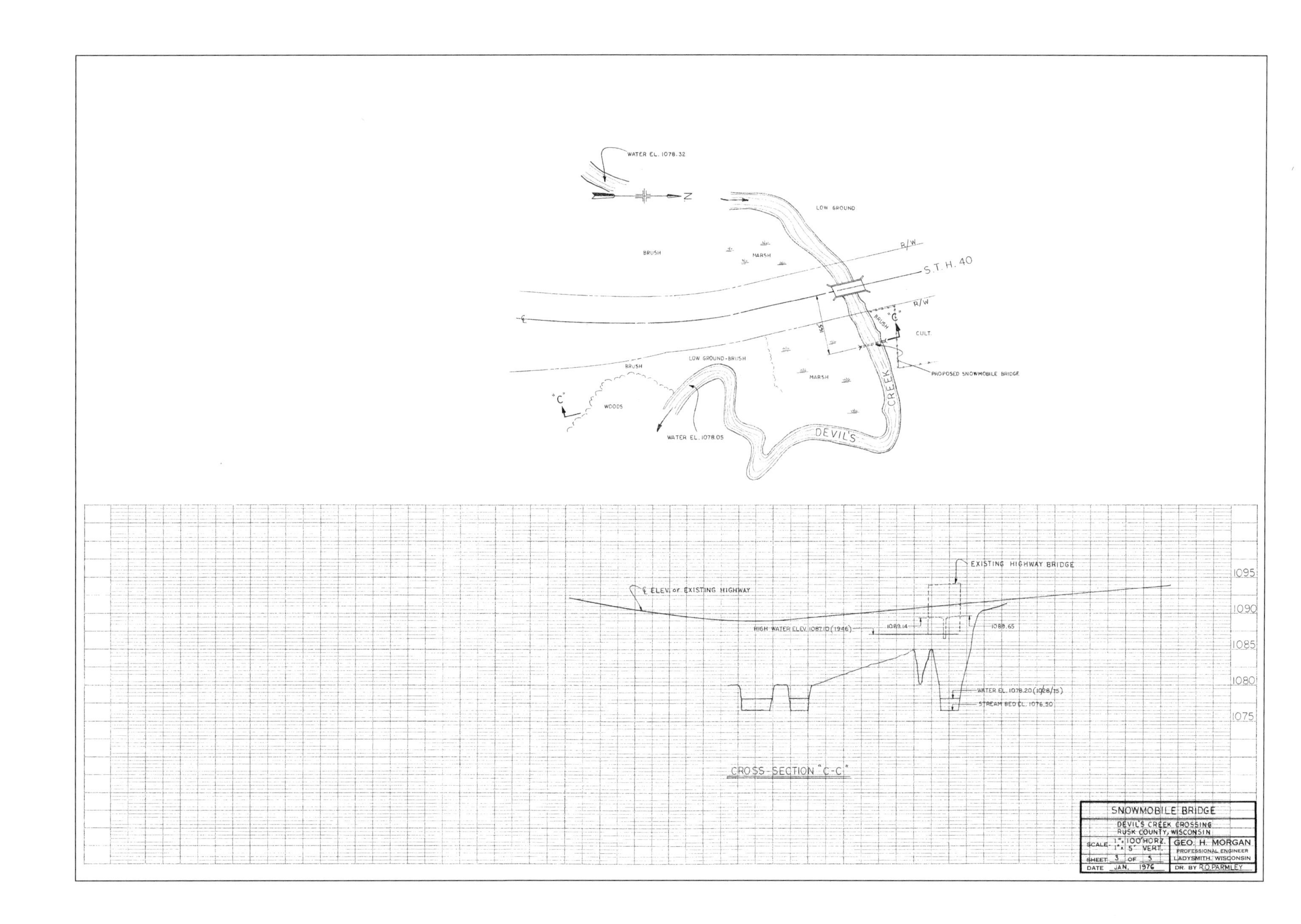
WATER EL. 1078.32
N
LOW GROUND
BRUSH
MARSH
R/W
S.T.H. 40
R/W
BRUSH
"C"
CULT.
PROPOSED SNOWMOBILE BRIDGE
LOW GROUND-BRUSH
BRUSH
MARSH
CREEK
"C"
WOODS
DEVIL'S
WATER EL. 1078.05
EXISTING HIGHWAY BRIDGE
℄ ELEV. of EXISTING HIGHWAY
HIGH WATER ELEV. 1087.10 (1946)
1089.14
1089.65
WATER EL. 1078.20 (10/28/75)
STREAM BED EL. 1076.50
1095
1090
1085
1080
1075
CROSS-SECTION "C-C"
SNOWMOBILE BRIDGE
DEVIL'S CREEK CROSSING
RUSK COUNTY, WISCONSIN
SCALE 1"=100' HORZ. 1"=5' VERT.
SHEET 3 OF 5
DATE JAN. 1976
GEO. H. MORGAN
PROFESSIONAL ENGINEER
LADYSMITH, WISCONSIN
DR. BY R.O. PARMLEY

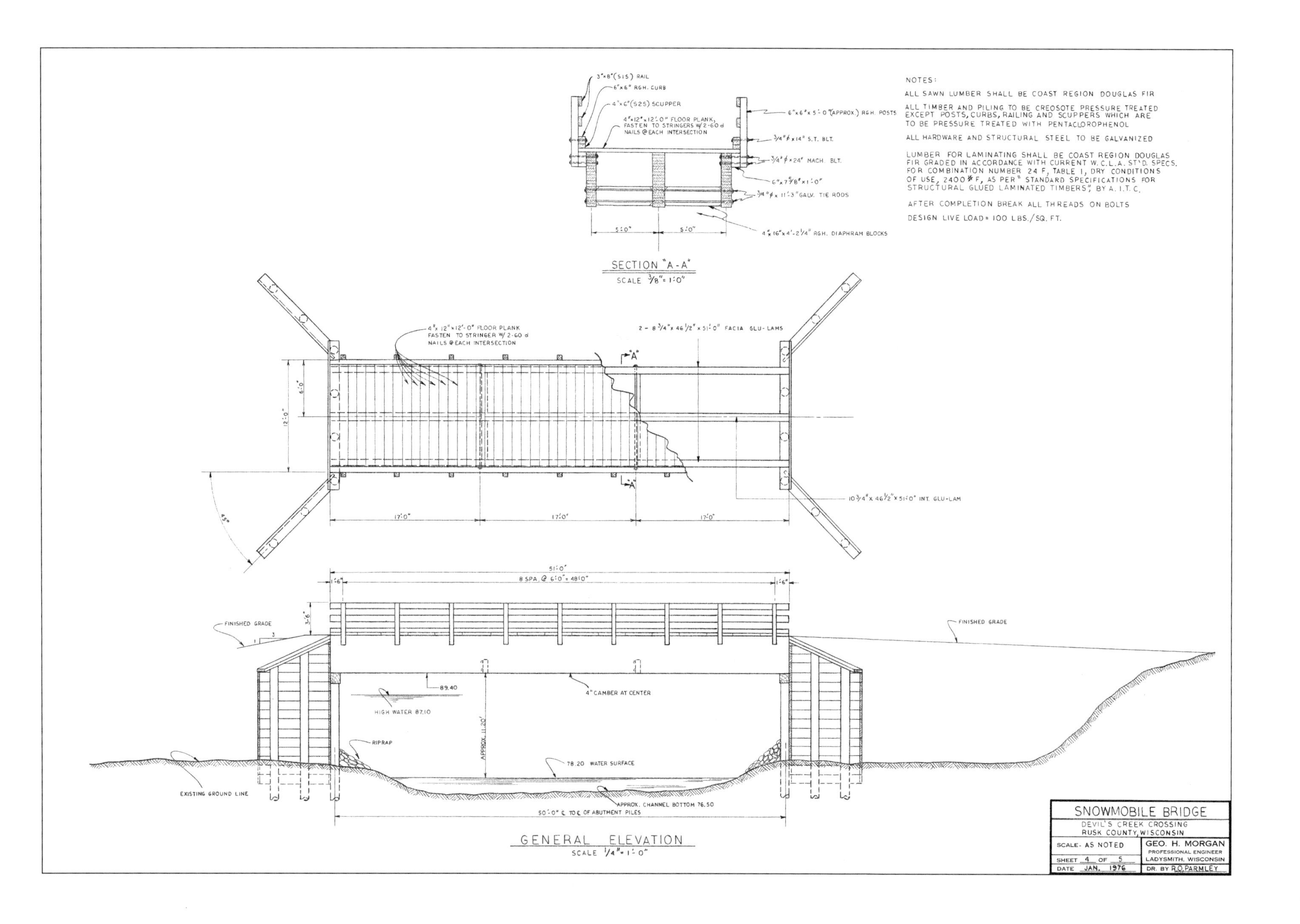
NOTES:
ALL SAWN LUMBER SHALL BE COAST REGION DOUGLAS FIR
ALL TIMBER AND PILING TO BE CREOSOTE PRESSURE TREATED EXCEPT POSTS, CURBS, RAILING AND SCUPPERS WHICH ARE TO BE PRESSURE TREATED WITH PENTACLOROPHENOL
ALL HARDWARE AND STRUCTURAL STEEL TO BE GALVANIZED
LUMBER FOR LAMINATING SHALL BE COAST REGION DOUGLAS FIR GRADED IN ACCORDANCE WITH CURRENT W.C.L.A. ST'D. SPECS. FOR COMBINATION NUMBER 24 F, TABLE I, DRY CONDITIONS OF USE, 2400# F, AS PER "STANDARD SPECIFICATIONS FOR STRUCTURAL GLUED LAMINATED TIMBERS", BY A.I.T.C.
AFTER COMPLETION BREAK ALL THREADS ON BOLTS
DESIGN LIVE LOAD = 100 LBS./SQ. FT.
SECTION "A-A"
SCALE 3/8" = 1'-0"
GENERAL ELEVATION
SCALE 1/4" = 1'-0"
FINISHED GRADE
EXISTING GROUND LINE
HIGH WATER 87.10
RIPRAP
78.20 WATER SURFACE
4" CAMBER AT CENTER
APPROX. CHANNEL BOTTOM 76.50
50'-0" ℄ TO ℄ OF ABUTMENT PILES
SNOWMOBILE BRIDGE
DEVIL'S CREEK CROSSING
RUSK COUNTY, WISCONSIN
SCALE- AS NOTED
GEO. H. MORGAN
PROFESSIONAL ENGINEER
LADYSMITH, WISCONSIN
SHEET 4 OF 5
DATE JAN. 1976
DR. BY R.O. PARMLEY

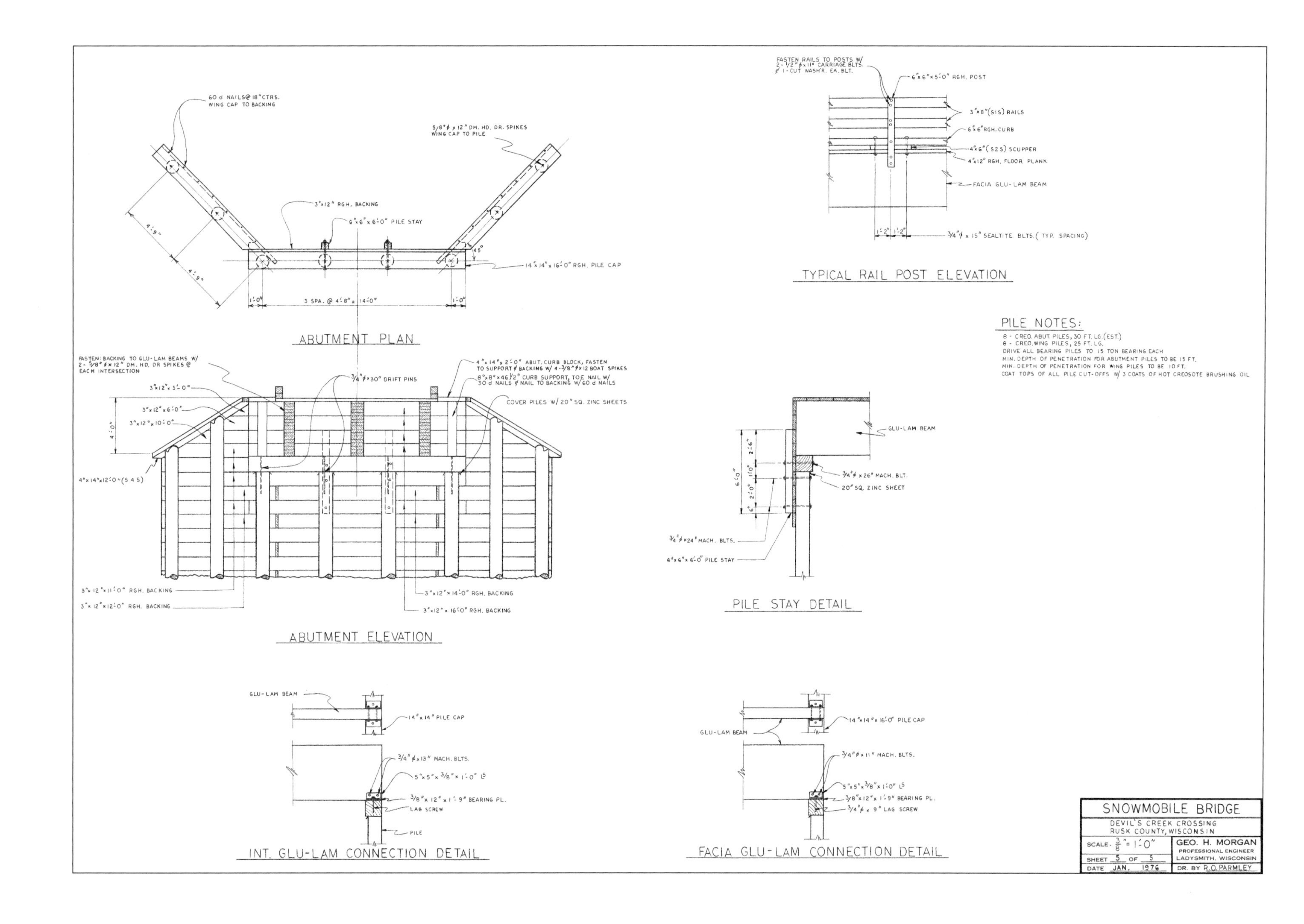
60 d NAILS@ 18" CTRS. WING CAP TO BACKING
5/8"ϕ x 12" DM. HD. DR. SPIKES WING CAP TO PILE
3"x12" RGH. BACKING
6"x6"x6'-0" PILE STAY
45°
14"x14"x16'-0" RGH. PILE CAP
4'-9"
4'-9"
1'-0"
3 SPA. @ 4'-8" = 14'-0"
1'-0"
ABUTMENT PLAN
FASTEN: BACKING TO GLU-LAM BEAMS W/ 2-5/8"ϕ x 12" DM. HD. DR SPIKES @ EACH INTERSECTION
3/4"ϕ x 30" DRIFT PINS
4"x14"x2'-0" ABUT. CURB BLOCK, FASTEN TO SUPPORT & BACKING W/ 4-3/8"ϕ x 12 BOAT SPIKES
8"x8"x46 1/2" CURB SUPPORT, TOE NAIL W/ 30 d NAILS & NAIL TO BACKING W/ 60 d NAILS
COVER PILES W/ 20" SQ. ZINC SHEETS
3"x12"x3'-0"
3"x12"x6'-0"
3"x12"x10'-0"
4'-0"
4"x14"x12'-0"(S 4 S)
3"x12"x11'-0" RGH. BACKING
3"x12"x12'-0" RGH. BACKING
3"x12"x14'-0" RGH. BACKING
3"x12"x16'-0" RGH. BACKING
ABUTMENT ELEVATION
FASTEN RAILS TO POSTS W/ 2-1/2"ϕ x 11" CARRIAGE BLTS. & 1-CUT WASH'R. EA. BLT.
6"x6"x5'-0" RGH. POST
3"x8"(S1S) RAILS
6"x6" RGH. CURB
4"x6"(S2S) SCUPPER
4"x12" RGH. FLOOR PLANK
FACIA GLU-LAM BEAM
1'-2" 1'-2"
3/4"ϕ x 15" SEALTITE BLTS. (TYP. SPACING)
TYPICAL RAIL POST ELEVATION
PILE NOTES:
8 - CREO. ABUT. PILES, 30 FT. LG. (EST.)
8 - CREO. WING PILES, 25 FT. LG.
DRIVE ALL BEARING PILES TO 15 TON BEARING EACH
MIN. DEPTH OF PENETRATION FOR ABUTMENT PILES TO BE 15 FT.
MIN. DEPTH OF PENETRATION FOR WING PILES TO BE 10 FT.
COAT TOPS OF ALL PILE CUT-OFFS W/ 3 COATS OF HOT CREOSOTE BRUSHING OIL
GLU-LAM BEAM
6'-0"
2'-6"
1'-0"
2'-0"
6"
3/4"ϕ x 26" MACH. BLT.
20" SQ. ZINC SHEET
3/4"ϕ x 24" MACH. BLTS.
6"x6"x6'-0" PILE STAY
PILE STAY DETAIL
GLU-LAM BEAM
14"x14" PILE CAP
3/4"ϕ x 13" MACH. BLTS.
5"x5"x3/8"x1'-0" LS
3/8"x12"x1'-9" BEARING PL.
LAG SCREW
PILE
INT. GLU-LAM CONNECTION DETAIL
14"x14"x16'-0" PILE CAP
GLU-LAM BEAM
3/4"ϕ x 11" MACH. BLTS.
5"x5"x3/8"x1'-0" LS
3/8"x12"x1'-9" BEARING PL.
3/4"ϕ x 9" LAG SCREW
FACIA GLU-LAM CONNECTION DETAIL
SNOWMOBILE BRIDGE
DEVIL'S CREEK CROSSING
RUSK COUNTY, WISCONSIN
SCALE - 3/8" = 1'-0"
SHEET 5 OF 5
DATE JAN. 1976
GEO. H. MORGAN
PROFESSIONAL ENGINEER
LADYSMITH, WISCONSIN
DR. BY R.O. PARMLEY

Design

AIRPORTS

Section 16

The following example of plans for an airport renovation project contains 67 sheets. Only the basic sheets and the larger number of cross section sheets with many typical standard details are presented, due to limited space in this sourcebook. However, the reader should refer to the title page and review the "index of sheets" to appreciate the full scope of the project plans.

TABLE OF CONTENTS

BLACK RIVER FALLS AREA AIRPORT
JACKSON COUNTY, WISCONSIN
A. I. P. 3-55-0008-01

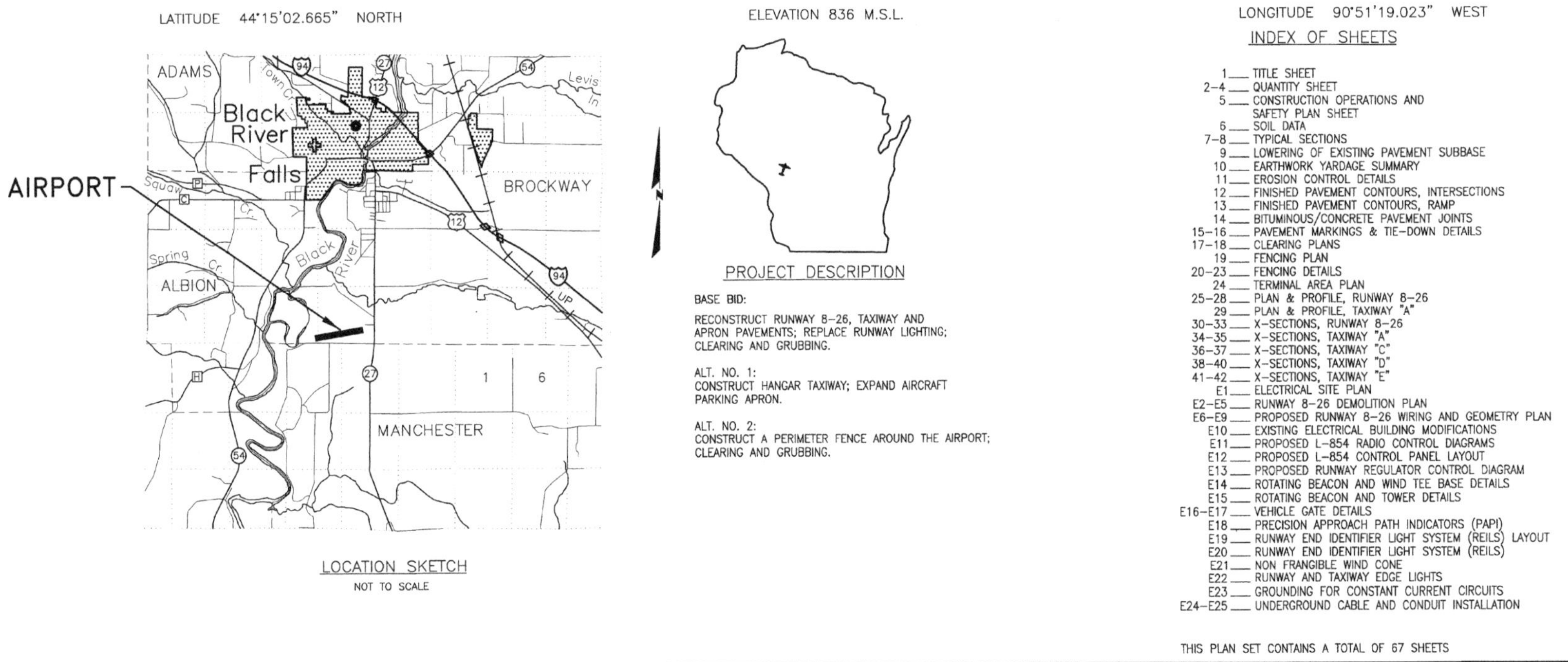

LATITUDE 44°15'02.665" NORTH
ELEVATION 836 M.S.L.
LONGITUDE 90°51'19.023" WEST
AIRPORT
ADAMS
Black River Falls
BROCKWAY
ALBION
MANCHESTER
Black River
LOCATION SKETCH
NOT TO SCALE
PROJECT DESCRIPTION
BASE BID:
RECONSTRUCT RUNWAY 8-26, TAXIWAY AND APRON PAVEMENTS; REPLACE RUNWAY LIGHTING; CLEARING AND GRUBBING.
ALT. NO. 1:
CONSTRUCT HANGAR TAXIWAY; EXPAND AIRCRAFT PARKING APRON.
ALT. NO. 2:
CONSTRUCT A PERIMETER FENCE AROUND THE AIRPORT; CLEARING AND GRUBBING.
INDEX OF SHEETS
1 TITLE SHEET
2-4 QUANTITY SHEET
5 CONSTRUCTION OPERATIONS AND SAFETY PLAN SHEET
6 SOIL DATA
7-8 TYPICAL SECTIONS
9 LOWERING OF EXISTING PAVEMENT SUBBASE
10 EARTHWORK YARDAGE SUMMARY
11 EROSION CONTROL DETAILS
12 FINISHED PAVEMENT CONTOURS, INTERSECTIONS
13 FINISHED PAVEMENT CONTOURS, RAMP
14 BITUMINOUS/CONCRETE PAVEMENT JOINTS
15-16 PAVEMENT MARKINGS & TIE-DOWN DETAILS
17-18 CLEARING PLANS
19 FENCING PLAN
20-23 FENCING DETAILS
24 TERMINAL AREA PLAN
25-28 PLAN & PROFILE, RUNWAY 8-26
29 PLAN & PROFILE, TAXIWAY "A"
30-33 X-SECTIONS, RUNWAY 8-26
34-35 X-SECTIONS, TAXIWAY "A"
36-37 X-SECTIONS, TAXIWAY "C"
38-40 X-SECTIONS, TAXIWAY "D"
41-42 X-SECTIONS, TAXIWAY "E"
E1 ELECTRICAL SITE PLAN
E2-E5 RUNWAY 8-26 DEMOLITION PLAN
E6-E9 PROPOSED RUNWAY 8-26 WIRING AND GEOMETRY PLAN
E10 EXISTING ELECTRICAL BUILDING MODIFICATIONS
E11 PROPOSED L-854 RADIO CONTROL DIAGRAMS
E12 PROPOSED L-854 CONTROL PANEL LAYOUT
E13 PROPOSED RUNWAY REGULATOR CONTROL DIAGRAM
E14 ROTATING BEACON AND WIND TEE BASE DETAILS
E15 ROTATING BEACON AND TOWER DETAILS
E16-E17 VEHICLE GATE DETAILS
E18 PRECISION APPROACH PATH INDICATORS (PAPI)
E19 RUNWAY END IDENTIFIER LIGHT SYSTEM (REILS) LAYOUT
E20 RUNWAY END IDENTIFIER LIGHT SYSTEM (REILS)
E21 NON FRANGIBLE WIND CONE
E22 RUNWAY AND TAXIWAY EDGE LIGHTS
E23 GROUNDING FOR CONSTANT CURRENT CIRCUITS
E24-E25 UNDERGROUND CABLE AND CONDUIT INSTALLATION
THIS PLAN SET CONTAINS A TOTAL OF 67 SHEETS

WISCONSIN DEPARTMENT OF TRANSPORTATION
BUREAU OF AERONAUTICS

DESIGN PHASE
APPROVED: ____ AIRPORT ENGINEERING SPECIALIST
DATE: ____
APPROVED: ____ P.E. CHIEF AIRPORT DEVELOPMENT ENGINEER
DATE: ____

RECORD DRAWING
APPROVED: ____ AIRPORT ENGINEERING SPECIALIST
DATE: ____

CONSULTANT, DESIGN:

CONSULTANT, RECORD DRAWINGS:
APPROVED:
RESIDENT ENGINEER
DATE:

BLACK RIVER FALLS AREA AIRPORT
JACKSON COUNTY, WISCONSIN

TITLE SHEET

A.I.P. 3-55-0008-01

DESIGN B.B.V. 5-02 FILE NO. ____
DRAWN M.R.E. 5-02 F. B. ____
CHECKED W.R. 5-02 SCALE: AS NOTED

COOPER ENGINEERING
310 WEST SOUTH STREET, P.O. BOX 230
RICE LAKE, WISCONSIN 54868-0230
TELEPHONE (715) 234-7008
FAX (715) 234-1025

PROJECT NO. 01552074 SHEET 1 OF 67

Source: Cooper Engineering, Co.

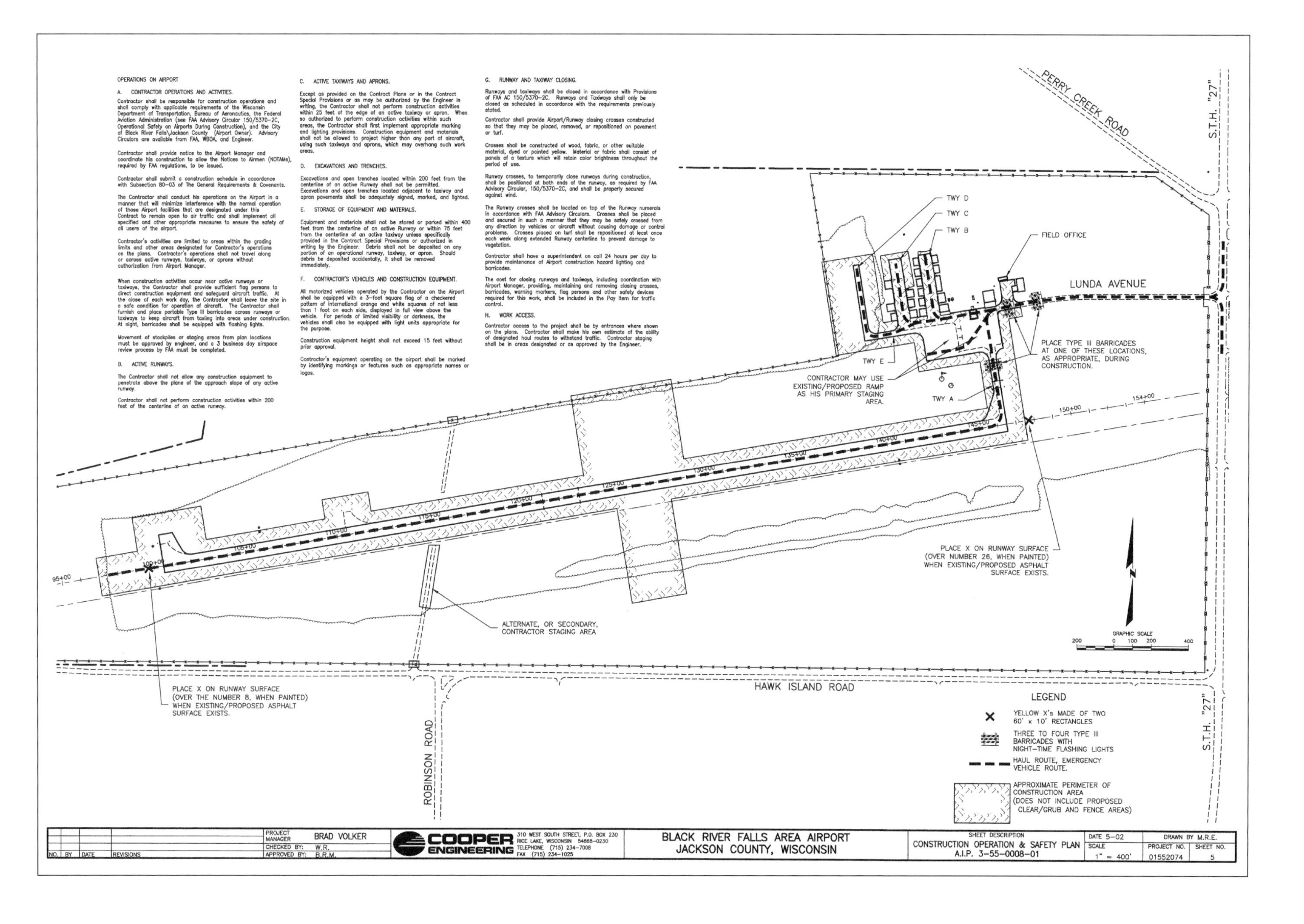
OPERATIONS ON AIRPORT
A. CONTRACTOR OPERATIONS AND ACTIVITIES.
Contractor shall be responsible for construction operations and shall comply with applicable requirements of the Wisconsin Department of Transportation, Bureau of Aeronautics, the Federal Aviation Administration (see FAA Advisory Circular 150/5370-2C, Operational Safety on Airports During Construction), and the City of Black River Falls\Jackson County (Airport Owner). Advisory Circulars are available from FAA, WBOA, and Engineer.
Contractor shall provide notice to the Airport Manager and coordinate his construction to allow the Notices to Airmen (NOTAMs), required by FAA regulations, to be issued.
Contractor shall submit a construction schedule in accordance with Subsection 80-03 of The General Requirements & Covenants.
The Contractor shall conduct his operations on the Airport in a manner that will minimize interference with the normal operation of those Airport facilities that are designated under this Contract to remain open to air traffic and shall implement all specified and other appropriate measures to ensure the safety of all users of the airport.
Contractor's activities are limited to areas within the grading limits and other areas designated for Contractor's operations on the plans. Contractor's operations shall not travel along or across active runways, taxiways, or aprons without authorization from Airport Manager.
When construction activities occur near active runways or taxiways, the Contractor shall provide sufficient flag persons to direct construction equipment and safeguard aircraft traffic. At the close of each work day, the Contractor shall leave the site in a safe condition for operation of aircraft. The Contractor shall furnish and place portable Type III barricades across runways or taxiways to keep aircraft from taxiing into areas under construction. At night, barricades shall be equipped with flashing lights.
Movement of stockpiles or staging areas from plan locations must be approved by engineer, and a 3 business day airspace review process by FAA must be completed.
B. ACTIVE RUNWAYS.
The Contractor shall not allow any construction equipment to penetrate above the plane of the approach slope of any active runway.
Contractor shall not perform construction activities within 200 feet of the centerline of an active runway.
C. ACTIVE TAXIWAYS AND APRONS.
Except as provided on the Contract Plans or in the Contract Special Provisions or as may be authorized by the Engineer in writing, the Contractor shall not perform construction activities within 25 feet of the edge of an active taxiway or apron. When so authorized to perform construction activities within such areas, the Contractor shall first implement appropriate marking and lighting provisions. Construction equipment and materials shall not be allowed to project higher than any part of aircraft, using such taxiways and aprons, which may overhang such work areas.
D. EXCAVATIONS AND TRENCHES.
Excavations and open trenches located within 200 feet from the centerline of an active Runway shall not be permitted. Excavations and open trenches located adjacent to taxiway and apron pavements shall be adequately signed, marked, and lighted.
E. STORAGE OF EQUIPMENT AND MATERIALS.
Equipment and materials shall not be stored or parked within 400 feet from the centerline of an active Runway or within 75 feet from the centerline of an active taxiway unless specifically provided in the Contract Special Provisions or authorized in writing by the Engineer. Debris shall not be deposited on any portion of an operational runway, taxiway, or apron. Should debris be deposited accidentally, it shall be removed immediately.
F. CONTRACTOR'S VEHICLES AND CONSTRUCTION EQUIPMENT.
All motorized vehicles operated by the Contractor on the Airport shall be equipped with a 3-foot square flag of a checkered pattern of international orange and white squares of not less than 1 foot on each side, displayed in full view above the vehicle. For periods of limited visibility or darkness, the vehicles shall also be equipped with light units appropriate for the purpose.
Construction equipment height shall not exceed 15 feet without prior approval.
Contractor's equipment operating on the airport shall be marked by identifying markings or features such as appropriate names or logos.
G. RUNWAY AND TAXIWAY CLOSING.
Runways and taxiways shall be closed in accordance with Provisions of FAA AC 150/5370-2C. Runways and Taxiways shall only be closed as scheduled in accordance with the requirements previously stated.
Contractor shall provide Airport/Runway closing crosses constructed so that they may be placed, removed, or repositioned on pavement or turf.
Crosses shall be constructed of wood, fabric, or other suitable material, dyed or painted yellow. Material or fabric shall consist of panels of a texture which will retain color brightness throughout the period of use.
Runway crosses, to temporarily close runways during construction, shall be positioned at both ends of the runway, as required by FAA Advisory Circular, 150/5370-2C, and shall be properly secured against wind.
The Runway crosses shall be located on top of the Runway numerals in accordance with FAA Advisory Circulars. Crosses shall be placed and secured in such a manner that they may be safely crossed from any direction by vehicles or aircraft without causing damage or control problems. Crosses placed on turf shall be repositioned at least once each week along extended Runway centerline to prevent damage to vegetation.
Contractor shall have a superintendent on call 24 hours per day to provide maintenance of Airport construction hazard lighting and barricades.
The cost for closing runways and taxiways, including coordination with Airport Manager, providing, maintaining and removing closing crosses, barricades, warning markers, flag persons and other safety devices required for this work, shall be included in the Pay Item for traffic control.
H. WORK ACCESS.
Contractor access to the project shall be by entrances where shown on the plans. Contractor shall make his own estimate of the ability of designated haul routes to withstand traffic. Contractor staging shall be in areas designated or as approved by the Engineer.
PERRY CREEK ROAD
S.T.H. "27"
TWY D
TWY C
TWY B
FIELD OFFICE
LUNDA AVENUE
PLACE TYPE III BARRICADES AT ONE OF THESE LOCATIONS, AS APPROPRIATE, DURING CONSTRUCTION.
TWY E
CONTRACTOR MAY USE EXISTING/PROPOSED RAMP AS HIS PRIMARY STAGING AREA.
TWY A
150+00
154+00
145+00
140+00
135+00
130+00
125+00
120+00
115+00
110+00
105+00
100+00
95+00
PLACE X ON RUNWAY SURFACE (OVER NUMBER 26, WHEN PAINTED) WHEN EXISTING/PROPOSED ASPHALT SURFACE EXISTS.
N
ALTERNATE, OR SECONDARY, CONTRACTOR STAGING AREA
GRAPHIC SCALE
200 0 100 200 400
PLACE X ON RUNWAY SURFACE (OVER THE NUMBER 8, WHEN PAINTED) WHEN EXISTING/PROPOSED ASPHALT SURFACE EXISTS.
HAWK ISLAND ROAD
ROBINSON ROAD
LEGEND
YELLOW X's MADE OF TWO 60' x 10' RECTANGLES
THREE TO FOUR TYPE III BARRICADES WITH NIGHT-TIME FLASHING LIGHTS
HAUL ROUTE, EMERGENCY VEHICLE ROUTE.
APPROXIMATE PERIMETER OF CONSTRUCTION AREA (DOES NOT INCLUDE PROPOSED CLEAR/GRUB AND FENCE AREAS)
S.T.H. "27"
NO. BY DATE REVISIONS
PROJECT MANAGER BRAD VOLKER
CHECKED BY: W.R.
APPROVED BY: B.R.M.
COOPER ENGINEERING
310 WEST SOUTH STREET, P.O. BOX 230 RICE LAKE, WISCONSIN 54868-0230 TELEPHONE (715) 234-7008 FAX (715) 234-1025
BLACK RIVER FALLS AREA AIRPORT
JACKSON COUNTY, WISCONSIN
SHEET DESCRIPTION
CONSTRUCTION OPERATION & SAFETY PLAN
A.I.P. 3-55-0008-01
DATE 5-02
SCALE 1" = 400'
DRAWN BY M.R.E.
PROJECT NO. 01552074
SHEET NO. 5

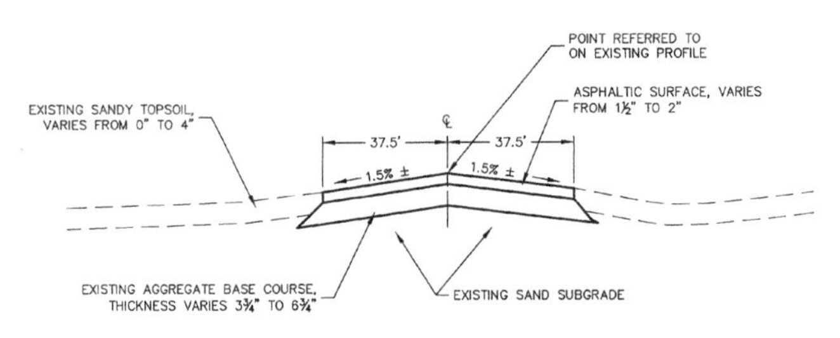

EXISTING TYPICAL SECTION
RUNWAY 8–26
NOT TO SCALE

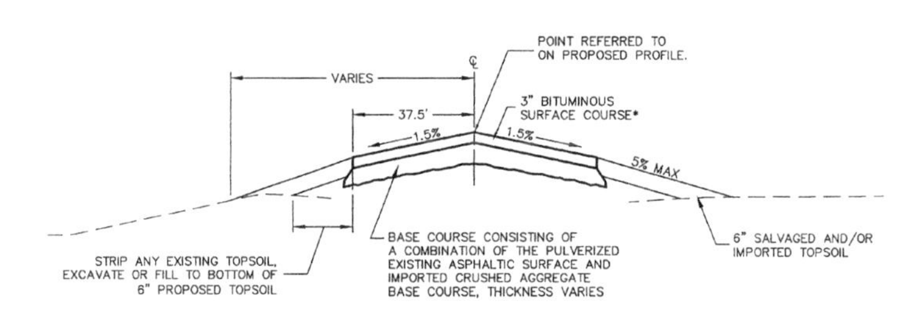

*SURFACE 1.5-INCH MAT (PG 64-28)
BINDER 1.5-INCH MAT (PG 58-28)

PROPOSED TYPICAL SECTION
RUNWAY 8–26
STA. 100+50 TO 146+50
NOT TO SCALE

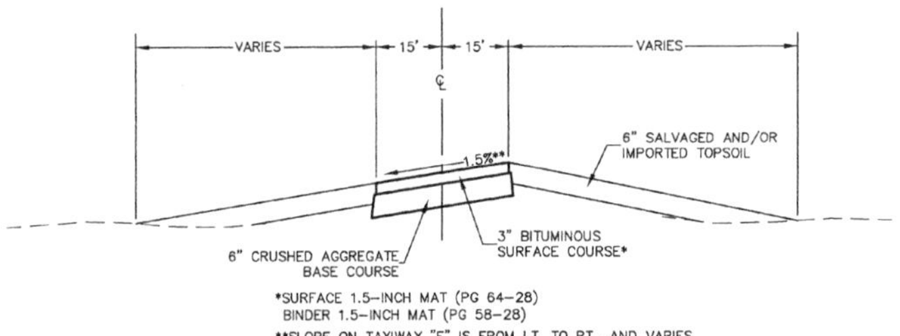

PROPOSED TYPICAL SECTION
TAXIWAY "D"
STA 79+85 TO STA 80+50
AND
TAXIWAY "E"
STA 49+85 TO STA 51+50±
NOT TO SCALE

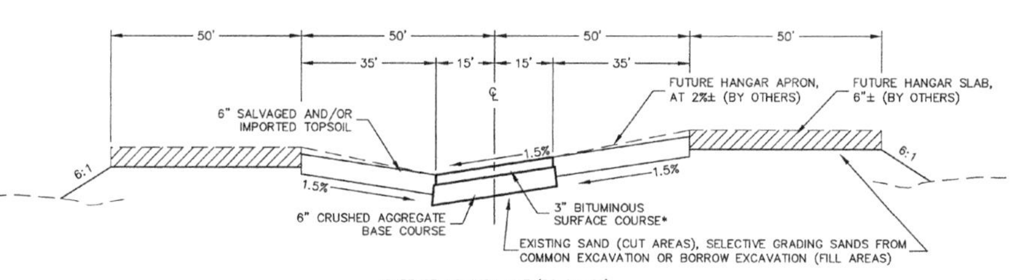

PROPOSED TYPICAL SECTION
TAXIWAY "D"
STA 80+50± TO STA 83+85
NOT TO SCALE

NO.	BY	DATE	REVISIONS	PROJECT MANAGER: BRAD VOLKER CHECKED BY: W.R. APPROVED BY: B.R.M.	COOPER ENGINEERING 310 WEST SOUTH STREET, P.O. BOX 230 RICE LAKE, WISCONSIN 54868-0230 TELEPHONE (715) 234-7008 FAX (715) 234-1025	BLACK RIVER FALLS AREA AIRPORT JACKSON COUNTY, WISCONSIN	SHEET DESCRIPTION TYPICAL SECTIONS A.I.P. 3-55-0008-01	DATE 5-02 SCALE AS NOTED	DRAWN BY M.R.E. PROJECT NO. 01552074	SHEET NO. 7

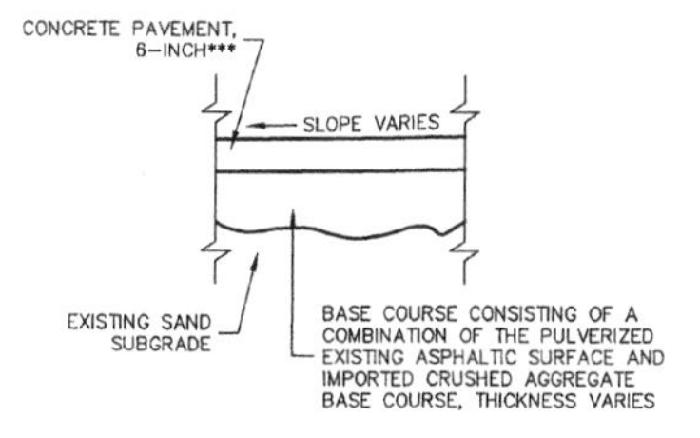

***SAW-CUT JOINTS AS SHOWN ON SHEET NO. 14. SAW-CUT DEPTH TO BE ¼ OF THE CONCRETE THICKNESS. (INCIDENTAL)

PROPOSED TYPICAL SECTION
CONCRETE FUELING PAD

NOT TO SCALE

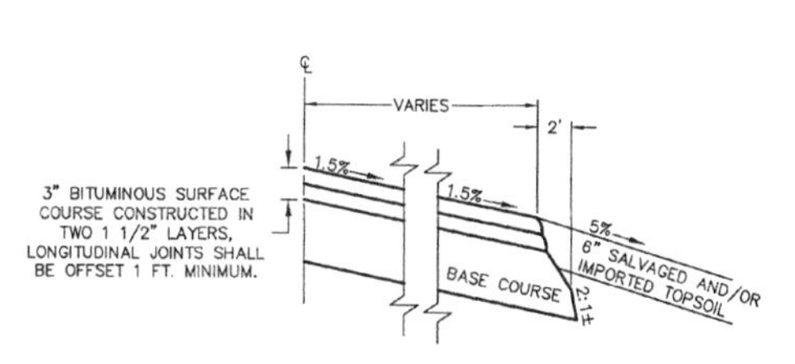

EDGE OF PAVEMENT DETAIL

NOT TO SCALE

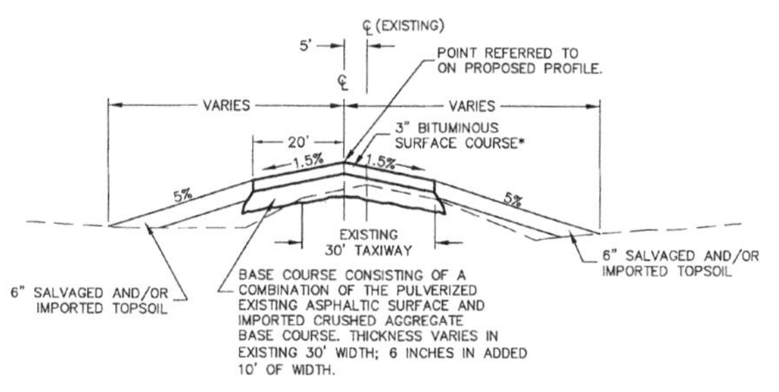

*SURFACE 1.5-INCH MAT (PG 64-28)
BINDER 1.5-INCH MAT (PG 58-28)

PROPOSED TYPICAL SECTION
TAXIWAY "A"

NOT TO SCALE

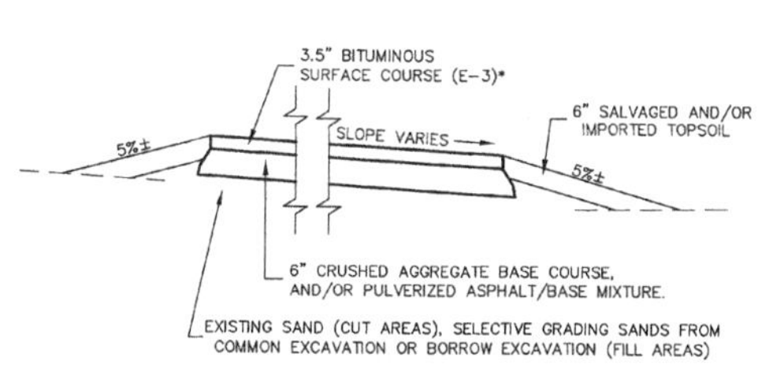

*SURFACE 1.5-INCH MAT (PG 64-28)
BINDER 1.5-INCH MAT (PG 58-28)

PROPOSED TYPICAL SECTION
AIRCRAFT PARKING APRON AND
HANGAR TAXIWAY EXPANSIONS

NOT TO SCALE

NO.	BY	DATE	REVISIONS	PROJECT MANAGER: BRAD VOLKER CHECKED BY: W.R. APPROVED BY: B.R.M.	COOPER ENGINEERING 310 WEST SOUTH STREET, P.O. BOX 230 RICE LAKE, WISCONSIN 54868-0230 TELEPHONE (715) 234-7008 FAX (715) 234-1025	BLACK RIVER FALLS AREA AIRPORT JACKSON COUNTY, WISCONSIN	SHEET DESCRIPTION TYPICAL SECTIONS A.I.P. 3-55-0008-01	DATE 5-02 SCALE AS NOTED	DRAWN BY M.R.E. PROJECT NO. 01552074 SHEET NO. 8

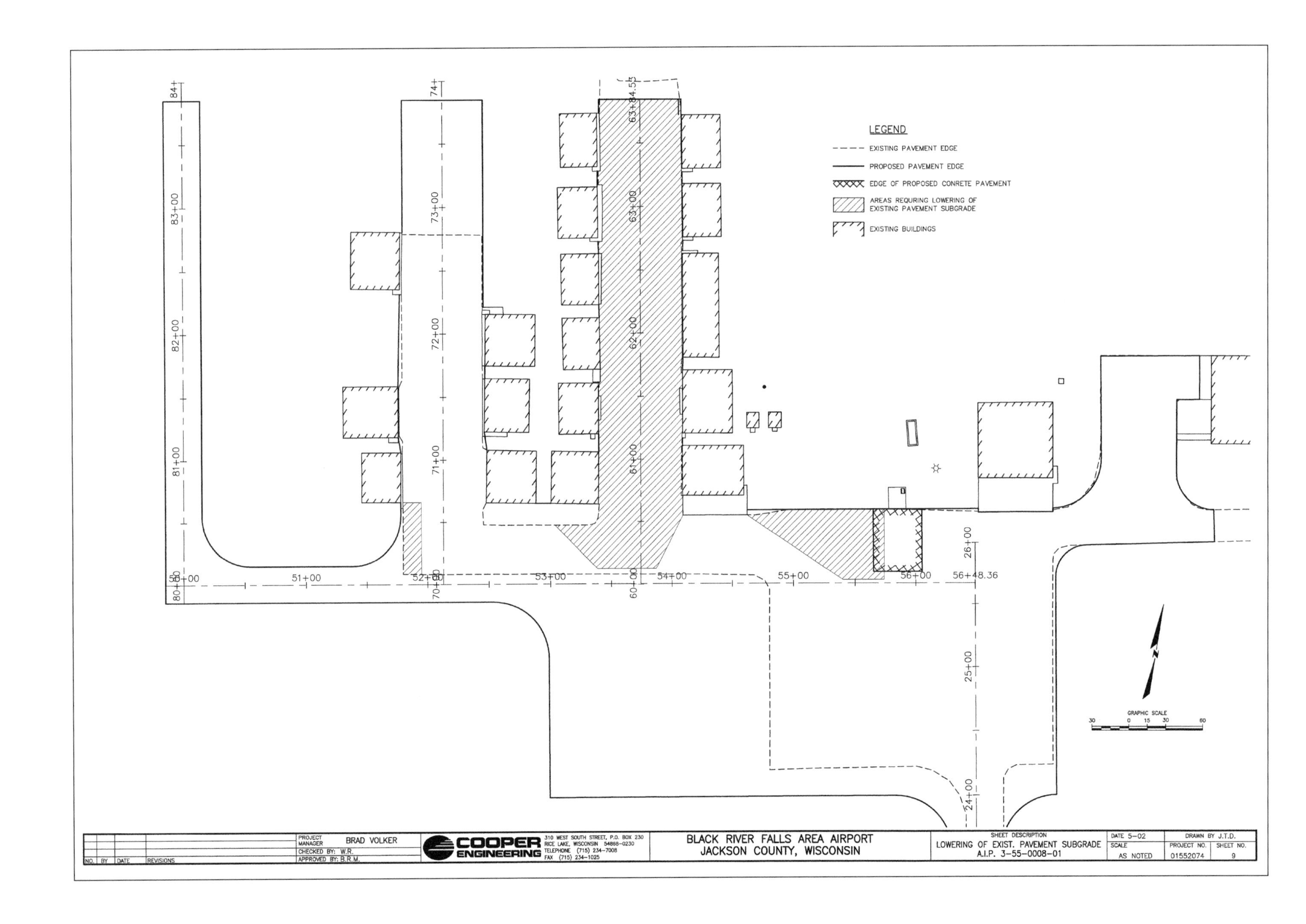
LEGEND
EXISTING PAVEMENT EDGE
PROPOSED PAVEMENT EDGE
EDGE OF PROPOSED CONRETE PAVEMENT
AREAS REQURING LOWERING OF EXISTING PAVEMENT SUBGRADE
EXISTING BUILDINGS
GRAPHIC SCALE
BRAD VOLKER
CHECKED BY: W.R.
APPROVED BY: B.R.M.
COOPER ENGINEERING
310 WEST SOUTH STREET, P.O. BOX 230
RICE LAKE, WISCONSIN 54868-0230
TELEPHONE (715) 234-7008
FAX (715) 234-1025
BLACK RIVER FALLS AREA AIRPORT
JACKSON COUNTY, WISCONSIN
SHEET DESCRIPTION
LOWERING OF EXIST. PAVEMENT SUBGRADE
A.I.P. 3-55-0008-01
DATE 5-02
SCALE AS NOTED
DRAWN BY J.T.D.
PROJECT NO. 01552074
SHEET NO. 9

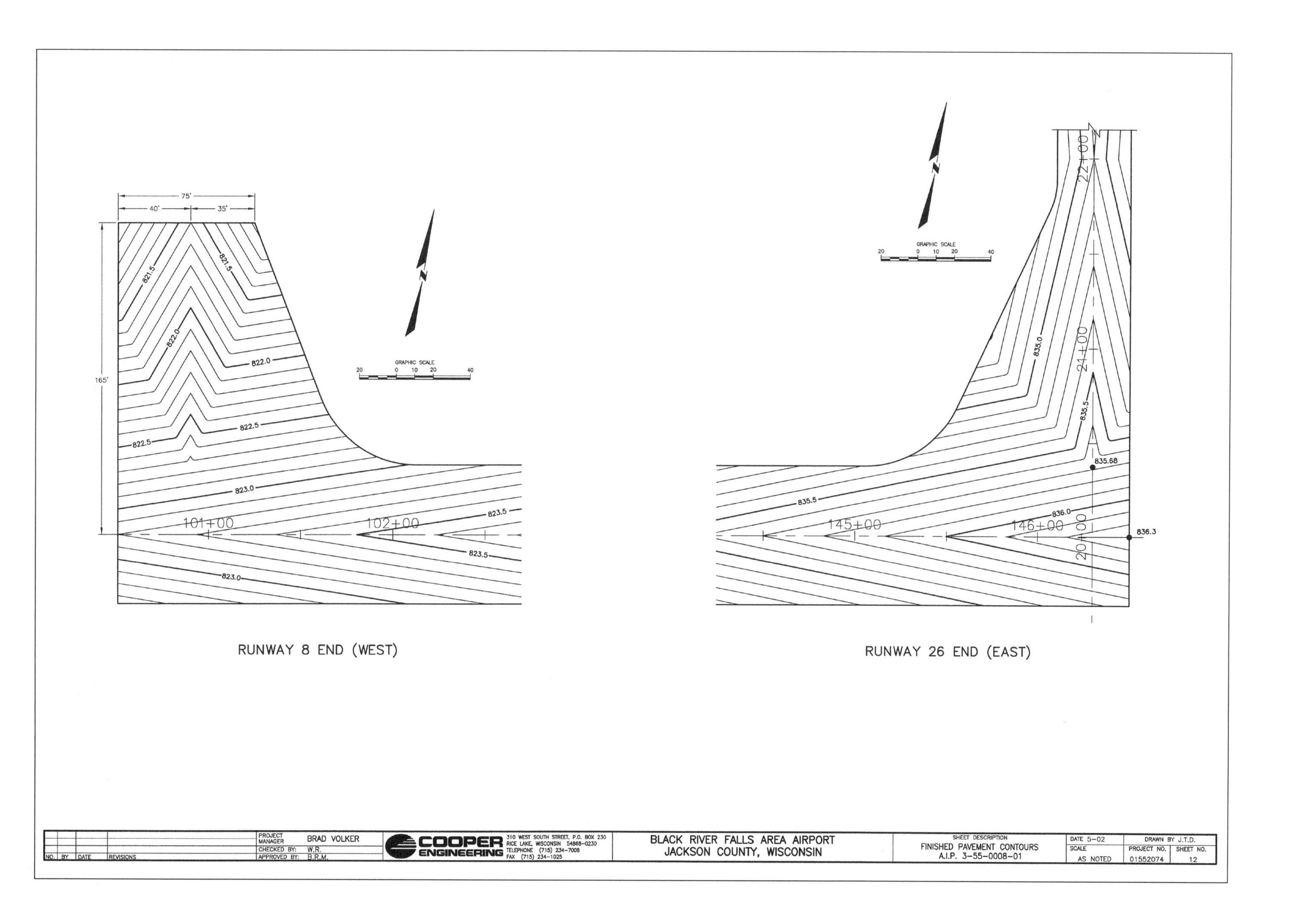
75'
40'
35'
165'
821.5
822.0
822.5
823.0
823.5
101+00
102+00
GRAPHIC SCALE
20 0 10 20 40
RUNWAY 8 END (WEST)
835.0
835.5
835.68
836.0
836.3
145+00
146+00
20+00
21+00
22+00
RUNWAY 26 END (EAST)
PROJECT MANAGER BRAD VOLKER
CHECKED BY: W.R.
APPROVED BY: B.R.M.
NO. BY DATE REVISIONS
COOPER ENGINEERING
310 WEST SOUTH STREET, P.O. BOX 230
RICE LAKE, WISCONSIN 54868-0230
TELEPHONE (715) 234-7008
FAX (715) 234-1025
BLACK RIVER FALLS AREA AIRPORT
JACKSON COUNTY, WISCONSIN
SHEET DESCRIPTION
FINISHED PAVEMENT CONTOURS
A.I.P. 3-55-0008-01
DATE 5-02
DRAWN BY J.T.D.
SCALE AS NOTED
PROJECT NO. 01552074
SHEET NO. 12

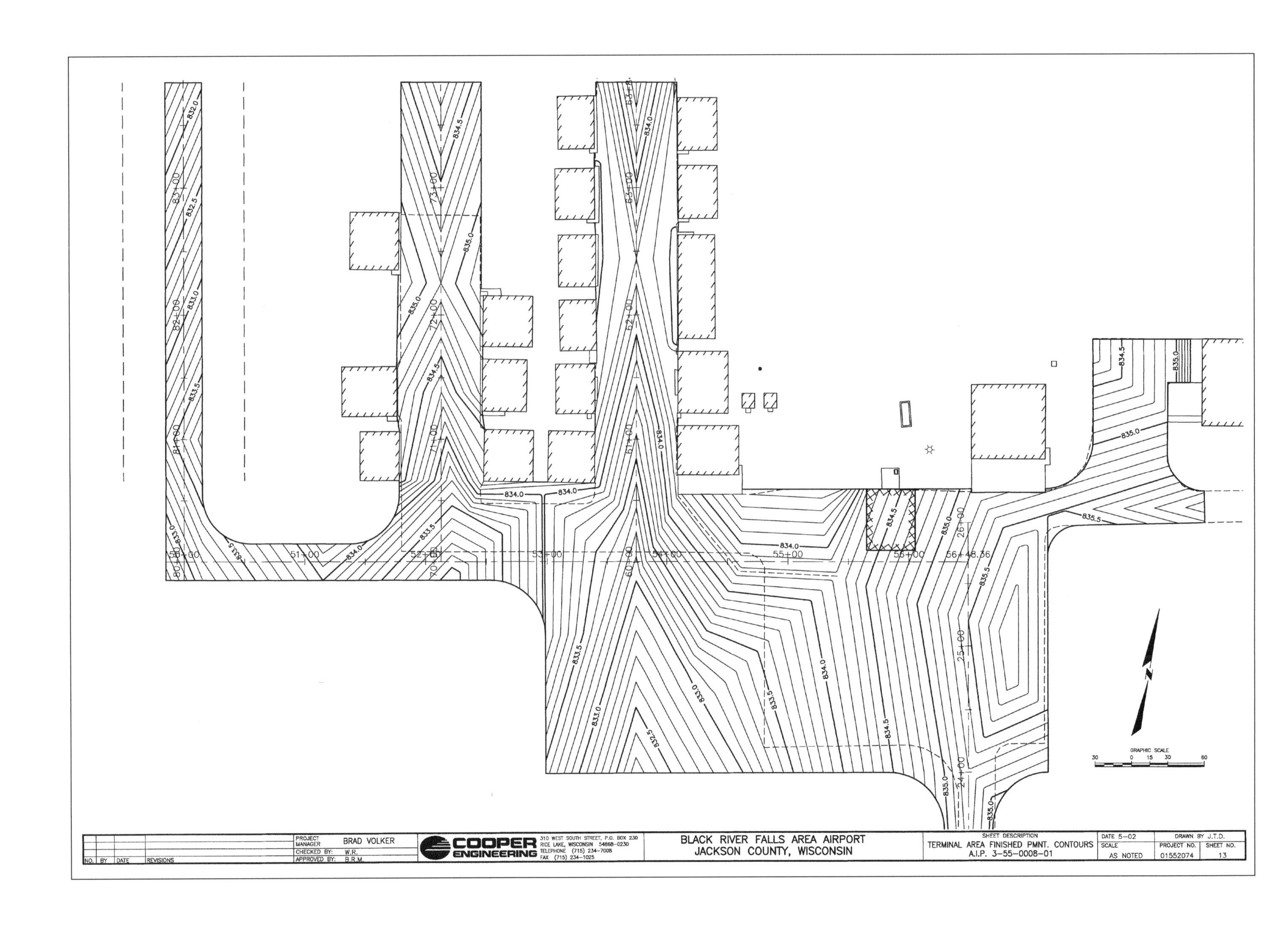

BLACK RIVER FALLS AREA AIRPORT
JACKSON COUNTY, WISCONSIN
SHEET DESCRIPTION
TERMINAL AREA FINISHED PMNT. CONTOURS
A.I.P. 3-55-0008-01
PROJECT MANAGER BRAD VOLKER
CHECKED BY: W.R.
APPROVED BY: B.R.M.
COOPER ENGINEERING
310 WEST SOUTH STREET, P.O. BOX 230
RICE LAKE, WISCONSIN 54868-0230
TELEPHONE (715) 234-7008
FAX (715) 234-1025
DATE 5-02
DRAWN BY J.T.D.
SCALE AS NOTED
PROJECT NO. 01552074
SHEET NO. 13
NO. BY DATE REVISIONS
GRAPHIC SCALE
30 0 15 30 60

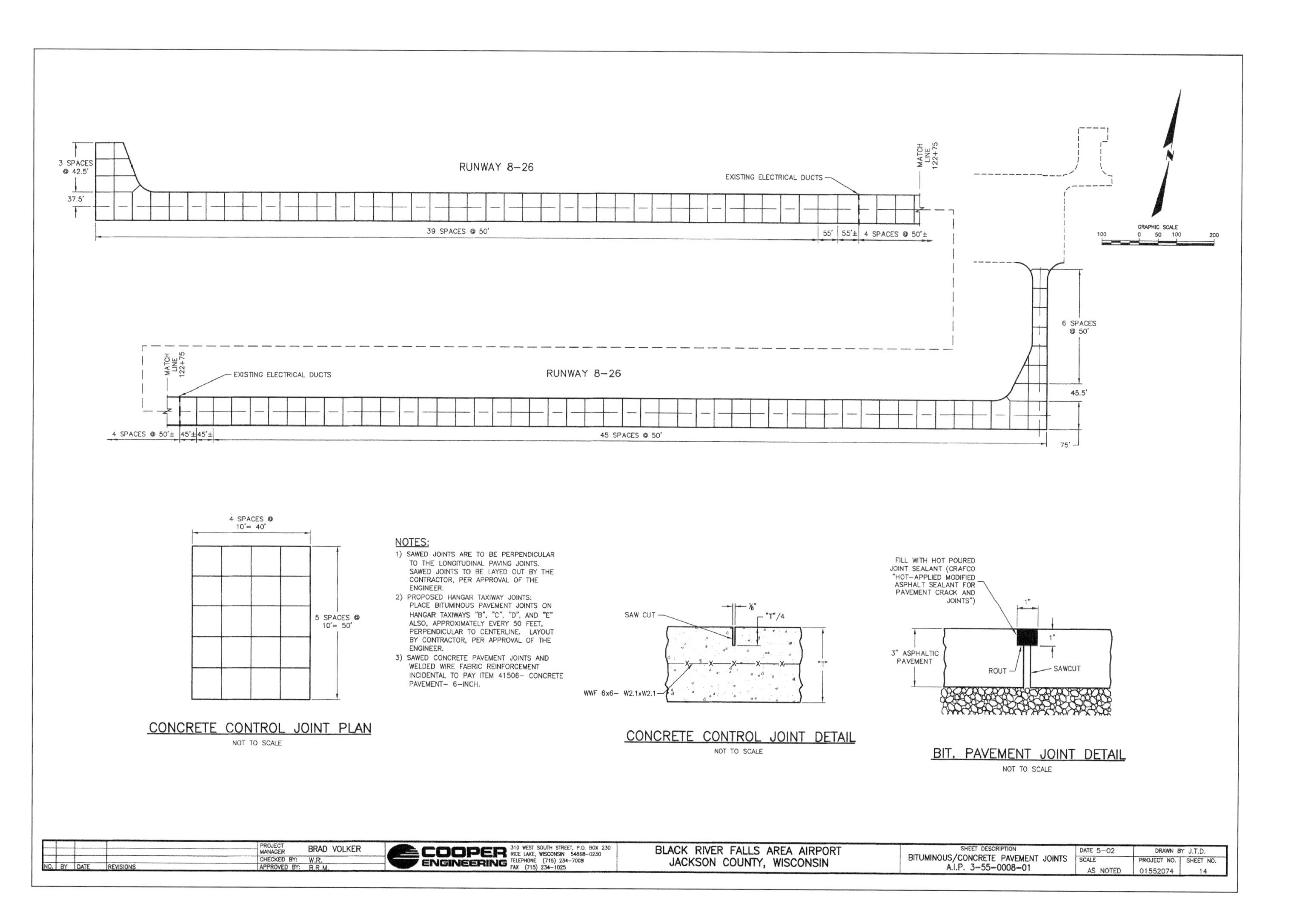

3 SPACES @ 42.5'
37.5'
RUNWAY 8–26
EXISTING ELECTRICAL DUCTS
MATCH LINE 122+75
39 SPACES @ 50'
55'
55'±
4 SPACES @ 50'±
GRAPHIC SCALE
100 0 50 100 200
6 SPACES @ 50'
45.5'
75'
4 SPACES @ 50'±
45'±
45'±
45 SPACES @ 50'
4 SPACES @ 10'= 40'
5 SPACES @ 10'= 50'
CONCRETE CONTROL JOINT PLAN
NOT TO SCALE
NOTES:
1) SAWED JOINTS ARE TO BE PERPENDICULAR TO THE LONGITUDINAL PAVING JOINTS. SAWED JOINTS TO BE LAYED OUT BY THE CONTRACTOR, PER APPROVAL OF THE ENGINEER.
2) PROPOSED HANGAR TAXIWAY JOINTS: PLACE BITUMINOUS PAVEMENT JOINTS ON HANGAR TAXIWAYS "B", "C", "D", AND "E" ALSO, APPROXIMATELY EVERY 50 FEET, PERPENDICULAR TO CENTERLINE. LAYOUT BY CONTRACTOR, PER APPROVAL OF THE ENGINEER.
3) SAWED CONCRETE PAVEMENT JOINTS AND WELDED WIRE FABRIC REINFORCEMENT INCIDENTAL TO PAY ITEM 41506– CONCRETE PAVEMENT– 6–INCH.
SAW CUT
⅛"
"T"/4
"T"
WWF 6x6– W2.1xW2.1
CONCRETE CONTROL JOINT DETAIL
NOT TO SCALE
FILL WITH HOT POURED JOINT SEALANT (CRAFCO "HOT–APPLIED MODIFIED ASPHALT SEALANT FOR PAVEMENT CRACK AND JOINTS")
1"
1"
3" ASPHALTIC PAVEMENT
ROUT
SAWCUT
BIT. PAVEMENT JOINT DETAIL
NOT TO SCALE
NO. BY DATE REVISIONS
PROJECT MANAGER BRAD VOLKER
CHECKED BY: W.R.
APPROVED BY: B.R.M.
COOPER ENGINEERING
310 WEST SOUTH STREET, P.O. BOX 230
RICE LAKE, WISCONSIN 54868–0230
TELEPHONE (715) 234–7008
FAX (715) 234–1025
BLACK RIVER FALLS AREA AIRPORT
JACKSON COUNTY, WISCONSIN
SHEET DESCRIPTION
BITUMINOUS/CONCRETE PAVEMENT JOINTS
A.I.P. 3–55–0008–01
DATE 5–02
SCALE AS NOTED
DRAWN BY J.T.D.
PROJECT NO. 01552074
SHEET NO. 14

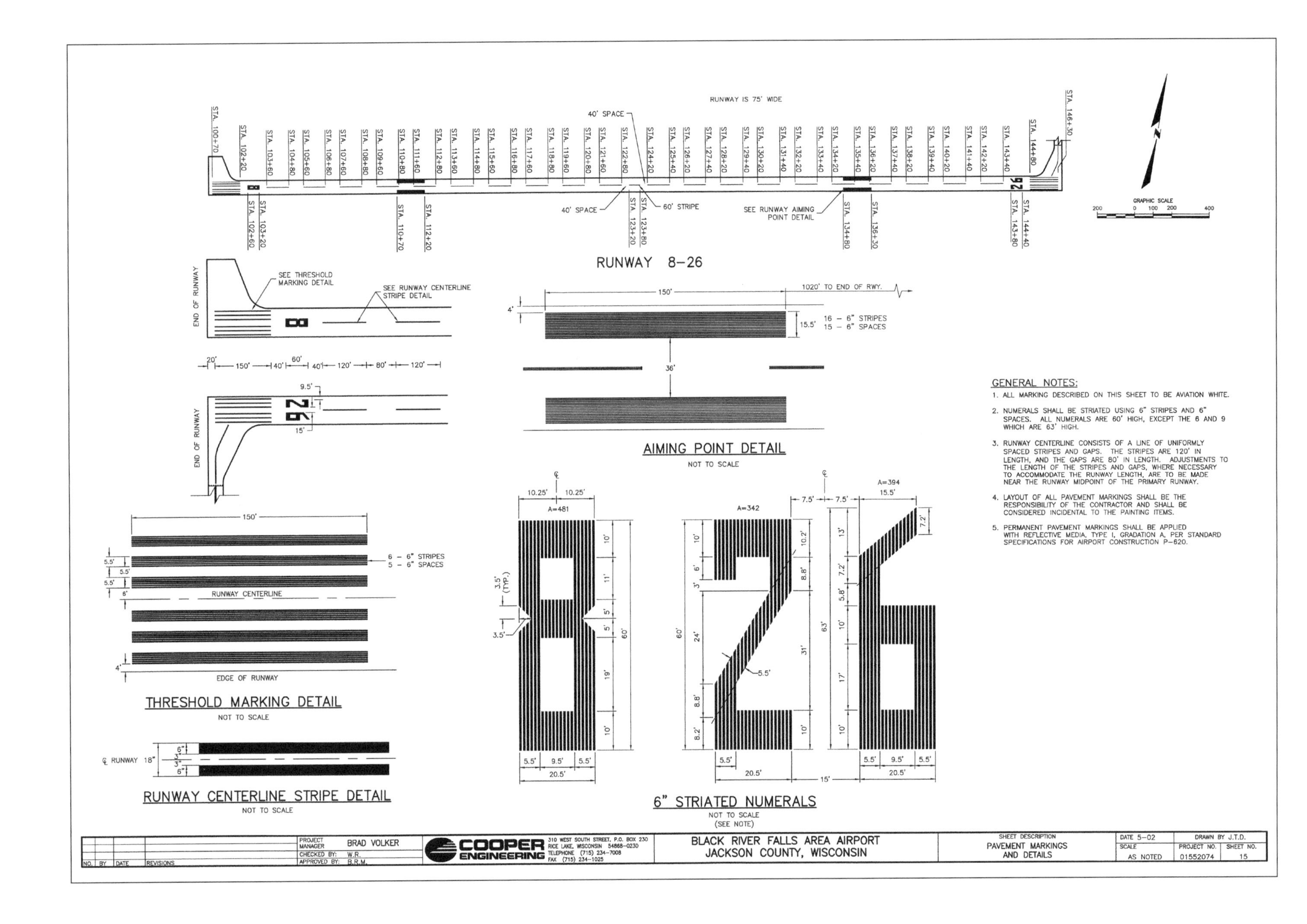
RUNWAY IS 75' WIDE
40' SPACE
60' STRIPE
SEE RUNWAY AIMING POINT DETAIL
GRAPHIC SCALE
RUNWAY 8-26
SEE THRESHOLD MARKING DETAIL
SEE RUNWAY CENTERLINE STRIPE DETAIL
END OF RUNWAY
1020' TO END OF RWY.
16 - 6" STRIPES
15 - 6" SPACES
AIMING POINT DETAIL
NOT TO SCALE
GENERAL NOTES:
1. ALL MARKING DESCRIBED ON THIS SHEET TO BE AVIATION WHITE.
2. NUMERALS SHALL BE STRIATED USING 6" STRIPES AND 6" SPACES. ALL NUMERALS ARE 60' HIGH, EXCEPT THE 6 AND 9 WHICH ARE 63' HIGH.
3. RUNWAY CENTERLINE CONSISTS OF A LINE OF UNIFORMLY SPACED STRIPES AND GAPS. THE STRIPES ARE 120' IN LENGTH, AND THE GAPS ARE 80' IN LENGTH. ADJUSTMENTS TO THE LENGTH OF THE STRIPES AND GAPS, WHERE NECESSARY TO ACCOMMODATE THE RUNWAY LENGTH, ARE TO BE MADE NEAR THE RUNWAY MIDPOINT OF THE PRIMARY RUNWAY.
4. LAYOUT OF ALL PAVEMENT MARKINGS SHALL BE THE RESPONSIBILITY OF THE CONTRACTOR AND SHALL BE CONSIDERED INCIDENTAL TO THE PAINTING ITEMS.
5. PERMANENT PAVEMENT MARKINGS SHALL BE APPLIED WITH REFLECTIVE MEDIA, TYPE I, GRADATION A, PER STANDARD SPECIFICATIONS FOR AIRPORT CONSTRUCTION P-620.
6 - 6" STRIPES
5 - 6" SPACES
RUNWAY CENTERLINE
EDGE OF RUNWAY
THRESHOLD MARKING DETAIL
NOT TO SCALE
RUNWAY CENTERLINE STRIPE DETAIL
NOT TO SCALE
6" STRIATED NUMERALS
NOT TO SCALE
(SEE NOTE)
PROJECT MANAGER BRAD VOLKER
CHECKED BY: W.R.
APPROVED BY: B.R.M.
COOPER ENGINEERING
310 WEST SOUTH STREET, P.O. BOX 230
RICE LAKE, WISCONSIN 54868-0230
TELEPHONE (715) 234-7008
FAX (715) 234-1025
BLACK RIVER FALLS AREA AIRPORT
JACKSON COUNTY, WISCONSIN
SHEET DESCRIPTION
PAVEMENT MARKINGS AND DETAILS
DATE 5-02
DRAWN BY J.T.D.
SCALE AS NOTED
PROJECT NO. 01552074
SHEET NO. 15
NO. BY DATE REVISIONS

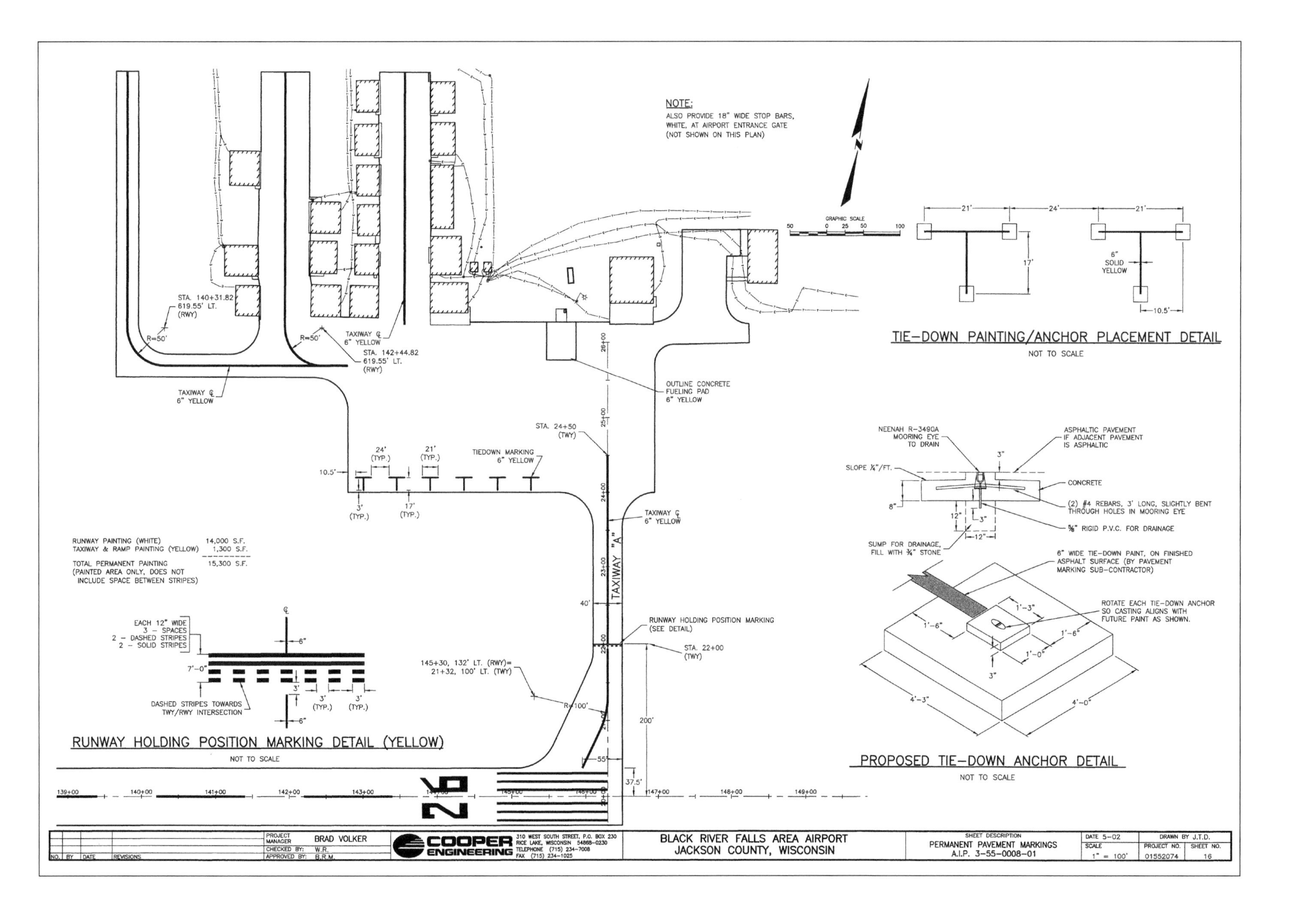
NOTE:
ALSO PROVIDE 18" WIDE STOP BARS, WHITE, AT AIRPORT ENTRANCE GATE (NOT SHOWN ON THIS PLAN)
GRAPHIC SCALE
TIE-DOWN PAINTING/ANCHOR PLACEMENT DETAIL
NOT TO SCALE
6" SOLID YELLOW
STA. 140+31.82 619.55' LT. (RWY)
R=50'
TAXIWAY ℄ 6" YELLOW
STA. 142+44.82 619.55' LT. (RWY)
OUTLINE CONCRETE FUELING PAD 6" YELLOW
STA. 24+50 (TWY)
TIEDOWN MARKING 6" YELLOW
RUNWAY PAINTING (WHITE) 14,000 S.F.
TAXIWAY & RAMP PAINTING (YELLOW) 1,300 S.F.
TOTAL PERMANENT PAINTING 15,300 S.F.
(PAINTED AREA ONLY, DOES NOT INCLUDE SPACE BETWEEN STRIPES)
TAXIWAY "A"
RUNWAY HOLDING POSITION MARKING (SEE DETAIL)
STA. 22+00 (TWY)
145+30, 132' LT. (RWY)= 21+32, 100' LT. (TWY)
EACH 12" WIDE
3 - SPACES
2 - DASHED STRIPES
2 - SOLID STRIPES
DASHED STRIPES TOWARDS TWY/RWY INTERSECTION
RUNWAY HOLDING POSITION MARKING DETAIL (YELLOW)
NOT TO SCALE
NEENAH R-3490A MOORING EYE TO DRAIN
SLOPE ¼"/FT.
ASPHALTIC PAVEMENT IF ADJACENT PAVEMENT IS ASPHALTIC
CONCRETE
(2) #4 REBARS, 3' LONG, SLIGHTLY BENT THROUGH HOLES IN MOORING EYE
⅝" RIGID P.V.C. FOR DRAINAGE
SUMP FOR DRAINAGE, FILL WITH ¾" STONE
6" WIDE TIE-DOWN PAINT, ON FINISHED ASPHALT SURFACE (BY PAVEMENT MARKING SUB-CONTRACTOR)
ROTATE EACH TIE-DOWN ANCHOR SO CASTING ALIGNS WITH FUTURE PAINT AS SHOWN.
PROPOSED TIE-DOWN ANCHOR DETAIL
NOT TO SCALE
PROJECT MANAGER BRAD VOLKER
CHECKED BY: W.R.
APPROVED BY: B.R.M.
COOPER ENGINEERING
310 WEST SOUTH STREET, P.O. BOX 230 RICE LAKE, WISCONSIN 54868-0230
TELEPHONE (715) 234-7008
FAX (715) 234-1025
BLACK RIVER FALLS AREA AIRPORT
JACKSON COUNTY, WISCONSIN
SHEET DESCRIPTION
PERMANENT PAVEMENT MARKINGS
A.I.P. 3-55-0008-01
DATE 5-02
DRAWN BY J.T.D.
SCALE 1" = 100'
PROJECT NO. 01552074
SHEET NO. 16

	CLEARING QUANTITY ESTIMATE SUMMARY				
	LOCATION DESCRIPTION	OWNER	ACRES TO CLEAR & GRUB	ACRES TO GRUB ONLY	INDIVIDUAL TREES TO CLEAR
BASE BID	LEFT OF RUNWAY CENTERLINE (NORTH), EXCEPT NORTH FENCELINE & PROPOSED HANGAR AREA	AIRPORT	8.4	7.3	–
	RIGHT (SOUTH) OF RUNWAY CENTERLINE EXCEPT FENCELINE ALONG S.T.H. "27" & HAWK ISLAND RD.	AIRPORT	17.0	11.7	–
	GROUP OF SMALL WHITE PINES, RIGHT (SOUTH), OF RUNWAY	AIRPORT	5.2	–	–
	EAST SIDE OF S.T.H. "27"	L. ROCHESTER	1.5*	–	–
	EAST SIDE OF S.T.H. "27"	K. LECHNER	1.9*	–	–
	EAST SIDE OF S.T.H. "27"	J. HEBERLAIN	0.4*	–	–
	UNDISTRIBUTED	AIRPORT	5.0	–	5.0±
ALT. NO. 1	PROPOSED HANGAR TAXIWAY AREA	AIRPORT	0.5	3.5	–
ALT. NO. 2	PROPOSED FENCELINE, WEST, NORTH & EAST OF HANGARS	AIRPORT	2.1	–	–
	PROPOSED FENCELINE, SOUTH SIDE OF RUNWAY, ALONG S.T.H. "27" AND HAWK ISLAND RD.	AIRPORT	7.1	–	–
	TOTALS		45.3	22.5	5±

* ESTIMATED ACREAGE FOR INFORMATION ONLY. CLEARING/GRUBBING/STUMP GRINDING ON THESE PRIVATE PROPERTIES BID LUMP SUM BY PROPERTY.

LEGEND

- TALLER TREES THAT ARE OUTSIDE THE PROPOSED CLEARING LIMITS THAT MAY HAVE TO BE INDIVIDUALLY HARVESTED OR TOPPED.
- PAVEMENT EDGE
- FAR PART 77 AIRSPACE FEATURE
- EDGE OF WOODS
- SURVEYED DECIDUOUS TREE
- SURVEYED CONIFEROUS TREE
- 875 HEIGHT LIMITATION ZONING ORDINANCE ELEVATION
- PROPOSED WOVEN WIRE FENCE
- PROPOSED CHAIN LINK FENCE
- PROPOSED CLEARING/GRUBBING
- GRUBBING ONLY (BRUSH REMOVAL INCIDENTAL)– MERCHANDISABLE TIMBER WAS PREVIOUSLY REMOVED BY OTHERS.

NO. BY DATE REVISIONS

PROJECT MANAGER BRAD VOLKER
CHECKED BY: W.R.
APPROVED BY: B.R.M.

COOPER ENGINEERING
310 WEST SOUTH STREET, P.O. BOX 230
RICE LAKE, WISCONSIN 54868–0230
TELEPHONE (715) 234–7008
FAX (715) 234–1025

BLACK RIVER FALLS AREA AIRPORT
JACKSON COUNTY, WISCONSIN

SHEET DESCRIPTION
CLEARING PLAN SHEET
A.I.P. 3–55–0008–01

DATE 5–02 | DRAWN BY M.R.E.
SCALE 1" = 600' | PROJECT NO. 01552074 | SHEET NO. 17

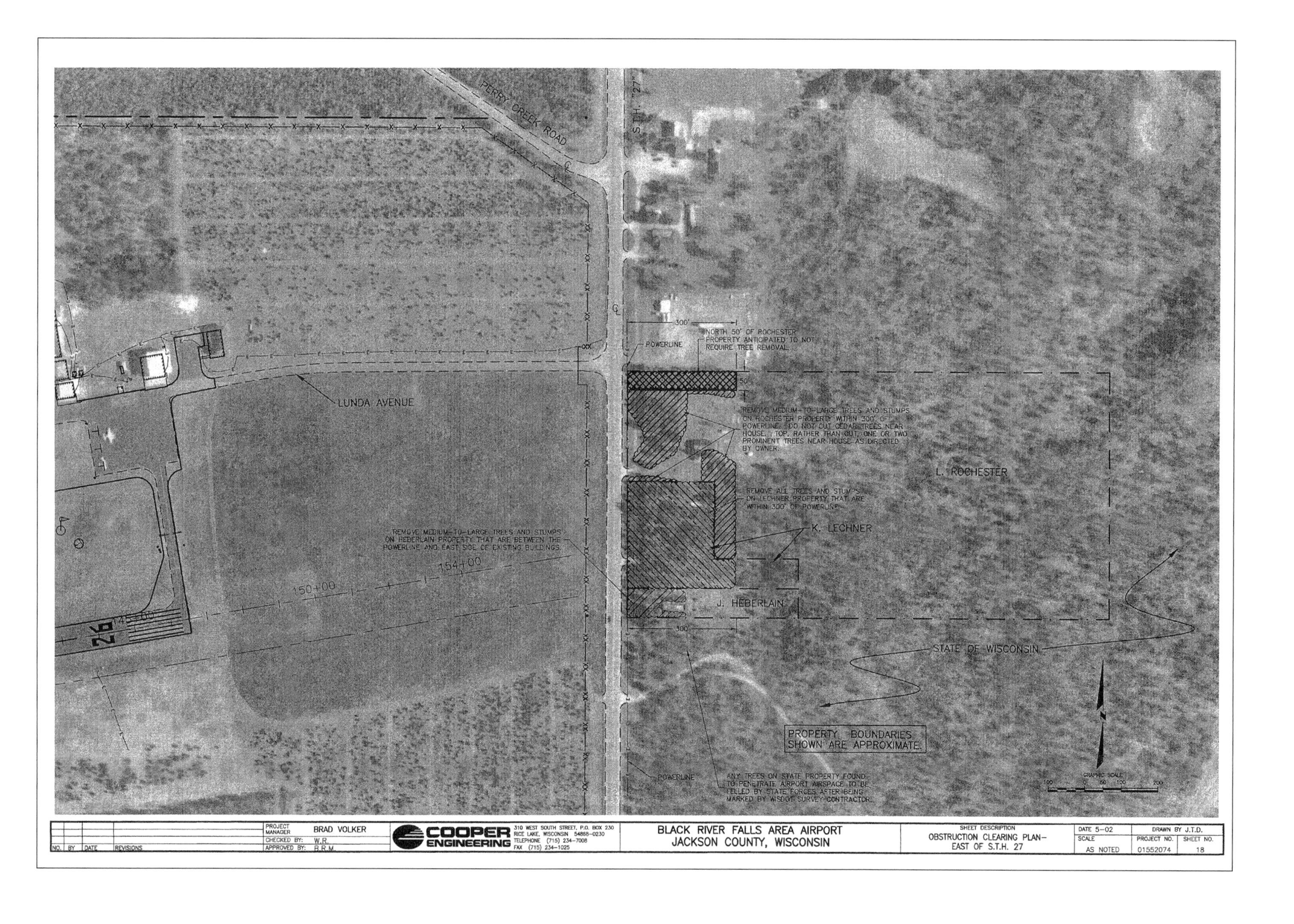
PERRY CREEK ROAD
S.T.H. "27"
LUNDA AVENUE
300'
NORTH 50' OF ROCHESTER PROPERTY ANTICIPATED TO NOT REQUIRE TREE REMOVAL.
POWERLINE
50'
REMOVE MEDIUM-TO-LARGE TREES AND STUMPS ON ROCHESTER PROPERTY WITHIN 300' OF POWERLINE. DO NOT CUT CEDAR TREES NEAR HOUSE. TOP, RATHER THAN CUT, ONE OR TWO PROMINENT TREES NEAR HOUSE AS DIRECTED BY OWNER.
L. ROCHESTER
REMOVE ALL TREES AND STUMPS ON LECHNER PROPERTY THAT ARE WITHIN 300' OF POWERLINE.
K. LECHNER
REMOVE MEDIUM-TO-LARGE TREES AND STUMPS ON HEBERLAIN PROPERTY THAT ARE BETWEEN THE POWERLINE AND EAST SIDE OF EXISTING BUILDINGS.
154+00
150+00
145+00
26
J. HEBERLAIN
300'
STATE OF WISCONSIN
PROPERTY BOUNDARIES SHOWN ARE APPROXIMATE.
POWERLINE
ANY TREES ON STATE PROPERTY FOUND TO PENETRATE AIRPORT AIRSPACE TO BE FELLED BY STATE FORCES AFTER BEING MARKED BY WISDOT SURVEY CONTRACTOR.
GRAPHIC SCALE
100 0 50 100 200
NO. BY DATE REVISIONS
PROJECT MANAGER BRAD VOLKER
CHECKED BY: W.R.
APPROVED BY: B.R.M.
COOPER ENGINEERING
310 WEST SOUTH STREET, P.O. BOX 230
RICE LAKE, WISCONSIN 54868-0230
TELEPHONE (715) 234-7008
FAX (715) 234-1025
BLACK RIVER FALLS AREA AIRPORT
JACKSON COUNTY, WISCONSIN
SHEET DESCRIPTION
OBSTRUCTION CLEARING PLAN–
EAST OF S.T.H. 27
DATE 5-02
SCALE AS NOTED
DRAWN BY J.T.D.
PROJECT NO. 01552074
SHEET NO. 18

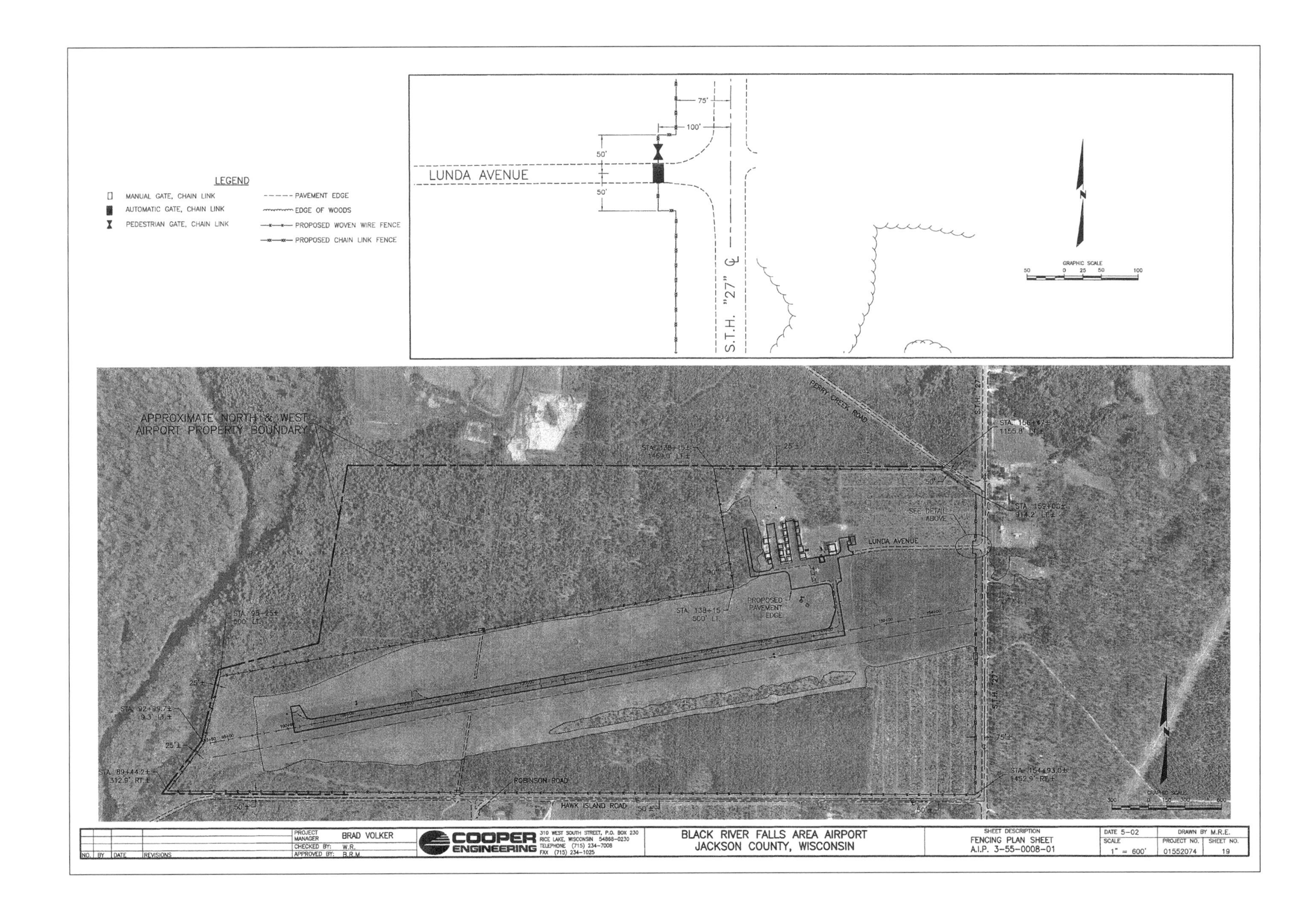
LEGEND
MANUAL GATE, CHAIN LINK
AUTOMATIC GATE, CHAIN LINK
PEDESTRIAN GATE, CHAIN LINK
PAVEMENT EDGE
EDGE OF WOODS
PROPOSED WOVEN WIRE FENCE
PROPOSED CHAIN LINK FENCE
LUNDA AVENUE
S.T.H. "27" ℄
APPROXIMATE NORTH & WEST AIRPORT PROPERTY BOUNDARY
PERRY CREEK ROAD
SEE DETAIL ABOVE
PROPOSED PAVEMENT EDGE
ROBINSON ROAD
HAWK ISLAND ROAD
PROJECT MANAGER BRAD VOLKER
CHECKED BY: W.R.
APPROVED BY: B.R.M.
COOPER ENGINEERING
310 WEST SOUTH STREET, P.O. BOX 230
RICE LAKE, WISCONSIN 54868-0230
TELEPHONE (715) 234-7008
FAX (715) 234-1025
BLACK RIVER FALLS AREA AIRPORT
JACKSON COUNTY, WISCONSIN
SHEET DESCRIPTION
FENCING PLAN SHEET
A.I.P. 3-55-0008-01
DATE 5-02
DRAWN BY M.R.E.
SCALE 1" = 600'
PROJECT NO. 01552074
SHEET NO. 19

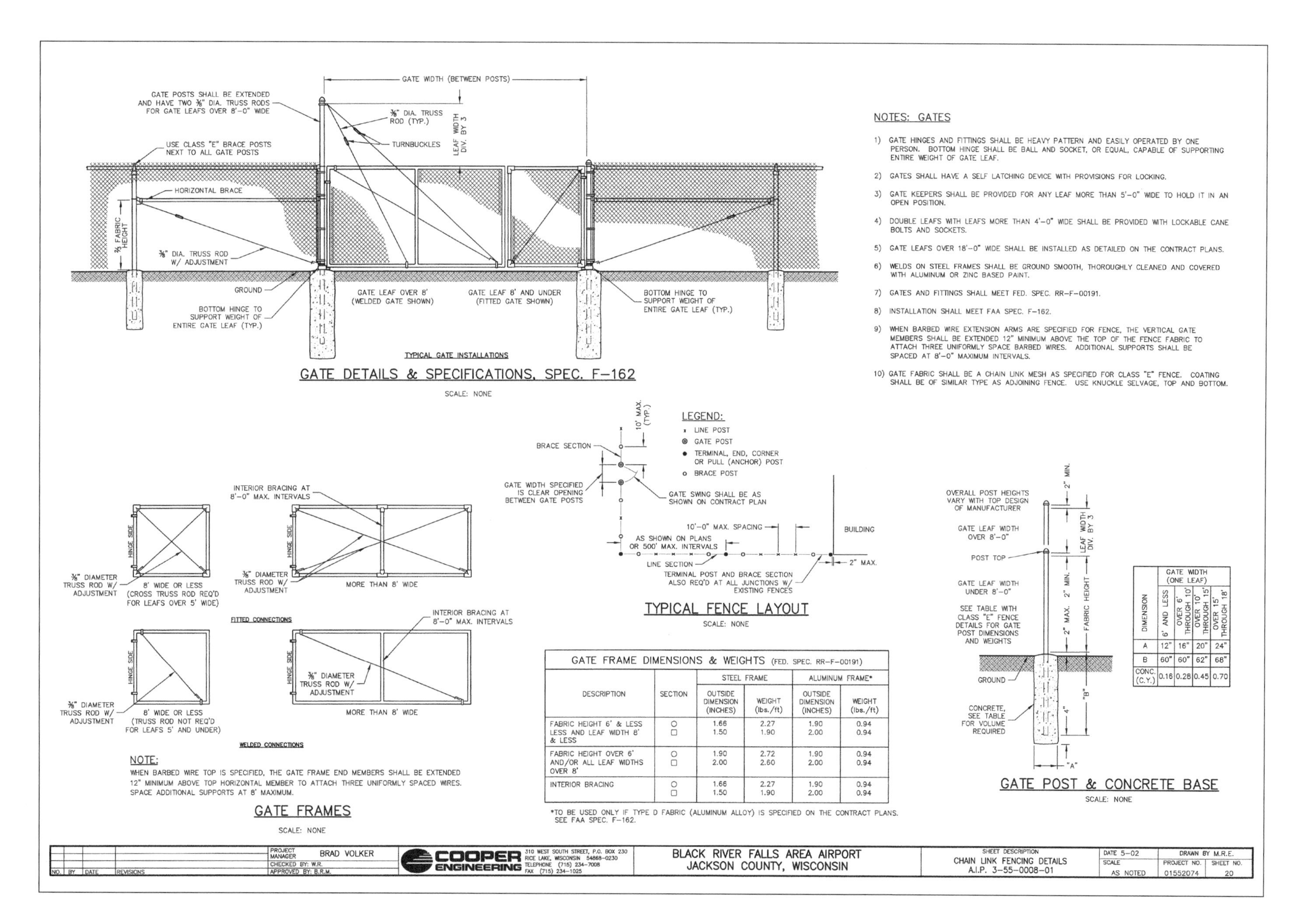

GATE FRAME DIMENSIONS & WEIGHTS (FED. SPEC. RR-F-00191)

DESCRIPTION	SECTION	STEEL FRAME: OUTSIDE DIMENSION (INCHES)	STEEL FRAME: WEIGHT (lbs./ft)	ALUMINUM FRAME*: OUTSIDE DIMENSION (INCHES)	ALUMINUM FRAME*: WEIGHT (lbs./ft)
FABRIC HEIGHT 6' & LESS LESS AND LEAF WIDTH 8' & LESS	○ □	1.66 1.50	2.27 1.90	1.90 2.00	0.94 0.94
FABRIC HEIGHT OVER 6' AND/OR ALL LEAF WIDTHS OVER 8'	○ □	1.90 2.00	2.72 2.60	1.90 2.00	0.94 0.94
INTERIOR BRACING	○ □	1.66 1.50	2.27 1.90	1.90 2.00	0.94 0.94

*TO BE USED ONLY IF TYPE D FABRIC (ALUMINUM ALLOY) IS SPECIFIED ON THE CONTRACT PLANS. SEE FAA SPEC. F-162.

DIMENSION	GATE WIDTH (ONE LEAF): 6' AND LESS	OVER 6' THROUGH 10'	OVER 10' THROUGH 15'	OVER 15' THROUGH 18'
A	12"	16"	20"	24"
B	60"	60"	62"	68"
CONC. (C.Y.)	0.16	0.28	0.45	0.70

CHAIN LINK FENCE MEMBERS DIMENSIONS & WEIGHTS (FED. SPEC. RR-F-00191)

DESCRIPTION	SECTION	STEEL FRAME		ALUMINUM FRAME*	
		OUTSIDE DIMENSION (INCHES)	WEIGHT (lbs./ft)	OUTSIDE DIMENSION (INCHES)	WEIGHT (lbs./ft)
CORNER, BRACE, END & PULL POSTS					
FABRIC HEIGHTS 6' & LESS	○	2.375	3.650	2.375	1.253
	□	2.000	3.650	2.500	1.253
FABRIC HEIGHTS OVER 6'	○	2.875	5.790	2.875	2.000
	□	2.500	5.700	3.000	2.000
ALL HEIGHTS	ROLL FORM.	3.5 X 3.5	5.100	–	–
GATE POSTS					
GATE LEAF 6' & LESS	○	2.875	5.790	2.875	2.000
	□	2.500	5.700	3.000	2.000
	ROLL FORM.	3.5 X 3.5	5.100	–	–
GATE LEAF WIDTH OVER 6' THRU 13'	○	4.000	9.100	4.000	3.000
GATE LEAF WIDTH OVER 13' THRU 18'	○	6.625	18.970	6.625	7.000
GATE LEAF WIDTH OVER 18'	○	8.625	24.700	8.625	10.500
LINE POSTS					
FABRIC HEIGHTS 6' & LESS	⊔	1.875 X 1.625	2.340	–	–
FABRIC HEIGHTS 8' & LESS	⊔	2.25 X 1.6875	2.730	–	–
FABRIC HEIGHTS 6' & LESS	○	1.900	2.720	1.900	0.940
FABRIC HEIGHTS OVER 6'	○	2.375	3.650	2.375	1.250
FABRIC HEIGHTS 8' & LESS	H	1.875 X 1.625 X 0.113	2.700	1.875 X 1.625 X 0.113	0.910
FABRIC HEIGHTS OVER 8'	H	2.250 X 1.950 X 0.143	4.100	2.250 X 1.950 X 0.143	1.250
RAILS & BRACES	○	1.660	1.806	1.660	0.786
	ROLL FORM.	1.625 X 1.250	1.350	–	–

*TO BE USED ONLY IF TYPE D FABRIC (ALUMINUM ALLOY) IS SPECIFIED ON THE CONTRACT PLANS. SEE FAA SPEC. F-162.

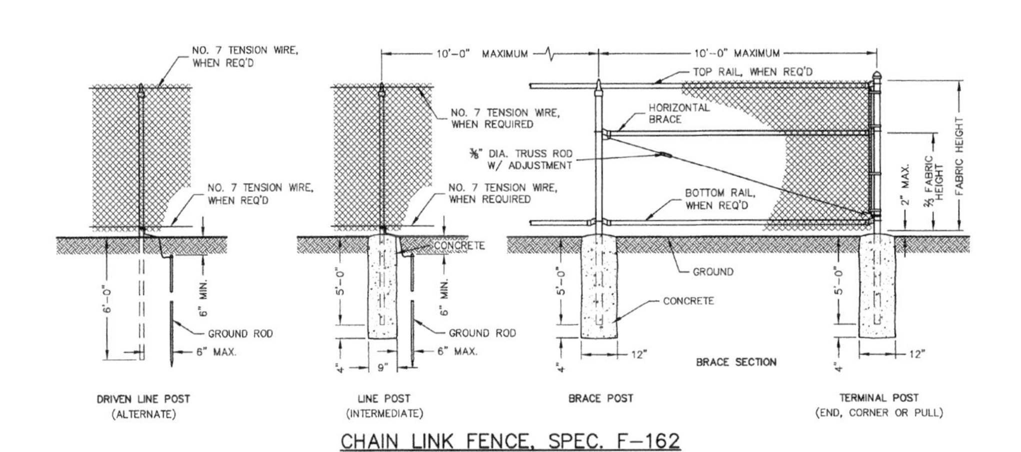

CHAIN LINK FENCE, SPEC. F-162

SCALE: NONE

REFER TO SHEETS E16 AND E17 FOR FURTHER DETAILS ON POWERED GATE. DETAILS ON SHEETS E16 AND E17 SHALL GOVERN IF DIFFERENCES ARE FOUND BETWEEN THOSE DETAILS AND THE DETAIL BELOW.

GATE WIDTH (BETWEEN POSTS)
TYPICAL CHAIN LINK FENCE (SHOWN W/ GATE & BRACE POSTS)
"OUTSIDE"
"INSIDE"
GATE FRAME
SLIDE GATE OPERATOR
DRIVE CHAIN
TYPICAL OPEN-ROLLER
PUSH BUTTON AT 5' ABOVE GROUND

PLAN VIEW

GATE POST
GATE FRAME
GATE POST
FENCE BRACE
GROUND
SLIDE GATE OPERATOR (SEE SPECIFICATIONS FOR INSTALLATION DETAILS)
MINIMUM #50 ROLLER DRIVE CHAIN

FRONT VIEW

OPEN-ROLLER POWER GATE DETAIL

SCALE: NONE

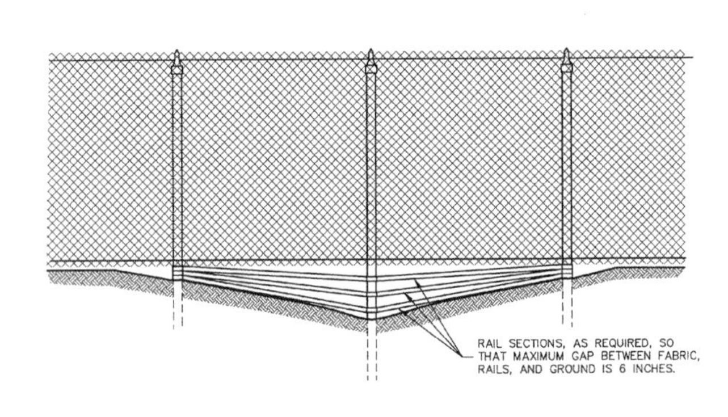

CHANNEL GUARD FOR MINOR DITCH CROSSINGS

SCALE: NONE

NO.	BY	DATE	REVISIONS

PROJECT MANAGER BRAD VOLKER
CHECKED BY: W.R.
APPROVED BY: B.R.M.

COOPER ENGINEERING
310 WEST SOUTH STREET, P.O. BOX 230
RICE LAKE, WISCONSIN 54868-0230
TELEPHONE (715) 234-7008
FAX (715) 234-1025

BLACK RIVER FALLS AREA AIRPORT
JACKSON COUNTY, WISCONSIN

SHEET DESCRIPTION: CHAIN LINK FENCING DETAILS A.I.P. 3-55-0008-01

DATE 5-02 | DRAWN BY M.R.E. | SCALE AS NOTED | PROJECT NO. 01552074 | SHEET NO. 21

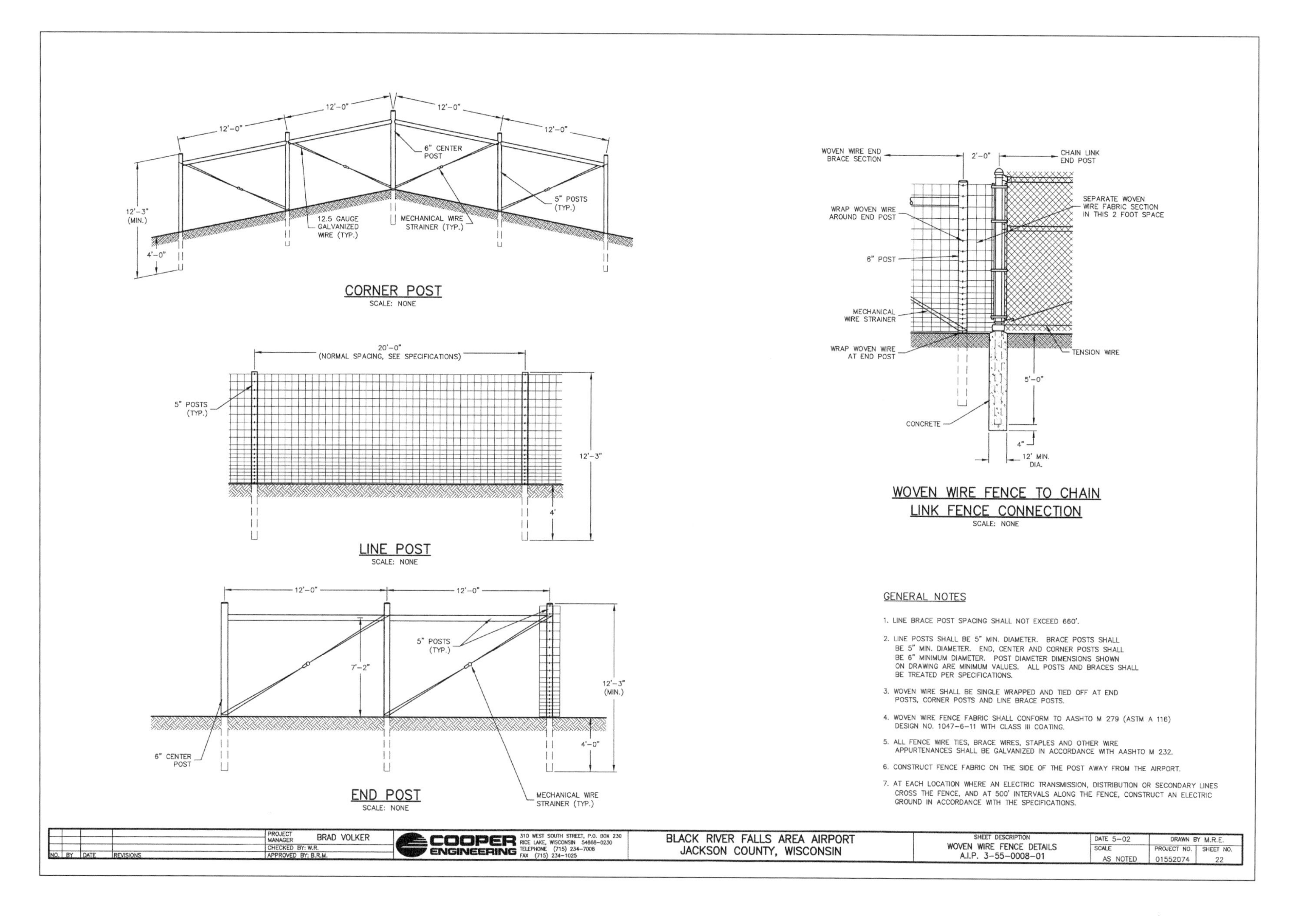
12'-0"
12'-0"
12'-0"
12'-0"
6" CENTER POST
5" POSTS (TYP.)
12'-3" (MIN.)
4'-0"
12.5 GAUGE GALVANIZED WIRE (TYP.)
MECHANICAL WIRE STRAINER (TYP.)
CORNER POST
SCALE: NONE
20'-0" (NORMAL SPACING, SEE SPECIFICATIONS)
5" POSTS (TYP.)
12'-3"
4'
LINE POST
SCALE: NONE
12'-0"
12'-0"
5" POSTS (TYP.)
7'-2"
12'-3" (MIN.)
4'-0"
6" CENTER POST
MECHANICAL WIRE STRAINER (TYP.)
END POST
SCALE: NONE
WOVEN WIRE END BRACE SECTION
2'-0"
CHAIN LINK END POST
SEPARATE WOVEN WIRE FABRIC SECTION IN THIS 2 FOOT SPACE
WRAP WOVEN WIRE AROUND END POST
6" POST
MECHANICAL WIRE STRAINER
WRAP WOVEN WIRE AT END POST
TENSION WIRE
5'-0"
CONCRETE
4"
12" MIN. DIA.
WOVEN WIRE FENCE TO CHAIN LINK FENCE CONNECTION
SCALE: NONE
GENERAL NOTES
1. LINE BRACE POST SPACING SHALL NOT EXCEED 660'.
2. LINE POSTS SHALL BE 5" MIN. DIAMETER. BRACE POSTS SHALL BE 5" MIN. DIAMETER. END, CENTER AND CORNER POSTS SHALL BE 6" MINIMUM DIAMETER. POST DIAMETER DIMENSIONS SHOWN ON DRAWING ARE MINIMUM VALUES. ALL POSTS AND BRACES SHALL BE TREATED PER SPECIFICATIONS.
3. WOVEN WIRE SHALL BE SINGLE WRAPPED AND TIED OFF AT END POSTS, CORNER POSTS AND LINE BRACE POSTS.
4. WOVEN WIRE FENCE FABRIC SHALL CONFORM TO AASHTO M 279 (ASTM A 116) DESIGN NO. 1047-6-11 WITH CLASS III COATING.
5. ALL FENCE WIRE TIES, BRACE WIRES, STAPLES AND OTHER WIRE APPURTENANCES SHALL BE GALVANIZED IN ACCORDANCE WITH AASHTO M 232.
6. CONSTRUCT FENCE FABRIC ON THE SIDE OF THE POST AWAY FROM THE AIRPORT.
7. AT EACH LOCATION WHERE AN ELECTRIC TRANSMISSION, DISTRIBUTION OR SECONDARY LINES CROSS THE FENCE, AND AT 500' INTERVALS ALONG THE FENCE, CONSTRUCT AN ELECTRIC GROUND IN ACCORDANCE WITH THE SPECIFICATIONS.
NO. BY DATE REVISIONS
PROJECT MANAGER BRAD VOLKER
CHECKED BY: W.R.
APPROVED BY: B.R.M.
COOPER ENGINEERING
310 WEST SOUTH STREET, P.O. BOX 230
RICE LAKE, WISCONSIN 54868-0230
TELEPHONE (715) 234-7008
FAX (715) 234-1025
BLACK RIVER FALLS AREA AIRPORT
JACKSON COUNTY, WISCONSIN
SHEET DESCRIPTION
WOVEN WIRE FENCE DETAILS
A.I.P. 3-55-0008-01
DATE 5-02
SCALE AS NOTED
DRAWN BY M.R.E.
PROJECT NO. 01552074
SHEET NO. 22

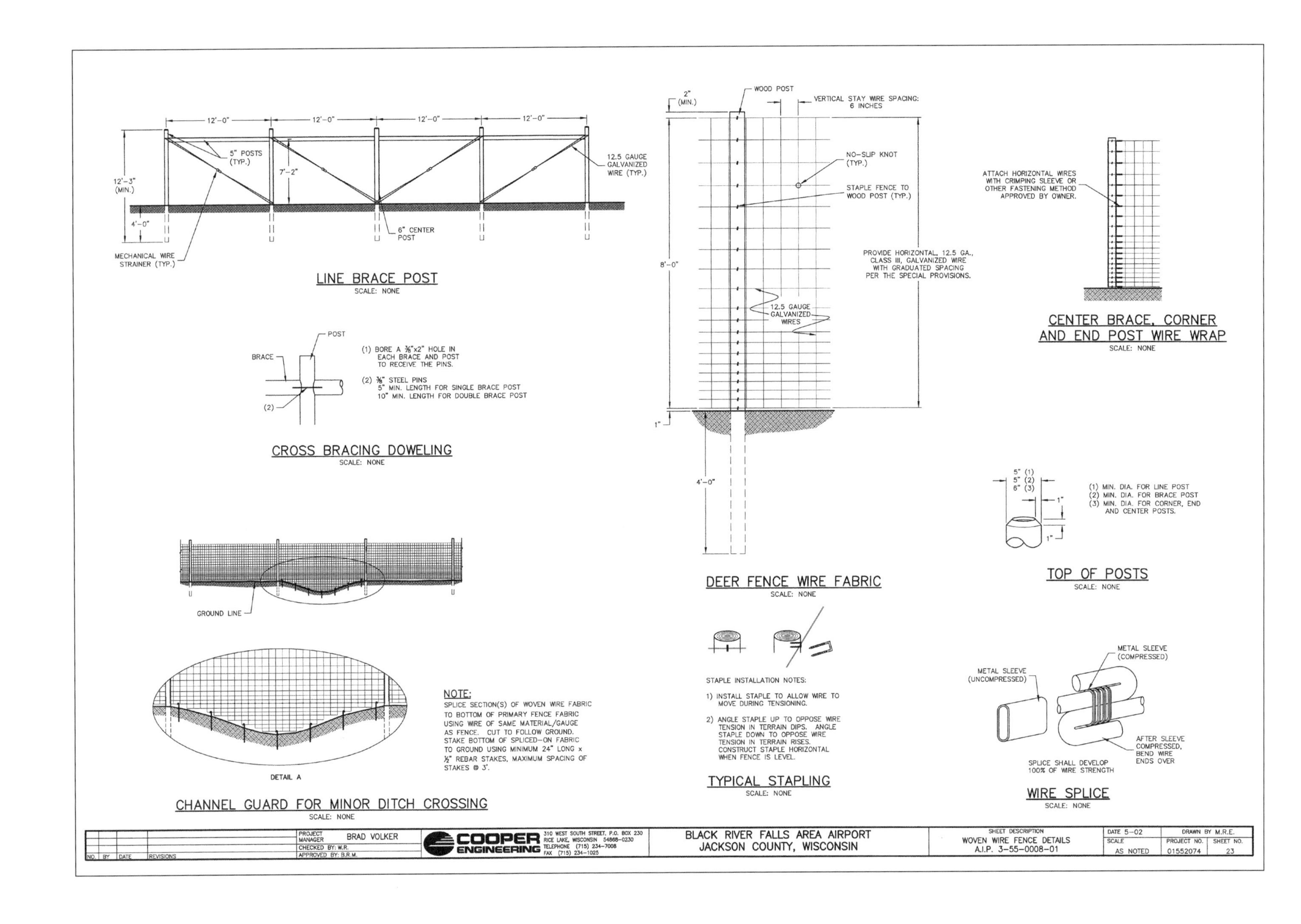

LINE BRACE POST
SCALE: NONE
12'-0"
5" POSTS (TYP.)
12.5 GAUGE GALVANIZED WIRE (TYP.)
12'-3" (MIN.)
7'-2"
4'-0"
6" CENTER POST
MECHANICAL WIRE STRAINER (TYP.)
CROSS BRACING DOWELING
SCALE: NONE
POST
BRACE
(2)
(1) BORE A ⅜"x2" HOLE IN EACH BRACE AND POST TO RECEIVE THE PINS.
(2) ⅜" STEEL PINS 5" MIN. LENGTH FOR SINGLE BRACE POST 10" MIN. LENGTH FOR DOUBLE BRACE POST
DEER FENCE WIRE FABRIC
SCALE: NONE
2" (MIN.)
WOOD POST
VERTICAL STAY WIRE SPACING: 6 INCHES
NO-SLIP KNOT (TYP.)
STAPLE FENCE TO WOOD POST (TYP.)
PROVIDE HORIZONTAL, 12.5 GA., CLASS III, GALVANIZED WIRE WITH GRADUATED SPACING PER THE SPECIAL PROVISIONS.
12.5 GAUGE GALVANIZED WIRES
8'-0"
1"
4'-0"
CENTER BRACE, CORNER AND END POST WIRE WRAP
SCALE: NONE
ATTACH HORIZONTAL WIRES WITH CRIMPING SLEEVE OR OTHER FASTENING METHOD APPROVED BY OWNER.
TOP OF POSTS
SCALE: NONE
5" (1)
5" (2)
6" (3)
1"
(1) MIN. DIA. FOR LINE POST
(2) MIN. DIA. FOR BRACE POST
(3) MIN. DIA. FOR CORNER, END AND CENTER POSTS.
GROUND LINE
DETAIL A
CHANNEL GUARD FOR MINOR DITCH CROSSING
SCALE: NONE
NOTE:
SPLICE SECTION(S) OF WOVEN WIRE FABRIC TO BOTTOM OF PRIMARY FENCE FABRIC USING WIRE OF SAME MATERIAL/GAUGE AS FENCE. CUT TO FOLLOW GROUND. STAKE BOTTOM OF SPLICED-ON FABRIC TO GROUND USING MINIMUM 24" LONG x ½" REBAR STAKES, MAXIMUM SPACING OF STAKES @ 3'.
STAPLE INSTALLATION NOTES:
1) INSTALL STAPLE TO ALLOW WIRE TO MOVE DURING TENSIONING.
2) ANGLE STAPLE UP TO OPPOSE WIRE TENSION IN TERRAIN DIPS. ANGLE STAPLE DOWN TO OPPOSE WIRE TENSION IN TERRAIN RISES. CONSTRUCT STAPLE HORIZONTAL WHEN FENCE IS LEVEL.
TYPICAL STAPLING
SCALE: NONE
METAL SLEEVE (UNCOMPRESSED)
METAL SLEEVE (COMPRESSED)
AFTER SLEEVE COMPRESSED, BEND WIRE ENDS OVER
SPLICE SHALL DEVELOP 100% OF WIRE STRENGTH
WIRE SPLICE
SCALE: NONE
NO. BY DATE REVISIONS
PROJECT MANAGER BRAD VOLKER
CHECKED BY: W.R.
APPROVED BY: B.R.M.
COOPER ENGINEERING
310 WEST SOUTH STREET, P.O. BOX 230 RICE LAKE, WISCONSIN 54868-0230 TELEPHONE (715) 234-7008 FAX (715) 234-1025
BLACK RIVER FALLS AREA AIRPORT
JACKSON COUNTY, WISCONSIN
SHEET DESCRIPTION
WOVEN WIRE FENCE DETAILS
A.I.P. 3-55-0008-01
DATE 5-02
DRAWN BY M.R.E.
SCALE AS NOTED
PROJECT NO. 01552074
SHEET NO. 23

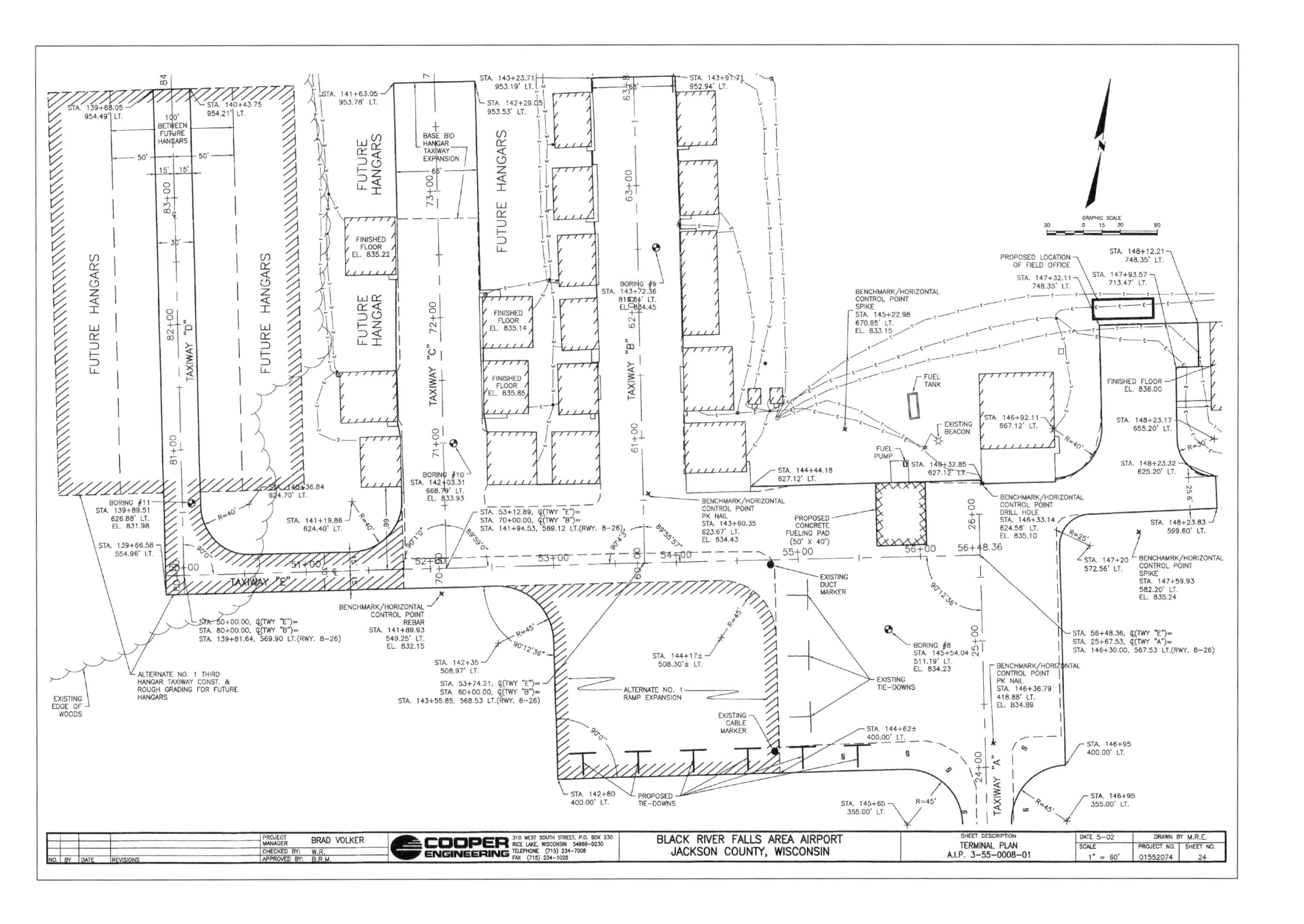
FUTURE HANGARS
TAXIWAY "D"
TAXIWAY "C"
TAXIWAY "B"
TAXIWAY "E"
TAXIWAY "A"
FUTURE HANGAR
BASE BID HANGAR TAXIWAY EXPANSION
PROPOSED CONCRETE FUELING PAD (50' X 40')
ALTERNATE NO. 1 RAMP EXPANSION
ALTERNATE NO. 1 THIRD HANGAR TAXIWAY CONST. & ROUGH GRADING FOR FUTURE HANGARS
EXISTING EDGE OF WOODS
EXISTING TIE-DOWNS
PROPOSED TIE-DOWNS
PROPOSED LOCATION OF FIELD OFFICE
FUEL TANK
FUEL PUMP
EXISTING BEACON
EXISTING DUCT MARKER
EXISTING CABLE MARKER
GRAPHIC SCALE
COOPER ENGINEERING
310 WEST SOUTH STREET, P.O. BOX 230
RICE LAKE, WISCONSIN 54868-0230
TELEPHONE (715) 234-7008
FAX (715) 234-1025
PROJECT MANAGER BRAD VOLKER
CHECKED BY: W.R.
APPROVED BY: B.R.M.
BLACK RIVER FALLS AREA AIRPORT
JACKSON COUNTY, WISCONSIN
SHEET DESCRIPTION
TERMINAL PLAN
A.I.P. 3-55-0008-01
DATE 5-02
SCALE 1" = 60'
DRAWN BY M.R.E.
PROJECT NO. 01552074
SHEET NO. 24
NO. BY DATE REVISIONS

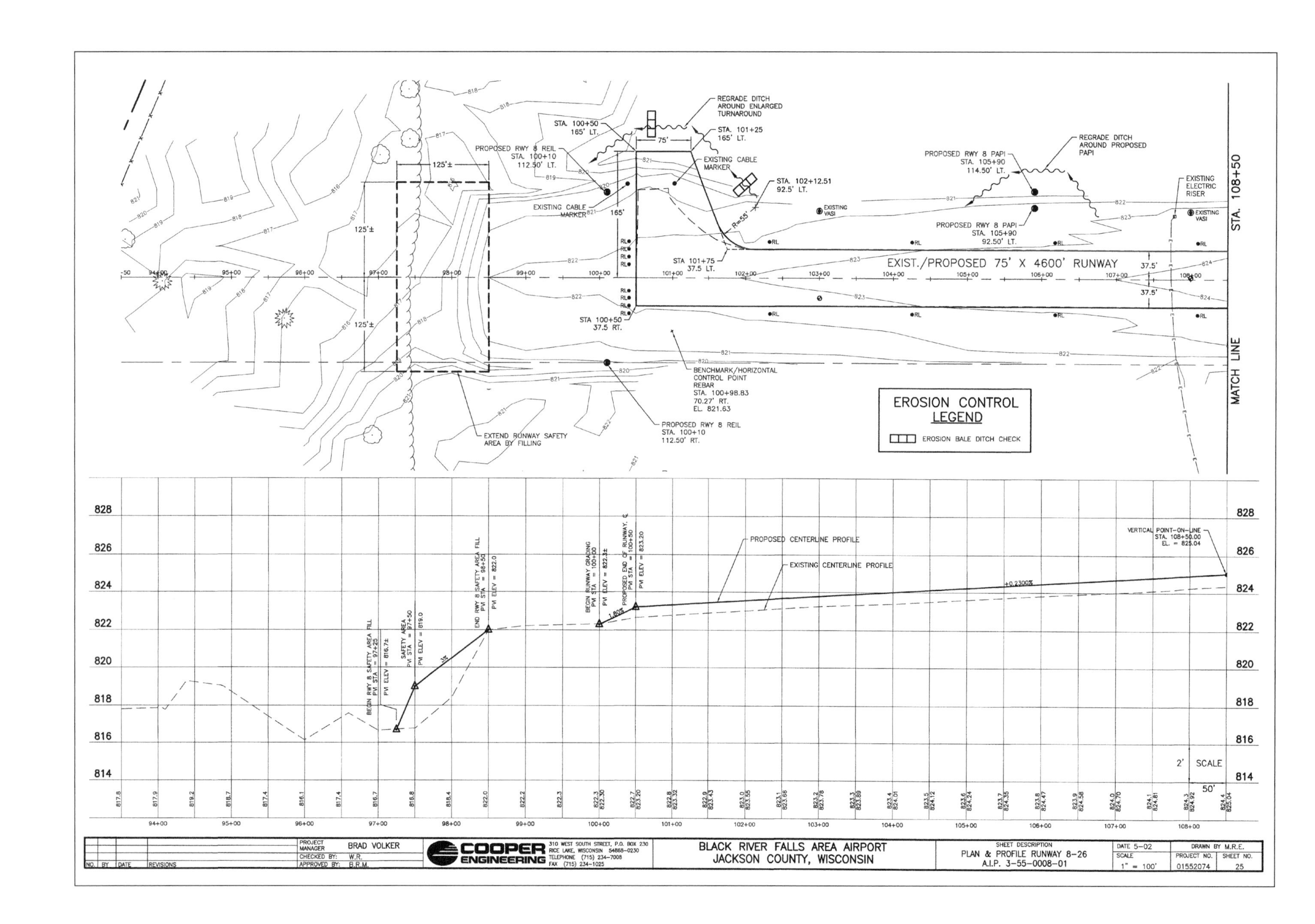
REGRADE DITCH AROUND ENLARGED TURNAROUND
STA. 100+50 165' LT.
STA. 101+25 165' LT.
PROPOSED RWY 8 REIL STA. 100+10 112.50' LT.
EXISTING CABLE MARKER
STA. 102+12.51 92.5' LT.
PROPOSED RWY 8 PAPI STA. 105+90 114.50' LT.
REGRADE DITCH AROUND PROPOSED PAPI
EXISTING ELECTRIC RISER
EXISTING VASI
PROPOSED RWY 8 PAPI STA. 105+90 92.50' LT.
STA 101+75 37.5 LT.
EXIST./PROPOSED 75' X 4600' RUNWAY
STA. 108+50
MATCH LINE
STA 100+50 37.5 RT.
BENCHMARK/HORIZONTAL CONTROL POINT REBAR STA. 100+98.83 70.27' RT. EL. 821.63
PROPOSED RWY 8 REIL STA. 100+10 112.50' RT.
EXTEND RUNWAY SAFETY AREA BY FILLING
EROSION CONTROL LEGEND
EROSION BALE DITCH CHECK
PROPOSED CENTERLINE PROFILE
EXISTING CENTERLINE PROFILE
VERTICAL POINT-ON-LINE STA. 108+50.00 EL. = 825.04
+0.2300%
BEGIN RWY 8 SAFETY AREA FILL PVI STA = 97+25 PVI ELEV = 816.7±
SAFETY AREA PVI STA = 97+50 PVI ELEV = 819.0
END RWY 8 SAFETY AREA FILL PVI STA = 98+50 PVI ELEV = 822.0
BEGIN RUNWAY GRADING PVI STA = 100+00 PVI ELEV = 822.3±
PROPOSED END OF RUNWAY PVI STA = 100+50 PVI ELEV = 823.20
2' SCALE 50'
PROJECT MANAGER BRAD VOLKER
CHECKED BY: W.R.
APPROVED BY: B.R.M.
COOPER ENGINEERING
310 WEST SOUTH STREET, P.O. BOX 230 RICE LAKE, WISCONSIN 54868-0230 TELEPHONE (715) 234-7008 FAX (715) 234-1025
BLACK RIVER FALLS AREA AIRPORT
JACKSON COUNTY, WISCONSIN
SHEET DESCRIPTION
PLAN & PROFILE RUNWAY 8-26
A.I.P. 3-55-0008-01
DATE 5-02
SCALE 1" = 100'
DRAWN BY M.R.E.
PROJECT NO. 01552074
SHEET NO. 25
NO. BY DATE REVISIONS

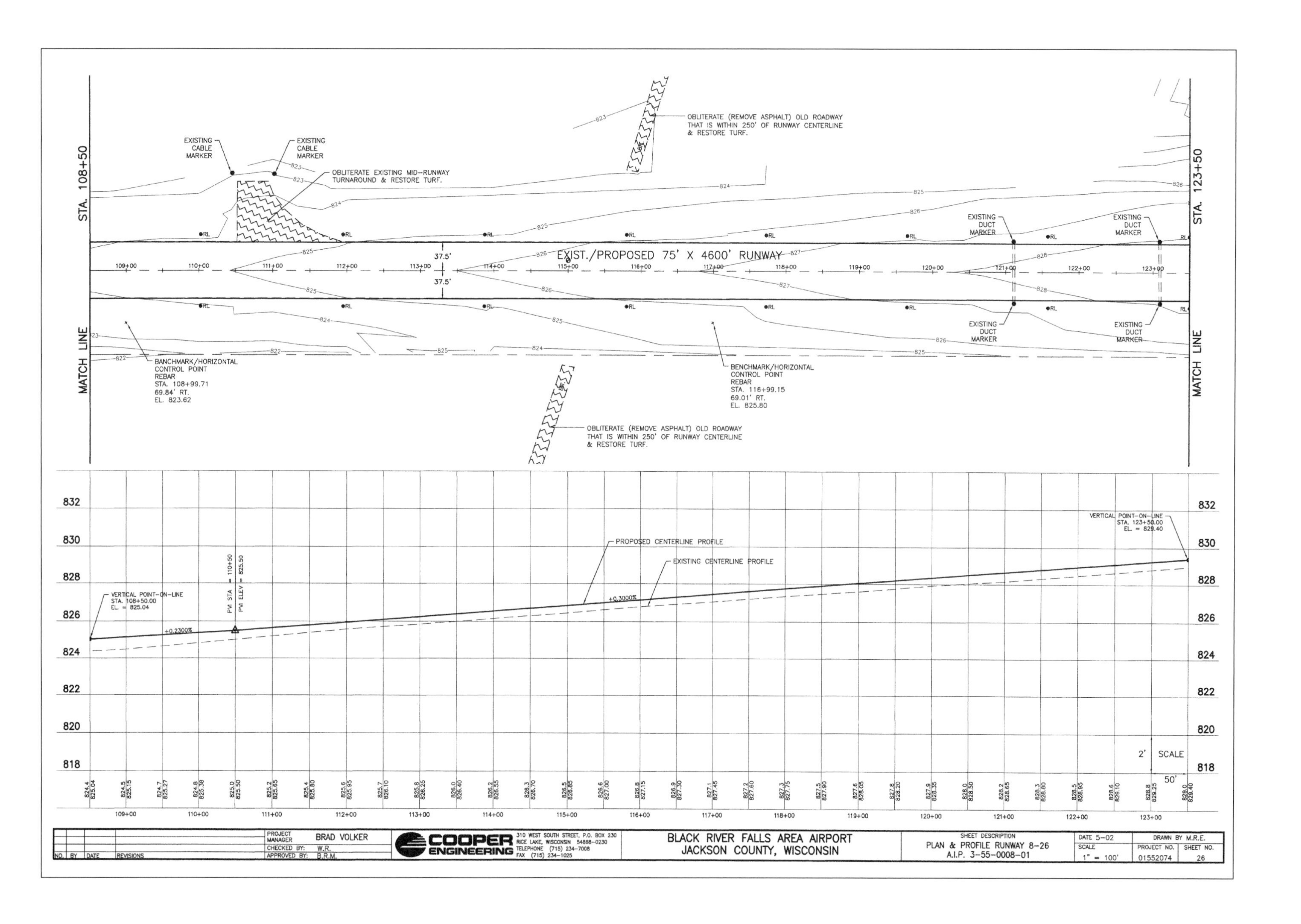
STA. 108+50
MATCH LINE
STA. 123+50
MATCH LINE
EXISTING CABLE MARKER
EXISTING CABLE MARKER
OBLITERATE EXISTING MID-RUNWAY TURNAROUND & RESTORE TURF.
OBLITERATE (REMOVE ASPHALT) OLD ROADWAY THAT IS WITHIN 250' OF RUNWAY CENTERLINE & RESTORE TURF.
EXISTING DUCT MARKER
EXIST./PROPOSED 75' X 4600' RUNWAY
37.5'
37.5'
BANCHMARK/HORIZONTAL CONTROL POINT REBAR STA. 108+99.71 69.84' RT. EL. 823.62
BENCHMARK/HORIZONTAL CONTROL POINT REBAR STA. 116+99.15 69.01' RT. EL. 825.80
VERTICAL POINT-ON-LINE STA. 108+50.00 EL. = 825.04
PVI STA = 110+50
PVI ELEV = 825.50
+0.2300%
PROPOSED CENTERLINE PROFILE
EXISTING CENTERLINE PROFILE
+0.3000%
VERTICAL POINT-ON-LINE STA. 123+50.00 EL. = 829.40
2' SCALE
50'
PROJECT MANAGER BRAD VOLKER
CHECKED BY: W.R.
APPROVED BY: B.R.M.
COOPER ENGINEERING
310 WEST SOUTH STREET, P.O. BOX 230 RICE LAKE, WISCONSIN 54868-0230 TELEPHONE (715) 234-7008 FAX (715) 234-1025
BLACK RIVER FALLS AREA AIRPORT
JACKSON COUNTY, WISCONSIN
SHEET DESCRIPTION
PLAN & PROFILE RUNWAY 8-26
A.I.P. 3-55-0008-01
DATE 5-02
SCALE 1" = 100'
DRAWN BY M.R.E.
PROJECT NO. 01552074
SHEET NO. 26
NO. BY DATE REVISIONS

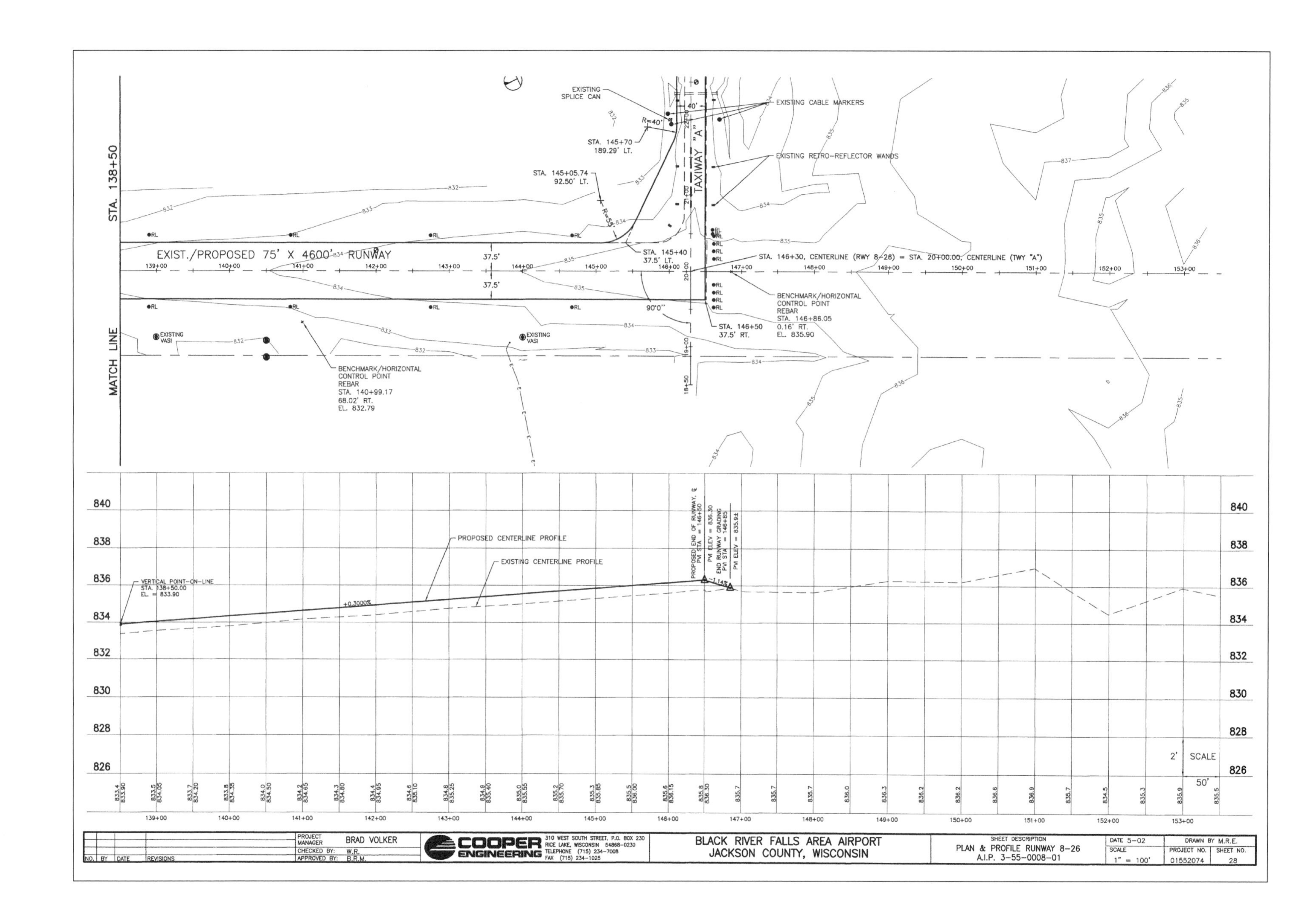
EXISTING SPLICE CAN
EXISTING CABLE MARKERS
EXISTING RETRO-REFLECTOR WANDS
TAXIWAY "A"
STA. 145+70 189.29' LT.
STA. 145+05.74 92.50' LT.
STA. 138+50
MATCH LINE
EXIST./PROPOSED 75' X 4600' RUNWAY
STA. 145+40 37.5' LT.
STA. 146+30, CENTERLINE (RWY 8-26) = STA. 20+00.00, CENTERLINE (TWY "A")
BENCHMARK/HORIZONTAL CONTROL POINT REBAR STA. 146+86.05 0.16' RT. EL. 835.90
STA. 146+50 37.5' RT.
EXISTING VASI
BENCHMARK/HORIZONTAL CONTROL POINT REBAR STA. 140+99.17 68.02' RT. EL. 832.79
PROPOSED CENTERLINE PROFILE
EXISTING CENTERLINE PROFILE
VERTICAL POINT-ON-LINE STA. 138+50.00 EL. 833.90
+0.3000%
PROPOSED END OF RUNWAY PVI STA = 146+50 PVI ELEV = 836.30
END RUNWAY GRADING PVI STA = 146+85 PVI ELEV = 835.9±
SCALE
BRAD VOLKER
COOPER ENGINEERING
310 WEST SOUTH STREET, P.O. BOX 230 RICE LAKE, WISCONSIN 54868-0230
BLACK RIVER FALLS AREA AIRPORT JACKSON COUNTY, WISCONSIN
PLAN & PROFILE RUNWAY 8-26 A.I.P. 3-55-0008-01

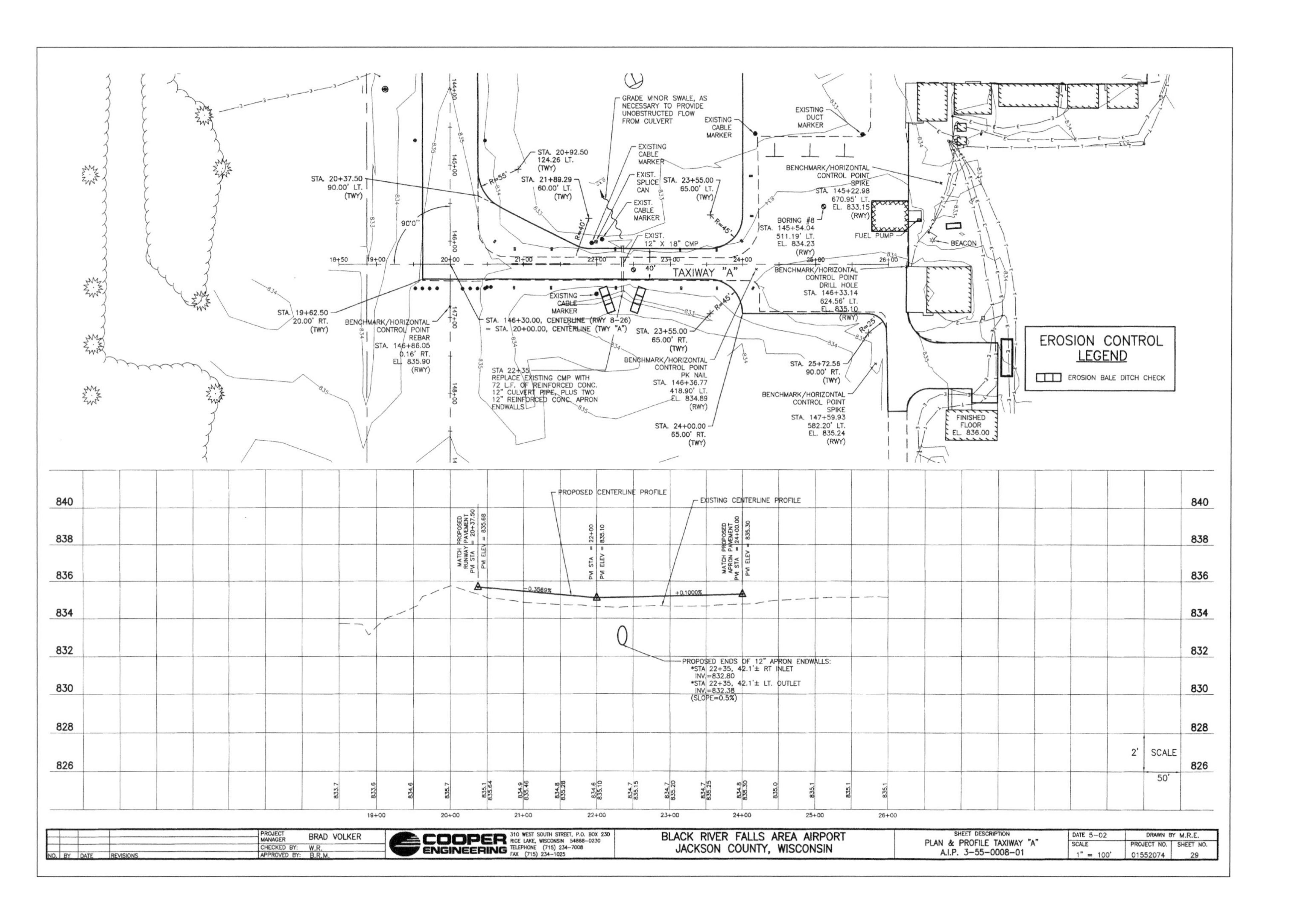
TAXIWAY "A"
EROSION CONTROL LEGEND
EROSION BALE DITCH CHECK
PROPOSED CENTERLINE PROFILE
EXISTING CENTERLINE PROFILE
PROPOSED ENDS OF 12" APRON ENDWALLS:
*STA 22+35, 42.1'± RT INLET
INV.=832.80
*STA 22+35, 42.1'± LT. OUTLET
INV.=832.38
(SLOPE=0.5%)
COOPER ENGINEERING
BLACK RIVER FALLS AREA AIRPORT
JACKSON COUNTY, WISCONSIN
PLAN & PROFILE TAXIWAY "A"
A.I.P. 3-55-0008-01

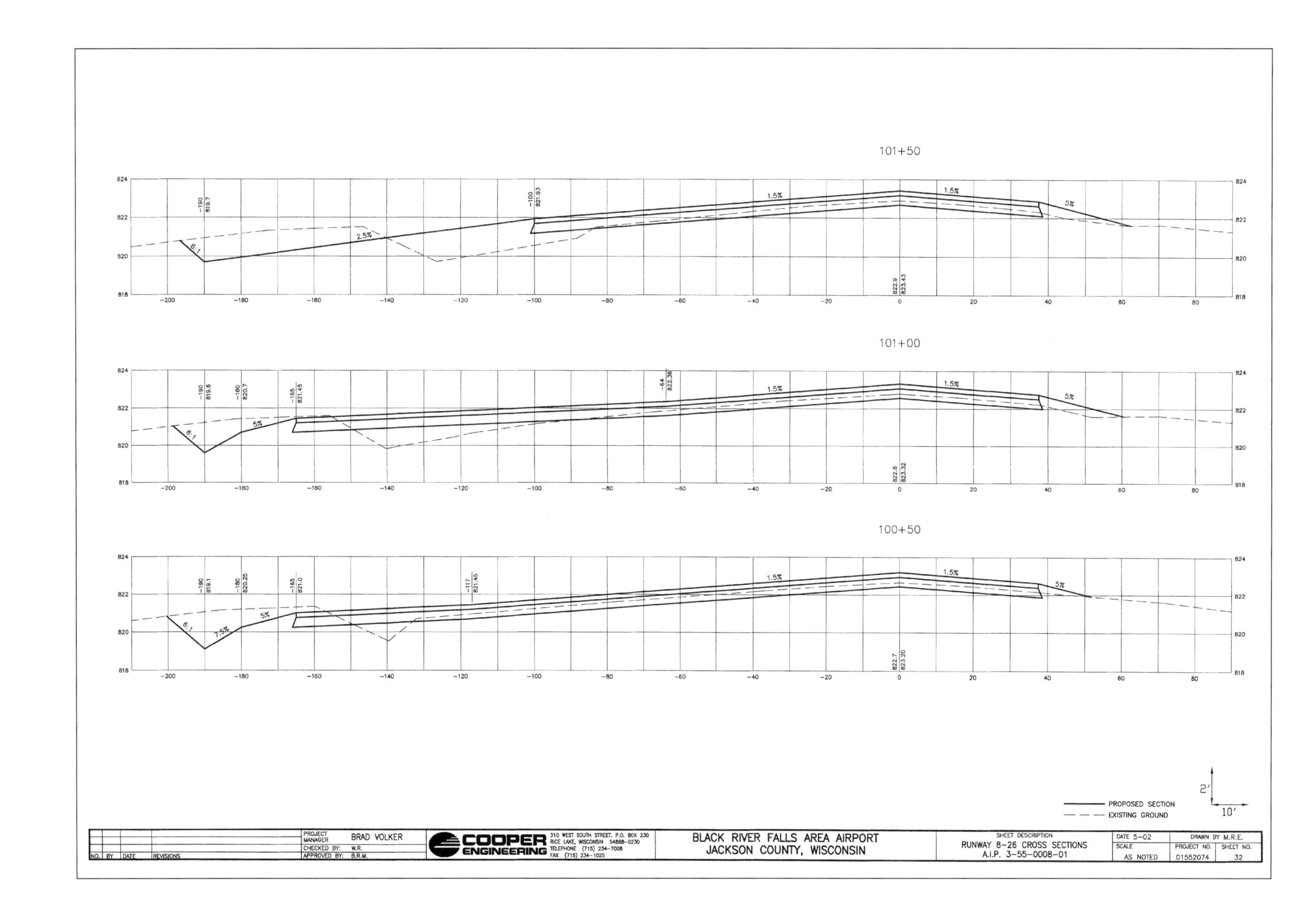
101+50
101+00
100+50
PROPOSED SECTION
EXISTING GROUND
PROJECT MANAGER BRAD VOLKER
CHECKED BY: W.R.
APPROVED BY: B.R.M.
COOPER ENGINEERING
310 WEST SOUTH STREET, P.O. BOX 230
RICE LAKE, WISCONSIN 54868-0230
TELEPHONE (715) 234-7008
FAX (715) 234-1025
BLACK RIVER FALLS AREA AIRPORT
JACKSON COUNTY, WISCONSIN
SHEET DESCRIPTION
RUNWAY 8-26 CROSS SECTIONS
A.I.P. 3-55-0008-01
DATE 5-02
SCALE AS NOTED
DRAWN BY M.R.E.
PROJECT NO. 01552074
SHEET NO. 32

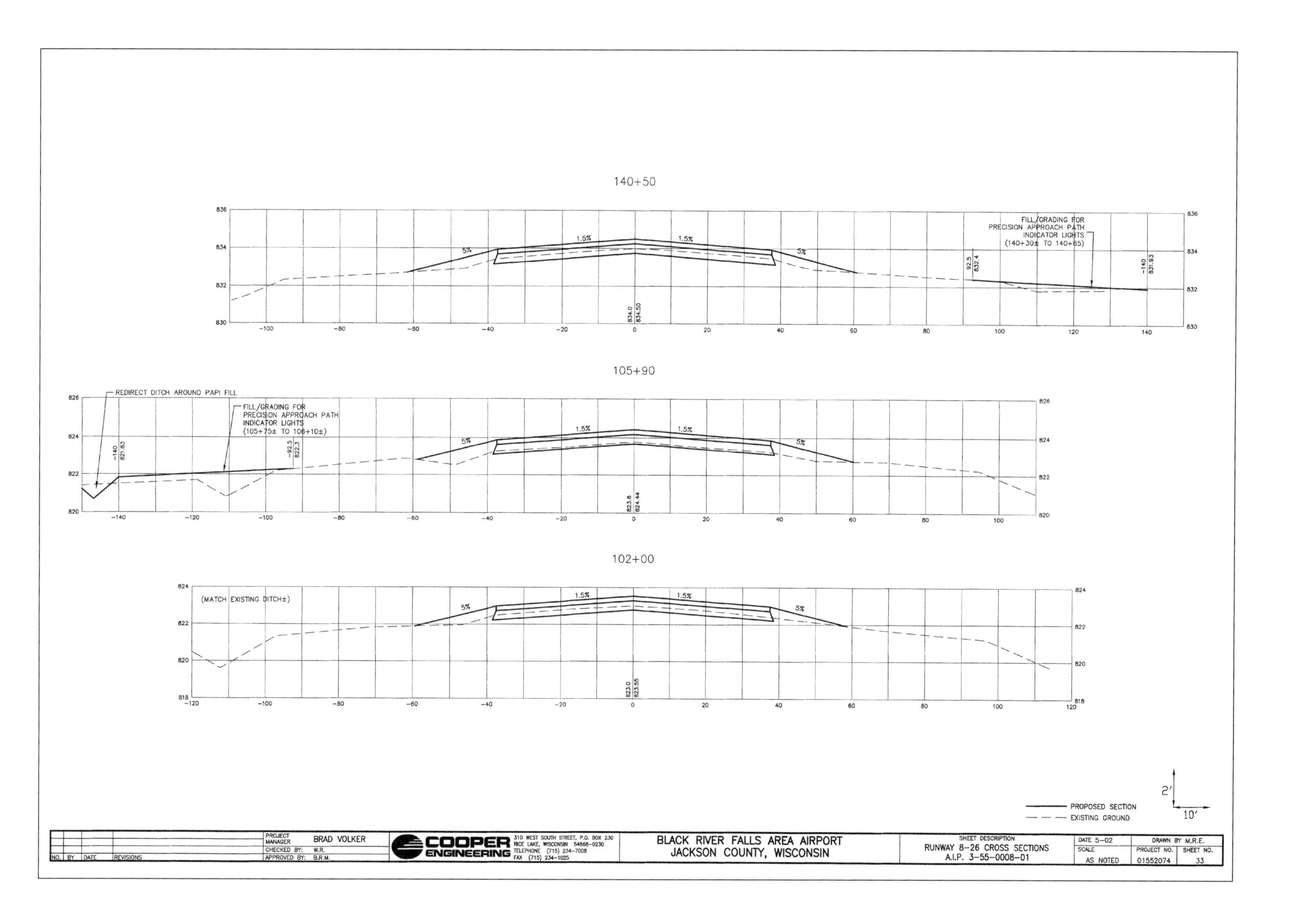
140+50
105+90
102+00
1.5%
5%
FILL/GRADING FOR PRECISION APPROACH PATH INDICATOR LIGHTS (140+30± TO 140+65)
REDIRECT DITCH AROUND PAPI FILL
FILL/GRADING FOR PRECISION APPROACH PATH INDICATOR LIGHTS (105+75± TO 106+10±)
(MATCH EXISTING DITCH±)
PROPOSED SECTION
EXISTING GROUND
2′
10′
PROJECT MANAGER BRAD VOLKER
CHECKED BY: W.R.
APPROVED BY: B.R.M.
NO. BY DATE REVISIONS
COOPER ENGINEERING
310 WEST SOUTH STREET, P.O. BOX 230 RICE LAKE, WISCONSIN 54868-0230 TELEPHONE (715) 234-7008 FAX (715) 234-1025
BLACK RIVER FALLS AREA AIRPORT
JACKSON COUNTY, WISCONSIN
SHEET DESCRIPTION
RUNWAY 8-26 CROSS SECTIONS
A.I.P. 3-55-0008-01
DATE 5-02
DRAWN BY M.R.E.
SCALE AS NOTED
PROJECT NO. 01552074
SHEET NO. 33

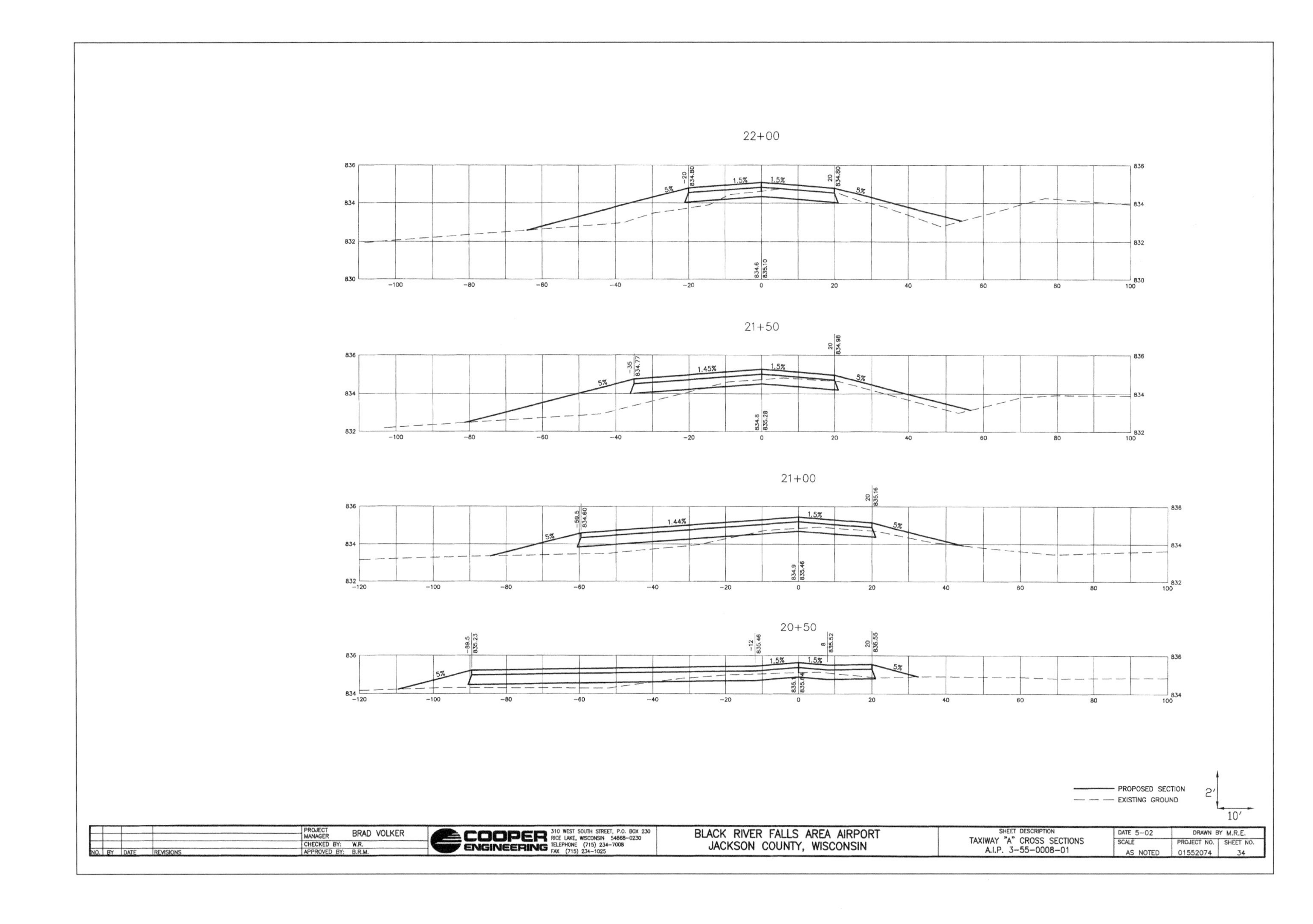
22+00
21+50
21+00
20+50
PROPOSED SECTION
EXISTING GROUND
2'
10'
NO. BY DATE REVISIONS
PROJECT MANAGER BRAD VOLKER
CHECKED BY: W.R.
APPROVED BY: B.R.M.
COOPER ENGINEERING
310 WEST SOUTH STREET, P.O. BOX 230
RICE LAKE, WISCONSIN 54868-0230
TELEPHONE (715) 234-7008
FAX (715) 234-1025
BLACK RIVER FALLS AREA AIRPORT
JACKSON COUNTY, WISCONSIN
SHEET DESCRIPTION
TAXIWAY "A" CROSS SECTIONS
A.I.P. 3-55-0008-01
DATE 5-02
SCALE AS NOTED
DRAWN BY M.R.E.
PROJECT NO. 01552074
SHEET NO. 34

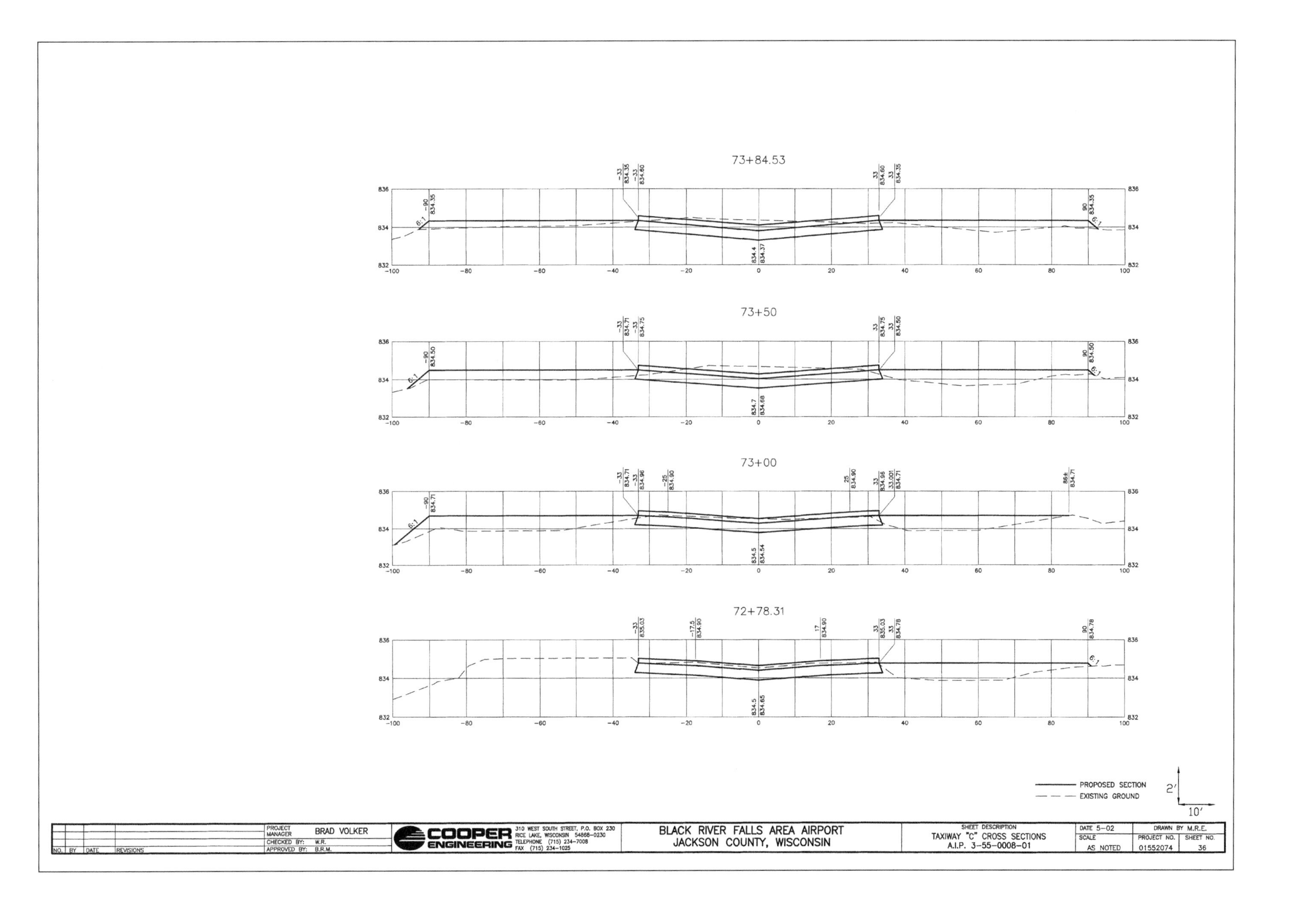
73+84.53
73+50
73+00
72+78.31
PROPOSED SECTION
EXISTING GROUND
2'
10'
PROJECT MANAGER BRAD VOLKER
CHECKED BY: W.R.
APPROVED BY: B.R.M.
COOPER ENGINEERING
310 WEST SOUTH STREET, P.O. BOX 230
RICE LAKE, WISCONSIN 54868-0230
TELEPHONE (715) 234-7008
FAX (715) 234-1025
BLACK RIVER FALLS AREA AIRPORT
JACKSON COUNTY, WISCONSIN
SHEET DESCRIPTION
TAXIWAY "C" CROSS SECTIONS
A.I.P. 3-55-0008-01
DATE 5-02
DRAWN BY M.R.E.
SCALE AS NOTED
PROJECT NO. 01552074
SHEET NO. 36
NO. BY DATE REVISIONS

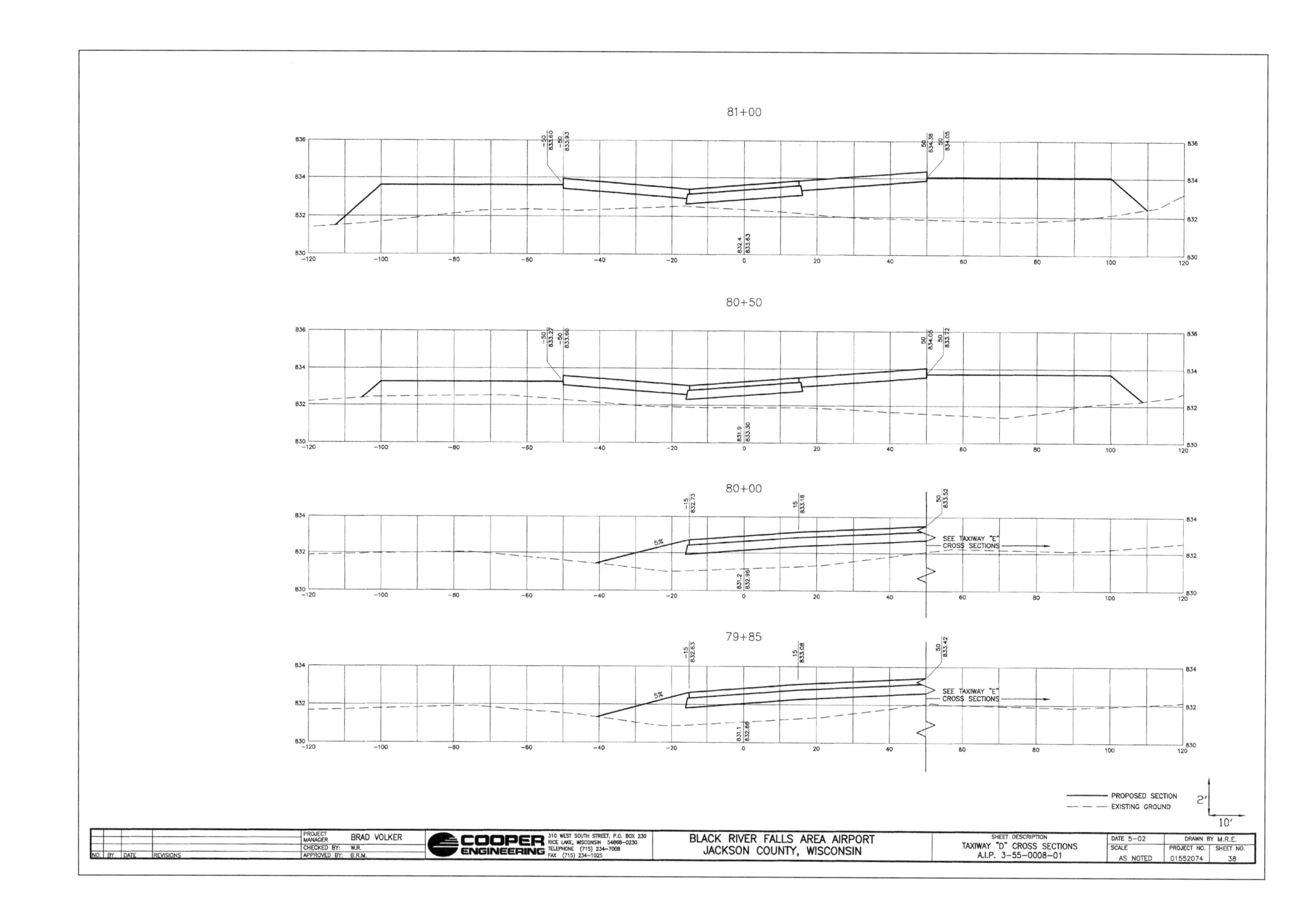
81+00
80+50
80+00
79+85
5%
SEE TAXIWAY "E" CROSS SECTIONS
PROPOSED SECTION
EXISTING GROUND
2'
10'
NO. BY DATE REVISIONS
PROJECT MANAGER BRAD VOLKER
CHECKED BY: W.R.
APPROVED BY: B.R.M.
COOPER ENGINEERING
310 WEST SOUTH STREET, P.O. BOX 230
RICE LAKE, WISCONSIN 54868-0230
TELEPHONE (715) 234-7008
FAX (715) 234-1025
BLACK RIVER FALLS AREA AIRPORT
JACKSON COUNTY, WISCONSIN
SHEET DESCRIPTION
TAXIWAY "D" CROSS SECTIONS
A.I.P. 3-55-0008-01
DATE 5-02
DRAWN BY M.R.E.
SCALE AS NOTED
PROJECT NO. 01552074
SHEET NO. 38

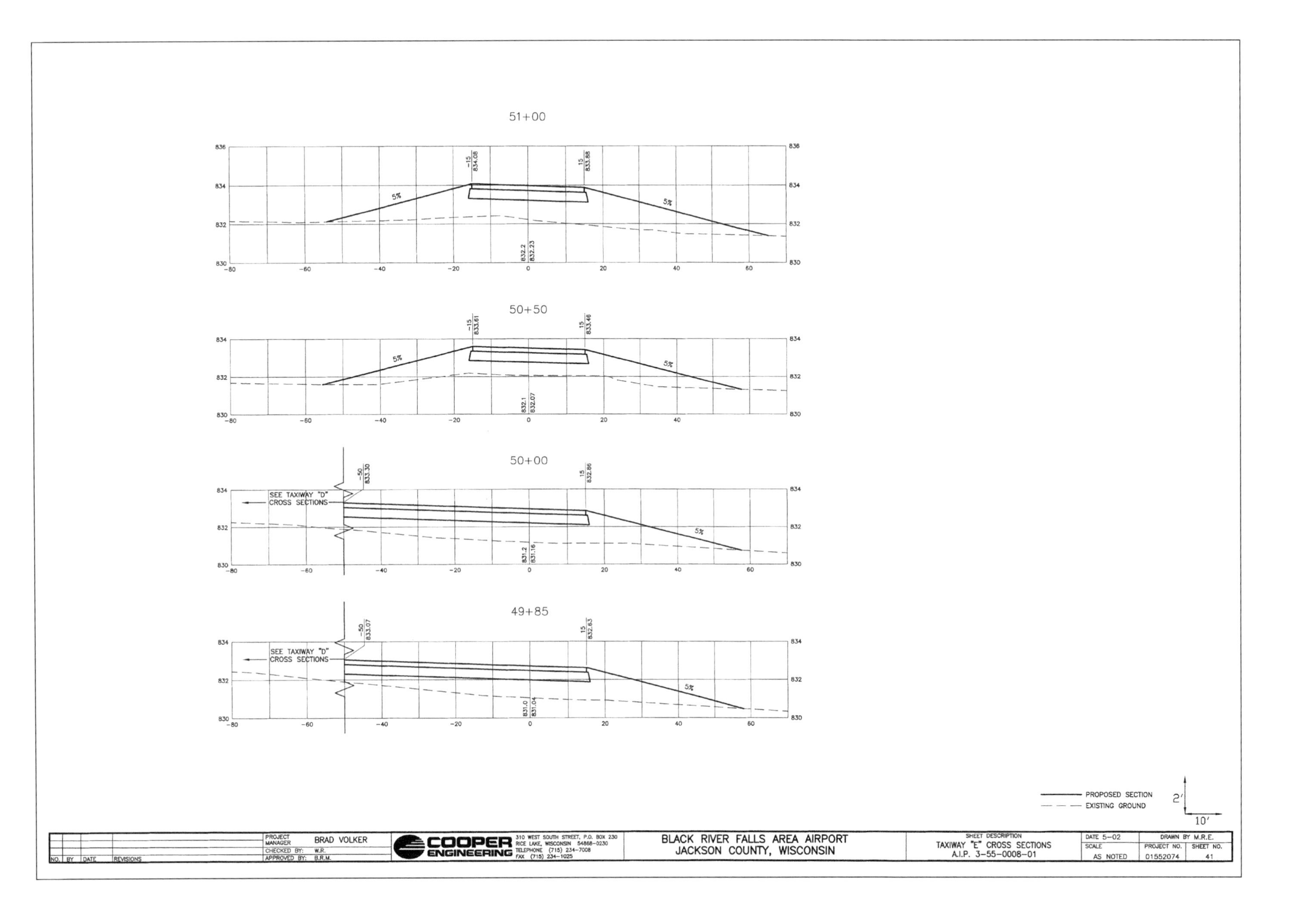
51+00
50+50
50+00
49+85
SEE TAXIWAY "D" CROSS SECTIONS
5%
PROPOSED SECTION
EXISTING GROUND
2'
10'
PROJECT MANAGER BRAD VOLKER
CHECKED BY: W.R.
APPROVED BY: B.R.M.
COOPER ENGINEERING
310 WEST SOUTH STREET, P.O. BOX 230
RICE LAKE, WISCONSIN 54868-0230
TELEPHONE (715) 234-7008
FAX (715) 234-1025
BLACK RIVER FALLS AREA AIRPORT
JACKSON COUNTY, WISCONSIN
SHEET DESCRIPTION
TAXIWAY "E" CROSS SECTIONS
A.I.P. 3-55-0008-01
DATE 5-02
DRAWN BY M.R.E.
SCALE AS NOTED
PROJECT NO. 01552074
SHEET NO. 41
NO. BY DATE REVISIONS

Design

ATHLETIC FACILITIES

Section 17

Athletic facilities can be a speciality area of practice, but occasionally the general practitioner is called upon to participate in these projects. Therefore, this section contains a number of examples for the reader's reference.

The first set of plans are for the retrofitting of an existing track and its conversion from a 1/4 mile sized to a metric 400 meter facility. The following three plans are at the same site, but are for expanded facilities. The final example represents a combination with racquetball courts as a component. This illustrates how an athletic facility may be only a segment of the total project.

TABLE OF CONTENTS

LADYSMITH HIGH SCHOOL ATHLETIC TRACK
-METRIC CONVERSION & REHABILITATION-

LADYSMITH / HAWKINS SCHOOL DISTRICT

PROJECT NO. 93-131

JUNE, 1994

(REVISED: MAR. '95)

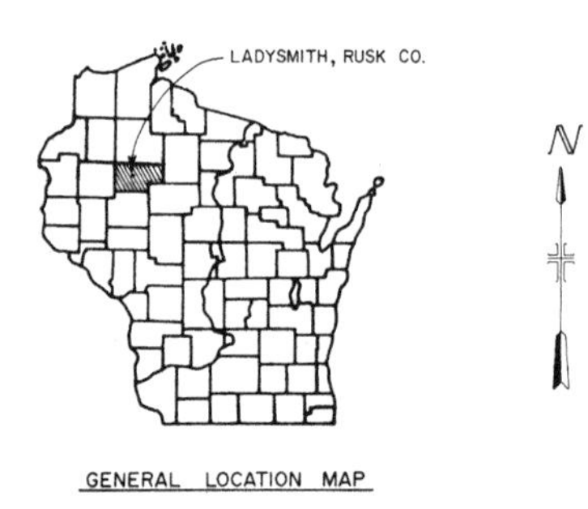

GENERAL LOCATION MAP

N

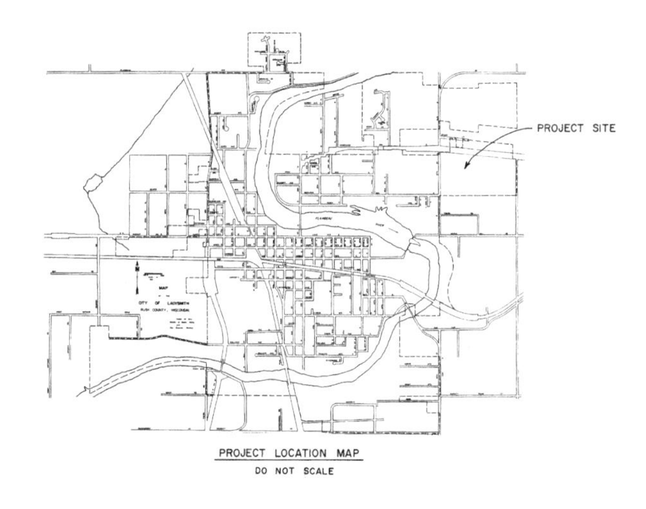

PROJECT LOCATION MAP

DO NOT SCALE

INDEX OF SHEETS

NO.	DESCRIPTION
1	TITLE SHEET W/ LOCATION MAPS
2	EXISTING ATHLETIC FACILITY LAYOUT
3	METRIC CONVERSION & REHABILITATION OF TRACK WITH CONSTRUCTION DETAILS
4	POLE VAULT & LONG/TRIPLE JUMP CONSTRUCTION DETAILS
5	TYPICAL EROSION CONTROL DETAILS **NOTE:** THIS SHEET NOT INCLUDED

NOTES

REFER TO ACCOMPANYING SPECIFICATIONS AND BIDDING DOCUMENTS FOR SPECIFIC WRITTEN DETAILS.

ENGINEER SHALL FIELD STAKE ALL BASIC HORIZONTAL AND VERTICAL CONTROLS.

-UTILITIES-

CONTACT: Ladysmith Director of Public Works (Water/Sewer)
Bill Christianson, - Director
120 W. Miner Avenue
Ladysmith, Wisconsin 54848
Telephone: (715) 532-2601

CONTACT: Northern States Power Company (Electrical)
711 W. 9th Street, N.
Ladysmith, Wisconsin 54848
Telephone: (715) 532-6226

CONTACT: Wisconsin Gas Company
Telephone: 1-800-242-8511 (Diggers Hotline)

CONTACT: Marcus Communications (TV Cable)
Brian Gustafson
219 W. Miner Avenue
Ladysmith, Wisconsin 54848
Telephone: (715) 532-6040

CONTACT: Ameritech (Telephone Company)
Telephone: 1-800-242-8511 (Diggers Hotline)

CONTACT: Bob Jenness (School Maintenance Supervisor)
Telephone: (715) 532-5277

PREPARED BY:
MORGAN & PARMLEY, LTD
CONSULTING ENGINEERS
LADYSMITH, WISCONSIN

Courtesy: Morgan & Parmley, Ltd.

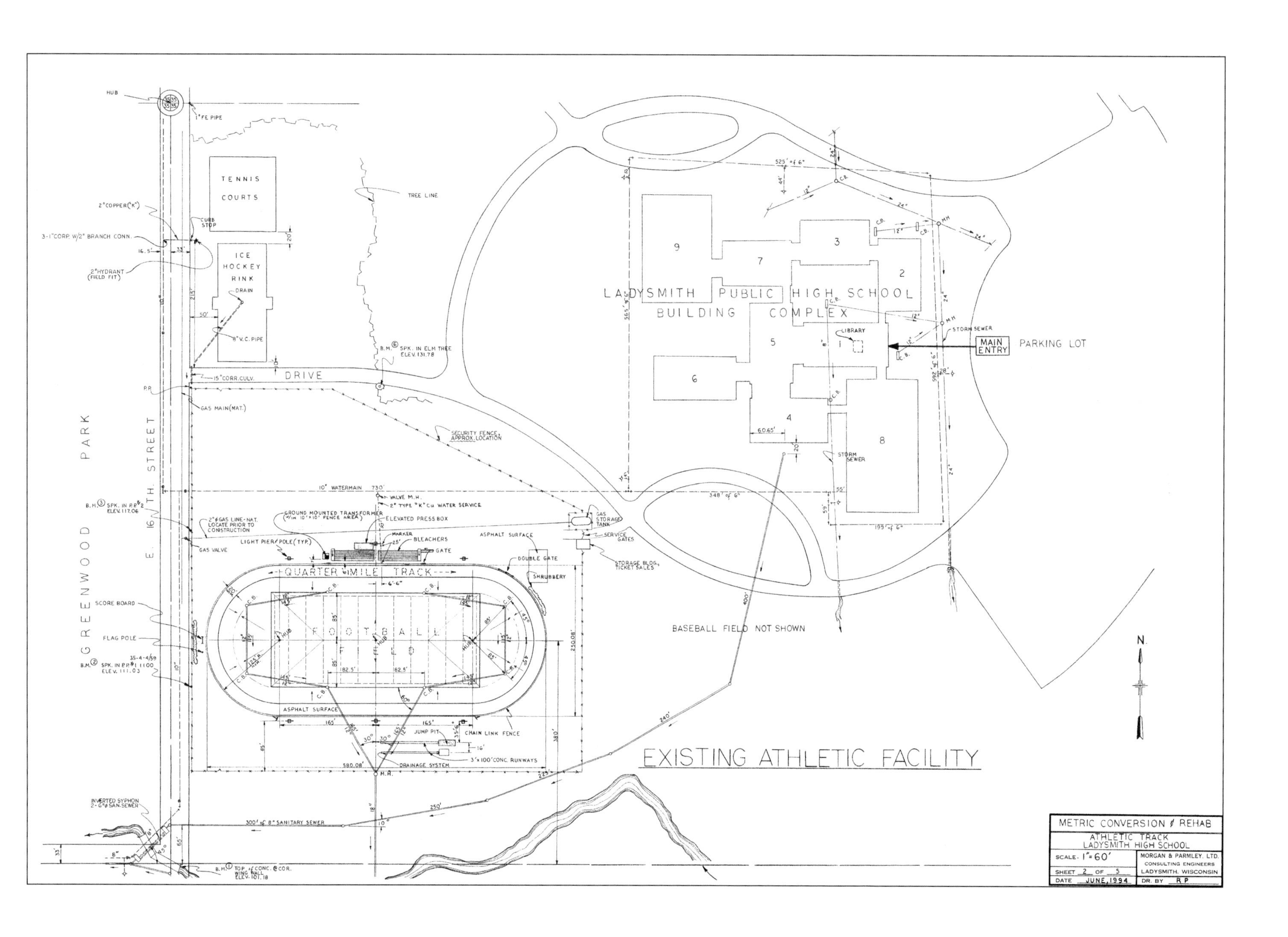
EXISTING ATHLETIC FACILITY
LADYSMITH PUBLIC HIGH SCHOOL
BUILDING COMPLEX
MAIN ENTRY
PARKING LOT
LIBRARY
TENNIS COURTS
ICE HOCKEY RINK
TREE LINE
DRIVE
SECURITY FENCE APPROX. LOCATION
QUARTER MILE TRACK
FOOTBALL FIELD
BASEBALL FIELD NOT SHOWN
GREENWOOD PARK
E. 16TH. STREET
ELEVATED PRESS BOX
BLEACHERS
GATE
DOUBLE GATE
SHRUBBERY
ASPHALT SURFACE
CHAIN LINK FENCE
JUMP PIT
3'x100' CONC. RUNWAYS
DRAINAGE SYSTEM
SCORE BOARD
FLAG POLE
LIGHT PIER/POLE (TYP.)
GAS STORAGE TANK
SERVICE GATES
STORAGE BLDG. TICKET SALES
STORM SEWER
10" WATERMAIN 730'
300' of 8" SANITARY SEWER
N
METRIC CONVERSION & REHAB
ATHLETIC TRACK
LADYSMITH HIGH SCHOOL
SCALE 1"=60'
SHEET 2 OF 5
DATE JUNE, 1994
MORGAN & PARMLEY, LTD.
CONSULTING ENGINEERS
LADYSMITH, WISCONSIN
DR. BY R.P.

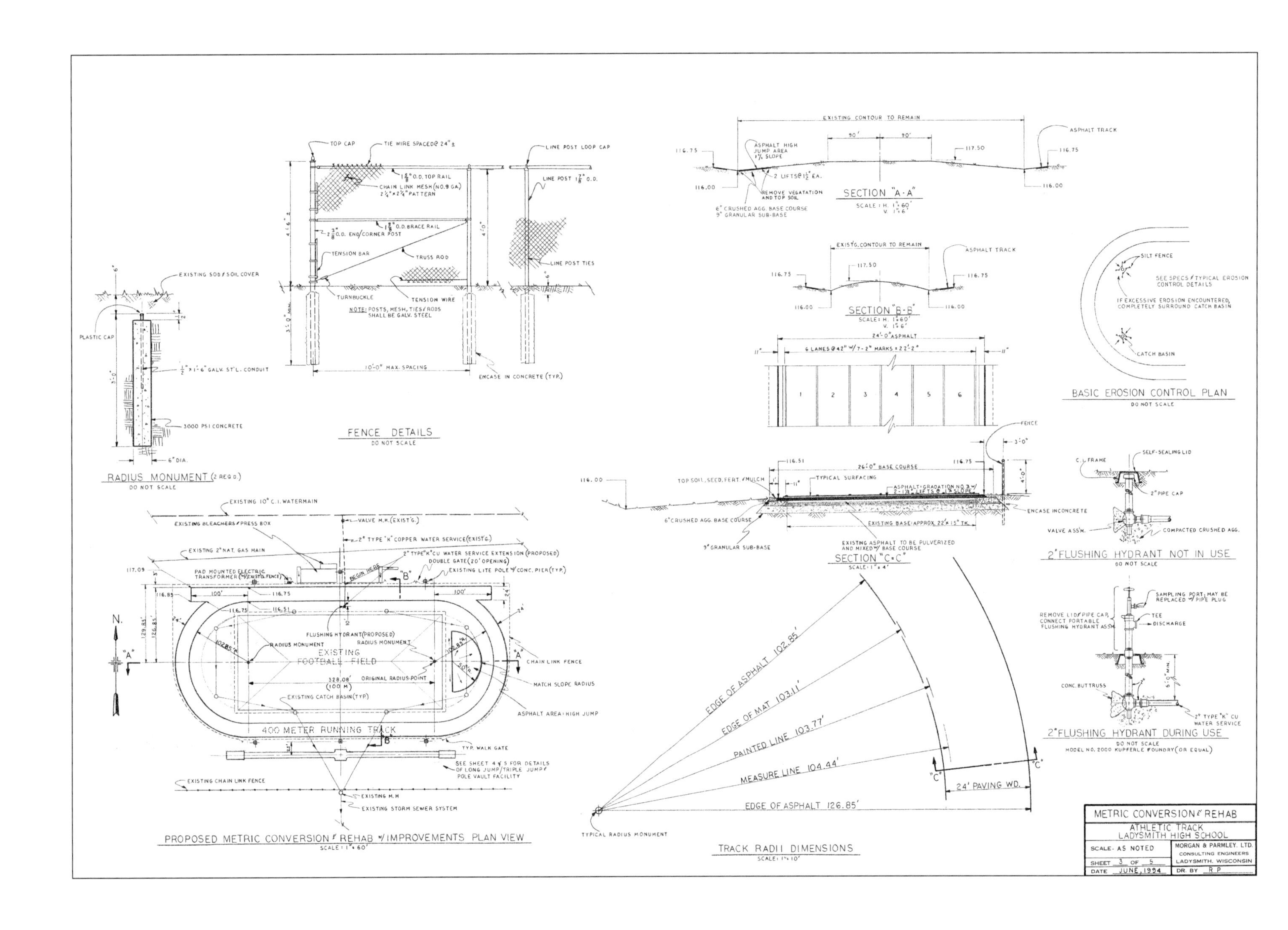
FENCE DETAILS
DO NOT SCALE
RADIUS MONUMENT (2 REQ'D)
DO NOT SCALE
SECTION "A-A"
SECTION "B-B"
SECTION "C-C"
BASIC EROSION CONTROL PLAN
DO NOT SCALE
2" FLUSHING HYDRANT NOT IN USE
DO NOT SCALE
2" FLUSHING HYDRANT DURING USE
DO NOT SCALE
PROPOSED METRIC CONVERSION & REHAB W/IMPROVEMENTS PLAN VIEW
SCALE: 1" = 60'
TRACK RADII DIMENSIONS
SCALE: 1" = 10'
EDGE OF ASPHALT 102.85'
EDGE OF MAT 103.11'
PAINTED LINE 103.77'
MEASURE LINE 104.44'
EDGE OF ASPHALT 126.85'
24' PAVING WD.
TYPICAL RADIUS MONUMENT
400 METER RUNNING TRACK
EXISTING FOOTBALL FIELD
METRIC CONVERSION & REHAB
ATHLETIC TRACK
LADYSMITH HIGH SCHOOL
SCALE - AS NOTED
SHEET 3 OF 5
DATE JUNE, 1994
MORGAN & PARMLEY, LTD.
CONSULTING ENGINEERS
LADYSMITH, WISCONSIN
DR. BY R.P.

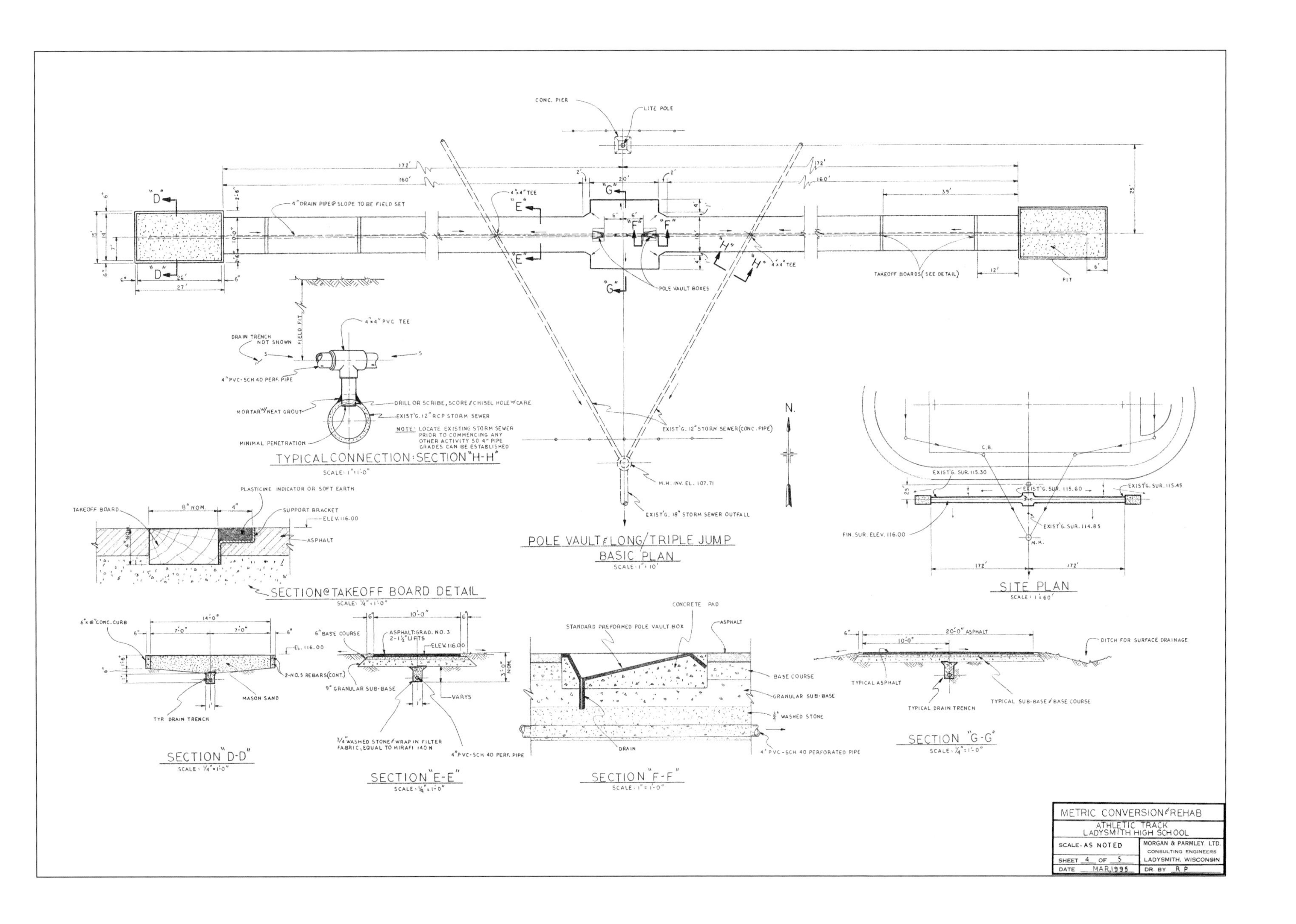
POLE VAULT & LONG/TRIPLE JUMP
BASIC PLAN
SCALE: 1" = 10'
TYPICAL CONNECTION: SECTION "H-H"
SCALE: 1" = 1'-0"
SECTION @ TAKEOFF BOARD DETAIL
SITE PLAN
SECTION "D-D"
SECTION "E-E"
SECTION "F-F"
SECTION "G-G"
METRIC CONVERSION & REHAB
ATHLETIC TRACK
LADYSMITH HIGH SCHOOL
SCALE: AS NOTED
MORGAN & PARMLEY, LTD.
CONSULTING ENGINEERS
LADYSMITH, WISCONSIN
SHEET 4 OF 5
DATE MAR 1995
DR. BY R P

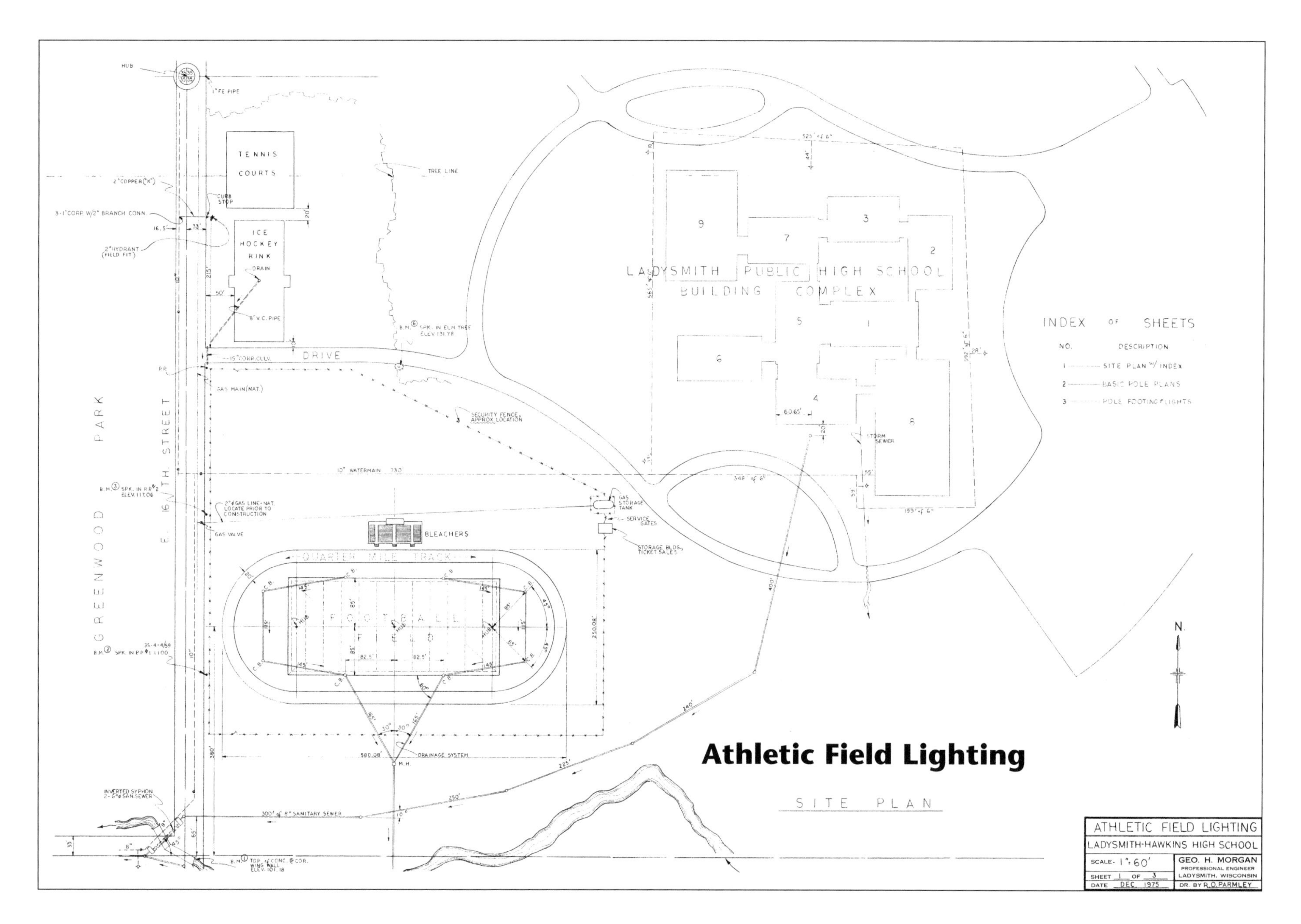

Courtesy: Morgan & Parmley, Ltd.

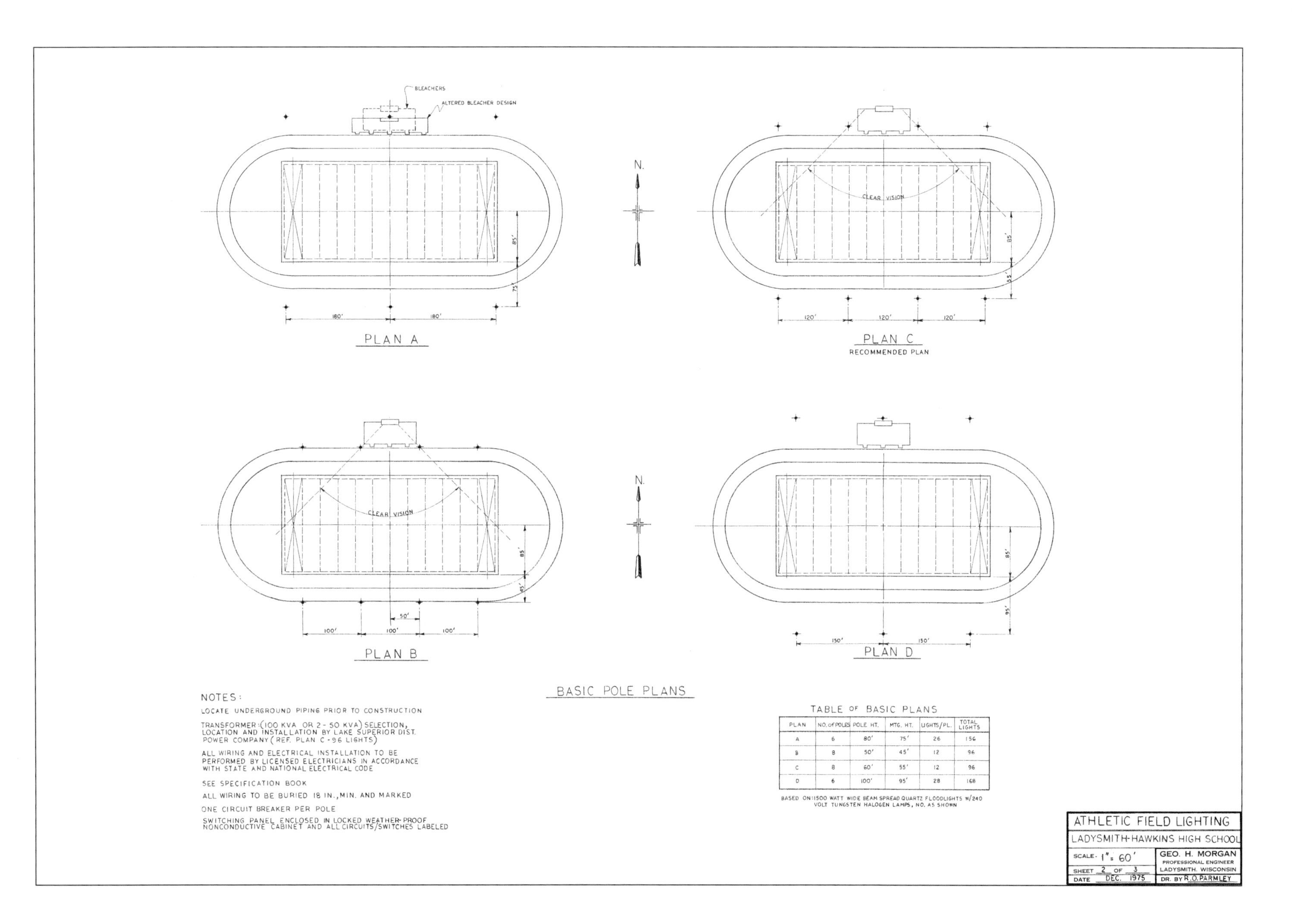
BLEACHERS
ALTERED BLEACHER DESIGN
N.
85'
75'
180'
180'
PLAN A
CLEAR VISION
85'
35'
120'
120'
120'
PLAN C
RECOMMENDED PLAN
CLEAR VISION
85'
45'
50'
100'
100'
100'
PLAN B
85'
95'
150'
150'
PLAN D
BASIC POLE PLANS
NOTES:
LOCATE UNDERGROUND PIPING PRIOR TO CONSTRUCTION
TRANSFORMER:(100 KVA OR 2 - 50 KVA) SELECTION, LOCATION AND INSTALLATION BY LAKE SUPERIOR DIST. POWER COMPANY (REF. PLAN C - 96 LIGHTS)
ALL WIRING AND ELECTRICAL INSTALLATION TO BE PERFORMED BY LICENSED ELECTRICIANS IN ACCORDANCE WITH STATE AND NATIONAL ELECTRICAL CODE
SEE SPECIFICATION BOOK
ALL WIRING TO BE BURIED 18 IN., MIN. AND MARKED
ONE CIRCUIT BREAKER PER POLE
SWITCHING PANEL ENCLOSED IN LOCKED WEATHER-PROOF NONCONDUCTIVE CABINET AND ALL CIRCUITS/SWITCHES LABELED
TABLE OF BASIC PLANS
BASED ON: 1500 WATT WIDE BEAM SPREAD QUARTZ FLOODLIGHTS W/240 VOLT TUNGSTEN HALOGEN LAMPS, NO. AS SHOWN
ATHLETIC FIELD LIGHTING
LADYSMITH-HAWKINS HIGH SCHOOL
SCALE- 1" = 60'
SHEET 2 OF 3
DATE DEC. 1975
GEO. H. MORGAN
PROFESSIONAL ENGINEER
LADYSMITH, WISCONSIN
DR. BY R.O. PARMLEY

PLAN	NO. of POLES	POLE HT.	MTG. HT.	LIGHTS/PL.	TOTAL LIGHTS
A	6	80'	75'	26	156
B	8	50'	45'	12	96
C	8	60'	55'	12	96
D	6	100'	95'	28	168

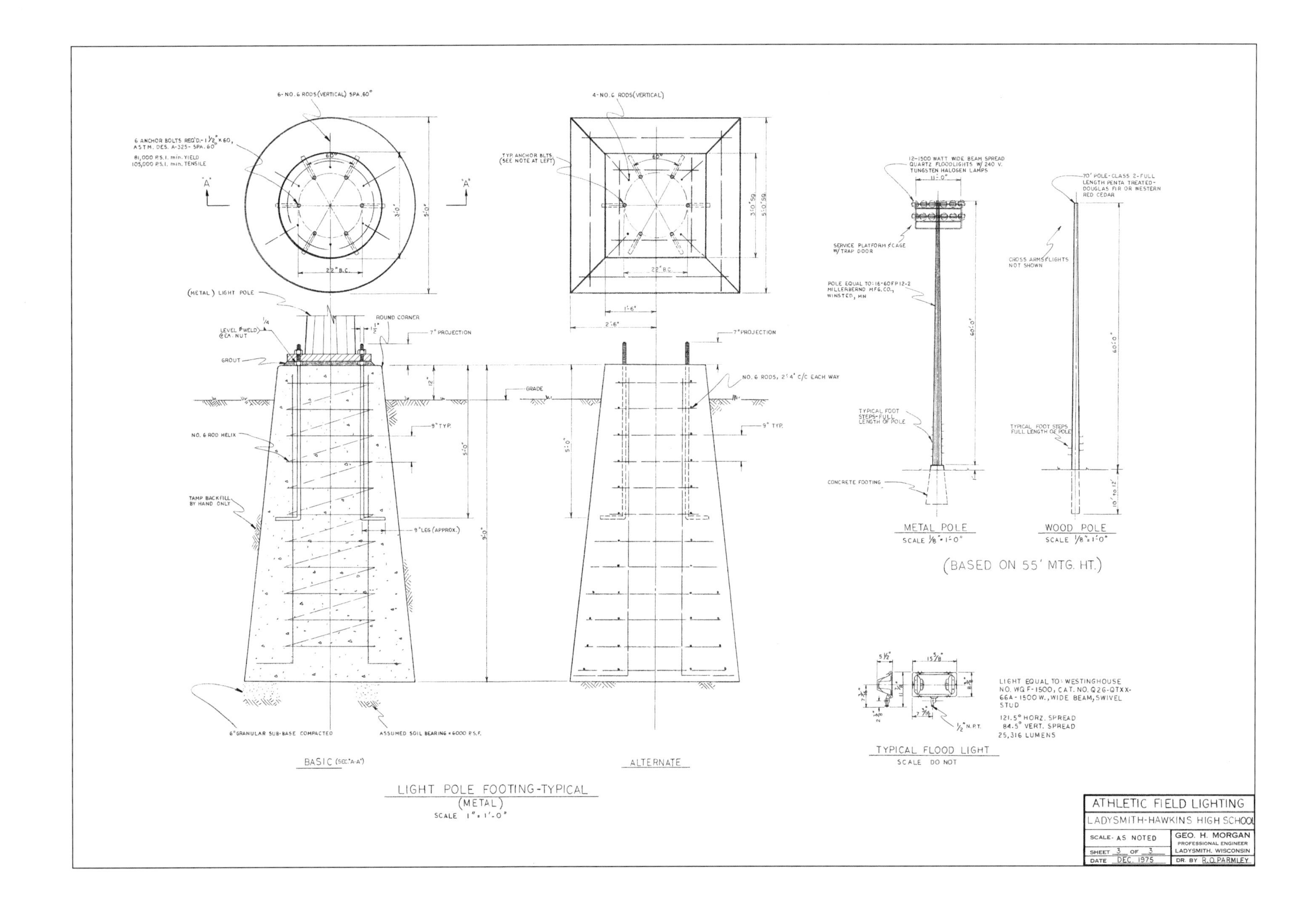
6-NO.6 RODS (VERTICAL) SPA.60°
6 ANCHOR BOLTS REQ'D.-1½"×60, ASTM. DES. A-325- SPA. 60°
81,000 P.S.I. min. YIELD
105,000 P.S.I. min. TENSILE
4-NO.6 RODS (VERTICAL)
TYP. ANCHOR BLTS. (SEE NOTE AT LEFT)
22" B.C.
(METAL) LIGHT POLE
ROUND CORNER
7" PROJECTION
LEVEL & WELD @ EA. NUT
GROUT
GRADE
NO. 6 RODS, 2'-4" C/C EACH WAY
9" TYP.
NO. 6 ROD HELIX
TAMP BACKFILL BY HAND ONLY
9" LEG (APPROX.)
6" GRANULAR SUB-BASE COMPACTED
ASSUMED SOIL BEARING = 6000 P.S.F.
BASIC (SEC. "A-A")
ALTERNATE
LIGHT POLE FOOTING-TYPICAL
(METAL)
SCALE 1" = 1'-0"
12-1500 WATT WIDE BEAM SPREAD QUARTZ FLOODLIGHTS W/ 240 V. TUNGSTEN HALOGEN LAMPS
SERVICE PLATFORM & CAGE W/ TRAP DOOR
POLE EQUAL TO: 16-60FP12-2 MILLERBERND MFG. CO., WINSTED, MN
TYPICAL FOOT STEPS-FULL LENGTH OF POLE
CONCRETE FOOTING
70' POLE-CLASS 2-FULL LENGTH PENTA TREATED-DOUGLAS FIR OR WESTERN RED CEDAR
CROSS ARMS & LIGHTS NOT SHOWN
TYPICAL FOOT STEPS FULL LENGTH OF POLE
METAL POLE
SCALE ⅛" = 1'-0"
WOOD POLE
SCALE ⅛" = 1'-0"
(BASED ON 55' MTG. HT.)
LIGHT EQUAL TO: WESTINGHOUSE NO. WQ F-1500, CAT. NO. Q26-QTXX-66A-1500 W., WIDE BEAM, SWIVEL STUD
121.5° HORZ. SPREAD
84.5° VERT. SPREAD
25,316 LUMENS
TYPICAL FLOOD LIGHT
SCALE DO NOT
ATHLETIC FIELD LIGHTING
LADYSMITH-HAWKINS HIGH SCHOOL
SCALE- AS NOTED
SHEET 3 OF 3
DATE DEC. 1975
GEO. H. MORGAN
PROFESSIONAL ENGINEER
LADYSMITH, WISCONSIN
DR. BY R.O. PARMLEY

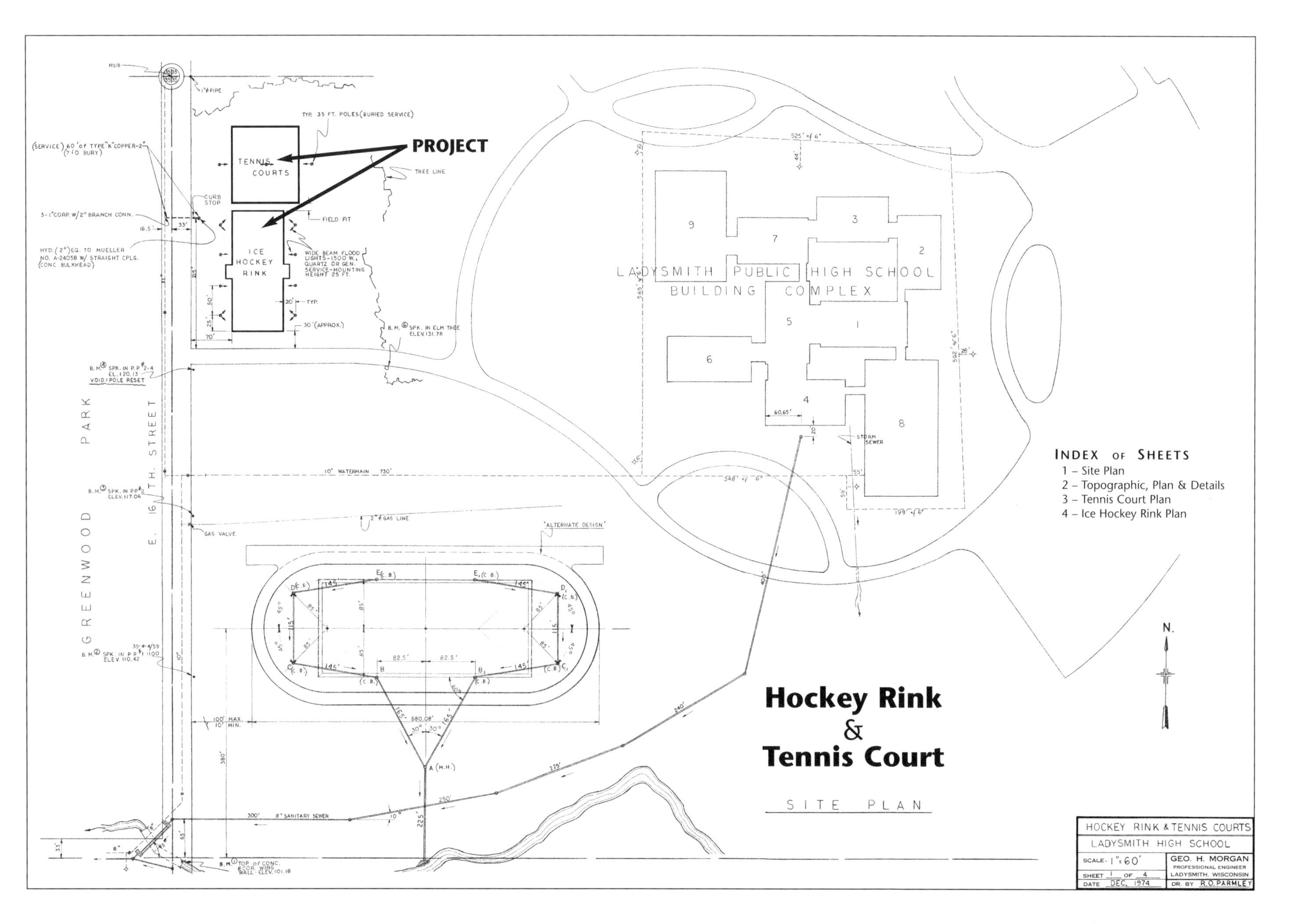

Courtesy: Morgan & Parmley, Ltd.

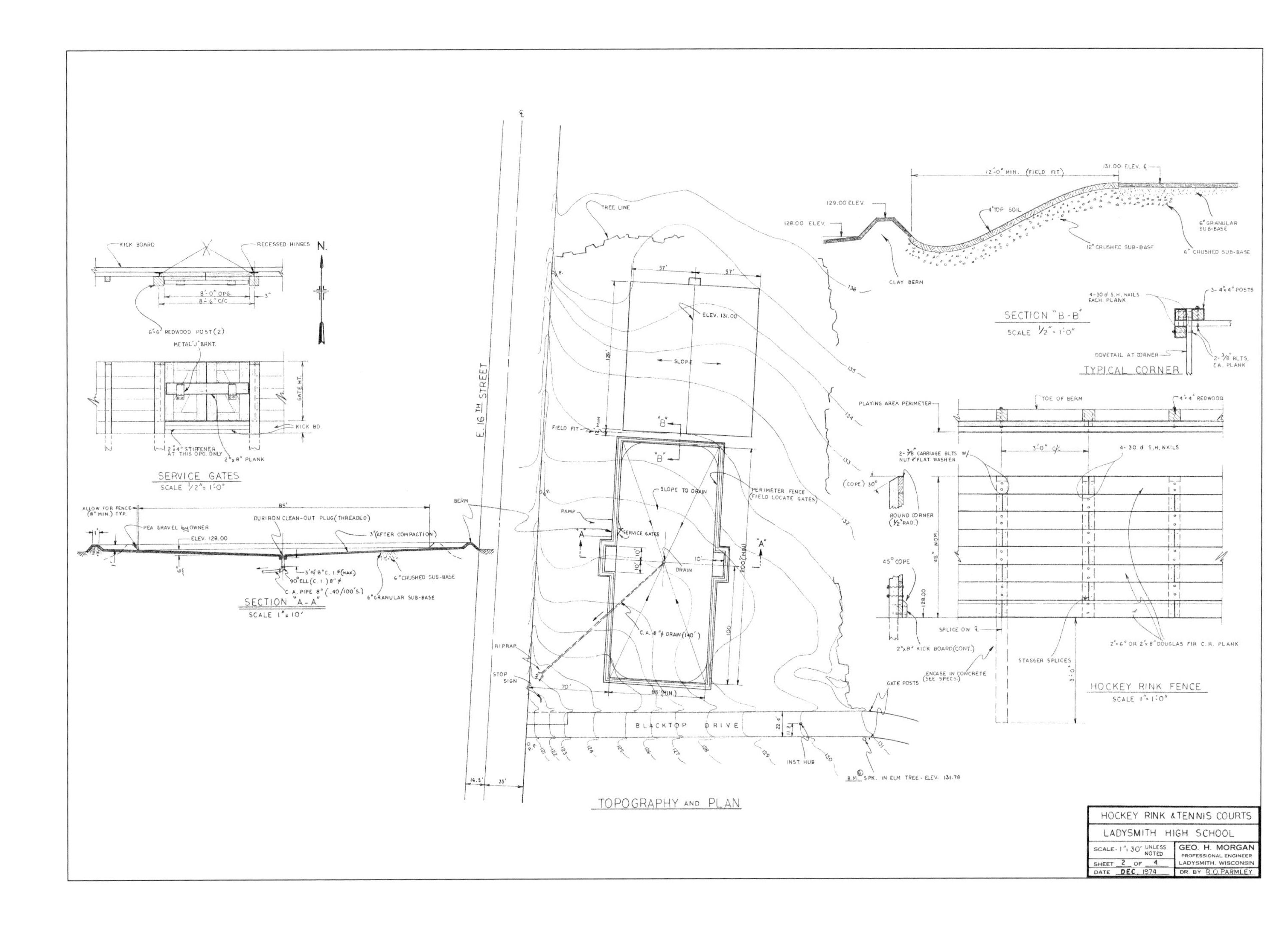

KICK BOARD
RECESSED HINGES
N.
8'-0" OPG.
8'-6" C/C
6"x6" REDWOOD POST (2)
METAL "J" BRKT.
GATE HT.
KICK BD.
2"x4" STIFFENER AT THIS OPG. ONLY
2"x8" PLANK
SERVICE GATES
SCALE 1/2" = 1'-0"
ALLOW FOR FENCE (8" MIN.) TYP.
85'
PEA GRAVEL by OWNER
ELEV. 128.00
DURIRON CLEAN-OUT PLUG (THREADED)
3" (AFTER COMPACTION)
BERM
6" CRUSHED SUB-BASE
6" GRANULAR SUB-BASE
90° ELL (C.I.) 8" ϕ
C.A. PIPE 8" (.40/100'S.)
SECTION "A-A"
SCALE 1" = 10'
E. 16TH STREET
TREE LINE
57'
57'
ELEV. 131.00
SLOPE
FIELD FIT
"B"
SLOPE TO DRAIN
PERIMETER FENCE (FIELD LOCATE GATES)
RAMP
SERVICE GATES
"A"
DRAIN
200 (MIN)
C.A. 8" ϕ DRAIN (140')
RIPRAP
STOP SIGN
70'
85' (MIN)
BLACKTOP DRIVE
INST. HUB
B.M. SPK. IN ELM TREE - ELEV. 131.78
16.5'
33'
TOPOGRAPHY AND PLAN
12'-0" MIN. (FIELD FIT)
131.00 ELEV.
129.00 ELEV.
128.00 ELEV.
4" TOP SOIL
6" GRANULAR SUB-BASE
12" CRUSHED SUB-BASE
6" CRUSHED SUB-BASE
CLAY BERM
SECTION "B-B"
SCALE 1/2" = 1'-0"
4-30d S.H. NAILS EACH PLANK
3- 4"x4" POSTS
DOVETAIL AT CORNER
2-3/8" BLTS. EA. PLANK
TYPICAL CORNER
PLAYING AREA PERIMETER
TOE OF BERM
4"x4" REDWOOD
3'-0" c/c
4-30 d S.H. NAILS
2-3/8" CARRIAGE BLTS W/ NUT & FLAT WASHER
(COPE) 30°
ROUND CORNER (1/2" RAD.)
45° COPE
48" NOM.
128.00
SPLICE ON ℄
2"x8" KICK BOARD (CONT.)
STAGGER SPLICES
2"x6" OR 2"x8" DOUGLAS FIR C.R. PLANK
ENCASE IN CONCRETE (SEE SPECS.)
GATE POSTS
HOCKEY RINK FENCE
SCALE 1" = 1'-0"
HOCKEY RINK & TENNIS COURTS
LADYSMITH HIGH SCHOOL
SCALE - 1" = 30' UNLESS NOTED
GEO. H. MORGAN
PROFESSIONAL ENGINEER
LADYSMITH, WISCONSIN
SHEET 2 OF 4
DATE DEC. 1974
DR. BY R.O. PARMLEY

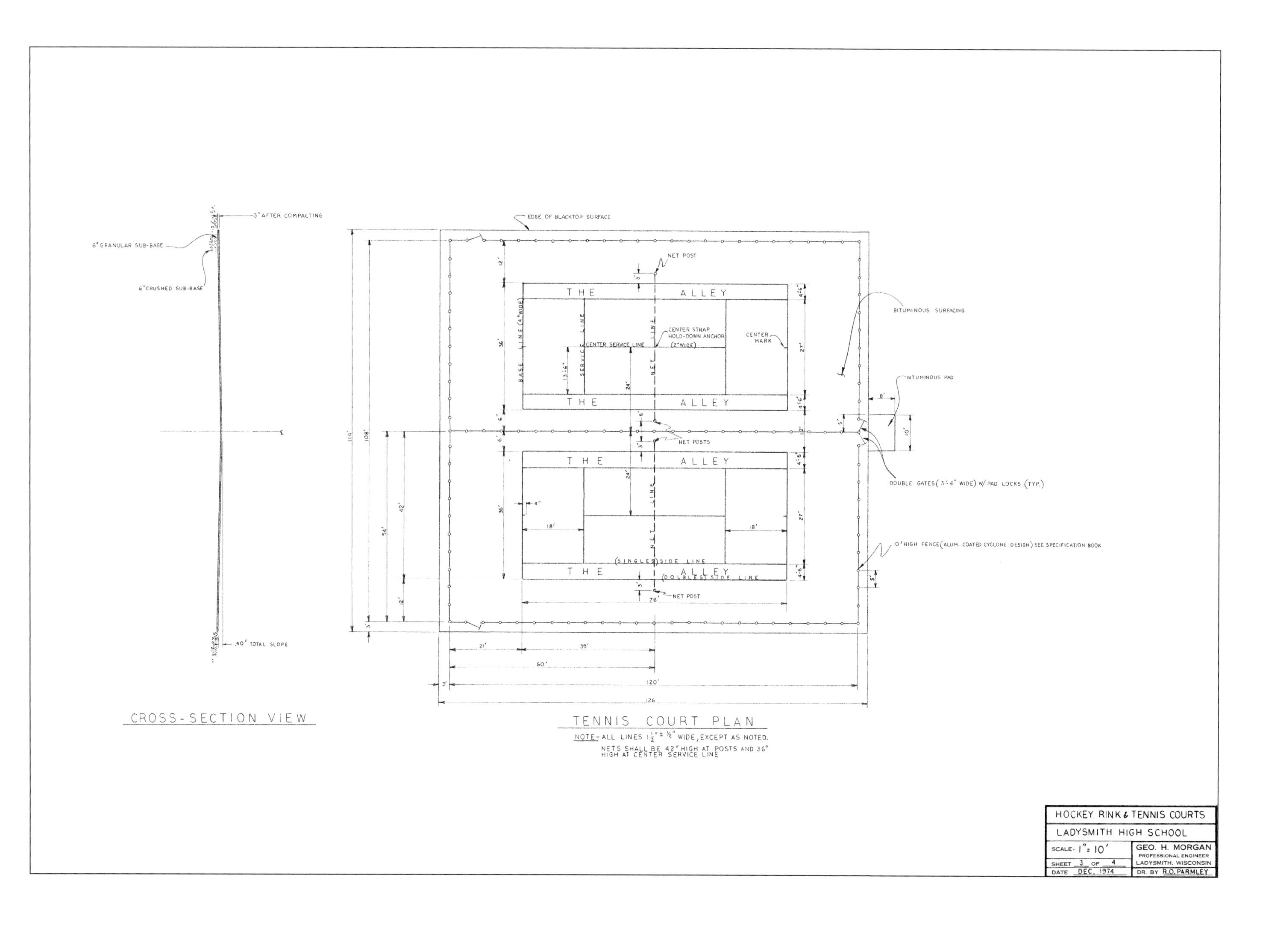
3" AFTER COMPACTING
6" GRANULAR SUB-BASE
6" CRUSHED SUB-BASE
.40' TOTAL SLOPE
CROSS-SECTION VIEW
EDGE OF BLACKTOP SURFACE
NET POST
THE ALLEY
CENTER STRAP HOLD-DOWN ANCHOR (2" WIDE)
CENTER MARK
CENTER SERVICE LINE
BASE LINE (4" WIDE)
SERVICE LINE
NET LINE
BITUMINOUS SURFACING
BITUMINOUS PAD
NET POSTS
DOUBLE GATES (3'-6" WIDE) W/ PAD LOCKS (TYP.)
10' HIGH FENCE (ALUM. COATED CYCLONE DESIGN) SEE SPECIFICATION BOOK
(SINGLES) SIDE LINE
(DOUBLES) SIDE LINE
TENNIS COURT PLAN
NOTE- ALL LINES 1½" ± ½" WIDE, EXCEPT AS NOTED.
NETS SHALL BE 42" HIGH AT POSTS AND 36" HIGH AT CENTER SERVICE LINE
HOCKEY RINK & TENNIS COURTS
LADYSMITH HIGH SCHOOL
SCALE- 1" = 10'
SHEET 3 OF 4
DATE DEC. 1974
GEO. H. MORGAN
PROFESSIONAL ENGINEER
LADYSMITH, WISCONSIN
DR. BY R.O. PARMLEY

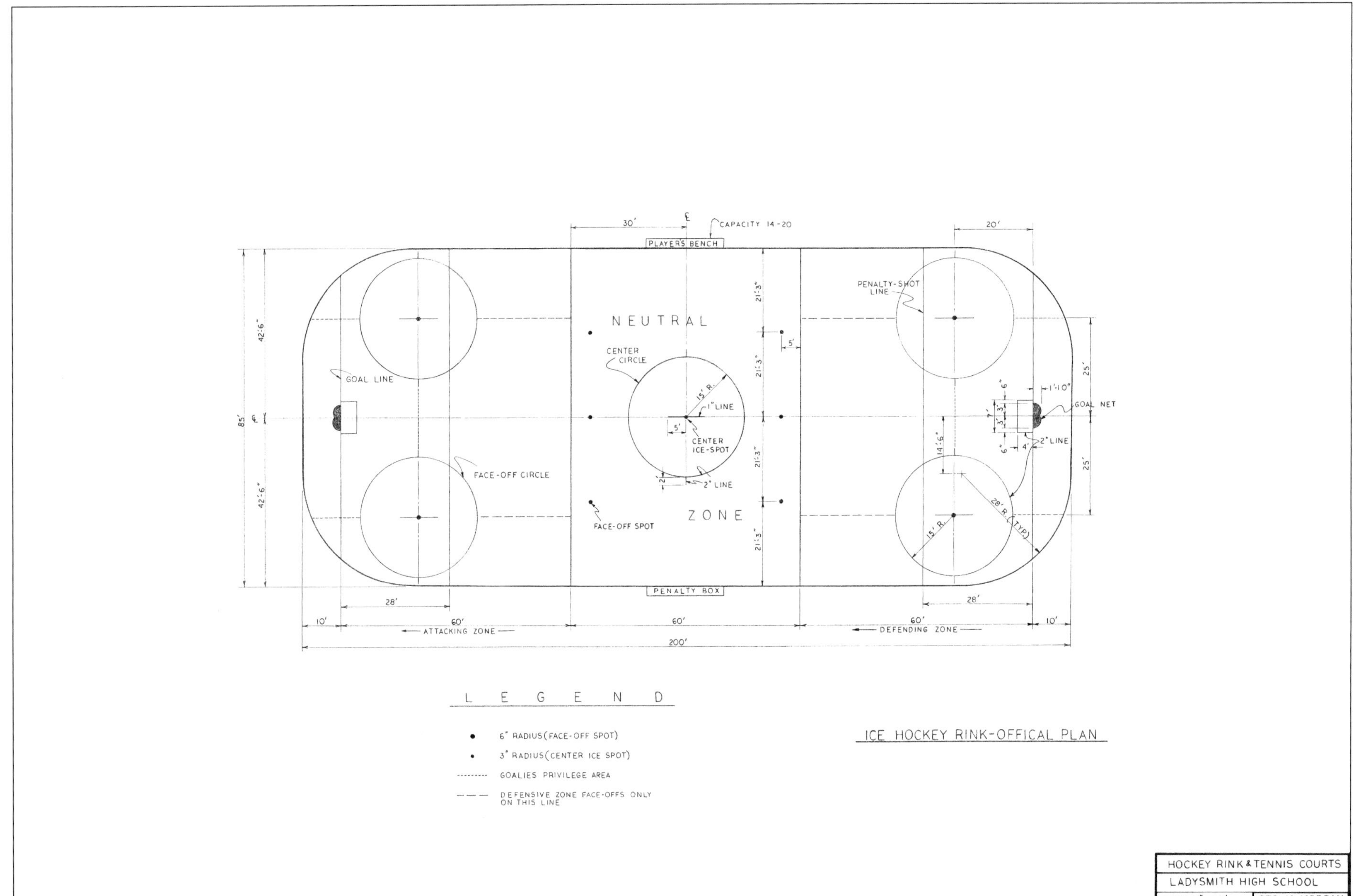

CAPACITY 14-20
PLAYERS BENCH
PENALTY BOX
NEUTRAL
ZONE
CENTER CIRCLE
15' R.
1" LINE
CENTER ICE-SPOT
2" LINE
FACE-OFF SPOT
FACE-OFF CIRCLE
GOAL LINE
PENALTY-SHOT LINE
GOAL NET
28' R. (TYP.)
15' R.
ATTACKING ZONE
DEFENDING ZONE
30'
20'
28'
10'
60'
200'
85'
42'-6"
21'-3"
25'
14'-6"
1'-10"
L E G E N D
6" RADIUS (FACE-OFF SPOT)
3" RADIUS (CENTER ICE SPOT)
GOALIES PRIVILEGE AREA
DEFENSIVE ZONE FACE-OFFS ONLY ON THIS LINE
ICE HOCKEY RINK-OFFICAL PLAN
HOCKEY RINK & TENNIS COURTS
LADYSMITH HIGH SCHOOL
SCALE- 1" = 10'
SHEET 4 OF 4
DATE DEC. 1974
GEO. H. MORGAN
PROFESSIONAL ENGINEER
LADYSMITH, WISCONSIN
DR. BY R.O. PARMLEY

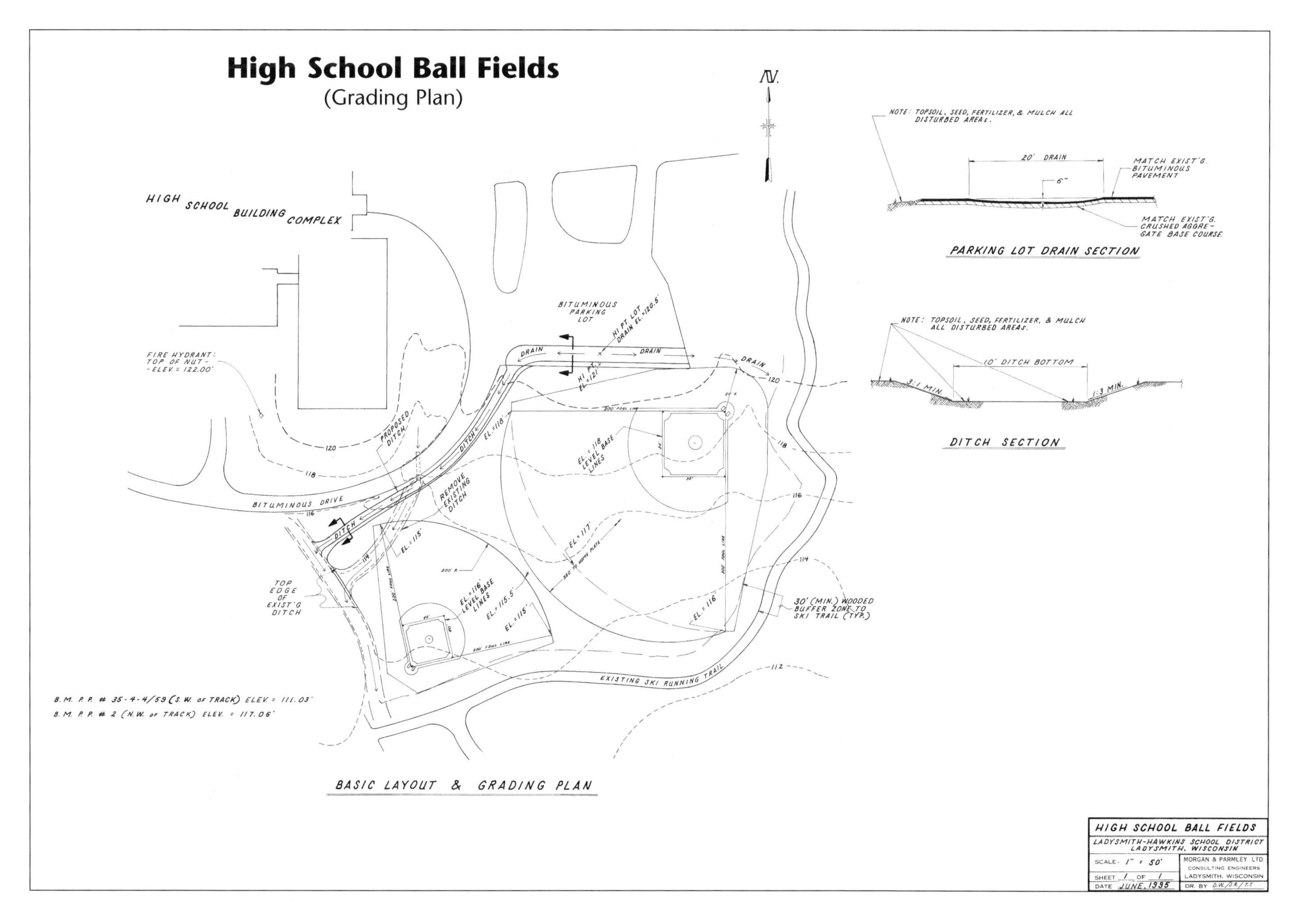

Courtesy: Morgan & Parmley, Ltd.

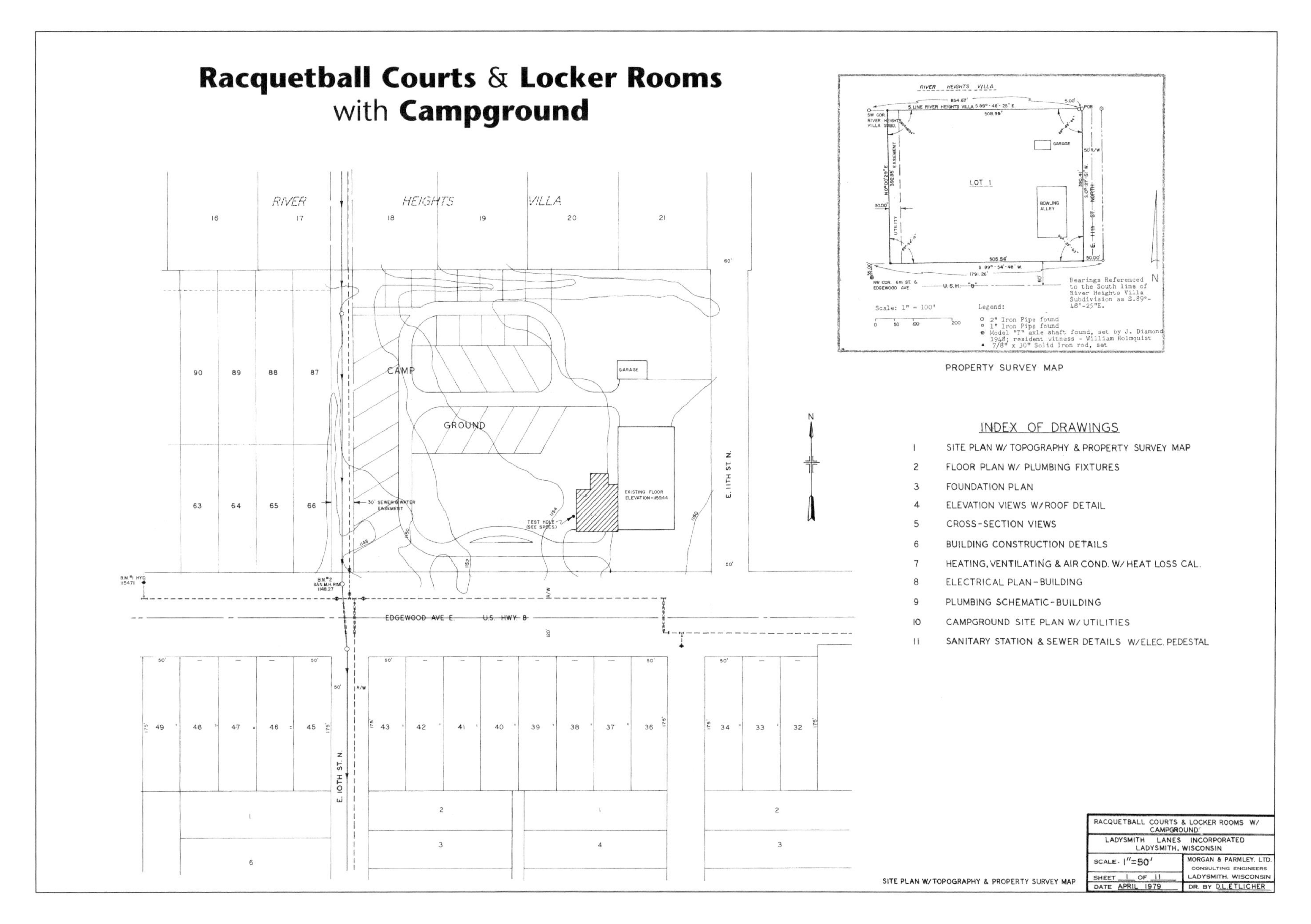

Courtesy: Morgan & Parmley, Ltd.

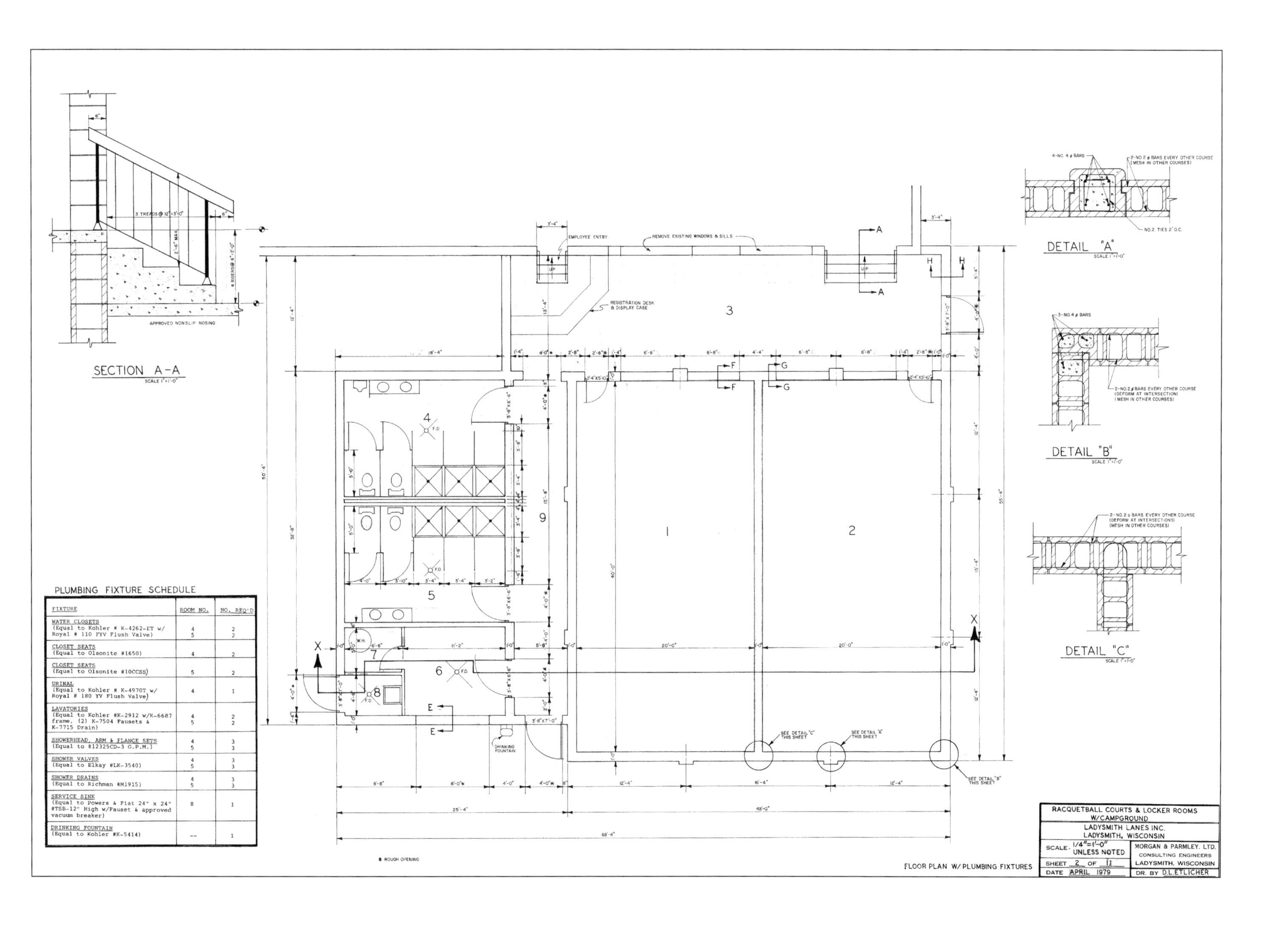

PLUMBING FIXTURE SCHEDULE

FIXTURE	ROOM NO.	NO. REQ'D
WATER CLOSETS (Equal to Kohler # K-4262-ET w/ Royal # 110 FYV Flush Valve)	4 5	2 2
CLOSET SEATS (Equal to Olsonite #1650)	4	2
CLOSET SEATS (Equal to Olsonite #10CCSS)	5	2
URINAL (Equal to Kohler # K-4970T w/ Royal # 180 YV Flush Valve)	4	1
LAVATORIES (Equal to Kohler #K-2912 w/K-6687 frame, (2) K-7504 Fausets & K-7715 Drain)	4 5	2 2
SHOWERHEAD, ARM & FLANGE SETS (Equal to #12325CD-3 G.P.M.)	4 5	3 3
SHOWER VALVES (Equal to Elkay #LK-3540)	4 5	3 3
SHOWER DRAINS (Equal to Richman #M1915)	4 5	3 3
SERVICE SINK (Equal to Powers & Fiat 24" x 24" #TSB-12" High w/Fauset & approved vacuum breaker)	8	1
DRINKING FOUNTAIN (Equal to Kohler #K-5414)	--	1

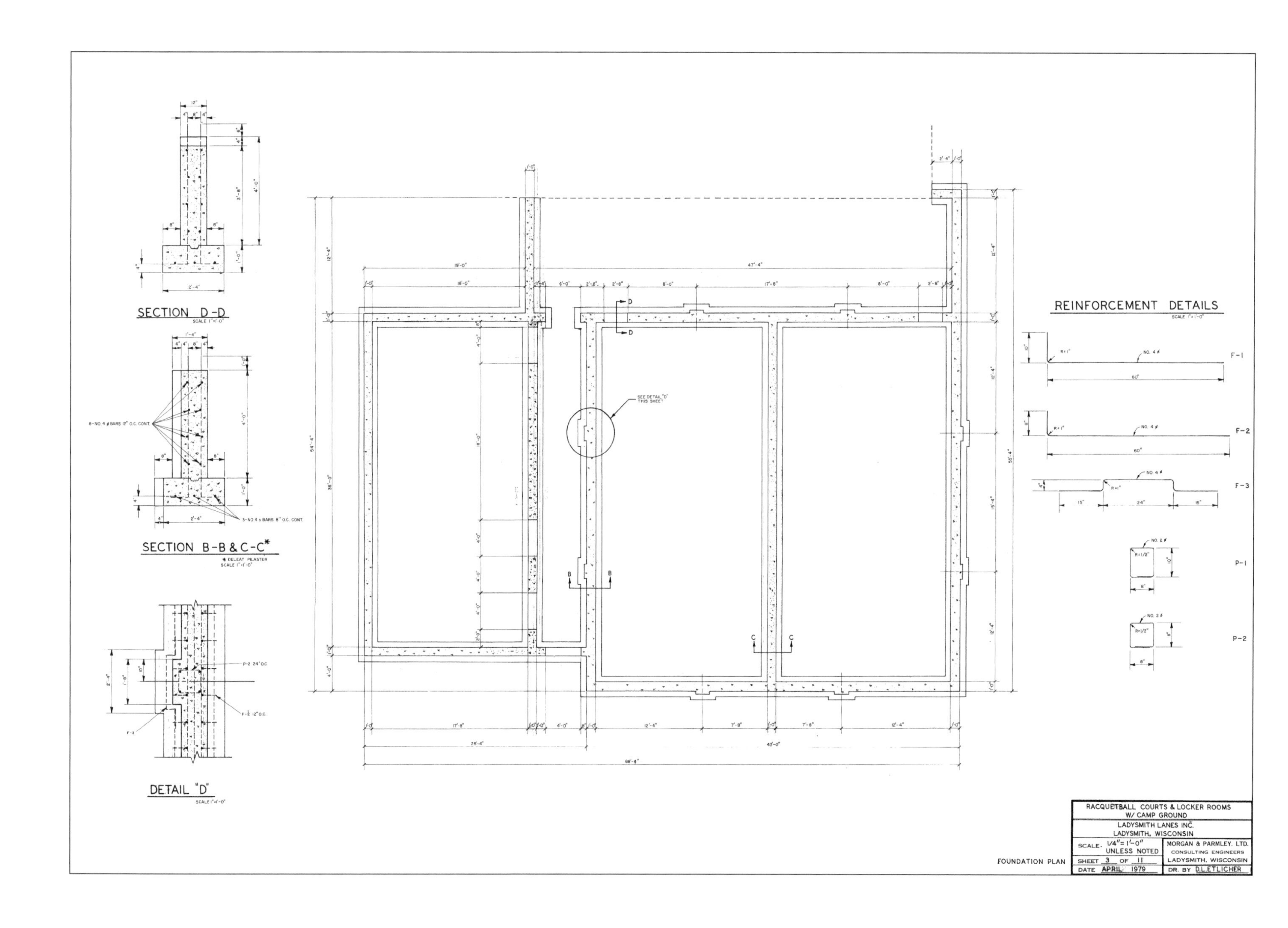
SECTION D-D
SECTION B-B & C-C*
DETAIL "D"
REINFORCEMENT DETAILS
F-1
F-2
F-3
P-1
P-2
SEE DETAIL "D" THIS SHEET
FOUNDATION PLAN
RACQUETBALL COURTS & LOCKER ROOMS
W/ CAMP GROUND
LADYSMITH LANES INC.
LADYSMITH, WISCONSIN
SCALE. 1/4"=1'-0" UNLESS NOTED
SHEET 3 OF 11
DATE APRIL 1979
MORGAN & PARMLEY, LTD.
CONSULTING ENGINEERS
LADYSMITH, WISCONSIN
DR. BY D.L.ETLICHER

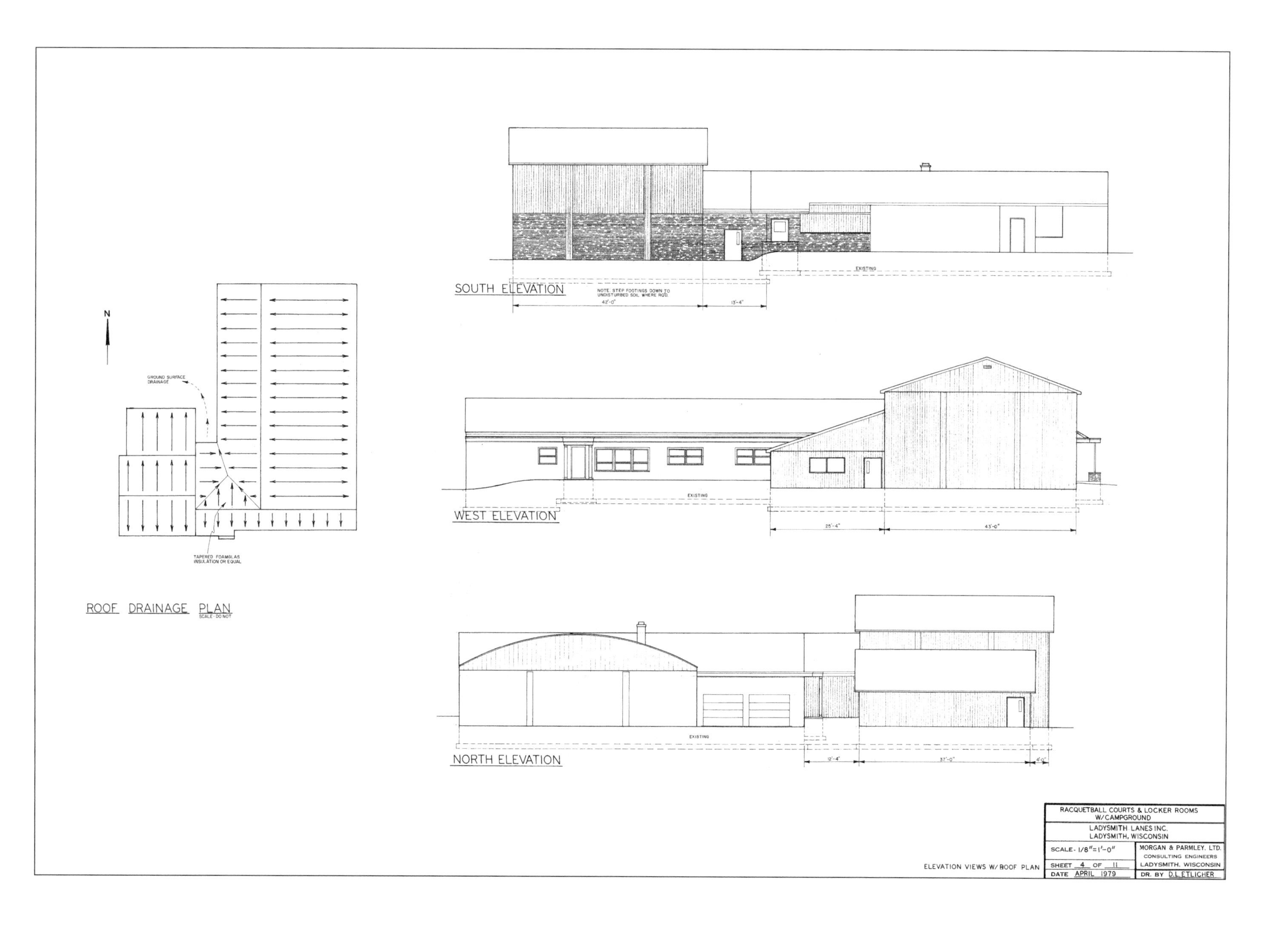
SOUTH ELEVATION
NOTE: STEP FOOTINGS DOWN TO UNDISTURBED SOIL WHERE RQ'D.
42'-0"
13'-4"
EXISTING
N
GROUND SURFACE DRAINAGE
TAPERED FOAMGLAS INSULATION OR EQUAL
ROOF DRAINAGE PLAN
SCALE - DO NOT
WEST ELEVATION
EXISTING
25'-4"
43'-0"
NORTH ELEVATION
EXISTING
12'-4"
37'-0"
4'-0"
ELEVATION VIEWS W/ ROOF PLAN
RACQUETBALL COURTS & LOCKER ROOMS W/CAMPGROUND
LADYSMITH LANES INC.
LADYSMITH, WISCONSIN
SCALE - 1/8"=1'-0"
MORGAN & PARMLEY, LTD.
CONSULTING ENGINEERS
LADYSMITH, WISCONSIN
SHEET 4 OF 11
DATE APRIL 1979
DR. BY D.L.ETLICHER

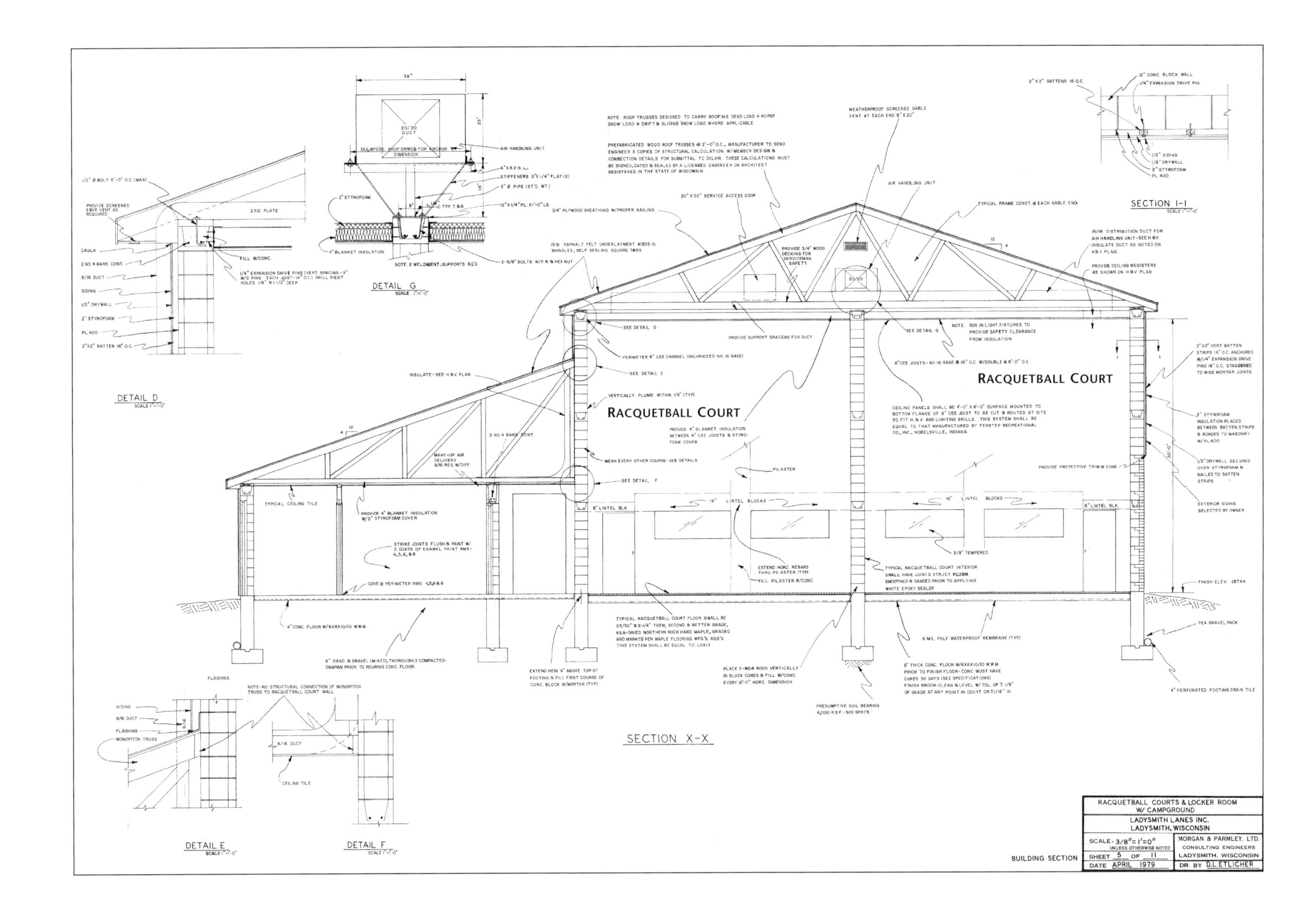
SECTION X-X
DETAIL D
DETAIL G
DETAIL E
DETAIL F
SECTION 1-1
RACQUETBALL COURT
RACQUETBALL COURT
RACQUETBALL COURTS & LOCKER ROOM W/ CAMPGROUND
LADYSMITH LANES INC.
LADYSMITH, WISCONSIN
SCALE - 3/8" = 1'=0" UNLESS OTHERWISE NOTED
SHEET 5 OF 11
DATE APRIL 1979
MORGAN & PARMLEY, LTD.
CONSULTING ENGINEERS
LADYSMITH, WISCONSIN
DR. BY D.L.EYLICHER
BUILDING SECTION

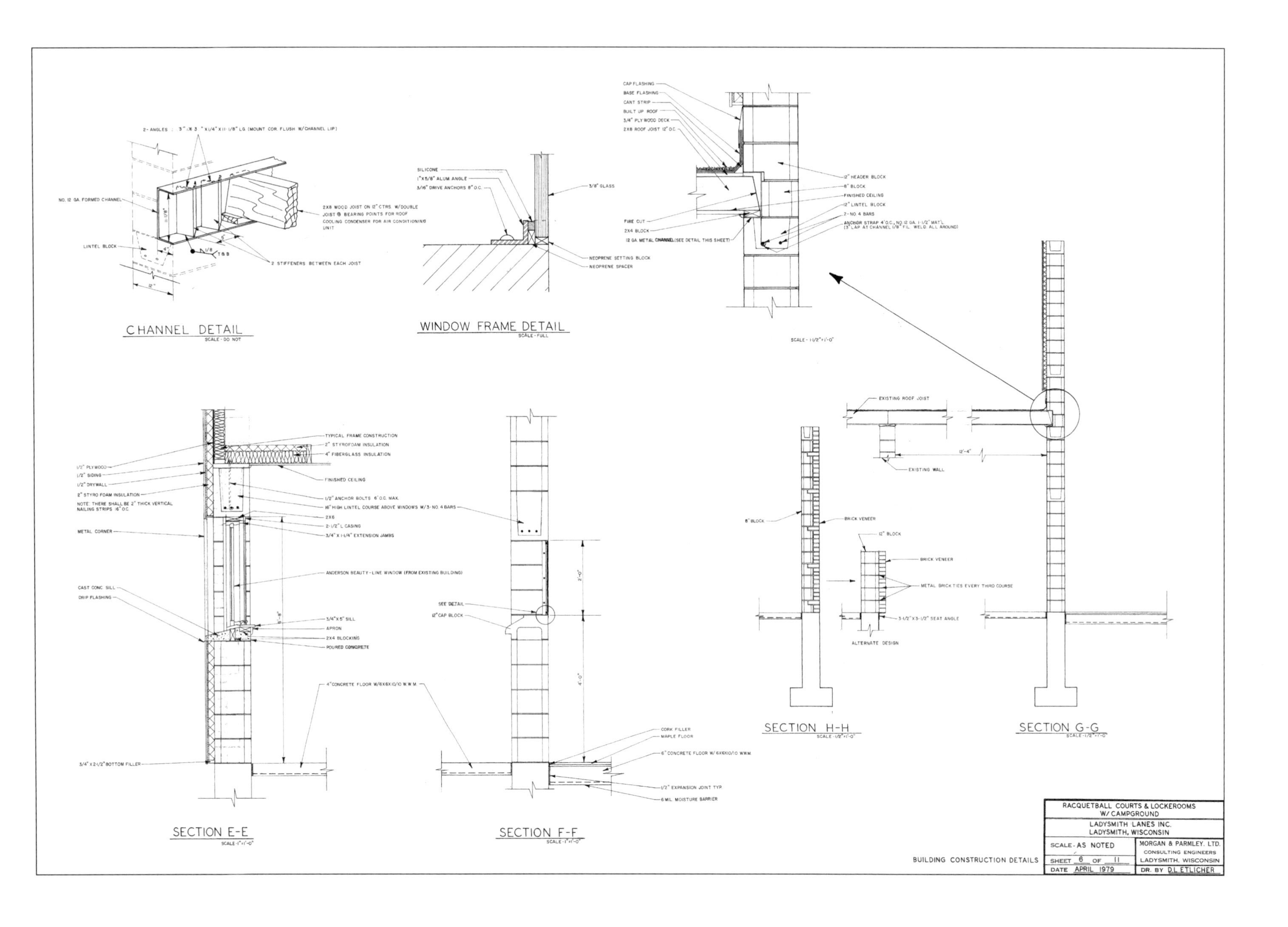
2-ANGLES 3" X 3" X 1/4" X 11-1/8" LG. (MOUNT COR. FLUSH W/CHANNEL LIP)
NO. 12 GA. FORMED CHANNEL
11-1/8"
2X8 WOOD JOIST ON 12" CTRS. W/DOUBLE JOIST @ BEARING POINTS FOR ROOF COOLING CONDENSER FOR AIR CONDITIONING UNIT
LINTEL BLOCK
4"
5"
1/8
T & B
2 STIFFENERS BETWEEN EACH JOIST
12"
CHANNEL DETAIL
SCALE - DO NOT
SILICONE
1"X5/8" ALUM. ANGLE
3/16" DRIVE ANCHORS 8" O.C.
3/8" GLASS
NEOPRENE SETTING BLOCK
NEOPRENE SPACER
WINDOW FRAME DETAIL
SCALE - FULL
CAP FLASHING
BASE FLASHING
CANT STRIP
BUILT UP ROOF
3/4" PLY WOOD DECK
2X8 ROOF JOIST 12" O.C.
12" HEADER BLOCK
8" BLOCK
FINISHED CEILING
12" LINTEL BLOCK
2-NO. 4 BARS
ANCHOR STRAP 4' O.C., NO. 12 GA. 1-1/2" MAT'L
(3" LAP AT CHANNEL 1/8" FIL. WELD. ALL AROUND)
FIRE CUT
2X4 BLOCK
12 GA. METAL CHANNEL (SEE DETAIL THIS SHEET)
SCALE - 1-1/2" = 1'-0"
EXISTING ROOF JOIST
12'-4"
EXISTING WALL
TYPICAL FRAME CONSTRUCTION
2" STYROFOAM INSULATION
4" FIBERGLASS INSULATION
1/2" PLYWOOD
1/2" SIDING
1/2" DRYWALL
2" STYRO FOAM INSULATION
NOTE: THERE SHALL BE 2" THICK VERTICAL NAILING STRIPS 16" O.C.
FINISHED CEILING
1/2" ANCHOR BOLTS 6' O.C. MAX.
16" HIGH LINTEL COURSE ABOVE WINDOWS W/3-NO. 4 BARS
2X6
2-1/2" L CASING
3/4" X 1-1/4" EXTENSION JAMBS
METAL CORNER
ANDERSON BEAUTY-LINE WINDOW (FROM EXISTING BUILDING)
CAST CONC. SILL
DRIP FLASHING
3/4"X5" SILL
APRON
2X4 BLOCKING
POURED CONCRETE
5'-8"
4" CONCRETE FLOOR W/6X6X10/10 W.W.M.
3/4" X 2-1/2" BOTTOM FILLER
SECTION E-E
SCALE - 1" = 1'-0"
SEE DETAIL
12" CAP BLOCK
2'-0"
4'-0"
CORK FILLER
MAPLE FLOOR
6" CONCRETE FLOOR W/ 6X6X10/10 W.W.M.
1/2" EXPANSION JOINT TYP.
6 MIL. MOISTURE BARRIER
SECTION F-F
SCALE - 1" = 1'-0"
8" BLOCK
BRICK VENEER
12" BLOCK
BRICK VENEER
METAL BRICK TIES EVERY THIRD COURSE
3-1/2" X 3-1/2" SEAT ANGLE
ALTERNATE DESIGN
SECTION H-H
SCALE - 1/2" = 1'-0"
SECTION G-G
SCALE - 1/2" = 1'-0"
BUILDING CONSTRUCTION DETAILS
RACQUETBALL COURTS & LOCKEROOMS
W/ CAMPGROUND
LADYSMITH LANES INC.
LADYSMITH, WISCONSIN
SCALE - AS NOTED
SHEET 6 OF 11
DATE APRIL 1979
MORGAN & PARMLEY, LTD.
CONSULTING ENGINEERS
LADYSMITH, WISCONSIN
DR. BY D.L. ETLICHER

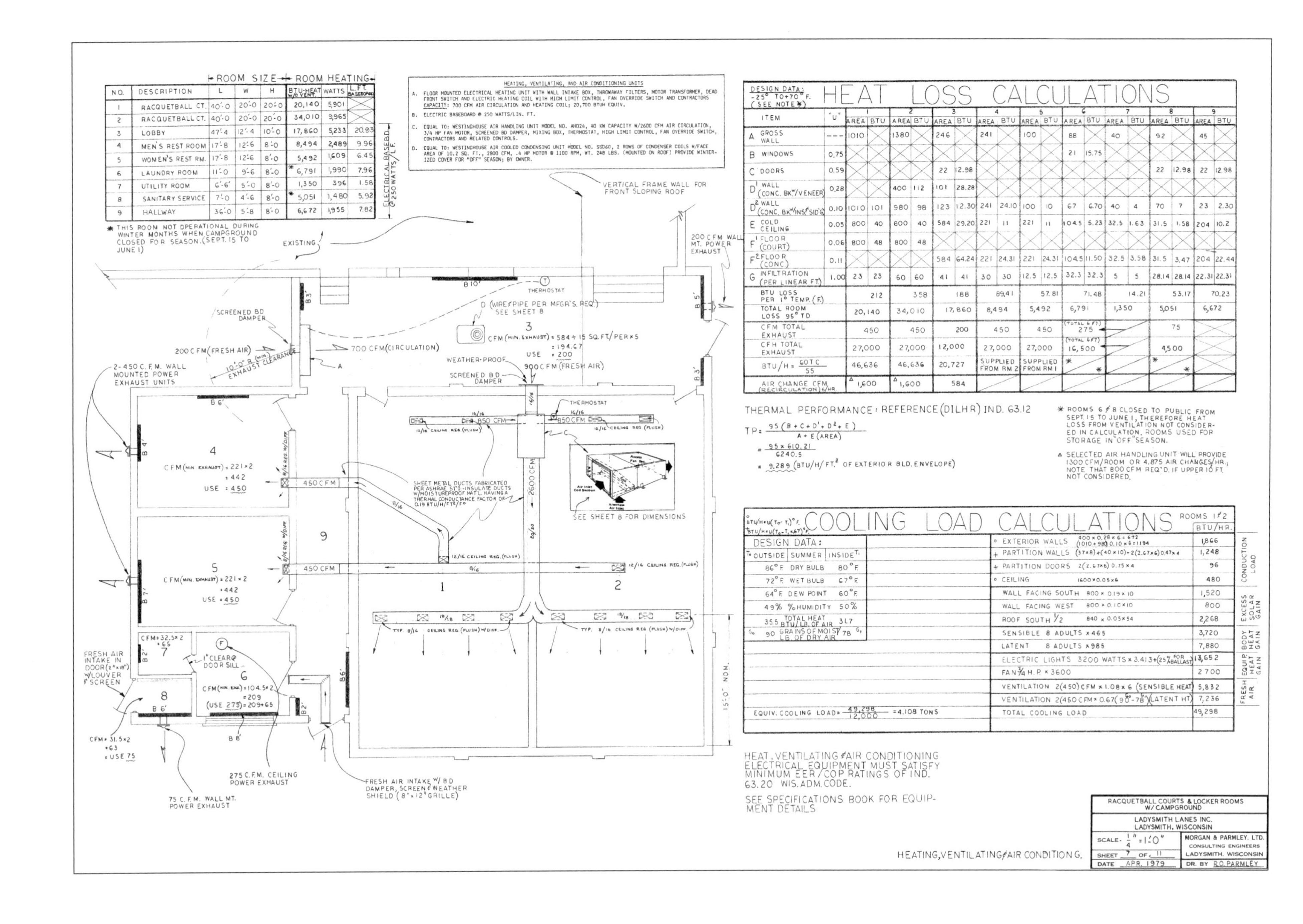

ROOM SIZE
ROOM HEATING
NO. DESCRIPTION L W H BTU-HEAT W/O VENT. WATTS L.F. BASEBOARD
1 RACQUETBALL CT. 40'-0 20'-0 20'-0 20,140 5,901
2 RACQUETBALL CT. 40'-0 20'-0 20'-0 34,010 9,965
3 LOBBY 47'-4 12'-4 10'-0 17,860 5,233 20.93
4 MEN'S REST ROOM 17'-8 12'-6 8'-0 8,494 2,489 9.96
5 WOMEN'S REST RM. 17'-8 12'-6 8'-0 5,492 1,609 6.45
6 LAUNDRY ROOM 11'-0 9'-6 8'-0 *6,791 1,990 7.96
7 UTILITY ROOM 6'-6' 5'-0 8'-0 1,350 396 1.58
8 SANITARY SERVICE 7'-0 4'-6 8'-0 *5,051 1,480 5.92
9 HALLWAY 36'-0 5'-8 8'-0 6,672 1,955 7.82
ELECTRICAL BASEBD. @ 250 WATTS/L.F.
* THIS ROOM NOT OPERATIONAL DURING WINTER MONTHS WHEN CAMPGROUND CLOSED FOR SEASON. (SEPT. 15 TO JUNE 1)
HEATING, VENTILATING, AND AIR CONDITIONING UNITS
A. FLOOR MOUNTED ELECTRICAL HEATING UNIT WITH WALL INTAKE BOX, THROWAWAY FILTERS, MOTOR TRANSFORMER, DEAD FRONT SWITCH AND ELECTRIC HEATING COIL WITH HIGH LIMIT CONTROL, FAN OVERRIDE SWITCH AND CONTRACTORS CAPACITY: 700 CFM AIR CIRCULATION AND HEATING COIL; 20,700 BTUH EQUIV.
B. ELECTRIC BASEBOARD @ 250 WATTS/LIN. FT.
C. EQUAL TO: WESTINGHOUSE AIR HANDLING UNIT MODEL NO. AHO24, 40 KW CAPACITY W/2600 CFM AIR CIRCULATION, 3/4 HP FAN MOTOR, SCREENED BD DAMPER, MIXING BOX, THERMOSTAT, HIGH LIMIT CONTROL, FAN OVERRIDE SWITCH, CONTRACTORS AND RELATED CONTROLS.
D. EQUAL TO: WESTINGHOUSE AIR COOLED CONDENSING UNIT MODEL NO. SSD60, 2 ROWS OF CONDENSER COILS W/FACE AREA OF 10.2 SQ. FT., 2800 CFM, .4 HP MOTOR @ 1100 RPM, WT. 248 LBS. (MOUNTED ON ROOF) PROVIDE WINTERIZED COVER FOR "OFF" SEASON; BY OWNER.
HEAT LOSS CALCULATIONS
DESIGN DATA: -25° TO +70° F. (SEE NOTE *)
BTU LOSS PER 1° TEMP. (F.) 212 358 188 89.41 57.81 71.48 14.21 53.17 70.23
TOTAL ROOM LOSS 95° TD 20,140 34,010 17,860 8,494 5,492 6,791 1,350 5,051 6,672
THERMAL PERFORMANCE: REFERENCE (DILHR) IND. 63.12
* ROOMS 6 & 8 CLOSED TO PUBLIC FROM SEPT. 15 TO JUNE 1, THEREFORE HEAT LOSS FROM VENTILATION NOT CONSIDERED IN CALCULATION. ROOMS USED FOR STORAGE IN "OFF" SEASON.
Δ SELECTED AIR HANDLING UNIT WILL PROVIDE 1300 CFM/ROOM OR 4.875 AIR CHANGES/HR.; NOTE THAT 800 CFM REQ'D. IF UPPER 10 FT. NOT CONSIDERED.
COOLING LOAD CALCULATIONS ROOMS 1 & 2
DESIGN DATA:
OUTSIDE SUMMER INSIDE
86° F DRY BULB 80° F
72° F WET BULB 67° F
64° F DEW POINT 60° F
49% % HUMIDITY 50%
35.5 TOTAL HEAT BTU/LB. OF AIR 31.7
90 GRAINS OF MOIST/LB. OF DRY AIR 78
EXTERIOR WALLS 1,866
PARTITION WALLS 1,248
PARTITION DOORS 96
CEILING 480
WALL FACING SOUTH 1,520
WALL FACING WEST 800
ROOF SOUTH 1/2 2,268
SENSIBLE 8 ADULTS x 465 3,720
LATENT 8 ADULTS x 985 7,880
ELECTRIC LIGHTS 13,652
FAN 3/4 H.P. x 3600 2700
VENTILATION 2(450) CFM x 1.08 x 6 (SENSIBLE HEAT) 5,832
VENTILATION 2(450 CFM x 0.67(90-78) (LATENT HT) 7,236
TOTAL COOLING LOAD 49,298
EQUIV. COOLING LOAD = 49,298 / 12,000 = 4.108 TONS
CONDUCTION LOAD
EXCESS SOLAR GAIN
BODY HEAT GAIN
EQUIP. HEAT GAIN
FRESH AIR
HEAT, VENTILATING & AIR CONDITIONING ELECTRICAL EQUIPMENT MUST SATISFY MINIMUM EER/COP RATINGS OF IND. 63.20 WIS. ADM. CODE.
SEE SPECIFICATIONS BOOK FOR EQUIPMENT DETAILS
VERTICAL FRAME WALL FOR FRONT SLOPING ROOF
EXISTING
200 CFM WALL MT. POWER EXHAUST
THERMOSTAT
D (WIRE & PIPE PER MFGR'S. REQ'). SEE SHEET 8
SCREENED BD DAMPER
200 CFM (FRESH AIR)
700 CFM (CIRCULATION)
2-450 C.F.M. WALL MOUNTED POWER EXHAUST UNITS
WEATHER-PROOF
900 CFM (FRESH AIR)
SEE SHEET 8 FOR DIMENSIONS
SHEET METAL DUCTS FABRICATED PER ASHRAE STD.-INSULATE DUCTS W/MOISTUREPROOF MAT'L. HAVING A THERMAL CONDUCTANCE FACTOR OF 0.19 BTU/H/FT²/F°
450 CFM
2600 CFM
FRESH AIR INTAKE IN DOOR (8"x18") W/LOUVER & SCREEN
1" CLEAR DOOR SILL
275 C.F.M. CEILING POWER EXHAUST
75 C.F.M. WALL MT. POWER EXHAUST
FRESH AIR INTAKE W/ BD DAMPER, SCREEN & WEATHER SHIELD (8"x12" GRILLE)
15'-0" NOM.
RACQUETBALL COURTS & LOCKER ROOMS W/ CAMPGROUND
LADYSMITH LANES INC. LADYSMITH, WISCONSIN
SCALE: 1/4" = 1'-0"
MORGAN & PARMLEY, LTD. CONSULTING ENGINEERS LADYSMITH, WISCONSIN
SHEET 7 OF 11
DATE APR. 1979
DR. BY R.O. PARMLEY
HEATING, VENTILATING & AIR CONDITIONING.

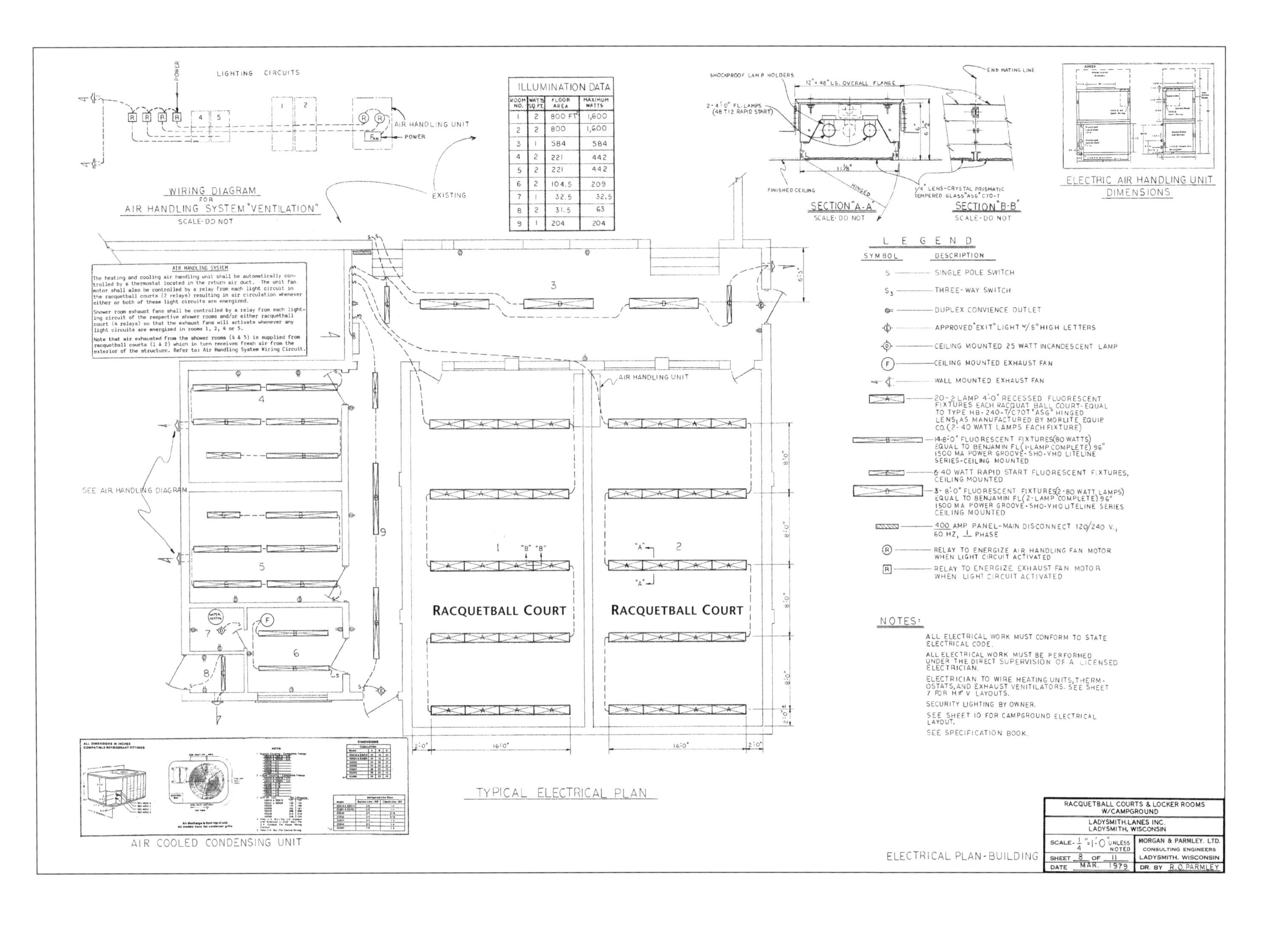
LIGHTING CIRCUITS
POWER
AIR HANDLING UNIT
POWER
WIRING DIAGRAM
FOR
AIR HANDLING SYSTEM "VENTILATION"
SCALE- DO NOT
EXISTING
ILLUMINATION DATA
ROOM NO. | WATTS SQ FT | FLOOR AREA | MAXIMUM WATTS
1 | 2 | 800 FT² | 1,600
2 | 2 | 800 | 1,600
3 | 1 | 584 | 584
4 | 2 | 221 | 442
5 | 2 | 221 | 442
6 | 2 | 104.5 | 209
7 | 1 | 32.5 | 32.5
8 | 2 | 31.5 | 63
9 | 1 | 204 | 204
SHOCKPROOF LAMP HOLDERS
12" x 48" LG. OVERALL FLANGE
2-4'-0" FL. LAMPS (48 T12 RAPID START)
END MATING LINE
FINISHED CEILING
HINGED
1/4" LENS-CRYSTAL PRISMATIC TEMPERED GLASS "ASG" C70-T
SECTION "A-A"
SCALE- DO NOT
SECTION "B-B"
SCALE- DO NOT
ELECTRIC AIR HANDLING UNIT DIMENSIONS
AIR HANDLING SYSTEM
The heating and cooling air handling unit shall be automatically controlled by a thermostat located in the return air duct. The unit fan motor shall also be controlled by a relay from each light circuit in the racquetball courts (2 relays) resulting in air circulation whenever either or both of these light circuits are energized.
Shower room exhaust fans shall be controlled by a relay from each lighting circuit of the respective shower rooms and/or either racquetball court (4 relays) so that the exhaust fans will activate whenever any light circuits are energized in rooms 1, 2, 4 or 5.
Note that air exhausted from the shower rooms (4 & 5) is supplied from racquetball courts (1 & 2) which in turn receives fresh air from the exterior of the structure. Refer to: Air Handling System Wiring Circuit.
SEE AIR HANDLING DIAGRAM
AIR HANDLING UNIT
RACQUETBALL COURT
RACQUETBALL COURT
TYPICAL ELECTRICAL PLAN
AIR COOLED CONDENSING UNIT
LEGEND
SYMBOL
DESCRIPTION
SINGLE POLE SWITCH
THREE-WAY SWITCH
DUPLEX CONVIENCE OUTLET
APPROVED "EXIT" LIGHT W/5" HIGH LETTERS
CEILING MOUNTED 25 WATT INCANDESCENT LAMP
CEILING MOUNTED EXHAUST FAN
WALL MOUNTED EXHAUST FAN
20-2 LAMP 4'-0" RECESSED FLUORESCENT FIXTURES EACH RACQUAT BALL COURT-EQUAL TO TYPE HB-240-T/C70T "ASG" HINGED LENS, AS MANUFACTURED BY MORLITE EQUIP. CO. (2-40 WATT LAMPS EACH FIXTURE)
14-8'-0" FLUORESCENT FIXTURES (80 WATTS) EQUAL TO BENJAMIN FL (1-LAMP COMPLETE) 96" 1500 MA POWER GROOVE-SHO-VHO LITELINE SERIES-CEILING MOUNTED
6-40 WATT RAPID START FLUORESCENT FIXTURES, CEILING MOUNTED
3-8'-0" FLUORESCENT FIXTURES (2-80 WATT LAMPS) EQUAL TO BENJAMIN FL (2-LAMP COMPLETE) 96" 1500 MA POWER GROOVE-SHO-VHO LITELINE SERIES CEILING MOUNTED
400 AMP PANEL-MAIN DISCONNECT 120/240 V., 60 HZ, 1 PHASE
RELAY TO ENERGIZE AIR HANDLING FAN MOTOR WHEN LIGHT CIRCUIT ACTIVATED
RELAY TO ENERGIZE EXHAUST FAN MOTOR WHEN LIGHT CIRCUIT ACTIVATED
NOTES:
ALL ELECTRICAL WORK MUST CONFORM TO STATE ELECTRICAL CODE.
ALL ELECTRICAL WORK MUST BE PERFORMED UNDER THE DIRECT SUPERVISION OF A LICENSED ELECTRICIAN.
ELECTRICIAN TO WIRE HEATING UNITS, THERMOSTATS, AND EXHAUST VENTILATORS. SEE SHEET 7 FOR H&V LAYOUTS.
SECURITY LIGHTING BY OWNER.
SEE SHEET 10 FOR CAMPGROUND ELECTRICAL LAYOUT.
SEE SPECIFICATION BOOK.
ELECTRICAL PLAN-BUILDING
RACQUETBALL COURTS & LOCKER ROOMS W/CAMPGROUND
LADYSMITH LANES INC.
LADYSMITH, WISCONSIN
SCALE- 1/4"=1'-0" UNLESS NOTED
SHEET 8 OF 11
DATE MAR. 1979
MORGAN & PARMLEY, LTD.
CONSULTING ENGINEERS
LADYSMITH, WISCONSIN
DR. BY R.O. PARMLEY

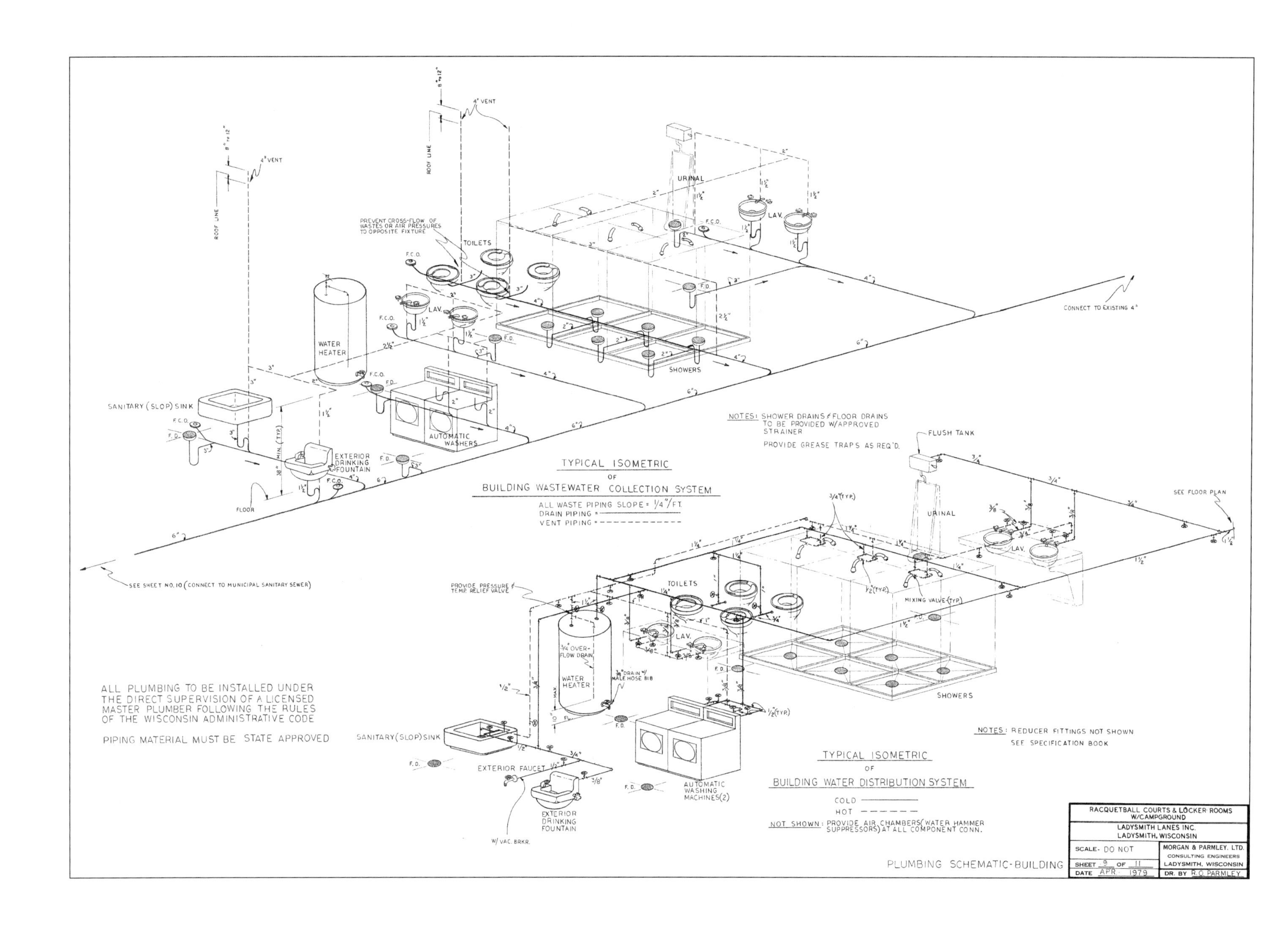
TYPICAL ISOMETRIC OF BUILDING WASTEWATER COLLECTION SYSTEM
ALL WASTE PIPING SLOPE = 1/4"/FT.
DRAIN PIPING =
VENT PIPING =
NOTES: SHOWER DRAINS & FLOOR DRAINS TO BE PROVIDED W/APPROVED STRAINER
PROVIDE GREASE TRAPS AS REQ'D.
SEE SHEET NO. 10 (CONNECT TO MUNICIPAL SANITARY SEWER)
CONNECT TO EXISTING 4"
PREVENT CROSS-FLOW OF WASTES OR AIR PRESSURES TO OPPOSITE FIXTURE
TOILETS
URINAL
LAV.
SHOWERS
WATER HEATER
SANITARY (SLOP) SINK
EXTERIOR DRINKING FOUNTAIN
AUTOMATIC WASHERS
ROOF LINE
4" VENT
FLOOR
TYPICAL ISOMETRIC OF BUILDING WATER DISTRIBUTION SYSTEM
COLD
HOT
NOT SHOWN: PROVIDE AIR CHAMBERS (WATER HAMMER SUPPRESSORS) AT ALL COMPONENT CONN.
NOTES: REDUCER FITTINGS NOT SHOWN
SEE SPECIFICATION BOOK
FLUSH TANK
SEE FLOOR PLAN
MIXING VALVE (TYP)
PROVIDE PRESSURE & TEMP. RELIEF VALVE
3/4" OVER-FLOW DRAIN
3/4" DRAIN W/ MALE HOSE BIB
EXTERIOR FAUCET
W/ VAC. BRKR.
AUTOMATIC WASHING MACHINES (2)
ALL PLUMBING TO BE INSTALLED UNDER THE DIRECT SUPERVISION OF A LICENSED MASTER PLUMBER FOLLOWING THE RULES OF THE WISCONSIN ADMINISTRATIVE CODE
PIPING MATERIAL MUST BE STATE APPROVED
PLUMBING SCHEMATIC-BUILDING
RACQUETBALL COURTS & LOCKER ROOMS W/CAMPGROUND
LADYSMITH LANES INC.
LADYSMITH, WISCONSIN
SCALE- DO NOT
SHEET 9 OF 11
DATE APR. 1979
MORGAN & PARMLEY, LTD.
CONSULTING ENGINEERS
LADYSMITH, WISCONSIN
DR. BY R.O. PARMLEY

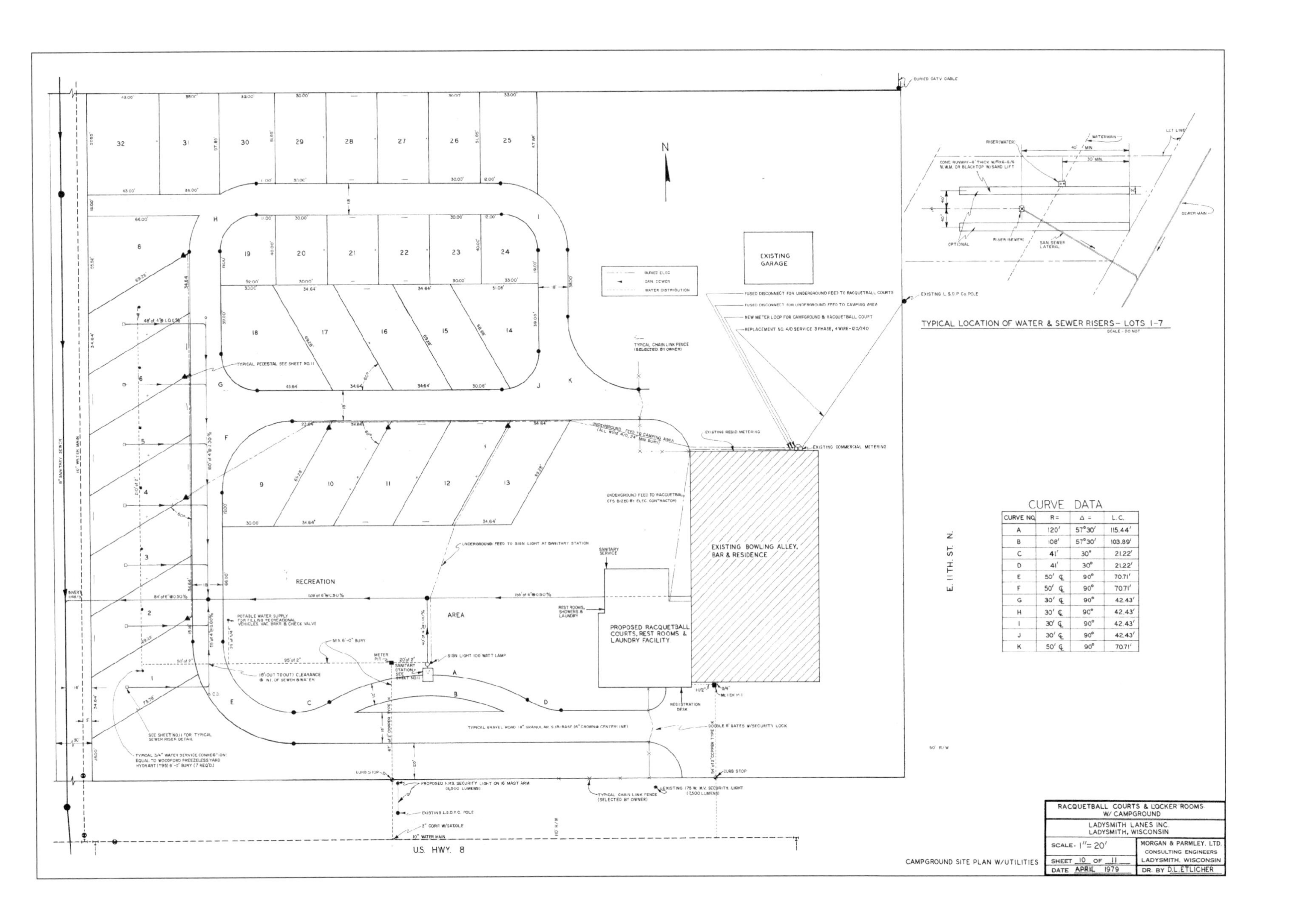
N
EXISTING GARAGE
TYPICAL LOCATION OF WATER & SEWER RISERS – LOTS 1-7
EXISTING BOWLING ALLEY, BAR & RESIDENCE
PROPOSED RACQUETBALL COURTS, REST ROOMS & LAUNDRY FACILITY
RECREATION
AREA
E. 11TH. ST. N.
U.S. HWY. 8
CURVE DATA
CURVE NO. | R= | Δ= | L.C.
A | 120' | 57°30' | 115.44'
B | 108' | 57°30' | 103.89'
C | 41' | 30° | 21.22'
D | 41' | 30° | 21.22'
E | 50' ℄ | 90° | 70.71'
F | 50' ℄ | 90° | 70.71'
G | 30' ℄ | 90° | 42.43'
H | 30' ℄ | 90° | 42.43'
I | 30' ℄ | 90° | 42.43'
J | 30' ℄ | 90° | 42.43'
K | 50' ℄ | 90° | 70.71'
CAMPGROUND SITE PLAN W/UTILITIES
RACQUETBALL COURTS & LOCKER ROOMS W/ CAMPGROUND
LADYSMITH LANES INC.
LADYSMITH, WISCONSIN
SCALE- 1" = 20'
SHEET 10 OF 11
DATE APRIL 1979
MORGAN & PARMLEY, LTD.
CONSULTING ENGINEERS
LADYSMITH, WISCONSIN
DR. BY D.L. ETLICHER

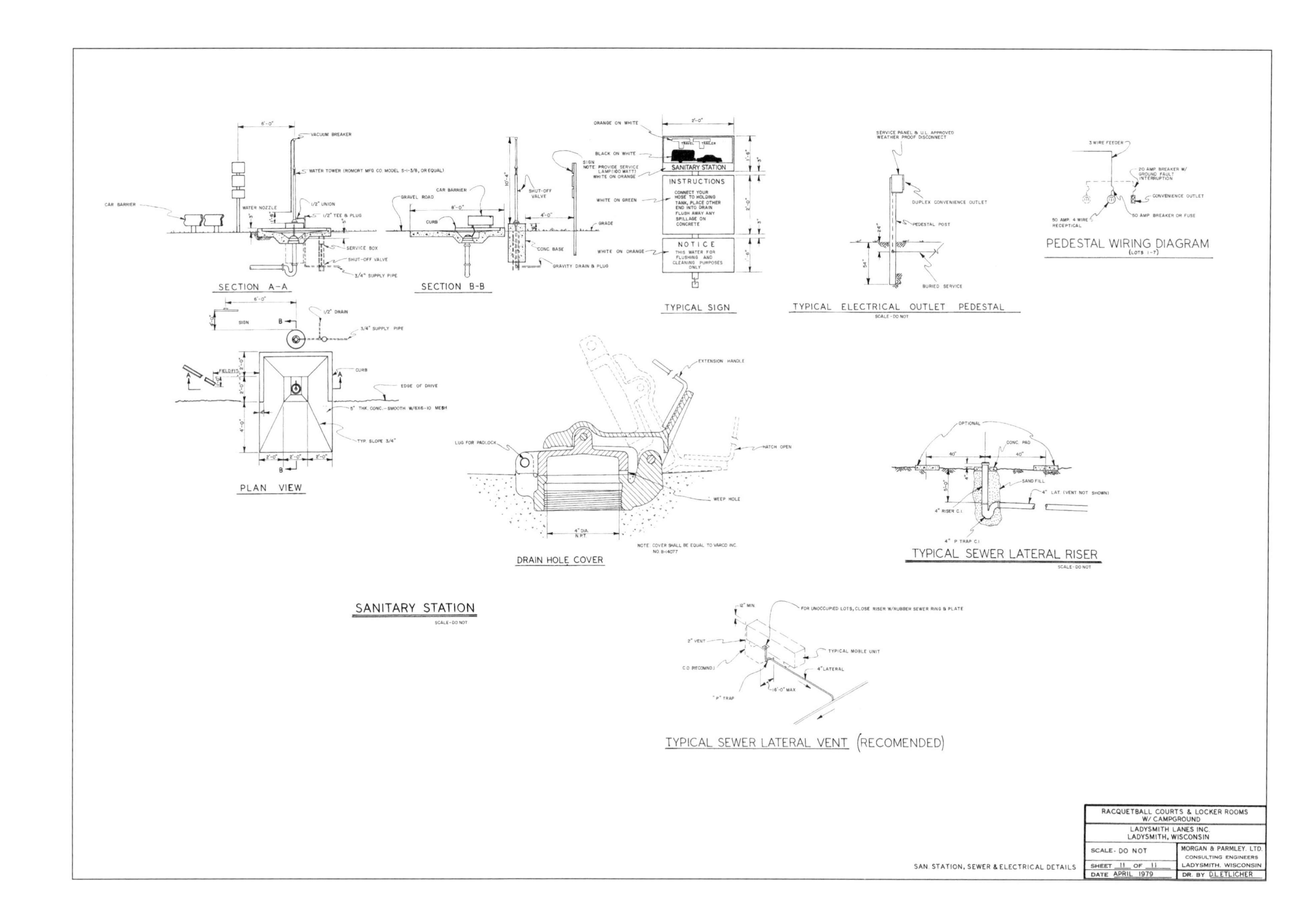

SECTION A-A
SECTION B-B
TYPICAL SIGN
TYPICAL ELECTRICAL OUTLET PEDESTAL
PEDESTAL WIRING DIAGRAM
PLAN VIEW
DRAIN HOLE COVER
TYPICAL SEWER LATERAL RISER
SANITARY STATION
TYPICAL SEWER LATERAL VENT (RECOMENDED)
SANITARY STATION
INSTRUCTIONS
CONNECT YOUR HOSE TO HOLDING TANK, PLACE OTHER END INTO DRAIN FLUSH AWAY ANY SPILLAGE ON CONCRETE
NOTICE
THIS WATER FOR FLUSHING AND CLEANING PURPOSES ONLY
RACQUETBALL COURTS & LOCKER ROOMS W/ CAMPGROUND
LADYSMITH LANES INC.
LADYSMITH, WISCONSIN
SCALE- DO NOT
MORGAN & PARMLEY, LTD.
CONSULTING ENGINEERS
LADYSMITH, WISCONSIN
SHEET 11 OF 11
DATE APRIL 1979
DR. BY D.L.ETLICHER
SAN. STATION, SEWER & ELECTRICAL DETAILS

Design

TRAILER COURTS & CAMPGROUNDS

Section 18

Trailer courts and campgrounds have been increasing at a rapid rate within the last three decades throughout the country. As the population increases and the desire for recreation in a rural setting becomes more popular, these types of facilities will increase.

The first set of plans detail a proposed trailer court that is adjacent to municipal sanitary sewer and watermain systems which will connect to these facilities. The second set of plans have access only to municipal sanitary sewer, therefore, a private water supply and distribution network is required, as illustrated on the plans. The third set of plans detail a rural campground without any municipal services, except for electrical service. The final set of example plans show an addition to an existing facility, which was the first set of plans presented in this section. Note that the name was changed from Parkview Mobile Home Park to Leatherman Trailer Court.

TABLE OF CONTENTS

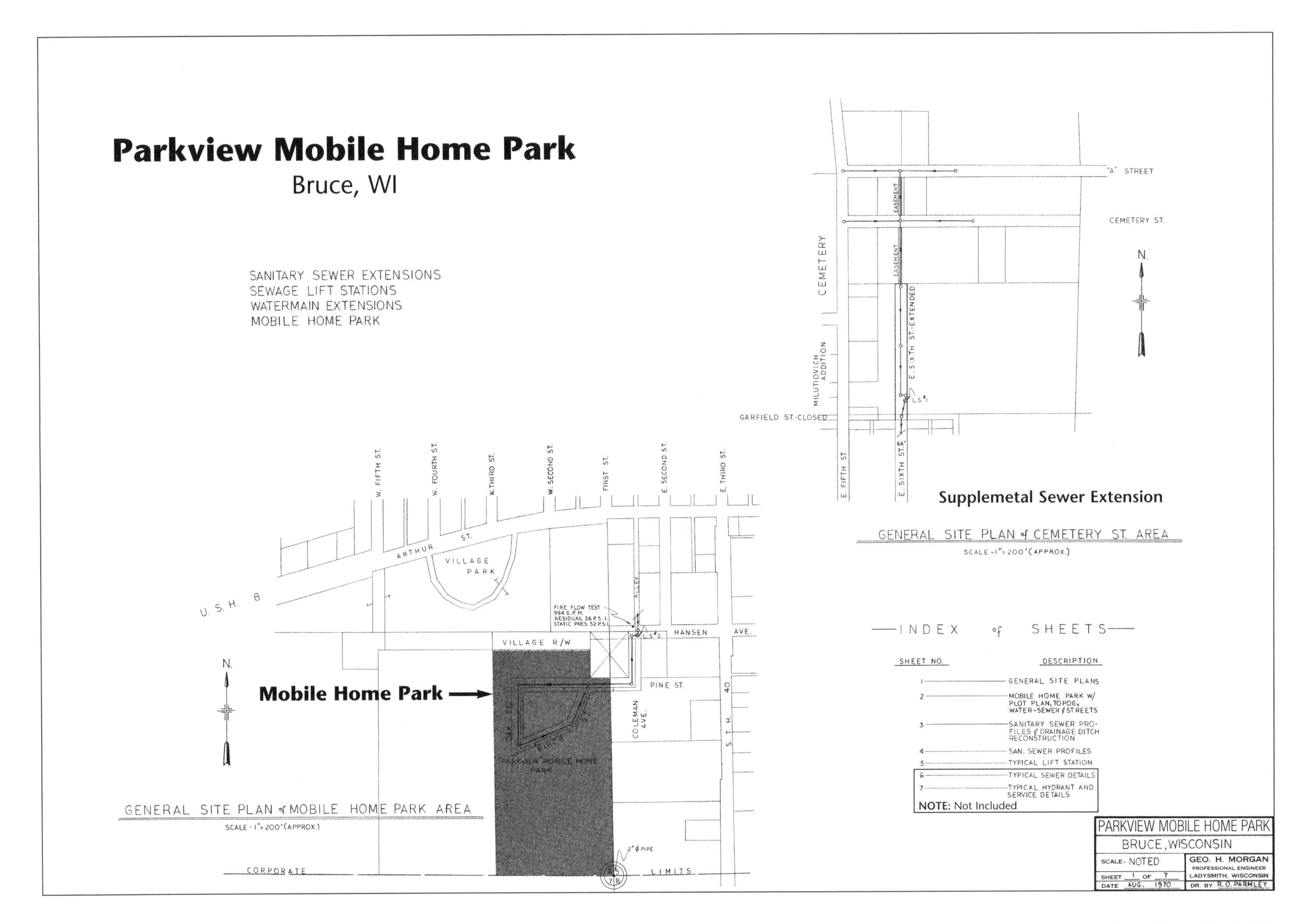

Courtesy: Morgan & Parmley, Ltd.

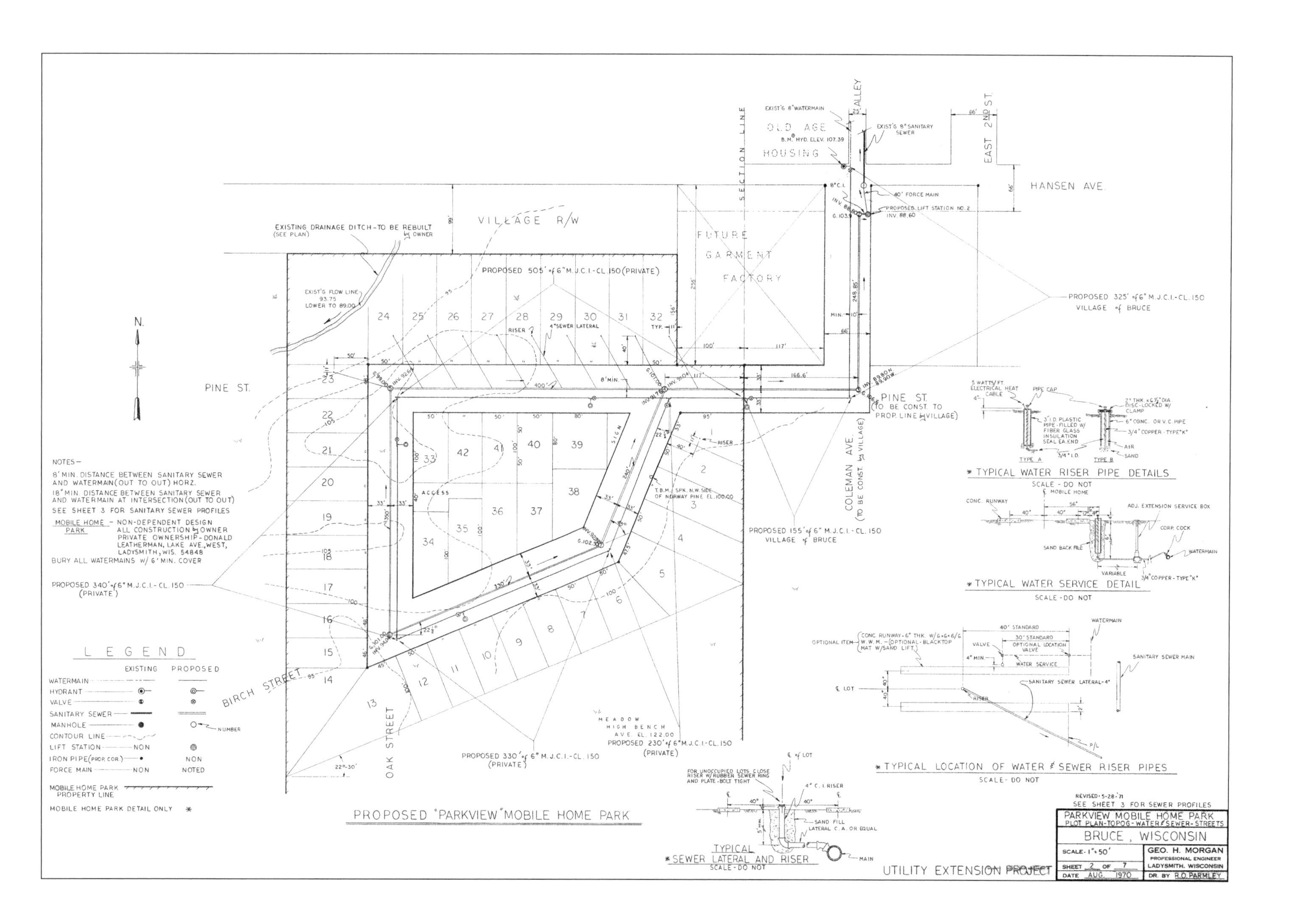
PROPOSED "PARKVIEW" MOBILE HOME PARK
PARKVIEW MOBILE HOME PARK
PLOT PLAN-TOPOG-WATER & SEWER-STREETS
BRUCE, WISCONSIN
SCALE- 1" = 50'
SHEET 2 OF 7
DATE AUG. 1970
GEO. H. MORGAN
PROFESSIONAL ENGINEER
LADYSMITH, WISCONSIN
DR. BY R.O. PARMLEY
REVISED: 5-28-71
SEE SHEET 3 FOR SEWER PROFILES
UTILITY EXTENSION PROJECT
NOTES –
8' MIN. DISTANCE BETWEEN SANITARY SEWER AND WATERMAIN (OUT TO OUT) HORZ.
18" MIN. DISTANCE BETWEEN SANITARY SEWER AND WATERMAIN AT INTERSECTION (OUT TO OUT)
SEE SHEET 3 FOR SANITARY SEWER PROFILES
MOBILE HOME PARK – NON-DEPENDENT DESIGN ALL CONSTRUCTION by OWNER PRIVATE OWNERSHIP - DONALD LEATHERMAN, LAKE AVE., WEST, LADYSMITH, WIS. 54848
BURY ALL WATERMAINS W/ 6' MIN. COVER
LEGEND
EXISTING PROPOSED
WATERMAIN
HYDRANT
VALVE
SANITARY SEWER
MANHOLE
NUMBER
CONTOUR LINE
LIFT STATION NON
IRON PIPE (PROP. COR.) NON
FORCE MAIN NON NOTED
MOBILE HOME PARK PROPERTY LINE
MOBILE HOME PARK DETAIL ONLY *
PINE ST.
BIRCH STREET
OAK STREET
COLEMAN AVE. (TO BE CONST. by VILLAGE)
PINE ST. (TO BE CONST. TO PROP. LINE by VILLAGE)
HANSEN AVE.
EAST 2ND ST.
ALLEY
SECTION LINE
OLD AGE HOUSING
B.M. HYD. ELEV. 107.39
FUTURE GARMENT FACTORY
VILLAGE R/W
EXISTING DRAINAGE DITCH - TO BE REBUILT (SEE PLAN) by OWNER
EXIST'G FLOW LINE 93.75 LOWER TO 89.00
EXIST'G 8" WATERMAIN
EXIST'G 8" SANITARY SEWER
8" FORCE MAIN
PROPOSED LIFT STATION NO. 2 INV. 88.60
PROPOSED 505' of 6" M.J.C.I.-CL.150 (PRIVATE)
PROPOSED 325' of 6" M.J.C.I.-CL.150 VILLAGE of BRUCE
PROPOSED 155' of 6" M.J.C.I.-CL.150 VILLAGE of BRUCE
PROPOSED 340' of 6" M.J.C.I.-CL.150 (PRIVATE)
PROPOSED 330' of 6" M.J.C.I.-CL.150 (PRIVATE)
PROPOSED 230' of 6" M.J.C.I.-CL.150 (PRIVATE)
MEADOW HIGH BENCH AVE. EL. 122.00
T.B.M. SPK. N.W. SIDE OF NORWAY PINE EL. 100.00
4" SEWER LATERAL
RISER
ACCESS
SIGN
*TYPICAL WATER RISER PIPE DETAILS
SCALE - DO NOT
5 WATTS/FT. ELECTRICAL HEAT CABLE
PIPE CAP
3" I.D. PLASTIC PIPE - FILLED W/ FIBER GLASS INSULATION SEAL EA. END
2" THK. x 6½" DIA. DISC - LOCKED W/ CLAMP
6" CONC. OR V.C. PIPE
3/4" COPPER - TYPE "K"
AIR
SAND
TYPE A
TYPE B
*TYPICAL WATER SERVICE DETAIL
SCALE - DO NOT
CONC. RUNWAY
MOBILE HOME
ADJ. EXTENSION SERVICE BOX
CORP. COCK
WATERMAIN
SAND BACK FILL
VARIABLE
3/4" COPPER - TYPE "K"
*TYPICAL LOCATION OF WATER & SEWER RISER PIPES
SCALE - DO NOT
OPTIONAL ITEM
CONC. RUNWAY - 6" THK. W/ 6+6+6/6 W.W.M. - (OPTIONAL - BLACKTOP MAT W/SAND LIFT)
40' STANDARD
30' STANDARD
OPTIONAL LOCATION VALVE
VALVE
WATER SERVICE
SANITARY SEWER MAIN
SANITARY SEWER LATERAL - 4"
RISER
LOT
TYPICAL
*SEWER LATERAL AND RISER
SCALE - DO NOT
FOR UNOCCUPIED LOTS, CLOSE RISER W/ RUBBER SEWER RING AND PLATE - BOLT TIGHT
4" C.I. RISER
SAND FILL LATERAL C.A. OR EQUAL
MAIN
N.

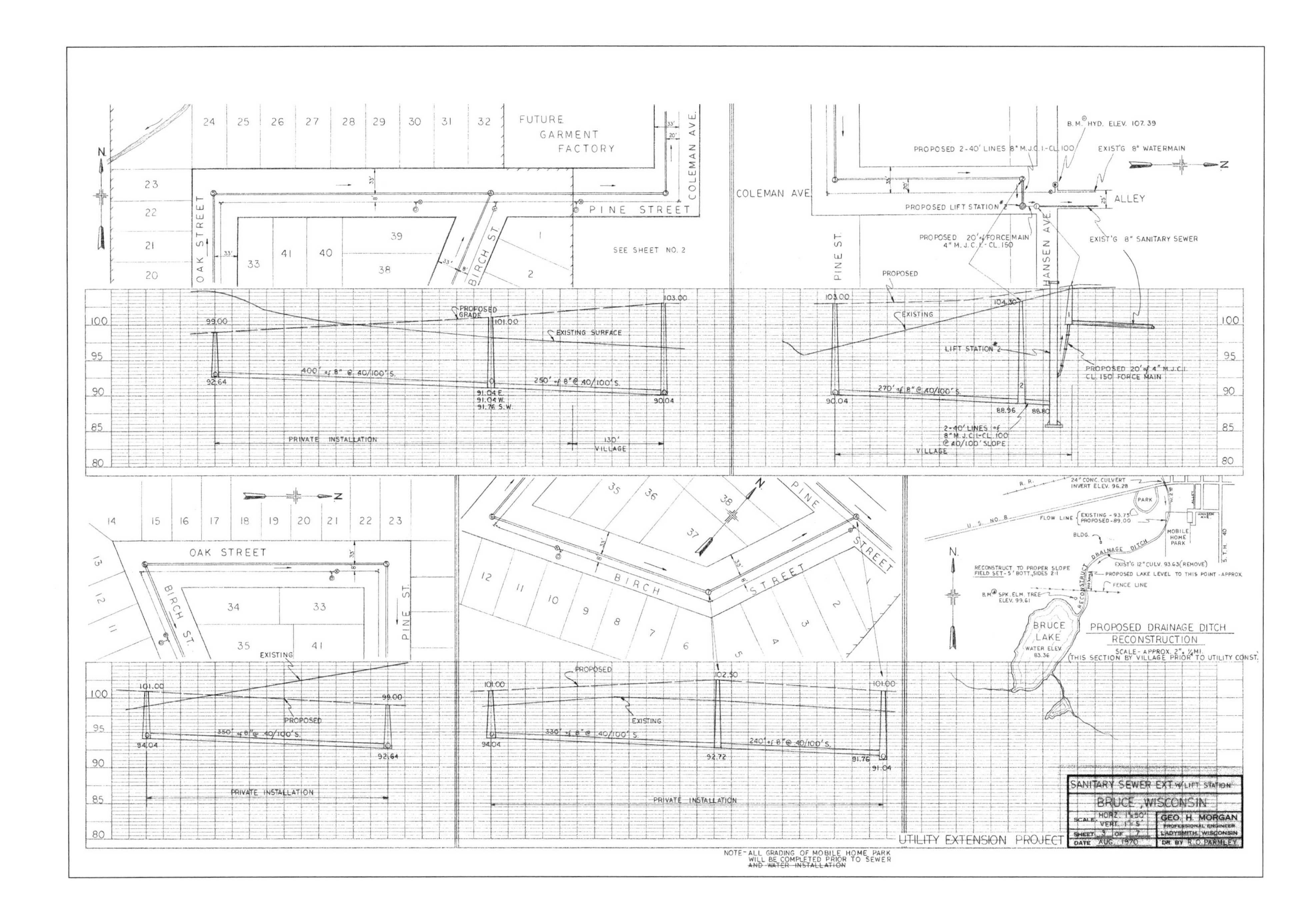

FUTURE GARMENT FACTORY
COLEMAN AVE.
PINE STREET
OAK STREET
BIRCH ST.
SEE SHEET NO. 2
PROPOSED GRADE
EXISTING SURFACE
400' of 8" @ .40/100'S.
250' of 8" @ .40/100'S.
PRIVATE INSTALLATION
130' VILLAGE
B.M. HYD. ELEV. 107.39
PROPOSED 2-40' LINES 8" M.J.C.I.-CL.100
EXIST'G 8" WATERMAIN
ALLEY
PROPOSED LIFT STATION #2
PROPOSED 20' of FORCE MAIN 4" M.J.C.I.-CL.150
EXIST'G 8" SANITARY SEWER
HANSEN AVE.
LIFT STATION #2
PROPOSED 20' of 4" M.J.C.I. CL.150 FORCE MAIN
270' of 8" @ .40/100'S.
2-40' LINES of 8" M.J.C.I.-CL.100 @ .40/100' SLOPE
VILLAGE
330' of 8" @ .40/100'S
240' of 8" @ .40/100'S.
350' of 8" @ .40/100'S.
24" CONC. CULVERT INVERT ELEV. 96.28
FLOW LINE EXISTING - 93.75 PROPOSED - 89.00
MOBILE HOME PARK
EXIST'G 12" CULV. 93.63 (REMOVE)
RECONSTRUCT TO PROPER SLOPE FIELD SET - 5' BOTT., SIDES 2:1
PROPOSED LAKE LEVEL TO THIS POINT - APPROX.
FENCE LINE
B.M. SPK. ELM TREE ELEV. 99.61
BRUCE LAKE WATER ELEV. 83.36
PROPOSED DRAINAGE DITCH RECONSTRUCTION
SCALE - APPROX. 2" = 1/2 MI.
(THIS SECTION BY VILLAGE PRIOR TO UTILITY CONST.)
SANITARY SEWER EXT. W/LIFT STATION
BRUCE, WISCONSIN
SCALE HORZ. 1"=50' VERT. 1"=5'
SHEET 3 OF 7
DATE AUG. 1970
GEO. H. MORGAN PROFESSIONAL ENGINEER LADYSMITH, WISCONSIN
DR. BY R.O. PARMLEY
UTILITY EXTENSION PROJECT
NOTE - ALL GRADING OF MOBILE HOME PARK WILL BE COMPLETED PRIOR TO SEWER AND WATER INSTALLATION

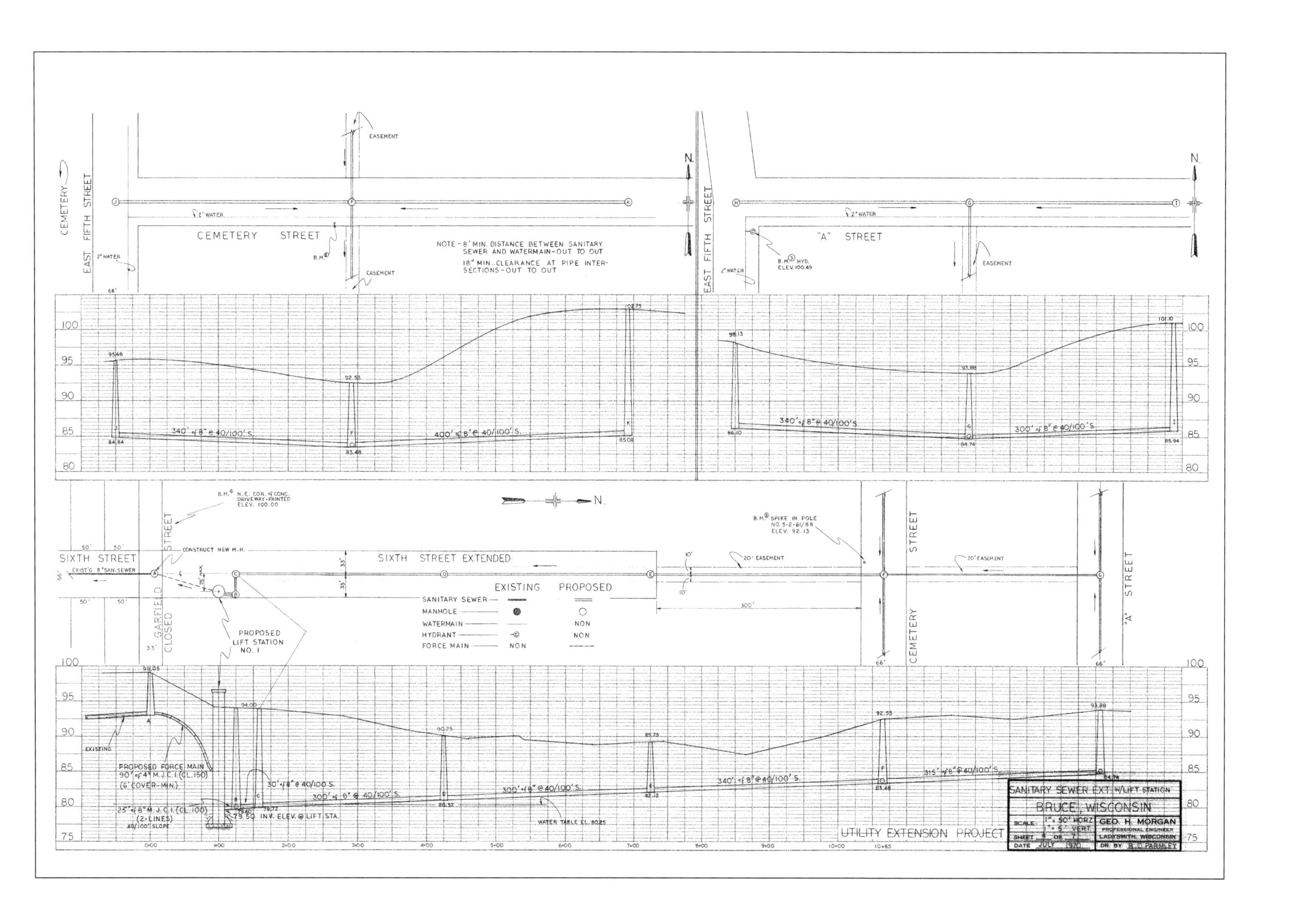
CEMETERY STREET
EAST FIFTH STREET
"A" STREET
NOTE - 8' MIN. DISTANCE BETWEEN SANITARY SEWER AND WATERMAIN-OUT TO OUT
18" MIN. CLEARANCE AT PIPE INTERSECTIONS-OUT TO OUT
EASEMENT
2" WATER
340' of 8" @ 40/100'S.
400' of 8" @ 40/100'S.
300' of 8" @ 40/100'S.
SIXTH STREET
SIXTH STREET EXTENDED
GARFIELD STREET CLOSED
CONSTRUCT NEW M.H.
PROPOSED LIFT STATION NO. 1
20' EASEMENT
EXISTING PROPOSED
SANITARY SEWER
MANHOLE
WATERMAIN NON
HYDRANT NON
FORCE MAIN NON
PROPOSED FORCE MAIN 90' of 4" M.J.C.I. (CL.150) (6' COVER-MIN.)
25' of 8" M.J.C.I. (CL.100) (2-LINES) 40/100' SLOPE
30' of 8" @ 40/100 S.
79.50 INV. ELEV. @ LIFT STA.
WATER TABLE EL. 80.25
315' of 8" @ 40/100'S.
SANITARY SEWER EXT. W/LIFT STATION
BRUCE, WISCONSIN
UTILITY EXTENSION PROJECT
GEO. H. MORGAN
PROFESSIONAL ENGINEER
LADYSMITH, WISCONSIN
DATE JULY 1970
DR. BY R.O. PARMLEY

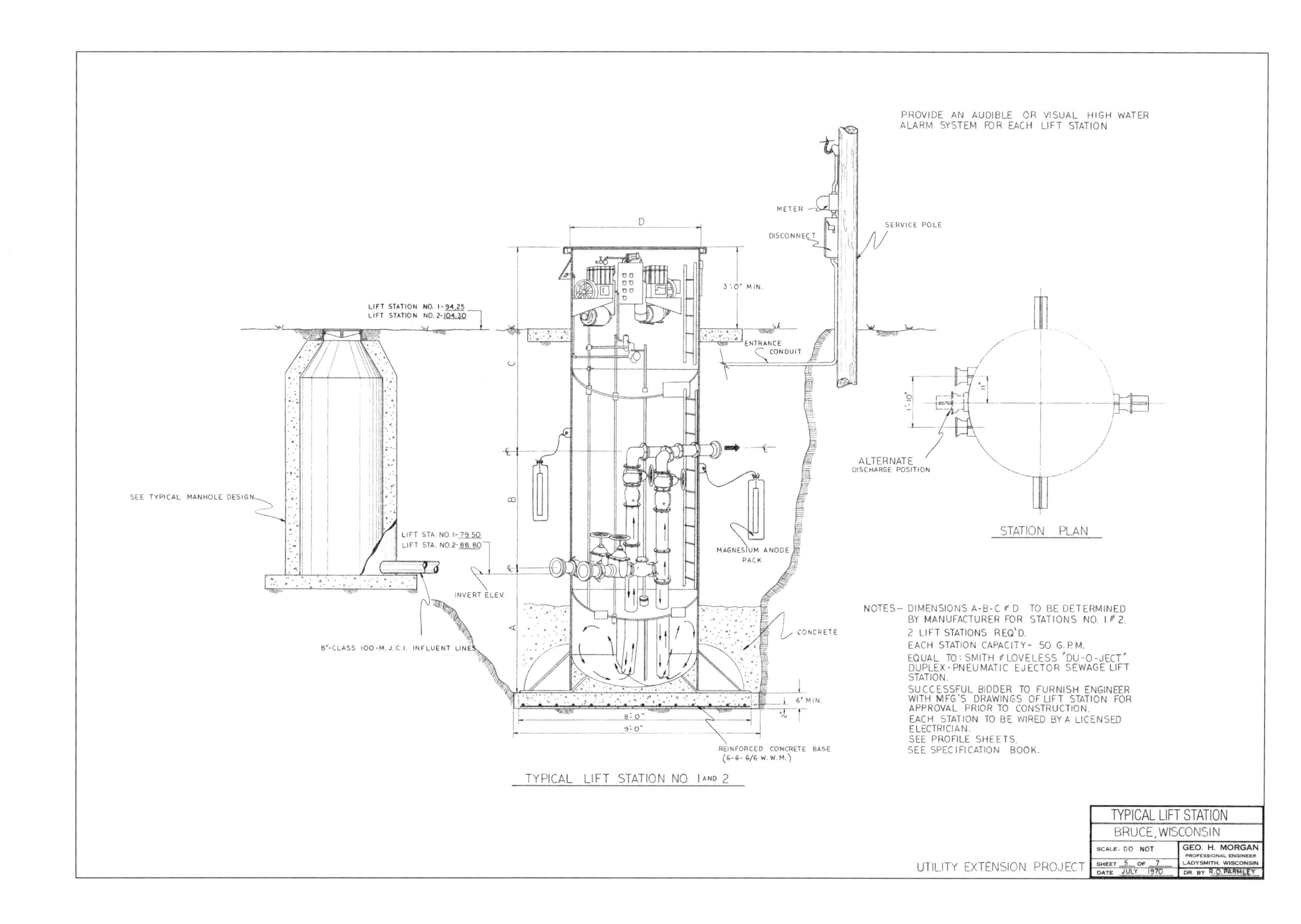

PROVIDE AN AUDIBLE OR VISUAL HIGH WATER ALARM SYSTEM FOR EACH LIFT STATION
METER
DISCONNECT
SERVICE POLE
D
3'-0" MIN.
LIFT STATION NO. 1-94.25
LIFT STATION NO. 2-104.30
ENTRANCE CONDUIT
C
B
A
SEE TYPICAL MANHOLE DESIGN
LIFT STA. NO. 1-79.50
LIFT STA. NO. 2-88.80
INVERT ELEV.
MAGNESIUM ANODE PACK
8"-CLASS 100-M. J. C. I. INFLUENT LINES
CONCRETE
6" MIN.
8'-0"
9'-0"
REINFORCED CONCRETE BASE (6-6- 6/6 W. W. M.)
TYPICAL LIFT STATION NO. 1 AND 2
1'-10"
11"
ALTERNATE DISCHARGE POSITION
STATION PLAN
NOTES– DIMENSIONS A-B-C & D TO BE DETERMINED BY MANUFACTURER FOR STATIONS NO. 1 & 2.
2 LIFT STATIONS REQ'D.
EACH STATION CAPACITY- 50 G.P.M.
EQUAL TO: SMITH & LOVELESS "DU-O-JECT" DUPLEX-PNEUMATIC EJECTOR SEWAGE LIFT STATION.
SUCCESSFUL BIDDER TO FURNISH ENGINEER WITH MFG'S DRAWINGS OF LIFT STATION FOR APPROVAL PRIOR TO CONSTRUCTION.
EACH STATION TO BE WIRED BY A LICENSED ELECTRICIAN.
SEE PROFILE SHEETS.
SEE SPECIFICATION BOOK.
TYPICAL LIFT STATION
BRUCE, WISCONSIN
UTILITY EXTENSION PROJECT
SCALE- DO NOT
SHEET 5 OF 7
DATE JULY 1970
GEO. H. MORGAN
PROFESSIONAL ENGINEER
LADYSMITH, WISCONSIN
DR. BY R.O. PARMLEY

SHELDON TRAILER PARK

NONDEPENDENT MOBILE HOME PARK - OWER: J. HAMMEN

SANITARY SEWER SYSTEM, WELL, WATER DISTRIBUTION SYSTEM, ELECTRICAL PLAN, SERVICE CONNECTIONS & PARK DESIGN

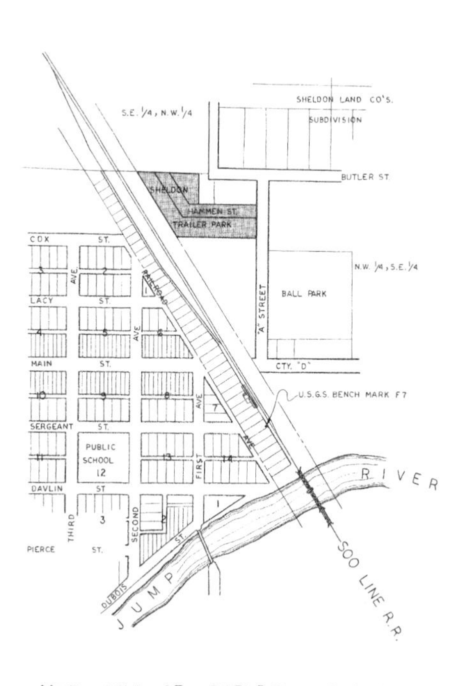

MAP of VILLAGE of SHELDON, WISCONSIN

INDEX of SHEETS

SHEET NO.	DESCRIPTION
1	TITLE SHEET, LOCATION MAP AND INDEX
2	SANITARY SEWER w/ PROFILES AND ISOMETRIC
3	WELL & WATER SYSTEM w/ TYP RISERS
4	ELECTRICAL PLAN
5	TYPICAL SEWER DETAILS

PROPOSED UTILITIES	
SHELDON TRAILER PARK SHELDON, WI	
SCALE- 1"=300'	GEO. H. MORGAN PROFESSIONAL ENGINEER LADYSMITH, WISCONSIN
SHEET 1 OF 5	
DATE OCT. 1975	DR. BY R.O. PARMLEY

Courtesy: Morgan & Parmley, Ltd.

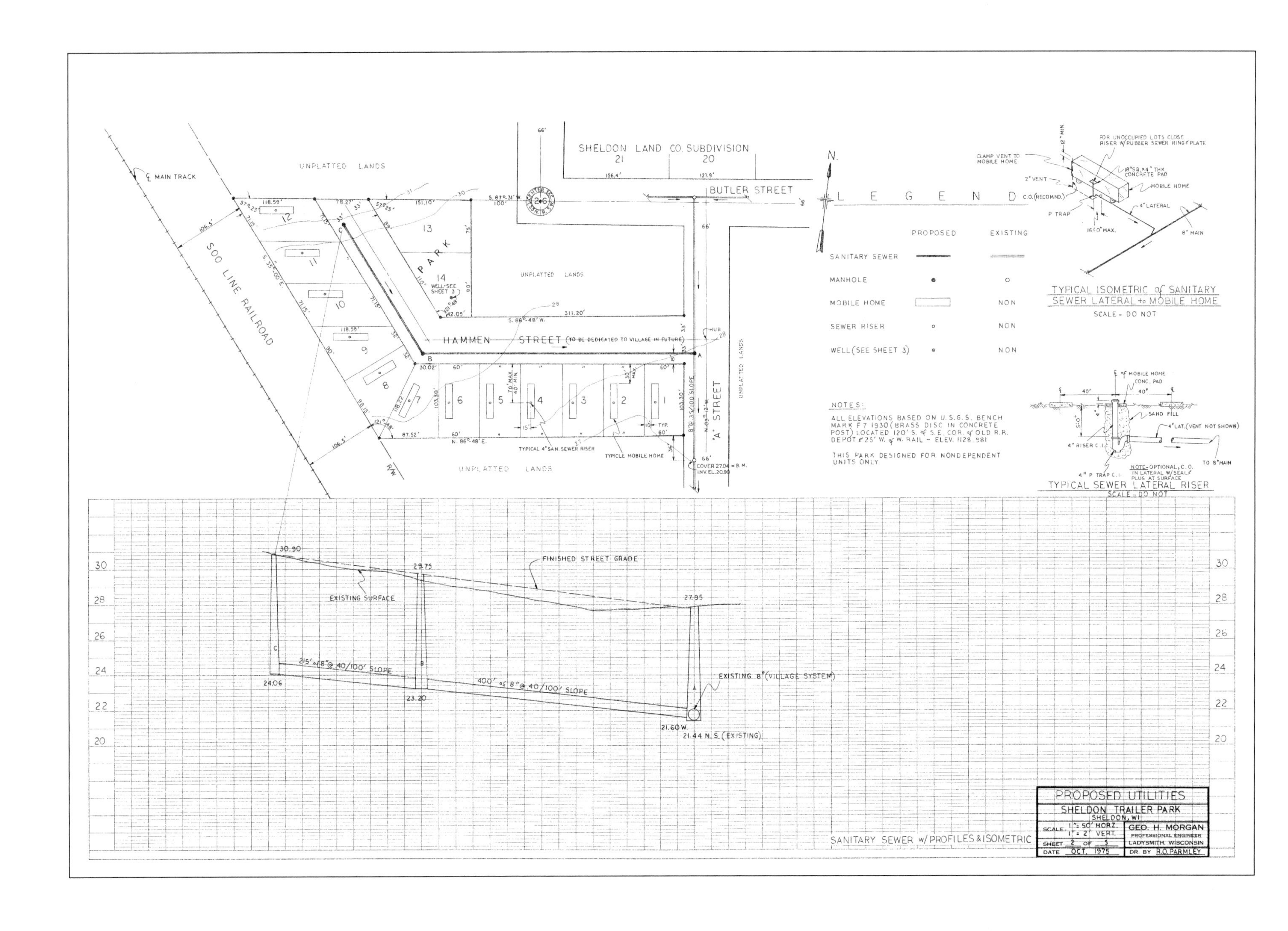
SHELDON LAND CO. SUBDIVISION
BUTLER STREET
UNPLATTED LANDS
SOO LINE RAILROAD
℄ MAIN TRACK
PARK
HAMMEN STREET (TO BE DEDICATED TO VILLAGE IN FUTURE)
"A" STREET
TYPICAL 4" SAN. SEWER RISER
TYPICLE MOBILE HOME
COVER 27.04 = B.M.
INV. EL. 20.90
L E G E N D
PROPOSED
EXISTING
SANITARY SEWER
MANHOLE
MOBILE HOME
SEWER RISER
WELL (SEE SHEET 3)
NON
NOTES:
ALL ELEVATIONS BASED ON U.S.G.S. BENCH MARK F7 1930 (BRASS DISC IN CONCRETE POST) LOCATED 120' S. of S.E. COR. of OLD R.R. DEPOT & 25' W. of W. RAIL - ELEV. 1128.981
THIS PARK DESIGNED FOR NONDEPENDENT UNITS ONLY
TYPICAL ISOMETRIC of SANITARY SEWER LATERAL to MOBILE HOME
SCALE - DO NOT
FOR UNOCCUPIED LOTS CLOSE RISER W/RUBBER SEWER RING & PLATE
CLAMP VENT TO MOBILE HOME
2" VENT
C.O. (RECOMND.)
P TRAP
18"SQ.x4" THK. CONCRETE PAD
MOBILE HOME
4" LATERAL
8" MAIN
TYPICAL SEWER LATERAL RISER
SCALE - DO NOT
SAND FILL
4" LAT. (VENT NOT SHOWN)
4" RISER C.I.
4" P TRAP C.I.
TO 8" MAIN
NOTE - OPTIONAL C.O. IN LATERAL W/SEAL & PLUG AT SURFACE
FINISHED STREET GRADE
EXISTING SURFACE
215' of 8" @ .40/100' SLOPE
400' of 8" @ .40/100' SLOPE
EXISTING 8" (VILLAGE SYSTEM)
30.90
29.75
27.95
24.06
23.20
21.60 W.
21.44 N.S. (EXISTING)
PROPOSED UTILITIES
SHELDON TRAILER PARK
SHELDON, WI.
SCALE: 1" = 50' HORZ. 1" = 2' VERT.
SHEET 2 OF 5
DATE OCT. 1975
GEO. H. MORGAN
PROFESSIONAL ENGINEER
LADYSMITH, WISCONSIN
DR. BY R.O. PARMLEY
SANITARY SEWER w/ PROFILES & ISOMETRIC

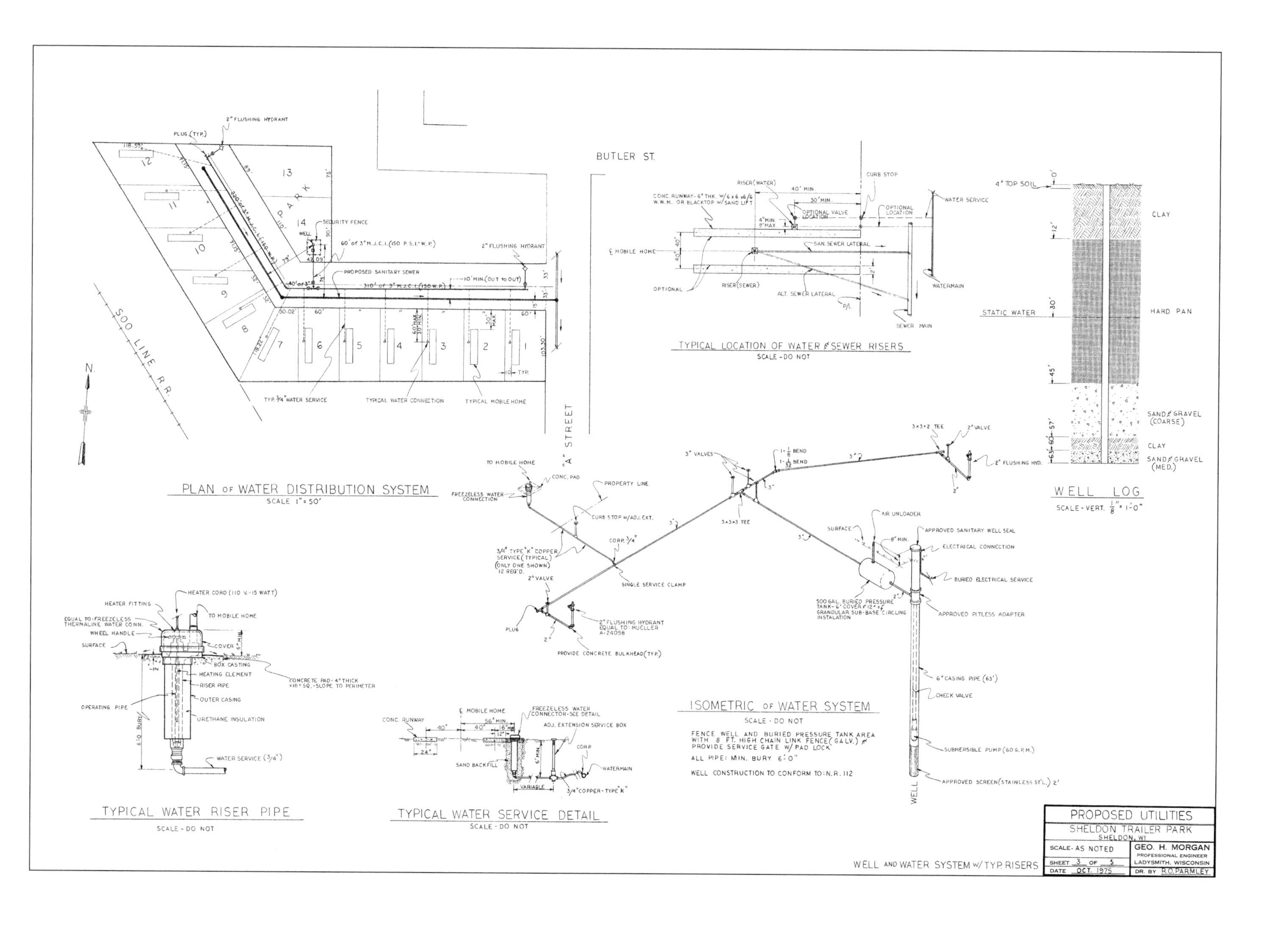
PLAN OF WATER DISTRIBUTION SYSTEM
SCALE 1" = 50'
BUTLER ST.
PARK
SOO LINE R.R.
"A" STREET
2" FLUSHING HYDRANT
PLUG (TYP.)
SECURITY FENCE
WELL
60' of 3" M.J.C.I. (150 P.S.I. W.P.)
PROPOSED SANITARY SEWER
310' of 3" M.J.C.I. (150 W.P.)
10' MIN. (OUT to OUT)
TYP. 3/4" WATER SERVICE
TYPICAL WATER CONNECTION
TYPICAL MOBILE HOME
TYPICAL LOCATION OF WATER & SEWER RISERS
SCALE - DO NOT
RISER (WATER)
CURB STOP
CONC. RUNWAY - 6" THK. W/ 6x6 x6/6 W.W.M. OR BLACKTOP W/ SAND LIFT
OPTIONAL VALVE LOCATION
WATER SERVICE
SAN. SEWER LATERAL
RISER (SEWER)
ALT. SEWER LATERAL
WATERMAIN
SEWER MAIN
OPTIONAL
STATIC WATER
WELL LOG
SCALE - VERT. 1/8" = 1'-0"
4" TOP SOIL
CLAY
HARD PAN
SAND & GRAVEL (COARSE)
SAND & GRAVEL (MED.)
ISOMETRIC OF WATER SYSTEM
SCALE - DO NOT
FENCE WELL AND BURIED PRESSURE TANK AREA WITH 8 FT. HIGH CHAIN LINK FENCE (GALV.) & PROVIDE SERVICE GATE W/ PAD LOCK
ALL PIPE: MIN. BURY 6'-0"
WELL CONSTRUCTION TO CONFORM TO: N.R. 112
3x3x2 TEE
2" VALVE
3" VALVES
3x3x3 TEE
2" FLUSHING HYD.
TO MOBILE HOME
CONC. PAD
PROPERTY LINE
FREEZELESS WATER CONNECTION
CURB STOP w/ADJ. EXT.
CORP. 3/4"
3/4" TYPE "K" COPPER SERVICE (TYPICAL) (ONLY ONE SHOWN) 12 REQ'D.
SINGLE SERVICE CLAMP
2" FLUSHING HYDRANT EQUAL TO: MUELLER A-24058
PROVIDE CONCRETE BULKHEAD (TYP.)
AIR UNLOADER
APPROVED SANITARY WELL SEAL
ELECTRICAL CONNECTION
BURIED ELECTRICAL SERVICE
500 GAL. BURIED PRESSURE TANK - 6' COVER & 12" of GRANDULAR SUB-BASE CIRCLING INSTALATION
APPROVED PITLESS ADAPTER
6" CASING PIPE (63')
CHECK VALVE
SUBMERSIBLE PUMP (60 G.P.M.)
APPROVED SCREEN (STAINLESS ST'L.) 2'
WELL
TYPICAL WATER RISER PIPE
SCALE - DO NOT
HEATER CORD (110 V. - 15 WATT)
HEATER FITTING
EQUAL TO: FREEZELESS THERMALINE WATER CONN.
WHEEL HANDLE
SURFACE
TO MOBILE HOME
COVER
BOX CASTING
HEATING ELEMENT
RISER PIPE
OUTER CASING
URETHANE INSULATION
OPERATING PIPE
CONCRETE PAD - 4" THICK x18" SQ. - SLOPE TO PERIMETER
WATER SERVICE (3/4")
TYPICAL WATER SERVICE DETAIL
SCALE - DO NOT
℄ MOBILE HOME
CONC. RUNWAY
FREEZELESS WATER CONNECTOR - SEE DETAIL
ADJ. EXTENSION SERVICE BOX
SAND BACKFILL
CORP.
WATERMAIN
VARIABLE
3/4" COPPER - TYPE "K"
WELL AND WATER SYSTEM W/ TYP. RISERS
PROPOSED UTILITIES
SHELDON TRAILER PARK
SHELDON, WI
SCALE - AS NOTED
SHEET 3 OF 5
DATE OCT. 1975
GEO. H. MORGAN
PROFESSIONAL ENGINEER
LADYSMITH, WISCONSIN
DR. BY R.O. PARMLEY

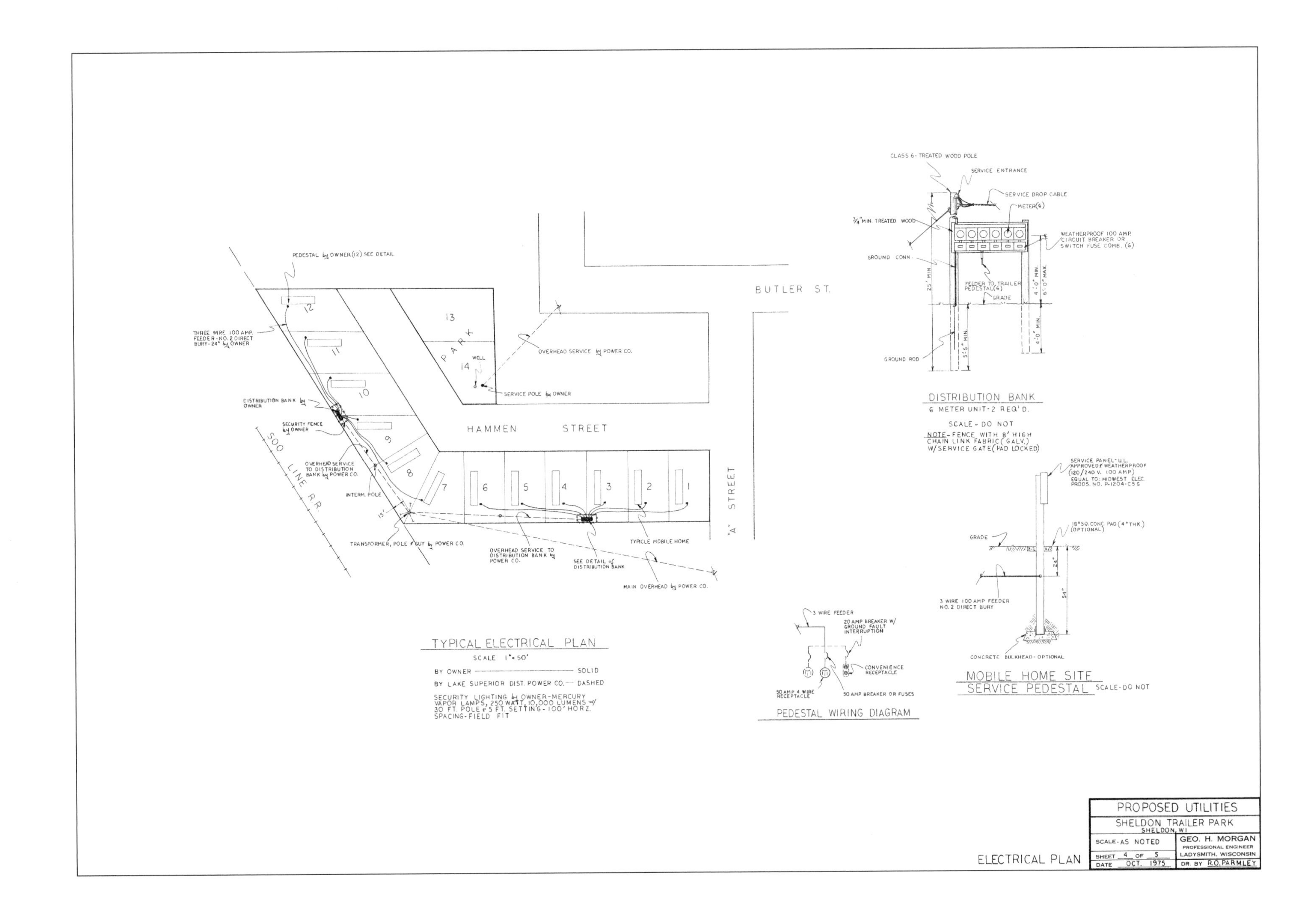

TYPICAL ELECTRICAL PLAN
SCALE 1"=50'
BY OWNER — SOLID
BY LAKE SUPERIOR DIST. POWER CO. — DASHED
SECURITY LIGHTING by OWNER-MERCURY VAPOR LAMPS, 250 WATT, 10,000 LUMENS w/ 30 FT. POLE & 5 FT. SETTING - 100' HORZ. SPACING - FIELD FIT
BUTLER ST.
HAMMEN STREET
"A" STREET
SOO LINE RR.
PARK
DISTRIBUTION BANK
6 METER UNIT-2 REQ'D.
SCALE - DO NOT
NOTE - FENCE WITH 8' HIGH CHAIN LINK FABRIC (GALV.) W/SERVICE GATE (PAD LOCKED)
PEDESTAL WIRING DIAGRAM
MOBILE HOME SITE SERVICE PEDESTAL SCALE-DO NOT
PROPOSED UTILITIES
SHELDON TRAILER PARK
SHELDON, WI
SCALE-AS NOTED
SHEET 4 OF 5
DATE OCT. 1975
GEO. H. MORGAN
PROFESSIONAL ENGINEER
LADYSMITH, WISCONSIN
DR. BY R.O. PARMLEY
ELECTRICAL PLAN

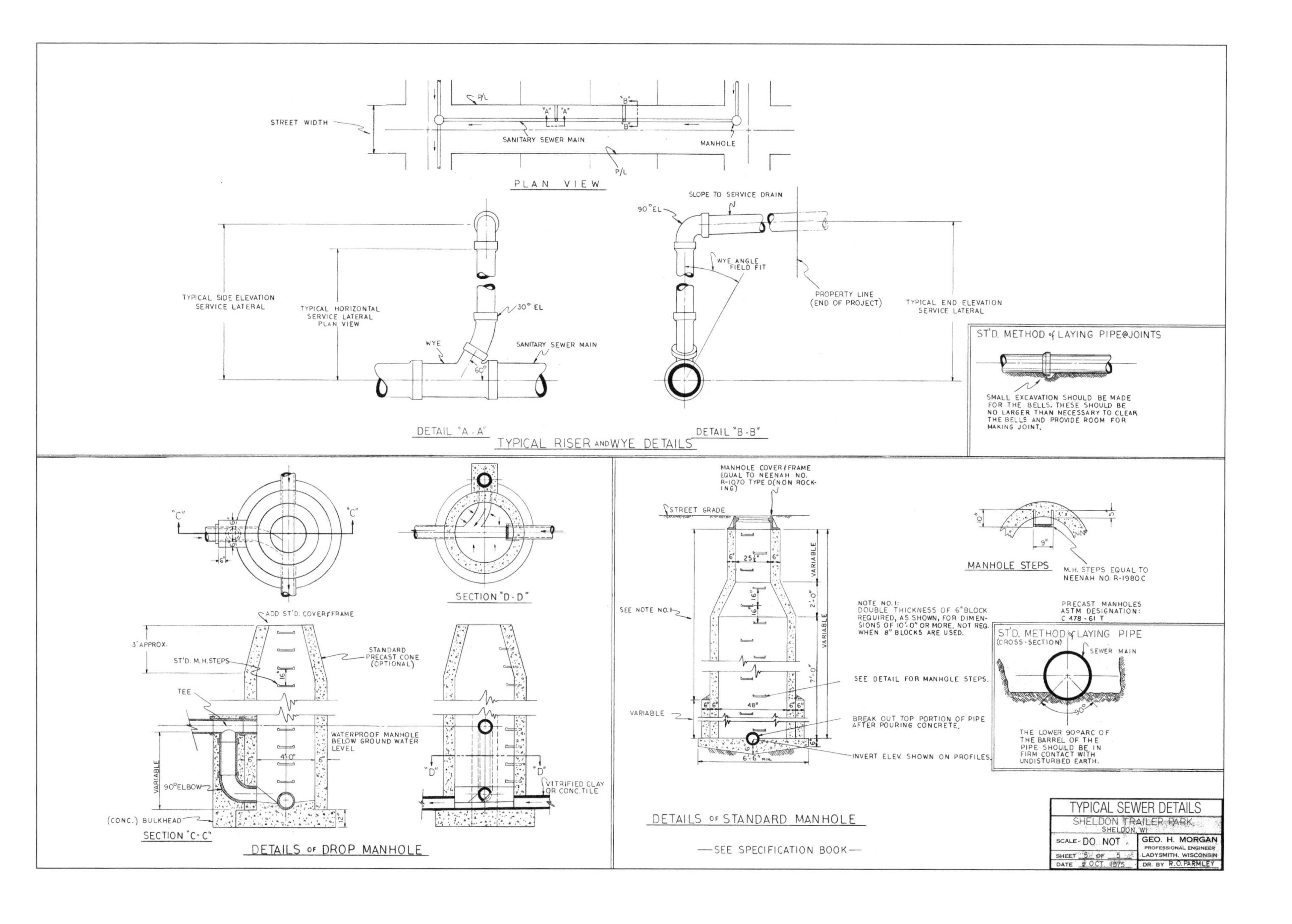
STREET WIDTH
P/L
"A" "A"
"B" "B"
SANITARY SEWER MAIN
MANHOLE
PLAN VIEW
SLOPE TO SERVICE DRAIN
90° EL
WYE ANGLE FIELD FIT
PROPERTY LINE (END OF PROJECT)
TYPICAL SIDE ELEVATION SERVICE LATERAL
TYPICAL HORIZONTAL SERVICE LATERAL PLAN VIEW
30° EL
WYE
SANITARY SEWER MAIN
60°
TYPICAL END ELEVATION SERVICE LATERAL
ST'D. METHOD of LAYING PIPE@JOINTS
SMALL EXCAVATION SHOULD BE MADE FOR THE BELLS. THESE SHOULD BE NO LARGER THAN NECESSARY TO CLEAR THE BELLS AND PROVIDE ROOM FOR MAKING JOINT.
DETAIL "A-A"
DETAIL "B-B"
TYPICAL RISER AND WYE DETAILS
"C"
"C"
SECTION "D-D"
ADD ST'D. COVER & FRAME
3' APPROX.
ST'D. M.H. STEPS
STANDARD PRECAST CONE (OPTIONAL)
TEE
WATERPROOF MANHOLE BELOW GROUND WATER LEVEL
4'-0"
VARIABLE
90° ELBOW
(CONC.) BULKHEAD
"D"
"D"
VITRIFIED CLAY OR CONC. TILE
SECTION "C-C"
DETAILS of DROP MANHOLE
MANHOLE COVER & FRAME EQUAL TO NEENAH NO. R-1070 TYPE D(NON ROCKING)
STREET GRADE
SEE NOTE NO.1
VARIABLE
2'-0"
7'-0"
48"
6'-6" MIN.
NOTE NO. 1: DOUBLE THICKNESS OF 6" BLOCK REQUIRED, AS SHOWN, FOR DIMENSIONS OF 10'-0" OR MORE. NOT REQ. WHEN 8" BLOCKS ARE USED.
SEE DETAIL FOR MANHOLE STEPS.
BREAK OUT TOP PORTION OF PIPE AFTER POURING CONCRETE.
INVERT ELEV. SHOWN ON PROFILES.
MANHOLE STEPS
M.H. STEPS EQUAL TO NEENAH NO. R-1980C
PRECAST MANHOLES ASTM DESIGNATION: C 478-61 T
ST'D. METHOD of LAYING PIPE (CROSS-SECTION)
SEWER MAIN
90°
THE LOWER 90° ARC OF THE BARREL OF THE PIPE SHOULD BE IN FIRM CONTACT WITH UNDISTURBED EARTH.
DETAILS of STANDARD MANHOLE
—SEE SPECIFICATION BOOK—
TYPICAL SEWER DETAILS
SHELDON TRAILER PARK
SHELDON, WI
SCALE- DO NOT
SHEET 5 OF 5
DATE 2 OCT 1975
GEO. H. MORGAN
PROFESSIONAL ENGINEER
LADYSMITH, WISCONSIN
DR. BY R.O. PARMLEY

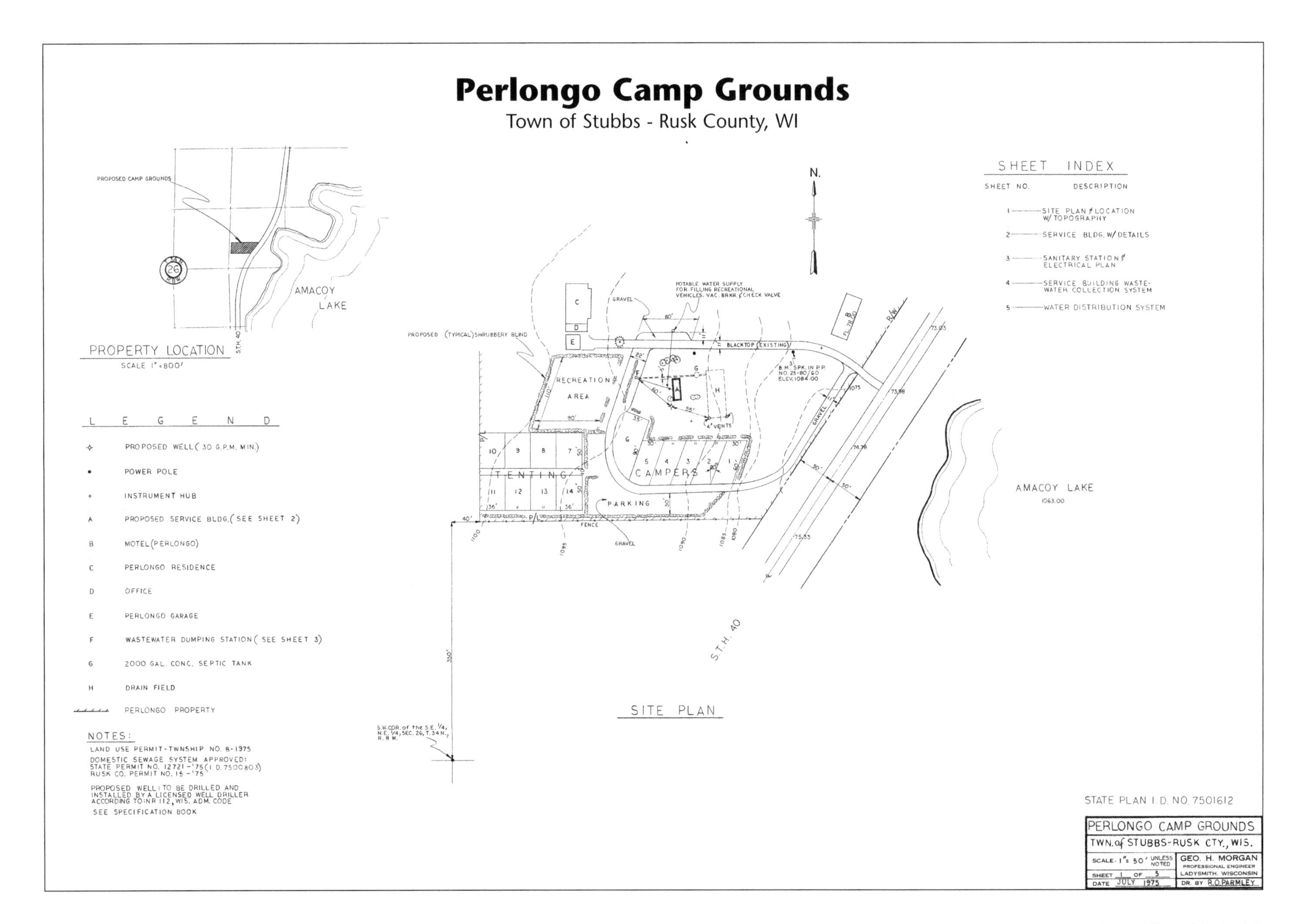

Courtesy: Morgan & Parmley, Ltd.

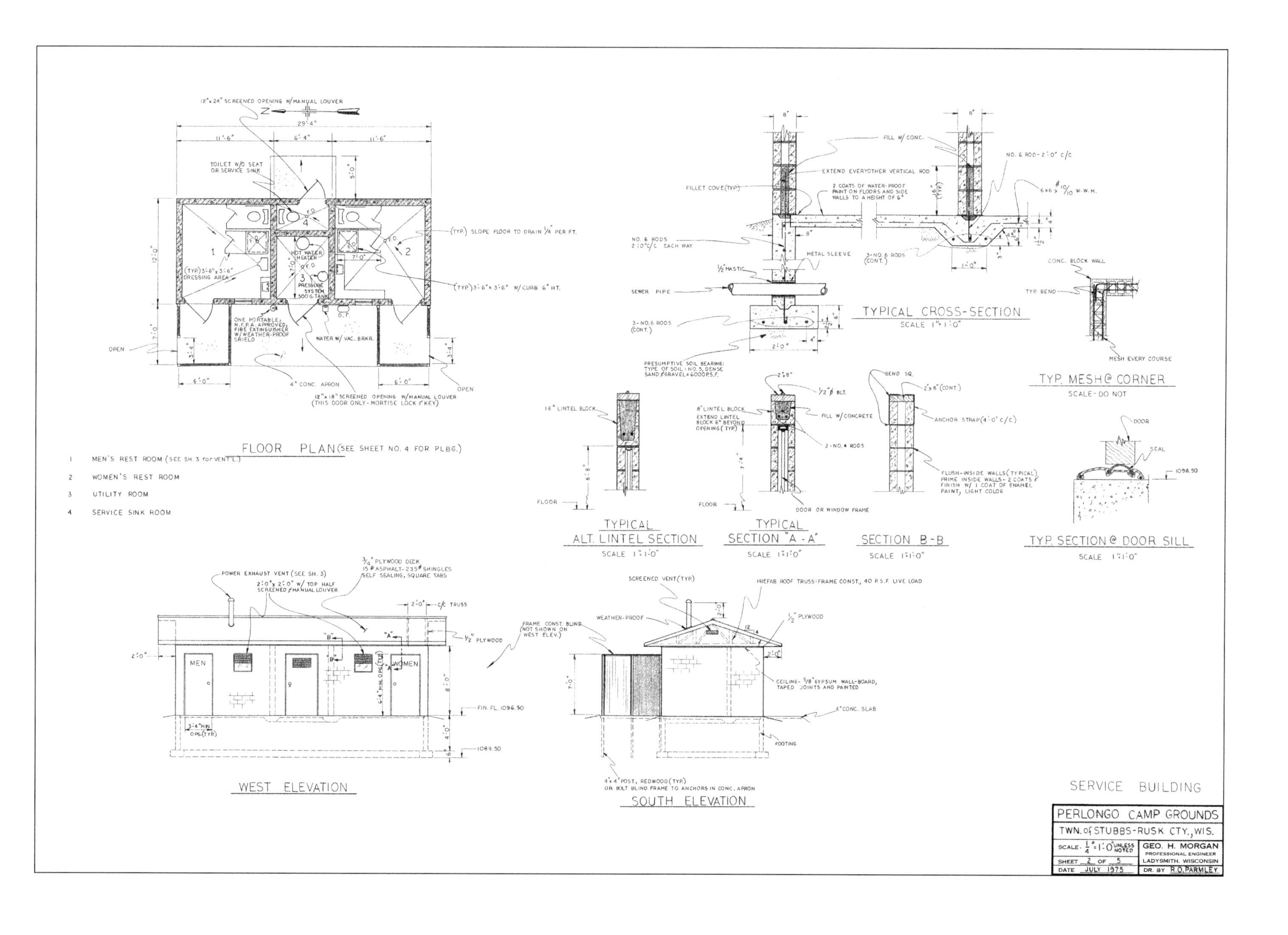
FLOOR PLAN (SEE SHEET NO. 4 FOR PLBG.)
1 MEN'S REST ROOM (SEE SH. 3 for VENT'L)
2 WOMEN'S REST ROOM
3 UTILITY ROOM
4 SERVICE SINK ROOM
12"x 24" SCREENED OPENING W/MANUAL LOUVER
TOILET W/O SEAT OR SERVICE SINK
(TYP) SLOPE FLOOR TO DRAIN 1/4" PER FT.
(TYP) 3'-6"x 3'-6" DRESSING AREA
HOT WATER HEATER
PRESSURE SYSTEM 500 G. TANK
(TYP) 3'-6"x 3'-6" W/CURB 6" HT.
ONE PORTABLE, N.F.P.A. APPROVED, FIRE EXTINGUISHER W/WEATHER-PROOF SHIELD
WATER W/ VAC. BRKR.
OPEN
4" CONC. APRON
12"x 18" SCREENED OPENING W/MANUAL LOUVER (THIS DOOR ONLY-MORTISE LOCK & KEY)
TYPICAL CROSS-SECTION
SCALE 1"=1'-0"
FILL W/ CONC.
EXTEND EVERYOTHER VERTICAL ROD
FILLET COVE (TYP)
2 COATS OF WATER-PROOF PAINT ON FLOORS AND SIDE WALLS TO A HEIGHT OF 6"
NO. 6 ROD-2'-0" C/C
NO. 6 RODS 2'-0" C/C EACH WAY
METAL SLEEVE
3-NO. 6 RODS (CONT.)
1/2" MASTIC
SEWER PIPE
3-NO. 6 RODS (CONT.)
PRESUMPTIVE SOIL BEARING: TYPE OF SOIL: NO. 5, DENSE SAND & GRAVEL = 6000 P.S.F.
CONC. BLOCK WALL
TYP. BEND
MESH EVERY COURSE
TYP. MESH @ CORNER
SCALE - DO NOT
16" LINTEL BLOCK
FLOOR
TYPICAL ALT. LINTEL SECTION
SCALE 1"=1'-0"
8" LINTEL BLOCK
EXTEND LINTEL BLOCK 6" BEYOND OPENING (TYP.)
1/2" ∅ BLT.
FILL W/ CONCRETE
2-NO. 4 RODS
DOOR OR WINDOW FRAME
TYPICAL SECTION "A-A"
SCALE 1"=1'-0"
BEND SQ.
2"x 8" (CONT.)
ANCHOR STRAP (4'-0" C/C)
FLUSH-INSIDE WALLS (TYPICAL), PRIME INSIDE WALLS - 2 COATS & FINISH W/ 1 COAT OF ENAMEL PAINT, LIGHT COLOR
SECTION B-B
SCALE 1"=1'-0"
DOOR
SEAL
1096.50
TYP. SECTION @ DOOR SILL
SCALE 1"=1'-0"
POWER EXHAUST VENT (SEE SH. 3)
2'-0"x 2'-0" W/ TOP HALF SCREENED & MANUAL LOUVER
3/4" PLYWOOD DECK
15# ASPHALT-235# SHINGLES SELF SEALING, SQUARE TABS
C/L TRUSS
1/2" PLYWOOD
MEN
WOMEN
FIN. FL. 1096.50
1089.50
3'-4" MIN. OPG. (TYP.)
WEST ELEVATION
FRAME CONST. BLIND (NOT SHOWN ON WEST ELEV.)
SCREENED VENT (TYP.)
WEATHER-PROOF
PREFAB ROOF TRUSS: FRAME CONST., 40 P.S.F. LIVE LOAD
1/2" PLYWOOD
CEILING - 5/8" GYPSUM WALL-BOARD, TAPED JOINTS AND PAINTED
4" CONC. SLAB
FOOTING
4"x 4" POST, REDWOOD (TYP.) OR BOLT BLIND FRAME TO ANCHORS IN CONC. APRON
SOUTH ELEVATION
SERVICE BUILDING
PERLONGO CAMP GROUNDS
TWN. of STUBBS-RUSK CTY., WIS.
SCALE - 1/4"=1'-0" UNLESS NOTED
GEO. H. MORGAN
PROFESSIONAL ENGINEER
LADYSMITH, WISCONSIN
SHEET 2 OF 5
DATE JULY 1975
DR. BY R.O. PARMLEY

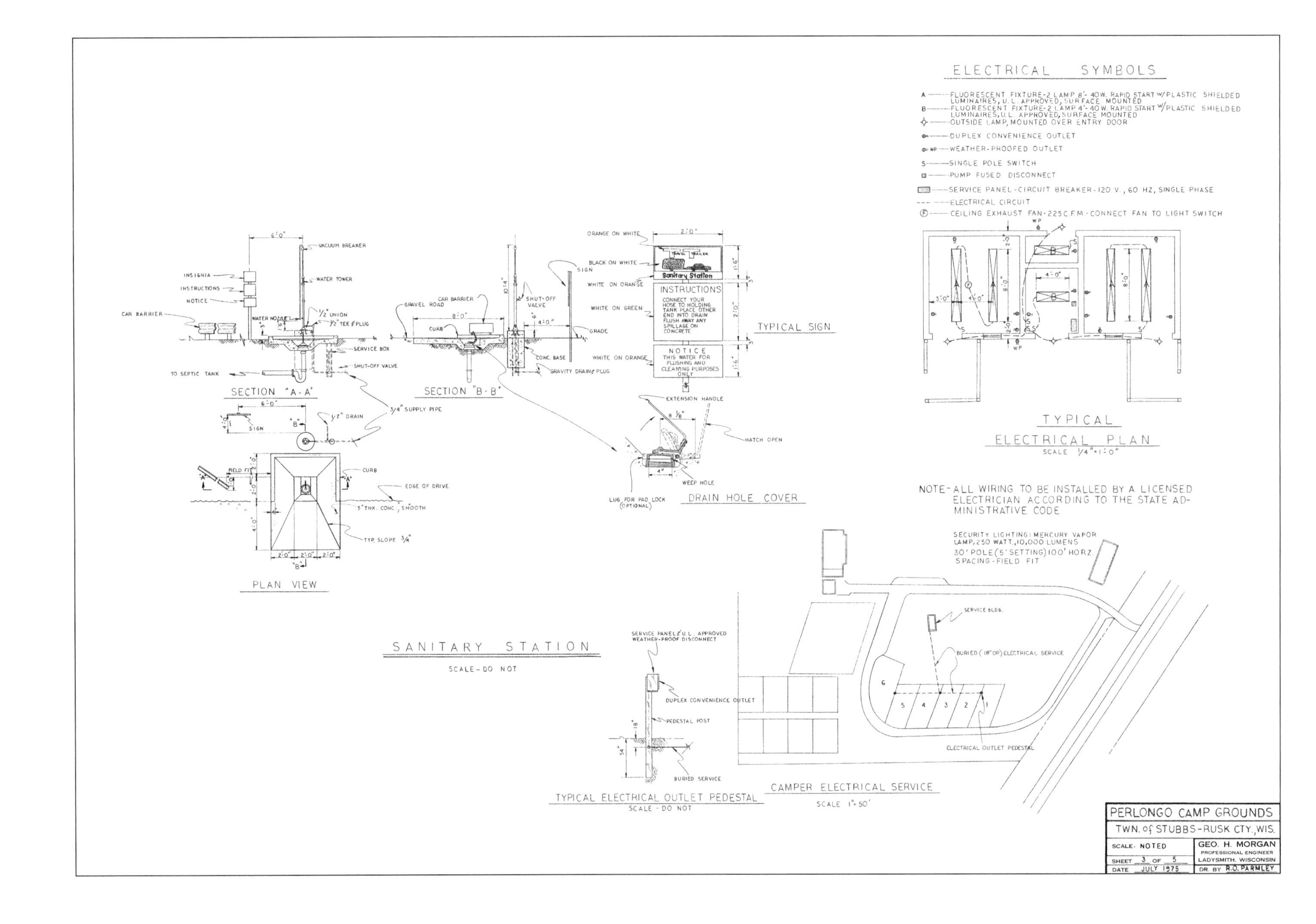
ELECTRICAL SYMBOLS
A — FLUORESCENT FIXTURE-2 LAMP 8'-40W. RAPID START W/PLASTIC SHIELDED LUMINAIRES, U.L. APPROVED, SURFACE MOUNTED
B — FLUORESCENT FIXTURE-2 LAMP 4'-40W. RAPID START W/PLASTIC SHIELDED LUMINAIRES, U.L. APPROVED, SURFACE MOUNTED
OUTSIDE LAMP, MOUNTED OVER ENTRY DOOR
DUPLEX CONVENIENCE OUTLET
WP — WEATHER-PROOFED OUTLET
S — SINGLE POLE SWITCH
PUMP FUSED DISCONNECT
SERVICE PANEL-CIRCUIT BREAKER-120 V., 60 HZ, SINGLE PHASE
ELECTRICAL CIRCUIT
F — CEILING EXHAUST FAN-225 C.F.M.-CONNECT FAN TO LIGHT SWITCH
TYPICAL ELECTRICAL PLAN
SCALE 1/4"=1'-0"
NOTE-ALL WIRING TO BE INSTALLED BY A LICENSED ELECTRICIAN ACCORDING TO THE STATE ADMINISTRATIVE CODE
SECURITY LIGHTING: MERCURY VAPOR LAMP, 250 WATT, 10,000 LUMENS
30' POLE (5' SETTING) 100' HORZ. SPACING-FIELD FIT
VACUUM BREAKER
WATER TOWER
INSIGNIA
INSTRUCTIONS
NOTICE
CAR BARRIER
WATER NOZZLE
1/2" UNION
1/2" TEE & PLUG
SERVICE BOX
SHUT-OFF VALVE
TO SEPTIC TANK
SECTION "A-A"
GRAVEL ROAD
CURB
SIGN
GRADE
CONC. BASE
GRAVITY DRAIN & PLUG
SECTION "B-B"
3/4" SUPPLY PIPE
1/2" DRAIN
FIELD FIT
EDGE OF DRIVE
5" THK. CONC., SMOOTH
TYP. SLOPE 3/4"
PLAN VIEW
ORANGE ON WHITE
BLACK ON WHITE
WHITE ON ORANGE
WHITE ON GREEN
TRAVEL TRAILER
Sanitary Station
INSTRUCTIONS
CONNECT YOUR HOSE TO HOLDING TANK PLACE OTHER END INTO DRAIN FLUSH AWAY ANY SPILLAGE ON CONCRETE
NOTICE
THIS WATER FOR FLUSHING AND CLEANING PURPOSES ONLY
TYPICAL SIGN
EXTENSION HANDLE
HATCH OPEN
WEEP HOLE
LUG FOR PAD LOCK (OPTIONAL)
DRAIN HOLE COVER
SANITARY STATION
SCALE-DO NOT
SERVICE PANEL & U.L. APPROVED WEATHER-PROOF DISCONNECT
DUPLEX CONVENIENCE OUTLET
PEDESTAL POST
BURIED SERVICE
TYPICAL ELECTRICAL OUTLET PEDESTAL
SCALE-DO NOT
SERVICE BLDG.
BURIED (18" DP) ELECTRICAL SERVICE
ELECTRICAL OUTLET PEDESTAL
CAMPER ELECTRICAL SERVICE
SCALE 1"=50'
PERLONGO CAMP GROUNDS
TWN. of STUBBS-RUSK CTY., WIS.
SCALE- NOTED
SHEET 3 OF 5
DATE JULY 1975
GEO. H. MORGAN
PROFESSIONAL ENGINEER
LADYSMITH, WISCONSIN
DR. BY R.O. PARMLEY

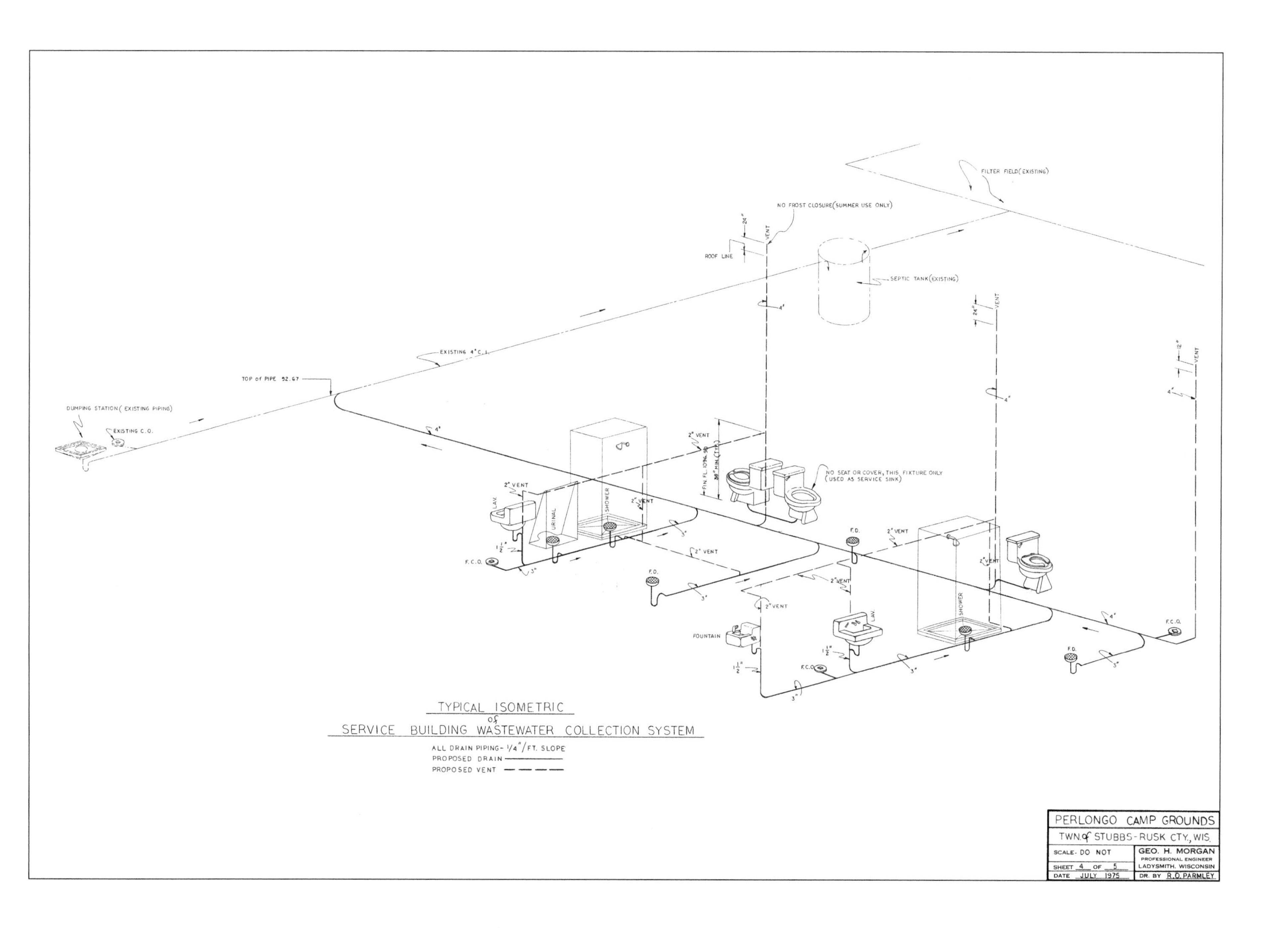

FILTER FIELD (EXISTING)
NO FROST CLOSURE (SUMMER USE ONLY)
ROOF LINE
VENT
SEPTIC TANK (EXISTING)
EXISTING 4" C.I.
TOP of PIPE 52.67
DUMPING STATION (EXISTING PIPING)
EXISTING C.O.
2" VENT
FIN. FL. 1096.90
38" MIN. (TYP.)
NO SEAT OR COVER, THIS FIXTURE ONLY (USED AS SERVICE SINK)
LAV.
URINAL
SHOWER
F.C.O.
F.D.
FOUNTAIN
TYPICAL ISOMETRIC
of
SERVICE BUILDING WASTEWATER COLLECTION SYSTEM
ALL DRAIN PIPING - 1/4"/FT. SLOPE
PROPOSED DRAIN
PROPOSED VENT
PERLONGO CAMP GROUNDS
TWN. of STUBBS - RUSK CTY, WIS.
SCALE - DO NOT
SHEET 4 OF 5
DATE JULY 1975
GEO. H. MORGAN
PROFESSIONAL ENGINEER
LADYSMITH, WISCONSIN
DR. BY R.O. PARMLEY

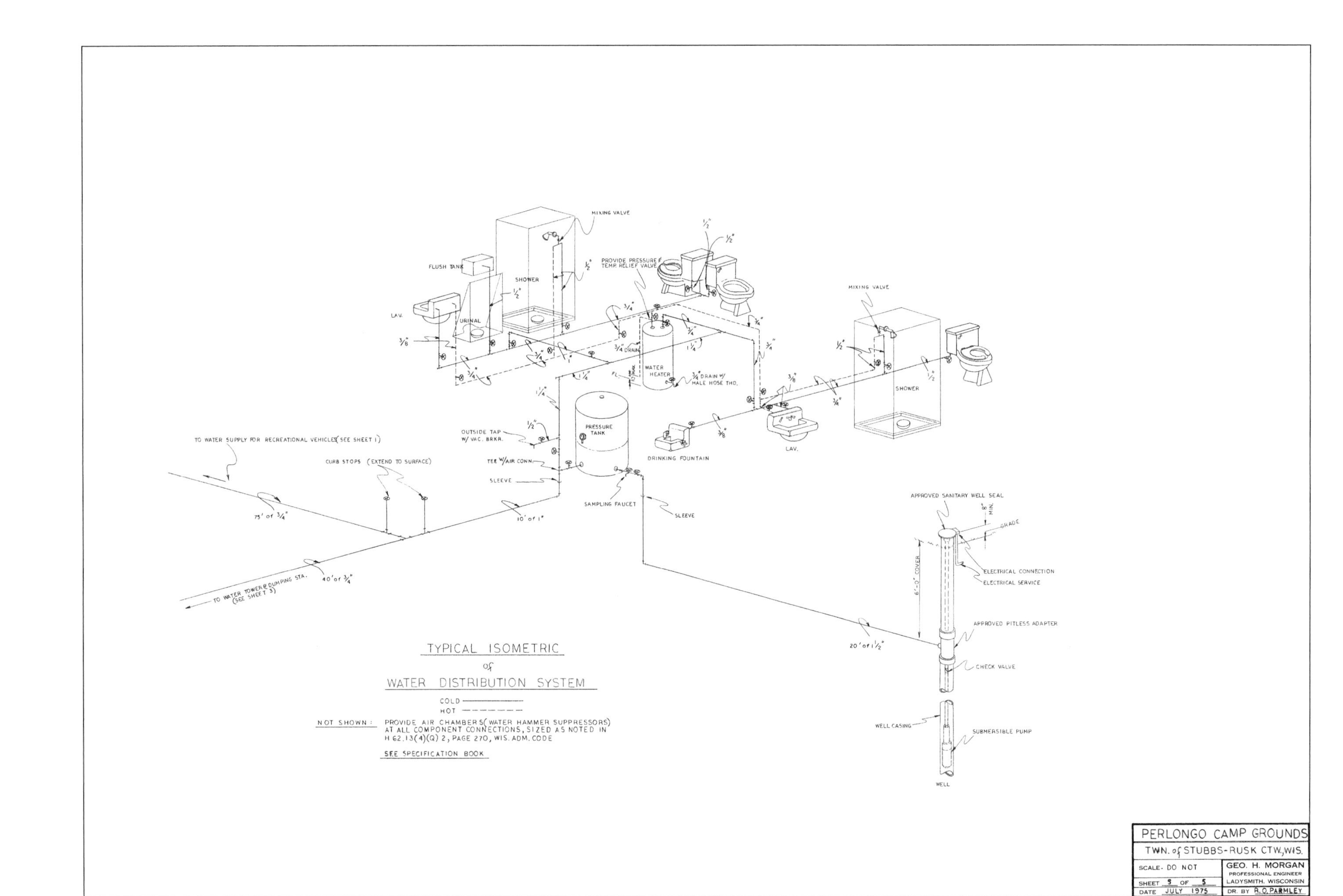
MIXING VALVE
FLUSH TANK
SHOWER
PROVIDE PRESSURE & TEMP. RELIEF VALVE
LAV.
URINAL
WATER HEATER
3/4" DRAIN W/ MALE HOSE THD.
PRESSURE TANK
OUTSIDE TAP W/ VAC. BRKR.
TEE W/AIR CONN.
SLEEVE
SAMPLING FAUCET
DRINKING FOUNTAIN
TO WATER SUPPLY FOR RECREATIONAL VEHICLES (SEE SHEET 1)
CURB STOPS (EXTEND TO SURFACE)
75' of 3/4"
10' of 1"
40' of 3/4"
TO WATER TOWER & DUMPING STA. (SEE SHEET 3)
20' of 1 1/2"
APPROVED SANITARY WELL SEAL
GRADE
6'-0" COVER
ELECTRICAL CONNECTION
ELECTRICAL SERVICE
APPROVED PITLESS ADAPTER
CHECK VALVE
WELL CASING
SUBMERSIBLE PUMP
WELL
TYPICAL ISOMETRIC
of
WATER DISTRIBUTION SYSTEM
COLD
HOT
NOT SHOWN: PROVIDE AIR CHAMBERS (WATER HAMMER SUPPRESSORS) AT ALL COMPONENT CONNECTIONS, SIZED AS NOTED IN H 62.13(4)(Q) 2, PAGE 270, WIS. ADM. CODE
SEE SPECIFICATION BOOK
PERLONGO CAMP GROUNDS
TWN. of STUBBS-RUSK CTW., WIS.
SCALE- DO NOT
SHEET 5 OF 5
DATE JULY 1975
GEO. H. MORGAN
PROFESSIONAL ENGINEER
LADYSMITH, WISCONSIN
DR. BY R.O. PARMLEY

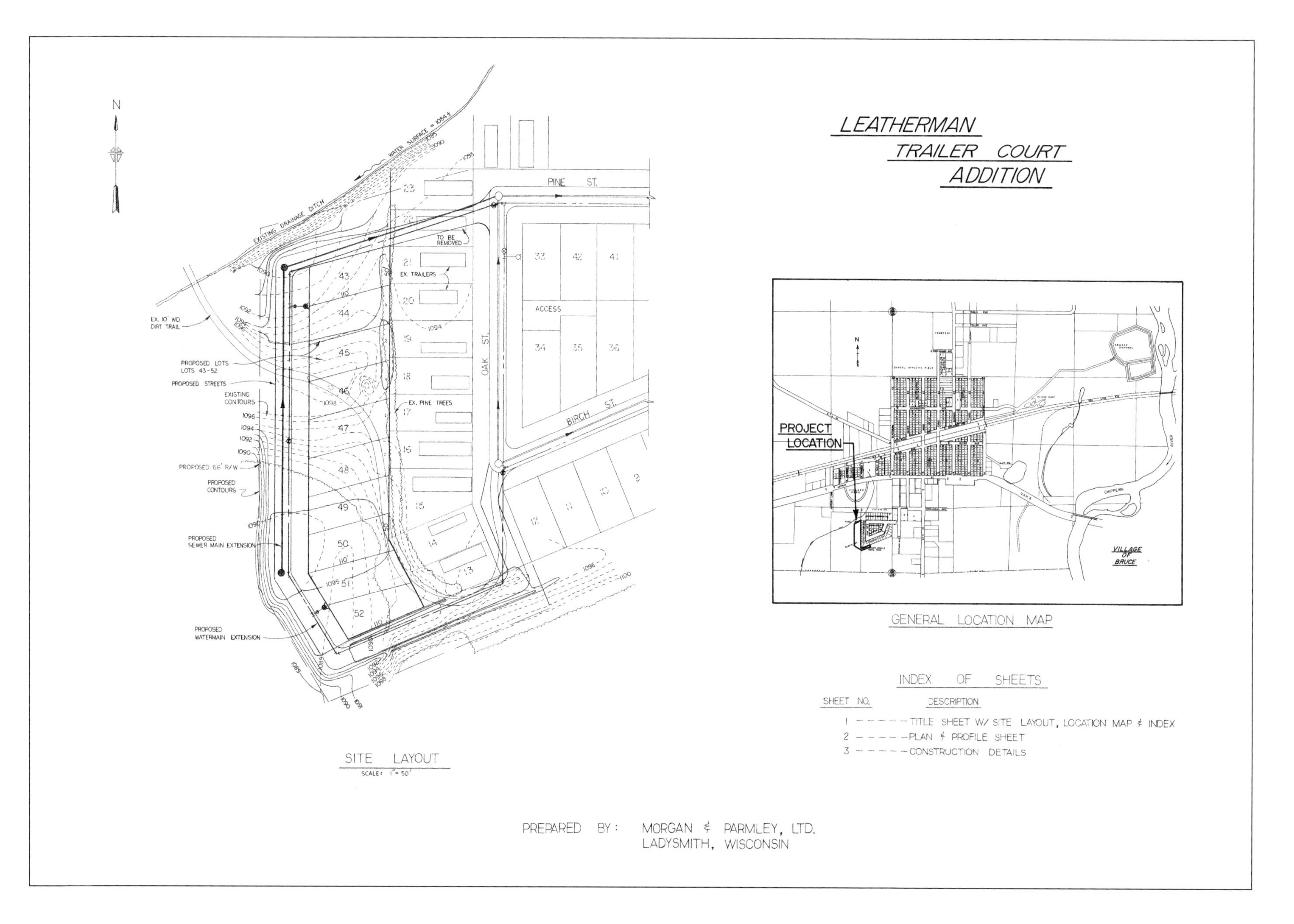

Courtesy: Morgan & Parmley, Ltd.

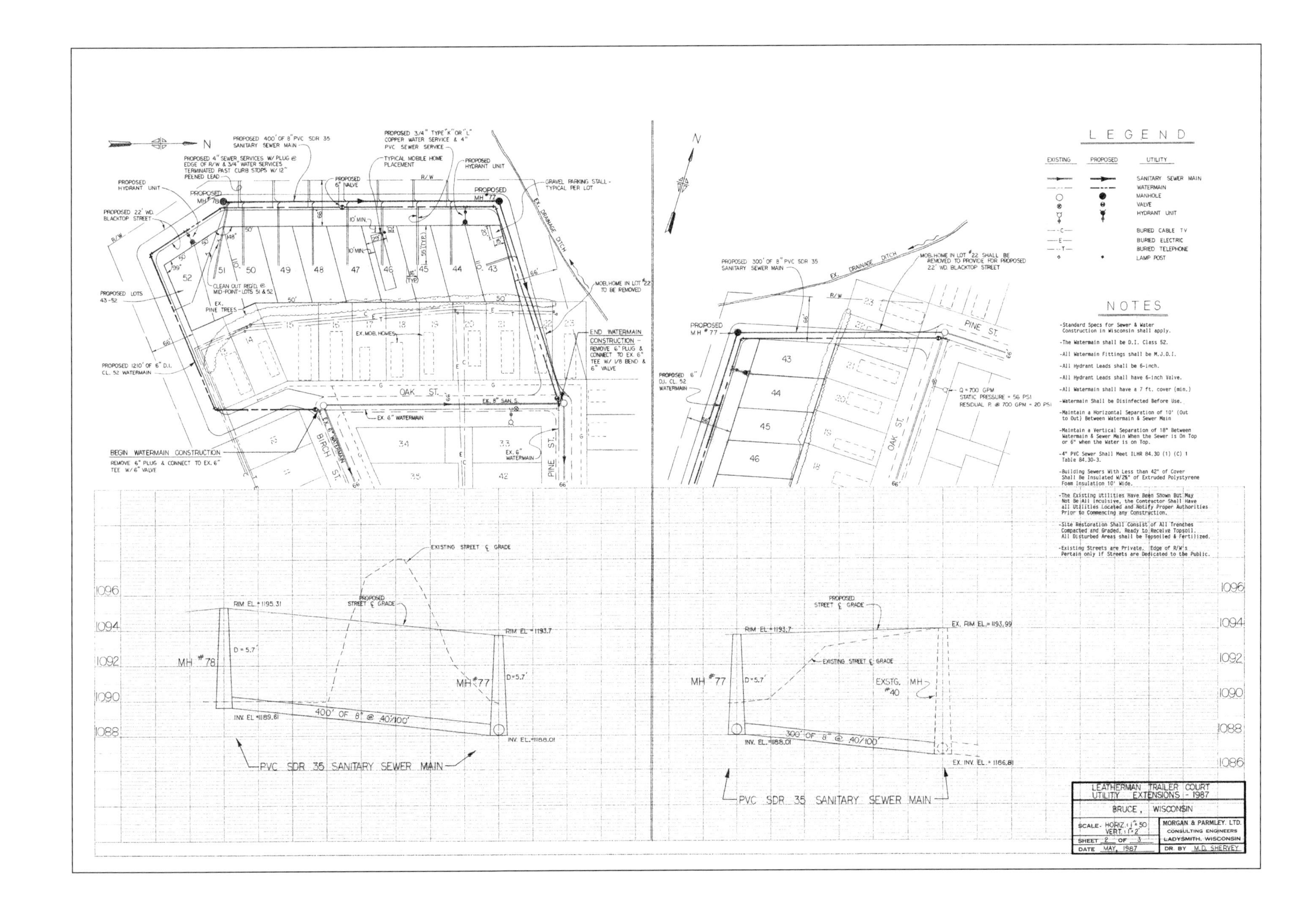
LEGEND
EXISTING
PROPOSED
UTILITY
SANITARY SEWER MAIN
WATERMAIN
MANHOLE
VALVE
HYDRANT UNIT
BURIED CABLE TV
BURIED ELECTRIC
BURIED TELEPHONE
LAMP POST
NOTES
-Standard Specs for Sewer & Water Construction in Wisconsin shall apply.
-The Watermain shall be D.I. Class 52.
-All Watermain Fittings shall be M.J.D.I.
-All Hydrant Leads shall be 6-inch.
-All Hydrant Leads shall have 6-inch Valve.
-All Watermain shall have a 7 ft. cover (min.)
-Watermain Shall be Disinfected Before Use.
-Maintain a Horizontal Separation of 10' (Out to Out) Between Watermain & Sewer Main
-Maintain a Vertical Separation of 18" Between Watermain & Sewer Main When the Sewer is On Top or 6" when the Water is on Top.
-4" PVC Sewer Shall Meet ILHR 84.30 (1) (C) 1 Table 84.30-3.
-Building Sewers With Less than 42" of Cover Shall Be Insulated W/2½" of Extruded Polystyrene Foam Insulation 10' Wide.
-The Existing Utilities Have Been Shown But May Not Be All Inculsive, the Contractor Shall Have all Utilities Located and Notify Proper Authorities Prior to Commencing any Construction.
-Site Restoration Shall Consist of All Trenches Compacted and Graded, Ready to Receive Topsoil. All Disturbed Areas shall be Topsoiled & Fertilized.
-Existing Streets are Private. Edge of R/W's Pertain only if Streets are Dedicated to the Public.
PROPOSED 400' OF 8" PVC SDR 35 SANITARY SEWER MAIN
PROPOSED 300' OF 8" PVC SDR 35 SANITARY SEWER MAIN
BEGIN WATERMAIN CONSTRUCTION
REMOVE 6" PLUG & CONNECT TO EX. 6" TEE W/ 6" VALVE
END WATERMAIN CONSTRUCTION
OAK ST.
BIRCH ST.
PINE ST.
EX. DRAINAGE DITCH
RIM EL.=1195.31
RIM EL.=1193.7
INV. EL.=1189.61
INV. EL.=1188.01
MH #78
MH #77
D=5.7'
400' OF 8" @ .40/100'
300' OF 8" @ .40/100'
EX. RIM EL.=1193.99
EX. INV. EL.=1186.81
EXSTG. MH #40
EXISTING STREET ℄ GRADE
PROPOSED STREET ℄ GRADE
PVC SDR 35 SANITARY SEWER MAIN
LEATHERMAN TRAILER COURT UTILITY EXTENSIONS - 1987
BRUCE, WISCONSIN
MORGAN & PARMLEY, LTD. CONSULTING ENGINEERS LADYSMITH, WISCONSIN
SHEET 2 OF 3
DATE MAY, 1987
DR. BY M.D. SHERVEY

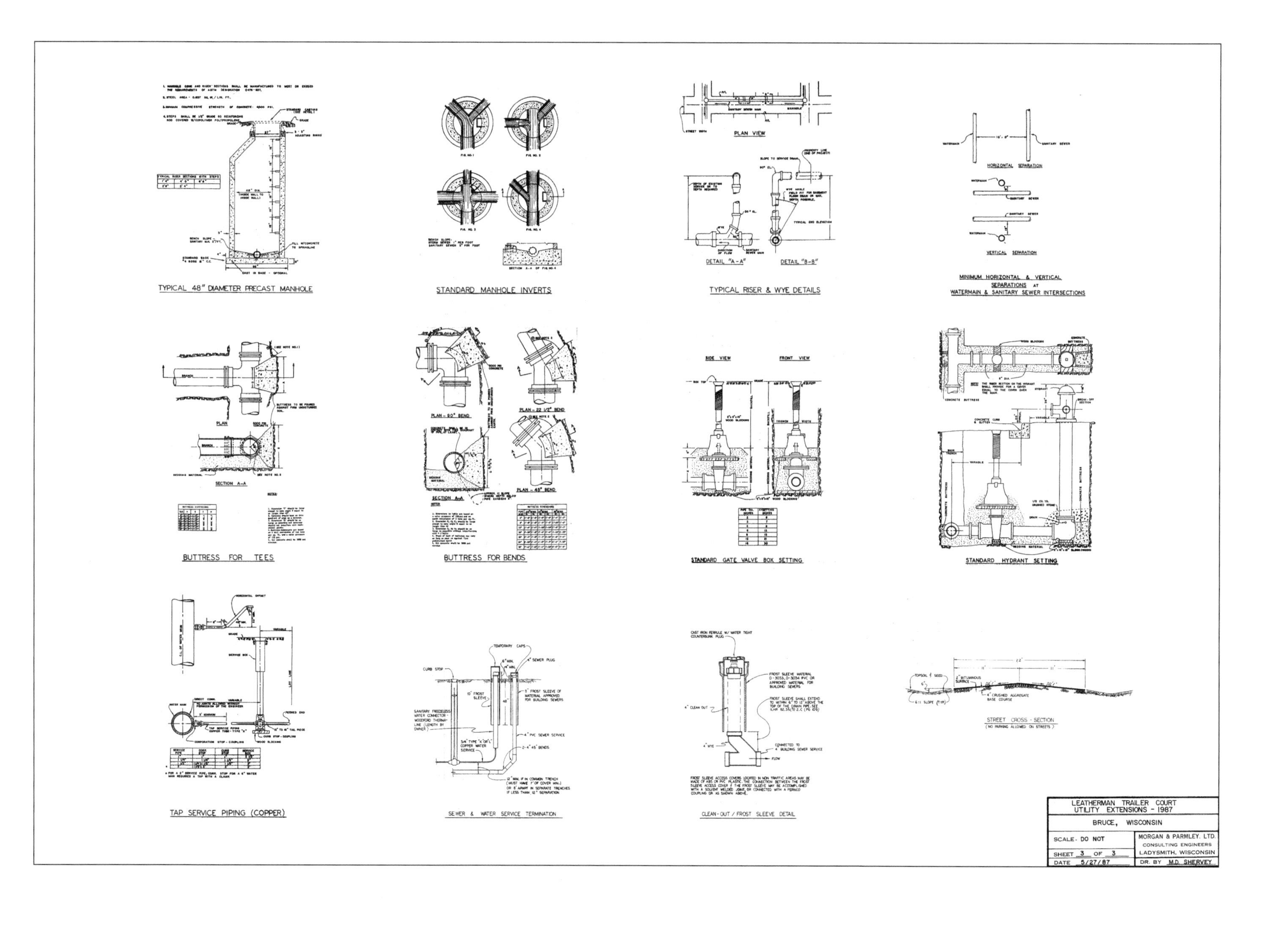
TYPICAL 48" DIAMETER PRECAST MANHOLE
STANDARD MANHOLE INVERTS
PLAN VIEW
DETAIL "A-A"
DETAIL "B-B"
TYPICAL RISER & WYE DETAILS
HORIZONTAL SEPARATION
VERTICAL SEPARATION
MINIMUM HORIZONTAL & VERTICAL SEPARATIONS AT WATERMAIN & SANITARY SEWER INTERSECTIONS
PLAN
SECTION A-A
BUTTRESS FOR TEES
PLAN - 90° BEND
PLAN - 22 1/2° BEND
PLAN - 45° BEND
BUTTRESS FOR BENDS
SIDE VIEW
FRONT VIEW
STANDARD GATE VALVE BOX SETTING
STANDARD HYDRANT SETTING
TAP SERVICE PIPING (COPPER)
TEMPORARY CAPS
CURB STOP
4" SEWER PLUG
SEWER & WATER SERVICE TERMINATION
CAST IRON FERRULE W/ WATER TIGHT COUNTERSUNK PLUG
4" CLEAN OUT
4" WYE
FLOW
CLEAN-OUT / FROST SLEEVE DETAIL
STREET CROSS - SECTION
(NO PARKING ALLOWED ON STREETS)
LEATHERMAN TRAILER COURT
UTILITY EXTENSIONS - 1987
BRUCE, WISCONSIN
SCALE - DO NOT
SHEET 3 OF 3
DATE 5/27/87
MORGAN & PARMLEY, LTD.
CONSULTING ENGINEERS
LADYSMITH, WISCONSIN
DR. BY M.D. SHERVEY

Design

RETROFITTING & REHABILITATION

Section 19

Retrofitting is the act of making modifications to an existing structure or system to upgrade its function or expand its operational capabilities.

Rehabilitation is the work required to restore a structure or system to its prior condition and often in the process make basic improvements.

The following pages show examples of plans for retrofitting and rehabilitation of structures and systems.

TABLE OF CONTENTS

Storage Structure for **Fly Ash Collector**

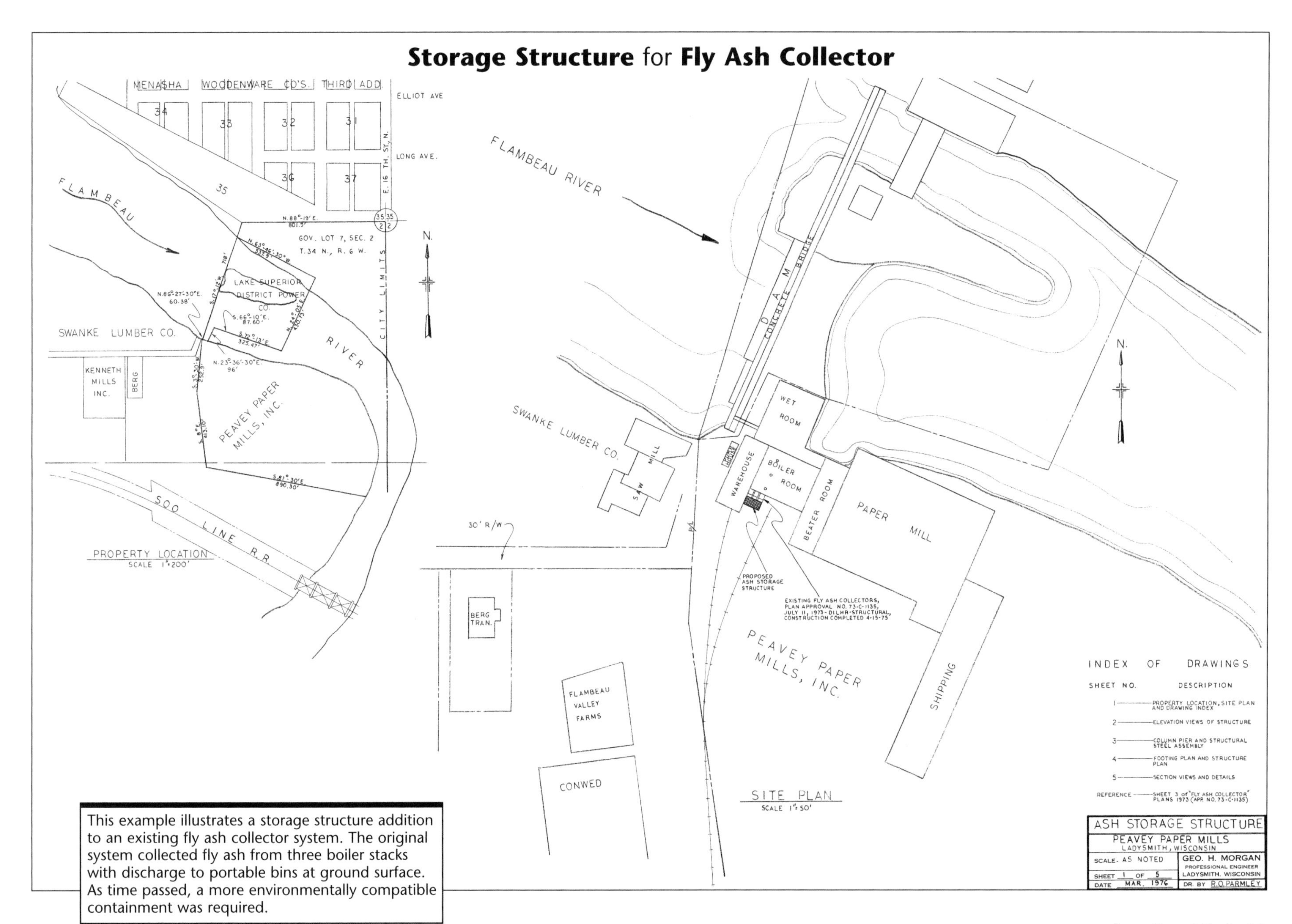

This example illustrates a storage structure addition to an existing fly ash collector system. The original system collected fly ash from three boiler stacks with discharge to portable bins at ground surface. As time passed, a more environmentally compatible containment was required.

Source: Morgan & Parmley, Ltd.

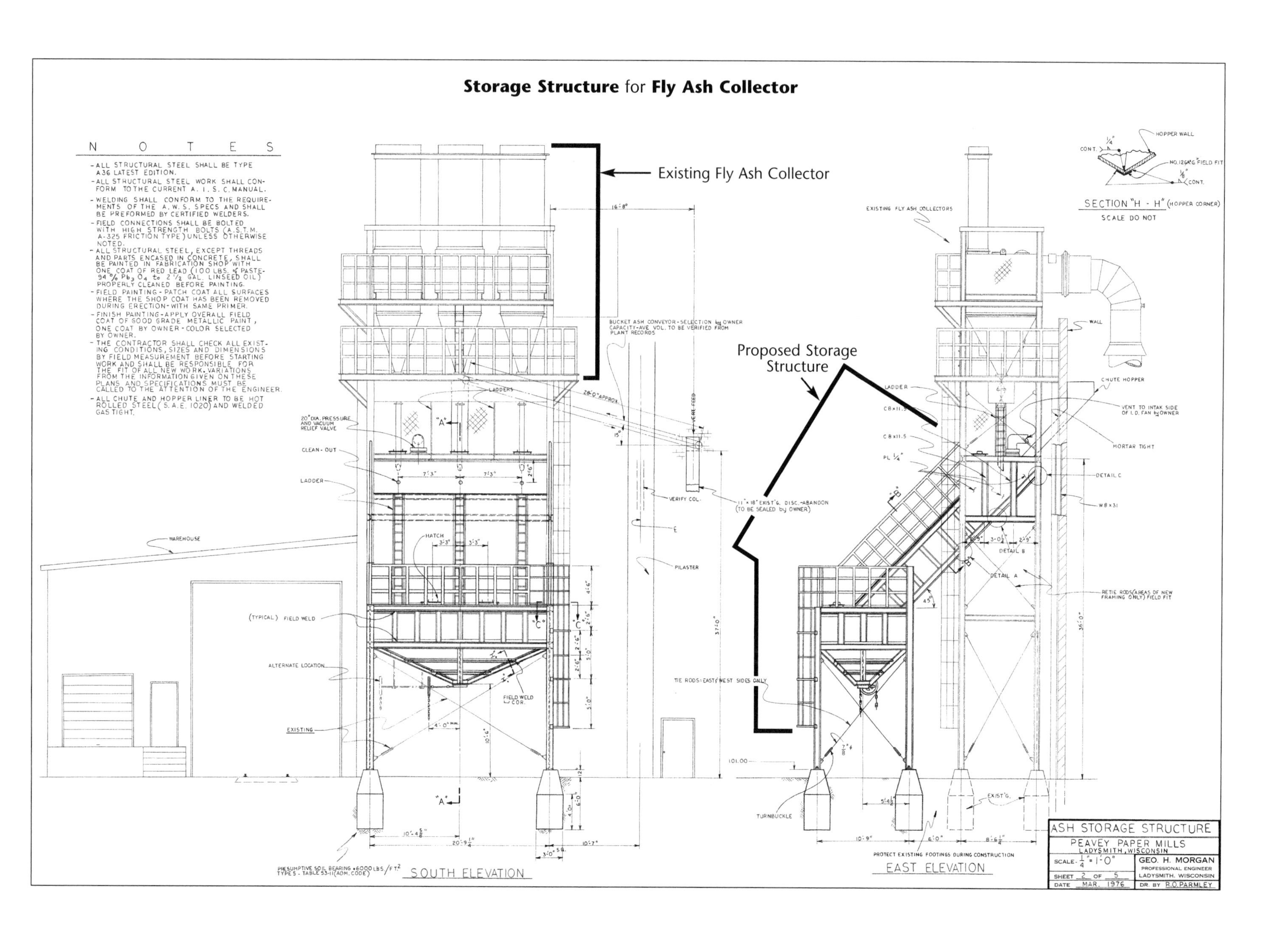
Storage Structure for Fly Ash Collector
Existing Fly Ash Collector
Proposed Storage Structure
NOTES
-ALL STRUCTURAL STEEL SHALL BE TYPE A36 LATEST EDITION.
-ALL STRUCTURAL STEEL WORK SHALL CONFORM TO THE CURRENT A. I. S. C. MANUAL.
-WELDING SHALL CONFORM TO THE REQUIREMENTS OF THE A. W. S. SPECS AND SHALL BE PREFORMED BY CERTIFIED WELDERS.
-FIELD CONNECTIONS SHALL BE BOLTED WITH HIGH STRENGTH BOLTS (A.S.T.M. A-325 FRICTION TYPE) UNLESS OTHERWISE NOTED.
-ALL STRUCTURAL STEEL, EXCEPT THREADS AND PARTS ENCASED IN CONCRETE, SHALL BE PAINTED IN FABRICATION SHOP WITH ONE COAT OF RED LEAD (100 LBS. of PASTE-94 % Pb3 O4 to 2 1/2 GAL. LINSEED OIL) PROPERLY CLEANED BEFORE PAINTING.
-FIELD PAINTING - PATCH COAT ALL SURFACES WHERE THE SHOP COAT HAS BEEN REMOVED DURING ERECTION - WITH SAME PRIMER.
-FINISH PAINTING - APPLY OVERALL FIELD COAT OF GOOD GRADE METALLIC PAINT, ONE COAT BY OWNER - COLOR SELECTED BY OWNER.
-THE CONTRACTOR SHALL CHECK ALL EXISTING CONDITIONS, SIZES AND DIMENSIONS BY FIELD MEASUREMENT BEFORE STARTING WORK AND SHALL BE RESPONSIBLE FOR THE FIT OF ALL NEW WORK. VARIATIONS FROM THE INFORMATION GIVEN ON THESE PLANS AND SPECIFICATIONS MUST BE CALLED TO THE ATTENTION OF THE ENGINEER.
-ALL CHUTE AND HOPPER LINER TO BE HOT ROLLED STEEL (S. A. E. 1020) AND WELDED GAS TIGHT.
WAREHOUSE
LADDERS
20" DIA. PRESSURE AND VACUUM RELIEF VALVE
CLEAN - OUT
LADDER
HATCH
(TYPICAL) FIELD WELD
ALTERNATE LOCATION
FIELD WELD
EXISTING
BUCKET ASH CONVEYOR - SELECTION by OWNER CAPACITY - AVE. VOL. TO BE VERIFIED FROM PLANT RECORDS
VERIFY COL.
PILASTER
TIE RODS: EAST/WEST SIDES ONLY
16'-8"
37'-0"
101.00
SOUTH ELEVATION
EXISTING FLY ASH COLLECTORS
LADDER
C8x11.5
C8x11.5
PL 1/4"
DETAIL B
DETAIL A
TURNBUCKLE
HOPPER WALL
CONT.
NO. 12 GA. FIELD FIT
SECTION "H - H" (HOPPER CORNER)
SCALE DO NOT
WALL
CHUTE HOPPER
VENT TO INTAK SIDE OF I.D. FAN by OWNER
MORTAR TIGHT
DETAIL C
W8x31
35'-0"
EXIST'G.
PROTECT EXISTING FOOTINGS DURING CONSTRUCTION
EAST ELEVATION
ASH STORAGE STRUCTURE
PEAVEY PAPER MILLS
LADYSMITH, WISCONSIN
SCALE: 1/4" = 1'-0"
GEO. H. MORGAN
PROFESSIONAL ENGINEER
LADYSMITH, WISCONSIN
SHEET 2 OF 5
DATE MAR. 1976
DR. BY R.O. PARMLEY

Storage Structure for **Fly Ash Collector**

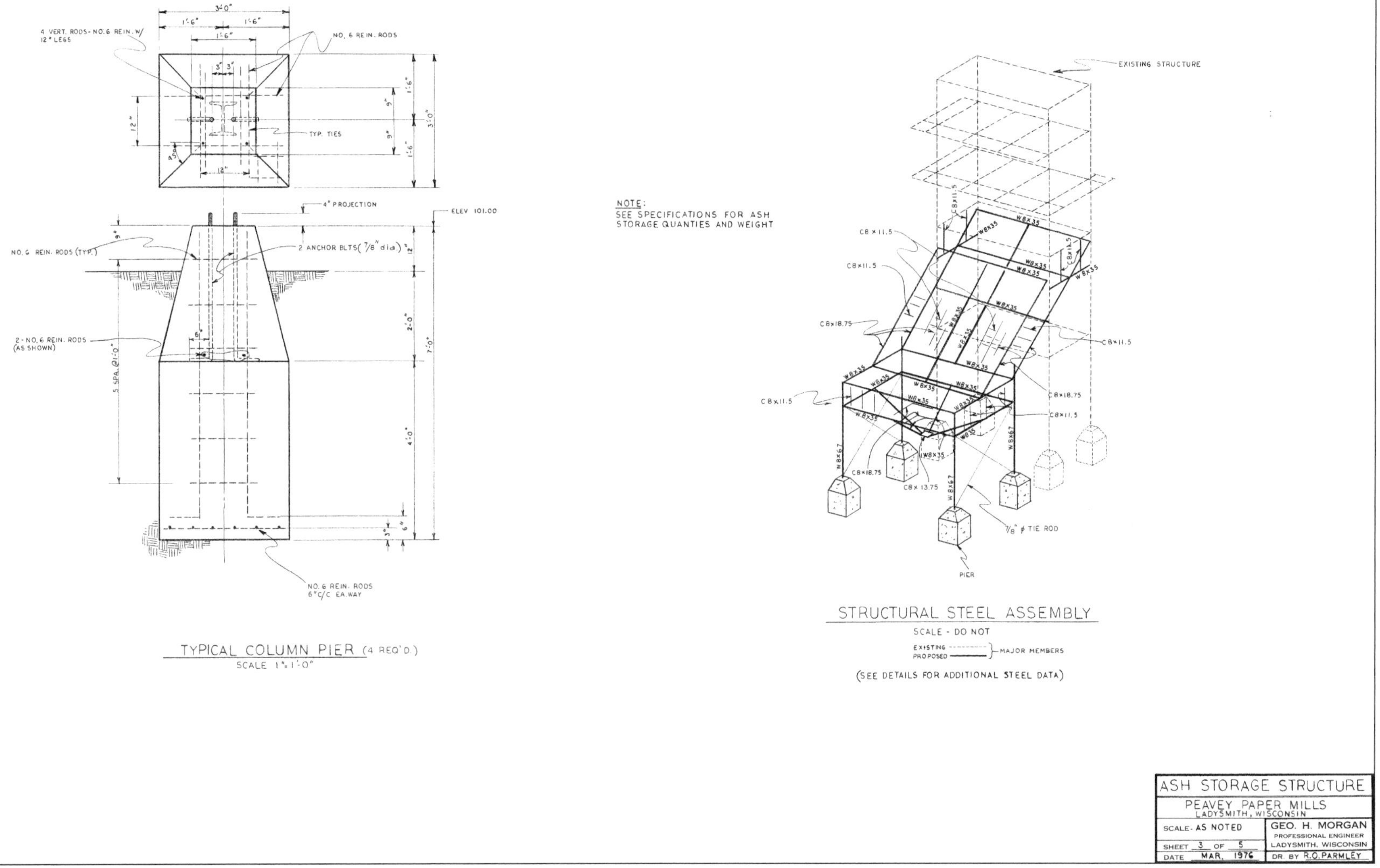
4 VERT. RODS - NO. 6 REIN. W/ 12" LEGS
NO. 6 REIN. RODS
TYP. TIES
4" PROJECTION
ELEV 101.00
NO. 6 REIN. RODS (TYP.)
2 ANCHOR BLTS (7/8" dia)
2 - NO. 6 REIN. RODS (AS SHOWN)
5 SPA. @ 1'-0"
NO. 6 REIN. RODS 6" C/C EA. WAY
TYPICAL COLUMN PIER (4 REQ'D.)
SCALE 1" = 1'-0"
NOTE:
SEE SPECIFICATIONS FOR ASH STORAGE QUANTIES AND WEIGHT
EXISTING STRUCTURE
7/8" ∅ TIE ROD
PIER
STRUCTURAL STEEL ASSEMBLY
SCALE - DO NOT
EXISTING
PROPOSED
MAJOR MEMBERS
(SEE DETAILS FOR ADDITIONAL STEEL DATA)
ASH STORAGE STRUCTURE
PEAVEY PAPER MILLS
LADYSMITH, WISCONSIN
SCALE - AS NOTED
GEO. H. MORGAN
PROFESSIONAL ENGINEER
LADYSMITH, WISCONSIN
SHEET 3 OF 5
DATE MAR. 1976
DR. BY R.O. PARMLEY

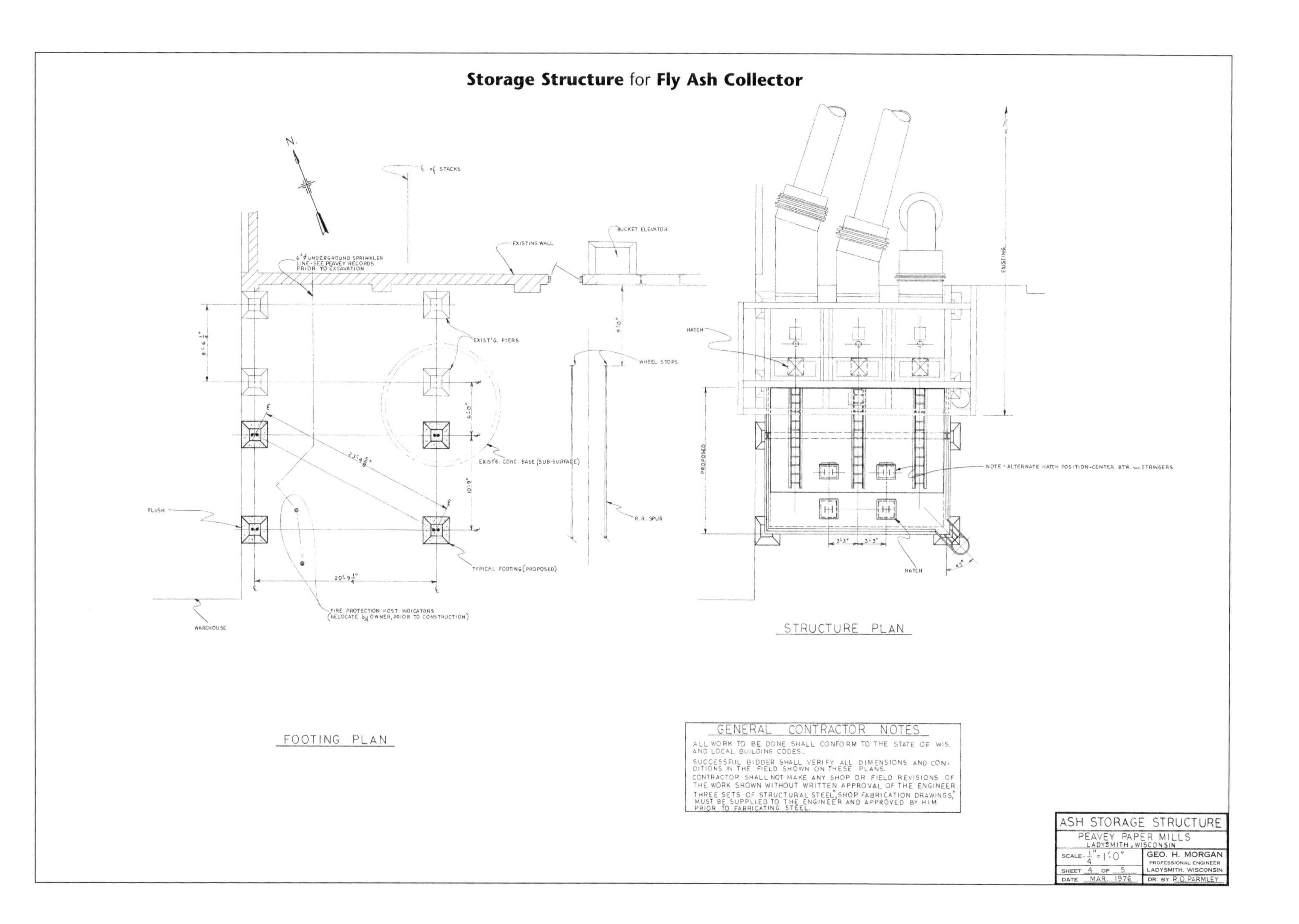
Storage Structure for Fly Ash Collector
FOOTING PLAN
STRUCTURE PLAN
GENERAL CONTRACTOR NOTES
ALL WORK TO BE DONE SHALL CONFORM TO THE STATE OF WIS. AND LOCAL BUILDING CODES.
SUCCESSFUL BIDDER SHALL VERIFY ALL DIMENSIONS AND CONDITIONS IN THE FIELD SHOWN ON THESE PLANS.
CONTRACTOR SHALL NOT MAKE ANY SHOP OR FIELD REVISIONS OF THE WORK SHOWN WITHOUT WRITTEN APPROVAL OF THE ENGINEER.
THREE SETS OF STRUCTURAL STEEL "SHOP FABRICATION DRAWINGS" MUST BE SUPPLIED TO THE ENGINEER AND APPROVED BY HIM PRIOR TO FABRICATING STEEL.
ASH STORAGE STRUCTURE
PEAVEY PAPER MILLS
LADYSMITH, WISCONSIN
SCALE - 1/4" = 1'-0"
GEO. H. MORGAN
PROFESSIONAL ENGINEER
LADYSMITH, WISCONSIN
SHEET 4 OF 5
DATE MAR. 1976
DR. BY R.O. PARMLEY
EXISTING WALL
BUCKET ELEVATOR
EXIST'G. PIERS
WHEEL STOPS
EXIST'G. CONC. BASE (SUB-SURFACE)
R.R. SPUR
TYPICAL FOOTING (PROPOSED)
FIRE PROTECTION POST INDICATORS (RELOCATE by OWNER, PRIOR TO CONSTRUCTION)
WAREHOUSE
FLUSH
HATCH
EXISTING
PROPOSED
NOTE - ALTERNATE HATCH POSITION - CENTER BTW. STRINGERS

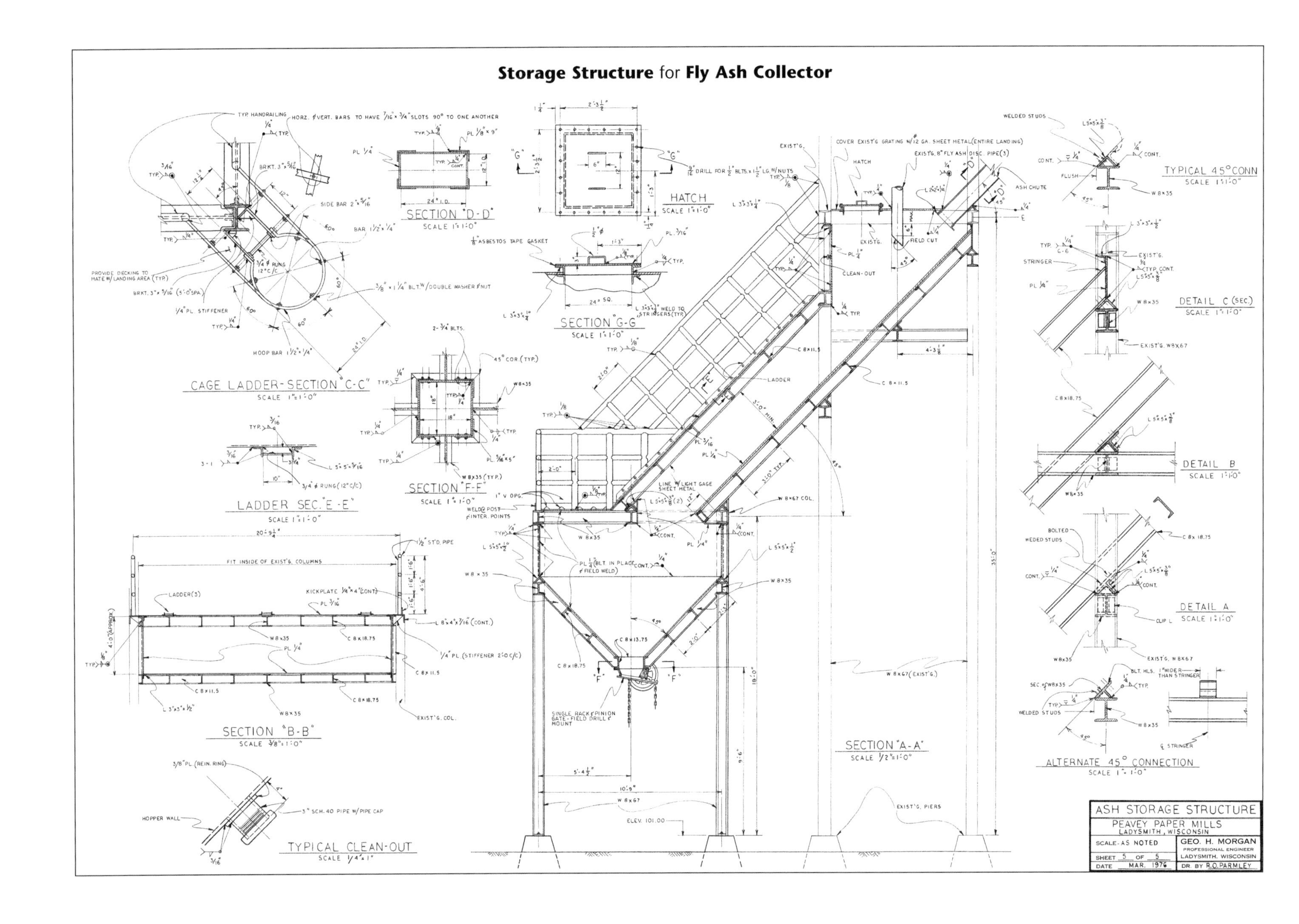
Storage Structure for Fly Ash Collector
CAGE LADDER-SECTION "C-C"
SCALE 1"=1'-0"
SECTION "D-D"
SCALE 1"=1'-0"
HATCH
SCALE 1"=1'-0"
SECTION "G-G"
SCALE 1"=1'-0"
TYPICAL 45° CONN
SCALE 1"=1'-0"
DETAIL C (SEC.)
SCALE 1"=1'-0"
LADDER SEC. "E-E"
SCALE 1"=1'-0"
SECTION "F-F"
SCALE 1"=1'-0"
DETAIL B
SCALE 1"=1'-0"
DETAIL A
SCALE 1"=1'-0"
SECTION "B-B"
SCALE 3/8"=1'-0"
SECTION "A-A"
SCALE 1/2"=1'-0"
ALTERNATE 45° CONNECTION
SCALE 1"=1'-0"
TYPICAL CLEAN-OUT
SCALE 1/4"=1"
ASH STORAGE STRUCTURE
PEAVEY PAPER MILLS
LADYSMITH, WISCONSIN
SCALE-AS NOTED
SHEET 5 OF 5
DATE MAR. 1976
GEO. H. MORGAN
PROFESSIONAL ENGINEER
LADYSMITH, WISCONSIN
DR. BY R.O. PARMLEY

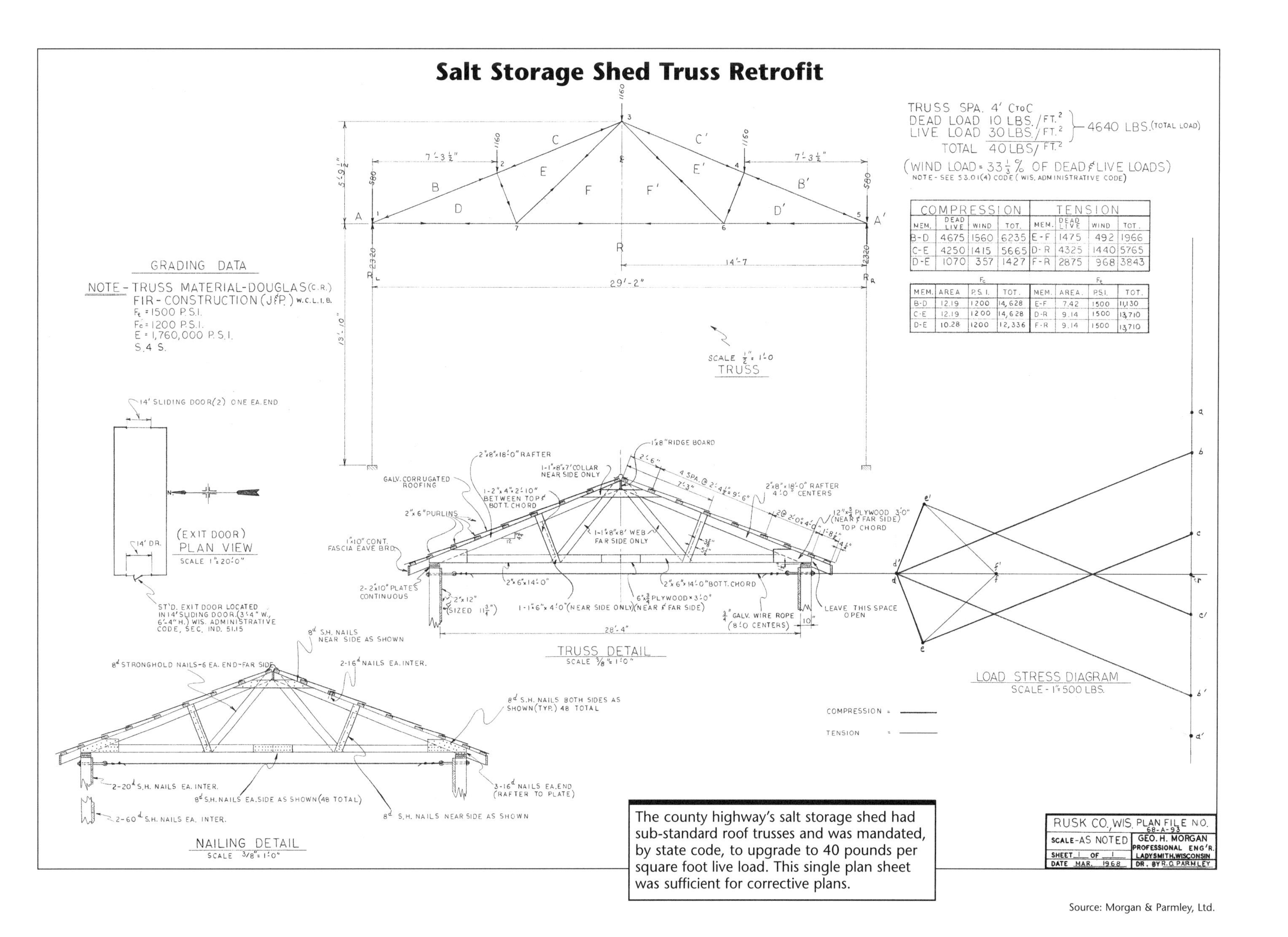

COMPRESSION				TENSION			
MEM.	DEAD LIVE	WIND	TOT.	MEM.	DEAD LIVE	WIND	TOT.
B-D	4675	1560	6235	E-F	1475	492	1966
C-E	4250	1415	5665	D-R	4325	1440	5765
D-E	1070	357	1427	F-R	2875	968	3843

Fc				Ft			
MEM.	AREA	P.S.I.	TOT.	MEM.	AREA	P.S.I.	TOT.
B-D	12.19	1200	14,628	E-F	7.42	1500	11,130
C-E	12.19	1200	14,628	D-R	9.14	1500	13,710
D-E	10.28	1200	12,336	F-R	9.14	1500	13,710

The county highway's salt storage shed had sub-standard roof trusses and was mandated, by state code, to upgrade to 40 pounds per square foot live load. This single plan sheet was sufficient for corrective plans.

Source: Morgan & Parmley, Ltd.

WWTF ADDITION
&
U.V. DISINFECTION FACILITY

E.D.A. PROJECT NO. 06-01-02571

VILLAGE
OF
GLEN FLORA, WISCONSIN
DEC., 1992

VILLAGE OF GLEN FLORA,
RUSK CO, WISCONSIN

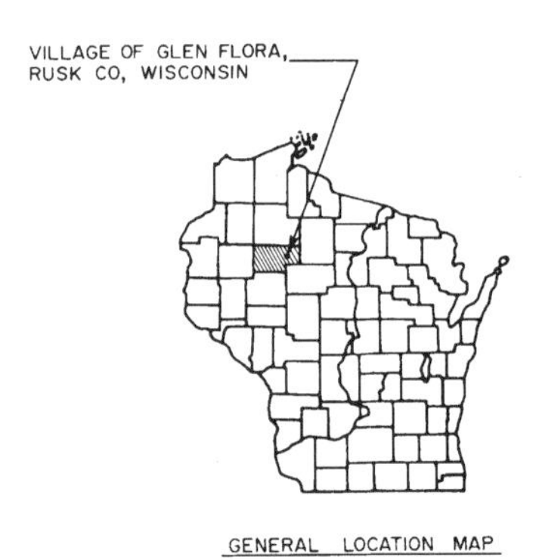

GENERAL LOCATION MAP

THESE PLANS REPRESENT ONLY 1 SECTION OF THE TOTAL PROJECT

THE TOTAL PROJECT EDA NO. 06-01-02571 IS PARTIALLY FUNDED BY A 60% GRANT FROM THE UNITED STATES ECONOMIC DEVELOPMENT ADMINISTRATION IN THE TOTAL GRANT AMOUNT OF $590,520

PROJECT SITE

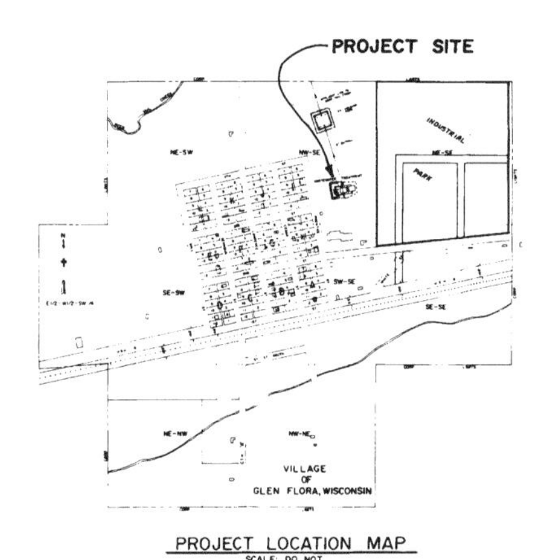

PROJECT LOCATION MAP
SCALE: DO NOT

UTILITIES

CONTACT – NORTHERN STATES POWER CO.
711 W. 9th Street, North
Ladysmith, WI. 54848

Attn: John Rymarkiewicz
715-532-6226

CONTACT – UNIVERSAL TELEPHONE CO.
OF NORTHERN WISCONSIN
Highway "8"
Hawkins, WI 54530

Attn: Jim Arquette
715-585-7707

CONTACT – DIGGERS HOTLINE

1-800 242-8511

CONTACT – VILLAGE OF GLEN FLORA WWCTF

Les Evjen, Operator
P.O. Box 253
Glen Flora, WI. 54526

715-322-5511

CONTACT – GLEN FLORA ELEMENTARY SCHOOL
SCHOOL WELL & PIPELINE
Larry Johnson, Custodian

715-322-5271

INDEX OF SHEETS

NO.	DESCRIPTION
1	TITLE SHEET, GENERAL LOCATION & INDEX
2	TREATMENT SYSTEM LAYOUT & HYDRAULIC PROFILE
3	FILTER BED NO. 5 & PIPING DETAILS
4	DOSING TANK & METERING MANHOLE MODIFICATIONS
5	U.V. DISINFECTION SYSTEM & BUILDING
6	TYPICAL EROSION CONTROL DETAILS

PREPARED BY:
MORGAN & PARMLEY LTD.
CONSULTING ENGINEERS
LADYSMITH, WISCONSIN

The wastewater treatment facility (WWTF) had to expand its hydraulic capacity and change its method of effluent disinfection from hazardous chlorination to the safe ultraviolet (U.V.) process.

Source: Morgan & Parmley, Ltd.

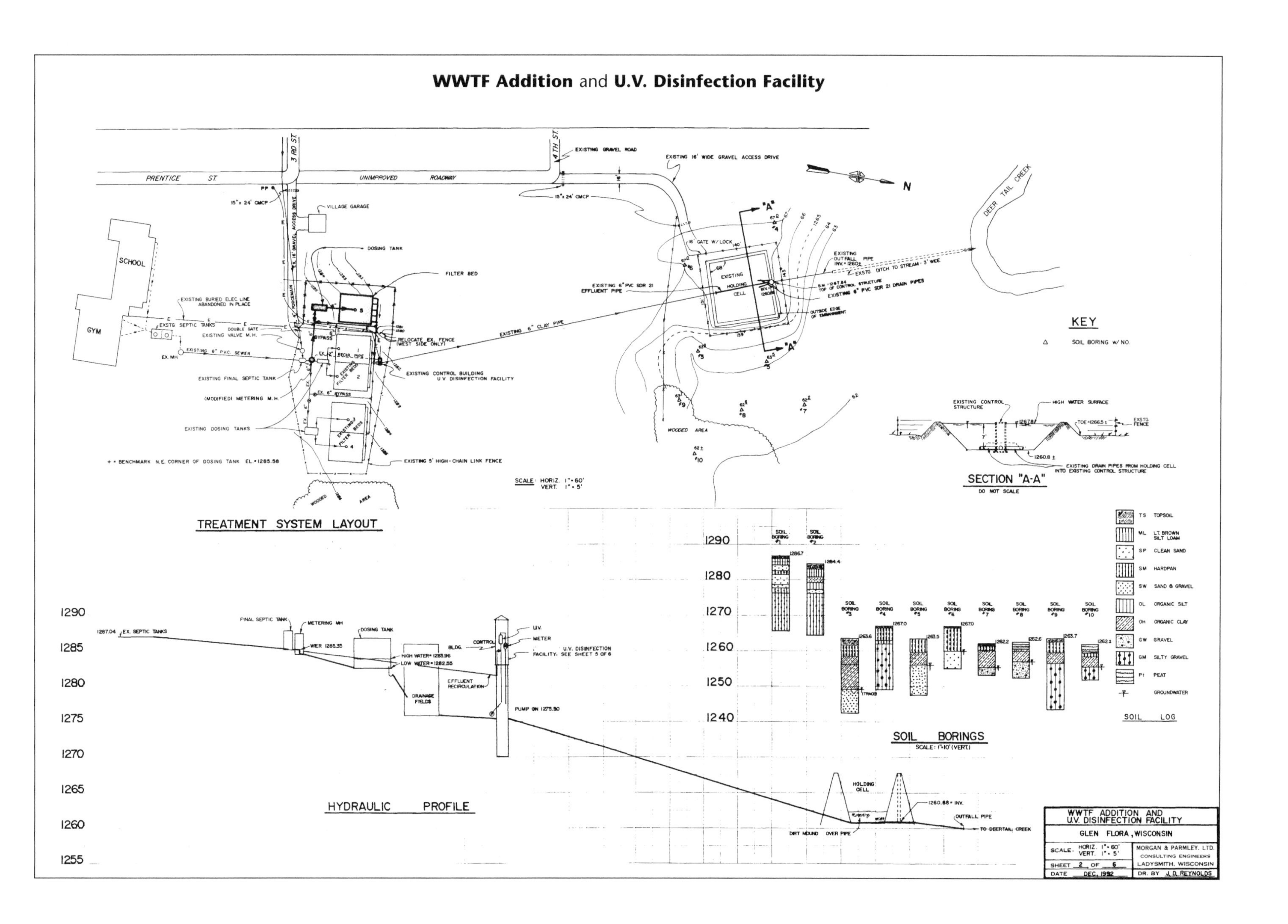
WWTF Addition and U.V. Disinfection Facility
TREATMENT SYSTEM LAYOUT
PRENTICE ST.
UNIMPROVED ROADWAY
3 RD ST.
4 TH ST.
EXISTING GRAVEL ROAD
EXISTING 16' WIDE GRAVEL ACCESS DRIVE
VILLAGE GARAGE
SCHOOL
GYM
DOSING TANK
FILTER BED
EXISTING BURIED ELEC LINE ABANDONED IN PLACE
EXSTG SEPTIC TANKS
DOUBLE GATE
EXISTING VALVE M.H.
EXISTING 6" P.V.C. SEWER
EX MH
EXISTING FINAL SEPTIC TANK
(MODIFIED) METERING M.H.
EXISTING DOSING TANKS
BENCHMARK N.E. CORNER OF DOSING TANK EL.=1285.58
EXISTING CONTROL BUILDING U.V. DISINFECTION FACILITY
RELOCATE EX. FENCE (WEST SIDE ONLY)
EXISTING 5' HIGH-CHAIN LINK FENCE
EXISTING 6" CLAY PIPE
EXISTING 6" PVC SDR 21 EFFLUENT PIPE
EXISTING HOLDING CELL
16' GATE W/ LOCK
EXISTING OUTFALL PIPE
EXSTG DITCH TO STREAM - 3' WIDE
EXISTING 6" PVC SDR 21 DRAIN PIPES
OUTSIDE EDGE OF EMBANKMENT
DEER TAIL CREEK
N
WOODED AREA
SCALE: HORIZ. 1" = 60' VERT. 1" = 5'
KEY
SOIL BORING W/ NO.
SECTION "A-A"
DO NOT SCALE
EXISTING CONTROL STRUCTURE
HIGH WATER SURFACE
EXSTG FENCE
EXISTING DRAIN PIPES FROM HOLDING CELL INTO EXISTING CONTROL STRUCTURE
HYDRAULIC PROFILE
EX. SEPTIC TANKS
FINAL SEPTIC TANK
METERING MH
DOSING TANK
WIER 1285.35
HIGH WATER = 1283.96
LOW WATER = 1282.55
DRAINAGE FIELDS
EFFLUENT RECIRCULATION
CONTROL BLDG.
U.V.
METER
U.V. DISINFECTION FACILITY. SEE SHEET 5 OF 6
PUMP ON 1275.50
HOLDING CELL
1260.88 = INV.
OUTFALL PIPE
DIRT MOUND OVER PIPE
TO DEERTAIL CREEK
SOIL BORINGS
SCALE: 1"=10' (VERT.)
SOIL LOG
TS TOPSOIL
ML LT. BROWN SILT LOAM
SP CLEAN SAND
SM HARDPAN
SW SAND & GRAVEL
OL ORGANIC SILT
OH ORGANIC CLAY
GW GRAVEL
GM SILTY GRAVEL
PT PEAT
GROUNDWATER
WWTF ADDITION AND U.V. DISINFECTION FACILITY
GLEN FLORA, WISCONSIN
SCALE: HORIZ. 1" = 60' VERT. 1" = 5'
MORGAN & PARMLEY, LTD. CONSULTING ENGINEERS LADYSMITH, WISCONSIN
SHEET 2 OF 6
DATE DEC. 1992
DR. BY J.D. REYNOLDS

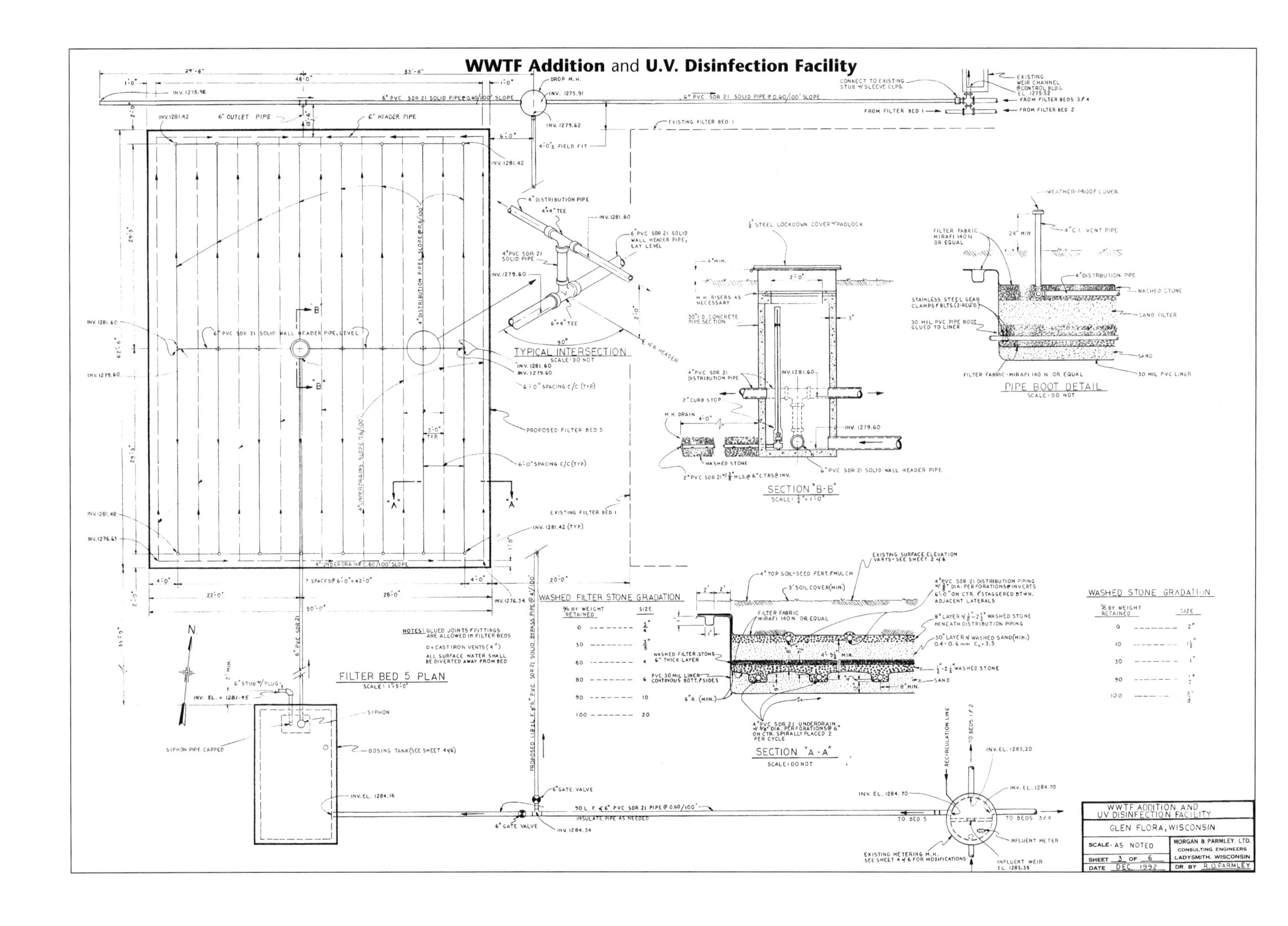

WWTF Addition and U.V. Disinfection Facility
DROP M.H.
INV. 1275.91
6" PVC SDR 21 SOLID PIPE @ 0.60/100' SLOPE
CONNECT TO EXISTING STUB W/ SLEEVE CLPG.
EXISTING WEIR CHANNEL @ CONTROL BLDG. EL. 1275.52
FROM FILTER BEDS 3 & 4
FROM FILTER BED 1
FROM FILTER BED 2
6" OUTLET PIPE
6" HEADER PIPE
INV. 1279.62
4'-0" ± FIELD FIT
EXISTING FILTER BED 1
INV. 1281.42
4" DISTRIBUTION PIPE
4"x4" TEE
INV. 1281.60
6" PVC SDR 21 SOLID WALL HEADER PIPE, LAY LEVEL
4" PVC SDR 21 SOLID PIPE
INV. 1279.60
6"x4" TEE
TYPICAL INTERSECTION
SCALE: DO NOT
6" PVC SDR 21 SOLID WALL HEADER PIPE, LEVEL
6'-0" SPACING C/C (TYP)
PROPOSED FILTER BED 5
4" DISTRIBUTION PIPES SLOPE @ 0.6/100'
4" UNDERDRAINS SLOPE 0.6/100'
4" UNDERDRAIN @ 0.60/100' SLOPE
7 SPACES @ 6'-0" = 42'-0"
INV. 1281.42 (TYP.)
INV. 1276.63
INV. 1276.34
NOTES: GLUED JOINTS & FITTINGS ARE ALLOWED IN FILTER BEDS
o = CAST IRON VENTS (4")
ALL SURFACE WATER SHALL BE DIVERTED AWAY FROM BED
FILTER BED 5 PLAN
SCALE: 1"=5'-0"
6" STUB W/ PLUG
INV. EL. = 1281.45
SIPHON
SIPHON PIPE CAPPED
DOSING TANK (SEE SHEET 4 of 6)
INV. EL. 1284.16
PROPOSED 118 L.F. of 6" PVC SDR 21 SOLID BYPASS PIPE @ 4/100'
6" GATE VALVE
90 L.F. of 6" PVC SDR 21 PIPE @ 0.60/100'
INSULATE PIPE AS NEEDED
INV. 1284.34
TO BED 5
TO BEDS 3 & 4
INV. EL. 1284.70
INV. EL. 1283.20
RECIRCULATION LINE
TO BEDS 1 & 2
INFLUENT METER
INFLUENT WEIR EL. 1283.35
EXISTING METERING M.H. SEE SHEET 4 of 6 FOR MODIFICATIONS
1/4" STEEL LOCKDOWN COVER W/ PADLOCK
M.H. RISERS AS NECESSARY
30" I.D. CONCRETE PIPE SECTION
4" PVC SDR 21 DISTRIBUTION PIPE
INV. 1281.60
2" CURB STOP
M.H. DRAIN
INV. 1279.60
WASHED STONE
6" PVC SDR 21 SOLID WALL HEADER PIPE
SECTION "B-B"
SCALE: 3/4" = 1'-0"
WEATHER-PROOF COVER
FILTER FABRIC MIRAFI 140N OR EQUAL
4" C.I. VENT PIPE
4" DISTRIBUTION PIPE
WASHED STONE
STAINLESS STEEL GEAR CLAMPS & BLTS. (2-REQ'D)
SAND FILTER
30 MIL PVC PIPE BOOT GLUED TO LINER
SAND
FILTER FABRIC-MIRAFI 140 N OR EQUAL
30 MIL PVC LINER
PIPE BOOT DETAIL
SCALE: DO NOT
EXISTING SURFACE ELEVATION VARYS - SEE SHEET 2 of 6
4" TOP SOIL-SEED FERT. & MULCH
3' SOIL COVER (MIN.)
4" PVC SDR 21 DISTRIBUTION PIPING
8" LAYER OF WASHED STONE BENEATH DISTRIBUTION PIPING
30" LAYER OF WASHED SAND (MIN.)
WASHED FILTER STONE 6" THICK LAYER
PVC 30 MIL LINER CONTINOUS BOTT. & SIDES
WASHED STONE
SAND
4" PVC SDR 21 UNDERDRAIN
SECTION "A-A"
SCALE: DO NOT
WASHED FILTER STONE GRADATION
% BY WEIGHT RETAINED | SIZE
0 | 3/4"
30 | 3/8"
60 | 4
80 | 6
90 | 10
100 | 20
WASHED STONE GRADATION
% BY WEIGHT RETAINED | SIZE
0 | 2"
10 | 1 1/2"
30 | 1"
90 | 1/2"
100 | 3/8"
WWTF ADDITION AND UV DISINFECTION FACILITY
GLEN FLORA, WISCONSIN
SCALE- AS NOTED
MORGAN & PARMLEY LTD. CONSULTING ENGINEERS LADYSMITH, WISCONSIN
SHEET 3 OF 6
DATE DEC. 1992
DR. BY R.O. FARMLEY

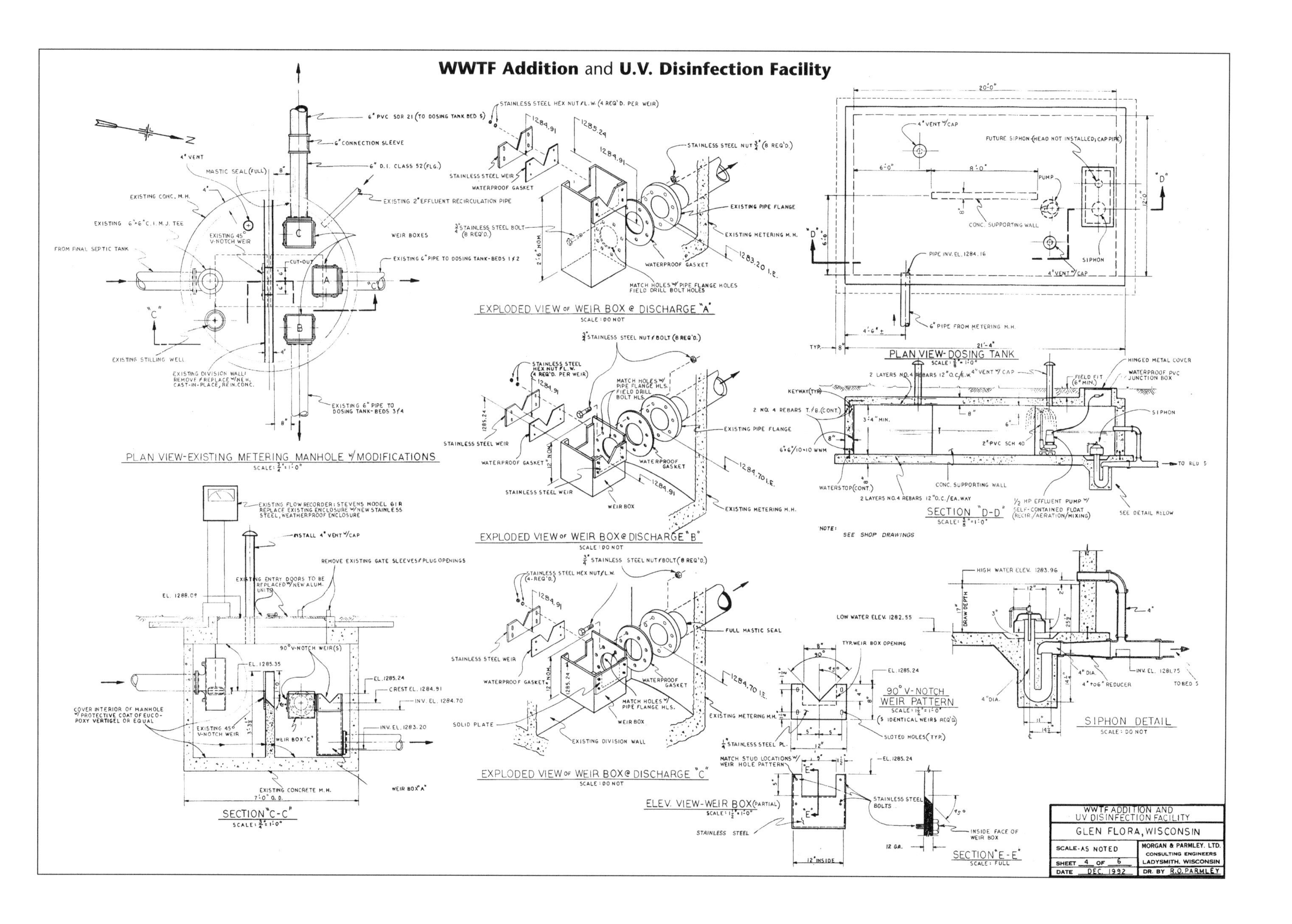

WWTF Addition and U.V. Disinfection Facility
PLAN VIEW-EXISTING METERING MANHOLE W/MODIFICATIONS
EXPLODED VIEW OF WEIR BOX @ DISCHARGE "A"
EXPLODED VIEW OF WEIR BOX @ DISCHARGE "B"
EXPLODED VIEW OF WEIR BOX @ DISCHARGE "C"
PLAN VIEW-DOSING TANK
SECTION "D-D"
SECTION "C-C"
90° V-NOTCH WEIR PATTERN
ELEV. VIEW-WEIR BOX (PARTIAL)
SECTION "E-E"
SIPHON DETAIL
WEIR BOXES
EXISTING METERING M.H.
EXISTING PIPE FLANGE
WATERPROOF GASKET
STAINLESS STEEL WEIR
NOTE: SEE SHOP DRAWINGS
WWTF ADDITION AND UV DISINFECTION FACILITY
GLEN FLORA, WISCONSIN
SCALE-AS NOTED
MORGAN & PARMLEY, LTD.
CONSULTING ENGINEERS
LADYSMITH, WISCONSIN
SHEET 4 OF 6
DATE DEC. 1992
DR. BY R.O. PARMLEY

WWTF Addition and U.V. Disinfection Facility

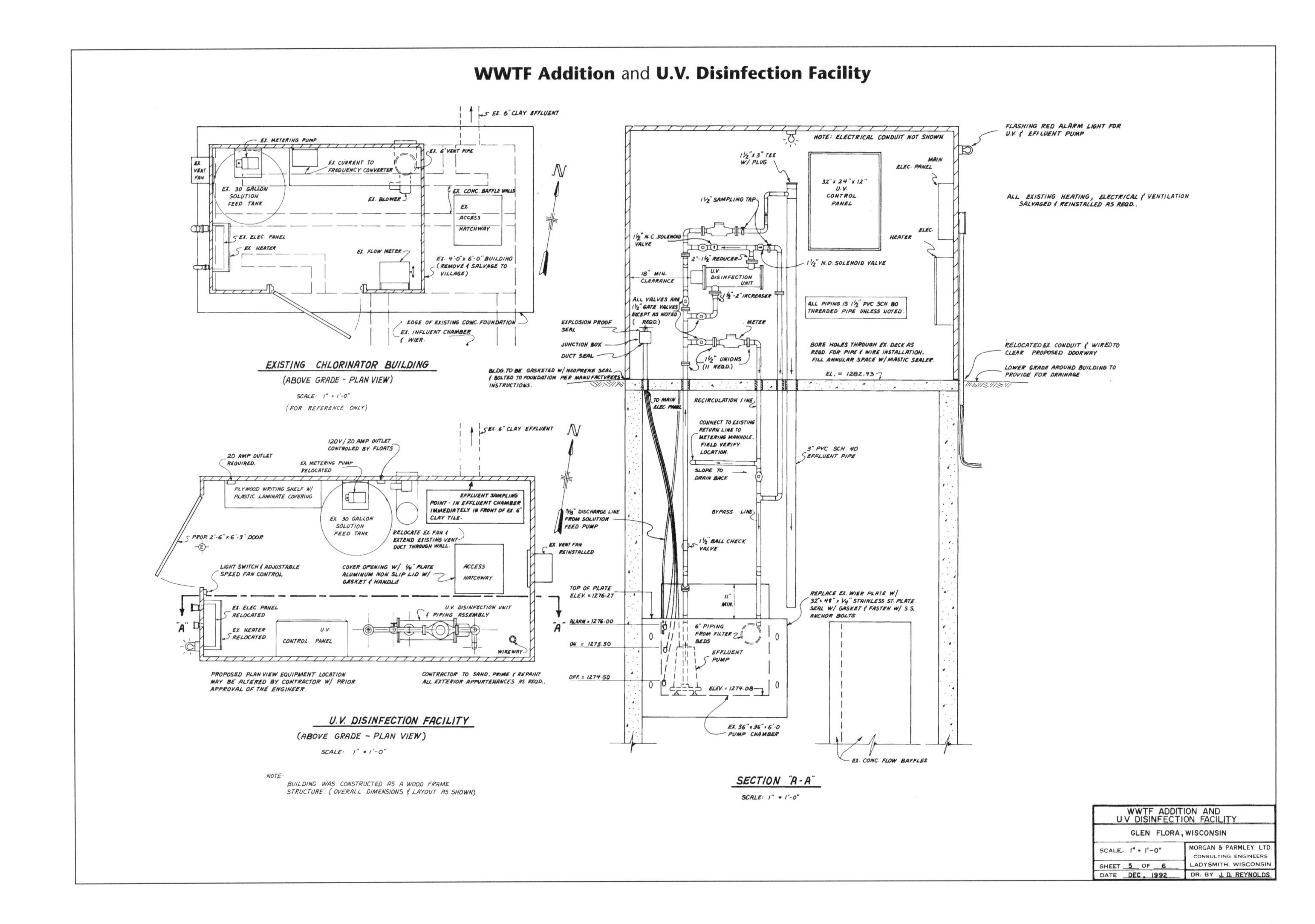

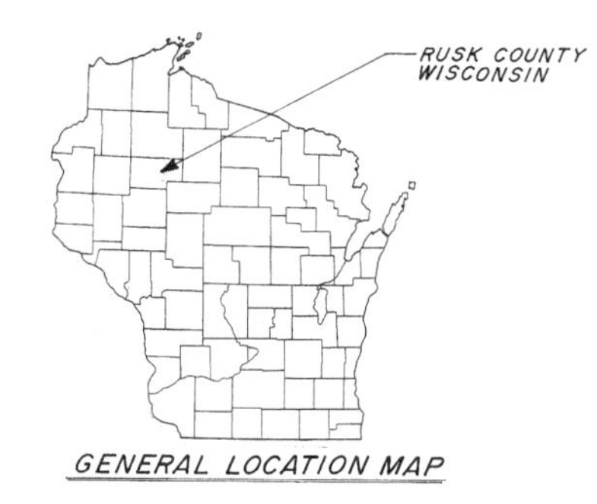

GENERAL LOCATION MAP

TRANSMISSION MAIN RE-ROUTE
RUSK COUNTY AIRPORT

VILLAGE

OF

TONY, WISCONSIN

MUNIPAL WATER SYSTEM

M. & P. PROJECT NO. 99-110

JUNE, 1999

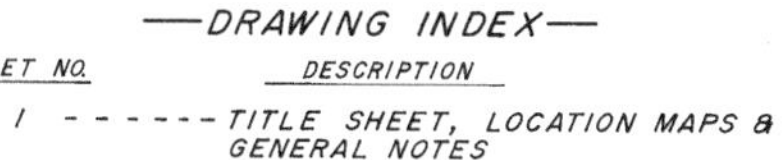

—DRAWING INDEX—
SHEET NO.
DESCRIPTION
1 ------- TITLE SHEET, LOCATION MAPS & GENERAL NOTES
2 ------ PROJECT SITE PLAN & CONSTRUCTION DETAILS
3 ------- EROSION CONTROL DETAILS w/ TYPICAL NOTES

GENERAL NOTES

1. ALL WATERMAIN SHALL BE DUCTILE IRON CL 52.
2. ALL WATERMAIN FITTINGS SHALL BE M.J.D.I. W / RESTRAINED JOINT LUGS AND MEET AWWA C153.
3. ALL HYDRANT LEADS SHALL BE 6-INCH DUCTILE IRON CL 52.
4. ALL HYDRANT LEADS SHALL HAVE A 6-INCH GATE VALVE.
5. HYDRANTS SHALL BE WATEROUS WB 67, 8.5 FT. BURY (4" PUMPER NOZZLE & 2-2½" NOZZLES), 22" BREAKOFF WITH THREADS MATCHING VILLAGE OF TONY PATTERN; I.E. CITY OF LADYSMITH FIRE DEPARTMENT.
6. ALL WATERMAIN SHALL HAVE AN 8 FT. BURY (MIN.) AND BE CABLE BONDED FOR CONTINUITY.
7. WATER SERVICES SHALL BE 1" TYPE K COPPER W / MUELLER OR FORD BRASS AND LOCATED BY OWNER.
8. ALL SERVICE CORPORATIONS SHALL HAVE A SADDLE CLAMP.
9. 8 FT. (MINIMUM) HORIZONTAL DISTANCE (℄ TO ℄) BETWEEN WATERMAIN AND SANITARY SEWER OR STORM SEWER.
10. 18-INCH (MINIMUM) VERTICAL DISTANCE (OUT TO OUT) BETWEEN WATERMAIN UNDER SANITARY SEWER OR STORM SEWER @ THEIR INTERSECTION.
11. 6-INCH (MINIMUM) VERTICAL DISTANCE (OUT TO OUT) BETWEEN WATERMAIN OVER SANITARY SEWER OR STORM SEWER @ THEIR INTERSECTION.
12. ALL ELEVATIONS BASED ON USGS DATUM.
13. ALL GATE VALVES IN TRANSMISSION MAIN SHALL HAVE A MAXIMUM SPACING OF 800 FEET.
14. EROSION CONTROL MEASURES SHALL BE IN PLACE BEFORE CONSTRUCTION BEINGS.
15. THE CONTRACTOR SHALL MINIMIZE EROSION DURING CONSTRUCTION USING GOOD CONSTRUCTION TECHNIQUES AND UTILIZING SILT FENCES AND BALE DITCH CHECKS. THE CONTRACTOR SHALL RELEASE RUNOFF FROM THE SITE IN A NUISANCE FREE MANNER.
16. THE CONTRACTOR IS RESPONSIBLE TO MAINTAIN THE SITE IN A SAFE CONDITION. THE SOLE RESPONSIBILITY FOR WARNING SIGNS, BARRICADES AND ALL ASPECTS OF SAFETY LIE WITH THE CONTRACTOR.
17. EXISTING UTILITIES SHOWN MAY NOT BE ALL INCLUSIVE, CONTRACTOR SHALL HAVE ALL BURIED UTILITIES LOCATED PRIOR TO COMMENCING CONSTRUCTION.
18. EXISTING WATER TRANSMISSION MAIN SHOWN ACCORDING TO EXISTING RECORDS, BUT LOCATION SHALL BE VERIFIED IN FIELD.
19. WATERMAIN & SERVICES SHALL BE INSULATED AT ALL STORM SEWER, CULVERT AND DITCH CROSSINGS WITH 2" THICK EXTRUDED POLYSTYRENE INSULATION BOARD.
20. PROPERTY CORNERS KNOWN TO EXIST SHALL NOT BE DISTURBED. DAMAGED CORNERS WILL BE REPLACED AT THE CONTRACTOR'S EXPENSE.
21. ALL WATERMAIN SHALL BE DISINFECTED BEFORE BEING PUT INTO USE.
22. ALL VALVE RISER CAPS SHALL BE SET TO GRADE SHOWN ON THE PLANS, UNLESS DIRECTED OTHERWISE.
23. CONSTRUCTION SHALL COMPLY WITH THE CONDITIONS OF DNR APPROVAL LETTER.
24. ALL DISTURBED AREAS SHALL BE SEEDED, FERTILIZED AND MULCHED, OR COVERED W / BASE COURSE AFTER CONSTRUCTION.
25. ALL DISTURBED AREAS SHALL BE RESTORED W / SEEDING AS FOLLOWS:
    - A) REPLACE ALL SALVAGED TOPSOIL
    - B) SEED MIXTURE: PERMANENT SEEDING - (LAWN TYPE TURF)
      - 35% KENTUCKY BLUEGRASS
      - 25% IMPROVED FINE PERENNIAL RYEGRASS
      - 15% CREEPING RED FESCUE
      - 15% IMPROVED HARD FESCUE
      - 10% WHITE CLOVER
    - C) SEED RATE: 2# / 1000 SQ. FT.
    - D) MULCH RATE: 3 TON / ACRE
    - E) PERMANENT SEEDING: ALLOWED TO SEPT. 7
      TEMPORARY SEEDING: SEPT. 8 THROUGH NOV. 10 - (4 BU. OATS / ACRE)
      DORMANT SEEDING: SEED PERM. SEEDING INTO TEMP. SEEDING AFTER NOV. 10
    - F) SEEDING MAY BE BROADCAST BUT MUST BE COVERED WITH A MAXIMUM OF ½" SOIL. MULCH MUST BE ANCHORED WITH A MULCH TILLER, TACKIFIER OR NETTING.
    - G) FERTILIZER: 500# / ACRE 20-10-10
26. CONTRACTOR SHALL DETERMINE CONSTRUCTION SEQUENCE IN ORDER TO MINIMIZE INSTALLATION CONFLICTS WITH PIPE CROSSINGS.
27. THERE SHALL BE NO DEVIATION FROM THE PLANS WITHOUT THE DESIGN ENGINEER'S APPROVAL.
28. CONSTRUCTION STAKES DESTROYED BY THE CONTRACTOR THAT NEED TO BE REPLACED, WILL BE CHARGED TO THE CONTRACTOR.
29. NO TREES ARE TO BE REMOVED WITHOUT THE APPROVAL OF THE ENGINEER.
30. CONTRACTOR SHALL COORDINATE AND SCHEDULE HIS WORK WITH VILLAGE & ENGINEER TO INSURE MINIMAL CONFLICT WITH WATER UTILITY DURING CONNECTION TO EXISTING TRANSMISSION MAIN. THEREFORE, THE CONTRACTOR SHALL SUPPLY A PLAN & SEQUENCE METHOD FOR TIE-IN, TESTING AND DISINFECTION FOR ENGINEER'S APPROVAL, PRIOR TO COMMENCING CONSTRUCTION.

REFERENCE SPECIFICATIONS*

*STANDARD SPECIFICATIONS FOR SEWER AND WATER CONSTRUCTION IN WISCONSIN (5TH EDITION) W / ADDENDUMS 1 & 2

*WISCONSIN CONSTRUCTION SITE BEST MANAGEMENT PRACTICE HANDBOOK

*MORGAN & PARMLEY, LTD. SUPPLEMENTAL SPECIFICATIONS

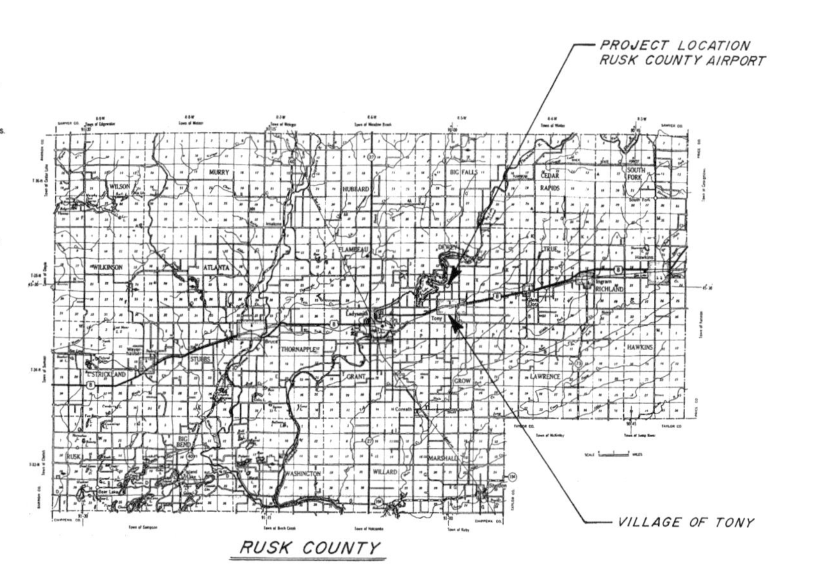

RUSK COUNTY

-UTILITIES-

SEWER & WATER: VILLAGE OF TONY
BILLY MECHELKE
N5297 LITTLE X RD.
TONY, WI 54563
715/532-3183 (W)
715/532-7046 (H)

ELECTRIC:
NORTHERN STATES POWER COMPANY
310 HICKORY HILL LANE
PHILLIPS, WISCONSIN 54555
ATTN: JOE PERKINS
715/836-1198

TELEPHONE:
CENTURYTEL
425 ELLINGSON AVENUE
P.O. BOX 78
HAWKINS, WI 54530
800/752-5637
ATTN: JAMES ARQUETTE

DIGGER'S HOTLINE:
800/242-8511

PREPARED BY:
MORGAN & PARMLEY, LTD.
PROFESSIONAL CONSULTING ENGINEERS
LADYSMITH, WISCONSIN

The airport initiated plans to construct a north-south cross runway which would intersect the existing potable water transmission main. This conflict needed to be resolved prior to constructing the new cross runway. The only logical solution to permanently protect the water supply was to reroute the transmission main, leaving the abandoned section in-place.

Source: Morgan & Parmley, Ltd.

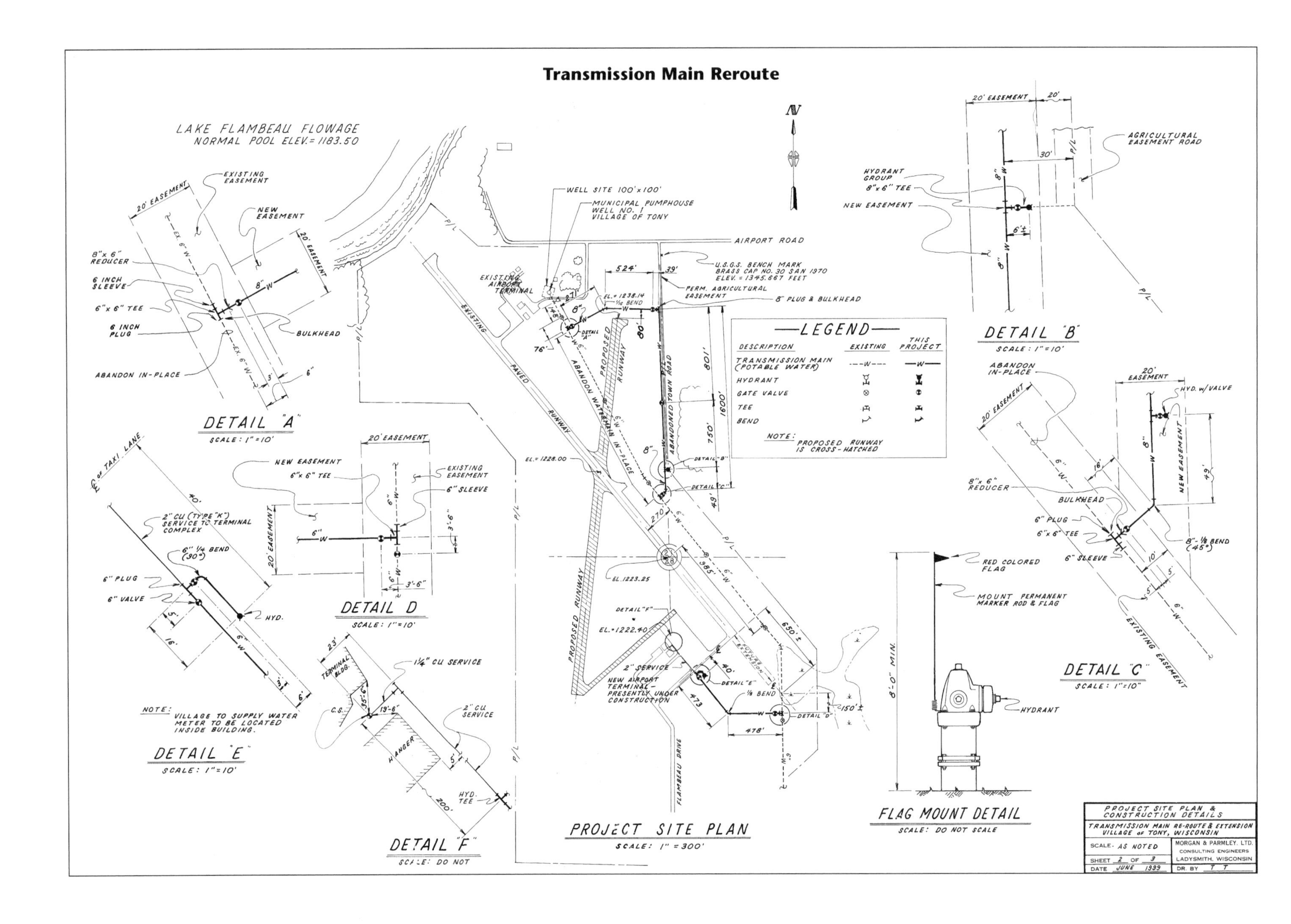
Transmission Main Reroute
LAKE FLAMBEAU FLOWAGE
NORMAL POOL ELEV.= 1183.50
DETAIL "A"
SCALE: 1"=10'
DETAIL "B"
SCALE: 1"=10'
DETAIL "C"
SCALE: 1"=10"
DETAIL D
SCALE: 1"=10'
DETAIL "E"
SCALE: 1"=10'
DETAIL "F"
PROJECT SITE PLAN
SCALE: 1" = 300'
FLAG MOUNT DETAIL
SCALE: DO NOT SCALE
LEGEND
DESCRIPTION
EXISTING
THIS PROJECT
TRANSMISSION MAIN (POTABLE WATER)
HYDRANT
GATE VALVE
TEE
BEND
NOTE: PROPOSED RUNWAY IS CROSS-HATCHED
WELL SITE 100'x100'
MUNICIPAL PUMPHOUSE WELL NO. 1 VILLAGE OF TONY
AIRPORT ROAD
U.S.G.S. BENCH MARK BRASS CAP NO. 30 SAN 1970 ELEV. = 1345.667 FEET
PERM. AGRICULTURAL EASEMENT
8" PLUG & BULKHEAD
NOTE: VILLAGE TO SUPPLY WATER METER TO BE LOCATED INSIDE BUILDING.
RED COLORED FLAG
MOUNT PERMANENT MARKER ROD & FLAG
HYDRANT
PROJECT SITE PLAN & CONSTRUCTION DETAILS
TRANSMISSION MAIN RE-ROUTE & EXTENSION VILLAGE of TONY, WISCONSIN
SCALE- AS NOTED
MORGAN & PARMLEY, LTD. CONSULTING ENGINEERS LADYSMITH, WISCONSIN
SHEET 2 OF 3
DATE JUNE 1999
DR. BY T T

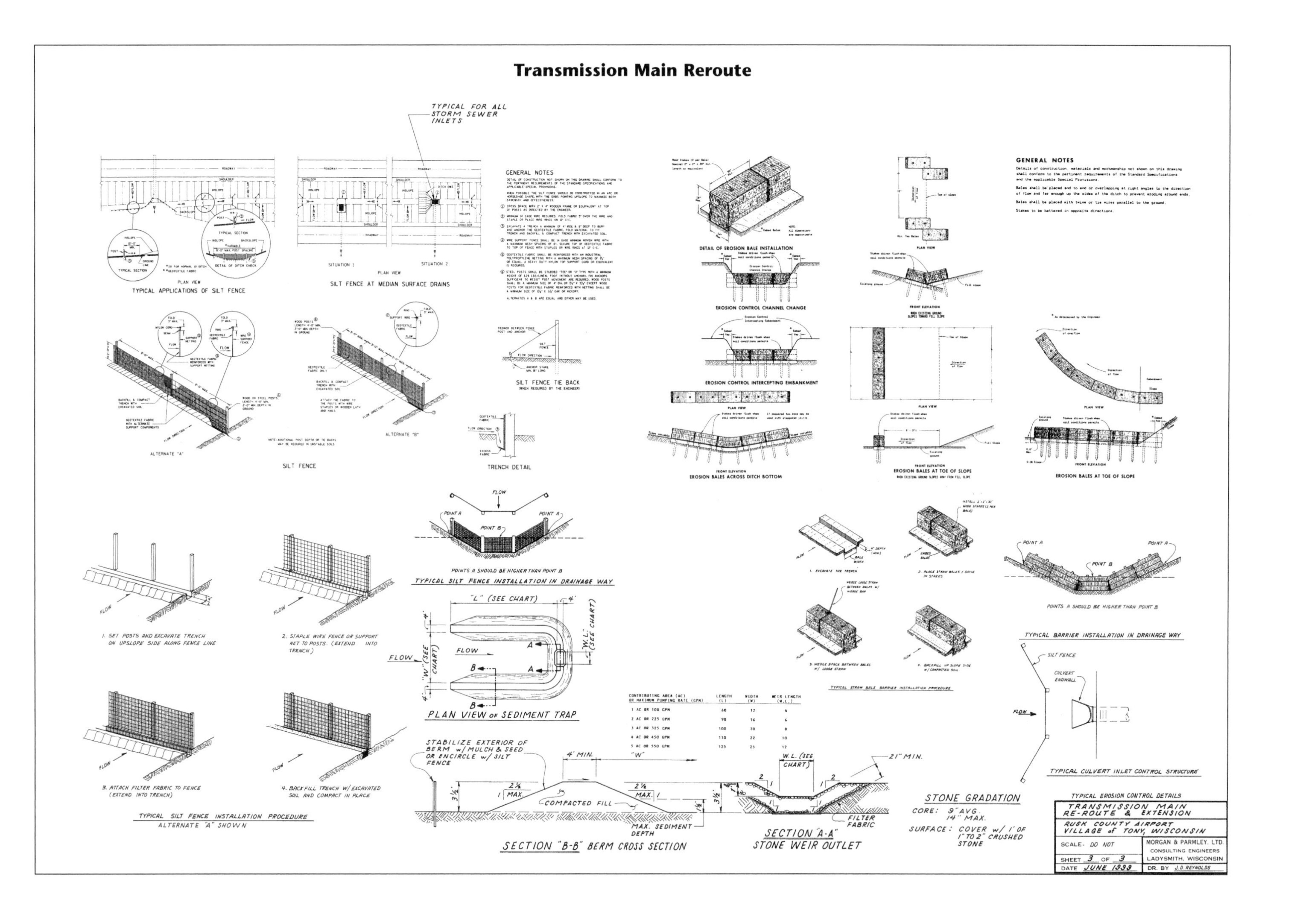

CONTRIBUTING AREA (AC) OR MAXIMUM PUMPING RATE (GPM)	LENGTH (L)	WIDTH (W)	WEIR LENGTH (W.L.)
1 AC OR 100 GPM	60	12	4
2 AC OR 225 GPM	90	16	6
3 AC OR 325 GPM	100	20	8
4 AC OR 450 GPM	110	22	10
5 AC OR 550 GPM	125	25	12

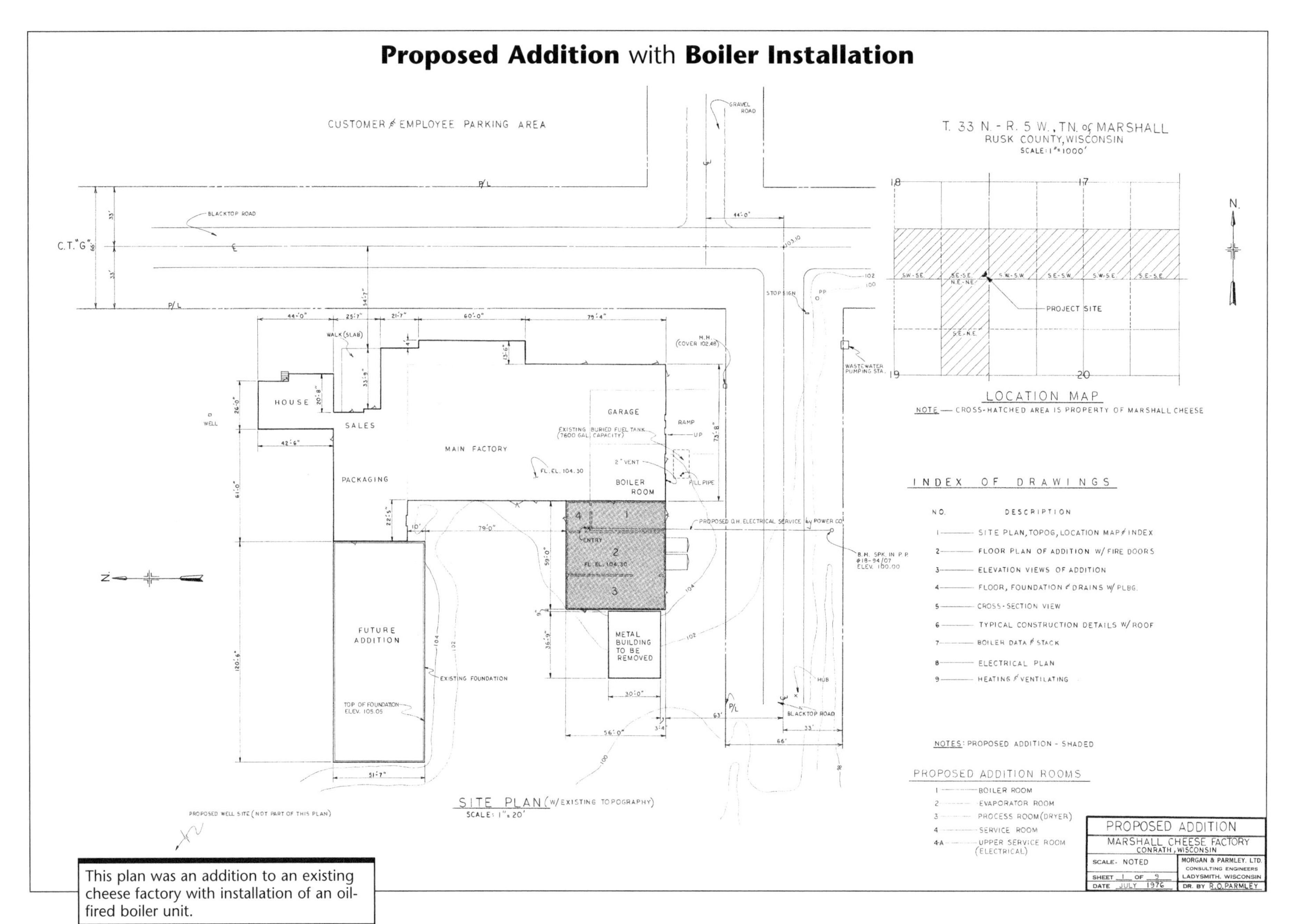

This plan was an addition to an existing cheese factory with installation of an oil-fired boiler unit.

Source: Morgan & Parmley, Ltd.

Proposed Addition with **Boiler Installation**

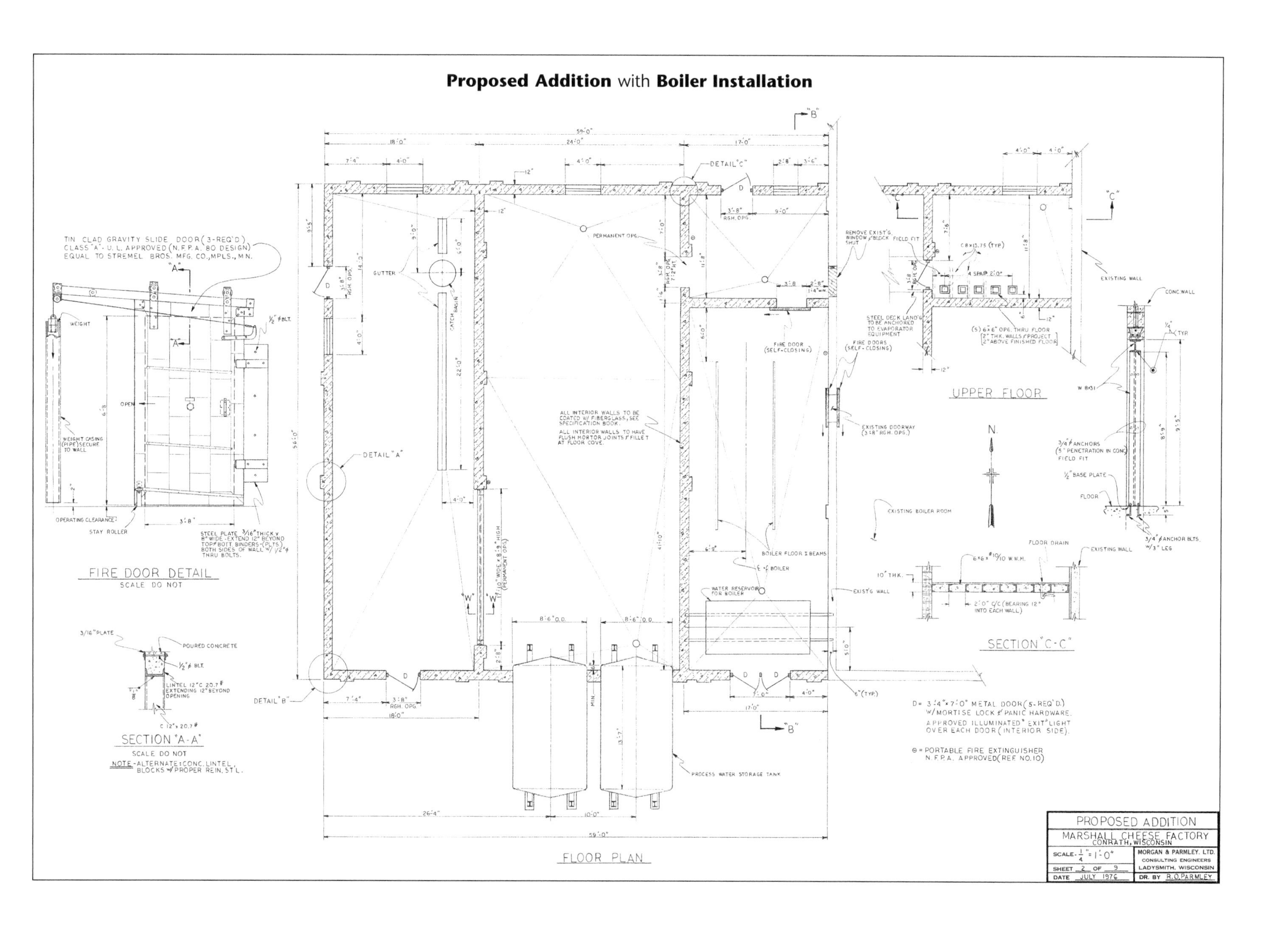

Proposed Addition with **Boiler Installation**

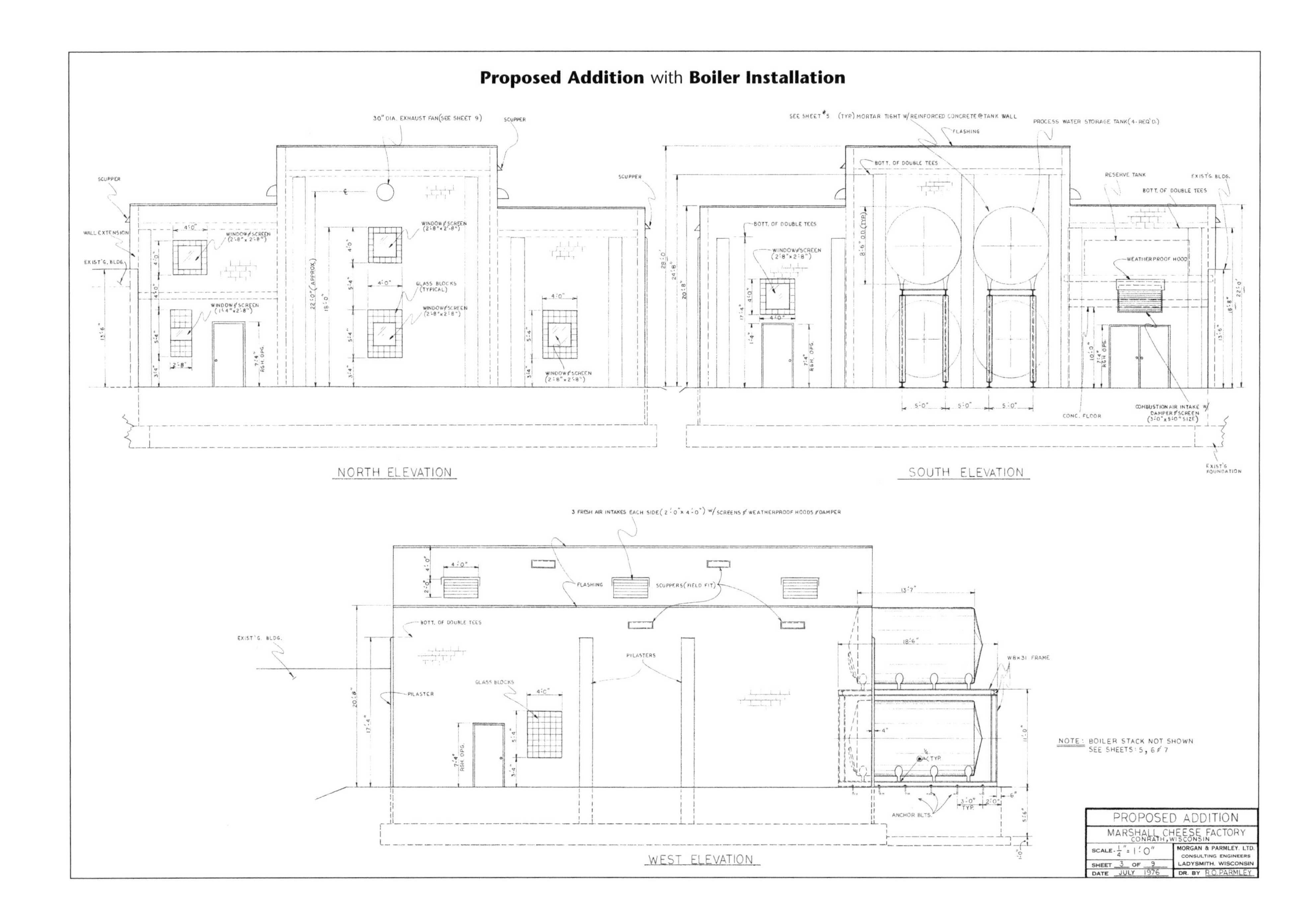

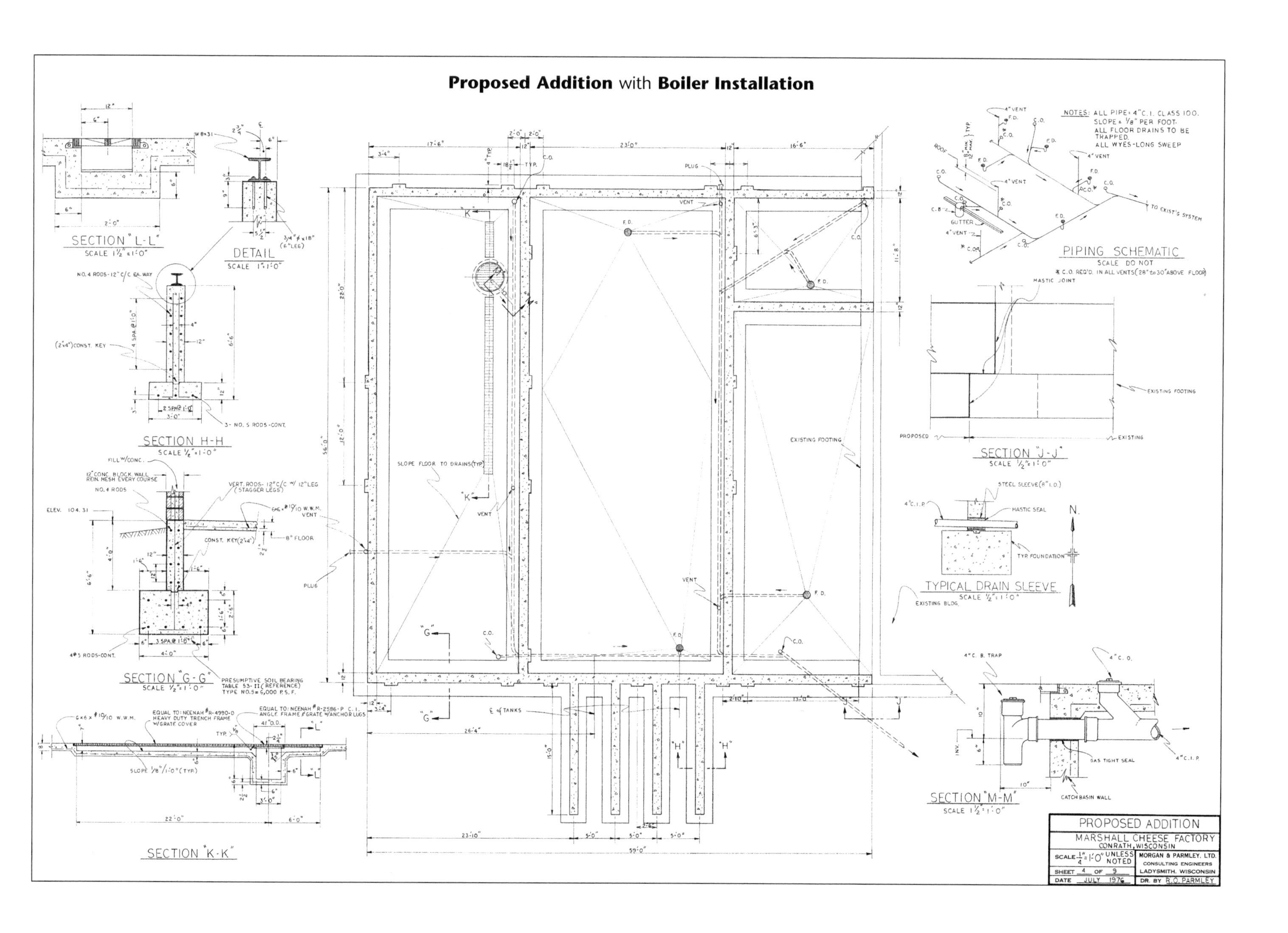
Proposed Addition with Boiler Installation
SECTION "L-L"
SCALE 1½"=1'-0"
DETAIL
SCALE 1"=1'-0"
SECTION H-H
SCALE ½"=1'-0"
NO. 4 RODS-12" C/C EA. WAY
(2"x4") CONST. KEY
3- NO. 5 RODS-CONT.
FILL W/CONC.
12" CONC. BLOCK WALL REIN. MESH EVERY COURSE
NO. 4 RODS
ELEV. 104.31
VERT. RODS- 12" C/C W/ 12" LEG (STAGGER LEGS)
CONST. KEY(2"x4")
8" FLOOR
4#5 RODS-CONT.
SECTION "G-G"
SCALE ½"=1'-0"
PRESUMPTIVE SOIL BEARING TABLE 53-II (REFERENCE) TYPE NO. 5=6,000 P.S.F.
EQUAL TO: NEENAH #R-4990-D HEAVY DUTY TRENCH FRAME W/GRATE COVER
EQUAL TO: NEENAH #R-2586-P C.I. ANGLE FRAME & GRATE W/ANCHOR LUGS
SLOPE ⅛"/1'-0" (TYP.)
SECTION "K-K"
SLOPE FLOOR TO DRAINS (TYP.)
VENT
PLUG
C.O.
F.D.
EXISTING FOOTING
℄ of TANKS
NOTES: ALL PIPE= 4" C.I. CLASS 100.
SLOPE = ⅛" PER FOOT.
ALL FLOOR DRAINS TO BE TRAPPED.
ALL WYES-LONG SWEEP
TO EXIST'G SYSTEM
PIPING SCHEMATIC
SCALE DO NOT
* C.O. REQ'D. IN ALL VENTS (28" to 30" ABOVE FLOOR)
MASTIC JOINT
EXISTING FOOTING
PROPOSED
EXISTING
SECTION "J-J"
SCALE ½"=1'-0"
STEEL SLEEVE (8" I.D.)
4" C.I.P.
MASTIC SEAL
TYP. FOUNDATION
TYPICAL DRAIN SLEEVE
SCALE ½"=1'-0"
EXISTING BLDG.
N.
4" C.B. TRAP
4" C.O.
GAS TIGHT SEAL
4" C.I.P.
CATCH BASIN WALL
SECTION "M-M"
SCALE 1½"=1'-0"
PROPOSED ADDITION
MARSHALL CHEESE FACTORY
CONRATH, WISCONSIN
SCALE ¼"=1'-0" UNLESS NOTED
MORGAN & PARMLEY, LTD.
CONSULTING ENGINEERS
LADYSMITH, WISCONSIN
SHEET 4 OF 9
DATE JULY 1976
DR. BY R.O. PARMLEY

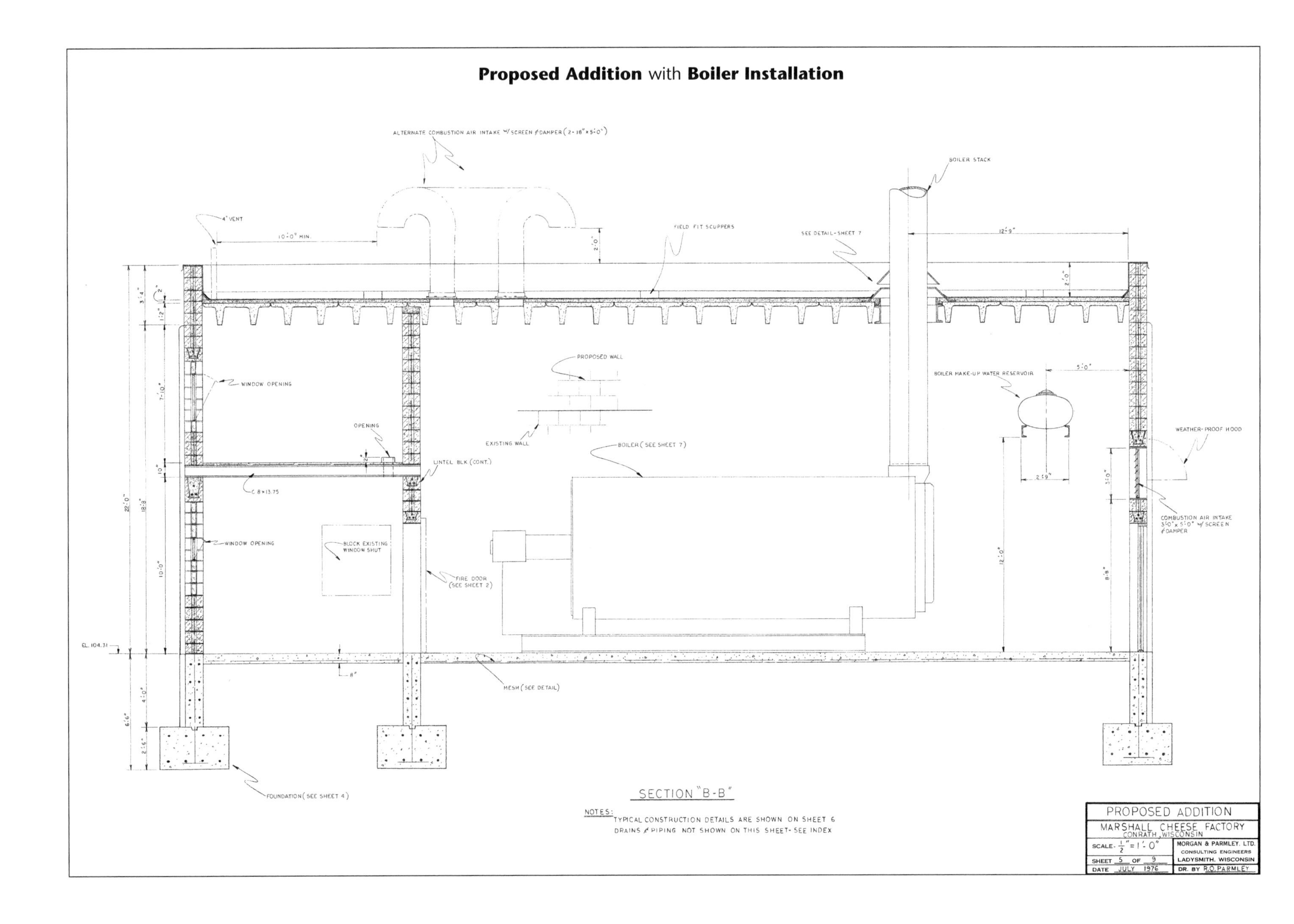
Proposed Addition with Boiler Installation
ALTERNATE COMBUSTION AIR INTAKE W/ SCREEN & DAMPER (2-18" x 5'-0")
BOILER STACK
4" VENT
10'-0" MIN.
FIELD FIT SCUPPERS
SEE DETAIL-SHEET 7
12'-9"
PROPOSED WALL
BOILER MAKE-UP WATER RESERVOIR
5'-0"
WINDOW OPENING
OPENING
EXISTING WALL
BOILER (SEE SHEET 7)
WEATHER-PROOF HOOD
LINTEL BLK (CONT.)
C 8 x 13.75
COMBUSTION AIR INTAKE 3'-0" x 5'-0" W/ SCREEN & DAMPER
WINDOW OPENING
BLOCK EXISTING WINDOW SHUT
FIRE DOOR (SEE SHEET 2)
EL. 104.31
MESH (SEE DETAIL)
FOUNDATION (SEE SHEET 4)
SECTION "B-B"
NOTES:
TYPICAL CONSTRUCTION DETAILS ARE SHOWN ON SHEET 6
DRAINS & PIPING NOT SHOWN ON THIS SHEET- SEE INDEX
PROPOSED ADDITION
MARSHALL CHEESE FACTORY
CONRATH, WISCONSIN
SCALE- 1/2" = 1'-0"
SHEET 5 OF 9
DATE JULY 1976
MORGAN & PARMLEY, LTD.
CONSULTING ENGINEERS
LADYSMITH, WISCONSIN
DR. BY R.O. PARMLEY

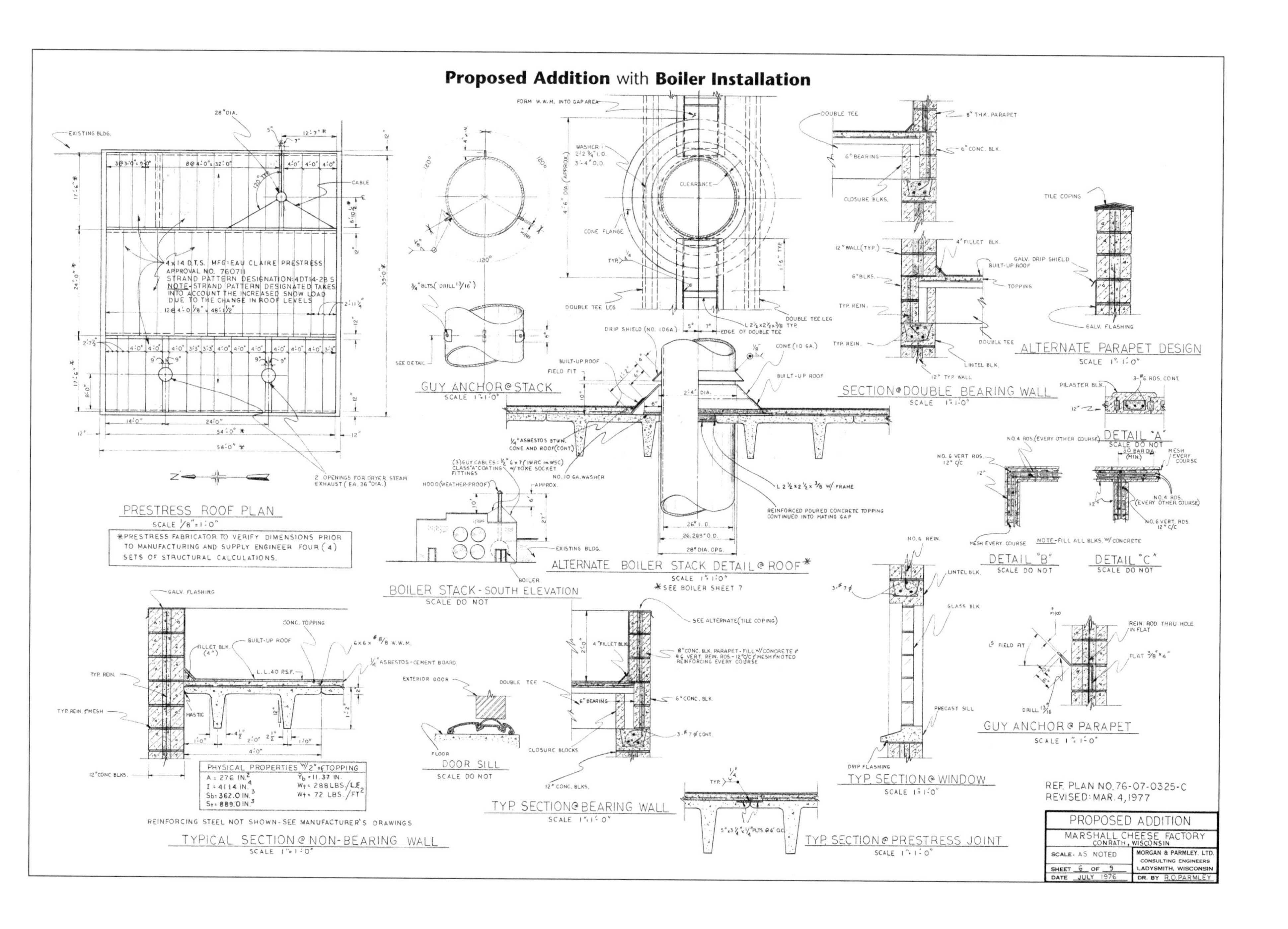
Proposed Addition with Boiler Installation
PRESTRESS ROOF PLAN
SCALE 1/8"=1'-0"
*PRESTRESS FABRICATOR TO VERIFY DIMENSIONS PRIOR TO MANUFACTURING AND SUPPLY ENGINEER FOUR (4) SETS OF STRUCTURAL CALCULATIONS.
GUY ANCHOR @ STACK
ALTERNATE BOILER STACK DETAIL @ ROOF*
*SEE BOILER SHEET 7
BOILER STACK - SOUTH ELEVATION
SECTION @ DOUBLE BEARING WALL
ALTERNATE PARAPET DESIGN
DETAIL "A"
DETAIL "B"
DETAIL "C"
TYPICAL SECTION @ NON-BEARING WALL
DOOR SILL
TYP. SECTION @ BEARING WALL
TYP. SECTION @ WINDOW
TYP. SECTION @ PRESTRESS JOINT
GUY ANCHOR @ PARAPET
REINFORCING STEEL NOT SHOWN - SEE MANUFACTURER'S DRAWINGS
REF. PLAN NO. 76-07-0325-C
REVISED: MAR. 4, 1977
PROPOSED ADDITION
MARSHALL CHEESE FACTORY
CONRATH, WISCONSIN
MORGAN & PARMLEY, LTD.
CONSULTING ENGINEERS
LADYSMITH, WISCONSIN
SHEET 6 OF 9
DATE JULY 1976
DR. BY R.O. PARMLEY

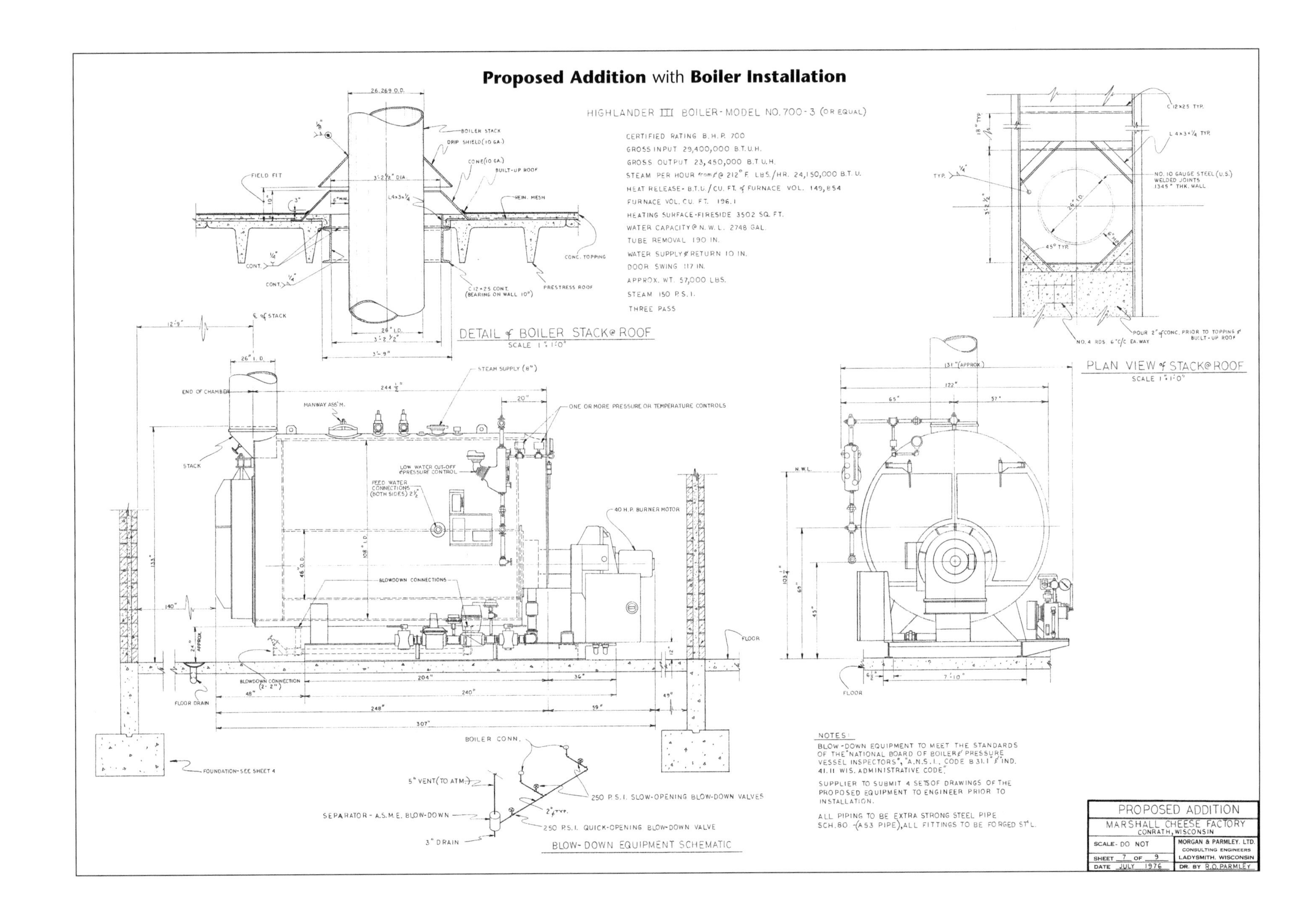
Proposed Addition with Boiler Installation
HIGHLANDER III BOILER-MODEL NO. 700-3 (OR EQUAL)
CERTIFIED RATING B.H.P. 700
GROSS INPUT 29,400,000 B.T.U.H.
GROSS OUTPUT 23,450,000 B.T.U.H.
STEAM PER HOUR from & @ 212° F. LBS./HR. 24,150,000 B.T.U.
HEAT RELEASE-B.T.U./CU. FT. of FURNACE VOL. 149,854
FURNACE VOL. CU. FT. 196.1
HEATING SURFACE-FIRESIDE 3502 SQ. FT.
WATER CAPACITY @ N.W.L. 2748 GAL.
TUBE REMOVAL 190 IN.
WATER SUPPLY & RETURN 10 IN.
DOOR SWING 117 IN.
APPROX. WT. 57,000 LBS.
STEAM 150 P.S.I.
THREE PASS
DETAIL of BOILER STACK @ ROOF
SCALE 1" = 1'-0"
PLAN VIEW of STACK @ ROOF
SCALE 1" = 1'-0"
BOILER STACK
DRIP SHIELD (10 GA.)
CONE (10 GA.)
BUILT-UP ROOF
FIELD FIT
REIN. MESH
CONC. TOPPING
PRESTRESS ROOF
C 12×25 CONT. (BEARING ON WALL 10")
C 12×25 TYP.
NO. 10 GAUGE STEEL (U.S.) WELDED JOINTS .1345" THK. WALL
POUR 2" of CONC. PRIOR TO TOPPING & BUILT-UP ROOF
NO. 4 RDS. 6" c/c EA. WAY
END OF CHAMBER
STEAM SUPPLY (8")
MANWAY ASS'M.
ONE OR MORE PRESSURE OR TEMPERATURE CONTROLS
STACK
LOW WATER CUT-OFF & PRESSURE CONTROL
FEED WATER CONNECTIONS (BOTH SIDES) 2½"
40 H.P. BURNER MOTOR
BLOWDOWN CONNECTIONS
BLOWDOWN CONNECTION (2-2")
FLOOR DRAIN
FLOOR
N.W.L.
FOUNDATION-SEE SHEET 4
BOILER CONN.
5" VENT (TO ATM.)
250 P.S.I. SLOW-OPENING BLOW-DOWN VALVES
SEPARATOR - A.S.M.E. BLOW-DOWN
250 P.S.I. QUICK-OPENING BLOW-DOWN VALVE
3" DRAIN
BLOW-DOWN EQUIPMENT SCHEMATIC
NOTES:
BLOW-DOWN EQUIPMENT TO MEET THE STANDARDS OF THE "NATIONAL BOARD OF BOILER & PRESSURE VESSEL INSPECTORS", "A.N.S.I. CODE B31.1" & IND. 41.11 WIS. ADMINISTRATIVE CODE.
SUPPLIER TO SUBMIT 4 SETS OF DRAWINGS OF THE PROPOSED EQUIPMENT TO ENGINEER PRIOR TO INSTALLATION.
ALL PIPING TO BE EXTRA STRONG STEEL PIPE SCH. 80 -(A53 PIPE), ALL FITTINGS TO BE FORGED ST'L.
PROPOSED ADDITION
MARSHALL CHEESE FACTORY
CONRATH, WISCONSIN
SCALE- DO NOT
SHEET 7 OF 9
DATE JULY 1976
MORGAN & PARMLEY, LTD.
CONSULTING ENGINEERS
LADYSMITH, WISCONSIN
DR. BY R.O. PARMLEY

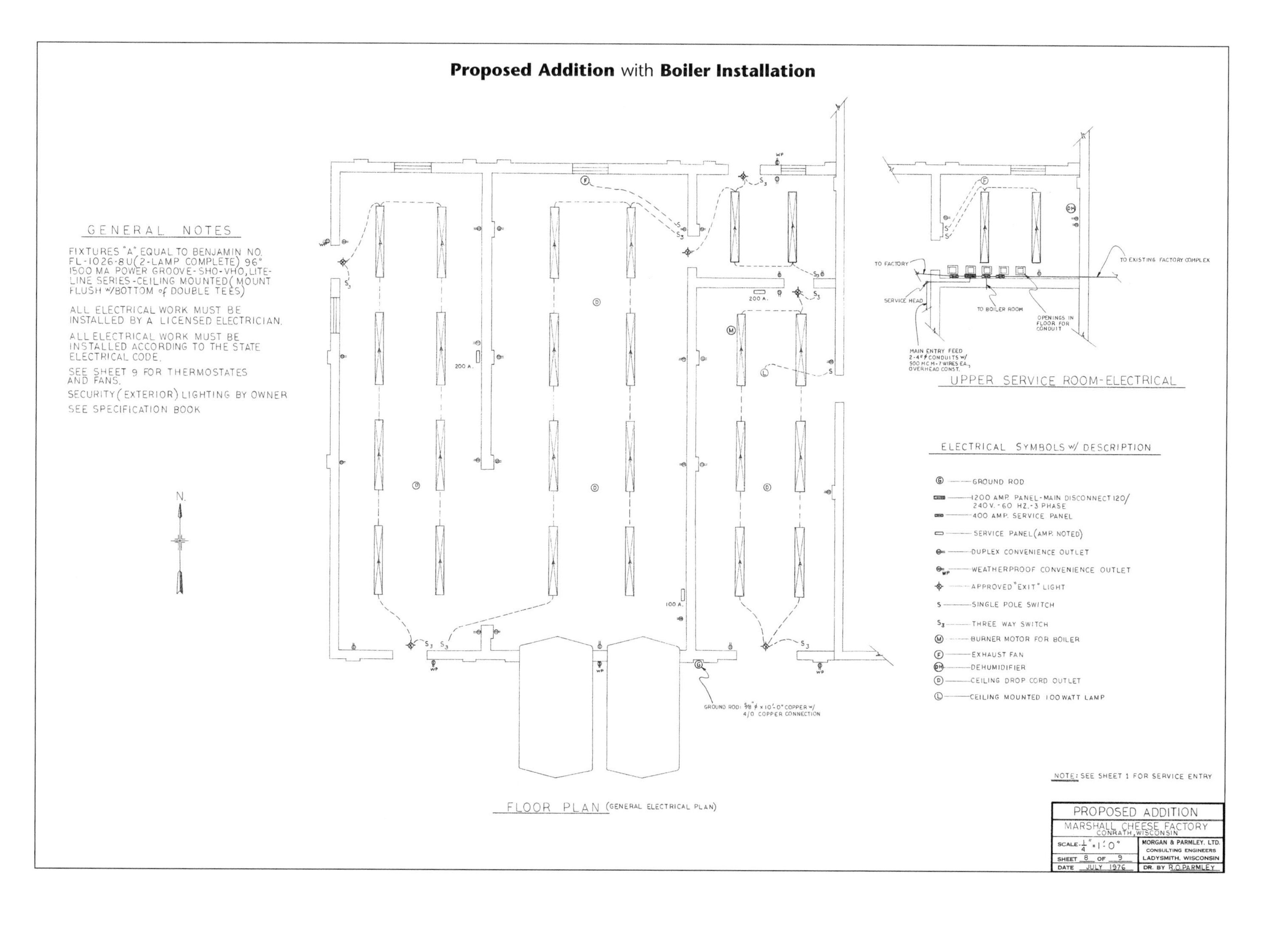
Proposed Addition with Boiler Installation
GENERAL NOTES
FIXTURES "A" EQUAL TO BENJAMIN NO. FL-1026-8U (2-LAMP COMPLETE) 96" 1500 MA POWER GROOVE-SHO-VHO, LITE-LINE SERIES-CEILING MOUNTED (MOUNT FLUSH w/BOTTOM of DOUBLE TEES)
ALL ELECTRICAL WORK MUST BE INSTALLED BY A LICENSED ELECTRICIAN.
ALL ELECTRICAL WORK MUST BE INSTALLED ACCORDING TO THE STATE ELECTRICAL CODE.
SEE SHEET 9 FOR THERMOSTATES AND FANS.
SECURITY (EXTERIOR) LIGHTING BY OWNER
SEE SPECIFICATION BOOK
N.
200 A.
100 A.
GROUND ROD: 5/8" ϕ x 10'-0" COPPER w/ 4/0 COPPER CONNECTION
FLOOR PLAN (GENERAL ELECTRICAL PLAN)
TO FACTORY
TO EXISTING FACTORY COMPLEX
SERVICE HEAD
TO BOILER ROOM
OPENINGS IN FLOOR FOR CONDUIT
MAIN ENTRY FEED 2-4" ϕ CONDUITS w/ 500 MCM-7 WIRES EA., OVERHEAD CONST.
UPPER SERVICE ROOM-ELECTRICAL
ELECTRICAL SYMBOLS w/ DESCRIPTION
GROUND ROD
1200 AMP. PANEL-MAIN DISCONNECT 120/240 V.-60 HZ.-3 PHASE
400 AMP. SERVICE PANEL
SERVICE PANEL (AMP. NOTED)
DUPLEX CONVENIENCE OUTLET
WEATHERPROOF CONVENIENCE OUTLET
APPROVED "EXIT" LIGHT
SINGLE POLE SWITCH
THREE WAY SWITCH
BURNER MOTOR FOR BOILER
EXHAUST FAN
DEHUMIDIFIER
CEILING DROP CORD OUTLET
CEILING MOUNTED 100 WATT LAMP
NOTE: SEE SHEET 1 FOR SERVICE ENTRY
PROPOSED ADDITION
MARSHALL CHEESE FACTORY
CONRATH, WISCONSIN
SCALE 1/4" = 1'-0"
SHEET 8 OF 9
DATE JULY 1976
MORGAN & PARMLEY, LTD.
CONSULTING ENGINEERS
LADYSMITH, WISCONSIN
DR. BY R.O. PARMLEY

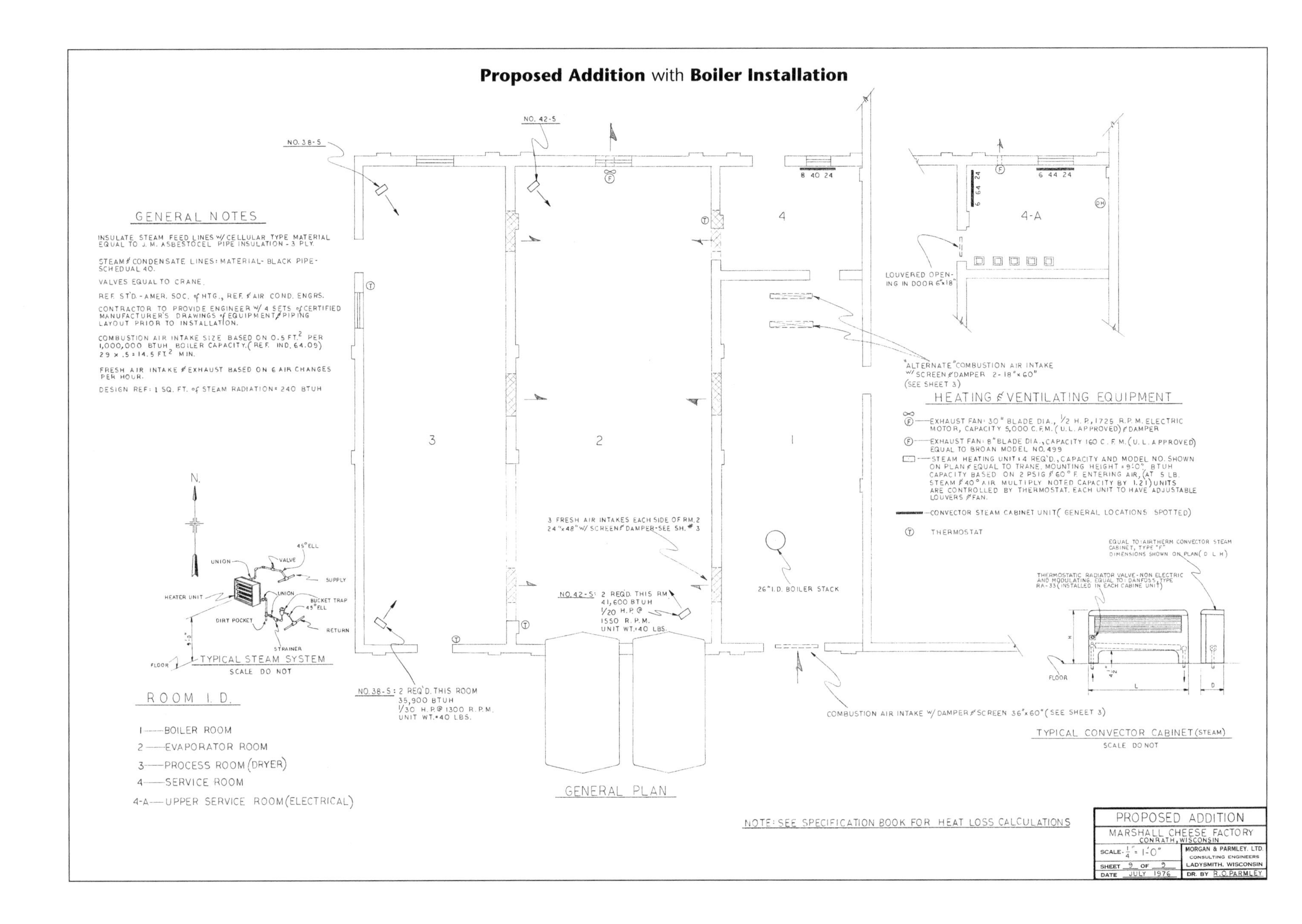
Proposed Addition with Boiler Installation
GENERAL NOTES
INSULATE STEAM FEED LINES W/ CELLULAR TYPE MATERIAL EQUAL TO J. M. ASBESTOCEL PIPE INSULATION - 3 PLY.
STEAM & CONDENSATE LINES: MATERIAL- BLACK PIPE- SCHEDUAL 40.
VALVES EQUAL TO CRANE.
REF. ST'D.- AMER. SOC. of HTG., REF. & AIR COND. ENGRS.
CONTRACTOR TO PROVIDE ENGINEER W/ 4 SETS of CERTIFIED MANUFACTURER'S DRAWINGS of EQUIPMENT & PIPING LAYOUT PRIOR TO INSTALLATION.
COMBUSTION AIR INTAKE SIZE BASED ON 0.5 FT.2 PER 1,000,000 BTUH BOILER CAPACITY. (REF. IND. 64.09) 29 x .5 = 14.5 FT.2 MIN.
FRESH AIR INTAKE & EXHAUST BASED ON 6 AIR CHANGES PER HOUR.
DESIGN REF: 1 SQ. FT. of STEAM RADIATION = 240 BTUH
NO. 38-S
NO. 42-S
8 40 24
6 44 24
6 64 24
4
4-A
DH
LOUVERED OPENING IN DOOR 6"x18"
"ALTERNATE" COMBUSTION AIR INTAKE W/ SCREEN & DAMPER 2- 18"x 60" (SEE SHEET 3)
HEATING & VENTILATING EQUIPMENT
EXHAUST FAN: 30" BLADE DIA., 1/2 H. P., 1725 R. P. M. ELECTRIC MOTOR, CAPACITY 5,000 C. F. M. (U. L. APPROVED) & DAMPER
EXHAUST FAN: 8" BLADE DIA., CAPACITY 160 C. F. M. (U. L. APPROVED) EQUAL TO BROAN MODEL NO. 499
STEAM HEATING UNIT: 4 REQ'D., CAPACITY AND MODEL NO. SHOWN ON PLAN & EQUAL TO TRANE. MOUNTING HEIGHT = 9'-0". BTUH CAPACITY BASED ON 2 PSIG & 60° F. ENTERING AIR, (AT 5 LB. STEAM & 40° AIR MULTIPLY NOTED CAPACITY BY 1.21) UNITS ARE CONTROLLED BY THERMOSTAT. EACH UNIT TO HAVE ADJUSTABLE LOUVERS & FAN.
CONVECTOR STEAM CABINET UNIT (GENERAL LOCATIONS SPOTTED)
THERMOSTAT
3
2
1
N.
45° ELL
UNION
VALVE
SUPPLY
HEATER UNIT
UNION
BUCKET TRAP
45° ELL
DIRT POCKET
RETURN
STRAINER
FLOOR
TYPICAL STEAM SYSTEM
SCALE DO NOT
3 FRESH AIR INTAKES EACH SIDE OF RM. 2 24"x48" W/ SCREEN & DAMPER- SEE SH. # 3
NO. 42-S: 2 REQ'D. THIS RM. 41,600 BTUH 1/20 H. P. @ 1550 R. P. M. UNIT WT. = 40 LBS.
26" I. D. BOILER STACK
EQUAL TO: AIRTHERM CONVECTOR STEAM CABINET, TYPE "F" DIMENSIONS SHOWN ON PLAN (D L H)
THERMOSTATIC RADIATOR VALVE- NON ELECTRIC AND MODULATING. EQUAL TO: DANFOSS, TYPE RA-33 (INSTALLED IN EACH CABINET UNIT)
FLOOR
L
D
ROOM I. D.
1 — BOILER ROOM
2 — EVAPORATOR ROOM
3 — PROCESS ROOM (DRYER)
4 — SERVICE ROOM
4-A — UPPER SERVICE ROOM (ELECTRICAL)
NO. 38-S: 2 REQ'D. THIS ROOM 35,900 BTUH 1/30 H. P. @ 1300 R. P. M. UNIT WT. = 40 LBS.
COMBUSTION AIR INTAKE W/ DAMPER & SCREEN 36"x 60" (SEE SHEET 3)
TYPICAL CONVECTOR CABINET (STEAM)
SCALE DO NOT
GENERAL PLAN
NOTE: SEE SPECIFICATION BOOK FOR HEAT LOSS CALCULATIONS
PROPOSED ADDITION
MARSHALL CHEESE FACTORY
CONRATH, WISCONSIN
SCALE- 1/4" = 1'-0"
SHEET 9 OF 9
DATE JULY 1976
MORGAN & PARMLEY, LTD.
CONSULTING ENGINEERS
LADYSMITH, WISCONSIN
DR. BY R. O. PARMLEY

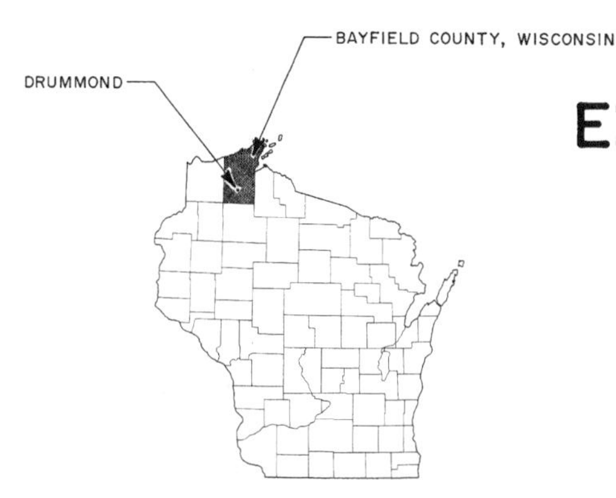

GENERAL LOCATION MAP

ELEVATED WATER TANK REHABILITATION

DRUMMOND SANITARY DISTRICT

DRUMMOND, WISCONSIN

M. & P. PROJECT NO. 99-157

APRIL 2000

DRUMMOND FIRE HALL

DRUMMOND TOWN HALL

EXISTING MUNICIPAL WELL & PUMPHOUSE

PROJECT SITE: EXISTING ELEVATED WATER TANK

MAP of DRUMMOND, WISCONSIN

SCALE 0' 200'

NOTES

REFER TO ACCOMPANYING SPECIFICATIONS AND REGULATORY APPROVALS FOR SPECIFIC DETAILS AND PROCEDURES.

REFER TO SECTION 01010-SUMMARY OF WORK-FOR PROPOSED WORK AND SPECIAL PROVISIONS.

PAINTING SYSTEMS AND REMOVAL / APPLICATION PROCEDURES SHALL BE CONSISTANT WITH CURRENT AWWA STANDARDS AND APPLICABLE REGULATORY RULES & GUIDLINES.

PRIOR TO PLACING FACILITY BACK INTO SERVICE, THE ENTIRE RESERVOIR FACILITY SHALL BE COMPLETELY DISINFECTED, AS OUTLINED IN THE CURRENT AWWA STANDARD C652. SAFE WATER SAMPLES SHALL BE OBTAINED FROM A CERTIFIED LAB BEFORE PLACING INTO SERVICE.

BEFORE TAKING ELEVATED TANK OFF THE DISTRIBUTION SYSTEM, CONTRACTOR SHALL GIVE THE WATERWORKS OPERATOR AND THE FIRE DEPARTMENT ADEQUATE ADVANCE NOTICE.

UTILITIES

MUNICIPAL WATER SYSTEM:

DRUMMOND SANITARY DISTRICT
P.O. BOX 43 — FRONT AVE.
DRUMMOND, WISCONSIN 54832
MARK JEROME: WATERWORKS OPERATOR
TELE. NO. 715 / 739-6741

TELEPHONE COMPANY:

CHEQUAMEGON TELEPHONE CO.
1st AVENUE – P. O. BOX 67
CABLE, WISCONSIN 54821
TELE. NO. 800 / 250-8927

ELECTRICAL SERVICE:

(A.K.A.) NORTHERN STATES POWER CO. (XCEL ENERGY)
16048 ELECTRIC AVENUE
HAYWARD, WISCONSIN 54843
KEN DISHER: FIELD ENGINEER
TELE. NO. 800 / 895 - 4999

FIRE DEPARTMENT

DRUMMOND FIRE & RESCUE
WADE SPEARS, CHIEF
TELE. NO. 715 / 739 - 6696

DIGGERS HOTLINE

TELE. NO. 800/242-8511

INDEX OF SHEETS

SHEET NO.	DESCRIPTION
1	TITLE SHEET, GENERAL LOCATION MAP, PROJECT LOCATION MAP, UTILITIES LIST, SHEET INDEX & NOTES
2	EXISTING TANK GEOMETRY AND CONDITIONS w/ REHABILITATION DETAILS
3	PROPOSED CIRCULATION PUMP SYSTEM AND RETROFITTING DETAILS

PREPARED BY:
MORGAN & PARMLEY, Ltd.
PROFESSIONAL CONSULTING ENGINEERS
115 W. 2nd STREET SOUTH
LADYSMITH, WISCONSIN 54848

A 20 year old elevated potable water storage tank required a major rehabilitation. The tank insulation had seriously deteriorated from ultraviolet rays, the paint contained lead, cadmium, and chromium, and there was need for some structural upgrades.

Source: Morgan & Parmley, Ltd.

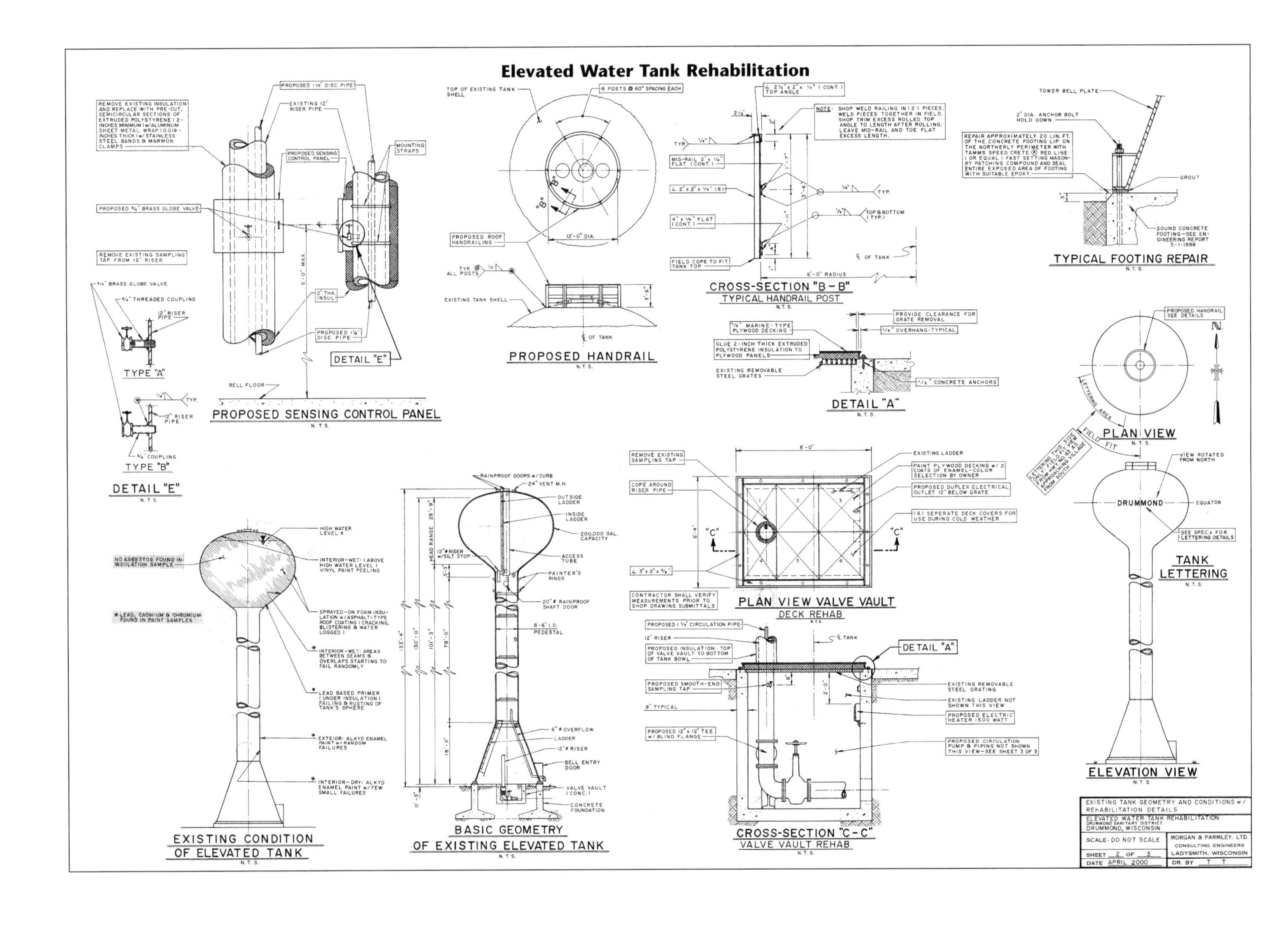

Elevated Water Tank Rehabilitation
PROPOSED SENSING CONTROL PANEL
DETAIL "E"
TYPE "A"
TYPE "B"
PROPOSED HANDRAIL
CROSS-SECTION "B-B"
TYPICAL HANDRAIL POST
TYPICAL FOOTING REPAIR
DETAIL "A"
PLAN VIEW
PLAN VIEW VALVE VAULT
DECK REHAB
CROSS-SECTION "C-C"
VALVE VAULT REHAB
TANK LETTERING
ELEVATION VIEW
EXISTING CONDITION OF ELEVATED TANK
BASIC GEOMETRY OF EXISTING ELEVATED TANK
DRUMMOND
EXISTING TANK GEOMETRY AND CONDITIONS w/ REHABILITATION DETAILS
ELEVATED WATER TANK REHABILITATION
DRUMMOND, WISCONSIN
SCALE - DO NOT SCALE
SHEET 2 OF 3
DATE APRIL 2000
MORGAN & PARMLEY, LTD.
CONSULTING ENGINEERS
LADYSMITH, WISCONSIN
DR. BY T T

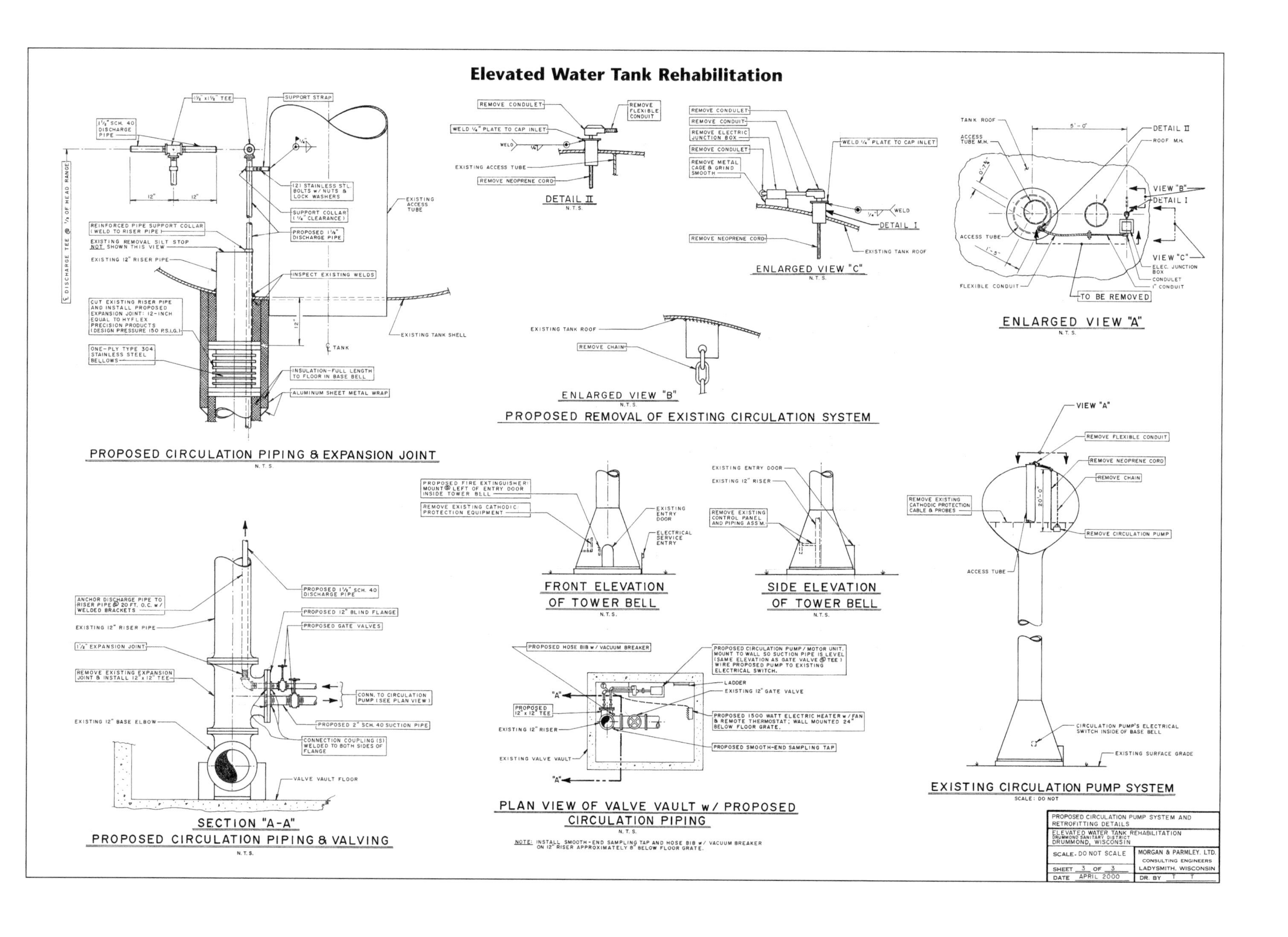
Elevated Water Tank Rehabilitation
1½" x 1½" TEE
SUPPORT STRAP
1½" SCH. 40 DISCHARGE PIPE
(2) STAINLESS STL. BOLTS w/ NUTS & LOCK WASHERS
SUPPORT COLLAR (¼" CLEARANCE)
PROPOSED 1½" DISCHARGE PIPE
EXISTING ACCESS TUBE
REINFORCED PIPE SUPPORT COLLAR (WELD TO RISER PIPE)
EXISTING REMOVAL SILT STOP NOT SHOWN THIS VIEW
EXISTING 12" RISER PIPE
INSPECT EXISTING WELDS
CUT EXISTING RISER PIPE AND INSTALL PROPOSED EXPANSION JOINT: 12-INCH EQUAL TO HYFLEX PRECISION PRODUCTS (DESIGN PRESSURE 150 P.S.I.G.)
ONE-PLY TYPE 304 STAINLESS STEEL BELLOWS
INSULATION-FULL LENGTH TO FLOOR IN BASE BELL
ALUMINUM SHEET METAL WRAP
EXISTING TANK SHELL
℄ TANK
℄ DISCHARGE TEE @ ⅓ OF HEAD RANGE
PROPOSED CIRCULATION PIPING & EXPANSION JOINT
N.T.S.
REMOVE CONDULET
REMOVE FLEXIBLE CONDUIT
WELD ¼" PLATE TO CAP INLET
WELD
EXISTING ACCESS TUBE
REMOVE NEOPRENE CORD
DETAIL II
N.T.S.
REMOVE CONDULET
REMOVE CONDUIT
REMOVE ELECTRIC JUNCTION BOX
REMOVE CONDULET
REMOVE METAL CAGE & GRIND SMOOTH
WELD ¼" PLATE TO CAP INLET
WELD
DETAIL I
REMOVE NEOPRENE CORD
EXISTING TANK ROOF
ENLARGED VIEW "C"
N.T.S.
EXISTING TANK ROOF
REMOVE CHAIN
ENLARGED VIEW "B"
N.T.S.
PROPOSED REMOVAL OF EXISTING CIRCULATION SYSTEM
TANK ROOF
ACCESS TUBE M.H.
5'-0"
DETAIL II
ROOF M.H.
VIEW "B"
DETAIL I
ACCESS TUBE
VIEW "C"
ELEC. JUNCTION BOX
CONDULET
1" CONDUIT
FLEXIBLE CONDUIT
TO BE REMOVED
ENLARGED VIEW "A"
N.T.S.
PROPOSED FIRE EXTINGUISHER; MOUNT @ LEFT OF ENTRY DOOR INSIDE TOWER BELL
REMOVE EXISTING CATHODIC PROTECTION EQUIPMENT
EXISTING ENTRY DOOR
ELECTRICAL SERVICE ENTRY
FRONT ELEVATION OF TOWER BELL
N.T.S.
EXISTING ENTRY DOOR
EXISTING 12" RISER
REMOVE EXISTING CONTROL PANEL AND PIPING ASS'M.
SIDE ELEVATION OF TOWER BELL
N.T.S.
VIEW "A"
REMOVE FLEXIBLE CONDUIT
REMOVE NEOPRENE CORD
REMOVE CHAIN
REMOVE EXISTING CATHODIC PROTECTION CABLE & PROBES
20'-0"
REMOVE CIRCULATION PUMP
ACCESS TUBE
CIRCULATION PUMP'S ELECTRICAL SWITCH INSIDE OF BASE BELL
EXISTING SURFACE GRADE
EXISTING CIRCULATION PUMP SYSTEM
SCALE: DO NOT
ANCHOR DISCHARGE PIPE TO RISER PIPE @ 20 FT. O.C. w/ WELDED BRACKETS
EXISTING 12" RISER PIPE
1½" EXPANSION JOINT
REMOVE EXISTING EXPANSION JOINT & INSTALL 12" x 12" TEE
EXISTING 12" BASE ELBOW
PROPOSED 1½" SCH. 40 DISCHARGE PIPE
PROPOSED 12" BLIND FLANGE
PROPOSED GATE VALVES
CONN. TO CIRCULATION PUMP (SEE PLAN VIEW)
PROPOSED 2" SCH. 40 SUCTION PIPE
CONNECTION COUPLING(S) WELDED TO BOTH SIDES OF FLANGE
VALVE VAULT FLOOR
SECTION "A-A"
PROPOSED CIRCULATION PIPING & VALVING
N.T.S.
PROPOSED HOSE BIB w/ VACUUM BREAKER
PROPOSED CIRCULATION PUMP/MOTOR UNIT. MOUNT TO WALL SO SUCTION PIPE IS LEVEL (SAME ELEVATION AS GATE VALVE @ TEE) WIRE PROPOSED PUMP TO EXISTING ELECTRICAL SWITCH.
LADDER
EXISTING 12" GATE VALVE
"A"
PROPOSED 12" x 12" TEE
EXISTING 12" RISER
PROPOSED 1500 WATT ELECTRIC HEATER w/ FAN & REMOTE THERMOSTAT; WALL MOUNTED 24" BELOW FLOOR GRATE.
PROPOSED SMOOTH-END SAMPLING TAP
EXISTING VALVE VAULT
PLAN VIEW OF VALVE VAULT w/ PROPOSED CIRCULATION PIPING
N.T.S.
NOTE: INSTALL SMOOTH-END SAMPLING TAP AND HOSE BIB w/ VACUUM BREAKER ON 12" RISER APPROXIMATELY 8" BELOW FLOOR GRATE.
PROPOSED CIRCULATION PUMP SYSTEM AND RETROFITTING DETAILS
ELEVATED WATER TANK REHABILITATION
DRUMMOND SANITARY DISTRICT
DRUMMOND, WISCONSIN
SCALE- DO NOT SCALE
MORGAN & PARMLEY, LTD.
CONSULTING ENGINEERS
LADYSMITH, WISCONSIN
SHEET 3 OF 3
DATE APRIL 2000
DR. BY T T

Design

SPECIALIZED PROJECTS

Section 20

Specialized projects are those projects that do not fit into established general areas of practice and most often are a "one-of-a-kind." Often they are hybrids that incorporate several basic disciplines into the mix to accomplish the goal of the special project. Structural, HVAC, electrical, plumbing, and piping systems are just a few of the components that may be called into play to complete the specialty. The wide variety of design skills may require sub-consultants working under the direction of the prime consulting engineer to round out the plans and specification. These conditions were present in the latter two plans displayed in this section, namely the rail site improvement project and the fish hatchery.

TABLE OF CONTENTS

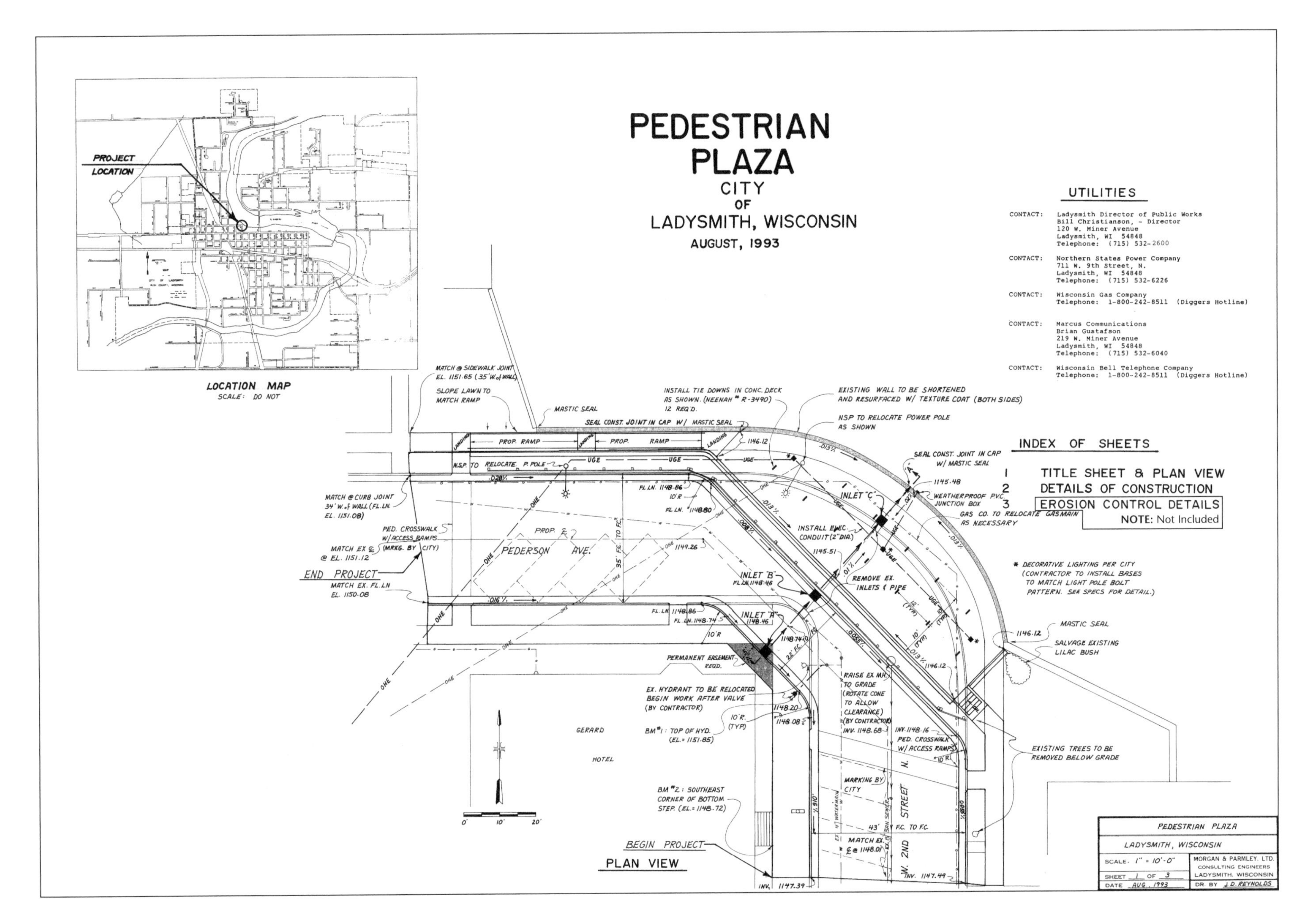

Courtesy: Morgan & Parmley, Ltd.

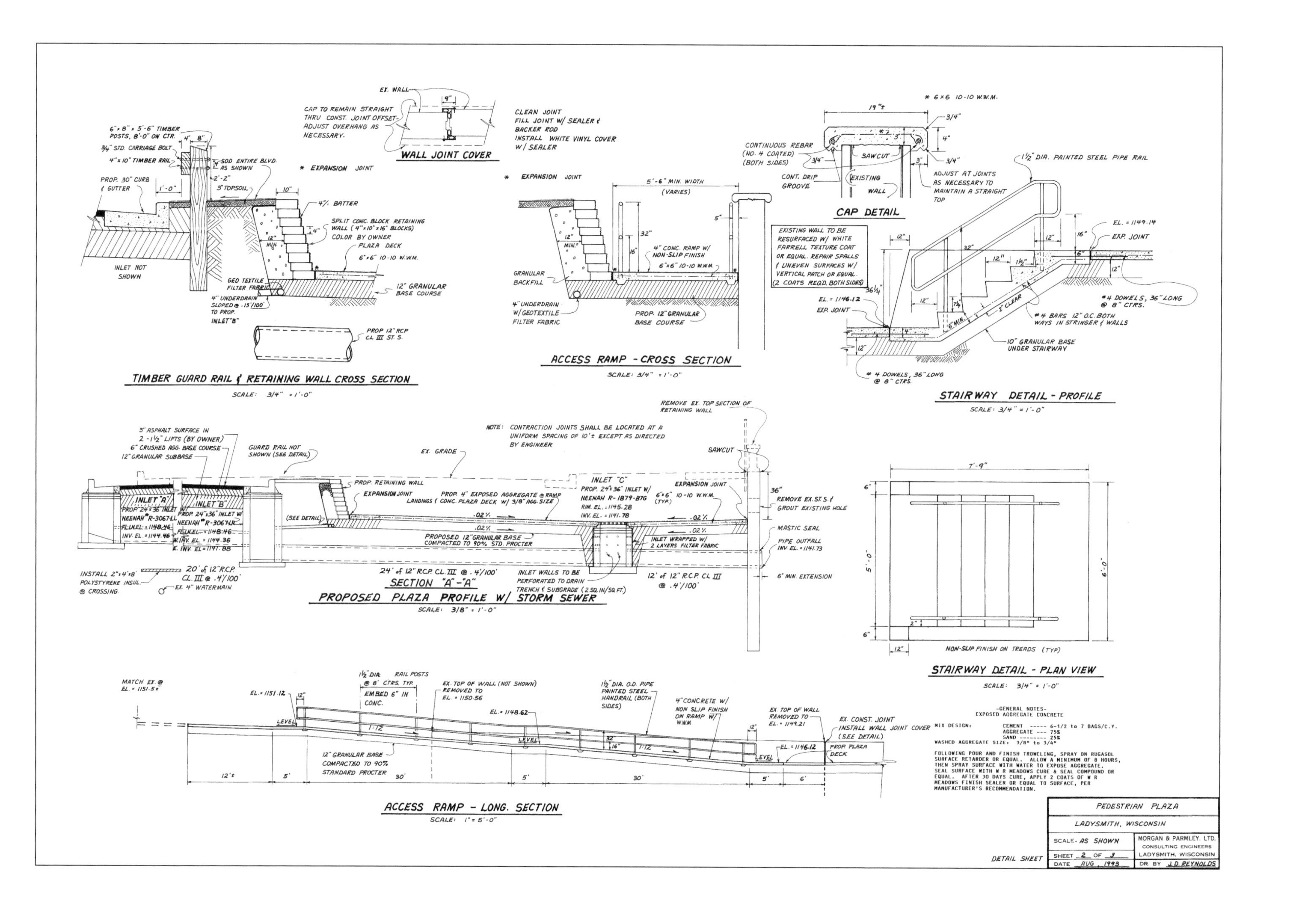
TIMBER GUARD RAIL & RETAINING WALL CROSS SECTION
SCALE: 3/4" = 1'-0"
WALL JOINT COVER
ACCESS RAMP - CROSS SECTION
SCALE: 3/4" = 1'-0"
CAP DETAIL
STAIRWAY DETAIL - PROFILE
SCALE: 3/4" = 1'-0"
NOTE: CONTRACTION JOINTS SHALL BE LOCATED AT A UNIFORM SPACING OF 10' ± EXCEPT AS DIRECTED BY ENGINEER
SECTION "A"-"A"
PROPOSED PLAZA PROFILE W/ STORM SEWER
SCALE: 3/8" = 1'-0"
STAIRWAY DETAIL - PLAN VIEW
SCALE: 3/4" = 1'-0"
ACCESS RAMP - LONG. SECTION
SCALE: 1" = 5'-0"
-GENERAL NOTES-
EXPOSED AGGREGATE CONCRETE
MIX DESIGN: CEMENT ----- 6-1/2 to 7 BAGS/C.Y.
AGGREGATE --- 75%
SAND -------- 25%
WASHED AGGREGATE SIZE: 3/8" to 3/4"
FOLLOWING POUR AND FINISH TROWELING, SPRAY ON RUGASOL SURFACE RETARDER OR EQUAL. ALLOW A MINIMUM OF 8 HOURS, THEN SPRAY SURFACE WITH WATER TO EXPOSE AGGREGATE. SEAL SURFACE WITH W R MEADOWS CURE & SEAL COMPOUND OR EQUAL. AFTER 30 DAYS CURE, APPLY 2 COATS OF W R MEADOWS FINISH SEALER OR EQUAL TO SURFACE, PER MANUFACTURER'S RECOMMENDATION.
PEDESTRIAN PLAZA
LADYSMITH, WISCONSIN
SCALE- AS SHOWN
SHEET 2 OF 3
DATE AUG. 1993
MORGAN & PARMLEY, LTD.
CONSULTING ENGINEERS
LADYSMITH, WISCONSIN
DR. BY J.D. REYNOLDS
DETAIL SHEET

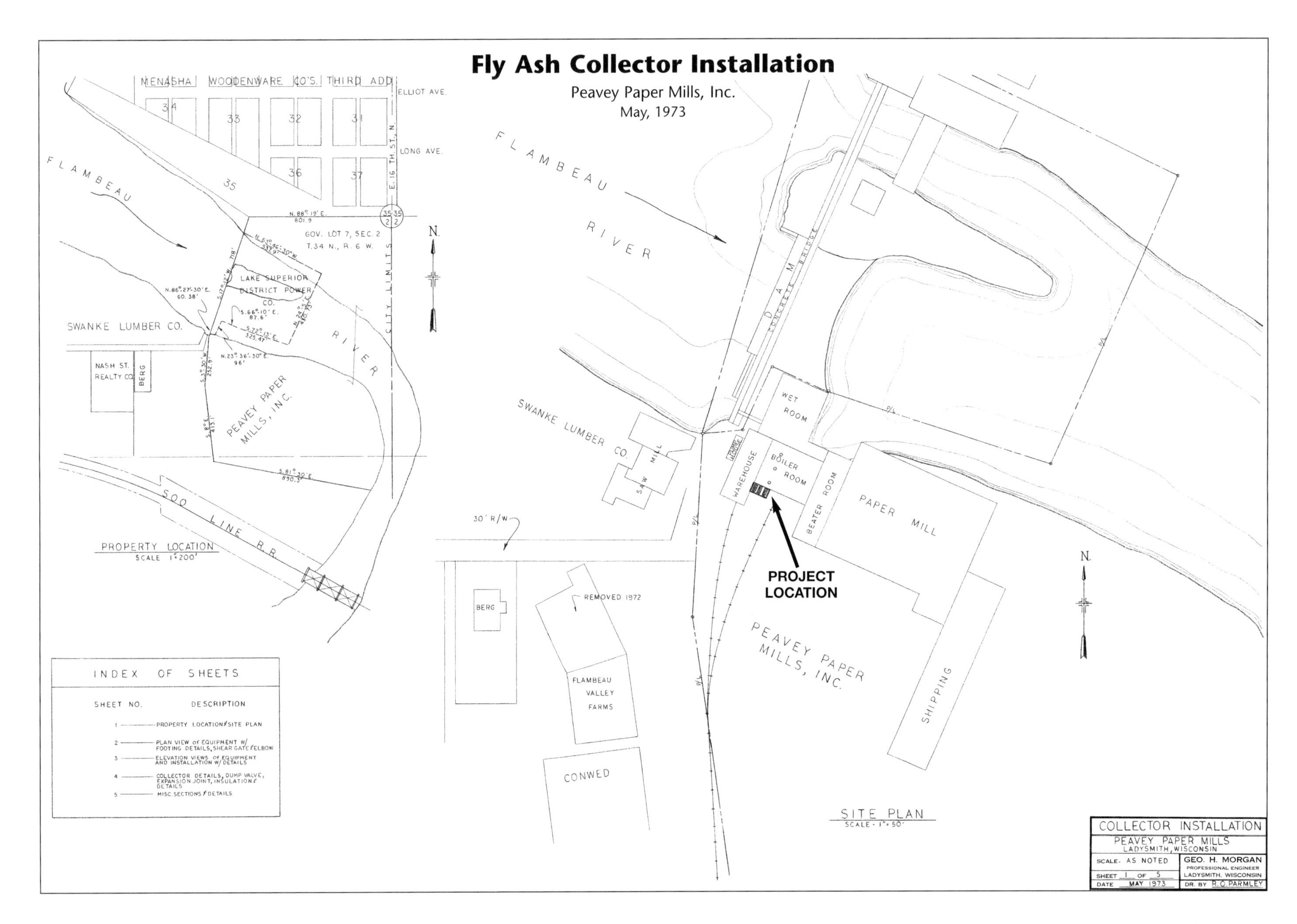

Courtesy: Morgan & Parmley, Ltd.

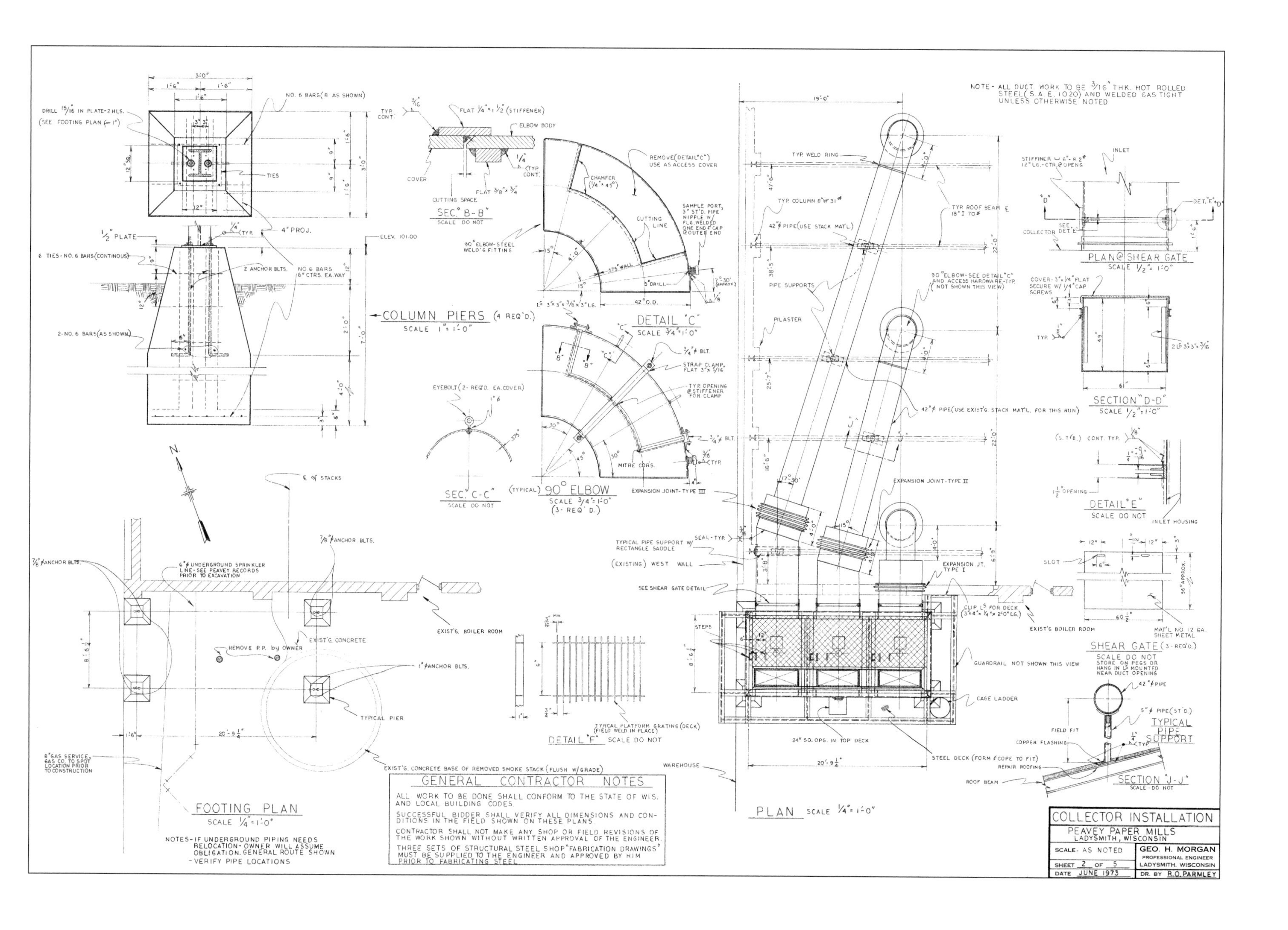
COLUMN PIERS (4 REQ'D.)
SCALE 1" = 1'-0"
SEC. "B-B"
SCALE DO NOT
DETAIL "C"
SCALE 3/4" = 1'-0"
SEC. "C-C"
SCALE DO NOT
(TYPICAL) 90° ELBOW
SCALE 3/4" = 1'-0"
(3-REQ'D.)
NOTE- ALL DUCT WORK TO BE 3/16" THK. HOT ROLLED STEEL (S.A.E. 1020) AND WELDED GAS TIGHT UNLESS OTHERWISE NOTED
PLAN @ SHEAR GATE
SCALE 1/2" = 1'-0"
SECTION "D-D"
SCALE 1/2" = 1'-0"
DETAIL "E"
SCALE DO NOT
SHEAR GATE (3-REQ'D.)
SCALE DO NOT
STORE ON PEGS OR HANG IN LS MOUNTED NEAR DUCT OPENING
TYPICAL PIPE SUPPORT
SECTION "J-J"
SCALE - DO NOT
FOOTING PLAN
SCALE 1/4" = 1'-0"
NOTES- IF UNDERGROUND PIPING NEEDS RELOCATION- OWNER WILL ASSUME OBLIGATION. GENERAL ROUTE SHOWN - VERIFY PIPE LOCATIONS
DETAIL "F" SCALE DO NOT
GENERAL CONTRACTOR NOTES
ALL WORK TO BE DONE SHALL CONFORM TO THE STATE OF WIS. AND LOCAL BUILDING CODES.
SUCCESSFUL BIDDER SHALL VERIFY ALL DIMENSIONS AND CONDITIONS IN THE FIELD SHOWN ON THESE PLANS.
CONTRACTOR SHALL NOT MAKE ANY SHOP OR FIELD REVISIONS OF THE WORK SHOWN WITHOUT WRITTEN APPROVAL OF THE ENGINEER
THREE SETS OF STRUCTURAL STEEL SHOP "FABRICATION DRAWINGS" MUST BE SUPPLIED TO THE ENGINEER AND APPROVED BY HIM PRIOR TO FABRICATING STEEL.
PLAN SCALE 1/4" = 1'-0"
COLLECTOR INSTALLATION
PEAVEY PAPER MILLS
LADYSMITH, WISCONSIN
SCALE- AS NOTED
SHEET 2 OF 5
DATE JUNE 1973
GEO. H. MORGAN
PROFESSIONAL ENGINEER
LADYSMITH, WISCONSIN
DR. BY R.O. PARMLEY

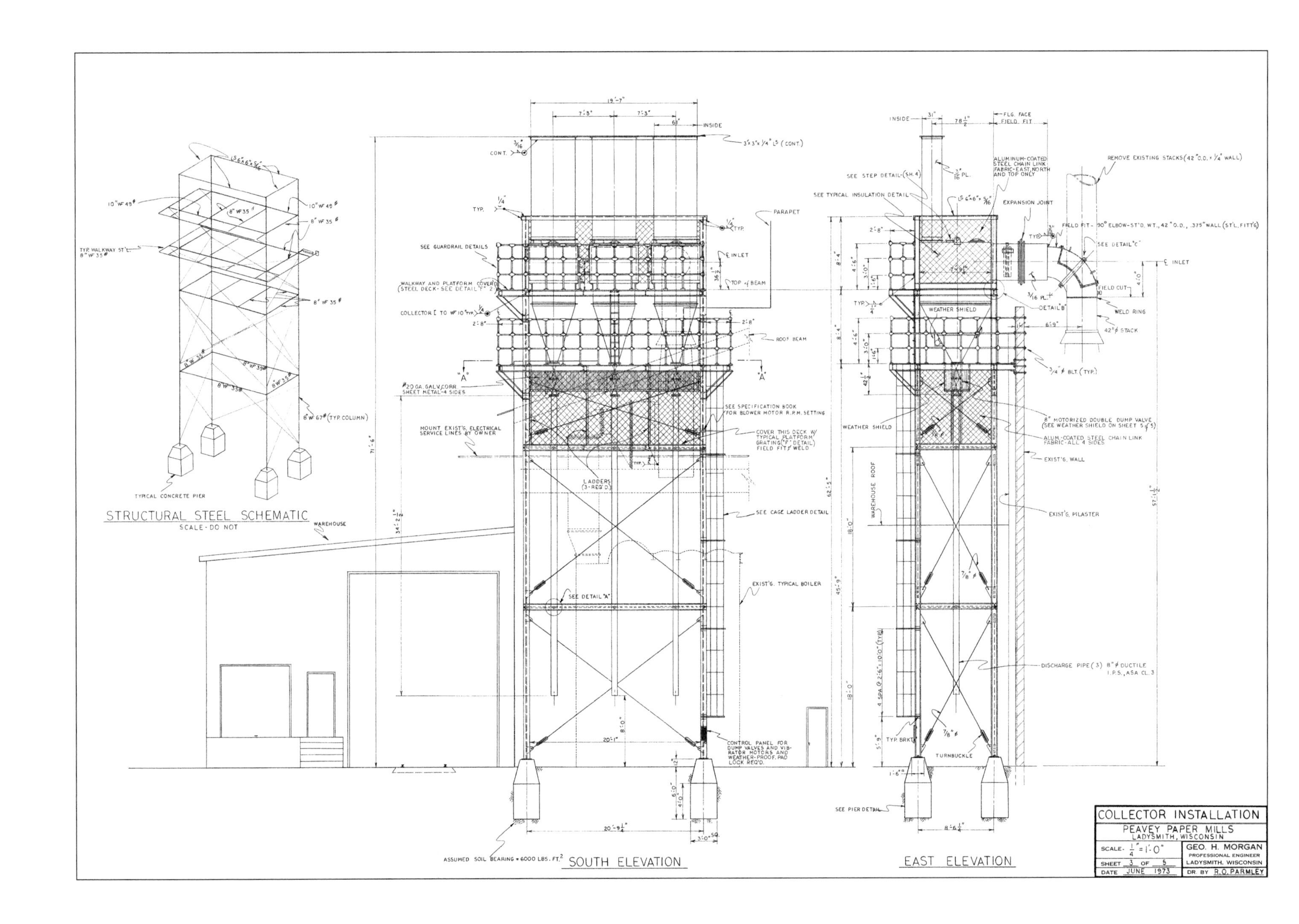
STRUCTURAL STEEL SCHEMATIC
SCALE - DO NOT
TYPICAL CONCRETE PIER
10" WF 49#
8" WF 35#
TYP. WALKWAY ST'L. 8" WF 35#
8" WF 67# (TYP. COLUMN)
WAREHOUSE
SOUTH ELEVATION
INSIDE
3"x3"x1/4" L's (CONT.)
PARAPET
SEE GUARDRAIL DETAILS
WALKWAY AND PLATFORM COVER (STEEL DECK - SEE DETAIL "F" 2)
COLLECTOR [TO WF 10" TYP.
ROOF BEAM
TOP of BEAM
#20 GA. GALV. CORR. SHEET METAL - 4 SIDES
SEE SPECIFICATION BOOK FOR BLOWER MOTOR R.P.M. SETTING
MOUNT EXIST'G. ELECTRICAL SERVICE LINES BY OWNER
COVER THIS DECK W/ TYPICAL PLATFORM GRATING ("F" DETAIL) FIELD FIT / WELD
LADDERS (3-REQ'D)
SEE CAGE LADDER DETAIL
EXIST'G. TYPICAL BOILER
SEE DETAIL "A"
CONTROL PANEL FOR DUMP VALVES AND VIBRATOR MOTORS AND WEATHER-PROOF, PAD LOCK REQ'D.
ASSUMED SOIL BEARING = 6000 LBS. FT.²
EAST ELEVATION
SEE STEP DETAIL - (SH. 4)
SEE TYPICAL INSULATION DETAIL
ALUMINUM-COATED STEEL CHAIN LINK FABRIC - EAST, NORTH AND TOP ONLY
REMOVE EXISTING STACKS (42" O.D. + 1/4" WALL)
EXPANSION JOINT
FIELD FIT - 90° ELBOW - ST'D. WT., 42" O.D., .375" WALL (ST'L. FITT'G)
SEE DETAIL "C"
INLET
FIELD CUT
WELD RING
42" ϕ STACK
WEATHER SHIELD
DETAIL "B"
3/4" ϕ BLT. (TYP.)
8" MOTORIZED DOUBLE DUMP VALVE (SEE WEATHER SHIELD ON SHEET 5 of 5)
ALUM.-COATED STEEL CHAIN LINK FABRIC - ALL 4 SIDES
EXIST'G. WALL
EXIST'G. PILASTER
WAREHOUSE ROOF
DISCHARGE PIPE (3) 8" ϕ DUCTILE I.P.S., ASA CL. 3
TYP. BRKT.
TURNBUCKLE
SEE PIER DETAIL
COLLECTOR INSTALLATION
PEAVEY PAPER MILLS
LADYSMITH, WISCONSIN
SCALE - 1/4" = 1'-0"
SHEET 3 OF 5
DATE JUNE 1973
GEO. H. MORGAN
PROFESSIONAL ENGINEER
LADYSMITH, WISCONSIN
DR. BY R.O. PARMLEY

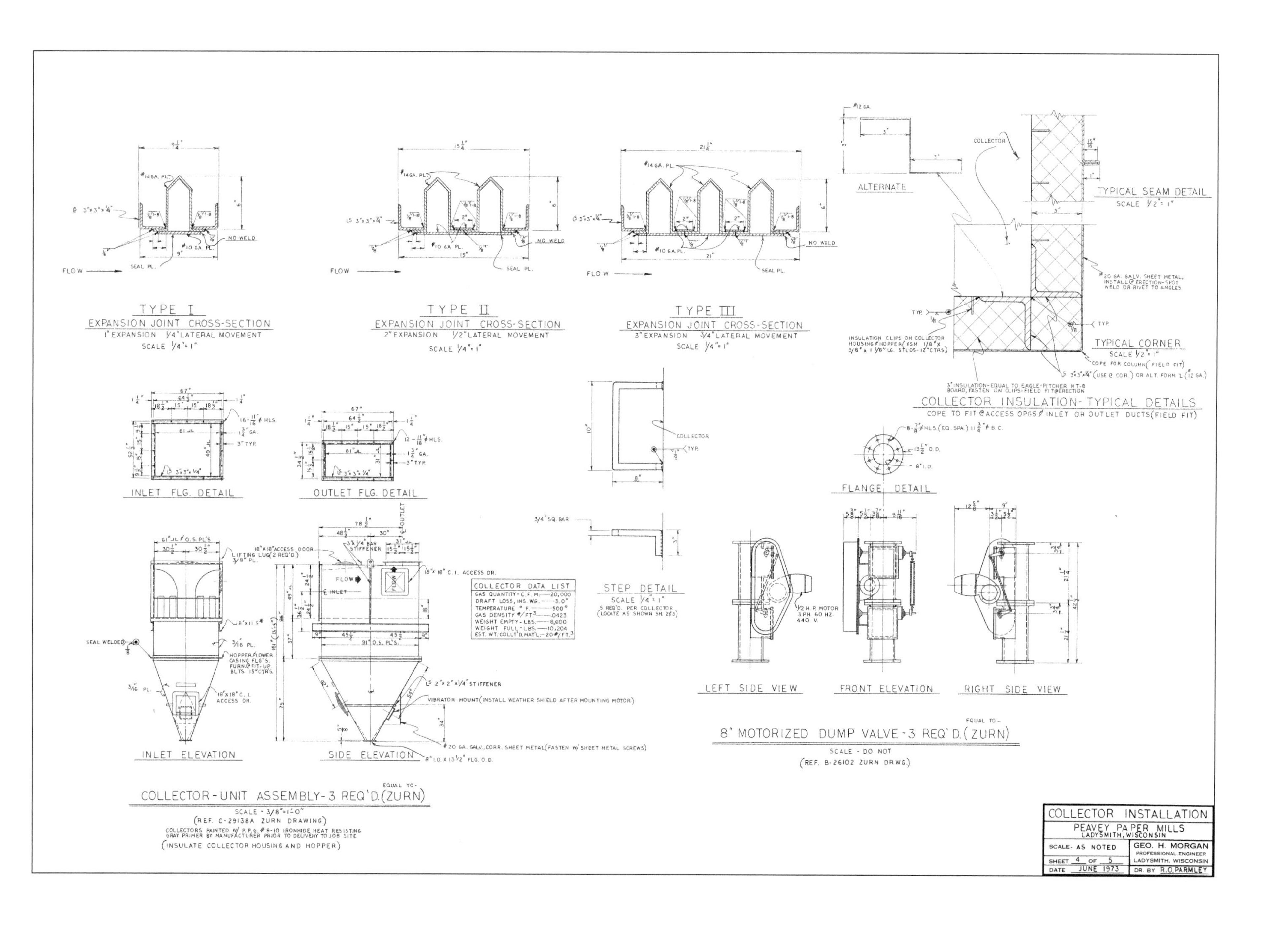
TYPE I
EXPANSION JOINT CROSS-SECTION
1" EXPANSION 1/4" LATERAL MOVEMENT
SCALE 1/4" = 1"
TYPE II
EXPANSION JOINT CROSS-SECTION
2" EXPANSION 1/2" LATERAL MOVEMENT
SCALE 1/4" = 1"
TYPE III
EXPANSION JOINT CROSS-SECTION
3" EXPANSION 3/4" LATERAL MOVEMENT
SCALE 1/4" = 1"
FLOW
NO WELD
SEAL PL.
#14 GA. PL.
#10 GA. PL.
ALTERNATE
COLLECTOR
TYPICAL SEAM DETAIL
SCALE 1/2" = 1"
20 GA. GALV. SHEET METAL, INSTALL @ ERECTION-SPOT WELD OR RIVET TO ANGLES
INSULATION CLIPS ON COLLECTOR HOUSING & HOPPER (KSH 1/8" x 3/8" x 1 1/8" LG. STUDS-12" CTRS)
TYPICAL CORNER
SCALE 1/2" = 1"
COPE FOR COLUMN (FIELD FIT)
3" INSULATION-EQUAL TO EAGLE-PITCHER M.T.-8 BOARD, FASTEN ON CLIPS-FIELD FIT @ ERECTION
COLLECTOR INSULATION-TYPICAL DETAILS
COPE TO FIT @ ACCESS OPGS. & INLET OR OUTLET DUCTS (FIELD FIT)
INLET FLG. DETAIL
OUTLET FLG. DETAIL
FLANGE DETAIL
STEP DETAIL
SCALE 1/4" = 1"
5 REQ'D. PER COLLECTOR (LOCATE AS SHOWN SH. 2/3)
3/4" SQ. BAR
COLLECTOR DATA LIST
GAS QUANTITY-C.F.M. — 20,000
DRAFT LOSS, INS. WG. — 3.0"
TEMPERATURE °F — 500°
GAS DENSITY #/FT.3 — .0423
WEIGHT EMPTY-LBS. — 8,600
WEIGHT FULL-LBS. — 10,204
EST. WT. COLLT'D. MAT'L. — 20#/FT.3
18" x 18" ACCESS DOOR
LIFTING LUG (2 REQ'D.) 3/8" PL.
18" x 18" C.I. ACCESS DR.
SEAL WELDED
HOPPER & LOWER CASING FLG'S. FURN. @ FIT-UP BLTS. 15" CTRS.
2" x 2" x 1/4" STIFFENER
VIBRATOR MOUNT (INSTALL WEATHER SHIELD AFTER MOUNTING MOTOR)
20 GA. GALV., CORR. SHEET METAL (FASTEN W/ SHEET METAL SCREWS)
8" I.D. x 13 1/2" FLG. O.D.
INLET ELEVATION
SIDE ELEVATION
COLLECTOR-UNIT ASSEMBLY-3 REQ'D. (ZURN)
EQUAL TO-
SCALE - 3/8" = 1'-0"
(REF. C-29138A ZURN DRAWING)
COLLECTORS PAINTED W/ P.P.G. #6-10 IRONHIDE HEAT RESISTING GRAY PRIMER BY MANUFACTURER PRIOR TO DELIVERY TO JOB SITE
(INSULATE COLLECTOR HOUSING AND HOPPER)
1/2 H.P. MOTOR 3 PH. 60 HZ. 440 V.
LEFT SIDE VIEW
FRONT ELEVATION
RIGHT SIDE VIEW
8" MOTORIZED DUMP VALVE-3 REQ'D. (ZURN)
EQUAL TO-
SCALE - DO NOT
(REF. B-26102 ZURN DRWG.)
COLLECTOR INSTALLATION
PEAVEY PAPER MILLS
LADYSMITH, WISCONSIN
SCALE- AS NOTED
SHEET 4 OF 5
DATE JUNE 1973
GEO. H. MORGAN
PROFESSIONAL ENGINEER
LADYSMITH, WISCONSIN
DR. BY R.O. PARMLEY

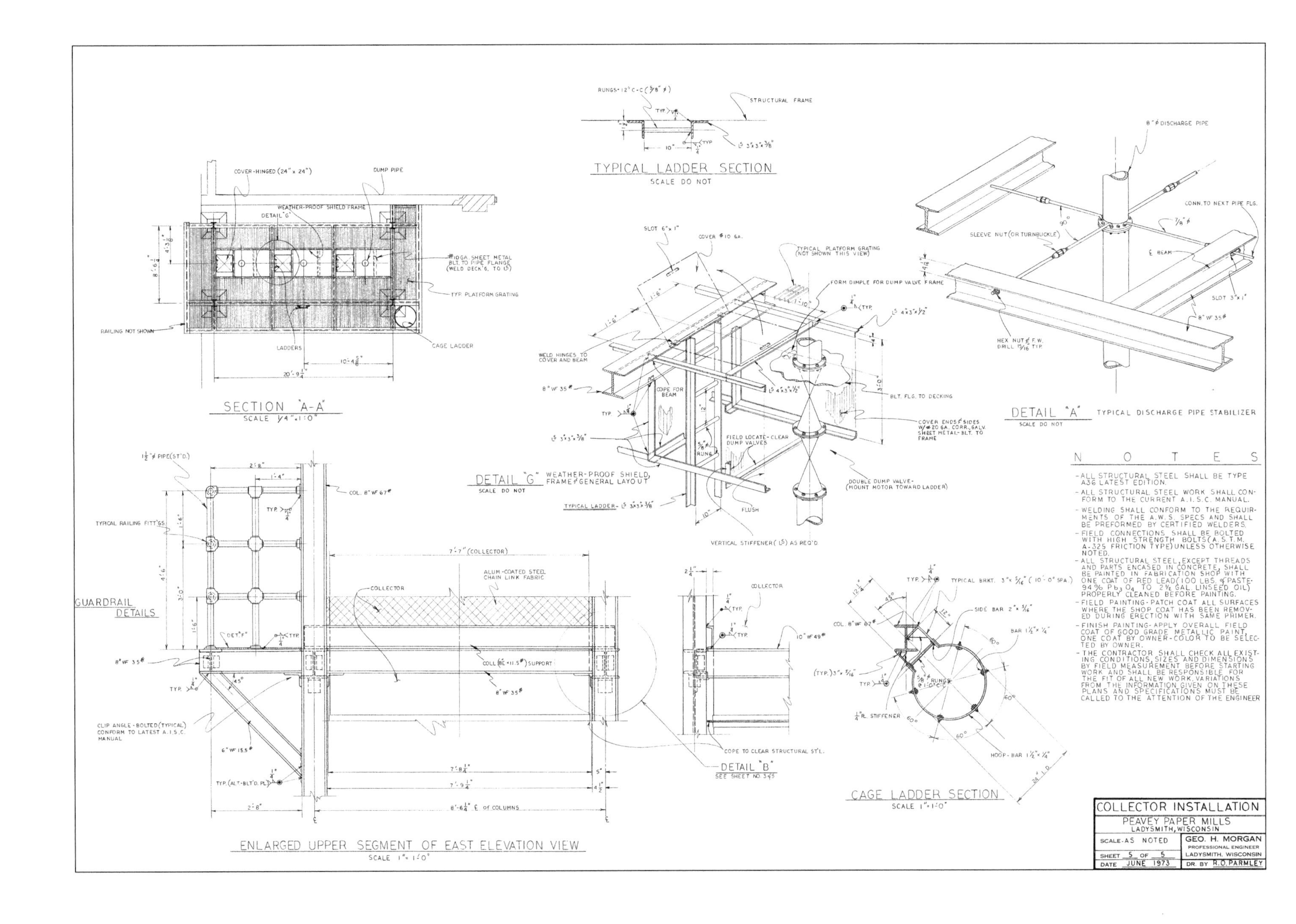

RUNGS-12" C-C (7/8" ϕ)
STRUCTURAL FRAME
TYPICAL LADDER SECTION
SCALE DO NOT
8" ϕ DISCHARGE PIPE
CONN. TO NEXT PIPE FLG.
SLEEVE NUT (OR TURNBUCKLE)
℄ BEAM
SLOT 3"x1"
8" WF 35#
HEX NUT ℄ F.W. DRILL 15/16 TYP.
DETAIL "A" TYPICAL DISCHARGE PIPE STABILIZER
SCALE DO NOT
COVER-HINGED (24" x 24")
DUMP PIPE
WEATHER-PROOF SHIELD FRAME
DETAIL "G"
#10GA SHEET METAL BLT. TO PIPE FLANGE (WELD DECK'G. TO L'S)
TYP. PLATFORM GRATING
RAILING NOT SHOWN
LADDERS
CAGE LADDER
SECTION "A-A"
SCALE 1/4"=1'-0"
SLOT 6"x1"
COVER #10 GA.
TYPICAL PLATFORM GRATING (NOT SHOWN THIS VIEW)
FORM DIMPLE FOR DUMP VALVE FRAME
WELD HINGES TO COVER AND BEAM
COPE FOR BEAM
BLT. FLG. TO DECKING
COVER ENDS & SIDES W/ #20 GA. CORR. GALV. SHEET METAL- BLT. TO FRAME
FIELD LOCATE-CLEAR DUMP VALVES
DOUBLE DUMP VALVE- (MOUNT MOTOR TOWARD LADDER)
FLUSH
VERTICAL STIFFENER (L'S) AS REQ'D
DETAIL "G" WEATHER-PROOF SHIELD, FRAME & GENERAL LAYOUT
SCALE DO NOT
TYPICAL LADDER- L 3x3x3/8"
NOTES
-ALL STRUCTURAL STEEL SHALL BE TYPE A36 LATEST EDITION.
-ALL STRUCTURAL STEEL WORK SHALL CONFORM TO THE CURRENT A.I.S.C. MANUAL.
-WELDING SHALL CONFORM TO THE REQUIREMENTS OF THE A.W.S. SPECS AND SHALL BE PREFORMED BY CERTIFIED WELDERS.
-FIELD CONNECTIONS SHALL BE BOLTED WITH HIGH STRENGTH BOLTS (A.S.T.M. A-325 FRICTION TYPE) UNLESS OTHERWISE NOTED.
-ALL STRUCTURAL STEEL, EXCEPT THREADS AND PARTS ENCASED IN CONCRETE, SHALL BE PAINTED IN FABRICATION SHOP WITH ONE COAT OF RED LEAD (100 LBS. OF PASTE-94% Pb3O4 TO 2½ GAL. LINSEED OIL) PROPERLY CLEANED BEFORE PAINTING.
-FIELD PAINTING-PATCH COAT ALL SURFACES WHERE THE SHOP COAT HAS BEEN REMOVED DURING ERECTION WITH SAME PRIMER.
-FINISH PAINTING-APPLY OVERALL FIELD COAT OF GOOD GRADE METALLIC PAINT, ONE COAT BY OWNER-COLOR TO BE SELECTED BY OWNER.
-THE CONTRACTOR SHALL CHECK ALL EXISTING CONDITIONS, SIZES AND DIMENSIONS BY FIELD MEASUREMENT BEFORE STARTING WORK AND SHALL BE RESPONSIBLE FOR THE FIT OF ALL NEW WORK. VARIATIONS FROM THE INFORMATION GIVEN ON THESE PLANS AND SPECIFICATIONS MUST BE CALLED TO THE ATTENTION OF THE ENGINEER
1½" ϕ PIPE (ST'D.)
COL. 8" WF 67#
TYPICAL RAILING FITT'GS
GUARDRAIL DETAILS
7'-7" (COLLECTOR)
ALUM.-COATED STEEL CHAIN LINK FABRIC
COLLECTOR
COLL. (℄ =11.5#) SUPPORT
8" WF 35#
CLIP ANGLE-BOLTED (TYPICAL) CONFORM TO LATEST A.I.S.C. MANUAL
6" WF 15.5#
8'-6¼" ℄ OF COLUMNS
ENLARGED UPPER SEGMENT OF EAST ELEVATION VIEW
SCALE 1"=1'-0"
10" WF 49#
COPE TO CLEAR STRUCTURAL ST'L.
DETAIL "B"
SEE SHEET NO. 3/5
TYPICAL BRKT. 3"x 5/16" (10'-0" SPA.)
SIDE BAR 2"x 5/16"
BAR 1½"x ¼"
COL. 8" WF 67#
¼" PL. STIFFENER
HOOP-BAR 1½"x ¼"
CAGE LADDER SECTION
SCALE 1"=1'-0"
COLLECTOR INSTALLATION
PEAVEY PAPER MILLS
LADYSMITH, WISCONSIN
SCALE-AS NOTED
SHEET 5 OF 5
DATE JUNE 1973
GEO. H. MORGAN
PROFESSIONAL ENGINEER
LADYSMITH, WISCONSIN
DR. BY R.O. PARMLEY

LADYSMITH RAIL SITE IMPROVEMENT PROJECT
CITY OF LADYSMITH INDUSTRIAL PARK

M & P PROJECT NO. 99–118

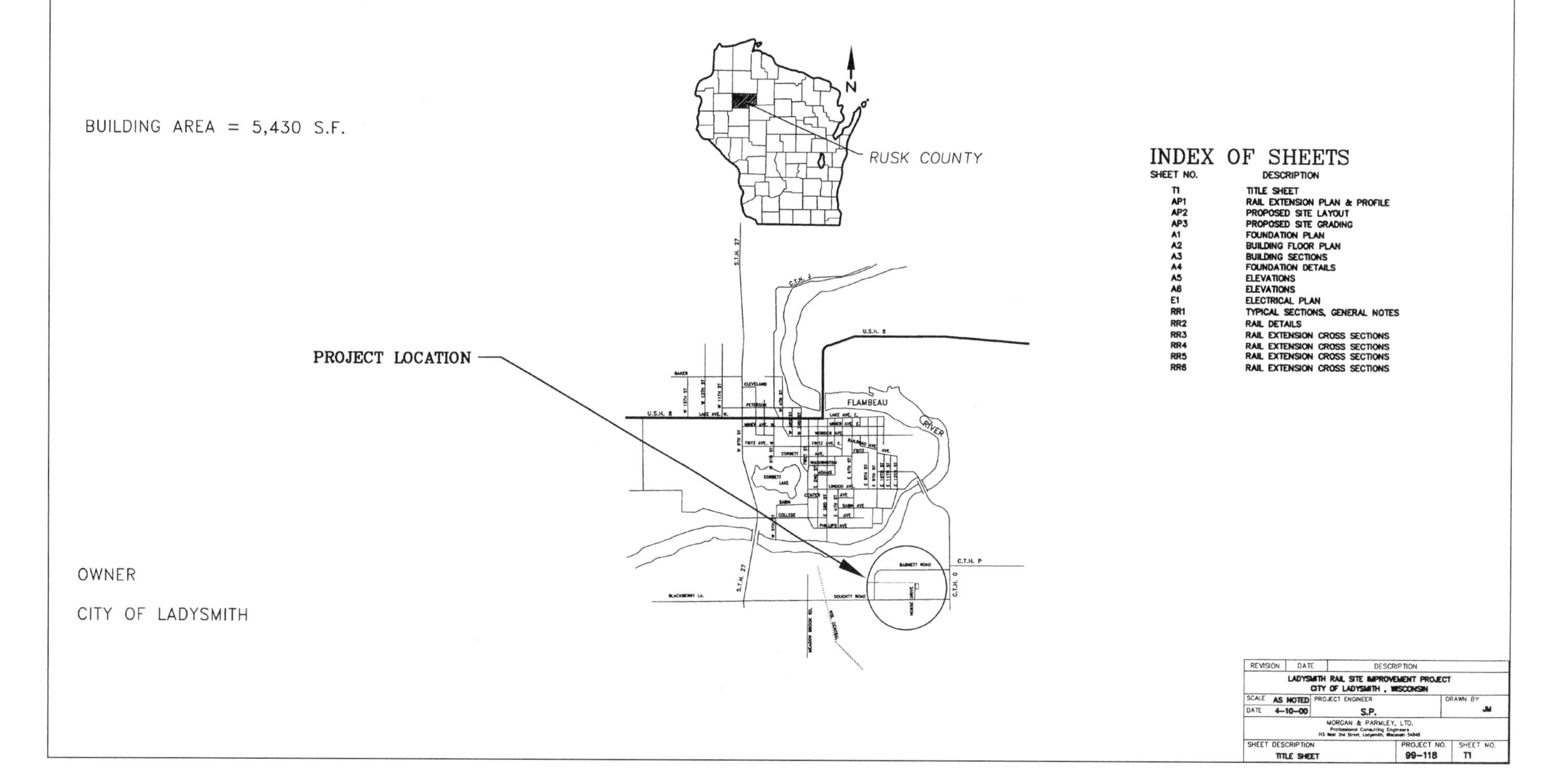

Courtesy: Morgan & Parmley, Ltd.

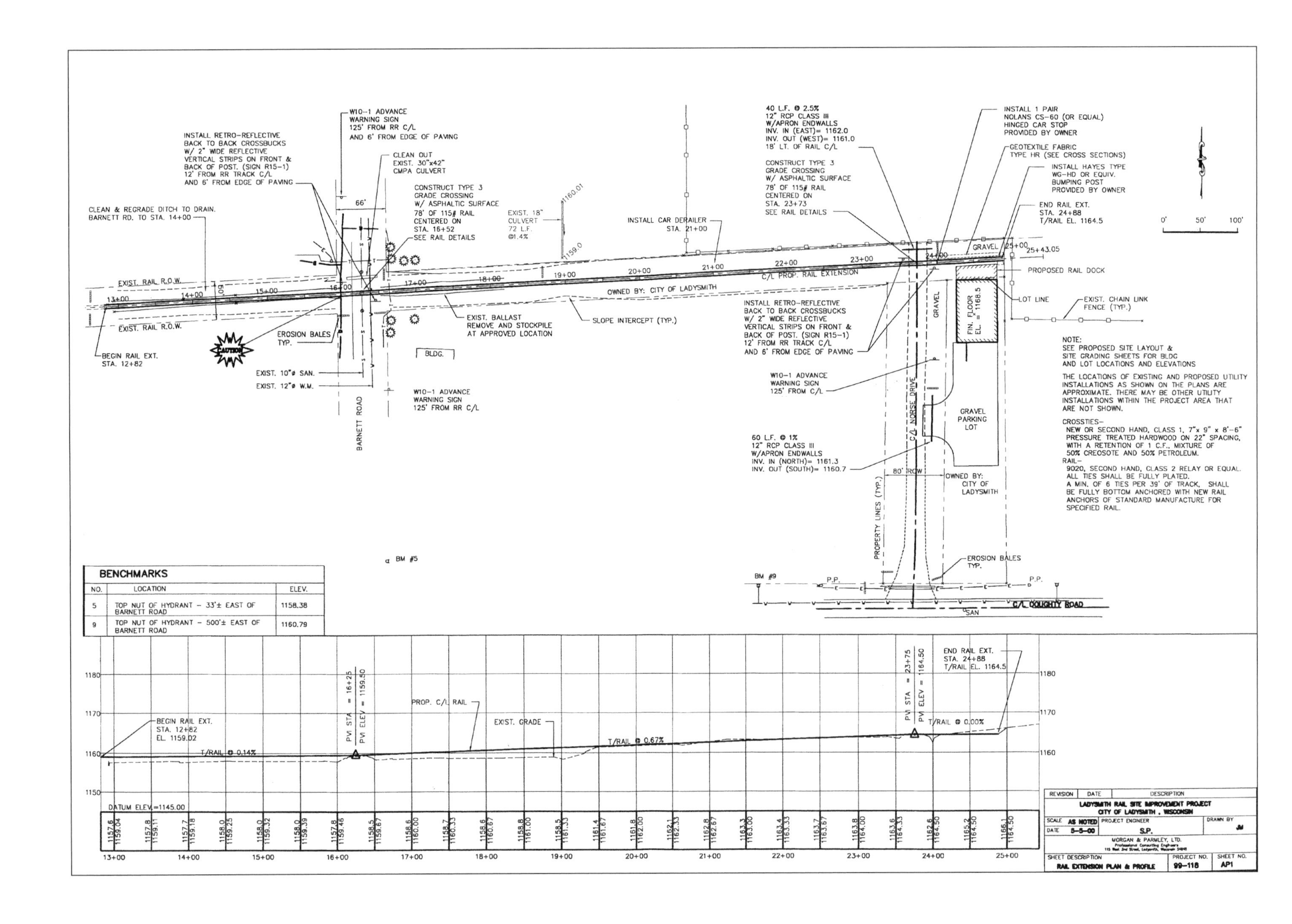
W10–1 ADVANCE WARNING SIGN 125' FROM RR C/L AND 6' FROM EDGE OF PAVING
INSTALL RETRO–REFLECTIVE BACK TO BACK CROSSBUCKS W/ 2" WIDE REFLECTIVE VERTICAL STRIPS ON FRONT & BACK OF POST. (SIGN R15–1) 12' FROM RR TRACK C/L AND 6' FROM EDGE OF PAVING
CLEAN OUT EXIST. 30"x42" CMPA CULVERT
CONSTRUCT TYPE 3 GRADE CROSSING W/ ASPHALTIC SURFACE 78' OF 115# RAIL CENTERED ON STA. 16+52 SEE RAIL DETAILS
EXIST. 18" CULVERT 72 L.F. @1.4%
CLEAN & REGRADE DITCH TO DRAIN. BARNETT RD. TO STA. 14+00
INSTALL CAR DERAILER STA. 21+00
40 L.F. @ 2.5% 12" RCP CLASS III W/APRON ENDWALLS INV. IN (EAST)= 1162.0 INV. OUT (WEST)= 1161.0 18' LT. OF RAIL C/L
CONSTRUCT TYPE 3 GRADE CROSSING W/ ASPHALTIC SURFACE 78' OF 115# RAIL CENTERED ON STA. 23+73 SEE RAIL DETAILS
INSTALL 1 PAIR NOLANS CS–60 (OR EQUAL) HINGED CAR STOP PROVIDED BY OWNER
GEOTEXTILE FABRIC TYPE HR (SEE CROSS SECTIONS)
INSTALL HAYES TYPE WG–HD OR EQUIV. BUMPING POST PROVIDED BY OWNER
END RAIL EXT. STA. 24+88 T/RAIL EL. 1164.5
0' 50' 100'
EXIST. RAIL R.O.W.
BEGIN RAIL EXT. STA. 12+82
EROSION BALES TYP.
EXIST. 10"ø SAN.
EXIST. 12"ø W.M.
BARNETT ROAD
BLDG.
EXIST. BALLAST REMOVE AND STOCKPILE AT APPROVED LOCATION
W10–1 ADVANCE WARNING SIGN 125' FROM RR C/L
SLOPE INTERCEPT (TYP.)
OWNED BY: CITY OF LADYSMITH
C/L PROP. RAIL EXTENSION
PROPOSED RAIL DOCK
LOT LINE
EXIST. CHAIN LINK FENCE (TYP.)
GRAVEL
FIN. FLOOR EL. = 1168.5
INSTALL RETRO–REFLECTIVE BACK TO BACK CROSSBUCKS W/ 2" WIDE REFLECTIVE VERTICAL STRIPS ON FRONT & BACK OF POST. (SIGN R15–1) 12' FROM RR TRACK C/L AND 6' FROM EDGE OF PAVING
W10–1 ADVANCE WARNING SIGN 125' FROM C/L
C/L NORSE DRIVE
GRAVEL PARKING LOT
60 L.F. @ 1% 12" RCP CLASS III W/APRON ENDWALLS INV. IN (NORTH)= 1161.3 INV. OUT (SOUTH)= 1160.7
80' ROW
OWNED BY: CITY OF LADYSMITH
PROPERTY LINES (TYP.)
EROSION BALES TYP.
BM #5
BM #9
C/L DOUGHTY ROAD
NOTE:
SEE PROPOSED SITE LAYOUT & SITE GRADING SHEETS FOR BLDG AND LOT LOCATIONS AND ELEVATIONS
THE LOCATIONS OF EXISTING AND PROPOSED UTILITY INSTALLATIONS AS SHOWN ON THE PLANS ARE APPROXIMATE. THERE MAY BE OTHER UTILITY INSTALLATIONS WITHIN THE PROJECT AREA THAT ARE NOT SHOWN.
CROSSTIES–
NEW OR SECOND HAND, CLASS 1, 7"x 9" x 8'–6" PRESSURE TREATED HARDWOOD ON 22" SPACING, WITH A RETENTION OF 1 C.F., MIXTURE OF 50% CREOSOTE AND 50% PETROLEUM.
RAIL–
9020, SECOND HAND, CLASS 2 RELAY OR EQUAL. ALL TIES SHALL BE FULLY PLATED. A MIN. OF 6 TIES PER 39' OF TRACK, SHALL BE FULLY BOTTOM ANCHORED WITH NEW RAIL ANCHORS OF STANDARD MANUFACTURE FOR SPECIFIED RAIL.
BENCHMARKS
NO. LOCATION ELEV.
5 TOP NUT OF HYDRANT – 33'± EAST OF BARNETT ROAD 1158.38
9 TOP NUT OF HYDRANT – 500'± EAST OF BARNETT ROAD 1160.79
BEGIN RAIL EXT. STA. 12+82 EL. 1159.02
T/RAIL @ 0.14%
PVI STA = 16+25
PVI ELEV = 1159.50
PROP. C/L RAIL
EXIST. GRADE
T/RAIL @ 0.67%
PVI STA = 23+75
PVI ELEV = 1164.50
T/RAIL @ 0.00%
END RAIL EXT. STA. 24+88 T/RAIL EL. 1164.5
DATUM ELEV.=1145.00
1180
1170
1160
1150
13+00 14+00 15+00 16+00 17+00 18+00 19+00 20+00 21+00 22+00 23+00 24+00 25+00
LADYSMITH RAIL SITE IMPROVEMENT PROJECT
CITY OF LADYSMITH, WISCONSIN
SCALE AS NOTED
DATE 5-5-00
PROJECT ENGINEER S.P.
MORGAN & PARMLEY, LTD.
SHEET DESCRIPTION RAIL EXTENSION PLAN & PROFILE
PROJECT NO. 99-118
SHEET NO. AP1

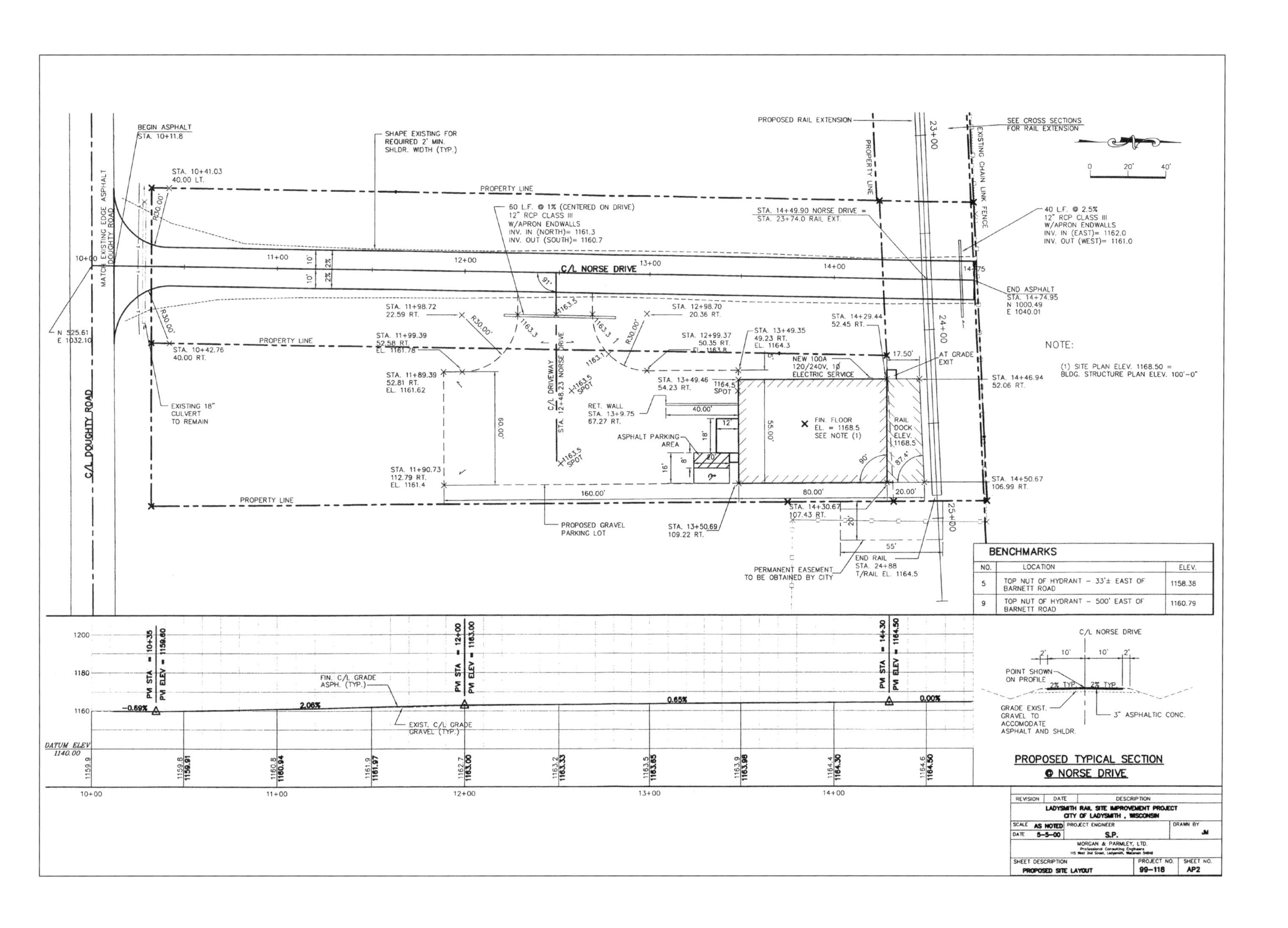
C/L NORSE DRIVE
C/L DOUGHTY ROAD
PROPERTY LINE
PROPOSED RAIL EXTENSION
SEE CROSS SECTIONS FOR RAIL EXTENSION
EXISTING CHAIN LINK FENCE
NOTE:
(1) SITE PLAN ELEV. 1168.50 = BLDG. STRUCTURE PLAN ELEV. 100'-0"
BENCHMARKS
NO. | LOCATION | ELEV.
5 | TOP NUT OF HYDRANT - 33'± EAST OF BARNETT ROAD | 1158.38
9 | TOP NUT OF HYDRANT - 500' EAST OF BARNETT ROAD | 1160.79
PROPOSED TYPICAL SECTION @ NORSE DRIVE
LADYSMITH RAIL SITE IMPROVEMENT PROJECT
CITY OF LADYSMITH, WISCONSIN
SCALE AS NOTED
DATE 5-5-00
PROJECT ENGINEER S.P.
MORGAN & PARMLEY, LTD.
SHEET DESCRIPTION PROPOSED SITE LAYOUT
PROJECT NO. 99-118
SHEET NO. AP2

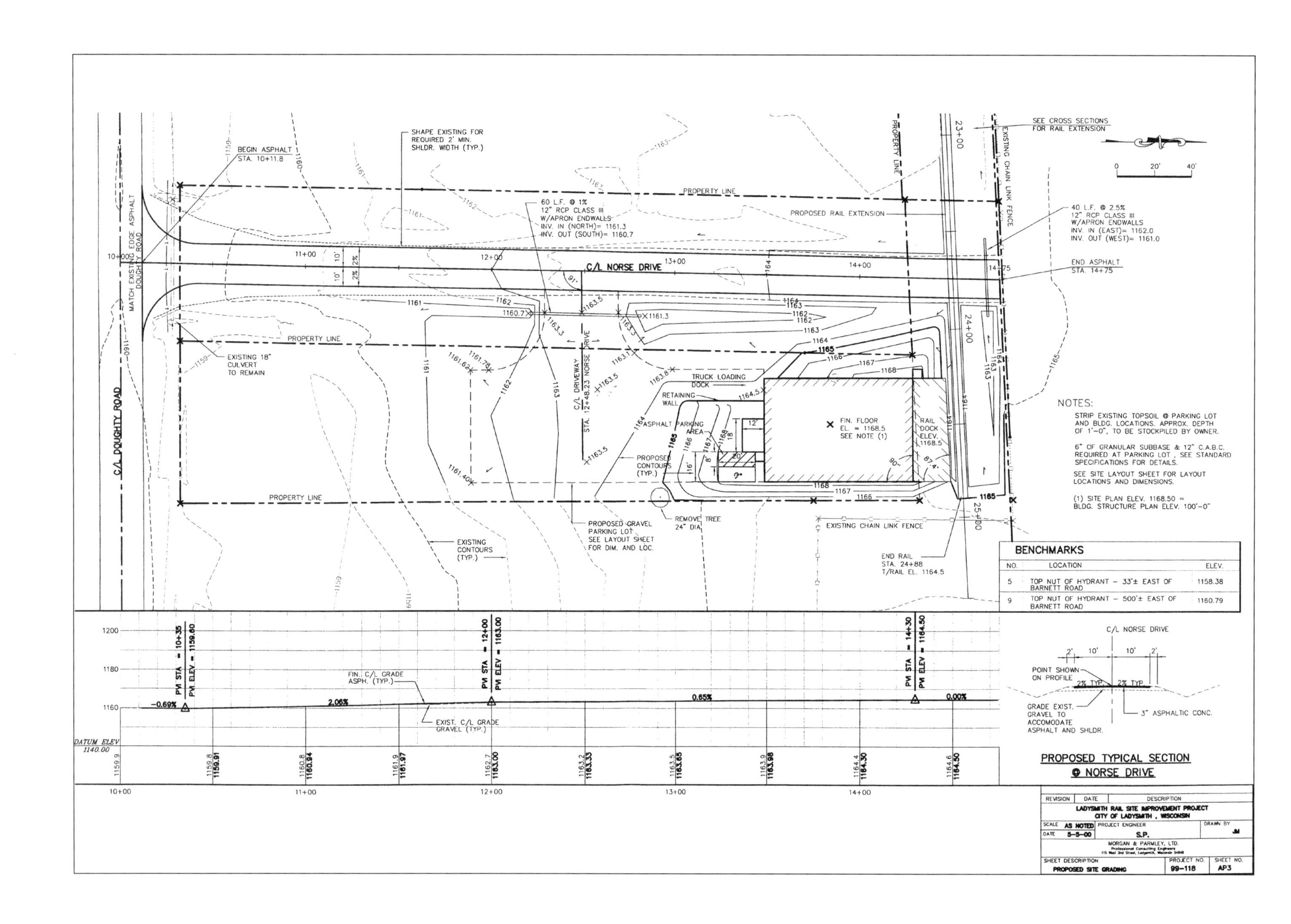

SEE CROSS SECTIONS FOR RAIL EXTENSION
SHAPE EXISTING FOR REQUIRED 2' MIN. SHLDR. WIDTH (TYP.)
BEGIN ASPHALT STA. 10+11.8
PROPERTY LINE
PROPOSED RAIL EXTENSION
60 L.F. @ 1% 12" RCP CLASS III W/APRON ENDWALLS INV. IN (NORTH)= 1161.3 INV. OUT (SOUTH)= 1160.7
40 L.F. @ 2.5% 12" RCP CLASS III W/APRON ENDWALLS INV. IN (EAST)= 1162.0 INV. OUT (WEST)= 1161.0
END ASPHALT STA. 14+75
C/L NORSE DRIVE
MATCH EXISTING EDGE ASPHALT DOUGHTY ROAD
C/L DOUGHTY ROAD
EXISTING 18" CULVERT TO REMAIN
C/L DRIVEWAY STA. 12+48.23 NORSE DRIVE
TRUCK LOADING DOCK
RETAINING WALL
ASPHALT PARKING AREA
PROPOSED CONTOURS (TYP.)
FIN. FLOOR EL. = 1168.5 SEE NOTE (1)
RAIL DOCK ELEV. 1168.5
REMOVE TREE 24" DIA.
EXISTING CHAIN LINK FENCE
PROPOSED GRAVEL PARKING LOT SEE LAYOUT SHEET FOR DIM. AND LOC.
EXISTING CONTOURS (TYP.)
END RAIL STA. 24+88 T/RAIL EL. 1164.5
NOTES:
STRIP EXISTING TOPSOIL @ PARKING LOT AND BLDG. LOCATIONS. APPROX. DEPTH OF 1'-0", TO BE STOCKPILED BY OWNER.
6" OF GRANULAR SUBBASE & 12" C.A.B.C. REQUIRED AT PARKING LOT , SEE STANDARD SPECIFICATIONS FOR DETAILS.
SEE SITE LAYOUT SHEET FOR LAYOUT LOCATIONS AND DIMENSIONS.
(1) SITE PLAN ELEV. 1168.50 = BLDG. STRUCTURE PLAN ELEV. 100'-0"
BENCHMARKS
NO. LOCATION ELEV.
5 TOP NUT OF HYDRANT - 33'± EAST OF BARNETT ROAD 1158.38
9 TOP NUT OF HYDRANT - 500'± EAST OF BARNETT ROAD 1160.79
PVI STA = 10+35 PVI ELEV = 1159.60
PVI STA = 12+00 PVI ELEV = 1163.00
PVI STA = 14+30 PVI ELEV = 1164.50
FIN. C/L GRADE ASPH. (TYP.)
EXIST. C/L GRADE GRAVEL (TYP.)
DATUM ELEV 1140.00
C/L NORSE DRIVE
POINT SHOWN ON PROFILE
GRADE EXIST. GRAVEL TO ACCOMODATE ASPHALT AND SHLDR.
3" ASPHALTIC CONC.
PROPOSED TYPICAL SECTION @ NORSE DRIVE
LADYSMITH RAIL SITE IMPROVEMENT PROJECT CITY OF LADYSMITH , WISCONSIN
SCALE AS NOTED
DATE 5-5-00
PROJECT ENGINEER S.P.
DRAWN BY JM
MORGAN & PARMLEY, LTD.
SHEET DESCRIPTION PROPOSED SITE GRADING
PROJECT NO. 99-118
SHEET NO. AP3

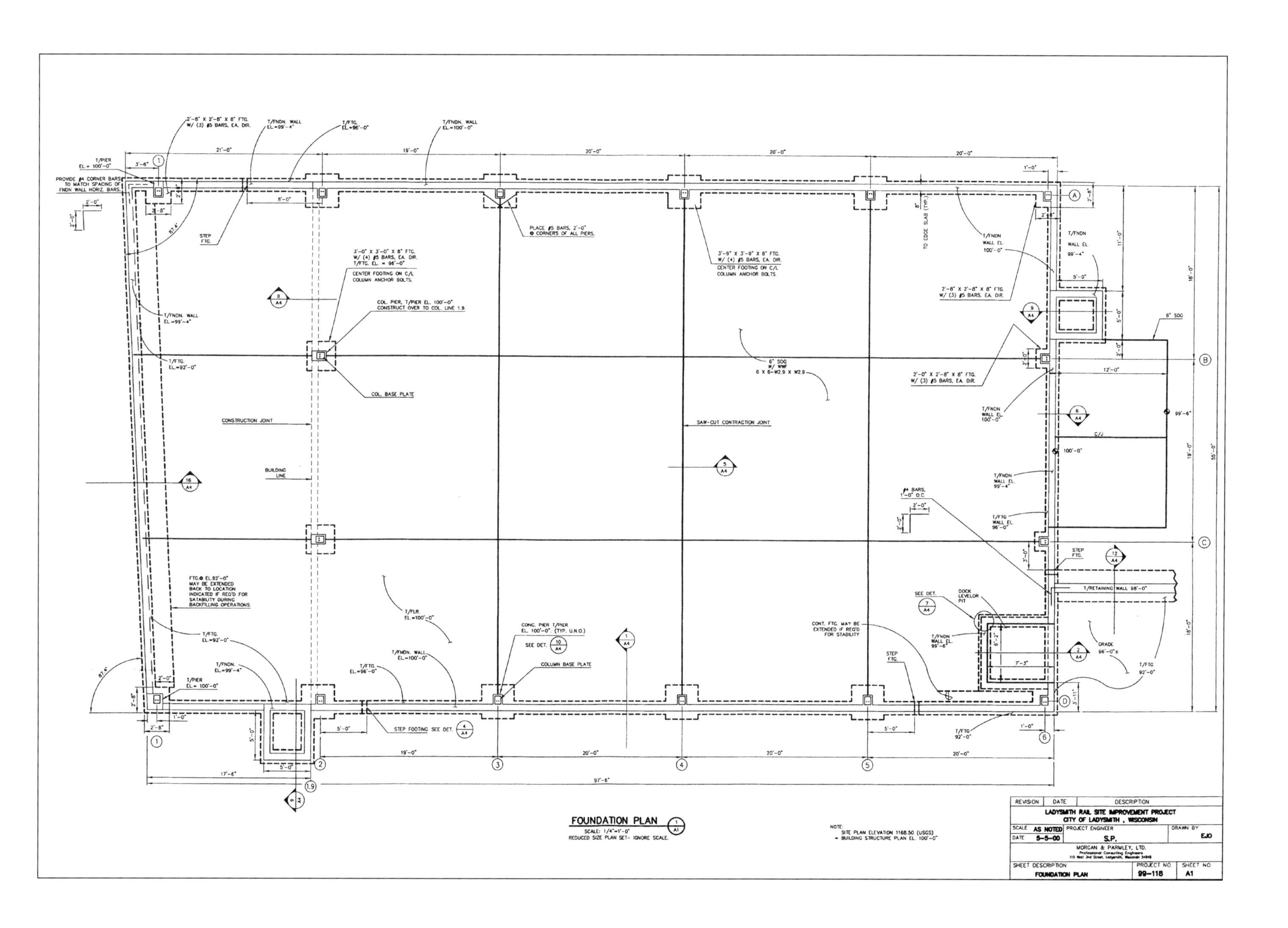
FOUNDATION PLAN
SCALE: 1/4"=1'-0"
REDUCED SIZE PLAN SET- IGNORE SCALE.
NOTE:
SITE PLAN ELEVATION 1168.50 (USGS)
= BUILDING STRUCTURE PLAN EL. 100'-0"
LADYSMITH RAIL SITE IMPROVEMENT PROJECT
CITY OF LADYSMITH, WISCONSIN
MORGAN & PARMLEY, LTD.
SHEET DESCRIPTION FOUNDATION PLAN
PROJECT NO. 99-118
SHEET NO. A1
DATE 5-5-00
SCALE AS NOTED
PROJECT ENGINEER S.P.
DRAWN BY EJO

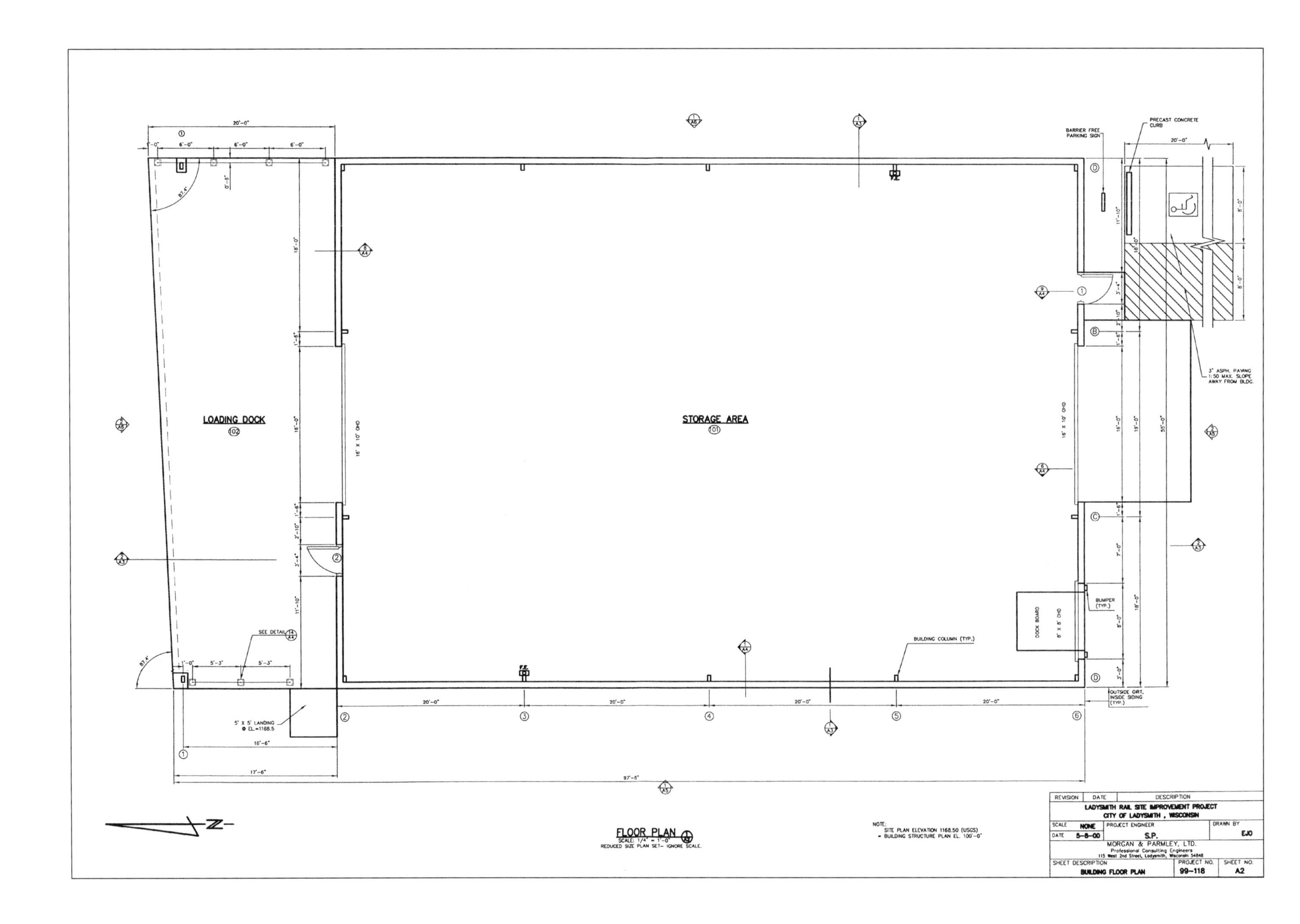
LOADING DOCK
102
STORAGE AREA
101
16' X 10' OHD
BARRIER FREE PARKING SIGN
PRECAST CONCRETE CURB
3" ASPH. PAVING 1:50 MAX. SLOPE AWAY FROM BLDG.
DOCK BOARD
8' X 8' OHD
BUMPER (TYP.)
BUILDING COLUMN (TYP.)
OUTSIDE GIRT, INSIDE SIDING (TYP.)
SEE DETAIL
5' X 5' LANDING @ EL.=1168.5
FLOOR PLAN
SCALE: 1/4" = 1'-0"
REDUCED SIZE PLAN SET- IGNORE SCALE.
NOTE:
SITE PLAN ELEVATION 1168.50 (USGS)
= BUILDING STRUCTURE PLAN EL. 100'-0"
REVISION
DATE
DESCRIPTION
LADYSMITH RAIL SITE IMPROVEMENT PROJECT
CITY OF LADYSMITH, WISCONSIN
SCALE NONE
DATE 5-8-00
PROJECT ENGINEER S.P.
DRAWN BY EJO
MORGAN & PARMLEY, LTD.
Professional Consulting Engineers
115 West 2nd Street, Ladysmith, Wisconsin 54848
SHEET DESCRIPTION BUILDING FLOOR PLAN
PROJECT NO. 99-118
SHEET NO. A2

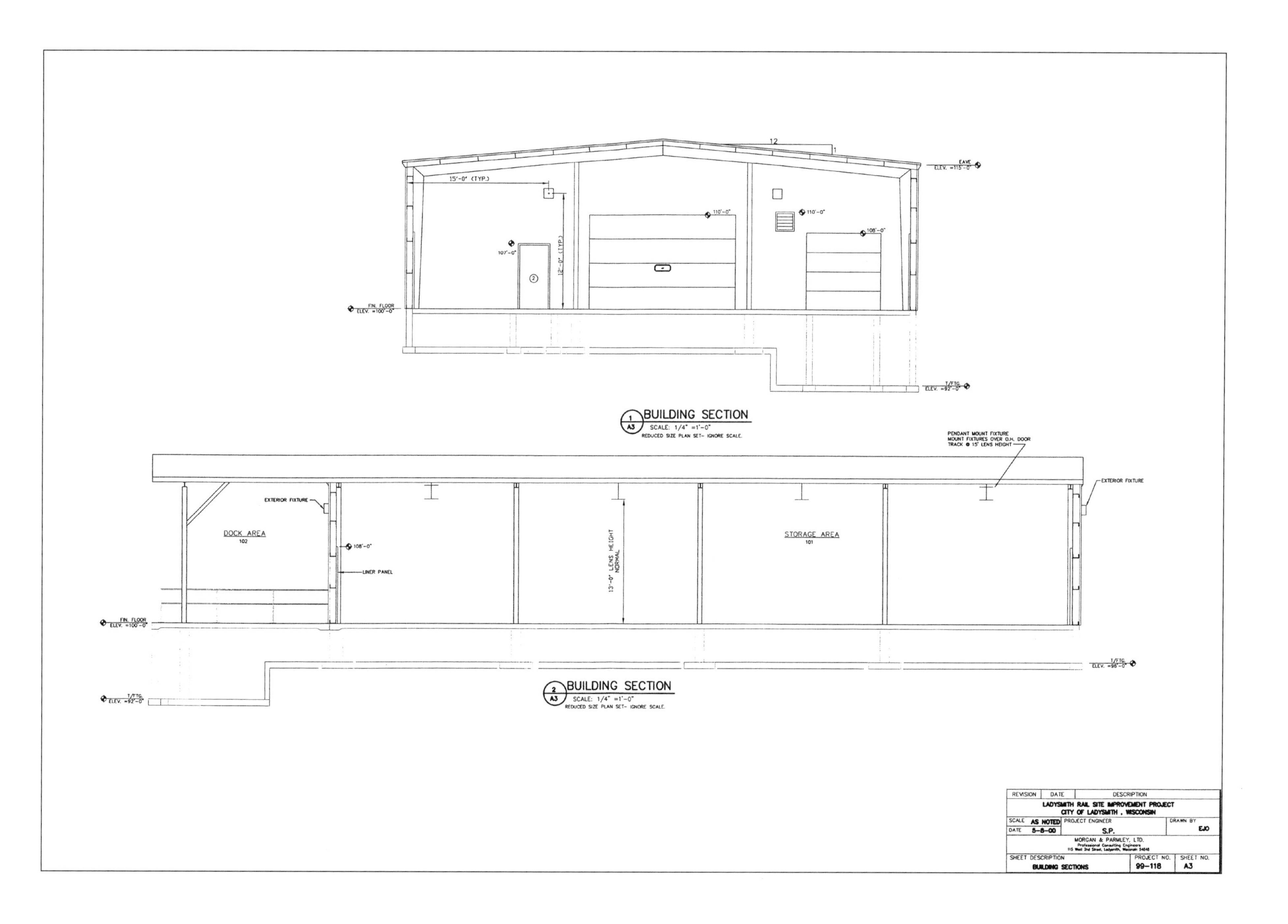
BUILDING SECTION
SCALE: 1/4" =1'-0"
REDUCED SIZE PLAN SET- IGNORE SCALE.
15'-0" (TYP.)
12'-0" (TYP.)
EAVE ELEV. =115'-0"
FIN. FLOOR ELEV. =100'-0"
T/FTG. ELEV. =92'-0"
110'-0"
108'-0"
107'-0"
PENDANT MOUNT FIXTURE MOUNT FIXTURES OVER O.H. DOOR TRACK @ 15' LENS HEIGHT
EXTERIOR FIXTURE
DOCK AREA
102
STORAGE AREA
101
LINER PANEL
13'-0" LENS HEIGHT NORMAL
T/FTG. ELEV. =98'-0"
LADYSMITH RAIL SITE IMPROVEMENT PROJECT
CITY OF LADYSMITH, WISCONSIN
SCALE AS NOTED
DATE 5-8-00
PROJECT ENGINEER S.P.
DRAWN BY EJO
MORGAN & PARMLEY, LTD.
Professional Consulting Engineers
SHEET DESCRIPTION BUILDING SECTIONS
PROJECT NO. 99-118
SHEET NO. A3

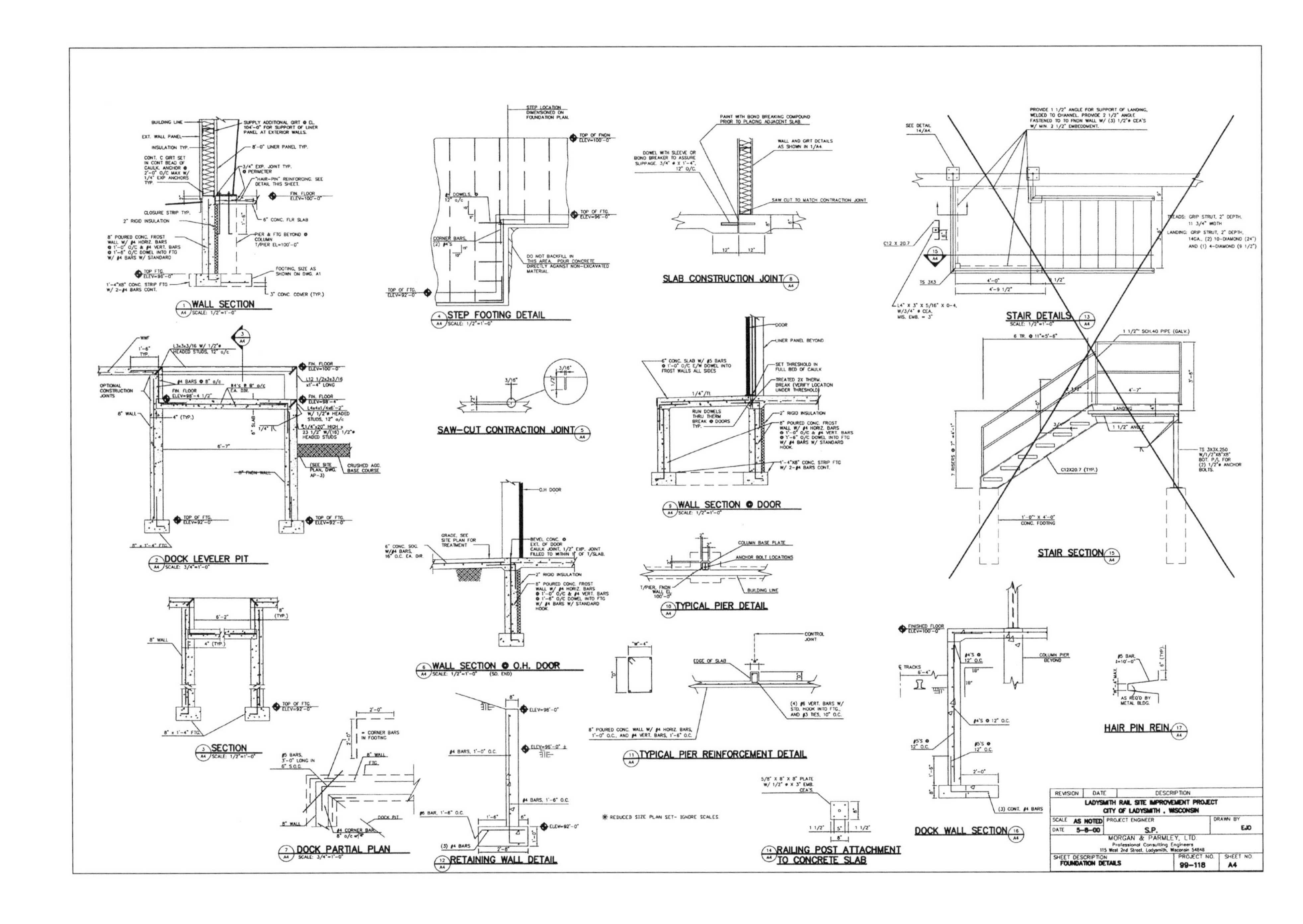
WALL SECTION
STEP FOOTING DETAIL
SLAB CONSTRUCTION JOINT
STAIR DETAILS
DOCK LEVELER PIT
SAW-CUT CONTRACTION JOINT
WALL SECTION @ DOOR
STAIR SECTION
TYPICAL PIER DETAIL
SECTION
WALL SECTION @ O.H. DOOR
TYPICAL PIER REINFORCEMENT DETAIL
HAIR PIN REIN.
DOCK PARTIAL PLAN
RETAINING WALL DETAIL
RAILING POST ATTACHMENT TO CONCRETE SLAB
DOCK WALL SECTION
LADYSMITH RAIL SITE IMPROVEMENT PROJECT
CITY OF LADYSMITH, WISCONSIN
MORGAN & PARMLEY, LTD.
Professional Consulting Engineers
FOUNDATION DETAILS
99-118
A4

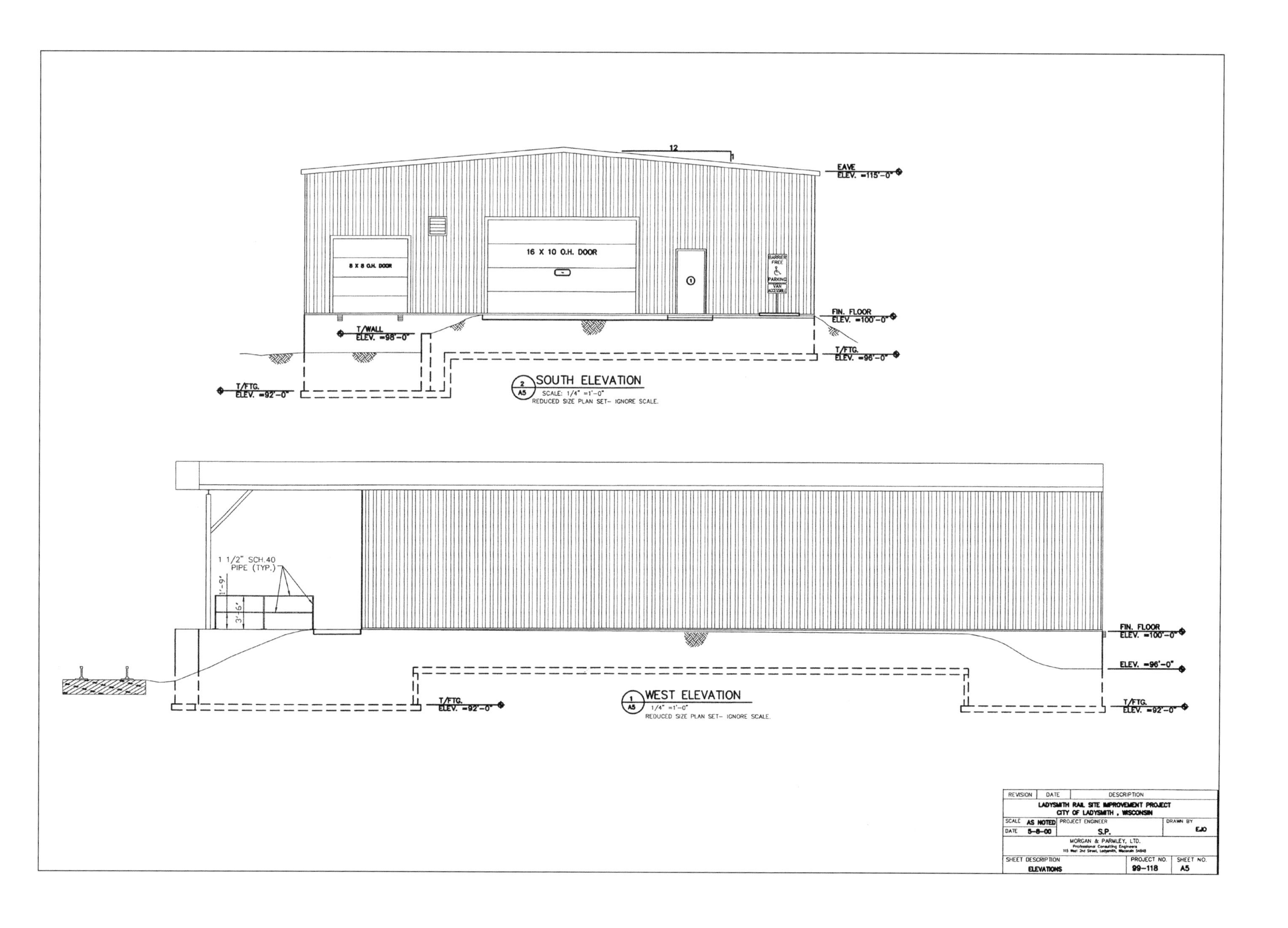
EAVE ELEV. =115'-0"
16 X 10 O.H. DOOR
8 X 8 O.H. DOOR
FIN. FLOOR ELEV. =100'-0"
T/WALL ELEV. =98'-0"
T/FTG. ELEV. =96'-0"
T/FTG. ELEV. =92'-0"
SOUTH ELEVATION
SCALE: 1/4" =1'-0"
REDUCED SIZE PLAN SET- IGNORE SCALE.
1 1/2" SCH.40 PIPE (TYP.)
FIN. FLOOR ELEV. =100'-0"
ELEV. =96'-0"
T/FTG. ELEV. =92'-0"
WEST ELEVATION
1/4" =1'-0"
REDUCED SIZE PLAN SET- IGNORE SCALE.
LADYSMITH RAIL SITE IMPROVEMENT PROJECT
CITY OF LADYSMITH, WISCONSIN
SCALE AS NOTED
DATE 5-8-00
PROJECT ENGINEER S.P.
DRAWN BY EJO
MORGAN & PARMLEY, LTD.
SHEET DESCRIPTION ELEVATIONS
PROJECT NO. 99-118
SHEET NO. A5

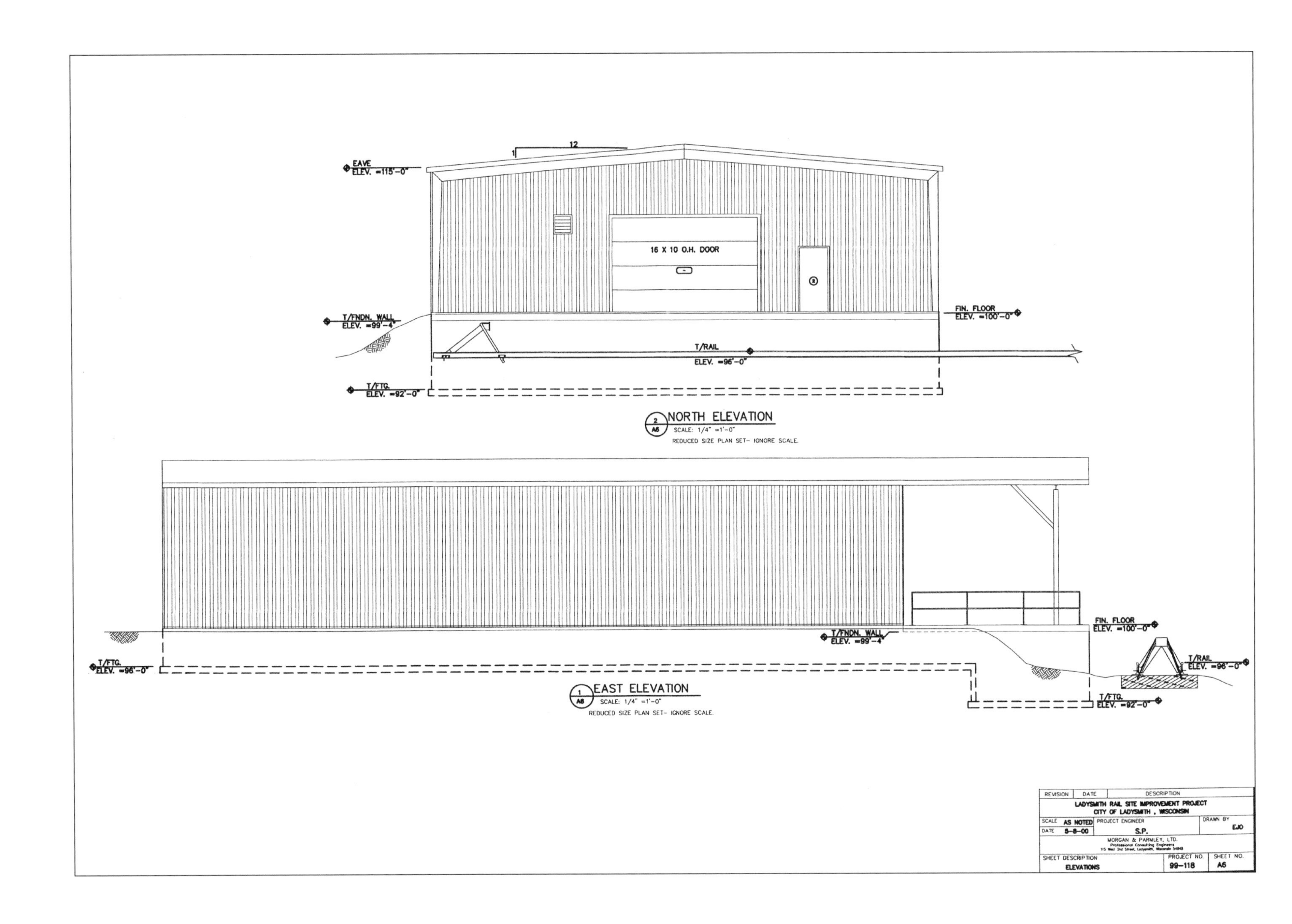
12
1
EAVE
ELEV. =115'-0"
16 X 10 O.H. DOOR
T/FNDN. WALL
ELEV. =99'-4"
FIN. FLOOR
ELEV. =100'-0"
T/RAIL
ELEV. =96'-0"
T/FTG.
ELEV. =92'-0"
2 A6 NORTH ELEVATION
SCALE: 1/4" =1'-0"
REDUCED SIZE PLAN SET- IGNORE SCALE.
T/FTG.
ELEV. =96'-0"
T/FNDN. WALL
ELEV. =99'-4"
FIN. FLOOR
ELEV. =100'-0"
T/RAIL
ELEV. =96'-0"
T/FTG.
ELEV. =92'-0"
1 A6 EAST ELEVATION
SCALE: 1/4" =1'-0"
REDUCED SIZE PLAN SET- IGNORE SCALE.
REVISION
DATE
DESCRIPTION
LADYSMITH RAIL SITE IMPROVEMENT PROJECT
CITY OF LADYSMITH, WISCONSIN
SCALE AS NOTED
DATE 5-8-00
PROJECT ENGINEER
S.P.
DRAWN BY
EJO
MORGAN & PARMLEY, LTD.
Professional Consulting Engineers
SHEET DESCRIPTION
ELEVATIONS
PROJECT NO.
99-118
SHEET NO.
A6

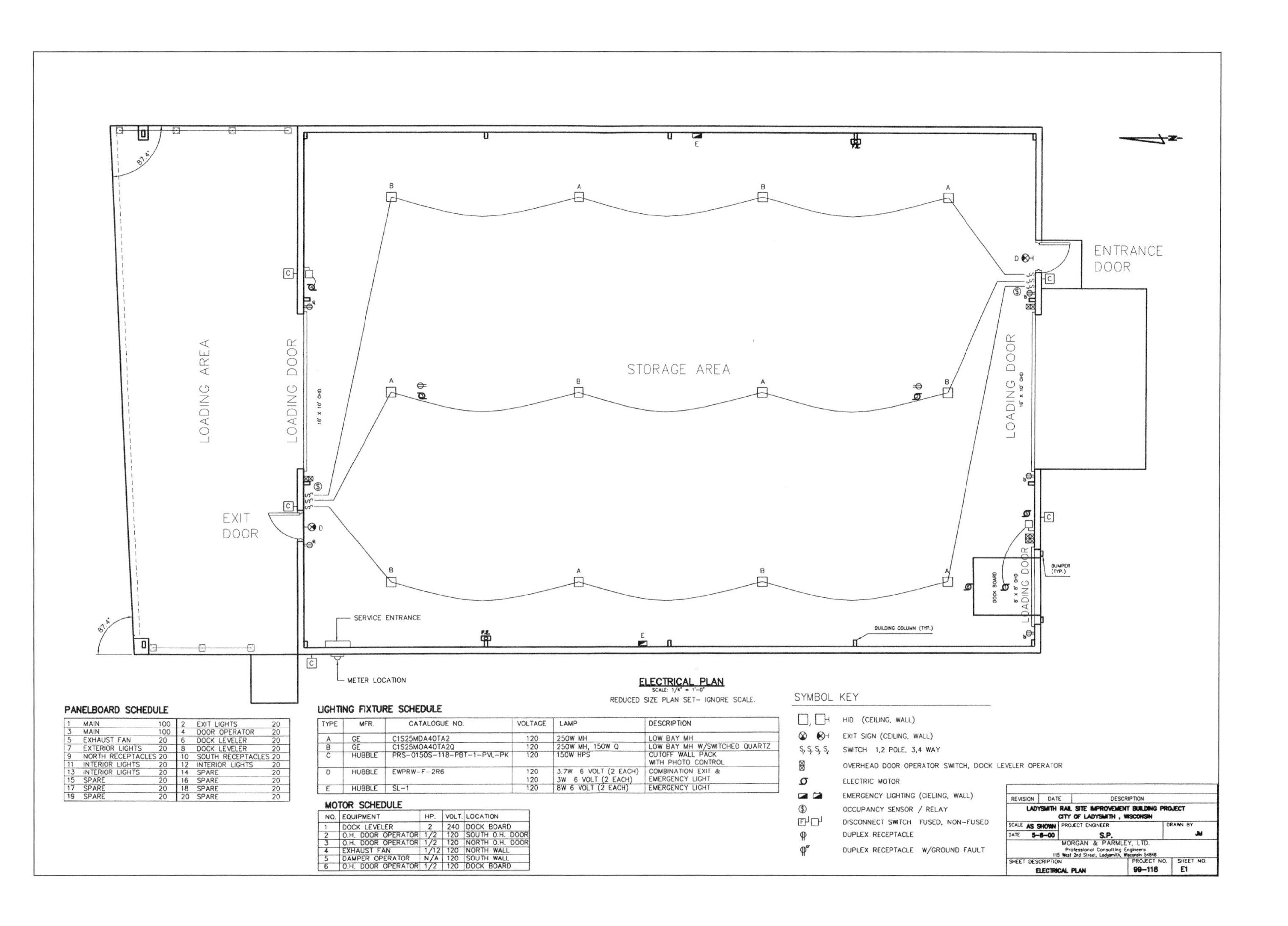

PANELBOARD SCHEDULE

1	MAIN	100	2	EXIT LIGHTS	20
3	MAIN	100	4	DOOR OPERATOR	20
5	EXHAUST FAN	20	6	DOCK LEVELER	20
7	EXTERIOR LIGHTS	20	8	DOCK LEVELER	20
9	NORTH RECEPTACLES	20	10	SOUTH RECEPTACLES	20
11	INTERIOR LIGHTS	20	12	INTERIOR LIGHTS	20
13	INTERIOR LIGHTS	20	14	SPARE	20
15	SPARE	20	16	SPARE	20
17	SPARE	20	18	SPARE	20
19	SPARE	20	20	SPARE	20

LIGHTING FIXTURE SCHEDULE

TYPE	MFR.	CATALOGUE NO.	VOLTAGE	LAMP	DESCRIPTION
A	GE	C1S25MDA40TA2	120	250W MH	LOW BAY MH
B	GE	C1S25MOA40TA2Q	120	250W MH, 150W Q	LOW BAY MH W/SWITCHED QUARTZ
C	HUBBLE	PRS-0150S-118-PBT-1-PVL-PK	120	150W HPS	CUTOFF WALL PACK WITH PHOTO CONTROL
D	HUBBLE	EWPRW-F-2R6	120 120	3.7W 6 VOLT (2 EACH) 3W 6 VOLT (2 EACH)	COMBINATION EXIT & EMERGENCY LIGHT
E	HUBBLE	SL-1	120	8W 6 VOLT (2 EACH)	EMERGENCY LIGHT

MOTOR SCHEDULE

NO.	EQUIPMENT	HP.	VOLT.	LOCATION
1	DOCK LEVELER	2	240	DOCK BOARD
2	O.H. DOOR OPERATOR	1/2	120	SOUTH O.H. DOOR
3	O.H. DOOR OPERATOR	1/2	120	NORTH O.H. DOOR
4	EXHAUST FAN	1/12	120	NORTH WALL
5	DAMPER OPERATOR	N/A	120	SOUTH WALL
6	O.H. DOOR OPERATOR	1/2	120	DOCK BOARD

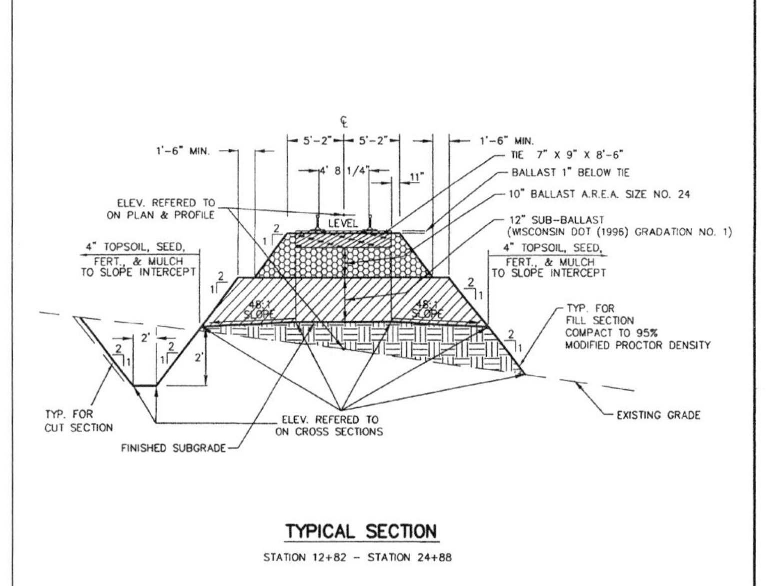

TYPICAL SECTION

STATION 12+82 – STATION 24+88

BULIDING DESIGN INFORMATION

Safe Soil Bearing Capacity = 2000 psf

Pre-engineered Metal Building Design Loading
(In accordance with Wisconsin Administrative Code, Chapter 53)

Roof Live Loading (Snow): 40 psf
Roof Collateral Live Load: 3 psf
Wind Loading:
Structure, wall components: 20 psf
Uplift pressure, normal to surface: 20 psf
Roof overhangs, eaves: 30 psf uplift

Building Design :
Metal building supplier to provide State of Wisconsin Department of Commerce, Safety and Buildings with building submittal for approval in accordance with requirements of the Administrative Code and Specification Section 13200. Plans and calculations shall be prepared by a professional engineer registered to practice in the State of Wisconsin.

Concrete & Concrete Reinforcing Design Stresses (See Specifications)

Thermal Performance ? Energy Calculations
Building is without space heating and therefor exempt from provisions of Administrative Code, Chapter 63.

GENERAL NOTES

ADDITIONAL NOTES ON FOLLOWING SHEETS

1. THE LOCATIONS OF EXISTING AND PROPOSED UTILITY INSTALLATIONS AS SHOWN ON THE PLANS ARE APPROXIMATE. THERE MAY BE OTHER UTILITY INSTALLATIONS WITHIN THE PROJECT AREA THAT ARE NOT SHOWN.
2. THE SEED MIXTURE TO BE USED ON DESIGNATED AREAS SHALL CONFORM TO THE PERTINENT REQUIREMENTS OF THE SPECIFICATIONS.
3. ALL DISTURBED AREAS OTHER THAN THE RAILROAD TRACK BED SHALL BE SEEDED, FERTILIZED AND MULCHED.
4. PRIOR TO THE PLACEMENT OF FILL MATERIAL; THE BOTTOM OF THE EXCAVATION SHALL BE EVALUATED BY A QUALIFIED SOILS ENGINEER TO DETERMINE IF THE EXPOSED SOILS OFFER SUFFICIENT BEARING CAPACITY FOR THE ANTICIPATED LOADS.
5. ALL COMPACTION SHALL BE 95 % OF MAXIMUM MODIFIED PROCTOR DENSITY (ASTM D-1557)
6. THE RAILROAD EXCAVATION AND EMBANKMENT WILL BE BROUGHT TO FINISHED GRADES BY THE EARTHWORK CONTRACTOR.
7. THE EARTHWORK CONTRACTOR WILL BE RESPONSIBLE FOR THE CONSTRUCTION OF ALL EMBANKMENT TO FINISHED SUBGRADE , INSTALLATION OF CULVERTS AND DITCHES AND RESTORATION OF ALL DISTURBED AREAS WITH TOPSOIL , SEED AND FERTILIZER , OR CRUSHED BASE COURSE AND PAVEMENT.
8. THE RAILROAD TRACK CONTRACTOR WILL BE RESPONSIBLE FOR THE PLACEMENT OF SUB-BALLAST , BALLAST , TIES , RAILS AND APPURTENANCES ON BASE PREPARED BY THE EARTHWORK CONTRACTOR.
9. WORK ZONE LIMITS SHALL BE 25' EACH SIDE OF THE RAIL CENTERLINE.
10. THE CONTRACTOR SHALL REMOVE AND STOCKPILE ALL TOPSOIL AND ORGANIC MATERIAL AFTER CLEARING & GRUBBING THE SITE. PRIOR TO PLACING ANY EMBANKMENT MATERIAL, THE CONTRACTOR SHALL SCARIFY THE TOP 6" OF MATERIAL AND COMPACT TO 95% MODIFIED DENSITY.
11. IN A CUT SECTION THE CONTRACTOR SHALL SCARIFY THE 6" OF MATERIAL BELOW THE SUBGRADE AND RECOMPACT TO 95% MODIFIED DENSITY.

MISC. QUANTITIES

4850 C.Y. .... COMMON EXCAVATION
465 C.Y. .... FILL MATERIAL
3830 S.Y. .... SALVAGED TOPSOIL, FERTILIZER, SEED, AND MULCH
90 C.Y. .... C.A.B.C. (SHOULDER MATERIAL – NORSE DRIVE) EXP. 30%
560 C.Y. .... C.A.B.C. (PARKING LOT) EXP. 30%
280 C.Y. .... GRANULAR FILL (PARKING LOT) EXP. 30%
875 C.Y. .... GRANULAR FILL (BLDG & DOCK) EXP. 30%
180 TONS. .... 3" SINGLE AGGREGATE ASPHALTIC PAVEMENT
100 L.F. .... 12" REINFORCED CONCRETE PIPE CLASS III
4 Ea. .... 12" REINFORCED CONCRETE PIPE CLASS III APRON ENDWALL
70 Ea. .... EROSION BALES

RAIL CONSTRUCTION

STATION 12+82 TO 24+88

2,100 L.F. .... RE9020 90 # SECOND HAND CLASS 2, RELAY RAIL (1.050 LF. OF TRACK)
312 L.F. .... RE11525 115 # SECOND HAND CLASS 2, RELAY RAIL (156' OF RAILROAD TRACK) (ROAD CROSSING)

675 Ea. .... NEW OR SECOND HAND RAILROAD TIES 7" X 9" X 8'-6" CLASS 1
850 C.Y. .... 2-1/2"-3/4" CLEAN STONE (A.R.E.A. SIZE NO. 24)(BALLAST)
50 S.Y. .... GEOTEXTILE FABRIC , TYPE HR
1302 C.Y. .... SUBBALLAST (WDOT GRADATION NO. 1)

1 EA. .... HAYS WG – HD (OR EQUAL) BUMPING POST (PROVIDED BY OWNER)
1 PAIR .... NOLAN'S MODEL CS-60 HINGED CAR STOPS (PROVIDED BY OWNER)

EARTHWORK SUMMARY

	CUT C.Y.	FILL C.Y.	NET WASTE C.Y.
BUILDING	950	215	109
RAIL	3900	250	3650
TOTALS	4850	465	3759

UTILITIES

UTILITIES WILL BE MOVED OR ADJUSTED BY THEIR OWNERS UNLESS OTHERWISE CALLED FOR ON THE PLANS.

*** CALL DIGGERS HOTLINE FOR LOCATION 1-800-242-8511**

THE UTILITY OWNERS ARE:

CONTACT: ELECTRIC
XCEL ENERGY
310 HICKORY HILL LANE
PHILLIPS , WISCONSIN 54555
ATTEN: JOE PERKINS
TELEPHONE: (715) 836-1198

CONTACT: NATURAL GAS
WISCONSIN GAS COMPANY
104 WEST SOUTH STREET
RICE LAKE , WISCONSIN 54868
ATTEN: DON WEDIN
TELEPHONE: (800) 925-2104

CONTACT: TELEPHONE
CENTURYTEL
425 ELLINGSON AVENUE
P.O. BOX 78
HAWKINS, WI 54530
ATTEN: JAMES ARQUETTE
TELEPHONE: (715) 585-7707

CONTACT: WATER & SEWER
LADYSMITH DEPT. OF PUBLIC WORKS
120 WEST MINOR AVENUE
P.O. BOX 431
LADYSMITH, WI 54848
ATTEN: BILL CHRISTIANSON, DIRECTOR
TELEPHONE: (715) 532-2601

CONTACT: CABLE TELEVISION
CHARTER COMMUNICATIONS
P.O. BOX 539
RICE LAKE, WISCONSIN
TELEPHONE: (800) 262-2578

REVISION	DATE	DESCRIPTION
LADYSMITH RAIL SITE IMPROVEMENT PROJECT CITY OF LADYSMITH , WISCONSIN		
SCALE AS NOTED DATE 4-24-00	PROJECT ENGINEER S.P.	DRAWN BY JM
MORGAN & PARMLEY, LTD. Professional Consulting Engineers 115 West 2nd Street, Ladysmith, Wisconsin 54848		
SHEET DESCRIPTION TYPICAL SECTIONS, GENERAL NOTES	PROJECT NO. 99-118	SHEET NO. RR1

GENERAL NOTES

DETAILS OF CONSTRUCTION , MATERIALS AND WORKMANSHIP NOT SHOWN ON THIS DRAWING SHALL CONFORM TO THE PERTINENT REQUIREMENTS OF THE STANDARD SPECIFICATIONS , THE APPLICABLE SPECIAL PROVISIONS, WISCONSIN DOT STANDARD SPECIFICATIONS AND FACILITIES DEVELOPEMENT MANUAL FOR TYPE 3 GRADE CROSSINGS. THE A.R.E.A DESIGN MANUAL AND WCL SPECIFICATIONS FOR GRADING AND CONSTRUCTION OF INDUSTRIAL TRACKS.

5/8"X12" LONG CAMRAIL TROX TRUSS WASHER HEAD TIMBER SCREWS EVERY OTHER TIE AND AT THE END OF EACH TIMBER.

CROSSING TO BE CONSTRUCTED OF 115 # NUMBER 2 RELAY RAIL AND SHALL BE FULLY TIE PLATED WITH FOUR SPIKES PER TIE PLATE AND FULLY BOX ANCHORED

3°52'49" AT NORSE DRIVE
2°07'29" AT BARNETT ROAD
SKEW
ALL TIES BETWEEN STATION 15+85 TO 16+65 AND STATION 16+35 TO 24+15 19 1/4" ON CENTER
COUNTERSUNK 5/8"X12" CAMRAIL TORX TRUSS WASHER HEAD TIMBER SCREW EVERY OTHER TIE
STATION 16+25 OR 23+75
CENTERLINE OF ROAD
EDGE OF GRAVEL
C
C
15' ALL ROADWAY CROWN REMOVED
ASPH. SURFACE

TYPE 3 GRADE CROSSING
BARNETT ROAD AND NORSE DRIVE CROSSINGS
CENTERLINE STATION 16+25 AND 23+75

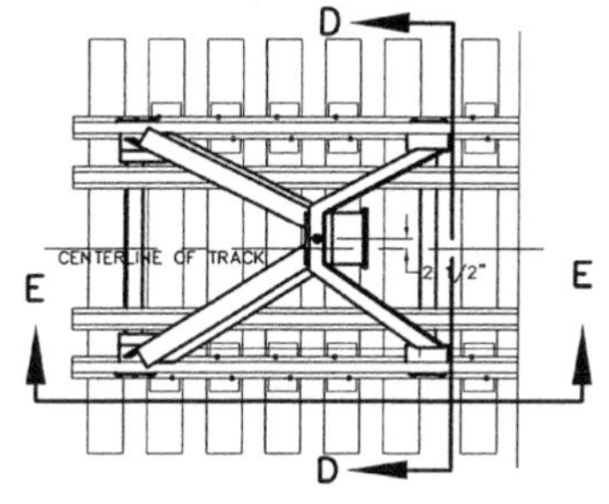

PLAN VIEW
BUMPING POST DETAIL
HAYES TYPE WG-HD OR EQUAL

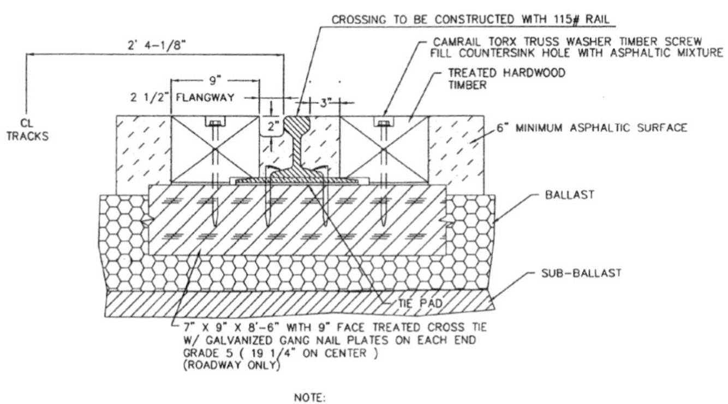

NOTE:
CROSSING BID TO INCLUDE RAIL, TIE PADS, TREATED HARDWOOD TIMBERS AND ALL OTHER NECESSARY ITEMS INCLUDING CROSSBUCKS AND ADVANCED WARNING SIGNS
CONTRACTOR TO INSTALL 6" ASPHALTIC SURFACE THROUGH CROSSING AND BETWEEN TREATED HARDWOOD TIMBERS AS NOTED ELSEWHERE

SECTION C – C

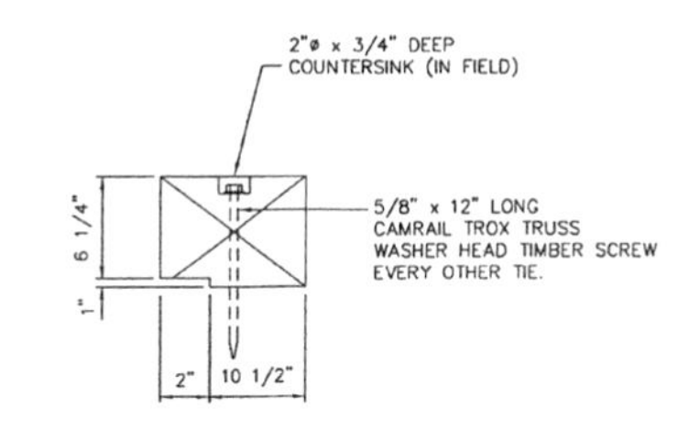

CROSSING PLANK DETAIL
FOR 9020 RAIL

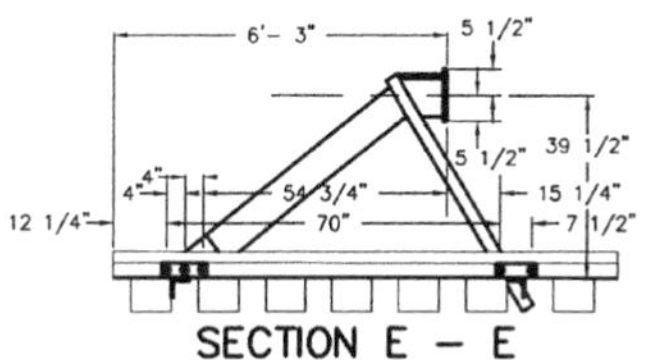

SECTION E – E

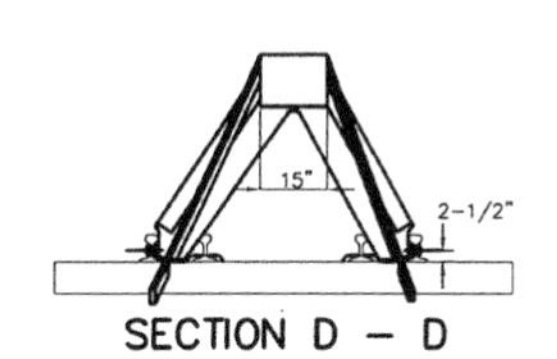

SECTION D – D

REVISION	DATE	DESCRIPTION
LADYSMITH RAIL SITE IMPROVEMENT PROJECT CITY OF LADYSMITH , WISCONSIN		
SCALE AS NOTED DATE 4-24-00	PROJECT ENGINEER S.P.	DRAWN BY JM
MORGAN & PARMLEY, LTD. Professional Consulting Engineers 115 West 2nd Street, Ladysmith, Wisconsin 54848		
SHEET DESCRIPTION RAIL DETAILS	PROJECT NO. 99-118	SHEET NO. RR2

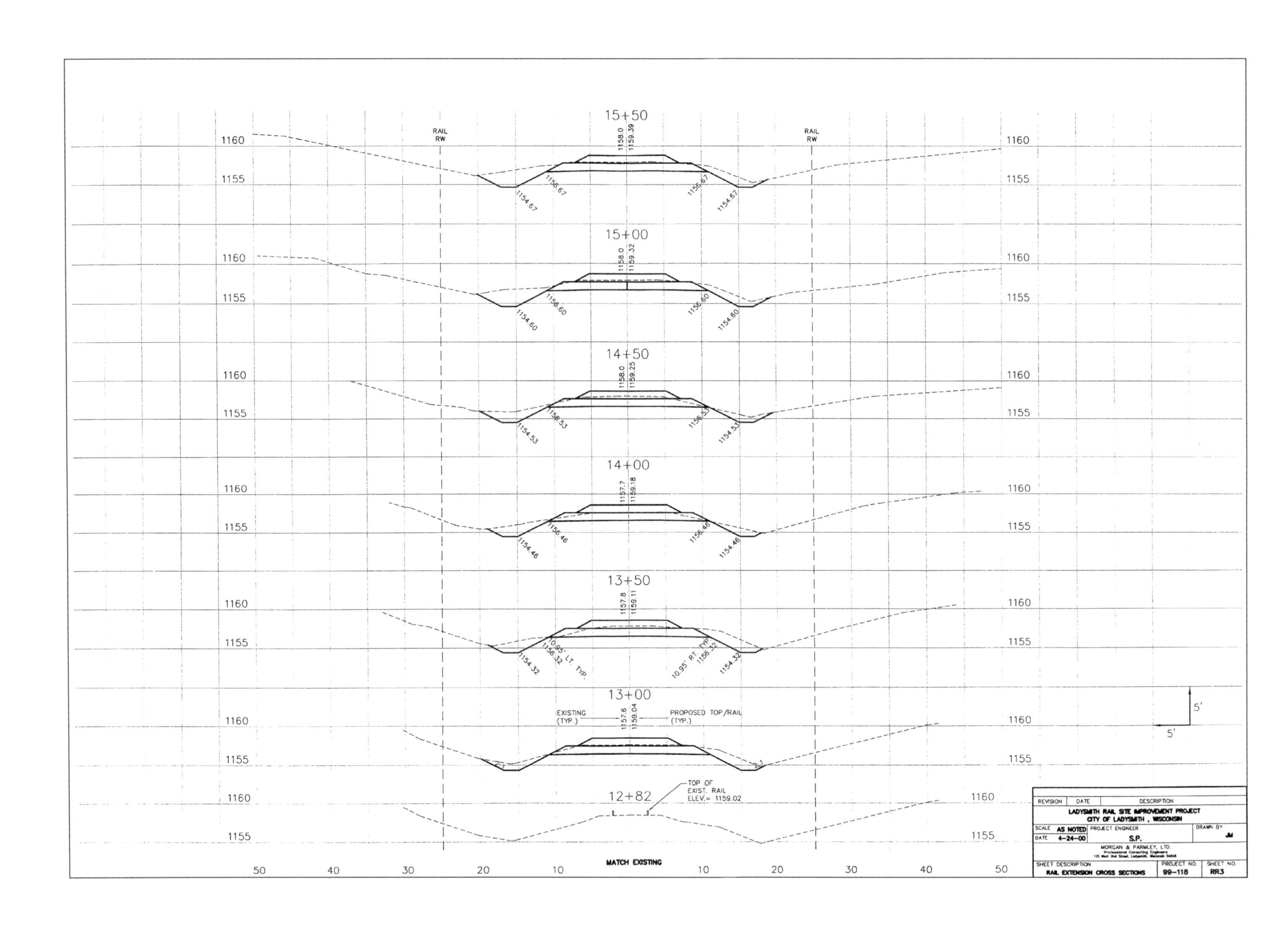
15+50
15+00
14+50
14+00
13+50
13+00
12+82
RAIL RW
EXISTING (TYP.)
PROPOSED TOP/RAIL (TYP.)
TOP OF EXIST. RAIL ELEV.= 1159.02
MATCH EXISTING
LADYSMITH RAIL SITE IMPROVEMENT PROJECT
CITY OF LADYSMITH, WISCONSIN
SCALE AS NOTED
DATE 4-24-00
PROJECT ENGINEER S.P.
MORGAN & PARMLEY, LTD.
SHEET DESCRIPTION RAIL EXTENSION CROSS SECTIONS
PROJECT NO. 99-118
SHEET NO. RR3

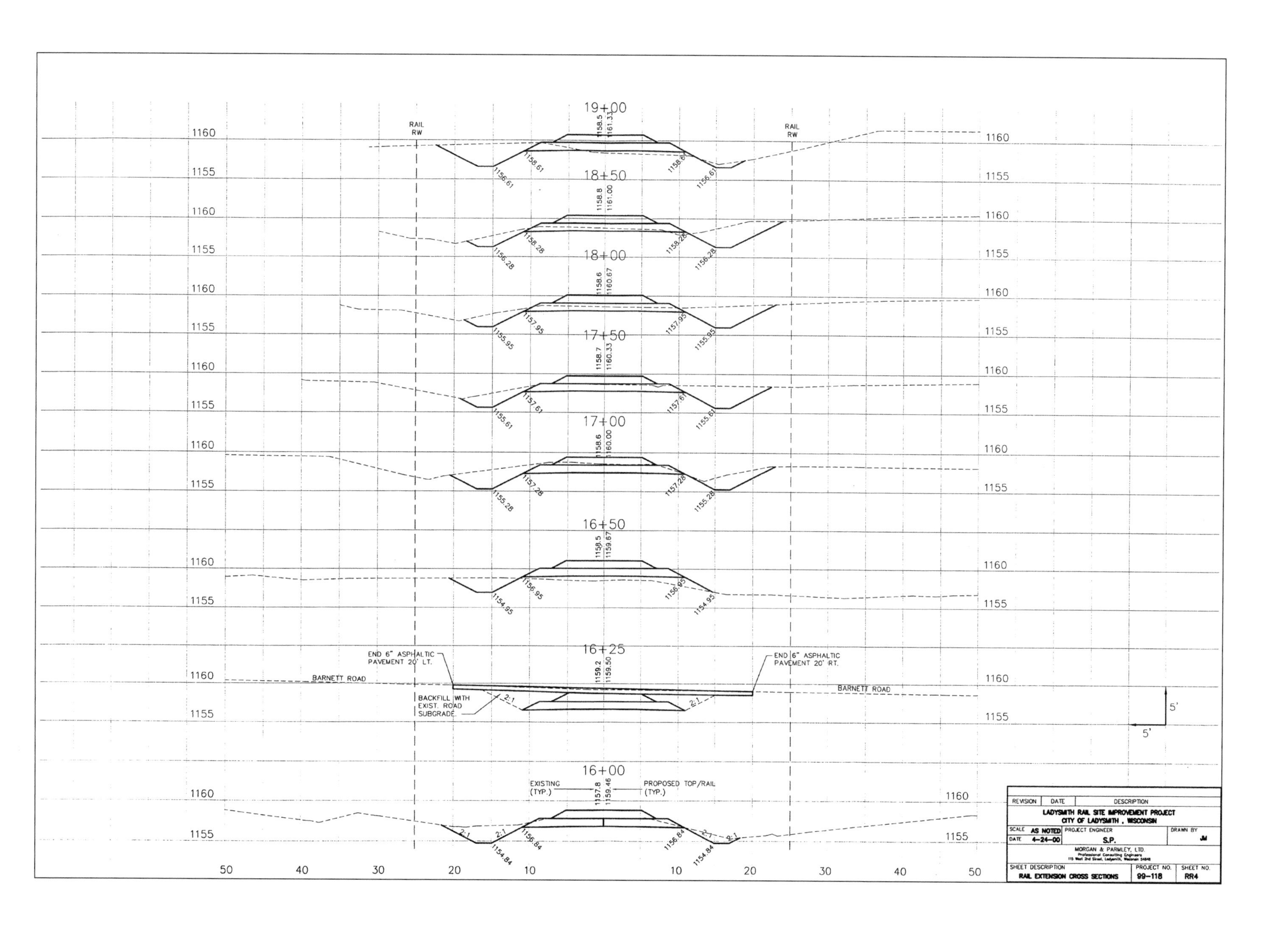
19+00
18+50
18+00
17+50
17+00
16+50
16+25
16+00
RAIL RW
RAIL RW
1160
1155
END 6" ASPHALTIC PAVEMENT 20' LT.
END 6" ASPHALTIC PAVEMENT 20' RT.
BARNETT ROAD
BARNETT ROAD
BACKFILL WITH EXIST. ROAD SUBGRADE.
EXISTING (TYP.)
PROPOSED TOP/RAIL (TYP.)
2:1
5'
5'
50
40
30
20
10
10
20
30
40
50
REVISION
DATE
DESCRIPTION
LADYSMITH RAIL SITE IMPROVEMENT PROJECT
CITY OF LADYSMITH, WISCONSIN
SCALE AS NOTED
DATE 4-24-00
PROJECT ENGINEER S.P.
DRAWN BY JM
MORGAN & PARMLEY, LTD.
Professional Consulting Engineers
SHEET DESCRIPTION RAIL EXTENSION CROSS SECTIONS
PROJECT NO. 99-118
SHEET NO. RR4

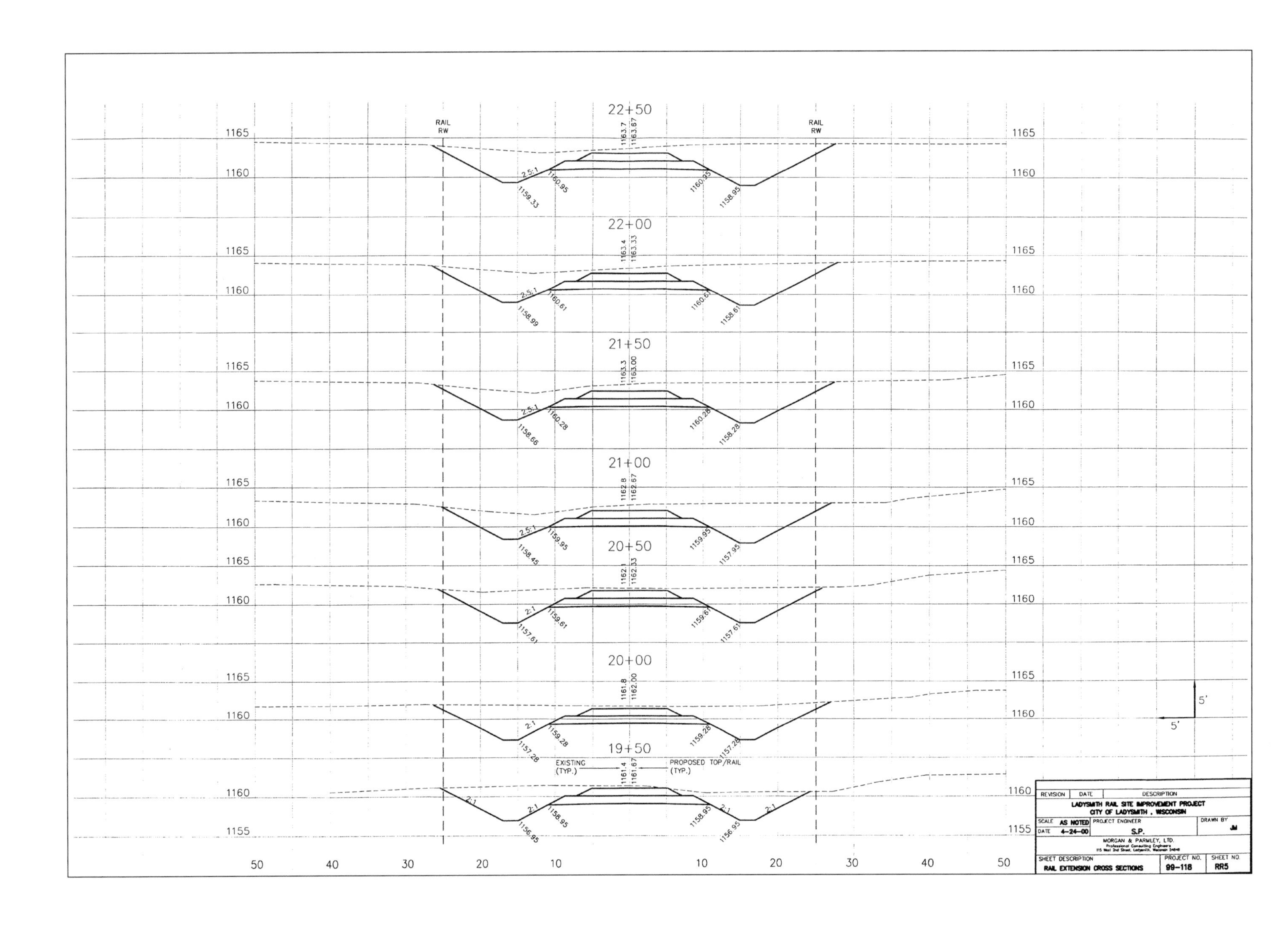
22+50
22+00
21+50
21+00
20+50
20+00
19+50
RAIL RW
EXISTING (TYP.)
PROPOSED TOP/RAIL (TYP.)
LADYSMITH RAIL SITE IMPROVEMENT PROJECT
CITY OF LADYSMITH, WISCONSIN
SCALE AS NOTED
DATE 4-24-00
PROJECT ENGINEER S.P.
DRAWN BY JM
MORGAN & PARMLEY, LTD.
SHEET DESCRIPTION RAIL EXTENSION CROSS SECTIONS
PROJECT NO. 99-118
SHEET NO. RR5

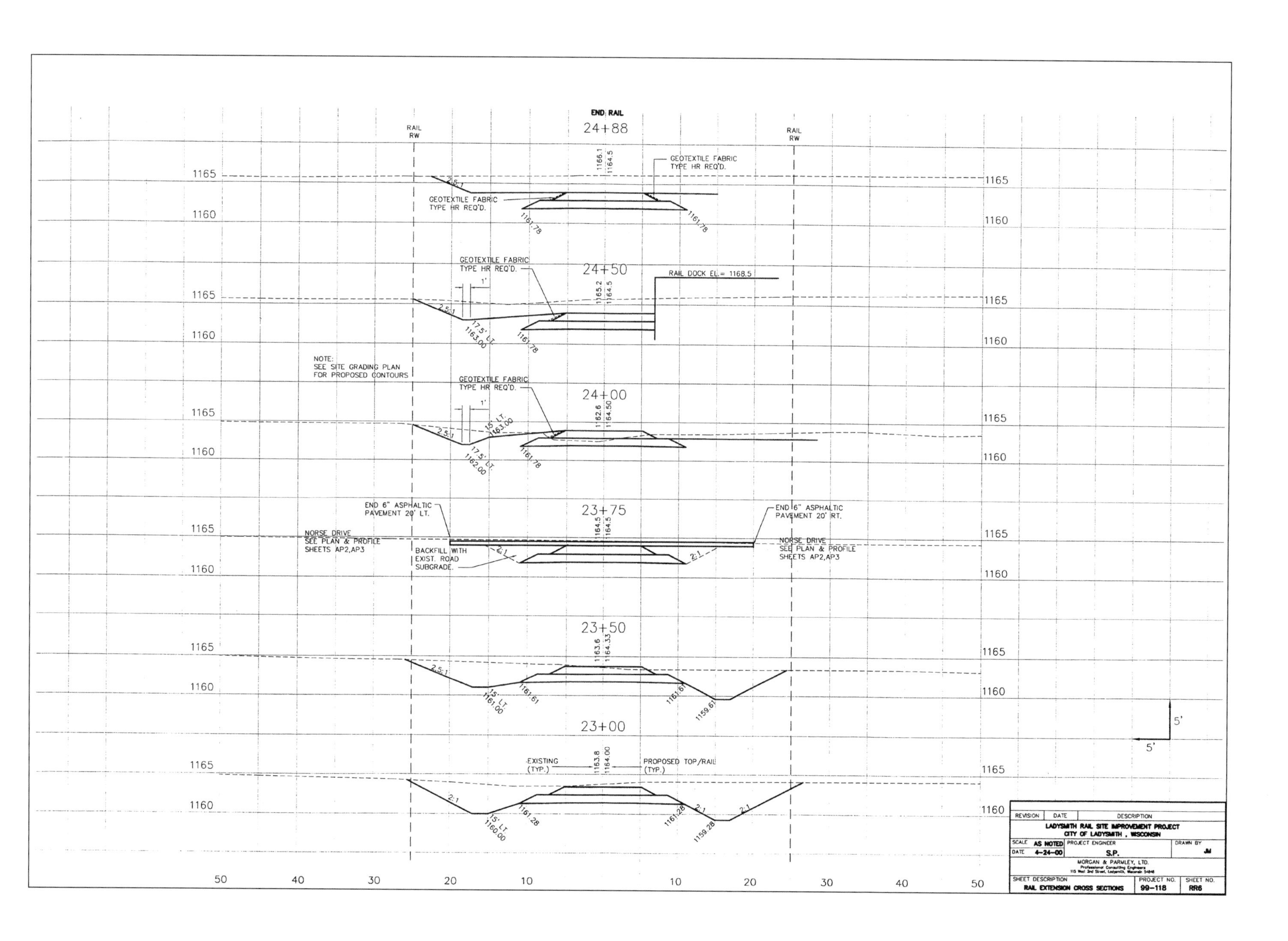
END RAIL
24+88
RAIL RW
GEOTEXTILE FABRIC TYPE HR REQ'D.
24+50
RAIL DOCK EL.= 1168.5
NOTE: SEE SITE GRADING PLAN FOR PROPOSED CONTOURS
24+00
END 6" ASPHALTIC PAVEMENT 20' LT.
END 6" ASPHALTIC PAVEMENT 20' RT.
23+75
NORSE DRIVE SEE PLAN & PROFILE SHEETS AP2,AP3
BACKFILL WITH EXIST. ROAD SUBGRADE.
23+50
23+00
EXISTING (TYP.)
PROPOSED TOP/RAIL (TYP.)
LADYSMITH RAIL SITE IMPROVEMENT PROJECT
CITY OF LADYSMITH , WISCONSIN
SCALE AS NOTED
DATE 4-24-00
PROJECT ENGINEER S.P.
DRAWN BY JM
MORGAN & PARMLEY, LTD.
SHEET DESCRIPTION RAIL EXTENSION CROSS SECTIONS
PROJECT NO. 99-118
SHEET NO. RR6

EDA Project No. 06-01-02846

WILLIAM J. POUPART TRIBAL FISH HATCHERY

FOR THE

LAC DU FLAMBEAU BAND OF LAKE SUPERIOR CHIPPEWA INDIANS OF WISCONSIN

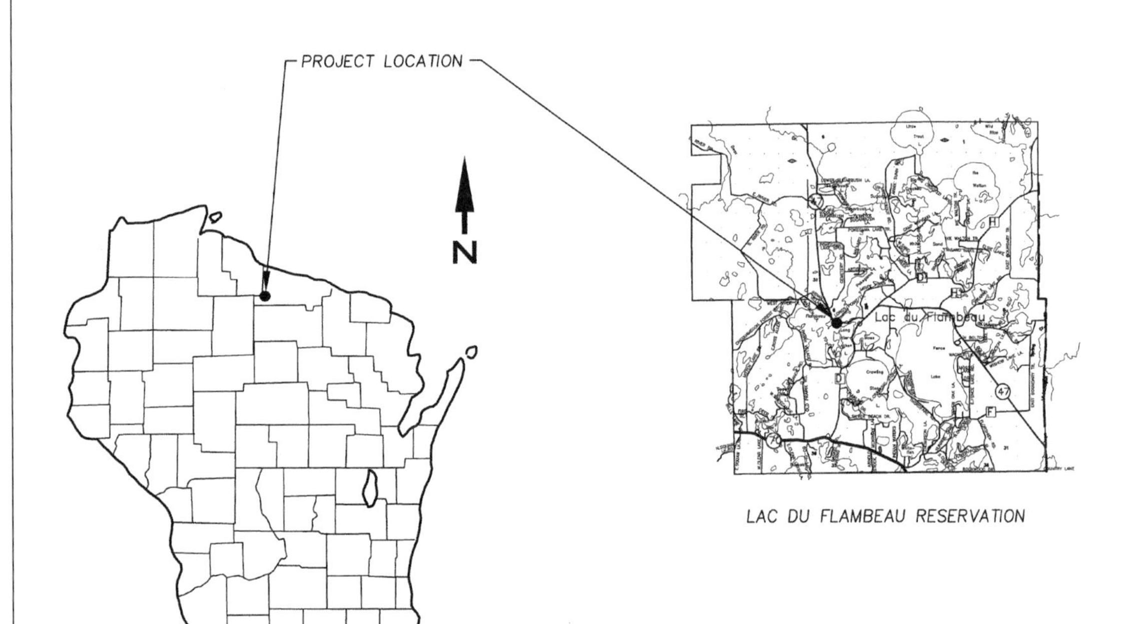

TABLE OF CONTENTS

G1	TITLE SHEET
G2	SITE PLAN
A1	NOT USED
A2	FIRST FLOOR PLAN, MEZZANINE FLOOR PLAN, NOTES
A3	BUILDING EXTERIOR ELEVATIONS
A4	BUILDING EXTERIOR ELEVATIONS
A5	BUILDING SECTION, WALL SECTION
A6	FOUNDATION PLAN AND DETAILS
A7	FOUNDATION DETAILS
A8	MEZZANINE FRAMING PLAN AND DETAILS
A9	ROOF FRAMING PLAN AND DETAILS
A10	ROOF FRAMING DETAILS
A11	ROOM FINISH SCHEDULE, DOOR SCHEDULE, DOOR AND FRAME TYPES
A12	DOOR AND WINDOW DETAILS, CABINET ELEVATION
M1	PROPOSED BUILDING PIPING PLAN
M2	PROPOSED BUILDING PIPING PLAN
M3	UNDER FLOOR PIPING PLAN
M4	JAR RACK PLANS, 60 JARS
M5	JAR RACK PLANS, 20 JARS
M6	TRENCH DRAIN/FLOOR SLOPE PLAN
P1	SITE UTILITY PLAN
P2	FIRST FLOOR PLAN
P3	PIPING SCHEMATICS
HV1	FIRST FLOOR & MEZZANINE FLOOR HEATING AND VENTILATION PLAN
HV2	SCHEDULES AND DETAILS
E1	ELECTRICAL SITE PLAN
E2	FIRST FLOOR/MEZZANINE LIGHTING PLAN
E3	FIRST FLOOR/MEZZANINE POWER PLAN
E4	ELECTRICAL RISER DIAGRAM/DETAILS
E5	INSTRUMENTATION AND CONTROL PLAN
E6	INSTRUMENTATION DETAILS, SCHEDULES AND SCADA OVERVIEW

REVISION	DATE	DESCRIPTION

WILLIAM J. POUPART TRIBAL FISH HATCHERY
LAC DU FLAMBEAU, WISCONSIN

SCALE NONE	PROJECT ENGINEER	DRAWN BY D.A.N.
DATE 7-30-98	BRUCE MARKGREN	

COOPER ENGINEERING COMPANY
310 WEST SOUTH STREET RICE LAKE, WISCONSIN

TELEPHONE	SHEET DESCRIPTION	PROJECT NO.	SHEET NO.
715-234-7008	TITLE SHEET	CE97008	G-1

Courtesy: Copper Engineering Co., Inc.

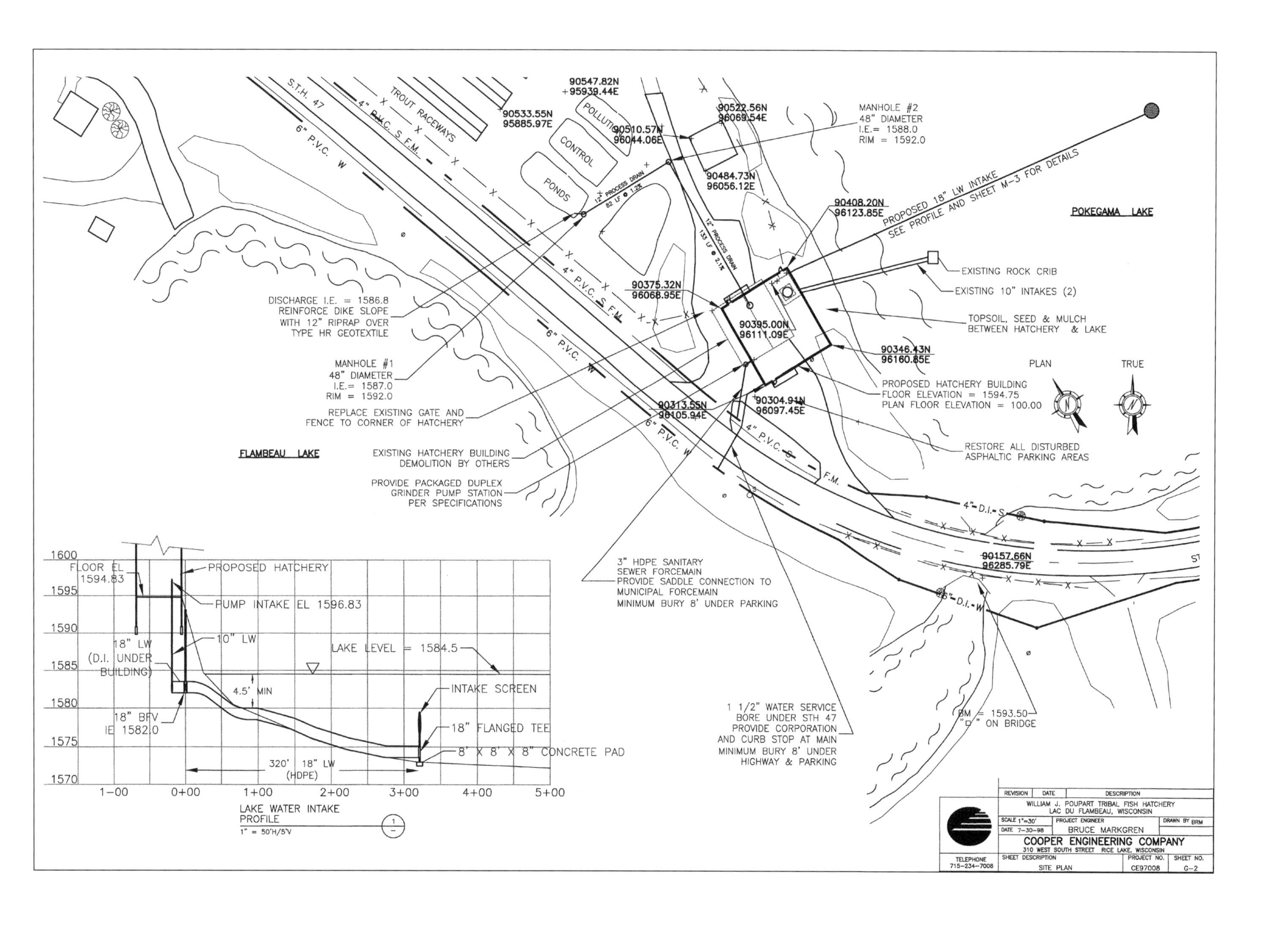
S.T.H. 47
4" P.V.C. S.F.M.
6" P.V.C. W
TROUT RACEWAYS
POLLUTION
CONTROL
PONDS
90547.82N
95939.44E
90533.55N
95885.97E
90510.57N
96044.06E
90522.56N
96069.54E
90484.73N
96056.12E
MANHOLE #2
48" DIAMETER
I.E.= 1588.0
RIM = 1592.0
12" PROCESS DRAIN
82 LF @ 1.2%
133 LF @ 2.1%
PROPOSED 18" LW INTAKE
SEE PROFILE AND SHEET M-3 FOR DETAILS
90408.20N
96123.85E
POKEGAMA LAKE
EXISTING ROCK CRIB
EXISTING 10" INTAKES (2)
TOPSOIL, SEED & MULCH
BETWEEN HATCHERY & LAKE
90375.32N
96068.95E
90395.00N
96111.09E
90346.43N
96160.85E
DISCHARGE I.E. = 1586.8
REINFORCE DIKE SLOPE
WITH 12" RIPRAP OVER
TYPE HR GEOTEXTILE
MANHOLE #1
48" DIAMETER
I.E.= 1587.0
RIM = 1592.0
REPLACE EXISTING GATE AND
FENCE TO CORNER OF HATCHERY
PLAN
TRUE
PROPOSED HATCHERY BUILDING
FLOOR ELEVATION = 1594.75
PLAN FLOOR ELEVATION = 100.00
90313.55N
96105.94E
90304.91N
96097.45E
RESTORE ALL DISTURBED
ASPHALTIC PARKING AREAS
FLAMBEAU LAKE
EXISTING HATCHERY BUILDING
DEMOLITION BY OTHERS
PROVIDE PACKAGED DUPLEX
GRINDER PUMP STATION
PER SPECIFICATIONS
4" P.V.C. S.F.M.
4" D.I. S
6" D.I. W
90157.66N
96285.79E
3" HDPE SANITARY
SEWER FORCEMAIN
PROVIDE SADDLE CONNECTION TO
MUNICIPAL FORCEMAIN
MINIMUM BURY 8' UNDER PARKING
1 1/2" WATER SERVICE
BORE UNDER STH 47
PROVIDE CORPORATION
AND CURB STOP AT MAIN
MINIMUM BURY 8' UNDER
HIGHWAY & PARKING
BM = 1593.50
"□" ON BRIDGE
1600
1595
1590
1585
1580
1575
1570
FLOOR EL
1594.83
PROPOSED HATCHERY
PUMP INTAKE EL 1596.83
10" LW
18" LW
(D.I. UNDER
BUILDING)
LAKE LEVEL = 1584.5
4.5' MIN
INTAKE SCREEN
18" BFV
IE 1582.0
18" FLANGED TEE
8' X 8' X 8" CONCRETE PAD
320' 18" LW
(HDPE)
1-00
0+00
1+00
2+00
3+00
4+00
5+00
LAKE WATER INTAKE
PROFILE
1" = 50'H/5'V
REVISION
DATE
DESCRIPTION
WILLIAM J. POUPART TRIBAL FISH HATCHERY
LAC DU FLAMBEAU, WISCONSIN
SCALE 1"=30'
PROJECT ENGINEER
DRAWN BY BRM
DATE 7-30-98
BRUCE MARKGREN
COOPER ENGINEERING COMPANY
310 WEST SOUTH STREET RICE LAKE, WISCONSIN
TELEPHONE
715-234-7008
SHEET DESCRIPTION
SITE PLAN
PROJECT NO.
CE97008
SHEET NO.
G-2

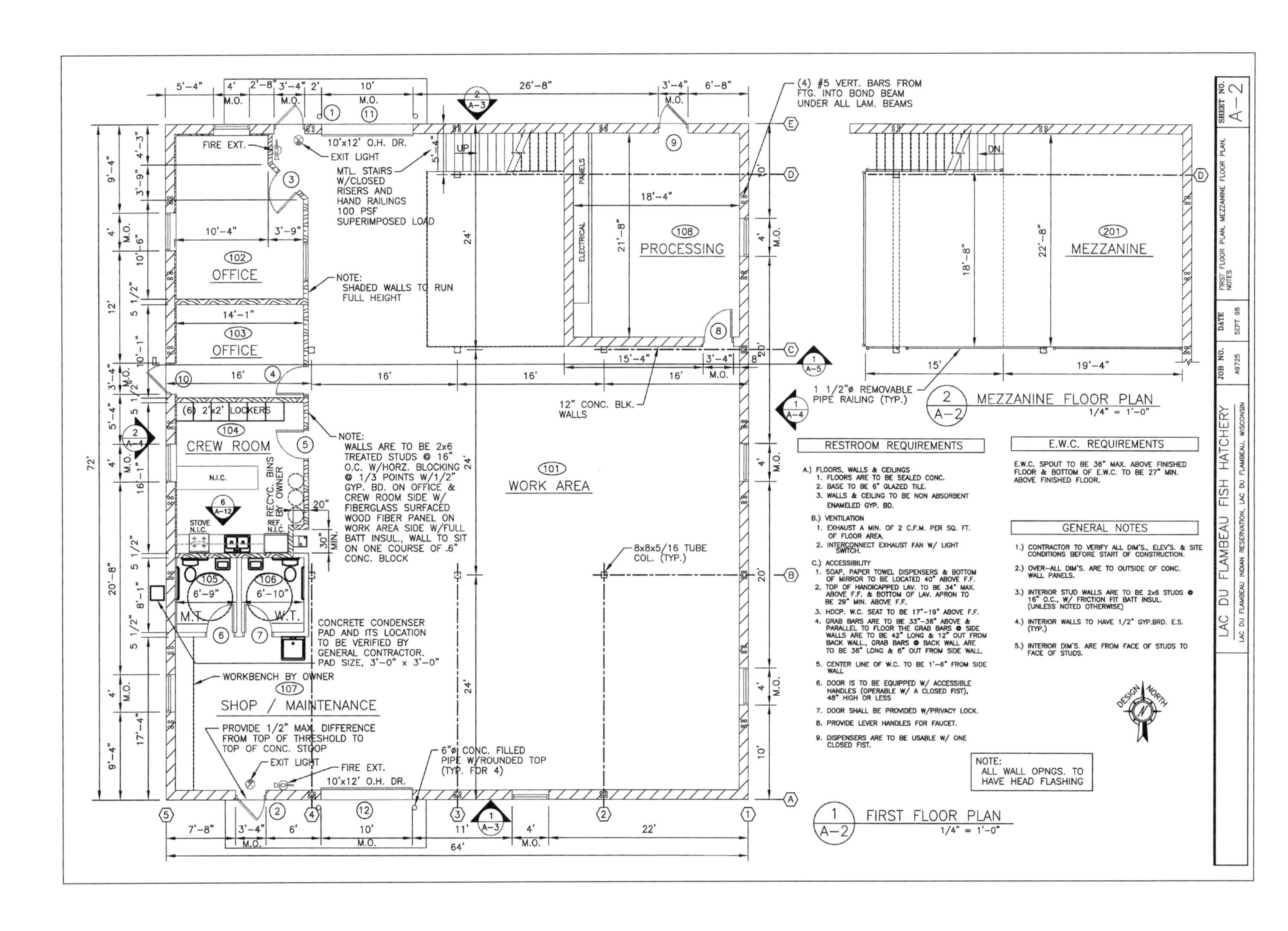

(4) #5 VERT. BARS FROM FTG. INTO BOND BEAM UNDER ALL LAM. BEAMS
10'x12' O.H. DR.
EXIT LIGHT
FIRE EXT.
MTL. STAIRS W/CLOSED RISERS AND HAND RAILINGS 100 PSF SUPERIMPOSED LOAD
NOTE: SHADED WALLS TO RUN FULL HEIGHT
102 OFFICE
103 OFFICE
(6) 2'x2' LOCKERS
104 CREW ROOM
N.I.C.
RECYC. BINS BY OWNER
STOVE N.I.C.
REF. N.I.C.
NOTE: WALLS ARE TO BE 2x6 TREATED STUDS @ 16" O.C. W/HORZ. BLOCKING @ 1/3 POINTS W/1/2" GYP. BD. ON OFFICE & CREW ROOM SIDE W/ FIBERGLASS SURFACED WOOD FIBER PANEL ON WORK AREA SIDE W/FULL BATT INSUL., WALL TO SIT ON ONE COURSE OF 6" CONC. BLOCK
105 M.T.
106 W.T.
CONCRETE CONDENSER PAD AND ITS LOCATION TO BE VERIFIED BY GENERAL CONTRACTOR. PAD SIZE, 3'-0" x 3'-0"
WORKBENCH BY OWNER
107 SHOP / MAINTENANCE
PROVIDE 1/2" MAX. DIFFERENCE FROM TOP OF THRESHOLD TO TOP OF CONC. STOOP
6"ø CONC. FILLED PIPE W/ROUNDED TOP (TYP. FOR 4)
101 WORK AREA
12" CONC. BLK. WALLS
8x8x5/16 TUBE COL. (TYP.)
PANELS
ELECTRICAL
108 PROCESSING
201 MEZZANINE
1 1/2"ø REMOVABLE PIPE RAILING (TYP.)
2 A-2 MEZZANINE FLOOR PLAN 1/4" = 1'-0"
1 A-2 FIRST FLOOR PLAN 1/4" = 1'-0"
RESTROOM REQUIREMENTS
A.) FLOORS, WALLS & CEILINGS
1. FLOORS ARE TO BE SEALED CONC.
2. BASE TO BE 6" GLAZED TILE.
3. WALLS & CEILING TO BE NON ABSORBENT ENAMELED GYP. BD.
B.) VENTILATION
1. EXHAUST A MIN. OF 2 C.F.M. PER SQ. FT. OF FLOOR AREA.
2. INTERCONNECT EXHAUST FAN W/ LIGHT SWITCH.
C.) ACCESSIBILITY
1. SOAP, PAPER TOWEL DISPENSERS & BOTTOM OF MIRROR TO BE LOCATED 40" ABOVE F.F.
2. TOP OF HANDICAPPED LAV. TO BE 34" MAX. ABOVE F.F. & BOTTOM OF LAV. APRON TO BE 29" MIN. ABOVE F.F.
3. HDCP. W.C. SEAT TO BE 17"-19" ABOVE F.F.
4. GRAB BARS ARE TO BE 33"-36" ABOVE & PARALLEL TO FLOOR THE GRAB BARS @ SIDE WALLS ARE TO BE 42" LONG & 12" OUT FROM BACK WALL., GRAB BARS @ BACK WALL ARE TO BE 36" LONG & 6" OUT FROM SIDE WALL.
5. CENTER LINE OF W.C. TO BE 1'-6" FROM SIDE WALL
6. DOOR IS TO BE EQUIPPED W/ ACCESSIBLE HANDLES (OPERABLE W/ A CLOSED FIST), 48" HIGH OR LESS
7. DOOR SHALL BE PROVIDED W/PRIVACY LOCK.
8. PROVIDE LEVER HANDLES FOR FAUCET.
9. DISPENSERS ARE TO BE USABLE W/ ONE CLOSED FIST.
E.W.C. REQUIREMENTS
E.W.C. SPOUT TO BE 36" MAX. ABOVE FINISHED FLOOR & BOTTOM OF E.W.C. TO BE 27" MIN. ABOVE FINISHED FLOOR.
GENERAL NOTES
1.) CONTRACTOR TO VERIFY ALL DIM'S., ELEV'S. & SITE CONDITIONS BEFORE START OF CONSTRUCTION.
2.) OVER-ALL DIM'S. ARE TO OUTSIDE OF CONC. WALL PANELS.
3.) INTERIOR STUD WALLS ARE TO BE 2x6 STUDS @ 16" O.C., W/ FRICTION FIT BATT INSUL. (UNLESS NOTED OTHERWISE)
4.) INTERIOR WALLS TO HAVE 1/2" GYP.BRD. E.S. (TYP.)
5.) INTERIOR DIM'S. ARE FROM FACE OF STUDS TO FACE OF STUDS.
NOTE: ALL WALL OPNGS. TO HAVE HEAD FLASHING
DESIGN NORTH
LAC DU FLAMBEAU FISH HATCHERY
LAC DU FLAMBEAU INDIAN RESERVATION, LAC DU FLAMBEAU, WISCONSIN
FIRST FLOOR PLAN, MEZZANINE FLOOR PLAN, NOTES
SHEET NO. A-2
DATE SEPT 98
JOB NO. A9725

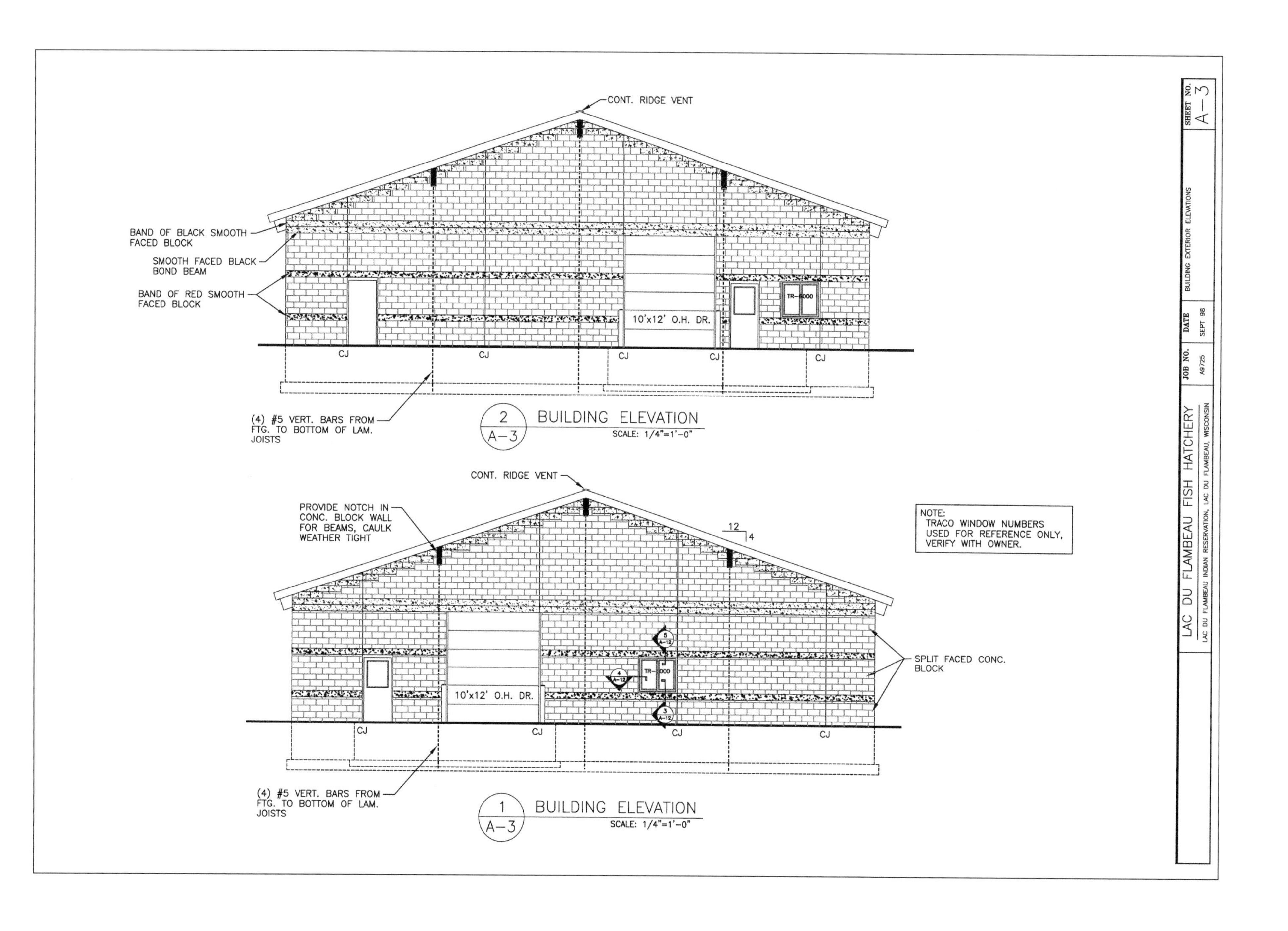
CONT. RIDGE VENT
BAND OF BLACK SMOOTH FACED BLOCK
SMOOTH FACED BLACK BOND BEAM
BAND OF RED SMOOTH FACED BLOCK
10'x12' O.H. DR.
TR-5000
CJ
(4) #5 VERT. BARS FROM FTG. TO BOTTOM OF LAM. JOISTS
2 A-3 BUILDING ELEVATION
SCALE: 1/4"=1'-0"
CONT. RIDGE VENT
PROVIDE NOTCH IN CONC. BLOCK WALL FOR BEAMS, CAULK WEATHER TIGHT
12
4
NOTE: TRACO WINDOW NUMBERS USED FOR REFERENCE ONLY, VERIFY WITH OWNER.
SPLIT FACED CONC. BLOCK
10'x12' O.H. DR.
TR-5000
CJ
(4) #5 VERT. BARS FROM FTG. TO BOTTOM OF LAM. JOISTS
1 A-3 BUILDING ELEVATION
SCALE: 1/4"=1'-0"
SHEET NO. A-3
BUILDING EXTERIOR ELEVATIONS
DATE SEPT 98
JOB NO. A9725
LAC DU FLAMBEAU FISH HATCHERY
LAC DU FLAMBEAU INDIAN RESERVATION, LAC DU FLAMBEAU, WISCONSIN

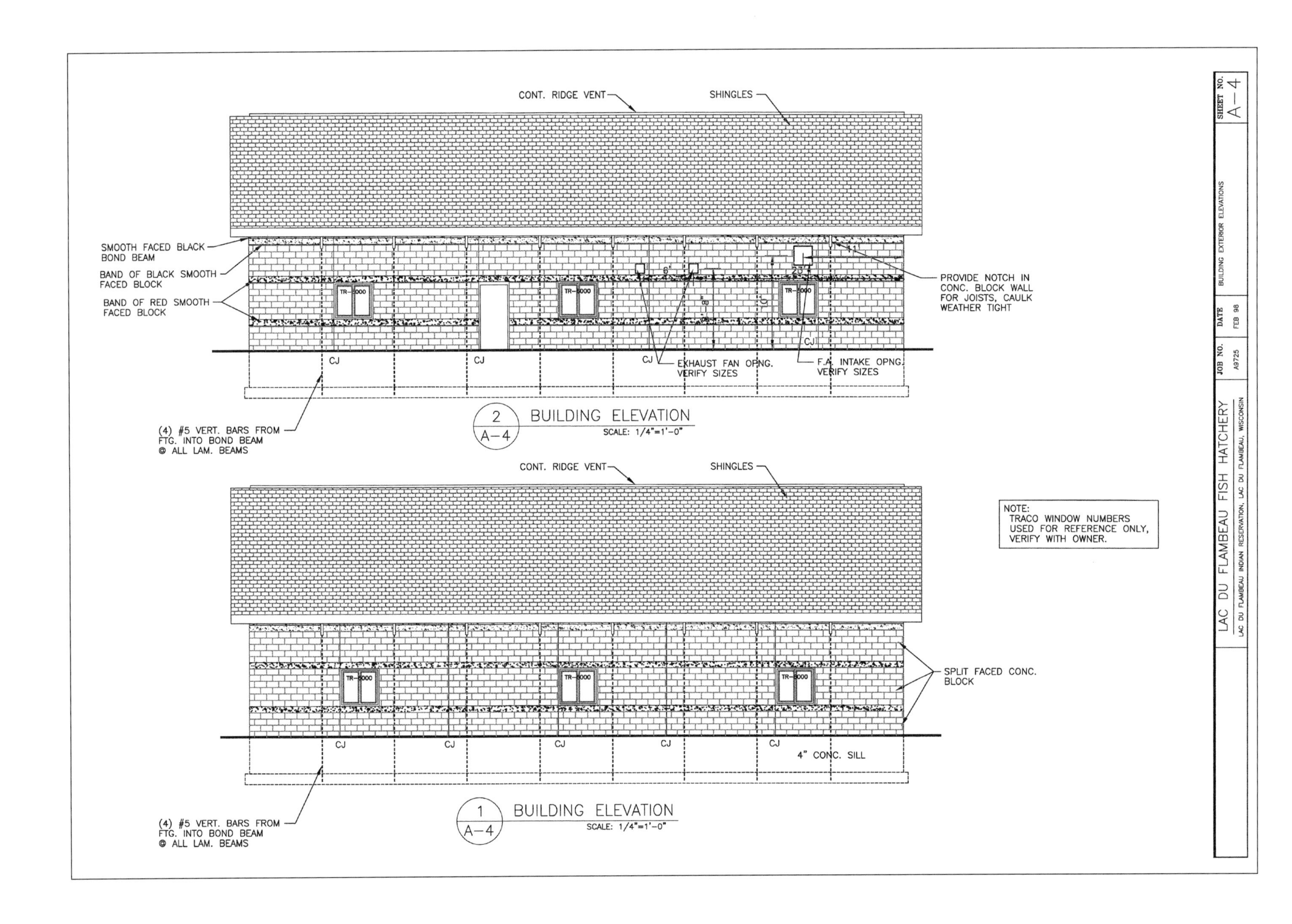
CONT. RIDGE VENT
SHINGLES
SMOOTH FACED BLACK BOND BEAM
BAND OF BLACK SMOOTH FACED BLOCK
BAND OF RED SMOOTH FACED BLOCK
TR-5000
CJ
PROVIDE NOTCH IN CONC. BLOCK WALL FOR JOISTS, CAULK WEATHER TIGHT
EXHAUST FAN OPNG. VERIFY SIZES
F.A. INTAKE OPNG. VERIFY SIZES
(4) #5 VERT. BARS FROM FTG. INTO BOND BEAM @ ALL LAM. BEAMS
2 A-4 BUILDING ELEVATION
SCALE: 1/4"=1'-0"
NOTE:
TRACO WINDOW NUMBERS USED FOR REFERENCE ONLY, VERIFY WITH OWNER.
SPLIT FACED CONC. BLOCK
4" CONC. SILL
1 A-4 BUILDING ELEVATION
SCALE: 1/4"=1'-0"
SHEET NO. A-4
BUILDING EXTERIOR ELEVATIONS
DATE FEB 98
JOB NO. A9725
LAC DU FLAMBEAU FISH HATCHERY
LAC DU FLAMBEAU INDIAN RESERVATION, LAC DU FLAMBEAU, WISCONSIN

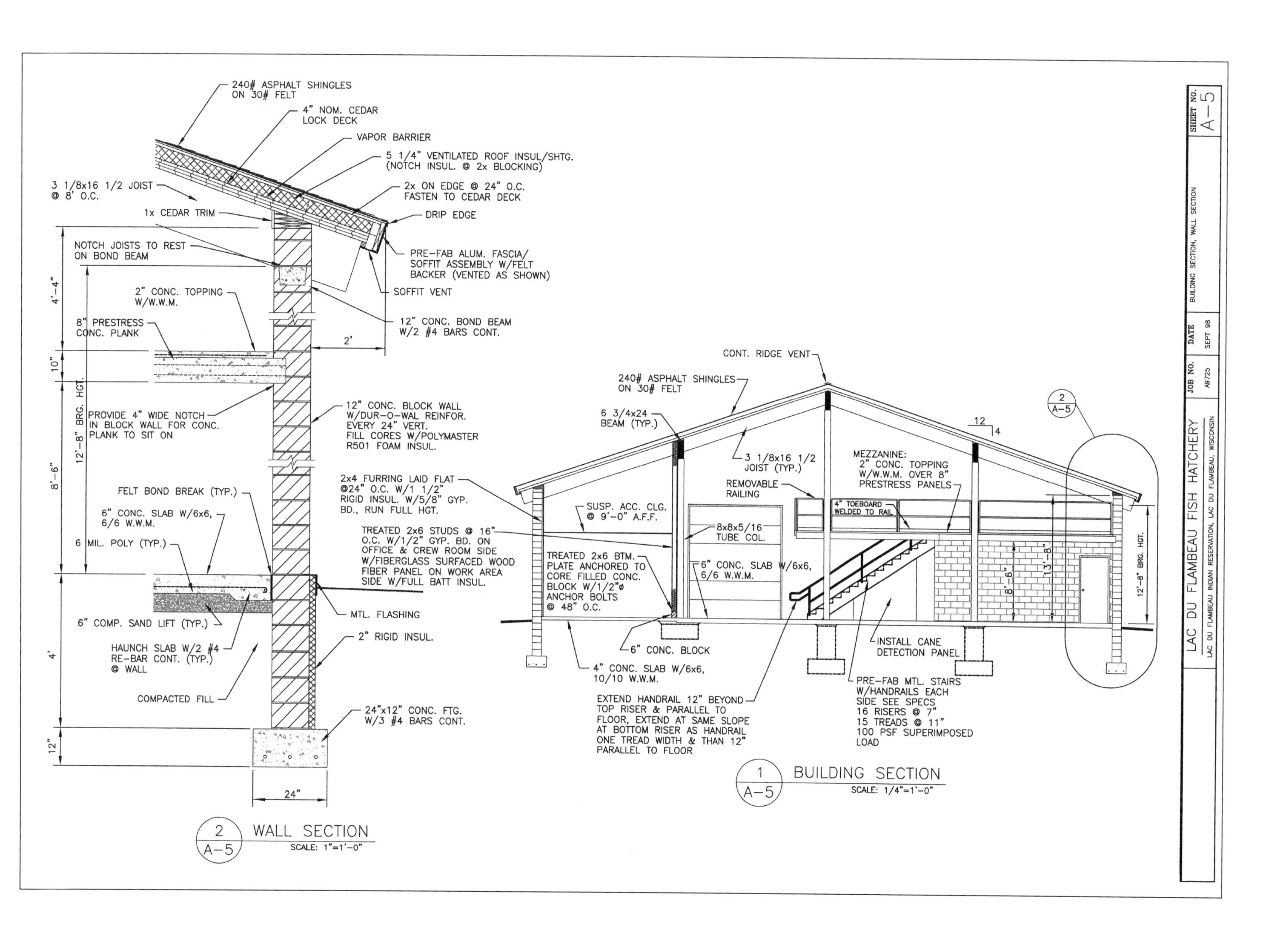
240# ASPHALT SHINGLES ON 30# FELT
4" NOM. CEDAR LOCK DECK
VAPOR BARRIER
5 1/4" VENTILATED ROOF INSUL/SHTG. (NOTCH INSUL. @ 2x BLOCKING)
3 1/8x16 1/2 JOIST @ 8' O.C.
2x ON EDGE @ 24" O.C. FASTEN TO CEDAR DECK
1x CEDAR TRIM
DRIP EDGE
NOTCH JOISTS TO REST ON BOND BEAM
PRE-FAB ALUM. FASCIA/ SOFFIT ASSEMBLY W/FELT BACKER (VENTED AS SHOWN)
SOFFIT VENT
2" CONC. TOPPING W/W.W.M.
12" CONC. BOND BEAM W/2 #4 BARS CONT.
8" PRESTRESS CONC. PLANK
2'
4'-4"
10"
12'-8" BRG. HGT.
8'-6"
4'
12"
PROVIDE 4" WIDE NOTCH IN BLOCK WALL FOR CONC. PLANK TO SIT ON
12" CONC. BLOCK WALL W/DUR-O-WAL REINFOR. EVERY 24" VERT. FILL CORES W/POLYMASTER R501 FOAM INSUL.
2x4 FURRING LAID FLAT @24" O.C. W/1 1/2" RIGID INSUL. W/5/8" GYP. BD., RUN FULL HGT.
FELT BOND BREAK (TYP.)
6" CONC. SLAB W/6x6, 6/6 W.W.M.
TREATED 2x6 STUDS @ 16" O.C. W/1/2" GYP. BD. ON OFFICE & CREW ROOM SIDE W/FIBERGLASS SURFACED WOOD FIBER PANEL ON WORK AREA SIDE W/FULL BATT INSUL.
6 MIL. POLY (TYP.)
MTL. FLASHING
6" COMP. SAND LIFT (TYP.)
2" RIGID INSUL.
HAUNCH SLAB W/2 #4 RE-BAR CONT. (TYP.) @ WALL
COMPACTED FILL
24"x12" CONC. FTG. W/3 #4 BARS CONT.
24"
2 A-5 WALL SECTION
SCALE: 1"=1'-0"
CONT. RIDGE VENT
240# ASPHALT SHINGLES ON 30# FELT
6 3/4x24 BEAM (TYP.)
12
4
2 A-5
3 1/8x16 1/2 JOIST (TYP.)
MEZZANINE: 2" CONC. TOPPING W/W.W.M. OVER 8" PRESTRESS PANELS
REMOVABLE RAILING
4" TOEBOARD WELDED TO RAIL
SUSP. ACC. CLG. @ 9'-0" A.F.F.
8x8x5/16 TUBE COL.
TREATED 2x6 BTM. PLATE ANCHORED TO CORE FILLED CONC. BLOCK W/1/2"ø ANCHOR BOLTS @ 48" O.C.
6" CONC. SLAB W/6x6, 6/6 W.W.M.
8'-6"
13'-8"
12'-8" BRG. HGT.
6" CONC. BLOCK
INSTALL CANE DETECTION PANEL
4" CONC. SLAB W/6x6, 10/10 W.W.M.
PRE-FAB MTL. STAIRS W/HANDRAILS EACH SIDE SEE SPECS 16 RISERS @ 7" 15 TREADS @ 11" 100 PSF SUPERIMPOSED LOAD
EXTEND HANDRAIL 12" BEYOND TOP RISER & PARALLEL TO FLOOR, EXTEND AT SAME SLOPE AT BOTTOM RISER AS HANDRAIL ONE TREAD WIDTH & THAN 12" PARALLEL TO FLOOR
1 A-5 BUILDING SECTION
SCALE: 1/4"=1'-0"
LAC DU FLAMBEAU FISH HATCHERY
LAC DU FLAMBEAU INDIAN RESERVATION, LAC DU FLAMBEAU, WISCONSIN
JOB NO. A9725
DATE SEPT 98
BUILDING SECTION, WALL SECTION
SHEET NO. A-5

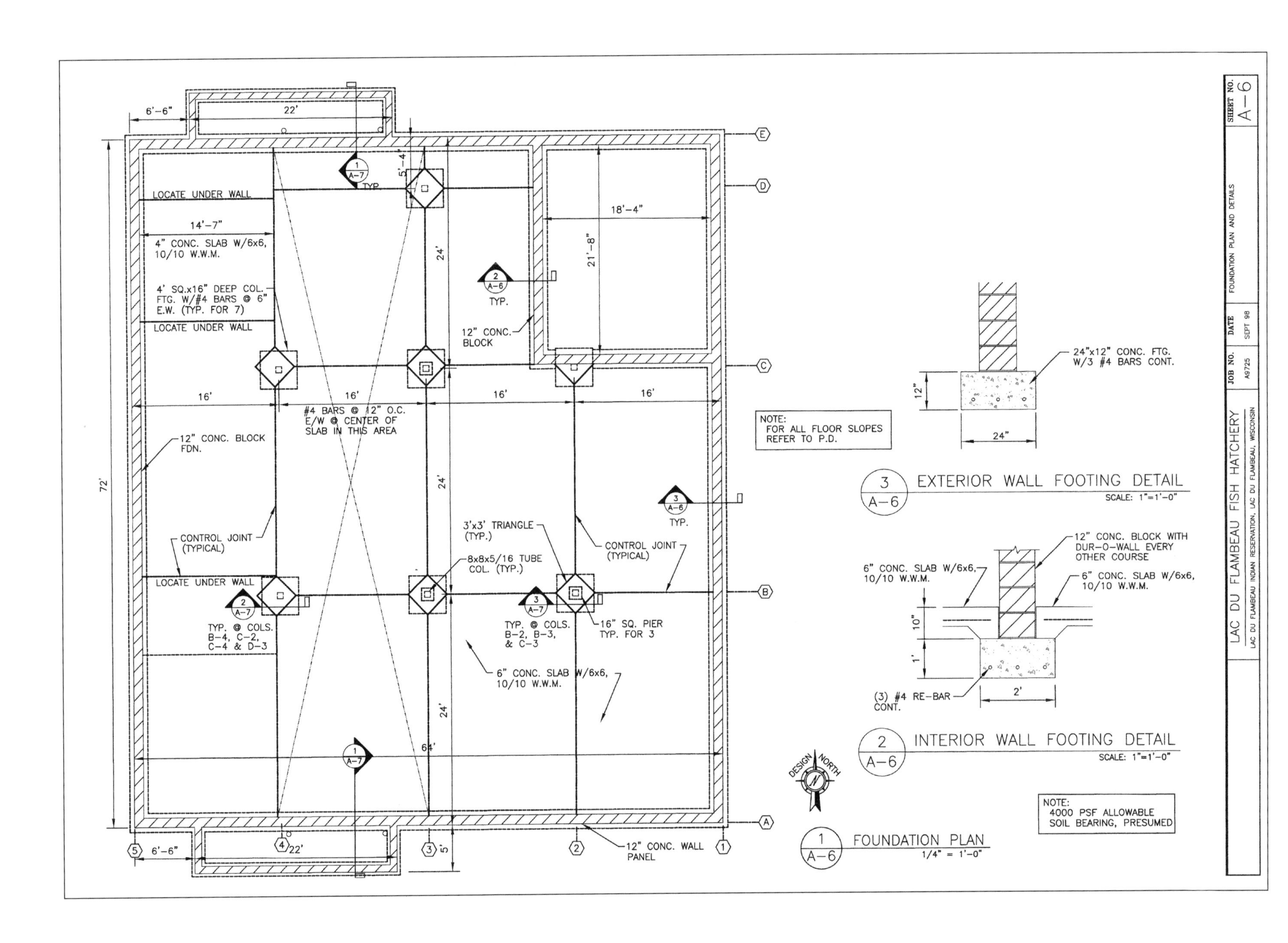

LOCATE UNDER WALL
14'-7"
4" CONC. SLAB W/6x6, 10/10 W.W.M.
4' SQ.x16" DEEP COL. FTG. W/#4 BARS @ 6" E.W. (TYP. FOR 7)
LOCATE UNDER WALL
12" CONC. BLOCK
18'-4"
21'-8"
#4 BARS @ 12" O.C. E/W @ CENTER OF SLAB IN THIS AREA
12" CONC. BLOCK FDN.
CONTROL JOINT (TYPICAL)
3'x3' TRIANGLE (TYP.)
8x8x5/16 TUBE COL. (TYP.)
CONTROL JOINT (TYPICAL)
LOCATE UNDER WALL
TYP. @ COLS. B-4, C-2, C-4 & D-3
TYP. @ COLS. B-2, B-3, & C-3
16" SQ. PIER TYP. FOR 3
6" CONC. SLAB W/6x6, 10/10 W.W.M.
12" CONC. WALL PANEL
NOTE: FOR ALL FLOOR SLOPES REFER TO P.D.
24"x12" CONC. FTG. W/3 #4 BARS CONT.
3 A-6 EXTERIOR WALL FOOTING DETAIL SCALE: 1"=1'-0"
12" CONC. BLOCK WITH DUR-O-WALL EVERY OTHER COURSE
6" CONC. SLAB W/6x6, 10/10 W.W.M.
(3) #4 RE-BAR CONT.
2 A-6 INTERIOR WALL FOOTING DETAIL SCALE: 1"=1'-0"
DESIGN NORTH
NOTE: 4000 PSF ALLOWABLE SOIL BEARING, PRESUMED
1 A-6 FOUNDATION PLAN 1/4" = 1'-0"
LAC DU FLAMBEAU FISH HATCHERY
LAC DU FLAMBEAU INDIAN RESERVATION, LAC DU FLAMBEAU, WISCONSIN
JOB NO. A9725
DATE SEPT 98
FOUNDATION PLAN AND DETAILS
SHEET NO. A-6

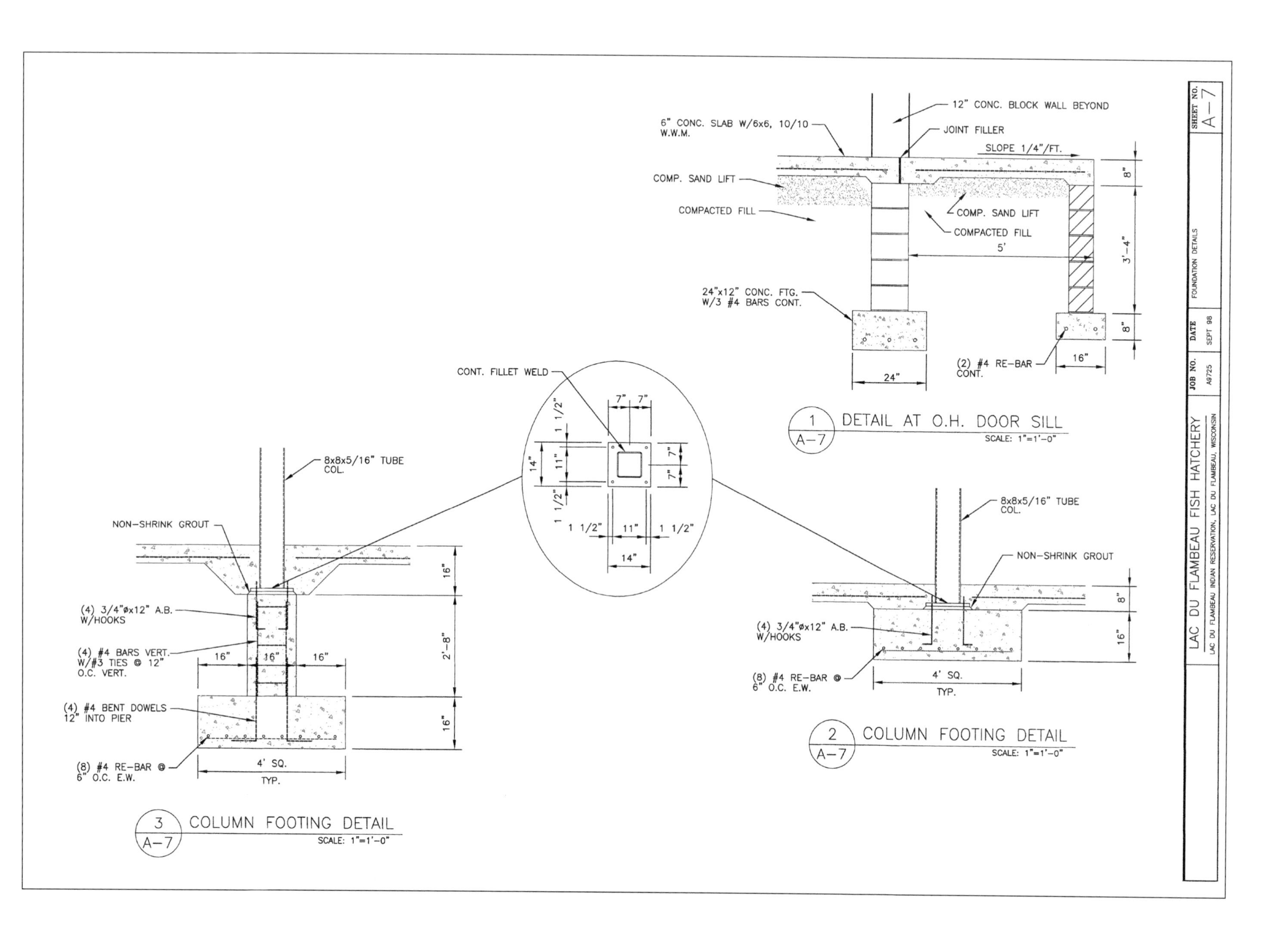
12" CONC. BLOCK WALL BEYOND
6" CONC. SLAB W/6x6, 10/10 W.W.M.
JOINT FILLER
SLOPE 1/4"/FT.
COMP. SAND LIFT
COMPACTED FILL
COMP. SAND LIFT
COMPACTED FILL
5'
8"
3'-4"
8"
24"x12" CONC. FTG. W/3 #4 BARS CONT.
(2) #4 RE-BAR CONT.
16"
24"
1 DETAIL AT O.H. DOOR SILL A-7
SCALE: 1"=1'-0"
CONT. FILLET WELD
7" 7"
1 1/2"
14"
11"
1 1/2"
7"
7"
1 1/2" 11" 1 1/2"
14"
8x8x5/16" TUBE COL.
NON-SHRINK GROUT
(4) 3/4"øx12" A.B. W/HOOKS
(4) #4 BARS VERT. W/#3 TIES @ 12" O.C. VERT.
16" 16" 16"
16"
2'-8"
16"
(4) #4 BENT DOWELS 12" INTO PIER
(8) #4 RE-BAR @ 6" O.C. E.W.
4' SQ.
TYP.
3 COLUMN FOOTING DETAIL A-7
SCALE: 1"=1'-0"
8x8x5/16" TUBE COL.
NON-SHRINK GROUT
(4) 3/4"øx12" A.B. W/HOOKS
(8) #4 RE-BAR @ 6" O.C. E.W.
8"
16"
4' SQ.
TYP.
2 COLUMN FOOTING DETAIL A-7
SCALE: 1"=1'-0"
SHEET NO. A-7
FOUNDATION DETAILS
DATE SEPT 98
JOB NO. A9725
LAC DU FLAMBEAU FISH HATCHERY
LAC DU FLAMBEAU INDIAN RESERVATION, LAC DU FLAMBEAU, WISCONSIN

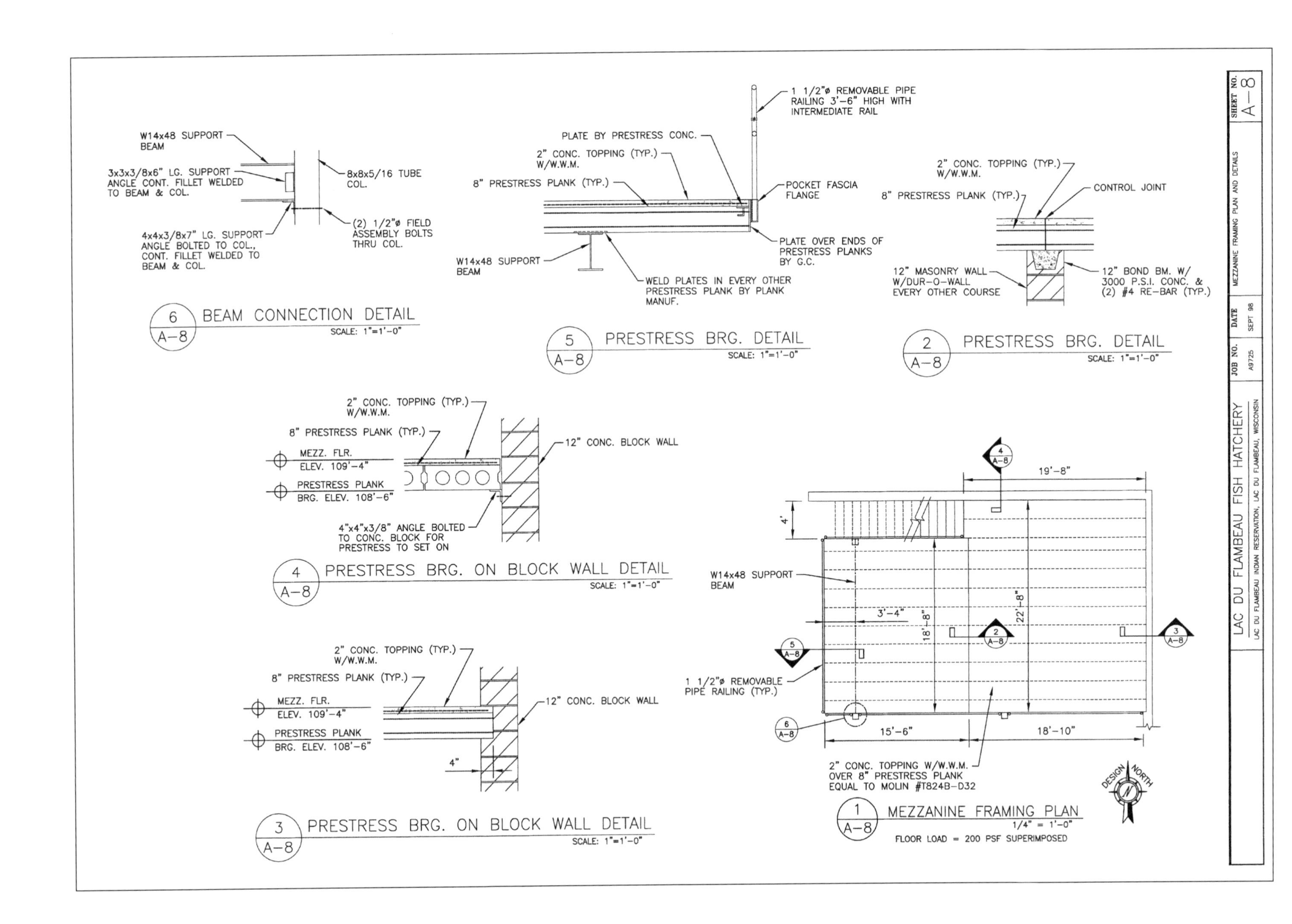
W14x48 SUPPORT BEAM
3x3x3/8x6" LG. SUPPORT ANGLE CONT. FILLET WELDED TO BEAM & COL.
8x8x5/16 TUBE COL.
4x4x3/8x7" LG. SUPPORT ANGLE BOLTED TO COL., CONT. FILLET WELDED TO BEAM & COL.
(2) 1/2"ø FIELD ASSEMBLY BOLTS THRU COL.
6 A-8 BEAM CONNECTION DETAIL
SCALE: 1"=1'-0"
1 1/2"ø REMOVABLE PIPE RAILING 3'-6" HIGH WITH INTERMEDIATE RAIL
PLATE BY PRESTRESS CONC.
2" CONC. TOPPING (TYP.) W/W.W.M.
8" PRESTRESS PLANK (TYP.)
POCKET FASCIA FLANGE
PLATE OVER ENDS OF PRESTRESS PLANKS BY G.C.
W14x48 SUPPORT BEAM
WELD PLATES IN EVERY OTHER PRESTRESS PLANK BY PLANK MANUF.
5 A-8 PRESTRESS BRG. DETAIL
SCALE: 1"=1'-0"
2" CONC. TOPPING (TYP.) W/W.W.M.
8" PRESTRESS PLANK (TYP.)
CONTROL JOINT
12" MASONRY WALL W/DUR-O-WALL EVERY OTHER COURSE
12" BOND BM. W/ 3000 P.S.I. CONC. & (2) #4 RE-BAR (TYP.)
2 A-8 PRESTRESS BRG. DETAIL
SCALE: 1"=1'-0"
2" CONC. TOPPING (TYP.) W/W.W.M.
8" PRESTRESS PLANK (TYP.)
MEZZ. FLR. ELEV. 109'-4"
PRESTRESS PLANK BRG. ELEV. 108'-6"
12" CONC. BLOCK WALL
4"x4"x3/8" ANGLE BOLTED TO CONC. BLOCK FOR PRESTRESS TO SET ON
4 A-8 PRESTRESS BRG. ON BLOCK WALL DETAIL
SCALE: 1"=1'-0"
2" CONC. TOPPING (TYP.) W/W.W.M.
8" PRESTRESS PLANK (TYP.)
MEZZ. FLR. ELEV. 109'-4"
PRESTRESS PLANK BRG. ELEV. 108'-6"
12" CONC. BLOCK WALL
4"
3 A-8 PRESTRESS BRG. ON BLOCK WALL DETAIL
SCALE: 1"=1'-0"
19'-8"
4'
W14x48 SUPPORT BEAM
3'-4"
18'-8"
22'-8"
1 1/2"ø REMOVABLE PIPE RAILING (TYP.)
15'-6"
18'-10"
2" CONC. TOPPING W/W.W.M. OVER 8" PRESTRESS PLANK EQUAL TO MOLIN #T824B-D32
DESIGN NORTH
1 A-8 MEZZANINE FRAMING PLAN
1/4" = 1'-0"
FLOOR LOAD = 200 PSF SUPERIMPOSED
LAC DU FLAMBEAU FISH HATCHERY
LAC DU FLAMBEAU INDIAN RESERVATION, LAC DU FLAMBEAU, WISCONSIN
JOB NO. A9725
DATE SEPT 98
MEZZANINE FRAMING PLAN AND DETAILS
SHEET NO. A-8

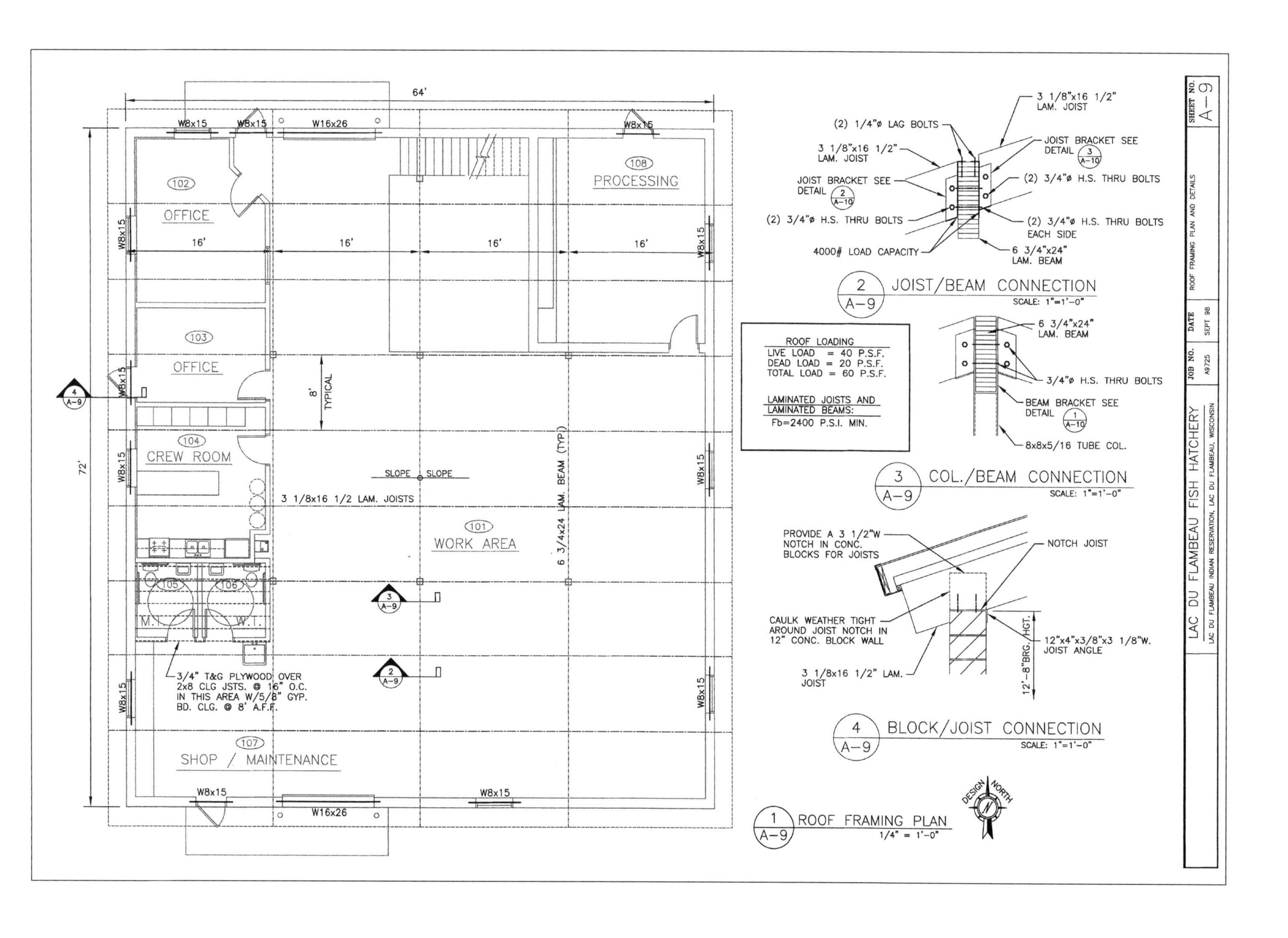
64'
72'
16'
16'
16'
16'
W8x15
W16x26
102
OFFICE
103
OFFICE
104
CREW ROOM
105
106
M.
W.
107
SHOP / MAINTENANCE
108
PROCESSING
101
WORK AREA
SLOPE
SLOPE
3 1/8x16 1/2 LAM. JOISTS
6 3/4x24 LAM. BEAM (TYP.)
8' TYPICAL
3/4" T&G PLYWOOD OVER 2x8 CLG JSTS. @ 16" O.C. IN THIS AREA W/5/8" GYP. BD. CLG. @ 8' A.F.F.
1 A-9 ROOF FRAMING PLAN 1/4" = 1'-0"
DESIGN NORTH
(2) 1/4"ø LAG BOLTS
3 1/8"x16 1/2" LAM. JOIST
JOIST BRACKET SEE DETAIL 3 A-10
JOIST BRACKET SEE DETAIL 2 A-10
(2) 3/4"ø H.S. THRU BOLTS
(2) 3/4"ø H.S. THRU BOLTS EACH SIDE
4000# LOAD CAPACITY
6 3/4"x24" LAM. BEAM
2 A-9 JOIST/BEAM CONNECTION SCALE: 1"=1'-0"
ROOF LOADING
LIVE LOAD = 40 P.S.F.
DEAD LOAD = 20 P.S.F.
TOTAL LOAD = 60 P.S.F.
LAMINATED JOISTS AND LAMINATED BEAMS:
Fb=2400 P.S.I. MIN.
6 3/4"x24" LAM. BEAM
3/4"ø H.S. THRU BOLTS
BEAM BRACKET SEE DETAIL 1 A-10
8x8x5/16 TUBE COL.
3 A-9 COL./BEAM CONNECTION SCALE: 1"=1'-0"
PROVIDE A 3 1/2"W NOTCH IN CONC. BLOCKS FOR JOISTS
NOTCH JOIST
CAULK WEATHER TIGHT AROUND JOIST NOTCH IN 12" CONC. BLOCK WALL
3 1/8x16 1/2" LAM. JOIST
12"x4"x3/8"x3 1/8"W. JOIST ANGLE
12'-8"BRG. HGT.
4 A-9 BLOCK/JOIST CONNECTION SCALE: 1"=1'-0"
LAC DU FLAMBEAU FISH HATCHERY
LAC DU FLAMBEAU INDIAN RESERVATION, LAC DU FLAMBEAU, WISCONSIN
JOB NO. A9725
DATE SEPT 98
ROOF FRAMING PLAN AND DETAILS
SHEET NO. A-9

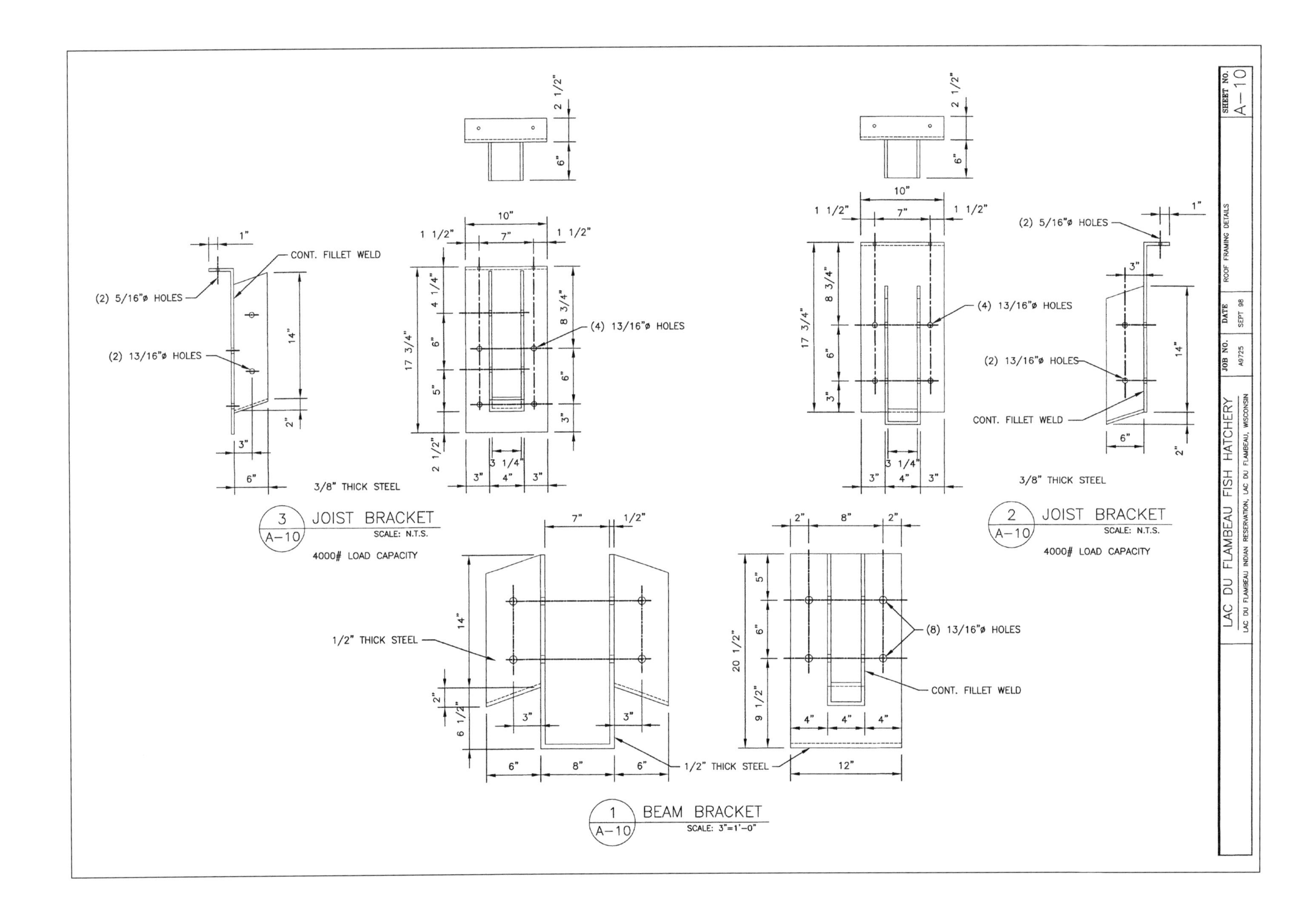
(2) 5/16"ø HOLES
CONT. FILLET WELD
(2) 13/16"ø HOLES
(4) 13/16"ø HOLES
3/8" THICK STEEL
3 JOIST BRACKET A-10
SCALE: N.T.S.
4000# LOAD CAPACITY
2 JOIST BRACKET A-10
SCALE: N.T.S.
4000# LOAD CAPACITY
1/2" THICK STEEL
(8) 13/16"ø HOLES
1 BEAM BRACKET A-10
SCALE: 3"=1'-0"
LAC DU FLAMBEAU FISH HATCHERY
LAC DU FLAMBEAU INDIAN RESERVATION, LAC DU FLAMBEAU, WISCONSIN
JOB NO. A9725
DATE SEPT 98
ROOF FRAMING DETAILS
SHEET NO. A-10

ROOM FINISH SCHEDULE										
ROOM NO.	ROOM NAME	FLOOR	BASE	MATERIALS					CEILING HEIGHT	NOTES
				WALLS				CEILING		
				NORTH	SOUTH	EAST	WEST			
101	WORK AREA	CONC.	–	PCB	PCB	PCB	F.P.	WD. DECK	VARIES	
102	OFFICE	CONC.	–	P.G.B.	P.G.B.	P.G.B.	P.G.B.	SUSP.	9'–0"	
103	OFFICE	CONC.	–	P.G.B.	P.G.B.	P.G.B.	P.G.B.	SUSP.	9'–0"	
104	CREW ROOM	CONC.	–	P.G.B.	P.G.B.	P.G.B.	P.G.B.	SUSP.	9'–0"	
105	MENS TOILET	CONC.	–	P.G.B.	P.G.B.	P.G.B.	P.G.B.	P.G.B.	8'–0"	
106	WOMENS TOILET	CONC.	–	P.G.B.	P.G.B.	P.G.B.	P.G.B.	P.G.B.	8'–0"	
107	SHOP/MAINTENANCE	CONC.	–	F.P.	PCB	–	PCB	WD. DECK	VARIES	
108	PROCESSING	CONC.	–	PCB	PCB	PCB	PCB	CONC. DECK	8'–6"	
201	MEZZANINE	CONC.	–	PCB	–	PCB	–	WD. DECK	VARIES	

ABBREVIATIONS

CONC.– SEALED CONCRETE
PCB– PAINTED CONC. BLOCK
P.G.B.– PAINTED GYPSUM BOARD
F.P.– FIBERGLASS SURFACED WOOD FIBER PANEL

DOOR & FRAME SCHEDULE																
DOOR NO.	DOOR							FRAME						FIRE LABEL	HDWR. GROUP	NOTES
	SIZE										DETAILS					
	WIDTH	HEIGHT	THICK	MATL.	TYPE	GLASS	U-CUT OR LOUVER	MATL.	ELEV.	DEPTH	HEAD	JAMB	SILL			
1	3'–0"	7'–0"	1–3/4"	H.M.	C	A	–	H.M.	2	5 3/4"	4"	2"	–	–	–	2
2	3'–0"	7'–0"	1–3/4"	H.M.	C	A	–	H.M.	2	5 3/4"	4"	2"	–	–	–	2
3	3'–0"	7'–0"	1–3/4"	H.M.	D	B	16"x8"	H.M.	1	5 3/4"	2"	2"	–	–	–	–
4	3'–0"	7'–0"	1–3/4"	H.M.	D	B	16"x8"	H.M.	1	5 3/4"	2"	2"	–	–	–	–
5	3'–0"	7'–0"	1–3/4"	H.M.	D	B	16"x8"	H.M.	1	5 3/4"	2"	2"	–	–	–	–
6	3'–0"	7'–0"	1–3/4"	H.M.	B	–	16"x8"	H.M.	1	5 3/4"	2"	2"	–	–	–	–
7	3'–0"	7'–0"	1–3/4"	H.M.	B	–	16"x8"	H.M.	1	5 3/4"	2"	2"	–	–	–	–
8	3'–0"	7'–0"	1–3/4"	H.M.	A	–	–	H.M.	1	5 3/4"	2"	2"	–	B	–	–
9	3'–0"	7'–0"	1–3/4"	H.M.	A	–	–	H.M.	2	5 3/4"	4"	2"	–	–	–	2
10	3'–0"	7'–0"	1–3/4"	H.M.	A	–	–	H.M.	2	5 3/4"	4"	2"	–	–	–	–
11	10'–0"	12'–0"	–	MTL.	E	–	–	–	–	–	–	–	–	–	–	1
11	10'–0"	12'–0"	–	MTL.	E	–	–	–	–	–	–	–	–	–	–	1

NOTES
1.) PROVIDE AUTOMATIC DOOR OPENER
2.) SEE DETAILS 1 & 2/A–12

GLASS TYPES FOR DOORS
A=1" INSULATED TEMPERED GLASS, CLEAR
B=1/4" SAFETY GLASS, CLEAR

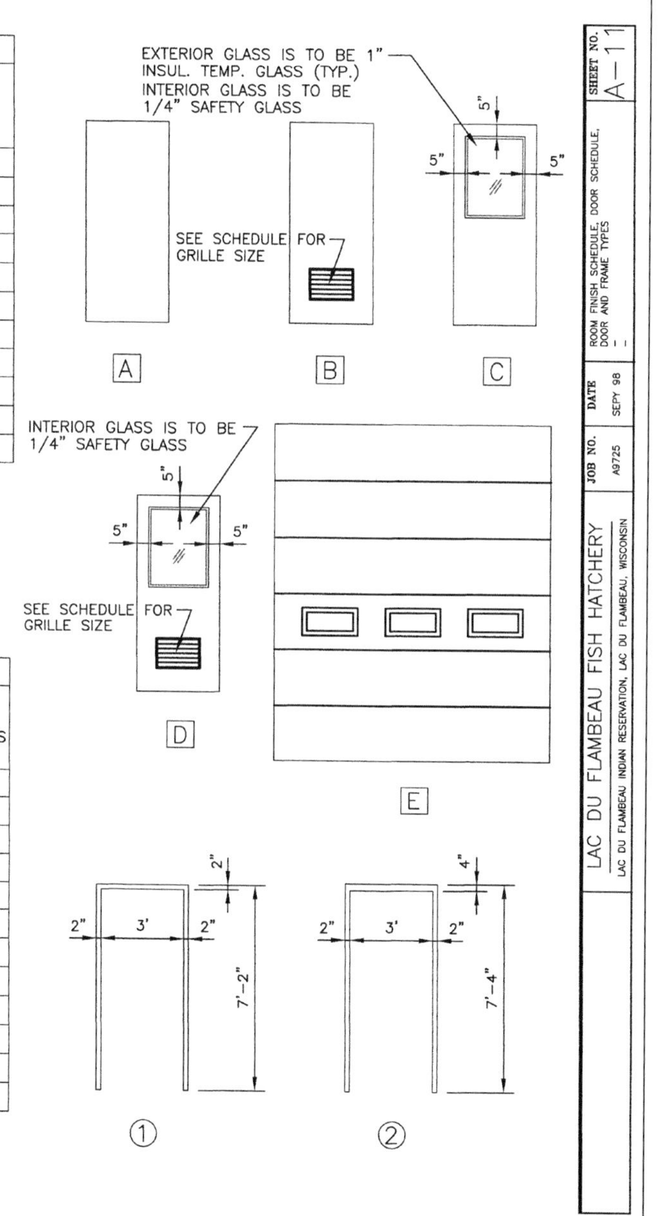

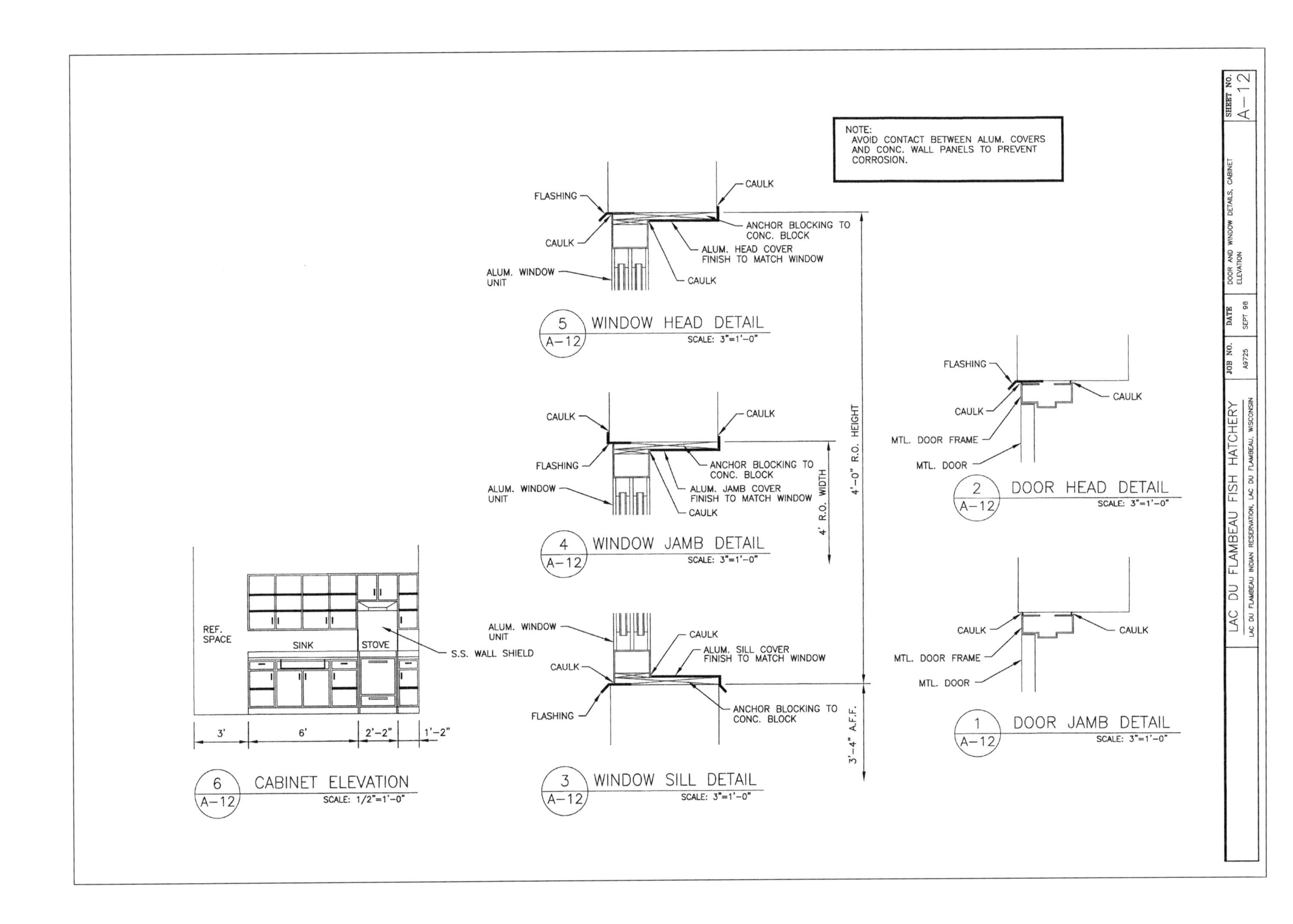

NOTE:
AVOID CONTACT BETWEEN ALUM. COVERS AND CONC. WALL PANELS TO PREVENT CORROSION.
FLASHING
CAULK
ANCHOR BLOCKING TO CONC. BLOCK
ALUM. HEAD COVER FINISH TO MATCH WINDOW
ALUM. WINDOW UNIT
5 A-12 WINDOW HEAD DETAIL
SCALE: 3"=1'-0"
ALUM. JAMB COVER FINISH TO MATCH WINDOW
4 A-12 WINDOW JAMB DETAIL
SCALE: 3"=1'-0"
4' R.O. WIDTH
4'-0" R.O. HEIGHT
3'-4" A.F.F.
ALUM. SILL COVER FINISH TO MATCH WINDOW
3 A-12 WINDOW SILL DETAIL
SCALE: 3"=1'-0"
MTL. DOOR FRAME
MTL. DOOR
2 A-12 DOOR HEAD DETAIL
SCALE: 3"=1'-0"
1 A-12 DOOR JAMB DETAIL
SCALE: 3"=1'-0"
REF. SPACE
SINK
STOVE
S.S. WALL SHIELD
3'
6'
2'-2"
1'-2"
6 A-12 CABINET ELEVATION
SCALE: 1/2"=1'-0"
SHEET NO. A-12
DOOR AND WINDOW DETAILS, CABINET ELEVATION
DATE SEPT 98
JOB NO. A9725
LAC DU FLAMBEAU FISH HATCHERY
LAC DU FLAMBEAU INDIAN RESERVATION, LAC DU FLAMBEAU, WISCONSIN

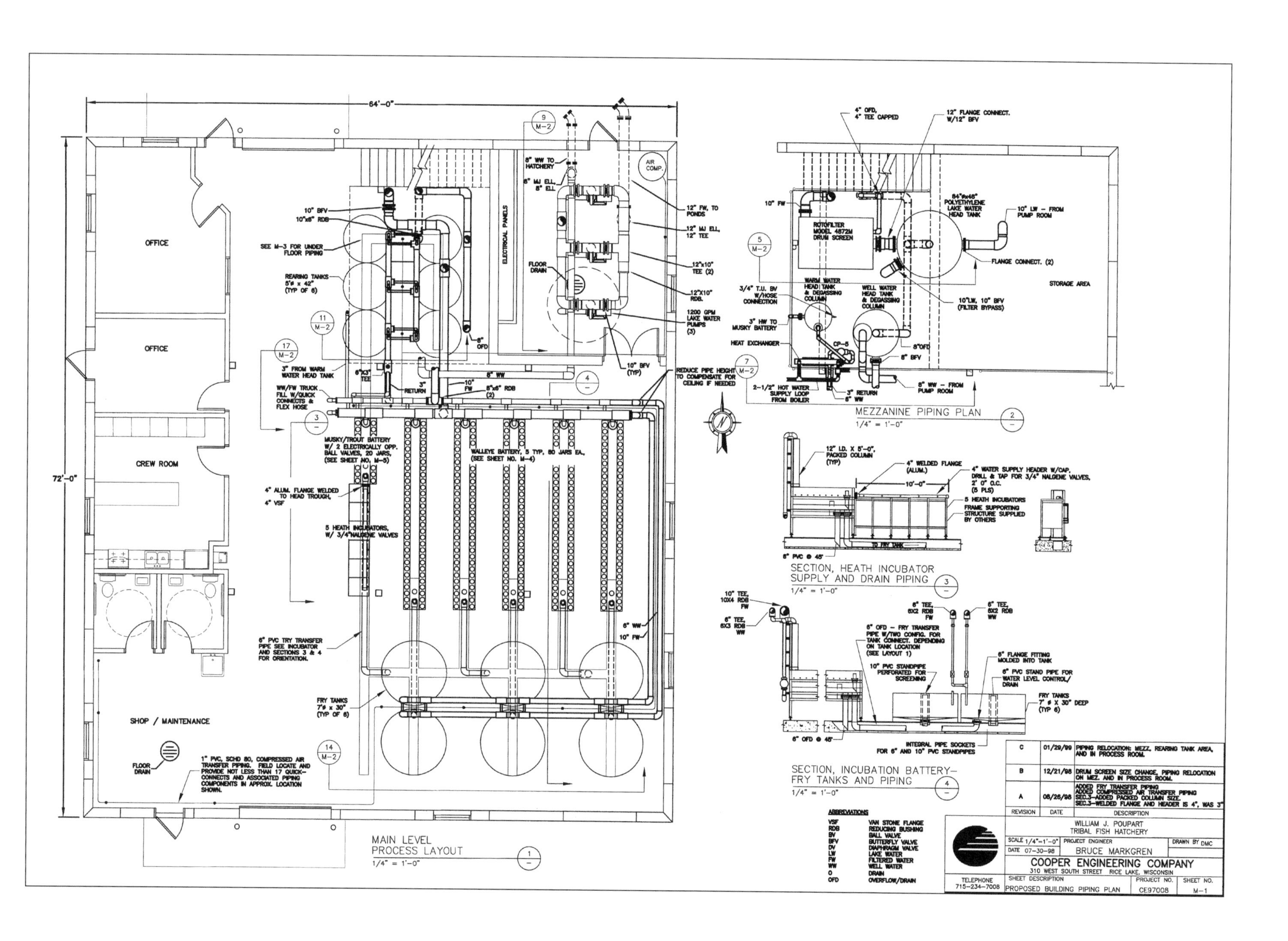
64'-0"
72'-0"
OFFICE
OFFICE
CREW ROOM
SHOP / MAINTENANCE
FLOOR DRAIN
ELECTRICAL PANELS
AIR COMP.
10" BFV
10"x8" RDB
SEE M-3 FOR UNDER FLOOR PIPING
REARING TANKS 5'ø x 42" (TYP OF 6)
3" FROM WARM WATER HEAD TANK
WW/FW TRUCK FILL W/QUICK CONNECTS & FLEX HOSE
MUSKY/TROUT BATTERY W/ 2 ELECTRICALLY OPP. BALL VALVES, 20 JARS, (SEE SHEET NO. M-5)
WALLEYE BATTERY, 5 TYP. 80 JARS EA., (SEE SHEET NO. M-4)
4" ALUM. FLANGE WELDED TO HEAD TROUGH, 4" VSF
5 HEATH INCUBATORS, W/ 3/4" NALGENE VALVES
6" PVC TRY TRANSFER PIPE SEE INCUBATOR AND SECTIONS 3 & 4 FOR ORIENTATION.
FRY TANKS 7'ø x 30" (TYP OF 6)
1" PVC, SCHD 80, COMPRESSED AIR TRANSFER PIPING. FIELD LOCATE AND PROVIDE NOT LESS THAN 17 QUICK-CONNECTS AND ASSOCIATED PIPING COMPONENTS IN APPROX. LOCATION SHOWN.
8" WW TO HATCHERY
8" MJ ELL, 8" ELL
FLOOR DRAIN
12" FW, TO PONDS
12" MJ ELL, 12" TEE
12"x10" TEE (2)
12"X10" RDB.
1200 GPM LAKE WATER PUMPS (3)
10" BFV (TYP)
REDUCE PIPE HEIGHT TO COMPENSATE FOR CEILING IF NEEDED
6"X3" TEE
RETURN
3" RETURN
10" FW
8" WW
8"x6" RDB (2)
6" WW
10" FW
8" OFD
MAIN LEVEL PROCESS LAYOUT
1/4" = 1'-0"
4" OFD, 4" TEE CAPPED
12" FLANGE CONNECT. W/12" BFV
10" FW
ROTOFILTER MODEL 4872M DRUM SCREEN
84"øx48" POLYETHYLENE LAKE WATER HEAD TANK
10" LW - FROM PUMP ROOM
FLANGE CONNECT. (2)
STORAGE AREA
3/4" T.U. BV W/HOSE CONNECTION
WARM WATER HEAD TANK & DEGASSING COLUMN
WELL WATER HEAD TANK & DEGASSING COLUMN
10"LW, 10" BFV (FILTER BYPASS)
3" HW TO MUSKY BATTERY
HEAT EXCHANGER
CP-5
8"OFD
8" BFV
8" WW - FROM PUMP ROOM
2-1/2" HOT WATER SUPPLY LOOP FROM BOILER
3" RETURN
8" WW
MEZZANINE PIPING PLAN
1/4" = 1'-0"
12" I.D. X 5'-0", PACKED COLUMN (TYP)
4" WELDED FLANGE (ALUM.)
10'-0"
4" WATER SUPPLY HEADER W/CAP, DRILL & TAP FOR 3/4" NALGENE VALVES, 2' 0" O.C. (5 PLS)
5 HEATH INCUBATORS
FRAME SUPPORTING STRUCTURE SUPPLIED BY OTHERS
TO FRY TANK
8" PVC @ 45°
SECTION, HEATH INCUBATOR SUPPLY AND DRAIN PIPING
1/4" = 1'-0"
10" TEE, 10X4 RDB FW
6" TEE, 6X3 RDB WW
6" TEE, 6X2 RDB FW
6" TEE, 6X2 RDB WW
6" OFD - FRY TRANSFER PIPE W/TWO CONFIG. FOR TANK CONNECT. DEPENDING ON TANK LOCATION (SEE LAYOUT 1)
10" PVC STANDPIPE PERFORATED FOR SCREENING
6" FLANGE FITTING MOLDED INTO TANK
6" PVC STAND PIPE FOR WATER LEVEL CONTROL/ DRAIN
FRY TANKS 7' ø X 30" DEEP (TYP 6)
6" OFD @ 45°
INTEGRAL PIPE SOCKETS FOR 6" AND 10" PVC STANDPIPES
SECTION, INCUBATION BATTERY-FRY TANKS AND PIPING
1/4" = 1'-0"
ABBREVIATIONS
VSF VAN STONE FLANGE
RDB REDUCING BUSHING
BV BALL VALVE
BFV BUTTERFLY VALVE
DV DIAPHRAGM VALVE
LW LAKE WATER
FW FILTERED WATER
WW WELL WATER
O DRAIN
OFD OVERFLOW/DRAIN
C 01/29/99 PIPING RELOCATION: MEZZ. REARING TANK AREA, AND IN PROCESS ROOM.
B 12/21/98 DRUM SCREEN SIZE CHANGE, PIPING RELOCATION ON MEZ. AND IN PROCESS ROOM.
A 08/26/98 ADDED FRY TRANSFER PIPING ADDED COMPRESSED AIR TRANSFER PIPING SEC.3-ADDED PACKED COLUMN SIZE. SEC.3-WELDED FLANGE AND HEADER IS 4", WAS 3"
REVISION DATE DESCRIPTION
WILLIAM J. POUPART TRIBAL FISH HATCHERY
SCALE 1/4"=1'-0"
PROJECT ENGINEER
DRAWN BY DMC
DATE 07-30-98
BRUCE MARKGREN
COOPER ENGINEERING COMPANY
310 WEST SOUTH STREET RICE LAKE, WISCONSIN
TELEPHONE 715-234-7008
SHEET DESCRIPTION PROPOSED BUILDING PIPING PLAN
PROJECT NO. CE97008
SHEET NO. M-1

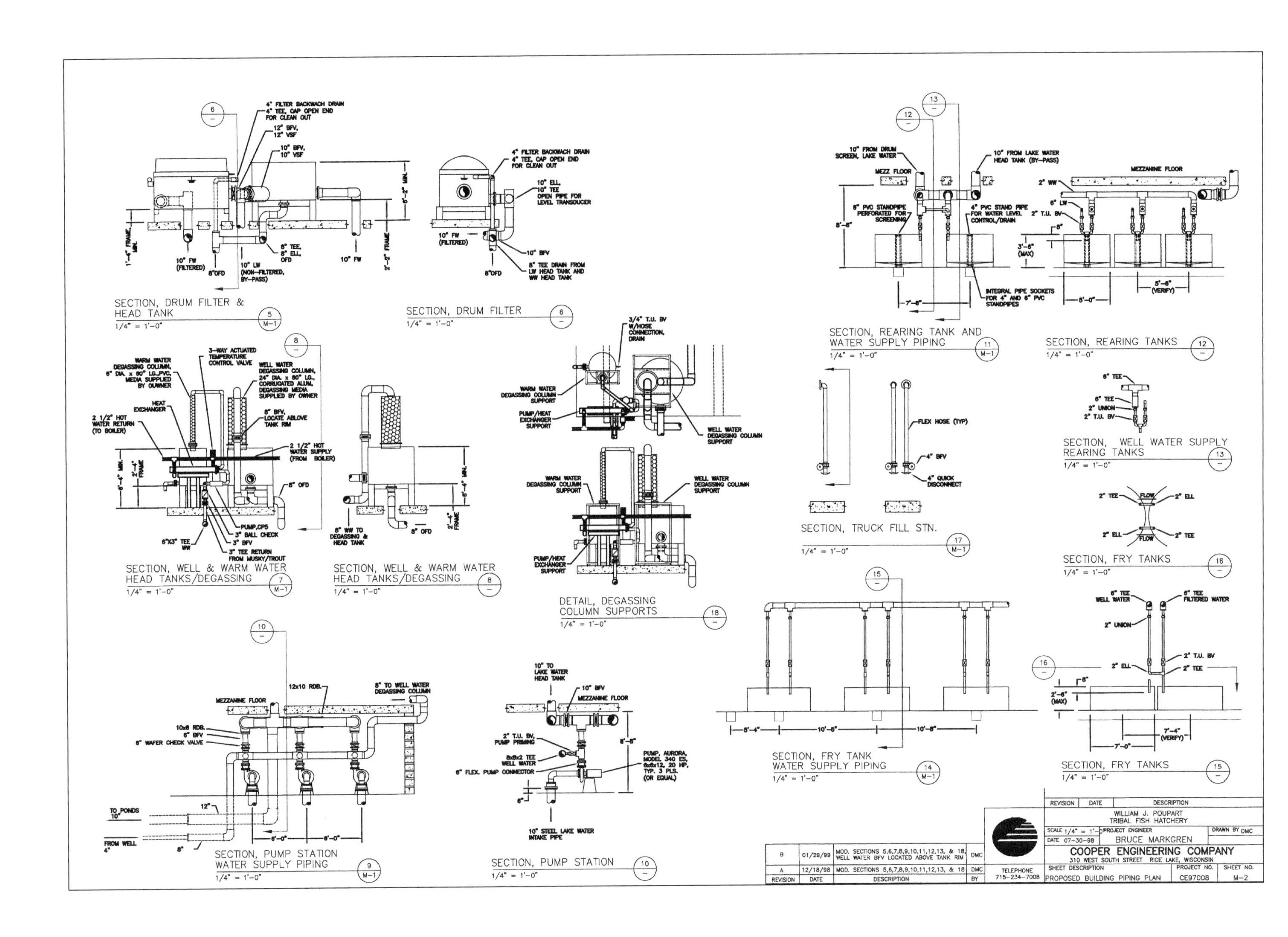
SECTION, DRUM FILTER & HEAD TANK
1/4" = 1'-0"
SECTION, DRUM FILTER
1/4" = 1'-0"
SECTION, REARING TANK AND WATER SUPPLY PIPING
1/4" = 1'-0"
SECTION, REARING TANKS
1/4" = 1'-0"
SECTION, WELL & WARM WATER HEAD TANKS/DEGASSING
1/4" = 1'-0"
SECTION, WELL & WARM WATER HEAD TANKS/DEGASSING
1/4" = 1'-0"
DETAIL, DEGASSING COLUMN SUPPORTS
1/4" = 1'-0"
SECTION, TRUCK FILL STN.
1/4" = 1'-0"
SECTION, WELL WATER SUPPLY REARING TANKS
1/4" = 1'-0"
SECTION, FRY TANKS
1/4" = 1'-0"
SECTION, FRY TANK WATER SUPPLY PIPING
1/4" = 1'-0"
SECTION, FRY TANKS
1/4" = 1'-0"
SECTION, PUMP STATION WATER SUPPLY PIPING
1/4" = 1'-0"
SECTION, PUMP STATION
1/4" = 1'-0"
4" FILTER BACKWASH DRAIN
4" TEE, CAP OPEN END FOR CLEAN OUT
10" FW (FILTERED)
10" LW (NON-FILTERED, BY-PASS)
MEZZANINE FLOOR
FLEX HOSE (TYP)
4" BFV
4" QUICK DISCONNECT
WARM WATER DEGASSING COLUMN SUPPORT
WELL WATER DEGASSING COLUMN SUPPORT
PUMP/HEAT EXCHANGER SUPPORT
10" STEEL LAKE WATER INTAKE PIPE
TO PONDS
FROM WELL
B 01/29/99 MOD. SECTIONS 5,6,7,8,9,10,11,12,13, & 18 WELL WATER BFV LOCATED ABOVE TANK RIM DMC
A 12/18/98 MOD. SECTIONS 5,6,7,8,9,10,11,12,13, & 18 DMC
REVISION DATE DESCRIPTION BY
TELEPHONE 715-234-7008
WILLIAM J. POUPART TRIBAL FISH HATCHERY
SCALE 1/4" = 1'-0"
DATE 07-30-98
PROJECT ENGINEER BRUCE MARKGREN
DRAWN BY DMC
COOPER ENGINEERING COMPANY
310 WEST SOUTH STREET RICE LAKE, WISCONSIN
SHEET DESCRIPTION PROPOSED BUILDING PIPING PLAN
PROJECT NO. CE97008
SHEET NO. M-2

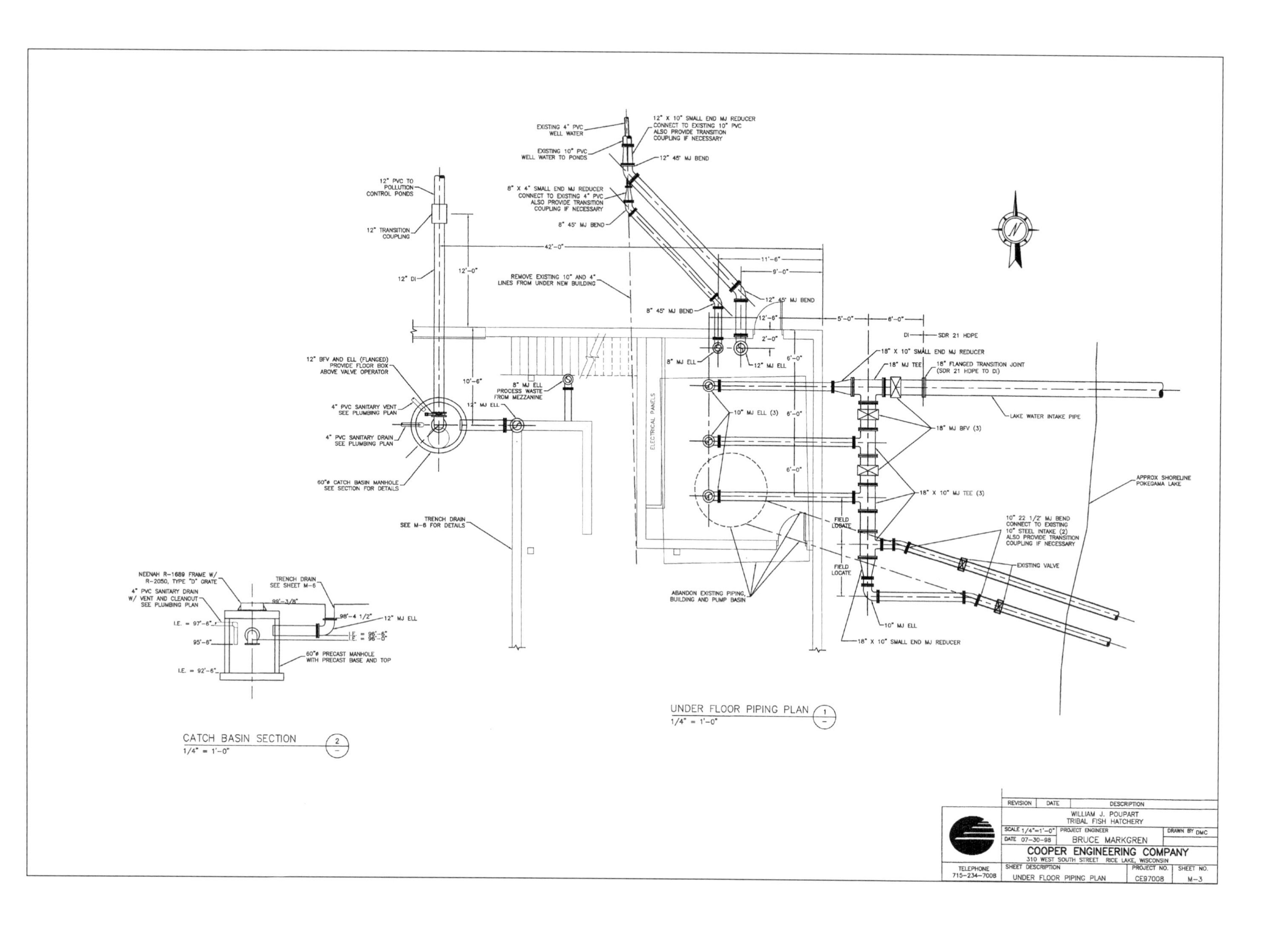
EXISTING 4" PVC WELL WATER
12" X 10" SMALL END MJ REDUCER CONNECT TO EXISTING 10" PVC ALSO PROVIDE TRANSITION COUPLING IF NECESSARY
EXISTING 10" PVC WELL WATER TO PONDS
12" 45' MJ BEND
12" PVC TO POLLUTION CONTROL PONDS
8" X 4" SMALL END MJ REDUCER CONNECT TO EXISTING 4" PVC ALSO PROVIDE TRANSITION COUPLING IF NECESSARY
8" 45' MJ BEND
12" TRANSITION COUPLING
42'-0"
11'-6"
9'-0"
12" DI
12'-0"
REMOVE EXISTING 10" AND 4" LINES FROM UNDER NEW BUILDING
12" 45' MJ BEND
8" 45' MJ BEND
12'-6"
5'-0"
6'-0"
DI
SDR 21 HDPE
2'-0"
12" BFV AND ELL (FLANGED) PROVIDE FLOOR BOX ABOVE VALVE OPERATOR
8" MJ ELL
12" MJ ELL
18" X 10" SMALL END MJ REDUCER
18" MJ TEE
18" FLANGED TRANSITION JOINT (SDR 21 HDPE TO DI)
10'-6"
8" MJ ELL PROCESS WASTE FROM MEZZANINE
4" PVC SANITARY VENT SEE PLUMBING PLAN
12" MJ ELL
ELECTRICAL PANELS
10" MJ ELL (3)
6'-0"
LAKE WATER INTAKE PIPE
18" MJ BFV (3)
4" PVC SANITARY DRAIN SEE PLUMBING PLAN
60"ø CATCH BASIN MANHOLE SEE SECTION FOR DETAILS
6'-0"
18" X 10" MJ TEE (3)
APPROX SHORELINE POKEGAMA LAKE
TRENCH DRAIN SEE M-6 FOR DETAILS
FIELD LOCATE
10" 22 1/2' MJ BEND CONNECT TO EXISTING 10" STEEL INTAKE (2) ALSO PROVIDE TRANSITION COUPLING IF NECESSARY
EXISTING VALVE
FIELD LOCATE
NEENAH R-1689 FRAME W/ R-2050, TYPE "D" GRATE
TRENCH DRAIN SEE SHEET M-6
4" PVC SANITARY DRAIN W/ VENT AND CLEANOUT SEE PLUMBING PLAN
99'-3/8"
ABANDON EXISTING PIPING, BUILDING AND PUMP BASIN
98'-4 1/2"
12" MJ ELL
I.E. = 97'-6"
I.E. = 96'-6"
I.E. = 96'-0"
95'-6"
10" MJ ELL
18" X 10" SMALL END MJ REDUCER
60"ø PRECAST MANHOLE WITH PRECAST BASE AND TOP
I.E. = 92'-6"
UNDER FLOOR PIPING PLAN
1/4" = 1'-0"
1
CATCH BASIN SECTION
1/4" = 1'-0"
2
REVISION
DATE
DESCRIPTION
WILLIAM J. POUPART TRIBAL FISH HATCHERY
SCALE 1/4"=1'-0"
PROJECT ENGINEER
DRAWN BY DMC
DATE 07-30-98
BRUCE MARKGREN
COOPER ENGINEERING COMPANY
310 WEST SOUTH STREET RICE LAKE, WISCONSIN
TELEPHONE 715-234-7008
SHEET DESCRIPTION
UNDER FLOOR PIPING PLAN
PROJECT NO.
CE97008
SHEET NO.
M-3

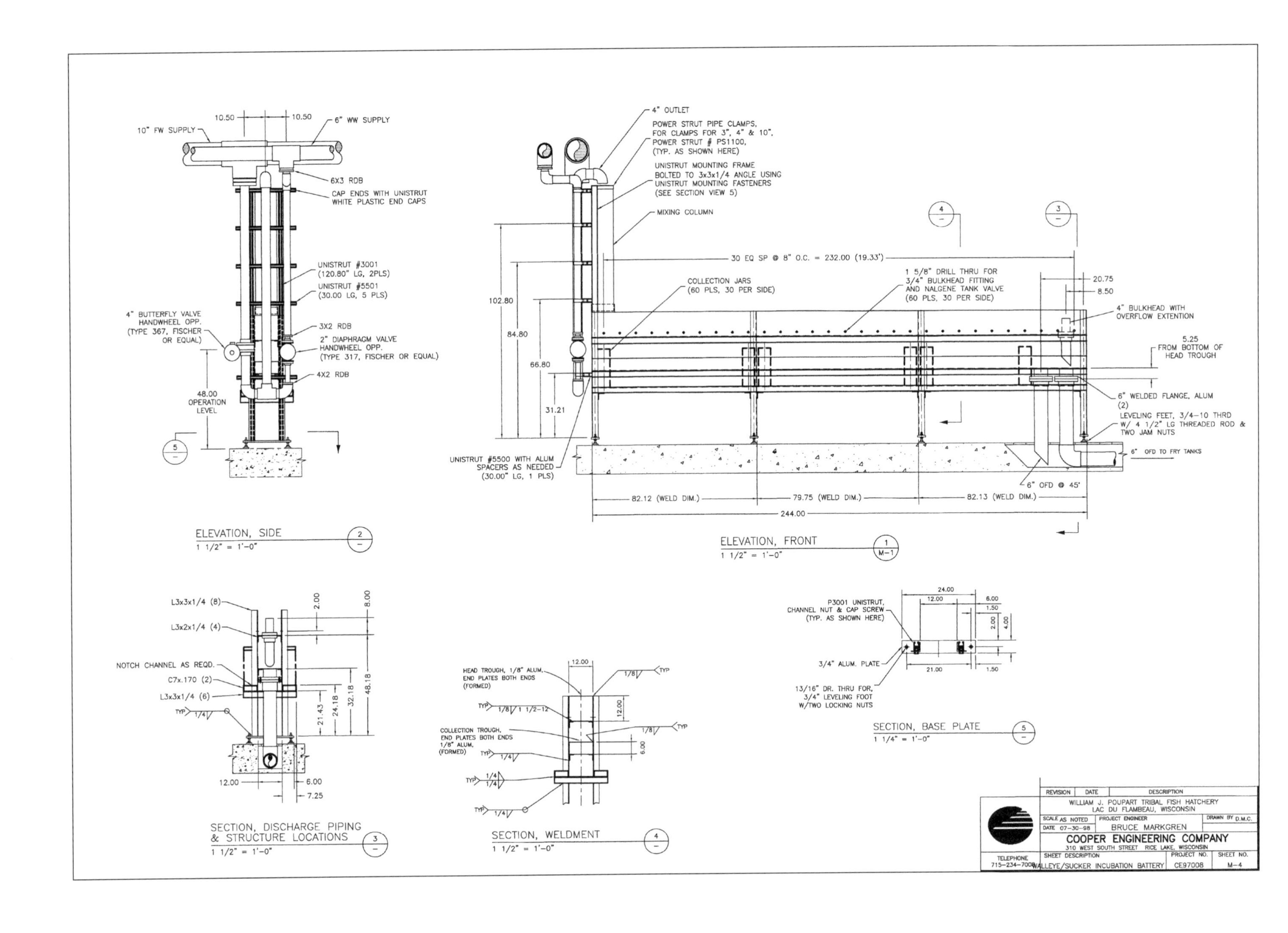
10" FW SUPPLY
10.50
10.50
6" WW SUPPLY
6X3 RDB
CAP ENDS WITH UNISTRUT WHITE PLASTIC END CAPS
UNISTRUT #3001 (120.80" LG, 2PLS)
UNISTRUT #5501 (30.00 LG, 5 PLS)
4" BUTTERFLY VALVE HANDWHEEL OPP. (TYPE 367, FISCHER OR EQUAL)
3X2 RDB
2" DIAPHRAGM VALVE HANDWHEEL OPP. (TYPE 317, FISCHER OR EQUAL)
4X2 RDB
48.00 OPERATION LEVEL
ELEVATION, SIDE
1 1/2" = 1'-0"
4" OUTLET
POWER STRUT PIPE CLAMPS, FOR CLAMPS FOR 3", 4" & 10", POWER STRUT # PS1100, (TYP. AS SHOWN HERE)
UNISTRUT MOUNTING FRAME BOLTED TO 3x3x1/4 ANGLE USING UNISTRUT MOUNTING FASTENERS (SEE SECTION VIEW 5)
MIXING COLUMN
30 EQ SP @ 8" O.C. = 232.00 (19.33')
COLLECTION JARS (60 PLS, 30 PER SIDE)
1 5/8" DRILL THRU FOR 3/4" BULKHEAD FITTING AND NALGENE TANK VALVE (60 PLS, 30 PER SIDE)
20.75
8.50
4" BULKHEAD WITH OVERFLOW EXTENTION
5.25 FROM BOTTOM OF HEAD TROUGH
102.80
84.80
66.80
31.21
6" WELDED FLANGE, ALUM (2)
LEVELING FEET, 3/4-10 THRD W/ 4 1/2" LG THREADED ROD & TWO JAM NUTS
6" OFD TO FRY TANKS
UNISTRUT #5500 WITH ALUM SPACERS AS NEEDED (30.00" LG, 1 PLS)
6" OFD @ 45'
82.12 (WELD DIM.)
79.75 (WELD DIM.)
82.13 (WELD DIM.)
244.00
ELEVATION, FRONT
1 1/2" = 1'-0"
L3x3x1/4 (8)
L3x2x1/4 (4)
NOTCH CHANNEL AS REQD.
C7x.170 (2)
L3x3x1/4 (6)
TYP 1/4
2.00
8.00
21.43
24.18
32.18
48.18
12.00
6.00
7.25
SECTION, DISCHARGE PIPING & STRUCTURE LOCATIONS
1 1/2" = 1'-0"
HEAD TROUGH, 1/8" ALUM, END PLATES BOTH ENDS (FORMED)
COLLECTION TROUGH, END PLATES BOTH ENDS 1/8" ALUM, (FORMED)
1/8 1 1/2-12
12.00
6.00
SECTION, WELDMENT
1 1/2" = 1'-0"
P3001 UNISTRUT, CHANNEL NUT & CAP SCREW (TYP. AS SHOWN HERE)
3/4" ALUM. PLATE
13/16" DR. THRU FOR, 3/4" LEVELING FOOT W/TWO LOCKING NUTS
24.00
12.00
6.00
1.50
2.00
4.00
21.00
SECTION, BASE PLATE
1 1/4" = 1'-0"
REVISION DATE DESCRIPTION
WILLIAM J. POUPART TRIBAL FISH HATCHERY
LAC DU FLAMBEAU, WISCONSIN
SCALE AS NOTED
DATE 07-30-98
PROJECT ENGINEER BRUCE MARKGREN
DRAWN BY D.M.C.
COOPER ENGINEERING COMPANY
310 WEST SOUTH STREET RICE LAKE, WISCONSIN
TELEPHONE 715-234-7008
SHEET DESCRIPTION WALLEYE/SUCKER INCUBATION BATTERY
PROJECT NO. CE97008
SHEET NO. M-4

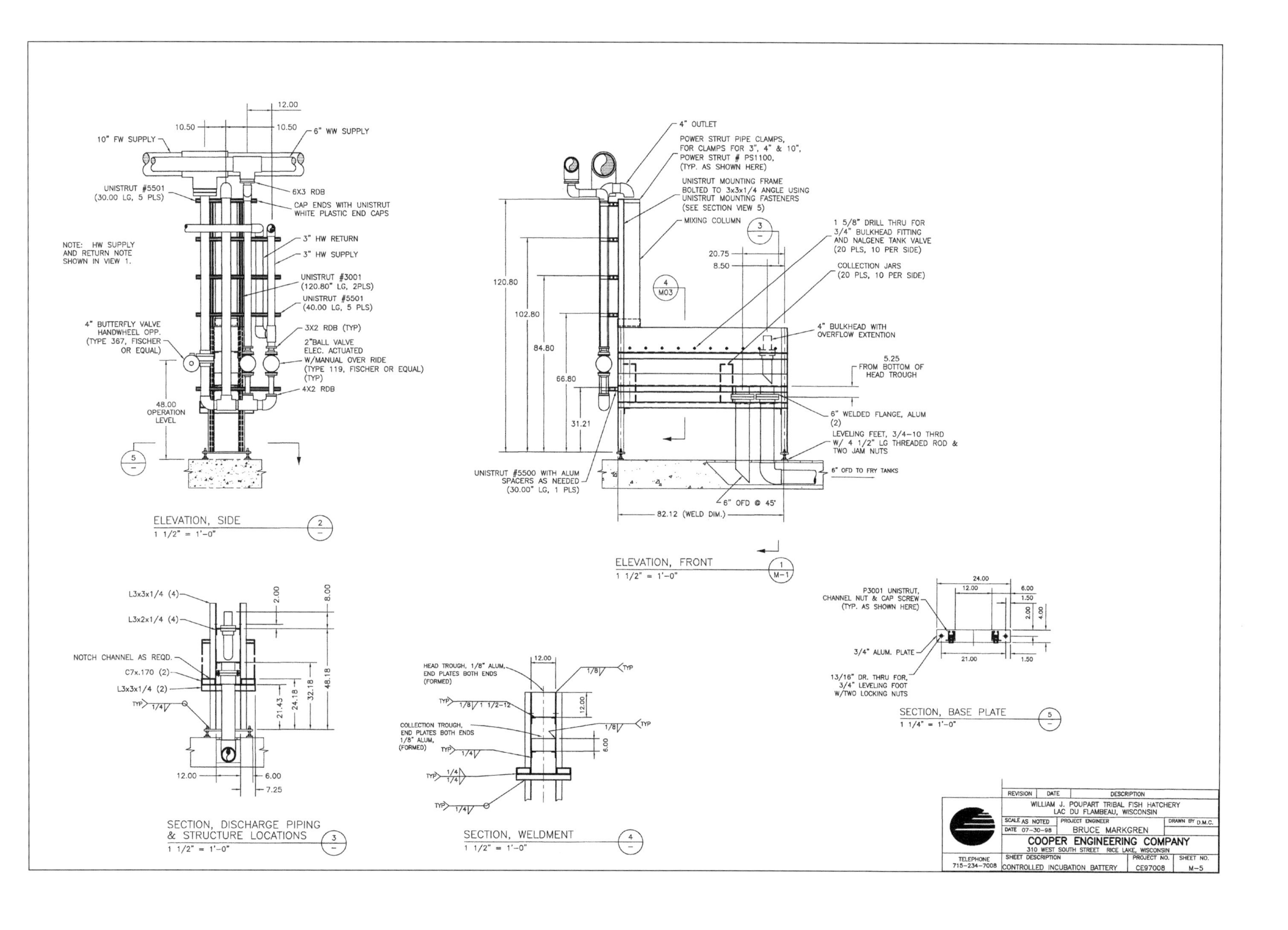

10" FW SUPPLY
10.50
12.00
10.50
6" WW SUPPLY
UNISTRUT #5501 (30.00 LG, 5 PLS)
6X3 RDB
CAP ENDS WITH UNISTRUT WHITE PLASTIC END CAPS
NOTE: HW SUPPLY AND RETURN NOTE SHOWN IN VIEW 1.
3" HW RETURN
3" HW SUPPLY
UNISTRUT #3001 (120.80" LG, 2PLS)
UNISTRUT #5501 (40.00 LG, 5 PLS)
4" BUTTERFLY VALVE HANDWHEEL OPP. (TYPE 367, FISCHER OR EQUAL)
3X2 RDB (TYP)
2"BALL VALVE ELEC. ACTUATED W/MANUAL OVER RIDE (TYPE 119, FISCHER OR EQUAL) (TYP)
4X2 RDB
48.00 OPERATION LEVEL
ELEVATION, SIDE
1 1/2" = 1'-0"
4" OUTLET
POWER STRUT PIPE CLAMPS, FOR CLAMPS FOR 3", 4" & 10", POWER STRUT # PS1100, (TYP. AS SHOWN HERE)
UNISTRUT MOUNTING FRAME BOLTED TO 3x3x1/4 ANGLE USING UNISTRUT MOUNTING FASTENERS (SEE SECTION VIEW 5)
MIXING COLUMN
1 5/8" DRILL THRU FOR 3/4" BULKHEAD FITTING AND NALGENE TANK VALVE (20 PLS, 10 PER SIDE)
COLLECTION JARS (20 PLS, 10 PER SIDE)
20.75
8.50
M03
120.80
102.80
84.80
66.80
31.21
4" BULKHEAD WITH OVERFLOW EXTENTION
5.25 FROM BOTTOM OF HEAD TROUGH
6" WELDED FLANGE, ALUM (2)
LEVELING FEET, 3/4-10 THRD W/ 4 1/2" LG THREADED ROD & TWO JAM NUTS
6" OFD TO FRY TANKS
UNISTRUT #5500 WITH ALUM SPACERS AS NEEDED (30.00" LG, 1 PLS)
6" OFD @ 45°
82.12 (WELD DIM.)
ELEVATION, FRONT
1 1/2" = 1'-0"
M-1
L3x3x1/4 (4)
L3x2x1/4 (4)
NOTCH CHANNEL AS REQD.
C7x.170 (2)
L3x3x1/4 (2)
TYP 1/4
2.00
8.00
21.43
24.18
32.18
48.18
12.00
6.00
7.25
SECTION, DISCHARGE PIPING & STRUCTURE LOCATIONS
1 1/2" = 1'-0"
HEAD TROUGH, 1/8" ALUM, END PLATES BOTH ENDS (FORMED)
12.00
1/8 TYP
TYP 1/8 1 1/2-12
12.00
COLLECTION TROUGH, END PLATES BOTH ENDS 1/8" ALUM, (FORMED)
TYP 1/4
1/8 TYP
6.00
TYP 1/4 1/4
TYP 1/4
SECTION, WELDMENT
1 1/2" = 1'-0"
24.00
12.00
6.00
1.50
2.00
4.00
P3001 UNISTRUT, CHANNEL NUT & CAP SCREW (TYP. AS SHOWN HERE)
3/4" ALUM. PLATE
21.00
1.50
13/16" DR. THRU FOR, 3/4" LEVELING FOOT W/TWO LOCKING NUTS
SECTION, BASE PLATE
1 1/4" = 1'-0"
REVISION
DATE
DESCRIPTION
WILLIAM J. POUPART TRIBAL FISH HATCHERY
LAC DU FLAMBEAU, WISCONSIN
SCALE AS NOTED
PROJECT ENGINEER
DRAWN BY D.M.C.
DATE 07-30-98
BRUCE MARKGREN
COOPER ENGINEERING COMPANY
310 WEST SOUTH STREET RICE LAKE, WISCONSIN
TELEPHONE 715-234-7008
SHEET DESCRIPTION
CONTROLLED INCUBATION BATTERY
PROJECT NO.
CE97008
SHEET NO.
M-5

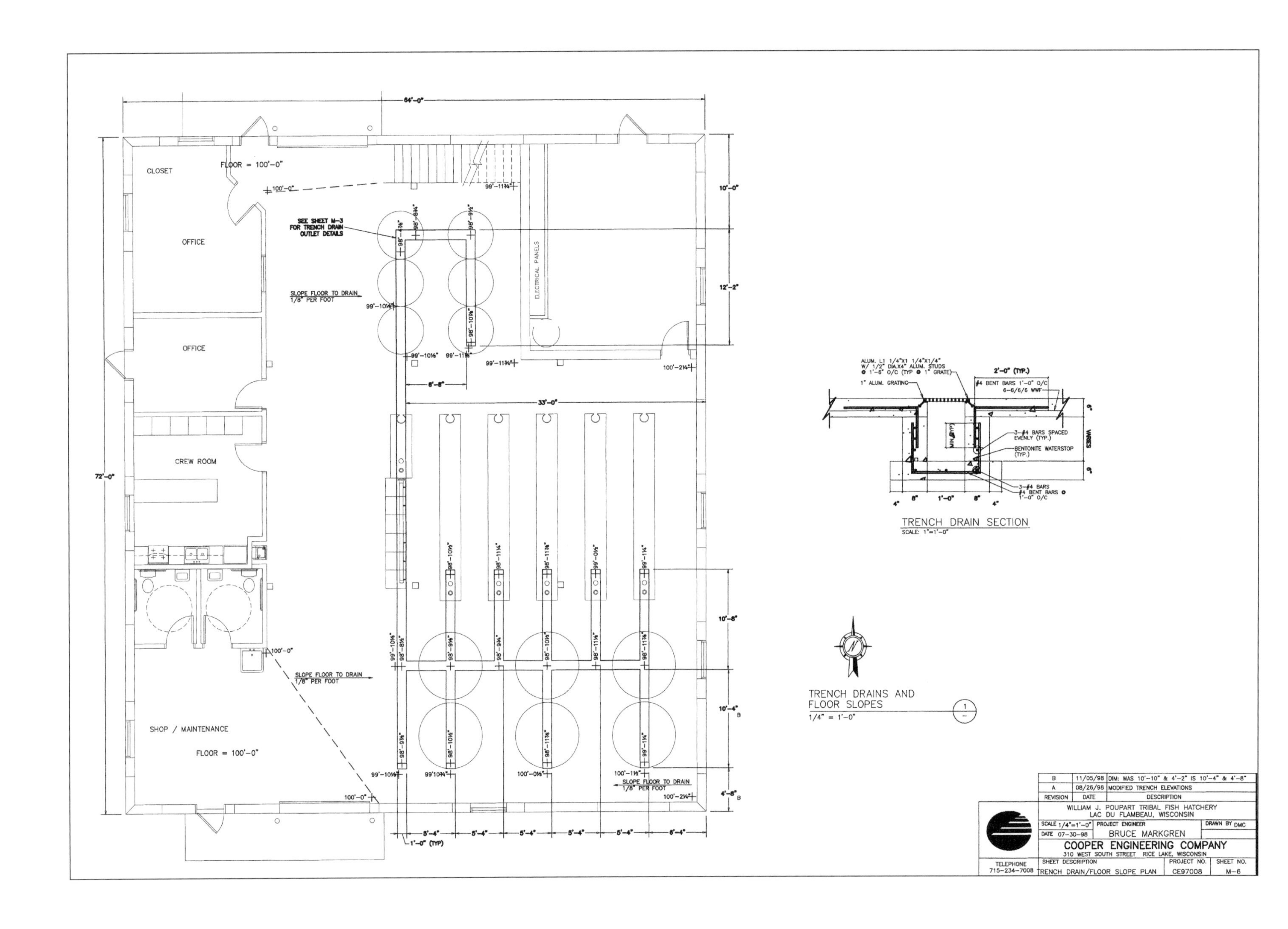

CLOSET
FLOOR = 100'-0"
OFFICE
OFFICE
CREW ROOM
SHOP / MAINTENANCE
FLOOR = 100'-0"
ELECTRICAL PANELS
SEE SHEET M-3 FOR TRENCH DRAIN OUTLET DETAILS
SLOPE FLOOR TO DRAIN 1/8" PER FOOT
64'-0"
72'-0"
33'-0"
TRENCH DRAIN SECTION
SCALE: 1"=1'-0"
TRENCH DRAINS AND FLOOR SLOPES
1/4" = 1'-0"
WILLIAM J. POUPART TRIBAL FISH HATCHERY
LAC DU FLAMBEAU, WISCONSIN
BRUCE MARKGREN
COOPER ENGINEERING COMPANY
310 WEST SOUTH STREET RICE LAKE, WISCONSIN
TELEPHONE 715-234-7008
TRENCH DRAIN/FLOOR SLOPE PLAN
CE97008
M-6

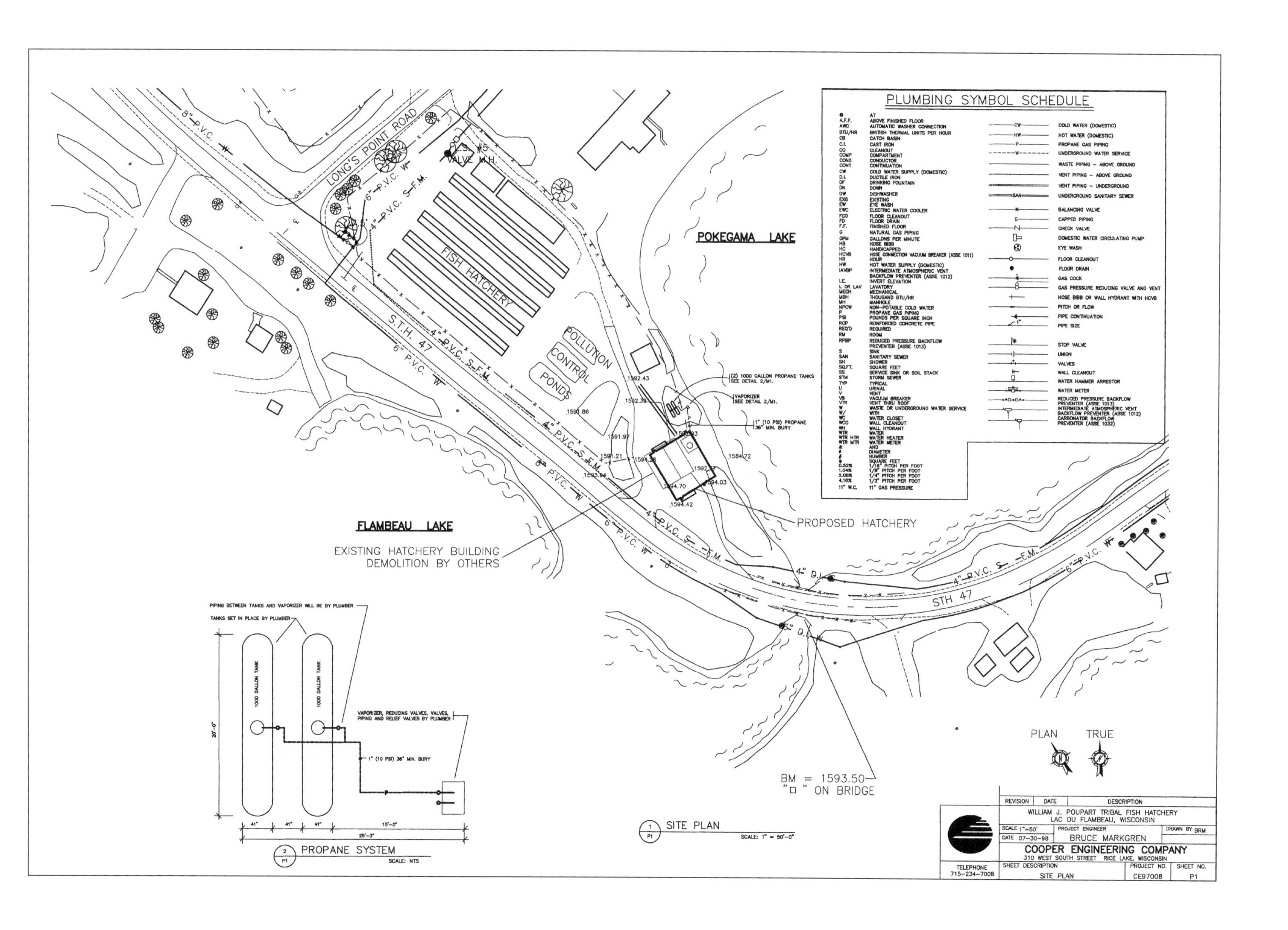
PLUMBING SYMBOL SCHEDULE
@ AT
A.F.F. ABOVE FINISHED FLOOR
AWC AUTOMATIC WASHER CONNECTION
BTU/HR BRITISH THERMAL UNITS PER HOUR
CB CATCH BASIN
C.I. CAST IRON
CO CLEANOUT
COMP COMPARTMENT
COND CONDUCTOR
CONT CONTINUATION
CW COLD WATER SUPPLY (DOMESTIC)
D.I. DUCTILE IRON
DF DRINKING FOUNTAIN
DN DOWN
DW DISHWASHER
EXG EXISTING
EW EYE WASH
EWC ELECTRIC WATER COOLER
FCO FLOOR CLEANOUT
FD FLOOR DRAIN
F.F. FINISHED FLOOR
G NATURAL GAS PIPING
GPM GALLONS PER MINUTE
HB HOSE BIBB
HC HANDICAPPED
HCVB HOSE CONNECTION VACUUM BREAKER (ASSE 1011)
HR HOUR
HW HOT WATER SUPPLY (DOMESTIC)
IAVBP INTERMEDIATE ATMOSPHERIC VENT BACKFLOW PREVENTER (ASSE 1012)
I.E. INVERT ELEVATION
L OR LAV LAVATORY
MECH MECHANICAL
MBH THOUSAND BTU/HR
MH MANHOLE
NPCW NON-POTABLE COLD WATER
P PROPANE GAS PIPING
PSI POUNDS PER SQUARE INCH
RCP REINFORCED CONCRETE PIPE
REQ'D REQUIRED
RM ROOM
RPBP REDUCED PRESSURE BACKFLOW PREVENTER (ASSE 1013)
S SINK
SAN SANITARY SEWER
SH SHOWER
SQ.FT. SQUARE FEET
SS SERVICE SINK OR SOIL STACK
STM STORM SEWER
TYP TYPICAL
U URINAL
V VENT
VB VACUUM BREAKER
VTR VENT THRU ROOF
W WASTE OR UNDERGROUND WATER SERVICE
W/ WITH
WC WATER CLOSET
WCO WALL CLEANOUT
WH WALL HYDRANT
WTR WATER
WTR HTR WATER HEATER
WTR MTR WATER METER
& AND
ø DIAMETER
NUMBER
SQUARE FEET
0.52% 1/16" PITCH PER FOOT
1.04% 1/8" PITCH PER FOOT
2.08% 1/4" PITCH PER FOOT
4.16% 1/2" PITCH PER FOOT
11" W.C. 11" GAS PRESSURE
CW COLD WATER (DOMESTIC)
HW HOT WATER (DOMESTIC)
P PROPANE GAS PIPING
W UNDERGROUND WATER SERVICE
WASTE PIPING – ABOVE GROUND
VENT PIPING – ABOVE GROUND
VENT PIPING – UNDERGROUND
SAN UNDERGROUND SANITARY SEWER
BALANCING VALVE
CAPPED PIPING
CHECK VALVE
DOMESTIC WATER CIRCULATING PUMP
EYE WASH
FLOOR CLEANOUT
FLOOR DRAIN
GAS COCK
GAS PRESSURE REDUCING VALVE AND VENT
HOSE BIBB OR WALL HYDRANT WITH HCVB
PITCH OR FLOW
PIPE CONTINUATION
1" PIPE SIZE
STOP VALVE
UNION
VALVES
WALL CLEANOUT
WATER HAMMER ARRESTOR
WATER METER
REDUCED PRESSURE BACKFLOW PREVENTER (ASSE 1013)
INTERMEDIATE ATMOSPHERIC VENT BACKFLOW PREVENTER (ASSE 1012)
CARBONATOR BACKFLOW PREVENTER (ASSE 1032)
8" P.V.C.
LONG'S POINT ROAD
6" P.V.C. W
4" P.V.C. S.F.M.
L.S. #5
VALVE M.H.
FISH HATCHERY
POKEGAMA LAKE
POLLUTION CONTROL PONDS
S.T.H. 47
4" P.V.C. S.F.M.
6" P.V.C. W
(2) 1000 GALLON PROPANE TANKS SEE DETAIL 2/M1.
VAPORIZER SEE DETAIL 2/M1.
1" (10 PSI) PROPANE 36" MIN. BURY
1592.43
1592.32
1591.86
1591.97
1591.21
1594.28
1593.94
1594.70
1594.42
1594.03
1584.72
FLAMBEAU LAKE
PROPOSED HATCHERY
EXISTING HATCHERY BUILDING DEMOLITION BY OTHERS
4" D.I.
6" D.I. W
STH 47
4" P.V.C. S.F.M.
6" P.V.C. W
BM = 1593.50
"□" ON BRIDGE
PLAN
TRUE
PIPING BETWEEN TANKS AND VAPORIZER WILL BE BY PLUMBER
TANKS SET IN PLACE BY PLUMBER
1000 GALLON TANK
1000 GALLON TANK
VAPORIZER, REDUCING VALVES, VALVES, PIPING AND RELIEF VALVES BY PLUMBER
1" (10 PSI) 36" MIN. BURY
20'-0"
41"
41"
41"
15'-0"
25'-3"
2 P1 PROPANE SYSTEM
SCALE: NTS
1 P1 SITE PLAN
SCALE: 1" = 50'-0"
REVISION
DATE
DESCRIPTION
WILLIAM J. POUPART TRIBAL FISH HATCHERY
LAC DU FLAMBEAU, WISCONSIN
SCALE 1"=50'
PROJECT ENGINEER
DRAWN BY BRM
DATE 07-30-98
BRUCE MARKGREN
COOPER ENGINEERING COMPANY
310 WEST SOUTH STREET RICE LAKE, WISCONSIN
TELEPHONE 715-234-7008
SHEET DESCRIPTION
SITE PLAN
PROJECT NO. CE97008
SHEET NO. P1

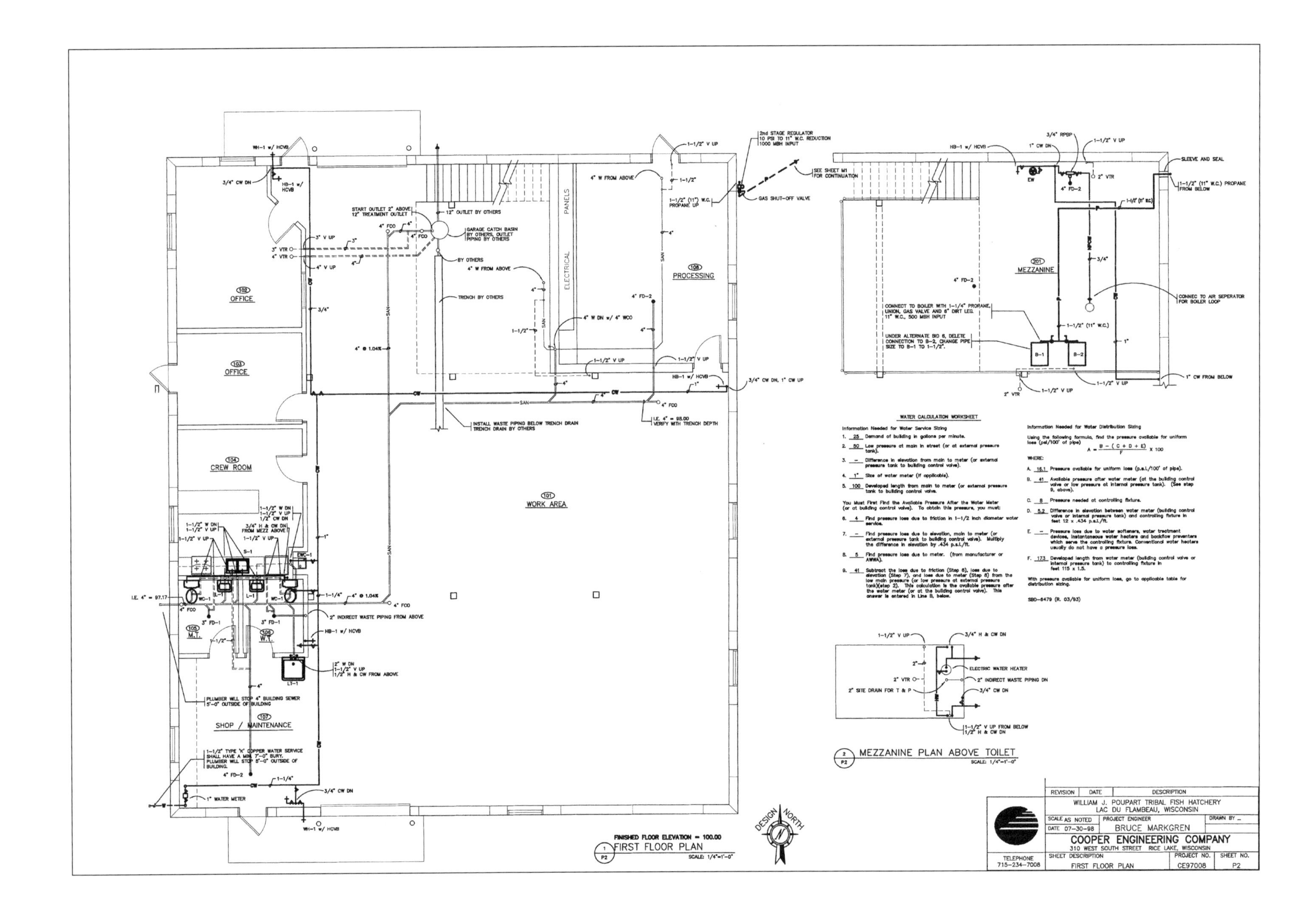
WH-1 w/ HCVB
3/4" CW DN
HB-1 w/ HCVB
START OUTLET 2" ABOVE
12" TREATMENT OUTLET
12" OUTLET BY OTHERS
GARAGE CATCH BASIN BY OTHERS, OUTLET PIPING BY OTHERS
4" FCO
3" V UP
3" VTR
4" VTR
4" V UP
BY OTHERS
4" W FROM ABOVE
TRENCH BY OTHERS
102
OFFICE
103
OFFICE
104
CREW ROOM
101
WORK AREA
108
PROCESSING
PANELS
ELECTRICAL
SAN
CW
4" @ 1.04%
4" W DN w/ 4" WCO
4" FD-2
1-1/2" V UP
1-1/2" (11") W.C. PROPANE UP
2nd STAGE REGULATOR
10 PSI TO 11" W.C. REDUCTION
1000 MBH INPUT
SEE SHEET M1 FOR CONTINUATION
GAS SHUT-OFF VALVE
3/4" CW DN, 1" CW UP
INSTALL WASTE PIPING BELOW TRENCH DRAIN
TRENCH DRAIN BY OTHERS
I.E. 4" = 98.00
VERIFY WITH TRENCH DEPTH
1-1/2" W DN
1-1/2" V UP
1/2" CW DN
3/4" H & CW DN FROM MEZZ ABOVE
S-1
EWC-1
WC-1
L-1
I.E. 4" = 97.17
3" FD-1
2" INDIRECT WASTE PIPING FROM ABOVE
HB-1 w/ HCVB
105
M.T.
106
W.T.
LT-1
2" W DN
1-1/2" V UP
1/2" H & CW FROM ABOVE
PLUMBER WILL STOP 4" BUILDING SEWER 5'-0" OUTSIDE OF BUILDING
107
SHOP / MAINTENANCE
1-1/2" TYPE "K" COPPER WATER SERVICE SHALL HAVE A MIN. 7'-0" BURY. PLUMBER WILL STOP 5'-0" OUTSIDE OF BUILDING.
4" FD-2
1-1/4"
3/4" CW DN
1" WATER METER
FINISHED FLOOR ELEVATION = 100.00
1
P2
FIRST FLOOR PLAN
SCALE: 1/4"=1'-0"
DESIGN NORTH
3/4" RPBP
1" CW DN
1-1/2" V UP
SLEEVE AND SEAL
1-1/2" (11" W.C.) PROPANE FROM BELOW
EW
4" FD-2
2" VTR
201
MEZZANINE
CONNEC TO AIR SEPERATOR FOR BOILER LOOP
CONNECT TO BOILER WITH 1-1/4" PROPANE, UNION, GAS VALVE AND 6" DIRT LEG. 11" W.C., 500 MBH INPUT
UNDER ALTERNATE BID 6, DELETE CONNECTION TO B-2, CHANGE PIPE SIZE TO B-1 TO 1-1/2".
B-1
B-2
1-1/2" (11" W.C.)
1" CW FROM BELOW
WATER CALCULATION WORKSHEET
Information Needed for Water Service Sizing
1. 25 Demand of building in gallons per minute.
2. 50 Low pressure at main in street (or at external pressure tank).
3. – Difference in elevation from main to meter (or external pressure tank to building control valve).
4. 1" Size of water meter (if applicable).
5. 100 Developed length from main to meter (or external pressure tank to building control valve.
You Must First Find the Available Pressure After the Water Meter (or at building control valve). To obtain this pressure, you must:
6. 4 Find pressure loss due to friction in 1-1/2 inch diameter water service.
7. – Find pressure loss due to elevation, main to meter (or external pressure tank to building control valve). Multiply the difference in elevation by .434 p.s.i./ft.
8. 5 Find pressure loss due to meter. (from manufacturer or AWWA).
9. 41 Subtract the loss due to friction (Step 6), loss due to elevation (Step 7), and loss due to meter (Step 8) from the low main pressure (or low pressure at external pressure tank)(step 2). This calculation is the available pressure after the water meter (or at the building control valve). This answer is entered in Line B, below.
Information Needed for Water Distribution Sizing
Using the following formula, find the pressure available for uniform loss (psi/100' of pipe)
A = (B − (C + D + E)) / F X 100
WHERE:
A. 16.1 Pressure available for uniform loss (p.s.i./100' of pipe).
B. 41 Available pressure after water meter (at the building control valve or low pressure at internal pressure tank). (See step 9, above).
C. 8 Pressure needed at controlling fixture.
D. 5.2 Difference in elevation between water meter (building control valve or internal pressure tank) and controlling fixture in feet 12 x .434 p.s.i./ft.
E. – Pressure loss due to water softeners, water treatment devices, instantaneous water heaters and backflow preventers which serve the controlling fixture. Conventional water heaters usually do not have a pressure loss.
F. 173 Developed length from water meter (building control valve or internal pressure tank) to controlling fixture in feet 115 x 1.5.
With pressure available for uniform loss, go to applicable table for distribution sizing.
SBD-6479 (R. 03/93)
3/4" H & CW DN
ELECTRIC WATER HEATER
2" INDIRECT WASTE PIPING DN
3/4" CW DN
2" VTR
2" SITE DRAIN FOR T & P
1-1/2" V UP FROM BELOW
1/2" H & CW DN
2
P2
MEZZANINE PLAN ABOVE TOILET
SCALE: 1/4"=1'-0"
REVISION
DATE
DESCRIPTION
WILLIAM J. POUPART TRIBAL FISH HATCHERY
LAC DU FLAMBEAU, WISCONSIN
SCALE AS NOTED
DATE 07-30-98
PROJECT ENGINEER
BRUCE MARKGREN
DRAWN BY _
COOPER ENGINEERING COMPANY
310 WEST SOUTH STREET RICE LAKE, WISCONSIN
TELEPHONE
715-234-7008
SHEET DESCRIPTION
FIRST FLOOR PLAN
PROJECT NO.
CE97008
SHEET NO.
P2

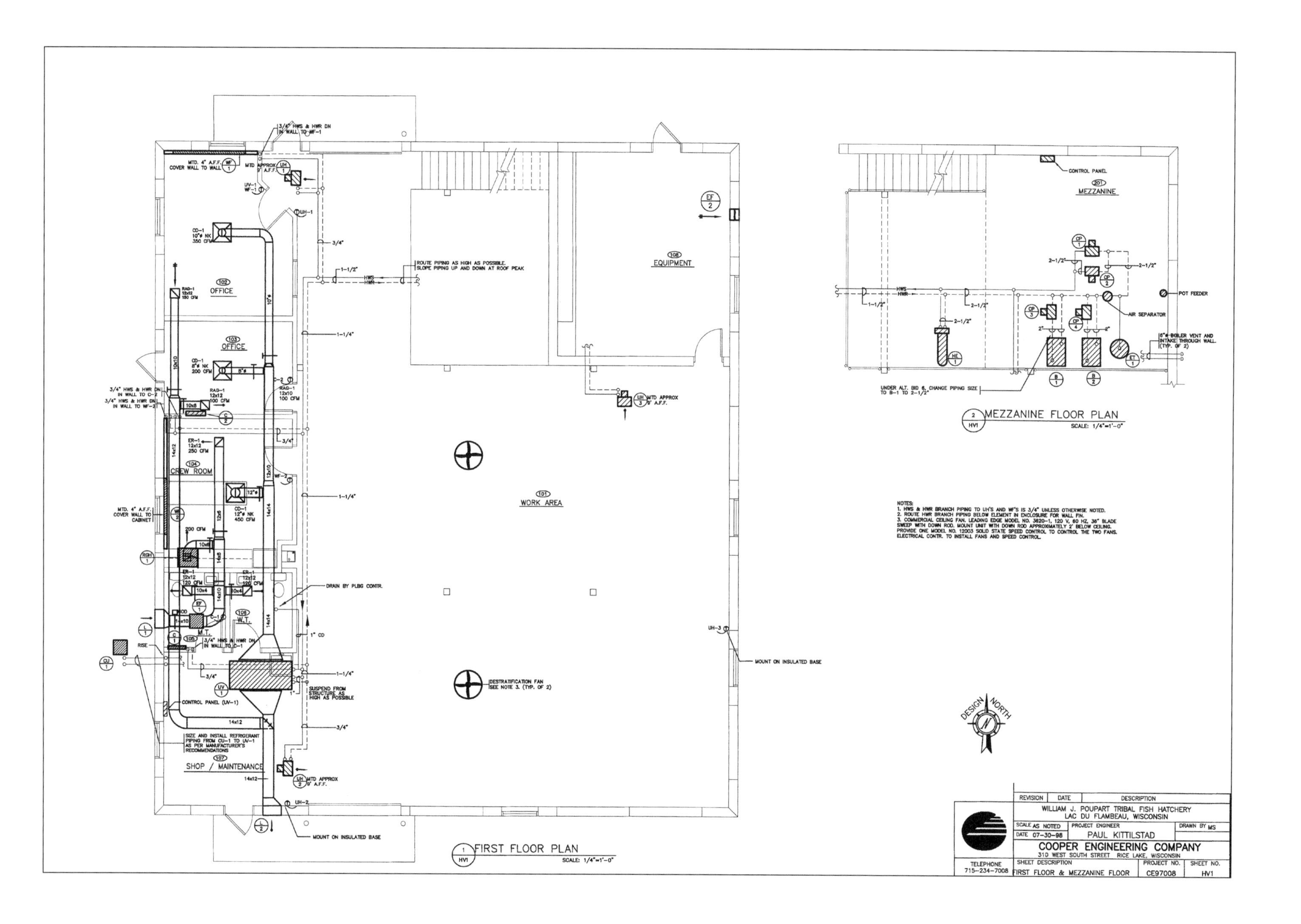
OFFICE
OFFICE
CREW ROOM
W.T.
M.T.
SHOP / MAINTENANCE
WORK AREA
EQUIPMENT
ROUTE PIPING AS HIGH AS POSSIBLE. SLOPE PIPING UP AND DOWN AT ROOF PEAK
DRAIN BY PLBG CONTR.
SUSPEND FROM STRUCTURE AS HIGH AS POSSIBLE
DESTRATIFICATION FAN SEE NOTE 3. (TYP. OF 2)
MOUNT ON INSULATED BASE
CONTROL PANEL (UV-1)
SIZE AND INSTALL REFRIGERANT PIPING FROM CU-1 TO UV-1 AS PER MANUFACTURER'S RECOMMENDATIONS
FIRST FLOOR PLAN
SCALE: 1/4"=1'-0"
CONTROL PANEL
MEZZANINE
POT FEEDER
AIR SEPARATOR
UNDER ALT. BID 6, CHANGE PIPING SIZE TO B-1 TO 2-1/2"
MEZZANINE FLOOR PLAN
SCALE: 1/4"=1'-0"
NOTES:
1. HWS & HWR BRANCH PIPING TO UH'S AND WF'S IS 3/4" UNLESS OTHERWISE NOTED.
2. ROUTE HWR BRANCH PIPING BELOW ELEMENT IN ENCLOSURE FOR WALL FIN.
3. COMMERCIAL CEILING FAN. LEADING EDGE MODEL NO. 3820-1, 120 V, 60 HZ, 36" BLADE SWEEP WITH DOWN ROD. MOUNT UNIT WITH DOWN ROD APPROXIMATELY 2' BELOW CEILING. PROVIDE ONE MODEL NO. 12003 SOLID STATE SPEED CONTROL TO CONTROL THE TWO FANS. ELECTRICAL CONTR. TO INSTALL FANS AND SPEED CONTROL.
DESIGN NORTH
REVISION
DATE
DESCRIPTION
WILLIAM J. POUPART TRIBAL FISH HATCHERY
LAC DU FLAMBEAU, WISCONSIN
SCALE AS NOTED
DATE 07-30-98
PROJECT ENGINEER
PAUL KITTILSTAD
DRAWN BY MS
COOPER ENGINEERING COMPANY
310 WEST SOUTH STREET RICE LAKE, WISCONSIN
TELEPHONE
715-234-7008
SHEET DESCRIPTION
FIRST FLOOR & MEZZANINE FLOOR
PROJECT NO.
CE97008
SHEET NO.
HV1

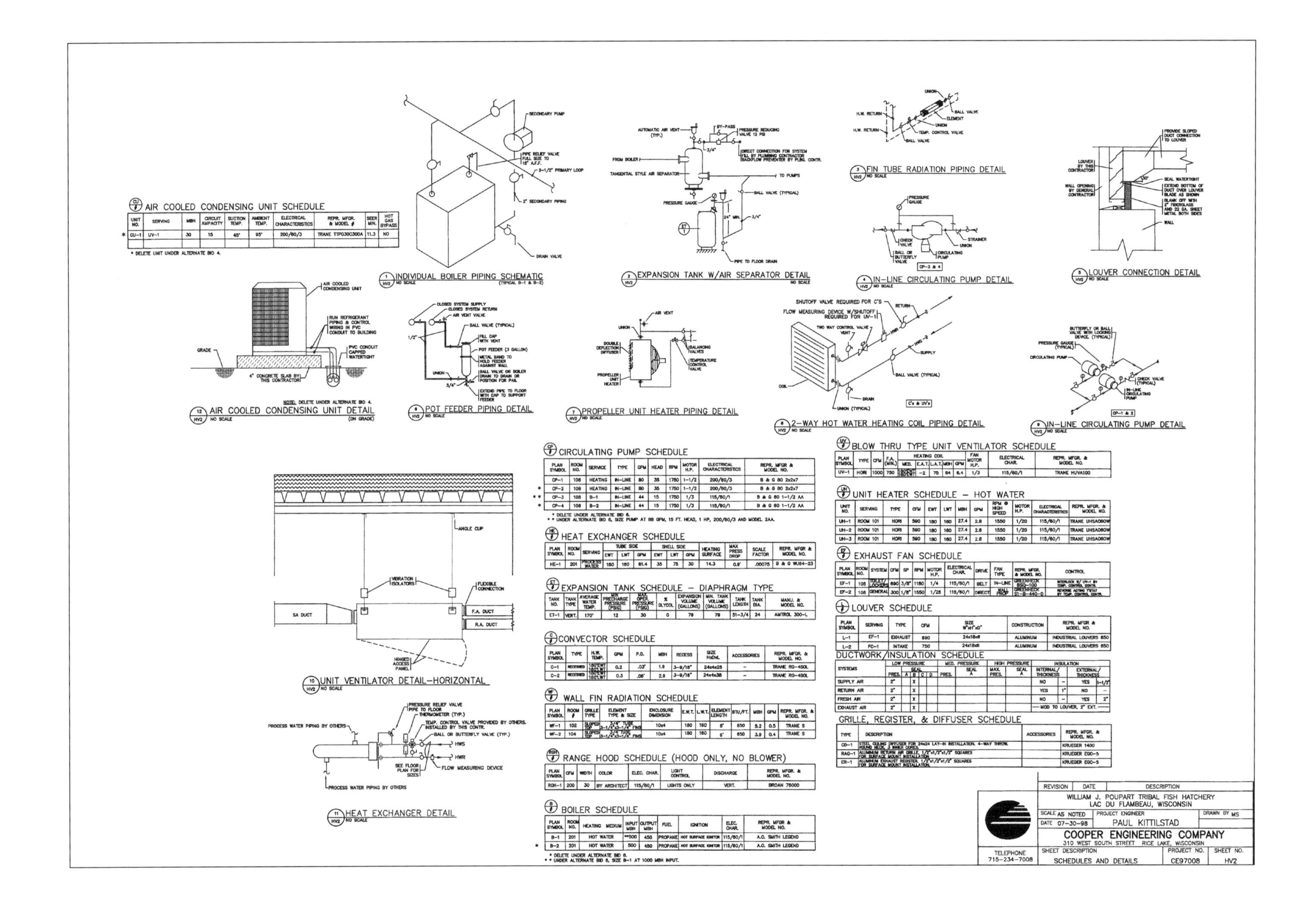

AIR COOLED CONDENSING UNIT SCHEDULE
1 INDIVIDUAL BOILER PIPING SCHEMATIC
HV2 NO SCALE
2 EXPANSION TANK W/AIR SEPARATOR DETAIL
HV2 NO SCALE
3 FIN TUBE RADIATION PIPING DETAIL
HV2 NO SCALE
4 IN-LINE CIRCULATING PUMP DETAIL
HV2 NO SCALE
5 LOUVER CONNECTION DETAIL
HV2 NO SCALE
12 AIR COOLED CONDENSING UNIT DETAIL
HV2 NO SCALE
6 POT FEEDER PIPING DETAIL
HV2 NO SCALE
7 PROPELLER UNIT HEATER PIPING DETAIL
HV2 NO SCALE
8 2-WAY HOT WATER HEATING COIL PIPING DETAIL
HV2 NO SCALE
9 IN-LINE CIRCULATING PUMP DETAIL
HV2 NO SCALE
CIRCULATING PUMP SCHEDULE
BLOW THRU TYPE UNIT VENTILATOR SCHEDULE
UNIT HEATER SCHEDULE - HOT WATER
HEAT EXCHANGER SCHEDULE
EXHAUST FAN SCHEDULE
EXPANSION TANK SCHEDULE - DIAPHRAGM TYPE
LOUVER SCHEDULE
CONVECTOR SCHEDULE
DUCTWORK/INSULATION SCHEDULE
10 UNIT VENTILATOR DETAIL-HORIZONTAL
HV2 NO SCALE
WALL FIN RADIATION SCHEDULE
GRILLE, REGISTER, & DIFFUSER SCHEDULE
RANGE HOOD SCHEDULE (HOOD ONLY, NO BLOWER)
11 HEAT EXCHANGER DETAIL
HV2 NO SCALE
BOILER SCHEDULE
REVISION
DATE
DESCRIPTION
WILLIAM J. POUPART TRIBAL FISH HATCHERY
LAC DU FLAMBEAU, WISCONSIN
SCALE AS NOTED
PROJECT ENGINEER
DRAWN BY MS
DATE 07-30-98
PAUL KITTILSTAD
COOPER ENGINEERING COMPANY
310 WEST SOUTH STREET RICE LAKE, WISCONSIN
TELEPHONE
715-234-7008
SHEET DESCRIPTION
SCHEDULES AND DETAILS
PROJECT NO.
CE97008
SHEET NO.
HV2

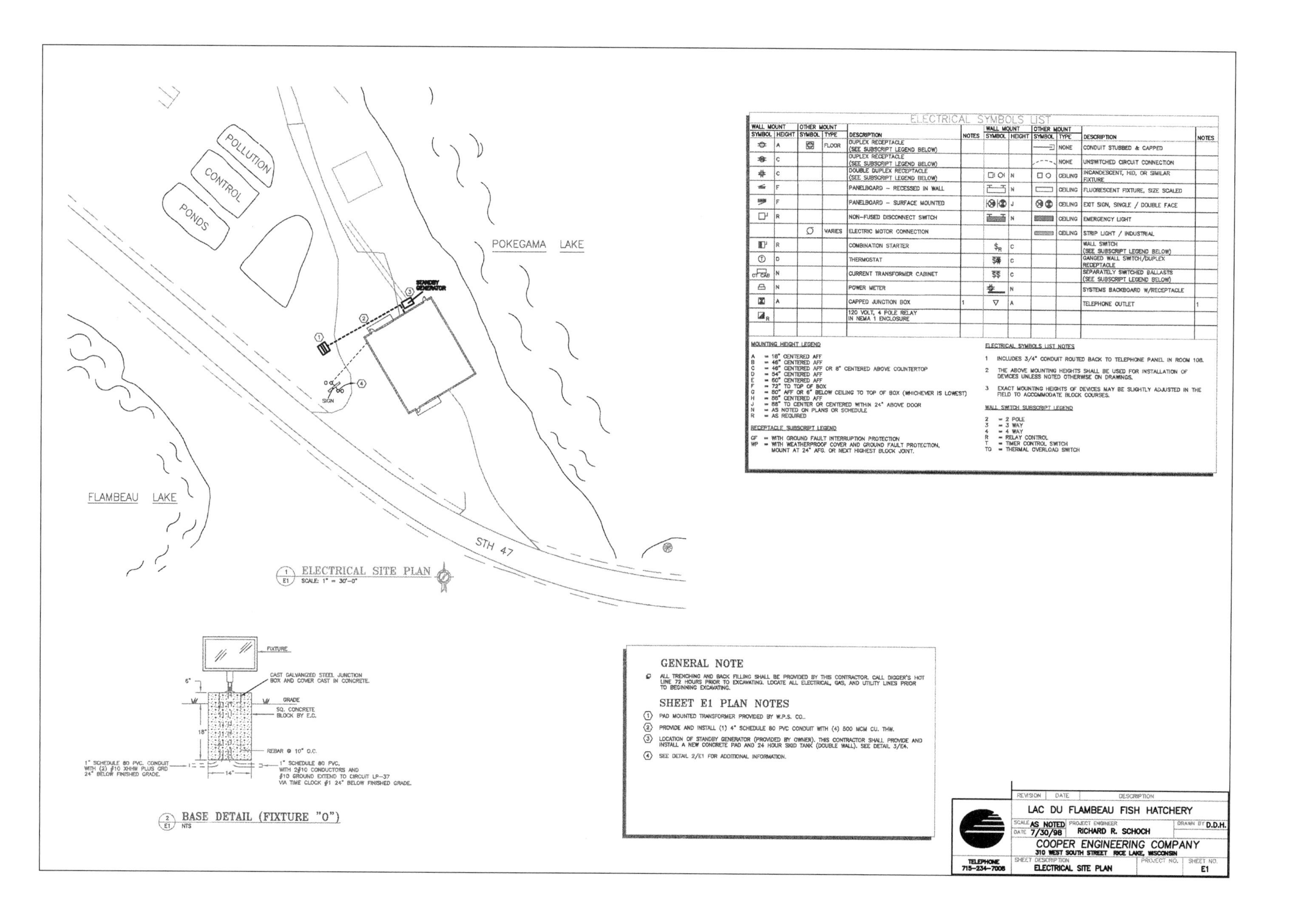
POLLUTION
CONTROL
PONDS
STANDBY GENERATOR
POKEGAMA LAKE
FLAMBEAU LAKE
SIGN
STH 47
ELECTRICAL SITE PLAN
SCALE: 1" = 30'-0"
ELECTRICAL SYMBOLS LIST
WALL MOUNT
OTHER MOUNT
SYMBOL
HEIGHT
TYPE
DESCRIPTION
NOTES
DUPLEX RECEPTACLE (SEE SUBSCRIPT LEGEND BELOW)
FLOOR
DOUBLE DUPLEX RECEPTACLE (SEE SUBSCRIPT LEGEND BELOW)
PANELBOARD – RECESSED IN WALL
PANELBOARD – SURFACE MOUNTED
NON–FUSED DISCONNECT SWITCH
VARIES
ELECTRIC MOTOR CONNECTION
COMBINATION STARTER
THERMOSTAT
CT CAB
CURRENT TRANSFORMER CABINET
POWER METER
CAPPED JUNCTION BOX
120 VOLT, 4 POLE RELAY IN NEMA 1 ENCLOSURE
NONE
CONDUIT STUBBED & CAPPED
UNSWITCHED CIRCUIT CONNECTION
CEILING
INCANDESCENT, HID, OR SIMILAR FIXTURE
FLUORESCENT FIXTURE, SIZE SCALED
EXIT SIGN, SINGLE / DOUBLE FACE
EMERGENCY LIGHT
STRIP LIGHT / INDUSTRIAL
WALL SWITCH (SEE SUBSCRIPT LEGEND BELOW)
GANGED WALL SWITCH/DUPLEX RECEPTACLE
SEPARATELY SWITCHED BALLASTS (SEE SUBSCRIPT LEGEND BELOW)
SYSTEMS BACKBOARD W/RECEPTACLE
TELEPHONE OUTLET
MOUNTING HEIGHT LEGEND
A = 18" CENTERED AFF
B = 46" CENTERED AFF
C = 46" CENTERED AFF OR 8" CENTERED ABOVE COUNTERTOP
D = 54" CENTERED AFF
E = 60" CENTERED AFF
F = 72" TO TOP OF BOX
G = 80" AFF OR 6" BELOW CEILING TO TOP OF BOX (WHICHEVER IS LOWEST)
H = 88" CENTERED AFF
J = 88" TO CENTER OR CENTERED WITHIN 24" ABOVE DOOR
N = AS NOTED ON PLANS OR SCHEDULE
R = AS REQUIRED
RECEPTACLE SUBSCRIPT LEGEND
GF = WITH GROUND FAULT INTERRUPTION PROTECTION
WP = WITH WEATHERPROOF COVER AND GROUND FAULT PROTECTION, MOUNT AT 24" AFG. OR NEXT HIGHEST BLOCK JOINT.
ELECTRICAL SYMBOLS LIST NOTES
1 INCLUDES 3/4" CONDUIT ROUTED BACK TO TELEPHONE PANEL IN ROOM 108.
2 THE ABOVE MOUNTING HEIGHTS SHALL BE USED FOR INSTALLATION OF DEVICES UNLESS NOTED OTHERWISE ON DRAWINGS.
3 EXACT MOUNTING HEIGHTS OF DEVICES MAY BE SLIGHTLY ADJUSTED IN THE FIELD TO ACCOMMODATE BLOCK COURSES.
WALL SWITCH SUBSCRIPT LEGEND
2 = 2 POLE
3 = 3 WAY
4 = 4 WAY
R = RELAY CONTROL
T = TIMER CONTROL SWITCH
TO = THERMAL OVERLOAD SWITCH
FIXTURE
CAST GALVANIZED STEEL JUNCTION BOX AND COVER CAST IN CONCRETE.
6"
GRADE
SQ. CONCRETE BLOCK BY E.C.
18"
REBAR @ 10" O.C.
1" SCHEDULE 80 PVC. CONDUIT WITH (2) #10 XHHW PLUS GRD 24" BELOW FINISHED GRADE.
14"
1" SCHEDULE 80 PVC, WITH 2#10 CONDUCTORS AND #10 GROUND EXTEND TO CIRCUIT LP-37 VIA TIME CLOCK #1 24" BELOW FINISHED GRADE.
BASE DETAIL (FIXTURE "O")
NTS
GENERAL NOTE
ALL TRENCHING AND BACK FILLING SHALL BE PROVIDED BY THIS CONTRACTOR. CALL DIGGER'S HOT LINE 72 HOURS PRIOR TO EXCAVATING. LOCATE ALL ELECTRICAL, GAS, AND UTILITY LINES PRIOR TO BEGINNING EXCAVATING.
SHEET E1 PLAN NOTES
1 PAD MOUNTED TRANSFORMER PROVIDED BY W.P.S. CO..
2 PROVIDE AND INSTALL (1) 4" SCHEDULE 80 PVC CONDUIT WITH (4) 500 MCM CU. THW.
3 LOCATION OF STANDBY GENERATOR (PROVIDED BY OWNER). THIS CONTRACTOR SHALL PROVIDE AND INSTALL A NEW CONCRETE PAD AND 24 HOUR SKID TANK (DOUBLE WALL). SEE DETAIL 3/E4.
4 SEE DETAIL 2/E1 FOR ADDITIONAL INFORMATION.
REVISION
DATE
DESCRIPTION
LAC DU FLAMBEAU FISH HATCHERY
SCALE AS NOTED
DATE 7/30/98
PROJECT ENGINEER
RICHARD R. SCHOCH
DRAWN BY D.D.H.
COOPER ENGINEERING COMPANY
310 WEST SOUTH STREET RICE LAKE, WISCONSIN
TELEPHONE 715-234-7008
SHEET DESCRIPTION
ELECTRICAL SITE PLAN
PROJECT NO.
SHEET NO.
E1

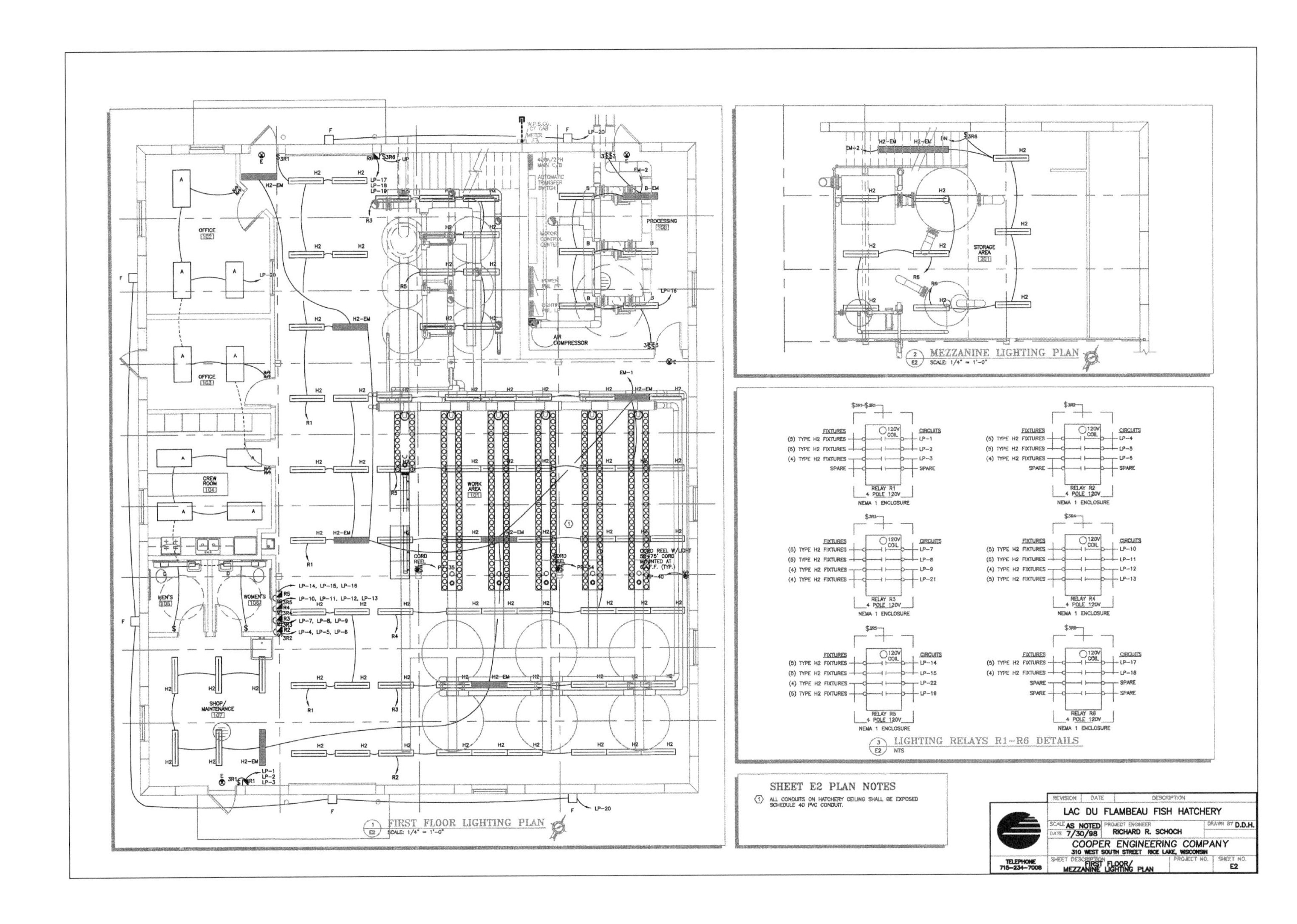
FIRST FLOOR LIGHTING PLAN
SCALE: 1/4" = 1'-0"
MEZZANINE LIGHTING PLAN
SCALE: 1/4" = 1'-0"
LIGHTING RELAYS R1-R6 DETAILS
NTS
SHEET E2 PLAN NOTES
1 ALL CONDUITS ON HATCHERY CEILING SHALL BE EXPOSED SCHEDULE 40 PVC CONDUIT.
LAC DU FLAMBEAU FISH HATCHERY
AS NOTED
7/30/98
RICHARD R. SCHOCH
D.D.H.
COOPER ENGINEERING COMPANY
310 WEST SOUTH STREET RICE LAKE, WISCONSIN
TELEPHONE 715-234-7008
FIRST FLOOR/ MEZZANINE LIGHTING PLAN
E2

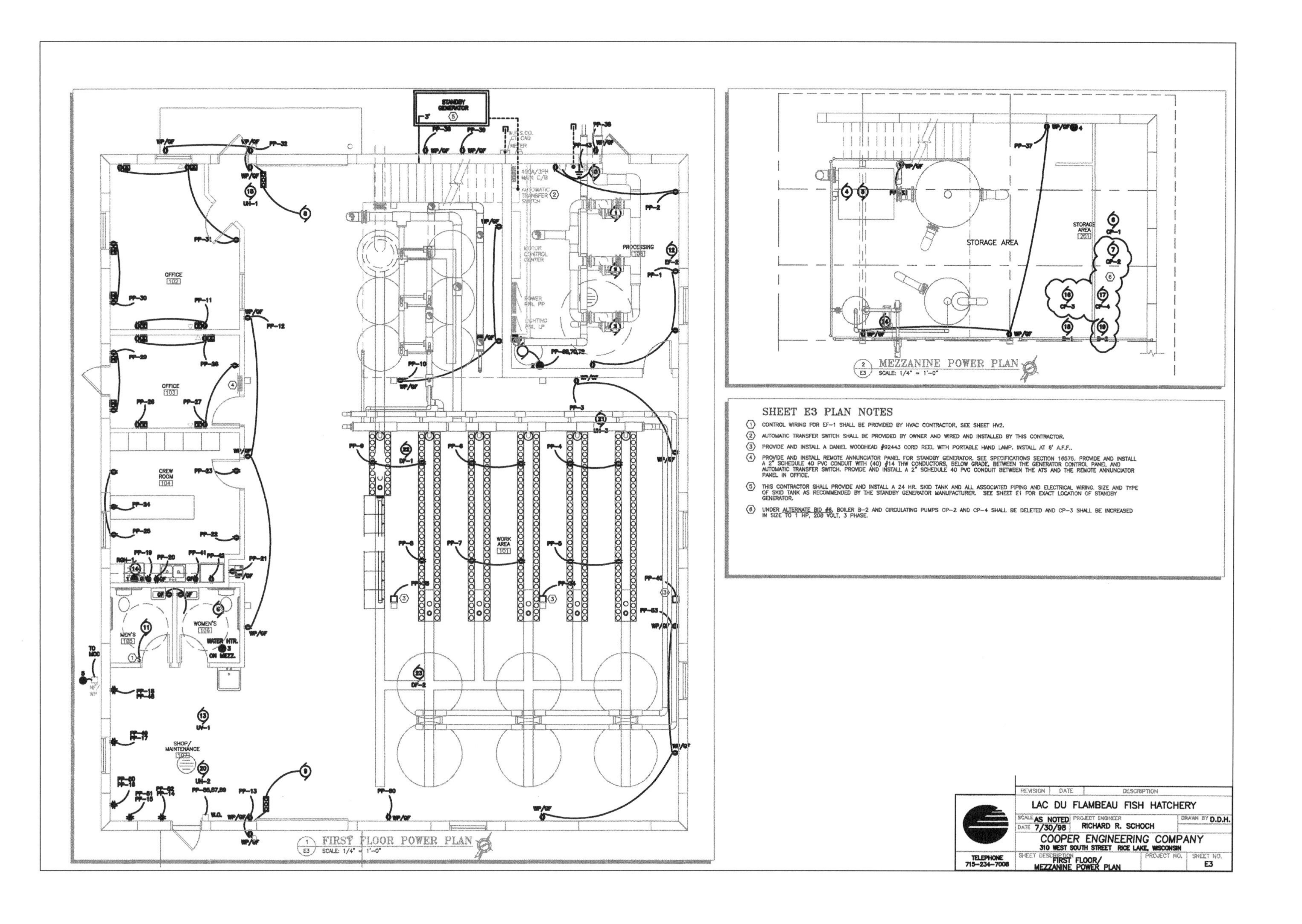
STANDBY GENERATOR
OFFICE 102
OFFICE 103
CREW ROOM 104
MEN'S 105
WOMEN'S 106
SHOP/ MAINTENANCE 107
PROCESSING 108
WORK AREA 101
MOTOR CONTROL CENTER
AUTOMATIC TRANSFER SWITCH
POWER PNL PP
LIGHTING PNL LP
WATER HTR. ON MEZZ.
TO MCC
FIRST FLOOR POWER PLAN
SCALE: 1/4" = 1'-0"
STORAGE AREA
STORAGE AREA 201
MEZZANINE POWER PLAN
SCALE: 1/4" = 1'-0"
SHEET E3 PLAN NOTES
1. CONTROL WIRING FOR EF-1 SHALL BE PROVIDED BY HVAC CONTRACTOR, SEE SHEET HV2.
2. AUTOMATIC TRANSFER SWITCH SHALL BE PROVIDED BY OWNER AND WIRED AND INSTALLED BY THIS CONTRACTOR.
3. PROVIDE AND INSTALL A DANIEL WOODHEAD #92443 CORD REEL WITH PORTABLE HAND LAMP. INSTALL AT 6' A.F.F..
4. PROVIDE AND INSTALL REMOTE ANNUNCIATOR PANEL FOR STANDBY GENERATOR. SEE SPECIFICATIONS SECTION 16575. PROVIDE AND INSTALL A 2" SCHEDULE 40 PVC CONDUIT WITH (40) #14 THW CONDUCTORS, BELOW GRADE, BETWEEN THE GENERATOR CONTROL PANEL AND AUTOMATIC TRANSFER SWITCH. PROVIDE AND INSTALL A 2" SCHEDULE 40 PVC CONDUIT BETWEEN THE ATS AND THE REMOTE ANNUNCIATOR PANEL IN OFFICE.
5. THIS CONTRACTOR SHALL PROVIDE AND INSTALL A 24 HR. SKID TANK AND ALL ASSOCIATED PIPING AND ELECTRICAL WIRING. SIZE AND TYPE OF SKID TANK AS RECOMMENDED BY THE STANDBY GENERATOR MANUFACTURER. SEE SHEET E1 FOR EXACT LOCATION OF STANDBY GENERATOR.
6. UNDER ALTERNATE BID #6, BOILER B-2 AND CIRCULATING PUMPS CP-2 AND CP-4 SHALL BE DELETED AND CP-3 SHALL BE INCREASED IN SIZE TO 1 HP, 208 VOLT, 3 PHASE.
REVISION
DATE
DESCRIPTION
LAC DU FLAMBEAU FISH HATCHERY
SCALE AS NOTED
DATE 7/30/98
PROJECT ENGINEER RICHARD R. SCHOCH
DRAWN BY D.D.H.
COOPER ENGINEERING COMPANY
310 WEST SOUTH STREET RICE LAKE, WISCONSIN
TELEPHONE 715-234-7008
SHEET DESCRIPTION FIRST FLOOR/ MEZZANINE POWER PLAN
PROJECT NO.
SHEET NO. E3

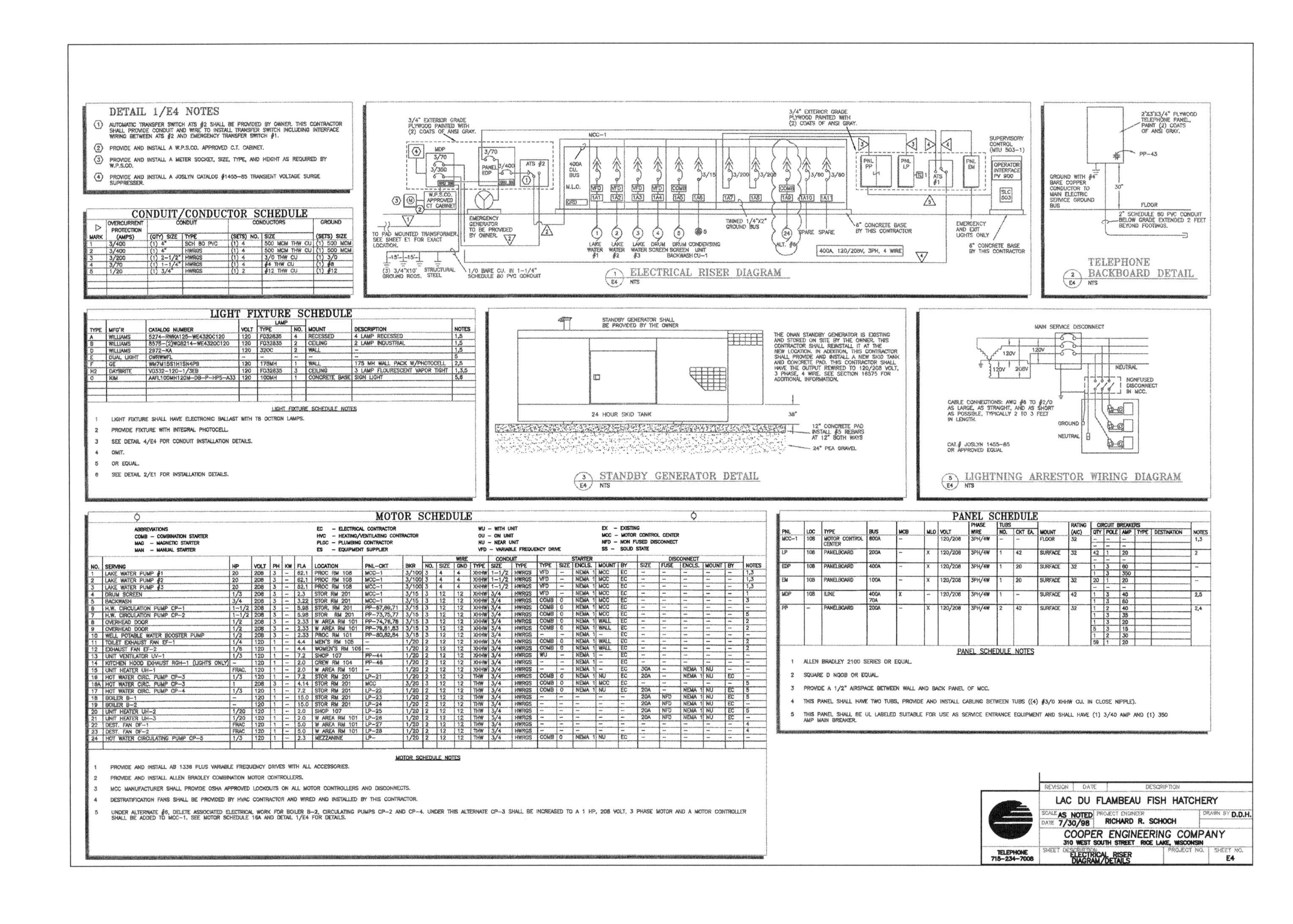
DETAIL 1/E4 NOTES
1 AUTOMATIC TRANSFER SWITCH ATS #2 SHALL BE PROVIDED BY OWNER. THIS CONTRACTOR SHALL PROVIDE CONDUIT AND WIRE TO INSTALL TRANSFER SWITCH INCLUDING INTERFACE WIRING BETWEEN ATS #2 AND EMERGENCY TRANSFER SWITCH #1.
2 PROVIDE AND INSTALL A W.P.S.CO. APPROVED C.T. CABINET.
3 PROVIDE AND INSTALL A METER SOCKET, SIZE, TYPE, AND HEIGHT AS REQUIRED BY W.P.S.CO.
4 PROVIDE AND INSTALL A JOSLYN CATALOG #1455-85 TRANSIENT VOLTAGE SURGE SUPPRESSER.
CONDUIT/CONDUCTOR SCHEDULE
ELECTRICAL RISER DIAGRAM
NTS
TELEPHONE BACKBOARD DETAIL
NTS
LIGHT FIXTURE SCHEDULE
LIGHT FIXTURE SCHEDULE NOTES
1 LIGHT FIXTURE SHALL HAVE ELECTRONIC BALLAST WITH T8 OCTRON LAMPS.
2 PROVIDE FIXTURE WITH INTEGRAL PHOTOCELL.
3 SEE DETAIL 4/E4 FOR CONDUIT INSTALLATION DETAILS.
4 OMIT.
5 OR EQUAL.
6 SEE DETAIL 2/E1 FOR INSTALLATION DETAILS.
STANDBY GENERATOR SHALL BE PROVIDED BY THE OWNER
24 HOUR SKID TANK
12" CONCRETE PAD INSTALL #5 REBARS AT 12" BOTH WAYS
24" PEA GRAVEL
STANDBY GENERATOR DETAIL
NTS
MAIN SERVICE DISCONNECT
NEUTRAL
NONFUSED DISCONNECT IN MCC.
GROUND
CAT.# JOSLYN 1455-85 OR APPROVED EQUAL
LIGHTNING ARRESTOR WIRING DIAGRAM
NTS
MOTOR SCHEDULE
MOTOR SCHEDULE NOTES
1 PROVIDE AND INSTALL AB 1336 PLUS VARIABLE FREQUENCY DRIVES WITH ALL ACCESSORIES.
2 PROVIDE AND INSTALL ALLEN BRADLEY COMBINATION MOTOR CONTROLLERS.
3 MCC MANUFACTURER SHALL PROVIDE OSHA APPROVED LOCKOUTS ON ALL MOTOR CONTROLLERS AND DISCONNECTS.
4 DESTRATIFICATION FANS SHALL BE PROVIDED BY HVAC CONTRACTOR AND WIRED AND INSTALLED BY THIS CONTRACTOR.
PANEL SCHEDULE
PANEL SCHEDULE NOTES
1 ALLEN BRADLEY 2100 SERIES OR EQUAL.
2 SQUARE D NQOB OR EQUAL.
3 PROVIDE A 1/2" AIRSPACE BETWEEN WALL AND BACK PANEL OF MCC.
LAC DU FLAMBEAU FISH HATCHERY
AS NOTED
7/30/98
RICHARD R. SCHOCH
D.D.H.
COOPER ENGINEERING COMPANY
310 WEST SOUTH STREET RICE LAKE, WISCONSIN
TELEPHONE 715-234-7008
ELECTRICAL RISER DIAGRAM/DETAILS
E4

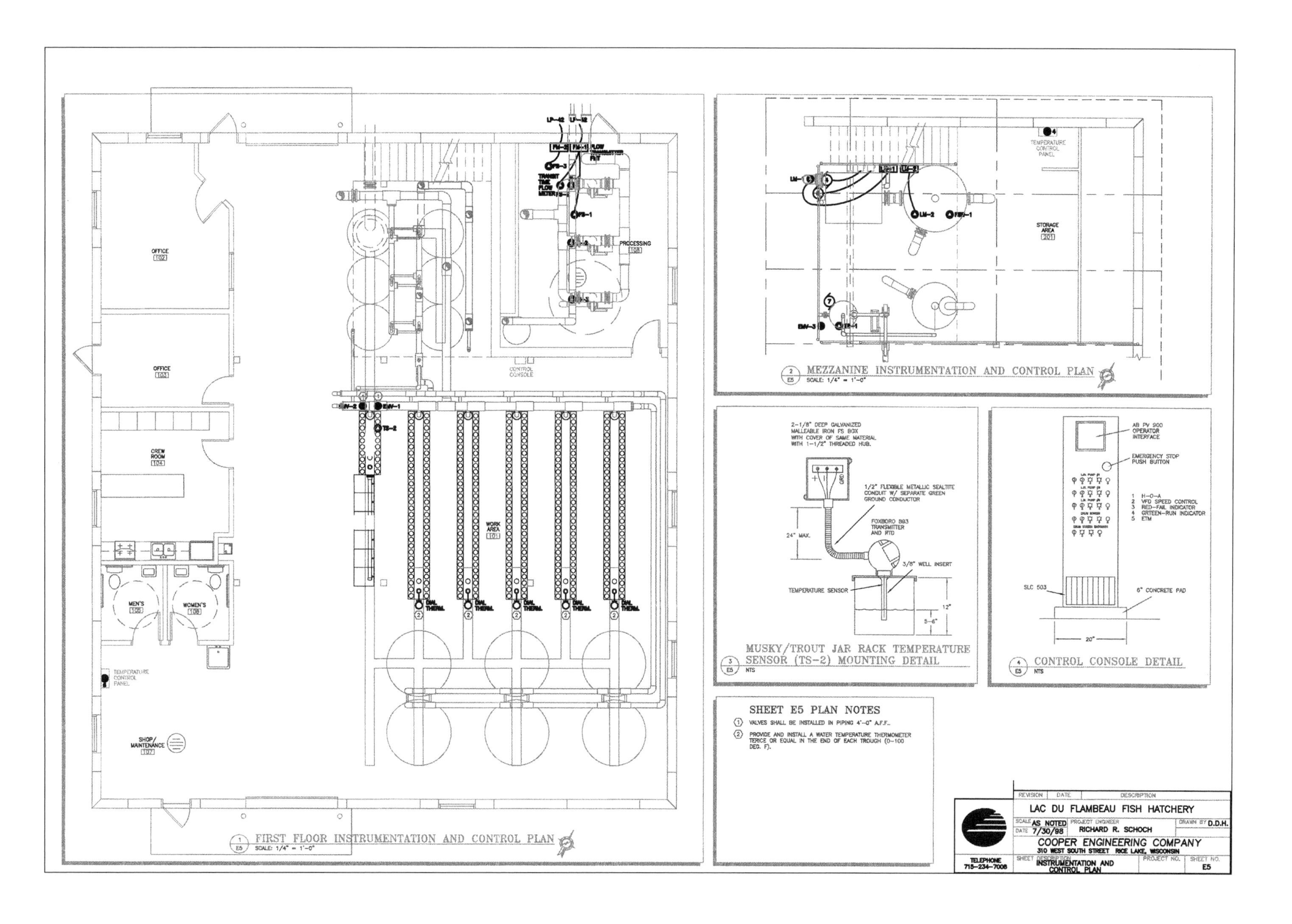
FIRST FLOOR INSTRUMENTATION AND CONTROL PLAN
SCALE: 1/4" = 1'-0"
MEZZANINE INSTRUMENTATION AND CONTROL PLAN
SCALE: 1/4" = 1'-0"
MUSKY/TROUT JAR RACK TEMPERATURE SENSOR (TS-2) MOUNTING DETAIL
NTS
CONTROL CONSOLE DETAIL
NTS
SHEET E5 PLAN NOTES
1 VALVES SHALL BE INSTALLED IN PIPING 4'-0" A.F.F..
2 PROVIDE AND INSTALL A WATER TEMPERATURE THERMOMETER TERICE OR EQUAL IN THE END OF EACH TROUGH (0-100 DEG. F).
LAC DU FLAMBEAU FISH HATCHERY
AS NOTED
7/30/98
RICHARD R. SCHOCH
D.D.H.
COOPER ENGINEERING COMPANY
310 WEST SOUTH STREET RICE LAKE, WISCONSIN
TELEPHONE 715-234-7008
INSTRUMENTATION AND CONTROL PLAN
E5

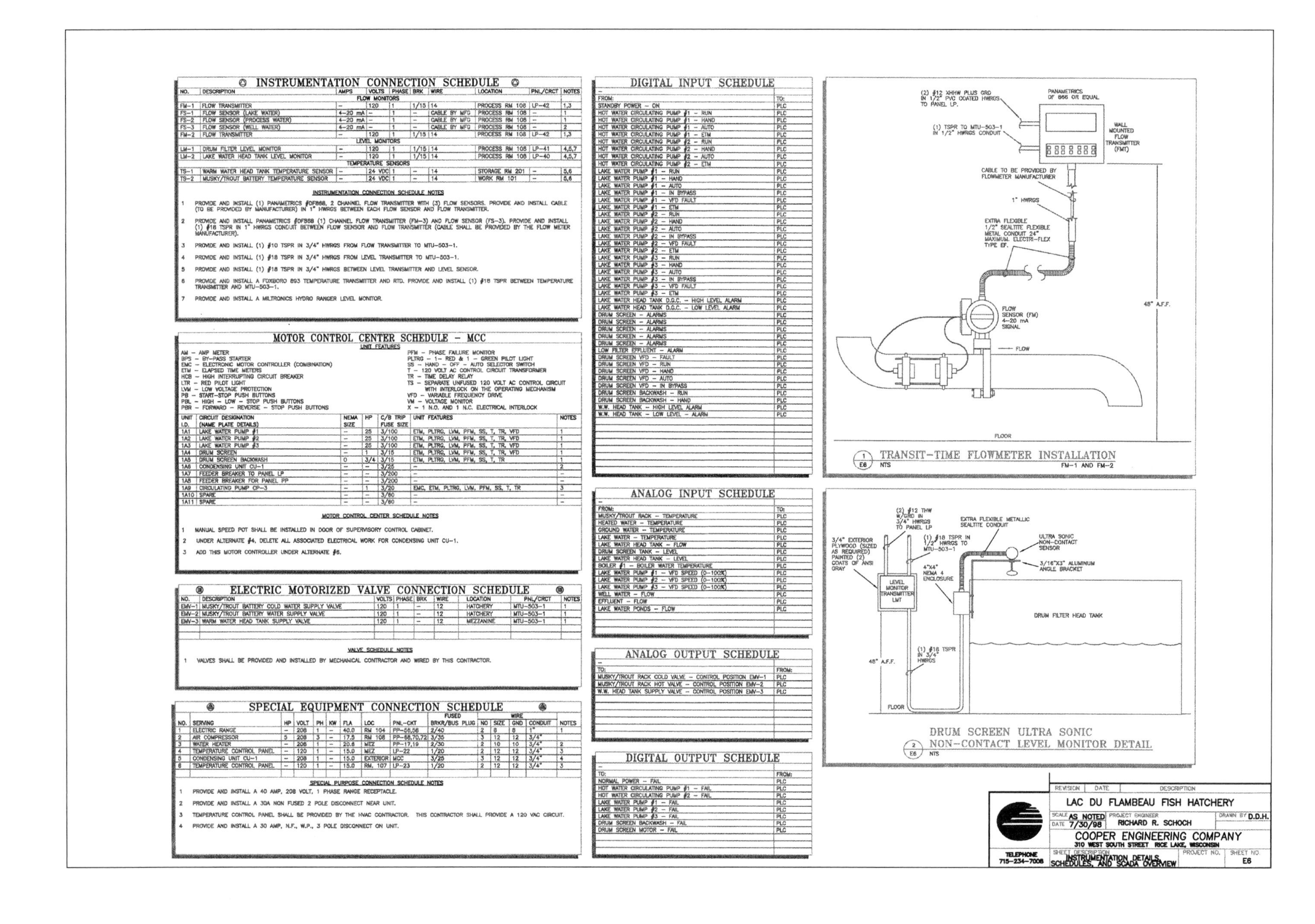
INSTRUMENTATION CONNECTION SCHEDULE
NO. DESCRIPTION AMPS VOLTS PHASE BRK WIRE LOCATION PNL/CRCT NOTES
FLOW MONITORS
FM-1 FLOW TRANSMITTER - 120 1 1/15 14 PROCESS RM 108 LP-42 1,3
FS-1 FLOW SENSOR (LAKE WATER) 4-20 mA - 1 - CABLE BY MFG PROCESS RM 108 - 1
FS-2 FLOW SENSOR (PROCESS WATER) 4-20 mA - 1 - CABLE BY MFG PROCESS RM 108 - 1
FS-3 FLOW SENSOR (WELL WATER) 4-20 mA - 1 - CABLE BY MFG PROCESS RM 108 - 2
FM-2 FLOW TRANSMITTER - 120 1 1/15 14 PROCESS RM 108 LP-42 1,3
LEVEL MONITORS
LM-1 DRUM FILTER LEVEL MONITOR - 120 1 1/15 14 PROCESS RM 108 LP-41 4,5,7
LM-2 LAKE WATER HEAD TANK LEVEL MONITOR - 120 1 1/15 14 PROCESS RM 108 LP-40 4,5,7
TEMPERATURE SENSORS
TS-1 WARM WATER HEAD TANK TEMPERATURE SENSOR - 24 VDC 1 - 14 STORAGE RM 201 - 5,6
TS-2 MUSKY/TROUT BATTERY TEMPERATURE SENSOR - 24 VDC 1 - 14 WORK RM 101 - 5,6
INSTRUMENTATION CONNECTION SCHEDULE NOTES
1 PROVIDE AND INSTALL (1) PANAMETRICS #DF868, 2 CHANNEL FLOW TRANSMITTER WITH (3) FLOW SENSORS. PROVIDE AND INSTALL CABLE (TO BE PROVIDED BY MANUFACTURER) IN 1" HWRGS BETWEEN EACH FLOW SENSOR AND FLOW TRANSMITTER.
2 PROVIDE AND INSTALL PANAMETRICS #DF868 (1) CHANNEL FLOW TRANSMITTER (FM-3) AND FLOW SENSOR (FS-3). PROVIDE AND INSTALL (1) #18 TSPR IN 1" HWRGS CONDUIT BETWEEN FLOW SENSOR AND FLOW TRANSMITTER (CABLE SHALL BE PROVIDED BY THE FLOW METER MANUFACTURER).
3 PROVIDE AND INSTALL (1) #10 TSPR IN 3/4" HWRGS FROM FLOW TRANSMITTER TO MTU-503-1.
4 PROVIDE AND INSTALL (1) #18 TSPR IN 3/4" HWRGS FROM LEVEL TRANSMITTER TO MTU-503-1.
5 PROVIDE AND INSTALL (1) #18 TSPR IN 3/4" HWRGS BETWEEN LEVEL TRANSMITTER AND LEVEL SENSOR.
6 PROVIDE AND INSTALL A FOXBORO 893 TEMPERATURE TRANSMITTER AND RTD. PROVIDE AND INSTALL (1) #18 TSPR BETWEEN TEMPERATURE TRANSMITTER AND MTU-503-1.
7 PROVIDE AND INSTALL A MILTRONICS HYDRO RANGER LEVEL MONITOR.
MOTOR CONTROL CENTER SCHEDULE - MCC
UNIT FEATURES
AM - AMP METER
BPS - BY-PASS STARTER
EMC - ELECTRONIC MOTOR CONTROLLER (COMBINATION)
ETM - ELAPSED TIME METERS
HCB - HIGH INTERRUPTING CIRCUIT BREAKER
LTR - RED PILOT LIGHT
LVM - LOW VOLTAGE PROTECTION
PB - START-STOP PUSH BUTTONS
PBL - HIGH - LOW - STOP PUSH BUTTONS
PBR - FORWARD - REVERSE - STOP PUSH BUTTONS
PFM - PHASE FAILURE MONITOR
PLTRG - 1- RED & 1 - GREEN PILOT LIGHT
SS - HAND - OFF - AUTO SELECTOR SWITCH
T - 120 VOLT AC CONTROL CIRCUIT TRANSFORMER
TR - TIME DELAY RELAY
TS - SEPARATE UNFUSED 120 VOLT AC CONTROL CIRCUIT WITH INTERLOCK ON THE OPERATING MECHANISM
VFD - VARIABLE FREQUENCY DRIVE
VM - VOLTAGE MONITOR
X - 1 N.O. AND 1 N.C. ELECTRICAL INTERLOCK
UNIT I.D. CIRCUIT DESIGNATION (NAME PLATE DETAILS) NEMA SIZE HP C/B TRIP FUSE SIZE UNIT FEATURES NOTES
1A1 LAKE WATER PUMP #1 - 25 3/100 ETM, PLTRG, LVM, PFM, SS, T, TR, VFD 1
1A2 LAKE WATER PUMP #2 - 25 3/100 ETM, PLTRG, LVM, PFM, SS, T, TR, VFD 1
1A3 LAKE WATER PUMP #3 - 25 3/100 ETM, PLTRG, LVM, PFM, SS, T, TR, VFD 1
1A4 DRUM SCREEN - 1 3/15 ETM, PLTRG, LVM, PFM, SS, T, TR, VFD 1
1A5 DRUM SCREEN BACKWASH 0 3/4 3/15 ETM, PLTRG, LVM, PFM, SS, T, TR 1
1A6 CONDENSING UNIT CU-1 - - 3/25 - 2
1A7 FEEDER BREAKER TO PANEL LP - - 3/200 - -
1A8 FEEDER BREAKER FOR PANEL PP - - 3/200 - -
1A9 CIRCULATING PUMP CP-3 - 1 3/20 EMC, ETM, PLTRG, LVM, PFM, SS, T, TR 3
1A10 SPARE - - 3/60 - -
1A11 SPARE - - 3/60 - -
MOTOR CONTROL CENTER SCHEDULE NOTES
1 MANUAL SPEED POT SHALL BE INSTALLED IN DOOR OF SUPERVISORY CONTROL CABINET.
2 UNDER ALTERNATE #4, DELETE ALL ASSOCIATED ELECTRICAL WORK FOR CONDENSING UNIT CU-1.
3 ADD THIS MOTOR CONTROLLER UNDER ALTERNATE #6.
ELECTRIC MOTORIZED VALVE CONNECTION SCHEDULE
NO. DESCRIPTION VOLTS PHASE BRK WIRE LOCATION PNL/CRCT NOTES
EMV-1 MUSKY/TROUT BATTERY COLD WATER SUPPLY VALVE 120 1 - 12 HATCHERY MTU-503-1 1
EMV-2 MUSKY/TROUT BATTERY WATER SUPPLY VALVE 120 1 - 12 HATCHERY MTU-503-1 1
EMV-3 WARM WATER HEAD TANK SUPPLY VALVE 120 1 - 12 MEZZANINE MTU-503-1 1
VALVE SCHEDULE NOTES
1 VALVES SHALL BE PROVIDED AND INSTALLED BY MECHANICAL CONTRACTOR AND WIRED BY THIS CONTRACTOR.
SPECIAL EQUIPMENT CONNECTION SCHEDULE
NO. SERVING HP VOLT PH KW FLA LOC PNL-CKT FUSED BRKR/BUS PLUG WIRE NO SIZE GND CONDUIT NOTES
1 ELECTRIC RANGE - 208 1 - 40.0 RM 104 PP-56,58 2/40 2 8 8 1" 1
2 AIR COMPRESSOR 5 208 3 - 17.5 RM 108 PP-68,70,72 3/35 3 12 12 3/4" -
3 WATER HEATER - 208 1 - 20.8 MEZ PP-17,19 2/30 2 10 10 3/4" 2
4 TEMPERATURE CONTROL PANEL - 120 1 - 15.0 MEZ LP-22 1/20 2 12 12 3/4" 3
5 CONDENSING UNIT CU-1 - 208 1 - 15.0 EXTERIOR MCC 3/25 3 12 12 3/4" 4
6 TEMPERATURE CONTROL PANEL - 120 1 - 15.0 RM. 107 LP-23 1/20 2 12 12 3/4" 3
SPECIAL PURPOSE CONNECTION SCHEDULE NOTES
1 PROVIDE AND INSTALL A 40 AMP, 208 VOLT, 1 PHASE RANGE RECEPTACLE.
2 PROVIDE AND INSTALL A 30A NON FUSED 2 POLE DISCONNECT NEAR UNIT.
3 TEMPERATURE CONTROL PANEL SHALL BE PROVIDED BY THE HVAC CONTRACTOR. THIS CONTRACTOR SHALL PROVIDE A 120 VAC CIRCUIT.
4 PROVIDE AND INSTALL A 30 AMP, N.F., W.P., 3 POLE DISCONNECT ON UNIT.
DIGITAL INPUT SCHEDULE
FROM: TO:
STANDBY POWER - ON PLC
HOT WATER CIRCULATING PUMP #1 - RUN PLC
HOT WATER CIRCULATING PUMP #1 - HAND PLC
HOT WATER CIRCULATING PUMP #1 - AUTO PLC
HOT WATER CIRCULATING PUMP #1 - ETM PLC
HOT WATER CIRCULATING PUMP #2 - RUN PLC
HOT WATER CIRCULATING PUMP #2 - HAND PLC
HOT WATER CIRCULATING PUMP #2 - AUTO PLC
HOT WATER CIRCULATING PUMP #2 - ETM PLC
LAKE WATER PUMP #1 - RUN PLC
LAKE WATER PUMP #1 - HAND PLC
LAKE WATER PUMP #1 - AUTO PLC
LAKE WATER PUMP #1 - IN BYPASS PLC
LAKE WATER PUMP #1 - VFD FAULT PLC
LAKE WATER PUMP #1 - ETM PLC
LAKE WATER PUMP #2 - RUN PLC
LAKE WATER PUMP #2 - HAND PLC
LAKE WATER PUMP #2 - AUTO PLC
LAKE WATER PUMP #2 - IN BYPASS PLC
LAKE WATER PUMP #2 - VFD FAULT PLC
LAKE WATER PUMP #2 - ETM PLC
LAKE WATER PUMP #3 - RUN PLC
LAKE WATER PUMP #3 - HAND PLC
LAKE WATER PUMP #3 - AUTO PLC
LAKE WATER PUMP #3 - IN BYPASS PLC
LAKE WATER PUMP #3 - VFD FAULT PLC
LAKE WATER PUMP #3 - ETM PLC
LAKE WATER HEAD TANK D.O.C. - HIGH LEVEL ALARM PLC
LAKE WATER HEAD TANK D.O.C. - LOW LEVEL ALARM PLC
DRUM SCREEN - ALARMS PLC
DRUM SCREEN - ALARMS PLC
DRUM SCREEN - ALARMS PLC
DRUM SCREEN - ALARMS PLC
DRUM SCREEN - ALARMS PLC
DRUM SCREEN - ALARMS PLC
LOW FILTER EFFLUENT - ALARM PLC
DRUM SCREEN VFD - FAULT PLC
DRUM SCREEN VFD - RUN PLC
DRUM SCREEN VFD - HAND PLC
DRUM SCREEN VFD - AUTO PLC
DRUM SCREEN VFD - IN BYPASS PLC
DRUM SCREEN BACKWASH - RUN PLC
DRUM SCREEN BACKWASH - HAND PLC
W.W. HEAD TANK - HIGH LEVEL ALARM PLC
W.W. HEAD TANK - LOW LEVEL - ALARM PLC
ANALOG INPUT SCHEDULE
FROM: TO:
MUSKY/TROUT RACK - TEMPERATURE PLC
HEATED WATER - TEMPERATURE PLC
GROUND WATER - TEMPERATURE PLC
LAKE WATER - TEMPERATURE PLC
LAKE WATER HEAD TANK - FLOW PLC
DRUM SCREEN TANK - LEVEL PLC
LAKE WATER HEAD TANK - LEVEL PLC
BOILER #1 - BOILER WATER TEMPERATURE PLC
LAKE WATER PUMP #1 - VFD SPEED (0-100%) PLC
LAKE WATER PUMP #2 - VFD SPEED (0-100%) PLC
LAKE WATER PUMP #3 - VFD SPEED (0-100%) PLC
WELL WATER - FLOW PLC
EFFLUENT - FLOW PLC
LAKE WATER PONDS - FLOW PLC
ANALOG OUTPUT SCHEDULE
TO: FROM:
MUSKY/TROUT RACK COLD VALVE - CONTROL POSITION EMV-1 PLC
MUSKY/TROUT RACK HOT VALVE - CONTROL POSITION EMV-2 PLC
W.W. HEAD TANK SUPPLY VALVE - CONTROL POSITION EMV-3 PLC
DIGITAL OUTPUT SCHEDULE
TO: FROM:
NORMAL POWER - FAIL PLC
HOT WATER CIRCULATING PUMP #1 - FAIL PLC
HOT WATER CIRCULATING PUMP #2 - FAIL PLC
LAKE WATER PUMP #1 - FAIL PLC
LAKE WATER PUMP #2 - FAIL PLC
LAKE WATER PUMP #3 - FAIL PLC
DRUM SCREEN BACKWASH - FAIL PLC
DRUM SCREEN MOTOR - FAIL PLC
(2) #12 XHHW PLUS GRD IN 1/2" PVC COATED HWRGS TO PANEL LP.
(1) TSPR TO MTU-503-1 IN 1/2" HWRGS CONDUIT
PANAMETRICS DF 868 OR EQUAL
WALL MOUNTED FLOW TRANSMITTER (FMT)
CABLE TO BE PROVIDED BY FLOWMETER MANUFACTURER
1" HWRGS
EXTRA FLEXIBLE 1/2" SEALTITE FLEXIBLE METAL CONDUIT 24" MAXIMUM. ELECTRI-FLEX TYPE EF.
FLOW SENSOR (FM) 4-20 mA SIGNAL
FLOW
48" A.F.F.
FLOOR
1 E6 TRANSIT-TIME FLOWMETER INSTALLATION NTS FM-1 AND FM-2
(2) #12 THW W/GRD IN 3/4" HWRGS TO PANEL LP
EXTRA FLEXIBLE METALLIC SEALTITE CONDUIT
3/4" EXTERIOR PLYWOOD (SIZED AS REQUIRED) PAINTED (2) COATS OF ANSI GRAY
(1) #18 TSPR IN 1/2" HWRGS TO MTU-503-1
ULTRA SONIC NON-CONTACT SENSOR
3/16"X3" ALUMINUM ANGLE BRACKET
4"X4" NEMA 4 ENCLOSURE
LEVEL MONITOR TRANSMITTER LMT
DRUM FILTER HEAD TANK
(1) #18 TSPR IN 3/4" HWRGS
48" A.F.F.
FLOOR
2 E6 DRUM SCREEN ULTRA SONIC NON-CONTACT LEVEL MONITOR DETAIL NTS
REVISION DATE DESCRIPTION
LAC DU FLAMBEAU FISH HATCHERY
SCALE AS NOTED PROJECT ENGINEER RICHARD R. SCHOCH DRAWN BY D.D.H.
DATE 7/30/98
COOPER ENGINEERING COMPANY
310 WEST SOUTH STREET RICE LAKE, WISCONSIN
TELEPHONE 715-234-7008
SHEET DESCRIPTION INSTRUMENTATION DETAILS, SCHEDULES, AND SCADA OVERVIEW
PROJECT NO.
SHEET NO. E6

Design

STANDARD DETAILS

Section 21

Every design professional should have an extensive file of typical or standard details that are repeatedly used with appropriate projects. These illustrations usually are time tested and have become key components in the project plans. Generally they are printed on 8 1/2" x 11" pages and inserted in the specification book rather than reproduced on the plan sheets.

The following pages represent a sampling of some standard details. They are shown here for reference only and should not be used as a final authority in a design until all applicable codes, rules and regulations have been properly satisfied.

TABLE OF CONTENTS

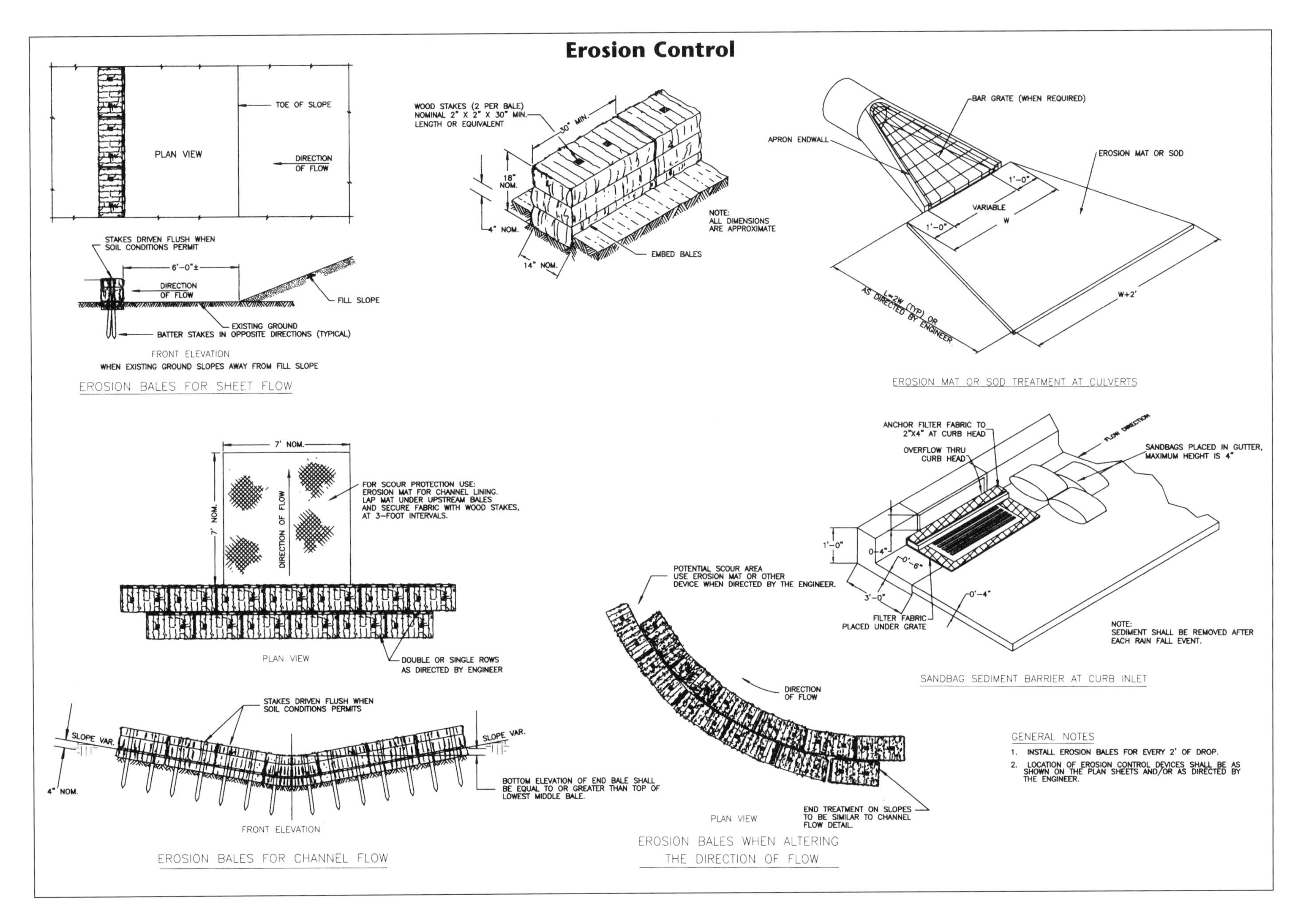

Source: Morgan & Parmley, Ltd.

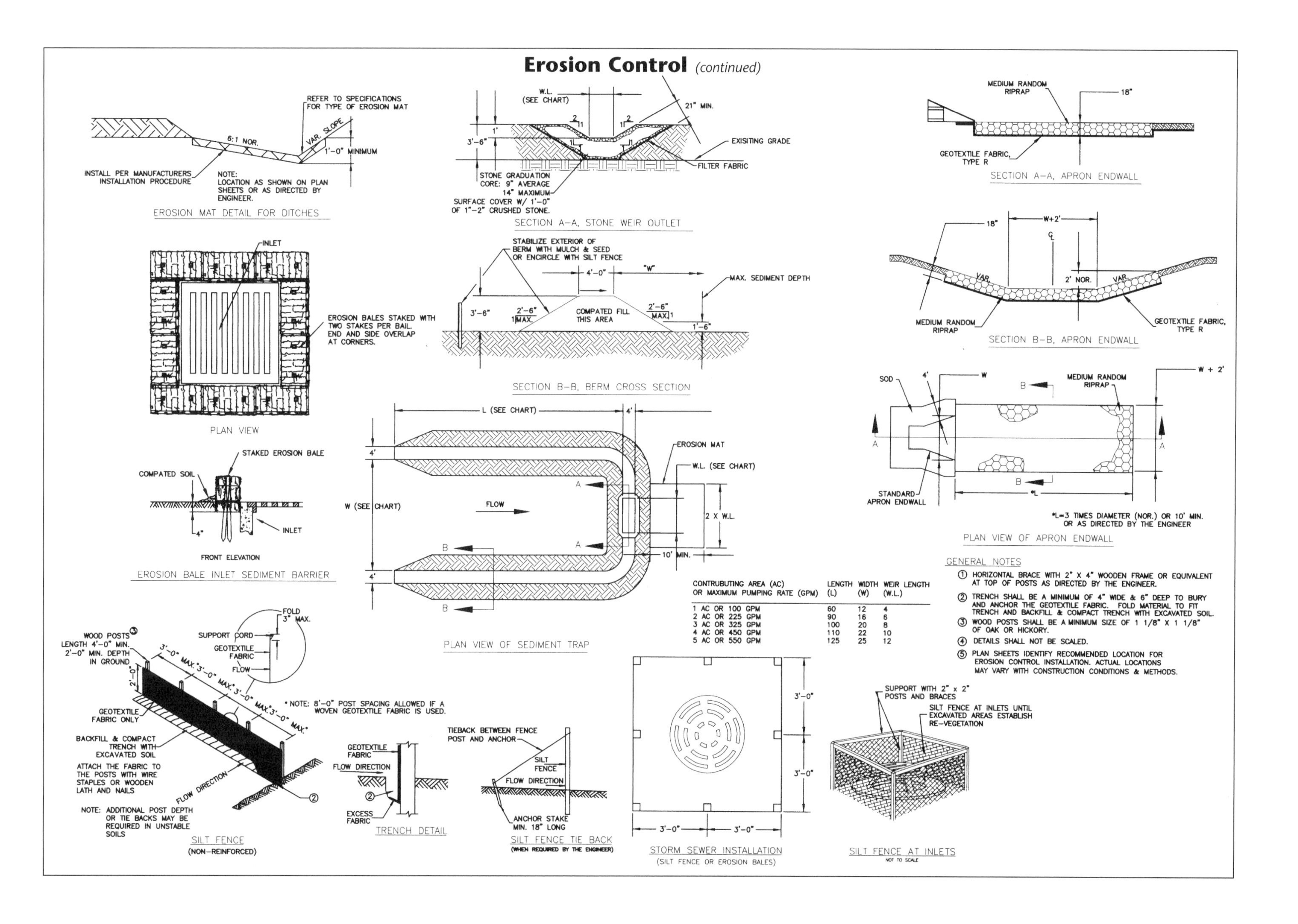

CONTRUBUTING AREA (AC) OR MAXIMUM PUMPING RATE (GPM)	LENGTH (L)	WIDTH (W)	WEIR LENGTH (W.L.)
1 AC OR 100 GPM	60	12	4
2 AC OR 225 GPM	90	16	6
3 AC OR 325 GPM	100	20	8
4 AC OR 450 GPM	110	22	10
5 AC OR 550 GPM	125	25	12

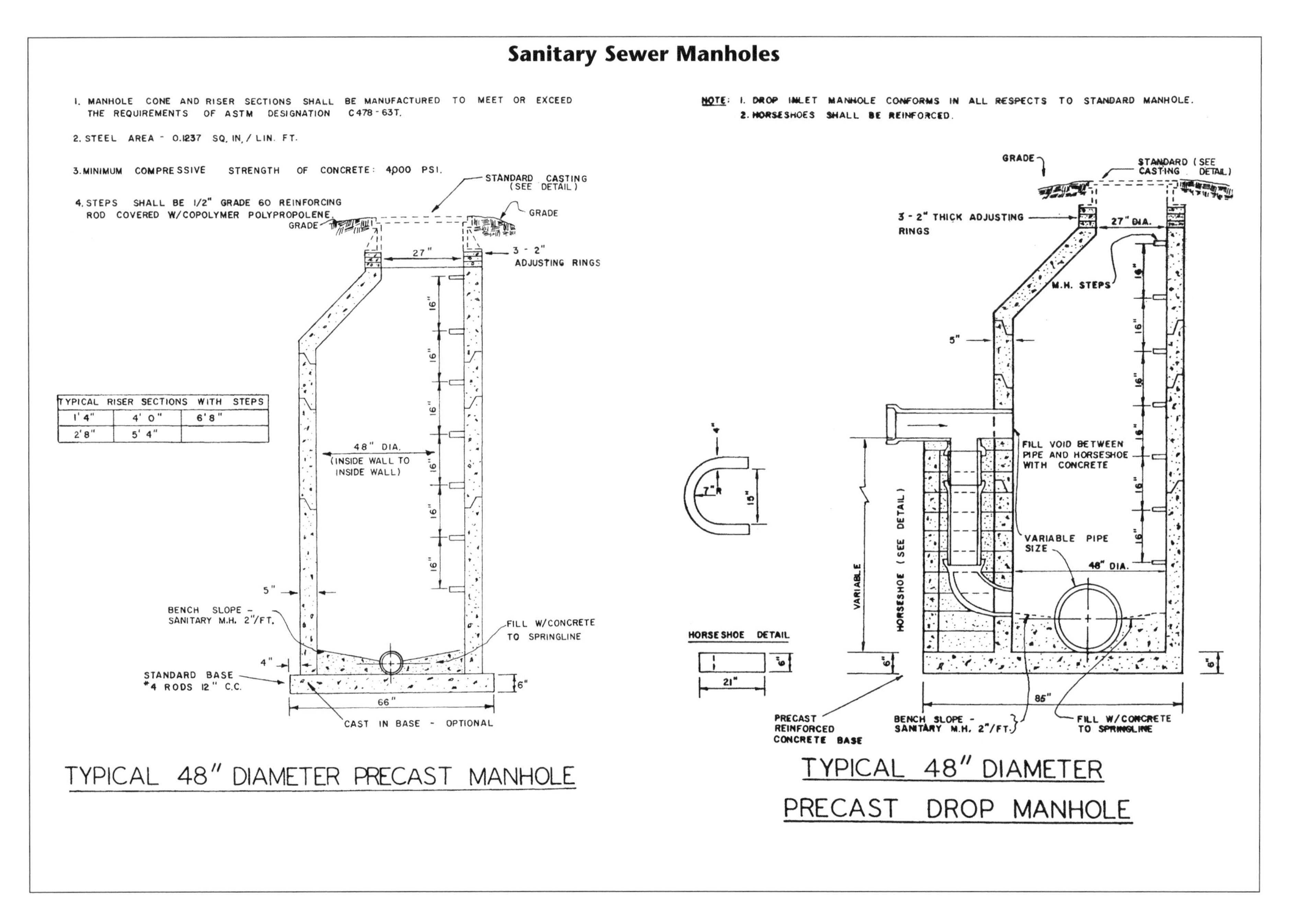
Sanitary Sewer Manholes
1. MANHOLE CONE AND RISER SECTIONS SHALL BE MANUFACTURED TO MEET OR EXCEED THE REQUIREMENTS OF ASTM DESIGNATION C478 - 63T.
2. STEEL AREA - 0.1237 SQ. IN./LIN. FT.
3. MINIMUM COMPRESSIVE STRENGTH OF CONCRETE: 4000 PSI.
4. STEPS SHALL BE 1/2" GRADE 60 REINFORCING ROD COVERED W/COPOLYMER POLYPROPOLENE.
STANDARD CASTING (SEE DETAIL)
GRADE
GRADE
27"
3 - 2" ADJUSTING RINGS
16"
48" DIA. (INSIDE WALL TO INSIDE WALL)
5"
BENCH SLOPE - SANITARY M.H. 2"/FT.
FILL W/CONCRETE TO SPRINGLINE
4"
STANDARD BASE #4 RODS 12" C.C.
6"
66"
CAST IN BASE - OPTIONAL
TYPICAL 48" DIAMETER PRECAST MANHOLE
NOTE: 1. DROP INLET MANHOLE CONFORMS IN ALL RESPECTS TO STANDARD MANHOLE.
2. HORSESHOES SHALL BE REINFORCED.
GRADE
STANDARD (SEE CASTING DETAIL)
3 - 2" THICK ADJUSTING RINGS
27" DIA.
M.H. STEPS
16"
5"
FILL VOID BETWEEN PIPE AND HORSESHOE WITH CONCRETE
VARIABLE PIPE SIZE
48" DIA.
4"
7"R
15"
VARIABLE
HORSESHOE (SEE DETAIL)
HORSESHOE DETAIL
6"
21"
6"
85"
PRECAST REINFORCED CONCRETE BASE
BENCH SLOPE - SANITARY M.H. 2"/FT.
FILL W/CONCRETE TO SPRINGLINE
TYPICAL 48" DIAMETER PRECAST DROP MANHOLE

TYPICAL RISER SECTIONS WITH STEPS		
1' 4"	4' 0"	6' 8"
2' 8"	5' 4"	

Source: Morgan & Parmley, Ltd.

Typical Flexible Manhole Sleeve

SLEEVE DIMENSIONS			
PIPE DIA.	"A"	"B"	"C"
4"	6"	14"	6"
6"	8 1/8"	16 1/8"	6 1/2"
8"	10 3/8"	18 3/8"	7 1/2"
10"	12 5/8"	20 5/8"	7 1/2"
12"	14 7/8"	22 7/8"	7 1/2"
15"	18 7/8"	26 7/8"	9"

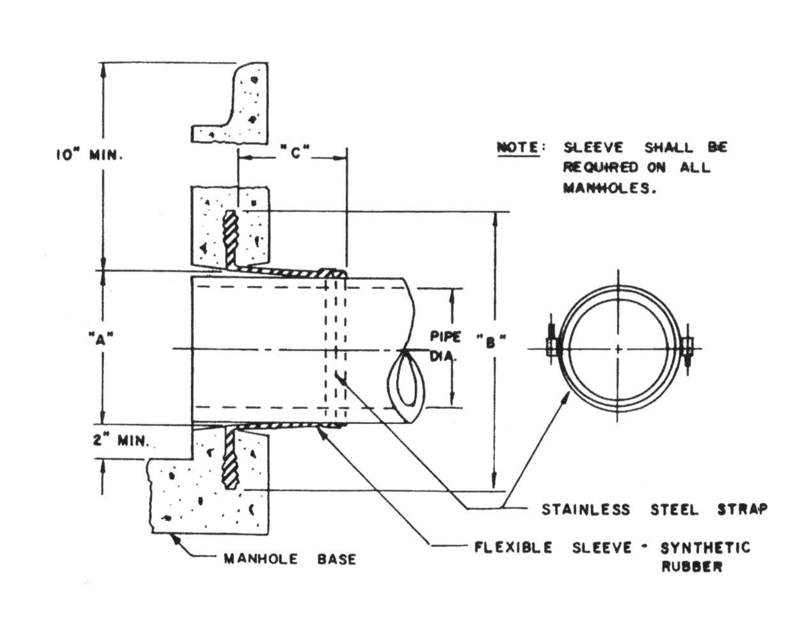

Source: Morgan & Parmley, Ltd.

Standard Section for Laying Pipe

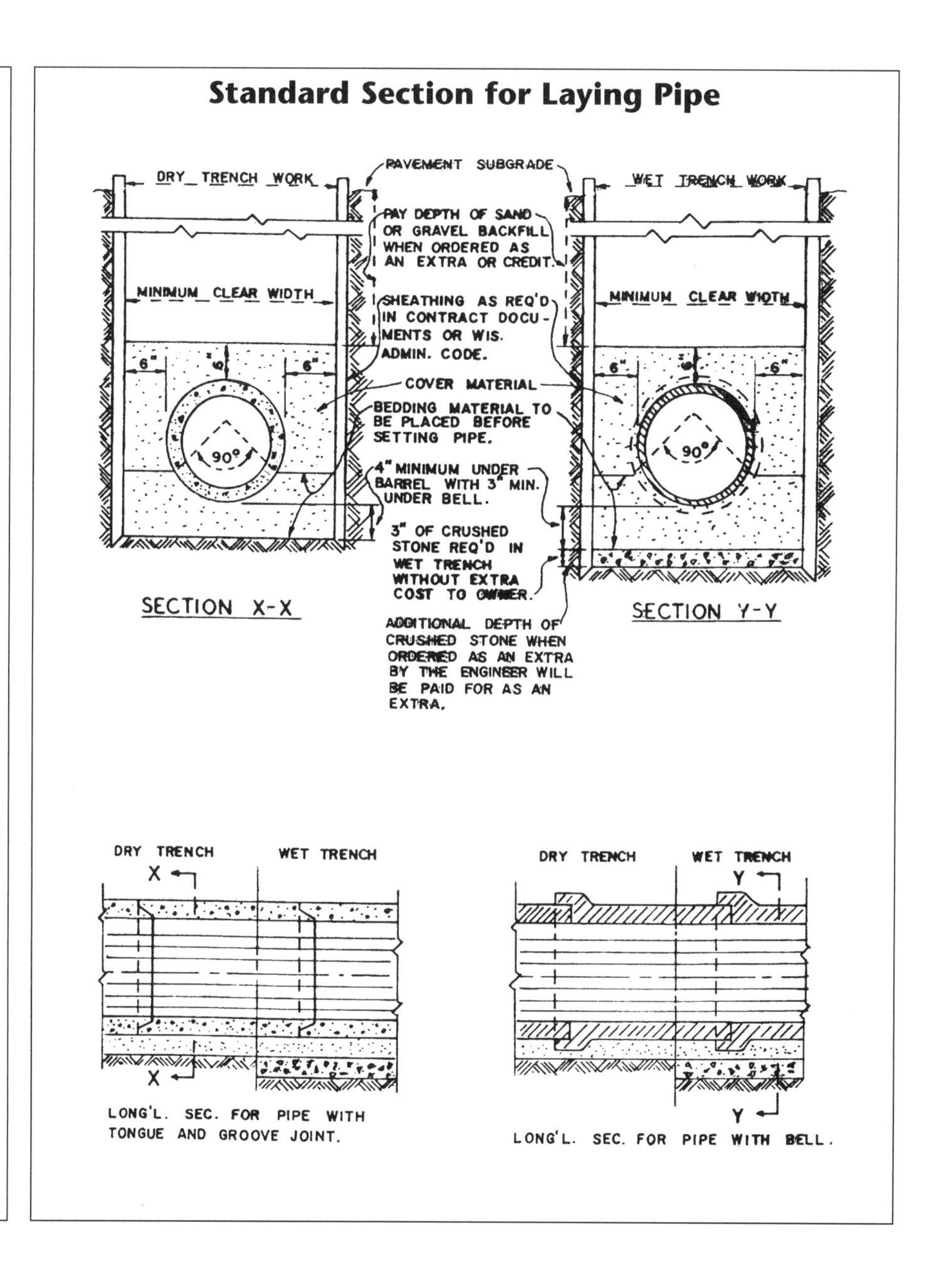

Source: Morgan & Parmley, Ltd.

Horizontal & Vertical Separations at Intersection

WATERMAIN

10'- 0"

SANITARY SEWER

HORIZONTAL SEPARATION

WATERMAIN

6"

SANITARY SEWER

SANITARY SEWER

18"

WATERMAIN

VERTICAL SEPARATION

MINIMUM HORIZONTAL & VERTICAL SEPARATIONS AT WATERMAIN & SANITARY SEWER INTERSECTIONS

Source: Morgan & Parmley, Ltd.

Common Trench Detail

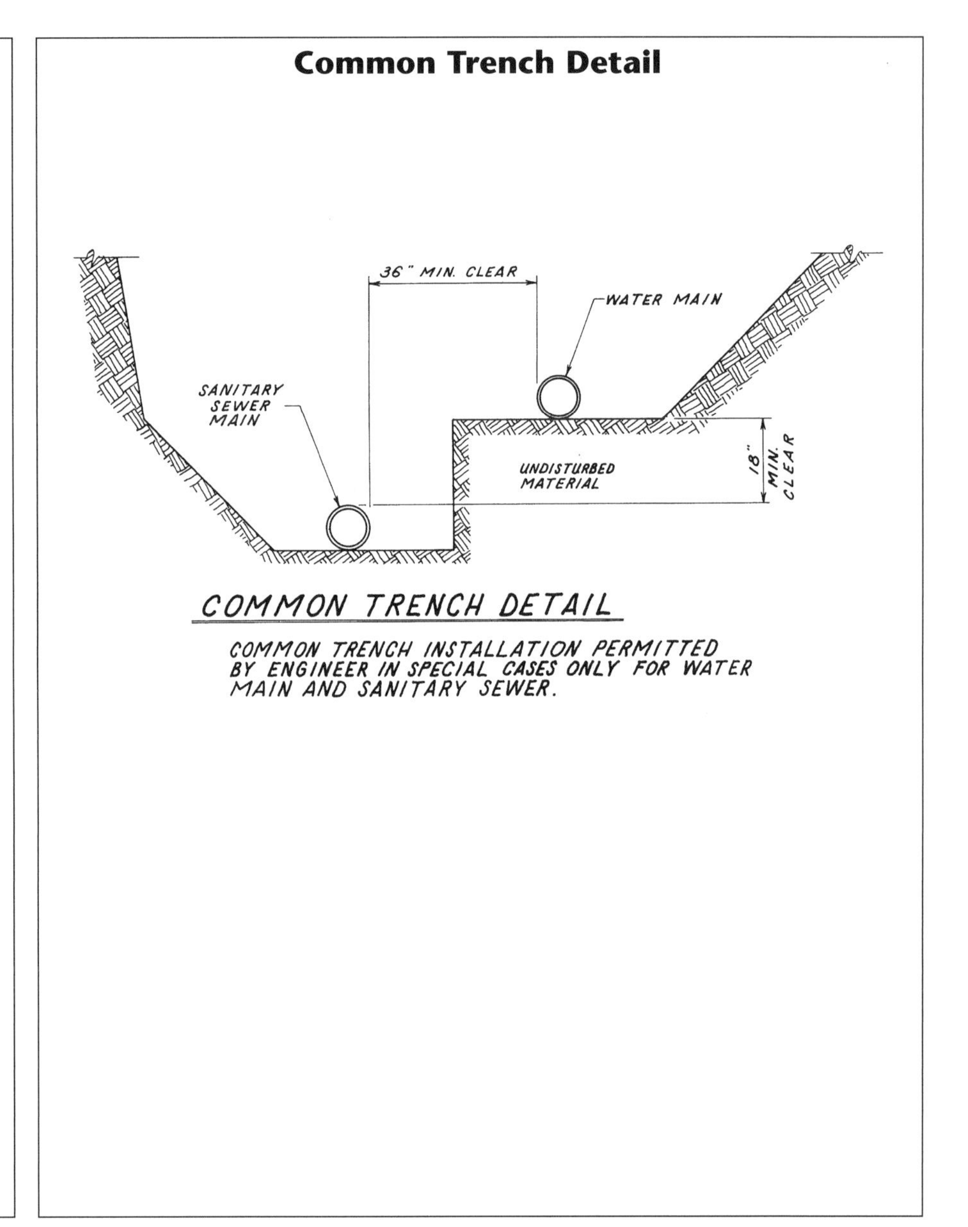

Source: Morgan & Parmley, Ltd.

Compacted Section for Laying Sewer Pipe

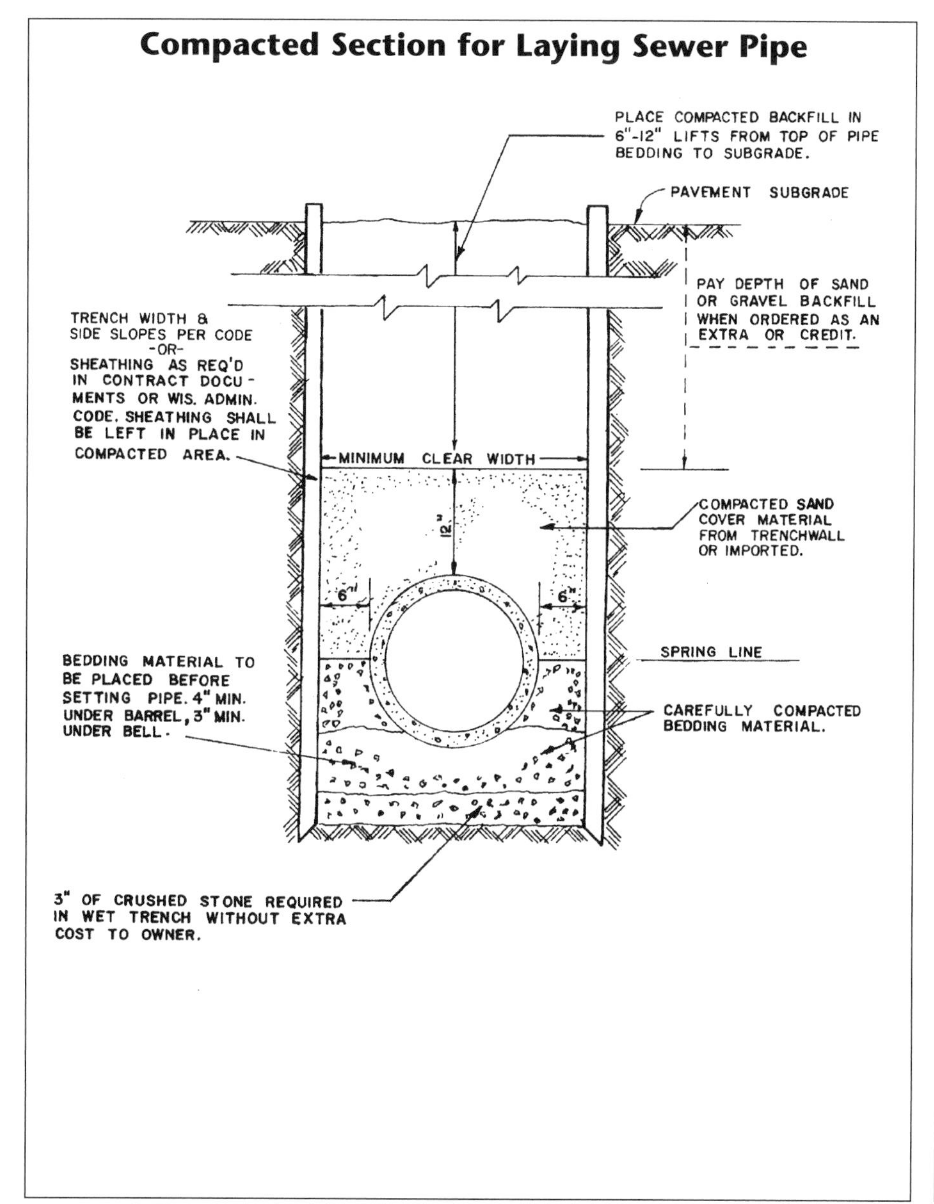

Source: Morgan & Parmley, Ltd.

Typical Riser & Wye Detail

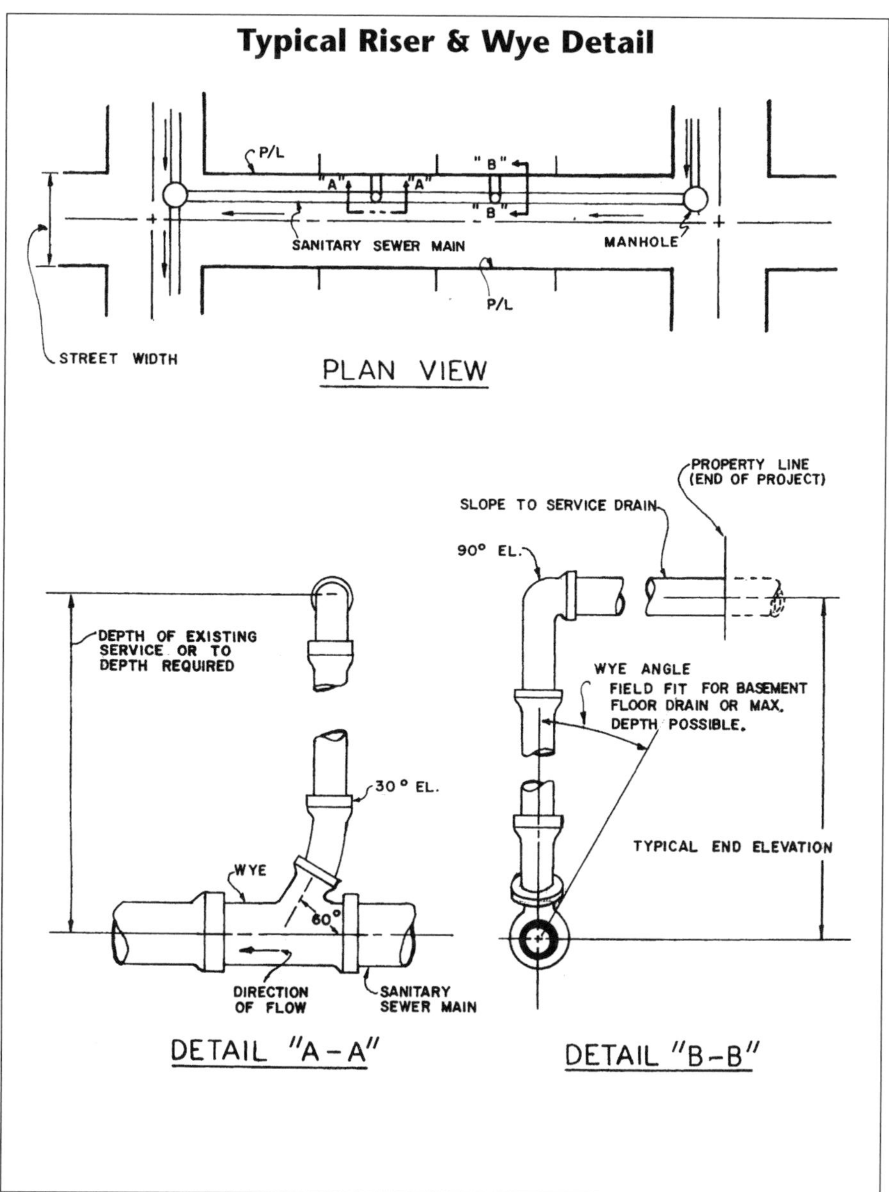

Source: Morgan & Parmley, Ltd.

Typical Hydrant Installation

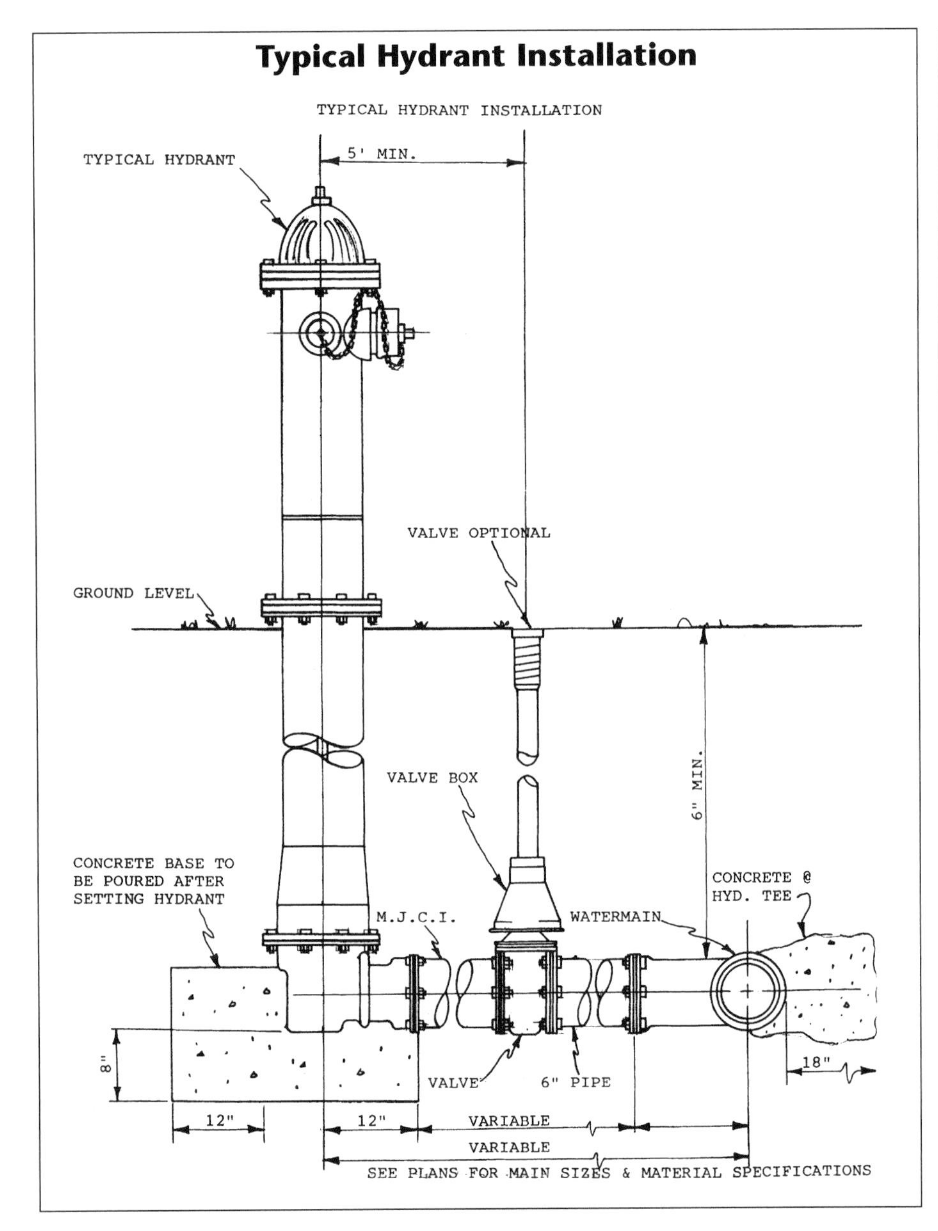

Source: Morgan & Parmley, Ltd.

Typical Dry Hydrant Installation

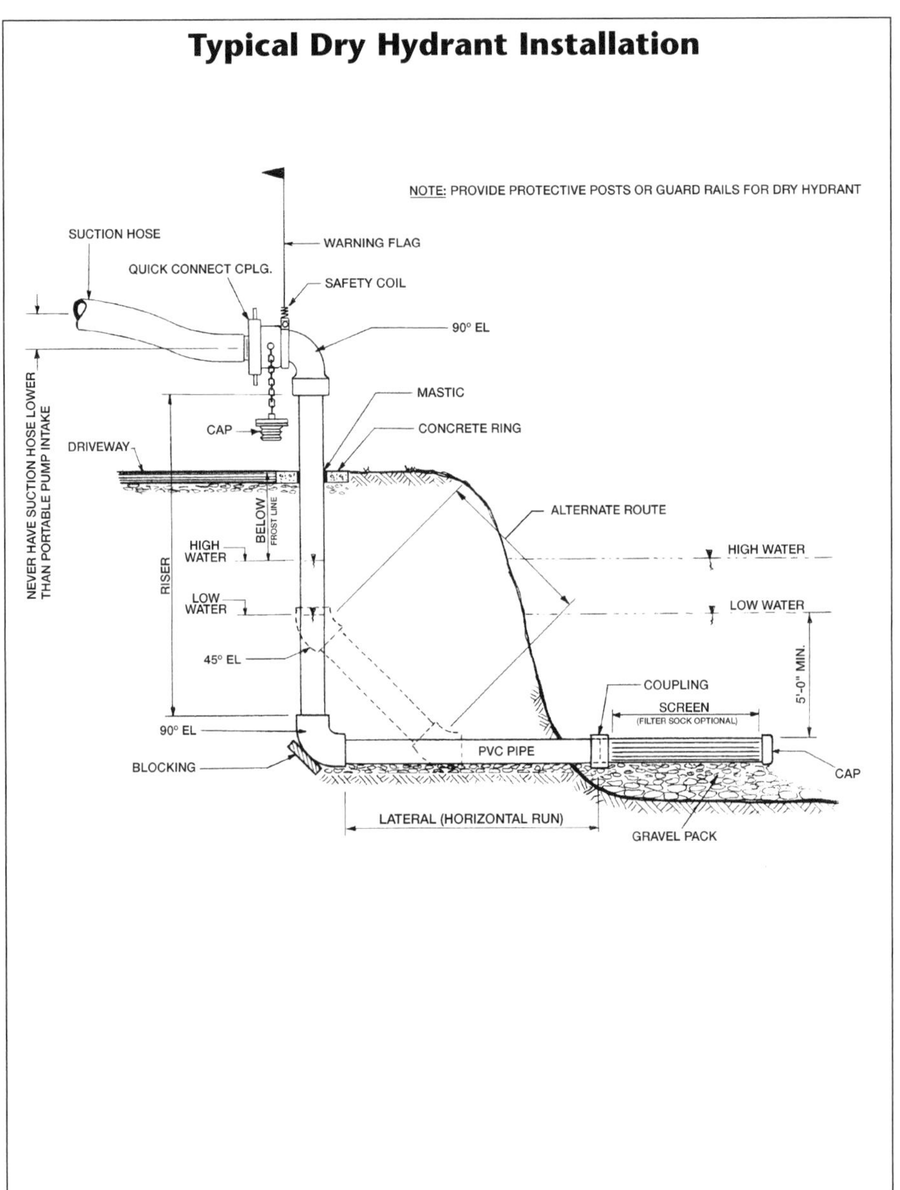

Source: *Hydraulics Field Manual*, 2/e
Robert O. Parmley, P.E.
McGraw-Hill © 2001

Gate Valve Settings

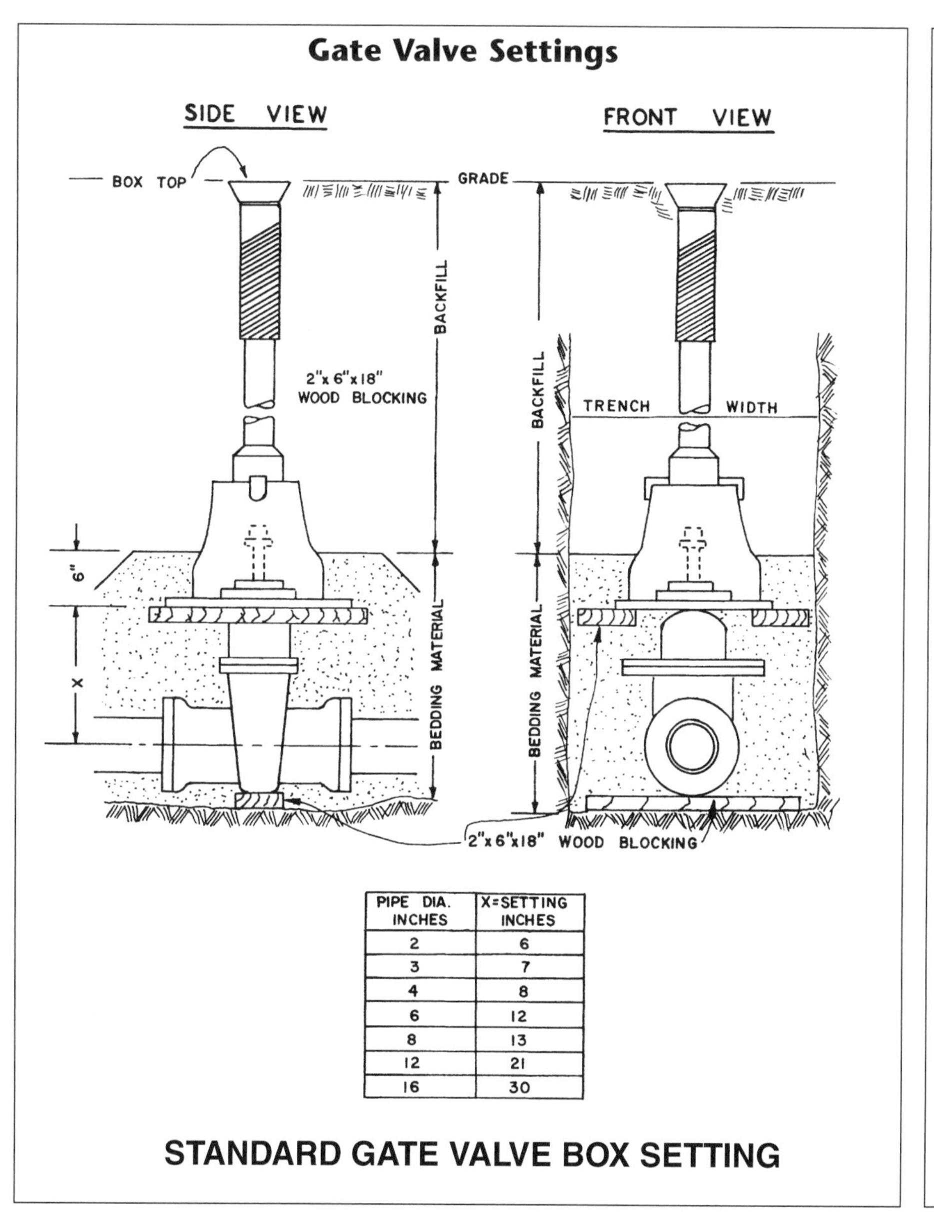

PIPE DIA. INCHES	X=SETTING INCHES
2	6
3	7
4	8
6	12
8	13
12	21
16	30

STANDARD GATE VALVE BOX SETTING

Gate Valve Settings *(Continued)*

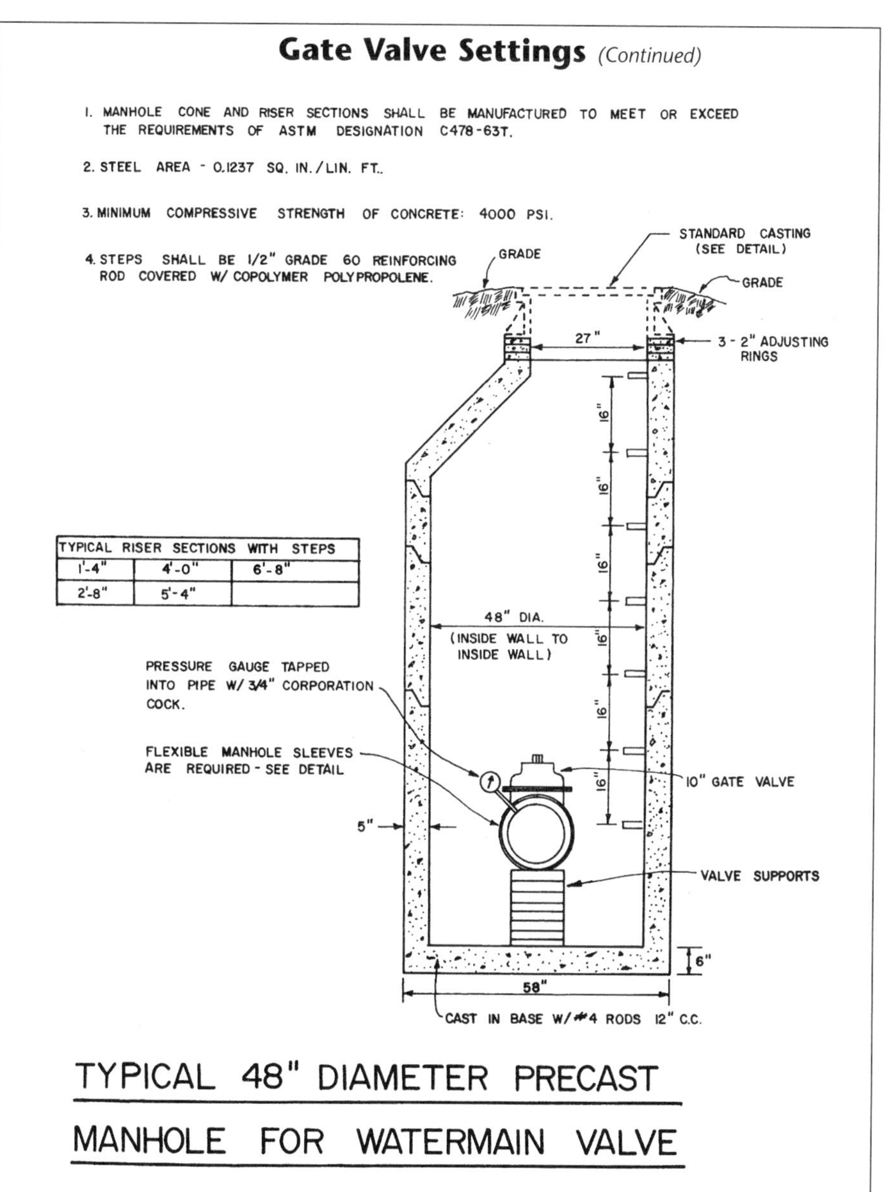

TYPICAL RISER SECTIONS WITH STEPS		
1'-4"	4'-0"	6'-8"
2'-8"	5'-4"	

TYPICAL 48" DIAMETER PRECAST MANHOLE FOR WATERMAIN VALVE

Source: Morgan & Parmley, Ltd.

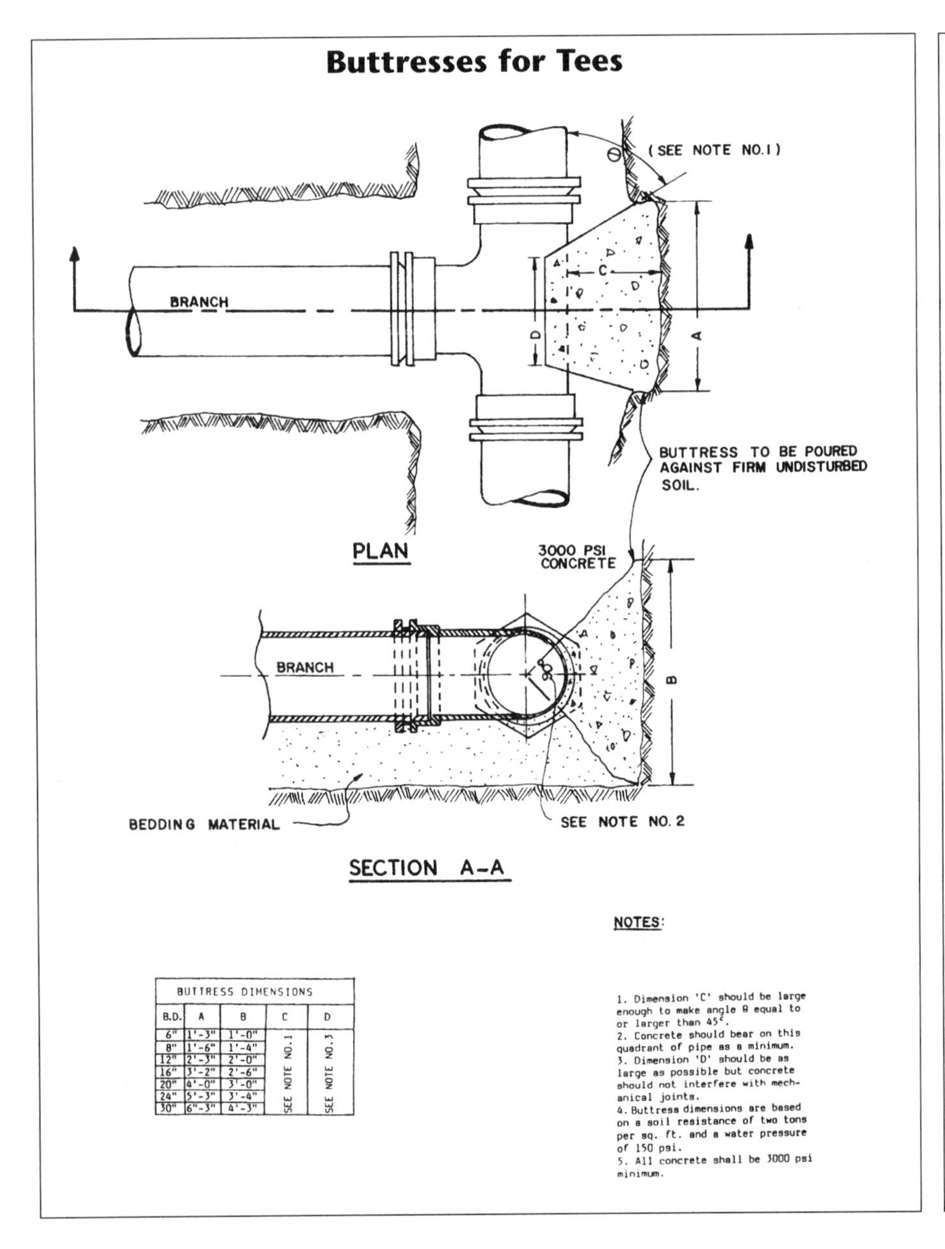

BUTTRESS DIMENSIONS				
B.D.	A	B	C	D
6"	1'-3"	1'-0"	SEE NOTE NO.1	SEE NOTE NO.3
8"	1'-6"	1'-4"		
12"	2'-3"	2'-0"		
16"	3'-2"	2'-6"		
20"	4'-0"	3'-0"		
24"	5'-3"	3'-4"		
30"	6"-3"	4'-3"		

NOTES:

1. Dimension 'C' should be large enough to make angle θ equal to or larger than 45°.
2. Concrete should bear on this quadrant of pipe as a minimum.
3. Dimension 'D' should be as large as possible but concrete should not interfere with mechanical joints.
4. Buttress dimensions are based on a soil resistance of two tons per sq. ft. and a water pressure of 150 psi.
5. All concrete shall be 3000 psi minimum.

Source: Morgan & Parmley, Ltd.

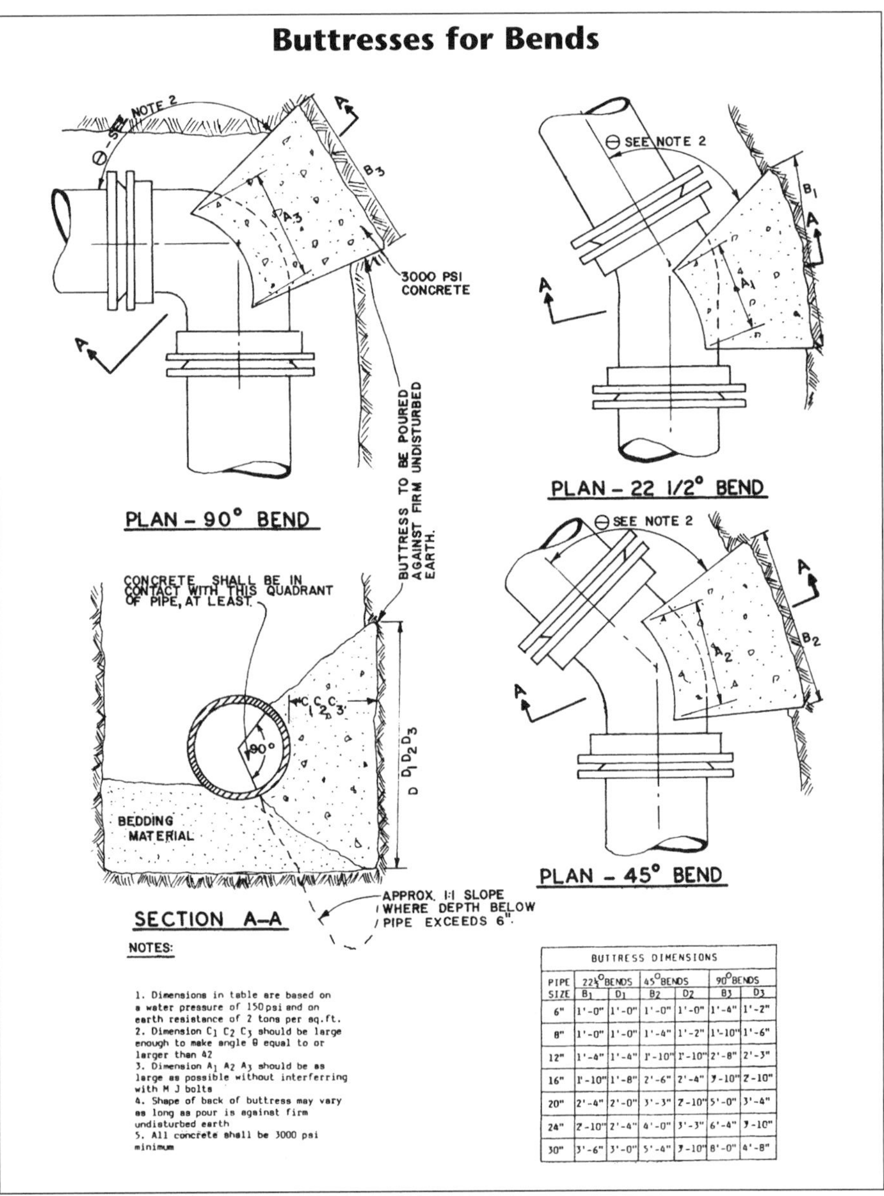

NOTES:

1. Dimensions in table are based on a water pressure of 150 psi and on earth resistance of 2 tons per sq.ft.
2. Dimension C_1 C_2 C_3 should be large enough to make angle θ equal to or larger then 42
3. Dimension A_1 A_2 A_3 should be as large as possible without interferring with M J bolts
4. Shape of back of buttress may vary as long as pour is against firm undisturbed earth
5. All concrete shall be 3000 psi minimum

BUTTRESS DIMENSIONS						
PIPE	22½° BENDS		45° BENDS		90° BENDS	
SIZE	B_1	D_1	B_2	D_2	B_3	D_3
6"	1'-0"	1'-0"	1'-0"	1'-0"	1'-4"	1'-2"
8"	1'-0"	1'-0"	1'-4"	1'-2"	1'-10"	1'-6"
12"	1'-4"	1'-4"	1'-10"	1'-10"	2'-8"	2'-3"
16"	1'-10"	1'-8"	2'-6"	2'-4"	3'-10"	2'-10"
20"	2'-4"	2'-0"	3'-3"	2'-10"	5'-0"	3'-4"
24"	2'-10"	2'-4"	4'-0"	3'-3"	6'-4"	3'-10"
30"	3'-6"	3'-0"	5'-4"	3'-10"	8'-0"	4'-8"

Source: Morgan & Parmley, Ltd.

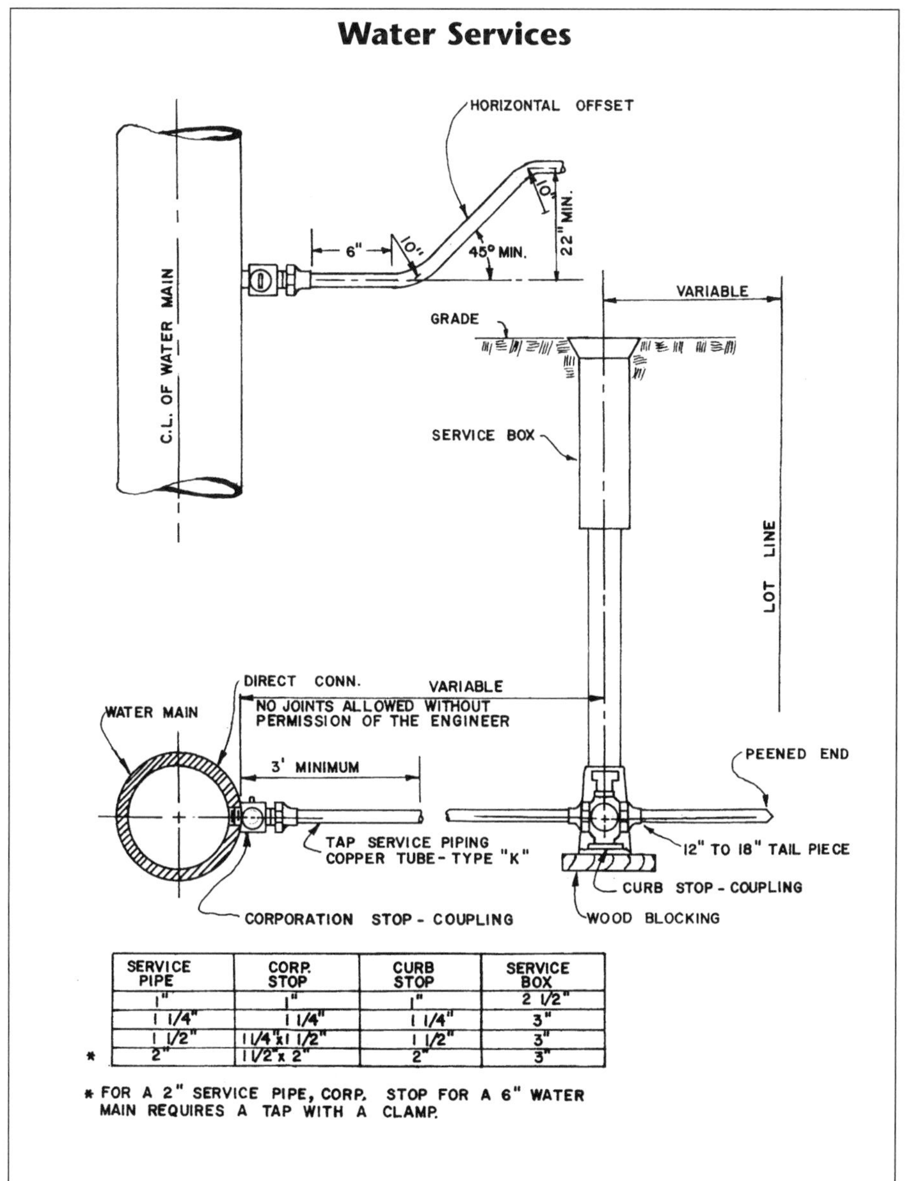

	SERVICE PIPE	CORP. STOP	CURB STOP	SERVICE BOX
	1"	1"	1"	2 1/2"
	1 1/4"	1 1/4"	1 1/4"	3"
	1 1/2"	1 1/4"x 1 1/2"	1 1/2"	3"
*	2"	1 1/2"x 2"	2"	3"

* FOR A 2" SERVICE PIPE, CORP. STOP FOR A 6" WATER MAIN REQUIRES A TAP WITH A CLAMP.

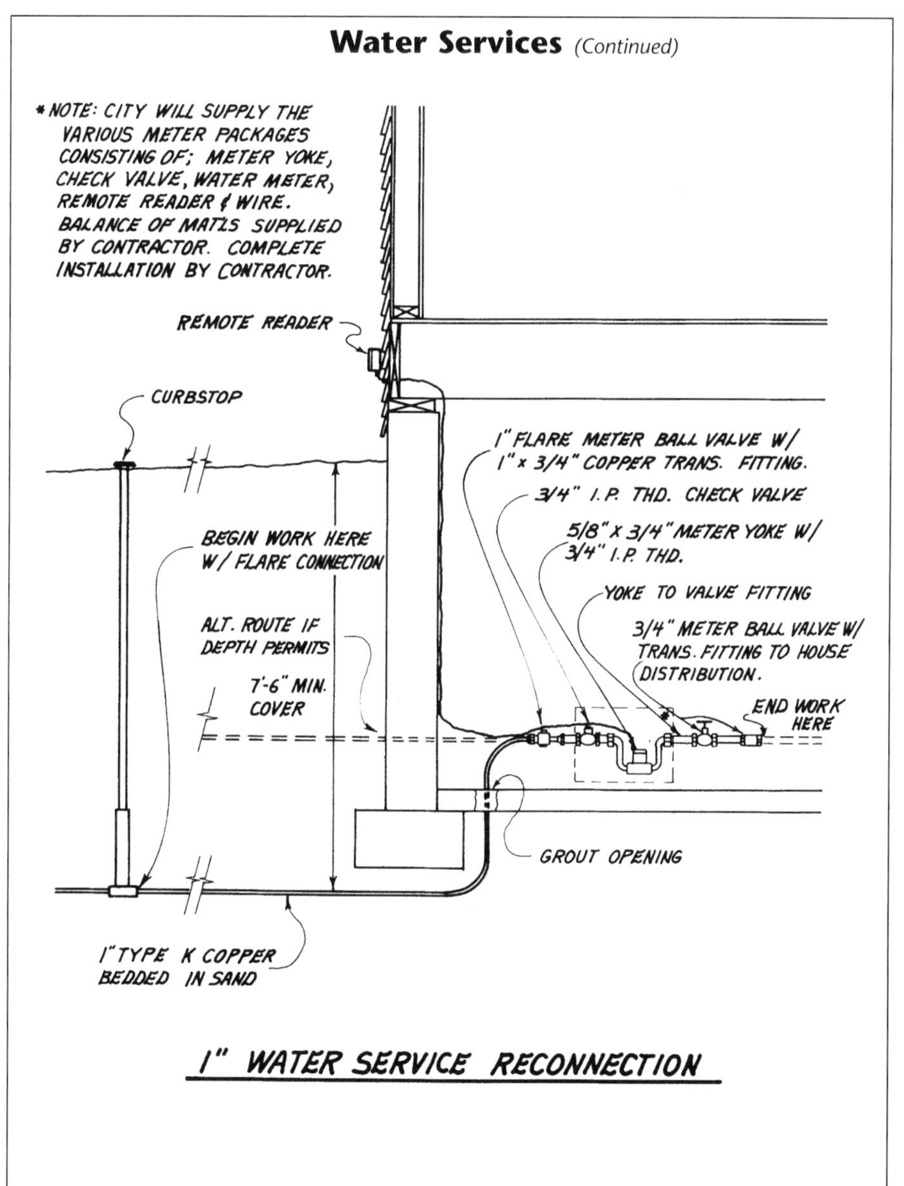

Source: Morgan & Parmley, Ltd.

Standard Watermain Trench Section

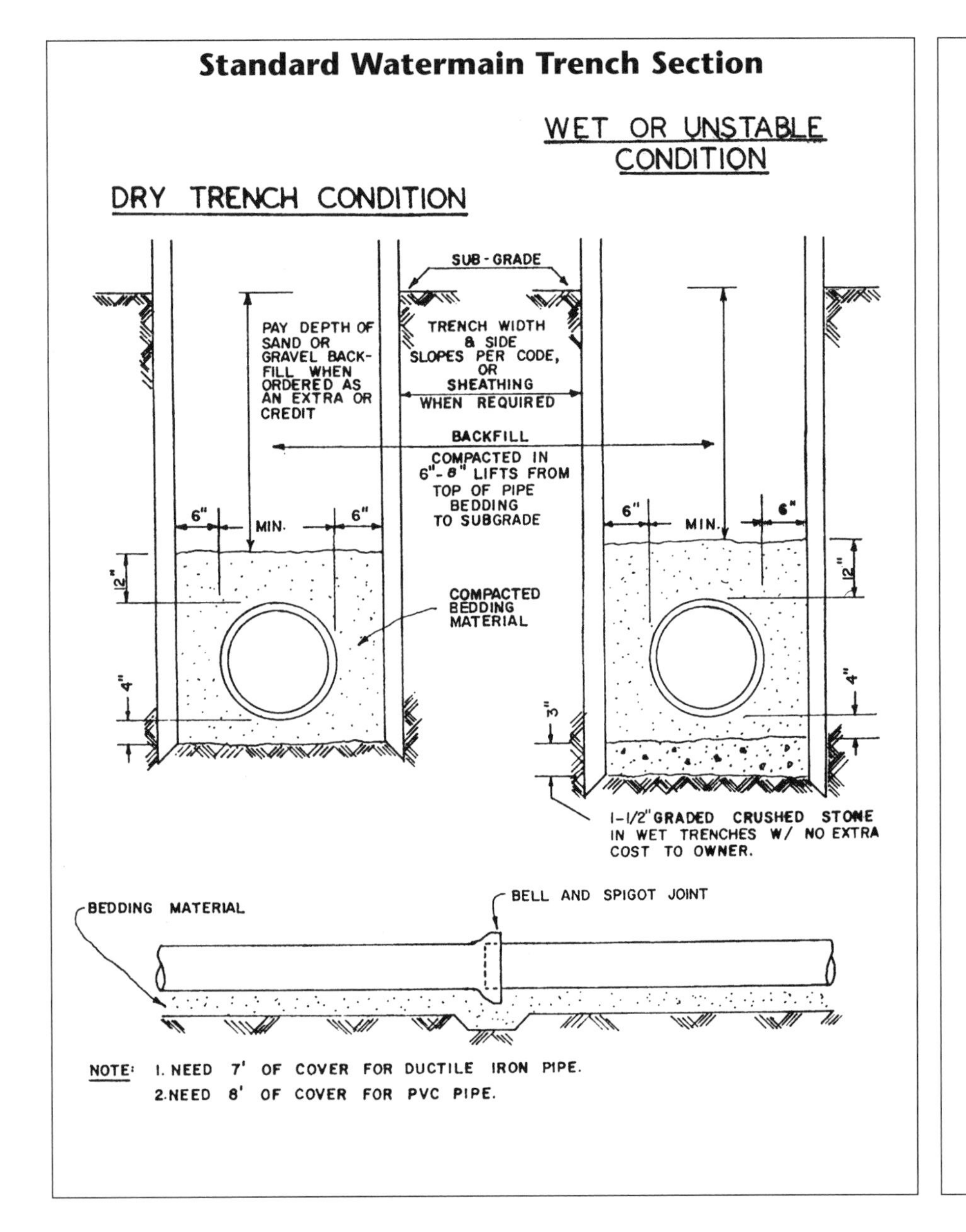

Source: Morgan & Parmley, Ltd.

Typical Installation Details for Water Pipe

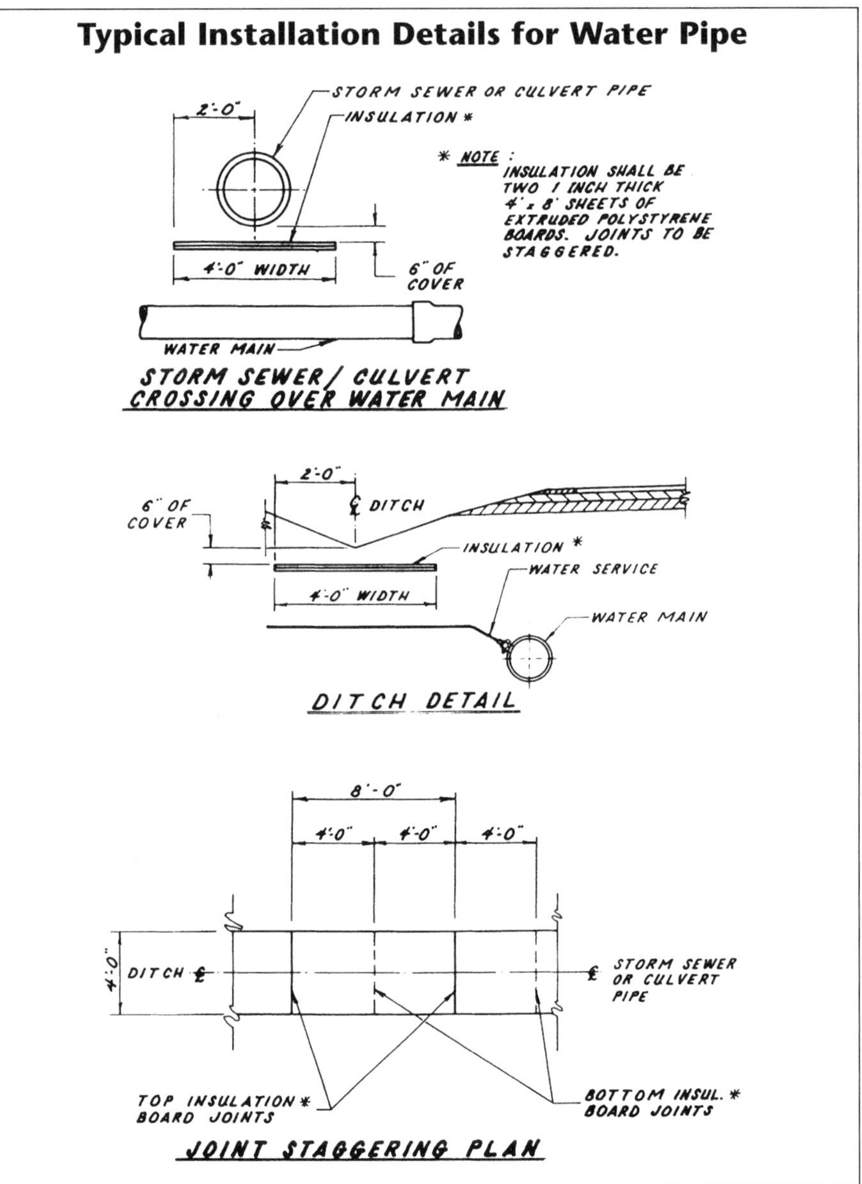

Source: Morgan & Parmley, Ltd.

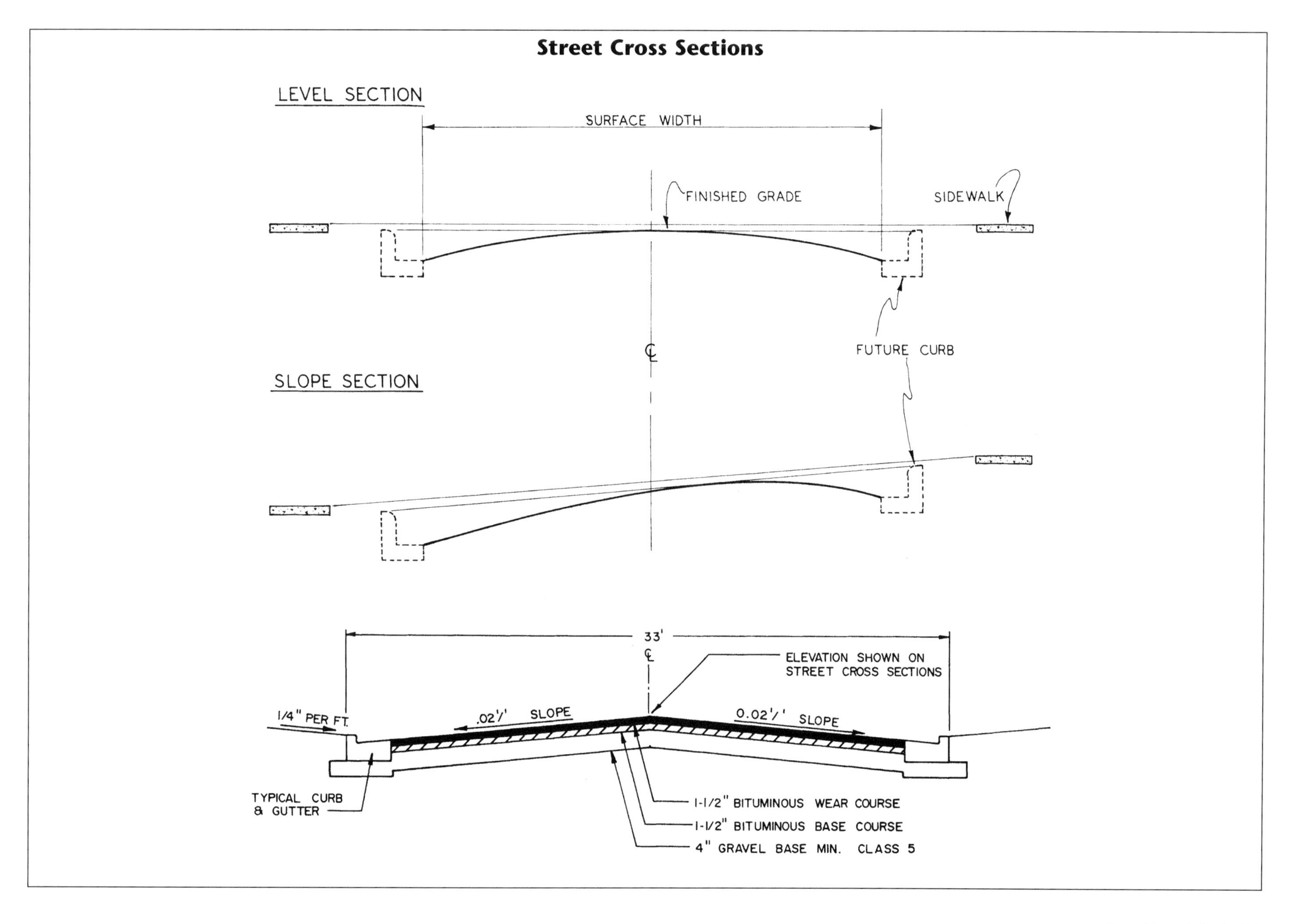
Street Cross Sections
LEVEL SECTION
SURFACE WIDTH
FINISHED GRADE
SIDEWALK
FUTURE CURB
SLOPE SECTION
33'
ELEVATION SHOWN ON
STREET CROSS SECTIONS
1/4" PER FT.
.02'/' SLOPE
0.02'/' SLOPE
TYPICAL CURB
& GUTTER
1-1/2" BITUMINOUS WEAR COURSE
1-1/2" BITUMINOUS BASE COURSE
4" GRAVEL BASE MIN. CLASS 5

Source: Morgan & Parmley, Ltd.

Typical Concrete Curb & Gutter

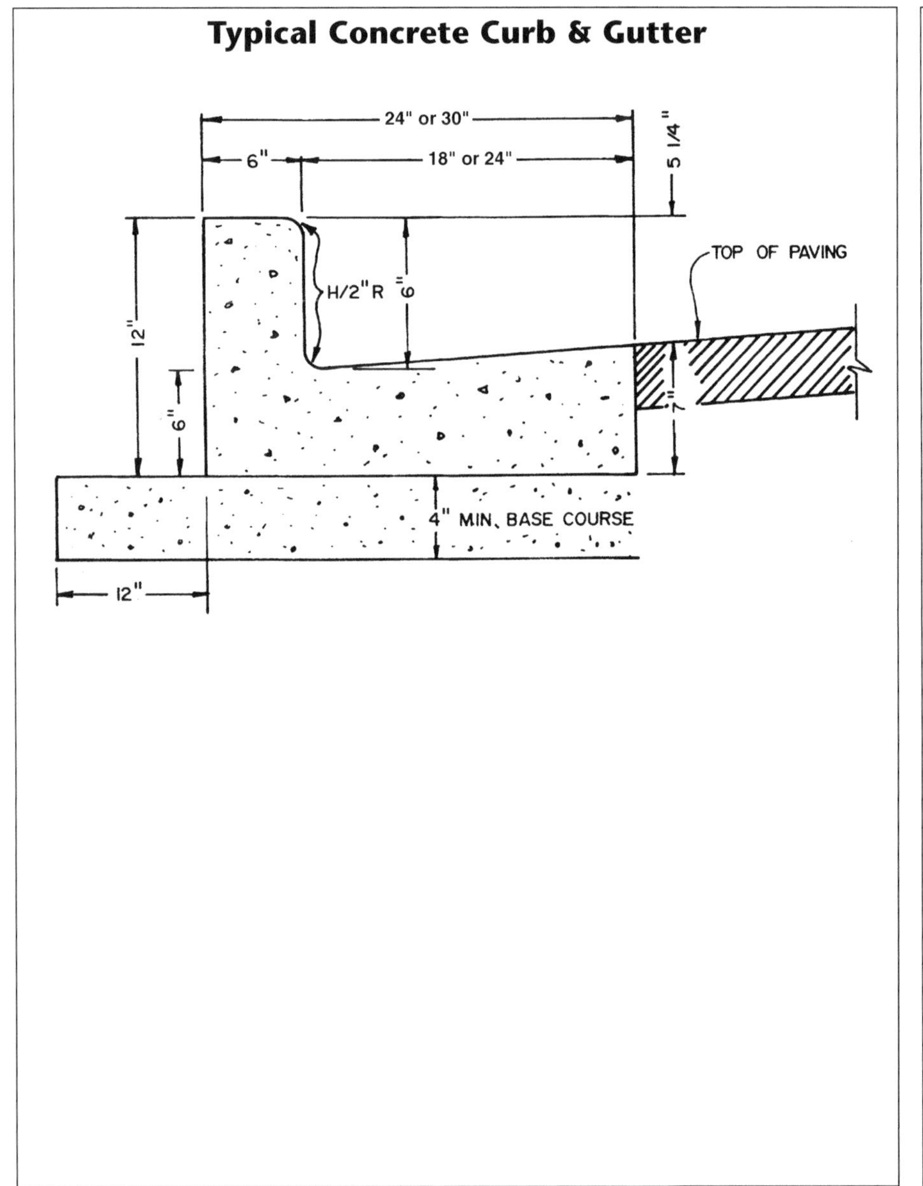

Source: Morgan & Parmley, Ltd.

Standard Catch Basin Detail

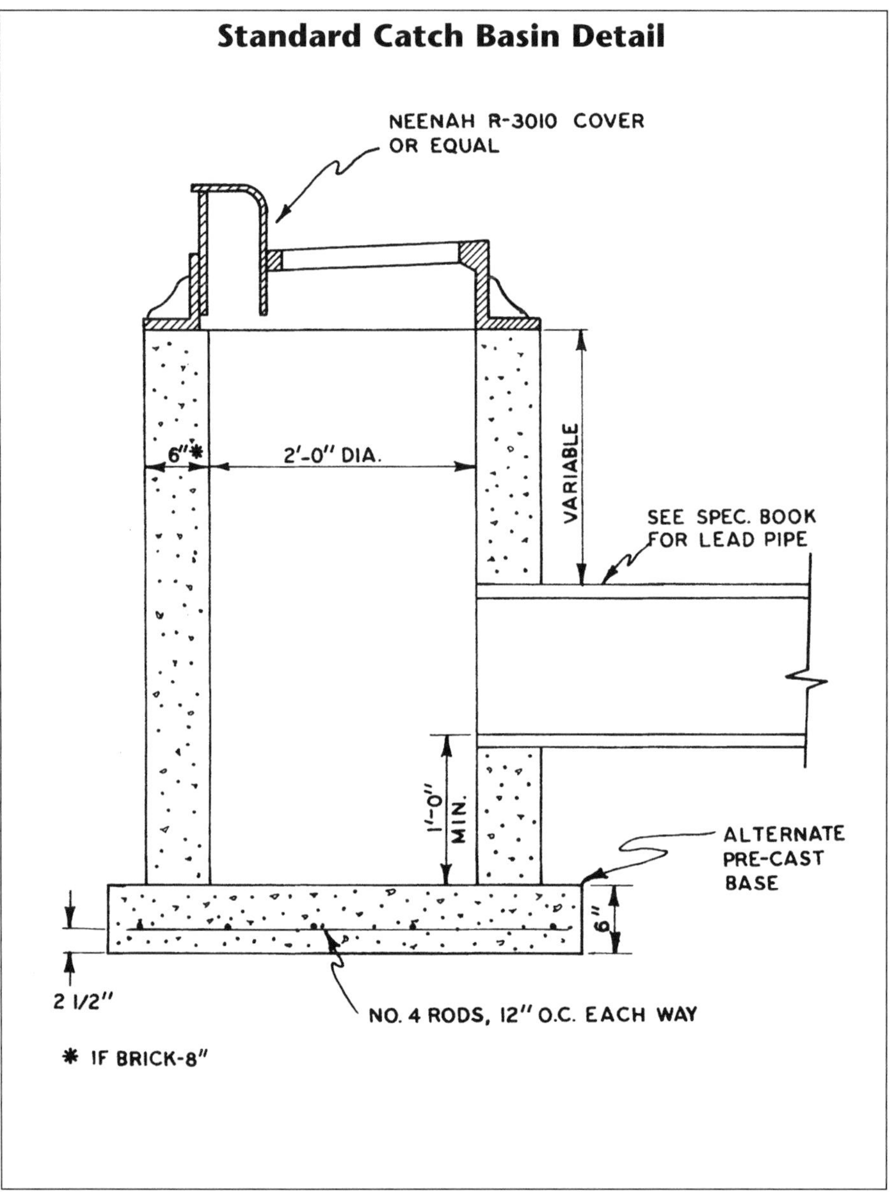

Source: Morgan & Parmley, Ltd.

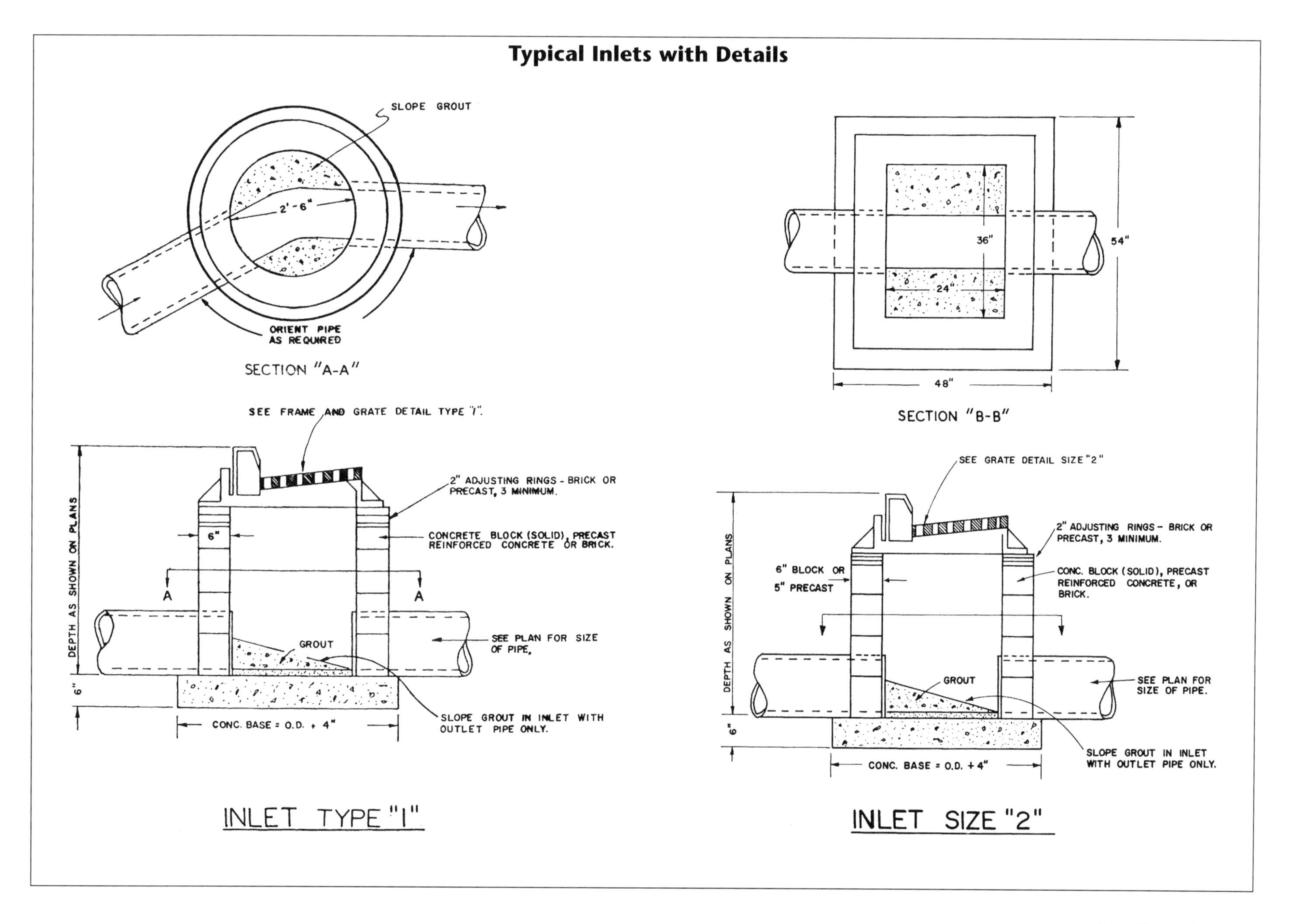
Typical Inlets with Details
SLOPE GROUT
2'-6"
ORIENT PIPE AS REQUIRED
SECTION "A-A"
36"
54"
24"
48"
SECTION "B-B"
SEE FRAME AND GRATE DETAIL TYPE "1".
2" ADJUSTING RINGS - BRICK OR PRECAST, 3 MINIMUM.
CONCRETE BLOCK (SOLID), PRECAST REINFORCED CONCRETE OR BRICK.
6"
A
A
DEPTH AS SHOWN ON PLANS
SEE PLAN FOR SIZE OF PIPE.
GROUT
SLOPE GROUT IN INLET WITH OUTLET PIPE ONLY.
CONC. BASE = O.D. + 4"
INLET TYPE "1"
SEE GRATE DETAIL SIZE "2"
2" ADJUSTING RINGS - BRICK OR PRECAST, 3 MINIMUM.
6" BLOCK OR
5" PRECAST
CONC. BLOCK (SOLID), PRECAST REINFORCED CONCRETE, OR BRICK.
SEE PLAN FOR SIZE OF PIPE.
GROUT
SLOPE GROUT IN INLET WITH OUTLET PIPE ONLY.
CONC. BASE = O.D. + 4"
INLET SIZE "2"

Source: Morgan & Parmley, Ltd.

Typical Inlets with Details *(Continued)*

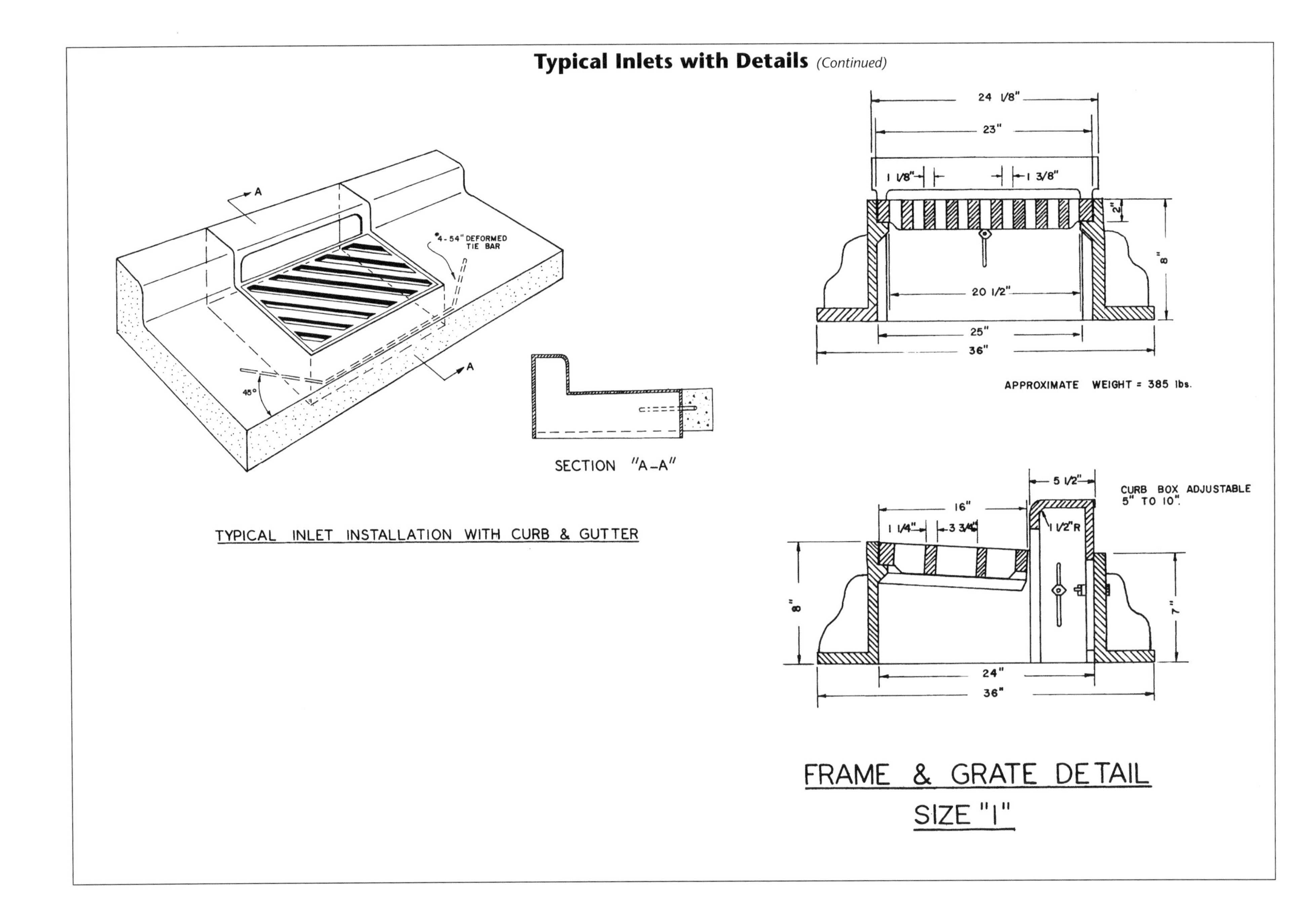

Typical Inlets with Details *(Continued)*

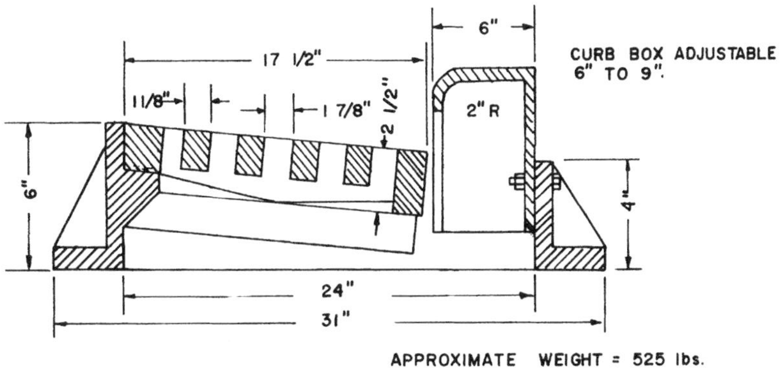

6"
17 1/2"
CURB BOX ADJUSTABLE
6" TO 9".
11/8"
1 7/8"
2 1/2"
2" R
6"
4"
24"
31"
APPROXIMATE WEIGHT = 525 lbs.

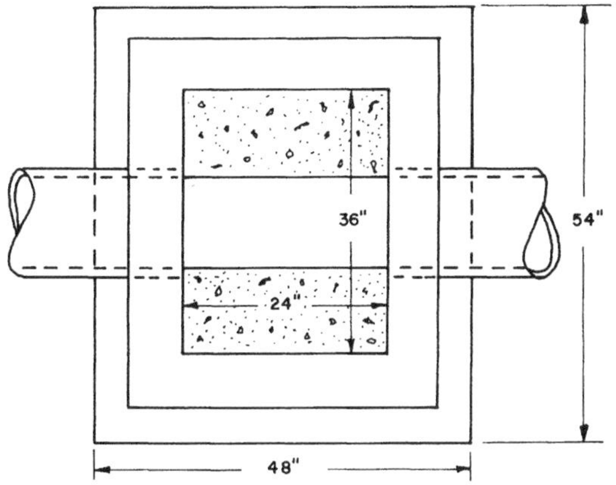

36"
54"
24"
48"
SECTION "B-B"

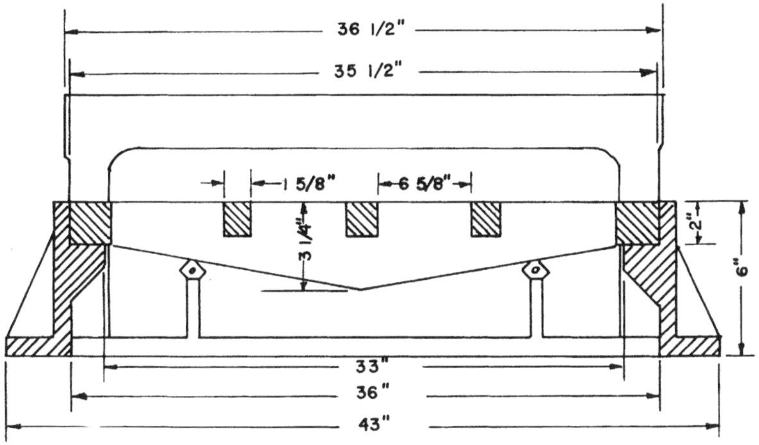

36 1/2"
35 1/2"
1 5/8"
6 5/8"
3 1/4"
2"
6"
33"
36"
43"

FRAME & GRATE DETAIL
SIZE "2"

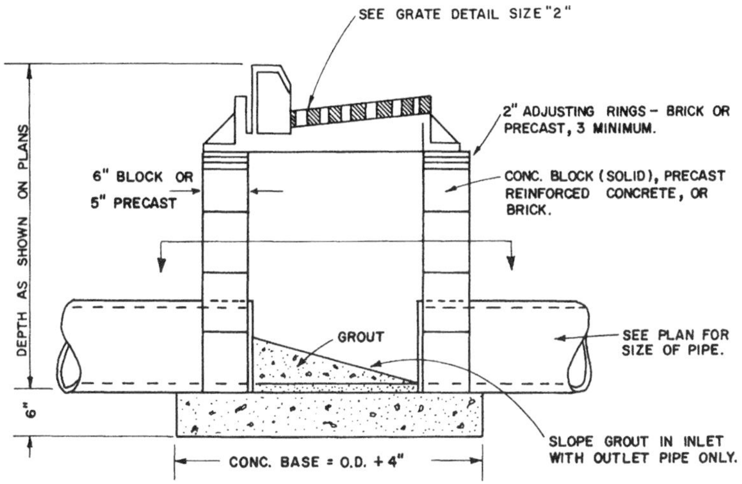

SEE GRATE DETAIL SIZE "2"
2" ADJUSTING RINGS – BRICK OR PRECAST, 3 MINIMUM.
6" BLOCK OR
5" PRECAST
CONC. BLOCK (SOLID), PRECAST REINFORCED CONCRETE, OR BRICK.
DEPTH AS SHOWN ON PLANS
GROUT
SEE PLAN FOR SIZE OF PIPE.
6"
CONC. BASE = O.D. + 4"
SLOPE GROUT IN INLET WITH OUTLET PIPE ONLY.

INLET SIZE "2"

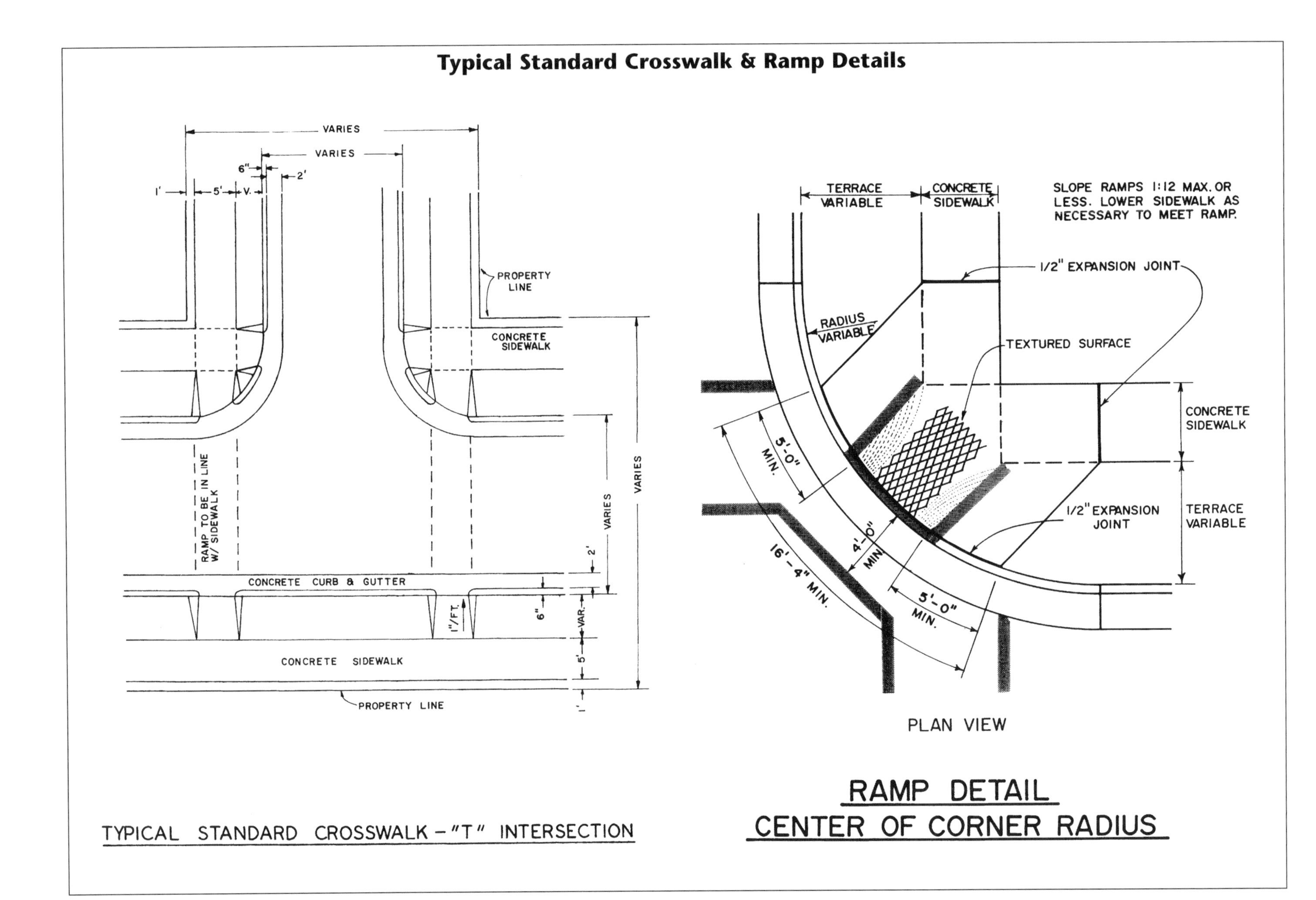

Source: Morgan & Parmley, Ltd.

Typical Driveway Approach

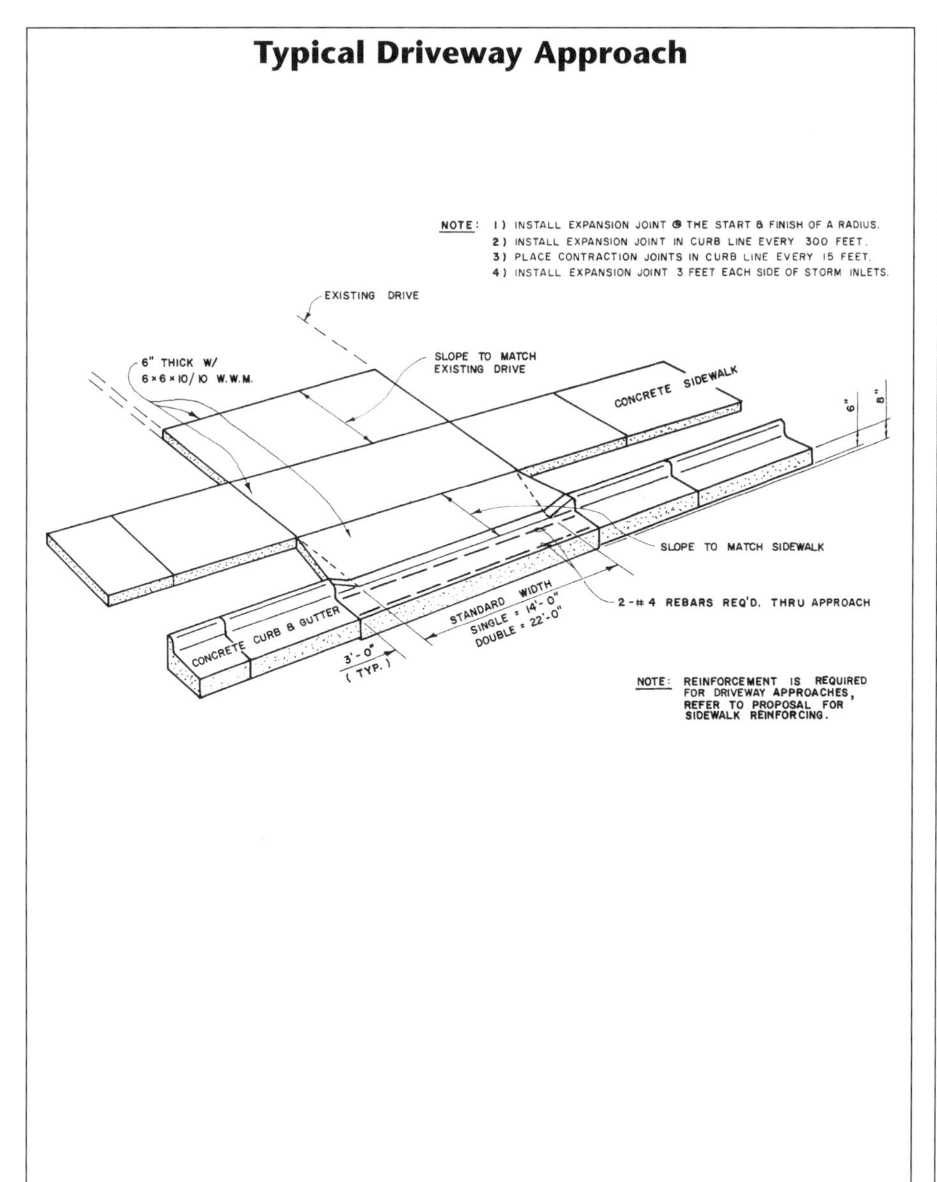

Source: Morgan & Parmley, Ltd.

Typical Sidewalk Opening for Trees

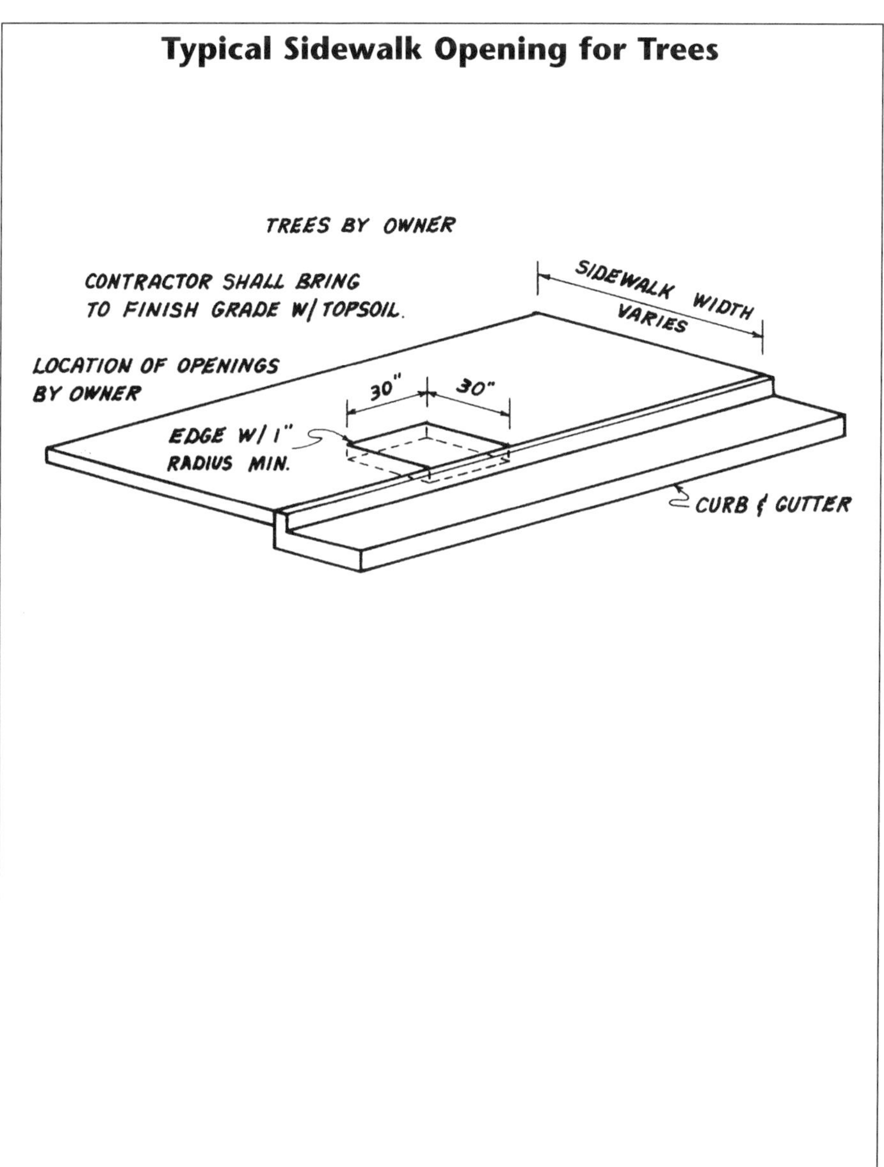

Source: Morgan & Parmley, Ltd.

Sidewalk Details

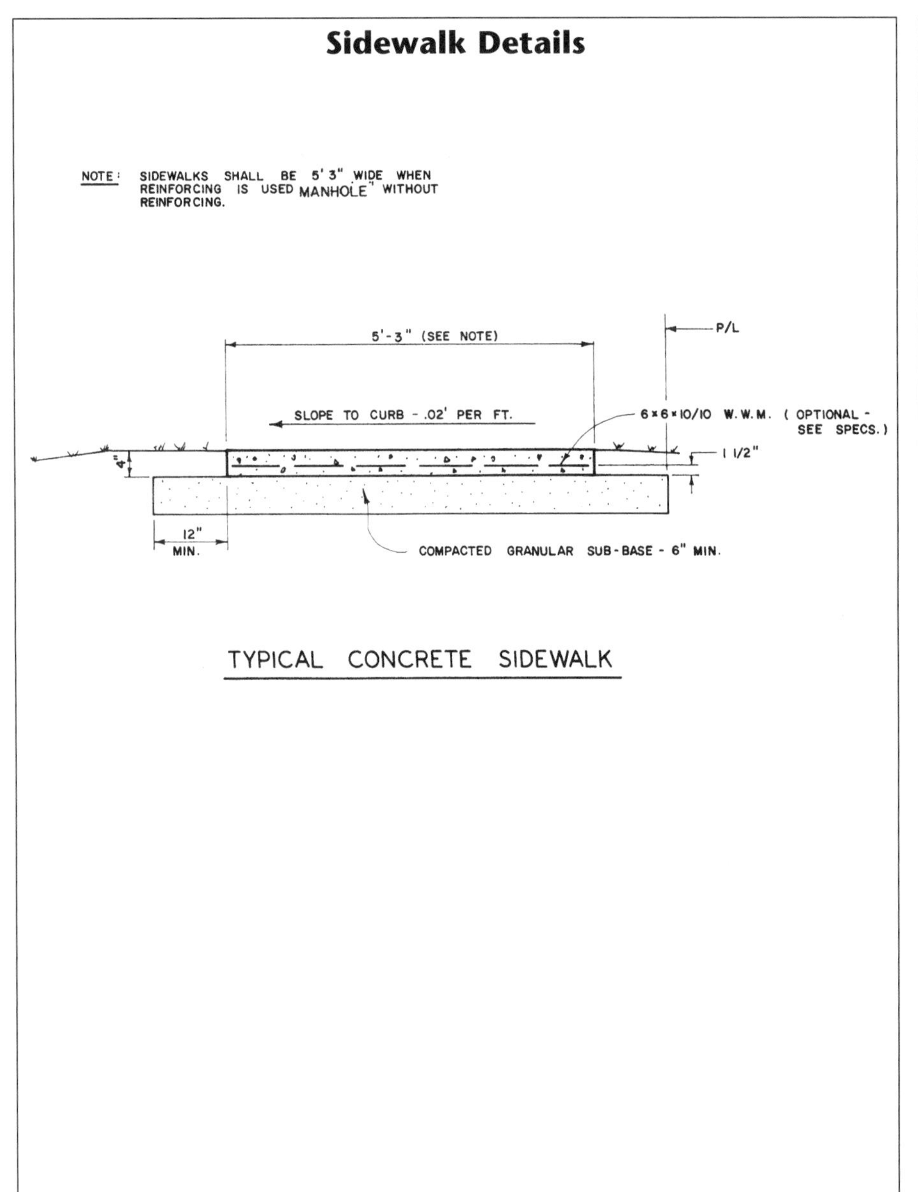

Sidewalk Details *(Continued)*

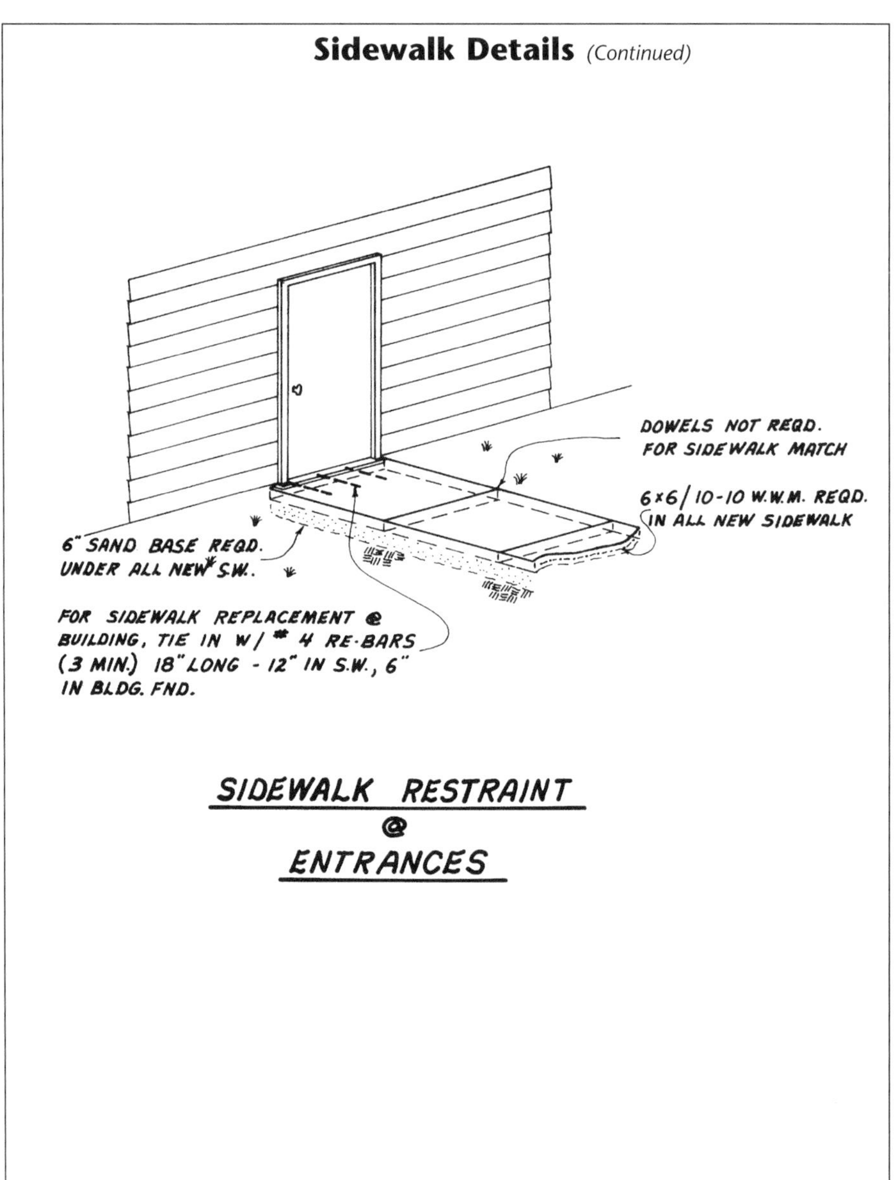

Source: Morgan & Parmley, Ltd.

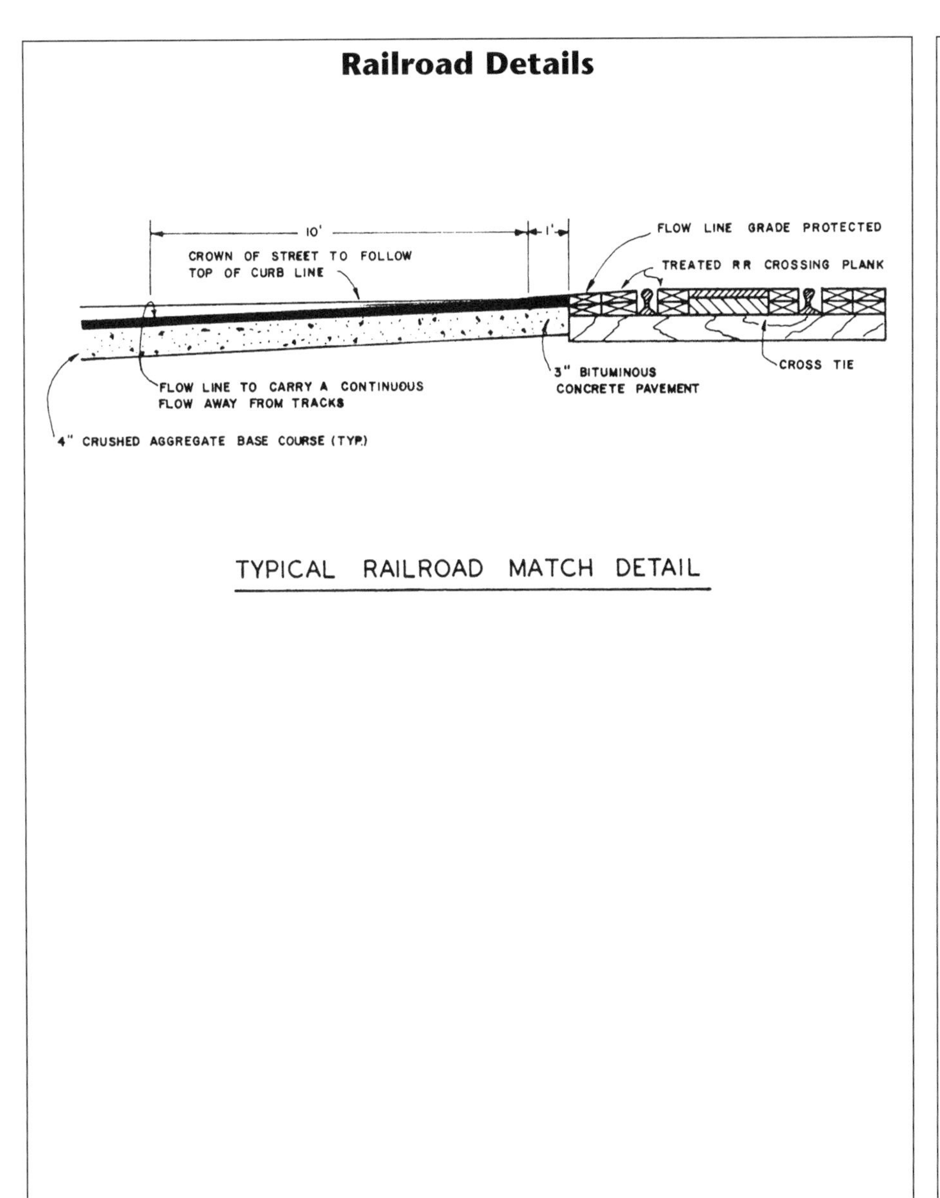

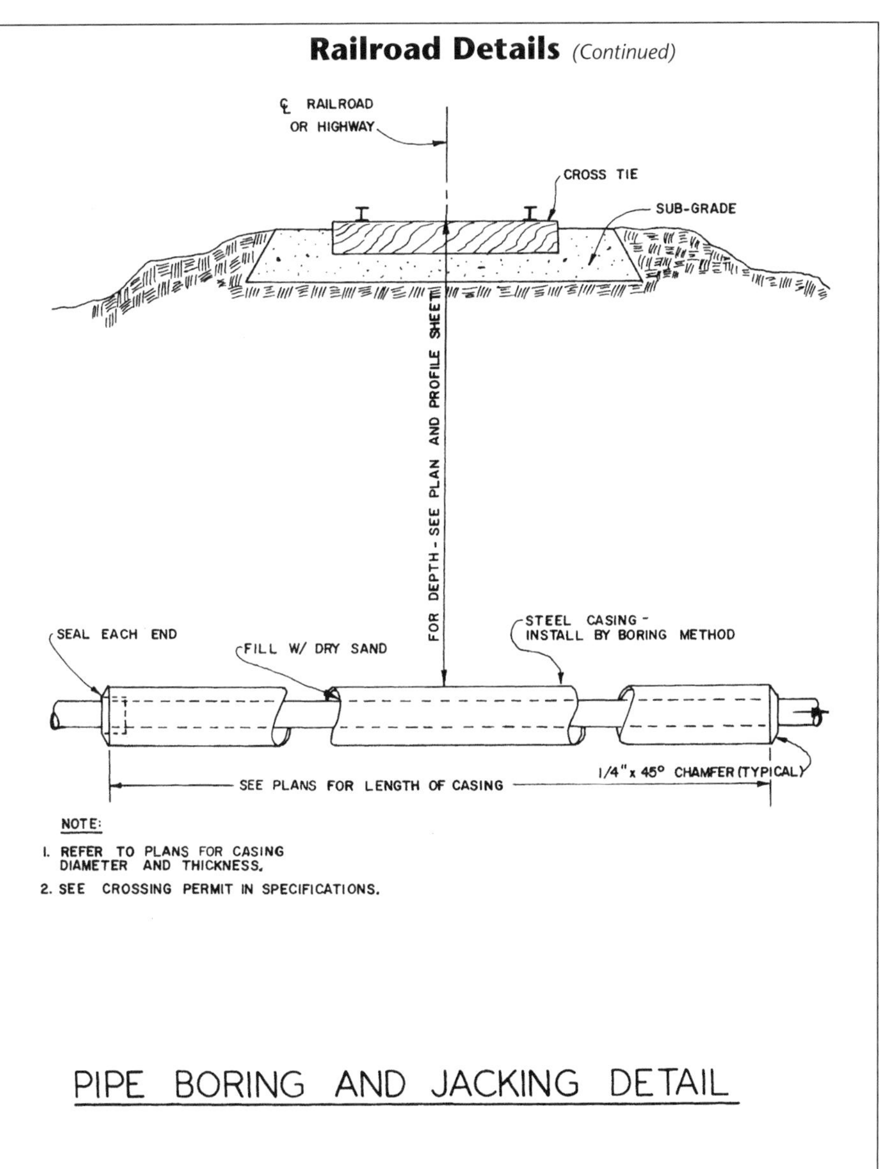

Source: Morgan & Parmley, Ltd.

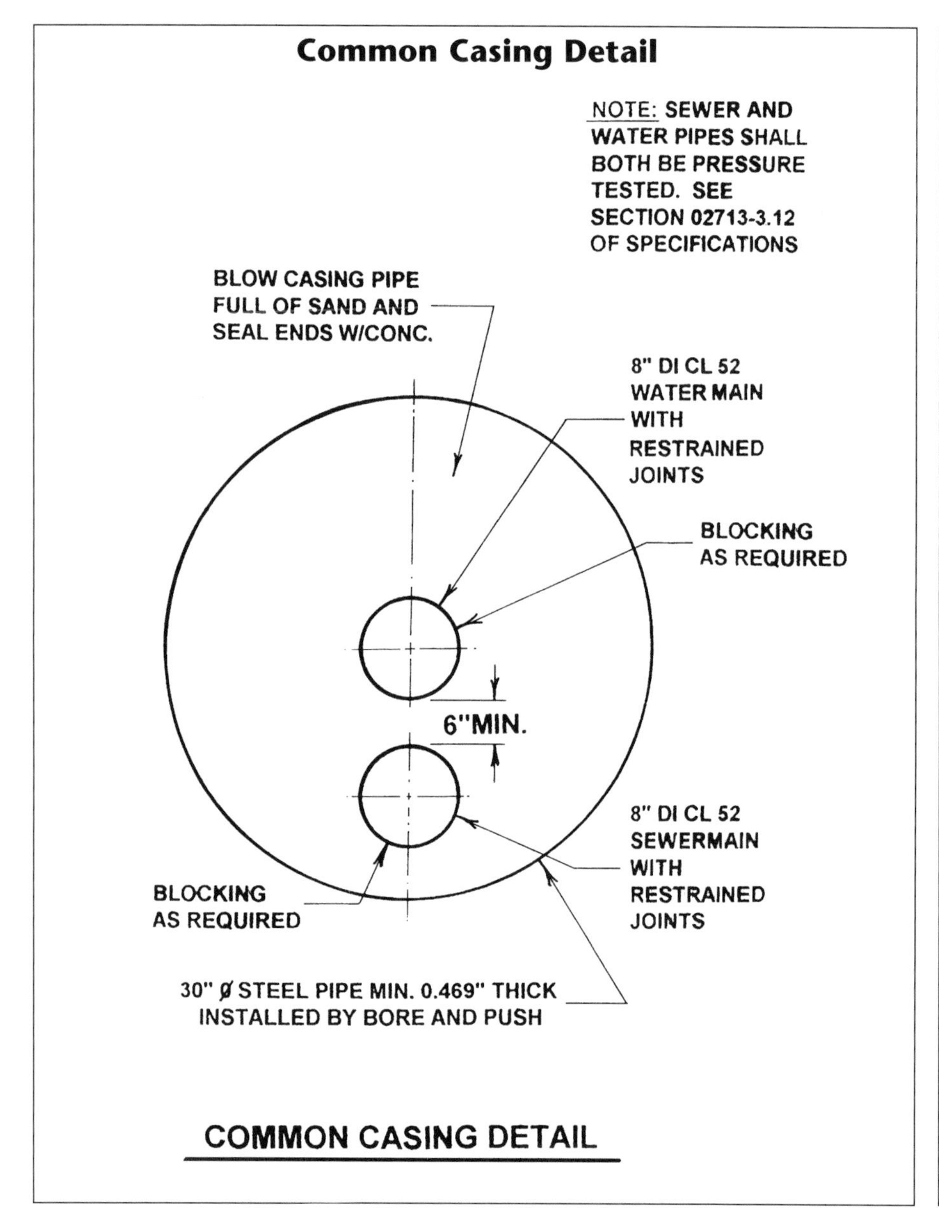
Common Casing Detail
NOTE: SEWER AND WATER PIPES SHALL BOTH BE PRESSURE TESTED. SEE SECTION 02713-3.12 OF SPECIFICATIONS
BLOW CASING PIPE FULL OF SAND AND SEAL ENDS W/CONC.
8" DI CL 52 WATER MAIN WITH RESTRAINED JOINTS
BLOCKING AS REQUIRED
6"MIN.
8" DI CL 52 SEWERMAIN WITH RESTRAINED JOINTS
BLOCKING AS REQUIRED
30" Ø STEEL PIPE MIN. 0.469" THICK INSTALLED BY BORE AND PUSH
COMMON CASING DETAIL

Source: Morgan & Parmley, Ltd.

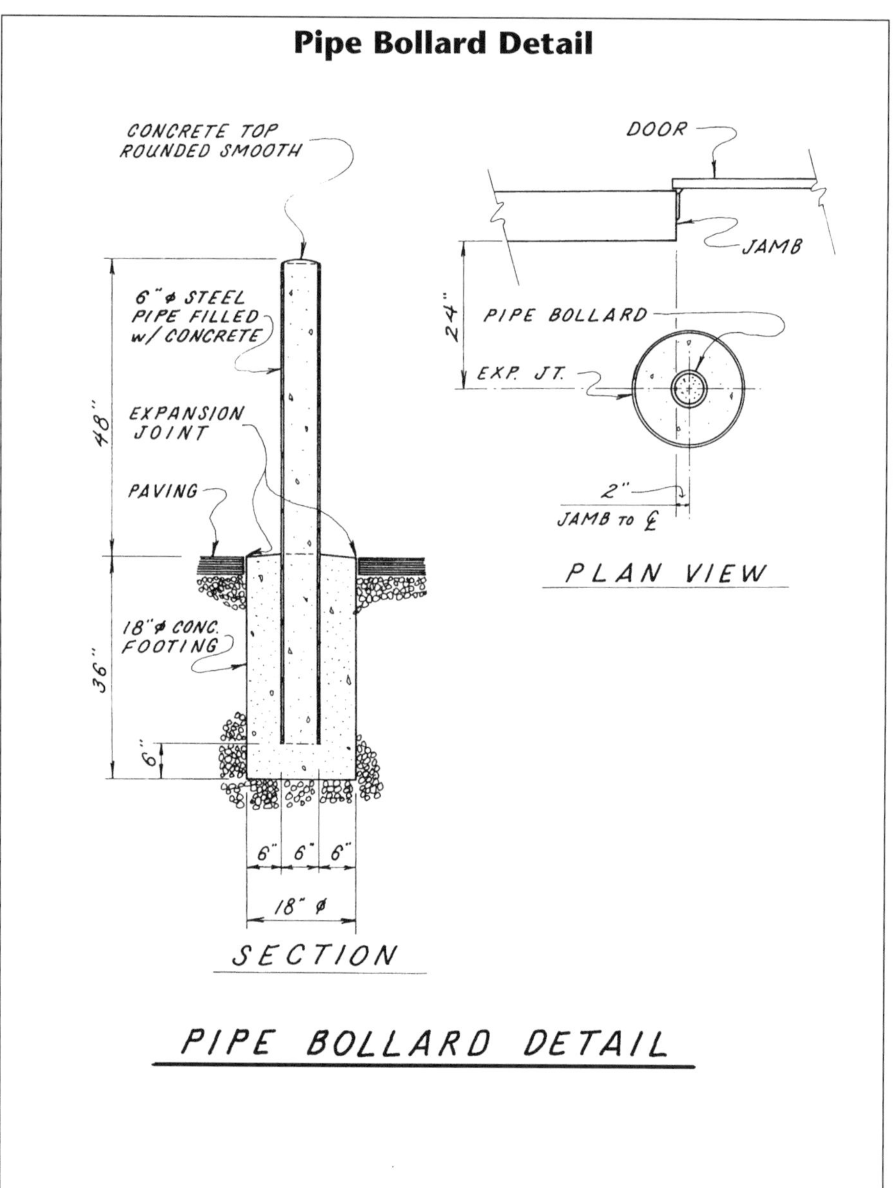
Pipe Bollard Detail
CONCRETE TOP ROUNDED SMOOTH
6" ø STEEL PIPE FILLED w/ CONCRETE
EXPANSION JOINT
PAVING
18" ø CONC. FOOTING
48"
36"
6"
6" 6" 6"
18" ø
SECTION
DOOR
JAMB
24"
PIPE BOLLARD
EXP. JT.
2"
JAMB TO ℄
PLAN VIEW
PIPE BOLLARD DETAIL

Source: Morgan & Parmley, Ltd.

Handicapped Details

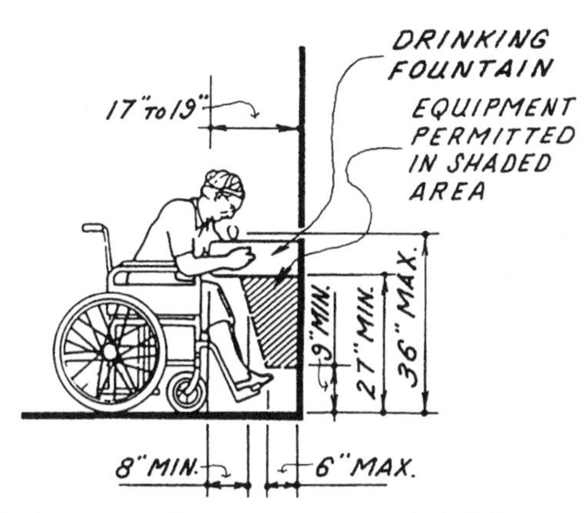

SPOUT HEIGHT & KNEE CLEARANCE

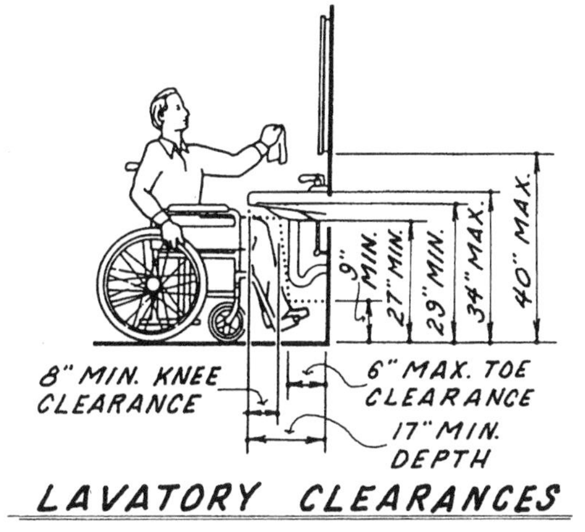

LAVATORY CLEARANCES

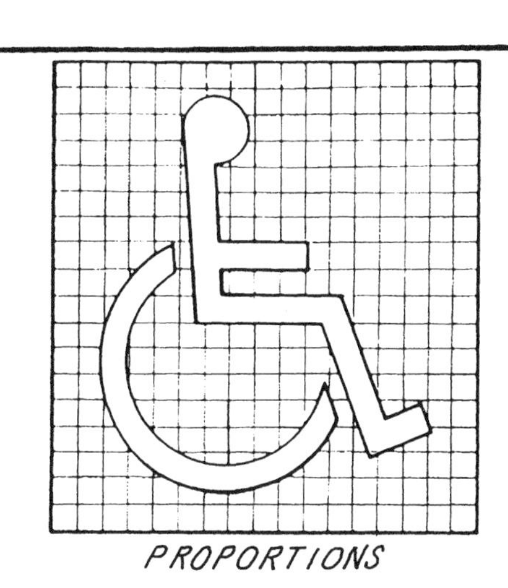

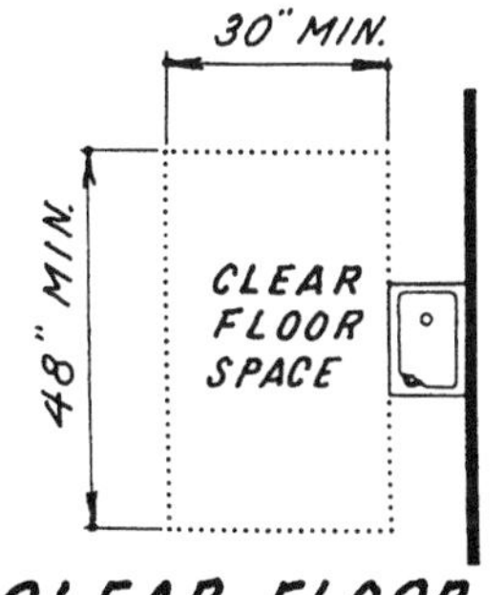

CLEAR FLOOR SPACE @ FREE-STANDING DRINKING FOUNTAIN

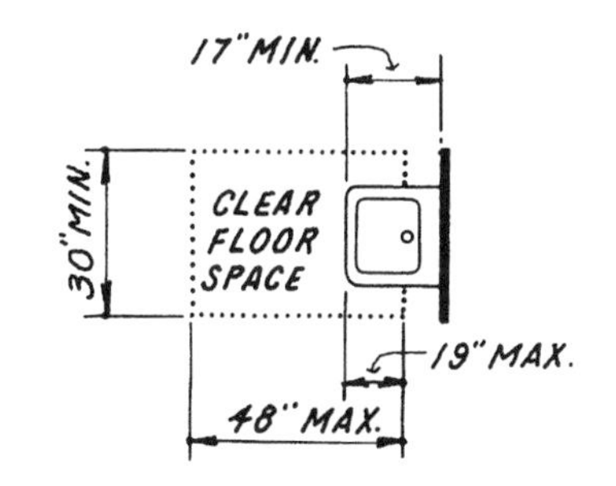

CLEAR FLOOR SPACE @ LAVATORIES

INTERNATIONAL SYMBOL OF ACCESSIBILITY

SIGNAGE DETAIL

Source: Morgan & Parmley, Ltd.

Plumbing Layouts

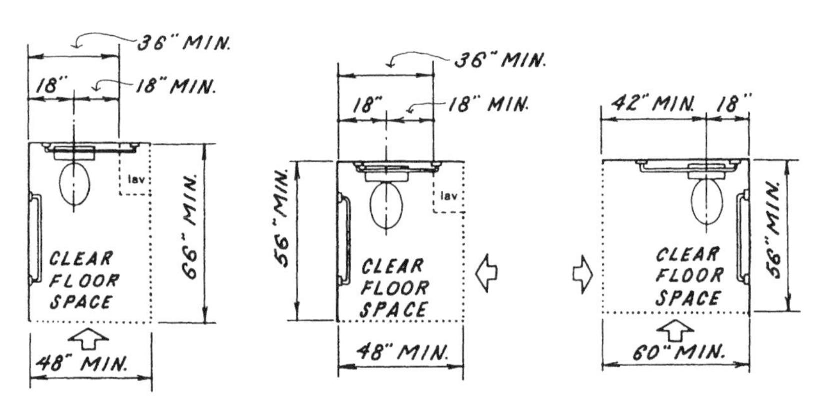

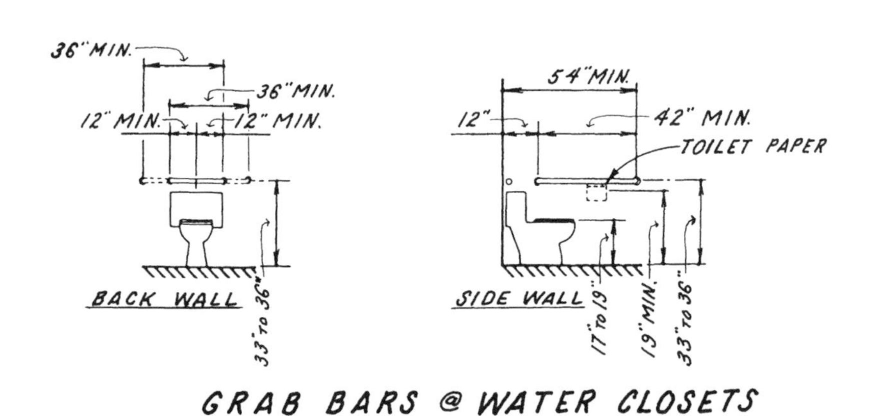

SHOWER SEAT DESIGN

16" MAX.
1½" MAX.
1½" MAX.
23" MAX.
15" MAX.
FULL DEPTH OF STALL

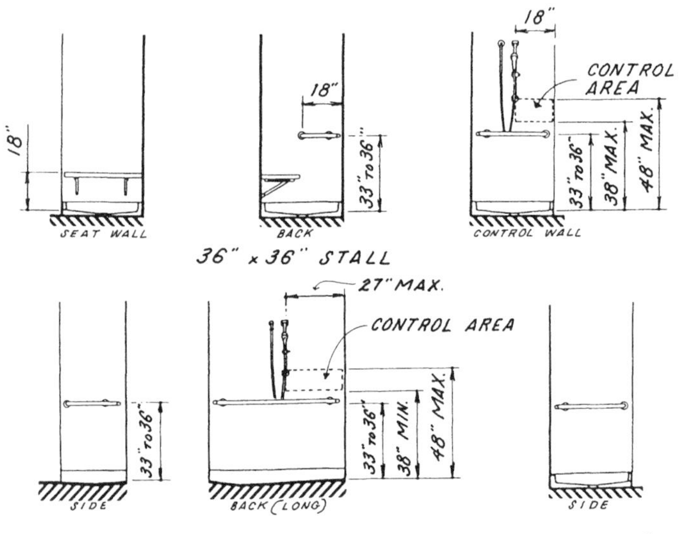

GRAB BARS SHOWER STALL

Source: Morgan & Parmley, Ltd.

Design

SPECIFICATIONS

Section 22

Specifications are the written description of the project design that clarifies specific portions of the plans that can not be graphically illustrated in the drawings.

This section commences with a brief description of the MasterFormat™ as developed by the Construction Specifications Institute (CSI). A table of contents from an actual project follows with selected specifications reproduced to give the reader a practical example of the technical writing style.

The selected sample specifications are sections from a complex project that includes retrofitting an existing potable water supply facility.

TABLE OF CONTENTS

Numbering Format

The Numbers and Titles used in this sub-section are from *MasterFormat*™ (1995 Edition) and is published by the Construction Specifications Institute (CSI), and Canada Construction Specifications (CSC), 2002. Reproduction of this material is by permission from CSI.

For those interested in a more in-depth explanation of MasterFormat™ and its use in the construction industry contact:

The Construction Specifications Institute (CSI)
99 Canal Center Plaza, Suite 300
Alexandria, VA 223-14
800-689-2900; 703-684-0300
CSINet URL: http://www.csinet.org

Over the past 30 years, the system of numbers and titles at the heart of MasterFormat has been used increasingly by the construction industry. Today, MasterFormat is the only system of organizing construction specifications in widespread use in the United States and Canada. And its system of organizing construction information into "Bidding Requirements," "Contracting Requirements," and 16 Divisions of products and activities is commonplace.

MasterFormat is adopted for use by the United States Department of Defense and is also used at state and municipal levels of government.

CSC promotes standardization of construction documents in Canada. Its efforts have been supported by the Royal Architectural Institute of Canada, the Association of Consulting Engineers of Canada, and the Canadian Construction Association. MasterFormat is also used by provincial and municipal governments to organize master specifications, data files, and cost analysis systems. MasterFormat is the basis of the titling and numbering system for the Canadian Federal Government's National Master Specification (NMS), which is used by all Canadian federal government agencies and a large portion of the private sector.

One of the most important things to remember when organizing a project manual is that specification sections do not have a hierarchical relationship to each other. Even though MasterFormat establishes a hierarchy for it's numbers and titles, specification sections are complementary with each other and with other contract documents. One specification section cannot "govern" another except to the extent that, as complementary documents, all specification sections potentially affect all other specification sections.

Another, important consideration is that "Introductory Information," "Bidding Requirements," and "Contracting Requirements" are not specifications and are not usually the responsibility of the specifier or design professional. Standard legal practice leans toward identifying these documents by name only, and not by number. MasterFormat numbers can be used to identify these documents in the project manual, but it is important to avoid referring to these documents as "sections" or otherwise confusing them with specifications. It is also not necessary to renumber preprinted documents if they are already identified. This includes the standard general conditions published by various professional societies.

Courtesy: CSI

Level Two Numbers and Titles

NOTE: See Master List of /Numbers 1995 Manual for complete system

INTRODUCTORY INFORMATION

00001 PROJECT TITLE PAGE
00005 CERTIFICATIONS PAGE
00007 SEALS PAGE
00010 TABLE OF CONTENTS
00015 LIST OF DRAWINGS
00020 LIST OF SCHEDULES

BIDDING REQUIREMENTS

00100 BID SOLICITATION
00200 INSTRUCTIONS TO BIDDERS
00300 INFORMATION AVAILABLE TO BIDDERS
00400 BID FORMS AND SUPPLEMENTS
00490 BIDDING ADDENDA

CONTRACTING REQUIREMENTS

00500 AGREEMENT
00600 BONDS AND CERTIFICATES
00700 GENERAL CONDITIONS
00800 SUPPLEMENTARY CONDITIONS
00900 ADDENDA AND MODIFICATIONS

FACILITIES AND SPACES

FACILITIES AND SPACES

SYSTEMS AND ASSEMBLIES

SYSTEMS AND ASSEMBLIES

CONSTRUCTION PRODUCTS AND ACTIVITIES

DIVISION 1 GENERAL REQUIREMENTS

01100 SUMMARY
01200 PRICE AND PAYMENT PROCEDURES
01300 ADMINISTRATIVE REQUIREMENTS
01400 QUALITY REQUIREMENTS
01500 TEMPORARY FACILITIES AND CONTROLS
01600 PRODUCT REQUIREMENTS
01700 EXECUTION REQUIREMENTS
01800 FACILITY OPERATION
01900 FACILITY DECOMMISSIONING

DIVISION 2 SITE CONSTRUCTION

02050 BASIC SITE MATERIALS AND METHODS
02100 SITE REMEDIATION
02200 SITE PREPARATION
02300 EARTHWORK
02400 TUNNELING, BORING, AND JACKING
02450 FOUNDATION AND LOAD-BEARING ELEMENTS
02500 UTILITY SERVICES
02600 DRAINAGE AND CONTAINMENT
02700 BASES, BALLASTS, PAVEMENTS, AND APPURTENANCES
02800 SITE IMPROVEMENTS AND AMENITIES
02900 PLANTING
02950 SITE RESTORATION AND REHABILITATION

DIVISION 3 CONCRETE

03050 BASIC CONCRETE MATERIALS AND METHODS
03100 CONCRETE FORMS AND ACCESSORIES
03200 CONCRETE REINFORCEMENT
03300 CAST-IN-PLACE CONCRETE
03400 PRECAST CONCRETE
03500 CEMENTITIOUS DECKS AND UNDERLAYMENT
03600 GROUTS
03700 MASS CONCRETE
03900 CONCRETE RESTORATION AND CLEANING

DIVISION 4 MASONRY

04050 BASIC MASONRY MATERIALS AND METHODS
04200 MASONRY UNITS
04400 STONE
04500 REFRACTORIES
04600 CORROSION-RESISTANT MASONRY
04700 SIMULATED MASONRY
04800 MASONRY ASSEMBLIES
04900 MASONRY RESTORATION AND CLEANING

DIVISION 5 METALS

05050 BASIC METAL MATERIALS AND METHODS
05100 STRUCTURAL METAL FRAMING
05200 METAL JOISTS
05300 METAL DECK
05400 COLD-FORMED METAL FRAMING
05500 METAL FABRICATIONS
05600 HYDRAULIC FABRICATIONS
05650 RAILROAD TRACK AND ACCESSORIES
05700 ORNAMENTAL METAL
05800 EXPANSION CONTROL
05900 METAL RESTORATION AND CLEANING

DIVISION 6 WOOD AND PLASTICS

06050 BASIC WOOD AND PLASTIC MATERIALS AND METHODS
06100 ROUGH CARPENTRY
06200 FINISH CARPENTRY
06400 ARCHITECTURAL WOODWORK
06500 STRUCTURAL PLASTICS
06600 PLASTIC FABRICATIONS
06900 WOOD AND PLASTIC RESTORATION AND CLEANING

Courtesy: CSI

DIVISION 7 THERMAL AND MOISTURE PROTECTION

07050 BASIC THERMAL AND MOISTURE PROTECTION MATERIALS AND METHODS
07100 DAMPPROOFING AND WATERPROOFING
07200 THERMAL PROTECTION
07300 SHINGLES, ROOF TILES, AND ROOF COVERINGS
07400 ROOFING AND SIDING PANELS
07500 MEMBRANE ROOFING
07600 FLASHING AND SHEET METAL
07700 ROOF SPECIALTIES AND ACCESSORIES
07800 FIRE AND SMOKE PROTECTION
07900 JOINT SEALERS

DIVISION 8 DOORS AND WINDOWS

08050 BASIC DOOR AND WINDOW MATERIALS AND METHODS
08100 METAL DOORS AND FRAMES
08200 WOOD AND PLASTIC DOORS
08300 SPECIALTY DOORS
08400 ENTRANCES AND STOREFRONTS
08500 WINDOWS
08600 SKYLIGHTS
08700 HARDWARE
08800 GLAZING
08900 GLAZED CURTAIN WALL

DIVISION 9 FINISHES

09050 BASIC FINISH MATERIALS AND METHODS
09100 METAL SUPPORT ASSEMBLIES
09200 PLASTER AND GYPSUM BOARD
09300 TILE
09400 TERRAZZO
09500 CEILINGS
09600 FLOORING
09700 WALL FINISHES
09800 ACOUSTICAL TREATMENT
09900 PAINTS AND COATINGS

DIVISION 10 SPECIALTIES

10100 VISUAL DISPLAY BOARDS
10150 COMPARTMENTS AND CUBICLES
10200 LOUVERS AND VENTS
10240 GRILLES AND SCREENS
10250 SERVICE WALLS
10260 WALL AND CORNER GUARDS
10270 ACCESS FLOORING
10290 PEST CONTROL
10300 FIREPLACES AND STOVES
10340 MANUFACTURED EXTERIOR SPECIALTIES
10350 FLAGPOLES
10400 IDENTIFICATION DEVICES
10450 PEDESTRIAN CONTROL DEVICES
10500 LOCKERS
10520 FIRE PROTECTION SPECIALTIES
10530 PROTECTIVE COVERS
10550 POSTAL SPECIALTIES
10600 PARTITIONS
10670 STORAGE SHELVING
10700 EXTERIOR PROTECTION
10750 TELEPHONE SPECIALTIES
10800 TOILET, BATH, AND LAUNDRY ACCESSORIES
10880 SCALES
10900 WARDROBE AND CLOSET SPECIALTIES

DIVISION 11 EQUIPMENT

11010 MAINTENANCE EQUIPMENT
11020 SECURITY AND VAULT EQUIPMENT
11030 TELLER AND SERVICE EQUIPMENT
11040 ECCLESIASTICAL EQUIPMENT
11050 LIBRARY EQUIPMENT
11060 THEATER AND STAGE EQUIPMENT
11070 INSTRUMENTAL EQUIPMENT
11080 REGISTRATION EQUIPMENT
11090 CHECKROOM EQUIPMENT
11100 MERCANTILE EQUIPMENT
11110 COMMERCIAL LAUNDRY AND DRY CLEANING EQUIPMENT
11120 VENDING EQUIPMENT
11130 AUDIO-VISUAL EQUIPMENT
11140 VEHICLE SERVICE EQUIPMENT
11150 PARKING CONTROL EQUIPMENT
11160 LOADING DOCK EQUIPMENT
11170 SOLID WASTE HANDLING EQUIPMENT
11190 DETENTION EQUIPMENT
11200 WATER SUPPLY AND TREATMENT EQUIPMENT
11280 HYDRAULIC GATES AND VALVES
11300 FLUID WASTE TREATMENT AND DISPOSAL EQUIPMENT
11400 FOOD SERVICE EQUIPMENT
11450 RESIDENTIAL EQUIPMENT
11460 UNIT KITCHENS
11470 DARKROOM EQUIPMENT
11480 ATHLETIC, RECREATIONAL, AND THERAPEUTIC EQUIPMENT
11500 INDUSTRIAL AND PROCESS EQUIPMENT
11600 LABORATORY EQUIPMENT
11650 PLANETARIUM EQUIPMENT
11660 OBSERVATORY EQUIPMENT
11680 OFFICE EQUIPMENT
11700 MEDICAL EQUIPMENT
11780 MORTUARY EQUIPMENT
11850 NAVIGATION EQUIPMENT
11870 AGRICULTURAL EQUIPMENT
11900 EXHIBIT EQUIPMENT

DIVISION 12 FURNISHINGS

12050 FABRICS
12100 ART
12300 MANUFACTURED CASEWORK
12400 FURNISHINGS AND ACCESSORIES
12500 FURNITURE
12600 MULTIPLE SEATING
12700 SYSTEMS FURNITURE
12800 INTERIOR PLANTS AND PLANTERS
12900 FURNISHINGS RESTORATION AND REPAIR

DIVISION 13 SPECIAL CONSTRUCTION

13010 AIR-SUPPORTED STRUCTURES
13020 BUILDING MODULES
13030 SPECIAL PURPOSE ROOMS
13080 SOUND, VIBRATION, AND SEISMIC CONTROL
13090 RADIATION PROTECTION
13100 LIGHTNING PROTECTION
13110 CATHODIC PROTECTION
13120 PRE-ENGINEERED STRUCTURES
13150 SWIMMING POOLS
13160 AQUARIUMS
13165 AQUATIC PARK FACILITIES
13170 TUBS AND POOLS
13175 ICE RINKS
13185 KENNELS AND ANIMAL SHELTERS
13190 SITE-CONSTRUCTED INCINERATORS
13200 STORAGE TANKS
13220 FILTER UNDERDRAINS AND MEDIA
13230 DIGESTER COVERS AND APPURTENANCES
13240 OXYGENATION SYSTEMS
13260 SLUDGE CONDITIONING SYSTEMS
13280 HAZARDOUS MATERIAL REMEDIATION
13400 MEASUREMENT AND CONTROL INSTRUMENTATION
13500 RECORDING INSTRUMENTATION
13550 TRANSPORTATION CONTROL INSTRUMENTATION
13600 SOLAR AND WIND ENERGY EQUIPMENT
13700 SECURITY ACCESS AND SURVEILLANCE
13800 BUILDING AUTOMATION AND CONTROL
13850 DETECTION AND ALARM
13900 FIRE SUPPRESSION

DIVISION 14 CONVEYING SYSTEMS

14100 DUMBWAITERS
14200 ELEVATORS
14300 ESCALATORS AND MOVING WALKS
14400 LIFTS
14500 MATERIAL HANDLING
14600 HOISTS AND CRANES
14700 TURNTABLES
14800 SCAFFOLDING
14900 TRANSPORTATION

DIVISION 15 MECHANICAL

15050 BASIC MECHANICAL MATERIALS AND METHODS
15100 BUILDING SERVICES PIPING
15200 PROCESS PIPING
15300 FIRE PROTECTION PIPING
15400 PLUMBING FIXTURES AND EQUIPMENT
15500 HEAT-GENERATION EQUIPMENT
15600 REFRIGERATION EQUIPMENT
15700 HEATING, VENTILATING, AND AIR CONDITIONING EQUIPMENT
15800 AIR DISTRIBUTION
15900 HVAC INSTRUMENTATION AND CONTROLS
15950 TESTING, ADJUSTING, AND BALANCING

DIVISION 16 ELECTRICAL

16050 BASIC ELECTRICAL MATERIALS AND METHODS
16100 WIRING METHODS
16200 ELECTRICAL POWER
16300 TRANSMISSION AND DISTRIBUTION
16400 LOW-VOLTAGE DISTRIBUTION
16500 LIGHTING
16700 COMMUNICATIONS
16800 SOUND AND VIDEO

SPECIFICATIONS
-BIDDING DOCUMENTS-
for
WATER SUPPLY, PUMPHOUSE, DISTRIBUTION SYSTEM & ELEVATED STORAGE RESERVOIR

Sample Table of Contents from Specifications Book

ii

* Represents specific specs that are reproduced.

Courtesy: Morgan & Parmley, Ltd.

NOTE: Specification Section Title

SUMMARY OF WORK
(W/SPECIAL PROVISIONS)

PART ONE - GENERAL

1.1 BACKGROUND

1.1.1 History of Project: The High School Complex in Tony, is owned and operated by the Flambeau School District. Due to historically poor potable water and continuous maintenance problems, the School Board retained Morgan & Parmley, Ltd. to solve their problem. On May 23, 1977, construction was completed for their new water supply well located on Rusk County Airport property. The pumphouse, pumping equipment and transmission main were placed in operation early in 1979. This facility has been in continuous use, serving the High School Complex, since that time with no significant operational problems.

On several occasions, the Village Board attempted to establish a municipal water system without success. However, their recent efforts have proven successful by obtaining funding through the Rural Economic and Community Development, USDA (formally, FmHA).

An agreement between the Flambeau School District and the Village of Tony was recently executed. The Village will purchase the existing school well, pumphouse, transmission main and related appurtenances. These components will be incorporated into their proposed (total) water system; as detailed in the accompanying Plans and Specifications.

The Village has established their public water utility under the rules and regulations of the Wisconsin Public Service Commission.

The Well Site Survey and Engineering Report have been approved by DNR.

1.2 SCOPE OF WORK

1.2.1 Description: The proposed Project consists of the following major items:

1-Rehab Existing Pumphouse
2-Vertical Turbine Pump w/Controls
3-Interior Piping & Controls
4-Right Angle Drive Mechanism
5-Chlorination Facility
6-Electrical Upgrade
7-Watermain Distribution System
8-Elevated Water Tank & Tower
9-Services & Meters

NOTE: Page Number

NOTE: Project Number

NOTE: Section Number

93-159 01010-1

Courtesy: Morgan & Parmley, Ltd.

The foregoing list is only major items. Refer to the accompanying Plans, applicable sections of the Specifications and the Proposal for a total concept.

1.2.2 Proposals: There will be two (2) Proposals. The first will be for all items, except the elevated water tank, tower and related components which will be bid separately.

1.3 SCHEDULE

1.3.1 Anticipated Construction Schedule:

Commence Advertising 11-13-1995
Bid Opening 12-14-1995
Notice of Award 12-20-1995
Pre-Construction Conference 1-15-1996
Notice to Proceed 1-15-1996
Complete Construction 11-10-1996

1.3.2 Highway Construction: All construction on U.S. Highway 8 shall be substantially completed by June 30, 1996.

1.4 HYDRANT PATTERN

1.4.1 Description: Since the Village of Tony is served by the Ladysmith Fire Department, it is mandatory that Tony's fire hydrant pattern be compatable with their equipment.

PART TWO - SPECIAL PROVISIONS

2.1 WAGE RATES

2.1.1 Wages: The Contractors and Sub-Contractors shall meet the requirements of all applicable official State Wage Rate Determinations for this Project. A copy is attached to these Specifications.

2.2 EQUAL EMPLOYMENT OPPORTUNITY

2.2.1 Compliance: The Contractors and Sub-Contractors shall comply with all current and applicable rules and regulations governing equal employment opportunities.

2.3 APPROVALS

2.3.1 General: The Contractors and Sub-Contractors shall comply with all applicable items contained in the regulatory approvals and permits for this project. Approvals and permits from DNR, PSC, FAA, DOT, SHPO, RR, Rusk County, and others are attached to this

93-159 01010-2

Specification Book under the heading; Supplemental Documents. It is your responsibility to understand and abide by the conditions contained in the documents.

2.4 EROSION CONTROL

2.4.1 Responsibility: It is the Contractor's responsibility to provide the specified erosion control for the project. Refer to Section 02770, sheet 15 of 15 of the Plans and DOT's Policy 96.55 and related sections.

2.5 TRAFFIC CONTROL

2.5.1 Responsibility: It is the Contractor's responsibility for traffic control. Refer to DOT's Policy 96.51 and related sections.

2.6 EXISTING UTILITIES

2.6.1 General: It is the responsibility of the Contractor to acquaint himself with the location of all underground structures which may be encountered or which may be affected by work under the Contract. The locations of any underground structures furnished, shown on the Plans or given on the site are based upon available records, but are not guaranteed to be complete or correct and are given only to assist the Contractor in making a determination of the location of all underground structures.

2.6.2 Utility Notification: The Contractor shall give notice in writing to all utilities that may be affected by Contractor's operations at least seventy-two (72) hours before starting work. The Contractor shall not hinder or interfere with any person in the protection of such work.

The Contractor is encouraged to utilize the "Call Before You Dig" telephone notification system.

2.6.3 Cable TV Notification: The Contractor shall notify Marcus Cable, 219 W. Miner Avenue, Ladysmtih, WI 54848,; Tele. No. 715/532-6040.

2.6.4 Electrical Utility Notification: The Contractor shall notify Northern States Power Co., 129 N. Lake Ave., Phillips, WI 54555; Atten: Joe Perkins; Tele. No. 715/836-1198.

2.6.5 Telephone Utility: The Contractor shall notify Ameritech @ 1-800-242-8511 (Diggers Hotline).

2.6.6 Village Sewer System: The Contractor shall notify the Village's Sanitary Sewer Utility Operator, Billy Mechelke, N5297 Little X, Tony, WI 54563, Tele. No. 715/532-7046.

93-159 01010-3

2.6.7 Emergency Facilities: The Contractor shall give written notice to the Fire Department, Rusk Co. Sheriff's Office and Ambulance Service, at least twenty-four (24) hours before closing off or in any way affecting through vehicular traffic on any street.

Rusk Co. Sheriff's Office 715/532-2200

Ladysmith Rural Fire Department 715/532-2186

Ambulance Service 715-532-2100

2.7 CONSTRUCTION SIGN

2.7.1 Construction Sign: One (1) construction sign shall be provided and erected at a location to be specified by the Engineer at the Pre-Construction Conference. Refer to Supplemental Specifications, page A for its size and design. The construction sign will be a bid item in the Proposal.

2.8 SURVEY MARKERS

2.8.1 General: No survey markers (e.g. DOT, County, USGS, Property Corners, etc.) located in the project area shall be disturbed, unless prior approval has been obtained. Replacement will be at the expense of the Contractor.

2.9 STORAGE OF MATERIALS

2.9.1 General: Contractor shall confine his apparatus, storage of materials and operations of his workmen to limits indicated by law, ordinances, permits or directions of the Engineer.

All materials stored on the site shall be protected from the elements by watertight coverings or shed with wood floors raised above the ground and protected from damage by Contractor's operations by barriers. The Owner assumes no responsibility for stored materials.

All materials shall be delivered in original sealed containers, with seals unbroken and with labels plainly indicating materials and directions for storage. Directions for storage shall be complied with unless specified to the contrary in the detailed Contract Specifications.

Damaged materials shall be immediately removed from the site. Stored material will be paid for in accordance with standard practice.

2.10 PROTECTING WORK

2.10.1 General: The Contractor shall protect existing improvements and Work installed by him from damage. Improvements and Work damaged shall be replaced by the Contractor. All replacement methods shall be approved by the Engineer.

2.11 LABOR LAWS

2.11.1 General: All Contractors and Subcontractors employed upon the Work shall be required to conform to applicable Federal and State labor laws and the various acts amendatory and supplementary thereto and to all other laws, ordinances and legal requirements applicable thereto.

All labor shall be performed in the best and most workmanlike manner by mechanics skilled in their respective trades. The standard of the Work required throughout shall be such grade as will bring results of the first class only.

2.12 TEMPORARY FACILITIES

2.12.1 Description: The Contractor shall provide his own temporary facilities such as electricity, toilets and heat. He shall also furnish, erect and maintain a weatherproof bulletin board for displaying wage rates, Equal Opportunity Requirement, etc. All temporary facilities shall be removed prior to job completion.

2.13 COOPERATION

2.13.1 General: The Contractors and Subcontractors shall cooperate and coordinate their Work with adjacent Work and shall give due notice to other Contractors and Subcontractors of intersecting Work to assure that all items are installed at an agreeable time and to facilitate general progress of the Work.

2.14 PROGRESS SCHEDULE

2.14.1 Description: The Contractor shall, within ten (10) days after the effective date of Notice to Proceed, furnish three (3) copies of a preliminary progress schedule covering his operations for the first 30 days.

The preliminary progress schedule shall be a bar graph or an arrow diagram showing the times he intends to commence and complete the various Work stages, operations and contract items planned to be started during the first 30 days.

The Contractor shall submit for approval by the Engineer within 30 days after the effective Notice to Proceed three (3) copies of a detailed bar graph or a graphic network diagram. The graph or diagram shall be accompanied by a brief written explanation of the proposed schedule and a list of activities.

The bar graph or graphic network diagram shall clearly depict the order and interdependencies of activities planned by the Contractor as well as activities by others which affects the Contractor's planning. For those activities lasting more than 30 days, either the estimated time for 25-50 and 75% completion or other significant milestone in the course of the activity shall be shown. In addition to the actual construction operations, the schedule shall show such items as submittal of samples and shop drawings, delivery of materials and equipment, construction in the area by other forces and other significant items related to the progress of construction. The graph or network diagram shall be printed or neatly and legibly drawn to a time scale.

The list of activities shall show for each activity the estimated duration and anticipated starting and finishing dates. Activities which are critical to complete the project in shortest time shall be identified. The cost of each activity shall be shown so the schedule can be cost loaded.

The Contractor shall submit with each payment request a copy of the schedule showing the current status of the project and the activities for which payment is being requested. If the project is behind schedule, a detailed report shall accompany the submittal stating why the project is behind schedule and what means are being taken to get the project back on schedule, including, if appropriate, contract time extensions.

The Schedule shall be revised:

a) When a Change Order significantly affects the contract completion date or the sequence of activities.

b) When progress of any critical activity falls significantly behind the scheduled progress.

c) When delay on a noncritical activity is of such magnitude as to make it become critical.

d) At any time the Contractor elects to change any sequence of activities affecting the shortest completion time.

The revised analysis shall be made in the same form and detail as the original submittal and shall be accompanied by an explanation of the reasons for the revisions.

2.15 SHOP DRAWINGS

2.15.1 Description: Shop drawings and descriptive data shall be required on all manufactured or fabricated items. Seven (7) copies of drawings and descriptive data shall be submitted for approval. (Two of these will be returned to the Contractor after their approval.)

Shop drawings shall be scheduled to be submitted for approval within 4 weeks after the effective Notice to Proceed.

2.16 SUPERVISION OF ERECTION FOR ALL EQUIPMENT

2.16.1 General: All items of structural, mechanical or electrical equipment whose cost delivered to the Owner exceeds $1,000 shall have included in the price bid the services of a qualified representative of the manufacturer for one (1) day to inspect equipment in operation and train the operator. For items whose cost exceeds $2,500, the bid price shall include not less than one (1) day during erection and not less than one (1) day operation inspection. Over $5,000, the bid price shall include not less than three (3) days during erection and two (2) days operation inspection. After each trip, the manufacturer is to submit a report to the Engineer covering the findings made during the inspection including changes and/or repairs, if any. Lesser time than specified may be for "make up" trips during the twelve (12) month warranty period. Trips made for the purpose of correcting defective materials or workmanship are to be made without charge and are not to be credited to the required number.

2.17 SCHOOL WATER SUPPLY

2.17.1 Service: It is mandatory that the High School Complex be provided potable water on a continuous basis during school hours. Therefore, the Contractor shall schedule his Work to avoid any interruption of water during school operation or normal activity.

2.18 EXISTING SHRUBBERY

2.18.1 Protection: It shall be the responsibility of the Contractor to protect all existing private shrubbery, trees, etc. within the construction area. If they are damaged, or destroyed, they shall be replaced to the owner's satisfaction at the Contractor's expense.

2.19 SAFETY

2.19.1 Basic: All work shall be done in accordance with OSHA standards and other specified regulations as noted throughout the Specifications.

2.20 PROJECT RECORD DOCUMENTS

2.20.1 Contractor's Responsibility: At the conclusion of the project, prior to final payment, the Contractor shall supply the Engineer with a legible record of all work as installed. Example: a map detailing watermain installed with location, lengths, tie-offs, valves, hydrants and appurtenances accurately shown so an "as-built" layout can be developed. This information can be marked on the appropriate sheet of the Plans with a permanent color marker.

2.21 OPERATION & MAINTENANCE DATA

2.21.1 O & M Data: The Contractors shall furnish to the Engineer three (3) complete unbound copies of complete operation and maintenance data for all equipment items. This data shall include a list of parts and a recommendation of spare parts that should be kept on hand.

2.22 TESTING

2.22.1 Testing & Equipment: All mechanical and electrical equipment furnished for this project shall be field tested to insure full compliance with the Specifications. Testing shall be performed by the Contractor and/or equipment supplier in the presence of the Engineer. If the field test shows that the equipment does not meet the Specifications, it shall be replaced with acceptable equipment at Contractor's expense.

Full payment for equipment will not be made until it has been field tested and accepted.

2.23 FINAL CLEANUP

2.23.1 General: The Contractor shall be responsible for cleaning up this Work and his waste and rubbish. It is the General Contractor's responsibility to completely clean up the inside and outside of all buildings and the total construction site at the completion of the project.

2.24 PRIVATE SERVICES

2.24.1 Description: Per specified bid items, the Contractor shall furnish and install service laterals from the distribution main to (and including) the curb stop. The Contractor shall also furnish and install the water meter including fittings and appurtenances within each respective user's structure.

Please be advised that section of the service lateral from the curb stop to the meter is the responsibility of each user and is not part of this construction contract.

2.25 PAYMENT TO CONTRACTOR

2.25.1 General: The Owner will make progress payments to the Contractors in accordance with the General Conditions. It is understood that grant reimbursement payments to the Owner are dependent upon compliance by the Owner with all the grant conditions identified in the grant request and grant agreement. The Owner shall be responsible for progress payments to the Contractor even when failure to comply with said grant conditions delays grant payments from the funding agency.

2.25.2 Final Payment: Final payment to the Contractor will not be made until the Owner and Engineer are satisfied that the facility is fully operational and all aspects of the Plans, Specifications and Construction Contract have been satisfied.

93-159 01010-8

2.26 FmHA NAME CHANGE

2.26.1 Clarification: Wherever the words "Farmers Home Administration (FmHA) or Rural Development Administration (RDA)" appear in the project manual, substitute the words "U.S. of America."

2.26 RAILROAD NOTIFICATION

2.26.1 General:

-END OF SECTION-

93-159 01010-9

DISINFECTION OF WATER STORAGE FACILITY

PART ONE - GENERAL

1.1 GENERAL

1.1.1 Work Included: The Work described in this Section shall consist of the disinfection of the total elevated water storage facility so it can safely store potable water for use by the Tony Municipal Water system.

1.1.2 Related Work Described Elsewhere:

1) Summary of Work 01010
2) Field Engineering & Inspection 01050
3) Applicable Standards 01085
4) Submittals & Substitutions 01300
5) Testing Laboratory Services 01410
6) Water Systems 02713
7) Painting 09901
8) Surface Preparation Elev. Storage Tank 09910
9) Painting Elevated Water Storage Facility 09920
10) Elevated Water Storage Facility 13410
Applicable Standards & Codes

1.1.3 Timing: The newly painted water storage facility shall be allowed to dry for a minimum of 7 days before being subjected to this disinfection process.

1.1.4 Holding Period: Chlorine solutions will remain in contact with the surfaces to be disinfected for a minimum of 24 hours.

1.1.5 Forms of Chlorine: Three (3) forms of chlorine are acceptable for disinfecting water storage facilities. The CONTRACTOR will select the form of chlorine he intends to use and so notify the ENGINEER, three (3) days before beginning the disinfection process.

1) Liquid chlorine will meet the requirements of AWWA B-301.

2) Sodium Hypochlorite will meet the requirements AWWA B-300.

3) Calcium Hypochlorite will meet the requirements of AWWA B-300.

1.1.6 Disinfection Methods: The CONTRACTOR must use one of the three (3) alternative disinfection methods specified in AWWA Standard C-652.

93-159 02673-1

CONTRACTOR shall notify the ENGINEER of the selected method at the same time he notifies him of the form of chlorine to be used.

1.1.7 Initial Fill: The initial filling of the elevated storage reservoir (tank) shall be accomplished by using water from the newly constructed distribution system. The expense for providing this water shall be borne by the Village of Tony.

1.1.8 Drainage: After the "holding period", the highly chlorinated water in the tank and riser shall be completely drained to waste, via the municipal sanitary sewer collection system at a rate that will not hydraulically and biologically upset the WWTF.

1.1.9 Refill: Following the complete draining of the tank and riser, as described in 1.1.8, the storage facility shall be refilled with potable water from the municipal distribution system at no expense to the Contractor. After refilling, samples of water should be taken from the tank and tested to demonstrate and record the good sanitary condition of the tank before it is placed into regular service.

The CONTRACTOR and ENGINEER will together drain, refill and sample the water in the tower. The ENGINEER will select the laboratory to analyze the collected water samples and the OWNER will pay the corresponding fee.

1.1.10 Unsuccessful Test: If the test for coliform organisms is positive, the tower and riser must be rechlorinated. The cost of rechlorination, water, chlorine, sample collection and sample analysis will be the responsibility of the CONTRACTOR. THE OWNER WILL CHARGE THE CONTRACTOR $1.15/1000 GALLONS OF WATER USED DURING ANY AND ALL RECHLORINATION WORK.

END OF SECTION

93-159 02673-2

Courtesy: Morgan & Parmley, Ltd.

DEEP WELL TURBINE PUMP

PART ONE - GENERAL

1.1 DESCRIPTION

1.1.1 Work included: This section of the Specifications applies to the furnishing and complete installation of one (1) deep well turbine water pump. Removal of the existing submersible pump and related appurtenances is also included in this Work.

1.1.2 Work location: The location of this pump installation is at Municipal Well No. 1, Rusk County Airport, as shown on the accompanying Plans.

1.1.3 Related work described elsewhere:

1) Summary of Work 01010
2) Field Engineering 01050
3) Submittals and Substitutions 01300
4) Testing & Laboratory Services 01410
5) Water Systems 02713
6) Right Angle Drive 11211
7) Chlorination Facility 11212
8) Internal Piping 11213
9) Pressure Switch & Surge Eliminator 11214
10) Water Supply Control System 13601
11) Electrical 16051

1.2 SUBMITTALS

1.2.1 Shop Drawings: Within thirty (30) days after Notice to Proceed, the Contractor shall submit tot he Engineer a complete manufacturer's specification and descriptive literature of the pump and related appurtenances.

1.2.2 DNR Form: The Contractor shall supply the Engineer with the necessary technical information required to complete Form 3300-226 *WELL PUMP SUBMITTAL CHECKLIST MUNICIPAL SYSTEMS.* A copy of this form is available to the Contractor upon request.

1.2.3 Operation & Maintenance Data: O & M data shall be supplied with illustrative literature for operator's Manual.

1.3 GUARANTEE

93-159 11210-1

1.3.1 Manufacturer: The manufacturer shall guarantee all materials and workmanship for a period of one (1) year from date of acceptance. All costs of field repair and/or replacement of components or devices to enforce the warranty will be borne by the Contractor or his representative for warranty period.

1.3.2 Contractor: The Contractor shall guarantee the total installation for one (1) year from date of acceptance, free of any expense to the Owner.

1.4 ACCEPTANCE

The pump and its installation will be accepted when it has been tested and demonstrates that it complies with the Specifications and design.

PART TWO - PRODUCTS

2.1 DEEP WELL TURBINE PUMP

2.1.1 Operating Conditions: The capacity of the well pump (for bidding purposes) shall be 200 G.P.M. against a total (field) dynamic head of 230 feet. To this head the manufacturer shall add the column friction loss, the pump friction loss and the line shaft bearing loss, as well as the motor loss. All computations for loss shall be as stated in Standards of the Hydraulic Institute and in the Standards of the current Edition of the National Association of Vertical Turbine Pump Manufacturers. Calculations shall accompany the shop drawing submittal. The speed of the pump and motor shall not exceed 1800 rpm. The pump shall be water lubricated with open shaft and rubber bearing.

2.1.2 Efficiency: Each Bidder shall submit with his Proposal a guaranteed wire to water efficiency of the pump he proposes to furnish when operating at the above total head and capacity, and shall submit performance curves showing head, capacity, efficiency, and horsepower demand from no delivery to maximum. The guarantee shall be stated in kilowatt hours electrical consumption per 1000 gallons of water produced, and the pump may be throttled during the acceptance test to develop the guaranteed head.

2.1.3 Bowl Assembly: The bowl assembly shall consist of close grain iron bowls, stainless steel impeller shaft, bronze enclosed type impellers, and bronze renewable wear rings. A discharge nozzle shall be furnished with the bowl assembly, which shall contain an extra long main bearing, and shall be bolted or screwed to the top bowl. Pumps without a discharge nozzle will not be acceptable.

2.1.4 Discharge Column: The discharge column shall be of standard weight 6" I.D. nominal steel pipe with heavy screw type couplings located not less than every 10 feet along column. Sufficient column shall be furnished to set the top of the bowls at the top of the screen and adequate suction tail pipe shall be located to place the suction entrance at a point three feet

93-159 11210-2

Courtesy: Morgan & Parmley, Ltd.

above the bottom of the screen. The line shaft shall be not less than 1" in diameter, and shall be composed of high carbon steel, and shall have adequately lubricated bearings located at least every 10 feet along the shaft. Water lubricated, Monel or Stainless Steel sleeves shall be located opposite rubber bearings fluted for water lubrication.

2.1.5 Pump Head: The pump head shall have a separate and removable cast iron base plate which can accommodate a one inch extension of the inner well casing above the top of the concrete pump foundation. The base plate shall be permanently grouted to the concrete pump base by the Contractor after the pump is installed and properly aligned. The pump head shall be large enough to accommodate a coupling in the motor drive shaft below the base of the motor and above the stuffing box. There shall be furnished and installed suitable protecting pipe guards to properly steady the protecting pipe and line shaft assembly. The Contractor shall furnish and install a pre-lubrication line, ½ inch solenoid valve for 110 volt operation, 90 second time delay relay and other electrical controls as needed to afford a 30 to 90 second prelubrication before each pump start-up. The pump shall have a 6-inch flanged above ground discharge drilled for 150 pound American Standard flanges.

The pump Contractor shall furnish and install all electrical equipment and controls required for the complete pump installation and chemical feed unit; commencing from the service entrance box as installed by the pumphouse subcontractor. Refer to technical Sections 11211, 11212, 11213, and 11214.

2.1.6 Electric Pump Motor: The Contractor shall furnish and install a vertical hollow shaft, part-winding, 240 volt, three phase, 60 Hz. drip-proof, 1800 rpm no load speed, motor of sufficient horsepower to operate the deep well pump at any head capacity condition, even at reduced heads, without overloading the name plate rating of the motor. The motor is to be manufactured by only a domestic manufacturer, such as G.E., Westinghouse, U.S., Allis-Chalmers, or equal.

2.1.7 Electrical Controls: This Contractor shall furnish with the pump at 240 volt, three phase, 60 Hz, part-winding magnetic starter in NEMA one enclosure equal to Square D, with Hand-Off-Automatic selector switch, together with fusible heavy duty disconnect switch. This Contractor shall also provide a time delay relay and pre-lubrication solenoid for operating the pre-lubrication line for up to 90 seconds prior to pump start. A Hand-Off-Automatic selector switch and fused disconnect switch shall be furnished and installed on the control circuit to the chemical feeder.

2.1.8 Pump Installation: Pump Sub-Contractor shall not sublet the installation of the pump; as it is intended that this Contractor shall be solely responsible for the satisfactory operation of the system. The Contractor shall provide an orifice and pressure gauge, and other testing equipment required to run a two-hour efficiency test upon completion of the installation. In the event the efficiency or performance of the pump does not meet his guarantee, the Contractor must make necessary modifications or replacements at his own expense.

2.1.9 Pump Control: The Pump Control System shall be the product of a manufacturer experienced, skilled and regularly engaged in the design and fabrication of this type of equipment, with similar installations within convenient inspection distance. Complete detail drawings and descriptive information shall be submitted to the Engineer for approval before fabrication. Refer to Section 13601.

2.1.10 Pressure Switch: The Contractor shall furnish and install pressure switch for sensing of main pressure for pump start/stop equipment. Refer to Specification 11214 for details.

2.1.11 Surge Eliminator: The Contractor shall furnish and install surge eliminator tank and related appurtenances. Refer to Specification 11214 for details.

2.1.12 Installation: The Contractor, through the manufacturer, shall provide the services of factory Engineers to inspect installation, supervise final adjustments and instruct the operating personnel in the use and maintenance of the equipment. This supervision shall be provided by the Contractor on the job site and is considered incidental to the contract.

2.1.13 Water Level Indicator: Contractor shall furnish and install a plastic airline with gage and tire pump to allow Waterworks Operator to measure water levels in well.

PART THREE - EXECUTION

3.1 GENERAL

3.1.1 Location: Refer to applicable drawings on the accompanying Plans for location and typical layouts.

3.1.2 Installation: All electrical Work shall be executed by a qualified electrician or under his direct supervision. All pump installation shall be executed by or under the direct supervision of fully qualified personnel.

3.1.3 Discrepancies: Where real discrepancies are found between the Plans and the Specifications they shall be brought to the attention of the Engineer as soon as found so that corrections can be made that will least affect the interests of the Owner and Contractor.

3.1.4 Final Connections: This Contractor shall be responsible for final connections of all electrical circuits at the control panel relative to pump installation and related appurtenances as defined.

3.1.5 Testing: This Contractor, together with a representative of the manufacturer, shall in cooperation with each other, check all circuitry before energizing.

3.1.6 Operation and Maintenance Manual Data: Spare parts Data and Shop Drawings requirements are specified in the Supplemental General Conditions. This Contractor is governed by those conditions.

As stated previously, workmanship and materials shall comply with Wisconsin and National Electric Codes, NEMA Standards and all equipment shall bear UL labels.

3.1.7 Labor for Testing: All labor, instruction, and equipment used for testing and fine tuning the electrical and mechanical systems shall be furnished by the Contractor and is considered incidental to the Contract.

3.2 ACCEPTANCE

3.2.1 Final Acceptance: By the Engineer and Owner, will follow the testing of the "total" pumping system and related equipment.

3.3 PAYMENT METHOD

3.3.1 Payment: Will be based upon the Proposal.

END OF SECTION

INTERIOR PIPING

PART ONE - GENERAL

1.1 DESCRIPTION

1.1.1 Work included: This Section of the Specifications applies to the furnishing and installation of interior pumphouse piping and shall commence at the discharge of the Deep Well Turbine Pump assembly and terminate at the 6-inch riser pipe, as illustrated on the accompanying Plans. The existing interior piping and related components shall be removed by Contractor. Disposal is the Contractor's responsibility.

1.1.2 Related Work described elsewhere:

1) Summary of Work 01010
2) Field Engineering 01050
3) Submittals & Substitutions 01300
4) Water Systems 02713
5) Deep Well Turbine Pump 11210
6) Right Angle Gear Drive 11211
7) Pumphouse Rehab 15050

1.2 SUBMITTALS

1.2.1 Shop Drawings: Within thirty (30) days after Notice to Proceed, the Contractor shall submit to the Engineer a piping layout of interior piping.

1.3 GUARANTEE

Interior piping and equipment shall be fully guaranteed for one (1) year after final acceptance for all material, workmanship, and installation. All costs for field repair and/or replacement shall be borne by Contractor for duration of warranty.

1.4 ACCEPTANCE

Interior piping system shall be accepted when it has been fully tested and demonstrated in conjunction with the "total" pumping system.

PART TWO - PRODUCT

2.1 Description: The Contractor shall furnish and install the following list of materials and equipment to fully complete the interior piping for connection of pumping system to the existing 6-inch riser pipe, as shown on the accompanying Plans. Material shall be ductile iron pipe

Courtesy: Morgan & Parmley, Ltd.

with flanged joints, 6-inch I.D. with a minimum working pressure of 250 p.s.i.

Automatic Air Relief System
Check Valve (6-inch)
Smooth Sampling Tap (½-inch)
Sleeves & Couplings (6-inch) (as needed)
Meter (6-inch)
O.S. & Y Gate Valve (6-inch)
6" x 6" x 6" Tee W/Blind Flange
Screened Well Vent
Floor Pipe Supports

This Work, also includes the removal of existing interior piping and components not needed for new facility.

The foregoing list includes major components. However, the list is not necessarily complete. No claim for extras will be allowed.

Refer to accompanying Plans for basic layout and technical Section 02713 of the Specifications.

PART THREE - EXECUTION

3.1 GENERAL

3.1.1 Location: Refer to accompanying Plans for details and location.

3.1.2 Installation: This Contractor shall provide experienced personnel to properly install the equipment.

3.1.3 Testing: The Contractor shall test this system in conjunction with testing of Deep Well Turbine Pump.

3.2 ACCEPTANCE

3.2.1 Final Acceptance: Final acceptance by the Engineer and Owner will follow the successful testing of the "total" pump facility.

3.3 PAYMENT METHOD

3.3.1 Payment: Payment method is based upon a lump sum as bid for the segment as listed in the Proposal.

END OF SECTION

ELEVATED WATER STORAGE RESERVOIR

PART ONE - GENERAL

1.1 BASE BID

1.1.1 Description: The base bid shall be for a Single Pedestal Sphere-Shaped elevated water storage reservoir; including all applicable appurtenances.

1.2 ALTERNATE BID

1.2.1 Description: The alternate bid shall be for a Multi-Leg Ellipsoidal-Shaped elevated water storage reservoir; including all applicable appurtenances.

1.3 PROJECT DESCRIPTION

1.3.1 Scope of Work: The storage facility basically consists of furnishing all materials, supplies, equipment, tools, labor and related skills necessary for fabricating, delivering, erecting, painting, testing and certifying one (1) 50,000 gallon elevated steel storage reservoir, complete with the specified accessories; ready for service to the Tony Water Utility. The elevated reservoir (tank) and tower will be fabricated and erected in strict accordance with all applicable regulatory codes, plan details, these general specifications, NFPA and AWWA Standards so it will meet the Wisconsin Department of Natural Resources approval requirements.

1.3.2 Related Work described elsewhere:

1) Summary of Work 01010
2) Field Engineering & Inspection 01050
3) Applicable Standards 01085
4) Submittals & Substitutions 01300
5) Testing Laboratory Services 01410
6) Temporary Facilities & Controls 01500
7) Contract Closeout 01700
8) Clearing & Grubbing 02110
9) Earthwork 02200
10) Topsoil & Landscaping 02480
11) Water Systems 02713
12) Erosion Control 02770
13) Concrete Testing & Sampling 03050
14) Concrete Formwork 03100
15) Concrete Reinforcement 03200
16) Cast-in-Place Concrete 03300
17) Structural Steel 05120

Courtesy: Morgan & Parmley, Ltd.

18) Structural Steel Erection 05121
19) Metal Fabrication 05500
20) Flashing & Sheet Metal 07600
21) Sealants & Calking 07920
22) Painting 09901
23) Surface Preparation Elevated Tank 09910
24) Painting Elevated Storage Tank 09920
25) Water Supply Control System 13601
26) Electrical 16051
27) Standard Specifications: .... 5th Edition
Sewer & Water Construction in Wisconsin
28) Applicable Standards NFPA & AWWA
29) Refer to Soil Boring Report (Supplemental Document I)

1.4 LOCATION

1.4.1 Site: The elevated water storage reservoir shall be constructed on the 100' x 100' lot located westerly of the intersection of Cedar Street and Central Avenue; as shown on the accompanying Plans. The Village holds a long term lease on this property.

1.4.2 Access: The Owner (Village of Tony) will provide suitable access to the site, via a twenty foot (20') wide easement from the adjacent municipal street.

1.5 SUBMITTALS

1.5.1 Shop Drawings: Within thirty (30) days after executing the Notice to Proceed document, the Contractor shall submit to the Engineer five (5) complete sets of fabrication shop drawings detailing the total facility; including design of the concrete foundation.

All drawings and design submittals shall be stamped and signed by a Professional Engineer currently registered in the State of Wisconsin.

1.5.2 Structural Calculations Accompanying the previously noted shop drawings, shall be five (5) completed sets of structural calculations certifying the structural integrity of the concrete foundation, elevated water storage reservoir and tower. These calculations shall be stamped and signed by a Professional Engineer currently registered in the State of Wisconsin.

1.6 GUARANTEE

1.6.1 Manufacturer: The manufacturing/fabricator shall fully guarantee all materials and workmanship for a period of one (1) year from date of final acceptance by Owner. All costs of field repair and/or replacement of any component or device to enforce the warranty will be completely borne by the Contractor or his representative for the warranty period.

1.6.2 Contractor: The Contractor shall fully guarantee the total installation for one (1) year from date of final acceptance by the Owner.

1.7 NOTIFICATION

1.7.1 Airport and General: The Contractor shall notify the manager of the Rusk County Airport, in writing, a minimum of five (5) working days, prior to commencing steel erection. The Contractor shall also, comply with any other applicable requirements specified in the regulatory approvals and permits; including FAA mandates.

PART TWO - PRODUCTS

2.1 GENERAL

2.1.1 Base-Bid Description: The reservoir (tank) and supporting structure will be of all-welded steel construction. The transition sections between the pedestal and tank, and between the pedestal and base shall be smooth curves.

2.1.2 Alternate - Bid Description: The reservoir (tank) and supporting structure will be of all-welded steel construction. The tank shall be supported on multiple tubular steel columns with a five (5) foot diameter dry riser. Access to the tank and valve vault shall be provided through the dry riser base, hereafter referred to as the bell. If the dry riser is to be used a support column, smooth curve transitions will be required between the dry riser and bell, and between the dry riser and tank.

2.2 DESIGN

2.2.1 Design Parameters: The following list of design parameters follows the basics outlined in the Forward of Section III, AWWA Standard D-100-84:

1) Capacity: 50,000 U.S. Gallons

2) Height to Overflow: 122 feet (nominal)

3) Type of Roof: Base Bid - Single Pedestal Sphere
Alternate Bid - Ellipsoidal

4) Head Range: 20 Feet (Nominal)

5) Riser Pipe: 8 inch diameter (steel)

6) Location: See Plans

7) Time for Completion: See Bidding Documents

8) Nearest Town: Within corporate limits of the Village of Tony, Wisconsin

9) Nearest Railroad Siding: Wisconsin Central, Ltd. Siding adjacent to construction site

10) Type of Road Near Site: Municipal street (asphalt pavement) w/access road graveled

11) Electrical Power: Northern States Power (NSP) Company, local office in Ladysmith, WI (5 miles west of Tony)

12) Compressed Air: Not available at site

13) Single Pedestal Design Supplement: Not required

14) Welded Joints: Detailed drawings are required to be submitted. Welds must be continuous and on all sides of overlaps.

15) Copper-Bearing Steel: Not required.

16) Type of Pipe & Fittings: Steel

17) Snow Load: 25 PSF Minimum

18) Wind Load: 100 MPH Minimum

19) Earthquake Load: None required

20) Corrosion Allowance: None required

21) Inspection & Painting Balcony: Not applicable

22) Increased Wind Load: Required to accommodate future paint removal, dust control apparatus (tarps, curtains, etc.) and resistance to the overturning forces.

23) Manholes, Ladders, Additional Accessories: See Item No. 30

24) Pipe Connections: An 8 inch diameter steel pipe will carry water from the bottom of the tank to a base elbow 8 feet below grade. The riser pipe will have an Expansion joint immediately above the base elbow. The riser pipe will be insulated to protect against freezing to -40°F (outside temp.)

93-159 13410-4

25) Removable Silt Stop: A removable silt stop is required, 12 inches high minimum on the riser pipe inside tank.

26) Overflow: A 6 inch diameter steel pipe over flow is required. The overflow piping will be fitted with an adequate anti-vortex entrance detail. The overflow piping will extend down the inside of the access tube and pedestal and discharge at a point perpendicular to and approximately 1 foot above ground level onto a concrete splash pad.

27) Roof Ladder: A roof ladder, handrail, or non-skid surface is not required for access to hatches or vents, provided however, that the access surface is sloped 2 inches in 12 inches or less.

28) Safety Devices: Safety devices will be installed on ladders in accordance with OSHA and the State of Wisconsin Safety codes.

29) Special Tank Vent: A special tank vent is required to handle pressure differential caused when pumping or withdrawing at a maximum rate of 750 GPM. The overflow piping will not be considered as a part of the venting system when designing this special vent. The vent will terminate in a "U"-Bend or mushroom cap constructed with the opening at lease 4 inches above the roof and covered with 4 mesh to the inch noncorrodible screen. The screen will be installed within the pipe or cap at a location protected from the environment that will insure fail-safe operation in the event that the insect screens frost over. The vent must be easily dismantled to remove the screens for cleaning.

30) Additional Accessories:

A) A 30" x 60" access door located in the supporting bell complete with locking device.

B) Ladders in bell, pedestal, access tube and tank, with safety devices to meet applicable State Safety Standards.

C) Steel condensate ceiling located at the junction of the pedestal cylinder and truncated cone complete with drain. The drain piping will extend down the inside of the supporting bell and discharge perpendicular to and approximately 1 foot above ground level onto the same concrete splash pad as the overflow piping.

93-159 13410-5

D) A manhole giving access to two painter's rings located at the top of the pedestal. There will be a platform inside the pedestal at this point. All manholes will be rainproof and will have a 4" high curb. The cover will overlap the curb by at least 2 inches.

E) Eight interior water proof light sockets and bulbs with conduit, wiring, and switch will be provided inside of the pedestal and access tube. There will be one light located at the top of the access tube, one light near the top of the midpoint of the supporting pedestal, one light above the condensate ceiling, three lights in the bell and one light in the valve vault. Two (2) - 110 volt grounded double convenience outlets will also be installed as directed by the ENGINEER; one in the bell and the other near the top of the access tube.

F) An access tube 36" in diameter will be provided from the top of the pedestal to the tank roof. A manway will be provided at the base of the tube for access to the ladder on the exterior of the tube (interior of the tank)

G) All vents, overflows, or other openings in the tank will be screened with 4 mesh to the inch noncorrodible screen. Details of this screening will be submitted to the ENGINEER for approval.

H) Roof Lugs - Sufficient lugs will be furnished and installed in all sections of the tank to permit lashing of painters' scaffolding and boatswain's chairs for painting all portions of the tank.

I) Riser Insulation - The 8 inch piping will be insulated with precut, semi-circular sections of Extruded Polystyrene (2-inches minimum). Masking tape will be used to hold insulation in place until a .019" thick aluminum sheet is wrapped around the insulation and held in place with stainless steel bands and marmon clamps.

J) An aircraft warning obstruction light, enclosed in an aviation red obstruction light globe, complete with photo-electric cell, conduit, and wire to an electrical breaker panel located in the base, shall be provided. Obstruction light equipment shall meet approval of the Federal Aviation Administration (FAA).

K) Circulation Pump - Furnish and install one (1) complete circulation pump assembly inside the tank. The assembly will be securely mounted to the exterior of the access tube running up through the tank. The assembly will be located in plain view, with the pump inlet approximately one-third (1/3) of the way up the tank bowl. The pump power cable will be enclosed in rigid steel conduit attached securely to the exterior of the access tube. The circulation pump will be Red Jacket 50 (½ H.P.) CNW (2-wire), 1 (208/230 V), CN (Corrosion-Proof NEMA), 9 (number of stages) BC (10 GPM), or approved equal. The pump discharge will be fitted with 1-1/4"galvanized steel discharge piping as shown on the construction plans. The pump power cables will terminate in an electrical junction box located near the top of the inside of the access tube.

L) Electrical Service - The CONTRACTOR will make arrangements to provide 120/240 volt, single phase permanent underground electrical power service to the base of the tank. Temporary power requirements necessary for construction will also be the CONTRACTOR's responsibility.

The CONTRACTOR will provide a meter socket and disconnect switch in a NEMA 3R enclosure suitable for the permanent service. These items will be located on the outside of the tank base. The CONTRACTOR will also provide for all necessary conduit and conductors necessary to connect the electrical service to the 200 amp minimum load control circuit breaker panel located inside the tank base. Separate electrical circuits will be provided for as follows:

1. Circulation Pump No. 1
2. Circulation Pump No. 2
3. Aircraft obstruction lighting system
4. Interior lights/Convenience Outlets
5. Interior convenience outlets.
6. Valve pit sump pump.

M) Expansion Joint - A stainless steel, bellows type expansion joint will be installed on the vertical riser pipe.

N) Sediment Drain - A 4 inch drain will be installed at the bottom of the tank bowl for draining sediment. The drain will discharge to the 6" overflow pipe as shown on the plans. A water tight, removable cap will be provided on the 4 inch drain inside the tank.

31. Butt-Joint Welds: Continuous with 100% joint penetration required.

32. Mill or Shop Inspection: Mill or shop inspection is not required by the OWNER.

33. Certified Written Report: At the conclusion of the work, the CONTRACTOR will submit a written report prepared by the CONTRACTOR's qualified personnel certifying that the work was inspected as listed in Section 11.2.1 of AWWA, D-100-84.

34. Radiographic Film & Test Segments: These items will become the property of the OWNER.

35. Type of Inspection: Inspection of welded - shell butt joints will be by radiographic testing in accordance with Section 11 of AWWA D-100-84.

36. Cleaning & Painting: See detailed specifications, Sections 09900 and 09910 plus manufacturer's recommendations. The tank must be painted prior to water testing and filling of the tank.

37. Soil Investigation: See Detailed Soils Report, Supplemental Document I. The top of the foundation will be 6 inches above finish grade.

38. Piling Compensation adjustment: Refer to the Standard General Conditions of the construction contract section on extra work.

39. Buoyancy Effect: Buoyancy effect shall be considered in the foundation design.

40. Concrete requirements: All concrete used during construction will be in accordance specifications located herein and as stipulated in other specific conditions with the in Section 3000 of the Specifications.

41. Earth Cover: There will be 8 feet of cover from finished ground level to the crown of the inlet and outlet pipe.

42. Seismic Design Specification Sheet: Not required to be completed.

43. Vertical Acceleration: Not required to be considered.

44. Local Seismic Data: Not available.

2.3 FOUNDATION

2.3.1 Description: The concrete foundation is included in the elevated tank Contract and will be designed by the CONTRACTOR (utilizing the services of a Registered Professional Engineer), in accordance with allowable soil bearing values and minimum depths previously provided. The foundation design must include the additional future (temporary) load resulting from paint removal/dust control apparatus (tarps, curtains, etc.) A concrete pipe vault, dimensions as shown on the Plans, will be constructed inside the pedestal base (or beneath the dry riser base; if alternate selected). Water main, extending from centerline of the tower to (and including) the 8" x 6" tee is included as a part of the elevated tank Contract. Water main includes one hydrant package and valve. An 8-inch blind flange shall be furnished and installed on the 8" x 6" tee which is the end of this section of the Work.

PART THREE - QUALITY ASSURANCE

3.1 EXPERIENCE

3.1.1 Contractor's Experience: Bids will be accepted only from experienced tank Contractors who are skilled and have furnished and erected at least five (5) similar facilities of equal, or greater, capacity within recent years. A letter shall accompany the Bid listing the five (5) such examples and their location with relevant details.

3.2 CONCEPTUAL DESIGN

3.2.1 Included with Bid submittal: A drawing showing the dimensions of the tank and tower, including the tank diameter, the height to lower and upper capacity levels, sizes of principal members, and thickness of plates in all parts of the tank and tower shall be submitted with the Bid. Also, the maximum wind gross moment and shear on the foundation system shall be identified.

3.3 QUALIFICATIONS

3.3.1 Welders: All welders employed on this project shall be qualified by ASME requirement standards.

3.3.2 Electricians: All electrical work shall be executed by experienced electricians.

3.4 WELDING

3.4.1 Welding supervision: The Contractor shall employ the services of a welding supervisor independent of the tank/tower erection foreman's jurisdiction.

3.4.2 Inspection: Welding inspection shall be in strict accordance with Section 11 of AWWA Specifications and shall be provided by the tank manufacturer.

3.5 DISCREPANCIES

3.5.1 Definition: Where real discrepancies are found between the Plans and the Specifications, they shall be brought to the attention of the Engineer as soon as they are discovered so that corrections can be made that will least affect the interest of the Owner and Contractor.

3.6 FINAL CONNECTIONS

3.6.1 General: This Contractor shall be responsible for final connection of all applicable electrical circuits and watermain.

3.7 TESTING

3.7.1 Labor, Material and Equipment: This Contractor shall furnish all labor, material, supplies and equipment necessary to test the facility.

3.7.2 Chlorination: This Contractor shall furnish all labor, material, supplies and equipment necessary to successfully disinfect the total facility, pursuant to applicable codes and DNR's approval.

The OWNER will furnish and dispose of all water required for initial testing and sterilization purposes.

3.8 ACCEPTANCE

3.8.1 Final Acceptance: Final acceptance, by the Owner, will follow the successful testing of the total facility and assurance that all other specified requirements have been satisfied.

3.9 PAYMENT

3.9.1 Method: Payment will be based upon the Proposal's bid items for this phase of the Work. Periodic partial payments will be made as work progresses utilizing customary practice. Final payment will not be made until the total facility has been accepted by the Engineer and Owner.

END OF SECTION

93-159 13410-10

WATER SUPPLY CONTROL SYSTEM

PART ONE - GENERAL

1.1 DESCRIPTION

1.1.1 Work Included: This section of the Specifications applies to the furnishing and installation of an integrated, solid state water supervisory control system with alarm monitoring of Tony's water system.

1.1.2 Water Location: Work sites are at the pumphouse and the elevated tank. Refer to the accompanying Plans for exact locations.

1.1.3 Related Work Described Elsewhere:

1) Summary of Work-Special Provisions 01010
2) Field Engineering 01050
3) Submittals & Substitutions 01300
4) Testing Laboratory Services 01410
5) Water Systems 02713
6) Deep Well Turbine Pump 11210
7) Elevated Water Storage Reservoir 13410

1.2 QUALITY ASSURANCE

1.2.1 General: The control system shall be the product of a manufacturer experienced, skilled, and regularly engaged in the design and fabrication of this type of equipment with similar installations within convenient inspection distance. The general design of the equipment shall be such that all working parts are readily accessible for inspection and repairs, easily duplicated and replaced, and each and every component suitable for the service required. Design, manufacture and testing of all components shall be in conformance with all requirements of governing and regulating agencies and all applicable industry codes.

1.2.2 Factory Supervision: The manufacturer shall provide the services of factory engineers to inspect the installation, supervise final adjustments, and instruct the operating personnel in the use and maintenance of the equipment. This supervision shall be incidental to the project and the cost included in the bid item.

1.2.3 Governing Codes & Inspections: The completed electrical installation shall meet all the requirements of the latest edition of the National Electric Code as well as state and local regulations as they may apply. This shall not be construed to permit a lower grade of construction where Plans and Specifications call for workmanship or materials in excess of code requirements.

93-159 13601-1

Courtesy: Morgan & Parmley, Ltd.

The work herein specified shall be subject to inspection and approved by authorized representatives of the National Board of Fire Underwriters, state and local governing authorities, and the Engineer.

The control system as specified herein, shall be constructed/manufactured by a UL certified #058 control manufacturing facility.

1.3 IDENTIFICATION

1.3.1 Labeling: All control devices and device enclosures shall be labeled with individual nameplates or legend plates.

Individual nameplates or legend plates shall be one of the following types:

1. Black laminated plastic or micarta with white cut letters.

2. Corrosion-resistant metal plates with engraved or raised letters and backfill.

1.4 ENVIRONMENTAL REQUIREMENTS

1.4.1 General: The Supervisory Control System shall function reliably within an operating temperature of -20^0 to $+70^0$ C (-40^0 to 158^0 F). Humidity shall be 0 to 95 percent non-condensing. Each supervisory or remote unit shall be furnished with integral lightning protection devices to protect against lightning induced transients on leased telephone and power lines.

The entire solid state Supervisory Control System shall provide a typical system accuracy and repeatability of better than $\pm$ ½% over a temperature range of 0^0 to $+50^0$ C ($+32^0$ to $+122^0$ F).

1.5 WARRANTY

1.5.1 General: The Contractor, thru the Manufacturer, shall warrant the Supervisory Control and Alarm System to be free of defective material and workmanship for a period of one (1) year from date of acceptance. The manufacturer shall be obligated to furnish and install, at no charge to the Village, replacement material items or unit proven defective within this warranty period. This warranty shall not be construed to cover lights, fuses, or other items normally consumed in service nor those items which have been damaged due to outside forces such as vandalism, lighting, etc. Service calls for Control System problems will not be considered part of the factory supervision.

1.6 SYSTEM OPERATION RESPONSIBILITY

1.6.1 General: The Contractor, thru the manufacturer, shall assume the total, undivided

responsibility for the correct operation of the Water Control System. This shall include everything required to make the system work, including additional control changes, if required.

The manufacturer shall maintain a service center with all necessary replacement parts/personnel. A certified letter shall be required to show compliance with the service center requirements.

PART TWO - PRODUCE

2.1 DESCRIPTION

2.1.1 Telemetry System: All control signals, status signals, alarm or variable analog data shall be transmitted and received between the tower and Well #1 utilizing an audio tone pulse frequency telemetry system operating on a low cost signal grade line or buried cable. The pulse frequency shall be directly proportional to the level in the tower and vary between 5-25 pulses per second. Each transmitter and receiver shall contain individual span and offset potentiometers for calibration and each unit shall contain a pulse indicating LED.

Audio tone telemetry system shall utilize 3 state FSK audio tone. Transmitter output shall be adjustable from -20 dbm to +3 dbm. Receiver sensitivity shall be adjustable to -40 dbm. Each transmitter/receiver unit shall be provided with a three stage surge suppression lighting protector consisting of gas tube arresters, varistors, and zener diodes.

2.1.2 LED Bargraph Indicator/Controller: The Control System shall sense the tower level over a calibrated range, display it on a 4" LED bar graph on the face of the controller, graphically display 8 level adjustments for automatic pump and alarm control in a coordinated arrangement with the level display and provide automatic operation of the pump and alarms as herein-after or otherwise described. Include:

A. 0-20 Ft. tower level range.
B. 40 Segment LED bar graph level display.
C. 40 Level adjustability of each 8 control levels utilizing gold plated adjusting pins.
D. "Raise-auto-lower" level simulation switch with spring return to "auto".
E. "Pump-up" control with eight level settings for four (4) demand levels.
F. Full-range adjustable dual settings for each control and alarm circuit, utilize gold plated adjusting pins to set levels.
G. LED indicator for each control and alarm circuit, or equal.
H. Power-up "wake-up" with pump and alarms off; held off for 18 seconds.
I. 20 Second time interval after 1 pump starts before another pump is allowed to start.
J. Rate of change limiting of analog level signal.
K. High and low Level alarm sensing.
L. Unit shall door mount on the outside of the enclosure.

The above controller format is desired for operating personnel because of the ease in comprehension.

2.1.3 Pump Control Module: The pump shall have a control module which combines the Hand-Off-Automatic switch with the pilot lights for "Required"-"Run"-"Failed". HOA switches shall be full size, rotary type units.

The pilot lights shall be LED type or equal and shall light when the pump is called for, and when it is running, or when it has failed.

2.1.4 Annunciator Panel:

A. An annunciator system shall be provided to announce a change of condition for all of the monitored status or alarm conditions.

B. The annunciator system shall consist of the required number of self-contained modules. The points shall be individually selectable for alarm annunciation or status indication. Individual points shall be grouped on the modules by location, category or function as designated by the Village to provide the operator with the easiest comprehension. Critical alarms shall be grouped on separate modules with an independent, easily identifiable audible alarm.

C. Alarm condition shall be annunciated by the ISA-1A sequence as follows:

1.

Condition	Light	Audible
Normal	Off	Off
Alarm	Flashing	On
Acknowledge	On	Off
Return to Normal	Off	Off

2. Silencing one alarm shall not inhibit sequence alarms.

3. All annunciator modules shall have a common acknowledge/silence push-button

4. A lamp test provision shall be included.

2.1.5 Elevated Tower Panel:

A. A solid state pressure responsive tower level/alarm transmitter system housed in a new NEMA 3R GALVANIZED enclosure with other components as hereinafter described shall be mounted on the Elevated Tower riser.

The enclosure shall incorporate a freezing protective dual electric heater with heavy duty adjustable thermostat to provide protection for the equipment located herein. ½" thick insulation of interior enclosure walls shall be provided. Provide the following major items of equipment:

1. Control breaker.
2. 120 VAC lighting protection as required.
3. Block and bleed valves.
4. 3½" Altitude gauge.
5. Solid state transducer to convert head range of the tower to an analog signal. Transducer to have stainless steel bellows and stainless steel internal parts to insure long life.
6. Surge quelling and suppression card. The process output of the transducer shall be fed into this card to remove electronically (not hydraulically) any irregularities in the process signal due to water pressure surges or friction losses. The rate of limit quelling or suppression shall be adjustable from 30 seconds to 180 seconds. Hydraulic pressure snubbers are not acceptable.
7. Telemetry transmitter as previously specified.

   Transmit:

   a. Elevated tower level data.
   b. Signal failure/Power failure.
   c. High water alarm.
   d. Low water alarm.
   e. Control panel low temperature.

8. Internal dual heaters to protect panel interior equipment at temperature of -40^0 Fahrenheit or wind chill of -85^0 Fahrenheit. Heaters to be screw-in type, Chromalox SCB150, or equal.
9. Low panel temperature alarm thermostat.
10. Enclosure shall mount on the elevated tower riser, with no tubing or piping exposed.

2.1.6 Supervisory Control Center at Well No. 1

A. The enclosure shall be a NEMA type 12 and all surfaces shall be phosphatized before painting and finished with rust-inhibiting base coat and a final coat of exterior grade baked enamel. All instrumentation and operator controls shall be mounted on the front panel.

B. The basic functions and components of the control center are to be as follows:

1. Provide a modular solid state type system with plug in type circuit cards.
2. Control breaker with 120 VAC lightning protection.
3. Telemetry receiver to receive tower level/alarm signals from the Elevated Tower (as previously specified).
4. Provide a seven-day circular chart electronic recorder (10") to record the water level in the Elevated Tower. Provide two year's supply of charts and ink.
5. Provide LED bargraph monitor/controller as previously specified for the Elevated Tower. This unit shall control the operation of the pumps. Control stages shall be as follows:

    1. High Level Alarm
    2. Lead pump on/off control
    3. Future pump on/off control
    4. Low Level Alarm

6. Provide a pump control module as previously specified with selector switch and pilot lights.
7. Provide an alarm annunciator as specified for the following alarms:

    A. Elevated Tower High Level
    B. Elevated Tower Low Level
    C. Elevated Tower Power/Signal Fail
    D. Elevated Tower Control Panel Low Temperature
    E. Well #1 Fail

Include audible horn/silence button. Any annunciator alarm shall activate an external 100 watt flashing red alarm light.

8. Provide space for future well pump control logic and future telephone dialer.

2.1.7 Discrepancies: Where real discrepancies are found between the Plans and Specifications they shall be brought to the attention of the Engineer as soon as found so that corrections can be made that will least affect the interests of the Owner and Contractor.

PART THREE - ACCEPTANCE

3.1 GENERAL

3.1.1 Testing: The Contractor, manufacturer's representative, Village personnel and Engineer shall be present during testing.

3.1.2 Instructions: The manufacturer's representative shall fully demonstrate, to Village personnel and Engineer, the correct operational procedures of the system.

3.1.3 O & M Data: The manufacturer shall provide copies of complete operation and maintenance data; plus complete parts catolog..

3.1.4. Final Acceptance: By the Engineer and Owner, will follow the testing of the new Well facility and Elevated Tower.

3.1.5 Payment: Will be based upon a lump sum as bid an listed in the Proposal.

-END OF SECTION-

Bidding Process

BIDDING DOCUMENTS

Section 23

The bidding process is a critical element of a construction project. It is extremely important that the design professional directly oversees the process to avoid any confusion or impropriety. All facets of the bidding process must be beyond suspect.

It must be noted that there are wide ranges of specific requirements from one project to the next. All applicable rules and regulations for each respective bidding process must adhere to local, state and/or federal requirements; i.e., labor standards, wage rates, nondiscrimination, funding conditions, etc. Therefore, the following pages reflect only a general overview of this most important process. The readers are reminded to fully understand all of the conditions and requirements of their respective project prior to commencing the bidding process. Federal agencies such as EPA, DOT, USDA Rural Development, EDA, etc. have very specific conditions for projects they fund, and the project engineer must fully comply with their respective rules.

The engineer may choose bidding documents prepared by professional societies such as the National Society of Professional Engineers or American Institute of Architects. In any event, the project engineer must exercise professional judgement when selecting and preparing the bidding documents.

NOTE: Documents marked with an asterisk (*) were jointly prepared and endorsed by:

Economic Development Administration, Department of Commerce
Environmental Protection Agency
Farmers Home Administration, Department of Agriculture
Department of Housing and Urban Development
American Consulting Engineer Council
American Public Works Association
American Society of Civil Engineers
Associated General Contractors of America
National Society of Professional Engineers
National Utility Contractors Association

TABLE OF CONTENTS

Advertisement for Bids

Project No. 99-157

Drummond Sanitary District, Drummond, WI
(OWNER)

Sealed bids for Elevated Water Storage Tank Rehabilitation will be received by the Sanitary District at Drummond Fire Hall Meeting Room (North of Town Hall) until 2:00 o'clock., P.M., Local Time, Feb. 8, 2001 and then at said location publicly opened and read aloud.

MAJOR ITEMS OF BID

Removal and Disposal of Insulation
Blasting: Paint Removal & Disposal
Steel Tank Repair
Handrailing Installation
Circulation Pump System
Water Supply Control System
Related Appurtenances
Total Tank Painting

The Information for Bidders, Pre-Qualification Statement, Form of Bid, Form of Contract, Plans, Specifications, and Forms of Bid Bond, Performance and Payment Bond, and other contract documents may be examined at the following:

Eau Claire Builders Exchange, Wausau Builders Exchange, Duluth Builders Exchange, F.W. Dodge - Minneapolis, Morgan & Parmley, Ltd., Ladysmith, Wisconsin

Copies may be obtained at the office of Morgan & Parmley, Ltd. located at 115 W. 2nd Street, S., Ladysmith, WI 54848, (715) 532-3721 upon payment of a $ 25.00 non refundable handling fee.

The Owner reserves the right to waive any informalities or to reject any or all bids. Letting is subject to Section 61.54, 61.55, 61.56 and to Section 66.29, Wisconsin Statutes.

Each Bidder must deposit with their Bid, security in the amount, form and subject to the conditions provided in the Information for Bidders.

Attention of Bidders is particularly called to the requirements as to conditions of employment to be observed and minimum wage rates to be paid under the contract, Section 3, Segregated Facility, Section 109 and E.O. 11246.

No Bidder may withdraw their Bid within 60 days after the actual date of the opening thereof.

Date: Jan. 2, 2001 James Unseth, Commission President

Source: Morgan & Parmley, Ltd.

Advertisement for Bids

____
Owner
____
Address
____

Separate sealed BIDS for the construction of (briefly describe nature, scope, and major elements of the work)______________________________

____

____

____

will be received by______________________________

at the office of ______________________________

until ______________, (Standard Time—Daylight Savings Time)______________,

__________, and then at said office publicly opened and read aloud.

The CONTRACT DOCUMENTS may be examined at the following locations:

____

____

____

____

Copies of the CONTRACT DOCUMENTS may be obtained at the office of ________ ______________located at ______________________________

upon payment of $______________ for each set.

Any BIDDER, upon returning the CONTRACT DOCUMENTS promptly and in good condition, will be refunded his payment, and any non-bidder upon so returning the CONTRACT DOCUMENTS will be refunded $______________.

______________________________ ______________________________
Date

Source: * (See page 23-1)

Information for Bidders

BIDS will be received by ______________________

(herein called the "OWNER"), at ______________________

until __________, ________, and then at said office publicly opened and read aloud.

Each BID must be submitted in a sealed envelope, addressed to __________ __________ at __________. Each sealed envelope containing a BID must be plainly marked on the outside as BID for __________ and the envelope should bear on the outside the name of the BIDDER, his address, his license number if applicable and the name of the project for which the BID is submitted. If forwarded by mail, the sealed envelope containing the BID must be enclosed in another envelope addressed to the OWNER at __________ __________.

All BIDS must be made on the required BID form. All blank spaces for BID prices must be filled in, in ink or typewritten, and the BID form must be fully completed and executed when submitted. Only one copy of the BID form is required.

The OWNER may waive any informalities or minor defects or reject any and all BIDS. Any BID may be withdrawn prior to the above scheduled time for the opening of BIDS or authorized postponement thereof. Any BID received after the time and date specified shall not be considered. No BIDDER may withdraw a BID within 60 days after the actual date of the opening thereof. Should there be reasons why the contract cannot be awarded within the specified period, the time may be extended by mutual agreement between the OWNER and the BIDDER.

BIDDERS must satisfy themselves of the accuracy of the estimated quantities in the BID Schedule by examination of the site and a review of the drawings and specifications including ADDENDA. After BIDS have been submitted, the BIDDER shall not assert that there was a misunderstanding concerning the quantities of WORK or of the nature of the WORK to be done.

The OWNER shall provide to BIDDERS prior to BIDDING, all information which is pertinent to, and delineates and describes, the land owned and rights-of-way acquired or to be acquired.

The CONTRACT DOCUMENTS contain the provisions required for the construction of the PROJECT. Information obtained from an officer, agent, or employee of the OWNER or any other person shall not affect the risks or obligations assumed by the CONTRACTOR or relieve him from fulfilling any of the conditions of the contract.

Each BID must be accompanied by a BID bond payable to the OWNER for five percent of the total amount of the BID. As soon as the BID prices have been compared, the OWNER will return the BONDS of all except the three lowest responsible BIDDERS. When the Agreement is executed the bonds of the two remaining unsuccessful BIDDERS will be returned. The BID BOND of the successful BIDDER will be retained until the payment BOND and performance BOND have been executed and approved, after which it will be returned. A certified check may be used in lieu of a BID BOND.

A performance BOND and a payment BOND, each in the amount of 100 percent of the CONTRACT PRICE, with a corporate surety approved by the OWNER, will be required for the faithful performance of the contract.

Attorneys-in-fact who sign BID BONDS or payment BONDS and performance BONDS must file with each BOND a certified and effective dated copy of their power of attorney.

The party to whom the contract is awarded will be required to execute the Agreement and obtain the performance BOND and payment BOND within ten (10) calendar days from the date when NOTICE OF AWARD is delivered to the BIDDER. The NOTICE OF AWARD shall be accompanied by the necessary Agreement and BOND forms. In case of failure of the BIDDER to execute the Agreement, the OWNER may at his option consider the BIDDER in default, in which case the BID BOND accompanying the proposal shall become the property of the OWNER.

The OWNER within ten (10) days of receipt of acceptable performance BOND, payment BOND and Agreement signed by the party to whom the Agreement was awarded shall sign the Agreement and return to such party an executed duplicate of the Agreement. Should the OWNER not execute the Agreement within such period, the BIDDER may by WRITTEN NOTICE withdraw his signed Agreement. Such notice of withdrawal shall be effective upon receipt of the notice by the OWNER.

The NOTICE TO PROCEED shall be issued within ten (10) days of the execution of the Agreement by the OWNER. Should there be reasons why the NOTICE TO PROCEED cannot be issued within such period, the time may be extended by mutual agreement between the OWNER and CONTRACTOR. If the NOTICE TO PROCEED has not been issued within the ten (10) day period or within the period mutually agreed upon, the CONTRACTOR may terminate the Agreement without further liability on the part of either party.

The OWNER may make such investigations as he deems necessary to determine the ability of the BIDDER to perform the WORK, and the BIDDER shall furnish to the OWNER all such information and data for this purpose as the OWNER may request. The OWNER reserves the right to reject any BID if the evidence submitted by, or investigation of, such BIDDER fails to satisfy the OWNER that such BIDDER is properly qualified to carry out the obligations of the Agreement and to complete the WORK contemplated therein.

A conditional or qualified BID will not be accepted.

Award will be made to the lowest responsible BIDDER.

All applicable laws, ordinances, and the rules and regulations of all authorities having jurisdiction over construction of the PROJECT shall apply to the contract throughout.

Each BIDDER is responsible for inspecting the site and for reading and being thoroughly familiar with the CONTRACT DOCUMENTS. The failure or omission of any BIDDER to do any of the foregoing shall in no way relieve any BIDDER from any obligation in respect to his BID.

Further, the BIDDER agrees to abide by the requirements under Executive Order No. 11246, as amended, including specifically the provisions of the equal opportunity clause set forth in the SUPPLEMENTAL GENERAL CONDITIONS.

The low BIDDER shall supply the names and addresses of major material SUPPLIERS and SUBCONTRACTORS when requested to do so by the OWNER.

Inspection trips for prospective BIDDERS will leave from the office of the __________ at __________

The ENGINEER is __________. His address is __________.

Source: * (See page 23-1)

Bid Bond

KNOW ALL MEN BY THESE PRESENTS, that we, the undersigned, ______________

____ as Principal, and

____ as Surety, are hereby

held and firmly bound unto ______________________________ as OWNER

in the penal sum of____

for the payment of which, well and truly to be made, we hereby jointly and severally bind ourselves, successors and assigns.

Signed, this ________________day of________________, 20__________.

The Condition of the above obligation is such that whereas the Principal has submitted

to ____a certain BID,

attached hereto and hereby made a part hereof to enter into a contract in writing, for the

____

____

____

NOW, THEREFORE,

(a) If said BID shall be rejected, or

(b) If said BID shall be accepted and the Principal shall execute and deliver a contract in the Form of Contract attached hereto (properly completed in accordance with said BID) and shall furnish a BOND for his faithful performance of said contract, and for the payment of all persons performing labor or furnishing materials in connection therewith, and shall in all other respects perform the agreement created by the acceptance of said BID,

then this obligation shall be void, otherwise the same shall remain in force and effect; it being expressly understood and agreed that the liability of the Surety for any and all claims hereunder shall, in no event, exceed the penal amount of this obligation as herein stated.

The Surety, for value received, hereby stipulates and agrees that the obligations of said Surety and its BOND shall be in no way impaired or affected by any extension of the time within which the OWNER may accept such BID; and said Surety does herby waive notice of any such extension.

IN WITNESS WHEREOF, the Principal and the Surety have hereunto set their hands and seals, and such of them as are corporations have caused their corporate seals to be hereto affixed and these presents to be signed by their proper officers, the day and year first set forth above.

______________________________ (L.S.)
Principal

____
Surety

By: ______________________________

IMPORTANT—Surety companies executing BONDS must appear on the Treasury Department's most current list (Circular 570 as amended) and be authorized to transact business in the state where the project is located.

Source: * (See page 23-1)

Owner's Instructions for Bonds and Insurance

Project: ______________________ Date: ______________________

______________________ Project No.: ______________________

Owner: ______________________

The Contractor is hereby instructed that the following bonds and insurance are required, by the Owner, as a condition of this contract. Proof of coverage is not required for bidding but is required as a part of the agreement.

1) Bid Security:

Not Required

Required in the amount of $__________ OR ________ percent of Total Bid

2) Performance and Payment Bond:

Not Required

Required in the amount of $__________ OR ________ percent of Total Bid

3) Contractors Liability Insurance:

A. Workers Compensation

1)	State	Statutory
2)	Applicable Federal	Statutory
3)	Employers Liability	$__________
4)	Benefits Required by Union Labor Contracts	As Applicable

B: Commercial General Liability including: Premises-Operations, Independent Contractors' Protective, Products and Completed Operations, Broad Form Property Damage, Personal Injury and Contracted.

1) Bodily Injury & Property Damage Combined Single Limit

$__________ Each Occurrence

$__________ Aggregate Products and Completed Operations

2) Products and completed Operations Insurance shall be maintained after final payment for:

One Year

Two Years

3) Property Damage Liability Insurance shall include:

Explosion

Collapse

Underground

____

4) Contractual Liability (Hold Harmless Coverage) for bodily injury and property damage combined single limit:

$__________ Each Occurrence

5) Personal injury with Employment Exclusions deleted:

$__________ Aggregate

C: Commercial Automobile Liability; owned, non-owned, hired:

Bodily Injury & Property Damage $__________ Accident

Combined Single Limit $__________ Aggregate

D: Aircraft Liability; owned and non-owned:

Bodily Injury & Property Damage $__________ Each Accident

Combined Single Limit $__________ Aggregate

E: Watercraft Liability; owned and non-owned:

Bodily Injury & Property Damage $__________ Each Accident

Combined Single Limit $__________ Aggregate

F: Property Insurance: Contractor shall purchase all Risk, completed Value Builders Risk Insurance.

$ Full Value

G: Other Insurance:
(If none so state)

Description	Amount

By: ______________________________ Date: ______________

Title: ______________________________

Source: Morgan & Parmley, Ltd.

General Conditions

1. *DEFINITIONS*

1.1 Wherever used in the CONTRACT DOCUMENTS, the following terms shall have the meanings indicated which shall be applicable to both the singular and plural thereof:

1.2 ADDENDA—Written or graphic instruments issued prior to the execution of the Agreement which modify or interpret the CONTRACT DOCUMENTS, DRAWINGS and SPECIFICATIONS, by additions, deletions, clarifications or corrections.

1.3 BID—The offer or proposal of the BIDDER submitted on the prescribed form setting forth the prices for the WORK to be performed.

1.4 BIDDER—Any person, firm or corporation submitting a BID for the WORK.

1.5 BONDS—Bid, Performance, and Payment Bonds and other instruments of security, furnished by the CONTRACTOR and his surety in accordance with the CONTRACT DOCUMENTS.

1.6 CHANGE ORDER—A written order to the CONTRACTOR authorizing an addition, deletion or revision in the WORK within the general scope of the CONTRACT DOCUMENTS, or authorizing an adjustment in the CONTRACT PRICE or CONTRACT TIME.

1.7 CONTRACT DOCUMENTS—The contract, including Advertisement For Bids, Information For Bidders, BID, Bid Bond, Agreement, Payment Bond, Performance Bond, NOTICE OF AWARD, NOTICE TO PROCEED, CHANGE ORDER, DRAWINGS, SPECIFICATIONS, and ADDENDA.

1.8 CONTRACT PRICE—The total monies payable to the CONTRACTOR under the terms and conditions of the CONTRACT DOCUMENTS.

1.9 CONTRACT TIME—The number of calendar days stated in the CONTRACT DOCUMENTS for the completion of the WORK.

1.10 CONTRACTOR—The person, firm or corporation with whom the OWNER has executed the Agreement.

1.11 DRAWINGS—The part of the CONTRACT DOCUMENTS which show the characteristics and scope of the WORK to be performed and which have been prepared or approved by the ENGINEER.

1.12 ENGINEER—The person, firm or corporation named as such in the CONTRACT DOCUMENTS.

1.13 FIELD ORDER—A written order effecting a change in the WORK not involving an adjustment in the CONTRACT PRICE or an extension of the CONTRACT TIME, issued by the ENGINEER to the CONTRACTOR during construction.

1.14 NOTICE OF AWARD—The written notice of the acceptance of the BID from the OWNER to the successful BIDDER.

1.15 NOTICE TO PROCEED—Written communication issued by the OWNER to the CONTRACTOR authorizing him to proceed with the WORK and establishing the date of commencement of the WORK.

1.16 OWNER—A public or quasi-public body or authority, corporation, association, partnership, or individual for whom the WORK is to be performed.

1.17 PROJECT—The undertaking to be performed as provided in the CONTRACT DOCUMENTS.

1.18 RESIDENT PROJECT REPRESENTATIVE—The authorized representative of the OWNER who is assigned to the PROJECT site or any part thereof.

1.19 SHOP DRAWINGS—All drawings, diagrams, illustrations, brochures, schedules and other data which are prepared by the CONTRACTOR, a SUBCONTRACTOR, manufacturer, SUPPLIER or distributor, which illustrate how specific portions of the WORK shall be fabricated or installed.

1.20 SPECIFICATIONS—A part of the CONTRACT DOCUMENTS consisting of written descriptions of a technical nature of materials, equipment, construction systems, standards and workmanship.

1.21 SUBCONTRACTOR—An individual, firm or corporation having a direct contract with the CONTRACTOR or with any other SUBCONTRACTOR for the performance of a part of the WORK at the site.

1.22 SUBSTANTIAL COMPLETION—That date as certified by the ENGINEER when the construction of the PROJECT or a specified part thereof is sufficiently completed, in accordance with the CONTRACT DOCUMENTS, so that the PROJECT or specified part can be utilized for the purposes for which it is intended.

1.23 SUPPLEMENTAL GENERAL CONDITIONS—Modifications to General Conditions required by a Federal agency for participation in the PROJECT and approved by the agency in writing prior to inclusion in the CONTRACT DOCUMENTS, or such requirements that may be imposed by applicable state laws.

1.24 SUPPLIER—Any person or organization who supplies materials or equipment for the WORK, including that fabricated to a special design, but who does not perform labor at the site.

1.25 WORK—All labor necessary to produce the construction required by the CONTRACT DOCUMENTS, and all materials and equipment incorporated or to be incorporated in the PROJECT.

1.26 WRITTEN NOTICE—Any notice to any party of the Agreement relative to any part of this Agreement in writing and considered delivered and the service thereof completed, when posted by certified or registered mail to the said party at his last given address, or delivered in person to said party or his authorized representative on the WORK.

2. *ADDITIONAL INSTRUCTIONS AND DETAIL DRAWINGS*

2.1 The CONTRACTOR may be furnished additional instructions and detail drawings, by the ENGINEER, as necessary to carry out the WORK required by the CONTRACT DOCUMENTS.

2.2 The additional drawings and instruction thus supplied will become a part of the CONTRACT DOCUMENTS. The CONTRACTOR shall carry out the WORK in accordance with the additional detail drawings and instructions.

3. *SCHEDULES, REPORTS AND RECORDS*

3.1 The CONTRACTOR shall submit to the OWNER such schedule of quantities and costs, progress schedules, payrolls, reports, estimates, records and other data where applicable as are required by the CONTRACT DOCUMENTS for the WORK to be performed.

3.2 Prior to the first partial payment estimate the CONTRACTOR shall submit construction progress schedules showing the order in which he proposes to carry on the WORK, including dates at which he will start the various parts of the WORK, estimated date of completion of each part and, as applicable:

3.2.1. The dates at which special detail drawings will be required; and

3.2.2 Respective dates for submission of SHOP DRAWINGS, the beginning of manufacture, the testing and the installation of materials, supplies and equipment.

3.3 The CONTRACTOR shall also submit a schedule of payments that he anticipates he will earn during the course of the WORK.

4. *DRAWINGS AND SPECIFICATIONS*

4.1 The intent of the DRAWINGS and SPECIFICATIONS is that the CONTRACTOR shall furnish all labor, materials, tools, equipment, and transportation necessary for the proper execution of the WORK in accordance with the CONTRACT DOCUMENTS and all incidental work necessary to complete the PROJECT in an acceptable manner, ready for use, occupancy or operation by the OWNER.

4.2 In case of conflict between the DRAWINGS and SPECIFICATIONS, the SPECIFICATIONS shall govern. Figure dimensions on DRAWINGS shall govern over scale dimensions, and detailed DRAWINGS shall govern over general DRAWINGS.

4.3 Any discrepancies found between the DRAWINGS and SPECIFICATIONS and site conditions or any inconsistencies or ambiguities in the DRAWINGS or SPECIFICATIONS shall be immediate y reported to the ENGINEER, in writing, who shall promptly correct such inconsistencies or ambiguities in writing. WORK done by the CONTRACTOR after his discovery of such discrepancies, inconsistencies or ambiguities shall be done at the CONTRACTOR'S risk.

5. *SHOP DRAWINGS*

5.1 The CONTRACTOR shall provide SHOP DRAWINGS as may be necessary for the prosecution of the WORK as required by the CONTRACT DOCUMENTS. The ENGINEER shall promptly review all SHOP DRAWINGS. The ENGINEER'S approval of any SHOP DRAWING shall not release the CONTRACTOR from responsibility for deviations from the CONTRACT DOCUMENTS. The approval of any SHOP DRAWING which substantially deviates from the requirement of the CONTRACT DOCUMENTS shall be evidenced by a CHANGE ORDER.

5.2 When submitted for the ENGINEER'S review, SHOP DRAWINGS shall bear the CONTRACTOR'S certification that he has reviewed, checked and approved the SHOP DRAWINGS and that they are in conformance with the requirements of the CONTRACT DOCUMENTS.

5.3 Portions of the WORK requiring a SHOP DRAWING or sample submission shall not begin until the SHOP DRAWING or submission has been approved by the ENGINEER. A copy of each approved SHOP DRAWING and each approved sample shall be kept in good order by the CONTRACTOR at the site and shall be available to the ENGINEER.

6. *MATERIALS, SERVICES AND FACILITIES*

6.1 It is understood that, except as otherwise specifically stated in the CONTRACT DOCUMENTS, the CONTRACTOR shall provide and pay for all materials, labor, tools, equipment, water, light, power, transportation, supervision, temporary construction of any nature, and all other services and facilities of any nature whatsoever necessary to execute, complete, and deliver the WORK within the specified time.

6.2 Materials and equipment shall be so stored as to insure the preservation of their quality and fitness for the WORK. Stored materials and equipment to be incorporated in the WORK shall be located so as to facilitate prompt inspection.

6.3 Manufactured articles, materials and equipment shall be applied, installed, connected, erected, used, cleaned and conditioned as directed by the manufacturer.

6.4 Materials, supplies and equipment shall be in accordance with samples submitted by the CONTRACTOR and approved by the ENGINEER.

6.5 Materials, supplies or equipment to be incorporated into the WORK shall not be purchased by the

Source: * (See page 23-1)

General Conditions *(continued)*

CONTRACTOR or the SUBCONTRACTOR subject to a chattel mortgage or under a conditional sale contract or other agreement by which an interest is retained by the seller.

7. *INSPECTION AND TESTING*

7.1 All materials and equipment used in the construction of the PROJECT shall be subject to adequate inspection and testing in accordance with generally accepted standards, as required and defined in the CONTRACT DOCUMENTS.

7.2 The OWNER shall provide all inspection and testing services not required by the CONTRACT DOCUMENTS.

7.3 The CONTRACTOR shall provide at his expense the testing and inspection services required by the CONTRACT DOCUMENTS.

7.4 If the CONTRACT DOCUMENTS, laws, ordinances, rules, regulations or orders of any public authority having jurisdiction require any WORK to specifically be inspected, tested, or approved by someone other than the CONTRACTOR, the CONTRACTOR will give the ENGINEER timely notice of readiness. The CONTRACTOR will then furnish the ENGINEER the required certificates of inspection, testing or approval.

7.5 Inspections, tests or approvals by the engineer or others shall not relieve the CONTRACTOR from his obligations to perform the WORK in accordance with the requirements of the CONTRACT DOCUMENTS.

7.6 The ENGINEER and his representatives will at all times have access to the WORK. In addition, authorized representatives and agents of any participating Federal or state agency shall be permitted to inspect all work, materials, payrolls, records of personnel, invoices of materials, and other relevant data and records. The CONTRACTOR will provide proper facilities for such access and observation of the WORK and also for any inspection, or testing thereof.

7.7 If any WORK is covered contrary to the written instructions of the ENGINEER it must, if requested by the ENGINEER, be uncovered for his observation and replaced at the CONTRACTOR'S expense.

7.8 If the ENGINEER considers it necessary or advisable that covered WORK be inspected or tested by others, the CONTRACTOR, at the ENGINEER'S request, will uncover, expose or otherwise make available for observation, inspection or testing as the ENGINEER may require, that portion of the WORK in question, furnishing all necessary labor, materials, tools, and equipment. If it is found that such WORK is defective, the CONTRACTOR will bear all the expenses of such uncovering, exposure, observation, inspection and testing and of satisfactory reconstruction. If, however, such WORK is not found to be defective, the CONTRACTOR will be allowed an increase in the CONTRACT PRICE or an extension of the CONTRACT TIME, or both, directly attributable to such uncovering, exposure, observation, inspection, testing and reconstruction and an appropriate CHANGE ORDER shall be issued.

8. *SUBSTITUTIONS*

8.1 Whenever a material, article or piece of equipment is identified on the DRAWINGS or SPECIFICATIONS by reference to brand name or catalogue number, it shall be understood that this is referenced for the purpose of defining the performance or other salient requirements and that other products of equal capacities, quality and function shall be considered. The CONTRACTOR may recommend the substitution of a material, article, or piece of equipment of equal substance and function for those referred to in the CONTRACT DOCUMENTS by reference to brand name or catalogue number, and if, in the opinion of the ENGINEER, such material, article, or piece of equipment is of equal substance and function to that specified, the ENGINEER may approve its substitution and use by the CONTRACTOR. Any cost differential shall be deductible from the CONTRACT PRICE and the CONTRACT DOCUMENTS shall be appropriately modified by CHANGE ORDER. The CONTRACTOR warrants that if substitutes are approved, no major changes in the function or general design of the PROJECT will result. Incidental changes or extra component parts required to accommodate the substitute will be made by the CONTRACTOR without a change in the CONTRACT PRICE or CONTRACT TIME.

9. *PATENTS*

9.1 The CONTRACTOR shall pay all applicable royalties and license fees. He shall defend all suits or claims for infringement of any patent rights and save the OWNER harmless from loss on account thereof, except that the OWNER shall be responsible for any such loss when a particular process, design, or the product of a particular manufacturer or manufacturers is specified, however if the CONTRACTOR has reason to believe that the design, process or product specified is an infringement of a patent, he shall be responsible for such loss unless he promptly gives such information to the ENGINEER.

10. *SURVEYS, PERMITS, REGULATIONS*

10.1 The OWNER shall furnish all boundary surveys and establish all base lines for locating the principal component parts of the WORK together with a suitable number of bench marks adjacent to the WORK as shown in the CONTRACT DOCUMENTS. From the information provided by the OWNER, unless otherwise specified in the CONTRACT DOCUMENTS, the CONTRACTOR shall develop and make all detail surveys needed for construction such as slope stakes, batter boards, stakes for pile locations and other working points, lines, elevations and cut sheets.

10.2 The CONTRACTOR shall carefully preserve bench marks, reference points and stakes and, in case of willful or careless destruction, he shall be charged with the resulting expense and shall be responsible for any mistakes that may be caused by their unnecessary loss or disturbance.

10.3 Permits and licenses of a temporary nature necessary for the prosecution of the WORK shall be secured and paid for by the CONTRACTOR unless otherwise stated in the SUPPLEMENTAL GENERAL CONDITIONS. Permits, licenses and easements for permanent structures or permanent changes in existing facilities shall be secured and paid for by the OWNER, unless otherwise specified. The CONTRACTOR shall give all notices and comply with all laws, ordinances, rules and regulations bearing on the conduct of the WORK as drawn and specified. If the CONTRACTOR observes that the CONTRACT DOCUMENTS are at variance therewith, he shall promptly notify the ENGINEER in writing, and any necessary changes shall be adjusted as provided in Section 13, CHANGES IN THE WORK.

11. *PROTECTION OF WORK, PROPERTY AND PERSONS*

11.1 The CONTRACTOR will be responsible for initiating, maintaining and supervising all safety precautions and programs in connection with the WORK. He will take all necessary precautions for the safety of, and will provide the necessary protection to prevent damage, injury or loss to all employees on the WORK and other persons who may be affected thereby, all the WORK and all materials or equipment to be incorporated therein, whether in storage on or off the site, and other property at the site or adjacent thereto, including trees, shrubs, lawns, walks, pavements, roadways, structures and utilities not designated for removal, relocation or replacement in the course of construction.

11.2 The CONTRACTOR will comply with all applicable laws, ordinances, rules, regulations and orders of any public body having jurisdiction. He will erect and maintain, as required by the conditions and progress of the WORK, all necessary safeguards for safety and protection. He will notify owners of adjacent utilities when prosecution of the WORK may affect them. The CONTRACTOR will remedy all damage, injury or loss to any property caused, directly or indirectly, in whole or in part, by the CONTRACTOR, any SUBCONTRACTOR or anyone directly or indirectly employed by any of them or anyone for whose acts any of them be liable, except damage or loss attributable to the fault of the CONTRACT DOCUMENTS or to the acts or omissions of the OWNER or the ENGINEER or anyone employed by either of them or anyone for whose acts either of them may be liable, and not attributable, directly or indirectly, in whole or in part, to the fault or negligence of the CONTRACTOR.

11.3 In emergencies affecting the safety of persons or the WORK or property at the site or adjacent thereto, the CONTRACTOR, without special instruction or authorization from the ENGINEER or OWNER, shall act to prevent threatened damage, injury or loss. He will give the ENGINEER prompt WRITTEN NOTICE of any significant changes in the WORK or deviations from the CONTRACT DOCUMENTS caused thereby, and a CHANGE ORDER shall thereupon be issued covering the changes and deviations involved.

12. *SUPERVISION BY CONTRACTOR*

12.1 The CONTRACTOR will supervise and direct the WORK. He will be solely responsible for the means, methods, techniques, sequences and procedures of construction. The CONTRACTOR will employ and maintain on the WORK a qualified supervisor or superintendent who shall have been designated in writing by the CONTRACTOR as the CONTRACTOR'S representative at the site. The supervisor shall have full authority to act on behalf of the CONTRACTOR and all communications given to the supervisor shall be as binding as if given to the CONTRACTOR. The supervisor shall be present on the site at all times as required to perform adequate supervision and coordination of the WORK.

13. *CHANGES IN THE WORK*

13.1 The OWNER may at any time, as the need arises, order changes within the scope of the WORK without invalidating the Agreement. If such changes increase or decrease the amount due under the CONTRACT DOCUMENTS, or in the time required for performance of the WORK, an equitable adjustment shall be authorized by CHANGE ORDER.

13.2 The ENGINEER, also, may at any time, by issuing a FIELD ORDER, make changes in the details of the WORK. The CONTRACTOR shall proceed with the performance of any changes in the WORK so ordered by the ENGINEER unless the CONTRACTOR believes that such FIELD ORDER entitles him to a change in CONTRACT PRICE or TIME, or both, in which event he shall give the ENGINEER WRITTEN NOTICE thereof within seven (7) days after the receipt of the ordered change. Thereafter the CONTRACTOR shall document the basis for the change in CONTRACT PRICE or TIME within thirty (30) days. The CONTRACTOR shall not execute such changes pending the receipt of an executed CHANGE ORDER or further instruction from the OWNER.

14. *CHANGES IN CONTRACT PRICE*

14.1 The CONTRACT PRICE may be changed only by a CHANGE ORDER. The value of any WORK covered by a CHANGE ORDER or of any claim for increase or decrease in the CONTRACT PRICE shall be determined by one or more of the following methods in the order of precedence listed below:

(a) Unit prices previously approved.

(b) An agreed lump sum.

(c) The actual cost for labor, direct overhead, materials, supplies, equipment, and other services necessary to complete the work. In addition there shall be added an amount to be agreed upon but not to exceed fifteen (15) percent of the actual cost of the WORK to cover the cost of general overhead and profit.

15. *TIME FOR COMPLETION AND LIQUIDATED DAMAGES*

15.1 The date of beginning and the time for completion of the WORK are essential conditions of the CONTRACT DOCUMENTS and the WORK embraced shall be commenced on a date specified in the NOTICE TO PROCEED.

15.2 The CONTRACTOR will proceed with the WORK at such rate of progress to insure full completion within the CONTRACT TIME. It is expressly understood and agreed, by and between the CONTRACTOR and the OWNER, that the CONTRACT TIME for the completion of the WORK described herein is a reasonable time, taking into consideration the average climatic and economic conditions and other factors prevailing in the locality of the WORK.

15.3 If the CONTRACTOR shall fail to complete the WORK within the CONTRACT TIME, or extension of time granted by the OWNER, then the CONTRACTOR will pay to the OWNER the amount for liquidated damages as specified in the BID for each calendar day that the CONTRACTOR shall be in default after the time stipulated in the CONTRACT DOCUMENTS.

15.4 The CONTRACTOR shall not be charged with liquidated damages or any excess cost when the delay in completion of the WORK is due to the following, and the CONTRACTOR has promptly given WRITTEN NOTICE of such delay to the OWNER or ENGINEER.

15.4.1 To any preference, priority or allocation

General Conditions *(continued)*

order duly issued by the OWNER.

15.4.2 To unforeseeable causes beyond the control and without the fault or negligence of the CONTRACTOR, including but not restricted to, acts of God, or of the public enemy, acts of the OWNER, acts of another CONTRACTOR in the performance of a contract with the OWNER, fires, floods, epidemics, quarantine restrictions, strikes, freight embargoes, and abnormal and unforeseeable weather; and

15.4.3 To any delays of SUBCONTRACTORS occasioned by any of the causes specified in paragraphs 15.4.1 and 15.4.2 of this article.

16. *CORRECTION OF WORK*

16.1 The CONTRACTOR shall promptly remove from the premises all WORK rejected by the ENGINEER for failure to comply with the CONTRACT DOCUMENTS, whether incorporated in the construction or not, and the CONTRACTOR shall promptly replace and re-execute the WORK in accordance with the CONTRACT DOCUMENTS and without expense to the OWNER and shall bear the expense of making good all WORK of other CONTRACTORS destroyed or damaged by such removal or replacement.

16.2 All removal and replacement WORK shall be done at the CONTRACTOR'S expense. If the CONTRACTOR does not take action to remove such rejected WORK within ten (10) days after receipt of WRITTEN NOTICE, the OWNER may remove such WORK and store the materials at the expense of the CONTRACTOR.

17. *SUBSURFACE CONDITIONS*

17.1 The CONTRACTOR shall promptly, and before such conditions are disturbed, except in the event of an emergency, notify the OWNER by WRITTEN NOTICE of:

17.1.1 Subsurface or latent physical conditions at the site differing materially from those indicated in the CONTRACT DOCUMENTS; or

17.1.2 Unknown physical conditions at the site, of an unusual nature, differing materially from those ordinarily encountered and generally recognized as inherent in WORK of the character provided for in the CONTRACT DOCUMENTS.

17.2 The OWNER shall promptly investigate the conditions, and if he finds that such conditions do so materially differ and cause an increase or decrease in the cost of, or in the time required for, performance of the WORK, an equitable adjustment shall be made and the CONTRACT DOCUMENTS shall be modified by a CHANGE ORDER. Any claim of the CONTRACTOR for adjustment hereunder shall not be allowed unless he has given the required WRITTEN NOTICE; provided that the OWNER may, if he determines the facts so justify, consider and adjust any such claims asserted before the date of final payment.

18. *SUSPENSION OF WORK, TERMINATION AND DELAY*

18.1 The OWNER may suspend the WORK or any portion thereof for a period of not more than ninety days or such further time as agreed upon by the CONTRACTOR, by WRITTEN NOTICE to the CONTRACTOR and the ENGINEER which notice shall fix the date on which WORK shall be resumed. The CONTRACTOR will resume that WORK on the date so fixed. The CONTRACTOR will be allowed an increase in the CONTRACT PRICE or an extension of the CONTRACT TIME, or both, directly attributable to any suspension.

18.2 If the CONTRACTOR is adjudged a bankrupt or insolvent, or if he makes a general assignment for the benefit of his creditors, or if a trustee or receiver is appointed for the CONTRACTOR or for any of his property, or if he files a petition to take advantage of any debtor's act, or to reorganize under the bankruptcy or applicable laws, or if he repeatedly fails to supply sufficient skilled workmen or suitable materials or equipment, or if he repeatedly fails to make prompt payments to SUBCONTRACTORS or for labor, materials or equipment or if he disregards laws, ordinances, rules, regulations or orders of any public body having jurisdiction of the WORK or if he disregards the authority of the ENGINEER, or if he otherwise violates any provision of the CONTRACT DOCUMENTS, then the OWNER may, without prejudice to any other right or remedy and after giving the CONTRACTOR and his surety a minimum of ten (10) days from delivery of a WRITTEN NOTICE, terminate the services of the CONTRACTOR and take possession of the PROJECT and of all materials, equipment, tools, construction equipment and machinery thereon owned by the CONTRACTOR, and finish the WORK by whatever method he may deem expedient. In such case the CONTRACTOR shall not be entitled to receive any further payment until the WORK is finished. If the unpaid balance of the CONTRACT PRICE exceeds the direct and indirect costs of completing the PROJECT, including compensation for additional professional services, such excess SHALL BE PAID TO THE CONTRACTOR. If such costs exceed such unpaid balance, the CONTRACTOR will pay the difference to the OWNER. Such costs incurred by the OWNER will be determined by the ENGINEER and incorporated in a CHANGE ORDER.

18.3 Where the CONTRACTOR'S services have been so terminated by the OWNER, said termination shall not affect any right of the OWNER against the CONTRACTOR then existing or which may thereafter accrue. Any retention or payment of monies by the OWNER due the CONTRACTOR will not release the CONTRACTOR from compliance with the CONTRACT DOCUMENTS.

18.4 After ten (10) days from delivery of a WRITTEN NOTICE to the CONTRACTOR and the ENGINEER, the OWNER may, without cause and without prejudice to any other right or remedy, elect to abandon the PROJECT and terminate the Contract. In such case, the CONTRACTOR shall be paid for all WORK executed and any expense sustained plus reasonable profit.

18.5 If, through no act or fault of the CONTRACTOR, the WORK is suspended for a period of more than ninety (90) days by the OWNER or under an order of court or other public authority, or the ENGINEER fails to act on any request for payment within thirty (30) days after it is submitted, or the OWNER fails to pay the CONTRACTOR substantially the sum approved by the ENGINEER or awarded by arbitrators within thirty (30) days of its approval and presentation, then the CONTRACTOR may, after ten (10) days from delivery of a WRITTEN NOTICE to the OWNER and the ENGINEER, terminate the CONTRACT and recover from the OWNER payment for all WORK executed and all expenses sustained. In addition and in lieu of terminating the CONTRACT, if the ENGINEER has failed to act on a request for payment or if the OWNER has failed to make any payment as aforesaid, the CONTRACTOR may upon ten (10) days written notice to the OWNER and the ENGINEER stop the WORK until he has been paid all amounts then due, in which event and upon resumption of the WORK, CHANGE ORDERS shall be issued for adjusting the CONTRACT PRICE or extending the CONTRACT TIME or both to compensate for the costs and delays attributable to the stoppage of the WORK.

18.6 If the performance of all or any portion of the WORK is suspended, delayed, or interrupted as a result of a failure of the OWNER or ENGINEER to act within the time specified in the CONTRACT DOCUMENTS, or if no time is specified, within a reasonable time, an adjustment in the CONTRACT PRICE or an extension of the CONTRACT TIME, or both, shall be made by CHANGE ORDER to compensate the CONTRACTOR for the costs and delays necessarily caused by the failure of the OWNER or ENGINEER.

19. *PAYMENTS TO CONTRACTOR*

19.1 At least ten (10) days before each progress payment falls due (but not more often than once a month), the CONTRACTOR will submit to the ENGINEER a partial payment estimate filled out and signed by the CONTRACTOR covering the WORK performed during the period covered by the partial payment estimate and supported by such data as the ENGINEER may reasonably require. If payment is requested on the basis of materials and equipment not incorporated in the WORK but delivered and suitably stored at or near the site, the partial payment estimate shall also be accompanied by such supporting data, satisfactory to the OWNER, as will establish the OWNER's title to the material and equipment and protect his interest therein, including applicable insurance. The ENGINEER will, within ten (10) days after receipt of each partial payment estimate, either indicate in writing his approval of payment and present the partial payment estimate to the OWNER, or return the partial payment estimate to the CONTRACTOR indicating in writing his reasons for refusing to approve payment. In the latter case, the CONTRACTOR may make the necessary corrections and resubmit the partial payment estimate. The OWNER will, within ten (10) days of presentation to him of an approved partial payment estimate, pay the CONTRACTOR a progress payment on the basis of the approved partial payment estimate. The OWNER shall retain ten (10) percent of the amount of each payment until final completion and acceptance of all work covered by the CONTRACT DOCUMENTS. The OWNER at any time, however, after fifty (50) percent of the WORK has been completed, if he finds that satisfactory progress is being made, shall reduce retainage to five (5%) percent on the current and remaining estimates. When the WORK is substantially complete (operational or beneficial occupancy), the retained amount may be further reduced below five (5) percent to only that amount necessary to assure completion. On completion and acceptance of a part of the WORK on which the price is stated separately in the CONTRACT DOCUMENTS, payment may be made in full, including retained percentages, less authorized deductions.

19.2 The request for payment may also include an allowance for the cost of such major materials and equipment which are suitably stored either at or near the site.

19.3 Prior to SUBSTANTIAL COMPLETION, the OWNER, with the approval of the ENGINEER and with the concurrence of the CONTRACTOR, may use any completed or substantially completed portions of the WORK. Such use shall not constitute an acceptance of such portions of the WORK.

19.4 The OWNER shall have the right to enter the premises for the purpose of doing work not covered by the CONTRACT DOCUMENTS. This provision shall not be construed as relieving the CONTRACTOR of the sole responsibility for the care and protection of the WORK, or the restoration of any damaged WORK except such as may be caused by agents or employees of the OWNER.

19.5 Upon completion and acceptance of the WORK, the ENGINEER shall issue a certificate attached to the final payment request that the WORK has been accepted by him under the conditions of the CONTRACT DOCUMENTS. The entire balance found to be due the CONTRACTOR, including the retained percentages, but except such sums as may be lawfully retained by the OWNER, shall be paid to the CONTRACTOR within thirty (30) days of completion and acceptance of the WORK.

19.6 The CONTRACTOR will indemnify and save the OWNER or the OWNER'S agents harmless from all claims growing out of the lawful demands of SUBCONTRACTORS, laborers, workmen, mechanics, materialmen, and furnishers of machinery and parts thereof, equipment, tools, and all supplies, incurred in the furtherance of the performance of the WORK. The CONTRACTOR shall, at the OWNER'S request, furnish satisfactory evidence that all obligations of the nature designated above have been paid, discharged, or waived. If the CONTRACTOR fails to do so the OWNER may, after having notified the CONTRACTOR, either pay unpaid bills or withhold from the CONTRACTOR'S unpaid compensation a sum of money deemed reasonably sufficient to pay any and all such lawful claims until satisfactory evidence is furnished that all liabilities have been fully discharged whereupon payment to the CONTRACTOR shall be resumed, in accordance with the terms of the CONTRACT DOCUMENTS, but in no event shall the provisions of this sentence be construed to impose any obligations upon the OWNER to either the CONTRACTOR, his Surety, or any third party. In paying any unpaid bills of the CONTRACTOR, any payment so made by the OWNER shall be considered as a payment made under the CONTRACT DOCUMENTS by the OWNER to the CONTRACTOR and the OWNER shall not be liable to the CONTRACTOR for any such payments made in good faith.

19.7 If the OWNER fails to make payment thirty (30) days after approval by the ENGINEER, in addition to other remedies available to the CONTRACTOR, there shall be added to each such payment interest at the maximum legal rate commencing on the first day after said payment is due and continuing until the payment is received by the CONTRACTOR.

General Conditions *(continued)*

20. *ACCEPTANCE OF FINAL PAYMENT AS RELEASE*

20.1 The acceptance by the CONTRACTOR of final payment shall be and shall operate as a release to the OWNER of all claims and all liability to the CONTRACTOR other than claims in stated amounts as may be specifically excepted by the CONTRACTOR for all things done or furnished in connection with this WORK and for every act and neglect of the OWNER and others relating to or arising out of this WORK. Any payment, however, final or otherwise, shall not release the CONTRACTOR or his sureties from any obligations under the CONTRACT DOCUMENTS or the Performance BOND and Payment BONDS.

21. *INSURANCE*

21.1 The CONTRACTOR shall purchase and maintain such insurance as will protect him from claims set forth below which may arise out of or result from the CONTRACTOR'S execution of the WORK, whether such execution be by himself or by any SUBCONTRACTOR or by anyone directly or indirectly employed by any of them, or by anyone for whose acts any of them may be liable:

21.1.1 Claims under workmen's compensation, disability benefit and other similar employee benefit acts;

21.1.2 Claims for damages because of bodily injury, occupational sickness or disease, or death of his employees;

21.1.3 Claims for damages because of bodily injury, sickness or disease, or death of any person other than his employees;

21.1.4 Claims for damages insured by usual personal injury liability coverage which are sustained (1) by any person as a result of an offense directly or indirectly related to the employment of such person by the CONTRACTOR, or (2) by any other person; and

21.1.5 Claims for damages because of injury to or destruction of tangible property, including loss of use resulting therefrom.

21.2 Certificates of Insurance acceptable to the OWNER shall be filed with the OWNER prior to commencement of the WORK. These Certificates shall contain a provision that coverages afforded under the policies will not be cancelled unless at least fifteen (15) days prior WRITTEN NOTICE has been given to the OWNER.

21.3 The CONTRACTOR shall procure and maintain, at his own expense, during the CONTRACT TIME, liability insurance as hereinafter specified;

21.3.1 CONTRACTOR'S General Public Liability and Property Damage Insurance including vehicle coverage issued to the CONTRACTOR and protecting him from all claims for personal injury, including death, and all claims for destruction of or damage to property, arising out of or in connection with any operations under the CONTRACT DOCUMENTS, whether such operations be by himself or by any SUBCONTRACTOR under him, or anyone directly or indirectly employed by the CONTRACTOR or by a SUBCONTRACTOR under him. Insurance shall be written with a limit of liability of not less than $500,000 for all damages arising out of bodily injury, including death, at any time resulting therefrom, sustained by any one person in any one accident; and a limit of liability of not less than $500,000 aggregate for any such damages sustained by two or more persons in any one accident. Insurance shall be written with a limit of liability of not less than $200,000 for all property damage sustained by any one person in any one accident; and a limit of liability of not less than $200,000 aggregate for any such damage sustained by two or more persons in any one accident.

21.3.2 The CONTRACTOR shall acquire and maintain, if applicable, Fire and Extended Coverage insurance upon the PROJECT to the full insurable value thereof for the benefit of the OWNER, the CONTRACTOR, and SUBCONTRACTORS as their interest may appear. This provision shall in no way release the CONTRACTOR or CONTRACTOR'S surety from obligations under the CONTRACT DOCUMENTS to fully complete the PROJECT.

21.4 The CONTRACTOR shall procure and maintain, at his own expense, during the CONTRACT TIME, in accordance with the provisions of the laws of the state in which the work is performed, Workmen's Compensation Insurance, including occupational disease provisions, for all of his employees at the site of the PROJECT and in case any work is sublet, the CONTRACTOR shall require such SUBCONTRACTOR similarly to provide Workmen's Compensation Insurance, including occupational disease provisions for all of the latter's employees unless such employees are covered by the protection afforded by the CONTRACTOR. In case any class of employees engaged in hazardous work under this contract at the site of the PROJECT is not protected under Workmen's Compensation statute, the CONTRACTOR shall provide, and shall cause each SUBCONTRACTOR to provide, adequate and suitable insurance for the protection of his employees not otherwise protected.

21.5 The CONTRACTOR shall secure, if applicable, "All Risk" type Builder's Risk Insurance for WORK to be performed. Unless specifically authorized by the OWNER, the amount of such insurance shall not be less than the CONTRACT PRICE totaled in the BID. The policy shall cover not less than the losses due to fire, explosion, hail, lightning, vandalism, malicious mischief, wind, collapse, riot, aircraft, and smoke during the CONTRACT TIME, and until the WORK is accepted by the OWNER. The policy shall name as the insured the CONTRACTOR, the ENGINEER, and the OWNER.

22. *CONTRACT SECURITY*

22.1 The CONTRACTOR shall within ten (10) days after the receipt of the NOTICE OF AWARD furnish the OWNER with a Performance Bond and a Payment Bond in penal sums equal to the amount of the CONTRACT PRICE, conditioned upon the performance by the CONTRACTOR of all undertakings, covenants, terms, conditions and agreements of the CONTRACT DOCUMENTS, and upon the prompt payment by the CONTRACTOR to all persons supplying labor and materials in the prosecution of the WORK provided by the CONTRACT DOCUMENTS. Such BONDS shall be executed by the CONTRACTOR and a corporate bonding company licensed to transact such business in the state in which the WORK is to be performed and named on the current list of "Surety Companies Acceptable on Federal Bonds" as published in the Treasury Department Circular Number 570. The expense of these BONDS shall be borne by the CONTRACTOR. If at any time a surety on any such BOND is declared a bankrupt or loses its right to do business in the state in which the WORK is to be performed or is removed from the list of Surety Companies accepted on Federal BONDS, CONTRACTOR shall within ten (10) days after notice from the OWNER to do so, substitute an acceptable BOND (or BONDS) in such form and sum and signed by such other surety or sureties as may be satisfactory to the OWNER. The premiums on such BOND shall be paid by the CONTRACTOR. No further payments shall be deemed due nor shall be made until the new surety or sureties shall have furnished an acceptable BOND to the OWNER.

23. *ASSIGNMENTS*

23.1 Neither the CONTRACTOR nor the OWNER shall sell, transfer, assign or otherwise dispose of the Contract or any portion thereof, or of his right, title or interest therein, or his obligations thereunder, without written consent of the other party.

24. *INDEMNIFICATION*

24.1 The CONTRACTOR will indemnify and hold harmless the OWNER and the ENGINEER and their agents and employees from and against all claims, damages, losses and expenses including attorney's fees arising out of or resulting from the performance of the WORK, provided that any such claims, damage, loss or expense is attributable to bodily injury, sickness, disease or death, or to injury to or destruction of tangible property including the loss of use resulting therefrom; and is caused in whole or in part by any negligent or willful act or omission of the CONTRACTOR, and SUBCONTRACTOR, anyone directly or indirectly employed by any of them or anyone for whose acts any of them may be liable.

24.2 In any and all claims against the OWNER or the ENGINEER, or any of their agents or employees, by any employee of the CONTRACTOR, any SUBCONTRACTOR, anyone directly or indirectly employed by any of them, or anyone for whose acts any of them may be liable, the indemnification obligation shall not be limited in any way by any limitation on the amount or type of damages, compensation or benefits payable by or for the CONTRACTOR or any SUBCONTRACTOR under workmen's compensation acts, disability benefit acts or other employee benefits acts.

24.3 The obligation of the CONTRACTOR under this paragraph shall not extend to the liability of the ENGINEER, his agents or employees arising out of the preparation or approval of maps, DRAWINGS, opinions, reports, surveys, CHANGE ORDERS, designs or SPECIFICATIONS.

25. *SEPARATE CONTRACTS*

25.1 The OWNER reserves the right to let other contracts in connection with this PROJECT. The CONTRACTOR shall afford other CONTRACTORS reasonable opportunity for the introduction and storage of their materials and the execution of their WORK, and shall properly connect and coordinate his WORK with theirs. If the proper execution or results of any part of the CONTRACTOR'S WORK depends upon the WORK of any other CONTRACTOR, the CONTRACTOR shall inspect and promptly report to the ENGINEER any defects in such WORK that render it unsuitable for such proper execution and results.

25.2 The OWNER may perform additional WORK related to the PROJECT by himself, or he may let other contracts containing provisions similar to these. The CONTRACTOR will afford the other CONTRACTORS who are parties to such Contracts (or the OWNER, if he is performing the additional WORK himself), reasonable opportunity for the introduction and storage of materials and equipment and the execution of WORK, and shall properly connect and coordinate his WORK with theirs.

25.3 If the performance of additional WORK by other CONTRACTORS or the OWNER is not noted in the CONTRACT DOCUMENTS prior to the execution of the CONTRACT, written notice thereof shall be given to the CONTRACTOR prior to starting any such additional WORK. If the CONTRACTOR believes that the performance of such additional WORK by the OWNER or others involves him in additional expense or entities him to an extension of the CONTRACT TIME, he may make a claim therefor as provided in Sections 14 and 15.

26. *SUBCONTRACTING*

26.1 The CONTRACTOR may utilize the services of specialty SUBCONTRACTORS on those parts of the WORK which, under normal contracting practices, are performed by specialty SUBCONTRACTORS.

26.2 The CONTRACTOR shall not award WORK to SUBCONTRACTOR(s), in excess of fifty (50%) percent of the CONTRACT PRICE, without prior written approval of the OWNER.

26.3 The CONTRACTOR shall be fully responsible to the OWNER for the acts and omissions of his SUBCONTRACTORS, and of persons either directly or indirectly employed by them, as he is for the acts and omissions of persons directly employed by him.

26.4 The CONTRACTOR shall cause appropriate provisions to be inserted in all subcontracts relative to the WORK to bind SUBCONTRACTORS to the CONTRACTOR by the terms of the CONTRACT DOCUMENTS insofar as applicable to the WORK of SUBCONTRACTORS and to give the CONTRACTOR the same power as regards terminating any subcontract that the OWNER may exercise over the CONTRACTOR under any provision of the CONTRACT DOCUMENTS.

26.5 Nothing contained in this CONTRACT shall create any contractual relation between any SUBCONTRACTOR and the OWNER.

27. *ENGINEER'S AUTHORITY*

27.1 The ENGINEER shall act as the OWNER'S representative during the construction period. He shall decide questions which may arise as to quality and acceptability of materials furnished and WORK performed. He shall interpret the intent of the CONTRACT DOCUMENTS in a fair and unbiased manner. The

General Conditions *(continued)*

ENGINEER will make visits to the site and determine if the WORK is proceeding in accordance with the CONTRACT DOCUMENTS.

27.2 The CONTRACTOR will be held strictly to the intent of the CONTRACT DOCUMENTS in regard to the quality of materials, workmanship and execution of the WORK. Inspections may be made at the factory or fabrication plant of the source of material supply.

27.3 The ENGINEER will not be responsible for the construction means, controls, techniques, sequences, procedures, or construction safety.

27.4 The ENGINEER shall promptly make decisions relative to interpretation of the CONTRACT DOCUMENTS.

28. *LAND AND RIGHTS-OF-WAY*

28.1 Prior to issuance of NOTICE TO PROCEED, the OWNER shall obtain all land and rights-of-way necessary for carrying out and for the completion of the WORK to be performed pursuant to the CONTRACT DOCUMENTS, unless otherwise mutually agreed.

28.2 The OWNER shall provide to the CONTRACTOR information which delineates and describes the lands owned and rights-of-way acquired.

28.3 The CONTRACTOR shall provide at his own expense and without liability to the OWNER any additional land and access thereto that the CONTRACTOR may desire for temporary construction facilities, or for storage of materials.

29. *GUARANTY*

29.1 The CONTRACTOR shall guarantee all materials and equipment furnished and WORK performed for a period of one (1) year from the date of SUBSTANTIAL COMPLETION. The CONTRACTOR warrants and guarantees for a period of one (1) year from the date of SUBSTANTIAL COMPLETION of the system that the completed system is free from all defects due to faulty materials or workmanship and the CONTRACTOR shall promptly make such corrections as may be necessary by reason of such defects including the repairs of any damage to other parts of the system resulting from such defects. The OWNER will give notice of observed defects with reasonable promptness. In the event that the CONTRACTOR should fail to make such repairs, adjustments, or other WORK that may be made necessary by such defects, the OWNER may do so and charge the CONTRACTOR the cost thereby incurred. The Performance BOND shall remain in full force and effect through the guarantee period.

30. *ARBITRATION*

30.1 All claims, disputes and other matters in question arising out of, or relating to, the CONTRACT DOCUMENTS or the breach thereof, except for claims which have been waived by the making and acceptance of final payment as provided by Section 20, shall be decided by arbitration in accordance with the Construction Industry Arbitration Rules of the American Arbitration Association. This agreement to arbitrate shall be specifically enforceable under the prevailing arbitration law. The award rendered by the arbitrators shall be final, and judgment may be entered upon it in any court having jurisdiction thereof.

30.2 Notice of the demand for arbitration shall be filed in writing with the other party to the CONTRACT DOCUMENTS and with the American Arbitration Association, and a copy shall be filed with the ENGINEER. Demand for arbitration shall in no event be made on any claim, dispute or other matter in question which would be barred by the applicable statute of limitations.

30.3 The CONTRACTOR will carry on the WORK and maintain the progress schedule during any arbitration proceedings, unless otherwise mutually agreed in writing.

31. *TAXES*

31.1 The CONTRACTOR will pay all sales, consumer, use and other similar taxes required by the law of the place where the WORK is performed.

Supplemental General Conditions

1. NOTICE TO UTILITIES. The Contractor shall give notice in writing to all utilities that may be affected by Contractor's operations at least seventy-two (72) hours before starting work. The Contractor shall not hinder or interfere with any person in the protection of such work.

   NOTICE TO FIRE AND POLICE. The Contractor shall give written notice to the Fire and Police Department of the municipality at least twenty-four (24) hours before closing off or in any way affecting through vehicular traffic on any street.

   NOTICE TO PUBLIC WORKS DEPARTMENT. The Contractor shall give notice in writing to the Village's Public Works Department at least seventy-two (72) hours before starting work.

2. PERMITS AND LICENSES. The Contractor shall procure all necessary permits and licenses, pay all charges and fees, and give all notices necessary and incidental to the due and lawful prosecution of the work unless otherwise specifically provided. Copies of all written notices and permits shall be submitted to the Engineer prior to the commencement of construction. All required right-of-way easements and property acquisition have been procured by the municipality.

3. LOCATION OF UNDERGROUND STRUCTURES. It is the responsibility of the Contractor to acquaint himself with the location of all underground structures which may be encountered or which may be affected by work under the Contract. The locations of any underground structures furnished, shown on the Plans or given on the site are based upon the available records, but are not guaranteed to be complete or correct and are given only to assist the Contractor in making a determination of the location of all underground structures.

4. ORDER OF PRIORITY OF DOCUMENTS. In addition to Section 4.2 of General Conditions, State Regulations govern over any other section followed by the Supplemental General Conditions, then General Conditions; then the technical Specifications in that order.

5. PROGRESS SCHEDULE. The Contractor shall, within ten (10) days after the effective date of Notice to Proceed, furnish three (3) copies of a preliminary progress schedule covering his operations for the first 30 days.

   The preliminary progress schedule shall be a bar graph or an arrow diagram showing the times he intends to commence and complete the various work stages, operations and contract items planned to be started during the first 30 days.

Source: Morgan & Parmley, Ltd.

Supplemental General Conditions *(continued)*

The Contractor shall submit for approval by the Engineer within 30 days after the effective Notice to Proceed three (3) copies of a detailed bar graph or a graphic network diagram. The graph or diagram shall be accompanied by a brief written explanation of the proposed schedule and a list of activities.

The bar graph or graphic network diagram shall clearly depict the order and interdependencies of activities planned by the contractor as well as activities by others which affects the contractor's planning. For those activities lasting more than 30 days, either the estimated time for 25-50 and 75% completion or other significant milestone in the course of the activity shall be shown. In addition to the actual construction operations, the schedule shall show such items as submittal of samples and shop drawings, delivery of materials and equipment, construction in the area by other forces and other significant items related to the progress of construction. The graph or network diagram shall be printed or neatly and legibly drawn to a time scale.

The list of activities shall show for each activity the estimated duration and anticipated starting and finishing dates. Activities which are critical to complete the project in shortest time shall be identified. The cost of each activity shall be shown so the schedule can be cost loaded.

The Contractor shall submit with each payment request a copy of the schedule showing the current status of the project and the activities for which payment is being requested. If the project is behind schedule, a detailed report shall accompany the submittal stating why the project is behind schedule and what means are being taken to get the project back on schedule, including, if appropriate, contract time extensions.

The Schedule shall be revised when:

(a) When a Change Order significantly affects the contract completion date or the sequence of activities.

(b) When progress of any critical activity falls significantly behind the scheduled progress.

(c) When delay on a noncritical activity is of such magnitude as to make it become critical.

(d) At any time the Contractor elects to change any sequence of activities affecting the shortest completion time.

The revised analysis shall be made in the same form and detail as the original submittal and shall be accompanied by an explanation of the reasons for the revisions.

6. SHOP DRAWINGS. Shop drawings and descriptive data shall be required on all manufactured or fabricated items. Seven (7) copies of drawings and descriptive data shall be submitted for approval. (Two of these will be returned to the Contractor after their approval.)

   Shop drawings shall be scheduled to be submitted for approval within 4 weeks after the effective Notice to Proceed.

7. OPERATION AND MAINTENANCE DATA. The Contractor shall furnish to the Engineer three (3) complete unbound copies of complete operation and maintenance data for all equipment items. This data shall include a list of parts and a recommendation of spare parts that should be kept on hand.

8. SUPERVISION OF ERECTION FOR ALL EQUIPMENT. All items of mechanical or electrical equipment whose cost delivered to the Owner exceeds $1,000.00 shall have included in the price bid the services of a qualified representative of the manufacturer for one (1) day to inspect equipment in operation and train the operator. For items whose cost exceeds $2,500.00, the bid price shall include not less than one (1) day during erection and not less than one (1) day operation inspection. Over $5,000.00, the bid price shall include not less than three (3) days during erection and two (2) days operation inspection. After each trip, the manufacturer is to submit a report to the Engineer covering the findings made during the inspection including changes and/or repairs, if any. Lesser time than specified may be needed. In this case, the Owner reserves the right to call for "make up" trips during the twelve (12) months following acceptance until time specified has been utilized. Unused time will lapse after the twelve (12) month period. Trips made for the purpose of correcting defective materials or workmansip are to be made without charge and are not to be credited to the required number.

9. TESTING OF EQUIPMENT. All mechanical and electrical equipment furnished for this project shall be field tested to insure full compliance with the Specifications. Testing shall be performed by the Contractor and/or equipment supplier in the presence of the Engineer. If the field test shows that the equipment does not meet the Specifications, it shall be replaced with acceptable equipment at Contractor's expense.

   Full payment for equipment will not be made until it has been field tested and accepted.

Supplemental General Conditions *(continued)*

10. STORAGE OF MATERIALS. Contractor shall confine his apparatus, storage of materials and operations of his workmen to limits indicated by law, ordinances, permits or directions of the Engineer.

    All materials stored on the site shall be protected from the elements by watertight coverings or shed with wood floors raised above the ground and protected from damage by Contractor's operations by barriers. The Owner assumes no responsibility for stored materials.

    All materials shall be delivered in original sealed containers, with seals unbroken and with labels plainly indicating materials and directions for storage. Directions for storage shall be complied with unless specified to the contrary in the detailed Contract Specifications.

    Damaged materials shall be immediately removed from the site. Stored material will be paid for in accordance with Section 19 of the General Conditions.

11. PROTECTING WORK. The Contractor shall protect existing improvements and work installed by him from damage. Improvements and work damaged shall be replaced by the Contractor. All replacement methods shall be approved by the Engineer.

12. COOPERATION. The Contractors and Subcontractors shall cooperate and corrdinate their work with adjacent work and shall give due notice to other Contracotrs and Subcontractors of intersecting work to assure that all items are installed at an agreeable time and to facilitate general progress of the work.

13. LABOR LAWS. All Contractors and Subcontractors employed upon the work shall be required to conform to Federal and State labor laws and the various acts amendatory and supplementary thereto and to all other laws, ordinances and legal requirements applicable thereto.

    All labor shall be performed in the best and most workmanlike manner by mechanics skilled in their respective trades. The standard of the work required throughout shall be such grade as will bring results of the first class only.

14. TEMPORARY FACILITIES. The Contractor shall provide his own temporary facilities such as electricity, toilets and heat. He shall also furnish, erect and maintain a weatherproof bulletin board for displaying wage rates, Equal Opportunity Requirement, etc. All temporary facilities shall be removed prior to job completion.

15. FINAL CLEANUP. The Contractor shall be responsible for cleaning up his work and his waste and rubbish. It is the General Contractor's responsibility to completely clean up the inside and outside of all buildings and the construction site at the completion of the project.

16. PAYMENT TO CONTRACTOR. The Owner will make progress payment to the Contractors in accordance with Section 19 of the General Conditions. It is understood that grant reimbursement payment to the Owner are dependent upon compliance by the Owner with all the grant conditions identified in the grant request and grant agreement. The Owner shall be responsible for progress payments to the Contractor even when failure to comply with said grant conditions delays grant payments from the funding agency.

17. SAFETY. All work shall be done in accordance with OSHA standards.

18. AMERICAN-MADE PRODUCTS. Federally funded projects prohibit foreign-made products. Therefore, contractor(s) and sub-contractors shall buy/use American-made products on this project.

19. EASEMENTS. Some of the work will be done outside of existing right of ways. The Village has obtained needed easements. All disturbed areas within the construction limits shall be restored. Easements are available for review at the office of the Village Clerk.

20. MINORITY REQUIREMENTS. The goals for minority and female participation are 0.6% and 6.9% respectively. These goals are for both contractors and subcontractors. If these goals cannot be met, the contractor must be able to document a good faith effort in attempting to meet these goals. These requirements are explained elsewhere in the contract documents and the contractor is required to comply with them.

Summary of Work

PART ONE - GENERAL (W/SPECIAL PROVISIONS)

1.1 BACKGROUND

1.1.1 History of Project: The High School Complex in Tony, is owned and operated by the Flambeau School District. Due to historically poor potable water and continuous maintenance problems, the School Board retained Morgan & Parmley, Ltd. to solve their problem. On May 23, 1977, construction was completed for their new water supply well located on Rusk County Airport property. The pumphouse, pumping equipment and transmission main were placed in operation early in 1979. This facility has been in continuous use, serving the High School Complex, since that time with no significant operational problems.

On several occasions, the Village Board attempted to establish a municipal water system without success. However, their recent efforts have proven successful by obtaining funding through the Rural Economic and Community Development, USDA (formally, FmHA).

An agreement between the Flambeau School District and the Village of Tony was recently executed. The Village will purchase the existing school well, pumphouse, transmission main and related appurtenances. These components will be incorporated into their proposed (total) water system; as detailed in the accompanying Plans and Specifications.

The Village has established their public water utility under the rules and regulations of the Wisconsin Public Service Commission.

The Well Site Survey and Engineering Report have been approved by DNR.

1.2 SCOPE OF WORK

1.2.1 Description: The proposed Project consists of the following major items:

1-Rehab Existing Pumphouse
2-Vertical Turbine Pump w/Controls
3-Interior Piping & Controls
4-Right Angle Drive Mechanism
5-Chlorination Facility
6-Electrical Upgrade
7-Watermain Distribution System
8-Elevated Water Tank & Tower
9-Services & Meters

93-159 01010-1

The foregoing list is only major items. Refer to the accompanying Plans, applicable sections of the Specifications and the Proposal for a total concept.

1.2.2 Proposals: There will be two (2) Proposals. The first will be for all items, except the elevated water tank, tower and related components which will be bid separately.

1.3 SCHEDULE

1.3.1 Anticipated Construction Schedule:

Commence Advertising	11-13-1995
Bid Opening	12-14-1995
Notice of Award	12-20-1995
Pre-Construction Conference	1-15-1996
Notice to Proceed	1-15-1996
Complete Construction	11-10-1996

1.3.2 Highway Construction: All construction on U.S. Highway 8 shall be substantially completed by June 30, 1996.

1.4 HYDRANT PATTERN

1.4.1 Description: Since the Village of Tony is served by the Ladysmith Fire Department, it is mandatory that Tony's fire hydrant pattern be compatable with their equipment.

PART TWO - SPECIAL PROVISIONS

2.1 WAGE RATES

2.1.1 Wages: The Contractors and Sub-Contractors shall meet the requirements of all applicable official State Wage Rate Determinations for this Project. A copy is attached to these Specifications.

2.2 EQUAL EMPLOYMENT OPPORTUNITY

2.2.1 Compliance: The Contractors and Sub-Contractors shall comply with all current and applicable rules and regulations governing equal employment opportunities.

2.3 APPROVALS

2.3.1 General: The Contractors and Sub-Contractors shall comply with all applicable items contained in the regulatory approvals and permits for this project. Approvals and permits from DNR, PSC, FAA, DOT, SHPO, RR, Rusk County, and others are attached to this

93-159 01010-2

Source: Morgan & Parmley, Ltd.

Summary of Work *(continued)*

Specification Book under the heading; Supplemental Documents. It is your responsibility to understand and abide by the conditions contained in the documents.

2.4 EROSION CONTROL

2.4.1 Responsibility: It is the Contractor's responsibility to provide the specified erosion control for the project. Refer to Section 02770, sheet 15 of 15 of the Plans and DOT's Policy 96.55 and related sections.

2.5 TRAFFIC CONTROL

2.5.1 Responsibility: It is the Contractor's responsibility for traffic control. Refer to DOT's Policy 96.51 and related sections.

2.6 EXISTING UTILITIES

2.6.1 General: It is the responsibility of the Contractor to acquaint himself with the location of all underground structures which may be encountered or which may be affected by work under the Contract. The locations of any underground structures furnished, shown on the Plans or given on the site are based upon available records, but are not guaranteed to be complete or correct and are given only to assist the Contractor in making a determination of the location of all underground structures.

2.6.2 Utility Notification: The Contractor shall give notice in writing to all utilities that may be affected by Contractor's operations at least seventy-two (72) hours before starting work. The Contractor shall not hinder or interfere with any person in the protection of such work.

The Contractor is encouraged to utilize the "Call Before You Dig" telephone notification system.

2.6.3 Cable TV Notification: The Contractor shall notify Marcus Cable, 219 W. Miner Avenue, Ladysmtih, WI 54848,; Tele. No. 715/532-6040.

2.6.4 Electrical Utility Notification: The Contractor shall notify Northern States Power Co., 129 N. Lake Ave., Phillips, WI 54555; Atten: Joe Perkins; Tele. No. 715/836-1198.

2.6.5 Telephone Utility: The Contractor shall notify Ameritech @ 1-800-242-8511 (Diggers Hotline).

2.6.6 Village Sewer System: The Contractor shall notify the Village's Sanitary Sewer Utility Operator, Billy Mechelke, N5297 Little X, Tony, WI 54563, Tele. No. 715/532-7046.

2.6.7 Emergency Facilities: The Contractor shall give written notice to the Fire Department, Rusk Co. Sheriff's Office and Ambulance Service, at least twenty-four (24) hours before closing off or in any way affecting through vehicular traffic on any street.

Rusk Co. Sheriff's Office 715/532-2200

Ladysmith Rural Fire Department 715/532-2186

Ambulance Service 715-532-2100

2.7 CONSTRUCTION SIGN

2.7.1 Construction Sign: One (1) construction sign shall be provided and erected at a location to be specified by the Engineer at the Pre-Construction Conference. Refer to Supplemental Specifications, page A for its size and design. The construction sign will be a bid item in the Proposal.

2.8 SURVEY MARKERS

2.8.1 General: No survey markers (e.g. DOT, County, USGS, Property Corners, etc.) located in the project area shall be disturbed, unless prior approval has been obtained. Replacement will be at the expense of the Contractor.

2.9 STORAGE OF MATERIALS

2.9.1 General: Contractor shall confine his apparatus, storage of materials and operations of his workmen to limits indicated by law, ordinances, permits or directions of the Engineer.

All materials stored on the site shall be protected from the elements by watertight coverings or shed with wood floors raised above the ground and protected from damage by Contractor's operations by barriers. The Owner assumes no responsibility for stored materials.

All materials shall be delivered in original sealed containers, with seals unbroken and with labels plainly indicating materials and directions for storage. Directions for storage shall be complied with unless specified to the contrary in the detailed Contract Specifications.

Damaged materials shall be immediately removed from the site. Stored material will be paid for in accordance with standard practice.

2.10 PROTECTING WORK

2.10.1 General: The Contractor shall protect existing improvements and Work installed by him from damage. Improvements and Work damaged shall be replaced by the Contractor. All replacement methods shall be approved by the Engineer.

2.11 LABOR LAWS

Summary of Work *(continued)*

2.11.1 General: All Contractors and Subcontractors employed upon the Work shall be required to conform to applicable Federal and State labor laws and the various acts amendatory and supplementary thereto and to all other laws, ordinances and legal requirements applicable thereto.

All labor shall be performed in the best and most workmanlike manner by mechanics skilled in their respective trades. The standard of the Work required throughout shall be such grade as will bring results of the first class only.

2.12 TEMPORARY FACILITIES

2.12.1 Description: The Contractor shall provide his own temporary facilities such as electricity, toilets and heat. He shall also furnish, erect and maintain a weatherproof bulletin board for displaying wage rates, Equal Opportunity Requirement, etc. All temporary facilities shall be removed prior to job completion.

2.13 COOPERATION

2.13.1 General: The Contractors and Subcontractors shall cooperate and coordinate their Work with adjacent Work and shall give due notice to other Contractors and Subcontractors of intersecting Work to assure that all items are installed at an agreeable time and to facilitate general progress of the Work.

2.14 PROGRESS SCHEDULE

2.14.1 Description: The Contractor shall, within ten (10) days after the effective date of Notice to Proceed, furnish three (3) copies of a preliminary progress schedule covering his operations for the first 30 days.

The preliminary progress schedule shall be a bar graph or an arrow diagram showing the times he intends to commence and complete the various Work stages, operations and contract items planned to be started during the first 30 days.

The Contractor shall submit for approval by the Engineer within 30 days after the effective Notice to Proceed three (3) copies of a detailed bar graph or a graphic network diagram. The graph or diagram shall be accompanied by a brief written explanation of the proposed schedule and a list of activities.

The bar graph or graphic network diagram shall clearly depict the order and interdependencies of activities planned by the Contractor as well as activities by others which affects the Contractor's planning. For those activities lasting more than 30 days, either the estimated time for 25-50 and 75% completion or other significant milestone in the course of the activity shall be shown. In addition to the actual construction operations, the schedule shall show such items as submittal of samples and shop drawings, delivery of materials and equipment, construction in the area by other

forces and other significant items related to the progress of construction. The graph or network diagram shall be printed or neatly and legibly drawn to a time scale.

The list of activities shall show for each activity the estimated duration and anticipated starting and finishing dates. Activities which are critical to complete the project in shortest time shall be identified. The cost of each activity shall be shown so the schedule can be cost loaded.

The Contractor shall submit with each payment request a copy of the schedule showing the current status of the project and the activities for which payment is being requested. If the project is behind schedule, a detailed report shall accompany the submittal stating why the project is behind schedule and what means are being taken to get the project back on schedule, including, if appropriate, contract time extensions.

The Schedule shall be revised:

a) When a Change Order significantly affects the contract completion date or the sequence of activities.

b) When progress of any critical activity falls significantly behind the scheduled progress.

c) When delay on a noncritical activity is of such magnitude as to make it become critical.

d) At any time the Contractor elects to change any sequence of activities affecting the shortest completion time.

The revised analysis shall be made in the same form and detail as the original submittal and shall be accompanied by an explanation of the reasons for the revisions.

2.15 SHOP DRAWINGS

2.15.1 Description: Shop drawings and descriptive data shall be required on all manufactured or fabricated items. Seven (7) copies of drawings and descriptive data shall be submitted for approval. (Two of these will be returned to the Contractor after their approval.)

Shop drawings shall be scheduled to be submitted for approval within 4 weeks after the effective Notice to Proceed.

2.16 SUPERVISION OF ERECTION FOR ALL EQUIPMENT

2.16.1 General: All items of structural, mechanical or electrical equipment whose cost delivered to the Owner exceeds $1,000 shall have included in the price bid the services of a qualified representative of the manufacturer for one (1) day to inspect equipment in operation and train the operator. For items whose cost exceeds $2,500, the bid price shall include not less than one (1) day during erection and not less than one (1) day operation inspection. Over $5,000, the

Summary of Work *(continued)*

bid price shall include not less than three (3) days during erection and two (2) days operation inspection. After each trip, the manufacturer is to submit a report to the Engineer covering the findings made during the inspection including changes and/or repairs, if any. Lesser time than specified may be for "make up" trips during the twelve (12) month warranty period. Trips made for the purpose of correcting defective materials or workmanship are to be made without charge and are not to be credited to the required number.

2.17 SCHOOL WATER SUPPLY

2.17.1 Service: It is mandatory that the High School Complex be provided potable water on a continuous basis during school hours. Therefore, the Contractor shall schedule his Work to avoid any interruption of water during school operation or normal activity.

2.18 EXISTING SHRUBBERY

2.18.1 Protection: It shall be the responsibility of the Contractor to protect all existing private shrubbery, trees, etc. within the construction area. If they are damaged, or destroyed, they shall be replaced to the owner's satisfaction at the Contractor's expense.

2.19 SAFETY

2.19.1 Basic: All work shall be done in accordance with OSHA standards and other specified regulations as noted throughout the Specifications.

2.20 PROJECT RECORD DOCUMENTS

2.20.1 Contractor's Responsibility: At the conclusion of the project, prior to final payment, the Contractor shall supply the Engineer with a legible record of all work as installed. Example: a map detailing watermain installed with location, lengths, tie-offs, valves, hydrants and appurtenances accurately shown so an "as-built" layout can be developed. This information can be marked on the appropriate sheet of the Plans with a permanent color marker.

2.21 OPERATION & MAINTENANCE DATA

2.21.1 O & M Data: The Contractors shall furnish to the Engineer three (3) complete unbound copies of complete operation and maintenance data for all equipment items. This data shall include a list of parts and a recommendation of spare parts that should be kept on hand.

2.22 TESTING

2.22.1 Testing & Equipment: All mechanical and electrical equipment furnished for this project shall be field tested to insure full compliance with the Specifications. Testing shall be performed by the Contractor and/or equipment supplier in the presence of the Engineer. If the

field test shows that the equipment does not meet the Specifications, it shall be replaced with acceptable equipment at Contractor's expense.

Full payment for equipment will not be made until it has been field tested and accepted.

2.23 FINAL CLEANUP

2.23.1 General: The Contractor shall be responsible for cleaning up this Work and his waste and rubbish. It is the General Contractor's responsibility to completely clean up the inside and outside of all buildings and the total construction site at the completion of the project.

2.24 PRIVATE SERVICES

2.24.1 Description: Per specified bid items, the Contractor shall furnish and install service laterals from the distribution main to (and including) the curb stop. The Contractor shall also furnish and install the water meter including fittings and appurtenances within each respective user's structure.

Please be advised that section of the service lateral from the curb stop to the meter is the responsibility of each user and is not part of this construction contract.

2.25 PAYMENT TO CONTRACTOR

2.25.1 General: The Owner will make progress payments to the Contractors in accordance with the General Conditions. It is understood that grant reimbursement payments to the Owner are dependent upon compliance by the Owner with all the grant conditions identified in the grant request and grant agreement. The Owner shall be responsible for progress payments to the Contractor even when failure to comply with said grant conditions delays grant payments from the funding agency.

2.25.2 Final Payment: Final payment to the Contractor will not be made until the Owner and Engineer are satisfied that the facility is fully operational and all aspects of the Plans, Specifications and Construction Contract have been satisfied.

2.26 FmHA NAME CHANGE

2.26.1 Clarification: Wherever the words "Farmers Home Administration (FmHA) or Rural Development Administration (RDA)" appear in the project manual, substitute the words "U.S. of America."

Summary of Work *(continued)*

2.26 RAILROAD NOTIFICATION

2.26.1 General:

The Railroad is not a member of Diggers Hotline, therefore; before construction can begin, you should schedule signal wire location, a railroad flagger and the inspector at least three (3) working days in advance; to do so, please call the following people:

Construction Inspection
Cooper Engineering Company, Inc.
Maurice E. Smith
Telephone No. (715) 234-7008
Fax No. (715) 234-1025

Signal Wire Location
Wisconsin Central Ltd.
Jill Hamilton
Telephone No. (715) 345-2524
Fax No. (715) 345-2543

Railroad Flagger
Wisconsin Central Ltd.
Jerry Sernau
Telephone No. (715) 345-2511
Fax No. (715) 345-2507

-END OF SECTION-

93-159 01010-9

Scope of Project

WATER SUPPLY, PUMPHOUSE, DISTRIBUTION
SYSTEM & ELEVATED STORAGE RESERVOIR

Village
of
Tony, Wisconsin

The Village of Tony will exercise their option and purchase the existing well, pumphouse and transmission main from the Flabmeau School District and take title to the related easements.

The project proposes to rehab the existing pumphouse, located on the Rusk County Airport, replace the existing submersible pump with a 200 gpm vertical turbine pump and retrofit the interior piping. A right angle gear drive mechanism will be installed to provide emergency pumping via a farm tractor's PTO. A standby chlorination facility will be installed in the pumphouse.

A 6-inch watermain distribution system with adequate valves and fire hydrants will be installed within the Village's service area and all customers currently connected to the sanitary sewer collection system will be provided service from the new water system. All services will be furnished with water meters and billed accordingly, pursuant to PSC rate scheduling.

An elevated water storage reservoir will be erected to provide adequate pressure in the distribution system and have a reserve available for fire protection.

The work is fully described in the Specifications bound herein and on the following sheets of the accompanying Plans:

Source: Morgan & Parmley, Ltd.

Plan Sheet Index

SHEET NO.	DESCRIPTION
1	Title Sheet, Location Maps, Index & Notes
2	General Project Site Map w/Tower Site
3	Existing Well & Pumphouse Site Plan
4	Existing Well Log w/Construction Details
5	Pumphouse Rehab w/Piping Retrofit & Chlorination
6	Pumphouse Electrical w/H & V Layouts
7	Existing Transmission Main w/Easement - CSM
8	Village Map of Platted Area (w/watermain network)
9	Distribution System Layout (w/topography)
10	User Location Map w/Listing
11	Railroad & Highway Crossing: Oak St.
12	Railroad & Highway Crossing: Walnut St. (CTH "I")
13	Elevated Water Storage Reservoir
14	Typical Fire Hydrant & Service Details
15	Typical Erosion Control Details

Bid Proposal "A" Format

Place:____________________

Date:____________________

Project: Water System

Project No.: 93-159

Proposal of ____

(Hereinafter called BIDDER), organized and existing under the laws of the State of __________
,

doing business as a ____.
(Corporation, Partnership, or Individual; delete non-applicable organization)

To the VILLAGE OF TONY, WISCONSIN:
(hereinafter called OWNER)

The BIDDER, in compliance with your Advertisement for Bids for the construction and installation of a MUNICIPAL WATER SYSTEM - CONTRACT A: Pumphouse Rehab, Pump, Right Angle Drive, Piping, Controls, Watermain Distribution System, Highway & R.R. Crossings, Meters, Services and Related Appurtenances; having examined the Plans and Specifications and related Documents prepared by Morgan & Parmley, Ltd., Consulting Engineers, and the site of the proposed Work and being familiar with all conditions surrounding the construction of the proposed project, including the availability of materials and labor, hereby proposes to furnish all labor, equipment, materials and supplies, and to construct the complete project in accordance with the Contract Documents within the time set forth therein and at the prices set forth below. Those prices are to cover all of the expenses incurred in performing the Work required under the Contract Documents, of which this Proposal is a part thereof.

The BIDDER hereby agrees to commence Work on the Contract on or before a date specified in the written "Notice to Proceed" of the Owner and to complete the project within 300 consecutive calendar days thereafter.

Source: Morgan & Parmley, Ltd.

Bid Proposal "A" Format *(continued)*

The BIDDER agrees to pay, as liquidated damages, the sum of$350 for each consecutive calendar day thereafter as provided in Section 15 of the General Conditions.

Bidder acknowledges receipt of the following Addendum ________________

BIDDER shall fill in Subcontractors and Material Supplier List and Similar Projects Lists attached hereto.

BIDDERS are advised that two (2) alternate pipe materials are acceptable for the distribution system (except where specified on the Plans). The alternatives are:

Ductile Iron - Class 52

PVC-DR-18

BIDDERS are required to complete all Bid Items.

-UNIT BID SCHEDULE-
for
PROPOSAL "A"
WATER SYSTEM - TONY, WISCONSIN

ITEM NO.	ITEM DESCRIPTION	UNIT	EST. QUAN.	UNIT PRICE	TOTAL
*1-A)	6" DISTRIBUTION MAIN: 6-Inch DI-Class 52 Watermain (150 psi W.P., 8'-0" Bury w/Cable Bond Connector @ all joints	L.F.	9,325	$	$
*1-B)	6" DISTRIBUTION MAIN: 6-Inch PVC-DR-18 Watermain (150 psi W.P., 8'-0" Bury)	L.F.	9,325		
*SPECIAL NOTE: BIDDER shall use the lowest priced alternate pipe material in Items 1 & 2 to determine the total Bid Price for Proposal A.					
*2-A)	8" DISTRIBUTION MAIN: 8-Inch DI-Class 52 Watermain (150 psi W.P., 8'-0" Bury w/Cable Bond Connector @ all joints	L.F.	625		
*2-B)	8" DISTRIBUTION MAIN: 8-Inch PVC-DR-18 Watermain (150 psi W.P., 8'-0" Bury)	L.F.	625		

ITEM NO.	ITEM DESCRIPTION	UNIT	EST. QUAN.	UNIT PRICE	TOTAL
3)	6" MAIN @ CROSSINGS: 6-Inch M.J.D.I., Class 52 Watermain (150 psi W.P., 8'-0" Bury) including sand bedding & sealing ends of casings	L.F.	285		
4)	8" MAIN @ CROSSINGS & TOWER: 8-Inch M.J.D.I., Class 52 Watermain (150 psi W.P., 8'-0" Bury) including sand bedding & sealing ends of casings	L.F.	390		
5)	6" VALVES: 6-Inch M.J. Gate Valves w/Three Piece Box @ 8'-0" Bury	Each	31		
6)	8" VALVES: 8-Inch M.J. Gate Valves w/Three Piece Box @ 8'-0" Bury	Each	4		
7)	6" X 6" CUT-IN-TEE: 6" x 6" x 6" M.J. Cut-In-Tee	Each	1		
8)	6" X 6" TEES: 6" x 6" x 6" M.J. Tees	Each	34		
9)	8" X 8" TEES: 8" x 8" x 8" M.J. Tees	Each	3		
10)	8" X 6" TEES: 8" X 8" X 6" M.J. Tees	Each	2		
11)	8" X 6" REDUCERS: 8" x 6" M.J. Reducers	Each	3		
12)	6" - 45° BENDS: 6-Inch M.J. 1/8 Bends	Each	2		
13)	8" - 45° BENDS: 8-Inch M.J. 1/8 Bends	Each	1		
14)	6" BLIND FLANGES: 6-Inch M.J. Blind Flanges	Each	10		
15)	8" BLIND FLANGES: 8-Inch M.J. Blind Flanges	Each	1		
16)	8" X 8" CROSS: 8-Inch x 8-Inch M.J. Cross	Each	1		
17)	HYDRANT GROUP: 6-Inch Hydrant (8'-0" Bury-Ladysmith Pattern) including Gate Valve & Box with 15 Ft. of 6" M.J.D.I. Lead (Does not include Tee)	Each	19		
18)	BEDDING-6" PIPE: Imported sand bedding, per Specs, for 6-Inch pipe	L.F.	500		

Bid Proposal "A" Format *(continued)*

ITEM NO.	ITEM DESCRIPTION	UNIT	EST. QUAN.	UNIT PRICE	TOTAL
19)	PIPE INSULATION: Extruded Polystyrene (4'-0" wide) Insulation, per Plans	L.F.	100		
20)	BEDDING - 8-INCH: Imported sand bedding, per Specs., for 8-Inch pipe.	L.F.	100		
21)	R.R. CASING - CTH "I": Bore & Jack 24" steel casing, per Plans & Permit (Watermain not included in this Item)	L.F.	92		
22)	R.R. CASING - OAK ST.: Bore & Jack 24" steel casing, per Plans & Permit (Watermain not included in this Item)	L.F.	65		
23)	HWY. CASING - CTH "I": Bore & Jack 24" steel casing, per Plans & Permit (Watermain not included in this Item)	L.F.	45		
24)	HWY. CASING - OAK ST.: Bore & Jack 24" steel casing, per Plans & Permit (Watermain not included in this Item)	L.F.	45		
25)	CONNECTION @ TANK: Connection of 8-Inch Watermain to Hydrant Tee @ End of Tank Yard Piping. (Remove & Salvage 8" Blind Flange)	L.S.	1		
26)	1" WATER SERVICE: 1-Inch Type K, Copper Service (Main to P/L) w/Corp., Saddle Clamp, Curb Stop and Box	Each	54		
27)	WATER METERS: ⅝" X ¾" Water Meter complete w/remote reader & connection fittings, per Specifications	Each	54		
28)	WATER METER (SCHOOL): 1½" Water Meter Complete w/connection fittings, per Specifications	Each	1		
29)	ACCESS DRIVEWAY: Construct and gravel Tower access driveway, per Plans	L.S.	1		
30)	CULVERT REPLACEMENT: 12-Inch CMCP, Complete with Gravel & Restoration	L.F.	180		

ITEM NO.	ITEM DESCRIPTION	UNIT	EST. QUAN.	TOTAL PRICE	TOTAL
31)	ABANDON CONNECTION: Disconnect School's existing 4-Inch cross-connection from abandoned well	L.S.	1		
32)	CURB & GUTTER REPLACEMENT: Replacement of concrete curb & gutter damaged during construction, including sub-base	L.F.	50		
33)	SIDEWALK REPLACEMENT: Replacement of concrete sidewalk damaged during construction; 4-Inch thick w/6 x 6 x 10/10 W.W.M. & 6-Inch sand lift	S.F.	800		
34)	BASE COURSE: Crushed aggregate base course; includes compaction and fine grading	C.Y.	800		
35)	SUB-BASE: Granular sub-base; includes compaction and fine grading	C.Y.	1,250		
36)	6" VALVE REPLACEMENT: Replacement of existing 6-Inch gate valve & box located in transmission main	Each	2		
37)	ASPHALT PAVING: Replacement of existing asphalt paving damaged during construction (2 - 1½" lifts); includes square cutting prior to excavation	S.F.	45,000		
38)	EROSION BALES: Includes stakes and installation, per Specifications	Each	50		
39)	SILT FENCE: Includes stakes and installation, per Specifications	L.F.	200		
40)	PUMPHOUSE REHAB: Complete rehabilitation of existing Pumphouse, per Plans & Specifications	L.S.	1		
41)	WELL PUMP: Deep Well Turbine Pump, complete per Plans & Specifications, including related appurtenances	L.S.	1		
42)	RIGHT ANGLE DRIVE: Right Angle Gear Drive mechanism; complete per Specifications including appurtenances and shafts (extension, combination and short)	L.S.	1		

Bid Proposal "A" Format *(continued)*

ITEM NO.	ITEM DESCRIPTION	UNIT	EST. QUAN.	UNIT PRICE	TOTAL
43)	INTERIOR PIPING: Complete interior piping and related components for Pumphouse, per Plans & Specifications	L.S.	1		
44)	CHLORINATION FACILITY: Chlorination Facility and related equipment, per Plans & Specifications	L.S.	1		
45)	ELECTRICAL: Complete electrical system for Pumphouse, per Plans & Specifications; including NSP application, entry, panels, general wiring, power roof ventilator, electrical heaters, lighting fixtures, control panels, and related Work not specifically Bid elsewhere	L.S.	1		
46)	SURGE ELIMINATOR: Surge eliminator and pressure switch; complete with related components (per Specifications)	L.S.	1		
47)	WATER CONTROL SYSTEM: Water Supply control system; complete per Specifications	L.S.	1		
48)	TRAFFIC CONTROL: Traffic control, per Specs. & Permit; including flagging personnel & signage	L.S.	1		
49)	MAIN DISINFECTION: Disinfection, flushing and testing of distribution system	L.S.	1		
50)	CONSTRUCTION SIGN: Provide one (1) project construction sign, per Specifications, and install at location designated by Engineer	L.S.	1		
51)	FIELD OFFICE: Provide construction field office at site during construction, per standard practice	L.S.	1		
52)	SITE RESTORATION: General restoration of construction site not specifically itemized elsewhere; including topsoil, seeding, fertilizing, sodding and mulching, as needed	L.S.	1		
53)	MOBILIZATION: Mobilization, bonds, insurance and other general project administration items	L.S.	1		

TOTAL BID: PROPOSAL A -- **$**

The BIDDER understands that the OWNER reserves the right to reject any and all Bids, to waive any informalities in the Bidding process and to accept the BID most advantageous to the OWNER.

The undersigned further agrees to furnish surety in a penal sum of equal to or greater than the aggregate of the Contract and on the form hereto attached, to the OWNER, for carrying out the Work for which this PROPOSAL is submitted.

The undersigned agrees to execute the CONTRACT within ten (10) days from the date of notice from the OWNER that the CONTRACT has been awarded.

Accompanying this PROPOSAL is a check (certified) or Bidder's Bond in the amount of five (5%) percent of the bid amounting to ______________________________ Dollars ($__________) as required in the Advertisement for Bids.

In submitting this Bid it is understood that the right is reserved by the Owner to reject any or all Bids and to waive any informalities in the Bidding.

The Bidder agrees that this Bid shall be good and may not be withdrawn for a period of 90 calendar days after the scheduled closing for receiving Bids.

FIRM'S NAME	OFFICIAL ADDRESS
____________________________	____________________________
____________________________	____________________________
by __________________________	____________________________
Title ________________________	____________________________
	Tele. No. (__)__________________

(Seal: If Bid by Corporation)

Attest:

____

Bid
Proposal "B" Format

Place:____________________

Date:____________________

Project: Water System

Project No.: 93-159

Proposal of ____

(Hereinafter called BIDDER), organized and existing under the laws of the State of __________

doing business as a ____.

(Corporation, Partnership, or Individual; delete non-applicable organization)

To the VILLAGE OF TONY, WISCONSIN:
(hereinafter called OWNER)

The BIDDER, in compliance with your Advertisement for Bids for the construction and installation of a MUNICIPAL WATER SYSTEM - CONTRACT B: Elevated Steel Water Tank, Tower, Foundation, Valve Vault, Yard Piping, and Related Appurtenances; having examined the Plans and Specifications and related Documents prepared by Morgan & Parmley, Ltd., Consulting Engineers, and the site of the proposed Work and being familiar with all conditions surrounding the construction of the proposed project, including the availability of materials and labor, hereby proposes to furnish all labor, equipment, materials and supplies, and to construct the complete project in accordance with the Contract Documents within the time set forth therein and at the prices set forth below. Those prices are to cover all of the expenses incurred in performing the Work required under the Contract Documents, of which this Proposal is a part thereof.

The BIDDER hereby agrees to commence Work on the Contract on or before a date specified in the written "Notice to Proceed" of the Owner and to complete the project within 300 consecutive calendar days thereafter.

The BIDDER agrees to pay, as liquidated damages, the sum of$350 for each consecutive calendar day thereafter as provided in Section 15 of the General Conditions.

Bidder acknowledges receipt of the following Addendum ______________________________

BIDDER shall fill in Subcontractors and Material Supplier List and Similar Projects List attached hereto.

BIDDERS are advised that two (2) types of Elevated Steel Water Storage Tanks are acceptable for use on this Project. The acceptable tank types are as follows:

BASE BID - Single Pedestal (Sphere) Elevated Storage Tank
ALTERNATE BID - Multi-Leg (Ellipsodal) Elevated Storage Tank

When evaluating Bids (Base Bid vs. Alternate Bid) a Cost Deduct Factor will be applied to the Total Amount Base Bid. This factor is a present worth cost of life-cycle paint costs for painting multi-leg tanks (Alternate Bid). The Cost Deduct Factor will be subtracted from the Base Bid Total Amount, and the resultant will be termed Equalized Base Bid.

COST DEDUCT FACTOR = $ 11,650

BIDDERS are required to complete the unit price schedule for both types of Elevated Storage Tanks.

Bid Proposal "B" Format *(continued)*

-UNIT BID SCHEDULE-
for
PROPOSAL "B"

WATER SYSTEM - TONY, WISCONSIN

SINGLE PEDESTAL (SPHERE) ELEVATED STORAGE TANK

ITEM NO.	ITEM DESCRIPTION	UNIT	EST. QUAN.	UNIT PRICE	TOTAL
1)	FOUNDATION: Concrete foundation & valve vault; complete w/related appurtenances	L.S.	1		$
2)	ELEVATED TANK: Fabricate & erect 50,000 gal. elevated steel water tank, tower & riser; complete w/accessories	L.S.	1		
3)	ELECTRICAL SYSTEM: Complete electrical system including service, panels, wiring, lighting, outlets, warning system and related components	L.S.	1		
4)	CIRCULATION PUMP: Duplex circulation pump system; complete per Specifications	L.S.	1		
5)	WELD TESTING: Radiographic weld testing with report	L.S.	1		
6)	PAINTING: Cleaning, surface preparation and painting of tank, tower and related components; including lettering of Village name on tank	L.S.	1		
7)	DISINFECTION: Complete disinfection of tank, riser & yard piping; including flushing	L.S.	1		
8)	YARD PIPING: 8-Inch, M.J.D.I. CL. 52 watermain connection from riser to hydrant tee; including hydrant, valve, 6-Inch lead, tee and blind flange	L.S.	1		
9)	SITE RESTORATION: Complete site restoration including grading, removal of construction debris, topsoil, seeding, fertilizing & mulching	L.S.	1		
10)	EROSION CONTROL: Erosion control of construction site, per Plans & Specifications	L.S.	1		
11)	MOBILIZATION: Mobilization, bonds, insurance and other general project administration items	L.S.	1		

TOTAL BASE BID (PROPOSAL B) ------------------------------------ $

COST DEDUCT FACTOR -- $ - (11,650)

EQUALIZED BASE BID --- $

-ALTERNATE BID-

ELLIPSOIDAL (MULTI-LEG) ELEVATED STORAGE TANK

ITEM NO.	ITEM DESCRIPTION	UNIT	EST. QUAN.	UNIT PRICE	TOTAL
1)	FOUNDATION: Concrete foundation & valve vault; complete w/related appurtenances	L.S.	1		
2)	ELEVATED TANK: Fabricate & erect 50,000 gal. elevated steel water tank, tower & riser; complete w/accessories	L.S.	1		
3)	ELECTRICAL SYSTEM: Complete electrical system including service, panels, wiring, lighting, outlets, warning system and related components	L.S.	1		
4)	CIRCULATION PUMP: Duplex circulation pump system; complete per Specifications	L.S.	1		
5)	WELD TESTING: Radiographic weld testing with report	L.S.	1		
6)	PAINTING: Cleaning, surface preparation and painting of tank, tower and related components; including lettering of Village name on tank	L.S.	1		
7)	DISINFECTION: Complete disinfection of tank, riser & yard piping; including flushing	L.S.	1		

Bid Proposal "B" Format *(continued)*

ITEM NO.	ITEM DESCRIPTION	UNIT	EST. QUAN.	UNIT PRICE	TOTAL
8)	YARD PIPING: 8-Inch, M.J.D.I. CL. 52 watermain connection from riser to hydrant tee; including hydrant, valve, 6-Inch lead, tee and blind flange	L.S.	1		
9)	SITE RESTORATION: Complete site restoration including grading, removal of construction debris, topsoil, seeding, fertilizing & mulching	L.S.	1		
10)	EROSION CONTROL: Erosion control of construction site, per Plans & Specifications	L.S.	1		
11)	MOBILIZATION: Mobilization, bonds, insurance and other general project administration items	L.S.	1		
TOTAL ALTERNATE BID (PROPOSAL B) ------------------------------ $					
***SPECIAL NOTE:** BIDDER shall use the least costly Bid (i.e. Base or Alternate) total for proposed Base Price.					
TOTAL BID: PROPOSAL B --- $					

The BIDDER understands that the OWNER reserves the right to reject any and all Bids, to waive any informalities in the Bidding process and to accept the BID most advantageous to the OWNER.

The undersigned further agrees to furnish surety in a penal sum of equal to or greater than the aggregate of the Contract and on the form hereto attached, to the OWNER, for carrying out the Work for which this PROPOSAL is submitted.

The undersigned agrees to execute the CONTRACT within ten (10) days from the date of notice from the OWNER that the CONTRACT has been awarded.

Accompanying this PROPOSAL is a check (certified) or Bidder's Bond in the amount of five (5%) percent of the bid amounting to ______________________________ Dollars ($____________) as required in the Advertisement for Bids.

In submitting this Bid it is understood that the right is reserved by the Owner to reject any or all Bids and to waive any informalities in the Bidding.

The Bidder agrees that this Bid shall be good and may not be withdrawn for a period of 90 calendar days after the scheduled closing for receiving Bids.

FIRM'S NAME

____

____

by ______________________________

Title ______________________________

OFFICIAL ADDRESS

____

____

____

____

Tele. No. (___)______________________

(Seal: If Bid by Corporation)

Attest:

____

Table of Contents

WATER SYSTEM
TONY, WISCONSIN

ii

Source: Morgan & Parmley, Ltd.

Table of Contents *(continued)*

Table of Contents *(continued)*

v

vi

Bidding Process

ADVERTISING & BID OPENING

Section 24

The rules and regulations of advertising for construction bids vary from state to state and from funding agency to funding agency. Therefore, the reader is advised to follow the specific mandates of the controlling agency and/or municipality when preparing the documents for this phase of the project.

There are many officially approved documents and formats that are in use today; all of which can not be included within this sourcebook. Therefore, as in the previous section, we have chosen to include a generic presentation in the following pages to guide the reader through this phase of a project.

The previous section (23) and this section have several interactive components and documents. Therefore, the reader should study both sections together to obtain a full understanding of the bidding process.

TABLE OF CONTENTS

Submittal Letter to Publisher

Engineer's Letterhead

Date: ____________

Publication Name, Address, etc.

Attn: Legal Advertisement Editor

RE: Project Name: ______________________________
Project Location: ______________________________
Project Number: ______________________________

Dear (*Editor*):

Please publish the enclosed ADVERTISEMENT for BIDS for subject project for (*Number*) consecutive weeks, commencing on (*Month/Day*), (*Year*) in your legal ads section. An affidavit of publication is required and requested to be submitted to our office within three (3) days prior to the Bid Opening which is (*Month/Day*), (*Year*).

If you have any questions concerning this submittal, do not hesitate to contact me.

Sincerely,

____

Project Engineer

Enc. Ad for Bids

C: (*Owner/Client*)

Editor's Note: Refer to Section 23 for samples of Ads for Bids.

Pre-Qualification Statement Format

PRE-QUALIFICATION STATEMENT

Submitted to ____ Date Filed ___________

Project: ____

NOTE: If the municipality, board, public body, or officer is not satisfied with the sufficiency of the answers to the questionnaire and financial statement, the bid may be rejected or disregarded or additional information may be required.

1. Name of Bidder ____

2. Bidder's Address ____

3. Direct any questions regarding information provided on this form to:
   ________________________________ at ____________________

4. Type of organization(check one): Corporation_____ Partnership______ Ind._____

5. When organized? ____

6. If a corporation, when and where incorporated ______________________
   ____

7. Attach a statement listing the corporate officers, partners or other principal members of your organization. Detail the background and experience of the principal members of your personnel, including the officers.

8. How many years has your organization been engaged in the contracting business under the present firm name? ______________________________

9. General description of work performed by your firm ____________________
   ____

10. Attach a list of contracts on hand, for both public and private construction, including for each contract: the class of work; the contract amount; the percent completed; the estimated completion date; and the name and address of the owner or contracting officer.

11. Has your organization or any officer or partner acting either in their own name or as an officer or partner of some other organization ever defaulted on a contract or failed to complete a construction contract for any reason during the past 5 years?

Source: Morgan & Parmley, Ltd.

Pre-Qualification Statement Format *(continued)*

12. Has your organization, any of its owners, a subsidiary or corporate parent, or any officer or director thereof, been subjected to any regulatory action or violation of law concerning construction projects or employees in the last 10 years? ____ If so, indicate:

    (a) The Date______________________ (b) Claimant ________________

    (c) Claimant's Mailing Address ______________________________ and

    (d) Attach a statement reciting the particulars of such violations(s).

13. Attach a list of the last 5 projects your organization has completed, including for each project: the class of work; the contract amount; the completion date; and the name, address and telephone number of the owner or contracting officer.

14. Attach a list of the major equipment which is available to your organization for the proposed work.

15. Attach a statement of your organization's experience in the construction or work similar in nature and importance to this project.

16. Credit available____.
    Attach a letter from your bank(s) or other financial institution(s) advising line of credit set up for your organization.

17. Name of Bonding Company and name, address and telephone number of agent:

    ____

    ____

18. Financial Statement (An official company Financial Statement may be attached in lieu of completing this Item)

    a. Cash $____________________

    b. Accounts Receivable $____________________

    c. Real estate equity $____________________

    d. Materials in stock $____________________

    e. Equipment, book value $____________________

    f. Furniture and fixtures, book value $____________________

    g. Other assets $____________________

Financial Statement (continued on next page)

TOTAL ASSETS ____________________

Liabilities

h. Accounts, notes & interest payable $____________________

i. Other liabilities ____________________

TOTAL LIABILITIES ____________________

NET WORTH $____________________

19. Additional information may be submitted if desired.

Dated at ____________________________ this ______ day of ________, _______.

Name of organization ____

By____

Title____

State of ______________________)
)
County of _____________________)

______________________________________ being duly sworn says that he/she is ____________________________________ of ______________________________
Title of Officer Name of Organization

and that the answers to the foregoing questions and all statements contained herein and in the attachments are true and correct.

Signed ____________________________________

Subscribed and sworn to before me this
_______ day of _____________, ______.

____
Notary Public, State of ____________
My Commission Is/Expires __________

NOTE: This statement must be filed with the Owner or the Engineer not later than five (5) days prior to the opening of bids. This law further required that the information given by you in this statement shall be kept strictly confidential.

Sub-Contractors & Materials Suppliers List

All BIDDERS (Contractors) are required to list their Subcontractors and Material Suppliers including the phase of construction they will execute.

Failure to complete this list may be considered cause for rejection of PROPOSAL.

Name of Subcontractor and/or Material Supplier w/Address	Trade, Work or Material

Source: Morgan & Parmley, Ltd.

List of Agents or Subcontractors Form

LIST OF AGENTS OR SUBCONTRACTORS

NAME ____________________ NAME ____________________
ADDRESS ____________________ ADDRESS ____________________
CITY, STATE, ZIP ____________________ CITY, STATE, ZIP ____________________
TELEPHONE # _(____)____________________ TELEPHONE # _(____)____________________

NAME ____________________ NAME ____________________
ADDRESS ____________________ ADDRESS ____________________
CITY, STATE, ZIP ____________________ CITY, STATE, ZIP ____________________
TELEPHONE # _(____)____________________ TELEPHONE # _(____)____________________

NAME ____________________ NAME ____________________
ADDRESS ____________________ ADDRESS ____________________
CITY, STATE, ZIP ____________________ CITY, STATE, ZIP ____________________
TELEPHONE # _(____)____________________ TELEPHONE # _(____)____________________

NAME ____________________ NAME ____________________
ADDRESS ____________________ ADDRESS ____________________
CITY, STATE, ZIP ____________________ CITY, STATE, ZIP ____________________
TELEPHONE # _(____)____________________ TELEPHONE # _(____)____________________

NAME ____________________ NAME ____________________
ADDRESS ____________________ ADDRESS ____________________
CITY, STATE, ZIP ____________________ CITY, STATE, ZIP ____________________
TELEPHONE # _(____)____________________ TELEPHONE # _(____)____________________

NAME ____________________ NAME ____________________
ADDRESS ____________________ ADDRESS ____________________
CITY, STATE, ZIP ____________________ CITY, STATE, ZIP ____________________
TELEPHONE # _(____)____________________ TELEPHONE # _(____)____________________

Source: Morgan & Parmley, Ltd.

Addendum Format

ADDENDUM NO. 1

LADYSMITH SITE IMPROVEMENT PROJECT

LADYSMITH INDUSTRIAL PARK
RUSK COUNTY, WISCONSIN

Bid Date: July 10, 2000, 1:00 p.m.

Date of Addendum: June 21, 2000

CHANGES TO THE SPECIFICATIONS:

1 The bids received date on the INFORMATION TO BIDDERS form is revised to reflect the date on the ADVERTISEMENT FOR BIDS – July 10, 2000, 1:00 p.m.

2 Line 17 of the BID form is revised to require completion of the project within 160 days of the commencement date. The 1250 foot rail spur and crossing shall be completed within 60 consecutive calendar days of the commencement date. Line 14 of the AGREEMENT is revised to reflect the same change – 120 calendar days becomes 160 calendar days.

3 Revise line 24 of Section 13200-3, Pre-Engineered Metal Buildings, as follows:

Wall panels shall be 26 gauge (min.) galvanized,...

END OF ADDENDUM NO. 1

Morgan & Parmley, Ltd.
115 W. Second Street South
Ladysmith, WI 54848

Receipt of this Addendum is hereby acknowledged:

By:______________________ Date:______________

Contractor:________________________________

PLEASE INCLUDE ADDENDUM IN YOUR PROPOSAL

By:______________________ Date:______________

Contractor:________________________________

PLEASE INCLUDE ADDENDUM IN YOUR PROPOSAL

Source: Morgan & Parmley, Ltd.

Plan Holders List Form

PLAN HOLDERS LIST

Project: ______________

Page 1 of ___

Bid Date:
Time:
Place:

Set No.	Contractor	Name & Address	Date Requested	Date Sent	Bid	Date Addendum Sent Out No. 1	No. 2	No. 3	No. 4	Plan Deposit
1										
2										
3										
4										
5										
6										
7										
8										
9										
10										
11										
12										
13										
14										

Source: Morgan & Parmley, Ltd.

Bid Opening Attendance Sign-Up Sheet

ATTENDANCE SIGN-UP SHEET

BID OPENING

for

Project: ____________________

Date: ____________________

NAME	ADDRESS	REPRESENTING

Bid Tabulation Format

Bid Date: ____________

Time: ____________

Place: ____________

BID TABULATION

Name of Project

Page __ of __

NOTE: All items must include description

No.	Item	CONTRACTOR	CONTRACTOR	CONTRACTOR
1				
2				
3				
4				
5				
6				
7				
8				
9				
10				
11				
12				
13				
14				
15				
16				
17				
18				
19				
20				
21				
22				
23				
24				
25				
26				
27				
28				
29				
30				
31				
32				
33				
34				
35				
	Project Total			

Source: Morgan & Parmley, Ltd.

Bid Tabulation Format *(continued)*

Bid Date: May 26, 2002
Time: 10:00 a.m.
Place: Ladysmith City Hall

Page 1 of 2

BID TABULATION
River Avenue Street Reconstruction
Ladysmith, Wisconsin

No.	Item	Unit Price	Quantity	CONTRACTOR Baughman Trucking & Excavating, LLC	Unit Price	Quantity	CONTRACTOR Russ Thompson Excavting, Inc.	Unit Price	Quantity	CONTRACTOR A-1 Excavating, Inc.
1	Unclassified Excavation	6.00	2090	12540.00	4.55	2090	9509.50	5.20	2090	10868.00
2	Unclassified Excavtation - Above Base Bid	6.00	100	600.00	4.55	100	455.00	5.20	100	520.00
3	Clearing & Grubbing	1200.00	1	1200.00	700	1	700.00	23.50	1	23.50
4	Sawing Pavement	1.50	200	300.00	2.5	200	500.00	2.00	200	400.00
5	6" Valve	1600.00	2	3200.00	1163.01	2	2326.02	1100.00	2	2200.00
6	Hydrant Units	2300.00	1	2300.00	3000	1	3000.00	1900.00	1	1900.00
7	2" Polystyrene Insulation	1.25	320	400.00	2	320	640.00	1.60	320	512.00
8	1" Water Services	500.00	3	1500.00	800	3	2400.00	580.00	3	1740.00
9	4' Sanitary Sewer Manholes	2000.00	5	10000.00	1750	5	8750.00	1560.00	5	7800.00
10	8" PVC Sewermain	18.00	880	15840.00	35	880	30800.00	31.00	880	27280.00
11	4" Sewer Services	400.00	18	7200.00	650	18	11700.00	660.00	18	11880.00
12	6" Sewer Services	450.00	2	900.00	700	2	1400.00	740.00	2	1480.00
13	12" Storm Sewer CL III	35.00	30	1050.00	24	30	720.00	28.00	30	840.00
14	12" Storm Sewer CL V	38.00	65	2470.00	26.25	65	1706.25	28.00	65	1820.00
15	10" D.I. CL 52 Storm Sewer	32.00	190	6080.00	27.35	190	5196.50	29.50	190	5605.00
16	4' Storm Sewer Manhole	900.00	3	2700.00	1400	3	4200.00	1465.00	3	4395.00
17	2' Diameter Storm Sewer Inlets	300.00	2	600.00	500	2	1000.00	850.00	2	1700.00
18	Type 3 Inlet	750.00	6	4500.00	950	6	5700.00	955.00	6	5730.00
19	Rip Rap	50.00	2	100.00	100	2	200.00	88.00	2	176.00
20	Imported Sand Bedding - 8" Pipe	2.00	880	1760.00	1.5	880	1320.00	0.50	880	440.00
21	Imported Sand Bedding - 12" Pipe	2.00	95	190.00	1.5	95	142.50	0.50	95	47.50
22	Imported Sand Bedding - 10" Pipe	2.00	160	320.00	1	160	160.00	0.50	160	80.00
23	Crushed Aggregate Base Course	8.25	2000	16500.00	9	2000	18000.00	8.20	2000	16400.00
24	Curb & Gutter 30-inch	7.25	2200	15950.00	7	2200	15400.00	7.00	2200	15400.00
25	Topsoil Delivered & Placed	18.00	200	3600.00	18	200	3600.00	17.80	200	3560.00
26	Seed	3.00	50	150.00	5.5	50	275.00	8.50	50	425.00
27	Fertilizer	20.00	3	60.00	35	3	105.00	50.00	3	150.00
28	Mulch	500.00	1	500.00	350	1	350.00	200.00	1	200.00
29	Silt Fence Installation & Maintenance	2.00	300	600.00	2	300	600.00	2.20	300	660.00
30	Erosion Bales Installation & Maintenance	4.00	20	80.00	10	20	200.00	8.00	20	160.00
31	4" Concrete Sidewalk	10.00	20	200.00	5	20	100.00	5.00	20	100.00
32	Geotextile Fabric	1.25	500	625.00	1.1	500	550.00	1.00	500	500.00
33	Sediment Trap	500.00	1	500.00	1000	1	1000.00	800.00	1	800.00
34	Asphalt	40.00	800	32000.00	36.6	800	29280.00	38.60	800	30880.00
35	Cut in Type 3 Flat Grate	500.00	1	500.00	750	1	750.00	875.00	1	875.00
	Project Total			**$147,015.00**			**$162,735.77**			**$157,547.00**

Source: Morgan & Parmley, Ltd.

Bid Evaluation & Method of Award

After the bid opening, each proposal received will be reviewed, evaluated and checked for accuracy and for compliance with the bidding documents. Any non-responsive bids can be rejected by the Owner. A bid may be considered non-responsive for any of the following reasons:

1. Failure to fully complete each and every bid item listed in the BID SCHEDULE for the proposal submitted.
2. Failure to comply with the conditions of the ADVERTISEMENT FOR BIDS.
3. The Bidder imposes conditions or qualifies his bid in any way.
4. The Proposal is unsigned and/or not attested.

If a Contract is awarded, it will be awarded to the lowest responsible Bidder. A responsible Bidder is one who meets the following requirements.:

1. Has adequate financial resources to complete the Proposal for which the Bid was received.
2. Has the necessary experience, organization and technical qualifications and the necessary equipment to perform the work required.
3. Is able to comply with the required performance schedule to meet the completion date.
4. Has a satisfactory record of performance, integrity, judgment and skills.
5. Is qualified and eligible to receive the award under the applicable laws and regulations.

Recommendation to Owner/Client Format

Engineer's Letterhead

Date: ______________

Owner/Client Name, Address, etc.

RE: Project Name: ____________________________
Project Number: ____________________________

Dear ______________:

Construction bids for subject project were opened *(Time & Date)* ______________.

Reference copies of the following documents are enclosed:

1) AFFIDAVIT of PUBLICATION
2) PLAN HOLDERS LIST
3) ATTENDANCE LIST @ BID OPENING
4) BID TABULATION
5) LOW BIDDER'S STATEMENT of QUALIFICATIONS

Based upon the foregoing data and the contractor's references, *(Name of Engineering Firm)* recommends that *(Name of Owner/Client)* award the construction contract to *(Low Bidder's Name)*; subject to concurrence from (if applicable) *(Grant Agency or similar funding source)*.

Please advise me when the award will be made so I can issue the Notice of Award to the contractor and schedule the Pre-Construction Conference.

Sincerely,

____
(Name of Project Engineer)

Enc.

C: *(Appropriate Officials)*

Source: Morgan & Parmley, Ltd.

Bidding Process

CONSTRUCTION CONTRACTS

Section 25

The construction contract, or agreement, is the legal document between the owner and contractor that specifically details the work to be performed and the price to be paid. This legally binding document is customarily bound within the bidding documents of the specifications book. Unlike many contracts, this agreement has many components which include the plans, specifications, special provisions, general conditions, supplemental general conditions and often the funding agency conditions.

Again, it must be pointed out to the reader that the controlling agency and/or municipality may require the use of their specific contractual documents. Therefore, it is prudent to obtain them early in the planning process to avoid potential legal entanglements. The following pages present only generic material, some of which the editor-in-chief has successfully used in may projects.

It, also, must be noted that many professional societies have bidding and construction documents that are in common use. Therefore, the design professional may want to use documents prepared by one of these organizations; namely NSPE or AIA, etc.

TABLE OF CONTENTS

Notice of Award

To: ________________ ________

____

____

____

____

PROJECT Description: ____

____

____

The OWNER has considered the BID submitted by you for the above described WORK in response to its Advertisement for Bids dated ________, ________, and Information for Bidders.

You are hereby notified that your BID has been accepted for items in the amount of $________.

You are required by the Information for Bidders to execute the Agreement and furnish the required CONTRACTOR'S Performance BOND, Payment BOND and certificates of insurance within ten (10) calendar days from the date of this Notice to you.

If you fail to execute said Agreement and to furnish said BONDS within ten (10) days from the date of this Notice, said OWNER will be entitled to consider all your rights arising out of the OWNER'S acceptance of your BID as abandoned and as a forfeiture of your BID BOND. The OWNER will be entitled to such other rights as may be granted by law.

You are required to return an acknowledged copy of this NOTICE OF AWARD to the OWNER.

Dated this ________ day of ____________, ________.

____
Owner

By ______________________________

Title ______________________________

ACCEPTANCE OF NOTICE

Receipt of the above NOTICE OF AWARD is hereby acknowledged

by ____,

this the __________________day of ______________________, ____

By ______________________________

Title ______________________________

Source: * (See page 23-1)

Agreement: Construction Contract

THIS AGREEMENT, made this ________day of________, ________, by and between ______________________________ (Name of Owner), (an individual), hereinafter called "OWNER"

and ______________________________ doing business as (an individual,) or (a partnership,) or (a corporation) hereinafter called "CONTRACTOR".

WITNESSETH: That for and in consideration of the payments and agreements hereinafter mentioned:

1. The CONTRACTOR will commence and complete the construction of

____

2. The CONTRACTOR will furnish all of the material, supplies, tools, equipment, labor and other services necessary for the construction and completion of the PROJECT described herein.

3. The CONTRACTOR will commence the work required by the CONTRACT DOCUMENTS within ________ calendar days after the date of the NOTICE TO PROCEED and will complete the same within ________ calendar days unless the period for completion is extended otherwise by the CONTRACT DOCUMENTS.

4. The CONTRACTOR agrees to perform all of the WORK described in the CONTRACT DOCUMENTS and comply with the terms therein for the sum of $ ________, or as shown in the BID schedule.

5. The term "CONTRACT DOCUMENTS" means and includes the following:

(A) Advertisement For BIDS

(B) Information For BIDDERS

(C) BID

(D) BID BOND

(E) Agreement

Source: * (See page 23-1)

Agreement: Construction Contract *(continued)*

(F) General Conditions

(G) SUPPLEMENTAL GENERAL CONDITIONS

(H) Payment BOND

(I) Performance BOND

(J) NOTICE OF AWARD

(K) NOTICE TO PROCEED

(L) CHANGE ORDER

(M) DRAWINGS prepared by ______________________
numbered __________ through __________, and dated __________,
____

(N) SPECIFICATIONS prepared or issued by ______________________
____,
dated __________, __________

(O) ADDENDA:

No. __________, dated __________, __________

No. __________, dated __________, __________

No. __________, dated __________, __________

No. __________, dated __________, __________

No. __________, dated __________, __________

No. __________, dated __________, __________

6. The OWNER will pay to the CONTRACTOR in the manner and at such times as set forth in the General Conditions such amounts as required by the CONTRACT DOCUMENTS.

7. This Agreement shall be binding upon all parties hereto and their respective heirs, executors, administrators, successors, and assigns.

IN WITNESS WHEREOF, the parties hereto have executed, or caused to be executed by their duly authorized officials, this Agreement in (__________ (Number of Copies)) each of which shall be deemed an original on the date first above written.

OWNER:

____

BY ______________________

Name ______________________ (Please Type)

Title ______________________

(SEAL)

ATTEST:

____

Name ______________________ (Please Type)

Title ______________________

CONTRACTOR:

____

BY ______________________

Name ______________________ (Please Type)

Address ______________________

____

(SEAL)

ATTEST:

____

Name ______________________ (Please Type)

____

Performance Bond Form

KNOW ALL MEN BY THESE PRESENTS: that

____
(Name of Contractor)

____
(Address of Contractor)

a ____________________, hereinafter called Principal, and
(Corporation, Partnership, or Individual)

____
(Name of Surety)

____
(Address of Surety)

hereinafter called Surety, are held and firmly bound unto ____________

____
(Name of owner)

____
(Address of Owner)

hereinafter called OWNER, in the penal sum of ____________

____________________Dollars, $(__________)

in lawful money of the United States, for the payment of which sum well and truly to be made, we bind ourselves, successors, and assigns, jointly and severally, firmly by these presents.

THE CONDITION OF THIS OBLIGATION is such that whereas, the Principal entered into a certain contract with the OWNER, dated the ________day of________, 19____, a copy of which is hereto attached and made a part hereof for the construction of:

____

____

____

____

NOW, THEREFORE, if the Principal shall well, truly and faithfully perform its duties, all the undertakings, covenants, terms, conditions, and agreements of said contract during the original term thereof, and any extensions thereof which may be granted by the OWNER, with or without notice to the Surety and during the one year guaranty period, and if he shall satisfy all claims and demands incurred under such contract, and shall fully indemnify and save harmless the OWNER from all costs and damages which it may suffer by reason of failure to do so, and shall reimburse and repay the OWNER all outlay and expense which the OWNER may incur in making good any default, then this obligation shall be void; otherwise to remain in full force and effect.

PROVIDED, FURTHER, that the said surety, for value received hereby stipulates and agrees that no change, extension of time, alteration or addition to the terms of the contract or to WORK to be performed thereunder or the SPECIFICATIONS accompanying the same shall in any wise affect its obligation on this BOND, and it does hereby waive notice of any such change, extension of time, alteration or addition to the terms of the contract or to the WORK or to the SPECIFICATIONS.

PROVIDED, FURTHER, that no final settlement between the OWNER and the CONTRACTOR shall abridge the right of any beneficiary hereunder, whose claim may be unsatisfied.

IN WITNESS WHEREOF, this instrument is executed in ______ (Number) counterparts, each one of which shall be deemed an original, this the ________day of ________ 19______.

ATTEST:

____
Principal

____
(Principal) Secretary

By ____________________________________(s)

(SEAL)

____
(Witness as to Principal)

____
(Address)

____
(Address)

____

____

____
Surety

ATTEST:

____
(Surety) Secretary

(SEAL)

____
Witness as to Surety

By ____________________________________
Attorney-in-Fact

____
(Address)

____
(Address)

____

____

NOTE: Date of BOND must not be prior to date of Contract.
If CONTRACTOR is Partnership, all partners should execute BOND.

IMPORTANT: Surety companies executing BONDS must appear on the Treasury Department's most current list (Circular 570 as amended) and be authorized to transact business in the state where the PROJECT is located.

Source: * (See page 23-1)

Payment Bond Form

KNOW ALL MEN BY THESE PRESENTS: that

____
(Name of Contractor)

____
(Address of Contractor)

a ______________________________, hereinafter called Principal,
(Corporation, Partnership or Individual)

and ______________________________
(Name of Surety)

____
(Address of Surety)

hereinafter called Surety, are held and firmly bound unto ______________________________

____
(Name of Owner)

____
(Address of Owner)

hereinafter called OWNER, in the penal sum of______________Dollars, $(__________)

in lawful money of the United States, for the payment of which sum well and truly to be made, we bind ourselves, successors, and assigns, jointly and severally, firmly by these presents.

THE CONDITION OF THIS OBLIGATION is such that whereas, the Principal entered into a certain contract with the OWNER, dated the ______________day of __________ 19________, a copy of which is hereto attached and made a part hereof for the construction of:

____

____

____

____

NOW, THEREFORE, if the Principal shall promptly make payment to all persons, firms, SUBCONTRACTORS, and corporations furnishing materials for or performing labor in the prosecution of the WORK provided for in such contract, and any authorized extension or modification thereof, including all amounts due for materials, lubricants, oil, gasoline, coal and coke, repairs on machinery, equipment and tools, consumed or used in connection with the construction of such WORK, and all insurance premiums on said WORK, and for all labor, performed in such WORK whether by SUBCONTRACTOR or otherwise, then this obligation shall be void; otherwise to remain in full force and effect.

PROVIDED, FURTHER, that the said Surety for value received hereby stipulates and agrees that no change, extension of time, alteration or addition to the terms of the contract or to the WORK to be performed thereunder or the SPECIFICATIONS accompanying the same shall in any wise affect its obligation on this BOND, and it does hereby waive notice of any such change, extension of time, alteration or addition to the terms of the contract or to the WORK or to the SPECIFICATIONS.

PROVIDED, FURTHER, that no final settlement between the OWNER and the CONTRACTOR shall abridge the right of any beneficiary hereunder, whose claim may be unsatisfied.

IN WITNESS WHEREOF, this instrument is executed in ________counterparts, each one of which shall be deemed an original, this the __________ day of __________ 19 ________.
(number)

ATTEST:

____
Principal

____
(Principal) Secretary

(SEAL)

By ______________________________(s)

____
(Address)

____

____
[Wit]ness as to Principal

____
(Address)

____

____
Surety

ATTEST:

By ______________________________
Attorney-in-Fact

____
Witness as to Surety

____
(Address)

____
(Address)

____

NOTE: Date of BOND must not be prior to date of Contract.
If CONTRACTOR is Partnership, all partners should execute BOND.

IMPORTANT: Surety companies executing BONDS must appear on the Treasury Department's most current list (Circular 570 as amended) and be authorized to transact business in the State where the PROJECT is located.

Source: * (See page 23-1)

Contractor's Certificate of Insurance Form

Name of Contractor: ______________________ Name of Owner: ______________________

Address: ______________________ Address: ______________________

______________________ ______________________

KIND OF INSURANCE	POLICY NO.	EXPIRATION DATE	LIMITS
Workmen's Compensation Employers' Liability			$ ______ Statutory Workmen's Compensation $ ______ One Accident and Aggregate Disease
Commercial General Liab. Bodily Injury & Property Damage			$ ______ Each Occurrence $ ______ Aggregate $ ______ Aggregate Completed Operations and Products
Personal Injury			$ ______ Aggregate
Comprehensive Auto Liab. Bodily Injury & Property Damage			$ ______ Each Occurrence $ ______ Aggregate
Other Property Insurance			$

		Yes	No
1.	Does Property Damage Liability Insurance shown include coverage for XC and U Hazards?	___	___
2.	Is Occurrence Basis Coverage provided under Property Damage Liability?	___	___
3.	Is Broad Form Property Damage Coverage provided for this Project?	___	___
4.	Does Personal Injury Liability Insurance include coverage for personal injury sustained by any person as a result of an offense directly or indirectly related to the employment of such person by the insured?	___	___
5.	Is Coverage provided for Contractual Liability (including indemnification provision) assumed by Insured?	___	___
6.	Does Coverage above apply to non-owned and hired automobiles?	___	___
7.	Is Occurrence Basis Coverage provided under Automobile Damage Liability?	___	___

Notice: The above named Owner shall be notified 15 days prior to the cancellation of any of the above coverages.

____
NAME OF INSURANCE COMPANY

____
ADDRESS

____
SIGNATURE OF AUTHORIZED REPRESENTATIVE & DATE

Sample Letter to Contractor

Engineer's Letterhead

Date:

Contractor's Name, Address, etc.

Attn: (*Name and Title of C.E.O.*)

RE: Project Name: ______________________
Project Number: ______________________

Dear (*Name of C.E.O.*):

You are hereby notified that the (*Owner/Client's Name*) has accepted your PROPOSAL for (*Name of Project*).

I have enclosed the following documents for you to execute:

1) NOTICE OF AWARD: (6 copies) Sign, date and return 5 copies to our office immediately. The extra copy is for your files.
2) AGREEMENT: (5 copies) Sign, date, seal and return all copies to our office.
3) PERFORMANCE BOND: (5 copies) Execute and return all copies to our office.
4) PAYMENT BOND: (5 copies) Execute and return all copies to our office.
5) CERTIFICATE OF INSURANCE: (5 copies) Execute and return all, copies to our office. (You may use your insurance company s form).
6) OWNER'S INSTRUCTIONS FOR BONDS & INSURANCE: (I copy for reference only)

After these documents have been received and reviewed, the Contract will be formally signed and sealed by the appropriate Officials. A copy of these documents, bound in the Specifications Book, will be delivered to you at the Preconstruction Conference.

If you have any questions concerning this material, please feel free to contact me.

Sincerely,

____
Project Engineer

Enc.

C: (*Owner/Client*)

Source: Morgan & Parmley, Ltd.

Attorney's Certification Form

CERTIFICATE OF OWNER'S ATTORNEY

I, the undsigned, ______________________, the duly authorized and acting legal representative of ______________ ______________, do hereby certify as follows:

I have examined the attached Contract(s) and Performance and Payment Bond(s) and the manner of execution thereof, and I am of the opinion that each of the aforesaid Agreements are adequate and have has been duly executed by the proper parties thereto acting through their duly authorized representatives; that said representatives have full power and authority to execute said Agreements on behalf of the respective parties named thereon; and that the forgoing Agreements constitute valid and legally binding obligations upon the parties executing the same in accordance with terms, conditions, and provisions thereof.

Date: ____________

____
Attorney

NOTE: Delete phrase "performance and payment bonds" when not applicable.

Change Order Form

Order No. ______________

Date: ______________

Agreement Date: ______________

NAME OF PROJECT: ______________________________

____

OWNER: ______________________________

CONTRACTOR: ______________________________

The following changes are hereby made to the CONTRACT DOCUMENTS:

Justification:

Change to CONTRACT PRICE:

Original CONTRACT PRICE $__________

Current CONTRACT PRICE adjusted by previous CHANGE ORDER $__________

The CONTRACT PRICE due to this CHANGE ORDER will be (increased) (decreased)

by: $______________

The new CONTRACT PRICE including this CHANGE ORDER will be $__________

Change to CONTRACT TIME:

The CONTRACT TIME will be (increased) (decreased) by________calendar days.

The date for completion of all work will be ____________(Date).

Approvals Required:
To be effective this Order must be approved by the Federal agency if it changes the scope or objective of the PROJECT, or as may otherwise be required by the SUPPLEMENTAL GENERAL CONDITIONS.

Requested by: ______________________________

Recommended by: ______________________________

Ordered by: ______________________________

Accepted by: ______________________________

Federal Agency Approval (where applicable) ______________________________

Source: * (See page 23-1)

Construction

PRE-CONSTRUCTION CONFERENCE

Section 26

The pre-construction conference is the launch pad where the project is officially ignited and is propelled into the physical construction phase. The construction contract is formally signed, copies distributed and an official notice to proceed document issued to the contractor.

At this meeting, all key players should be assembled and the major elements of the project discussed, as well as pertinent technical questions answered. All special provisions should be properly addressed. In essence, the rules, regulations, format, schedules and related parameters of the project should be thoroughly reviewed. The importance of the pre-construction conference can not be overemphasized. It is the bridge from the planning stage of the project to the final, physical product.

Many times a groundbreaking ceremony is desired for a high profile project. This ceremony may be incorporated into the pre-construction conference, but is recommended that it be held following the technical briefing.

TABLE OF CONTENTS

Format of Notice

NOTICE
of
PRE-CONSTRUCTION CONFERENCE

Date: ______________

To: *(List Contractors, Utility Co. Representatives, Contractors, Regulatory Agency Staff, Municipal Officials, Engineering Staff and Other appropriate individuals who should be notified)*

RE: *(Name of Construction Project, Name of Owner and Project No.)*

Memo:

Please be advised that a PRE-CONSTRUCTION CONFERENCE for subject project will be held on ________________, ________________, ________,
(Day Name) *(Month)* *(Day)*

____________ @ ______________, in the ________________________,
(Year) *(Time)* *(Location)*

________________________, ________. (A location map is enclosed.)
(City) *(State)*

If the prime contractor wants their sub-contractors present at this meeting, it is their responsibility to notify them.

CONSTRUCTION CONTRACTS with a NOTICE TO PROCEED will be issued at this meeting.

Sincerely,

____

(Name of Engineer with Title)

Enc. Location Map

Sample Agenda

Page 1 of 2

-AGENDA-
PRE-CONSTRUCTION CONFERENCE
for
Wastewater Collection & Treatment Facility
Catawba-Kennan Joint Sewage Commission
M & P Project No. 98-106
CDBG Project No. PF FY00-0089

DATE: January 30, 2002

TIME: 1:00 p.m.

LOCATION: Community Center, Kennan, Wisconsin

I. **REGISTRATION:**
   A. Sign Attendance List

II. **INTRODUCTION:**
   A. Welcome & General Overview of Project (M & P, Ltd.)
   B. Sewage Commission Officials
   C. Morgan & Parmley, Ltd. Staff
   D. Rural Development (USDA) Representatives
   E. Grant Administrator
   F. Contractors' Representatives
   G. Subcontractors' Representatives
   H. DNR Representatives
   I. WDOT Representatives
   J. Price Co. Highway Commission Representatives
   K. Town of Catawba Representatives
   L. Town of Kennan Representatives
   M. Utilities (Electric, Telephone, Etc.)
   N. Others

III. **CONSTRUCTION CONTRACTS:**
   A. Commission to Sign Contracts
   B. Rural Development to Approve Contracts
   C. Review: Notice to Proceed (Each Contract)
   D. Contractor (2) to Sign Notice to Proceed
   E. Distribute Contract (2)
      1-Contractor
      2-Commission
      3-Rural Development
      4-Grant Administrator
      5-Engineer

IV. **USDA-RURAL DEVELOPMENT:**
   A. Project Specialist
   B. Reference Pages 233 thru 250
   C. Project Sign

Source: Morgan & Parmley, Ltd.

Page 2 of 2

V. **CDBG REGULATIONS:**
- A. Grant Administrator
- B. Refer to Supplemental CDBG Agenda

VI. **SCOPE OF WORK:**
- A. Review Section 01010 (Special Provisions)
- B. Timetable (Construction Schedule)
- C. Shop Drawing Process
- D. Construction Area Video Taped (Fall 2001 by M & P, Ltd.)
- E. Existing Utilities (Contact Diggers Hotline)
- F. List of Key Contacts

VII. **PLANS & SPECIFICATIONS:**
- A. General Review
- B. Specific Items & Questions

VIII. **PAY REQUEST:**
- A. RD Form (pg. 42 thru 44)
- B. Submit to Engineer by 3rd of Month (3 copies)
- C. Payroll Records (State & Federal)
- D. Payments will be processed monthly, in accordance with the "Progress Schedule of Construction". These schedules, cumulatively, form the basis of the Sewage Commission's outlay management program.

IX. **APPROVALS:**
- A. Contractors to comply with all applicable conditions of each respective approval/permit. Copies of these regulatory approvals and permits are contained in the Specifications & Contracts.

X. **GENERAL QUESTIONS & COMMENTS:**
- A. Overview-Supplemental
- B. Contractors' Representatives
- C. Utilities Representatives
- D. Ground Breaking Ceremony (Schedule)
- E. Other Related Items
- F. Closing Remarks

Sample Attendance Signature List

Page __ of __

ATTENDANCE SIGNATURE LIST
for
PRE-CONSTRUCTION CONFERENCE

Wastewater Collection & Treatment Facility
Catawba-Kennan Joint Sewage Commission
M & P Project No. 98-106
CDBG Project No. PF FY00-0089
January 30, 2002

NAME	REPRESENTING

Source: Morgan & Parmley, Ltd.

Generic Notice to Proceed

NOTICE TO PROCEED

To: ______________________ Date: ______________________

______________________ Project: ______________________

______________________ ______________________

______________________ ______________________

______________________ ______________________

You are hereby notified to commence WORK in accordance with the Agreement dated __________, ______, on or before __________, ______, and you are to complete the WORK within ________ consecutive calendar days thereafter. The date of completion of all WORK is therefore __________, ______.

____
Owner

By ______________________

Title ______________________

ACCEPTANCE OF NOTICE

Receipt of the above NOTICE TO PROCEED is hereby acknowledged by ____

______________________,

this the ______________________ day

of ______________________, ____

____

By ______________________

Title ______________________

Source: * (See page 23-1)

Generic Format for Construction Schedule

WASTEWATER COLLECTION & TREATMENT FACILITY

Project No. ______________________

January 30, 2002

BID ITEM	MAY 5-11	MAY 12-18	MAY 19-25	MAY/JUNE 26-1	JUNE 2-8	JUNE 9-15	JUNE 16-22	JUNE 23-29	JUNE/JULY 30-6	JULY 7-13	JULY 14-20	JULY 21-27	JULY/AUG 28-3	AUG 4-10	AUG 11-17
1) *(Name)*															
2) *(Name)*															
3) *(Name)*															
4) *(Name)*															
5) *(Name)*															
6) *(Name)*															
7) *(Name)*															
8) *(Name)*															
9) *(Name)*															

NOTE: 1) Restoration to be completed as installations progress.

Format for Application for Payment

Page 1 of 2

APPLICATION FOR PAYMENT

Project: ______________________________

Owner: ______________ Contractor: ______________

Date: ______________ Pay Request No.: ______________

From (Date): ______________ To (Date): ______________

1. Original Contract Amount .... $ ________
2. Net change by Change Orders .... $ ________

   Change Order No. Amount (+ or −)

   NET CHANGE $ ________

3. Contract Sum to Date .... $ ________
4. Total Completed and Stored to Date .... $ ________
   (Total from Page ___)
5. Retainage @ ____%. .... $ ________
6. Total Earned less Retainage .... $ ________
7. Less Previous Certificates for Payment .... $ ________
8. Current Payment Due .... $ ________
9. Balance to Finish, Plus Retainage .... $ ________

Contractor: ______________ Date: ______________

Certified By: ______________ Date: ______________

Accepted By: ______________ Date: ______________

Authorized By: ______________ Date: ______________
(When Applicable)

Page 2 of 2

SUMMARY OF COMPLETED WORK

Item No.	Description of Work	Contract Amount	Previous Application	This Application	Total Completed and Stored to Date	Balance to Finish
TOTAL						

Source: Morgan & Parmley, Ltd.

Payroll Record Format

PAYROLL

NAME OF CONTRACTOR____ or SUBCONTRACTOR____		ADDRESS	
PAYROLL NO.	FOR WEEK ENDING	PROJECT AND LOCATION	PROJECT NO.

(1) NAME, ADDRESS AND SOCIAL SECURITY NUMBER OF EMPLOYEE	(2) # OF EXEMP-TIONS	(3) WORK CLASSIFICATION	OT OR ST	(4) DAY AND DATE							(5) TOTAL HOURS	(6) RATE OF PAY	(7) Gross Amount Earned	(8) DEDUCTIONS						(9) NET WAGES PAID FOR WEEK
				Hours Worked Each Day										FICA	WITH HOLDING TAX			OTHER	TOTAL DEDUCTIONS	
			O																	
			S																	
			O																	
			S																	
			O																	
			S																	
			O																	
			S																	
			O																	
			S																	
			O																	
			S																	
			O																	
			S																	

Construction

Shop Drawings

Section 27

Plans and specifications describe the proposed project and detail the various components. However, it is good practice to specify various components in such a way as to avoid "cold speccing" so that more than one supplier can satisfy the specifications, thus ensuring competitive bidding and meet budgetary restrictions. Therefore, it is the engineer's duty to review all shop drawings for the project to ensure that the contractor, through their fabricator/supplier, complies with the plans and specifications.

Table of Contents

General Process

Many project components and equipment are not fabricated at the project site. Therefore, shop drawings are prepared for these specific items and submitted to the engineer for review, prior to their manufacture. It is the engineer's duty to review these drawings and note corrections where necessary so the final project will comply with the project plans and specifications.

It is extremely important that copies of all shop drawings be kept for the engineer's permanent file. Additionally, the project inspector must have copies at his disposal, as components arrive at the site, to check and verify that the actual products match the approved, items.

It is also wise to retain a complete set of all shop drawings to assist in preparing the O & M Manual at the close of construction.

The following samples of shop drawings have been selected from actual projects and are reproduced here to give the reader a general feel for their content since it is impossible and certainly impractical in this sourcebook to present the full range of shop drawings within the allotted pages. The reader should understand that electrical components versus structural steel fabrication differ widely in their respective graphic presentations. Often a manufacturer, such as one who produces items that have various models, will submit copies of their technical catalog with specific pages marked which indicate the selected item that meets the specification called out on the project plan. Usually, this is considered an acceptable submittal. This method, of course, is significantly different from the detailing process required to fabricate structural steel for a bridge. In any event, the reader must be aware of the wide range of shop drawing techniques in use today. Remember, it is the purpose of shop drawings to ensure that the detailed component will meet the specifications of the project plans.

Sample: Well Pump Shop Drawing

DEEP WELL TURBINE PUMP
ELECTRIC MOTOR DRIVEN

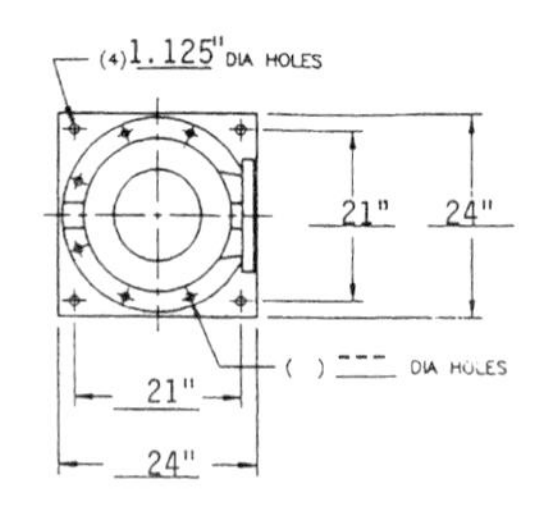

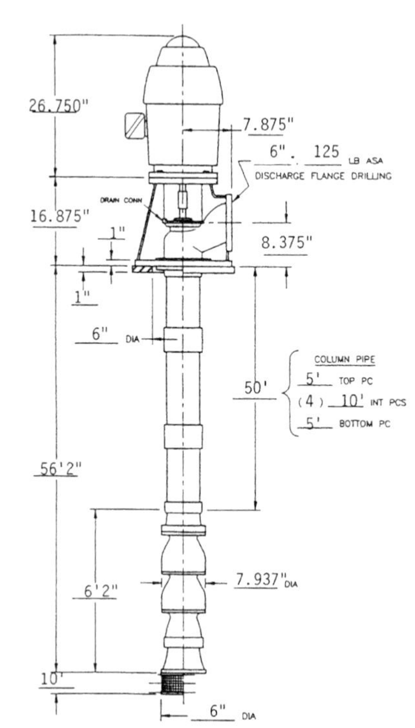

MATERIAL OF CONSTRUCTION			
Bowl	C.I. 30	Impeller	Bronze
Bowl Shaft	416	Shaft Coupling	C-1137
Bowl Bearings	Bronze	Shaft Bearings	Bronze
Strainer	N/A	Bowl W/R	Bronze
Impeller W/R	Bronze	Column Pipe	A120 B
Lineshaft	C-1045	Packing	Composition
Base Plate	Steel	Head	C.I. 30
PUMP			
Type	8 KC	Discharge Head	TR6B
Suction	6" N.P.T.	Discharge	6"
Lineshaft	C-1045	Column	6"
Lubrication	Product	Model	
Stage	9	GPM	200
TDH	221	Trim	6.182'
RPM	1770	BHP	14.36
MOTOR			
Make	U.S.	Type	AUE
Enclosure	W.P.I.	NRR	Yes
SRC	N/A	HP	15
RPM	1770	Phase	3
Hertz	60	Voltage	230/460
Frame No.	254 TP	Type Coupling	Bolted
OTHER SPECIFICATIONS			
Drawing No.		Serial No.	
Fluid	Well Water	Spec.Gravity	1
Viscosity		Temperature	
PH		No.Units Required	1

Customer		
Address		
City	ST	ZIP
Tel. ( )	Fax. ( )	
Rep.		
Supplier		
Salesman		

Source: Morgan & Parmley, Ltd.

Sample: Shop Drawings for Elevated Water Storage Tank

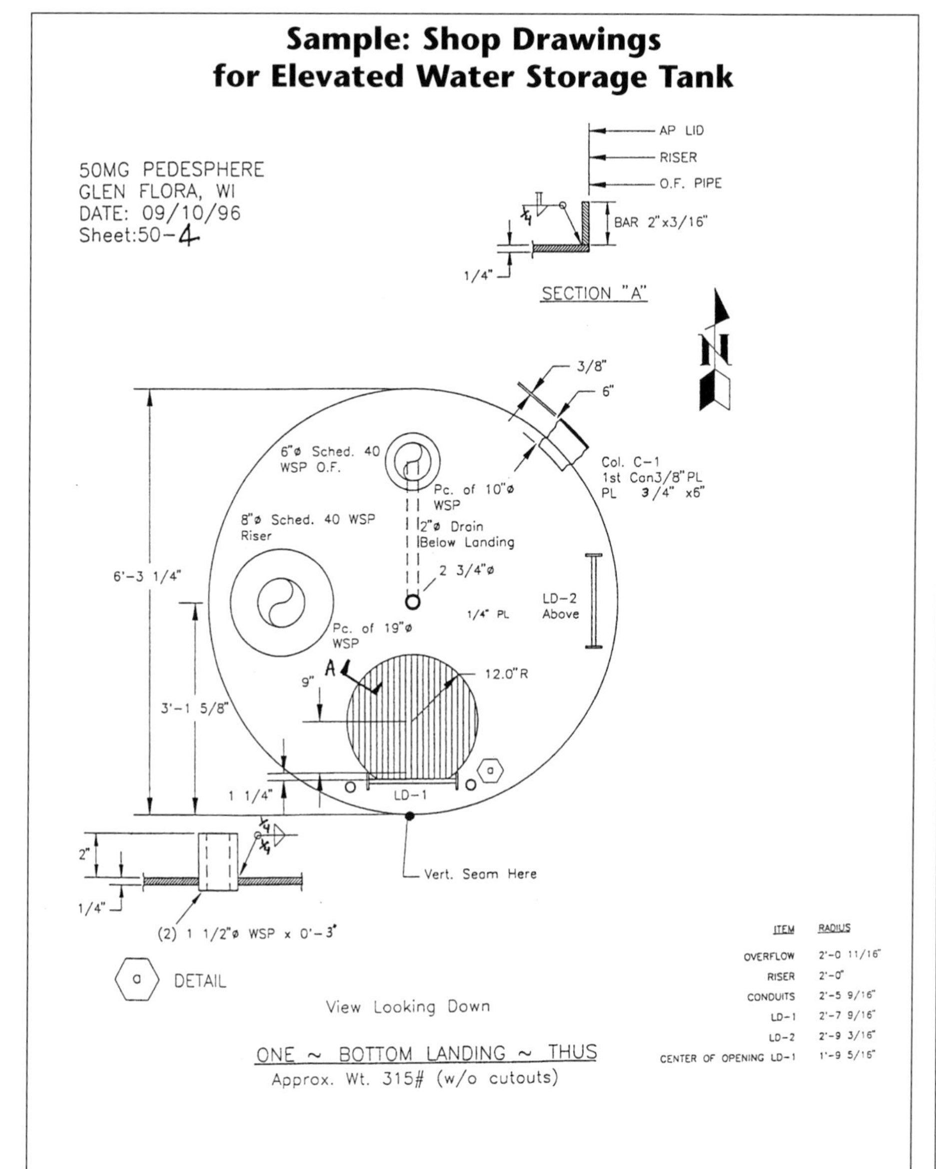

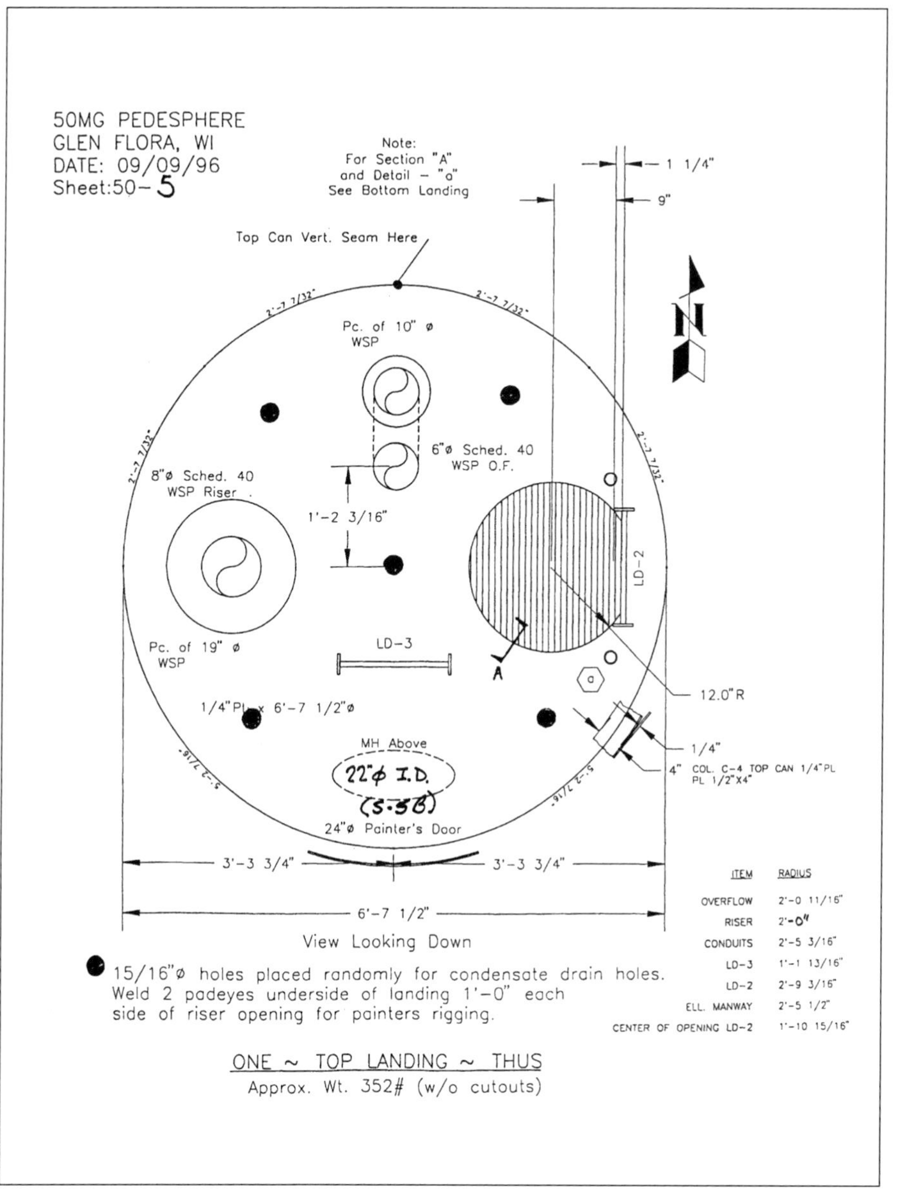

Source: Maguire Iron, Inc.

Sample: Shop Drawings for Elevated Water Storage Tank *(continued)*

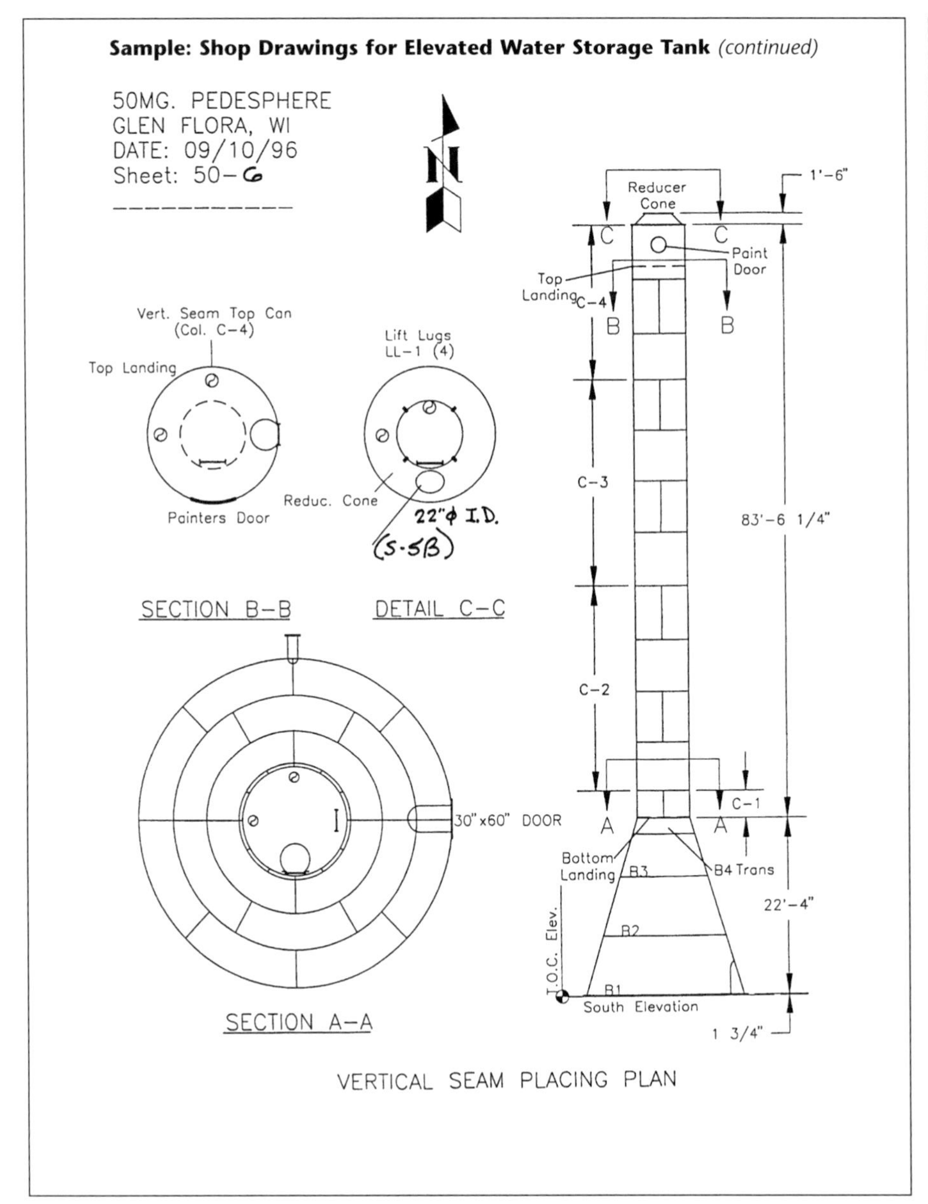

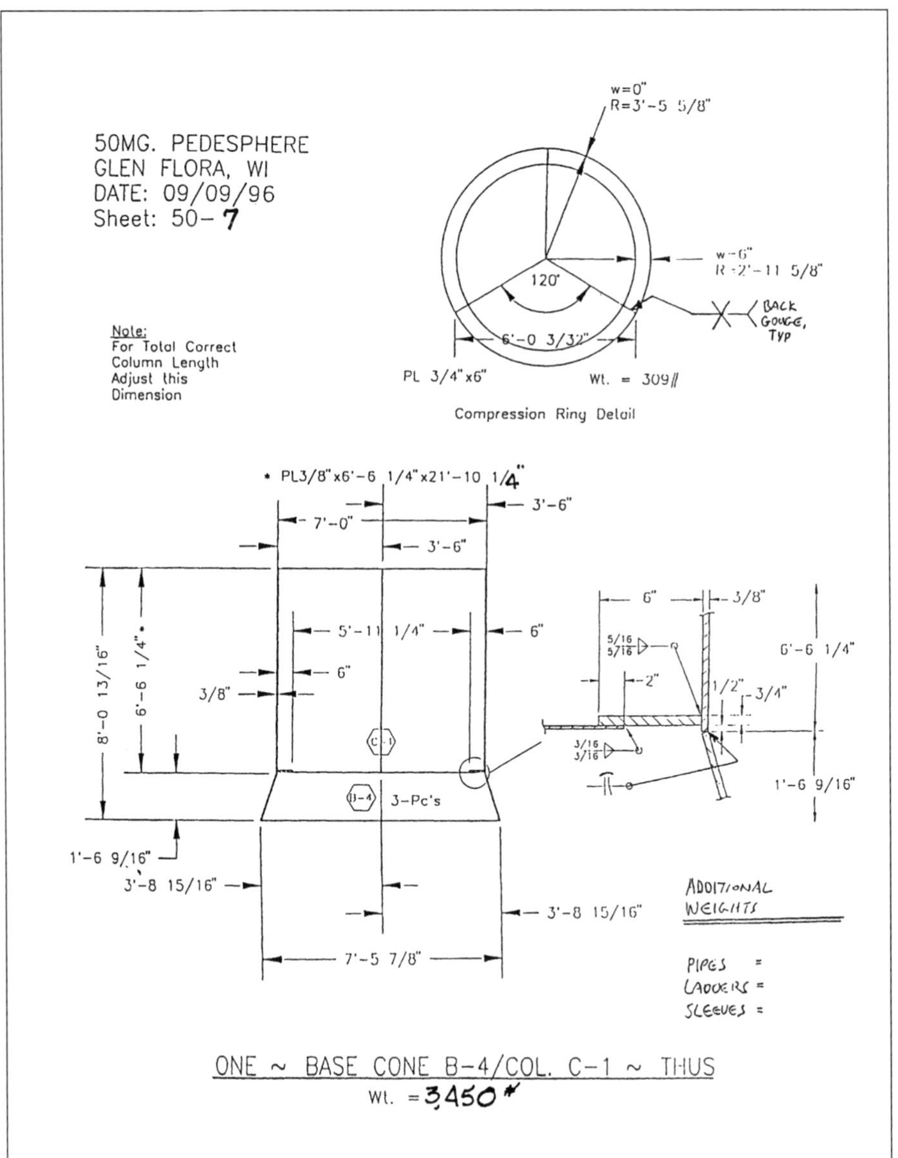

Sample: Shop Drawings for Elevated Water Storage Tank *(continued)*

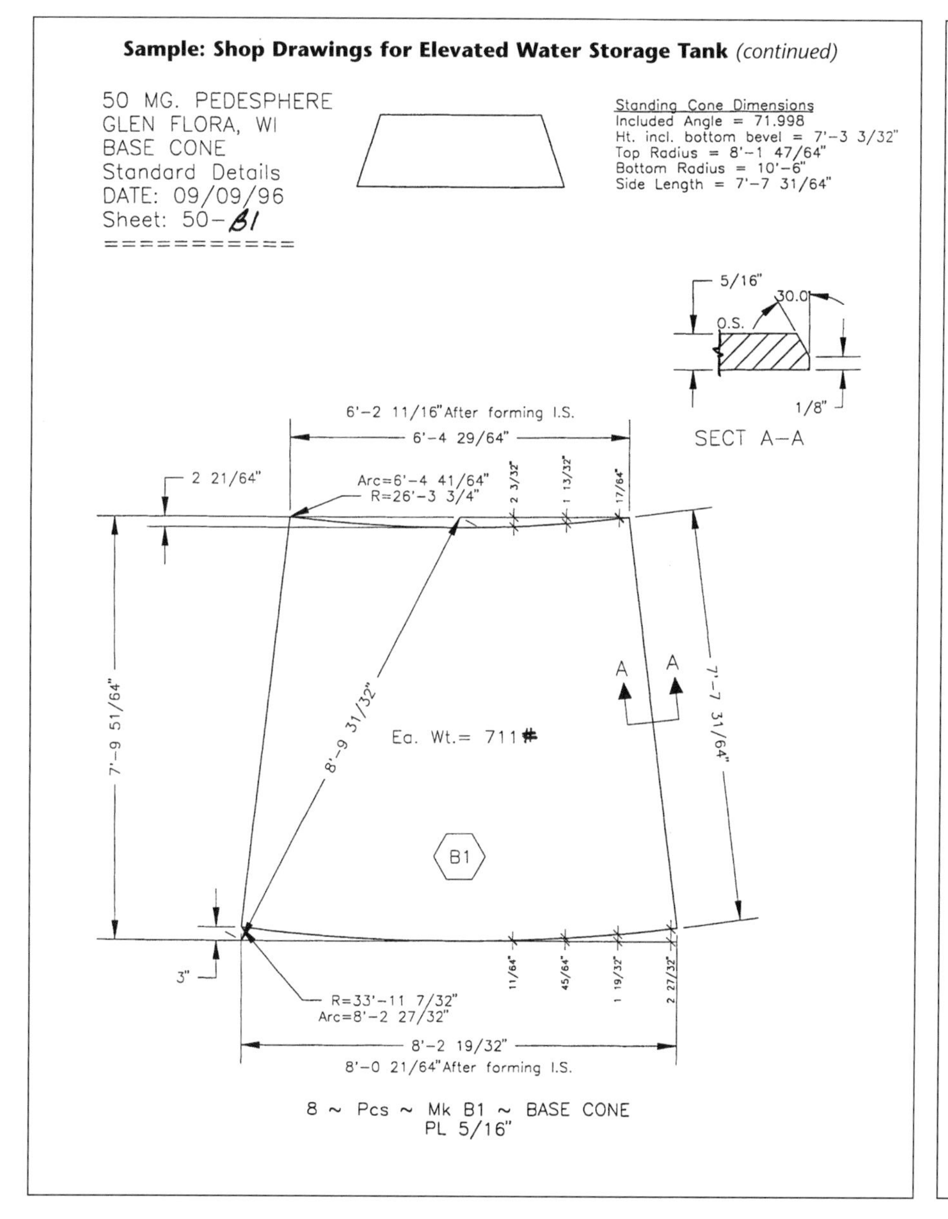

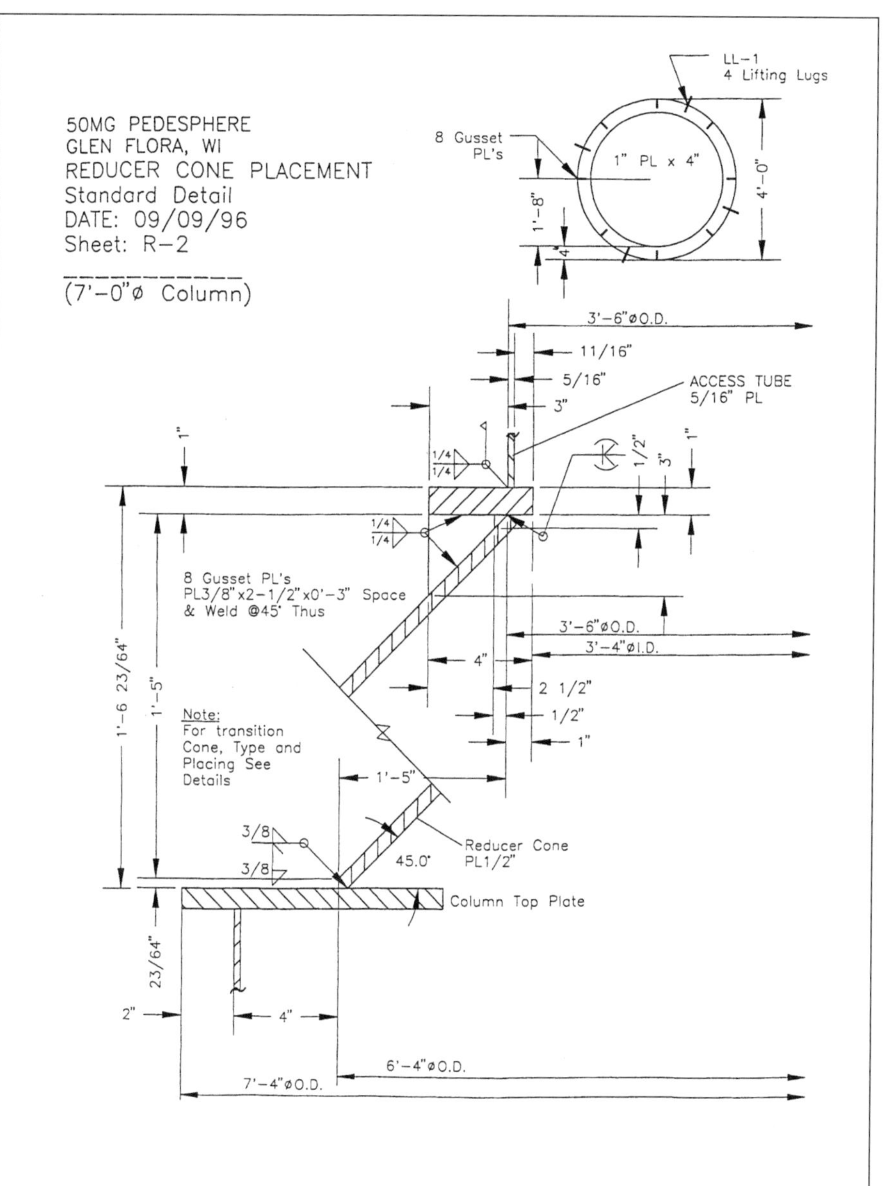

Engineer's Review Stamp

Review by the Engineer shall not be construed as a complete check of the submitted material nor shall it relieve the contractor from his responsibility to comply with the specifications.

Initials __________ Date ________ Reviewed: no exceptions noted

Initials __________ Date ________ Reviewed: exceptions noted

Typical Method of Certifying Review

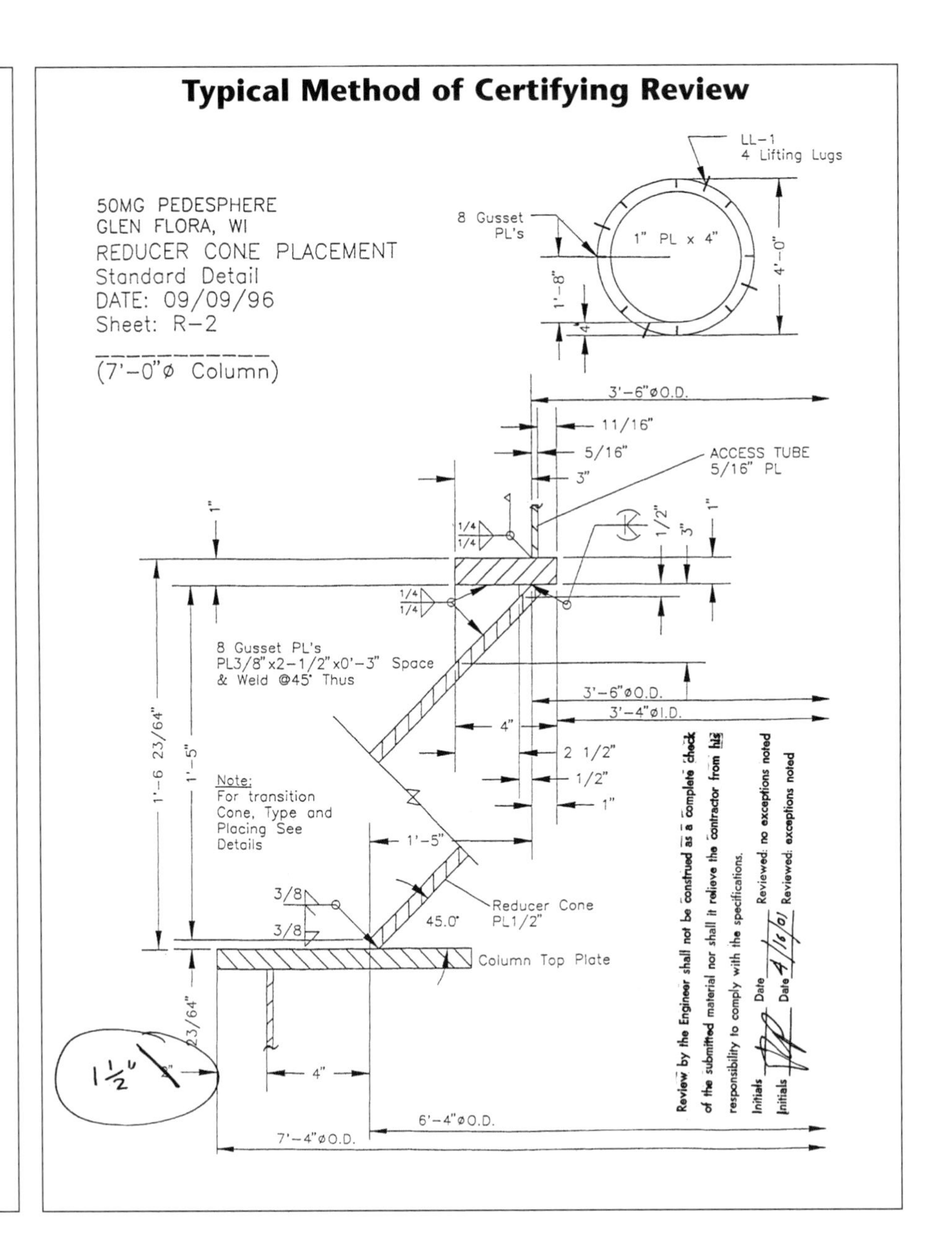

Source: Morgan & Parmley, Ltd. & Maguire Iron, Inc.

Typical Method of Certifying Review *(continued)*

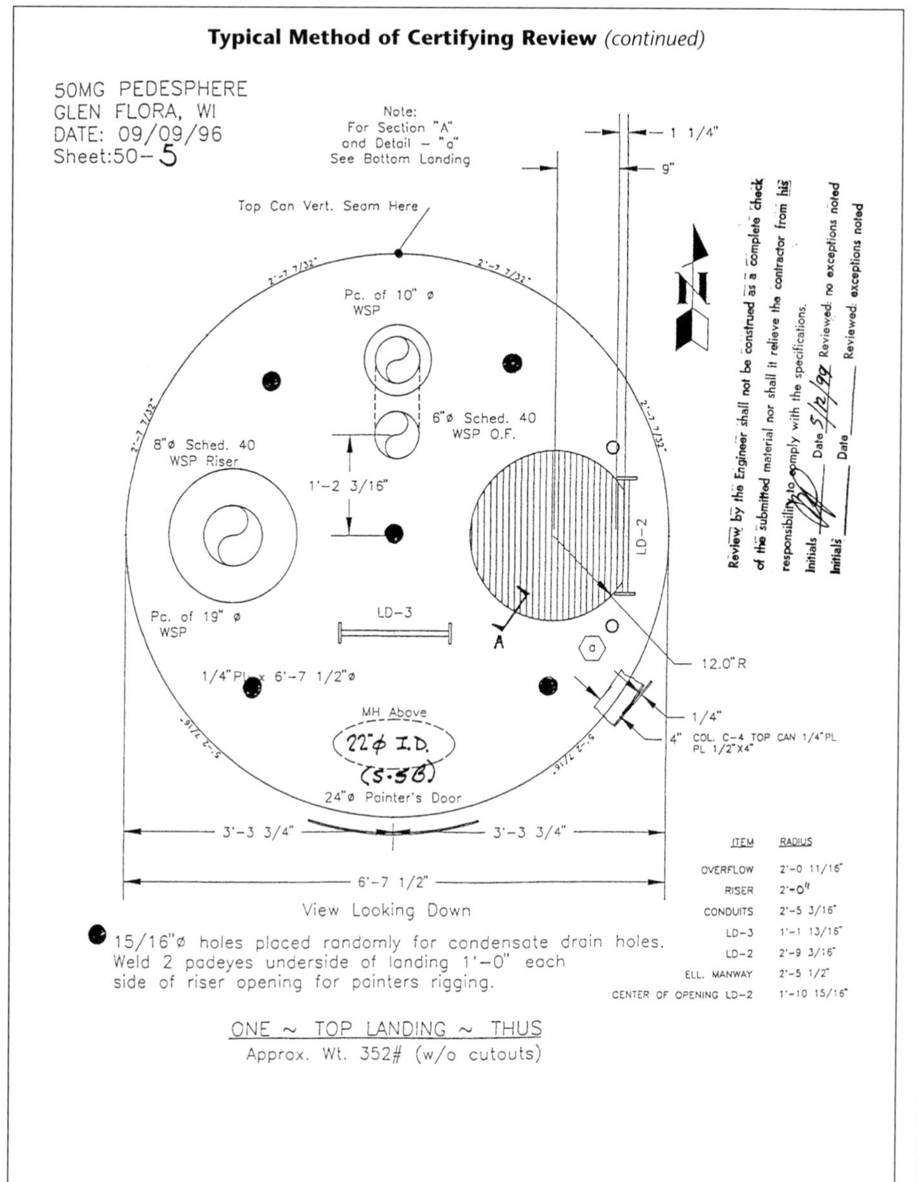

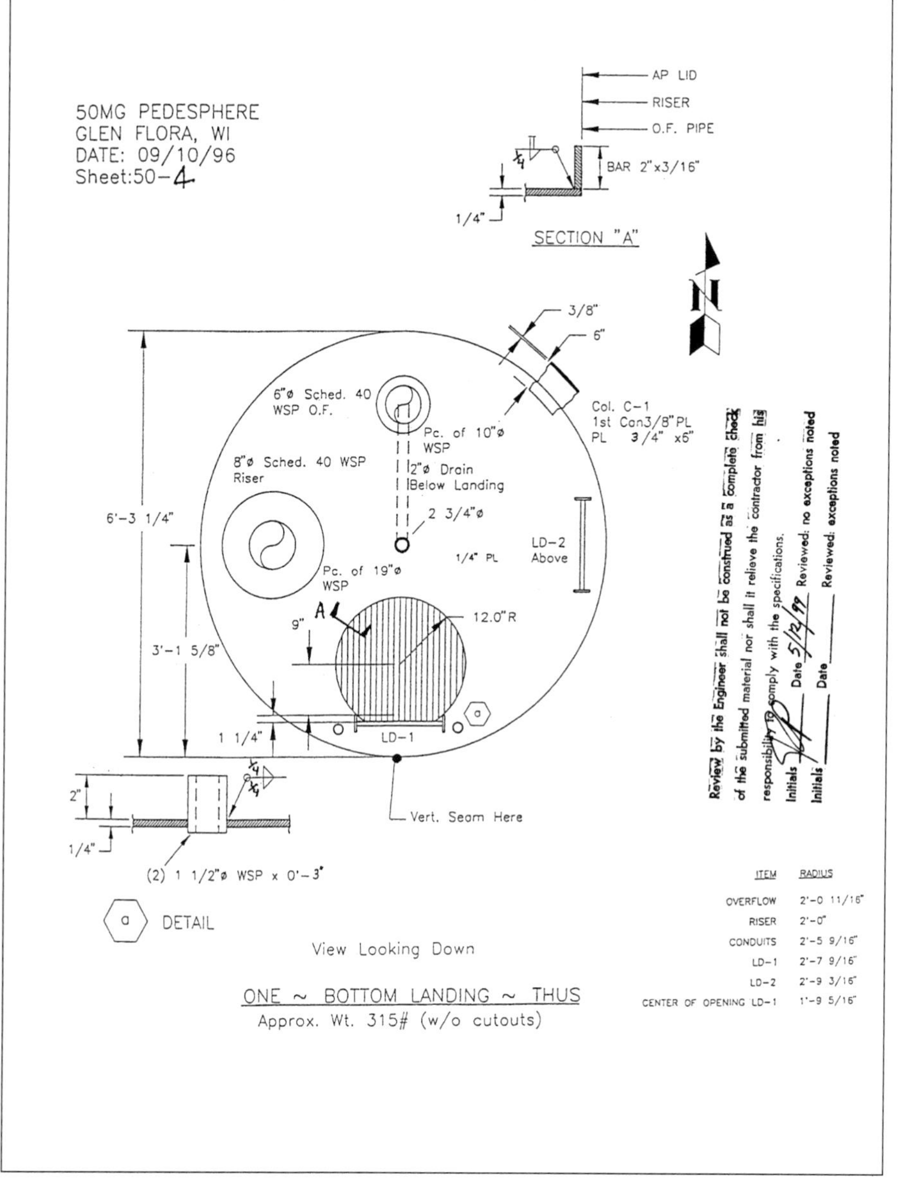

Construction

SAFETY

Section 28

Safety on the construction site must never be underestimated. It is paramount that adequate safety measures be properly in place before commencing any physical activity. Constant awareness and adequate barricades to protect the public and protection of working personnel are critical measures.

It is the intent of this section to make note of this important component of construction and draw the reader's attention to some basic fundamentals. We have included some reference information for general conditions. However, this material is not remotely all-exclusive and the reader is directed to consult applicable regulatory agencies, professional organizations, local, state & federal codes and OSHA regulations for relevant rules for each specific construction project. Additionally, there are some excellent books written on the subject. Some are general in nature and some describe specific types of construction.

The language of safety must be understood by all regardless of experience, age, background or ethnic origin. Safety at a construction site is an everyday concern and certainly requires constant vigilance. Thus, safety is an integral part of any construction activity and must be incorporated into the fabric of executing the project.

It is strongly recommended that regular safety meetings be held throughout the construction phase for all personnel. A constant reminder of the safety procedures will avoid many problems. Success of accident prevention requires teamwork.

First aid training for construction workers should be a prerequisite for employment, especially in some of the more dangerous projects.

TABLE OF CONTENTS

Potential Effects of Oxygen-Deficient Atmospheres

Oxygen Content (% by Volume)	Effects and Symptoms (At Atmospheric Pressure)
19.5%	Minimum permissible oxygen level.
15-19%	Decreased ability to work strenuously. May impair coordination and may induce early symptoms in persons with coronary, pulmonary, or circulatory problems.
12-14%	Respiration increases in exertion, pulse up, impaired coordination, perception, judgment.
10-12%	Respiration further increases in rate and depth, poor judgment, lips blue.
8-10%	Mental failure, fainting, unconsciousness, ashen face, blueness of lips, nausea, and vomiting.
6-8%	8 minutes, 100% fatal; 6 minutes, 50% fatal; 4–5 minutes, recovery with treatment.
4-6%	Coma in 40 seconds, convulsions, respiration ceases, death.

These values are approximate and vary as to the individual's state of health and his physical activities.

Source: Scott Instruments

Potential Effects of Hydrogen Sulfide Exposure

PPM*	Effects and Symptoms	Time
10	Permissible Exposure Level	8 Hours
50–100	Mild Eye Irritation, Mild Respiratory Irritation	1 Hour
200-300	Marked Eye Irritation, Marked Respiratory Irritation	1 Hour
500-700	Unconsciousness, Death	½-1 Hour
1000 or More	Unconsciousness, Death	Minutes

These values are approximate and vary as to the individual's state of health and his physical activity.

Source: Scott Instruments

Potential Effects of Carbon Monoxide Exposure

PPM*	Effects and Symptoms	Time
50	Permissible Exposure Level	8 Hours
200	Slight Headache, Discomfort	3 Hours
400	Headache, Discomfort	2 Hours
600	Headache, Discomfort	1 Hour
1000-2000	Confusion, Headache, Nausea	2 Hours
1000-2000	Tendency to Stagger	1½ Hours
1000-2000	Slight Palpitation of the Heart	30 Min.
2000-2500	Unconsciousness	30 Min.
4000	Fatal	Less Than 1 Hour

These values are approximate and vary as to the individual's state of health and his physical activity.

Source: Scott Instruments

Potentially Dangerous Areas

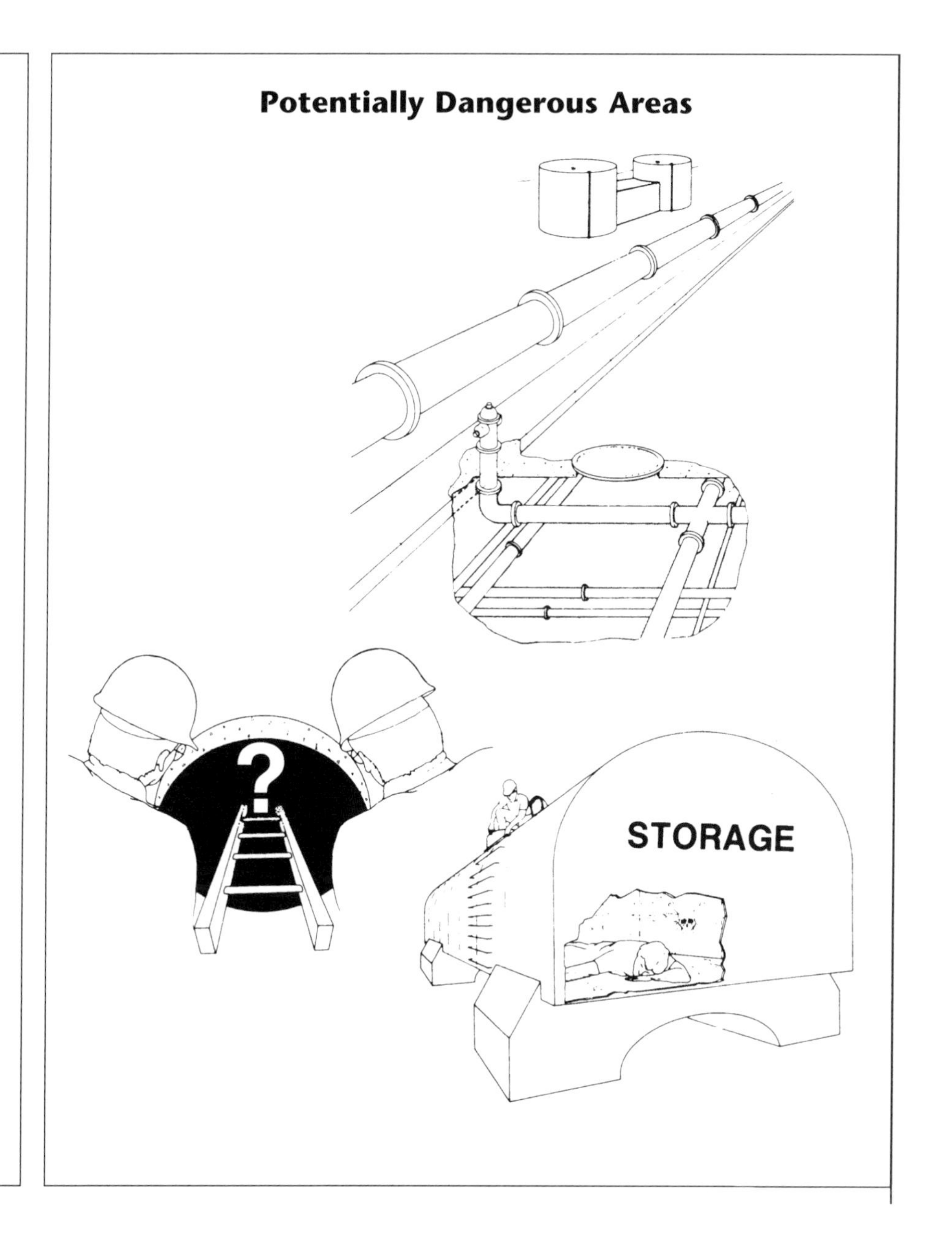

Source: Scott Instruments

Potentially Dangerous Areas *Continued*

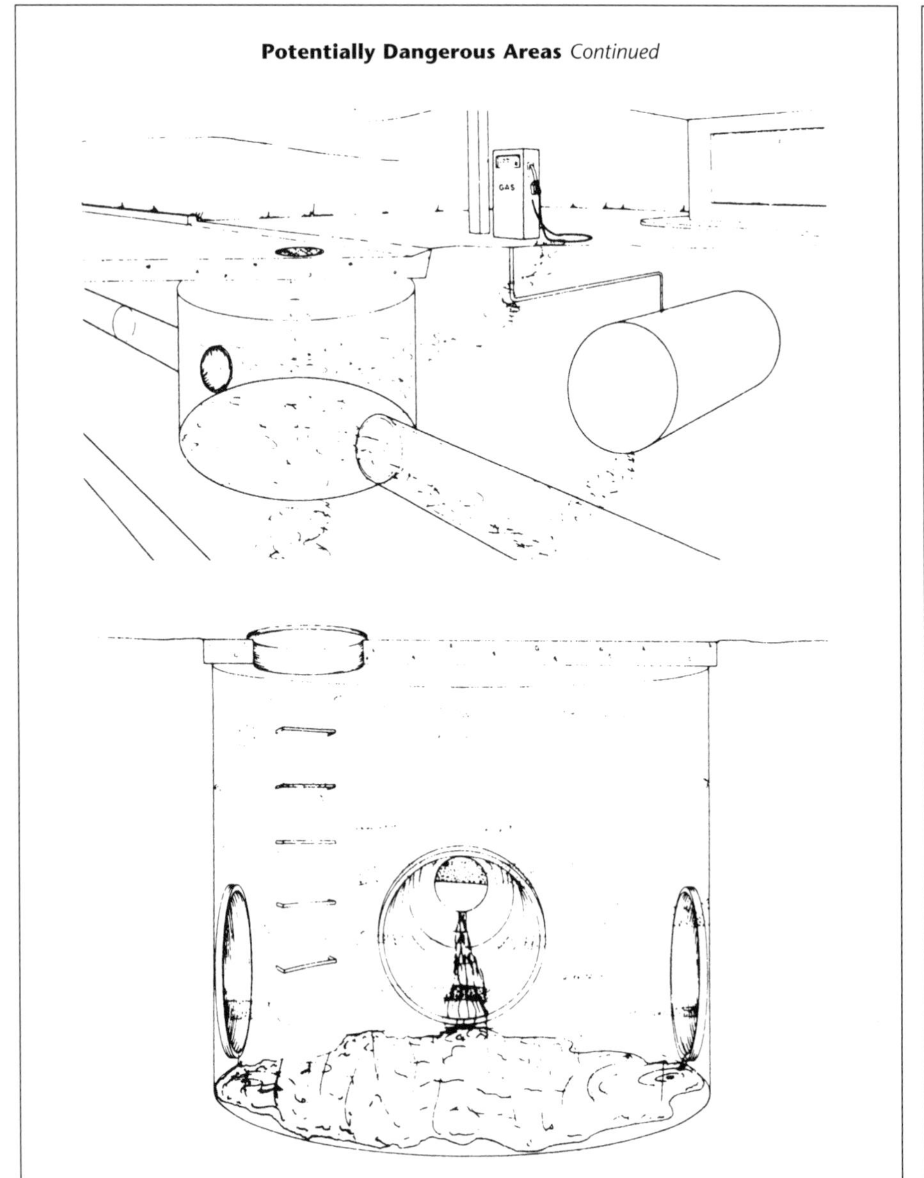

Monitor Atmosphere Before Entry

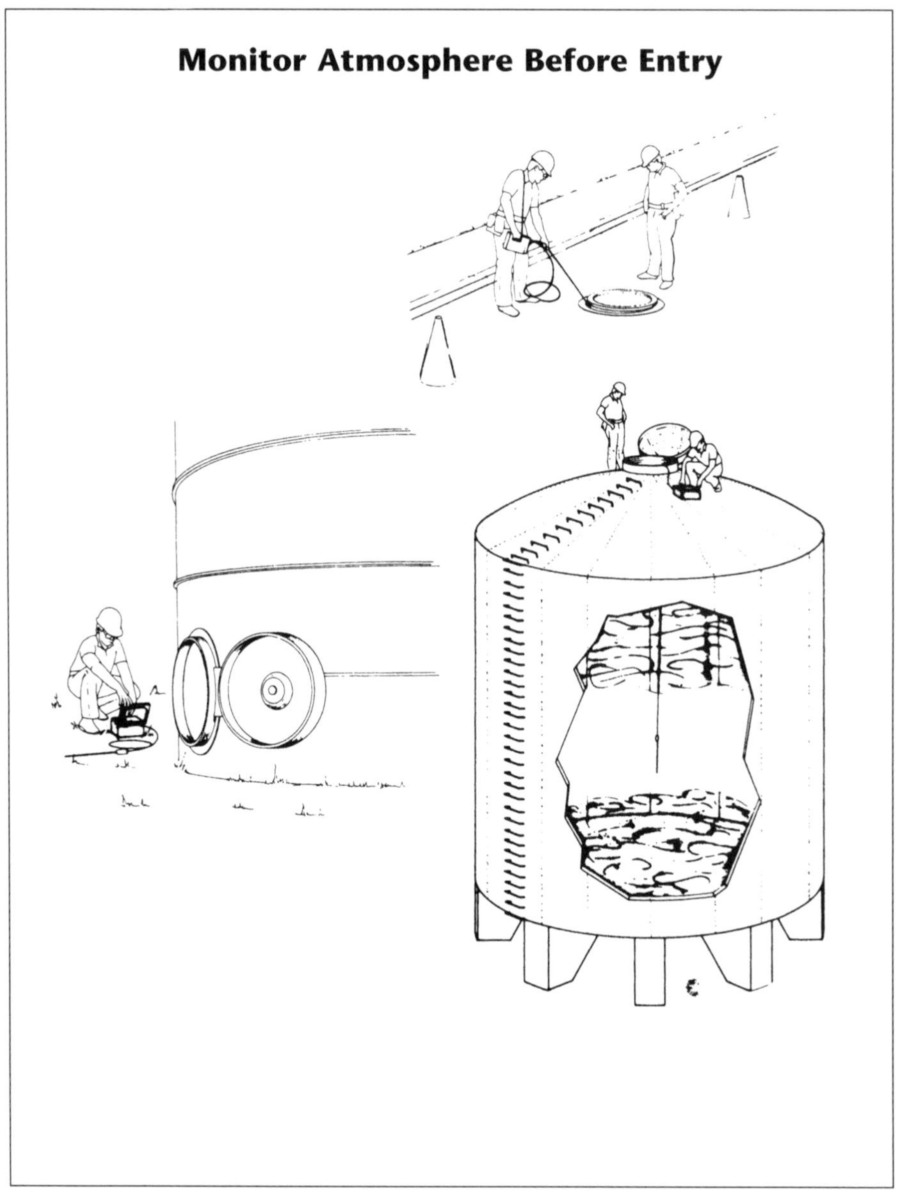

Source: Scott Instruments

Monitor Work Area During Construction

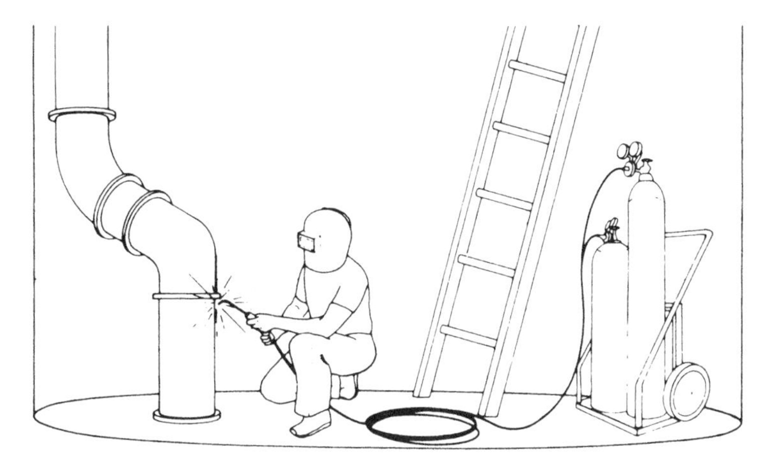

Source: Scott Instruments

Basic Equipment for Confined Entry

1 Gas meter
2 DO meter
3 PH meter
4 First-aid kit
5 Thermometers
6 Blower
7 Generator
8 Safety harness
9 Fire extinguishers
10 Tripod and winch
11 Lanyard with spreader bar
12 Airline respirator
13 Traffic cones
14 Safety vests
15 Air tank
16 Rope ladder
17 Cell phone
18 Protective coveralls
19 Rubber boots and gloves
20 Safety glasses or goggles

Source: R. O. Parmley, Field Engineer's Manual, 3/e ©2002 Published by McGraw-Hill and reproduced with permission of the McGraw-Hill Companies.

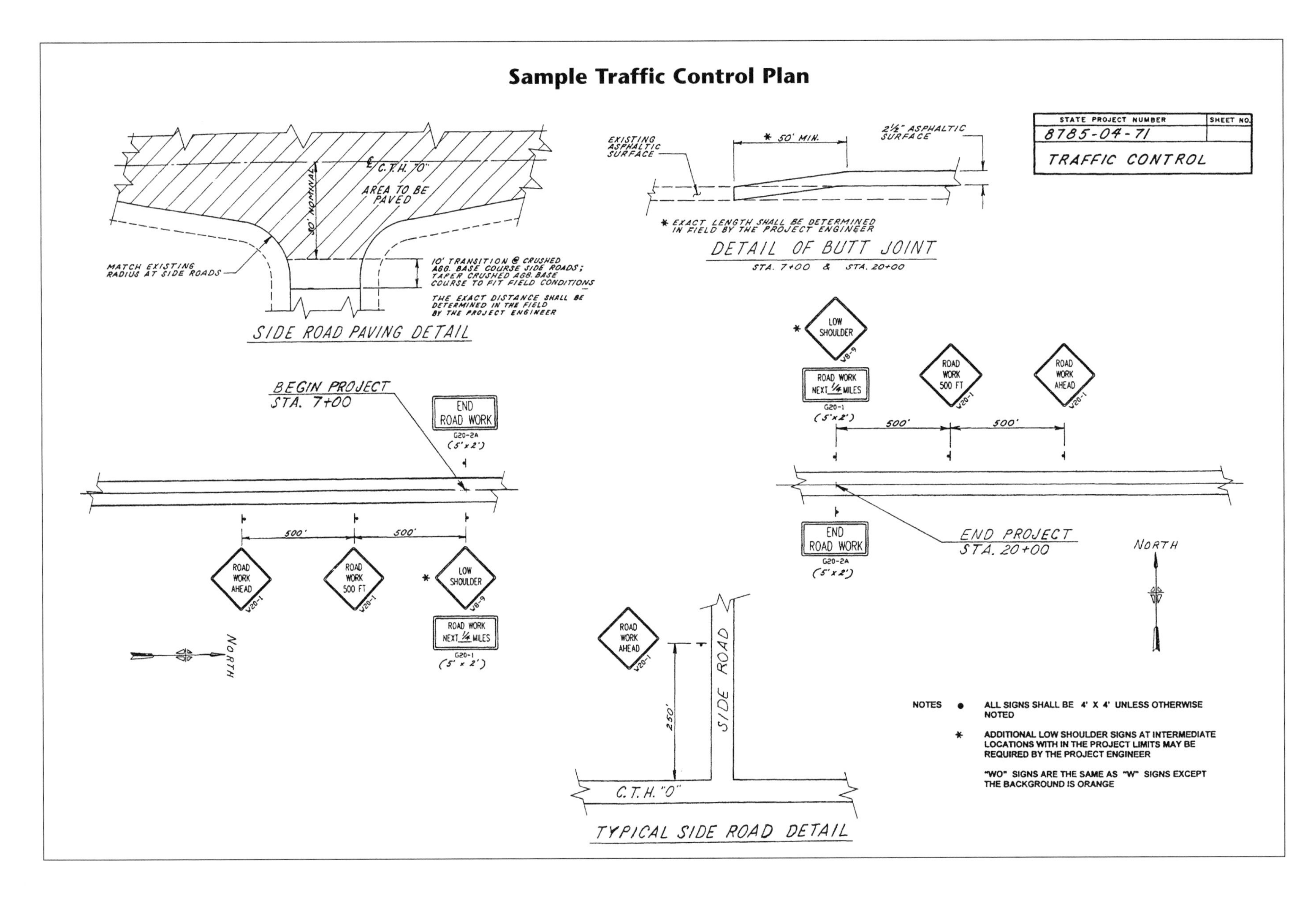

Source: Morgan & Parmley, Ltd.

Crane Signals

From: Pittsburgh DesMoines Steel Co., Pittsburgh, PA

First Aid Suggestions

Wear a Med-A-Lert tag at all times if you have any allergies to a particular medication or have a disease, e.g., diabetes, that requires special attention.

Store the majority of items listed in the table below in a dust and waterproof container. Label the container "First Aid Kit." Place a list of contents on the inside of the first aid kit. Replenish items after each use. The kit should be carried in field vehicle; i.e., inspector's truck and survey crew's van.

Quantity	*Item*	*Use*
12	4 x 4 Compresses	Dressings for wounds and burns
12	2 x 2 Compresses	
1 box	Assorted Band-aids	
3	40" Triangle bandages	Slings; to hold dressings in place
10 yards	1" Roller gauze or flexible gauze	Slings, open wounds, or dry dressings for burns
10 yards	2" Roller gauze or flexible gauze	
10 yards	4" Roller gauze or flexible gauze	
1	2" Roll elastic bandage	For sprains; to hold dressings in place
5 yards	1" Adhesive tape	To hold dressings in place
5 yards	2" Adhesive tape	
2	Small bath towels or large Pampers	Dressings for large wounds or burns
2	Eye pads	For eye injuries
1	Small bar of soap or bottle of liquid soap	Antiseptic for cleansing wounds
1	Single-edge razor blade, individually wrapped	For snake bite
1	Small container with needle, thread, and safety pins	To remove slivers; do emergency repair; fasten dressings
1	Flashlight with extra batteries	
1	Book matches	To sterilize needle (above)
1	Clean sheet, individually wrapped	For dressing large burns; splinting
1	Blanket	For transporting or shock
1	Cup and spoon	For giving fluids or for flushing wounds
1	First-Aid Book	
1	Tweezers	For removing foreign objects, insect stingers
1	Pocket knife	To cut splints
1	Scissors	To cut bandages
1	Small package salt	For shock, heat exhaustion (to be mixed with water)
1	Small package soda	
1	8"-long Tourniquet twist stick	As a last resort in bleeding; for snake bite
2	Small plastic bag	Container for ice or snow; for sealing chest wounds
1	1 Bottle antiseptic (Consult physician on selection and follow his or her recommendation.)	For general use

Note: The kit should include the latest publications of the American Safety & Health Institute (Emergency First Aid Information; CPR, Bleeding, Snake Bite, Etc.)

Construction

Inspection & Testing

Section 29

Inspectors of any construction activity are key personnel and are necessary for a successful project. They are the eyes and ears of the design professional, project engineer and owner at the job site. Their attention to details and monitoring can ensure that the project is constructed according to the plans and accurate logs are recorded.

Each project will have its specific areas that need testing. The inspector usually participates in this process and witnesses the events. If sampling is required, the inspector is responsible for the records and train of custody.

At the conclusion of construction, the inspector submits the log book, notes, photos and samples to the project engineer for the permanent project file.

Table of Contents

Inspector's Qualifications

The Inspector is the "eyes and ears" of the design professional (Architect or Engineer) at the job site. The inspector is the one with possibly the toughest job in construction, having the assignment that begins and ends with seeing that the physical construction produces a final result as described on the plans and in the specifications.

Ideally, an inspector should possess a broad range of technical education and be a well-seasoned technician in the area of construction to be monitored.

Inspector's Responsibility

Each project has its own specific rules and regulations, depending upon the organization and regulatory agencies involved. Generally, the supervising Professional Engineer or Architect will provide guidelines detailing the mandatory responsibilities and limit of authority of the Inspector. However, there are some basic responsibilities that all Inspectors should be apprised of that pertain to all projects. They are as follows:

1. Understand the importance of the Inspector's responsibility to attain the highest quality construction.
2. Possess good communicative skills and the technical knowledge to fully understand the project's plans and specifications.
3. Possess sound, practical judgment with good writing and sketching skills.
4. Understand the extent of Inspector's responsibilities.
5. Acceptance of Inspector's responsibilities.
6. Do not overstep the boundary of Inspector's authority.
7. Understand supervisor's (A/E) attitude and goals.
8. Be aware of responsibility to accurately record all significant events into Inspector's daily log, including all relevant verbal instructions and applicable conversations.
9. At all times, the Inspector should conduct oneself in a professional manner, demonstrating courtesy, patience and cooperation with all parties to insure a team effort for successful project completion.

Inspector's Authority

The Inspector has the authority and instruction to point out deviations from the plans and specification to the contractor. However, the Inspector does not have corresponding authority to approve any changes.

Working under specific guidelines and ruled by the legal portions of the specifications, the Inspector fulfills a key segment of the construction phase as he or she watches over each day's work; faithfully recording all activities for the record.

Inspector's Basic Duties

Basic duties that are generally applicable to all Inspectors are as follows:

1. Thoroughly review plans and specifications, prior to commencing construction. Report all errors, omissions, discrepancies and deficiencies to project A/E. Any portion that needs further explanation should be clarified by A/E well in advance of construction.
2. Secure a complete set of construction plans, specifications and contract documents for personal, on-the-job use. Record all applicable field measurements in "red" on plans so that "as builts" can be prepared at job completion.
3. Obtain complete file of all shop drawings and compare them with actual equipment and supplies delivered to job site, prior to actual installation.
4. Keep a neat and well-organized log book that accurately lists all applicable project activities encountered.

5. Always be alert to potential claims or situations of future dispute. When a possible claim or dispute is germinating, notify the project A/E and accurately record all pertinent facts in daily log book.
6. Keep abreast of any conditions that might delay the official progress schedule. Periodically report to project A/E any conditions that could affect the timely completion of construction.
7. Where applicable, monitor the labor standards and verify that the official wage rates, work hours and payroll records are being enforced.
8. Keep duplicate listing of on-site materials and monitor the storage facilities with attention to safety provisions, heating, ventilating and protection from severe weather.
9. Assist project A/E in reviewing the contractor's pay requests to insure that the work for which payment is requested has actually been executed in accordance with the approved plans and specifications.
10. Have in your immediate possession copies of all project permits, approvals, easements, cut sheets and other documents necessary for project to orderly proceed.
11. Take adequate photographs at all critical stages. Use Polaroid camera to insure that photograph records key areas. Keep a photo log and use a video camera with sound recorder if needed. Always remember that a photograph, supplemented with written/sketched notes is invaluable for future use. Never hesitate to take a picture, if in doubt.
12. Use a tape recorder liberally. You can always transcribe this verbal data into your log book later.
13. Always be fair and impartial, with the project as your prime objective.
14. Never assume anything! When in doubt, consult the plans, specifications, bidding documents, contract, shop drawings, files, approvals, permits or other relevant documents, bearing in mind that the project A/E must be consulted if the question can not be found in the project documents.

Inspector's Guides

There are several excellent manuals in print on construction inspection and it is recommended that anyone who is engaged in such activity should obtain one. Space is limited in this Sourcebook to fully describe and totally detail the complete and complex range of construction inspection. Therefore, specific guidelines are not included or implied.

Safety

The project site's construction area should be as orderly and safe as possible to ensure the safety of workers. Good housekeeping practices are always of paramount importance. This is where an Inspector's observation and general policing can be of great help.

On major construction jobs, safety is not part of the Inspector's responsibility. Generally, a Safety Engineer is assigned these duties with ample staff to enforce OSHA standards and similar regulatory rules. However, the Inspector certainly is a key person to alert the proper responsible individual of any potentially dangerous condition. Many personal-injury accidents are the direct result of careless handling of debris and a tactful Inspector's alertness and pressure can improve conditions in the work area.

Additionally, a skilled Inspector will be mindful of the requirement that all buried utilities be properly marked prior to any physical excavation activity. Utilizing regional telephone "hot-line" resources have proven extremely successful in recent years.

Event Calendar – (Basic Engineering Tasks Summary)

Item No.	Description	Projected Dates
1	Define project scope	
2	Client authorizes engineers to begin	
3	Initial conference: client/engineer	
4	Research records and files	
5	Field survey and reconnaissance	
6	Video photograph project site	
7	Buried utilities located and flagged	
8	Research property survey records	
9	Soil borings, test drilling, and subsurface investigation	
10	Committee meeting—refine project scope	
11	Reduce field notes	
12	Preliminary budget—construction cost estimates	
13	Committee meeting—review and approve preliminary concept	
14	Preliminary budget—construction cost estimates	
15	Preliminary plans and specifications	
16	Client meeting—review and approve preliminary P & S	
17	Easement survey and legal documents (if required)	
18	Final plans and specifications	
19	Engineering report (summary of work)	
20	Proposal and bidding documents	
21	Obtain concurrence from client for final P & S	
22	Submit plans and specifications to regulatory agencies	

Item No.	Description	Projected Dates
23	Obtain approvals from regulatory agencies	
24	Department of Natural Resources	
25	Public Service Commission	
26	Department of Transportation	
27	Corps of Engineers (if required)	
28	Private utility company approvals	
29	Water utility	
30	Sewer utility	
31	Railroad company	
32	Gas company	
33	Cable TV company	
34	Electrical company	
35	Telephone company	
36	Public hearing (assessments to property owners) (if required)	
37	Organize bidding process (commence)	
38	Obtain wage rates (state and/or federal)	
39	Advertise for bids	
40	Supply plans and specs to Builders Exchange	
41	Send plans and specifications to contractors	
42	Prebid meeting	
43	Bid opening	
44	Minutes, bid tabulation, and analysis of proposals	

Event Calendar – (Basic Engineering Tasks Summary) *Continued*

Item No.	Description	Projected Dates
45	Recommendation to client	
46	Client gives notice of award	
47	Prepare construction contract documents	
48	Secure client attorney's certification of construction contract	
49	Obtain client resolution to sign construction contract	
50	Supervise signing of construction contract	
51	Preconstruction conference	
52	Client signs notice to proceed	
53	Shop drawing review	
54	Construction staking	
55	General inspection services, photos, daily log, etc.	
56	Construction records	
57	Contractor payment requests review and certification	
58	Status reports (periodic)	
59	Inspection and testing of installation	
60	Prepare punch list	
61	Final inspection and certification	
62	Preparation of construction record drawings (as built)	
63	Final payment request review and certification	

Note: The projected dates are targets and may vary as a result of conditions beyond the control of the client, engineer, and contractor(s).

Inspector's Equipment List

- Briefcase
- Plans and specifications with contract documents
- Project log book
- Video camera
- 35-mm camera with spare film
- Polaroid camera with spare film
- Project file
- List of key project personnel
- Event calendar
- Project milestone list
- Proper personal ID
- A/E scales
- Magnetic locator
- Divining rods
- Magnifying glass
- Keys for project locks on restricted area
- Clipboard
- Table, pencils, markers, pen, etc.
- Tape recorder
- Carpenter's hand level
- 6-ft folding ruler (fractional/decimal)
- Cloth tape (100-ft English/metric)
- Electronic calculator (slide rule design)
- Measuring wheel (metric/English)
- First aid kit
- Watch with second hand
- Safety glasses
- Safety cones
- Specific testing equipment
- Rain gear
- Orange vest
- Hard hat
- Duct tape
- Paper toweling
- Tool kit
- Coveralls
- Rubber boots
- Rubber gloves
- Surveying equipment (if required)
- Cellular telephone
- Spray paint (red, green, blue, and white)
- Confined entry safety equipment (if applicable)
- Containers for collecting samples
- Traffic cones
- Flags
- Surveyor's ribbon
- Field vehicle
- Two-way radios

Miscellaneous Nomenclature

A/E	Architect/engineer
BB	Distance back-to-back
BC	Bottom chord
BF	Board feet
BM	Bench mark
BOD	Biochemical oxygen demand
BS	Back sight
C	Celsius
CC	Distance center-to-center
CDF	Creosoted Douglas fir (pressure-treated)
CF	Cubic feet
CFS	Cubic feet per second
CL	Centerline
COD	Chemical oxygen demand
CY	Cubic yard
D	Diameter
EL	Elevation
F	Fahrenheit
GPC/D	Gallons per capita per day
GPD	Gallons per day
GPM	Gallons per minutes
HI	Height of instrument
HL	Head loss or heat loss
HP	Horsepower
HWM	High-water mark
IE	Invert elevation
J	Joule
KIP	1000 lb
L	Liter
M	Meter
MGD	Millions of gallons per day
MH	Manhole
PC	Point of curve
PI	Point of intersection
PL	Property line
P & S	Plans and specifications
PT	Point of tangent
Q	Quantity
R	Radius
RC	Reinforced concrete
RW	Redwood
R/W	Right of way
S	Slope (grade)
SF	Square feet
SS	Suspended solids
STA	Station
SY	Square yards
T	Tons
TC	Top chord
TSS	Total suspended solids
UF	Untreated fir
V	Volt
VC	Vertical clearance
W	Watt
WL	Water level

Source: R. O. Parmley, Field Engineer's Manual, 3/e ©2002 Published by McGraw-Hill and reproduced with permission of the McGraw-Hill Companies.

Collecting Concrete Samples

Basic: If the sample is shoveled from the forms, do not take it from the over-watered concrete which collects on the surface of the concrete mass. If the sample is taken from the discharge stream of the concrete truck or mixer, take it at three different intervals, but not near the beginning or end.

The sample should weigh about 100 pounds. Carry it in two buckets to the place the cylinders are to be made and stored. Combine and remix the samples with a shovel in a wheelbarrow, buggy, or metal pan to insure uniformity before filling the molds.

Cylinder Casting: Use only steel, plastic or parafined paper molds, 6 inches in diameter by 12 inches long, with base plates or bottoms. Place the molds on a smooth, firm, level surface and fill with three equal layers of concrete, rodding each layer 25 times with a 5/8-inch round rod. Be sure to penetrate the previous layer slightly. After rodding, tap the sides of the mold to close any voids. The third layer should contain an excess which can be struck off smooth and level after rodding. Three cylinders are normally made for testing.

Fill out the data sheet thoroughly describing mix and placement. Then attach an envelope containing a copy of the data sheet to the side of each molded cylinder and cover with a plastic bag to prevent moisture loss.

Test Cylinder Handling: Do not remove or disturb the cylinders for 24 hours. Keep them in a protected area where the temperature remains between 60-80 degrees F. Then ship the cylinders to the testing laboratory packed in wet sawdust to keep them moist and protected.

Slump Test: Remix the concrete on a nonabsorbent surface. Fill the cone in three layers of equal volume and rod each layer 25 times with a 5/8-inch round rod. Clean away the excess concrete around the base before lifting the cone. Be very careful that the slump cone is lifted vertically. The distance in inches that the concrete sinks after the cone is lifted is termed the slump. Measure this with a ruler. If the molded concrete falls over, disregard this test and start over again.

Use a truncated cone with a 4-inch top diameter, 8-inch bottom diameter and 12-inch height, made of 16-gage sheet metal. Be sure the slump cone is clean and prewetted.

Source: R. O. Parmley, Field Engineer's Manual, 3/e ©2002 Published by McGraw-Hill and reproduced with permission of the McGraw-Hill Companies.

Sieve Analysis for Mixture of Fine & Coarse Aggregates

Project and/or Project No. Morgan & Parmley / 02LAB		Deposit Identification Filter Sand		
Contract	County Rusk			Specifications
Contractor and/or Producer	Sample No. 1	☐ Crushed Stone	☐ Base Course	☐ Grade 2
		☐ Crushed Gravel	X Other	X Other
Materials accepted at: ☐ Belt Stockpile ☐ Roadway	Date: 7/23/02	☐ Blend	Filter Sand	Special
Sampled at Location (pit, grid, station, etc.)	Time	Comments		

MOISTURE CONTENT
Weight of Sample (moist): 1041
Weight of Sample (dry): 1011
Moisture Loss: 30
% Moisture: 3.0%

Weight of Total Sample (dry, unwashed) 1011
Weight of R-4 (dry, Unwashed) 1011 = 1.000
Weight of P-4 (dry, unwashed) = 0.000

R-4 MATERIAL Washed: ☑ Yes ☐ No				P-4 MATERIAL Washed: ☐ Yes ☐ No Wt. = (minimum 500 grams)			TOTAL MATERIALS (% Passing)					
Sieve	Weight Retained	% Retained	% Pass (C)	Weight Retained	% Retained	% Pass (D)	R-4 (A*C)	P-4 (B*D)	Dry Sieved	Corr. Factor	Corr. or Washed Results	Spec.
3/8"	0	0.0	100.0								100.0	100
4	246	24.3	75.7								75.7*	77-100
8	574	56.8	43.2								43.2*	53-100
16	847	83.8	16.2								16.2	15-80
30	946	93.6	6.4								6.4	3-50
50	989	97.8	2.2								2.2*	0-1
100	1003	99.2	0.8								0.8	0-1
200	1006	99.5	0.5								0.5	0-1
pan	2											

R-4 FRACTURE COUNT Fracture Particles: Total Particles: % Fracture:	PLASTICITY CHECK Can P-40 be rolled into 1/8" thread when moist? ☐ Yes ☐ No	Weight/c.y. =
TEST RESULTS ☐ Passed X Failed	Morgan & Parmley Sampled by: Project Engineer: Morgan & Parmley Tested by: B.Walker	Date Tested: 7/24/02

Source: Morgan & Parmley, Ltd.

Report Form of Concrete Compressive Strength

REPORT OF CONCRETE COMPRESSIVE STRENGTH

Project:	Contractor:
Client:	
Date:	CEC File No.:

FIELD DATA

Location of Placement:		Specified Strength:	
		Supplier:	
Method of Placement:		Mix Number:	
Date Placed:		Cement (lbs):	
Time of Tests:		Fly Ash (lbs):	
Slump (in):		Water (lbs):	
Air Content (%):		Fine Aggregate (lbs):	
Concrete Temperature:		Coarse Aggregate #1 (lbs):	
Weather Conditions:		Coarse Aggregate #2 (lbs):	
Date Received in Lab:		Admixture #1 (oz):	
Field Data Submitted by:		Admixture #2 (oz):	

LABORATORY RESULTS

Test Method: ASTM C39-86

Set No.	Break Type	Age (days)	Load (lbs)	Strength (psi)

COMMENTS

☐ Cylinder cast by CEC	☐ Cylinders picked up by CEC	☐ Cylinders meet project specifications
☐ Cylinders cast by Contractors or Architects Representative	☐ Cylinders delivered to CEC laboratory	☐ Cylinders do not meet project specifications

Source: Cooper Engineering Co., Inc.

Nuclear Testing Field Data Sheet

CLIENT:	PROJECT NAME:
CLIENTS PROJECT #:	COOPER PROJECT #:
CONTRACTOR:	
TESTING METHOD:	GAUGE #:
DATE:	TEST PERFORMED BY:

Test Number	Station, Offset Northing, Easting Grid Lines-x,y Other (circle one)	Approx. Elevation	Dry Density (pcf)	Moisture (percent)	Wet Density (pcf)	Maximum Density (pcf)	Proctor Sample Number	Relative Compaction (percent)	Spec's Met?

Source: Cooper Engineering Co., Inc.

Nuclear Testing Data Spreadsheet

CLIENT:	PROJECT NAME:
CLIENT'S PROJECT #:	COOPER PROJECT #:
CONTRACTOR:	DATE:
TESTING METHOD:	GAUGE #:
LOT #'s:	TESTS PERFORMED BY:

Date Placed	Date Tested	Test Number	Station, Offset	Lower or Upper Course	Mainline or Shoulder	Maximum Density (pcf)	Density Count	Wet Density (pcf)	% Maximum Density
Lot #: ☐ This lot meets project specifications.			Average % Maximum Density =			Specified Density = ☐ This lot does not meet project specifications.			
Lot #: ☐ This lot meets project specifications.			Average % Maximum Density =			Specified Density = ☐ This lot does not meet project specifications.			
Lot #: ☐ This lot meets project specifications.			Average % Maximum Density =			Specified Density = ☐ This lot does not meet project specifications.			
Lot #: ☐ This lot meets project specifications.			Average % Maximum Density =			Specified Density = ☐ This lot does not meet project specifications.			

Source: Cooper Engineering Co., Inc.

Glossary of Crane and Derrick Terms

A frame: See Gantry and Jib mast. Also, a frame sometimes used with derricks, particularly barge-mounted, to support the boom foot and the topping lift.

Anti two-block device: A mechanism which, when activated, disengages all crane functions whose movement can cause the lower load block to strike the upper load block or boom head sheaves.

Articulated jib: A tower crane jib that in general has a pivot point somewhere in its middle area; also called pivoted luffing jib.

Axis of rotation: The vertical line about which a crane or derrick swings; also called center of rotation (obsolete) and swing axis.

Back-hitch gantry: See Gantry.

Ballast: Weight added to a crane base to create additional stability; it does not rotate when the crane swings.

Barrel: The lagging or body part of a rope drum in a drum hoist.

Base mounting: The structure forming the lowest element of a crane or derrick; it transmits loads to the ground or other supporting surface. For mobile cranes this is synonymous with carrier or crawler mounting. For tower cranes, the term includes a travel base, knee frame base, or fixed base (footing).

Base section: The lowermost section of a telescopic boom; it does not telescope but contains the boom foot pin mountings and the boom-hoist cylinder super end mountings.

Basic boom: The minimum length of sectional latticed boom that can be mounted and operated, usually consisting of a boom base and tip section only.

Bogie: An assembly of two or more axles arranged to permit both vertical wheel displacement and an equalization of loading on the wheels.

Boom: A crane or derrick member used to project the upper end of the hoisting tackle in reach or in a combination of height and reach; also called jib (European).

Boom angle: The angle between the horizontal and the longitudinal centerline of the boom, boom base, or base section.

Boom base: The lowermost section of a sectional latticed boom having the attachment or boom foot pins mounted at its lower end; also called boom butt or butt section.

Boom butt: See Boom base.

Boom foot mast: A component of some mobile-crane boom suspensions. It consists of a frame hinged at or near the boom foot that serves to increase the height of the inboard end of the fixed-boom suspension ropes, thereby increasing the angle those ropes make with the boom while being itself controlled by the boom-hoist ropes. Its purpose is to decrease the axial compressive force on the boom; also called hi-light gantry.

Boom guy line: See Pendant.

Boom head: The portion of a boom that houses the upper load sheaves.

Boom hoist: The rope drum(s), drive(s), and reeving controlling the derricking motion of the boom.

Boom-hoist cylinder: Hydraulic ram used instead of a rope boom suspension; the most common means of derricking telescopic booms.

Booming in (out): See Derricking.

Boom inserts: Center sections of a sectional latticed boom, usually having A four chords parallel.

Boom point: See Boom tip section.

Boom stay: See Pendant.

Boom stop: A device intended to limit the maximum angle to which the boom should be derricked.

Boom suspension: A system of ropes and fittings, either fixed or variable in length, that supports the boom and controls the boom angle.

Boom tip section: The uppermost section of a sectional latticed boom, which usually includes the weldment mounting the upper load sheaves as an integral part; also called boom point, head section, or tapered tip.

Bridle: See Floating harness.

Bull pole: A pole, generally of steel pipe, mounted to project laterally from the base of a derrick mast. It is used to swing the derrick manually.

Bull wheel: A horizontally mounted circular frame fixed to the base of a derrick mast to receive and guide the ropes used for swinging.

Butt section: See Boom base.

Carbody: That part of a crawler crane base mounting which carries the upperstructure and to which the crawler side frames are attached.

Source: H. I. Shapiro et al, Cranes and Derricks, 2d ed., McGraw-Hill, New York, ©1991. Reprinted with permission of the McGraw-Hill Companies.

Carrier: A wheeled chassis that is the base mounting, for mobile truck and rough terrain cranes.

Center of rotation (obsolete): See Axis of rotation.

Cheek weights: Overhauling weights attached to the side plates of a lower load block.

Climbing frame: A supplemented structure placed on, and forms a sleeve around, a tower crane mast; it is used in top-climbing the crane.

Climbing ladder: A steel member with crossbars (used in pairs) suspended from a climbing frame and used as jacking support points when some tower cranes climb.

Climbing schedule: A diagram or chart giving information for coordinating the periodic raising (climbing or jumping) of a tower crane with the increasing height of the building structure as the work progresses.

Counter jib: A horizontal member of a tower crane on which the counterweights and usually the hoisting machinery are mounted; also called counterweight jib.

Counterweight: Weights added to a crane upperstructure to create additional stability. They rotate with the crane as it swings.

Counterweight jib: See Counter jib.

Crawler frames: Part of the base mounting of a crawler crane attached to the car body and supporting the crawler threads, the track rollers, and the drive and idler sprockets. Crawler frames transmit crane weight and operational loadings to the ground; also called side frames.

Cribbing: Timber mats, steel plates, or structural members placed under mobile-crane tracks or outrigger floats to reduce the unit bearing pressure on the supporting surface below.

Crossover points: Points of rope contact where one layer of rope on a rope drum crosses over the previous layer.

Dead end: The point of fastening of one rope end in a running rope system, the other (live) end being fastened at the rope drum.

Deadman: An object or structure, either existing or built for the purpose, used as anchorage for a guy rope.

Derricking: Changing the boom angle by varying the length of the boom suspension ropes; also called luffing, booming in (out), or topping.

Dog: A pawl used in conjunction with a ratchet built into one flange of a rope drum to lock the drum from rotation in the spooling-out direction; also, one of a set of projecting lugs that support the weight of a tower crane.

Dogged off: The condition of a rope drum when its dog is engaged.

Drift: The vertical clearance between the top of a lifted load and the lifting crane hook when in its highest position; a measure of the gap available for slings and other rigging or for manipulating the load.

Drifting: Pulling a suspended load laterally to change its horizontal position.

Drum hoist: A hoisting mechanism incorporating one or more rope drums; also called hoist, winch, or hoisting engine.

Duty cycle work: Steady work at a fairly constant short cycle time with fairly constant loading levels for one or more daily shifts.

EOT crane: Electric overhead traveling crane.

Expendable base: For static-mounted tower cranes, a style of bottom mast section that is cast into the concrete footing block. All or part of this mast section is lost to future installations.

Extension cylinder: Hydraulic ram used to extend a section of a telescopic boom; the most common but not the only means for power-extending boom sections.

Fall: See Parts of line.

Flange point: The point of contact between the rope and the drum flange where the rope changes layers on a rope drum.

Fleet angle: The angle the rope leading onto a rope drum makes with the line perpendicular to the drum rotating axis when the lead rope is making a wrap against a flange.

Fleeting sheave: Sheave mounted on a shaft parallel to the rope-drum shaft and arranged so that it can slide laterally as the rope spools, permitting close sheave placement without excessive fleet angle.

Float: An outrigger pan that distributes the load from the outrigger to the supporting surface or to cribbing placed beneath it; part of the crane's outrigger support system.

Floating harness: A frame, forming part of the boom suspension, supporting sheaves for the live suspension ropes and attached to the fixed suspension ropes (pendants); also called bridle, spreader, upper spreader, live spreader, or spreader bar.

Fly section: On a telescopic boom, the outermost powered telescoping section.

Footblock: A steel weldment or assembly serving as the base mounting for a guy derrick, gin pole, or Chicago boom derrick.

Free fall: Lowering the hook (load) or boom by gravity; the lowering speed is controlled only by a retarding device such as a brake.

Frequency of vibration: Number of vibration cycles that will occur in a unit of time, usually expressed in hertz (cycles per second); also called frequency.

Front end attachments: Optional load-supporting members for use on mobile cranes, e.g., taper tip boom, hammerhead boom, guy derrick attachment, and tower attachment.

Gantry: A structure, fixed or adjustable in height, forming part of the upperstructure of a crane, to which the lower spreader (carrying the live boom-suspension ropes) is anchored; also called A-frame gantry, A frame, or back-hitch gantry.

Gate block: See Snatch block.

Gooseneck boom: A boom with an upper section projecting at an angle to the longitudinal centerline of the lower section.

Gudgeon pin: The pin at the top of a derrick mast forming a pivot for the spider or for the mast of a stiffleg derrick.

Guy rope: A fixed-length supporting rope intended to maintain a nominally fixed distance between the two points of attachment; also called stay rope or pendant.

Hammerhead boom: A boom tip arrangement in which both the boom suspension and the hoist ropes are greatly offset from the boom longitudinal centerline to provide increased load clearance.

Head section: See Boom tip section,

Hog line: See Pendant, Intermediate suspension.

Hoist: To lift an object; the mechanism used for lifting.

Horse: See Jib mast.

Impact: Increase in vertical-load effect from dynamic causes.

In-service wind: Wind encountered while a crane is working; usually used to define the maximum permissible level of wind pressure or velocity at the site before the crane must be taken out of service.

Intermediate suspension: An additional set of boom-suspension lines attached to the boom at some point between the main suspension attachment and the boom foot. On mobile cranes it is used to reduce boom elastic deflection during erection; on horizontal jib tower cranes it is used as part of the primary support; also called midpoint suspension, midpoint hitch, intermediate hitch, or intermediate hog line.

Jib: In American practice, an extension to the boom mounted at the boom tip, in line with the boom longitudinal axis or offset to it. It is equipped with its own suspension ropes made fast to a mast at the boom tip, which in turn is supported by guy ropes or a derricking system. (Europeans call this a fly jib and use jib to refer to a boom.)

Jib mast: A short strut or frame mounted on the boom head to provide a means for attachment of the jib support ropes; also called jib strut, rooster, horse, or A frame.

Jib strut: See Jib mast.

Jumping: Raising a tower crane or guy derrick from one operating elevation to the next in concert with the completion of additional floors of the building.

Lagging: Removable shells (optional) for use on a rope drum to produce change in line pull or line speed.

Latticed boom: A boom constructed of four longitudinal comer members, called chords, assembled with transverse and/or diagonal members, called lacings, to form a trusswork in two directions. The chords carry the axial boom forces and bending moments, while the lacings resist the shears.

Layer: A series of wraps of wire rope around a rope drum barrel, extending full from flange to flange.

Level luffing: An automatic arrangement whereby the crane or derrick hook does not significantly change elevation as the boom derricks.

Line pull: The pulling force attainable in a rope leading off a rope drum or lagging at a particular pitch diameter (number of layers).

Line speed: The speed attainable in a rope leading off a rope drum or lagging at a particular pitch diameter (number of layers).

Live spreader: See Floating harness.

Load jib: See Saddle jib.

Load radius: See Radius.

Loading: An external agency that induces force in members of a structure. It may be in the form of a direct physical entity (e.g., weight superimposed on the structure or pressure applied by wind) or in the form of an abstract condition (e.g., inertial effects associated with motion.

Lower spreader: A frame, forming part of the boom suspension, supporting sheaves for the live suspension ropes and attached to the gantry or upper structure.

Luffing: See Derricking.

Luffing jib: A tower crane boom that is raised and lowered about a pivot to move the hook radially, that is, to change its working radius.

Lumped masses: A concept used to simplify mathematical analysis, whereby distributed or discrete masses are replaced by aggregates of mass concentrated at a point (or points) where the resultant of the inertia forces associated with the actual masses seems to act.

Machine resisting moment: The moment of the dead weight of the crane or derrick, less boom weight, about the tipping fulcrum; hence, the moment that resists overturning; also called machine moment or stabilizing moment.

Manual insert: An optional nonpowered section in a telescopic boom forming the outermost boom section when provided and usable either fully extended or fully retracted.

Mast: An essentially vertical load-bearing component of a crane or derrick; the tower of a tower crane. See also Boom foot mast and Jib mast.

Mast cap: See Spider.

Midpoint suspension: See Intermediate suspension.

Midsection: On a telescopic boom, the intermediate powered telescoping section(s) mounted between the base and fly sections.

OET crane: See EOT crane.

Offset angle: The angle between the longitudinal centerline of a jib and the longitudinal centerline of the boom on which it is mounted.

Operating radius: See Radius.

Operating sectors: Portions of a horizontal circle about the axis of rotation of a mobile crane providing the limits of zones where over-the-side, over-the-rear, and over-the-front ratings are applicable.

Out-of-service wind: The wind speed or pressure that an upright inoperative crane is exposed to; usually used to define the maximum level of wind that the inoperative crane is designed to safely sustain.

Outriggers: Extendible arms attached to a crane base mounting, which include means for relieving the wheels (crawlers) of crane weight; used to increase stability.

Overhauling weight: Weight added to a load fall to overcome resistance and permit unspooling at the rope drum when no live load is being supported; also called headache ball; see also Cheek weights.

Overturning moment: The moment of the load plus the boom weight about the tipping fulcrum. Wind and dynamic effects can be included when appropriate.

Parking track: For rail-mounted traveling cranes, a section of track supported so that it is capable of sustaining storm-induced bogie loads; it is provided with storm anchorages when required.

Parts of line: A number of running ropes supporting a load or force; also called pails or falls.

Paying out: Adding slack to a line or relieving load on a line by letting (spooling) out rope.

Pendant: A fixed-length rope forming part of the boom-suspension system; also called boom guy line, hog line, boom stay, standing line, or stay rope.

Pitch diameter: The diameter of a sheave or rope drum measured at the centerline of the rope; tread diameter plus rope diameter.

Pivoted luffing jib: See Articulated jib.

Preventer: In rigging practice, a means, usually comprising but not limited to a wire rope, for preventing an unwanted movement or occurrence or acting as a saving device when an anchorage or attachment fails.

Radius, load (operating): Nominally, the horizontal distance from the axis of rotation to the center of gravity of a lifted load. In mobile-crane practice, this is more specifically defined as the horizontal distance from the projection to the ground of the axis of rotation before loading to the center of a loaded but vertical hoist line.

Range diagram: A diagram showing an elevation view of a crane with circular arcs marked off to show the luffing path of the tip for all boom and jib lengths and radial lines marking boom angles. A vertical scale indicates height above ground, while a horizontal scale is marked with operating radii. The diagram can be used to determine lift heights, clearance of the load from the boom, and clearances for lifts over obstructions.

Reach: Distance from the axis of rotation of a crane or derrick; sometimes used synonymously with radius.

Recurrence period: The interval of time between occurrences of repeating events; the statistically expected interval of time between occurrences, such as storms, of a given magnitude.

Reeving diagram: A diagram showing the path of the rope through a system of sheaves (blocks).

Revolving superstructure: See Upperstructure.

Rooster: Vernacular term for one or more struts at the top of a boom or mast, such as a jib strut, a tower-crane top tower, or the struts at the top of the mast of a mobile-crane tower attachment.

Root diameter: See Tread diameter.

Rope drum: That part of a drum hoist which consists of a rotating cylinder with side flanges on which hoisting rope is spooled in or out (wrapped).

Rotation-resistant rope: A wire rope consisting of an inner layer of strand laid in one direction covered by a layer of strand laid in the opposite direction.

Running line: A rope that moves over sheaves or drums.

Saddle jib: The horizontal live-load-supporting member of a hammerhead-type tower crane having the load falls supported from a trolley that traverses the jib; also called load jib.

Sheave: A wheel or pulley with a circumferential groove designed for a particular size of wire rope; used to change the direction of a running rope.

Side frames: See Crawler frames.

Side guys: Ropes supporting the flanks of a boom or mast to prevent lateral motion or lateral instability.

Side loading: A loading applied at any angle to the vertical plane of the boom.

Sill: One of the horizontal stationary members of a stiff-leg derrick, it secures the vertically diagonal stifflegs.

Slewing: See Swing.

Snatch block: A single- or double-sheave block arranged so that one or both check plates can be opened, permitting the block to be reeved without having to use a free rope end; also called gate block.

Spider: A fitting mounted to a pivot (gudgeon pin) at the top of a derrick mast, providing attachment points for guy ropes; also called mast cap.

Spreader: See Floating harness, Lower spreader.

Spreader bar: See Floating harness.

Stabilizers: Devices for increasing stability of a crane; they are attached to the crane base mounting but are incapable of relieving the wheels (crawlers) of crane weight.

Stabilizing moment: See Machine resisting moment.

Standing line: A fixed-length line that supports loads without being spooled on or off a drum; a line of which both ends are dead; also called guy line, stay rope, or pendant.

Static base: Tower-crane support (base mounting) where the crane mast is set on or into a foundation.

Stay rope: See Guy rope, Pendant.

Strand: A group of wires helically wound together forming all or part of a wire rope.

Strength factor: Failure load (or stress) divided by allowable working load (or stress).

Structural competence: The ability of the equipment and its components to support the stresses imposed by operating loads without the stresses exceeding specified limits.

Superstructure: See Upperstructure.

Swing: A crane or derrick function wherein the boom or load-supporting member rotates about a vertical axis (axis of rotation); also called slewing.

Swing axis: See Axis of rotation.

Tackle: An assembly of ropes and sheaves designed for pulling.

Tagline: A rope (usually fiber) attached to the load and used for controlling load spin or alignment from the ground. Also, for clamshell operations, a wire rope used to retard rotation and pendulum action of the bucket.

Tailing crane: In a multimachine operation in which a long object is erected from a horizontal starting position to a vertical final position, the crane controlling the base end of the object.

Taking up: The process of removing slack from a line or drawing (spooling) in on a line; loading a line by drawing in on it.

Tapered tip: See Boom tip section.

Telescoping: A process whereby the height of a traveling or freestanding tower crane is increased by adding sections at the top of the outer tower and then raising the inner tower. There are cranes that are telescoped by adding to the inner tower from below.

Tipping fulcrum: The horizontal line about which a crane or derrick will rotate should it overturn; the point(s) on which the entire weight of a crane or derrick will be imposed during tipping.

Tipping load: The load for a particular operating radius that brings the crane or derrick to the point of incipient overturning.

Top climbing: A method of raising a tower crane by adding mast sections; with the crane balanced, a climbing frame lifts the crane upperworks to permit insertion of a new tower section beneath the turntable. The process is repeated as required.

Top tower: A tower mounted above the jibs of some tower cranes providing means for attachment of the pendants; also called rooster or tower head.

Topping: See Derricking.

Topping lift: See Boom hoist.

Tower: Tower crane mast; on some tower cranes there is a top tower used to support the jib and counterjib pendants.

Tower attachment: A combination of vertical mast and luffing boom mounted to the front end of a mobile crane in place of a conventional boom.

Tower head: See Top tower.

Transit: Movement or transport of a crane from one jobsite to another.

Travel: Movement of a mobile or wheel-mounted crane about a jobsite under its own power.

Travel base: The base mounting for a wheel-mounted (traveling) tower crane.

Tread diameter: The diameter of a sheave or grooved rope drum measured at the base of the groove; the diameter of a smooth barrel on a rope drum.

Trolley: A carriage carrying the hook block for radial movement along the lower chords of a horizontally mounted tower crane jib; in some tower cranes, a counterweight trolley allows the counterweights to be moved radially to modulate their backward moment in proportion to load hook radius.

Two blocking: Excessive taking in of tackle rope, which causes the blocks to make contact with each other.

Upper: See Upperstructure.

Upper spreader: See floating harness.

Upperstructure: On a mobile crane, the entire rotating structure less the front-end attachment; also called upper, superstructure, or revolving superstructure.

Vangs (vang lines): Side lines reeved to a derrick boom and used to swing the boom.

Walking beam: A bogie member whose lone axis is nominally horizontal and parallel to the direction of travel; it is pivoted at its center and mounts a wheel, wheel pair, axle, or the center pivot of another walking beam at each end. Its purpose is to permit wheel oscillation, thus equalizing wheel loading during passage over travel-path irregularities.

Weathervane: To swing with the wind when out of service so as to expose a minimal area to the wind.

Whip line: A secondary or auxiliary hoist line.

Winch: See Drum hoist.

Wrap: One circumferential turn of wire rope around a rope-drum barrel.

Construction

CONSTRUCTION STAKING

Section 30

Since the common usage of the laser, construction staking has become much easier, but certainly not less important.

In times past, batter boards with parallel strings and range poles were used to transfer grade to sewer pipe laying crews. Now a (cold) laser beam, set at proper slope, is used from manhole to manhole. Additionally, a rotating laser beam can be used on roadways, building sites and a wide variety of construction projects. However, most projects do require physical staking to establish corners, centerlines, base lines, manhole locations, curb & gutter alignment and similar control points.

The following pages contain some typical examples of construction staking and include the historical method used before lasers became the norm.

It is a foregone conclusion that only some basic material can be included on this subject within the covers of this sourcebook. Therefore, the reader should refer to manuals and handbooks that are devoted entirely to surveying, leveling and construction staking.

TABLE OF CONTENTS

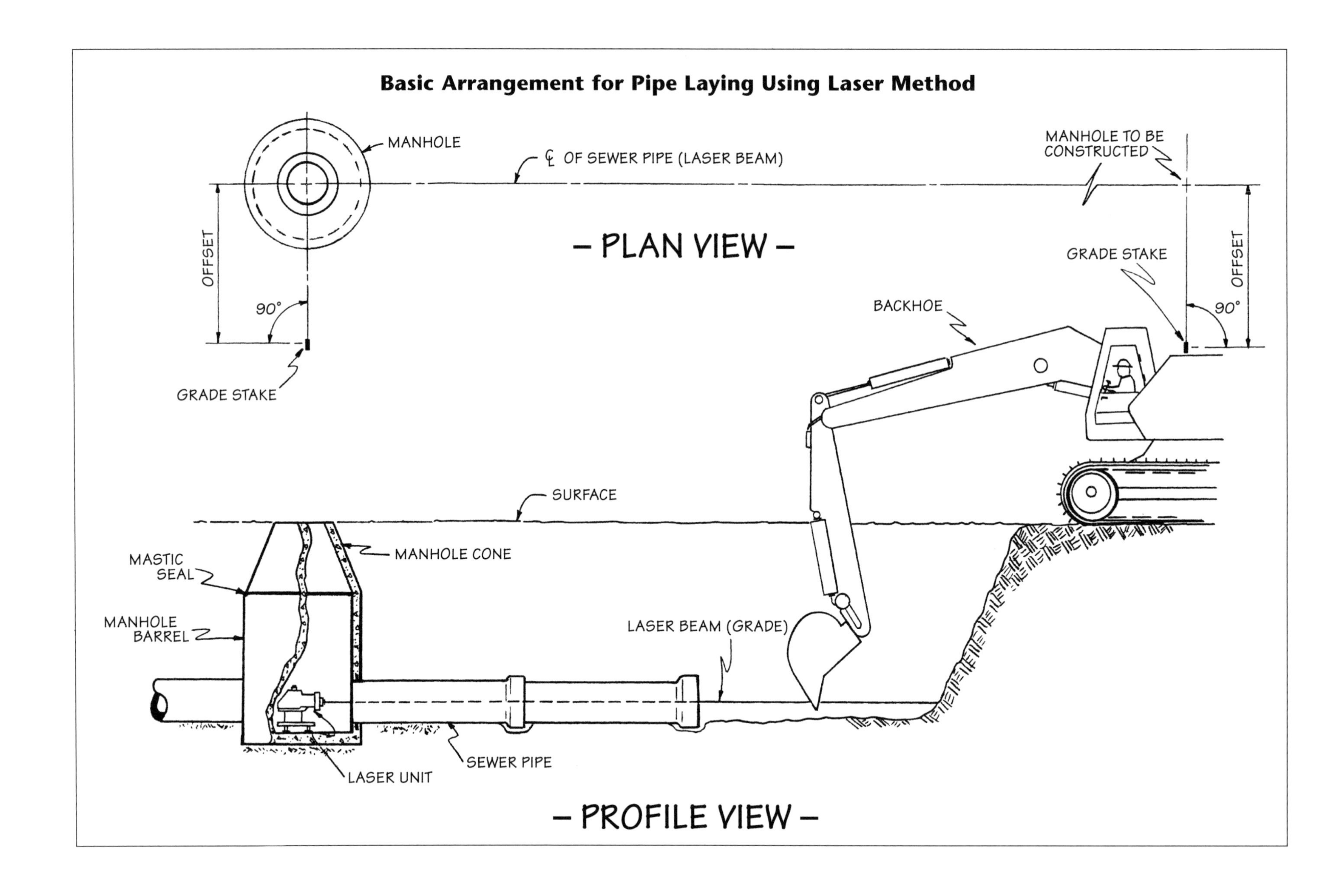
Basic Arrangement for Pipe Laying Using Laser Method
MANHOLE
℄ OF SEWER PIPE (LASER BEAM)
MANHOLE TO BE CONSTRUCTED
– PLAN VIEW –
OFFSET
90°
GRADE STAKE
BACKHOE
SURFACE
MASTIC SEAL
MANHOLE CONE
MANHOLE BARREL
LASER BEAM (GRADE)
SEWER PIPE
LASER UNIT
– PROFILE VIEW –

Typical Cut Sheet Form

PROJECT No. -												
PROJECT NAME -												
STA.	MARK No.	(INLET / E.W. / M.H.)	STAKE ELEV.	FLOW LINE	CUT	PIPE INVERT	CUT	OFFSET	FINISHED CENTER LINE	CUT/ FILL	LOCATION	NOTES

Source: Morgan & Parmley, Ltd.

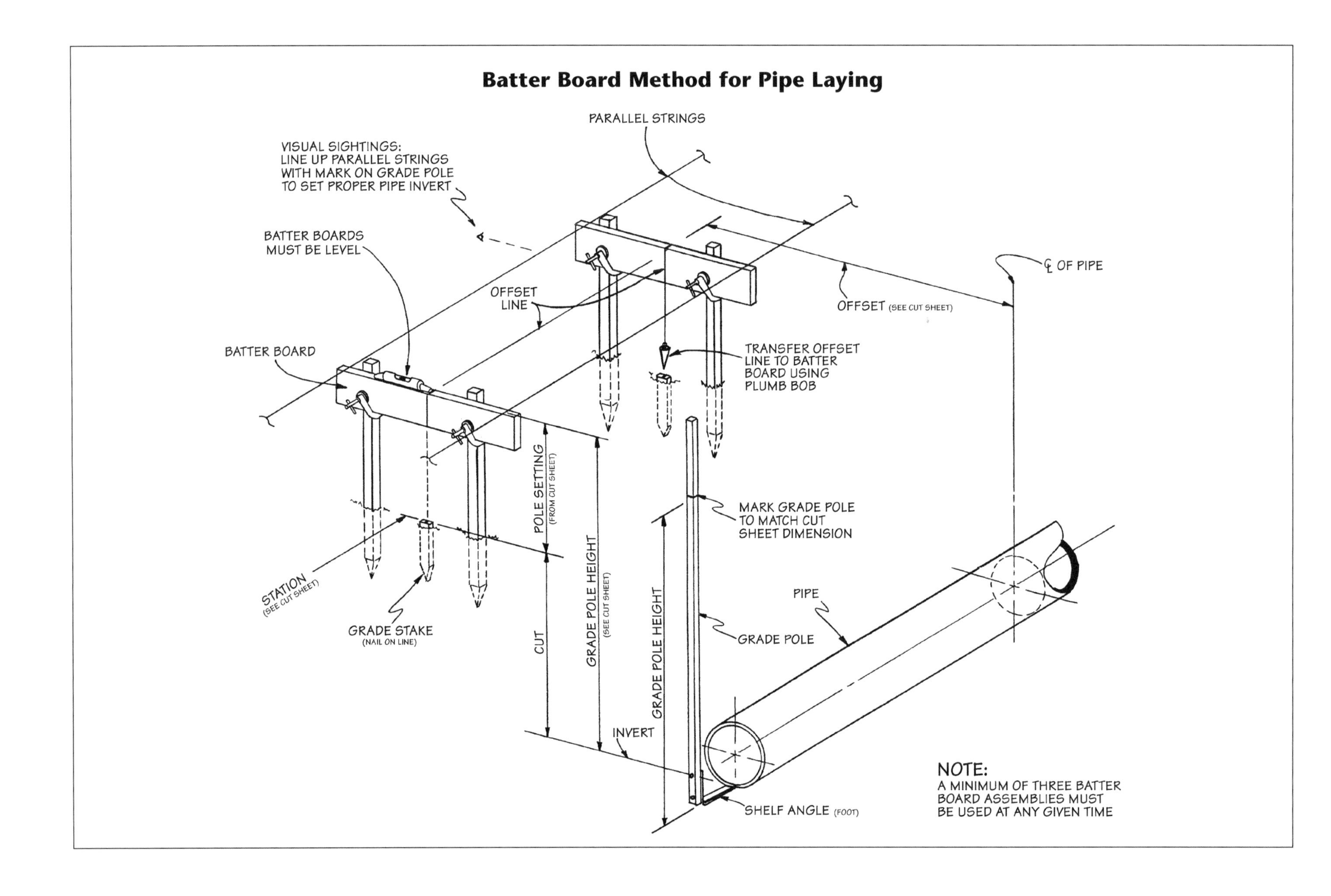
Batter Board Method for Pipe Laying
PARALLEL STRINGS
VISUAL SIGHTINGS:
LINE UP PARALLEL STRINGS
WITH MARK ON GRADE POLE
TO SET PROPER PIPE INVERT
BATTER BOARDS
MUST BE LEVEL
OFFSET
LINE
BATTER BOARD
℄ OF PIPE
OFFSET (SEE CUT SHEET)
TRANSFER OFFSET
LINE TO BATTER
BOARD USING
PLUMB BOB
POLE SETTING
(FROM CUT SHEET)
GRADE POLE HEIGHT
(SEE CUT SHEET)
MARK GRADE POLE
TO MATCH CUT
SHEET DIMENSION
STATION
(SEE CUT SHEET)
GRADE STAKE
(NAIL ON LINE)
PIPE
GRADE POLE
CUT
GRADE POLE HEIGHT
INVERT
SHELF ANGLE (FOOT)
NOTE:
A MINIMUM OF THREE BATTER
BOARD ASSEMBLIES MUST
BE USED AT ANY GIVEN TIME

Typical Field Book Pages: Sewer Grade Stakes

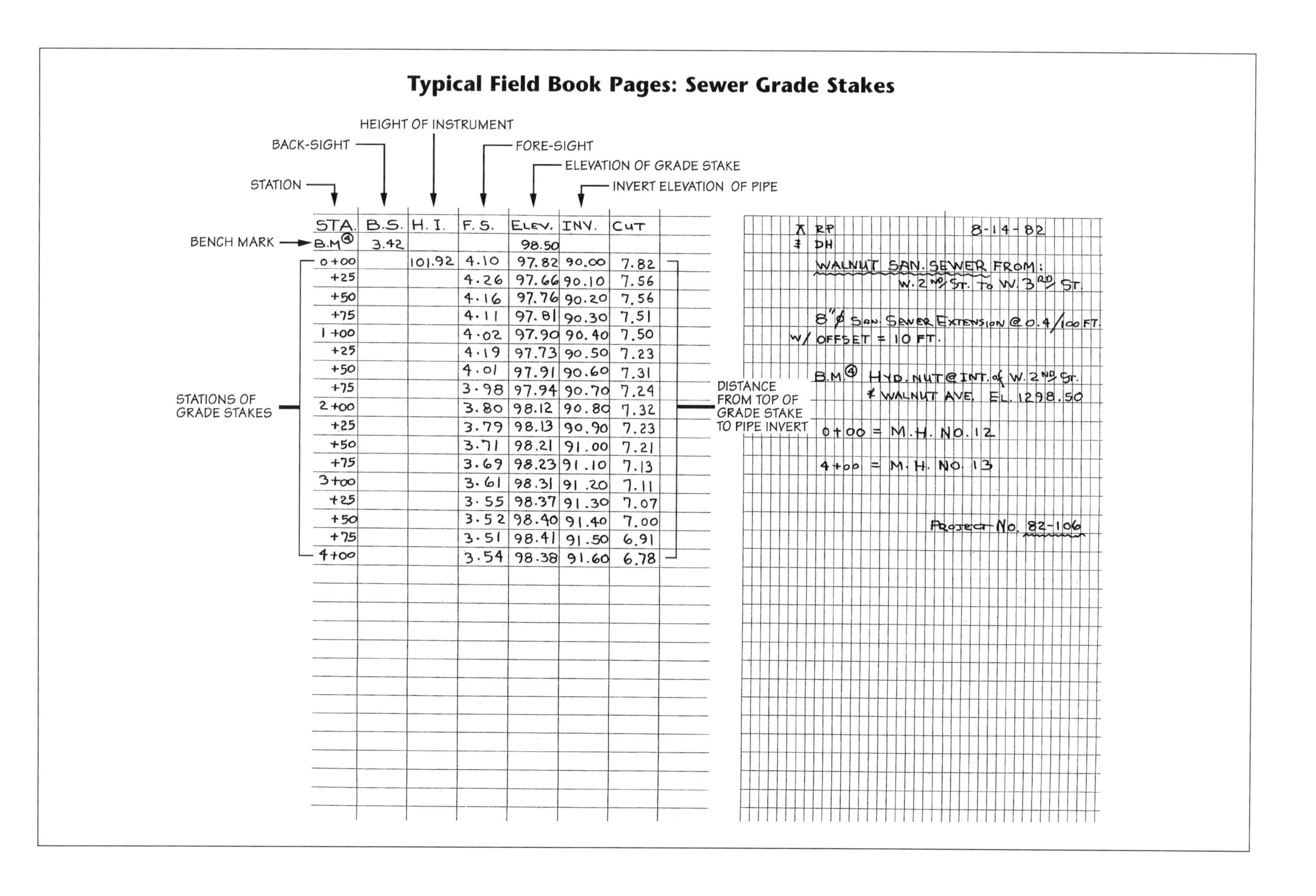

STA.	B.S.	H. I.	F. S.	ELEV.	INV.	CUT
B.M.④	3.42			98.50		
0+00		101.92	4.10	97.82	90.00	7.82
+25			4.26	97.66	90.10	7.56
+50			4.16	97.76	90.20	7.56
+75			4.11	97.81	90.30	7.51
1+00			4.02	97.90	90.40	7.50
+25			4.19	97.73	90.50	7.23
+50			4.01	97.91	90.60	7.31
+75			3.98	97.94	90.70	7.24
2+00			3.80	98.12	90.80	7.32
+25			3.79	98.13	90.90	7.23
+50			3.71	98.21	91.00	7.21
+75			3.69	98.23	91.10	7.13
3+00			3.61	98.31	91.20	7.11
+25			3.55	98.37	91.30	7.07
+50			3.52	98.40	91.40	7.00
+75			3.51	98.41	91.50	6.91
4+00			3.54	98.38	91.60	6.78

Typical Grade Sheet for Batter Board Method

GRADE SHEET

DATE: AUGUST 14, 1982 PAGE 1 OF 1

PROJECT: SANITARY SEWER EXTENSION NO. 82-106 (8-INCH)

LOCATION: WALNUT AVENUE

FROM: W. 2ND STREET TO: W. 3RD STREET

SLOPE: 0.40'/100 FT. OFFSET: 10 FT.

BENCH MARK ELEVATION: 1298.50 LOCATION: HYD. NUT @ WALNUT & W. 2ND

STATION	STAKE ELEVATION	GRADE ELEVATION	FILL	CUT	GRADE POLE 9 FT.	NOTES
0+00	97.82	90.00		7.82	1.18	M.H. NO.12
+25	97.66	90.10		7.56	1.44	
+50	97.76	90.20		7.56	1.44	
+75	97.81	90.30		7.51	1.49	
1+00	97.90	90.40		7.50	1.50	
+25	97.73	90.50		7.23	1.77	
+50	97.91	90.60		7.31	1.69	
+75	97.94	90.70		7.24	1.76	
2+00	98.12	90.80		7.32	1.68	
+25	98.13	90.90		7.23	1.77	
+50	98.21	91.00		7.21	1.79	
+75	98.23	91.10		7.13	1.87	
3+00	98.31	91.20		7.11	1.89	
+25	98.37	91.30		7.07	1.93	
+50	98.40	91.40		7.00	2.00	
+75	98.41	91.50		6.91	2.09	
4+00	98.38	91.60		6.78	2.22	M.H. 13

GRADE SHEET

DATE: PAGE ___ OF ___

PROJECT:

LOCATION:

FROM: TO:

SLOPE: OFFSET:

BENCH MARK ELEVATION: LOCATION:

STATION	STAKE ELEVATION	GRADE ELEVATION	FILL	CUT	GRADE POLE	NOTES

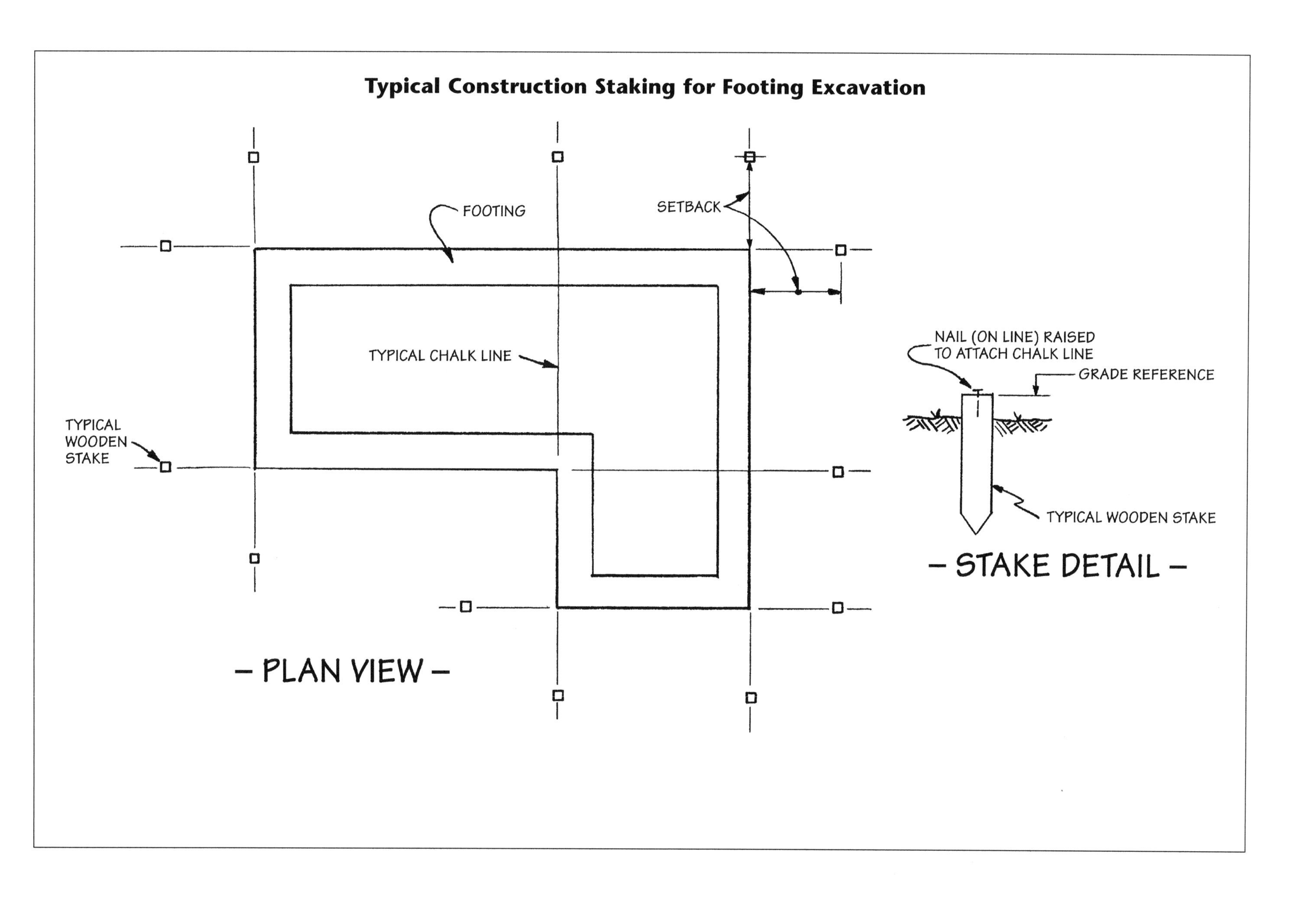
Typical Construction Staking for Footing Excavation
FOOTING
SETBACK
TYPICAL CHALK LINE
TYPICAL WOODEN STAKE
– PLAN VIEW –
NAIL (ON LINE) RAISED TO ATTACH CHALK LINE
GRADE REFERENCE
TYPICAL WOODEN STAKE
– STAKE DETAIL –

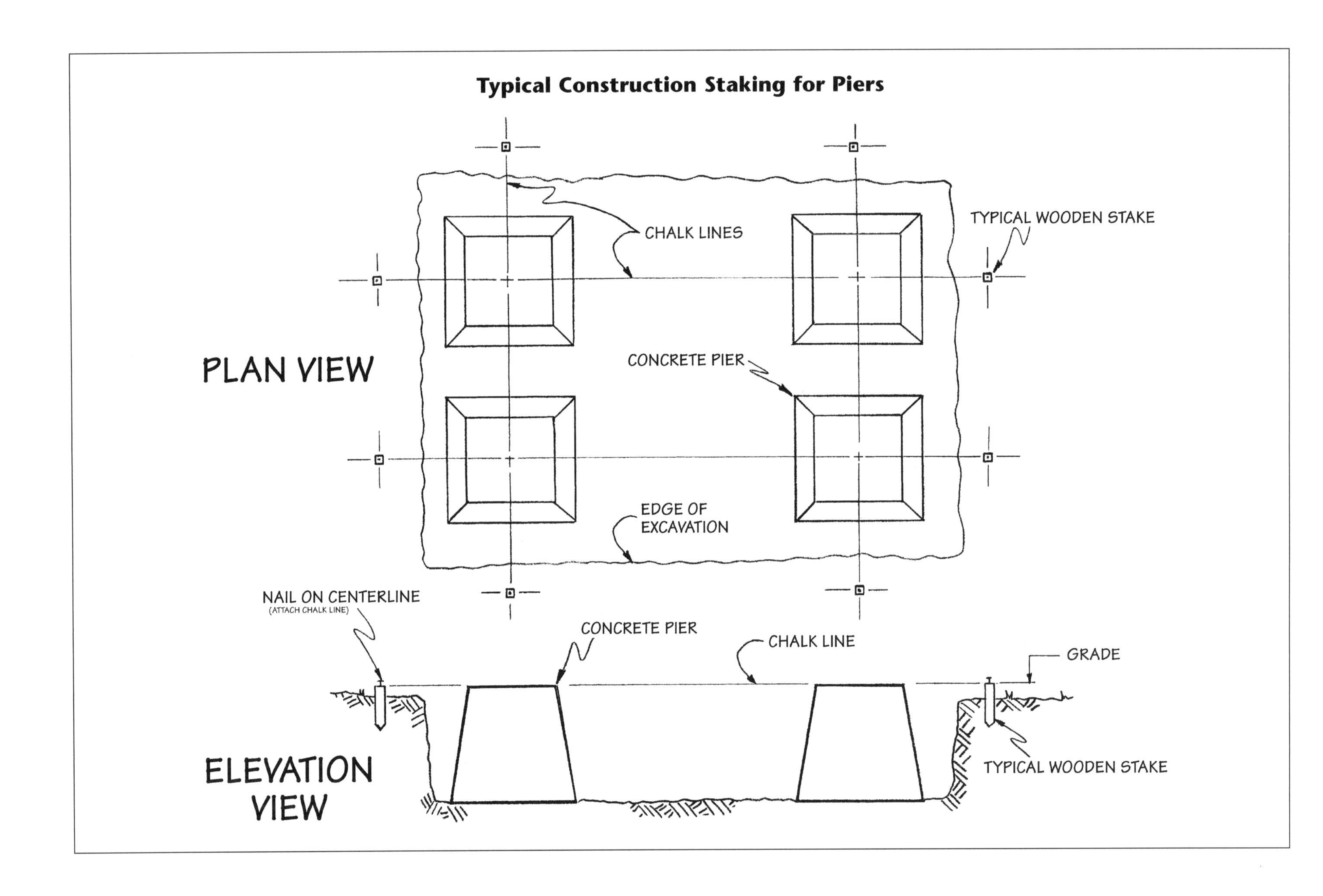
Typical Construction Staking for Piers
CHALK LINES
TYPICAL WOODEN STAKE
PLAN VIEW
CONCRETE PIER
EDGE OF EXCAVATION
NAIL ON CENTERLINE
(ATTACH CHALK LINE)
CONCRETE PIER
CHALK LINE
GRADE
TYPICAL WOODEN STAKE
ELEVATION VIEW

Construction

Close-out

Section 31

The last 5% of construction can sometimes be the most frustrating period of a project. While the construction is substantially complete and facility is operational, there are numerous minor items still to be completed to satisfy the plans and specifications. The contractor's crew and the subcontractors have all but vacated the site. The inspector has repeatedly pursued the foreman to finish the remaining punch list items without success. It now becomes the project engineer's responsibility to formally issue deadlines and represent the owner's interest to ensure that all project details have been properly completed. Most projects are completed in an orderly manner and close-outs progress in a timely manner. However, in that rare circumstance, the owner usually has the right to use the retainage that was held back from previous payments to complete the work, provided that this action is executed according the applicable laws.

The following pages contain some sample material and generic documents that should be helpful in finalizing a project and is normally called project "Close-out."

Table of Contents

Punch List Format

(Sample)

MORGAN and PARMLEY, Ltd.
Professional Consulting Engineers
115 West 2nd Street, S.
LADYSMITH, WI 54848
Phone 715-532-3721
FAX 715-532-5305

DATE:

-MEMO-

TO: (Name of Contractor)

RE: (Name & Number of Project)

PUNCH LIST:

1) ISCO Representative 2nd Training Session (3.3 – 15960)

2) Supply 3 hole paper punch (2.2.2 – 11600)

3) Supply 3 drawer, metal file cabinet (2.2.3 – 11600)

4) Supply UV spare parts (2.3.10 – 16601)

5) Complete telemetry @ lift station (2.8 – 02606) w/demo & testing

6) Furnish & install cap for emergency riser @ lift station

7) Install neutral (4th wire) from CT cabinet to control panel in service building (per spec & Xcel Energy)

8) Replace faulty control for Blower #2 and adjust properly.

9) Resolve timer sequence problem for effluent pump to insure UV bulbs submergence in weir channel and replace check valve & piping (alignment may be problem).

10) Install small rip rap @ outfall for balance & clean channel to river

11) Furnish 3 complete sets of O & M material for:

A-Lift Station; pumps, controls & electrical panel
B-Telemetry System @ Lift Station
C-Air Conditioner @ Service Building
D-Electrical equipment @ Service Building
  *Toe Heaters
  *Base Board Heaters
  *Power Ventilators
  *Suspended Heaters
E-UV (Ultraviolet Disinfection System)

Page 2

12) Furnish executed forms:

A-Consent of Surety Co. to final payment (Pg. 57)

B-Contractor's Affidavit of Release of Liens (Pg. 58)

C-Contractor's Affidavit of Debts & Claims (Pg. 59)

13) Written document for (3 yr.) extended warranty of structural integrity of service building.

14) Add gravel around air valves (flush w/surface) at south end of cell No. 1; i.e. north of parking lot.

15) Fill (seed, fert. & mulch) four (4) mud puddles at fence corners.

16) Shim exit door (east: Blower Room) Does not swing free.

17) Replace all window screens: all screen frames are bent.

18) Seal/Fix wall plate & clean-out in rest room.

19) Ceiling cove trim has shrunk: tighten and caulk to keep insulation from leaking; entire building.

20) Fix insulation leak @ existing fan in rest room.

21) Fix broken handle on lid to Valve Vault chamber "D" (effluent/UV pump area).

22) Replace all UV bulbs or guarantee their longevity: much damage was evidenced during failure of channel to keep bulbs submerged.

23) East fascia on Service Building wrinkled; must be fixed, but could wait until spring.

24) Provide staff gate for UV unit; reference specs 16601, Section 2.3.6, Item 5.

25) Provide 2-inch Gate Valve (for throttling) on effluent pump discharge, per Plans.

Sincerely,

Robert O. Parmley, P.E.
Project Engineer

Source: Morgan & Parmley, Ltd.

Consent of Surety Company to Final Payment

PROJECT: ______________________

____

Owner: ______________________

Contractor: ______________________

Address: ______________________

____

Contract Date: ______________________

In accordance with the provisions of the Contract between the Owner and the Contractor.

______________________, SURETY COMPANY,

on bond of

______________________, CONTRACTOR,

hereby approves of the final payment to the Contractor, and agrees that final payment to the Contractor shall not relieve the Surety Company of any of its obligations to

______________________, OWNER,

as set forth in the said Surety Company's bond.

IN WITNESS WHEREOF,
the Surety Company has hereunto set its hand this _______ day of ______________, 20___.

____
Surety Company

____
Signature of Authorized Representative

Attest:
(Seal):

____
Title

Source: Morgan & Parmley, Ltd.

Contractor's Affidavit of Debts and Claims

PROJECT: ______________________

____

Owner: ______________________

Contractor: ______________________

Address: ______________________

____

Contract Date: ______________________

State:

County of:

The undersigned, pursuant to Section 19.6 of the General Conditions of the Contract, hereby certifieS that, except as listed below, he has paid in full or has otherwise satisfied all obligations for all materials and equipment furnished, for all work, labor, and services performed, and for all known indebtedness and claims against the Contractor for damages arising in any manner in connection with the performance of the Contract referenced above for which the Owner or his property might in any way be held responsible.

EXCEPTIONS: (If none, write "None".)

CONTRACTOR:

Address:

BY:

Subscribed and sworn to before me

this ____ day of ___________, 20__

Notary Public:

My Commission Expires:

Source: Morgan & Parmley, Ltd.

Contractor's Affidavit of Release of Liens

PROJECT: ______________________

____

Owner: ______________________

Contractor: ______________________

Address: ______________________

____

Contract Date: ______________________

State of:

County of:

The undersigned, pursuant to Section 19.6 of the General Conditions of the Contract, hereby certifies that to the best of his knowledge, information and belief, except as listed below, the Releases or Waivers of Lien attached hereto include the Contractor, all Subcontractors, all suppliers of materials and equipment, and all performers of Work, labor or services who have or may have liens against any property of the Owner arising in any manner out of the performance of the Contract referenced above.

EXCEPTIONS: (if none, write "None".)

ATTACHMENTS:

CONTRACTOR:

Address:

BY:

Subscribed and sworn before me this ____ day of ______________, 20__

Notary Public:

My Commission Expires:

Source: Morgan & Parmley, Ltd.

Certificate of Substantial Completion

OWNER'S Project No. ____________ ENGINEER'S Project No. __________

Project Name: ______________________

CONTRACTOR ______________________

Contract Amount ______________ Contract Date ______________

This Certificate of Substantial Completion applies to all Work under the Contract Documents or to the following specified parts thereof:

To: ______________________
OWNER

And To ______________________
CONTRACTOR

The Work to which this Certificate applies has been inspected by authorized representatives of OWNER, CONTRACTOR and ENGINEER, and that Work is hereby declared to be substantially complete in accordance with the Contract Documents on:

____
Date of Substantial Completion

A Punch List of items to be completed or corrected is attached hereto. This list may not be all-inclusive, and the failure to include an item does not alter the responsibility of CONTRACTOR to complete all the Work in accordance with the Contract Documents. The items in the Punch List shall be completed or corrected by CONTRACTOR within _____ consecutive days of the above date of Substantial Completion.

Source: Morgan & Parmley, Ltd.

O & M Manual

(Sample Front Material & Table of Contents)

M & P Project No. 98-109

VOLUME I

-OPERATION & MAINTENANCE MANUAL-

WASTEWATER TREATMENT

FACILITY

Village

of

Sheldon, Wisconsin

Prepared by:

MORGAN & PARMLEY, LTD.
Professional Consulting Engineers
115 West 2nd Street, South
Ladysmith, Wisconsin 54848

PREFACE

The Sheldon Wastewater Treatment Facility and Sewer Collection System is owned and operated by the Village of Sheldon. The Village Board provides all necessary administrative and supervisory controls, including all monies to insure proper operation and maintenance (O & M) of the complete sewerage facility.

The following is a suggested list of management responsibilities:

1. Establish staff requirements, prepare job descriptions, develop organizational charts and assign personnel as required.
2. Provide operational personnel with sufficient funds to properly operate and maintain the treatment facility and collection system.
3. Ensure that operational personnel are paid a salary to commensurate with their level of responsibility.
4. Provide good working conditions, safety equipment and proper tools for the operational personnel.
5. Establish operator training programs.
6. Make periodic inspections of the treatment facility to discuss mutual problems with Operators and to observe operational practices.
7. Maintain good public relations and create an atmosphere that will make the Operators feel that they can bring all problems to management's attention.
8. Plan for future expansions and modifications to the sewage collection system and treatment facility.

-i-

Source: Morgan & Parmley, Ltd.

(Continued)

Sound water management goals, relative to wastewater treatment, should have the following primary objectives:

1. Provide an adequate level of waste treatment to meet applicable water quality standards and BPWTT (Best Practicable Wastewater Treatment Technology) requirements.
2. Provide adequate capacity in both conveyance and treatment facilities to handle both present and anticipated future flows.
3. Allow for expansion of treatment facilities in the future, should growth necessitate such expansion.

These three primary objectives have been addressed by the Village Board and achieved by the study, analysis, planning and construction of their new Wastewater Treatment Facility. Financing for Facility Planning, Design and Construction was supported by a grant-loan package from the USDA-Rural Development.

Volume I of the O & M Manual contains a detailed description of the total facility with related technical data for successful operation and maintenance. Volume II is devoted to suppliers' literature and manufacturers' technical material for all individual components and equipment.

It is recommended that the WWTP Operator insert additional material and add notes as he judges appropriate to assist him in fully personalize this manual.

We wish you GOOD LUCK and if you need any assistance, please do not hesitate to contact us: Morgan & Parmley, Ltd. Our telephone number is 532-3721 and our FAX number is 532-5305.

-ii-

VOLUME I

Sheldon WWTF O & M Manual

-TABLE OF CONTENTS-

-iii-

(Continued)

(Continued)

-vi-

*These publications are supplied separately.

LIST OF DRAWINGS, TABLES, FORMS AND CHARTS

-vii-

(Continued)

Typical Construction Record Drawing

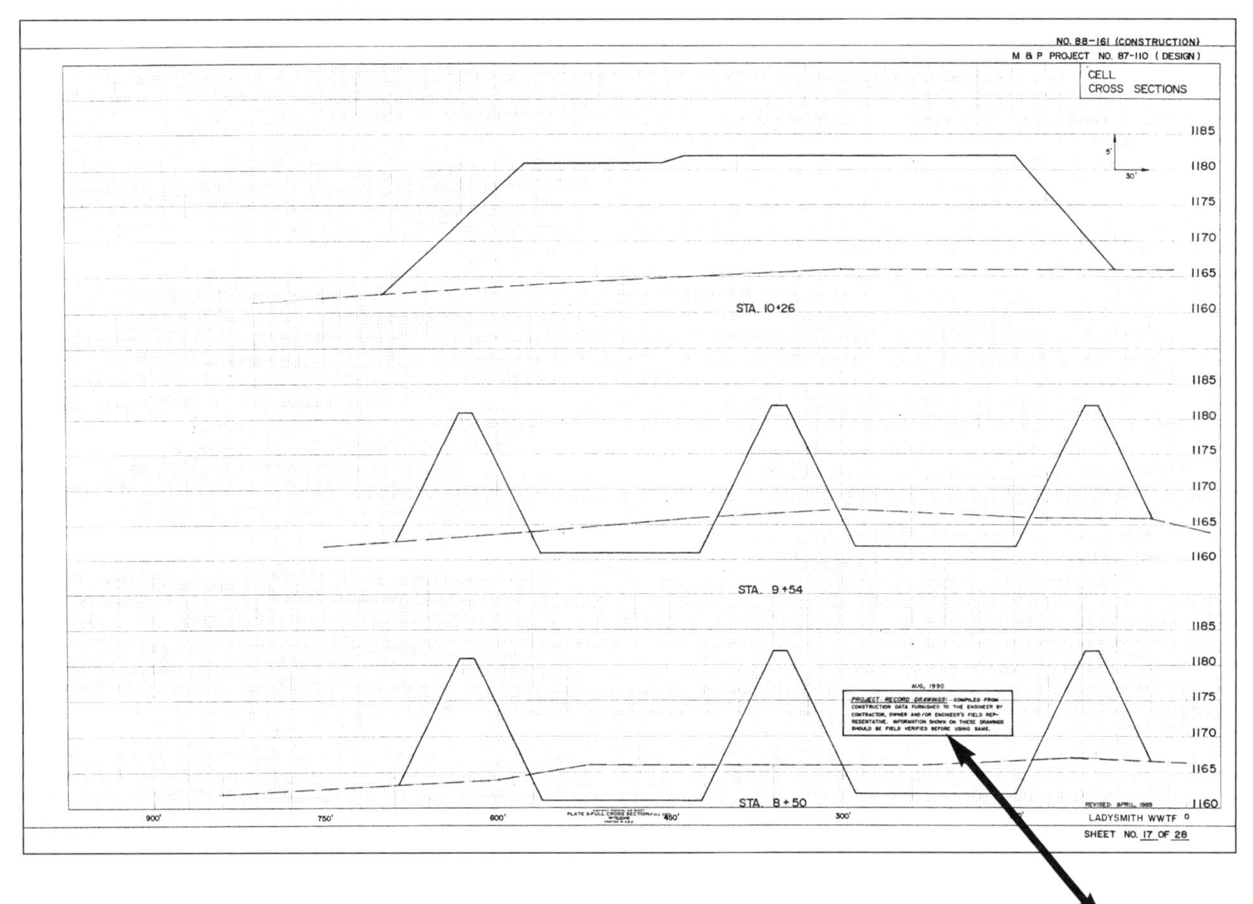

PROJECT RECORD DRAWINGS: COMPILED FROM CONSTRUCTION DATA FURNISHED TO THE ENGINEER BY CONTRACTOR, OWNER AND/OR ENGINEER'S FIELD REPRESENTATIVE. INFORMATION SHOWN ON THESE DRAWINGS SHOULD BE FIELD VERIFIED BEFORE USING SAME.

Source: Morgan & Parmley, Ltd.

Engineer's Certificate of Approval

PROJECT NAME: __________

PROJECT NO: __________

I, ____________________________, a registered Professional Engineer in the State of ___________, and project engineer representing ______________________, do hereby certify that I have inspected ____ and find the same accomplished according to the Specifications and/or duly authorized Change Orders to the prime contract of ____ .

I do approve of the above referred to improvements and recommend acceptance of this work.

Signature: ____________________________________
Engineer

Title: ____

Date: ____

....

As an authorized representative of the Owner ______________________________,
I do hereby accept the improvements referred herein this ____day of ______________, ______.

The warranty period begins ____________________ and ends ___________________

Signature:_____________________________________
Representative

Title: ____

Date: ____

Source: Morgan & Parmley, Ltd.

Supplemental

TECHNICAL REFERENCE

Section 32

All proficient engineers should have an extensive reference file geared to their specific area of interest and professional discipline. While this final section does not propose to replace or duplicate such a file, the Editor-in-Chief has inserted some hard-to-find and perhaps obscure material that may be of future value to the reader. This section concludes with a general discussion of the basics and format of metric measurement, followed by an in-depth tabulation of conversion factors for English to metric.

TABLE OF CONTENTS

Weight & Volume

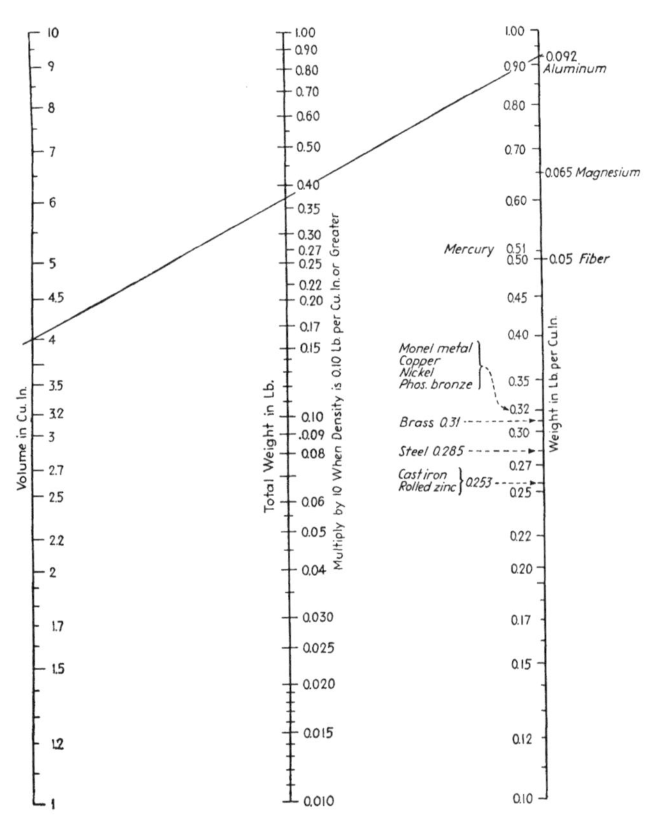

Draw a straight line through the two known points. The answer will be found at the intersection of this line with the third scale.

Example: 4 cu. in. of aluminum weighs 0.37 lb.

Volumes in Horizontal Round Tanks with Flat Ends

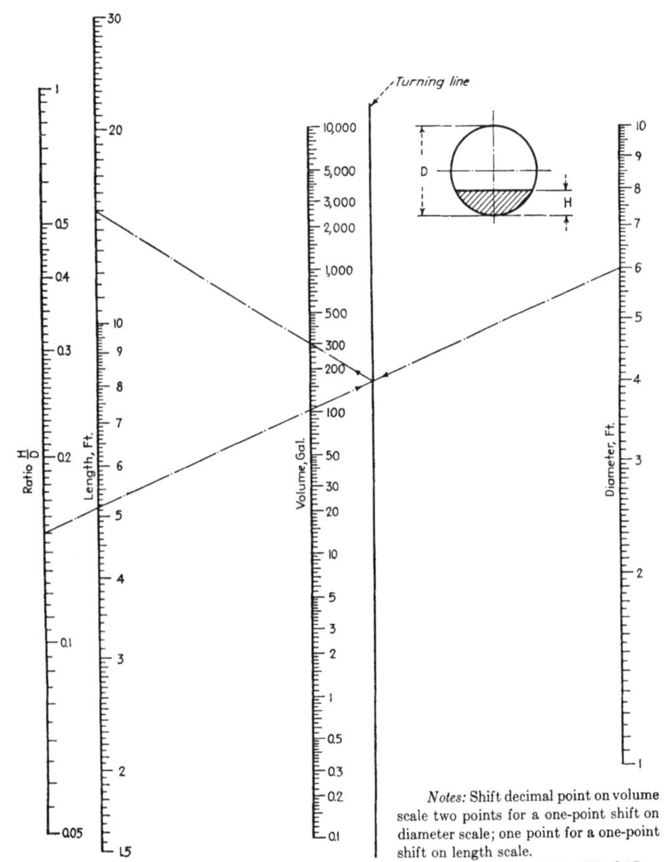

Notes: Shift decimal point on volume scale two points for a one-point shift on diameter scale; one point for a one-point shift on length scale.

Example: Tank is 6 ft. in diameter and 15 ft. long. $H = 0.9$ ft. $H/D = 0.15$. Join 0.15 on H/D scale with 6 on diameter scale. From point of intersection with turning line, draw line to 15 ft. on the length scale. The volume scale shows 300 gal. If D had been 0.6 ft., H 0.09 ft., and length the same, the answer would be 3.00 gal.

Find How Much Horsepower to Pump Liquids

The charts on these two pages will make it easier to design pumps and other equipment for handling liquid flowing in pipes.

First, it is often helpful to know the theoretical horsepower required to raise the liquid to various heads. This is obtained from the horsepower-gpm charts. They are plotted for a 10-ft head of water: For other heads, multiply hp by $H/10$ where H is the revised head; for other liquids, multiply by the corresponding specific gravity.

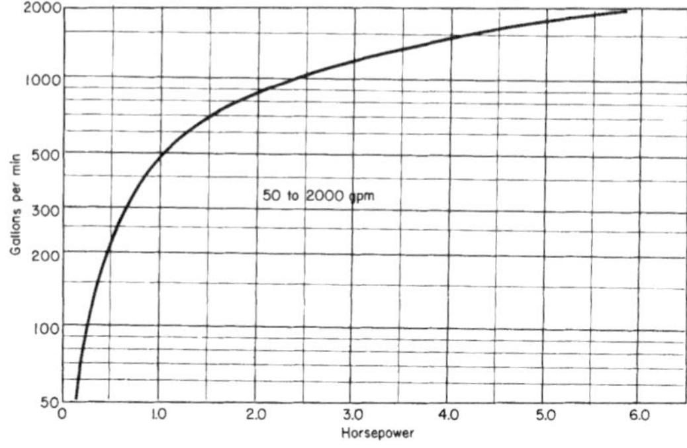

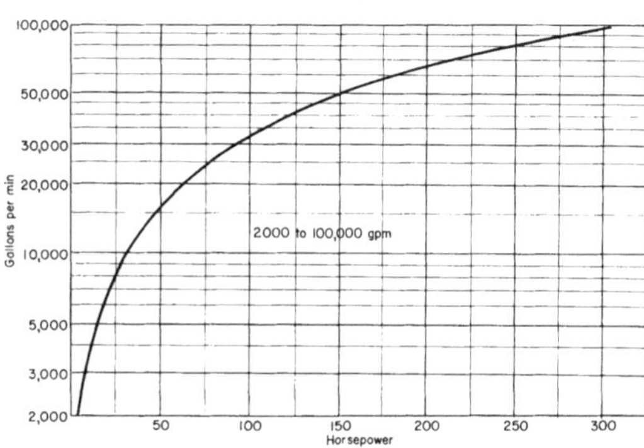

Hp-gpm Chart . . . shows how much hp is required to pump water against a 10-ft head. Full-pipe flow is assumed.

SYMBOLS

G = flow rate, gpm
H_l = head loss, ft
L = length of pipe, ft
P = pressure, psi
R = Reynold's number = $0.0833\ Vd/\nu$
V = velocity, fps
d = pipe dia, in.
f = friction factor
w = fluid density, lb per cu ft
r = relative roughness of pipe = ϵ/d
ϵ = effective height of roughness particle, in.
ν = kinematic viscosity, ft²/sec

Next, for practical results friction losses must be accounted for. These vary and should be known for each individual case.

Much used in liquid-flow calculations is the Darcy formula

$$H_l = f\frac{6LV^2}{d32.16}$$

which can be modified, if velocity is in gpm, to

$$H_l = 0.0312f\frac{LG^2}{d^5}w$$

or for head loss in psi units

$$P = 0.000217f\frac{LG^2}{d^5}w$$

Practical values of f vary from about 0.01 to 0.06 depending on pipe smoothness and dia. For laminar flow, $f = 64.4/R$. The flow chart gives f for various values of R and pipe roughness. Values ϵ for various pipes are: 0.00006 in. for smooth drawn tubing; 0.0018 in. for wrought iron; 0.01 in. for cast iron. Curves for relative roughness values of 0.0005 to 0.01 are plotted. Most of these lie in the transition zone between laminar flow and complete turbulence.

Fluid-flow Chart . . .

gives friction for various pipe conditions and values of Reynold's Number.

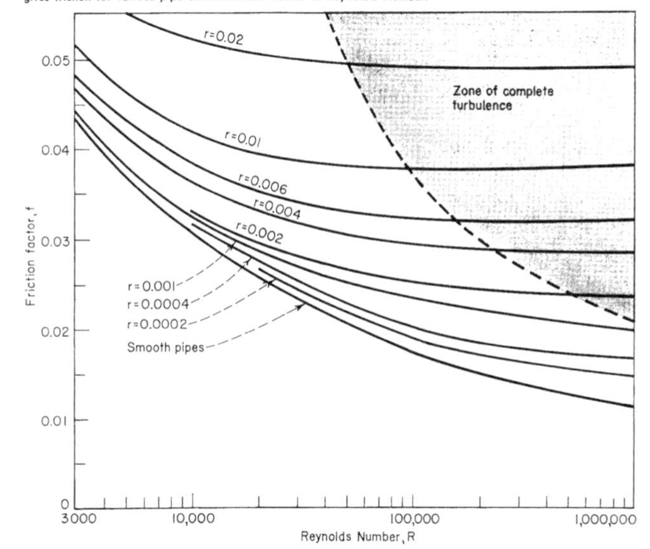

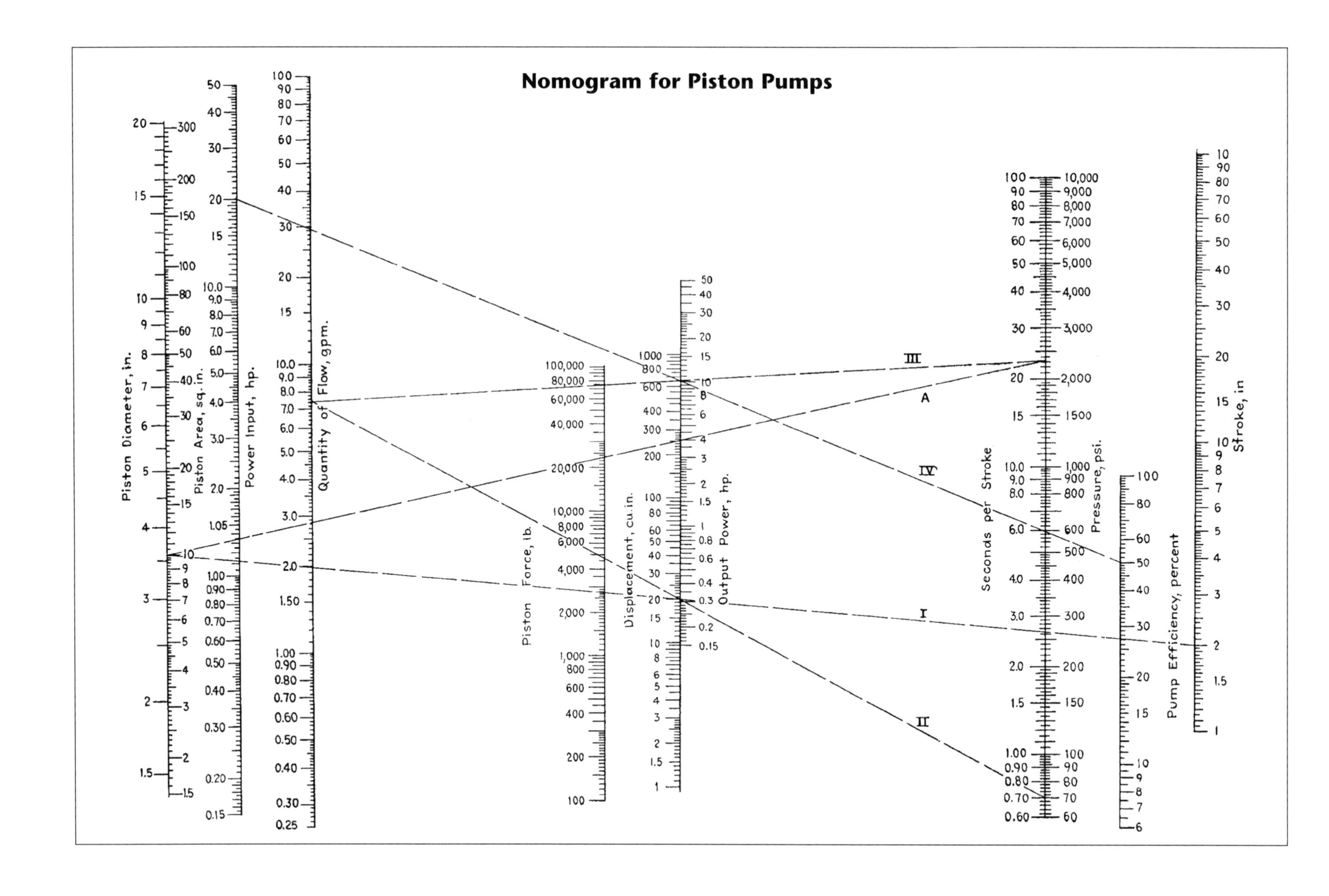

Nomogram for Piston Pumps
Piston Diameter, in.
Piston Area, sq. in.
Power Input, hp.
Quantity of Flow, gpm.
Piston Force, lb.
Displacement, cu. in.
Output Power, hp.
Seconds per Stroke
Pressure, psi.
Pump Efficiency, percent
Stroke, in
I
II
III
IV
A

Find the Length of Open & Closed Belts

The following formulas give the answers (see the illustrations for notation):

Open length, $L = \pi D + (\tan\theta - \theta)(D - d)$

Closed length, $L = (D + d)[\pi + (\tan\theta - \theta)]$

You can find θ from (for open belts): $\cos\theta = (D - d)/2C$; (for closed belts) $\cos\theta = (D + d)/2C$.

When you want to find the **center distance of belt drives,** however, it is much quicker if you have a table that gives you $y = \cos\theta$ in terms of $x = (\tan\theta) - \theta$. **Sidney Kravitz,** of Picatinny Arsenal, has compiled such a table. Now, all you need do to find C is first calculate $x = [L/(D + d)] - \pi$ for open drives.

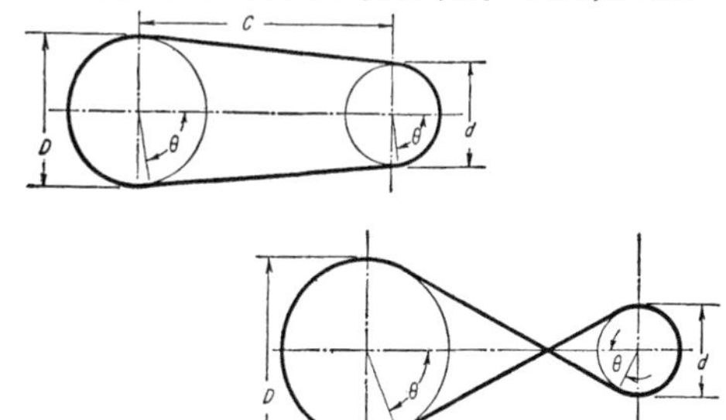

y values[1]

x	0.00	0.01	0.02	0.03	0.04	0.05	0.06	0.07	0.08	0.09
0.0	1.00000	0.95332	0.92731	0.90626	0.88804	0.87175	0.85690	0.84318	0.83039	0.81839
0.1	.80705	.79630	.78606	.77629	.76693	.75795	.74931	.74098	.73295	.72518
0.2	.71767	.71038	.70332	.69646	.68980	.68332	.67701	.67086	.66487	.65902
0.3	.65331	.64774	.64230	.63698	.63177	.62668	.62169	.61681	.61202	.60733
0.4	.60274	.59822	.59380	.58946	.58520	.58101	.57690	.57286	.56889	.56499
0.5	0.56116	0.55738	0.55367	0.55002	0.54643	0.54289	0.53941	0.53598	0.53260	0.52927
0.6	.52600	.52277	.51958	.51645	.51336	.51031	.50730	.50433	.50141	.49852
0.7	.49567	.49286	.49009	.48735	.48465	.48198	.47935	.47675	.47417	.47164
0.8	.46913	.46665	.46420	.46179	.45940	.45703	.45470	.45239	.45011	.44785
0.9	.44562	.44342	.44123	.43908	.43694	.43483	.43274	.43068	.42863	.42661
	.0	.1	.2	.3	.4	.5	.6	.7	.8	.9
1	0.42461	0.40568	0.38850	0.37284	0.35848	0.34526	0.33304	0.32170	0.31115	0.30130
2	.29208	.28344	.27531	.26766	.26043	.25359	.24712	.24098	.23515	.22960
3	.22431	.21926	.21445	.20984	.20544	.20121	.19717	.19328	.18955	.18596
4	.18251	.17918	.17598	.17289	.16991	.16703	.16424	.16156	.15895	.15644
5	0.15400	0.15163	0.14935	0.14712	0.14497	0.14287	0.14084	0.13886	0.13694	0.13508
6	.13326	.13149	.12977	.12810	.12646	.12487	.12332	.12181	.12033	.11889
7	.11748	.11611	.11477	.11346	.11217	.11092	.10970	.10850	.10733	.10618
8	.10506	.10396	.10289	.10183	.10080	.09979	.09880	.09783	.09688	.09594
9	.09503	.09413	.09325	.09238	.09153	.09070	.08988	.08908	.08829	.08751
	0	1	2	3	4	5	6	7	8	9
10	0.08675	0.07980	0.07389	0.06879	0.06436	0.06046	0.05701	0.05393	0.05116	0.04867
20	.04641	.04435	.04246	.04073	.03914	.03766	.03629	.03502	.03384	.03273
30	.03169	.03072	.02980	.02894	.02812	.02735	.02663	.02594	.02528	.02466
40	.02406	.02350	.02296	.02244	.02195	.02148	.02103	.02059	.02018	.01978
50	0.01939	0.01903	0.01867	0.01833	0.01800	0.01768	0.01737	0.01708	0.01679	0.01651
60	.01624	.01598	.01573	.01549	.01525	.01502	.01480	.01459	.01438	.01417
70	.01397	.01378	.01359	.01341	.01323	.01310	.01289	.01273	.01257	.01241
80	.01226	.01211	.01197	.01183	.01169	.01155	.01142	.01129	.01117	.01104
90	.01092	.01080	.01069	.01057	.01046	.01036	.01025	.01015	.01004	.00994
100	0.00985 (see note below for x > 100)									

[1] If $x = (\tan\psi) - \psi$; then $y = \cos\psi$.

If $x > 100$, calculate C from $C = \frac{L}{2} - \frac{\pi}{4}(D + d)$ for both open and closed belts.

$$x = \frac{L}{D + d} \quad \text{for closed drives}$$

Then

$$C = \frac{D - d}{2y} \quad \text{for open drives}$$

$$C = \frac{D + d}{2y} \quad \text{for closed drives}$$

Example: $L = 60.0$, $D = 15.0$, $d = 10.0$, $x = (L - \pi D)/(D - d) = 2.575$, $y = 0.24874$ by linear interpolation in the table. $C = (D - d)/2y = 10.051$.

• • • •

Morton P. Matthew's letter on fractional derivatives (*PE*—July 22 '63, p 105) drew several interesting comments from readers. Here's what Professor Komkov of the University of Utah had to say on the subject. He pointed out that the question raised by Mr Matthew is well known in mathematics, but very little publicized.

"The definition of fractional derivatives goes back to Abel, who developed around 1840 this fascinating little formula:

$$D^s(f) = \frac{d^s f(x)}{dx^s} = \frac{1}{\Gamma(-s)} \int_0^{\bullet} (\xi - t)^{(-s-1)} f(t)\, dt$$

($\Gamma(n)$ is the Euler's Gamma Function).

An elementary proof of this formula is given for example in Courant's *Differential and Integral Calculus,* Part II, page 340. Abel claimed that the formula works for all real values of S, although there is no guarantee that the range of values obtained can be bounded. For a negative S Abel's operator D^s becomes an integral operator:

$$\iota^s(f(x)) = D^{-s}(f(x)) = \frac{1}{\Gamma(s)} \int_0^{\bullet} (\xi - t)^{s-1} f(t)\, dt.$$

All results quoted by Mr Matthew may be easily obtained by application of Abel's Formula.

"There exists a generalization to partial differential equations of the fractional derivative. This is the so-called Riesz Operator. In one dimensional case it becomes Abel's derivative of fractional order.

"Details of the Riesz technique are explained, for example, in Chapter 10 of *Partial Differential Equations* by Duff. Unfortunately I know of no textbook which devotes more than a few pages to the subject of fractional derivatives. However, there exists a large number of papers on the subject in mathematical journals. I remember reading one by Professor John Barrett in the Pacific Journal of Mathematics (I think it was 1947) which discussed the equation:

$$\frac{d^s y}{dx^s} + \iota y = 0 \quad \text{where } 1 \leq s \leq 2$$

"There are some interesting applications in engineering and science for this theory. I was interested some years ago in formulation of elasticity equations for some plastics. I have never completed that investigation but I have established that in some cases, the behavior of plastics may be better simulated by assuming stress-strain relationship to be of the type:

$$\epsilon_{ij} = C_{ijkl} \frac{d^s \epsilon^{kl}}{dt^s}$$

where s is some number between 0 and 1, than by the usual assumption of linear superposition of Hooke's law and Newtonian Fluid properties. In case of some rubbers s worked to be close to 0.7."

Length of Material for 90 Degree Bends

As shown in Fig. 1, when a sheet or flat bar is bent, the position of the neutral plane with respect to the outer and inner surfaces will depend on the ratio of the radius of bend to the thickness of the bar or sheet. For a sharp corner, the neutral plane will lie one-third the distance from the inner to the outer surface. As the radius of the bend is increased, the neutral plane shifts until it reaches a position midway between the inner and outer surfaces. This factor should be taken into consideration when calculating the developed length of material required for formed pieces.

The table on the following pages gives the developed length of the material in the 90-deg. bend. The following formulas were used to calculate the quantities given in the table, the radius of the bend being measured as the distance from the center of curvature to the inner surface of the bend.

1. For a sharp corner and for any radius of bend up to T, the thickness of the sheet, the developed length L for a 90-deg. bend will be

$$L = 1.5708\left(R + \frac{T}{3}\right)$$

2. For any radius of bend greater than $2T$, the length L for a 90-deg. bend will be

$$L = 1.5708\left(R + \frac{T}{2}\right)$$

3. For any radius of bend between $1T$ and $2T$, the value of L as given in the table was found by interpolation.

The developed length L of the material in any bend other than 90 deg. can be obtained from the following formulas:

1. For a sharp corner or a radius up to T:

$$L = 0.0175\left(R + \frac{T}{3}\right) \times \text{degrees of bend}$$

2. For a radius of $2T$ or more:

$$L = 0.0175\left(R + \frac{T}{2}\right) \times \text{degrees of bend}$$

Fig. 1.

For double bends as shown in Fig. 2, if $R_1 + R_2$ is greater than B:

$$X = \sqrt{2B(R_1 + R_2 - B/2)}$$

With R_1, R_2, and B known:

$$\cos A = \frac{R_1 + R_2 - B}{R_1 + R_2}$$

$$L = 0.0175(R_1 + R_2)A$$

where A is in degrees and L is the developed length.

If $R_1 + R_2$ is less than B, as in Fig. 3,

$$Y = B \operatorname{cosec} A - (R_1 + R_2)(\operatorname{cosec} A - \operatorname{cotan} A)$$

Fig. 2.

Fig. 3.

The value of X when B is greater than $R_1 + R_2$ will be

$$X = B \cot A + (R_1 + R_2)(\operatorname{cosec} A - \operatorname{cotan} A)$$

The total developed length L required for the material in the straight section plus that in the two arcs will be

$$L = Y + 0.0175(R_1 + R_2)A$$

To simplify the calculations, the table on this page gives the equations for X, Y, and the developed length for various common angles of bend. The table on following pages gives L for values of R and T for 90-deg. bends.

EQUATIONS FOR X, Y, AND DEVELOPED LENGTHS

Angle A, deg.	X	Y	Developed length
15	$3.732B + 0.132(R_1 + R_2)$	$3.864B - 0.132(R_1 + R_2)$	$3.864B + 0.130(R_1 + R_2)$
22½	$2.414B + 0.199(R_1 + R_2)$	$2.613B - 0.199(R_1 + R_2)$	$2.613B + 0.194(R_1 + R_2)$
30	$1.732B + 0.268(R_1 + R_2)$	$2.000B - 0.268(R_1 + R_2)$	$2.000B + 0.256(R_1 + R_2)$
45	$B + 0.414(R_1 + R_2)$	$1.414B - 0.414(R_1 + R_2)$	$1.414B + 0.371(R_1 + R_2)$
60	$0.577(B + R_1 + R_2)$	$1.155B - 0.577(R_1 + R_2)$	$1.155B + 0.470(R_1 + R_2)$
67½	$0.414B + 0.668(R_1 + R_2)$	$1.082B - 0.668(R_1 + R_2)$	$1.082B + 0.510(R_1 + R_2)$
75	$0.268B + 0.767(R_1 + R_2)$	$1.035B - 0.767(R_1 + R_2)$	$1.035B + 0.542(R_1 + R_2)$
90	$R_1 + R_2$	$B - R_1 - R_2$	$B + 0.571(R_1 + R_2)$

Chordal Height & Length of Chord

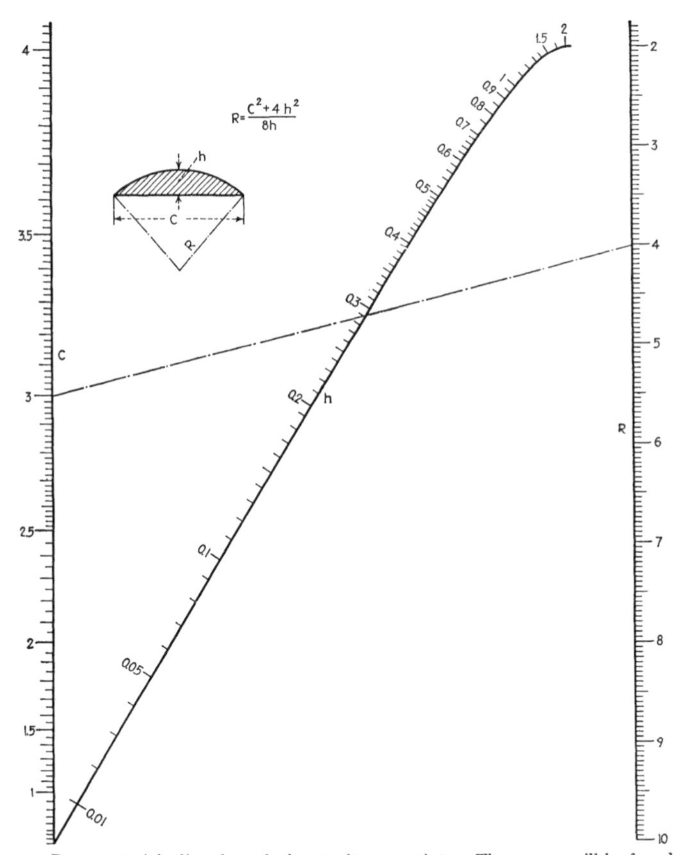

Draw a straight line through the two known points. The answer will be found at the intersection of this line with the third scale.

Example: Length of chord is 3 in., and radius of circle is 4 in. The height h of the chord is 0.29 in.

Forces in Toggle Joint with Equal Arms

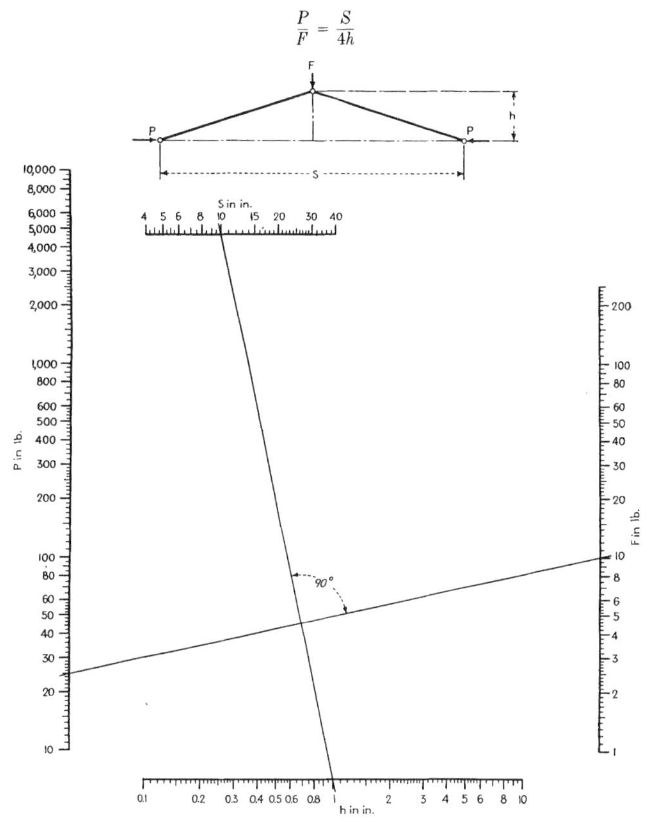

Example: Use mutually perpendicular lines drawn on tracing cloth or celluloid. In the example given for $S = 10$ in. and $h = 1$ in., a force F of 10 lb. exerts pressures P of 25 lb. each.

Moment of Inertia of a Prism About the Axis aa

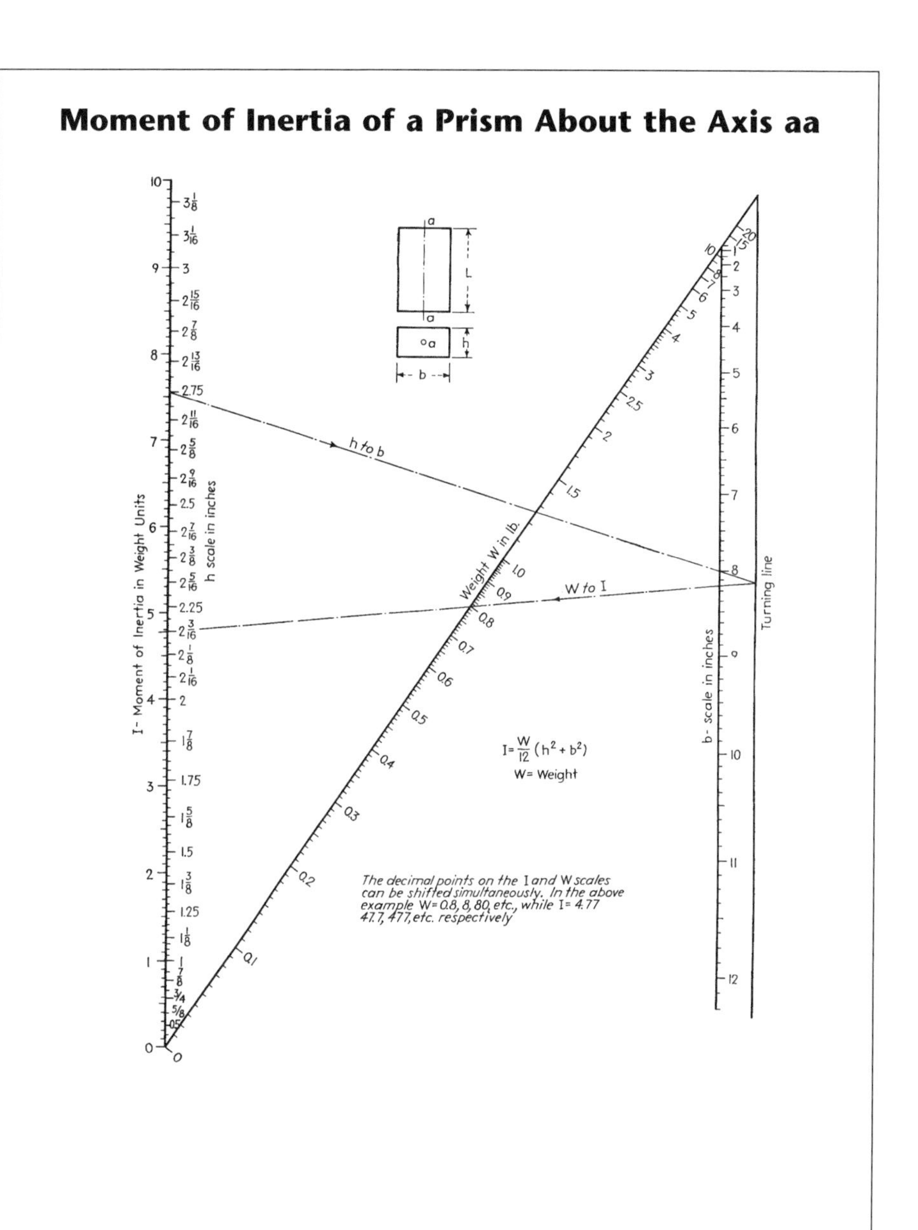

Chart for Transferring Moment of Inertia

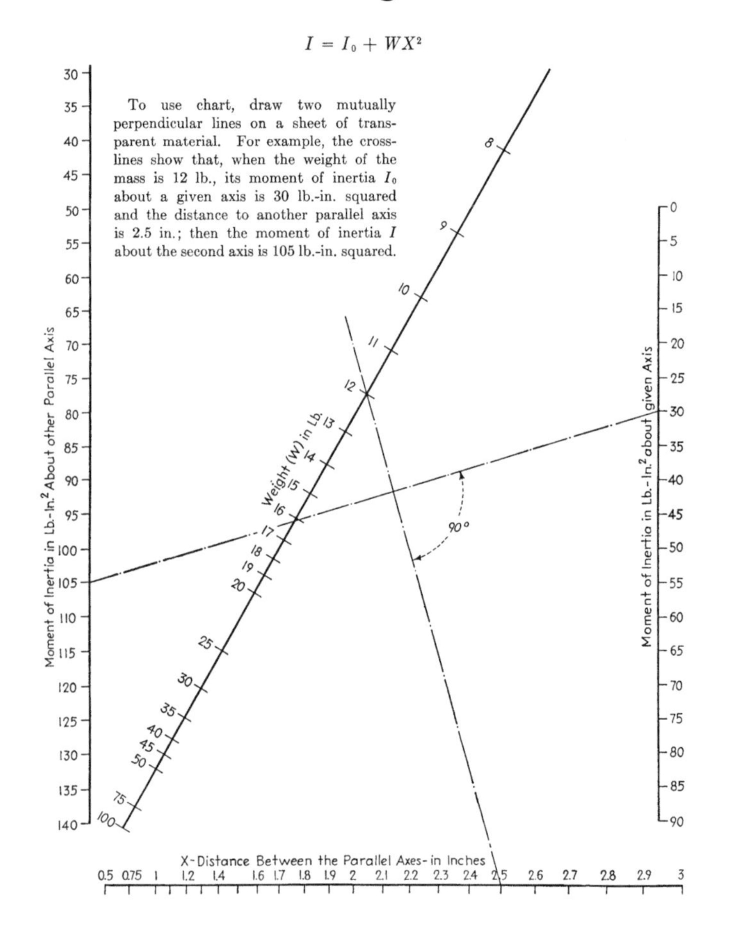

I = I₀ + WX²
To use chart, draw two mutually perpendicular lines on a sheet of transparent material. For example, the cross-lines show that, when the weight of the mass is 12 lb., its moment of inertia I₀ about a given axis is 30 lb.-in. squared and the distance to another parallel axis is 2.5 in.; then the moment of inertia I about the second axis is 105 lb.-in. squared.
Moment of Inertia in Lb.-In.² About other Parallel Axis
Weight (W) in Lb.
Moment of Inertia in Lb.-In.² about given Axis
X-Distance Between the Parallel Axes-in Inches
90°

Accelerated Linear Motion

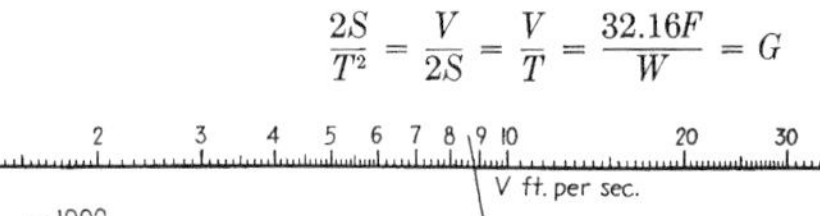

$$\frac{2S}{T^2} = \frac{V}{2S} = \frac{V}{T} = \frac{32.16F}{W} = G$$

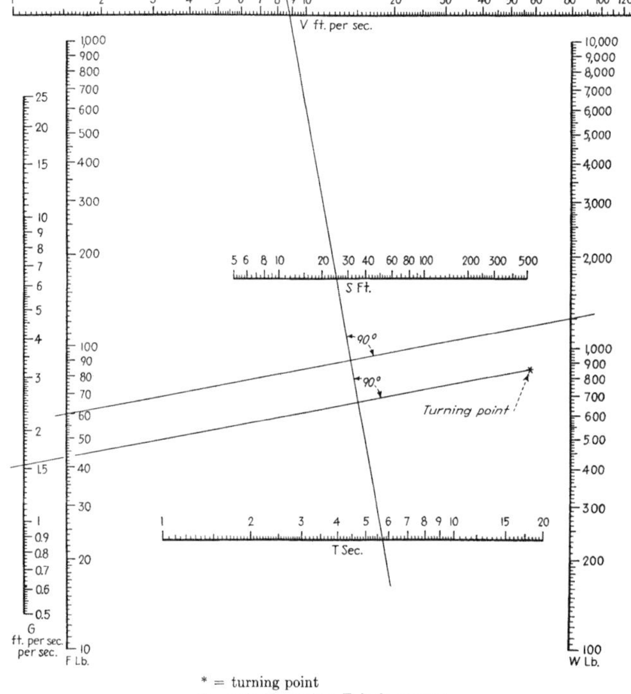

* = turning point
V = velocity at time T, in ft. per sec.
S = distance passed through, in ft.
T = time during which force acts, in sec.
F = accelerating force, in lb.
W = weight of moving body, in lb.
G = constant acceleration, in ft. per sec.

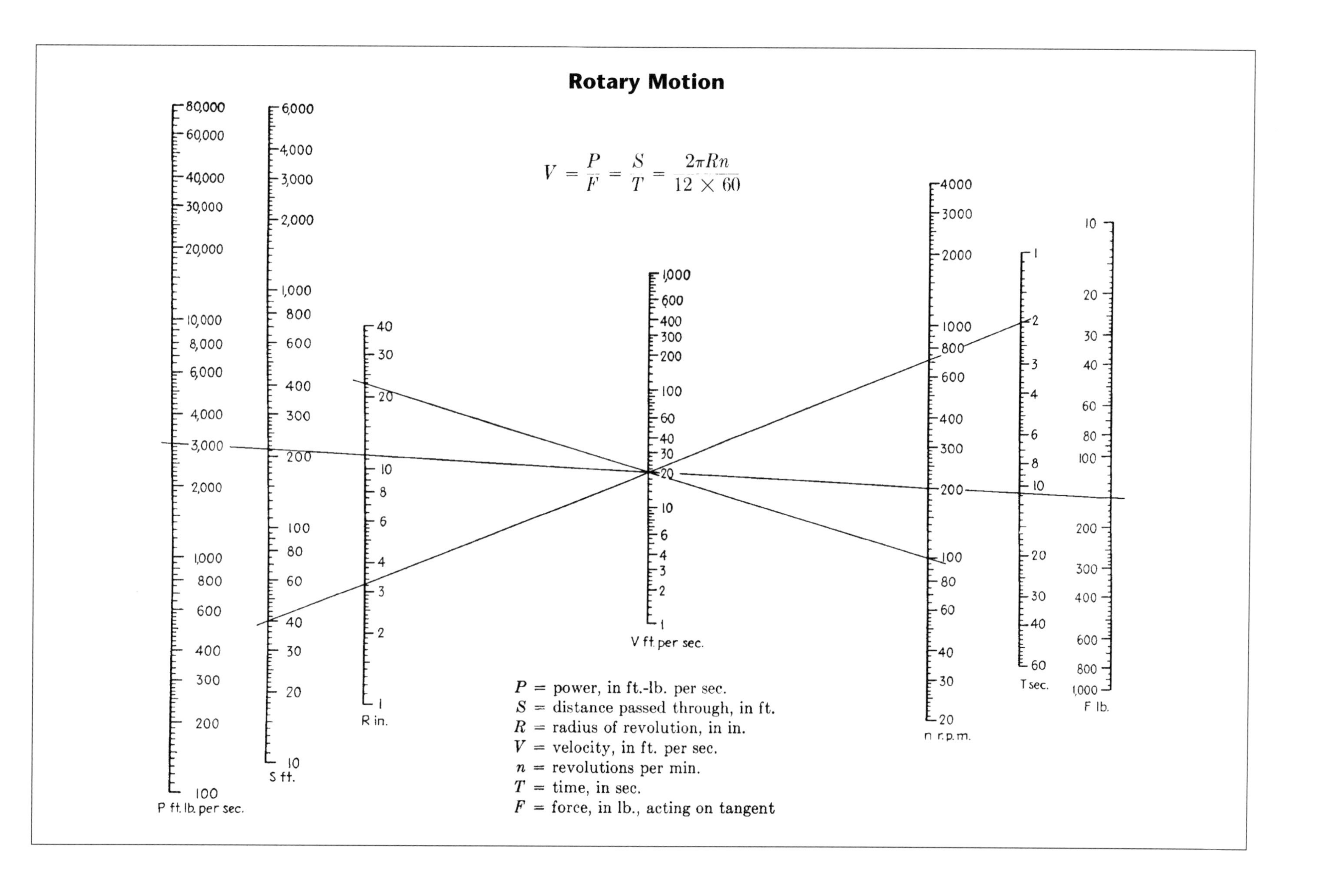
Rotary Motion
$$V = \frac{P}{F} = \frac{S}{T} = \frac{2\pi R n}{12 \times 60}$$
P ft. lb. per sec.
S ft.
R in.
V ft. per sec.
n r.p.m.
T sec.
F lb.
P = power, in ft.-lb. per sec.
S = distance passed through, in ft.
R = radius of revolution, in in.
V = velocity, in ft. per sec.
n = revolutions per min.
T = time, in sec.
F = force, in lb., acting on tangent

Radii of Gyration for Rotating Bodies

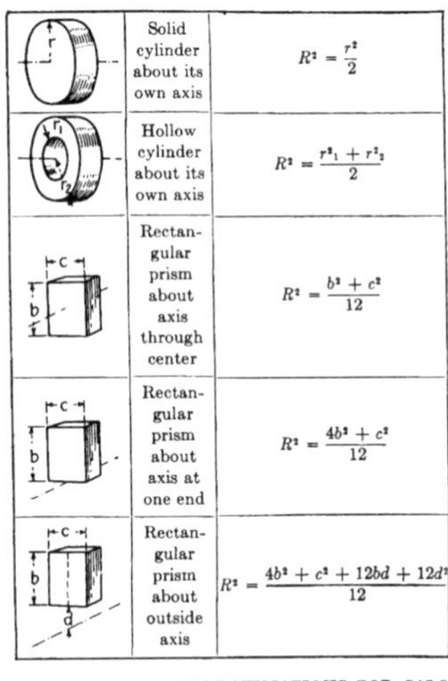

	Solid cylinder about its own axis	$R^2 = \frac{r^2}{2}$
	Hollow cylinder about its own axis	$R^2 = \frac{r_1^2 + r_2^2}{2}$
	Rectangular prism about axis through center	$R^2 = \frac{b^2 + c^2}{12}$
	Rectangular prism about axis at one end	$R^2 = \frac{4b^2 + c^2}{12}$
	Rectangular prism about outside axis	$R^2 = \frac{4b^2 + c^2 + 12bd + 12d^2}{12}$

	Cylinder about axis through center	$R^2 = \frac{l + 3r^2}{12}$
	Cylinder about axis at one end	$R^2 = \frac{4l^2 + 3r^2}{12}$
	Cylinder about outside axis	$R^2 = \frac{4l^2 + 3r^2 + 12dl + 12d^2}{12}$

Any body about axis outside its center of gravity
$R_1^2 = R_0^2 + d^2$
where R_0 = radius of gyration about axis through center of gravity
R_1 = radius of gyration about any other parallel axis
d = distance between center of gravity and axis of rotation

APPROXIMATIONS FOR CALCULATING MOMENTS OF INERTIA

NAME OF PART	MOMENT OF INERTIA
Flywheels (not applicable to belt pulleys)	Moment of inertia equal to 1.08 to 1.15 times that of rim alone
Flywheel (based on total weight and outside diameter)	Moment of inertia equal to two-thirds of that of total weight concentrated at the outer circumference
Spur or helical gears (teeth alone)	Moment of inertia of teeth equal to 40 per cent of that of a hollow cylinder of the limiting dimensions
Spur or helical gears (rim alone)	Figured as a hollow cylinder of same limiting dimensions
Spur or helical gears (total moment of inertia)	Equal to 1.25 times the sum of that of teeth plus rim
Spur or helical gears (with only weight and pitch diameter known)	Moment of inertia considered equal to 0.60 times the moment of inertia of the total weight concentrated at the pitch circle
Motor armature (based on total weight and outside diameter)	Multiply outer radius of armature by following factors to obtain radius of gyration: Large slow-speed motor .... 0.75–0.85 Medium speed d-c or induction motor ..... 0.70–0.80 Mill-type motor 0.60–0.65

WR^2 OF SYMMETRICAL BODIES

For computing WR^2 of rotating masses of weight per unit volume ρ, by resolving the body into elemental shapes.

Note: ρ in pounds per cubic inch and dimensions in inches give WR^2 in lb.-in. squared.

1. Weights per Unit Volume of Materials.

MATERIAL	WEIGHT, LB. PER CU. IN.
Cast iron	0.260
Cast-iron castings of heavy section i.e., flywheel rims	0.250
Steel	0.283
Bronze	0.319
Lead	0.410
Copper	0.318

2. Cylinder, about Axis Lengthwise through the Center of Gravity.

$$\text{Volume} = \frac{\pi}{4}L(D_1^2 - D_2^2)$$

(a) For any material:

$$WR^2 = \frac{\pi}{32}\rho L(D_1^4 - D_2^4)$$

where ρ is the weight per unit volume.

(b) For cast iron:

$$WR^2 = \frac{L(D_1^4 - D_2^4)}{39.2}$$

(c) For cast iron (heavy sections):

$$WR^2 = \frac{L(D_1^4 - D_2^4)}{40.75}$$

(d) For steel:

$$WR^2 = \frac{L(D_1^4 - D_2^4)}{36.0}$$

3. Cylinder, about an Axis Parallel to the Axis through Center of Gravity.

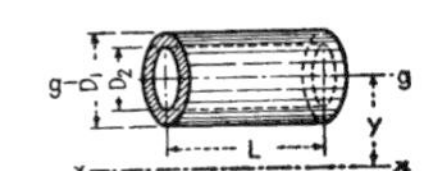

$$\text{Volume} = \frac{\pi}{4}L(D_1^2 - D_2^2)$$

(a) For any material:

$$WR^2_{x-x} = \frac{\pi}{4}\rho L(D_1^2 - D_2^2)\left(\frac{D_1^2 + D_2^2}{8} + y^2\right)$$

(b) For steel:

$$WR^2_{x-x} = \frac{(D_1^2 - D_2^2)L}{4.50}\left(\frac{D_1^2 + D_2^2}{8} + y^2\right)$$

4. Solid Cylinder, Rotated about an Axis Parallel to a Line that Passes through the Center of Gravity and Is Perpendicular to the Center Line.

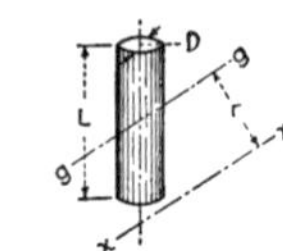

$$\text{Volume} = \frac{\pi}{4}D^2L$$

(a) For any material:

$$WR^2_{x-x} = \frac{\pi}{4}D^2L\rho\left(\frac{L^2}{12} + \frac{D^2}{16} + r^2\right)$$

(b) For steel:

$$WR^2_{x-x} = \frac{D^2L}{4.50}\left(\frac{L^2}{12} + \frac{D^2}{16} + r^2\right)$$

Radii of Gyration for Rotating Bodies *Continued*

5. Rod of Rectangular or Elliptical Section, Rotated about an Axis Perpendicular to and Passing through the Center Line.

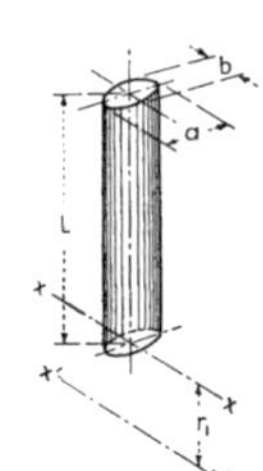

For rectangular cross sections:

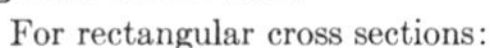

$$K_1 = 1/12; \qquad K_2 = 1$$

For elliptical cross sections:

$$K_1 = \frac{\pi}{64}; \qquad K_2 = \frac{\pi}{4}$$

$$\text{Volume} = K_2 abL$$

(*a*) For any material:

$$WR^2_{x'-x'} = \rho abL\left\{K_2\left[\frac{L^2}{3} + r_1(r_1 + L)\right] + K_1 a^2\right\}$$

(*b*) For a cast-iron rod of elliptical section ($\rho = 0.260$):

$$WR^2_{x'-x'} = \frac{abL}{4.90}\left[\frac{L^2}{3} + r_1(r_1 + L) + \frac{a^2}{16}\right]$$

6. Elliptical Cylinder, about an Axis Parallel to the Axis through the Center of Gravity.

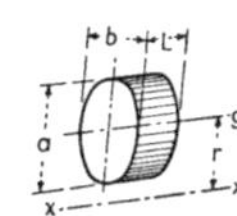

$$\text{Volume} = \frac{\pi}{4} abL$$

(*a*) For any material:

$$WR^2_{x-x} = \rho \frac{\pi}{4} abL\left(\frac{a^2 + b^2}{16} + r^2\right)$$

(*b*) For steel:

$$WR^2_{x-x} = \frac{abL}{4.50}\left(\frac{a^2 + b^2}{16} + r^2\right)$$

7. Cylinder with Frustum of a Cone Removed.

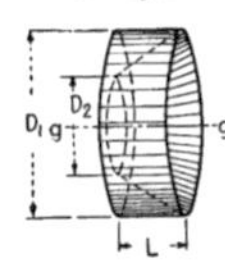

$$\text{Volume} = \frac{\pi L}{2(D_1 - D_2)}\left[\frac{1}{3}(D^3_1 - D^3_2) - \frac{D^2}{2}(D^2_1 - D^2_2)\right]$$

$$WR^2_{g-g} = \frac{\pi\rho L}{8(D_1 - D_2)}\left[\frac{1}{5}(D^5_1 - D^5_2) - \frac{D_2}{4}(D^4_1 - D^4_2)\right]$$

8. Frustum of a Cone with a Cylinder Removed.

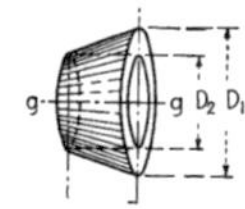

$$\text{Volume} = \frac{\pi L}{2(D_1 - D_2)}\left[\frac{D_1}{2}(D^2_1 - D^2_2) - \frac{1}{3}(D^3_1 - D^3_2)\right]$$

$$WR^2_{g-g} = \frac{\pi\rho L}{8(D_1 - D_2)}\left[\frac{D_1}{4}(D^4_1 - D^4_2) - \frac{1}{5}(D^5_1 - D^5_2)\right]$$

9. Solid Frustum of a Cone.

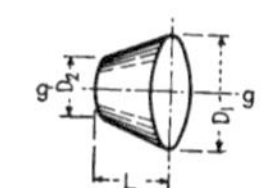

$$\text{Volume} = \frac{\pi L}{12}\frac{(D^3_1 - D^3_2)}{(D_1 - D_2)}$$

$$WR^2_{g-g} = \frac{\pi\rho L}{160}\frac{(D^5_1 - D^5_2)}{(D_1 - D_2)}$$

10. Chamfer Cut from Rectangular Prism Having One End Turned about a Center.

Distance to center of gravity, where $A = R_2/R_1$ and $B = C/2R_1$

$$r_x = \frac{jR^3_1B}{\text{volume} \times (1 - A)}\left[\frac{1}{3}(A^3 - 3A + 2) + \frac{B^2}{3}\left(1 - A - A\log_e\frac{1}{A}\right) + \frac{3}{40}\frac{B^4}{A}(A^2 - 2A + 1) + \frac{5}{672}\frac{B^6}{A^3}(3A^4_1 - 4A^3 + 1)\cdots\right]$$

$$\text{Volume} = \frac{jR^2_1B}{(1 - A)}\left\{(A^2 - 2A + 1) + \frac{B^2}{3}\left[\log_e\frac{1}{A} - (1 - A)\right] + \frac{1}{40}\frac{B^4}{A^2}(2A^3 - 3A + 1) + \frac{1}{224}\frac{B^6}{A^4}(4A^5 - 5A^4 + 1) + \cdots\right\}$$

$$WR^2_{x-x} = -\frac{\rho jR^4_1B}{6(1 - A)}\left\{(A^4 - 4A + 3) + B^2(A^2 - 2A + 1) + \frac{9}{10}B^4\left[\log_e\frac{1}{A} - (1 - A)\right] + \frac{5}{56}\frac{B^6}{A^2}(2A^3 - 3A^2 + 1) + \cdots\right\}$$

11. Complete Torus.

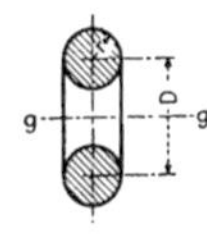

$$\text{Volume} = \pi^2Dr^2$$

$$WR^2_{g-g} = \frac{\pi^2\rho Dr^2}{4}(D^2 + 3r^2)$$

12. Outside Part of a Torus.

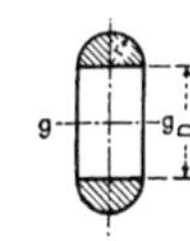

$$\text{Volume} = 2\pi r^2\left(\frac{\pi D}{4} + \frac{2}{3}r\right)$$

$$WR^2_{g-g} = \pi\rho r^2\left[\frac{D^2}{4}\left(\frac{\pi D}{2} + 4r\right) + r^2\left(\frac{3\pi}{8}D + \frac{8}{15}r\right)\right]$$

Radii of Gyration for Rotating Bodies *Continued*

13. Inside Part of a Torus.

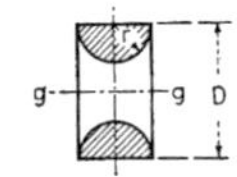

$$\text{Volume} = 2\pi r^2 \left(\frac{\pi D}{4} - \frac{2}{3} r\right)$$

$$WR^2_{g-g} = \pi \rho r^2 \left[\frac{D^2}{4}\left(\frac{\pi D}{2} - 4r\right) + r^2\left(\frac{3\pi}{8} D - \frac{8}{15} r\right)\right]$$

14. Circular Segment about an Axis through Center of Circle.

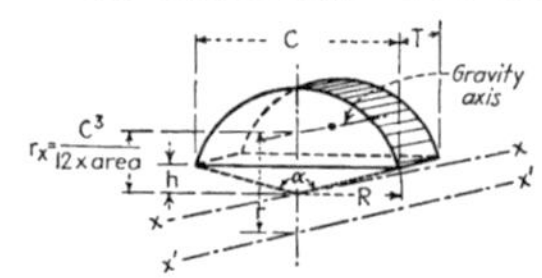

$$\alpha = 2 \sin^{-1} \frac{C}{2R} \text{ deg.}$$

$$\text{Area} = \frac{R^2\alpha}{114.59} - \frac{C}{2}\sqrt{R^2 - \frac{C^2}{4}}$$

(*a*) Any material:

$$WR^2_{x-x} = \rho T \left[\frac{R^4\alpha}{229.2} - \frac{1}{6}\left(3R^2 - \frac{C^2}{2}\right)\frac{C}{2}\sqrt{R^2 - \frac{C^2}{4}}\right]$$

(*b*) For steel:

$$WR^2_{x-x} = \frac{T}{3.534}\left[\frac{R^4\alpha}{229.2} - \frac{1}{6}\left(3R^2 - \frac{C^2}{2}\right)\frac{C}{2}\sqrt{R^2 - \frac{C^2}{4}}\right]$$

15. Circular Segment about Any Axis Parallel to an Axis through the Center of the Circles. (Refer to 14 for Figure.)

$$WR^2_{x'-x'} = WR^2_{x-x} + \text{weight } (r^2 - r^2_x)$$

16. Rectangular Prism about an Axis Parallel to the Axis through the Center of Gravity.

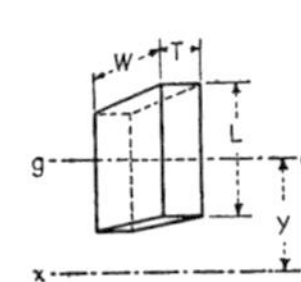

$$\text{Volume} = WLT$$

(*a*) For any material:

$$WR^2_{x-x} = \rho WLT\left(\frac{W^2 + L^2}{12} + y^2\right)$$

(*b*) For steel:

$$WR^2_{x-x} = \frac{WLT}{3.534}\left(\frac{W^2 + L^2}{12} + y^2\right)$$

17. Isosceles Triangular Prism, Rotated about an Axis through Its Vertex.

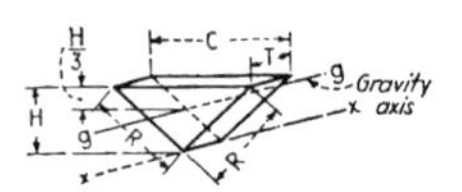

$$\text{Volume} = \frac{CHT}{2}$$

$$WR^2_{x-x} = \frac{\rho CHT}{2}\left(\frac{R^2}{2} - \frac{C^2}{12}\right)$$

18. Isosceles Triangular Prism, Rotated about Any Axis Parallel to an Axis through the Vertex.

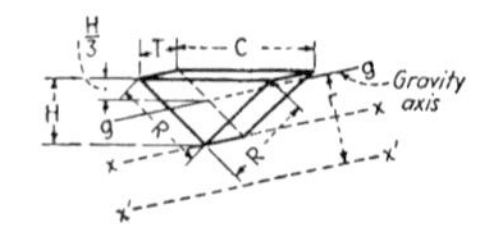

$$\text{Volume} = \frac{CHT}{2}$$

$$WR^2_{x'-x'} = \frac{\rho CHT}{2}\left(\frac{R^2}{2} - \frac{C^2}{12} - \frac{4}{9}H^2 + r^2\right)$$

19. Prism with Square Cross Section and Cylinder Removed, along Axis through Center of Gravity of Square.

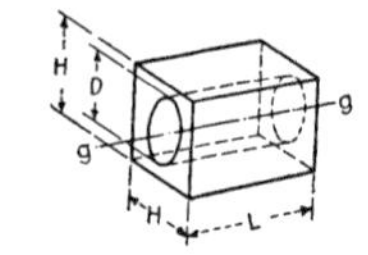

$$\text{Volume} = L\left(H^2 - \frac{\pi D^2}{4}\right)$$

$$WR^2_{g-g} = \frac{\pi \rho L}{32}(1.697H^4 - D^4)$$

20. Any Body about an Axis Parallel to the Gravity Axis, When WR^2 about the Gravity Axis Is Known.

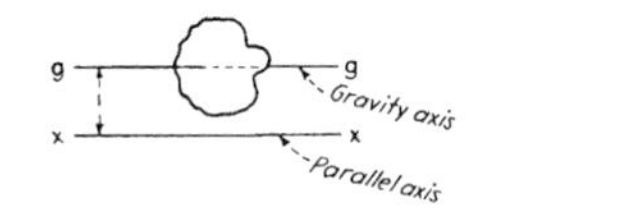

$$WR^2_{x-x} = WR^2_{g-g} + \text{weight} \times r^2$$

21. WR^2 of a Piston, Effective at the Cylinder Center Line, about the Crankshaft Center Line.

$$WR^2 = r^2 W_p \left(\frac{1}{2} + \frac{r^2}{8L^2}\right)$$

where r = crank radius
L = center-to-center length of connecting rod
W_p = weight of complete piston, rings, and pin

Radii of Gyration for Rotating Bodies *Continued*

22. WR^2 of a Connecting Rod, Effective at the Cylinder Center Line, about the Crankshaft Center Line.

$$WR^2 = r^2\left[W_1 + W_2\left(\frac{1}{2} + \frac{r^2}{8L^2}\right)\right]$$

where r = crank radius
L = center-to-center length of connecting rod
W_1 = weight of the lower or rotating part of the rod = $[W_R(L - L_1)]/L$
W_2 = weight of the upper or reciprocating part of the rod = $W_R L_1/L$
W_R = $W_1 + W_2$, the weight of the complete rod
L_1 = distance from the center line of the crank-pin to the center of gravity of the connecting rod

23. Mass Geared to a Shaft.—The equivalent flywheel effect at the shaft in question is

$$WR^2 = h^2(WR^2)'$$

where h = gear ratio $= \frac{\text{r.p.m. of mass geared to shaft}}{\text{r.p.m. of shaft}}$
$(WR^2)'$ = flywheel effect of the body in question about its own axis of rotation

24. Mass Geared to Main Shaft and Connected by a Flexible Shaft.—The effect of the mass $(WR^2)'$ at the position of the driving gear on the main shaft is

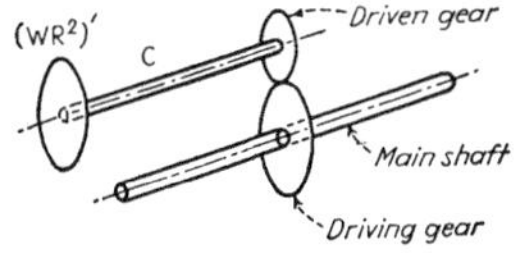

$$WR^2 = \frac{h^2(WR^2)'}{1 - \frac{(WR^2)'f^2}{9.775C}}$$

where h = gear ratio $= \frac{\text{r.p.m. of driven gear}}{\text{r.p.m. of driving gear}}$
$(WR^2)'$ = flywheel effect of geared-on mass
f = natural torsional frequency of the shafting system, in vibrations per sec.
C = torsional rigidity of flexible connecting shaft, in pound-inches per radian

25. Belted Drives.—The equivalent flywheel effect of the driven mass at the driving shaft is

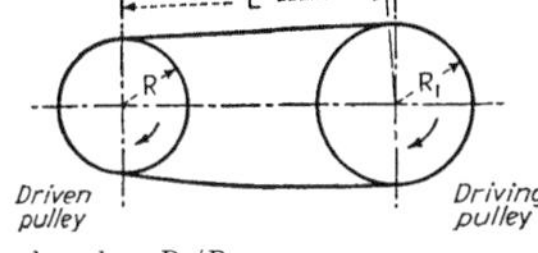

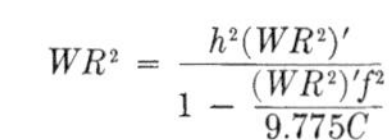

$$WR^2 = \frac{h^2(WR^2)'}{1 - \frac{(WR^2)'f^2}{9.775C}}$$

where $h = R_1/R$ $= \frac{\text{r.p.m. of pulley belted to shaft}}{\text{r.p.m. of shaft}}$
$(WR^2)'$ = flywheel effect of the driven body about its own axis of rotation
f = natural torsional frequency of the system, in vibrations per sec.
$C = R^2AE/L$
A = cross-sectional area of belt, in sq. in.
E = modulus of elasticity of belt material in tension, in lb. per sq. in.
R = radius of driven pulley, in in.
L = length of tight part of belt which is clear of the pulley, in in.

26. Effect of the Flexibility of Flywheel Spokes on WR^2 of Rim.—The effective WR^2 of the rim is

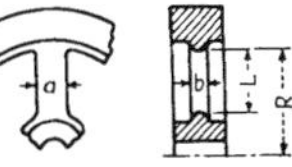

$$WR^2 = \frac{(WR^2)'}{1 - \frac{(WR^2)'f^2}{9.775C}}$$

where $(WR^2)'$ = flywheel effect of the rim
f = natural torsional frequency of the system of which the flywheel is a member, in vibrations per sec.
C = torque required to move the rim through one radian relative to the hub

$$C = \frac{12gEka^3bR}{L^2}\left(\frac{L}{3R} + \frac{R}{L} - 1\right)$$

where g = number of spokes
E = bending modulus of elasticity of the spoke material
$k = \pi/64$ for elliptical, and $k = 1/12$ for rectangular section spokes
All dimensions are in inches.

For cast-iron spokes of elliptical section:

$$E = 15 \times 10^6 \text{ lb. per sq. in.}$$

$$C = \frac{ga^3bR \times 10^6}{0.1132L^2}\left(\frac{L}{3R} + \frac{R}{L} - 1\right)\frac{\text{lb.-in.}}{\text{radians}}$$

Note: It is found by comparative calculations that with spokes of moderate taper very little error is involved in assuming the spoke to be straight and using cross section at mid-point for area calculation.

TYPICAL EXAMPLE

The flywheel shown below is used in a Diesel engine installation. It is required to determine effective WR^2 for calculation of one of the natural frequencies of torsional vibration. The anticipated natural frequency of the system is 56.4 vibrations per sec.

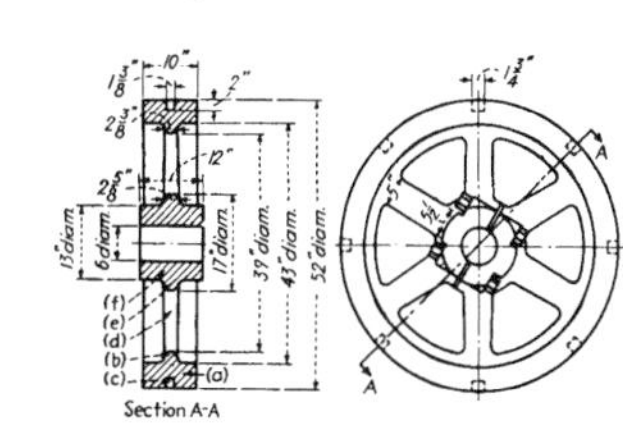

Note: Since the beads at the ends of the spokes comprise but a small part of the flywheel WR^2, very little error will result in assuming them to be of rectangular cross section. Also, because of the effect of the clamping bolts, the outer hub will be considered a square equal to the diameter. The spokes will be assumed straight and of mid-point cross section.

Part of fly wheel	Formula	WR^2
(a)	2c	$\frac{10[(52)^4 - (43)^4]}{40.75} = 955{,}300$
(b)	2b	$\frac{2.375[(43)^4 - (39)^4]}{39.2} = 67{,}000$
(c)	16a neglecting $\left(\frac{W^2 + L^2}{12}\right)$	$-0.250 \times 1.75 \times 2 \times 1.375(25)^2 \times 8 = -6{,}000$
		Total for rim = 1,016,300 lb.-in. squared
(d)	5b	$6 \times \frac{5.25 \times 2.5 \times 11}{4.90}\left[\frac{(11)^2}{3} + 8.5(8.5+11) + \frac{(5.25)^2}{16}\right] = 36{,}800$
(e)	2b	$\frac{2.625[(17)^4 - (13)^4]}{39.2} = 3{,}700$
(f)	19	$\frac{\pi \times 0.250 \times 12}{32}[1.697 \times (13)^4 - (6)^4] = 13{,}900$
		Total for remainder of flywheel = 54,400 lb.-in. squared

From formula (26)

$$C = \frac{6 \times (5.25)^3 \times 2.5 \times 19.5 \times 10^6}{0.1132 \times (11)^2}\left(\frac{11}{3 \times 19.5} + \frac{19.5}{11} - 1\right) = 2{,}970 \times 10^6 \frac{\text{lb.-in.}}{\text{radians}}$$

and

$$WR^2 = \frac{1{,}016{,}300}{1 - \frac{1{,}016{,}300 \times (56.4)^2}{9.775 \times 2{,}970 \times 10^6}} + 54{,}400 = \underline{1{,}197{,}000} \text{ lb.-in. squared}$$

Basic SI Units

Quantity	*Unit*
length	meter (m)
mass	kilogram (kg)
time	second (s)
electric current	ampere (A)
temperature (thermodynamic)	kelvin (K)
amount of substance	mole (mol)
luminous intensity	candela (cd)

Source: R. O. Parmley, Field Engineer's Manual, 3/e ©2002 Published by McGraw-Hill and reproduced with permission of the McGraw-Hill Companies.

Prefixes for SI Units

Multiple and submultiple	*Prefix*	*Symbol*
1,000,000,000,000 = 10^{12}	tera	T
1,000,000,000 = 10^{9}	giga	G
1,000,000 = 10^{6}	mega	M
1,000 = 10^{3}	kilo	k
100 = 10^{2}	hecto	h
10 = 10	deka	da
0.1 = 10^{-1}	deci	d
0.01 = 10^{-2}	centi	c
0.001 = 10^{-3}	milli	m
0.000 001 = 10^{-6}	micro	μ
0.000 000 001 = 10^{-9}	nano	n
0.000 000 000 001 = 10^{-12}	pico	p
0.000 000 000 000 001 = 10^{-15}	femto	f
0.000 000 000 000 000 001 = 10^{-18}	atto	a

Source: R. O. Parmley, Field Engineer's Manual, 3/e ©2002 Published by McGraw-Hill and reproduced with permission of the McGraw-Hill Companies.

SI Derived Units in Civil & Mechanical Engineering

Quantity	*Usual units*	*Symbol*
water usage	liter per person day	l/person·d
runoff	cubic meter per square kilometer	m^3/km^2
precipitation	millimeter per hour	mm/h
river flow	cubic meter per second	m^3/s
discharge	cubic meter per day	m^3/d
hydraulic load per unit area, e.g., filtration rates	cubic meter per square meter day	$m^3/m^2 \cdot d$
concentration	gram per cubic meter	g/m^3
	milligram per liter	mg/l
BOD produced	kilogram per day	kg/d
BOD loading	kilogram per cubic meter day	$kg/m^3 \cdot d$
hydraulic load per unit volume, e.g., biological filters	cubic meter per cubic meter day	$m^3/m^3 \cdot d$
sludge treatment	square meter per person	m^2/person
flow in pipes, channels, etc., peak demands; surface water runoff	cubic meters per second	m^3/s
flow in small pipes; demands of sanitary fittings; pumping rates	liters per second	l/s
water demand; water supply	cubic meters per day	m^3/d
force	kilonewtons	kN
moment of force	newton meter	Nm
	kilonewton meter	kNm

Source: *McGraw-Hill Metrication Manual.* Copyright © 1971 by McGraw-Hill, Inc. New York. (Used with permission of the McGraw-Hill Co.)

Derived Units of the International System

Quantity	*Name of unit*	*Unit symbol or abbreviation, where differing from basic form*	*Unit expressed in terms of basic or supplementary units**
area	square meter		m^2
volume	cubic meter		m^3
frequency	hertz, cycle per second†	Hz	s^{-1}
density	kilogram per cubic meter		kg/m^3
velocity	meter per second		m/s
angular velocity	radian per second		rad/s
acceleration	meter per second squared		m/s^2
angular acceleration	radian per second squared		rad/s^2
volumetric flow rate	cubic meter per second		m^3/s
force	newton	N	$kg \cdot m/s^2$
surface tension	newton per meter, joule per square meter	N/m, J/m^2	kg/s^2
pressure	newton per square meter, pascal†	N/m^2, Pa†	$kg/m \cdot s^2$
viscosity, dynamic	newton-second per square meter, poiseuille†	N s/m^2, Pl†	$kg/m \cdot s$
viscosity, kinematic	meter squared per second		m^2/s
work, torque, energy, quantity of heat	joule, newton-meter, watt-second	J, N·m, W·s	$kg \cdot m^2/s^2$
power, heat flux	watt, joule per second	W, J/s	$kg \cdot m^2/s^3$
heat flux density	watt per square meter	W/m^2	kg/s^3
volumetric heat release rate	watt per cubic meter	W/m^3	$kg/m \cdot s^3$
heat transfer coefficient	watt per square meter degree	$W/m^2 \cdot deg$	$kg/s^3 \cdot deg$
heat capacity (specific)	joule per kilogram degree	J/kg·deg	$m^2/s^2 \cdot deg$
capacity rate	watt per degree	W/deg	$kg \cdot m^2/s^3 \cdot deg$
thermal conductivity	watt per meter degree	W/m·deg, $\frac{Jm}{s \cdot m^2 \cdot deg}$	$kg \cdot m/s^3 \cdot deg$
quantity of electricity	coulomb	C	A·s
electromotive force	volt	V, W/A	$kg \cdot m^2/A \cdot s^3$
electric field strength	volt per meter		V/m
electric resistance	ohm	Ω, V/A	$kg \cdot m^2/A^2 \cdot s^3$
electric conductivity	ampere per volt meter	A/V·m	$A^2 s^3/kg \cdot m^3$
electric capacitance	farad	F, A·s/V	$A^3 s^1/kg \cdot m^2$
magnetic flux	weber	Wb, V·s	$kg \cdot m^2/A \cdot s^2$
inductance	henry	H, V·s/A	$kg \cdot m^2/A^2 s^2$
magnetic permeability	henry per meter	H/m	$kg \cdot m/A^2 s^2$
magnetic flux density	tesla, weber per square meter	T, Wb/m^2	$kg/A \cdot s^2$
magnetic field strength	ampere per meter		A/m
magnetomotive force	ampere		A
luminous flux	lumen	lm	cd sr
luminance	candela per square meter		cd/m^2
illumination	lux, lumen per square meter	lx, lm/m^2	$cd \cdot sr/m^2$

*Supplementary units are plane angle, radian (rad); solid angle, steradian (sr).
†Not used in all countries.

Source: *McGraw-Hill Metrication Manual.* Copyright © 1971 by McGraw-Hill, Inc. New York. (Used with permission of the McGraw-Hill Co.)

Conversion Factors as Exact Numerical Multiples of SI Units

The following tables express the definitions of various units of measure as exact numerical multiples of coherent SI units, and provide multiplying factors for converting numbers and miscellaneous units to corresponding new numbers and SI units.

The first two digits of each numerical entry represent a power of 10. An asterisk follows each number which expresses an exact definition. For example, the entry "–02 2.54*" expresses the fact that 1 inch = 2.54×10^{-2} meter, exactly, by definition. Most of the definitions are extracted from National Bureau of Standards documents. Numbers not followed by an asterisk are only approximate representations of definitions or are the results of physical measurements. The conversion factors are listed alphabetically and by physical quantity.

The Listing by Physical Quantity includes only relationships which are frequently encountered and deliberately omits the great multiplicity of combinations of units which are used for more specialized purposes. Conversion factors for combinations of units are easily generated from numbers given in the Alphabetical Listing by the technique of direct substitution or by other well-known rules for manipulating units. These rules are adequately discussed in many science and engineering textbooks and are not repeated here.

ALPHABETICAL LISTING

To convert from	*to*	*multiply by*
abampere	ampere	+01 1.00*
abcoulomb	coulomb	+01 1.00*
abfarad	farad	+09 1.00*
abhenry	henry	−09 1.00*
abmho	mho	+09 1.00*
abohm	ohm	−09 1.00*
abvolt	volt	−08 1.00*
acre	meter^2	+03 4.046 856 422 4*
ampere (international of 1948)	ampere	−01 9.998 35
angstrom	meter	−10 1.00*
are	meter^2	+02 1.00*
astronomical unit	meter	+11 1.495 978 9
atmosphere	newton/meter^2	+05 1.013 25*
bar	newton/meter^2	+05 1.00*
barn	meter^2	−28 1.00*
barrel (petroleum, 42 gallons)	meter^3	−01 1.589 873
barye	newton/meter^2	−01 1.00*
British thermal unit (ISO/TC 12)	joule	+03 1.055 06
British thermal unit (International Steam Table)	joule	+03 1.055 04
British thermal unit (mean)	joule	+03 1.055 87
British thermal unit (thermochemical)	joule	+03 1.054 350 264 488
British thermal unit (39° F)	joule	+03 1.059 67
British thermal unit (60° F)	joule	+03 1.054 68
bushel (U.S.)	meter^3	−02 3.523 907 016 688*
cable	meter	+02 2.194 56*
caliber	meter	−04 2.54*
calorie (International Steam Table)	joule	+00 4.1868
calorie (mean)	joule	+00 4.190 02
calorie (thermochemical)	joule	+00 4.184*
calorie (15° C)	joule	+00 4.185 80
calorie (20° C)	joule	+00 4.181 90
calorie (kilogram, International Steam Table)	joule	+03 4.1868
calorie (kilogram, mean)	joule	+03 4.190 02
calorie (kilogram, thermochemical)	joule	+03 4.184*
carat (metric)	kilogram	−04 2.00*
Celsius (temperature)	kelvin	$t_K = t_C + 273.15$
centimeter of mercury (0° C)	newton/meter^2	+03 1.333 22
centimeter of water (4° C)	newton/meter^2	+01 9.806 38
chain (engineer or ramden)	meter	+01 3.048*
chain (surveyor or gunter)	meter	+01 2.011 68*
circular mil	meter^2	−10 5.067 074 8
cord	meter^3	+00 3.624 556 3
coulomb (international of 1948)	coulomb	−01 9.998 35
cubit	meter	−01 4.572*
cup	meter^3	−04 2.365 882 365*
curie	disintegration/second	+10 3.70*
day (mean solar)	second (mean solar)	+04 8.64*
day (sidereal)	second (mean solar)	+04 8.616 409 0
degree (angle)	radian	−02 1.745 329 251 994 3
denier (international)	kilogram/meter	−07 1.00*
dram (avoirdupois)	kilogram	−03 1.771 845 195 312 5*
dram (troy or apothecary)	kilogram	−03 3.887 934 6*
dram (U.S. fluid)	meter^3	−06 3.696 691 195 312 5*
dyne	newton	−05 1.00*
electron volt	joule	−19 1.602 10
erg	joule	−07 1.00*
Fahrenheit (temperature)	kelvin	$t_K = (5/9)(t_F + 459.67)$
Fahrenheit (temperature)	Celsius	$t_C = (5/9)(t_F - 32)$

Source: *McGraw-Hill Metrication Manual.*

Conversion Factors as Exact Numerical Multiples of SI Units *(Continued)*

To convert from	*to*	*multiply by*
farad (international of 1948)	farad	−01 9.995 05
faraday (based on carbon 12)	coulomb	+04 9.648 70
faraday (chemical)	coulomb	+04 9.649 57
faraday (physical)	coulomb	+04 9.652 19
fathom	meter	+00 1.828 8*
fermi (femtometer)	meter	−15 1.00*
fluid ounce (U.S.)	meter^3	−05 2.957 352 956 25*
foot	meter	−01 3.048*
foot (U.S. survey)	meter	+00 1200/3937*
foot (U.S. survey)	meter	−01 3.048 006 096
foot of water (39.2° F)	newton/meter^2	+03 2.988 98
foot-candle	lumen/meter^2	+01 1.076 391 0
foot-lambert	candela/meter^2	+00 3.426 259
furlong	meter	+02 2.011 68*
gal (galileo)	meter/second^2	−02 1.00*
gallon (U.K. liquid)	meter^3	−03 4.546 087
gallon (U.S. dry)	meter^3	−03 4.404 883 770 86*
gallon (U.S. liquid)	meter^3	−03 3.785 411 784*
gamma	tesla	−09 1.00*
gauss	tesla	−04 1.00*
gilbert	ampere turn	−01 7.957 747 2
gill (U.K.)	meter^3	−04 1.420 652
gill (U.S.)	meter^3	−04 1.182 941 2
grad	degree (angular)	−01 9.00*
grad	radian	−02 1.570 796 3
grain	kilogram	−05 6.479 891*
gram	kilogram	−03 1.00*
hand	meter	−01 1.016*
hectare	meter^2	+04 1.00*
henry (international of 1948)	henry	+00 1.000 495
hogshead (U.S.)	meter^3	−01 2.384 809 423 92*
horsepower (550 foot lbf/second)	watt	+02 7.456 998 7
horsepower (boiler)	watt	+03 9.809 50
horsepower (electric)	watt	+02 7.46*
horsepower (metric)	watt	+02 7.354 99
horsepower (U.K.)	watt	+02 7.457
horsepower (water)	watt	+02 7.460 43
hour (mean solar)	second (mean solar)	+03 3.60*
hour (sidereal)	second (mean solar)	+03 3.590 170 4
hundredweight (long)	kilogram	+01 5.080 234 544*
hundredweight (short)	kilogram	+01 4.535 923 7*
inch	meter	−02 2.54*
inch of mercury (32° F)	newton/meter^2	+03 3.386 389
inch of mercury (60° F)	newton/meter^2	+03 3.376 85
inch of water (39.2° F)	newton/meter^2	+02 2.490 82
inch of water (60° F)	newton/meter^2	+02 2.4884
joule (international of 1948)	joule	+00 1.000 165
kayser	1/meter	+02 1.00*
kilocalorie (International Steam Table)	joule	+03 4.186 74
kilocalorie (mean)	joule	+03 4.190 02
kilocalorie (thermochemical)	joule	+03 4.184*
kilogram mass	kilogram	+00 1.00*
kilogram force (kgf)	newton	+00 9.806 65*
kilopond force	newton	+00 9.806 65*
kip	newton	+03 4.448 221 615 260 5*
knot (international)	meter/second	−01 5.144 444 444
lambert	candela/meter^2	+04 1/π*
lambert	candela/meter^2	+03 3.183 098 8
langley	joule/meter^2	+04 4.184*
lbf (pound force, avoirdupois)	newton	+00 4.448 221 615 260 5*
lbm (pound mass, avoirdupois)	kilogram	−01 4.535 923 7*
league (British nautical)	meter	+03 5.559 552*
league (international nautical)	meter	+03 5.556*
league (statute)	meter	+03 4.828 032*
light year	meter	+15 9.460 55
link (engineer or ramden)	meter	−01 3.048*
link (surveyor or gunter)	meter	−01 2.011 68*
liter	meter^3	−03 1.00*
lux	lumen/meter^2	+00 1.00*
maxwell	weber	−08 1.00*
meter	wavelengths Kr 86	+06 1.650 763 73*
micron	meter	−06 1.00*
mil	meter	−05 2.54*
mile (U.S. statute)	meter	+03 1.609 344*
mile (U.K. nautical)	meter	+03 1.853 184*
mile (international nautical)	meter	+03 1.852*
mile (U.S. nautical)	meter	+03 1.852*
millibar	newton/meter^2	+02 1.00*
millimeter of mercury (0° C)	newton/meter^2	+02 1.333 224
minute (angle)	radian	−04 2.908 882 086 66
minute (mean solar)	second (mean solar)	+01 6.00*
minute (sidereal)	second (mean solar)	+01 5.983 617 4
month (mean calendar)	second (mean solar)	+06 2.628*
nautical mile (international)	meter	+03 1.852*
nautical mile (U.S.)	meter	+03 1.852*
nautical mile (U.K.)	meter	+03 1.853 184*
oersted	ampere/meter	+01 7.957 747 2
ohm (international of 1948)	ohm	+00 1.000 495
ounce force (avoirdupois)	newton	−01 2.780 138 5
ounce mass (avoirdupois)	kilogram	−02 2.834 952 312 5*
ounce mass (troy or apothecary)	kilogram	−02 3.110 347 68*
ounce (U.S. fluid)	meter^3	−05 2.957 352 956 25*
pace	meter	−01 7.62*
parsec	meter	+16 3.083 74
pascal	newton/meter^2	+00 1.00*
peck (U.S.)	meter^3	−03 8.809 767 541 72*
pennyweight	kilogram	−03 1.555 173 84*
perch	meter	+00 5.0292*
phot	lumen/meter^2	+04 1.00
pica (printers)	meter	−03 4.217 517 6*
pint (U.S. dry)	meter^3	−04 5.506 104 713 575*
pint (U.S. liquid)	meter^3	−04 4.731 764 73*
point (printers)	meter	−04 3.514 598*
poise	newton second/meter^2	−01 1.00*
pole	meter	+00 5.0292*
pound force (lbf avoirdupois)	newton	+00 4.448 221 615 260 5*

Conversion Factors as Exact Numerical Multiples of SI Units *(Continued)*

To convert from	*to*	*multiply by*
pound mass (lbm avoirdupois)	kilogram	−01 4.535 923 7*
pound mass (troy or apothecary)	kilogram	−01 3.732 417 216*
poundal	newton	−01 1.382 549 543 76*
quart (U.S. dry)	meter³	−03 1.101 220 942 715*
quart (U.S. liquid)	meter³	−04 9.463 529 5
rad (radiation dose absorbed)	joule/kilogram	−02 1.00*
Rankine (temperature)	kelvin	$t_K=(5/9)t_R$
rayleigh (rate of photon emission)	1/second meter²	+10 1.00*
rhe	meter²/newton second	+01 1.00*
rod	meter	+00 5.0292*
roentgen	coulomb/kilogram	−04 2.579 76*
rutherford	disintegration/second	+06 1.00*
second (angle)	radian	−06 4.848 136 811
second (ephemeris)	second	+00 1.000 000 000
second (mean solar)	second (ephemeris)	Consult American Ephemeris and Nautical Almanac
second (sidereal)	second (mean solar)	−01 9.972 695 7
section	meter²	+06 2.589 988 110 336*
scruple (apothecary)	kilogram	−03 1.295 978 2*
shake	second	−08 1.00
skein	meter	+02 1.097 28*
slug	kilogram	+01 1.459 390 29
span	meter	−01 2.286*
statampere	ampere	−10 3.335 640
statcoulomb	coulomb	−10 3.335 640
statfarad	farad	−12 1.112 650
stathenry	henry	+11 8.987 554
statmho	mho	−12 1.112 650
statohm	ohm	+11 8.987 554
statute mile (U.S.)	meter	+03 1.609 344*
statvolt	volt	+02 2.997 925
stere	meter³	+00 1.00*
stilb	candela/meter²	+04 1.00
stoke	meter²/second	−04 1.00*
tablespoon	meter³	−05 1.478 676 478 125*
teaspoon	meter³	−06 4.928 921 593 75*
ton (assay)	kilogram	−02 2.916 666 6
ton (long)	kilogram	+03 1.016 046 908 8*
ton (metric)	kilogram	+03 1.00*
ton (nuclear equivalent of TNT)	joule	+09 4.20
ton (register)	meter³	+00 2.831 684 659 2*
ton (short, 2000 pound)	kilogram	+02 9.071 847 4*
tonne	kilogram	+03 1.00*
torr (0° C)	newton/meter²	+02 1.333 22
township	meter²	+07 9.323 957 2
unit pole	weber	−07 1.256 637
volt (international of 1948)	volt	+00 1.000 330
watt (international of 1948)	watt	+00 1.000 165
yard	meter	−01 9.144*
year (calendar)	second (mean solar)	+07 3.1536*
year (sidereal)	second (mean solar)	+07 3.155 815 0
year (tropical)	second (mean solar)	+07 3.155 692 6
year 1900, tropical, Jan., day 0, hour 12	second (ephemeris)	+07 3.155 692 597 47*
year 1900, tropical, Jan., day 0, hour 12	second	+07 3.155 692 597 47

Listing of Conversion Factors by Physical Quantity

To convert from	*to*	*multiply by*
ACCELERATION		
foot/second^2	meter/second^2	−01 3.048*
free fall, standard	meter/second^2	+00 9.806 65*
gal (galileo)	meter/second^2	−02 1.00*
inch/second^2	meter/second^2	−02 2.54*
AREA		
acre	meter^2	+03 4.046 856 422 4*
are	meter^2	+02 1.00*
barn	meter^2	−28 1.00*
circular mil	meter^2	−10 5.067 074 8
foot^2	meter^2	−02 9.290 304*
hectare	meter^2	+04 1.00*
inch^2	meter^2	−04 6.4516*
mile^2 (U.S. statute)	meter^2	+06 2.589 988 110 336*
section	meter^2	+06 2.589 988 110 336*
township	meter^2	+07 9.323 957 2
yard^2	meter^2	−01 8.361 273 6*
DENSITY		
gram/centimeter^3	kilogram/meter^3	+03 1.00*
lbm/inch^3	kilogram/meter^3	+04 2.767 990 5
lbm/foot^3	kilogram/meter^3	+01 1.601 846 3
slug/foot^3	kilogram/meter^3	+02 5.153 79
ENERGY		
British thermal unit (ISO/TC 12)	joule	+03 1.055 06
British thermal unit (International Steam Table)	joule	+03 1.055 04
British thermal unit (mean)	joule	+03 1.055 87
British thermal unit (thermochemical)	joule	+03 1.054 350 264 488
British thermal unit (39° F)	joule	+03 1.059 67
British thermal unit (60° F)	joule	+03 1.054 68
calorie (International Steam Table)	joule	+00 4.1868
calorie (mean)	joule	+00 4.190 02
calorie (thermochemical)	joule	+00 4.184*
calorie (15° C)	joule	+00 4.185 80
calorie (20° C)	joule	+00 4.181 90
calorie (kilogram, International Steam Table)	joule	+03 4.1868
calorie (kilogram, mean)	joule	+03 4.190 02
calorie (kilogram, thermochemical)	joule	+03 4.184*
electron volt	joule	−19 1.602 10
erg	joule	−07 1.00*
foot lbf	joule	+00 1.355 817 9
foot poundal	joule	−02 4.214 011 0
joule (international of 1948)	joule	+00 1.000 165
kilocalorie (International Steam Table)	joule	+03 4.1868
kilocalorie (mean)	joule	+03 4.190 02
kilocalorie (thermochemical)	joule	+03 4.184*
kilowatt hour	joule	+06 3.60*
kilowatt hour (international of 1948)	joule	+06 3.600 59
ton (nuclear equivalent of TNT)	joule	+09 4.20
watt hour	joule	+03 3.60*
ENERGY/AREA TIME		
Btu (thermochemical)/foot^2 second	watt/meter^2	+04 1.134 893 1
Btu (thermochemical)/foot^2 minute	watt/meter^2	+02 1.891 488 5
Btu (thermochemical)/foot^2 hour	watt/meter^2	+00 3.152 480 8
Btu (thermochemical)/inch^2 second	watt/meter^2	+06 1.634 246 2
calorie (thermochemical)/cm^2 minute	watt/meter^2	+02 6.973 333 3
erg/centimeter^2 second	watt/meter^2	−03 1.00*
watt/centimeter^2	watt/meter^2	+04 1.00*
FORCE		
dyne	newton	−05 1.00*
kilogram force (kgf)	newton	+00 9.806 65*
kilopond force	newton	+00 9.806 65*
kip	newton	+03 4.448 221 615 260 5*
lbf (pound force, avoirdupois)	newton	+00 4.448 221 615 260 5*
ounce force (avoirdupois)	newton	−01 2.780 138 5
pound force, lbf (avoirdupois)	newton	+00 4.448 221 615 260 5*
poundal	newton	−01 1.382 549 543 76*
LENGTH		
angstrom	meter	−10 1.00*
astronomical unit	meter	+11 1.495 978 9
cable	meter	+02 2.194 56*
caliber	meter	−04 2.54*
chain (surveyor or gunter)	meter	+01 2.011 68*
chain (engineer or ramden)	meter	+01 3.048*
cubit	meter	−01 4.572*
fathom	meter	+00 1.8288*
fermi (femtometer)	meter	−15 1.00*
foot	meter	−01 3.048*
foot (U.S. survey)	meter	+00 1200/3937*
foot (U.S. survey)	meter	−01 3.048 006 096
furlong	meter	+02 2.011 68*
hand	meter	−01 1.016*
inch	meter	−02 2.54*
league (U.K. nautical)	meter	+03 5.559 552*
league (international nautical)	meter	+03 5.556*
league (statute)	meter	+03 4.828 032*
light year	meter	+15 9.460 55
link (engineer or ramden)	meter	−01 3.048*
link (surveyor or gunter)	meter	−01 2.011 68*
meter	wavelengths Kr 86	+06 1.650 763 73*
micron	meter	−06 1.00*
mil	meter	−05 2.54*
mile (U.S. statute)	meter	+03 1.609 344*
mile (U.K. nautical)	meter	+03 1.853 184*
mile (international nautical)	meter	+03 1.852*
mile (U.S. nautical)	meter	+03 1.852*
nautical mile (U.K.)	meter	+03 1.853 184*
nautical mile (international)	meter	+03 1.852*
nautical mile (U.S.)	meter	+03 1.852*

Source: *McGraw-Hill Metrication Manual.* Copyright © 1971 by McGraw-Hill, Inc. New York. (Used with permission of the McGraw-Hill Co.)

Listing of Conversion Factors by Physical Quantity *(Continued)*

To convert from	*to*	*multiply by*
pace	meter	−01 7.62*
parsec	meter	+16 3.083 74
perch	meter	+00 5.0292*
pica (printers)	meter	−03 4.217 517 6*
point (printers)	meter	−04 3.514 598*
pole	meter	+00 5.0292*
rod	meter	+00 5.0292*
skein	meter	+02 1.097 28*
span	meter	−01 2.286*
statute mile (U.S.)	meter	+03 1.609 344*
yard	meter	−01 9.144*

MASS

To convert from	*to*	*multiply by*
carat (metric)	kilogram	−04 2.00*
dram (avoirdupois)	kilogram	−03 1.771 845 195 312 5*
dram (troy or apothecary)	kilogram	−03 3.887 934 6*
grain	kilogram	−05 6.479 891*
gram	kilogram	−03 1.00*
hundredweight (long)	kilogram	+01 5.080 234 544*
hundredweight (short)	kilogram	+01 4.535 923 7*
kgf second^2 meter (mass)	kilogram	+00 9.806 65*
kilogram mass	kilogram	+00 1.00*
lbm (pound mass, avoirdupois)	kilogram	−01 4.535 923 7*
ounce mass (avoirdupois)	kilogram	−02 2.834 952 312 5*
ounce mass (troy or apothecary)	kilogram	−02 3.110 347 68*
pennyweight	kilogram	−03 1.555 173 84*
pound mass, lbm (avoirdupois)	kilogram	−01 4.535 923 7*
pound mass (troy or apothecary)	kilogram	−01 3.732 417 216*
scruple (apothecary)	kilogram	−03 1.295 978 2*
slug	kilogram	+01 1.459 390 29
ton (assay)	kilogram	−02 2.916 666 6
ton (long)	kilogram	+03 1.016 046 908 8*
ton (metric)	kilogram	+03 1.00*
ton (short, 2000 pound)	kilogram	+02 9.071 847 4*
tonne	kilogram	+03 1.00*

POWER

To convert from	*to*	*multiply by*
Btu (thermochemical)/second	watt	+03 1.054 350 264 488
Btu (thermochemical)/minute	watt	+01 1.757 250 4
calorie (thermochemical)/second	watt	+00 4.184*
calorie (thermochemical)/minute	watt	−02 6.973 333 3
foot lbf/hour	watt	−04 3.766 161 0
foot lbf/minute	watt	−02 2.259 696 6
foot lbf/second	watt	+00 1.355 817 9
horsepower (550 foot lbf/second)	watt	+02 7.456 998 7
horsepower (boiler)	watt	+03 9.809 50
horsepower (electric)	watt	+02 7.46*
horsepower (metric)	watt	+02 7.354 99
horsepower (U.K.)	watt	+02 7.457
horsepower (water)	watt	+02 7.460 43
kilocalorie (thermochemical)/minute	watt	+01 6.973 333 3
kilocalorie (thermochemical)/second	watt	+03 4.184*
watt (international of 1948)	watt	+00 1.000 165

PRESSURE

To convert from	*to*	*multiply by*
atmosphere	newton/meter^2	+05 1.013 25*
bar	newton/meter^2	+05 1.00*
barye	newton/meter^2	−01 1.00*
centimeter of mercury (0° C)	newton/meter^2	+03 1.333 22
centimeter of water (4° C)	newton/meter^2	+01 9.806 38
dyne/centimeter^2	newton/meter^2	−01 1.00*
foot of water (39.2° F)	newton/meter^2	+03 2.988 98
inch of mercury (32° F)	newton/meter^2	+03 3.386 389
inch of mercury (60° F)	newton/meter^2	+03 3.376 85
inch of water (39.2° F)	newton/meter^2	+02 2.490 82
inch of water (60° F)	newton/meter^2	+02 2.4884
kgf/centimeter^2	newton/meter^2	+04 9.806 65*
kgf/meter^2	newton/meter^2	+00 9.806 65*
lbf/foot^2	newton/meter^2	+01 4.788 025 8
lbf/inch^2 (psi)	newton/meter^2	+03 6.894 757 2
millibar	newton/meter^2	+02 1.00*
millimeter of mercury (0° C)	newton/meter^2	+02 1.333 224
pascal	newton/meter^2	+00 1.00*
psi (lbf/inch^2)	newton/meter^2	+03 6.894 757 2
torr (0° C)	newton/meter^2	+02 1.333 22

SPEED

To convert from	*to*	*multiply by*
foot/hour	meter/second	−05 8.466 666 6
foot/minute	meter/second	−03 5.08*
foot/second	meter/second	−01 3.048*
inch/second	meter/second	−02 2.54*
kilometer/hour	meter/second	−01 2.777 777 8
knot (international)	meter/second	−01 5.144 444 444
mile/hour (U.S. statute)	meter/second	−01 4.4704*
mile/minute (U.S. statute)	meter/second	+01 2.682 24*
mile/second (U.S. statute)	meter/second	+03 1.609 344*

TEMPERATURE

To convert from	*to*	*multiply by*
Celsius	kelvin	$t_K = t_C + 273.15$
Fahrenheit	kelvin	$t_K = (5/9)(t_F + 459.67)$
Fahrenheit	Celsius	$t_C = (5/9)(t_F - 32)$
Rankine	kelvin	$t_K = (5/9)t_R$

TIME

To convert from	*to*	*multiply by*
day (mean solar)	second (mean solar)	+04 8.64*
day (sidereal)	second (mean solar)	+04 8.616 409 0
hour (mean solar)	second (mean solar)	+03 3.60*
hour (sidereal)	second (mean solar)	+03 3.590 170 4
minute (mean solar)	second (mean solar)	+01 6.00*
minute (sidereal)	second (mean solar)	+01 5.983 617 4
month (mean calendar)	second (mean solar)	+06 2.628*
second (ephemeris)	second	+00 1.000 000 000
second (mean solar)	second (ephemeris)	Consult American Ephemeris and Nautical Almanac
second (sidereal)	second (mean solar)	−01 9.972 695 7
year (calendar)	second (mean solar)	+07 3.1536*
year (sidereal)	second (mean solar)	+07 3.155 815 0
year (tropical)	second (mean solar)	+07 3.155 692 6
year 1900, tropical, Jan., day 0, hour 12	second (ephemeris)	+07 3.155 692 597 47*
year 1900, tropical, Jan., day 0, hour 12	second	+07 3.155 692 597 47

Listing of Conversion Factors by Physical Quantity *(Continued)*

To convert from	*to*	*multiply by*
	VISCOSITY	
centistoke	meter²/second	−06 1.00*
stoke	meter²/second	−04 1.00*
foot²/second	meter²/second	−02 9.290 304*
centipoise	newton second/meter²	−03 1.00*
lbm/foot second	newton second/meter²	+00 1.488 163 9
lbf second/foot²	newton second/meter²	+01 4.788 025 8
poise	newton second/meter²	−01 1.00*
poundal second/foot²	newton second/meter²	+00 1.488 163 9
slug/foot second	newton second/meter²	+01 4.788 025 8
rhe	meter²/newton second	+01 1.00*
	VOLUME	
acre foot	meter³	+03 1.233 481 9
barrel (petroleum, 42 gallons)	meter³	−01 1.589 873
board foot	meter³	−03 2.359 737 216*
bushel (U.S.)	meter³	−02 3.523 907 016 688*
cord	meter³	+00 3.624 556 3
cup	meter³	−04 2.365 882 365*
dram (U.S. fluid)	meter³	−06 3.696 691 195 312 5*
fluid ounce (U.S.)	meter³	−05 2.957 352 956 25*
foot³	meter³	−02 2.831 684 659 2*
gallon (U.K. liquid)	meter³	−03 4.546 087
gallon (U.S. dry)	meter³	−03 4.404 883 770 86*
gallon (U.S. liquid)	meter³	−03 3.785 411 784*
gill (U.K.)	meter³	−04 1.420 652
gill (U.S.)	meter³	−04 1.182 941 2
hogshead (U.S.)	meter³	−01 2.384 809 423 92*
inch³	meter³	−05 1.638 706 4*
liter	meter³	−03 1.00*
ounce (U.S. fluid)	meter³	−05 2.957 352 956 25*
peck (U.S.)	meter³	−03 8.809 767 541 72*
pint (U.S. dry)	meter³	−04 5.506 104 713 575*
pint (U.S. liquid)	meter³	−04 4.731 764 73*
quart (U.S. dry)	meter³	−03 1.101 220 942 715*
quart (U.S. liquid)	meter³	−04 9.463 529 5
stere	meter³	+00 1.00*
tablespoon	meter³	−05 1.478 676 478 125*
teaspoon	meter³	−06 4.928 921 593 75*
ton (register)	meter³	+00 2.831 684 659 2*
yard³	meter³	−01 7.645 548 579 84*

INDEX

NOTE: Subjects are followed by double numbers to indicate each page reference. The (first) bold-face number indicates the section; the (second) light-face number is the page number of that section.

U

V

W

Y

Z

ABOUT THE EDITOR-IN-CHIEF

Robert O. Parmley, P.E., CMfgE, CSI, is Co-Founder, President, and Principal Consulting Engineer of Morgan & Parmley, Ltd., Professional Consulting Engineers, Ladysmith, Wisconsin. He is also a member of the National Society of Professional Engineers, the American Society of Civil Engineers, the American Society of Mechanical Engineers, the Construction Specifications Institute, the American Design Drafting Association, the American Society of Heating, Refrigerating, and Air-Conditioning Engineers, and the Society of Manufacturing Engineers, and is listed in the AAES *Who's Who in Engineering*. Mr. Parmley holds a BSME and a MSCE from Columbia Pacific University and is a registered professional engineer in Wisconsin, California, and Canada. He is also a certified manufacturing engineer under SME's national certification program and a certified wastewater treatment plant operator in Wisconsin. In a career covering four decades, Mr. Parmley has worked on the design and construction supervision of a wide variety of structures, systems, and machines—from dams and bridges to municipal sewage treatment facilities and public water projects. The author of over 40 technical articles published in leading professional journals, he is also the Editor-in-Chief of the *Illustrated Sourcebook of Mechanical Components*; the *HVAC Field Manual*; the *Hydraulics Field Manual*, now in its Second Edition; the *HVAC Design Data Sourcebook*; the *Mechanical Components Handbook*; the *Standard Handbook of Fastening & Joining*, now in its Third Edition; and the *Field Engineer's Manual*, now in its Third Edition, all published by McGraw-Hill.